Lecture Notes in Mathematics

Volume 2373

Editors-in-Chief

Jean-Michel Morel, City University of Hong Kong, Kowloon Tong, China

Bernard Teissier, IMJ-PRG, Paris, France

Series Editors

Karin Baur, University of Leeds, Leeds, UK

Michel Brion, UGA, Grenoble, France

Rupert Frank, LMU, Munich, Germany

Annette Huber, Albert Ludwig University, Freiburg, Germany

Davar Khoshnevisan, The University of Utah, Salt Lake City, UT, USA

Ioannis Kontoyiannis, University of Cambridge, Cambridge, UK

Angela Kunoth, University of Cologne, Cologne, Germany

Ariane Mézard, IMJ-PRG, Paris, France

Mark Podolskij, University of Luxembourg, Esch-sur-Alzette, Luxembourg

Mark Policott, Mathematics Institute, University of Warwick, Coventry, UK

László Székelyhidi, MPI for Mathematics in the Sciences, Leipzig, Germany

Gabriele Vezzosi, UniFI, Florence, Italy

Anna Wienhard, MPI for Mathematics in the Sciences, Leipzig, Germany

This series reports on new developments in all areas of mathematics and their applications - quickly, informally and at a high level. Mathematical texts analysing new developments in modelling and numerical simulation are welcome. The type of material considered for publication includes:

1. Research monographs
2. Lectures on a new field or presentations of a new angle in a classical field
3. Summer schools and intensive courses on topics of current research.

Texts which are out of print but still in demand may also be considered if they fall within these categories. The timeliness of a manuscript is sometimes more important than its form, which may be preliminary or tentative. Please visit the LNM Editorial Policy (https://drive.google.com/file/d/1MOg4TbwOSokRnFJ3ZR3ciEeKs9hOnNX_/view?usp=sharing)

Titles from this series are indexed by Scopus, Web of Science, Mathematical Reviews, and zbMATH.

Simon Felten

Global Logarithmic Deformation Theory

 Springer

Simon Felten (iD)
Mathematical Institute
University of Oxford
Oxford, UK

ISSN 0075-8434   ISSN 1617-9692   (electronic)
Lecture Notes in Mathematics
ISBN 978-3-031-98750-2   ISBN 978-3-031-98751-9   (eBook)
https://doi.org/10.1007/978-3-031-98751-9

Mathematics Subject Classification: 14D15, 14F99, 14J32, 14-02, 14B10

This work was supported by Deutsche Forschungsgemeinschaft (FE 2102/1-1) and Royal Society (RF\ERE\231095).

This Springer imprint is published by the registered company Springer Nature Switzerland AG
The registered company address is: Gewerbestrasse 11, 6330 Cham, Switzerland

If disposing of this product, please recycle the paper.

To Elisabeth and Thomas

Preface

Background in the Theory of Semistable Degenerations

The most basic objects of interest in this monograph are degenerating families of algebraic varieties $f : X \to S$, where $S/\mathbb{C}$ is a smooth curve, and where we have a degenerate fiber X_0 over a distinguished point $s_0 \in S$.

An early insight into the structure of degenerating families is the *Semistable Reduction Theorem* of [174]. It guarantees that, after a finite base change $\pi : S' \to S$ with $\pi^{-1}(s_0) = \{s_0'\}$, the family $f : X \to S$ can be brought into semistable form, i.e., there is a semistable degeneration $f' : X' \to S'$ which is equal to $X \times_S S' \to S'$ over $S' \setminus \{s_0'\}$. Furthermore, the irreducible components of X_0' are smooth. As a consequence, we can assume that we have a semistable degeneration as long as we are primarily interested in the behavior of the smooth fibers around s_0. Here, a *semistable degeneration* is one which is locally of the form $t \mapsto z_1 \cdot \ldots \cdot z_r$.

When $f : X \to S$ is a proper semistable degeneration, then the de Rham complex $\Omega^\bullet_{X/S}(\log X_0)$ of differential forms with *logarithmic poles* in X_0 is well-behaved: Its pieces $\Omega^i_{X/S}(\log X_0)$ are locally free on X, and the Hodge bundles $R^j f_* \Omega^i_{X/S}(\log X_0)$ are locally free on S. These and similar features make semistable degenerations more accessible than more general degenerations.

Let us switch to the complex-analytic picture. When $f : X \to \Delta$ is a proper analytic semistable degeneration over a small disk Δ such that each fiber X_t for $t \neq 0$ is a Calabi–Yau manifold, then we might be interested in also having $\omega_X \cong \mathcal{O}_X$ for the total space. For degenerating families of Calabi–Yau *surfaces*, this has been established by Kulikov in [183] and Persson and Pinkham in [228]: If $f : X \to \Delta$ is a semistable degeneration of Calabi–Yau surfaces, then there is another semistable degeneration $f' : X' \to \Delta$ with $\omega_{X'} \cong \mathcal{O}_{X'}$ and such that f' is equal to f over $\Delta \setminus \{0\}$. The family $f' : X' \to \Delta$ is called a *Kulikov model* of the degeneration.

One may now attempt to classify all Kulikov models, say with the additional property that all irreducible components of the central fiber are smooth. A list of candidates for the central fibers has been provided by Kulikov in [183] and Persson in [227], see also [97]. A good way to understand this list is as follows: First, X_0 must be a *normal crossing space*, i.e., be locally of the form $z_1 \cdot \ldots \cdot z_r = 0$. Since

$f: X \to \Delta$ is a Kulikov model, we have $\omega_{X_0}^\circ \cong O_{X_0}$ for the normalized dualizing sheaf. Furthermore, it is an insight of Friedman from [87] that any central fiber X_0 of a semistable degeneration is d-semistable, i.e., $\mathcal{T}_{X_0}^1 = \mathcal{E}xt^1(\Omega_{X_0}^1, O_{X_0})$ is isomorphic to O_{D_0}, where $D_0 \subset X_0$ is the non-normal locus. These two properties on a normal crossing space give strong restrictions on the geometry of X_0, which we may use to classify the possible X_0. Given a candidate X_0 for the central fiber, we are then left with the problem of determining whether X_0 actually is the central fiber of some semistable degeneration $f: X \to \Delta$, or, put more optimistically, of constructing such a semistable degeneration $f: X \to \Delta$.

Above, a Calabi–Yau surface is a proper analytic surface Y such that $\omega_Y \cong O_Y$. They can be classified into K3 surfaces, Abelian surfaces, and primary Kodaira surfaces. The list of possible central fibers can be subdivided accordingly. For an X_0 of K3 type, Friedman showed in [87] that X_0 is the central fiber of a semistable degeneration $f: X \to \Delta$ with $\omega_X \cong O_X$ and X_t a K3 surface for $t \neq 0$. The proof uses a combination of classical methods in flat deformation theory and Steenbrink's limiting mixed Hodge structure of a semistable degeneration from [261].

The condition $\omega_{X_0}^\circ \cong O_{X_0}$ and d-semistability are necessary for the existence of a semistable degeneration not only for surfaces but in all dimensions. In [173], Kawamata and Namikawa considered the question of the existence of a semistable degeneration when X_0 is a d-semistable and Calabi–Yau normal crossing space of dimension ≥ 3. Under some mild additional assumptions, they showed that such an X_0 is indeed the central fiber of a semistable degeneration.

Kawamata and Namikawa use an early variant of *logarithmic geometry* in their article. From a modern point of view, d-semistability allows them to endow X_0 with a *semistable log smooth structure* over the standard log point. Semistable degenerations to X_0 then correspond with *log smooth deformations*, and some of the good behavior of semistable degenerations can be explained as a manifestation of the smoothness hidden in the log structure. The Calabi–Yau condition suggests that the existence of the semistable degeneration to X_0 results from an unobstructedness principle similar to the Bogomolov–Tian–Todorov theorem for compact Kähler manifolds.

Origins in the Gross–Siebert Program

Mirror symmetry is the observation, first brought forward by physicists, that Calabi–Yau manifolds apparently come in pairs with dual properties. Among the many works inspired by this observation is the *toric Gross–Siebert mirror construction* of [124, 125], which obtains candidates for mirror pairs by smoothing combinatorially constructed pairs of reducible toric schemes with trivial dualizing sheaf. These latter *toric Calabi–Yau schemes* are locally isomorphic to central fibers of log smooth degenerations, so one may expect to deform them via a generalization of Kawamata–Namikawa's result to log smooth spaces which are not necessarily semistable.

When Gross and Siebert attempted to do so, their approach failed for two reasons. The first reason is that their toric Calabi–Yau schemes X_0 are not d-semistable, i.e., they do not carry a global log smooth structure. One can only select a closed subset

$Z_0 \subset X_0$ of codimension ≥ 2 and a log smooth structure on the complement $U_0 = X_0 \setminus Z_0$. We call this phenomenon the presence of *log singularities*.

Luckily, when the dual intersection complex $(B, \mathscr{P})$ is simple, one can choose Z_0 and the log smooth structure on U_0 canonically. The canonical choices give rise to the *toric log Calabi–Yau spaces* of [124]. Then, the log singularities have explicit local models in the form of certain toric morphisms of affine toric varieties. This allows Gross and Siebert to control the local behavior of the log singularities and work with them as though they were smooth, although they are not even smooth from the logarithmic perspective.

The second reason why Gross and Siebert's initial approach failed is that, at the time, no general method to show the existence of global logarithmic deformations was available. For example, the approach of Kawamata–Namikawa uses the T^1-lifting criterion and necessitates that

$$H^2(X_0, \Theta^1_{X_0/S_0}) \to \mathrm{Ext}^2(\Omega^1_{X_0}, O_{X_0})$$

is injective, where $\Omega^1_{X_0}$ are the (non-logarithmic) differential forms, and $\Theta^1_{X_0/S_0}$ is the relative log tangent sheaf. This is not clear in the generality of the Gross–Siebert construction. To circumvent the issue, Gross and Siebert developed the *scattering algorithm* of [126], which constructs the smoothing of X_0 explicitly.

When X_0 is a toric log Calabi–Yau space with simple dual intersection complex $(B, \mathscr{P})$ such that moreover the (total outer) monodromy polytopes $\check{\Delta}(\tau)$ are standard simplices, Gross and Siebert proved in [125] that the associated log Hodge–de Rham spectral sequence degenerates at E_1. The proof relies on the explicit combinatorial construction of X_0. However, it was expected that this is not so much a consequence of the explicit description as it is a consequence of the nature of the log singularities of X_0. After all, the log Hodge–de Rham spectral sequence of a log smooth X_0 degenerates at E_1, a fact that has been known since the inception of log geometry in [162]. This problem was given to the author by his doctoral adviser Helge Ruddat. As it turned out, the E_1-degeneration is indeed true for all proper log spaces whose log singularities are locally given by toric varieties. Furthermore, this does not only hold over the standard log point but also over thickenings of the base. We published these results in our joint work [77] with Ruddat and Matej Filip.

We considered the Hodge–de Rham degeneration as a necessary step toward showing the existence of a degeneration abstractly (without the need to construct it explicitly), but while we had now generalized the Hodge–de Rham degeneration, this still left us with the problem of showing that imposing the log Calabi–Yau condition is actually sufficient to conclude the existence of a degeneration.

Meanwhile, Chan, Leung, and Ma considered the deformation problem for toric log Calabi–Yau spaces in terms of log polyvector fields and the log de Rham complex. By adapting methods of Katzarkov, Kontsevich, and Pantev from [171] and Terilla from [266], they discovered a homological argument to show that the log polyvector fields and the log de Rham complex can be lifted order by order. In particular, this applies to the structure sheaf, thus yielding the desired degeneration without the need of an explicit construction. They published their results in [38]. The argument of Chan–Leung–Ma is the breakthrough that was missing when Gross and Siebert first attempted to deform toric log Calabi–Yau spaces.

The preprint version of [38] was already available in 2019. We incorporated their results in [77] to conclude the existence of degenerations to a wide variety of spaces X_0. In particular, we showed that every proper normal crossing scheme X_0 with trivial dualizing sheaf and $\mathcal{T}^1_{X_0} = \mathcal{E}xt^1(\Omega^1_{X_0}, O_{X_0})$ globally generated is the central fiber of some degeneration of smooth manifolds, significantly strengthening the result of Kawamata–Namikawa from 1994.

About This Monograph I

The author undertook a first step toward integrating F. Kato's log smooth deformation theory of [160], the results of Gross and Siebert, and the argument of Chan–Leung–Ma into a systematic logarithmic deformation theory with the article [74], where he proves that the log smooth (no log singularities here) deformation problem is equivalent to the lifting problem for log polyvector fields and the log de Rham complex. Together with further insights of Chan–Leung–Ma, this shows that the log smooth deformation functor is controlled by a *curved* Lie algebra (called predg Lie algebra in [74]) rather than a dg Lie algebra. This latter fact is also one of the reasons why previous attempts at understanding log deformations failed.

In the joint work with Andrea Petracci [78], we gave a proof of the logarithmic Bogomolov–Tian–Todorov theorem over the trivial log point by showing homotopy Abelianity of the controlling dg Lie algebra (the curvature is zero in this case), and we deduced the general case from a weaker statement that follows more directly from Chan–Leung–Ma than the actual logarithmic Bogomolov–Tian–Todorov theorem. We restricted ourselves to the log smooth case in [78].

This monograph finally provides the systematic logarithmic deformation theory. Its core is the very careful treatment of the Chan–Leung–Ma argument (inspired by Katzarkov–Kontsevich–Pantev and Terilla) in Part I. The argument is purely homological (or homotopical if the reader likes) and independent of logarithmic geometry. The central results are the First Abstract Unobstructedness Theorem 1.98 and the Second Abstract Unobstructedness Theorem 1.101. Our treatment clarifies the assumptions necessary for the argument of Chan–Leung–Ma and presents the result in a form suitable for applications beyond logarithmic geometry.

In Part II, we expound the logarithmic geometry used in this monograph. We give a general introduction to logarithmic geometry and lay the foundations for the language of *generically log smooth families*, our chosen framework to study log singularities and their deformations. We also give an elementary introduction to toroidal crossing spaces, which are intermediate between schemes and log schemes.

Part III develops the deformation theory of generically log smooth families and provides the tools to connect logarithmic deformation problems with the Abstract Unobstructedness Theorems. This is a more general version of [74] and, in parts, crucially relies on ideas of Chan–Leung–Ma.

The final Part IV develops applications in logarithmic geometry. Here, we prove the main theorem of this monograph, the Log Toroidal Bogomolov–Tian–Todorov

Theorem 1.103. As an application, we (re-)prove the Smoothing Theorem 1.128 for normal crossing spaces in a slightly stronger form than we did in [77]. In particular, we treat the Fano case much more carefully than we did in [77].

The Log Toroidal Bogomolov–Tian–Todorov Theorem also holds for deformations of pairs $(X_0, \mathcal{L}_0)$, where $\mathcal{L}_0$ is a line bundle on the log Calabi–Yau X_0, see Theorem 1.120. While the classical version has been known, the logarithmic version was previously unknown. Theorem 1.120 allows us in particular to deform an ample line bundle on X_0 together with X_0 and therefore obtain a projective deformation.

About the Introduction

The introduction to this monograph is exceptionally long. It grew out of the experiences with many introductory talks on the subject of logarithmic deformation theory, and it can be considered as a self-contained elementary course in global logarithmic deformation theory. We gently guide the reader through all the main notions and results of the present monograph. That said, we recommend reading the introduction and then going to the main text whenever more details are desired.

About This Monograph II

This monograph is titled *Global Logarithmic Deformation Theory*. With the choice of the adjective "global," we emphasize that the main results in this monograph are about the passage from given local logarithmic deformations to a global logarithmic deformation. Nonetheless, we also lay the foundations for local logarithmic deformation theory with the basic definitions fixed and many examples of log singularities discussed. In particular, Chap. 14 contains the necessary theory (originally developed in [125]) to show that, in some situations, local deformations can be chosen consistently. However, much is still unknown about the local deformation theory of log singularities, and we hope to come back to this question in the future.

With this monograph, we hope to make logarithmic deformation theory and in particular its use in the Gross–Siebert program more complete and more accessible, and to facilitate further research into degenerations of varieties, compactification of their moduli, and classification of the components of moduli spaces. In particular, we wish to popularize Gerstenhaber algebras and calculi—our main technical tool in the Abstract Unobstructedness Theorems, whose usefulness in log deformation theory was discovered by Chan, Leung, and Ma—as a tool to study algebraic varieties. These calculi encode the deformation theory and cohomology of algebraic varieties. Moreover, they are closely related to Hodge structures as well as the approach of Barannikov–Kontsevich to the mirror symmetry B-model.

Oxford, UK   Simon Felten
February 2025

Acknowledgments The oldest parts of this monograph have been written while the author was a doctoral student at the Johannes Gutenberg University of Mainz. In particular, large parts of the material in Chap. 8 have been adapted from the author's unpublished[1] doctoral thesis. Some but not all of the material also appeared in abridged form in [77]. In this context, the author thanks JGU Mainz for its hospitality, Carl Zeiss Stiftung for financial support, and Studienstiftung for further support. He thanks his former doctoral adviser Helge Ruddat for many discussions on most aspects of the theory presented here, more recently in particular on toroidal crossing spaces and smoothings.

Some of the older materials in this monograph, in particular in Chap. 9, have been written while the author visited the Institut des Hautes Études Scientifiques in Bures-sur-Yvette. In this context, he thanks IHÉS for its hospitality and Maxim Kontsevich for the invitation.

Most of the material has been written while the author was at Columbia University as a Walter Benjamin research fellow of the German Research Council (DFG). He thanks his postdoctoral mentor Aise Johan de Jong for many discussions on most of the topics covered in this monograph as well as Columbia University for its hospitality and the DFG for financial support through the grant FE 2102/1-1.

The manuscript was finished while the author was a postdoctoral research associate at the University of Oxford. He also thanks his postdoctoral adviser Lukas Brantner for further discussions and suggestions. The author thanks the University of Oxford for its hospitality and the Royal Society for financial support through Brantner's grant URF\R1\211075.

The author thanks the three anonymous referees for their positive feedback and constructive comments. In the order assigned by the publisher, he particularly thanks the second referee for the proposed structural improvements and the third referee for their exceptionally careful reading of the manuscript and numerous suggestions, which have been very helpful in further improving the manuscript.

The author thanks Richard Thomas for discussion around his example of an obstructed vector bundle of rank 2, see Example 11.14; Luc Illusie for clarifications about the early history of log structures; Andrea Petracci and Andrés David Gómez Villegas for comments on the draft of this monograph; Sándor Kovács for helpful comments on the notion of semi–log canonical singularities; Bruno Vallette for pointing out many important references in operads and homotopical algebra; and Alessio Corti, Matej Filip, Robert Friedman, John Terilla, and Matthias Zach for discussions on various aspects of the material in this monograph.

While they played no direct role in writing this monograph, it directly depends on the work of Kwokwai Chan, Mark Gross, Fumiharu Kato, Conan Nai Chung Leung, Ziming Nikolas Ma, Arthur Ogus, and Bernd Siebert. Among these, the author thanks Fumiharu Kato for discussions on the author's earlier and related

---

[1] The thesis is permanently available at Gutenberg Open Science, https://doi.org/10.25358/openscience-5658.

article [76], and Arthur Ogus and Bernd Siebert for their support of and interest in the author's research.

The author thanks Taro Sano for bringing up the question about unobstructedness of line bundles on log Calabi–Yau spaces. His question largely guided the author to the present form of the material in this monograph.

Guide for the Reader

We discuss prerequisites, the structure of the monograph, and the dependencies of the material. Furthermore, we point out those results in this monograph which are new. This guide is best understood after reading the introduction.

Conventions

Unless specified otherwise, we work over a field $\mathbf{k}$ of characteristic 0 which is not necessarily algebraically closed. Double complexes $(C^{\bullet,\bullet}, \partial, \bar{\partial})$ satisfy $\partial\bar{\partial} + \bar{\partial}\partial = 0$.

Prerequisites

Throughout the monograph, we assume that the reader is familiar with algebraic geometry at the level of Hartshorne's [138] (and some standard topics not discussed there), and logarithmic geometry at the level of Ogus' [222], especially Chapters III and IV. For the novice, we give an overview of those concepts in log geometry which are most important for this monograph in Chap. 7. It also contains some advice on further study in log geometry. A basic understanding of complex geometry as in Grauert's and Remmert's [115] is helpful but not needed except in Chap. 15. We also assume acquaintance with infinitesimal deformation theory as in Schlessinger's short article [251] and the books [254] and [139], but most of the material there is not necessary for this monograph. The reader not acquainted with F. Kato's log smooth deformation theory may want to read [160] first; we give a summary of the theory in Sect. 7.6. We do not assume the reader familiar with the theory of log toroidal families, which we have developed in [77] and the author's doctoral thesis [75], in that we provide a comprehensive review in Chap. 8. However, we do not repeat the proofs of the main results here. Prior knowledge of the Gross–Siebert program like [124–126] is helpful as well, in particular to put results in context, but we do not assume that the reader is familiar with these works.

Structure of This Monograph

In Chap. 1, we start this monograph with a comprehensive introduction which gives motivation, background information, and an overview over the main concepts and results. Thereafter, we discuss the related research literature in Chap. 2.

The main body of this monograph is divided into four parts. In Part I, we define the algebraic structures (such as Lie algebras) which we consider in this monograph, and we prove the abstract unobstructedness theorems for the deformation functors associated with Batalin–Vilkovisky algebras and calculi. This is the engine of our theory. In Part II, we review the basic theory of log geometry and families of singular log schemes. We also give a comprehensive treatment of the notion of toroidal crossing space. This is preparation for geometric applications. In Part III, we elaborate on the transition from a geometric deformation problem to a deformation problem of Gerstenhaber algebras and calculi. This is the connector between the abstract machinery, our engine, and the geometric problems we are actually interested in. In Part IV, we have assembled various materials needed to apply the theory, such as that systems of deformations exist in practice, and that the sufficient conditions of the abstract theorems are satisfied in the geometric situation. Finally, in Part IV, we see what our machine can do. The appendices contain a variety of materials used throughout the monograph, particularly those for which we could not find an appropriate reference.

Figure 1 shows the (approximate) dependencies of the chapters of the four main parts and the introduction. As the reader can see, it is mathematically correct to follow the monograph in the given order of the chapters. However, this order is mainly chosen in order to group related topics together. The reader may find it more enlightening to follow a different path through the chapters, for example

$$1 - 7 - 8 - 3 - 4 - 10 - 12 - 5 - 13 - 6 - 9 - 14 - 15 - 17 - 11 - 16 - 18,$$

which requires only minor forward references (use of concepts that are not yet defined when following this order).

Part I: Abstract Unobstructedness Theorems

In Chap. 3, we define the algebraic structures which are the main subject of our study. In particular, we define Gerstenhaber and Batalin–Vilkovisky calculi as well as their bigraded and curved counterparts.

In Chap. 4, we study gauge transforms of our structures in some generality.

In Chap. 5, we study the classical extended Maurer–Cartan equation of a curved Lie algebra, in particular a curved Gerstenhaber calculus, as well as the semi-classical extended Maurer–Cartan equation of a curved Batalin–Vilkovisky calculus. We show that solutions of these equations parametrize the different ways of turning the curved structures into a differential graded one by modifying the predifferential $\bar{\partial}$. Furthermore, we define perfectness and quasi-perfectness for curved Gerstenhaber calculi and related structures.

In Chap. 6, we prove the two abstract unobstructedness theorems Theorems 1.98 and 1.101 by following the argument in [38]. In doing so, we also define a quantum extended Maurer–Cartan equation and study liftings of its solutions first.

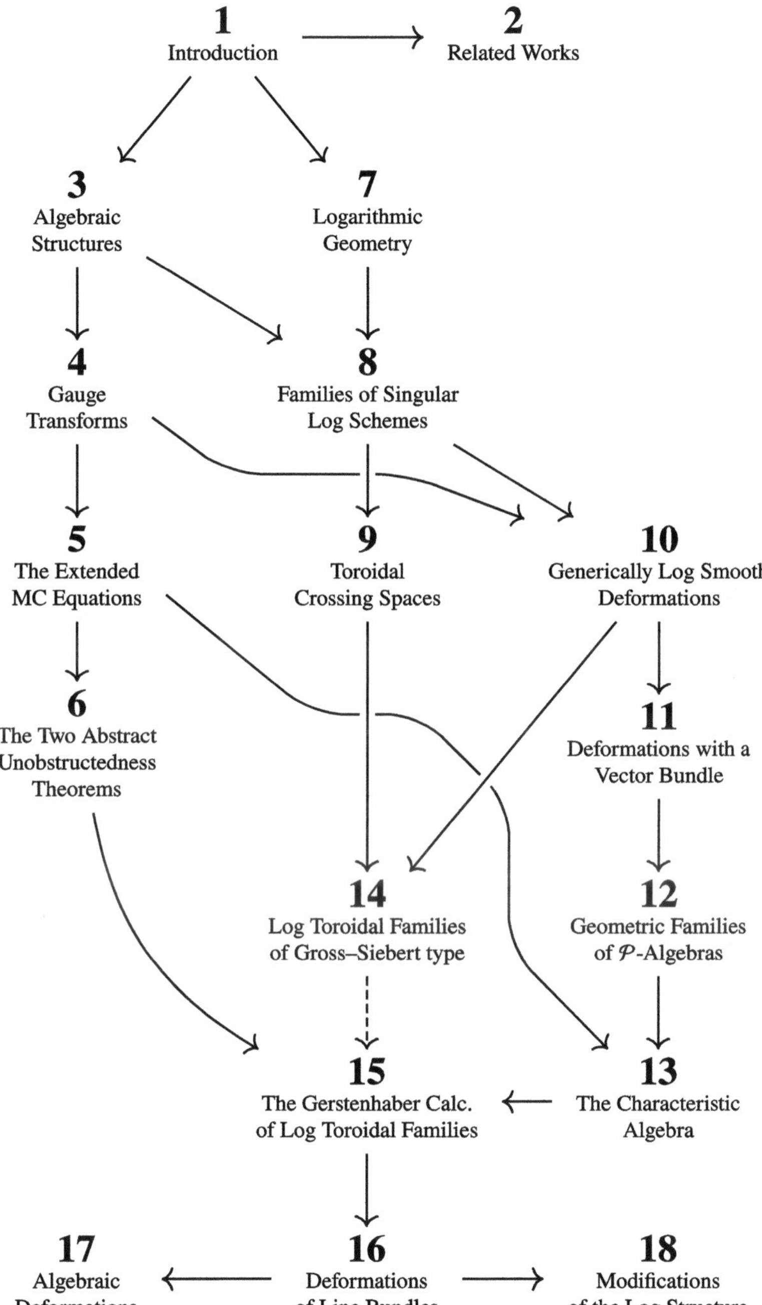

Fig. 1 The dependencies of the chapters

Part II: Logarithmic Geometry

In Chap. 7, we give a short introduction to logarithmic geometry with a focus on the topics which are most important for this monograph.

In Chap. 8, we discuss generically log smooth families and log toroidal families. It is a combination of the proofs of some minor new results and a review of already established results.

In Chap. 9, we discuss the notion of a toroidal crossing space $(V, \mathcal{P}, \bar{\rho})$, which goes back to [253], was essentially developed in [124], and was briefly defined in [77]. This chapter seems to be the first introductory discussion of toroidal crossing spaces available, and some of the results seem to have not yet been stated anywhere else (for example that V has semi–log canonical singularities).

Part III: Global Deformation Theory

In Chap. 10, we review the basic theory of generically log smooth deformations, in particular their automorphisms. We give the definition of a system of deformations $\mathscr{D}$ for $f_0 \colon X_0 \to S_0$. We also define enhanced generically log smooth families and their systems of deformations.

In Chap. 11, we review deformations of a pairs consisting of an (enhanced) generically log smooth family $f_0 \colon X_0 \to S_0$ and a vector bundle $\mathcal{E}_0$ on X_0. We briefly investigate the relationship between deformations of $(X_0, \mathcal{E}_0)$, deformations of $(X_0, \det(\mathcal{E}_0))$, and deformations of X_0.

In Chap. 12, we introduce a deformation theory of geometric families of algebraic structures. The most important example of this is obtained by replacing a generically log smooth deformation $f_A \colon X_A \to S_A$ with the deformation on the level of two-sided Gerstenhaber calculi. In doing so, we may forget about the log structures and just keep the Gerstenhaber calculi.

In Chap. 13, we give the very important construction of a characteristic curved algebra associated with a deformation problem of algebraic structures. This construction essentially relies on the Thom–Whitney resolution. It allows to form a single algebraic structure which controls the deformation problem and, in the case of Gerstenhaber calculi, computes the Hodge numbers. It is likely that this structure is closely tied to other important invariants of $f_0 \colon X_0 \to S_0$ as well, like Hodge structures, a Frobenius structure similar to [19],[2] or even Gromov–Witten invariants of the mirror, although the characteristic curved algebra alone is certainly not sufficient to compute these invariants.

Part IV: Applications

In Chap. 14, we discuss log toroidal families of (elementary) Gross–Siebert type, whose singularities are controlled by local models discussed by Gross–Siebert. For the reader's convenience, we also discuss the proof of the important fact that any two deformations which are log toroidal of elementary Gross–Siebert type (*divisorial* deformations in the terminology of Gross–Siebert) are locally isomorphic. The main

---

[2] Cf. also to [38, §6.2], where a semi-infinite Hodge structure is constructed.

point of this chapter is that log toroidal deformations of elementary Gross–Siebert type form a system of deformations as defined in Chap. 10, the most important example that we currently know.

In Chap. 15, we show that the characteristic curved Gerstenhaber calculus is indeed perfect if $f_0 \colon X_0 \to S_0$ is proper and log toroidal, and all deformations in a system of deformations $\mathscr{D}$ are log toroidal as well. The argument in this chapter relies on analytification, so we work over $\mathbf{k} = \mathbb{C}$. Through the Lefschetz principle, the result should be true over all fields $\mathbf{k}$ of characteristic 0. With the results of Chap. 15, we have everything together to show that log toroidal deformations of elementary Gross–Siebert type are unobstructed in the proper log Calabi–Yau case.

In Chap. 16, we study deformations of pairs $f_0 \colon (X_0, \mathcal{L}_0) \to S_0$ with a line bundle $\mathcal{L}_0$ on X_0. This yields the unobstructedness theorems for pairs.

In Chap. 17, we apply the results of Chap. 16 to an ample line bundle on a projective $f_0 \colon X_0 \to S_0$ over the standard log point $S_0 = \mathrm{Spec}(\mathbb{N} \to \mathbb{C})$ to enhance a formal deformation to an algebraic deformation over $\mathrm{Spec}\mathbb{C}[\![t]\!]$. We briefly discuss the geometric properties of such an algebraic deformation.

In Chap. 18, we present our theory of modifications $g \colon X(s) \to S$ of log morphisms $f \colon X \to S$, which often allows us to turn log Fano spaces into log Calabi–Yau spaces in a well-behaved way. As an application, we discuss the smoothing of normal crossing spaces, which is the main result of our article [77].

Appendices

In Appendix A, we review the basic definitions and theory of deformation functors, mostly to fix notation. We use Manetti's notion of a deformation functor from [199] but call the concept a *neat deformation functor* to distinguish it from the classical concept, which needs to satisfy one axiom less. This additional axiom guarantees a well-behaved obstruction theory, which we do not need in this monograph but anticipate to use in future applications.

In Appendix B, given a morphism $f \colon X \to S$, we study the relationship between that a coherent sheaf $\mathcal{F}$ on X is reflexive, and that its fibers $\mathcal{F}_s$ are reflexive.

In Appendix C, we introduce the notion of a *multilinear differential operator* of finite order

$$\mu \colon \ \mathcal{F}_1 \times \ldots \times \mathcal{F}_n \to \mathcal{G},$$

generalizing the case of differential operators $D \colon \mathcal{F} \to \mathcal{G}$ between two quasi-coherent sheaves $\mathcal{F}$ and $\mathcal{G}$. As it turns out, this theory is quite well-behaved. Our primary interest is in the Schouten–Nijenhuis bracket $[-, -]$ and the Lie derivative $\mathcal{L}_-(-)$ as bilinear differential operators.

In Appendix D, we prove the criterion for E_1-degeneration of spectral sequences over $\mathbf{k}$, which is alluded to in [171, Defn. 4.13]. We also give a more precise criterion for partial degeneration at E_1, and we study degeneration of spectral sequences over Artinian local rings A.

In Appendix E, we study the analytification of schemes X of finite type over $\mathbb{C}$. In particular, we show that all differential operators of first order between quasi-coherent algebraic sheaves can be analytified.

In Appendix F, we provide the code that we used in analyzing Example 10.5.

What Is New in This Monograph?

Many results in this monograph have been known before, in some form or another. Here is what is new in this monograph:

(1) The precise definitions of a Gerstenhaber calculus and Batalin–Vilkovisky calculus, in all their variants.

(2) The Second Abstract Unobstructedness Theorem 5.24.

(3) The criterion for partial degenerations of spectral sequences in Proposition D.2.

(4) The unobstructedness results for line bundles on log toroidal log Calabi–Yau spaces in Chap. 16; in particular, the deformations of the line bundle claimed in [126, Rem. 1.32] exist indeed, albeit that we have a different proof than the one announced there.

(5) As a consequence of the unobstructedness of a space with a line bundle, the existence of actual algebraic families over $\mathrm{Spec}\,\mathbf{k}[\![t]\!]$ in the case where the central fiber is a projective log Calabi–Yau space, not only formal families.

(6) The precise formulation of changing the log structure of a log Fano variety to obtain a log Calabi–Yau one in Chap. 18; a weaker and less precise version of this was given in [77].

(7) A theory of multilinear differential operators in algebraic geometry, given in Appendix C.

(8) A weakening of the notion of generically log smooth family to the notion of *enhanced generically log smooth family*, which has better stability properties under base change.

(9) The first comprehensive presentation of the theory of toroidal crossing spaces in our sense, given in Chap. 9; this also includes a number of new results such as the computation of the dualizing sheaf of a simple toroidal crossing space in Theorem 9.66.

(10) An interpretation of the sheaf $\mathcal{B}$ occurring in [125, §2.2], given in Chap. 14.

Contents

Part I   Abstract Unobstructedness Theorems

Part II   Logarithmic Geometry

Symbols

$\mathbf{k}$	A field of characteristic 0, not necessarily algebraically closed
Q	A sharp toric monoid, often $Q = \mathbb{N}$
Λ	A complete local Noetherian $\mathbf{k}$-algebra with residue field $\mathbf{k}$; often $\mathbf{k}[\![Q]\!]$ for a sharp toric monoid Q
$\mathbf{Art}$	The category of Artinian local $\mathbf{k}$-algebras with residue field $\mathbf{k}$; see Definition 1.26
$\mathbf{Art}_\Lambda$	The category of Artinian local Λ-algebras with residue field $\mathbf{k}$; see Definition A.1
$\mathbf{Art}_\Lambda^\circ$	The full subcategory of objects of $\mathbf{Art}_\Lambda$ with trivial Λ-algebra structure; considered in Sect. 1.8.3; this category is isomorphic (not only equivalent) to $\mathbf{Art}$
$\mathbf{Art}_Q$	The category of Artinian local $\mathbf{k}[\![Q]\!]$-algebras with residue field $\mathbf{k}$
A_k	The quotient $\Lambda/\mathfrak{m}_\Lambda^{k+1}$ of the complete local Noetherian ring Λ, usually of $\Lambda = \mathbf{k}[\![Q]\!]$
A_ε	The Artinian local Λ-algebra $\mathbf{k}[\varepsilon]/(\varepsilon^2)$, with the trivial Λ-algebra structure
S_0	The log point $\mathrm{Spec}(Q \to \mathbf{k})$ for a sharp toric monoid Q
S_k	The punctual log scheme $\mathrm{Spec}(Q \to A_k)$ for $A_k = \mathbf{k}[\![Q]\!]/\mathfrak{m}_Q^{k+1}$
S_ε	The punctual log scheme $\mathrm{Spec}(Q \to A_\varepsilon)$ for $\Lambda = \mathbf{k}[\![Q]\!]$
ω_V°	The normalized dualizing sheaf of the scheme $V/\mathbf{k}$; see also Definition 1.6
$(X \supset V)$	A scheme germ; see Sect. 1.1
$\mathrm{Def}_V^{\mathrm{lt}}$	The locally trivial deformation functor; see Sect. 1.5
$\mu_{d;r}^{\mathrm{nc}}$	The local model of a semistable degeneration; see Sect. 1.5
$\mathbb{T}_V^1$	$= \mathrm{Ext}^1(\Omega_V^1, \mathcal{O}_V)$, global Ext group on the scheme $V/\mathbf{k}$
$\mathrm{Def}_V^{\mathrm{ss}}$	The semistable deformation functor; see Definition 1.51
(X, U)	A partial log scheme; see Definition 1.75
X^{sze}	The log scheme $(\mathrm{Spec}\,\mathbf{k}[x, y, z, t]/(xy - tz)|\{t = 0\})$; see Example 1.79

$\mathscr{C}$	A class of log singularities; see Definition 1.84
$\mathscr{C}^{\text{anc}}$	The class of almost normal crossing log singularities; see Example 1.86
$\mathscr{C}_Q^{\text{tor}}$	The class of toroidal log singularities; see Example 1.87
$\mathscr{C}^{\text{GS}}, \mathscr{C}^{\text{eGS}}, \mathscr{C}^{\text{sGS}}$	Classes of log singularities of Gross–Siebert type; see Example 1.88 and the references therein
$\mathcal{P}$	An algebraic structure such as a Lie algebra or a Gerstenhaber algebra; "algebraic structures" in this sense are a naive notion of a colored operad; see page 141
$\mathcal{P}^{\text{bg}}$	Associated bigraded pre-algebraic structure of $\mathcal{P}$; see Sect. 3.6
$\mathcal{P}^{\text{crv}}$	Associated curved pre-algebraic structure of $\mathcal{P}$; see Sect. 3.6
$\mathfrak{Coh}(X/S)$	Context of coherent O_X-modules, flat over S, for a flat morphism $f: X \to S$; see Example 3.2
$\mathfrak{Coh}'(X/S)$	Context of all coherent O_X-modules for a flat morphism $f: X \to S$; see Example 3.2
$\mathfrak{Comp}(\Lambda)$	Context of flat and complete Λ-modules for a complete local Noetherian ring Λ with residue field $\mathbf{k}$; see Example 3.2
$\exp_\theta$	Gauge transform defined by θ, an (infinitesimal) automorphism of an algebraic structure such as a Gerstenhaber algebra; see Definition 4.3
$\theta \odot \xi$	Baker–Campbell–Hausdorff product of two elements in a Lie algebra; see Lemma 4.4
$T(x)$	The formal power series expansion of $(e^x - 1)/x$; see Lemma and Definition 5.3
$\exp_\theta * \phi$	The gauge action; see Lemma and Definition 5.3
$\text{Def}(L^\bullet, -)$	The deformation functor of the curved Lie algebra $L^\bullet$; see Sect. 5.1.1
$\text{Def}(E^{\bullet,\bullet}, -)$	The deformation functor of the curved algebra $E^{\bullet,\bullet}$; see Sect. 5.1.2
$\text{MC}(E^{\bullet,\bullet}, -)$	The Maurer–Cartan functor of the curved algebra $E^{\bullet,\bullet}$; see Sect. 5.1.2
$\text{SMC}(G^{\bullet,\bullet}, -)$	The semi-classical Maurer–Cartan functor of the curved Batalin–Vilkovisky algebra $G^{\bullet,\bullet}$; see Definition 5.12
$\text{SDef}(G^{\bullet,\bullet}, -)$	Gauge equivalence classes of elements in $\text{SMC}(G^{\bullet,\bullet}, -)$; see Eq. (5.2) on page 189
$\check{d}$	$\bar{\partial} + \Delta + (\ell + y) \wedge (-)$; see Lemma 5.20
d	$\partial + \bar{\partial} + \ell \lrcorner (-)$; see Sect. 5.4
$\text{QMC}(G^{\bullet,\bullet}, -)$	The quantum extended Maurer–Cartan functor of the curved Batalin–Vilkovisky algebra $G^{\bullet,\bullet}$; see Sect. 6.2
P^*	The subgroup of units in the monoid P
P^{gp}	The Grothendieck group of the monoid P
$\overline{P}$	The quotient P/P^* for the monoid P

LSch	The category of log schemes. We denote the category of coherent log schemes by $\mathbf{LSch}^{\mathrm{coh}}$ etc.
$\mathcal{M}_{(X\mid D)}$	The compactifying log structure defined by $D \subset X$
$\mathbb{A}_P$	The affine toric variety $\mathrm{Spec}\,\mathbb{Z}[P]$ for a toric monoid P, considered as a scheme or a log scheme with the trivial log structure; also over a field $\mathbf{k}$ instead of over $\mathbb{Z}$
A_P	The log scheme $\mathrm{Spec}(P \to \mathbb{Z}[P])$ for a toric monoid P; also over a field $\mathbf{k}$ instead of over $\mathbb{Z}$
$A_\theta : A_P \to A_Q$	The morphism of log schemes associated with a monoid homomorphism $\theta : Q \to P$
LD_{X_0/S_0}	The log smooth deformation functor of the log smooth morphism $f_0 : X_0 \to S_0$; see Sect. 7.6
Exp	The map $\Theta^1 \to \mathcal{A}ut$ which maps (log) derivations to infinitesimal automorphisms, or a more general version; see Sects. 7.6 and 10.2
Z	The log singular locus of the generically log smooth family $f : (X, U) \to S$, given as $X \setminus U$; see Definition 8.2
U°	The strict locus of the generically log smooth family $f : X \to S$, a dense open subset of U; see Sect. 8
$\mathcal{W}^\bullet_{X/S}$	The reflexive log de Rham complex of the generically log smooth family $f : (X, U) \to S$, obtained as $j_*\Omega^\bullet_{U/S}$; see Definition 8.3
$\Theta^1_{X/S}$	The sheaf of relative log derivations of the generically log smooth family $f : (X, U) \to S$, endowed with the usual Lie bracket; see Sect. 8.2
$\mathcal{V}^\bullet_{X/S}$	The reflexive log polyvector fields of the generically log smooth family $f : X \to S$, considered as a Gerstenhaber algebra in negative degrees and with the negative Schouten–Nijenhuis bracket. In particular, $\mathcal{V}^{-1}_{X/S}$ are the log derivations with the negative Lie bracket; see Sect. 8.2
$\mathcal{V}\mathcal{W}^\bullet_{X/S}$	The two-sided Gerstenhaber calculus of reflexive log differential forms and log polyvector fields of the generically log smooth family $f : (X, U) \to S$; see Proposition 8.5
V_F	The closed subscheme $\mathrm{Spec}\,\mathbf{k}[F] \subset \mathrm{Spec}\,\mathbf{k}[P] = \mathbb{A}_P$ for a face $F \subseteq P$ in a toric monoid; used primarily in Sect. 8.3
U_F	The open subscheme $\mathrm{Spec}\,\mathbf{k}[P_F] \subseteq \mathrm{Spec}\,\mathbf{k}[P] = \mathbb{A}_P$ for a face $F \subseteq P$ in a toric monoid; used primarily in Sect. 8.3
$\mathcal{T}_{X/S}$	The relative classical derivation of the morphism of schemes $f : X \to S$; often opposed to the relative log derivations $\Theta^1_{X/S}$
$\mathcal{T}^1_{X/S}$	The classical first relative tangent sheaf of the morphism of schemes $f : X \to S$, computed from the cotangent complex; primarily used for the map $\eta : \mathcal{LS}_V \to \mathcal{T}^1_V$ constructed in Sect. 9.12

$U(\sigma)$	The affine toric variety $\mathrm{Spec}\,\mathbf{k}[P_\sigma]$ constructed from a lattice polytope σ. A local model for log smooth, saturated, and vertical morphisms to the standard log point; see Construction 9.2
$V(\sigma)$	The central fiber of $U(\sigma) \to \mathbb{A}^1$; see Construction 9.2
$\widehat{\mathcal{L}}_V$	Sheaf of equivalence classes of u-integral log structure with specified ghost sheaf; see Definition 9.19
$(V, \mathcal{P}, \bar\rho)$	A toroidal crossing space, or more generally, a space with a pre-ghost structure; see Definition 9.26 and Definition 9.36
$\widehat{\mathcal{LS}}_V$	Sheaf of isomorphism classes of log structures with specified ghost sheaf, together with a morphism of log schemes to S_0; see Definition 9.29
$\mathcal{LS}_V$	On a toroidal crossing space $(V, \mathcal{P}, \bar\rho)$, the sheaf of log smooth log morphisms to the standard log point S_0; see Definition 9.41
$\mathcal{U}_m V$	On a toroidal crossing space V, the open locus where the rank of the ghost sheaf is $\leq m + 1$; see Sect. 9.6
$C_m V$	On a toroidal crossing space V, the closed locus where the rank of the ghost sheaf is $\geq m + 1$; see Sect. 9.6
$\mathcal{S}_m V$	On a toroidal crossing space V, the strata where the rank of the ghost sheaf is $= m + 1$; see Sect. 9.6
(V, Z, s)	A well-adjusted triple, consisting of a toroidal crossing space V, a closed subset $Z \subset V$, and a section $s \in \mathcal{LS}_V(V \setminus Z)$, satisfying certain conditions; see Definition 9.92
$\mathrm{LD}^{\mathrm{gen}}_{X_0/S_0}$	The generically log smooth deformation functor of a generically log smooth family $f_0 : (X_0, U_0) \to S_0$ over $S_0 = \mathrm{Spec}(Q \to \mathbf{k})$; see Definition 10.1
Log	Map from infinitesimal automorphisms to log derivations, inverse of Exp; see Sect. 10.2
$\mathcal{A}^\bullet_{X/S}$	The distinguished log de Rham complex of the enhanced generically log smooth family $f : X \to S$, endowed with a map $\varpi^\bullet : \mathcal{A}^\bullet_{X/S} \to \mathcal{W}^\bullet_{X/S}$ to the reflexive log de Rham complex; see Definition 10.8
$\mathcal{G}^\bullet_{X/S}$	The distinguished Gerstenhaber algebra of log polyvector fields of the enhanced generically log smooth family $f : X \to S$, endowed with a map $\varpi^\bullet : \mathcal{G}^\bullet_{X/S} \to \mathcal{V}^\bullet_{X/S}$ to the reflexive log polyvector fields; see Definition 10.8
$\mathcal{GC}^\bullet_{X/S}$	The two-sided Gerstenhaber calculus of distinguished log differential forms and log polyvector fields of the enhanced generically log smooth family $f : X \to S$, endowed with a map $\varpi^\bullet : \mathcal{GC}^\bullet_{X/S} \to \mathcal{VW}^\bullet_{X/S}$; see Definition 10.8
$\Gamma^1_{X/S}$	The distinguished relative log derivations of the enhanced generically log smooth family $f : X \to S$, endowed with the usual Lie bracket and with a map $\Gamma^1_{X/S} \to \Theta^1_{X/S}$ to the usual relative log derivations

$\mathscr{D}$	A system of deformations $V_{k;\alpha} \to S_k$; see Definition 10.15. Also an enhanced system of deformations, see Definition 10.21
$\mathcal{V} = \{V_\alpha\}_\alpha$	An open cover of the total space X_0 of a generically log smooth family $f_0 \colon X_0 \to S_0$, usually affine, used to define local models of infinitesimal deformations; see Definition 10.15
$V_{\alpha;k}$	An infinitesimal generically log smooth deformation of $V_\alpha = V_{\alpha;0}$ over S_k, usually as part of a system of deformations $\mathscr{D}$
$\mathrm{LD}^{\mathscr{D}}_{X_0/S_0}$	The functor of generically log smooth deformations of type $\mathscr{D}$. This is a subfunctor of $\mathrm{LD}^{\mathrm{gen}}_{X_0/S_0}$ which classifies only those generically log smooth deformations which are isomorphic to the ones prescribed in the system of deformations $\mathscr{D}$; see Definition 10.15
$\mathrm{ELD}^{\mathscr{D}}_{X_0/S_0}$	The functor of enhanced generically log smooth deformations of type $\mathscr{D}$. These are deformations of the enhanced generically log smooth family $f_0 \colon X_0 \to S_0$ which are locally isomorphic to the deformations prescribed in $\mathscr{D}$; see Definition 10.21
$\mathrm{LD}^{\mathrm{gen}}_{X_0/S_0}(\mathcal{E}_0)$	Generically log smooth deformations of $f_0 \colon X_0 \to S_0$ together with a deformation of the vector bundle $\mathcal{E}_0$; see Definition 11.1
$\Theta^1_{X/S}(\mathcal{E})$	Derivations of the family $f \colon (X, \mathcal{E}) \to S$ with a vector bundle; see Definition 11.3
$\mathrm{LD}^{\mathscr{D}}_{X_0/S_0}(\mathcal{E}_0)$	Generically log smooth deformations of type $\mathscr{D}$ together with a deformation of the vector bundle $\mathcal{E}_0$; see Definition 11.10
$\mathrm{ELD}^{\mathscr{D}}_{X_0/S_0}(\mathcal{E}_0)$	Enhanced generically log smooth deformations of type $\mathscr{D}$ together with a deformation of the vector bundle $\mathcal{E}_0$; see Definition 11.12
$\mathcal{E}^\bullet$	The sheaves of modules in a geometric family of $\mathcal{P}$-algebras; see Definition 12.1
$\mathrm{GDef}^{\mathscr{D}}(\mathcal{E}_0^\bullet, -)$	The functor of geometric deformations of $\mathcal{P}$-algebras of type $\mathscr{D}$; see Definition 12.15
$\mathscr{D}^{\mathrm{gc}}$	System of deformations of two-sided Gerstenhaber calculi associated with a system of deformations $\mathscr{D}$ of generically log smooth families. Denotes also its enhanced version; see Sect. 12.4
$\mathcal{U} = \{U_i\}_i$	An open cover, usually affine, of a scheme X, used to compute (Čech) cohomology on X
$\mathrm{TW}^\bullet(\mathcal{F})$	The Thom–Whitney resolution of the sheaf $\mathcal{F}$; see Sect. 13.1
$\bar{\partial}$	The horizontal or second differential of a double complex; often coming from the Thom–Whitney resolution; also often only a predifferential, i.e., $\bar{\partial}^2 \neq 0$
$\mathrm{TWD}^{\mathscr{D}}(\mathcal{E}_0^\bullet, -)$	The deformation functor of bigraded Thom–Whitney deformations together with a differential $\bar{\partial}$; see Definition 13.29
$E^{\bullet,\bullet}_{X_0/\Lambda}$	The characteristic Λ-linear $\mathcal{P}^{crv}$-pre-algebra; see Definition 13.33

$PV^{\bullet,\bullet}_{X_0/\Lambda}$	The characteristic curved Gerstenhaber algebra of $f_0 : X_0 \to S_0$ with a system of deformations $\mathscr{D}$; see Sect. 13.5
$DR^{\bullet,\bullet}_{X_0/\Lambda}$	The characteristic curved de Rham complex; see Sect. 13.5
$L^{\bullet}_{X_0/\Lambda}$	The curved Lie algebra controlling the deformation functor of $f_0 \colon X_0 \to S_0$; see Sect. 13.5
$L(\ell; s)$	The scheme $\mathrm{Spec}\,\mathbf{k}[x, y, z, t]/(xy - t^\ell z) \times \mathbb{A}^s$ as a family over $\mathbb{A}^1_t$, used as a local model for unisingular deformations; see Sect. 14.3
$\mathrm{LD}^{\mathrm{uni}}_{X_0/S_0}$	Unisingular deformation functor; see Definition 14.23
$\mathscr{B}_{X_0/S_0}$	The cokernel of the inclusion $\Theta^1_{X_0/S_0} \to \mathcal{T}_{X_0/S_0}$ for a generically log smooth family $f_0 \colon X_0 \to S_0$, considered in the context of unisingular deformations; see Sect. 14.3
$K \subset Q$	Ideal in the monoid Q; considered in Chap. 15
A_K	$\mathbf{k}[Q]/\mathbf{k}[K]$ for the monoid ideal $K \subset Q$
$\mathbf{0}$	The trivial log point $\mathrm{Spec}(0 \to \mathbf{k})$; considered in Chap. 15
S_K	$\mathrm{Spec}(Q \to A_K)$, thick log point; $K \subset Q$ is a monoid ideal
$\Omega^{\bullet}_{S_K/\mathbf{0}}$	Log de Rham complex of S_K over $\mathbf{0}$; see Sect. 15.1
$\mathcal{W}^{\bullet}_{X_K/\mathbf{0}}$	Reflexive absolute de Rham complex, over $\mathbf{0}$ instead of S_K; see Sect. 15.1
$\mathcal{DA}^{\bullet,\bullet}_k$	Curved absolute de Rham complex over S_k, absolute analog of $\mathcal{DR}^{\bullet,\bullet}_k$; see Sect. 15.2
$\mathscr{D}(s_0)$	Modified system of deformations for the modified generically log smooth family $f_0 \colon X_0(s_0) \to S_0$; see Sect. 18.6
$\mathrm{ELD}^{\mathscr{D}}_{X_0/S_0}(\mathcal{L}_0, s_0)$	Deformation functor which does not only capture deformations of a line bundle $\mathcal{L}_0$ but also deformations of a section $s_0 \in \mathcal{L}_0$; see Sect. 18.6

Overview of Formulae in Curved Algebras

We give an overview of our definitions of the various curved algebras. For the bigraded case, just omit the predifferential, and for the singly graded case, take the total degree. Λ is a complete local Noetherian $\mathbf{k}$-algebra with residue field $\mathbf{k}$. Our convention is to denote the (pre-)differential always by $\bar{\partial}$. It is usually not obtained from any (hypothetical) operator ∂ via conjugation.

Λ-Linear Curved Lie Algebras

- For every $\mathbf{i} \geq \mathbf{0}$, the object L^i is a **flat** Λ-module which is **complete** with respect to the $\mathfrak{m}_\Lambda$-adic topology; if $\theta \in L^i$, then its **total degree** is $|\theta| = i - 1$;
- Λ-bilinear **Lie bracket**

$$[-, -]: L^i \times L^j \to L^{i+j}$$

such that

$$[\theta, \xi] = -(-1)^{(|\theta|+1)(|\xi|+1)}[\xi, \theta];$$

- **Jacobi identity**

$$[\theta, [\xi, \eta]] = [[\theta, \xi], \eta] + (-1)^{(|\theta|+1)(|\xi|+1)}[\xi, [\theta, \eta]];$$

- Λ-linear **predifferential** $\bar{\partial}: L^i \to L^{i+1}$ with

$$\bar{\partial}[\theta, \xi] = [\bar{\partial}\theta, \xi] + (-1)^{|\theta|+1}[\theta, \bar{\partial}\xi];$$

- $\ell \in \mathfrak{m}_\Lambda \cdot L^2$ with $\bar{\partial}^2(\theta) = [\ell, \theta]$ and $\bar{\partial}(\ell) = 0$.

Λ-Linear Curved Lie–Rinehart Algebras

- For every $\mathbf{i} \geq \mathbf{0}$, the objects F^i and T^i are **flat** Λ-modules which are **complete** with respect to the $\mathfrak{m}_\Lambda$-adic topology; if $a \in F^i$, then its **total degree** is $|a| = i$; if $\theta \in T^i$, then its **total degree** is $|\theta| = i - 1$;
- Λ-bilinear **product** $-\wedge-\colon F^i \times F^j \to F^{i+j}$ and a **unit element** $1 \in F^0$ with

$$a \wedge (b \wedge c) = (a \wedge b) \wedge c, \quad a \wedge b = (-1)^{|a||b|} b \wedge a, \quad 1 \wedge a = a;$$

- Λ-bilinear **multiplication** $*\colon F^i \times T^j \to T^{i+j}$ with

$$(a \wedge b) * \theta = a * (b * \theta), \quad 1 * \theta = \theta;$$

- Λ-bilinear **derivative** $\nabla^F \colon T^i \times F^j \to F^{i+j}$ with

$$a * \nabla^F_\theta(b) = \nabla^F_{a*\theta}(b);$$

- Λ-bilinear **Lie bracket** $[-,-] = \nabla^T \colon T^i \times T^j \to T^{i+j}$ with

$$[\theta, \xi] = -(-1)^{(|\theta|+1)(|\xi|+1)}[\xi, \theta];$$

- For $P = F, T$, the two **Jacobi identities**

$$\nabla^P_{[\theta,\xi]}(p) = \nabla^P_\theta \nabla^P_\xi(p) - (-1)^{(|\theta|+1)(|\xi|+1)} \nabla^P_\xi \nabla^P_\theta(p);$$

- for $P = F, T$, the two **odd Poisson identities** here, we have $*^F = \wedge$ and $*^T = *$)

$$\nabla^P_\theta(a *^P p) = \nabla_\theta(a) *^P p + (-1)^{(|\theta|+1)|a|} a *^P \nabla^P_\theta(p);$$

- Two Λ-linear **predifferentials** $\bar{\partial}\colon F^i \to F^{i+1}$ and $\bar{\partial}\colon T^i \to T^{i+1}$ satisfying the four **derivation rules**

$$\bar{\partial}\nabla^P_\theta(p) = \nabla^P_{\bar{\partial}\theta}(p) + (-1)^{|\theta|+1}\nabla^P_\theta(\bar{\partial}p)$$

and

$$\bar{\partial}(a *^P p) = \bar{\partial}(a) *^P p + (-1)^{|a|} a *^P \bar{\partial}p;$$

- $\ell \in \mathfrak{m}_\Lambda$ with

$$\bar{\partial}^2(a) = \nabla^F_\ell(a) \quad \text{and} \quad \bar{\partial}^2(\theta) = \nabla^T_\ell(\theta) = [\ell, \theta]$$

as well as $\bar{\partial}(\ell) = 0$.

Λ-Linear Lie–Rinehart Pairs

A Λ-linear Lie–Rinehart algebra plus:

- For every $\mathbf{i} \geq \mathbf{0}$, the object E^i is a **flat** Λ-module which is **complete** with respect to the $\mathfrak{m}_\Lambda$-adic topology; if $e \in E^i$, then its **total degree** is $|e| = i$;
- Λ-bilinear **multiplication** $*^E : F^i \times E^j \to E^{i+j}$ with

$$(a \wedge b) *^E e = a *^E (b *^E e)$$

and $1 *^E e = e$;
- Λ-bilinear **derivative** $\nabla^E : T^i \times E^j \to E^{i+j}$ with

$$a *^E \nabla_\theta^E(e) = \nabla_{a*\theta}^E(e);$$

- **Jacobi identity**

$$\nabla_{[\theta,\xi]}^E(e) = \nabla_\theta^E \nabla_\xi^E(e) - (-1)^{(|\theta|+1)(|\xi|+1)} \nabla_\xi^E \nabla_\theta^E(e);$$

- **Derivation rule**

$$\nabla_\theta^E(a *^E e) = \nabla_\theta^F(a) *^E e + (-1)^{(|\theta|+1)|a|} a *^E \nabla_\theta^E(e);$$

- Λ-linear **predifferential** $\bar\partial : E^i \to E^{i+1}$ satisfying the **derivation rules**

$$\bar\partial \nabla_\theta^E(e) = \nabla_{\bar\partial\theta}^T(e) + (-1)^{|\theta|+1} \nabla_\theta^E(\bar\partial e)$$

and

$$\bar\partial(a *^E e) = \bar\partial(a) *^E e + (-1)^{|a|} a *^E \bar\partial(e)$$

as well as $\bar\partial^2(e) = \nabla_\ell^E(e)$.

Λ-Linear Curved Gerstenhaber Algebras and Calculi

Λ-linear Curved Gerstenhaber *Algebras*

- For every $-\mathbf{d} \leq \mathbf{p} \leq 0$ and $\mathbf{q} \geq \mathbf{0}$, the object $G^{p,q}$ is a **flat** Λ-module which is **complete** with respect to the $\mathfrak{m}_\Lambda$-adic topology; if $\theta \in G^{p,q}$, then its **total degree** is $|\theta| = p + q$;
- Λ-bilinear $\wedge$-**product** $-\wedge-: G^{p,q} \times G^{p',q'} \to G^{p+p',q+q'}$ and $\mathbf{1} \in G^{0,0}$ such that

$$\theta \wedge (\xi \wedge \eta) = (\theta \wedge \xi) \wedge \eta; \qquad \theta \wedge \xi = (-1)^{|\theta||\xi|} \xi \wedge \theta; \qquad 1 \wedge \theta = \theta;$$

  unadorned powers $\theta^n := \theta \wedge \ldots \wedge \theta$;
- Λ-bilinear **Lie bracket** $[-, -]: G^{p,q} \times G^{p',q'} \to G^{p+p'+1,q+q'}$ such that

$$[\theta, \xi] = -(-1)^{(|\theta|+1)(|\xi|+1)}[\xi, \theta]; \qquad [\theta, 1] = 0;$$

Jacobi identity

$$[\theta, [\xi, \eta]] = [[\theta, \xi], \eta] + (-1)^{(|\theta|+1)(|\xi|+1)}[\xi, [\theta, \eta]];$$

in particular, for φ with $|\varphi|$ even: $\quad [\varphi, [\varphi, \xi]] = [\tfrac{1}{2}[\varphi, \varphi], \xi];$
- **Odd Poisson identities**

$$[\theta, \xi \wedge \eta] = [\theta, \xi] \wedge \eta + (-1)^{(|\theta|+1)|\xi|} \xi \wedge [\theta, \eta],$$

$$[\theta \wedge \xi, \eta] = \theta \wedge [\xi, \eta] + (-1)^{(|\eta|+1)|\xi|}[\theta, \eta] \wedge \xi;$$

in particular, for φ with $|\varphi|$ even, we have $[\varphi^n, \theta] = n[\varphi, \theta] \wedge \varphi^{n-1}$ for $n \geq 1$;
- Λ-linear **predifferential** $\bar{\partial}: G^{p,q} \to G^{p,q+1}$ with $\quad \bar{\partial}(1) = 0 \quad$ and **derivation rules**

$$\bar{\partial}[\theta, \xi] = [\bar{\partial}\theta, \xi] + (-1)^{|\theta|+1}[\theta, \bar{\partial}\xi] \quad \text{and} \quad \bar{\partial}(\theta \wedge \xi) = \bar{\partial}\theta \wedge \xi + (-1)^{|\theta|}\theta \wedge \bar{\partial}\xi;$$

in particular, $\bar{\partial}(\varphi^n) = n \cdot \bar{\partial}\varphi \wedge \varphi^{n-1}$ for $|\varphi|$ even;
- $\ell \in \mathfrak{m}_\Lambda \cdot G^{-1,2} \quad$ with $\quad \bar{\partial}^2(\theta) = [\ell, \theta]; \qquad \bar{\partial}(\ell) = 0; \qquad \ell \wedge \ell = 0.$

Λ-linear Curved *One-sided* Gerstenhaber Calculi

A Λ-linear curved Gerstenhaber algebra plus:

- For every $\mathbf{0} \leq \mathbf{i} \leq \mathbf{d}$ and $\mathbf{j} \geq \mathbf{0}$, the object $A^{i,j}$ is a **flat** Λ-module which is **complete** with respect to the $\mathfrak{m}_\Lambda$-adic topology; if $\alpha \in A^{i,j}$, then its **total degree** is $|\alpha| = i + j$;

- Λ-bilinear $\wedge$-**product** $-\wedge-:\ A^{i,j} \times A^{i',j'} \to A^{i+i',j+j'}$ and $\mathbf{1} \in A^{0,0}$ such that

$$(\alpha \wedge \beta) \wedge \gamma = \alpha \wedge (\beta \wedge \gamma); \quad \alpha \wedge \beta = (-1)^{|\alpha||\beta|}\beta \wedge \alpha; \quad 1 \wedge \alpha = 1;$$

- Λ-linear **de Rham differential** $\partial:\ A^{i,j} \to A^{i+1,j}$ with $\partial^2 = 0, \quad \partial(1) = 0$, and the **derivation rule**

$$\partial(\alpha \wedge \beta) = \partial(\alpha) \wedge \beta + (-1)^{|\alpha|}\alpha \wedge \partial(\beta);$$

- Λ-bilinear **contraction map** $\lrcorner:\ G^{p,q} \times A^{i,j} \to A^{p+i,q+j}$ such that

$$1 \lrcorner \alpha = \alpha \quad \text{and} \quad (\theta \wedge \xi) \lrcorner \alpha = \theta \lrcorner (\xi \lrcorner \alpha);$$

- Λ-bilinear **Lie derivative** $\mathcal{L}_-(-):\ G^{p,q} \times A^{i,j} \to A^{p+i+1,q+j}$ such that

$$\mathcal{L}_{[\theta,\xi]}(\alpha) = \mathcal{L}_\theta(\mathcal{L}_\xi(\alpha)) - (-1)^{(|\theta|+1)(|\xi|+1)}\mathcal{L}_\xi(\mathcal{L}_\theta(\alpha)) \quad \text{and} \quad \mathcal{L}_1(\alpha) = 0;$$

- **Mixed Leibniz rule**

$$\theta \lrcorner \mathcal{L}_\xi(\alpha) = (-1)^{|\xi|+1}([\theta,\xi] \lrcorner \alpha) + (-1)^{|\theta|(|\xi|+1)}\mathcal{L}_\xi(\theta \lrcorner \alpha)$$

$$\Leftrightarrow \quad [\theta,\xi] \lrcorner \alpha = (-1)^{|\xi|+1}\theta \lrcorner \mathcal{L}_\xi(\alpha) - (-1)^{(|\theta|+1)(|\xi|+1)}\mathcal{L}_\xi(\theta \lrcorner \alpha)$$

$$\Leftrightarrow \quad \mathcal{L}_\theta(\xi \lrcorner \alpha) = (-1)^{(|\theta|+1)|\xi|}\xi \lrcorner \mathcal{L}_\theta(\alpha) + [\theta,\xi] \lrcorner \alpha;$$

- **Lie–Rinehart homotopy formula**

$$(-1)^{|\theta|}\mathcal{L}_\theta(\alpha) = \partial(\theta \lrcorner \alpha) - (-1)^{|\theta|}(\theta \lrcorner \partial\alpha);$$

consequently,

$$\mathcal{L}_{\theta \wedge \xi}(\alpha) = (-1)^{|\xi|}\mathcal{L}_\theta(\xi \lrcorner \alpha) + \theta \lrcorner \mathcal{L}_\xi(\alpha);$$

- For φ with $|\varphi|$ even, $\mathcal{L}_{\varphi^n}(\alpha) = n\varphi^{n-1} \lrcorner \mathcal{L}_\varphi(\alpha) + \frac{n(n-1)}{2}([\varphi,\varphi] \wedge \varphi^{n-2}) \lrcorner \alpha$ for $n \geq 2$;
- For $\theta \in G^{0,q}$, the identity $\theta \lrcorner \alpha = (\theta \lrcorner 1) \wedge \alpha$ holds;
- For $\theta \in G^{-1,q}$, the identity

$$\theta \lrcorner (\alpha \wedge \beta) = (\theta \lrcorner \alpha) \wedge \beta + (-1)^{|\alpha||\theta|}\alpha \wedge (\theta \lrcorner \beta)$$

holds; consequently, we have for $\theta \in G^{-1,q}$

$$\mathcal{L}_\theta(\alpha \wedge \beta) = \mathcal{L}_\theta(\alpha) \wedge \beta + (-1)^{|\alpha|(|\theta|+1)}\alpha \wedge \mathcal{L}_\theta(\beta);$$

- Λ-linear **predifferential** $\bar{\partial}\colon A^{i,j} \to A^{i,j+1}$ with $\bar{\partial}^2(\alpha) = \mathcal{L}_\ell(\alpha)$ and $\partial\bar{\partial} + \bar{\partial}\partial = 0$;
- **Derivation rules**

$$\bar{\partial}(\alpha \wedge \beta) = \bar{\partial}\alpha \wedge \beta + (-1)^{|\alpha|}\alpha \wedge \bar{\partial}\beta \quad \text{and} \quad \bar{\partial}(\theta \rightharpoonup \alpha) = (\bar{\partial}\theta) \rightharpoonup \alpha + (-1)^{|\theta|}\theta \rightharpoonup \bar{\partial}\alpha;$$

- Λ-linear isomorphism $\lambda\colon G^{0,q} \to A^{0,q}$, $\theta \mapsto \theta \rightharpoonup 1$, with $\lambda(\theta \wedge \xi) = \lambda(\theta) \wedge \lambda(\xi)$.

Λ-linear Curved *Two-sided* Gerstenhaber Calculi

A Λ-linear curved one-sided Gerstenhaber **calculus** (next page) plus:

- Λ-bilinear **left contraction** $\vdash\colon G^{p,q} \times A^{i,j} \to G^{p+i,q+j}$ satisfying $\theta \vdash 1 = \theta$ and
$$\theta \vdash (\alpha \wedge \beta) = (\theta \vdash \alpha) \vdash \beta;$$
- For $\alpha \in A^{0,j}$, the identity $\theta \vdash \alpha = \theta \wedge (1 \vdash \alpha)$;
- For $\alpha \in A^{1,j}$, the identity

$$(\theta \wedge \xi) \vdash \alpha = (-1)^{|\xi||\alpha|}(\theta \vdash \alpha) \wedge \xi + \theta \wedge (\xi \vdash \alpha);$$

- For $\theta \in G^{-i,q}$ and $\alpha \in A^{i,j}$, the identity $\lambda(\theta \vdash \alpha) = (-1)^i\, \theta \rightharpoonup \alpha$;
- For $\omega \in A^{d,j}$, the identity $(\theta \vdash \alpha) \rightharpoonup \omega = (-1)^{|\alpha||\omega|}(\theta \rightharpoonup \omega) \wedge \alpha$;
- For $\omega^\vee \in G^{-d,q}$, the identity $\omega^\vee \vdash (\theta \rightharpoonup \alpha) = (-1)^{|\omega^\vee||\theta|}\, \theta \wedge (\omega^\vee \vdash \alpha)$;
- For $\theta \in G^{-1,q}$, the **special left mixed Leibniz rule**

$$[\theta, \xi \vdash \alpha] = [\theta, \xi] \vdash \alpha + (-1)^{(|\theta|+1)|\xi|}\, \xi \vdash \mathcal{L}_\theta(\alpha);$$

- Derivation rule $\bar{\partial}(\theta \vdash \alpha) = \bar{\partial}\theta \vdash \alpha + (-1)^{|\theta|}\, \theta \vdash \bar{\partial}\alpha.$

Λ-Linear Curved Batalin–Vilkovisky Algebras and Calculi

Λ-linear Curved Batalin–Vilkovisky *Algebras*

- For every $-\mathbf{d} \leq \mathbf{p} \leq \mathbf{0}$ and $\mathbf{q} \geq \mathbf{0}$, the object $G^{p,q}$ is a **flat** Λ-module which is **complete** with respect to the $\mathfrak{m}_\Lambda$-adic topology; if $\theta \in G^{p,q}$, then its **total degree** is $|\theta| = p + q$;
- Λ-bilinear $\wedge$-**product** $-\wedge-: G^{p,q} \times G^{p',q'} \to G^{p+p',q+q'}$ and $\mathbf{1} \in G^{0,0}$ s.t.

$$\theta \wedge (\xi \wedge \eta) = (\theta \wedge \xi) \wedge \eta; \qquad \theta \wedge \xi = (-1)^{|\theta||\xi|}\xi \wedge \theta; \qquad 1 \wedge \theta = \theta;$$

unadorned powers $\theta^n := \theta \wedge \ldots \wedge \theta$;
- Λ-bilinear **Lie bracket** $[-,-]: G^{p,q} \times G^{p',q'} \to G^{p+p'+1,q+q'}$ such that

$$[\theta,\xi] = -(-1)^{(|\theta|+1)(|\xi|+1)}[\xi,\theta]; \qquad [\theta,1] = 0;$$

Jacobi identity

$$[\theta,[\xi,\eta]] = [[\theta,\xi],\eta] + (-1)^{(|\theta|+1)(|\xi|+1)}[\xi,[\theta,\eta]];$$

in particular, for φ with $|\varphi|$ even: $\quad [\varphi,[\varphi,\xi]] = [\tfrac{1}{2}[\varphi,\varphi],\xi]$;
- **odd Poisson identities**

$$[\theta,\xi \wedge \eta] = [\theta,\xi] \wedge \eta + (-1)^{(|\theta|+1)|\xi|}\xi \wedge [\theta,\eta] \quad \text{and}$$

$$[\theta \wedge \xi,\eta] = \theta \wedge [\xi,\eta] + (-1)^{(|\eta|+1)|\xi|}[\theta,\eta] \wedge \xi;$$

in particular, for φ with $|\varphi|$ even, we have $[\varphi^n,\theta] = n[\varphi,\theta] \wedge \varphi^{n-1}$ for $n \geq 1$;
- Λ-linear **predifferential** $\bar{\partial}: G^{p,q} \to G^{p,q+1}$ w/ $\bar{\partial}(1) = 0$ and **derivation rules**

$$\bar{\partial}[\theta,\xi] = [\bar{\partial}\theta,\xi] + (-1)^{|\theta|+1}[\theta,\bar{\partial}\xi] \quad \text{and} \quad \bar{\partial}(\theta \wedge \xi) = \bar{\partial}\theta \wedge \xi + (-1)^{|\theta|}\theta \wedge \bar{\partial}\xi;$$

in particular, $\bar{\partial}(\varphi^n) = n \cdot \bar{\partial}\varphi \wedge \varphi^{n-1}$ for $|\varphi|$ even;
- $\ell \in \mathfrak{m}_\Lambda \cdot G^{-1,2}$ with $\bar{\partial}^2(\theta) = [\ell,\theta]$; $\quad \bar{\partial}(\ell) = 0$; $\quad \ell \wedge \ell = 0$;
- Λ-linear **Batalin–Vilkovisky operator** $\Delta: G^{p,q} \to G^{p+1,q}$ with

$$\Delta(1) = 0; \qquad \Delta^2 = 0; \qquad \Delta[\theta,\xi] = [\Delta(\theta),\xi] + (-1)^{|\theta|+1}[\theta,\Delta(\xi)],$$

and the **Bogomolov–Tian–Todorov formula**

$$(-1)^{|\theta|}[\theta,\xi] = \Delta(\theta \wedge \xi) - \Delta(\theta) \wedge \xi - (-1)^{|\theta|}\theta \wedge \Delta(\xi);$$

for φ with $|\varphi|$ even, we have $\Delta(\varphi^n) = n\Delta(\varphi) \wedge \varphi^{n-1} + \frac{n(n-1)}{2}[\varphi,\varphi] \wedge \varphi^{n-2}$ for $n \geq 2$;
- $y \in \mathfrak{m}_\Lambda \cdot G^{0,1}$ with $\bar{\partial}\Delta(\theta) + \Delta(\bar{\partial}\theta) = [y,\theta]$ and $\bar{\partial}y + \Delta\ell = 0$ and

$$y \wedge y = 0; \quad y \wedge \ell + \ell \wedge y = 0; \quad [\ell,y] + [y,\ell] = 0.$$

Λ-linear Curved *One-sided* Batalin–Vilkovisky Calculi

A Λ-linear curved Batalin–Vilkovisky algebra plus:

- For every $0 \leq i \leq d$ and $j \geq 0$, the object $A^{i,j}$ is a **flat** Λ-module which is **complete** with respect to the $\mathfrak{m}_\Lambda$-adic topology; if $\alpha \in A^{i,j}$, then its **total degree** is $|\alpha| = i + j$;
- Λ-bilinear $\wedge$-**product** $-\wedge-\colon A^{i,j} \times A^{i',j'} \to A^{i+i',j+j'}$ and $1 \in A^{0,0}$ such that

$$(\alpha \wedge \beta) \wedge \gamma = \alpha \wedge (\beta \wedge \gamma); \quad \alpha \wedge \beta = (-1)^{|\alpha||\beta|}\beta \wedge \alpha; \quad 1 \wedge \alpha = 1;$$

- Λ-linear **de Rham differential** $\partial\colon A^{i,j} \to A^{i+1,j}$ with $\partial^2 = 0$, $\partial(1) = 0$, and the **derivation rule**

$$\partial(\alpha \wedge \beta) = \partial(\alpha) \wedge \beta + (-1)^{|\alpha|}\alpha \wedge \partial(\beta);$$

- Λ-bilinear **contraction map** $\lrcorner\colon G^{p,q} \times A^{i,j} \to A^{p+i,q+j}$ such that

$$1 \lrcorner \alpha = \alpha \quad \text{and} \quad (\theta \wedge \xi) \lrcorner \alpha = \theta \lrcorner (\xi \lrcorner \alpha);$$

- Λ-bilinear **Lie derivative** $\mathcal{L}_-(-)\colon G^{p,q} \times A^{i,j} \to A^{p+i+1,q+j}$ such that

$$\mathcal{L}_{[\theta,\xi]}(\alpha) = \mathcal{L}_\theta(\mathcal{L}_\xi(\alpha)) - (-1)^{(|\theta|+1)(|\xi|+1)}\mathcal{L}_\xi(\mathcal{L}_\theta(\alpha)) \quad \text{and} \quad \mathcal{L}_1(\alpha) = 0;$$

- **Mixed Leibniz rule**

$$\theta \lrcorner \mathcal{L}_\xi(\alpha) = (-1)^{|\xi|+1}([\theta,\xi] \lrcorner \alpha) + (-1)^{|\theta|(|\xi|+1)}\mathcal{L}_\xi(\theta \lrcorner \alpha)$$

- **Lie–Rinehart homotopy formula**

$$(-1)^{|\theta|}\mathcal{L}_\theta(\alpha) = \partial(\theta \lrcorner \alpha) - (-1)^{|\theta|}(\theta \lrcorner \partial\alpha);$$

consequently,

$$\mathcal{L}_{\theta \wedge \xi}(\alpha) = (-1)^{|\xi|}\mathcal{L}_\theta(\xi \lrcorner \alpha) + \theta \lrcorner \mathcal{L}_\xi(\alpha);$$

- For φ with $|\varphi|$ even, $\mathcal{L}_{\varphi^n}(\alpha) = n\varphi^{n-1} \lrcorner \mathcal{L}_\varphi(\alpha) + \frac{n(n-1)}{2}([\varphi,\varphi] \wedge \varphi^{n-2}) \lrcorner \alpha$ for $n \geq 2$;
- For $\theta \in G^{0,q}$, the identity $\theta \lrcorner \alpha = (\theta \lrcorner 1) \wedge \alpha$ holds;
- For $\theta \in G^{-1,q}$, the identity

$$\theta \lrcorner (\alpha \wedge \beta) = (\theta \lrcorner \alpha) \wedge \beta + (-1)^{|\alpha||\theta|}\alpha \wedge (\theta \lrcorner \beta)$$

holds; consequently, we have for $\theta \in G^{-1,q}$

$$\mathcal{L}_\theta(\alpha \wedge \beta) = \mathcal{L}_\theta(\alpha) \wedge \beta + (-1)^{|\alpha|(|\theta|+1)}\alpha \wedge \mathcal{L}_\theta(\beta);$$

- Λ-linear **predifferential** $\bar{\partial}\colon A^{i,j} \to A^{i,j+1}$ with $\bar{\partial}^2(\alpha) = \mathcal{L}_\ell(\alpha)$ and $\partial\bar{\partial} + \bar{\partial}\partial = 0$;
- **Derivation rules**

$$\bar{\partial}(\alpha \wedge \beta) = \bar{\partial}\alpha \wedge \beta + (-1)^{|\alpha|}\alpha \wedge \bar{\partial}\beta \quad \text{and} \quad \bar{\partial}(\theta \lrcorner \alpha) = (\bar{\partial}\theta) \lrcorner \alpha + (-1)^{|\theta|}\theta \lrcorner \bar{\partial}\alpha;$$

- **Volume element** $\omega \in A^{d,0}$ inducing isomorphisms of Λ-modules

$$\kappa\colon G^{p,q} \to A^{p+d,q}, \quad \theta \mapsto \theta \lrcorner \omega, \quad \text{and} \quad \upsilon\colon A^{0,q} \to A^{d,q}, \quad \alpha \mapsto \alpha \wedge \omega;$$

- $\Delta(\theta) \lrcorner \omega = \partial(\theta \lrcorner \omega)$ and $y \lrcorner \omega = \bar{\partial}\omega$;
- Λ-linear isomorphism $\lambda\colon G^{0,q} \to A^{0,q}$, $\theta \mapsto \theta \lrcorner 1$, with $\lambda(\theta \wedge \xi) = \lambda(\theta) \wedge \lambda(\xi)$.

A Λ-linear curved **two-sided BV calculus** is a Λ-bilinear curved one-sided BV calculus which is at the same time a Λ-linear curved two-sided Gerstenhaber calculus.

List of Figures

Chapter 1
Introduction

We give a self-contained introduction into logarithmic deformation theory. We explain how the construction of Calabi–Yau manifolds and in particular the toric Gross–Siebert mirror construction motivate logarithmic deformation theory. We give an overview over the basic concepts and main results discussed in this monograph and outline some of the arguments. In particular, we illustrate the concept of singularities in logarithmic geometry with many examples, and we explain how the logarithmic version of the Bogomolov–Tian–Todorov theorem allows us to smooth reducible schemes. At the end of this chapter, we give an outlook on questions that remain unaddressed.

1.1   A Plan to Construct Calabi–Yau Manifolds

In this section, we present our central motivation for logarithmic deformation theory, namely constructing smooth Calabi–Yau varieties by deforming degenerate ones.

1.1.1   Calabi–Yau Varieties

Let $Y/\mathbb{C}$ be a normal algebraic variety of dimension d, and let $j\colon U \subseteq Y$ be the smooth locus. We write $\omega_Y := j_* \Omega_U^d$ for the canonical sheaf.

A Noetherian local ring R with residue field $\mathbf{k}$ and of dimension d is *Gorenstein* if and only if $\mathrm{Ext}_R^d(\mathbf{k}, R) \cong \mathbf{k}$ and $\mathrm{Ext}_R^i(\mathbf{k}, R) = 0$ for $i \neq d$, and a locally Noetherian scheme Y is *Gorenstein* if each local ring $O_{Y,y}$ is Gorenstein. For example, every locally Noetherian scheme with local complete intersection singularities is Gorenstein. When the normal variety Y is Gorenstein, then ω_Y is a line bundle.

S. Felten, *Global Logarithmic Deformation Theory*, Lecture Notes
in Mathematics 2373, https://doi.org/10.1007/978-3-031-98751-9_1

One often classifies algebraic varieties according to properties of ω_Y, and the following is the case of our primary interest in this monograph.

Definition 1.1 (Calabi–Yau Varieties) A *Calabi–Yau variety* is a connected proper normal Gorenstein algebraic variety $Y/\mathbb{C}$ whose canonical bundle ω_Y is isomorphic to the structure sheaf O_Y. A *volume form* is a global section $\omega \in H^0(Y, \omega_Y)$ such that the induced map $O_Y \to \omega_Y$, $1 \mapsto \omega$, is an isomorphism. A Calabi–Yau variety Y is *strict* if

$$H^i(Y, O_Y) = 0$$

for $1 \leq i \leq d - 1$, where $d = \dim(Y)$. An *algebraic Calabi–Yau manifold* is a smooth Calabi–Yau variety.

Many authors require a Calabi–Yau variety to be strict. However, in the sense of our definition, for example Abelian varieties are Calabi–Yau varieties.

Example 1.2 The Fermat quartic surface

$$K = \{X_0^4 + X_1^4 + X_2^4 + X_3^4 = 0\} \subset \mathbb{P}^3$$

is a strict algebraic Calabi–Yau manifold. $\Diamond$

Example 1.3 The quartic surface

$$K' = \{X_0^4 + X_1^4 + X_2^4 + X_3^4 = (X_0 + X_1 + X_2 + X_3)^4\} \subset \mathbb{P}^3$$

is a strict Calabi–Yau variety with six A_3-singularities, i.e., with six singularities which are étale locally isomorphic to $xy = z^4$. $\Diamond$

Example 1.4 The Fermat quintic threefold

$$Q = \{X_0^5 + X_1^5 + X_2^5 + X_3^5 + X_4^5 = 0\} \subset \mathbb{P}^4$$

is a strict algebraic Calabi–Yau manifold. $\Diamond$

1.1.2 Another Perspective on the Canonical Sheaf

For a normal algebraic variety $Y/\mathbb{C}$, we have defined above the canonical sheaf as $\omega_Y = j_* \Omega_U^d$. This sheaf can also be understood in terms of the *dualizing complex*, which gives us an alternative perspective which generalizes to certain non-normal schemes.

For a general locally Noetherian scheme Y, a *dualizing complex* is an object $\omega_Y^\bullet \in \mathbf{D}_{qc}(O_Y)$ which has finite injective dimension, such that each $\mathcal{H}^i(\omega_Y^\bullet)$ is coherent, and such that $O_Y \to \mathcal{R}\mathcal{H}om(\omega_Y^\bullet, \omega_Y^\bullet)$ is a quasi-isomorphism. Such an $\omega_Y^\bullet$ does not exist in general, and if it exists, it is not unique. Nonetheless, it exists very often; for example, if Y is Gorenstein, then O_Y is a dualizing complex.

Let $f: X \to S$ be a separated morphism of finite type between Noetherian schemes. Then we have a functor

$$f^!: \ \mathbf{D}^+_{\mathrm{qc}}(O_S) \to \mathbf{D}^+_{\mathrm{qc}}(O_X),$$

which is right adjoint to Rf_* when f is proper, and which is given by f^* when f is an open immersion. When $\omega_S^\bullet$ is a dualizing complex for S, then $f^!\omega_S^\bullet$ is a dualizing complex for X. This dualizing complex is called *normalized* relative to $(S, \omega_S^\bullet)$. Unlike a general dualizing complex, a dualizing complex normalized relative to $(S, \omega_S^\bullet)$ is unique up to canonical isomorphism.

Definition 1.5 (Dualizing Complex) Let $f: Y \to \operatorname{Spec}\mathbb{C}$ be separated and of finite type, and endow $\operatorname{Spec}\mathbb{C}$ with the dualizing complex $O_{\operatorname{Spec}\mathbb{C}}$. Then *the dualizing complex* of $Y/\mathbb{C}$ is $\omega_Y^\bullet := f^!O_{\operatorname{Spec}\mathbb{C}}$.

A Noetherian local ring R is *Cohen–Macaulay* if $\operatorname{depth}(R) = \dim(R)$, and a locally Noetherian scheme Y is *Cohen–Macaulay* if $O_{Y,y}$ is Cohen–Macaulay for every point $y \in Y$. When $f: Y \to \operatorname{Spec}\mathbb{C}$ is separated and of finite type, and when Y is Cohen–Macaulay and equidimensional of dimension d, then $\omega_Y^\bullet$ is concentrated in degree d.

Definition 1.6 (Dualizing Sheaf) Let $Y/\mathbb{C}$ be separated, of finite type, Cohen–Macaulay, and equidimensional of dimension d. Then the *dualizing sheaf* is $\omega_Y^\circ := \mathcal{H}^{-d}(\omega_Y^\bullet)$.

In this situation, we have an isomorphism $\omega_Y^\bullet \cong \omega_Y^\circ[d]$ in $\mathbf{D}_{\mathrm{qc}}(O_Y)$.

Example 1.7 Let $Y/\mathbb{C}$ be proper, Cohen–Macaulay, and equidimensional of dimension d. By Authors [267, 0FVZ], we have functorial isomorphisms $\operatorname{Ext}^{d-i}_Y(\mathcal{F}, \omega_Y^\circ) \cong \operatorname{Hom}_\mathbb{C}(H^i(Y, \mathcal{F}), \mathbb{C})$ for every quasi-coherent sheaf $\mathcal{F}$. In particular, ω_Y° is a dualizing sheaf in the sense of [138, III, §7]. $\Diamond$

Let $Y/\mathbb{C}$ be separated, of finite type, Cohen–Macaulay, equidimensional of dimension d, and normal. In this case, we have an isomorphism $\omega_Y^\circ \cong \omega_Y$ with the canonical sheaf $\omega_Y = j_*\Omega_U^d$, giving another perspective on ω_Y. Now Y is Gorenstein if and only if ω_Y° is a line bundle.

1.1.3 Constructing Calabi–Yau Varieties

In general, it is not so easy to construct Calabi–Yau varieties. Among the hypersurfaces of degree n in $\mathbb{P}^{d+1}$, only $n = d + 2$ yields a Calabi–Yau variety. Some additional Calabi–Yau varieties can be constructed as complete intersections in $\mathbb{P}^m$ or as complete intersections in products of projective spaces. Many more constructions of Calabi–Yau varieties as subvarieties of some ambient space are nicely summarized by He in [142].

We may extend the definition of a Calabi–Yau variety to certain non-normal schemes. For the purpose of our discussion, we first fix the following notion.

Definition 1.8 (Degenerate Schemes) A *degenerate scheme* is a separated scheme $V/\mathbb{C}$ of finite type which is reduced, connected, Cohen–Macaulay, and equidimensional of dimension d.

While the canonical sheaf $\omega_Y = j_*\Omega^d_U$ is defined only for normal varieties $Y/\mathbb{C}$, every degenerate scheme $V/\mathbb{C}$ admits a dualizing sheaf ω°_V in the sense of Definition 1.6, and ω°_V is nothing but ω_V if V is normal. Thus, we obtain the following natural generalization of Calabi–Yau varieties.

Definition 1.9 (Calabi–Yau Schemes) A *Calabi–Yau scheme* is a proper Gorenstein scheme $V/\mathbb{C}$, equidimensional of dimension d, whose dualizing sheaf ω°_V is isomorphic to O_V.

Unlike Calabi–Yau varieties, Calabi–Yau schemes can be constructed by gluing simpler schemes along closed subschemes.

Example 1.10 Consider the (index) category $\mathcal{I}$ whose objects are subsets $S \subseteq \{0, 1, 2, 3\}$ with $1 \le |S| \le 3$, and whose morphisms are given by inclusions of subsets. For one-element sets $S = \{i\}$, consider $F(S) = H_i = \operatorname{Proj} \mathbb{C}[S \setminus \{i\}] \cong \mathbb{P}^2$. For two-element subsets $S = \{i, j\}$, consider $F(S) = L_{ij} = \operatorname{Proj} \mathbb{C}[S \setminus \{i, j\}] \cong \mathbb{P}^1$. For three-element subsets $S = \{i, j, k\}$, consider $F(S) = T_{ijk} = \operatorname{Proj} \mathbb{C}[S \setminus \{i, j, k\}] \cong \mathbb{P}^0$. For an inclusion $S \subseteq S'$, the quotient map of polynomial rings induces a closed immersion $F(S') \subset F(S)$ in the opposite direction. This defines a functor $F \colon \mathcal{I}^{\mathrm{op}} \to \mathbf{Sch}$. The colimit $H = \operatorname{colim}(F)$ exists in the category of schemes. Explicitly, and far more elementarily in this example, we can construct H as $\{X_0 X_1 X_2 X_3 = 0\} \subset \mathbb{P}^3$. Here, we recover each H_i as an irreducible component $H_i = \{X_i = 0\}$, each L_{ij} as the intersection of two irreducible components, and each T_{ijk} as the intersection of three irreducible components. The construction as a colimit rather than a hypersurface shows that we do not need the ambient space in the first place to construct H.

See Fig. 1.1 for an illustration of the functor $F \colon \mathcal{I}^{\mathrm{op}} \to \mathbf{Sch}$ and its colimit H.

To see that $H = \bigcup_{i=0}^3 H_i$ is a Calabi–Yau scheme, recall that the dualizing sheaf of a hypersurface $i \colon H \subset P = \mathbb{P}^3$ can be computed as $i_*\omega^\circ_H \cong \mathcal{E}xt^1(i_*O_H, \omega_P)$. Let $F = X_0 X_1 X_2 X_3 \in O_P(4)$. Then the short exact sequence

$$0 \to O_P(-4) \xrightarrow{F^\vee} O_P \to i_*O_H \to 0$$

yields a short exact sequence

$$0 \to \omega_P \otimes O_X \xrightarrow{\mathrm{id} \otimes F} \omega_P \otimes O_P(4) \to i_*\omega^\circ_H \to 0.$$

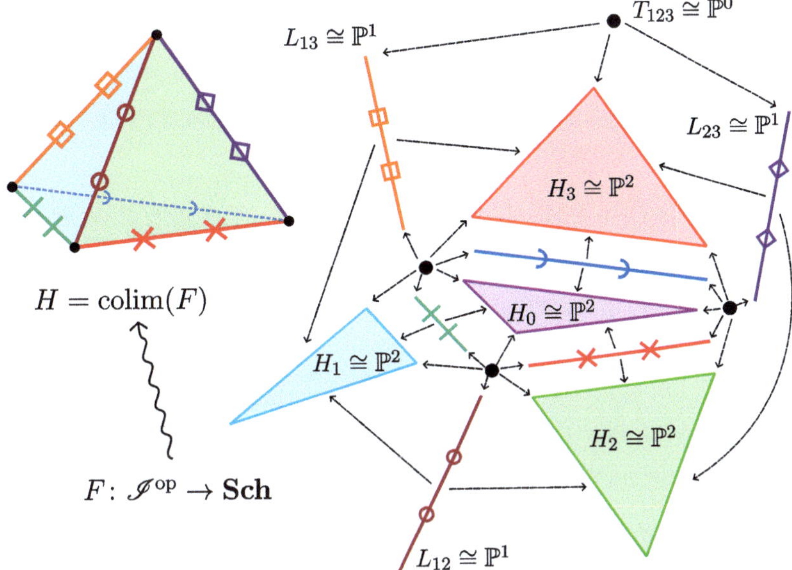

Fig. 1.1 Example 1.10. The degenerate scheme H is obtained as a colimit over the diagram $F \colon \mathscr{I}^{\mathrm{op}} \to \mathbf{Sch}$ of projective spaces with closed immersions between them. © Simon Felten 2025. All rights reserved

Since $\omega_P \cong O_P(-4)$, this yields $i_*\omega_H^\circ \cong i_*O_H$.   ◇

Given that Calabi–Yau schemes can be constructed by gluing irreducible components along closed subvarieties, we obtain the following strategy to construct Calabi–Yau varieties.

(A) The first step is to construct a degenerate Calabi–Yau scheme $V = \bigcup_i V_i$ by gluing the irreducible components V_i along closed subvarieties.

(B) The second step is to construct a flat and proper morphism $f \colon X \to S$ to a smooth curve S together with a marked $\mathbb{C}$-valued point $s_0 \in S$ and a Cartesian diagram

$$
\begin{array}{ccc}
V & \xrightarrow{\;c\;} & X \\
\downarrow & & \downarrow{\scriptstyle f} \\
\{s_0\} & \xrightarrow{\;b\;} & S
\end{array}
$$

such that every fiber $f^{-1}(s)$ in some neighborhood $S' \subseteq S$ of $s_0 \in S$ is a Calabi–Yau variety.

We call a family $f\colon X \to S$ as in Step (B) a *proper algebraic one-parameter normaling* of V. If the fibers $f^{-1}(s)$ in a neighborhood of $s_0 \in S$ are smooth, then we say that $f\colon X \to S$ is a *proper algebraic one-parameter smoothing* of V.

Example 1.11 For the degenerate scheme, we take $H = \{X_0 X_1 X_2 X_3 = 0\} \subset \mathbb{P}^3$ from Example 1.10. Now consider

$$E = \{T_0(X_0^4 + X_1^4 + X_2^4 + X_3^4) - T_1 X_0 X_1 X_2 X_3 = 0\} \subset \mathbb{P}^1 \times \mathbb{P}^3.$$

Here, T_0 and T_1 are the coordinates on $\mathbb{P}^1$. The first projection defines a flat projective family $f\colon E \to \mathbb{P}^1$ such that all but finitely many fibers are smooth. We have $H \cong f^{-1}(0)$ for $0 := [0 : 1]$ as a special fiber. The smooth fibers are smooth quartic surfaces and hence Calabi–Yau varieties. See Fig. 1.2 for an illustration.   ◇

Example 1.12 We take again $H = \{X_0 X_1 X_2 X_3 = 0\} \subset \mathbb{P}^3$ for the degenerate scheme, but for the total space of the family, we take

$$E' = \{T_0(X_0 + X_1 + X_2 + X_3)^4 - T_1 X_0 X_1 X_2 X_3 = 0\} \subset \mathbb{P}^1 \times \mathbb{P}^3.$$

The first projection defines a flat projective family $f'\colon E' \to \mathbb{P}^1$ with $f'^{-1}(0) \cong H$. All other fibers in a neighborhood of $0 \in \mathbb{P}^1$ are normal but not smooth Calabi–Yau varieties. The fiber over $1 \in \mathbb{P}^1$ is the singular quartic surface from Example 1.3.   ◇

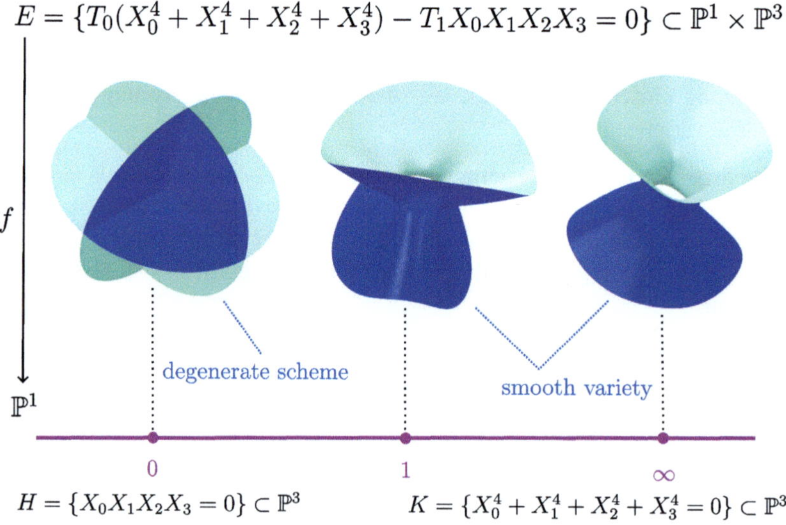

Fig. 1.2 Example 1.11. A family of algebraic varieties degenerating to $H = \{X_0 X_1 X_2 X_3 = 0\}$

1.1.4   The Language of Algebraic Deformations

Proper algebraic one-parameter normalings are examples of *deformations*. As a preparation for our further discussion, we define the concept of a deformation formally.

A *scheme germ* is a closed immersion $b\colon S_0 \to S$ of schemes. We write a scheme germ as $(S \supset S_0)$. If $S_0 = \{s_0\}$ is a reduced point, we also write $(S \ni s_0)$. A *premorphism* of scheme germs $f\colon (X \supset X_0) \to (S \supset S_0)$ is a triple (f_0, X', f), where $f_0\colon X_0 \to S_0$ is a morphism of schemes, $X' \subseteq X$ is an open subset with $X_0 \subset X'$, and $f\colon X' \to S$ is a morphism of schemes which fits into a commutative diagram

$$
\begin{array}{ccccc}
X_0 & \xrightarrow{\ b_X\ } & X' & \xrightarrow{\ \subseteq\ } & X \\
\big\downarrow{\scriptstyle f_0} & & \big\downarrow{\scriptstyle f} & & \\
S_0 & \xrightarrow[\ b_S\ ]{} & S & &
\end{array}
$$

of schemes. Two premorphisms (f_{01}, X_1', f_1) and (f_{02}, X_2', f_2) are *equivalent* if $f_{01} = f_{02}$, and if there is an open subset $U \subseteq X_1' \cap X_2'$ with $X_0 \subset U$ such that $f_1|_U = f_2|_U$. A *morphism* of scheme germs is an equivalence class of premorphisms.

A premorphism of scheme germs $(f_0, X', f)\colon (X \supset X_0) \to (S \supset S_0)$ is *Cartesian* if the above diagram is Cartesian, and a morphism of scheme germs is *Cartesian* if it can be represented by a Cartesian premorphism.

The category of scheme germs has fiber products.

Let **P** be a property of schemes. We say that a scheme germ $(S \supset S_0)$ has **P** if there is some open subset $S' \subseteq S$ with $S_0 \subset S'$ such that S' has **P**.

Let **P** be a property of morphisms of schemes with the following properties:

(a) an open immersion has **P**;
(b) **P** is stable under composition and base change.

Then, we say that a Cartesian premorphism $f\colon (X \supset X_0) \to (S \supset S_0)$ has **P** if there is some open subset $U \subseteq X$ with $X_0 \subset U$ such that f_0 and $f|_U\colon U \to S$ have **P**. A Cartesian morphism has **P** if one (equivalently: every) representing premorphism has **P**.

We say that a Cartesian premorphism $f\colon (X \supset X_0) \to (S \supset S_0)$ is a *closed immersion* respectively *proper* if there is some open subset $V \subseteq S$ with $S_0 \subset V$ and some open subset $U \subseteq X$ with $X_0 \subset U$ such that $U \subseteq f^{-1}(V)$ and $f\colon U \to V$ is a closed immersion respectively proper. A Cartesian morphism is a closed immersion/proper if one (equivalently: every) representing premorphism is a closed immersion/proper.

We are now ready to define a deformation.

Definition 1.13 (Algebraic Deformations) Let $V/\mathbb{C}$ be a degenerate scheme. An *algebraic deformation* of V consists of a scheme $S/\mathbb{C}$ together with a $\mathbb{C}$-valued point $s_0 \in S$, giving rise to a scheme germ $(S \ni s_0)$, a closed immersion $c\colon V \to X$ of $\mathbb{C}$-schemes, giving rise to a scheme germ $(X \supset V)$, and a flat morphism of scheme germs $f\colon (X \supset V) \to (S \ni s_0)$. A *morphism* between algebraic deformations is a Cartesian diagram

$$
\begin{array}{ccc}
(X_1 \supset V) & \xrightarrow{\;\gamma\;} & (X_2 \supset V) \\
\downarrow & & \downarrow \\
(S_1 \ni s_0) & \xrightarrow{\;\beta\;} & (S_2 \ni s_0)
\end{array}
$$

of scheme germs. If $V/\mathbb{C}$ is proper, then an algebraic deformation $f\colon (X \supset V) \to (S \ni s_0)$ is *proper* if it is proper as a morphism of scheme germs.

In this language, we can now redefine one-parameter normalings and smoothings.

Definition 1.14 (Algebraic One-Parameter Deformations) Let $V/\mathbb{C}$ be a degenerate scheme. An *algebraic one-parameter deformation* is an algebraic deformation $f\colon (X \supset V) \to (S \ni s_0)$ where S is a smooth curve. An *algebraic one-parameter normaling* is an algebraic one-parameter deformation such that there is an open subset $S' \subseteq S$ with $s_0 \in S'$ such that $f^{-1}(s)$ is normal for every $s \in S' \setminus \{s_0\}$. An *algebraic one-parameter smoothing* is an algebraic one-parameter deformation such that there is some open subset $S' \subseteq S$ with $s_0 \in S'$ such that $f^{-1}(s)$ is smooth for every $s \in S' \setminus \{s_0\}$.

In this definition, no condition ensures that $f^{-1}(s)$ is Calabi–Yau in the case where V is Calabi–Yau—something which we have required in Step (B). However, we have the following simple criterion which is sufficient in practice.

Lemma 1.15 *Let $V/\mathbb{C}$ be a Calabi–Yau scheme, and let $f\colon (X \supset V) \to (S \ni s_0)$ be a proper algebraic one-parameter normaling. Let $b\colon \{s_0\} \to S$ and $c\colon V \to X$ be the closed immersions.*

(1) f is a Gorenstein morphism of scheme germs.

Let $\omega_{X/S}^{\circ}$ be the relative dualizing sheaf, and assume that $b^* f_* \omega_{X/S}^{\circ} \to (f_0)_* \omega_V^{\circ}$ is surjective.

(2) There is an open subset $S' \subseteq S$ with $s_0 \in S'$ such that $f^{-1}(s)$ is a Calabi–Yau variety for every $s \in S'$.

1.1.5   Analytic and Formal Analytic Deformations

Constructing algebraic one-parameter normalings is a difficult problem which we do not solve in this monograph. Instead, we construct *analytic one-parameter normalings*.

Analytic deformations can be defined analogously to algebraic deformations. A *complex space germ* is a closed immersion $b\colon S_0 \to S$ of complex spaces, and the definition of premorphisms and morphisms is as expected. The full subcategory of complex space germs $(S \supset S_0)$ where S_0 is a reduced point is somewhat simpler than in the algebraic case. Namely, the map $(S \ni s_0) \mapsto O_{S,s_0}$ defines a contravariant equivalence with the category of *analytic local rings*, i.e., local $\mathbb{C}$-algebras A with residue field $\mathbb{C}$ which are quotients of the stalk $O_{\mathbb{C}^n,0}$ for some $n \geq 0$. An *analytic deformation* is nothing but a flat Cartesian morphism $f\colon (X \supset V) \to (S \ni 0)$ of complex space germs. Unlike in the algebraic case, an analytic deformation is automatically proper if V is proper.

The reason why it is easier to construct analytic one-parameter normalings is the close relation between analytic deformations and *formal analytic deformations*.

Definition 1.16 (Formal Analytic Deformations) Let V be a complex space, and let $(S \ni 0)$ be a complex space germ corresponding to an analytic local ring $A = O_{S,0}$. Then a *formal analytic deformation* over $(S \ni 0)$ is a sequence of analytic deformations

$$
\begin{array}{ccccccccc}
(V \supset V) & \longrightarrow & (X_1 \supset V) & \longrightarrow & (X_2 \supset V) & \longrightarrow & (X_3 \supset V) & \longrightarrow & \cdots \\
\downarrow{\scriptstyle f_0} & & \downarrow{\scriptstyle f_1} & & \downarrow{\scriptstyle f_2} & & \downarrow{\scriptstyle f_3} & & \\
(S_0 \ni 0) & \longrightarrow & (S_1 \ni 0) & \longrightarrow & (S_2 \ni 0) & \longrightarrow & (S_3 \ni 0) & \longrightarrow & \cdots,
\end{array}
$$

where $S_k = \operatorname{Spec} A/\mathfrak{m}_A^{k+1}$, which is a locally ringed space that is both a scheme and a complex space.

Every analytic deformation $f\colon (X \supset V) \to (S \ni 0)$ gives rise to a formal analytic deformation by base change along $(S_k \ni 0) \to (S \ni 0)$. The partial converse—if V is compact—relies on two results: on the existence of the *miniversal analytic deformation* and on the *Artin approximation theorem*.

We start with the miniversal analytic deformation. A somewhat simpler concept is a *universal analytic deformation*. Let $\Phi\colon (M \supset V) \to (D \ni 0)$ be a proper analytic deformation. When $\beta\colon (S \ni 0) \to (D \ni 0)$ is a morphism of complex space germs, then we can form the pull-back $(M \supset V) \times_{(D \ni 0)} (S \ni 0)$ and obtain an analytic deformation over $(S \ni 0)$. The deformation Φ is called *universal* if, for every analytic deformation $f\colon (X \supset V) \to (S \ni 0)$, there is a *unique* morphism $(S \ni 0) \to (D \ni 0)$ such that f is isomorphic to $\Phi \times_{(D \ni 0)} (S \ni 0)$. In general, V does not admit a universal analytic deformation, but a slightly weaker deformation exists, the *miniversal* analytic deformation.

Theorem 1.17 *Let V be a compact complex space. Then there is a proper analytic deformation $\Phi\colon (M \supset V) \to (D \ni 0)$ with the following properties:*

(a) *(Completeness) Let $f\colon (X \supset V) \to (S \ni 0)$ be a proper analytic deformation. Then there is a morphism of analytic deformations $(X/S) \to (M/D)$.*

(b) *(Versality) Let $(\tilde{\gamma}/\tilde{\beta})\colon (X_a/S_a) \to (X_b/S_b)$ be a morphism between proper analytic deformations of V such that $\tilde{\beta}\colon (S_a \ni 0) \to (S_b \ni 0)$ is a closed immersion, and let $(\gamma_a/\beta_a)\colon (X_a/S_a) \to (M/D)$ be a morphism of analytic deformations. Then there is a morphism of analytic deformations $(\gamma_b/\beta_b)\colon (X_b/S_b) \to (M/D)$ such that $(\gamma_b/\beta_b) \circ (\tilde{\gamma}/\tilde{\beta}) = (\gamma_a/\beta_a)$.*

(c) *(Effectivity) Let $(X_1 \supset V) \to (S_1 \ni 0)$ be a proper analytic deformation over $S_1 = \operatorname{Spec}\mathbb{C}[t]/(t^2)$. Let $(\gamma_1/\beta_1)\colon (X_1/S_1) \to (M/D)$ and $(\gamma_2/\beta_2)\colon (X_1/S_1) \to (M/D)$ be two morphisms of deformations. Then $\beta_1 = \beta_2$ (though not necessarily $\gamma_1 = \gamma_2$).*

Proof For a complex manifold Y, this has been first proven by Kuranishi in [184]. The case of complex spaces has been achieved by Douady in [65], by Grauert in [112], and by Forster and Knorr in [86]. More precisely, [65, 86, 112] show the existence of a complete and effective deformation (M/D). On the other-hand side, the infinitesimal deformation functor Def_V has a hull given by a formal deformation $(X_k/S_k)_{k \geq 0}$. Formal versality of $(X_k/S_k)_{k \geq 0}$ together with completeness of (M/D) and effectivity of both shows that the hull of Def_V is isomorphic to $\widehat{O}_{D,0}$, and that (M/D) is formally versal, i.e., versal for closed immersions of infinitesimal deformations. Then [27, Thm. 7.1] shows that (M/D) is not only formally versal but versal. $\qquad\square$

When $\Phi\colon (M \supset V) \to (D \ni 0)$ is a miniversal proper analytic deformation, and when $(f_k\colon (X_k \supset V) \to (S_k \ni 0))_{k \geq 0}$ is a formal analytic deformation, then inductive application of the versality property yields a sequence of maps

$$\beta_k\colon \quad (S_k \ni 0) \to (D \ni 0)$$

such that $\beta_\ell \circ b_{k\ell} = \beta_k$, where $b_{k\ell}\colon (S_k \ni 0) \to (S_\ell \ni 0)$ is the obvious closed immersion. On the level of analytic local rings, this corresponds to a sequence of maps $\bar{u}_k\colon B \to A_k$, where we write $B = O_{D,0}$, $A = O_{S,0}$, and $A_k = A/\mathfrak{m}_A^{k+1}$, and thus to a map $\bar{u}\colon B \to \widehat{A}$. The *Artin approximation theorem* states that this map can be approximated by a map $u\colon B \to A$ of analytic local rings, and thus the formal analytic family $(f_k)_{k \geq 0}$ can be approximated by the base change of $\Phi\colon (M \supset V) \to (D \ni 0)$ along the map $(S \ni 0) \to (D \ni 0)$ induced by $u\colon B \to A$.

Theorem 1.18 *Let A and B be analytic local rings. Let $A \to A_0$ be a non-zero quotient, and let $u_0\colon B \to A_0$ be a homomorphism. Let $\tilde{u}_0\colon B \to \widehat{A}_0$ be the map obtained by composition with the canonical map $A_0 \to \widehat{A}_0$ to the completion in the maximal ideal, and assume that we have a local homomorphism $\bar{u}\colon B \to \widehat{A}$ of $\mathbb{C}$-algebras such that the diagram*

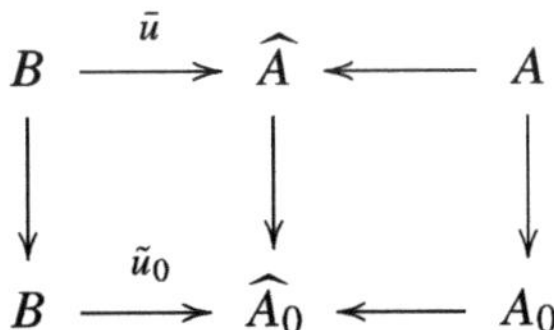

commutes. Let $c \geq 0$. Then there is a local homomorphism $u \colon B \to A$ of $\mathbb{C}$-algebras such that $\tilde{u} \equiv \bar{u} \mod \mathfrak{m}_A^{c+1}$, where $\tilde{u} \colon B \to \widehat{A}$ is the map obtained from $u \colon B \to A$ by composition with $A \to \widehat{A}$.

Proof This is [14, Thm. 1.5a (i)]. $\square$

Corollary 1.19 *Let V be a compact complex space. Consider a formal analytic deformation $(f_k \colon (X_k \supset V) \to (S_k \ni 0))_{k \geq 0}$, and let $c \geq 0$. Then there is a proper analytic one-parameter deformation $f \colon (X \supset V) \to (\Delta \ni 0)$ such that the base change of f along $(S_c \ni 0) \to (\Delta \ni 0)$ is isomorphic to $f_c \colon (X_c \supset V) \to (S_c \ni 0)$.*

Note that Corollary 1.19 does not state the existence of an analytic deformation which induces the given formal analytic deformation for all $k \geq 0$. Such a thing may not exist. Also note that it is not sufficient to have only $f_c \colon (X_c \supset V) \to (S_c \ni 0)$ given at the beginning. Such a thing may not extend to $(S \ni 0)$.

1.1.6   *Formal Analytic One-Parameter Normalings and Smoothings*

Let $(S \ni 0) = (\Delta \ni 0)$ be a disk. We have seen how to pass from a formal analytic deformation over $(\Delta \ni 0)$ to an analytic deformation $f \colon (X \supset V) \to (\Delta \ni 0)$. However, we want f to be a normaling or a smoothing, not just a deformation. For example, $V \times \Delta$ is an analytic one-parameter deformation but not a normaling/smoothing unless V is normal/smooth. Nonetheless, under suitable conditions on the formal deformation, the analytic deformation is a normaling/smoothing. Let us first introduce some terminology.

Definition 1.20 (Juxta-Smoothness)   Let $f \colon (X \supset V) \to (\Delta \ni 0)$ be an analytic one-parameter deformation. A point $v \in V$ is *juxta-smooth* (respectively *juxta-normal*) if v admits an open neighborhood $W \subseteq X$ such that, for every $t \in \Delta \setminus \{0\}$, the fiber $W \cap f^{-1}(t)$ is smooth (respectively normal).[1]

Juxta-smoothness can be detected infinitesimally.

---

[1] The preposition "juxta-" means "next to" in a spatial sense.

Lemma 1.21 *Let* $f: (X \supset V) \to (\Delta \ni 0)$ *be an analytic one-parameter deformation, and assume that all fibers* $f^{-1}(t)$ *are equidimensional of dimension* d. *Then* $v \in V$ *is juxta-smooth if and only if there is some* $m \geq 0$ *and some open neighborhood* $v \in W \subseteq X$ *such that*

$$t^m \in \mathrm{Fitt}_d(\Omega^1_{W_m/S_m})$$

for the d-*th Fitting ideal of the relative differential forms* $\Omega^1_{W_m/S_m}$, *where the map* $f_m: W_m \to S_m$ *is the base change of* $W \to \Delta$ *to* $S_m = \mathrm{Spec}\, \mathbb{C}[t]/(t^{m+1})$.

Proof The algebraic version can be found for example in [267, 0E7T]. Let us briefly recall the argument. A point $x \in X_t = f^{-1}(t)$ is smooth if and only if $\Omega^1_{X_t}$ is locally free of rank $\dim(X_t, x)$. Since X_t is equidimensional of dimension d, we have to check if $\Omega^1_{X_t}$ is locally free of rank d, with d independent of the point. This is the case if and only if $\mathrm{Fitt}_d(\Omega^1_{X_t})_x = O_{X_t,x}$ at $x \in X_t$. Because the formation of Fitt_d commutes with pull-back, this is equivalent to $\mathrm{Fitt}_d(\Omega^1_{X/\Delta})_x = O_{X,x}$. Therefore, we have $\mathrm{Sing}(f) = V(I)$, where $V(I)$ is the locus cut out by the ideal $I := \mathrm{Fitt}_d(\Omega^1_{X/\Delta})$. Now $v \in V$ is juxta-smooth if and only if $W \cap V(I) \subseteq V$, which is equivalent to $\mathrm{Rad}(I)|_W \supseteq (t)|_W$ for the radical ideal. This, in turn, is equivalent to the stated condition.   □

In the following global variant, we add the Cohen–Macaulay condition on the central fiber to drop the equidimensionality condition on the nearby fibers.

Corollary 1.22 *Let* V *be a compact complex space which is Cohen–Macaulay and equidimensional of dimension* d. *Let* $f: (X \supset V) \to (\Delta \ni 0)$ *be a proper analytic one-parameter deformation. Then* f *is a smoothing if and only if there is some* $m \geq 0$ *such that*

$$t^m \in \mathrm{Fitt}_d(\Omega^1_{X_m/S_m}).$$

Proof Since V is Cohen–Macaulay and equidimensional, so is X in a neighborhood of V, and then the fibers are also equidimensional of dimension d. Thus, we can apply Lemma 1.21. Since V is compact, we can find a single m which works for all points $v \in V$.   □

This motivates the following definition.

Definition 1.23 (Formal Smoothings) Let V be a compact complex space which is Cohen–Macaulay and equidimensional of dimension d, and consider a formal analytic one-parameter deformation $(f_k: (X_k \supset V) \to (S_k \ni 0))_{k \geq 0}$. Then $(f_k)_{k \geq 0}$ is a *formal analytic one-parameter smoothing* if there is some $m \geq 0$ such that $t^m \in \mathrm{Fitt}_d(\Omega^1_{X_m/S_m})$. A *formal algebraic one-parameter smoothings* are defined analogously.

We now discuss the less well known version for normalings. For an analytic one-parameter deformation f, let $\mathrm{JuxSm}(f) \subseteq V$ be the locus of juxta-smooth points, and note that its complement is a closed analytic subset.

Lemma 1.24 *Let V be a compact complex space which is Cohen–Macaulay and equidimensional of dimension d, and let $f \colon (X \supset V) \to (\Delta \ni 0)$ be a proper analytic one-parameter deformation. Then f is a normaling if and only if the complement of $\mathrm{JuxSm}(f)$ has codimension ≥ 2.*

Proof We discuss the elementary proof since we did not find an appropriate reference. The total space X is Cohen–Macaulay and equidimensional of dimension $d+1$ in a neighborhood of V. Thus, all fibers sufficiently close to $t = 0$ are Cohen–Macaulay and equidimensional of dimension d.

Let $A^\circ = \mathrm{Sing}(f) \cap \{t \neq 0\}$, a locally closed subset in the analytic topology of X, and let $A = \overline{A^\circ}$ be its closure in the analytic topology of X. The irreducible components of A are the analytic closures of the irreducible components of A°. Let $A_0 := A \cap V$. When $v \in V \setminus A_0$, then v is clearly juxta-smooth. The converse is also true. Namely, assume that $v \in A_0$ would be a juxta-smooth point. Then we have an Euclidean open subset $v \in W \subseteq V$ with $(W \setminus V) \cap A = \varnothing$. In other words, we have $W \cap A = W \cap A_0$. By the identity theorem for analytic sets in [114, p. 19], A_0 must contain one of the irreducible components of A. Since $A = \overline{A^\circ}$, this is impossible. Thus, we have $\mathrm{JuxSm}(f) = V \setminus A_0$.

Let us first assume that f is a normaling. After shrinking Δ, we can assume that all fibers X_t for $t \neq 0$ are normal. By Greuel et al. [116, Thm. 1.96], we have $\dim(\mathrm{Sing}(X_t)) \leq d - 2$. Let $\varnothing \neq D \subset A_0$ be an irreducible closed subset. Then D is contained in an irreducible component of A, and hence there is an irreducible component $B^\circ \subset A^\circ$ such that $D \subset B := \overline{B^\circ}$. Its image $f(B) \subset \Delta$ is closed and irreducible, so it is either a point or $= \Delta$. Since $B^\circ \subset \{t \neq 0\}$, we have $f(B) \neq \{0\}$, and since $\overline{B^\circ} \cap V \neq \varnothing$, the image is not only a single point in $\Delta \setminus \{0\}$. Thus, we have $f(B) = \Delta$. In particular, $f \in O_B$ is a non-zero divisor, and $B \to \Delta$ is flat by Greuel et al. [116, Prop. 1.85]. This implies $\dim(B, x) = \dim(B_{f(x)}, x) + 1$ for every $x \in B^\circ$. Since $B_{f(x)} \subset \mathrm{Sing}(X_{f(x)})$, we have $\dim(B_{f(x)}, x) \leq d - 2$ and hence $\dim(B, x) \leq d - 1$. Because B is reduced and irreducible, its dimension is constant, so we have $\dim(B, v) \leq d - 1$ for every $v \in D$, and hence $\dim(D, v) \leq \dim(B_0, v) \leq d - 2$ by flatness. This proves $\dim(A_0) \leq d - 2$.

Let us now assume that $\dim(A_0) \leq d - 2$. Again, let B° be an irreducible component of A° with analytic closure $B = \overline{B^\circ}$. As before, we can assume that $B \to \Delta$ is flat so that $\dim(B) = \dim(B \cap V) + 1 \leq d - 1$. But then flatness implies $\dim(B_{f(x)}, x) \leq d - 2$ for every $x \in B^\circ$. Since f is proper, we can assume that, after shrinking Δ, every irreducible component of A intersects V. Thus, we have $\dim(A_{f(x)}, x) \leq d - 2$ for every $x \in A^\circ$. Because $X_{f(x)}$ is Cohen–Macaulay, this implies that $X_{f(x)}$ is normal by Fischer [85, p. 120, Cor. 2]. $\qquad\square$

Definition 1.25 (Formal Normalings) Let V be a compact complex space which is Cohen–Macaulay and equidimensional of dimension d, and consider a formal analytic one-parameter deformation $(f_k \colon (X_k \supset V) \to (S_k \ni 0))_{k \geq 0}$. Then $(f_k)_{k \geq 0}$

is a *formal analytic one-parameter normaling* if there is some $m \geq 0$ such that $t^m \in \mathrm{Fitt}_d(\Omega^1_{X_m/S_m})$ holds on $V \setminus A_0$ for some closed analytic subset $A_0 \subset V$ of dimension $\leq d - 2$. A *formal algebraic one-parameter normaling* is defined analogously.

Given a formal analytic one-parameter normaling, Corollary 1.19 now allows to construct an analytic one-parameter normaling over Δ.

1.1.7 Constructing a Formal Analytic One-Parameter Smoothing

It remains how to construct a formal analytic one-parameter normaling or smoothing when a Calabi–Yau scheme $V/\mathbb{C}$ is given. The starting point is the *Bogomolov–Tian–Todorov theorem* for smooth Calabi–Yau manifolds $Y/\mathbb{C}$. To formulate it, we need the following category.

Definition 1.26 (Art) An *Artinian local $\mathbb{C}$-algebra with residue field $\mathbb{C}$* is a triple (A, e_A, p_A), where A is an Artinian local ring, $e_A : \mathbb{C} \to A$ endows A with a $\mathbb{C}$-algebra structure, and $p_A : A \to \mathbb{C}$ identifies the residue field with $\mathbb{C}$; in particular, we have $p_A \circ e_A = \mathrm{id}_\mathbb{C}$. A *homomorphism* between Artinian local $\mathbb{C}$-algebras with residue field $\mathbb{C}$ is a local homomorphism $\phi : B \to A$ of $\mathbb{C}$-algebras which is compatible with the maps $p_B : B \to \mathbb{C}$ and $p_A : A \to \mathbb{C}$. We denote the resulting category by **Art**.

Let $A \in \mathbf{Art}$. Then $\mathrm{Spec}\, A$ is, as a topological space, just a point. It defines a scheme germ $(S_A \ni 0) := (\mathrm{Spec}\, A \supset \mathrm{Spec}\, \mathbb{C})$, and its analytification is equal to S_A as a germ of locally ringed spaces. If $(Y_A \supset Y) \to (S_A \ni 0)$ is an algebraic or analytic deformation, then $|Y_A| = |Y|$ as topological spaces. Thus, deformations over S_A are particularly simple deformations: they modify only the structure sheaf but not the underlying topological space.

The classical Bogomolov–Tian–Todorov theorem concerns deformations of Calabi–Yau manifolds over scheme germs of the form $(S_A \ni 0)$.

Theorem 1.27 *Let Y be an algebraic Calabi–Yau manifold, let $A \in \mathbf{Art}$, and let $(Y_A \supset Y) \to (S_A \ni 0)$ be an algebraic deformation. Let $B \to A$ be a surjective homomorphism in* **Art**, *giving rise to a closed immersion $(S_A \ni 0) \to (S_B \ni 0)$ of scheme germs. Then there is an algebraic deformation $(Y_B \supset Y) \to (S_B \ni 0)$ together with a morphism of deformations $(\gamma/\beta) : (Y_A/S_A) \to (Y_B/S_B)$.*

The same theorem is also true for an analytic Calabi–Yau manifold Y if the $\partial\bar{\partial}$-lemma holds for Y. The theorem allows us to construct formal algebraic or analytic one-parameter deformations by extending a given deformation indefinitely along $(S_k \ni 0) \to (S_{k+1} \ni 0)$.

Example 1.28 Let $V/\mathbb{C}$ be a Calabi–Yau variety of dimension 3 which has at most A_1-singularities. Then the statement of Theorem 1.27 holds for algebraic deformations of V by Friedman and Laza [96, Cor. 1.5]. Thus, if $f_1 \colon (X_1 \supset V) \to (S_1 \ni 0)$ is a deformation such that $t \in \mathrm{Fitt}_3(\Omega^1_{X_1/S_1})$, then V admits a formal algebraic one-parameter smoothing and hence an analytic one-parameter smoothing. $\Diamond$

For many if not most degenerate Calabi–Yau schemes V, the statement of Theorem 1.27 does not hold. However, when we endow V with an additional structure, namely a *logarithmic structure*, then the analog of Theorem 1.27 becomes true for logarithmic deformations. This is the main topic of this monograph.

1.2 Normal Crossing Schemes

The simplest, best-known, and most-studied degenerate schemes are *normal crossing schemes*. We elaborate on the geometry of normal crossing schemes here to help the reader build intuition about degenerate schemes. At the end of the section, we explain how to compute the dualizing sheaf ω_V° of a normal crossing scheme.

First, we define normal crossing *divisors*, which is the slightly older notion.

Definition 1.29 (Normal Crossing Divisors) Let $Y/\mathbb{C}$ be smooth of dimension $d + 1$, and let $D \subset Y$ be a closed subscheme.

(1) $D \subset Y$ is a *normal crossing divisor* if every geometric point $\bar{y} \in D$ has an étale neighborhood $p \colon W \to Y$ which admits an étale morphism $h \colon W \to \mathbb{A}^{d+1}$ such that $p^{-1}(D) = h^{-1}(\{z_0 \cdot \ldots \cdot z_r = 0\})$ for some $0 \le r \le d$ which may depend on $\bar{y}$.
(2) A normal crossing divisor $D \subset Y$ is *simple* if every irreducible component of D is smooth.
(3) A simple normal crossing divisor $D \subset Y$ is *strictly simple* if the intersection of any set of irreducible components of D is connected.[2]

Example 1.30 We collect some examples and non-examples of normal crossing divisors. See also Fig. 1.3

(1) $\{X_1 X_2 = 0\} \subset \mathbb{P}^2$ is a strictly simple normal crossing divisor.
(2) $\{X_1(X_2^2 - X_0 X_1) = 0\} \subset \mathbb{P}^2$ is a simple normal crossing divisor but not strictly simple.
(3) $\{X_0 X_2^2 - X_1^2(X_1 + 1) = 0\} \subset \mathbb{P}^2$ is a normal crossing divisor but not simple.
(4) $\{X_0 X_2^2 - X_1^3 = 0\} \subset \mathbb{P}^2$ is not a normal crossing divisor. $\Diamond$

A normal crossing *scheme* is then a degenerate scheme V which is locally isomorphic to a normal crossing divisor.

---

[2] The terminology "strictly simple" is taken from [259].

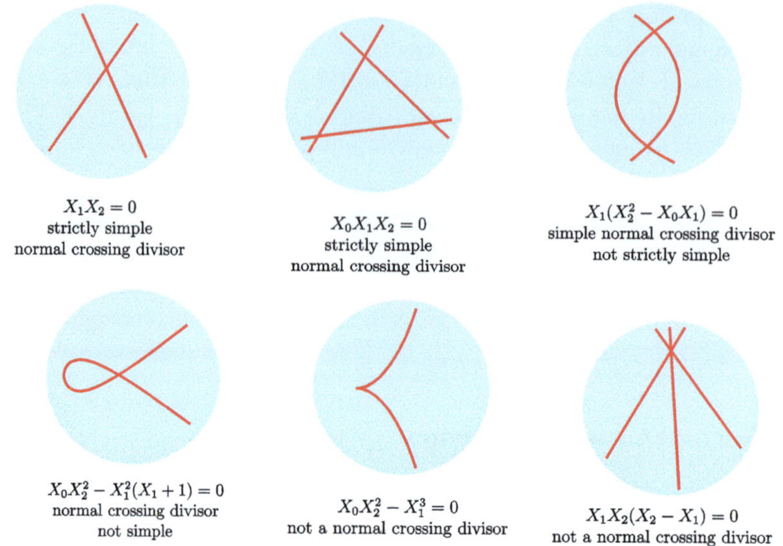

Fig. 1.3 Some examples and non-examples of normal crossing divisors. © Simon Felten 2025. All rights reserved

Definition 1.31 (Normal Crossing Schemes) Let $V/\mathbb{C}$ be a separated scheme of finite type, equidimensional of dimension d.

(1) V is a *normal crossing scheme* if every geometric point $\bar{v} \in V$ has an étale neighborhood W which admits an étale morphism

$$h \colon \ W \to \{z_0 \cdot \ldots \cdot z_r = 0\} \subset \mathbb{A}^{d+1}$$

for some $0 \leq r \leq d$, where r is allowed to depend on $\bar{v}$.
(2) A normal crossing scheme V is *simple* if all irreducible components are smooth.
(3) A simple normal crossing scheme V is *strictly simple* if the intersection of any set of irreducible components is connected.

We say that $\{z_0 \cdot \ldots \cdot z_r = 0\} \subset \mathbb{A}^{n+1}$ is a *local model* for the singularities of V. Since a normal crossing scheme has hypersurface singularities, it is Gorenstein.

Example 1.32 We collect some additional examples of normal crossing schemes. See also Fig. 1.4.

(1) The scheme $\{X_1 X_2 X_3 = 0\} \subset \mathbb{P}^3$ is a strictly simple normal crossing scheme.
(2) The scheme $\{X_1(X_2^2 - X_0 X_1)Y_0 Y_1 = 0\} \subset \mathbb{P}^2 \times \mathbb{P}^1$ is a simple normal crossing scheme but not strictly simple.
(3) The scheme $\{X_0 X_1 X_2 Y_0 Y_1 = 0\} \subset \mathbb{P}^2 \times \mathbb{P}^1$ is a strictly simple normal crossing scheme.   ◊

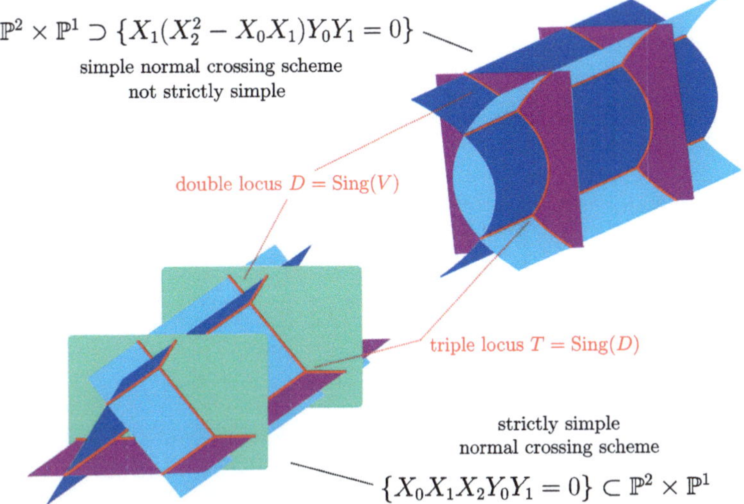

Fig. 1.4 Two examples of normal crossing schemes. © Simon Felten 2025. All rights reserved

1.2.1 The Stratification and the Dual Intersection Complex

Let V be a normal crossing scheme of dimension d. Then we obtain a stratification as follows. First, we set $C_0V = V$, and then inductively $C_{m+1}V = \mathrm{Sing}(C_mV)$, the singular locus with its reduced induced scheme structure. In particular, $C_mV \subset V$ is a closed subscheme. The special case $D = C_1V = \mathrm{Sing}(V)$ is called the *double locus*, and the special case $T = C_2V = \mathrm{Sing}(D)$ is called the *triple locus*. Then we set $S_mV = C_mV \setminus C_{m+1}V$. By construction, S_mV is non-singular. An *open stratum* is then a connected component of some S_mV, and a *closed stratum* is the closure of an open stratum. We denote the set of connected components of S_mV by $[S_mV]$. For an open stratum $S^\circ \in [S_mV]$, we write $S = \overline{S}^\circ$ for the associated closed stratum, and $\partial S = S \setminus S^\circ$ for the boundary. We denote the set of closed strata which are closures of open strata in $[S_mV]$ by $[C_mV]$. The formation of strata commutes with étale morphisms, so we see from the local model $\{z_0 \cdot \ldots \cdot z_r = 0\} \subset \mathbb{A}^{n+1}$ that $\dim(S^\circ) = d - m$ for an open stratum $S^\circ \in [S_mV]$. Furthermore, V has exactly $m+1$ local irreducible components around a point $v \in S^\circ$, hence the names "double locus" and "triple locus."

Assume that V is simple, and let $S^\circ \in [S_mV]$ be an open stratum. Let I be the set of the $m + 1$ irreducible components of V which contain S°. Then $\bigcap_{V_i \in I} V_i$ is smooth, and $S = \overline{S}^\circ$ is a connected component of $\bigcap_{V_i \in I} V_i$. Furthermore, $\partial S \subset S$ is a simple normal crossing divisor.

For a simple normal crossing scheme V, we encode the combinatorics of the strata in a topological space Γ_V, the *dual intersection complex* of V. First, we introduce a partial order on the set of open strata $[S_\bullet V]$ by declaring $S^\circ \preceq T^\circ$

if and only if $S^\circ \subseteq \overline{T}^\circ$. This turns $[S_\bullet V]$ into a category. For $S^\circ \in [S_m V]$, let $\mathrm{Comp}(S^\circ)$ be the set of irreducible components of V which contain S°. Since V is simple, we have $|\mathrm{Comp}(S^\circ)| = m + 1$ for $S^\circ \in [S_m V]$. Furthermore, if $S^\circ \preceq T^\circ$, then we have $\mathrm{Comp}(S^\circ) \supseteq \mathrm{Comp}(T^\circ)$. For each $S^\circ \in [S_m V]$, we take an $(m + 1)$-simplex $\sigma(S^\circ)$, and we choose a bijection between $\mathrm{Comp}(S^\circ)$ and the set of vertices of $\sigma(S^\circ)$. For every $S^\circ \preceq T^\circ$, we have a closed immersion $\sigma(T^\circ) \subset \sigma(S^\circ)$ as a face according to the identifications of vertices with irreducible components of V. This defines a functor $\sigma \colon [S_\bullet V]^{\mathrm{op}} \to \mathbf{Top}$. The dual intersection complex is defined as the colimit $\Gamma_V := \mathrm{colim}(\sigma)$.

1.2.2 The Dualizing Sheaf

For a simple normal crossing scheme V, we can compute the dualizing sheaf ω_V° in the sense of Definition 1.6 as follows. For each open stratum $S^\circ \in [S_m V]$, we have a smooth scheme S with a simple normal crossing divisor ∂S. We set

$$\omega_{(S,\partial S)} := \omega_S \otimes_{O_S} O_S(\partial S).$$

For any inclusion $i \colon S \subset T$, there is an isomorphism $i^* \omega_{(T,\partial T)} \to \omega_{(S,\partial S)}$. The correct global choice of all these isomorphisms is surprisingly subtle and will be addressed in Sect. 9.9.[3] We then obtain an $[S_\bullet V]^{\mathrm{op}}$-indexed diagram of coherent sheaves on V. The dualizing sheaf ω_V° is now isomorphic to the limit of the diagram, and furthermore, we have $\omega_V^\circ|_S \cong \omega_{(S,\partial S)}$ for every closed stratum S.

Let V be a Calabi–Yau simple normal crossing scheme. The explicit description of ω_V° shows that $\omega_{(S,\partial S)} \cong O_S$ for every $S^\circ \in [S_\bullet V]$. In this case, we say that $(S, \partial S)$ is a *log Calabi–Yau pair*.[4] Conversely, let V be a proper normal crossing scheme such that $\omega_{V_i} \otimes O_{V_i}(D_i) \cong O_{V_i}$ for every irreducible component V_i of V, and assume that $H^0(S, O_S) \cong \mathbb{C}$ for every closed stratum S. We obtain a diagram

$$F \colon [S_\bullet V]^{\mathrm{op}} \to \mathbf{Vect}_\mathbb{C}, \quad S^\circ \mapsto H^0(S, \omega_{(S,\partial S)}).$$

Then $H^0(V, \omega_V^\circ) \cong \lim(F)$. By construction, the homomorphisms $F(T^\circ) \to F(S^\circ)$ are isomorphisms. Since V is connected, the limit is either isomorphic to $\mathbb{C}$

---

[3] For a glimpse on why this is subtle, note that the sign of the residue map on the level of the de Rham complex for a simple normal crossing divisor depends on a chosen order of the components, see for example [230, § 4.2]. The ultimate source of the subtlety is that the identification between $\omega_V^\circ|_S$ and $\omega_{(S,\partial S)}$ is not canonical but relies on trivializing a constant sheaf with stalk $\mathbb{Z}$ which does not have a canonical generator.

[4] This is in fact true in the sense of log structures, but it can be just considered as terminology at this point.

or 0. In the first case, any non-zero global section of ω_V° does not vanish anywhere, and we have $\omega_V^\circ \cong O_V$.

Example 1.33 Let $H = \bigcup_{i=0}^{3} H_i$ be the normal crossing scheme of Example 1.10. Then each $(H_i, \partial H_i)$ is a log Calabi–Yau pair because $\omega_{\mathbb{P}^2} \cong O_{\mathbb{P}^2}(-3)$. Let

$$F: \quad [S_\bullet H]^{\mathrm{op}} \to \mathbf{Vect}_{\mathbb{C}}$$

be the functor as above. Let $0 \neq \lambda_0 \in F(H_0^\circ)$. Since $F(H_0^\circ) \to F(L_{0i}^\circ)$ and $F(H_i^\circ) \to F(L_{0i}^\circ)$ are isomorphisms for $1 \leq i \leq 3$, we can choose an element $0 \neq \lambda_1 \in F(H_i^\circ)$ which has the same image as λ_0 in $F(L_{0i}^\circ)$. Since λ_i and λ_j have both the same image as λ_0 in $F(T_{0ij}^\circ)$, they have the same image in $F(T_{0ij}^\circ)$ and hence in $F(L_{ij}^\circ)$. Finally, all three λ_i have the same image in $F(T_{123}^\circ)$ because they have pairwise the same image in $F(L_{ij}^\circ)$. Thus, we find $\omega_H^\circ \cong O_H$ in the colimit description.

The same argument also shows that $\{X_1(X_2^2 - X_0 X_1)Y_0 Y_1 = 0\} \subset \mathbb{P}^2 \times \mathbb{P}^1$ and $\{X_0 X_1 X_2 Y_0 Y_1 = 0\} \subset \mathbb{P}^2 \times \mathbb{P}^1$ are Calabi–Yau. The argument cannot be applied to most one-dimensional normal crossing schemes because it relies on the existence of sufficiently many triple intersections. $\Diamond$

Example 1.34 The normal crossing scheme $\{X_1 X_2 X_3 = 0\} \subset \mathbb{P}^3$ is not Calabi–Yau because $(\mathbb{P}^2, \{X_1 X_2 = 0\})$ is not a log Calabi–Yau pair. $\Diamond$

1.3 Infinitesimal Deformations and Functors of Artin Rings

We have reduced the problem of constructing an analytic one-parameter smoothing of V to constructing a formal analytic one-parameter smoothing, so it is sufficient to study deformations over $(S_A \ni 0)$ for $A \in \mathbf{Art}$, the category of Artinian local $\mathbb{C}$-algebras with residue field $\mathbb{C}$ from Definition 1.26. Such a deformation is called an *infinitesimal deformation*. When $f_A \colon (X_A \supset V) \to (S_A \ni 0)$ is an infinitesimal deformation, then we have $|S_A| = \{0\}$ and $|X_A| = |V|$ for the underlying topological spaces. Let us spell out the simpler definition of deformations in the infinitesimal case.

Definition 1.35 (Infinitesimal Deformations) Let $V/\mathbb{C}$ be a scheme. Let $A \in \mathbf{Art}$, and let $S_A = \mathrm{Spec}\, A$. An *infinitesimal algebraic deformation* is a Cartesian square

$$
\begin{array}{ccc}
V & \xrightarrow{\;c_A\;} & X_A \\[4pt]
{\scriptstyle f_0}\big\downarrow & & \big\downarrow{\scriptstyle f_A} \\[4pt]
\mathrm{Spec}\,\mathbb{C} & \xrightarrow{\;b_A\;} & S_A
\end{array}
$$

in the category of schemes where $f_A \colon X_A \to S_A$ is flat. When $f_A \colon X_A \to S_A$ and $f_B \colon X_B \to S_B$ are two infinitesimal flat deformations, then a *morphism* (γ/β) of deformations $(X_A/S_A) \to (X_B/S_B)$ is a Cartesian diagram

$$
\begin{array}{ccc}
X_A & \xrightarrow{\ \gamma\ } & X_B \\
{\scriptstyle f_A} \downarrow & & \downarrow {\scriptstyle f_B} \\
S_A & \xrightarrow{\ \beta\ } & S_B
\end{array}
$$

of schemes such that $\beta \circ b_A = b_B$ and $\gamma \circ c_A = c_B$. *Infinitesimal analytic deformations* and their morphisms are defined analogously.

If $V/\mathbb{C}$ is proper, so is $f_A \colon X_A \to S_A$.

We collect the isomorphism classes of infinitesimal deformations in the *flat deformation functor*

$$
\mathrm{Def}_V \colon \quad \mathbf{Art} \to \mathbf{Set}
$$

which maps $A \in \mathbf{Art}$ to the set of isomorphism classes of infinitesimal algebraic or analytic deformations of V. In general, a functor $F \colon \mathbf{Art} \to \mathbf{Set}$ such that $F(\mathbb{C})$ consists of a single element is called a *functor of Artin rings*.

We can now reformulate Theorem 1.27 in terms of Def_Y: When Y is an algebraic Calabi–Yau manifold, then Def_Y is *unobstructed* in the sense of the following definition.

Definition 1.36 (Unobstructedness I) Let $F \colon \mathbf{Art} \to \mathbf{Set}$ be a functor of Artin rings. Then F is *unobstructed* if, for every surjective homomorphism $\phi \colon B \to A$ in $\mathbf{Art}$, the induced map $F(B) \to F(A)$ is surjective.

1.4 Deformation Theory of Smooth Varieties

Before we embark on studying the deformation theory of degenerate schemes V, we review the deformation theory of smooth varieties Y. This will be our guide for studying deformations of degenerate schemes later on.

1.4.1 Local Rigidity of Smooth Deformations

The first fundamental property of infinitesimal deformations of complex manifolds Y is that they are *locally rigid*. More precisely, assume that Y is Stein (an analog of being affine), let $A \in \mathbf{Art}$, and let

$$
f_A \colon \ (Y_A \supset Y) \to (S_A \ni 0) \quad \text{and} \quad f'_A \colon \ (Y'_A \supset A) \to (S_A \ni 0)
$$

be two deformations. Then f_A and f'_A are isomorphic as deformations. Since $Y \times S_A$ is a deformation, called the *trivial deformation*, this implies that every infinitesimal deformation is locally isomorphic to $Y \times S_A$—they are also *locally trivial*.

The reason for local rigidity is as follows. Let $f : X \to S$ be a smooth morphism of complex spaces. As in the algebraic case, $f : X \to S$ has the *infinitesimal lifting property*: Whenever we have a solid commutative diagram

$$
\begin{array}{ccc}
T & \xrightarrow{\ c\ } & X \\
{\scriptstyle i}\downarrow & {\scriptstyle h}\nearrow\ {\scriptstyle b} & \downarrow{\scriptstyle f} \\
T' & \xrightarrow[\ \ \ ]{} & S
\end{array}
$$

of complex spaces where $i : T \to T'$ is a first-order thickening, i.e., $\mathcal{I}^2_{T/T'} = 0$ for the ideal sheaf $\mathcal{I}_{T/T'}$ of the closed immersion $i : T \to T'$, and where T is Stein, then there is a map $h : T' \to X$ making the diagram commutative.

1.4.2 Infinitesimal Automorphisms

The second fundamental property of infinitesimal deformations of complex manifolds is that their automorphisms are controlled by the holomorphic tangent sheaf $\mathcal{T}_Y$. Let us consider a morphism

$$
\begin{array}{ccccc}
Y & \xrightarrow{\ c_A\ } & Y_A & \xrightarrow{\ \gamma\ } & Y_B \\
{\scriptstyle f_0}\downarrow & & {\scriptstyle f_A}\downarrow & & \downarrow{\scriptstyle f_B} \\
\mathrm{Spec}\,\mathbb{C} & \xrightarrow[\ b_A\ ]{} & S_A & \xrightarrow[\ \beta\ ]{} & S_B
\end{array}
$$

between infinitesimal analytic deformations, where $\beta : S_A \to S_B$ is a closed immersion. An *infinitesimal automorphism* is an automorphism $\chi : Y_B \to Y_B$ such that $f_B \circ \chi = f_B$ and $\chi \circ \gamma = \gamma$. Since $|Y_B| = |Y_A| = |Y|$ as topological spaces, such an automorphism is given by a B-linear automorphism $\phi : \mathcal{O}_{Y_B} \to \mathcal{O}_{Y_B}$ of sheaves of rings such that $\gamma^* \circ \phi = \gamma^*$ for $\gamma^* : \mathcal{O}_{Y_B} \to \mathcal{O}_{Y_A}$. Such automorphisms are in one-to-one correspondence with B-linear derivations $D : \mathcal{O}_{Y_B} \to \mathcal{I}_{Y_A/Y_B}$ into the ideal sheaf of the embedding $Y_A \subset Y_B$ via

$$
D \mapsto \left(a \mapsto \sum_{n=0}^{\infty} \frac{D^n(a)}{n!} = a + D(a) + \frac{1}{2}D(D(a)) + \ldots \right).
$$

When $I_{A/B}$ is the kernel of $B \twoheadrightarrow A$, then these derivations are classified by

$$I_{A/B} \cdot \mathcal{T}_{Y_B/S_B},$$

where $\mathcal{T}_{Y_B/S_B}$ is the relative tangent sheaf. In particular, if either $I_{A/B} \cdot \mathfrak{m}_B = 0$ or $Y_B = Y \times S_B$ is a trivial deformation, then the infinitesimal automorphisms are classified by $I_{A/B} \otimes_{\mathbb{C}} \mathcal{T}_Y$.

1.4.3 Dolbeault Resolutions

We review the Dolbeault complex and define a Dolbeault resolution for any coherent analytic sheaf.

Let Y be a complex manifold. Then we have the *Dolbeault resolution*

$$
\begin{array}{ccccccccc}
& & \vdots & & \vdots & & \vdots & & \vdots \\
& & \uparrow & & \uparrow & & \uparrow & & \partial\uparrow \\
0 & \longrightarrow & \Omega_Y^2 & \longrightarrow & \mathcal{A}_Y^{2,0} & \xrightarrow{\bar\partial} & \mathcal{A}_Y^{2,1} & \xrightarrow{\bar\partial} & \mathcal{A}_Y^{2,2} & \longrightarrow \cdots \\
& & \partial\uparrow & & \partial\uparrow & & \partial\uparrow & & \partial\uparrow \\
0 & \longrightarrow & \Omega_Y^1 & \longrightarrow & \mathcal{A}_Y^{1,0} & \xrightarrow{\bar\partial} & \mathcal{A}_Y^{1,1} & \xrightarrow{\bar\partial} & \mathcal{A}_Y^{1,2} & \longrightarrow \cdots \\
& & \partial\uparrow & & \partial\uparrow & & \partial\uparrow & & \partial\uparrow \\
0 & \longrightarrow & O_Y & \longrightarrow & \mathcal{A}_Y^{0,0} & \xrightarrow{\bar\partial} & \mathcal{A}_Y^{0,1} & \xrightarrow{\bar\partial} & \mathcal{A}_Y^{0,2} & \longrightarrow \cdots
\end{array}
$$

of the holomorphic de Rham complex $(\Omega_Y^\bullet, \partial)$. Here, our convention is that $d = \partial + \bar\partial$ and $\partial\bar\partial + \bar\partial\partial = 0$, and we shall always follow this convention for double complexes.

The Dolbeault resolution gives the resolution $O_Y \to (\mathcal{A}_Y^{0,\bullet}, \bar\partial)$ of the structure sheaf. The differential $\bar\partial$ is O_Y-linear, and the resolution $\Omega_Y^i \to (\mathcal{A}_Y^{i,\bullet}, \bar\partial)$ is obtained from $O_Y \to (\mathcal{A}_Y^{0,\bullet}, \bar\partial)$ by the tensor product functor $(-) \otimes_{O_Y} \Omega_Y^i$ via the isomorphism

$$\mathcal{A}_Y^{0,j} \otimes_{O_Y} \Omega_Y^i \to \mathcal{A}_Y^{i,j}, \quad \beta \otimes \alpha \mapsto \beta \wedge \alpha.$$

The order of the two factors is important to obtain the correct signs.

Given this presentation of $\Omega_Y^i \to (\mathcal{A}_Y^{i,\bullet}, \bar\partial)$, we define the *Dolbeault resolution* of any coherent analytic sheaf $\mathcal{F}$ as

$$0 \to \mathcal{F} \to \mathcal{A}_Y^{0,0} \otimes_{O_Y} \mathcal{F} \to \mathcal{A}_Y^{0,1} \otimes_{O_Y} \mathcal{F} \to \mathcal{A}_Y^{0,2} \otimes_{O_Y} \mathcal{F} \to \cdots$$

1.4.4 A DG Lie Algebra Controlling Def_Y

The combination of local rigidity and the fact that $\mathcal{T}_Y$ encodes automorphisms allows us to describe Def_Y algebraically. Let us write $\mathcal{KS}_Y^\bullet := \mathcal{A}_Y^{0,\bullet} \otimes_{O_Y} \mathcal{T}_Y$ for the Dolbeault resolution of the holomorphic tangent sheaf $\mathcal{T}_Y$. The letters stand for "Kodaira–Spencer." The Lie bracket on $\mathcal{T}_Y$ can be extended to an operation

$$[-,-]\colon\ \mathcal{KS}_Y^{q_1} \times \mathcal{KS}_Y^{q_2} \to \mathcal{KS}_Y^{q_1+q_2},$$

which turns $\mathcal{KS}_Y^\bullet$ into a sheaf of *differential graded Lie algebras*. For compatibility with our sign conventions at other places, our convention here is to use $[-,-] = (-1) \cdot [-,-]_{\mathrm{L}}$, where $[-,-]_{\mathrm{L}}$ is the usual Lie bracket.

Definition 1.37 (DG Lie Algebras) Let R be a ring. A *differential graded Lie algebra* is a graded R-module $L^\bullet$ together with an R-linear map $\bar\partial\colon L^q \to L^{q+1}$ such that $\bar\partial^2 = 0$, and together with an R-bilinear map $[-,-]\colon L^{q_1} \times L^{q_2} \to L^{q_1+q_2}$ which satisfies

$$[\theta, \xi] = -(-1)^{\hat\theta \cdot \hat\xi}[\xi, \theta],$$

the Jacobi identity

$$[\theta, [\xi, \eta]] = [[\theta, \xi], \eta] + (-1)^{\hat\theta \cdot \hat\xi}[\xi, [\theta, \eta]],$$

and $\bar\partial[\theta, \xi] = [\bar\partial\theta, \xi] + (-1)^{\hat\theta}[\theta, \bar\partial\xi]$, where we write $\hat\theta = q$ if $\theta \in L^q$.[5]

For $A \in \mathbf{Art}$, we form the tensor product of our two resolutions with A:

$$0 \longrightarrow A \otimes_{\mathbb{C}} O_Y \longrightarrow A \otimes_{\mathbb{C}} \mathcal{A}_Y^{0,0} \overset{\bar\partial}{\longrightarrow} A \otimes_{\mathbb{C}} \mathcal{A}_Y^{0,1} \overset{\bar\partial}{\longrightarrow} A \otimes_{\mathbb{C}} \mathcal{A}_Y^{0,2} \longrightarrow \cdots$$

$$0 \longrightarrow A \otimes_{\mathbb{C}} \mathcal{T}_Y \longrightarrow A \otimes_{\mathbb{C}} \mathcal{KS}_Y^0 \overset{\bar\partial}{\longrightarrow} A \otimes_{\mathbb{C}} \mathcal{KS}_Y^1 \overset{\bar\partial}{\longrightarrow} A \otimes_{\mathbb{C}} \mathcal{KS}_Y^2 \longrightarrow \cdots$$

The structure sheaf of $Y \times S_A$ is nothing but $A \otimes_{\mathbb{C}} O_Y$, and thus $O_{Y \times S_A}$ is obtained as the kernel of $\bar\partial\colon A \otimes_{\mathbb{C}} \mathcal{A}_Y^{0,0} \to A \otimes_{\mathbb{C}} \mathcal{A}_Y^{0,1}$.

We also have an operation

$$[-,-]\colon\ \mathcal{KS}_Y^{q_1} \times \mathcal{A}_Y^{0,q_2} \to \mathcal{A}_Y^{0,q_1+q_2}.$$

---

[5] Our later degree convention will be $|\theta| = q - 1$ for $\theta \in L^q$, which explains the different signs in the definition in comparison with Definition 3.3.

Thus, if $\phi \in \Gamma(Y, \mathfrak{m}_A \otimes_{\mathbb{C}} \mathcal{KS}_Y^1)$, then

$$\bar{\partial}_\phi: \quad A \otimes_{\mathbb{C}} \mathcal{A}_Y^{0,q} \to A \otimes_{\mathbb{C}} \mathcal{A}_Y^{0,q+1}, \quad a \mapsto \bar{\partial}(a) + [\phi, a],$$

is an operation with the property that the induced operation on

$$\mathcal{A}_Y^{0,\bullet} = (A \otimes_{\mathbb{C}} \mathcal{A}_Y^{0,\bullet})/(\mathfrak{m}_A \otimes_{\mathbb{C}} \mathcal{A}_Y^{0,\bullet})$$

is the original Dolbeault differential. A direct computation shows that

$$O_Y(A; \phi) := \ker(\bar{\partial}_\phi: A \otimes_{\mathbb{C}} \mathcal{A}_Y^{0,0} \to A \otimes_{\mathbb{C}} \mathcal{A}_Y^{0,1})$$

is always a sheaf of rings on Y. It is in fact a sheaf of A-algebras equipped with a ring homomorphism $O_Y(A; \phi) \to O_Y$, so this is a candidate for an infinitesimal analytic deformation of Y.

Not all $O_Y(A; \phi)$ are different. Let $\theta \in \Gamma(Y, \mathfrak{m}_A \otimes_{\mathbb{C}} \mathcal{KS}_Y^0)$. Then we obtain an automorphism

$$\exp_\theta: \quad A \otimes_{\mathbb{C}} \mathcal{A}_Y^{0,q} \to A \otimes_{\mathbb{C}} \mathcal{A}_Y^{0,q}, \quad a \mapsto \sum_{n=0}^{\infty} \frac{([\theta, -])^n(a)}{n!} = a + [\theta, a] + \dots,$$

of sheaves of rings which induces the identity on $\mathcal{A}_Y^{0,q}$. Under this automorphism, $\phi \in \Gamma(Y, \mathfrak{m}_A \otimes_{\mathbb{C}} \mathcal{KS}_Y^1)$ corresponds to

$$\phi' = \exp_\theta * \phi = \phi + \sum_{n=0}^{\infty} \frac{([\theta, -])^n}{(n+1)!}([\theta, \phi] - \bar{\partial}\theta)$$

$$= \phi + ([\theta, \phi] - \bar{\partial}\theta) + \frac{1}{2}([\theta, [\theta, \phi] - \bar{\partial}\theta]) + \dots,$$

i.e., $\exp_\theta \circ \bar{\partial}_\phi = \bar{\partial}_{\phi'} \circ \exp_\theta$. In particular, we obtain an isomorphism

$$\exp_\theta: \quad O_Y(A; \phi) \to O_Y(A; \phi').$$

We say that $\exp_\theta$ is a *gauge transform*, and that ϕ and ϕ' are *gauge equivalent*.

A direct computation shows that

$$\bar{\partial}_\phi^2(a) = \left[\bar{\partial}(\phi) + \frac{1}{2}[\phi, \phi],\ a\right],$$

and one may prove that $\bar{\partial}_\phi^2 = 0$ is equivalent to $\bar{\partial}(\phi) + \frac{1}{2}[\phi, \phi] = 0$. This is called the *Maurer–Cartan equation*.

If $\bar{\partial}_\phi^2 = 0$, one can construct local gauge transforms which transform ϕ to 0. Thus, in this case, $O_Y(A; \phi)$ is locally isomorphic to $A \otimes_{\mathbb{C}} O_Y$ and therefore defines a deformation over S_A. Thus, we obtain a well-defined map from gauge equivalence classes of solutions of the Maurer–Cartan equation to isomorphism classes of infinitesimal deformations. This map is bijective.

Theorem 1.38 *Let Y be a complex manifold, and let $A \in$ **Art***.

(1) *Assume that $\phi \in \Gamma(Y, \mathfrak{m}_A \otimes_{\mathbb{C}} \mathcal{KS}_Y^1)$ solves the Maurer–Cartan equation. Then $O_Y(A; \phi)$ defines a deformation of Y over S_A.*
(2) *Let ϕ be a solution of the Maurer–Cartan equation, and let $\phi' = \exp_\theta * \phi$ for some $\theta \in \Gamma(Y, \mathfrak{m}_A \otimes_{\mathbb{C}} \mathcal{KS}_Y^0)$. Then $O_Y(A; \phi) \cong O_Y(A; \phi')$, and the deformations are isomorphic.*
(3) *If $O_Y(A; \phi)$ and $O_Y(A; \phi')$ define isomorphic deformations, then ϕ and ϕ' are gauge equivalent.*
(4) *If $f_A \colon (Y_A \supset Y) \to (S_A \ni 0)$ is a deformation, then there is a Maurer–Cartan element ϕ such that $O_Y(A; \phi) \cong O_{Y_A}$.*

Definition 1.39 (Kodaira–Spencer DG Lie Algebra) Let Y be a compact complex manifold. Then the *Kodaira–Spencer dg Lie algebra* is given by

$$K S_Y^\bullet := \Gamma(Y, \mathcal{KS}_Y^\bullet).$$

Definition 1.40 (Associated Deformation Functor) Let $(L^\bullet, [-, -], \bar{\partial})$ be a dg Lie algebra over $\mathbb{C}$. Then the *deformation functor associated with $L^\bullet$* is given by

$$\mathrm{Def}(L^\bullet, A) := \left\{ \phi \in \mathfrak{m}_A \otimes_{\mathbb{C}} L^1 \;\middle|\; \bar{\partial}(\phi) + \frac{1}{2}[\phi, \phi] = 0 \right\} /(\text{gauge equivalence})$$

for $A \in$ **Art**.

Thus, we may reformulate Theorem 1.38 as that we have an isomorphism

$$\mathrm{Def}_Y \cong \mathrm{Def}(K S_Y^\bullet, -)$$

of functors of Artin rings for a complex manifold Y. We say that the Kodaira–Spencer dg Lie algebra $K S_Y^\bullet$ *controls* Def_Y.

Even though we have not presented a proof of Theorem 1.38, note that it relies crucially on our two fundamental facts:

(1) $O_Y(A; \phi)$ is always locally trivial, so we need local rigidity to obtain all deformations.
(2) Infinitesimal automorphisms must be controlled by $\mathcal{T}_Y$ for its Dolbeault resolution to control when $O_Y(A; \phi)$ and $O_Y(A; \phi')$ are isomorphic.

1.4.5   A Proof of the Bogomolov–Tian–Todorov Theorem

There are now several proofs of Theorem 1.27 available. We present the proof of
Goldman–Millson from [109] since it allows us to introduce some terminology and
structures of later interest.

For the rest of the section, let Y be a compact Kähler Calabi–Yau manifold of
dimension d. More generally, it is sufficient to assume that the $\partial\bar{\partial}$-lemma holds for
Y.

Let $\psi\colon L^{\bullet}\to M^{\bullet}$ be a homomorphism of dg Lie algebras. This induces a natural
transformation

$$\mathrm{Def}(\psi)\colon\quad \mathrm{Def}(L^{\bullet},-)\Rightarrow\mathrm{Def}(M^{\bullet},-)$$

between functors of Artin rings. If ψ is a quasi-isomorphism of underlying cochain
complexes, then $\mathrm{Def}(\psi)$ is a natural isomorphism. We will show Theorem 1.27 by
simplifying $KS_Y^{\bullet}$.

We define $\Theta_Y^{-p}:=\bigwedge_{\mathcal{O}_Y}^{p}\mathcal{T}_Y$ for $0\le p\le d$. This is called the *sheaf of polyvector
fields*. Note our convention to put $\Theta_Y^{\bullet}$ in negative degrees.

The sheaves of polyvector fields $\Theta_Y^{\bullet}$ come with two operations: the *wedge
product*

$$-\wedge-\colon\quad \Theta_Y^{p_1}\times\Theta_Y^{p_2}\to\Theta_Y^{p_1+p_2}$$

and the *Schouten–Nijenhuis bracket*

$$[-,-]\colon\quad \Theta_Y^{p_1}\times\Theta_Y^{p_2}\to\Theta_Y^{p_1+p_2+1}.$$

The latter is an extension of the Lie bracket. With these two operations, $\Theta_Y^{\bullet}$ is a sheaf
of *Gerstenhaber algebras*. A precise definition can be found in Definition 3.10. Just
as in the case of the Lie bracket, our convention is to use $[-,-]=(-1)\cdot[-,-]_{\mathrm{SN}}$,
where $[-,-]_{\mathrm{SN}}$ is the usual Schouten–Nijenhuis bracket.

We apply the Dolbeault resolution from above to every Θ_Y^{p} and obtain a
differential bigraded Gerstenhaber algebra

$$
\begin{array}{ccccccc}
\mathcal{G}_Y^{0,0} & \xrightarrow{\bar{\partial}} & \mathcal{G}_Y^{0,1} & \xrightarrow{\bar{\partial}} & \mathcal{G}_Y^{0,2} & \longrightarrow & \cdots \\
\uparrow\Delta & & \uparrow\Delta & & \uparrow\Delta & & \\
\mathcal{G}_Y^{-1,0} & \xrightarrow{\bar{\partial}} & \mathcal{G}_Y^{-1,1} & \xrightarrow{\bar{\partial}} & \mathcal{G}_Y^{-1,2} & \longrightarrow & \cdots \\
\uparrow\Delta & & \uparrow\Delta & & \uparrow\Delta & & \\
\cdots & & \cdots & & \cdots & & \\
\uparrow\Delta & & \uparrow\Delta & & \uparrow\Delta & & \\
\mathcal{G}_Y^{-d,0} & \xrightarrow{\bar{\partial}} & \mathcal{G}_Y^{-d,1} & \xrightarrow{\bar{\partial}} & \mathcal{G}_Y^{-d,2} & \longrightarrow & \cdots
\end{array}
$$

which has only the solid horizontal differentials, not the dashed vertical ones. Here, we have $\mathcal{G}_Y^{0,q} = \mathcal{A}_Y^{0,q}$ and $\mathcal{G}_Y^{-1,q} = \mathcal{KS}_Y^q$ in the above notation. When $\theta \in \mathcal{G}_Y^{p,q}$, then its *degree* is $|\theta| = p + q$. A precise definition of a differential bigraded Gerstenhaber algebra can be found in Definition 3.10.

To construct the dashed vertical operations, we choose a volume form $\omega \in \Gamma(Y, \Omega_Y^d)$. Contraction with ω then induces isomorphisms

$$\kappa\colon \ \Theta_Y^p \to \Omega_Y^{p+d}, \quad \theta \mapsto \theta \lrcorner\, \omega,$$

for $-d \leq p \leq 0$, and therefore isomorphisms $\mathcal{G}_Y^{p,q} \cong \mathcal{A}_Y^{p+d,q}$ via the tensor product functor $\mathcal{A}_Y^{0,q} \otimes_{O_Y} (-)$. Now the *Batalin–Vilkovisky operator*

$$\Delta\colon \ \mathcal{G}_Y^{p,q} \to \mathcal{G}_Y^{p+1,q}, \quad \theta \mapsto \kappa^{-1} \circ \partial \circ \kappa(\theta),$$

is induced by the de Rham differential $\partial\colon \mathcal{A}_Y^{p+d,q} \to \mathcal{A}_Y^{p+d+1,q}$. This turns $\mathcal{G}_Y^{\bullet,\bullet}$ into a *differential bigraded Batalin–Vilkovisky algebra*, whose precise definition can be found in Definition 3.24.

By construction, the contraction isomorphism κ is compatible with the Dolbeault differential $\bar{\partial}$, so the double complex $(\mathcal{G}_Y^{\bullet,\bullet}, \Delta, \bar{\partial})$ is isomorphic to the double complex $(\mathcal{A}_Y^{\bullet,\bullet}, \partial, \bar{\partial})$. In particular, both satisfy the $\partial\bar{\partial}$-lemma on global sections.

Let $PV_Y^{p,q} := \Gamma(Y, \mathcal{G}_Y^{p,q})$ be the global sections. We define

$$C_Y^q := \ker(\Delta\colon PV_Y^{-1,q} \to PV_Y^{0,q}).$$

The *Bogomolov–Tian–Todorov formula*

$$(-1)^{|\theta|}[\theta, \xi] = \Delta(\theta \wedge \xi) - \Delta(\theta) \wedge \xi - (-1)^{|\theta|}\theta \wedge \Delta(\xi)$$

shows that $C_Y^\bullet \subseteq KS_Y^\bullet$ is a dg Lie subalgebra, and the $\partial\bar{\partial}$-lemma shows that this inclusion is a quasi-isomorphism. Let

$$Q_Y^q := C_Y^q / \mathrm{im}(\Delta\colon PV_Y^{-2,q} \to PV_Y^{-1,q}).$$

Now the Bogomolov–Tian–Todorov formula shows that $[-,-]$ descends to a well-defined Lie bracket $[-,-]_Q$ on $Q_Y^\bullet$. Thus, the quotient map $C_Y^\bullet \to Q_Y^\bullet$ is also a homomorphism of dg Lie algebras. Again, the $\partial\bar{\partial}$-lemma shows that it is a quasi-isomorphism. A third application of the Bogomolov–Tian–Todorov formula shows that $[-,-]_Q = 0$, so the bracket on $Q_Y^\bullet$ is actually trivial.

In summary, we have isomorphisms

$$\mathrm{Def}_Y \cong \mathrm{Def}(KS_Y^\bullet, -) \cong \mathrm{Def}(C_Y^\bullet, -) \cong \mathrm{Def}(Q_Y^\bullet, -)$$

of functors of Artin rings.

A dg Lie algebra $L^\bullet$ with $[-, -] = 0$ is called *Abelian*. If $L^\bullet$ is an Abelian dg Lie algebra, the Maurer–Cartan equation simplifies to $\bar\partial(\phi) = 0$, and the gauge action simplifies to $\exp_\theta * \phi = \phi - \bar\partial(\theta)$. Thus, we have

$$\mathrm{Def}(L^\bullet, A) = \mathfrak{m}_A \otimes_\mathbb{C} H^1(L^\bullet, \bar\partial),$$

and $\mathrm{Def}(L^\bullet, -)$ is unobstructed in the sense of Definition 1.36. This concludes the proof of Theorem 1.27 for compact Kähler Calabi–Yau manifolds Y.

Definition 1.41 (Homotopy Abelianity)   Let $L^\bullet$ be a dg Lie algebra over $\mathbb{C}$. Then $L^\bullet$ is *homotopy Abelian* if it is quasi-isomorphic to an Abelian dg Lie algebra.

1.5   Semistable Smoothings

In this section, we introduce semistable smoothings as a specific type of smoothings. We discuss the interplay between semistable smoothings and first-order deformations, leading to the definition of d-semistability. We give an example $\widetilde{H}$ of a d-semistable normal crossing scheme, we review Friedman's computation of the base of the miniversal deformation of $\widetilde{H}$, and we give an explicit example of a semistable smoothing of $\widetilde{H}$. At the end of the section, we introduce the semistable deformation functor.

1.5.1   Restricting the Local Structure of Deformations

We have seen above that infinitesimal deformations of a smooth variety Y are locally rigid, and this fact plays an important role in the above proof of the Bogomolov–Tian–Todorov Theorem 1.27. In contrast, degenerate schemes V have already locally many deformations.

Example 1.42   The degenerate scheme $V = \mathrm{Spec}\,\mathbb{C}[x, y, z]/(xy)$ has many deformations:

(1) $\mathrm{Spec}\,\mathbb{C}[x, y, z, t]/(xy) \to \mathbb{A}^1_t$, a trivial deformation;
(2) $\mathrm{Spec}\,\mathbb{C}[x, y, z, t]/(xy - t) \to \mathbb{A}^1_t$, a smoothing;
(3) $\mathrm{Spec}\,\mathbb{C}[x, y, z, t]/(xy - tz) \to \mathbb{A}^1_t$, another smoothing;
(4) $\mathrm{Spec}\,\mathbb{C}[x, y, z, t]/(xy - t^2z) \to \mathbb{A}^1_t$, yet another smoothing;
(5) $\mathrm{Spec}\,\mathbb{C}[x, y, z, t]/(xy - tz^2) \to \mathbb{A}^1_t$, a normaling;
(6) $\mathrm{Spec}\,\mathbb{C}[x, y, z, t]/(xy - tz^3) \to \mathbb{A}^1_t$, another normaling. $\Diamond$

Since our primary goal is not to understand all deformations but to construct a smoothing of a proper degenerate scheme V, it is useful to restrict our attention to some particularly simple local deformations of V, at least as a starting point.

The simplest deformations are locally trivial deformations, which are locally isomorphic to $V \times S_A$. They form a well-behaved functor of Artin rings

$$\mathrm{Def}_V^{\mathrm{lt}}\colon \quad \mathbf{Art} \to \mathbf{Set},$$

but they are not useful to construct a smoothing.

Let us now focus on normal crossing schemes V, which we have discussed in Sect. 1.2. We have defined a normal crossing scheme as being locally isomorphic to $\{z_0 \cdot \ldots \cdot z_r = 0\} \subset \mathbb{A}^{d+1}$. The fact that this scheme is the central fiber of the smoothing

$$\mu_{d;r}^{\mathrm{nc}}\colon \quad \mathbb{A}^{d+1} \to \mathbb{A}_t^1, \quad t \mapsto z_0 \cdot \ldots \cdot z_r,$$

suggests to consider smoothings of V which are locally isomorphic to the *local model* $\mu_{d;r}^{\mathrm{nc}}$. For an illustration of the special case $\mu_{2;2}^{\mathrm{nc}}$, see Fig. 1.5.

Definition 1.43 (Semistable Deformations) Let $V/\mathbb{C}$ be a normal crossing scheme. An algebraic one-parameter deformation $f\colon (X \supset V) \to (S \ni 0)$ is *semistable* if one (equivalently: every) representative has the following property: Every geometric point $\bar{x} \in X$ which is contained in $V \subset X$ admits two parameters $0 \le r \le d$ and an étale neighborhood $p\colon W \to X$ which fits into a commutative diagram

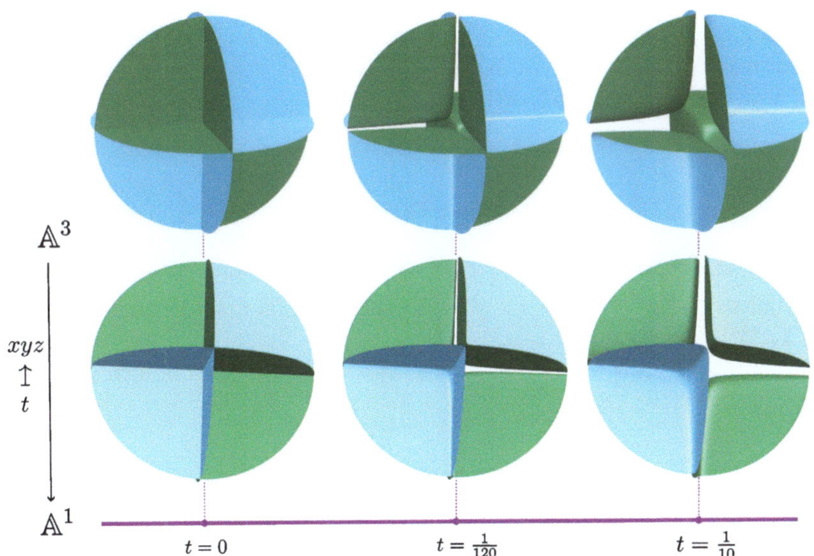

Fig. 1.5 Two perspectives on the local model $xyz = t$. © Simon Felten 2025. All rights reserved

$$
\begin{array}{ccccc}
X & \xleftarrow{\ p\ } & W & \xrightarrow{\ h\ } & \mathbb{A}^{d+1} \\
\downarrow{\scriptstyle f} & & \downarrow & & \downarrow{\scriptstyle \mu^{\mathrm{nc}}_{d;r}} \\
S & \longleftarrow & \widetilde{S} & \longrightarrow & \mathbb{A}^1_t
\end{array}
$$

where all horizontal maps are étale. Semistable analytic one-parameter deformations are defined analogously.

1.5.2   Criteria for Semistability

First, we prove the following criterion, which seems to be well-known, but for which we do not know a reference.

Lemma 1.44 *Let $V/\mathbb{C}$ be a normal crossing scheme, and consider an algebraic one-parameter deformation $f\colon (X \supset V) \to (S \ni 0)$. Then f is semistable if and only if $(X \supset V)$ is a smooth scheme germ.*

Proof If f is semistable, then X is smooth in a neighborhood of V. For the converse, let us assume that V is affine, and that X is smooth. Let $\bar{v} \to V$ be a geometric point. Since V is a normal crossing scheme, we can find an affine étale neighborhood $p_0\colon W_0 \to V$ such that all irreducible components of W_0 are smooth. As in Lemma 17.15, we can find an étale morphism $p\colon W \to X$ such that the induced map over $0 \in S$ is $p_0\colon W_0 \to V$. Since p is étale, also W is smooth. The claim now follows from [267, 0BIA]. $\qquad\qquad\Box$

Smoothness of $(X \supset V)$ can be detected on $X_1 = X \times_S S_1$ for an embedding

$$
S_1 = \operatorname{Spec}\mathbb{C}[t]/(t^2) \to S
$$

induced by an isomorphism $\widehat{O}_{S,0} \cong \mathbb{C}[\![t]\!]$. In particular, semistability of the deformation $f\colon (X \supset V) \to (S \ni 0)$ depends only on the *first-order deformation* $f_1\colon (X_1 \supset V) \to (S_1 \ni 0)$.

When $f_1\colon (X_1 \supset V) \to (S_1 \ni 0)$ is a first-order deformation, then we obtain an extension

$$
0 \to O_V \xrightarrow{\ 1 \mapsto c^* dt\ } c^* \Omega^1_{X_1/\mathbb{C}} \to \Omega^1_V \to 0,
$$

where $c\colon V \to X_1$ is the inclusion. It is a well-known fact that this induces a bijection between the set of isomorphism classes of first-order deformations and the set

$$
\mathbb{T}^1_V := \operatorname{Ext}^1(\Omega^1_V, O_V)
$$

of isomorphism classes of extensions, cf. [254, Rem. 2.4.4]. However, we can apply this construction also on every affine open subset of V and obtain a global section

$$\eta(f_1) \in \mathcal{T}_V^1 := \mathcal{E}xt^1(\Omega_V^1, O_V).$$

This is the image of the class in $\mathbb{T}_V^1$ under the canonical map $\gamma \colon \mathbb{T}_V^1 \to H^0(V, \mathcal{T}_V^1)$.

One may show by a direct computation in the local model $\{z_0 \cdot \ldots \cdot z_r = 0\} \subset \mathbb{A}^{n+1}$ that $\mathcal{T}_V^1$ is supported on $D = \mathrm{Sing}(V)$, and that $\mathcal{T}_V^1$ is furthermore a line bundle on D. If $f \colon (X \supset V) \to (S \ni 0)$ is an algebraic one-parameter deformation, then $(X \supset V)$ is smooth in a $\mathbb{C}$-valued point $v \in V \subset X$ if and only if either

(i) $v \in V \setminus D$ is a smooth point of the central fiber, or
(ii) $\eta(f_1) \in H^0(V, \mathcal{T}_V^1)$, considered as a section of a line bundle on D, does not have a zero at v.

Thus, $f \colon (X \supset V) \to (S \ni 0)$ is semistable if and only if $O_D \to \mathcal{T}_V^1$, $1 \mapsto \eta(f_1)$, is an isomorphism.

The necessary condition $\mathcal{T}_V^1 \cong O_D$ for the existence of a semistable smoothing was first described by Friedman in [87].

Definition 1.45 (*d*-**Semistability**) Let $V/\mathbb{C}$ be a normal crossing scheme, and let $D = \mathrm{Sing}(V)$. Then V is *d-semistable* if there is an isomorphism $\mathcal{T}_V^1 \cong O_D$ of coherent sheaves on V.

1.5.3 Computing $\mathcal{T}_V^1$

For a simple normal crossing scheme V, we compute $\mathcal{T}_V^1$ in terms of the geometry of V. For each irreducible component $V_i \in [C_0 V]$, we have the ideal sheaf $\mathcal{I}_{V_i/V} \subseteq O_V$ of the closed embedding $V_i \subset V$. One may show in the local model that $\mathcal{I}_{V_i/V} \otimes O_D$ is a line bundle on D. Then Friedman defines in [87]

$$O_D(-V) := \bigotimes_{V_i \in [C_0 V]} (\mathcal{I}_{V_i/V} \otimes O_D).$$

Now $\mathcal{T}_V^1$ is isomorphic to the dual $O_D(V) := \mathcal{H}om_D(O_D(-V), O_D)$. Since [87] does not give an explicit global isomorphism and we did not find it in the literature, we provide it here, based on the arguments in [87].

Lemma 1.46 *Let V be a normal crossing scheme. Then $O_D(V) \cong \mathcal{T}_V^1$.*

Proof Let $s \in \mathcal{T}_V^1$ be a local section corresponding to an extension

$$0 \to t \cdot O_V \to O_{X_1} \to O_V \to 0.$$

Let $a_1 \otimes \ldots \otimes a_c \in \mathcal{I}_{V_1/V} \otimes \ldots \otimes \mathcal{I}_{V_c/V} =: \mathcal{A}$. Choose lifts $\hat{a}_1, \ldots, \hat{a}_c \in O_{X_1}$. Then their product $\hat{a}_1 \cdot \ldots \cdot \hat{a}_c$ is contained in $t \cdot O_V$, and hence defines a local section of O_V. This section depends on our choice of lifts $\hat{a}_i$, but its image under the map $O_V \to O_D$ is independent of the choices of the lifts. Thus, we obtain a map $\beta \colon \mathcal{A} \times \mathcal{T}_V^1 \to O_D$. One may show that β is O_V-bilinear so that we obtain an induced map $b \colon \mathcal{A} \otimes_{O_V} \mathcal{T}_V^1 \to O_D$. Since $\mathcal{A} \otimes_{O_V} O_D \cong O_D(-V)$, we obtain a map $O_D(-V) \otimes_{O_D} \mathcal{T}_V^1 \to O_D$. A computation in the local model shows that this is an isomorphism. Hence $\mathcal{T}_V^1 \cong O_D(V)$. $\qquad\square$

In practice, it is sometimes useful to compute $O_D(V)|_B$ for a closed stratum $B \in [C_1 V]$ of codimension 1. Let $V_i \in [C_0 V]$ be an irreducible component. We distinguish two cases:

(i) $B \subset V_i$: Let $V_j \in [C_0 V]$ be the other irreducible component with $B \subset V_j$. Then the restriction map $\mathcal{I}_{V_i/V} \to \mathcal{I}_{B/V_j}$ induces an isomorphism

$$(\mathcal{I}_{V_i/V} \otimes O_D)|_B \cong \mathcal{I}_{B/V_j}/\mathcal{I}_{B/V_j}^2 = \mathcal{N}_{B/V_j}^{\vee}$$

with the conormal bundle of $B \subset V_j$.

(ii) $B \not\subset V_i$: In this case, $V_i \cap B \subset B$ is a smooth divisor. The restriction map $\mathcal{I}_{V_i/V} \to \mathcal{I}_{V_i \cap B/B}$ induces an isomorphism

$$(\mathcal{I}_{V_i/V} \otimes O_D)|_B \cong \mathcal{I}_{V_i \cap B/B}.$$

Now $O_D(-V)$ is the tensor product of these sheaves.

Example 1.47 We consider $V = H = \bigcup_{i=0}^{3} H_i$ from Example 1.10. The double locus decomposes into the union $D = \bigcup_{j=1}^{6} L_j$ of six lines $L_j \cong \mathbb{P}^1$. Let $L = H_0 \cap H_1$ be one of them. Then we have $\mathcal{I}_{H_0/H}|_L \cong \mathcal{N}_{L/H_1}^{\vee} \cong O_L(-1)$ and $\mathcal{I}_{H_1/H}|_L \cong \mathcal{N}_{L/H_0}^{\vee} \cong O_L(-1)$. Furthermore, we have $\mathcal{I}_{H_2/H}|_L \cong \mathcal{I}_{H_2 \cap L/L} \cong O_L(-1)$ and $\mathcal{I}_{H_3/H}|_L \cong \mathcal{I}_{H_3 \cap L/L} \cong O_L(-1)$. Thus, we have $O_D(H) \cong O_L(4)$, and H is not d-semistable. In particular, H does not admit a semistable smoothing. $\qquad\diamond$

1.5.4 An Example of a Semistable Smoothing

We have just seen in Example 1.47 that H is not d-semistable, but we can modify its geometry slightly and obtain a d-semistable normal crossing scheme $\widetilde{H}$.

Example 1.48 We construct a d-semistable modification $\widetilde{H}$ of H. Let $T = \mathrm{Sing}(D) \subset D$ be the triple locus of H, consisting of four isolated points. Then we choose four distinct points in $L_j \setminus T$ for each line L_j in D, and we denote the set of chosen points by Z. For each $p \in Z$, we choose one of the precisely two irreducible components of H which meet in p, and we denote the set of points

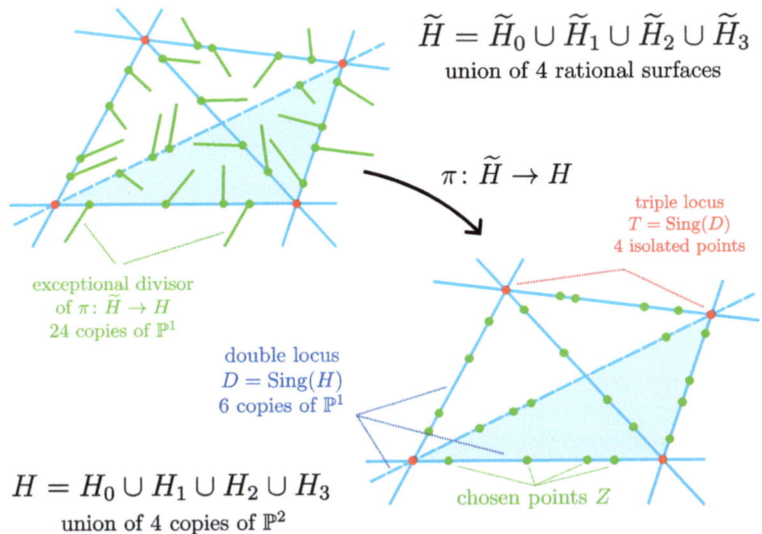

Fig. 1.6 Example 1.48. On the right, we have the normal crossing scheme H, which is not d-semistable. We choose 24 points on the double locus as indicated in the figure to produce a resolution of non-d-semistability $\pi \colon \widetilde{H} \to H$, which we see on the left. The resolution π is an isomorphism outside the 24 points; over each of them, the fiber is a copy of $\mathbb{P}^1$. © Simon Felten 2025. All rights reserved

for which H_i has been chosen by Z_i. Then we blow each H_i up in Z_i and obtain rational surfaces $\widetilde{H}_i$. We glue them along the strict transforms of the lines L_j to obtain a proper normal crossing space $\widetilde{H} = \bigcup_{i=0}^{3} \widetilde{H}_i$ and a proper birational map $\pi \colon \widetilde{H} \to H$. The situation is depicted in Fig. 1.6. We have

$$\mathcal{T}^1_{\widetilde{H}} \cong \mathcal{O}_{\widetilde{D}}$$

for $\widetilde{D} = \mathrm{Sing}(\widetilde{H})$ so that $\widetilde{H}$ is d-semistable. We still have a trivial dualizing sheaf $\omega_{\widetilde{H}} \cong \mathcal{O}_{\widetilde{H}}$ so that we may say that $\pi \colon \widetilde{H} \to H$ is a crepant resolution of the failure to be d-semistable.

Since $\widetilde{D} = \mathrm{Sing}(\widetilde{H})$ is connected, we have

$$\dim H^0(\widetilde{H}, \mathcal{T}^1_{\widetilde{H}}) = 1.$$

By Stevens [263, Thm. 3.4], the dimension of $\mathbb{T}^1_{\widetilde{H}}$ is 23 if $\widetilde{H}$ is in (-1)-form, which is the case if we have blown up each component of H in two points of each of the six lines in D. In any case, (-1)-form or not, the map

$$\gamma \colon \ \mathbb{T}^1_{\widetilde{H}} \to H^0(\widetilde{H}, \mathcal{T}^1_{\widetilde{H}})$$

is surjective, so a semistable deformation exists at least up to first order.

Friedman computed in [87, Thm. 5.10] the base $(B \ni 0)$ of the miniversal family of $\widetilde{H}$ and similar normal crossing Calabi–Yau surfaces. As complex space germs, there is a closed embedding $(B \ni 0) \to (\mathbb{T}^1_{\widetilde{H}} \ni 0)$. The base B decomposes as $(B \ni 0) = (B_1 \ni 0) \cup (B_2 \ni 0)$, where $(B_1 \ni 0)$ and $(B_2 \ni 0)$ are germs of smooth subvarieties of $(\mathbb{T}^1_{\widetilde{H}} \ni 0)$. The first component $(B_1 \ni 0) \subset (\mathbb{T}^1_{\widetilde{H}} \ni 0)$ is a divisor and corresponds to locally trivial deformations of $\widetilde{H}$. Its dimension is 22 if $\widetilde{H}$ is in (-1)-form. The second component $(B_2 \ni 0)$ is of dimension 20, and points in $B_2 \setminus B_1$ correspond to smooth K3 surfaces. The two components $(B_1 \ni 0)$ and $(B_2 \ni 0)$ meet transversally along a 19-dimensional subvariety of $(B_2 \ni 0)$ which corresponds to d-semistable locally trivial deformations of $\widetilde{H}$. Most curves $(\Delta \ni 0) \to (B_2 \ni 0)$ give rise to a semistable smoothing by base change of the miniversal deformation. This was the first major existence result for semistable smoothings, showing that every case in the Kulikov classification of semistable degenerations of K3 surfaces actually occurs. $\Diamond$

We now give an explicit semistable smoothing of $\widetilde{H}$.

Example 1.49 Consider the family $f : E \to \mathbb{P}^1$ of Example 1.11. The subset

$$Z := D \cap \{X_0^4 + X_1^4 + X_2^4 + X_3^4 = 0\} \subset H \subset \mathbb{P}^3$$

of the central fiber H consists of 24 points, four on each line $L_j \subset D$, and none of them is contained in the triple locus T. The set Z is the singular locus of the germ $(E \supset H)$, and the deformation $f : (E \supset H) \to (\mathbb{P}^1 \ni 0)$ is semistable outside Z. In the points $p \in Z$, the irreducible components H_i of H are not Cartier divisors but only Weil divisors. For each $p \in Z$, we choose a component $H_i \ni p$ and blow $(E \supset H)$ up in the subscheme $H_i \subset E$ locally around $p \in Z$. Since H_i is a Cartier divisor outside Z, this modifies E only over $p \in Z$, and thus we can choose the component independently for each $p \in Z$ and obtain a resolution of singularities $(\widetilde{E} \supset \widetilde{H}) \to (E \supset H)$. This gives rise to a semistable smoothing $g : (\widetilde{E} \supset \widetilde{H}) \to (\mathbb{P}^1 \ni 0)$. $\Diamond$

1.5.5 *The Semistable Deformation Functor*

So far, we have only considered semistable deformations over curves. The Cartesian diagram

$$
\begin{array}{ccc}
\operatorname{Spec} \mathbb{C}[x, y, z]/(xy) & \longrightarrow & \operatorname{Spec} \mathbb{C}[x, y] \\
\downarrow & & \downarrow {\scriptstyle t \mapsto xy} \\
\operatorname{Spec} \mathbb{C}[z] & \xrightarrow{\; z \mapsto 0 \;} & \operatorname{Spec} \mathbb{C}[t]
\end{array}
$$

shows that semistability is not stable under base change, so it is not obvious what a semistable deformation over a general base should be. The solution is to enrich the category **Art** of base spaces of deformations with information about a map to $\mathbb{A}_t^1$ to compare a given deformation with the local model $\mu_{d;r}^{\mathrm{nc}}$.

Definition 1.50 (Art$_{\mathbb{C}[\![t]\!]}$) An *Artinian local $\mathbb{C}[\![t]\!]$-algebra with residue field $\mathbb{C}$ is* a triple (A, p_A, e_A) consisting of an Artinian local ring A, a surjective local homomorphism $p_A \colon A \to \mathbb{C}$, and a local homomorphism $e_A \colon \mathbb{C}[\![t]\!] \to A$ such that $p_A \circ e_A$ equals the surjection $\mathbb{C}[\![t]\!] \to \mathbb{C}$. A *homomorphism* $(B, p_B, e_B) \to (A, p_A, e_A)$ is a local ring homomorphism $\phi \colon B \to A$ such that $p_A \circ \phi = p_B$ and $\phi \circ e_B = e_A$. We denote the category of Artinian local $\mathbb{C}[\![t]\!]$-algebras with residue field $\mathbb{C}$ by **Art$_{\mathbb{C}[\![t]\!]}$**.

Definition 1.51 (Infinitesimal Semistable Deformations) Let $V/\mathbb{C}$ be a normal crossing scheme. Let $A \in \textbf{Art}_{\mathbb{C}[\![t]\!]}$, and let $S_A = \operatorname{Spec} A$. An algebraic deformation $f_A \colon (X_A \supset V) \to (S_A \ni 0)$ is *semistable* if every geometric point $\bar{x} \in X_A$ admits two parameters $0 \le r \le d$ and an étale neighborhood $p \colon W \to X_A$ which fits into a commutative diagram

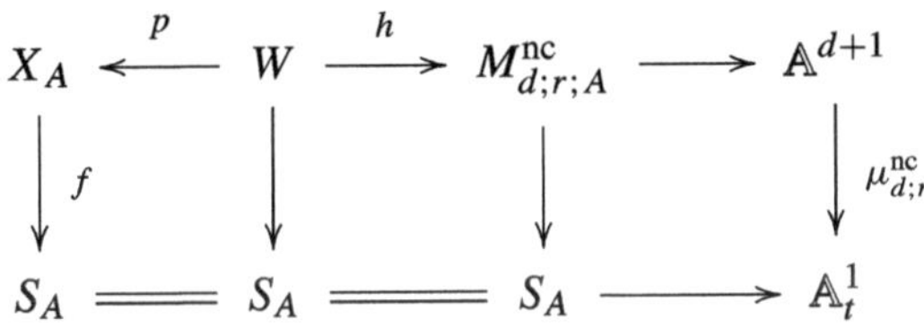

where h is étale, the right square is Cartesian, and $S_A \to \mathbb{A}_t^1$ is the morphism given by $e_A \colon \mathbb{C}[\![t]\!] \to A$. Isomorphism classes of semistable deformations form the *semistable deformation functor*

$$\operatorname{Def}_V^{\mathrm{ss}} \colon \quad \textbf{Art}_{\mathbb{C}[\![t]\!]} \to \textbf{Set}.$$

Semistable analytic deformations over $(S_A \ni 0)$ are defined analogously.

Let $f \colon (X \supset V) \to (S \ni 0)$ be a semistable one-parameter deformation. Let $(S_A \ni 0) \to (S \ni 0)$ be a morphism. Then the base change $f_A \colon X_A = X \times_S S_A \to S_A$ is a semistable deformation, where the $\mathbb{C}[\![t]\!]$-algebra structure on A comes from the choice of an isomorphism $\widehat{\mathcal{O}}_{S,0} \cong \mathbb{C}[\![t]\!]$.

Let V be a proper normal crossing scheme. Let $S_k = \operatorname{Spec} \mathbb{C}[t]/(t^{k+1})$, and assume that a formal semistable deformation $(f_k \colon (X_k \supset V) \to (S_k \ni 0))_{k \ge 0}$ is given. Then $(f_k)_{k \ge 0}$ is a formal algebraic one-parameter smoothing in the sense of Definition 1.23. Thus, unlike in the case of Def_V, formal smoothings (as opposed to mere formal deformations) can be constructed by analyzing the semistable deformation functor $\operatorname{Def}_V^{\mathrm{ss}}$, at least in principle.

Semistable deformations are not locally rigid.

Example 1.52 Let $V = \operatorname{Spec} \mathbb{C}[x, y]/(xy)$. Then

$$f_1\colon \quad \operatorname{Spec} \mathbb{C}[x, y, t]/(xy-t, t^2) \to S_1, \qquad f_1'\colon \quad \operatorname{Spec} \mathbb{C}[x, y, t]/(xy+t, t^2) \to S_1$$

are two non-isomorphic first-order deformations. Namely, we have

$$\eta(f_1) \neq \eta(f_1') \in \mathcal{T}_V^1 \cong \mathbb{C}.$$

The two deformations become isomorphic only when we allow the isomorphism to act non-trivially on the central fiber V. $\qquad\qquad\qquad\qquad\qquad\qquad\qquad\qquad\qquad \Diamond$

1.6   Logarithmic Algebraic Geometry

Let V be a normal crossing scheme. The passage from all deformations to semistable deformations has already greatly reduced the plethora of deformations under consideration, but we have just seen in Example 1.52 that semistable deformations are still not locally rigid. The solution offered by logarithmic algebraic geometry is to turn non-isomorphic semistable deformations into deformations of *different things*. This is achieved by endowing V with an additional structure, namely a *logarithmic structure*. In Chap. 7, the reader can find a somewhat more detailed short introduction into the log geometry used in this monograph. The standard reference for basic facts in log geometry is now Ogus' book [222]. At the beginning of Chap. 7, we give some advice on how to read [222]. In Sect. 2 (on works related to this monograph), the reader can find some comments on the history of log geometry.

1.6.1   Logarithmic Structures

Logarithmic algebraic geometry studies schemes endowed with log structures. All monoids are commutative and have a unit.

Definition 1.53 (Log Schemes) Let Y be a scheme. A *log structure* is a homomorphism of sheaves of monoids

$$\alpha\colon \quad (\mathcal{M}, \cdot) \to (\mathcal{O}_Y, \cdot)$$

in the étale topology of Y such that $\alpha^{-1}(\mathcal{O}_Y^*) = \mathcal{O}_Y^*$. A *homomorphism* between log structures $(\mathcal{M}_1, \alpha_1)$ and $(\mathcal{M}_2, \alpha_2)$ is a map $\phi\colon \mathcal{M}_1 \to \mathcal{M}_2$ of sheaves of monoids such that $\alpha_2 \circ \phi = \alpha_1$. A *log scheme* is a scheme Y together with a log structure $(\mathcal{M}, \alpha)$. A *morphism* between log schemes $(Y_1, \mathcal{M}_1, \alpha)$ and $(Y_2, \mathcal{M}_2, \alpha)$ consists of a morphism of schemes $(f, f^\sharp)\colon (Y_1, \mathcal{O}_{Y_1}) \to (Y_2, \mathcal{O}_{Y_2})$ together with a homomorphism of sheaves of monoids $f^\flat \colon \mathcal{M}_2 \to f_* \mathcal{M}_1$ which is compatible with $f^\sharp \colon O_{Y_2} \to f_* O_{Y_1}$.

The difference between $\mathcal{M}$ and O_Y is measured by the *characteristic sheaf* or *ghost sheaf* $\overline{\mathcal{M}}$. This is a somewhat crude combinatorial invariant of $\mathcal{M}$, which is nonetheless very important.

Definition 1.54 (Ghost Sheaves) Let $(Y, \mathcal{M})$ be a log scheme. Then the *ghost sheaf* is the quotient

$$\overline{\mathcal{M}} := \mathcal{M}/O_Y^*.$$

We have two fundamental constructions of log schemes. The simpler one starts with a closed embedding $C \subset Y$.

Construction 1.55 (Compactifying Log Structures) Let Y be a scheme, and let $C \subset Y$ a closed subscheme. We can define the *compactifying* (also called *divisorial*) log structure as follows. The sheaf of monoids $\mathcal{M}_{(Y|C)}$ is given by

$$\Gamma(W, \mathcal{M}_{(Y|C)}) = \{a \in \Gamma(W, O_Y) \mid a|_{W \setminus C} \text{ is invertible}\}$$

for open subsets $W \subseteq Y$, and by the multiplication of functions as the monoid operation. Together with the obvious monoid homomorphism $\alpha \colon \mathcal{M}_{(Y|C)} \to O_Y$, this gives a log structure on Y. The resulting log scheme is denoted by $(Y|C)$. $\Diamond$

Example 1.56 Let $C = \varnothing \subset Y$. Then $\mathcal{M}_{(Y|\varnothing)} = O_Y^*$ is the *trivial log structure*. $\Diamond$

The second fundamental construction starts with a ring R and a homomorphism of monoids $\beta \colon P \to (R, \cdot)$, or more generally with a scheme Y and a homomorphism of monoids $\beta \colon P \to \Gamma(Y, O_Y)$.

Construction 1.57 (Charts and the Spectrum) Let Y be a scheme, let P be a monoid, and let $\beta \colon P \to \Gamma(Y, O_Y)$ be a monoid homomorphism. Consider the push-out diagram

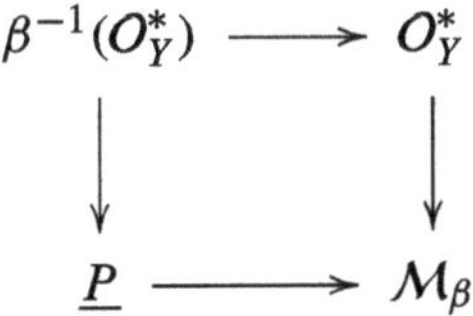

in the category of sheaves of monoids, where $\underline{P}$ is the constant sheaf in the étale topology with stalk P. Then, the map $\alpha \colon \mathcal{M}_\beta \to O_Y$ induced by the universal property of push-out squares is a log structure. If $Y = \operatorname{Spec} R$, then we write $\operatorname{Spec}(P \to R)$ for $(Y, \mathcal{M}_\beta)$, and we say that $\operatorname{Spec}(P \to R)$ is the *spectrum* of $\beta \colon P \to R$. We say that $\beta \colon P \to \Gamma(Y, O_Y)$ is a *chart* for the log structure $\mathcal{M}_\beta$. $\Diamond$

A log scheme $(Y, \mathcal{M}_Y)$ which is étale locally isomorphic to some $\mathrm{Spec}(P \to R)$ is called *quasi-coherent*. If P can be chosen to be finitely generated, then X is called *coherent*. Here, a monoid P is *finitely generated* if there is a surjection $\mathbb{N}^r \to P$ for some $r \geq 0$.

Example 1.58 Choose $0 \leq r \leq d$. Let $R = \mathbb{C}[z_0, \ldots, z_d]$, let $Y = \mathbb{A}^{d+1} = \mathrm{Spec}\, R$, and let $C = H_{d;r}^{\mathrm{nc}} = \{z_0 \cdot \ldots \cdot z_r = 0\} \subset Y$. Then we obtain a log scheme $(Y|C) = (\mathbb{A}^{d+1}|H_{d;r}^{\mathrm{nc}})$. Let now $P = \bigoplus_{i=0}^{r} \mathbb{N} \cdot e_i$, and consider $\beta \colon P \to R$, $e_i \mapsto z_i$. This gives rise to a log scheme $\mathrm{Spec}(P \to R)$. For $0 \leq i \leq r$, we have $z_i \in \mathcal{M}_{(Y|C)} \subset O_Y$, and thus we obtain an induced map $\phi \colon \underline{P} \to \mathcal{M}_{(Y|C)}$, $e_i \mapsto z_i$. The construction of the log structure $\mathcal{M}_\beta$ on $\mathrm{Spec}(P \to R)$ yields a homomorphism $\varphi \colon \mathcal{M}_\beta \to \mathcal{M}_{(Y|C)}$ of log structures, which is an isomorphism. In particular, $(\mathbb{A}^{d+1}|H_{d;r}^{\mathrm{nc}})$ is coherent. $\Diamond$

Whenever we have a one-parameter deformation $f \colon (X \supset V) \to (S \ni 0)$, the construction of the compactifying log structure allows us to consider f as a morphism of log schemes $(X|V) \to (S|\{0\})$ after choosing representatives of the germs.

Example 1.59 Consider the semistable smoothing $g \colon \widetilde{E} \to \mathbb{P}^1$ from Example 1.49. We endow $\mathbb{P}^1$ with the compactifying log structure defined by $\{0\} \subset \mathbb{P}^1$, and we endow $\widetilde{E}$ with the compactifying log structure defined by $\widetilde{H} = g^{-1}(0) \subset \widetilde{X}$. This turns $g \colon \widetilde{E} \to \mathbb{P}^1$ into a morphism of log schemes $g \colon (\widetilde{E}|\widetilde{H}) \to (\mathbb{P}^1|\{0\})$. Near $\widetilde{H}$, it is étale locally isomorphic to the morphism of log schemes $(\mathbb{A}^{d+1}|H_{d;r}^{\mathrm{nc}}) \to (\mathbb{A}_t^1|\{0\})$. $\Diamond$

The simplest morphisms of log schemes are *strict* morphisms, which are defined by the following universal property.

Definition 1.60 (Strictness) Let $(f, f^\sharp, f^\flat) \colon (Y_1, \mathcal{M}_1) \to (Y_2, \mathcal{M}_2)$ be a morphism of log schemes. Then f is *strict* if the following condition holds: Let $(Z, \mathcal{M}_Z)$ be another log scheme, let $(h, h^\sharp, h^\flat) \colon (Z, \mathcal{M}_Z) \to (Y_2, \mathcal{M}_2)$ be a morphism of log schemes, and let $(g, g^\sharp) \colon Z \to Y_1$ be a morphism of schemes such that $(h, h^\sharp) = (f, f^\sharp) \circ (g, g^\sharp)$. Then there is a unique $g^\flat \colon \mathcal{M}_1 \to h_* \mathcal{M}_Z$ such that $(g, g^\sharp, g^\flat)$ is a morphism of log schemes, and such that $(h, h^\sharp, h^\flat) = (f, f^\sharp, f^\flat) \circ (g, g^\sharp, g^\flat)$.

When $f \colon Y_1 \to Y_2$ is a morphism of schemes, and when $\mathcal{M}_2$ is a log structure on Y_2, then there is always a log structure $\mathcal{M}_1$ on Y_1 which admits a strict morphism $(f, f^\flat) \colon (Y_1, \mathcal{M}_1) \to (Y_2, \mathcal{M}_2)$. Here, $\mathcal{M}_1$ and $f^\flat$ are unique up to unique isomorphism. We write $\mathcal{M}_1 = f_{\log}^* \mathcal{M}_2$. This is called the *pull-back* of the log structure $\mathcal{M}_2$ along $f \colon Y_1 \to Y_2$.

1.6.2 Integral and Saturated Log Structures

In logarithmic geometry, we usually assume that our log structures are *fine and saturated*, and we will stick to that convention throughout the monograph. Monoids and log structures with these properties are particularly simple and well-behaved. Here, we give some brief indications about the definition of these properties. More details can be found in Chap. 7 and of course in Ogus' [222].

The inclusion of the category of Abelian groups into the category of monoids has a left adjoint, given by a universal map $P \to P^{\mathrm{gp}}$ into an Abelian group. A monoid P is *integral* if and only if the cancellation property holds, i.e., if $p + r = q + r$ implies $p = q$ for any three elements $p, q, r \in P$. This is equivalent to $P \to P^{\mathrm{gp}}$ being injective. An integral monoid P is *saturated* if, for every $n \geq 1$ and $p \in P^{\mathrm{gp}}$ with $np = p + \ldots + p \in P \subseteq P^{\mathrm{gp}}$, we already have $p \in P$. A log structure $\alpha \colon \mathcal{M} \to \mathcal{O}_Y$ is *integral* respectively *saturated* if this holds for all (geometric) stalks $\mathcal{M}_{\bar{y}}$. A log structure is *fine* if it is coherent and integral, and it is *fine and saturated* if it is coherent, integral, and saturated.

1.6.3 Fiber Products and Saturated Morphisms

The category of log schemes has fiber products. For two morphisms of log schemes $f \colon (X, \mathcal{M}) \to (S, \mathcal{N})$ and $b \colon (T, \mathcal{N}') \to (S, \mathcal{N})$, the underlying scheme of their fiber product $(X, \mathcal{M}) \times_{(S,\mathcal{N})} (T, \mathcal{N}')$ is precisely the fiber product $X \times_S T$ of schemes. The construction of the log structure on the fiber product is based on the push-out of monoids.

When $\theta_1 \colon Q \to P_1$ and $\theta_2 \colon Q \to P_2$ are two homomorphisms between integral or saturated monoids, then their push-out $P_1 \oplus_Q P_2$ may not be integral respectively saturated. Consequently, fiber products of fine and saturated log schemes may not be fine and saturated. Most often in log geometry, this situation is remedied by passing to the associated fine and saturated log scheme, a construction which does not preserve the underlying scheme. However, this solution is unsuitable for our purposes since we need the underlying scheme of the fiber product to be the fiber product of underlying schemes. Instead, we restrict our attention to *integral* and *saturated* homomorphisms of monoids respectively morphisms of log schemes. When $f \colon (X, \mathcal{M}) \to (S, \mathcal{N})$ is a saturated morphism, then the fiber product along $b \colon (T, \mathcal{N}') \to (S, \mathcal{N})$ is saturated for any morphism from a fine and saturated log scheme $(T, \mathcal{N}')$. More details can be found in Definition 7.9 and the surrounding discussion.

1.6.4 Logarithmic Derivations and Differential Forms

Let $f\colon (X, M) \to (S, N)$ be a morphism of log schemes. The logarithmic analog of a derivation with values in an O_X-module $\mathcal{E}$ is a pair (D, Δ), where $\Delta\colon O_X \to \mathcal{E}$ is an $f^{-1}O_S$-linear derivation, and where $\Delta\colon (M, \cdot) \to (\mathcal{E}, +)$ is a monoid homomorphism such that $D(\alpha(m)) = \alpha(m) \cdot \Delta(m)$ for every local section $m \in M$, and $\Delta(f^{\flat}(n)) = 0$ for every local section $n \in f^{-1}N$. There is a universal log derivation

$$(d, \delta)\colon \quad (O_X, M) \to \Omega^1_{(X, M)/(S, N)}.$$

The target is called the sheaf of *log differential forms*. We write

$$\Theta^1_{(X, M)/(S, N)} = \mathcal{H}om(\Omega^1_{(X, M)/(S, N)}, O_X)$$

for the sheaf of log derivations with values in O_X. This is a sheaf of Lie algebras which controls infinitesimal automorphisms.

1.6.5 Logarithmic Smoothness

A morphism $f\colon (X, M) \to (S, N)$ of log schemes is *formally log smooth* if it has the infinitesimal lifting property in the category of log schemes. More precisely, when we have a solid diagram of log schemes

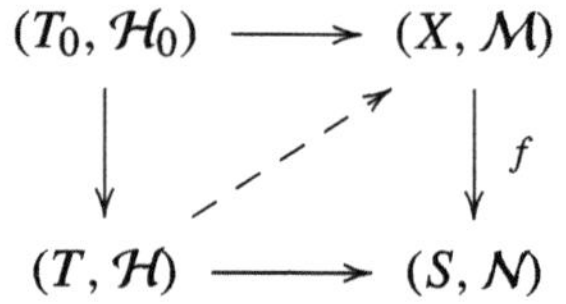

$$
\begin{array}{ccc}
(T_0, \mathcal{H}_0) & \longrightarrow & (X, M) \\
\downarrow & \nearrow & \downarrow{\scriptstyle f} \\
(T, \mathcal{H}) & \longrightarrow & (S, N)
\end{array}
$$

where $T_0 \to T$ is a first-order thickening of log schemes (see Definition 7.49), and where T_0 is an affine scheme, then there is a dashed arrow making the diagram commutative. A morphism of log schemes $f\colon (X, M) \to (S, M)$ is *log smooth* if it is formally log smooth, if M and N are coherent, and if $f\colon X \to S$ is locally of finite presentation.

Example 1.61 Let $f\colon X \to S$ be a morphism of schemes. Then $f\colon (X|\varnothing) \to (S|\varnothing)$ is log smooth if and only if $f\colon X \to S$ is smooth. $\Diamond$

When $f\colon (X, M) \to (S, N)$ is log smooth, then $\Omega^1_{(X, M)/(S, N)}$ is locally free.

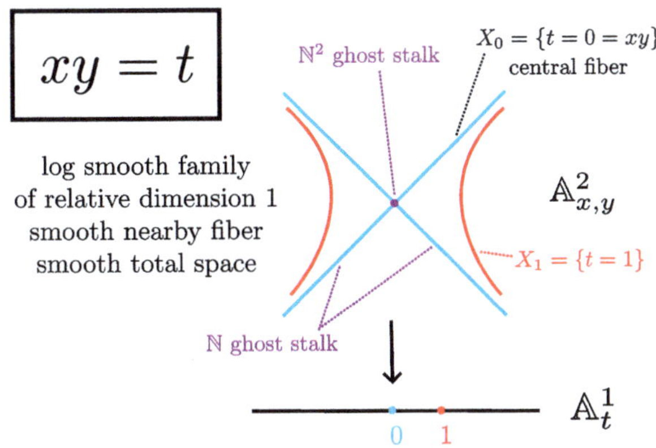

Fig. 1.7 Example 1.62. © Simon Felten 2025. All rights reserved

By K. Kato's toroidal characterization of log smoothness, every saturated log smooth morphism $f\colon (X, \mathcal{M}) \to (S, \mathcal{N})$ is étale locally isomorphic to the base change of some toric morphism

$$(\mathbb{A}_P, \mathcal{M}_P) \times \mathbb{A}^r \to (\mathbb{A}_P, \mathcal{M}_P) \xrightarrow{A_\theta} (\mathbb{A}_Q, \mathcal{M}_Q)$$

of toric varieties, where $(\mathbb{A}_P, \mathcal{M}_P) = \mathrm{Spec}(P \to \mathbb{C}[P])$ and $(\mathbb{A}_Q, \mathcal{M}_Q) = \mathrm{Spec}(Q \to \mathbb{C}[Q])$, where $\theta\colon Q \to P$ is a saturated monoid homomorphism, and where $\mathbb{A}^r$ carries the trivial log structure. The log structures on $\mathbb{A}_P$ and $\mathbb{A}_Q$ coincide with the compactifying log structures defined by the full toric boundaries $D_P \subset \mathbb{A}_P$ and $D_Q \subset \mathbb{A}_Q$.[6] Conversely, this construction allows us to give many explicit examples of log smooth morphisms.

Example 1.62 (xy = t) Let $Q = \mathbb{N}$, let $P = \mathbb{N}^2$, and let $\theta(1) = (1, 1)$. Then the log smooth morphism A_θ is given by

$$f\colon \left(\mathrm{Spec}\,\mathbb{C}[x, y] \big| \{xy = 0\} \right) \to \left(\mathbb{A}^1_t \big| \{0\} \right), \quad t \mapsto xy.$$

This map is log smooth of relative dimension 1. The nearby fiber is smooth, as is the total space. The central fiber consists of two copies of $\mathbb{A}^1$, intersecting in a point. The ghost stalk at this point is $\mathbb{N}^2$ while the ghost stalk at every other point is $\mathbb{N}$. This example is of course nothing but our local model $\mu^{\mathrm{nc}}_{1;1}$ for semistable smoothings. See Fig. 1.7 for an illustration. $\diamondsuit$

Example 1.63 (xy = $\mathbf{t^k}$) Let $k \geq 1$. Let $Q = \mathbb{N}$, let

$$P = P_k = (\mathbb{N}\,\bar{x} \oplus \mathbb{N}\,\bar{y} \oplus \mathbb{N}\,\bar{w})/(\bar{x} + \bar{y} = k \cdot \bar{w})$$

---

[6] We will later write $A_P := (\mathbb{A}_P, \mathcal{M}_P)$ while using the notation $\mathbb{A}_P$ for the underlying scheme.

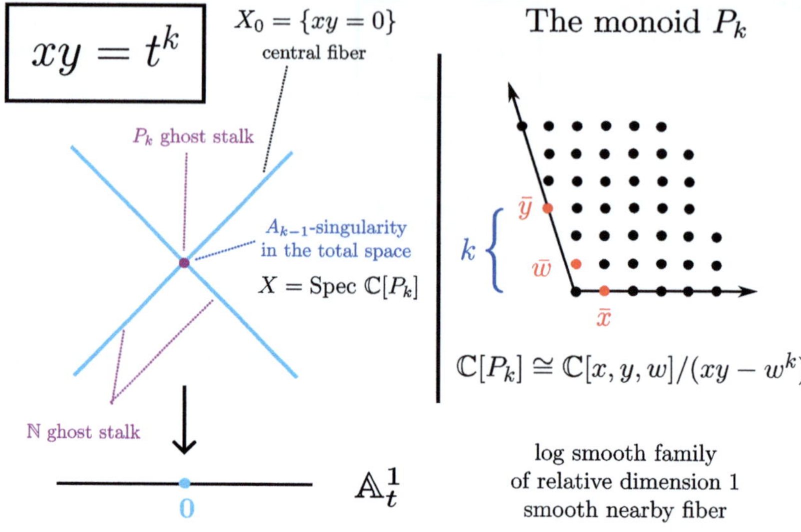

Fig. 1.8 Example 1.63. On the left, we have in blue the central fiber of the log smooth morphism given by $xy = t^k$. This central fiber is a normal crossing scheme with a single double point in purple. In this double point, the stalk of the ghost sheaf is given by P_k. Everywhere else on the central fiber, the stalk of the ghost sheaf is $\mathbb{N}$. On the right, we see P_k as a submonoid of $\mathbb{Z}^2$. © Simon Felten 2025. All rights reserved

be the monoid depicted in Fig. 1.8, and let $\theta(1) = \bar{w}$. Then the log smooth morphism A_θ is given by

$$f\colon \left(\mathrm{Spec}\ \mathbb{C}[x, y, w]/(xy - w^k)\,\big|\,\{w = 0\}\right) \to \left(\mathbb{A}^1_t\,\big|\,\{0\}\right), \quad t \mapsto w.$$

In particular, $f^{-1}(\{t = 0\}) = \{w = 0\}$ is the full toric boundary of $X = \mathrm{Spec}\ \mathbb{C}[P_k]$. While the total space in the previous Example 1.62 is smooth, now the total space has an A_{k-1}-singularity. The stalk of the ghost sheaf $\overline{\mathcal{M}}_X$ at the intersection of the two components of the central fiber X_0 is a copy of the monoid P_k. ◊

Example 1.64 (xyz = t) Let $Q = \mathbb{N}$, let $P = \mathbb{N}^3$, and let $\theta(1) = (1, 1, 1)$. Then A_θ is given by

$$f\colon \left(\mathrm{Spec}\ \mathbb{C}[x, y, z]\,\big|\,\{xyz = 0\}\right) \to \left(\mathbb{A}^1_t\,\big|\,\{0\}\right), \quad t \mapsto xyz.$$

This map is log smooth of relative dimension 2. It is the semistable smoothing $\mu^{\mathrm{nc}}_{2;2}$ whose central fiber $X_0 = H^{\mathrm{nc}}_{2;2}$ has three components intersecting in a single point. The stalk of the ghost sheaf $\overline{\mathcal{M}}_X$ at the triple intersection point is $\mathbb{N}^3$. On the double locus, it is $\mathbb{N}^2$, and in the interior of the irreducible components, it is $\mathbb{N}$. The general fiber of the family is smooth. For an illustration, see Fig. 1.9. ◊

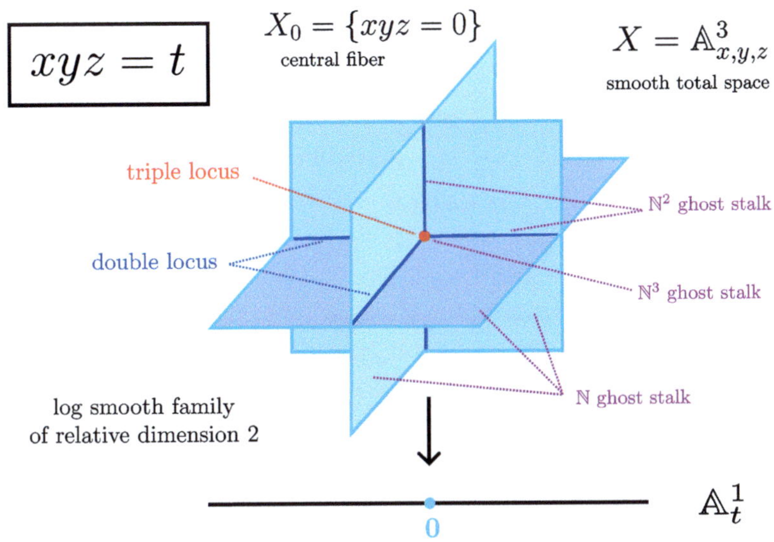

Fig. 1.9 Example 1.64. © Simon Felten 2025. All rights reserved

Example 1.65 ($z_0 \cdot \ldots \cdot z_r = t$) For $d \geq 0$ and $0 \leq r \leq d$, let $Q = \mathbb{N}$, let $P = \mathbb{N}^{r+1}$, and let $\theta(1) = (1, \ldots, 1)$. Then $(\mathbb{A}_P | D_P) \times \mathbb{A}^{d-r} \to (\mathbb{A}_Q | D_Q)$ is nothing but the local model $\mu^{\mathrm{nc}}_{d;r} \colon (\mathbb{A}^{d+1} | H^{\mathrm{nc}}_{d;r}) \to (\mathbb{A}^1_t | \{0\})$ of semistable smoothings. $\diamond$

In particular, the morphism of log schemes $g \colon (\widetilde{E} | \widetilde{H}) \to (\mathbb{P}^1 | \{0\})$ from Example 1.59 is log smooth in a neighborhood of $0 \in \mathbb{P}^1$ since $g \colon (\widetilde{E} \supset \widetilde{H}) \to (\mathbb{P}^1 \ni 0)$ is semistable.

Example 1.66 ($xy = w^k$, $zw = t$) This is a generalization of the semistable smoothing $\mu^{\mathrm{nc}}_{2;2}$. Fix $k \geq 1$. Let $Q = \mathbb{N}$, let $P = P_k \oplus \mathbb{N}\bar{z}$ with the monoid P_k from Example 1.63, and let $\theta(1) = \bar{z} + \bar{w}$. Then the log smooth morphism A_θ is given by

$$f \colon \left(\mathrm{Spec}\, \mathbb{C}[x, y, z, w]/(xy - w^k) \,|\, \{zw = 0\}\right) \to \left(\mathbb{A}^1_t \,|\, \{0\}\right), \quad t \mapsto zw.$$

The total space has a cA_{k-1}-singularity in the sense that we have an A_{k-1}-singularity multiplied with $\mathbb{A}^1$. The central fiber X_0 has three irreducible components; two of them, $V_{yz} = \{x = w = 0\} = \mathrm{Spec}\, \mathbb{C}[y, z]$ and $V_{xz} = \{y = w = 0\} = \mathrm{Spec}\, \mathbb{C}[x, z]$, are smooth; the remaining component $V_{xy} = \{z = 0\} = \mathrm{Spec}\, \mathbb{C}[x, y, w]/(xy - w^k)$ has an A_{k-1}-singularity which lies in the triple locus of X_0. The singular locus of the total space is then $V_{yz} \cap V_{xz} = \mathrm{Spec}\, \mathbb{C}[z]$. The stalk of the ghost sheaf $\overline{\mathcal{M}}_X$ at the single triple intersection point is $P_k \oplus \mathbb{N}$. Along $V_{yz} \cap V_{xz}$, this is Example 1.63; in particular, the ghost stalk is P_k. Along the other two lines of the double locus, this is $xy = t$, hence the stalk of the ghost sheaf $\overline{\mathcal{M}}_X$ is $\mathbb{N}^2$. For an illustration, see Fig. 1.10. $\diamond$

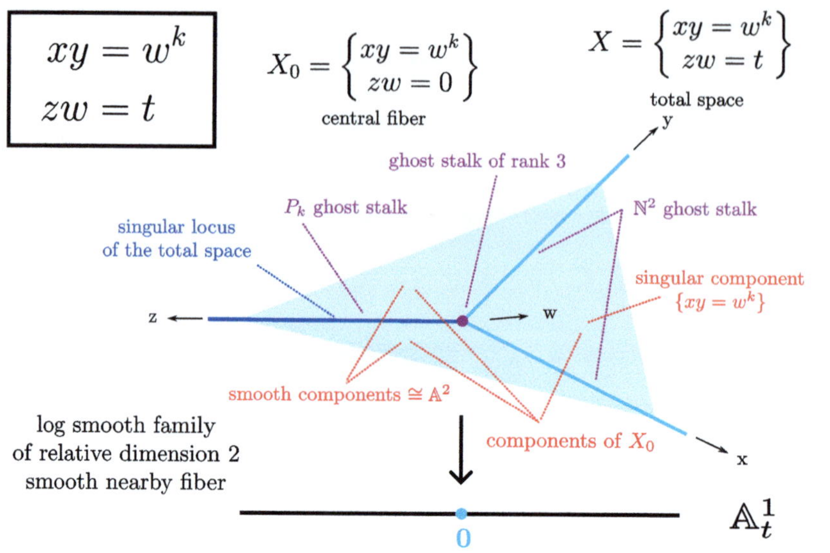

Fig. 1.10 Example 1.66. We see in light blue the three irreducible components of the central fiber X_0 of the log smooth morphism from Example 1.66. Two of them are smooth and intersect in the dark blue component of the double locus on the left. The third irreducible component of X_0 on the right is singular. It intersects the other two components of X_0 in the middle blue lines. © Simon Felten 2025. All rights reserved

1.6.6 Semistable Log Smooth Structures

We consider the log morphism $g \colon (\widetilde{E}|\widetilde{H}) \to (\mathbb{P}^1|\{0\})$. Let $S_0 \subset \mathbb{P}^1$ be the point $0 \in \mathbb{P}^1$, considered as a closed subscheme. Since the target $(\mathbb{P}^1|\{0\})$ carries a log structure, we can pull it back and obtain a log structure $\alpha \colon \mathcal{N}_0 \to O_{S_0}$ on S_0 such that $(S_0, \mathcal{N}_0) \to (\mathbb{P}^1|\{0\})$ is strict in the sense of Definition 1.60. We say that $(S_0, \mathcal{N}_0)$ is the *standard log point*. Explicitly, we have

$$\Gamma(S_0, \mathcal{N}_0) = \mathbb{C}^* \oplus \mathbb{N}, \quad \alpha(\lambda, n) = \lambda \cdot 0^n,$$

where $0^0 = 1$ and $0^n = 0$ for $n \geq 1$. In particular, $(S_0, \mathcal{N}_0)$ is not a trivial log point since the trivial log structure on S_0 is given by $\mathbb{C}^* \to \mathbb{C}$. We can pull back the log structure from $(\widetilde{E}|\widetilde{H})$ to the fiber $\widetilde{H}$ as well, thus constructing a log morphism $g_0 \colon (\widetilde{H}, \mathcal{M}_0) \to (S_0, \mathcal{N}_0)$. It fits into a Cartesian diagram

$$
\begin{array}{ccc}
(\widetilde{H}, \mathcal{M}_0) & \longrightarrow & (\widetilde{E}|\widetilde{H}) \\
\downarrow{\scriptstyle g_0} & & \downarrow{\scriptstyle g} \\
(S_0, \mathcal{N}_0) & \longrightarrow & (\mathbb{P}^1|\{0\})
\end{array}
$$

in the category of log schemes. The property of being log smooth is stable under base change. In other words, the log morphism $g_0 \colon (\widetilde{H}, \mathcal{M}_0) \to (S_0, \mathcal{N}_0)$ is log smooth. In this way, we have enhanced our original central fiber $\widetilde{H}$ with a *log smooth structure*.

Definition 1.67 (Log Smooth Structures I) Let $V/\mathbb{C}$ be a degenerate scheme. A *log smooth structure* consists of a log structure $\alpha \colon \mathcal{M}_0 \to \mathcal{O}_V$ together with a morphism of log schemes $f_0 \colon (V, \mathcal{M}_0) \to (S_0, \mathcal{N}_0)$ which is log smooth. A log smooth structure is *semistable* if $f_0 \colon (V, \mathcal{M}_0) \to (S_0, \mathcal{N}_0)$ is étale locally isomorphic to the central fiber of some $\mu_{d;r}^{\mathrm{nc}} \colon (\mathbb{A}^{d+1} | H_{d;r}^{\mathrm{nc}}) \to (\mathbb{A}_t^1 | \{0\})$.

The reader may want to compare this definition with its variant Definition 1.143 in the context of toroidal crossing schemes.

Note that V is necessarily a normal crossing scheme if it admits a semistable log smooth structure.

Different semistable smoothings may induce different log smooth structures on the central fiber. This explains how the two non-isomorphic semistable first-order deformations of Example 1.52 can be considered as "deformations of different things."

Example 1.68 Let $X = \mathbb{A}^2 = \operatorname{Spec} \mathbb{C}[x, y]$, and let us consider the two morphisms $f \colon X \to \mathbb{A}_t^1$, $t \mapsto xy$, and $f' \colon X \to \mathbb{A}^1$, $t \mapsto -xy$. Each of them induces a log smooth morphism $(X | V) \to (\mathbb{A}_t^1 | \{0\})$, where $V = \{xy = 0\} \subset X$. In particular, we obtain two log smooth morphisms $f_0, f_0' \colon (V, \mathcal{M}_0) \to (S_0, \mathcal{N}_0)$, where $\mathcal{M}_0 = i_{\log}^* \mathcal{M}_{(X|V)}$ for the inclusion $i \colon V \to X$. These two log smooth morphisms are not equal because they map $\tau = (0, 1) \in \Gamma(S_0, \mathcal{N}_0)$ to two different sections of $\Gamma(V, \mathcal{M}_0)$. In this way, the two semistable deformations in Example 1.52 can be considered as deformations of different logarithmic schemes over $(S_0, \mathcal{N}_0)$. $\Diamond$

Semistable log smooth structures admit a simple classification. The key observation is that they do not have any non-trivial (local) automorphisms: namely,

$$V \supseteq W \mapsto \{\text{semistable log smooth structures on } W\}/\{\text{isomorphisms}\}$$

is a sheaf $\mathcal{LS}_V$ on V. We will compute $\mathcal{LS}_V$ after introducing log smooth deformation theory.

1.6.7 *Log Smooth Deformation Theory*

We have constructed a semistable log smooth structure $g_0 \colon (\widetilde{H}, \mathcal{M}_0) \to (S_0, \mathcal{N}_0)$ on $\widetilde{H}$ by means of the semistable smoothing $g \colon (\widetilde{E} \supset \widetilde{H}) \to (\mathbb{P}^1 \ni 0)$. From a somewhat different angle, imagine that we do not already have the semistable smoothing g, but that we have constructed a semistable log smooth structure $g_0 \colon (\widetilde{H}, \mathcal{M}_0) \to (S_0, \mathcal{N}_0)$ on $\widetilde{H}$ by analyzing $\mathcal{LS}_{\widetilde{H}}$. We may then try to deform $\widetilde{H}$ as a log smooth morphism rather than as a scheme to construct a smoothing.

The concept of log smooth deformations goes back to Fumiharu Kato's article [160]. To define log smooth deformations, we first need a thickening of the log point $(S_0, \mathcal{N}_0)$. They are readily obtained from Artinian local $\mathbb{C}[\![t]\!]$-algebras $A \in \mathbf{Art}_{\mathbb{C}[\![t]\!]}$. Namely, each A admits a morphism

$$S_A = \operatorname{Spec} A \to \mathbb{P}^1$$

induced by the $\mathbb{C}[\![t]\!]$-algebra structure on A, and then we can endow S_A with the pull-back log structure $\alpha \colon \mathcal{N}_A \to O_{S_A}$. Note that the log structure $\mathcal{N}_A$ depends on the $\mathbb{C}[\![t]\!]$-algebra structure, not just on the underlying $\mathbb{C}$-algebra structure. For example,

$$(S_1 = \operatorname{Spec} \mathbb{C}[t]/(t^2), \ \mathcal{N}_1) \quad \text{and} \quad (S_\varepsilon = \operatorname{Spec} \mathbb{C}[\varepsilon]/(\varepsilon^2), \ \mathcal{N}_\varepsilon)$$

are not isomorphic as log schemes.

Given $A \in \mathbf{Art}_{\mathbb{C}[\![t]\!]}$, we can construct $(S_A, \mathcal{N}_A)$ more directly as the spectrum $\operatorname{Spec}(\mathbb{N} \to A, \ 1 \mapsto t)$ in the sense of Construction 1.57. This allows us to generalize to other log points than the standard log point: Let Q be a *sharp toric monoid*, i.e., $Q = \sigma^\vee \cap \mathbb{Z}^n$ is the set of lattice points of the dual $\sigma^\vee$ of some strongly convex rational polyhedral cone $\sigma \subseteq \mathbb{R}^n$, such as one usually considers in toric geometry. Then we obtain the complete local Noetherian ring $\mathbb{C}[\![Q]\!]$ by completing the monoid algebra $\mathbb{C}[Q]$ in the maximal ideal $\mathbb{C}[Q \setminus \{0\}]$. Every Artinian local $\mathbb{C}[\![Q]\!]$-algebra A with residue field $\mathbb{C}$ (defined in analogy with Definition 1.50) determines a log scheme $(S_A, \mathcal{N}_A) = \operatorname{Spec}(Q \to A)$ which is a thickening of the log point $(S_0, \mathcal{N}_0) = \operatorname{Spec}(Q \to \mathbb{C})$. The case $Q = \mathbb{N}$ with $\mathbb{C}[\![\mathbb{N}]\!] \cong \mathbb{C}[\![t]\!]$ is by far the most important, so the reader may restrict their attention to this case.

Definition 1.69 (Log Smooth Deformations) Let Q be a sharp toric monoid, let $(S_0, \mathcal{N}_0) = \operatorname{Spec}(Q \to \mathbb{C})$, and let $f_0 \colon (X_0, \mathcal{M}_0) \to (S_0, \mathcal{N}_0)$ be a separated and saturated log smooth morphism. Let $A \in \mathbf{Art}_{\mathbb{C}[\![Q]\!]}$. Then a *log smooth deformation* is a Cartesian diagram

$$
\begin{array}{ccc}
(V, \mathcal{M}_0) & \longrightarrow & (X_A, \mathcal{M}_A) \\
\downarrow {\scriptstyle f_0} & & \downarrow {\scriptstyle f_A} \\
(S_0, \mathcal{N}_0) & \longrightarrow & (S_A, \mathcal{N}_A)
\end{array}
$$

in the category of log schemes such that $f_A \colon (X_A, \mathcal{M}_A) \to (S_A, \mathcal{N}_A)$ is log smooth. A *morphism* between log smooth deformations is defined in analogy with a morphism between algebraic deformations, see Definition 1.35. Isomorphism classes of log smooth deformations form the *log smooth deformation functor*

$$\mathrm{LD}_{X_0/S_0} \colon \ \mathbf{Art}_{\mathbb{C}[\![Q]\!]} \to \mathbf{Set}.$$

Example 1.70 Consider $g\colon (\widetilde{E}|\widetilde{H}) \to (\mathbb{P}^1|\{0\})$, and let $A \in \mathbf{Art}_{\mathbb{C}[\![t]\!]}$. Then we can form the Cartesian square

$$
\begin{array}{ccc}
(\widetilde{E}_A, \mathcal{M}_A) & \longrightarrow & (\widetilde{E}|\widetilde{H}) \\
\downarrow{\scriptstyle g_A} & & \downarrow{\scriptstyle g} \\
(S_A, \mathcal{N}_A) & \longrightarrow & (\mathbb{P}^1|\{0\})
\end{array}
$$

in the category of log schemes by pulling back the log structure on $(\widetilde{E}|\widetilde{H})$ along $\widetilde{E}_A \to \widetilde{E}$. This is a log smooth deformation of $g_0\colon (\widetilde{H}, \mathcal{M}_0) \to (S_0, \mathcal{N}_0)$. $\diamond$

The infinitesimal lifting property shows that log smooth deformations are locally rigid: Let $f_0\colon (X_0, \mathcal{M}_0) \to (S_0, \mathcal{N}_0)$ be log smooth with X_0 affine, and let $f_A\colon (X_A, \mathcal{M}_A) \to (S_A, \mathcal{N}_A)$ and $f_A'\colon (X_A', \mathcal{M}_A') \to (S_A, \mathcal{N}_A)$ be two log smooth deformations. Then f_A and f_A' are isomorphic.

For infinitesimal algebraic or analytic deformations, local rigidity implies local triviality because every deformation must be locally isomorphic to the trivial deformation $V \times S_A$. This implication does no longer hold for log smooth deformations: In order to form the trivial deformation $X_0 \times S_A$, we need a map $S_A \to S_0$, but not every $(S_A, \mathcal{N}_A)$ admits a log morphism to $(S_0, \mathcal{N}_0)$. For example, $(S_1, \mathcal{N}_1)$ defined above does not admit a morphism to $(S_0, \mathcal{N}_0)$. This explains why the underlying flat deformation of a log smooth deformation does not need to be locally trivial although log smooth deformations are locally rigid.

Example 1.71 Consider the log morphism $\mu_{1;1}^{\mathrm{nc}}\colon (\mathbb{A}^2|H_{1;1}^{\mathrm{nc}}) \to (\mathbb{A}_t^1|\{0\})$. Then:

(1) $(\operatorname{Spec}\mathbb{C}[x, y, t]/(xy - t, t^2), \mathcal{M}_1) \to (S_1, \mathcal{N}_1)$ is a log smooth deformation, and it is a first-order smoothing.
(2) $(\operatorname{Spec}\mathbb{C}[x, y, \varepsilon]/(xy, \varepsilon^2), \mathcal{M}_\varepsilon) \to (S_\varepsilon, \mathcal{N}_\varepsilon)$ is a log smooth deformation whose underlying flat deformation is trivial. $\diamond$

Although some $(S_A, \mathcal{N}_A)$ do not admit a notion of a trivial deformation over it, a log smooth deformation always exists locally. The reason is that every log smooth map $f_0\colon (X_0, \mathcal{M}_0) \to (S_0, \mathcal{N}_0)$ is locally in the smooth topology isomorphic to the central fiber of

$$
A_\theta\colon \quad (\mathbb{A}_P, \mathcal{M}_P) := \operatorname{Spec}(P \to \mathbb{C}[P]) \to \operatorname{Spec}(Q \to \mathbb{C}[Q]) =: (\mathbb{A}_Q, \mathcal{M}_Q)
$$

for some monoid homomorphism $\theta\colon Q \to P$. Then an étale local log smooth deformation is given by the base change of A_θ along $(S_A, \mathcal{N}_A) \to (\mathbb{A}_Q, \mathcal{M}_Q)$.

Automorphisms of log smooth deformations are controlled by the sheaf of logarithmic derivations $\Theta^1_{X_0/S_0}$ introduced above. Let $\phi\colon B \to A$ be a first-order thickening in $\mathbf{Art}_{\mathbb{C}[\![Q]\!]}$, let $I \subset B$ be the kernel, and consider a morphism

$$(X_0, \mathcal{M}_0) \longrightarrow (X_A, \mathcal{M}_A) \longrightarrow (X_B, \mathcal{M}_B)$$

$$\downarrow f_0 \qquad\qquad \downarrow f_A \qquad\qquad \downarrow f_B$$

$$(S_0, \mathcal{N}_0) \longrightarrow (S_A, \mathcal{N}_A) \longrightarrow (S_B, \mathcal{N}_B)$$

of log smooth deformations. Then infinitesimal automorphisms of f_B which fix f_A are in one-to-one correspondence with sections of the sheaf $\Theta^1_{X_0/S_0} \otimes_{\mathbb{C}} I$ via the formulae in Sect. 10.2. As a consequence, when $\phi\colon B \to A$ and $f_A\colon (X_A, \mathcal{M}_A) \to (S_A, \mathcal{N}_A)$ are given, the obstruction to the existence of a lift to $(S_B, \mathcal{N}_B)$ lies in

$$H^2(X_0, \Theta^1_{X_0/S_0} \otimes_{\mathbb{C}} I),$$

and the space of lifts is a (pseudo-)torsor under

$$H^1(X_0, \Theta^1_{X_0/S_0} \otimes_{\mathbb{C}} I).$$

In particular, if X_0 is affine, then there is a unique log smooth deformation over $(S_A, \mathcal{N}_A)$ for every $A \in \mathbf{Art}_{\mathbb{C}[\![Q]\!]}$. It is unique up to non-unique isomorphism.

1.6.8 The Logarithmic Bogomolov–Tian–Todorov Theorem

In Theorem 1.27, we have seen that flat deformations of algebraic Calabi–Yau manifolds are unobstructed. If LD_{X_0/S_0} is unobstructed as well, then we can construct log smooth deformations. To define the concept of unobstructedness for LD_{X_0/S_0} is straightforward.

Definition 1.72 (Unobstructedness II) Let $F\colon \mathbf{Art}_{\mathbb{C}[\![Q]\!]} \to \mathbf{Set}$ be a functor of Artin rings. Then F is *unobstructed* if, for every surjection $\phi\colon B \to A$ in $\mathbf{Art}_{\mathbb{C}[\![Q]\!]}$, the map $F(\phi)\colon F(B) \to F(A)$ is surjective.

When $f\colon (X, \mathcal{M}) \to (S, \mathcal{N})$ is log smooth, and if the sheaf of log differential forms $\Omega^1_{X/S}$ is locally free of some *constant* rank $d \geq 1$, then we can form the *log canonical sheaf* as

$$\omega_{X/S} := \Omega^d_{X/S}.$$

We then say that $f_0\colon (X_0, \mathcal{M}_0) \to (S_0, \mathcal{N}_0)$ is *log Calabi–Yau* if $\omega_{X_0/S_0} \cong O_{X_0}$. With this definition, the logarithmic Bogomolov–Tian–Todorov theorem becomes the following.

Theorem 1.73 (Log Smooth Bogomolov–Tian–Todorov Theorem) *Let Q be a sharp toric monoid, let $(S_0, \mathcal{N}_0) = \mathrm{Spec}(Q \to \mathbb{C})$, and let $f_0\colon (X_0, \mathcal{M}_0) \to$*

$(S_0, \mathcal{N}_0)$ *be saturated, log smooth, and proper. Assume that f_0 is log Calabi–Yau. Then* LD_{X_0/S_0} *is unobstructed.*

In the main text, this is Theorem 15.2. We gave this theorem also as [78, Thm. 1.1] in our joint work with Petracci. A method that leads to the proof of the logarithmic Bogomolov–Tian–Todorov theorem was provided in 2019 by Chan, Leung, and Ma in [38], building on earlier work of Kontsevich, Manetti, and others. We will say more about the proof in Sect. 1.8 below.

Let now $Q = \mathbb{N}$, and assume that $f_0\colon (X_0, \mathcal{M}_0) \to (S_0, \mathcal{N}_0)$ is *vertical* in the sense of Definition 7.14. For example, $\mu_{d;r}^{\mathrm{nc}}$ is vertical. Then the log canonical sheaf ω_{X_0/S_0} is isomorphic to the dualizing sheaf $\omega_{X_0}^\circ$. Thus, in the vertical situation, the log Calabi–Yau condition can be checked on the underlying scheme.

Assume again that f_0 is vertical. Let $(S_k, \mathcal{N}_k) = \mathrm{Spec}(\mathbb{N} \to \mathbb{C}[t]/(t^{k+1}))$, and let $(f_k\colon (X_k, \mathcal{M}_k) \to (S_k, \mathcal{N}_k))_{k\geq 0}$ be a formal log smooth deformation. Then $(f_k)_{k\geq 0}$ is a formal smoothing. Thus, Theorem 1.73 allows us to construct a formal one-parameter smoothing, and hence also an analytic one-parameter smoothing if V is a Calabi–Yau scheme in the sense of Definition 1.9 which carries a vertical and saturated log smooth structure.

1.6.9 Comparison Between Log Smooth Deformations and Semistable Deformations

Let V be a normal crossing scheme, and let $f_0\colon (V, \mathcal{M}_0) \to (S_0, \mathcal{N}_0)$ be a semistable log smooth structure. Let $f_A\colon (X_A, \mathcal{M}_A) \to (S_A, \mathcal{N}_A)$ be a log smooth deformation. Local rigidity of log smooth deformations shows that the underlying deformation $f_A\colon (X_A \supset V) \to (S_A \ni 0)$ is semistable. Thus, we obtain a natural transformation

$$\mathrm{LD}_{V/S_0} \Rightarrow \mathrm{Def}_V^{\mathrm{ss}}.$$

In other words, choosing a semistable log smooth structure on V provides us with refined information about semistable deformations.

1.6.10 Computation of $\mathcal{LS}_V$

As promised above, we compute $\mathcal{LS}_V$ for a normal crossing scheme V.

Construction 1.74 Let $f_0\colon (V, \mathcal{M}_0) \to (S_0, \mathcal{N}_0)$ be a semistable log smooth structure on V. On an affine open subset $W \subseteq V$, log smooth deformations are unique up to non-unique isomorphism. Thus, we can form a log smooth deformation over $(S_1, \mathcal{N}_1)$. This gives rise to a flat deformation $f_1\colon (W_1 \supset W) \to (S_1 \ni 0)$

and thus defines a section $\eta(f_1) \in \mathcal{T}^1_W = \mathcal{E}xt^1(\Omega^1_W, \mathcal{O}_W)$. We can apply this construction on any affine open subset $W \subseteq V$ and obtain a map

$$\eta\colon \ \mathcal{LS}_V \to \mathcal{T}^1_V.$$

This map is injective, and its image is the sheaf $(\mathcal{T}^1_V)^*$ of local generators of $\mathcal{T}^1_V$. $\Diamond$

In particular, a global semistable log smooth structure on V defines a trivialization $\mathcal{T}^1_V \cong \mathcal{O}_D$, where $D = \mathrm{Sing}(V)$ is the double locus. We obtain a new interpretation of d-semistability: It is necessary to obtain a global semistable log smooth structure, which would be induced by any semistable smoothing.

The map $\eta\colon \mathcal{LS}_V \to \mathcal{T}^1_V$ allows us to construct log smooth structures: Any trivialization $\mathcal{T}^1_V \cong \mathcal{O}_D$ induces a semistable log smooth structure on V.

The equivalence between the existence of semistable log smooth structures and d-semistability was first formulated by Kawamata and Namikawa in [173, Prop. 1.1].

1.7 Generically Log Smooth Families

In Example 1.10, we have constructed the normal crossing scheme $H = \bigcup_{i=0}^3 H_i$ by gluing its irreducible components. We have then turned our attention to semistable smoothings but observed in Example 1.47 that H is not d-semistable so that H does not admit a semistable smoothing. To obtain a semistable smoothing, we have constructed a resolution $\pi\colon \widetilde{H} \to H$ of the failure to be d-semistable in Example 1.48, and in Example 1.49, we have provided an explicit semistable smoothing $g\colon (\widetilde{E} \supset \widetilde{H}) \to (\mathbb{P}^1 \ni 0)$ of $\widetilde{H}$. This semistable smoothing can be interpreted as a log morphism $g\colon (\widetilde{E}|\widetilde{H}) \to (\mathbb{P}^1|\{0\})$ which is log smooth in a neighborhood of $0 \in \mathbb{P}^1$, and our plan is to use logarithmic deformation theory in great generality to construct smoothings of degenerate schemes V once they are endowed with a log smooth structure in the sense of Definition 1.67.

The explicit semistable smoothing $g\colon (\widetilde{E} \supset \widetilde{H}) \to (\mathbb{P}^1 \ni 0)$ was obtained by a birational modification of the—in terms of defining equations—simpler smoothing $f\colon (E \supset H) \to (\mathbb{P}^1 \ni 0)$. Just as $g\colon \widetilde{E} \to \mathbb{P}^1$, we can also turn $f\colon E \to \mathbb{P}^1$ into a log morphism $f\colon (E|H) \to (\mathbb{P}^1|\{0\})$ by means of the compactifying log structures. However, this log morphism is not log smooth at the 24 points of Z, the singular locus of the germ $(E \supset H)$ defined in Example 1.49. Thus, the smoothing $f\colon (E \supset H) \to (\mathbb{P}^1 \ni 0)$ cannot be constructed by means of log smooth deformation theory—it is simply not log smooth.

The log singularities of $f\colon (E|H) \to (\mathbb{P}^1|\{0\})$ are even worse than being merely not log smooth: The log structure $\mathcal{M}_{(E|H)}$ is not quasi-coherent around $p \in Z$, and even the sheaf of logarithmic differential forms

$$\Omega^1_{(E|H)/(\mathbb{P}^1|\{0\})}$$

is not quasi-coherent, as was observed by Gross and Siebert in [125, Ex. 1.11].

We would like to have a theory that also constructs the smoothing $f\colon (E \supset H) \to (\mathbb{P}^1 \ni 0)$. This would allow us to smooth H directly without constructing the resolution $\pi\colon \widetilde{H} \to H$ of non-d-semistability. To this end, we need a notion of log morphism that captures log morphisms like $f\colon (E|H) \to (\mathbb{P}^1|\{0\})$ while circumventing the pathological behavior of the log structure in Z. Our solution is simple: We record Z as the *log singular locus* and drop the log structure on Z, only retaining it on $U = E \setminus Z$. We then say that $f\colon (E, U|H) \to (\mathbb{P}^1|\{0\})$ is a *generically log smooth family*, even though it is log smooth in relative codimension *one*, which is more than our usual understanding of "generically." First, we introduce the following auxiliary notion.

Definition 1.75 (Partial Log Schemes)

(1) A *partial log scheme* $(X, U, \mathcal{M})$ consists of a scheme (X, O_X), an open subset $j\colon U \subseteq X$, and a log structure $\alpha\colon \mathcal{M} \to O_U$ on U.
(2) A *morphism* between partial log schemes $(X, U_X, \mathcal{M})$ and $(S, U_S, \mathcal{N})$ consists of a morphism of schemes $(f, f^\sharp)\colon (X, O_X) \to (S, O_S)$ such that $U_X \subseteq f^{-1}(U_S)$ together with a map $f^\flat\colon (c|_{U_X})^{-1}\mathcal{N} \to \mathcal{M}$ of sheaves of monoids such that

$$(f|_{U_X}, f^\sharp|_{U_X}, f^\flat)\colon \quad (U_X, O_{U_X}, \mathcal{M}) \to (U_S, O_{U_S}, \mathcal{N})$$

is a morphism of log schemes.
(3) A morphism $f\colon (X, U_X, \mathcal{M}) \to (S, U_S, \mathcal{N})$ is *accurate* if $U_X = f^{-1}(U_S)$.

Remark 1.76 When talking about properties of partial log schemes and their morphisms, we use the following conventions:

(1) Let **P** be a property of schemes. Then $(X, U, \mathcal{M})$ has **P** if X has **P**.
(2) Let **P** be a property of log schemes. Then $(X, U, \mathcal{M})$ has **P** if $(U, \mathcal{M})$ has **P**.
(3) Let **P** be a property of scheme morphisms. A morphism of partial log schemes $f\colon (X, U_X, \mathcal{M}) \to (S, U_S, \mathcal{N})$ has **P** if $f\colon X \to S$ has **P**.
(4) Let **P** be a property of log morphisms. A morphism of partial log schemes $f\colon (X, U_X, \mathcal{M}) \to (S, U_S, \mathcal{N})$ has **P** if $f\colon (U_X, \mathcal{M}) \to (U_S, \mathcal{N})$ has **P**.

For example, $p\colon (X', U', \mathcal{M}') \to (X, U, \mathcal{M})$ is *accurate strict étale* if $p\colon X' \to X$ is étale, $(U', \mathcal{M}') \to (U, \mathcal{M})$ is strict, and $U' = p^{-1}(U)$. $\Diamond$

We consider every log scheme $(X, \mathcal{M})$ as a partial log scheme by virtue of $U := X$. In this case, we simply write $(X, \mathcal{M})$ instead of $(X, X, \mathcal{M})$.

The category of partial log schemes has fiber products. For two morphisms $f\colon (X, U_X) \to (S, U_S)$ and $b\colon (T, U_T) \to (S, U_S)$, the fiber product is given by the scheme $(Y, O_Y) = (X, O_X) \times_{(S, O_S)} (T, O_T)$ and the log scheme

$$(U_Y, \mathcal{M}_{U_Y}) = (U_X, \mathcal{M}_{U_X}) \times_{(U_S, \mathcal{M}_{U_S})} (U_T, \mathcal{M}_{U_T}).$$

Definition 1.77 (Generically Log Smooth Families) A *generically log smooth family* is a morphism of partial log schemes $f\colon (X, U, \mathcal{M}) \to (S, \mathcal{N})$ such that:

(a) $(S, \mathcal{N})$ is a locally Noetherian fine and saturated log scheme;
(b) $f \colon X \to S$ is flat, separated, and of finite type; its fibers satisfy Serre's condition (S_2), are geometrically reduced, and are equidimensional of some fixed dimension $d \geq 1$;
(c) the complement $Z = X \setminus U \subset X$ of $U \subseteq X$ has codimension ≥ 2 in every fiber;
(d) $(U, \mathcal{M})$ is a fine and saturated log scheme;
(e) $f \colon (U, \mathcal{M}) \to (S, \mathcal{N})$ is saturated and log smooth.

A *morphism* between generically log smooth families $g \colon (Y, V, \mathcal{M}_V) \to (T, \mathcal{N}_T)$ and $f \colon (X, U, \mathcal{M}_U) \to (S, \mathcal{N}_S)$ is a commutative diagram

$$
\begin{array}{ccc}
(Y, V, \mathcal{M}_V) & \xrightarrow{\ c\ } & (X, U, \mathcal{M}_U) \\
\big\downarrow{\scriptstyle g} & & \big\downarrow{\scriptstyle f} \\
(T, \mathcal{N}_T) & \xrightarrow{\ b\ } & (S, \mathcal{N}_S)
\end{array}
$$

of partial log schemes. A morphism of generically log smooth families is *accurate* if $V = c^{-1}(U)$.

Additional discussion of generically log smooth families and some motivation for the choices we made in this definition is given in the introduction to Chap. 8 and in Sect. 8.1. There, we also briefly discuss the related theory of *relatively log smooth families*, which is a useful but less general approach to log singularities due to Nakayama and Ogus, introduced in [212].

We think of generically log smooth families as logarithmic analogs of normal algebraic varieties. Namely, an algebraic variety Y is normal if and only if it is regular in codimension 1 and satisfies Serre's condition (S_2). In a generically log smooth family, regularity in codimension 1 is replaced with log smoothness outside a specified subset of codimension 2, and the (S_2)-condition is retained.

Let $j \colon U \to X$ be the open immersion of $U = X \setminus Z$. A consequence of and our main motivation for the (S_2)-condition is that it implies $j_* \mathcal{O}_U = \mathcal{O}_X$. This implies that $j_* \mathcal{E}$ is coherent and reflexive for every vector bundle $\mathcal{E}$ on U. We are mainly interested in the *reflexive log de Rham complex*

$$
\mathcal{W}^{\bullet}_{X/S} := j_* \Omega^{\bullet}_{U/S}
$$

and in the *reflexive log polyvector fields*

$$
\mathcal{V}^{p}_{X/S} := j_* \bigwedge_{\mathcal{O}_U}^{-p} \Theta^1_{U/S},
$$

where $p \leq 0$ and $\Theta^1_{U/S}$ is the sheaf of relative log derivations of $f \colon U \to S$. In situations where the log structure $\mathcal{M}$ is actually given on X rather than only on U, the sheaves $\mathcal{W}^i_{X/S}$ serve as a *coherent replacement* for $\Omega^i_{(X,\mathcal{M})/(S,\mathcal{N})}$, which may not be coherent as is the case for $f: (E|H) \to (\mathbb{P}^1|\{0\})$.

Example 1.158 below shows that the formation of $\mathcal{W}^i_{X/S}$ does not always commute with base change along $(T, \mathcal{N}_T) \to (S, \mathcal{N})$. If it commutes, then we say that $f: (X, U, \mathcal{M}) \to (S, \mathcal{N})$ has the *base change property*.

Definition 1.78 (Base Change Property) Let $f: (X, U, \mathcal{M}_X) \to (S, \mathcal{N}_S)$ be a generically log smooth family. Then f has the *base change property* if, for all morphisms of locally Noetherian coherent log schemes $b: (T, \mathcal{N}_T) \to (S, \mathcal{N}_S)$, the pull-back $c^*\mathcal{W}^i_{X/S}$ along the map $c: Y = X \times_S T \to X$ is reflexive.

Replacing differential forms with their reflexive hull to improve their properties has a long tradition. For toric varieties, this has been already done by Danilov in [57], and for families of log schemes, this trick has been used by Gross and Siebert in [125].

Another consequence of $j_*O_U = O_X$ is that $j_*\mathcal{M} \to j_*O_U = O_X$ is a log structure. However, this log structure does not seem to be very well behaved (for example, it coincides with $\mathcal{M}_{(E|H)}$ in that example), and its precise nature is unclear to us; it plays no role in our analysis of generically log smooth families.

1.7.1 Log Toroidal Families

Generically log smooth families without further restrictions are often too general to say much about them. In practice, we usually impose additional restrictions on the log singularities. *Log toroidality* is the most important such restriction.

As mentioned earlier, K. Kato's toroidal characterization of log smoothness ascertains that every saturated log smooth morphism is étale locally isomorphic to base changes of log smooth morphisms of the form $(\mathbb{A}_P|D_P) \times \mathbb{A}^r \to (\mathbb{A}_Q|D_Q)$ for saturated homomorphisms $\theta: Q \to P$ of sharp toric monoids. We define *log toroidal families* by a variation of this construction.

For every saturated and injective homomorphism of sharp toric monoids $\theta: Q \to P$, we obtain a morphism of toric varieties $f: \mathbb{A}_P \to \mathbb{A}_Q$. We endow $\mathbb{A}_Q$ with the compactifying log structure defined by the full toric boundary $D_Q \subset \mathbb{A}_Q$. When we endow $\mathbb{A}_P$ with the compactifying log structure defined by the full toric boundary D_P, we obtain a log smooth morphism. However, we can also choose any other toric boundary divisor $f^{-1}(D_Q) \subset D \subset D_P$ to define a log morphism $f: (\mathbb{A}_P|D) \to (\mathbb{A}_Q|D_Q)$. This is what we call an *elementary log toroidal family*. In Sect. 8.3, we construct an open subset $U_{P/Q} \subseteq \mathbb{A}_P$, only depending on $\theta: Q \to P$, such that $f: (\mathbb{A}_P|D) \to (\mathbb{A}_Q|D_Q)$ is log smooth on $U_{P/Q}$ for every D, and which turns f into a generically log smooth family. Details on elementary log toroidal families can be found in the aforementioned Sect. 8.3.

A *log toroidal family* is then a generically log smooth family which is étale locally isomorphic to elementary log toroidal families. We make this precise in Definition 8.27.

Example 1.79 (xy = tz) This example is the local model for the log singularities of the smoothing $f\colon (E, U|H) \to (\mathbb{P}^1|\{0\})$, as we will see in Example 1.80. We let

$$f\colon \ (X^{\mathrm{sze}}|X_0^{\mathrm{sze}}) = \big(\operatorname{Spec}\mathbb{C}[x, y, z, t]/(xy - tz)\big|\{t = 0\}\big) \to \big(\mathbb{A}_t^1\big|\{0\}\big), \quad t \mapsto t.$$

The central fiber X_0^{sze} has two smooth irreducible components. The log structure has an isolated singularity in $Z = \{x = y = z = t = 0\}$, the A_1-singularity of the total space. This is the simplest example of a generically log smooth family which is not log smooth. For an illustration of how the smooth fibers degenerate to the central fiber, see Fig. 1.11. For an illustration of the central fiber and its log-geometric aspects, see Fig. 1.12.

On $X_0 = X_0^{\mathrm{sze}}$, we have $\mathcal{LS}_{X_0} \hookrightarrow \mathcal{T}_{X_0}^1 = \mathbb{C}[z]$ for the sheaf of semistable log smooth structures. The section defining the log structure under consideration is $s = z \in \Gamma(X_0 \setminus Z, \mathcal{LS}_{X_0})$. We use the superscript "sze" because $s = z$ has a simple zero in Z. Outside Z, the ghost stalk is $\mathbb{N}^2$ on the double locus $\{x = y = 0\} = \operatorname{Sing}(X_0^{\mathrm{sze}})$ of the central fiber, but in 0, the ghost stalk is only $\mathbb{N}$.

The sheaf of log differential forms $\Omega^1_{X^{\mathrm{sze}}/\mathbb{A}^1}$, defined via the usual universal property, is not coherent. In fact, it is neither quasi-coherent nor of finite type. See [125, Ex. 1.11] and the author's master thesis [73, Thm. 4.7] for details. This shows that the log structure is *not coherent* in Z, i.e., the map is not only not log smooth, but there is no chart (as defined in Construction 1.57) for the log structure at all. This motivates both that we work with a coherent replacement $\mathcal{W}^1_{X^{\mathrm{sze}}/\mathbb{A}^1}$ of $\Omega^1_{X^{\mathrm{sze}}/\mathbb{A}^1}$ (see below) and that we ignore the log structure in Z, just working with the log structure on the log smooth locus $U = X^{\mathrm{sze}} \setminus Z$.

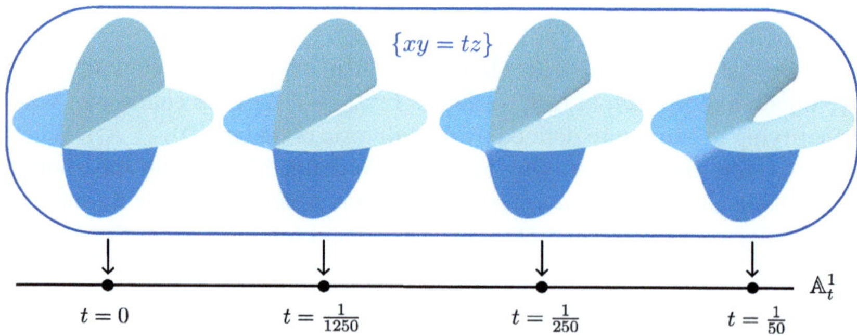

Fig. 1.11 Example 1.79: various fibers. In the family $f\colon X = \operatorname{Spec}\mathbf{k}[x, y, z, t]/(xy - tz) \to \mathbb{A}_t^1$, the smooth fibers X_t have a saddle point whose "curvature" (as measured in the picture) increases as $t \to 0$. The family f is not semistable in the limit point of the saddle points. This is the log singularity in the central fiber X_0. © Simon Felten 2025. All rights reserved

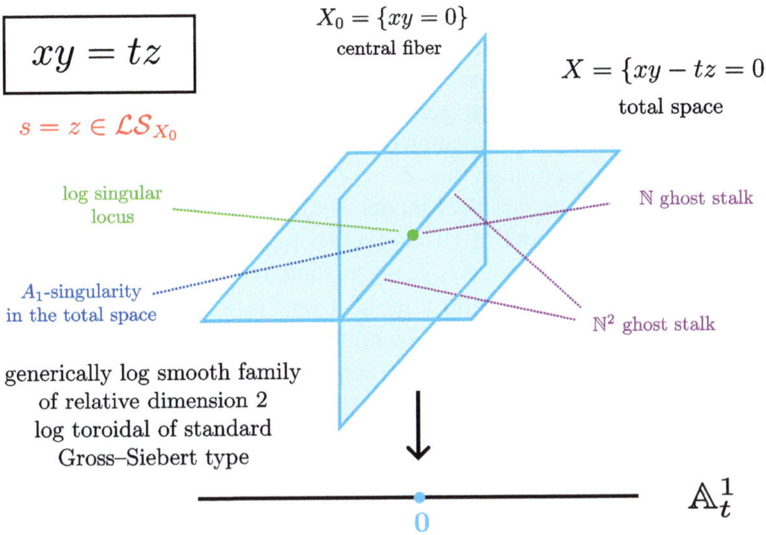

Fig. 1.12 Example 1.79: the central fiber. © Simon Felten 2025. All rights reserved

The generically log smooth family f is log toroidal. First, we set $Q = \mathbb{N}$. Next, let $N = \mathbb{Z}$. Consider the two functions

$$\check{\psi}_0 = \check{\psi}_1 \colon \ N \to \mathbb{Z}, \quad n \mapsto -\inf\{0, n\},$$

which define a monoid

$$P^{\mathrm{sze}} := P = \{(n, a_0, a_1) \mid a_i \geq \check{\psi}_i(n), \ i = 0, 1\} \subseteq N \oplus \mathbb{Z} \oplus \mathbb{Z}.$$

This monoid is generated by

$$\bar{x} = (1, 0, 0), \quad \bar{y} = (-1, 1, 1), \quad \bar{t} = (0, 1, 0), \quad \bar{z} = (0, 0, 1),$$

which induces the isomorphism $\mathbb{C}[x, y, z, t]/(xy - tz) \cong \mathbb{C}[P^{\mathrm{sze}}]$. Thus, X^{sze} is an affine toric variety with four toric boundary divisors $\{x = z = 0\}$, $\{x = t = 0\}$, $\{y = z = 0\}$, and $\{y = t = 0\}$. Our morphism f is the toric morphism given by $1 \mapsto \bar{t}$ on the level of monoids. For the log structure on X^{sze}, we have chosen $\{t = 0\} = \{x = t = 0\} \cup \{y = t = 0\}$ as part of the toric boundary. So far, f is an instance of our construction of elementary log toroidal families in Sect. 8.3. However, we have $U_{P/Q} = X^{\mathrm{sze}} \setminus \{x = y = z = 0\}$ for the open subset of (8.3) on page so that $U_{P/Q} \subsetneq X^{\mathrm{sze}} \setminus Z$. Therefore, f is log toroidal but not an elementary log toroidal family.

Instead of being an elementary log toroidal family, the family f is a (log toroidal) local model of standard Gross–Siebert type in the sense of Definitions 14.3 and

14.12. To see this, we take $\tau = [0, 1] \subseteq \mathbb{R}^1$ in the construction in Sect. 14.1; we set $q = 1$ and $\Delta_1 = \tau$. Then we obtain $\check{\psi}_0$, $\check{\psi}_1$, and P as above so that the local model of Gross–Siebert type associated with τ, Δ_1 has the same underlying morphism of schemes as f. The general fiber of f is smooth so that we have $Z^* = \mathrm{Sing}(\{t \neq 0\}) = \varnothing$ in the notation of Sect. 14.1. One may check that $Z^{\|} = \{0\}$ in the notation of that section so that $Z = Z^{\|} \cup Z^* = \{0\}$. Therefore, f has the same log singular locus as the local model of Gross–Siebert type. But then, f is isomorphic to the local model of Gross–Siebert type as a generically log smooth family. That $f : (X^{\mathrm{sze}}|X_0^{\mathrm{sze}}) \to (\mathbb{A}_t^1|\{0\})$ is of *standard* Gross–Siebert type follows from Lemma 14.8 since the general fiber of $f : X^{\mathrm{sze}} \to \mathbb{A}_t^1$ is smooth. Note also that this example is a special case of Example 14.5. For further discussion of this example and a depiction of the monoid P, see Example 8.10 and Fig. 8.1. $\Diamond$

Example 1.80 Let us now check that the previous Example 1.79 indeed provides the local model for the log singularities of $f : (E, U|H) \to (\mathbb{P}^1|\{0\})$. It is sufficient to show that f is étale locally isomorphic to

$$\mathrm{Spec}\,\tilde{R} = \mathrm{Spec}\,\mathbb{C}[\tilde{x}, \tilde{y}, \tilde{z}, t]/(\tilde{x}\tilde{y} - t\tilde{z}) \to \mathrm{Spec}\,\mathbb{C}[t].$$

Let us consider $X_3 \neq 0$ so that the family is locally given by $\mathrm{Spec}\,R \to \mathbb{A}_t^1$ for

$$R = \mathbb{C}[x, y, z, t]/(t(x^4 + y^4 + z^4 + 1) - xyz).$$

Now we consider the morphism $\tilde{R}[\tilde{u}, \tilde{v}] \to R$ given by

$$t \mapsto t, \quad \tilde{x} \mapsto x, \quad \tilde{y} \mapsto yz, \quad \tilde{z} \mapsto x^4 + y^4 + z^4 + 1, \quad \tilde{u} \mapsto z, \quad \tilde{v} \mapsto y.$$

This map is surjective with kernel $(f_1, f_2) = (\tilde{u}^4 + \tilde{v}^4 + \tilde{x}^4 + 1 - \tilde{z}, \tilde{u}\tilde{v} - \tilde{y})$. Thus, when we localize R in the image $h = 4(z^4 - y^4)$ of

$$\det \begin{pmatrix} \frac{\partial f_1}{\partial \tilde{u}} & \frac{\partial f_1}{\partial \tilde{v}} \\ \frac{\partial f_2}{\partial \tilde{u}} & \frac{\partial f_2}{\partial \tilde{v}} \end{pmatrix} = 4(\tilde{u}^4 - \tilde{v}^4),$$

then the resulting map $\mathrm{Spec}\,R_h \to \mathrm{Spec}\,\tilde{R}$ is étale. We have $Z \cap \{t = x = y = 0\} \subset \{h \neq 0\}$ so that $\mathrm{Spec}\,R_h$ is a neighborhood of those points in Z, and $\mathrm{Spec}\,\tilde{R} \to \mathbb{A}_t^1$ is a local model. The other log singular points are analogous. $\Diamond$

We will encounter more examples of log toroidal families throughout the introduction.

1.7.2   From Degenerate Schemes to Generically Log Smooth Families

To use the deformation theory of generically log smooth families to construct a smoothing, we would like to endow the degenerate scheme V with the structure of a generically log smooth family over the standard log point $(S_0, \mathcal{N}_0) = \mathrm{Spec}(\mathbb{N} \to \mathbb{C})$.

Example 1.81 We consider the normal crossing scheme H. Construction 1.74 gives us an injection $\eta \colon \mathcal{LS}_H \to \mathcal{T}_H^1$ so that we can construct a semistable log smooth structure on H as a section of $\mathcal{T}_H^1$. Now $\mathcal{T}_H^1$ is not isomorphic to $\mathcal{O}_D$ so that we cannot construct a global section of $\mathcal{LS}_V$, i.e., no global semistable log smooth structure. Nonetheless, $\mathcal{T}_H^1$ has many global sections—they just do not lie in the image of η. For example, the first-order deformation $f_1 \colon E_1 \to S_1$ induced by base change of $f \colon E \to \mathbb{P}^1$ defines a section $\sigma \in \Gamma(H, \mathcal{T}_H^1)$. The locus where $\sigma \in (\mathcal{T}_H^1)^* \subseteq \mathcal{T}_H^1$ is precisely $U_0 = H \setminus Z$. This gives rise to a section $s \in \Gamma(U_0, \mathcal{LS}_H)$ with $\eta(s) = \sigma|_{U_0}$, and s is exactly the semistable log smooth structure induced by the log smooth morphism $(U|H) \to (\mathbb{P}^1|\{0\})$. Thus, we can turn H into a generically log smooth family $f_0 \colon (H, U_0, \mathcal{R}_0) \to (S_0, \mathcal{N}_0)$ by means of a section $\sigma \in \Gamma(H, \mathcal{T}_H^1)$. $\diamond$

Example 1.82 ($xy = tz^{-1}$) Not every structure of a generically log smooth family that we can construct by means of sections of $\mathcal{LS}_V$ is well-behaved. For a simple yet pathological example, we take

$$X_0^{\mathrm{spo}} = V := \mathrm{Spec}\, \mathbb{C}[x, y, z]/(xy)$$

for the underlying scheme as in Example 1.79. This is a normal crossing scheme, and hence we have the sheaf $\mathcal{LS}_V$ of semistable log smooth structures. We take $z^{-1} \in \Gamma(V \setminus \{0\}, \mathcal{LS}_V)$ to define a log structure outside $x = y = z = 0$. In plain terms, we take the log structure from $xy = tz^{-1}$ on $\{z \neq 0\}$, and we take the log structure induced by pull-back from the base $(S_0, \mathcal{N}_0)$ on both $\{x \neq 0\}$ and $\{y \neq 0\}$. This gives rise to a generically log smooth family with log singular locus a single point $Z = \{0\}$. There is no generically log smooth deformation of this family over $(S_1, \mathcal{N}_1) = \mathrm{Spec}(\mathbb{N} \to \mathbb{C}[t]/(t^2))$. Namely, a flat deformation over S_1 defines a class in $\mathrm{Ext}^1(\Omega_V^1, \mathcal{O}_V) = \mathbb{C}[z]$ whose restriction to $\{z \neq 0\}$ is the underlying flat deformation of the log smooth deformation; this is $z^{-1} \in \mathbb{C}[z]_z$, which does not extend to the whole of V. We use the superscript "spo" because $z^{-1} \in \mathcal{LS}_V$ has a simple pole in Z. For an illustration, see Fig. 1.13. $\diamond$

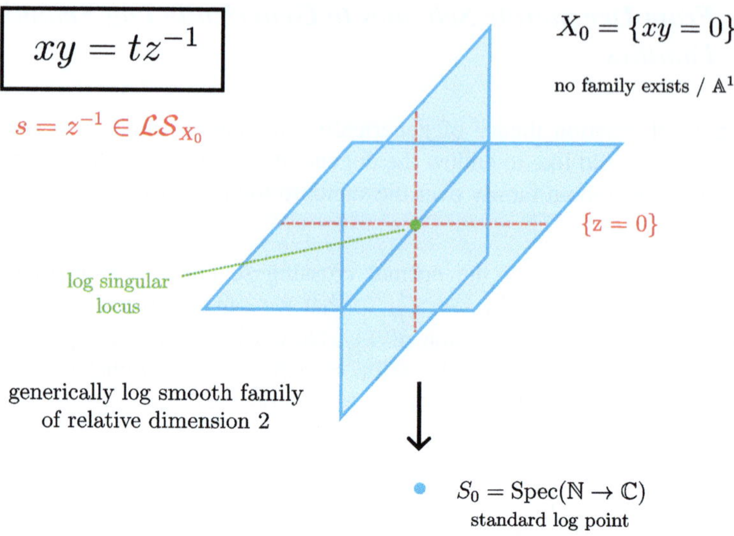

Fig. 1.13 Example 1.82. © Simon Felten 2025. All rights reserved

1.7.3 Deformations of Generically Log Smooth Families

Let $(S_0, \mathcal{N}_0)$ be the standard log point, and let $f_0 \colon (X_0, U_0, \mathcal{M}_0) \to (S_0, \mathcal{N}_0)$ be a generically log smooth family. It is straightforward to define *generically log smooth deformations* over $(S_A, \mathcal{N}_A)$ for $A \in \mathbf{Art}_{\mathbb{C}[\![t]\!]}$, and this gives rise to the *generically log smooth deformation functor*

$$\mathrm{LD}^{\mathrm{gen}}_{X_0/S_0} \colon \quad \mathbf{Art}_{\mathbb{C}[\![t]\!]} \to \mathbf{Set}.$$

Because generically log smooth families are the analog of normal varieties, such deformations are far from being locally rigid.

Example 1.83 ($xy = tzw$ and $xy = tzw + t^2$) This is a benign example of the local ambiguity of generically log smooth deformations, a more pathological example can be found in [125, Ex. 2.9]. First, we discuss the family

$$f \colon \quad (X|X_0) = \big(\mathrm{Spec}\, \mathbb{C}[x, y, z, w, t]/(xy - tzw) \big| \{t = 0\} \big) \to \big(\mathbb{A}^1_t \big| \{0\} \big), \quad t \mapsto t,$$

in some detail. It is illustrated in Fig. 1.14. The central fiber

$$X_0 = \mathrm{Spec}\, \mathbb{C}[x, y, z, w]/(xy)$$

has two smooth irreducible components. The log singular locus Z is given by $Z \cap X_0 = \{zw = 0\}$ inside the double locus $\mathbb{A}^2_{z,w} = \mathrm{Sing}(X_0)$, i.e., it consists of two

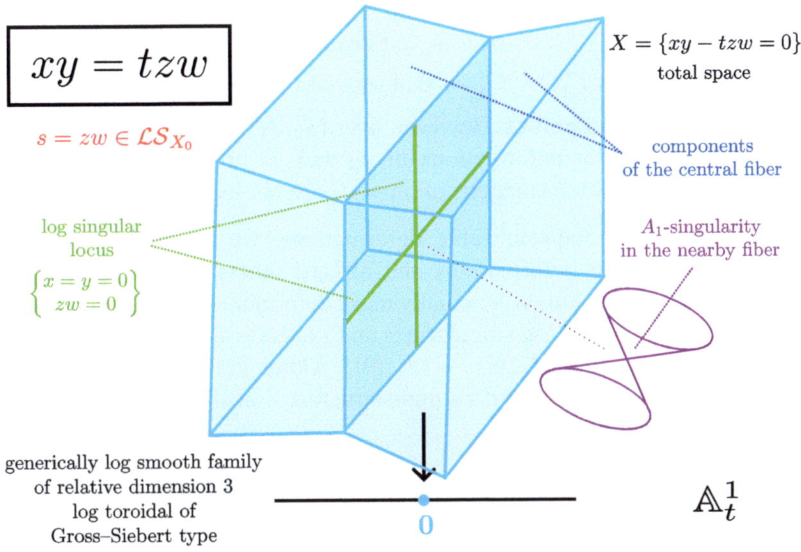

Fig. 1.14 Example 1.83. © Simon Felten 2025. All rights reserved

lines intersecting transversely in a point. Outside this point, the log singularity is of the form $xy = tz$ as in Example 1.79.

We discuss log toroidality. The family $f : X \to \mathbb{A}^1$ is identical with the family $L(1; 1, 1) \to \mathbb{A}^1$ in Example 14.5. Thus, it is a local model of Gross–Siebert type (in the sense of Definition 14.3) but not a local model of elementary Gross–Siebert type[7] since Δ_+ is a square and hence not an elementary simplex. The general fiber has a 3-dimensional A_1-singularity; a log toroidal family of elementary Gross–Siebert type of relative dimension 3 would have a smooth general fiber since, in this case, singularities in the general fiber can occur only in codimension 4.

Next, we consider the family

$$f' : \quad (X'|X_0') = \big(\operatorname{Spec} \mathbb{C}[x, y, z, w, t]/(xy - tzw - t^2)\big|\{t = 0\} \big) \to \big(\mathbb{A}^1_t \big| \{0\} \big).$$

Unlike f, this family is smooth over $t \neq 0$. Moreover, it is log smooth outside $Z' = \{zw = 0\} \subset \mathbb{A}^2_{z,w} = \operatorname{Sing}(X_0')$ so that f' is a generically log smooth family. One may show that f' does not have the base change property so that it is not log toroidal.

Note that $X_0 = X_0'$ as schemes, and that $Z \cap X_0 = Z' \cap X_0'$ under this identification. Around $X_0 \setminus Z$ respectively $X_0' \setminus Z'$, both families are semistable

---

[7] Do not confuse *elementary log toroidal families* in the sense of Definition 8.14 with *log toroidal families of elementary Gross–Siebert type* in the sense of Definition 14.19. In the former, "elementary" refers to the fact that they are basic building blocks while, in the latter, "elementary" refers to the fact that the associated polytope Δ_+ is an elementary simplex.

so that the log structures on the central fibers are given by sections s, s' of $\mathcal{LS}_{X_0} = \mathcal{LS}_{X_0'}$. Since $f_1 = f_1'$ over $S_1 = \operatorname{Spec}\mathbb{C}[t]/(t^2)$ as flat deformations, both sections are the same in $\mathcal{T}^1_{X_0} = \mathcal{T}^1_{X_0'}$ so that we find $s = s'$. Thus, we have $f_0 = f_0'$ as generically log smooth families. However, over $(S_2, \mathcal{N}_2) = \operatorname{Spec}(\mathbb{N} \to \mathbb{C}[t]/(t^3))$, even the underlying flat deformations of f_2 and f_2' are not isomorphic. Thus, generically log smooth deformations of $f_0 = f_0'$ are not locally rigid. $\Diamond$

As in the case of flat/semistable deformations, we are led to select some generically log smooth deformations and discard others. Unlike in the case of semistable deformations, this will actually make deformations locally rigid. We have two approaches: classes of log singularities and systems of deformations.

Let us go back to $f \colon (E, U|H) \to (\mathbb{P}^1|\{0\})$. Although the log singularities of f are not quasi-coherent, they have a simple structure. Locally in the étale topology, each of them is isomorphic to

$$(M, M \setminus \{0\}|M_0) = (\operatorname{Spec}\mathbb{C}[x, y, z, t]/(xy - tz), M \setminus \{0\}|\{t = 0\}) \to (\mathbb{A}^1_t|\{0\}),$$

where $0 \in M$ is the point given by $x = y = z = t = 0$. Thus, we may insist that an admissible generically log smooth deformation of the central fiber $f_0 \colon (H, U_0, \mathcal{R}_0) \to (S_0, \mathcal{N}_0)$ of f is étale locally isomorphic to

$$(M_A, M_A \setminus \{0\}, Q_A) = (M, M \setminus \{0\}|M_0) \times_{(\mathbb{A}^1_t|\{0\})} (S_A, \mathcal{N}_A) \to (S_A, \mathcal{N}_A).$$

In [125], Gross and Siebert have called such a deformation a *divisorial* deformation. In the example $f_0 \colon (H, U_0, \mathcal{R}_0) \to (S_0, \mathcal{N}_0)$, divisorial deformations are locally rigid, which is a special case of the proof of [125, Thm. 2.11]. This observation motivates the following concept, which is analogous to semistable deformations.

Definition 1.84 (Classes of Log Singularities) Let Q be a sharp toric monoid, and let $\mathbb{A}_Q = \operatorname{Spec}\mathbb{C}[Q]$ with full toric boundary $D_Q \subset \mathbb{A}_Q$.

(1) A *class of log singularities* is a set $\mathscr{C}$ of generically log smooth families

$$\mu \colon (M_\mu, U_\mu, \mathcal{K}_\mu) \to (\mathbb{A}_Q|D_Q).$$

(2) Let $A \in \mathbf{Art}_{\mathbb{C}[\![Q]\!]}$ and $(S_A, \mathcal{N}_A) = \operatorname{Spec}(Q \to A)$. A generically log smooth family $f_A \colon (X_A, U_A, \mathcal{M}_A) \to (S_A, \mathcal{N}_A)$ is *of class $\mathscr{C}$* if every geometric point $\bar{x} \in X_A$ admits an étale neighborhood $p \colon W_A \to X_A$ which fits into a commutative diagram

$$
\begin{array}{ccccccc}
(X_A, U_A, \mathcal{M}_A) & \xleftarrow{\ p\ } & (W_A, U_A', \mathcal{M}_A') & \xrightarrow{\ h\ } & (M_{\mu;A}, U_{\mu;A}, \mathcal{K}_{\mu;A}) & \xrightarrow{\ c\ } & (M_\mu, U_\mu, \mathcal{K}_\mu) \\
\ \downarrow{\scriptstyle f_A} & & \downarrow & & \downarrow & & \downarrow{\scriptstyle \mu} \\
(S_A, \mathcal{N}_A) & = & (S_A, \mathcal{N}_A) & = & (S_A, \mathcal{N}_A) & \xrightarrow{\ b\ } & (\mathbb{A}_Q|D_Q)
\end{array}
$$

of morphisms between partial log schemes where p and h are strict and étale, where $\mu \in \mathscr{C}$, and where the right-hand square is Cartesian. The family f_A is *accurately of class* $\mathscr{C}$ if p and h can be chosen to be accurate, i.e., $p^{-1}(U_A) = U'_A$ and $h^{-1}(U_{\mu;A}) = U'_A$.

(3) The class of log singularities $\mathscr{C}$ has the *base change property* if every $\mu \in \mathscr{C}$ has the base change property from Definition 1.78.

(4) The class of log singularities $\mathscr{C}$ is *log Gorenstein* if every μ is log Gorenstein, i.e., if $\mathcal{W}^d_{M_\mu/\mathbb{A}_Q}$ is a line bundle for d the relative dimension of μ.

(5) The class of log singularities $\mathscr{C}$ is *locally rigid* if any two deformations accurately of class $\mathscr{C}$ of any *affine* generically log smooth family are isomorphic.

Remark 1.85 We abstain from collecting deformations accurately of class $\mathscr{C}$ in a functor of Artin rings since it is in general not clear that this is a deformation functor. (Consider Schlessinger's condition (H_1) in Definition A.4 and two deformations with different local models μ_1 and μ_2.) $\Diamond$

Example 1.86 For $Q = \mathbb{N}$, we construct the class $\mathscr{C}^{\mathrm{anc}}$ of *almost normal crossing* families. First, we take the semistable local models $\mu^{\mathrm{nc}}_{d;r} \colon (\mathbb{A}^{d+1}|H^{\mathrm{nc}}_{d;r}) \to (\mathbb{A}^1_t|\{0\})$. Then we add the local models

$$\mu^{\mathrm{anc}}_{d;r} \colon \quad \left(\operatorname{Spec} \mathbb{C}[z_0, \ldots, z_{d-1}, u, t]/(z_0 \cdot \ldots \cdot z_r - tu), U^{\mathrm{anc}}_{d;r}\big|\{t = 0\}\right) \to \left(\mathbb{A}^1_t\big|\{0\}\right)$$

for every $0 \leq r \leq d - 1$. In Construction 18.46, we will see that $\mu^{\mathrm{anc}}_{d;r}$ is a local model of standard Gross–Siebert type. In particular, we choose the log singular locus to be $Z = Z^{\|} \cup Z^*$ in the notation of Sect. 14.1. The nearby fibers of $\mu^{\mathrm{anc}}_{d;r}$ are smooth so that formal generically log smooth one-parameter deformations are formal smoothings. The family $f_0 \colon (H, U_0, \mathcal{R}_0) \to (S_0, \mathcal{N}_0)$ is accurately of class $\mathscr{C}^{\mathrm{anc}}$. We discuss more details of $\mathscr{C}^{\mathrm{anc}}$ in Construction 18.46. $\Diamond$

Example 1.87 Fix a sharp toric monoid Q. Then the class $\mathscr{C}^{\mathrm{tor}}_Q$ of *toroidal log singularities* is given by all elementary log toroidal families

$$f \colon \quad (\mathbb{A}_P, U_{P/Q}|D) \to (\mathbb{A}_Q|D_Q)$$

over Q. For $A \in \mathbf{Art}_{\mathbb{C}[\![Q]\!]}$, we say that a generically log smooth family over $(S_A, \mathcal{N}_A)$ is *log toroidal with respect to* $(S_A, \mathcal{N}_A) \to (\mathbb{A}_Q|D_Q)$ if it is (possibly non-accurately) of class $\mathscr{C}^{\mathrm{tor}}_Q$. Families of class $\mathscr{C}^{\mathrm{anc}}$ are also of class $\mathscr{C}^{\mathrm{tor}}_{\mathbb{N}}$, but we do not have $\mathscr{C}^{\mathrm{anc}} \subset \mathscr{C}^{\mathrm{tor}}_{\mathbb{N}}$ since the chosen log smooth loci do not match. $\Diamond$

Example 1.88 In Definition 1.148, we will encounter three classes of log singularities

$$\mathscr{C}^{\mathrm{sGS}} \subset \mathscr{C}^{\mathrm{eGS}} \subset \mathscr{C}^{\mathrm{GS}}$$

which arise in the toric mirror construction of Gross and Siebert in [124, 125]. We have $\mathscr{C}^{\mathrm{anc}} \subset \mathscr{C}^{\mathrm{sGS}}$, as we will see in Construction 18.46. With Proposition 14.36, we will see that the class $\mathscr{C}^{\text{eGS}}$ is locally rigid. Then also the classes $\mathscr{C}^{\text{sGS}}$ and $\mathscr{C}^{\text{anc}}$ are locally rigid.   ◊

Instead of specifying a class of log singularities, we can also choose local deformations directly and admit only such deformations which are isomorphic to our choices. This leads to the notion of a *system of deformations*.

Definition 1.89 (Systems of Deformations)   Let $f_0 \colon (X_0, U_0, \mathcal{M}_0) \to (S_0, \mathcal{N}_0)$ be a generically log smooth family. A *system of deformations* $\mathscr{D}$ consists of the following data:

(a)   an open cover $\mathcal{V} = \{V_\alpha\}_\alpha$ of X_0;
(b)   a generically log smooth deformation $f_{\alpha;A} \colon (V_{\alpha;A}, U_{\alpha;A}, \mathcal{M}_{\alpha;A}) \to (S_A, \mathcal{N}_A)$
      for each V_α and $A \in \mathbf{Art}_{\mathbb{C}[\![t]\!]}$; we assume that $f_{\alpha;A}$ has the base change property;
(c)   isomorphisms

$$\psi_{\alpha\beta;A} \colon \quad (V_{\alpha;A}, U_{\alpha;A}, \mathcal{M}_{\alpha;A})|_{\alpha\beta} \cong (V_{\beta;A}, U_{\beta;A}, \mathcal{M}_{\beta;A})|_{\alpha\beta}$$

of generically log smooth deformations, where we write $(-)|_{\alpha\beta}$ for $(-)|_{V_\alpha \cap V_\beta}$;
(d)   for every $B \to A$ in $\mathbf{Art}_{\mathbb{C}[\![t]\!]}$, restriction maps

$$\rho_{\alpha;BA} \colon \quad (V_{\alpha;A}, U_{\alpha;A}, \mathcal{M}_{\alpha;A}) \to (V_{\alpha;B}, U_{\alpha;B}, \mathcal{M}_{\alpha;B})$$

which induce an isomorphism

$$(V_{\alpha;B}, U_{\alpha;B}, \mathcal{M}_{\alpha;B}) \times_{(S_B, \mathcal{N}_B)} (S_A, \mathcal{N}_A) \cong (V_{\alpha;A}, U_{\alpha;A}, \mathcal{M}_{\alpha;A}).$$

A generically log smooth deformation $f_A \colon (X_A, U_A, \mathcal{M}_A) \to (S_A, \mathcal{N}_A)$ is *of type $\mathscr{D}$* if $f_A|_W$ is isomorphic to $f_{\alpha;A}|_W$ for every affine open subset $W \subseteq V_\alpha$. Isomorphism classes of generically log smooth deformations of type $\mathscr{D}$ are collected in the deformation functor

$$\mathrm{LD}^{\mathscr{D}}_{X_0/S_0} \colon \quad \mathbf{Art}_{\mathbb{C}[\![t]\!]} \to \mathbf{Set}.$$

A slightly more precise version of the definition with additional technical assumptions is given in Definition 10.15.

Example 1.90   Consider $f \colon (E, U|H) \to (\mathbb{P}^1_t | \{0\})$. Let $V_\alpha = H$ for a single index α, and let $V_{\alpha;A} = (E_A, U_A, \mathcal{R}_A) \to (S_A, \mathcal{N}_A)$ be the base change of f along the obvious map $(S_A, \mathcal{N}_A) \to (\mathbb{P}^1 | \{0\})$. They form a system of deformations.   ◊

Deformations accurately of class $\mathscr{C}^{\text{eGS}}$ (and hence also those accurately of class $\mathscr{C}^{\text{sGS}}$ or $\mathscr{C}^{\text{anc}}$, see Example 1.88) are locally rigid. In this case, we can construct a system of deformations as follows.

Construction 1.91   Let $\mathscr{C}$ be a class of log singularities which is locally rigid in the sense of Definition 1.84. Let $f_0 \colon (X_0, U_0, \mathcal{M}_0) \to (S_0, \mathcal{N}_0)$ be a generically log smooth family accurately of class $\mathscr{C}$. We form a system of deformations $\mathscr{D}$ as follows. Let $\mathcal{V} = \{V_\alpha\}_\alpha$ be an affine open cover of X_0. Deformations accurately of class $\mathscr{C}$ exist étale locally by the assumption that f_0 is accurately of class $\mathscr{C}$, and they are locally rigid. Then on the affine open $V_\alpha \subseteq X_0$, deformations accurately of class $\mathscr{C}$ exist and are unique up to non-unique isomorphism.[8] Choose one such deformation $f_{\alpha;A} : (V_{\alpha;A}, U_{\alpha;A}, \mathcal{M}_{\alpha;A}) \to (S_A, N_A)$ for every V_α and every $A \in$ $\mathbf{Art}_{\mathbb{C}[\![t]\!]}$. Now a generically log smooth deformation of f_0 is accurately of class $\mathscr{C}$ if and only if it is of type $\mathscr{D}$. $\Diamond$

This construction allows us to pass from the class-of-log-singularities picture, which is more suitable to study properties of a log singularity, to the system-of-deformations picture, which is more suitable to study the deformation theory of a given $f_0 : (X_0, U_0, \mathcal{M}_0) \to (S_0, N_0)$.

1.8   Unobstructedness of the Logarithmic Deformation Functor

We have seen in Theorem 1.38 that infinitesimal deformations of a complex manifold Y are controlled by its Kodaira–Spencer dg Lie algebra $K S_Y^\bullet$ via gauge equivalence classes of solutions of the Maurer–Cartan equation. This in turn was a crucial ingredient in the proof of the Bogomolov–Tian–Todorov theorem which we have presented in Sect. 1.4.

Fix a sharp toric monoid Q, let $(S_0, N_0) = \mathrm{Spec}(Q \to \mathbb{C})$ be a log point, and let $f_0 : (X_0, U_0, \mathcal{M}_0) \to (S_0, N_0)$ be a generically log smooth family of relative dimension $d \geq 1$. Let $\mathscr{D}$ be a system of deformations in the sense of Definition 1.89. Let us assume that f_0 is *log Gorenstein*, i.e., $\omega_{X_0/S_0} = j_* W^d_{U_0/S_0}$ is a line bundle, and that $\mathscr{D}$ has the base change property in the sense of Definition 1.78. In this section, we present a logarithmic analog of Theorem 1.38, and then we proceed to an overview over the proof of the logarithmic Bogomolov–Tian–Todorov theorem in the case where f_0 is proper and log Calabi–Yau.

1.8.1   From DG Lie Algebras to Curved Lie Algebras

Definition 1.37 provides us with a notion of dg Lie algebra over $\mathbb{C}[\![Q]\!]$, and any such dg Lie algebra $L^\bullet$ gives rise to a functor

$$\mathrm{Def}(L^\bullet, -) : \quad \mathbf{Art}_{\mathbb{C}[\![Q]\!]} \to \mathbf{Set}$$

---

[8] This depends on the additional assumptions that $\mathscr{C}$ has the base change property and that f_0 is log Gorenstein.

by taking gauge equivalence classes of solutions of the Maurer–Cartan equation $\bar{\partial}\phi + \frac{1}{2}[\phi, \phi] = 0$ in $\mathfrak{m}_A \cdot L^1$. Clearly, $\phi = 0$ is a solution, and we find

$$\mathrm{Def}(L^{\bullet}, A) \neq \varnothing$$

for all $A \in \mathbf{Art}_{\mathbb{C}[\![Q]\!]}$. Furthermore, these solutions are compatible with restrictions, so we have a formal family of solutions. Therefore, if there is a dg Lie algebra $L^{\bullet}$ and an isomorphism

$$\mathrm{LD}_{X_0/S_0} \cong \mathrm{Def}(L^{\bullet}, -)$$

of functors of Artin rings, then $f_0 \colon (X_0, U_0, \mathcal{M}_0) \to (S_0, \mathcal{N}_0)$ admits a formal generically log smooth deformation of type $\mathscr{D}$. Rather unsurprisingly, such a dg Lie algebra does not always exist.

Example 1.92 This is an example which was constructed by Persson and Pinkham in [229, pp. 485–486] to show that not every d-semistable normal crossing complex space admits a semistable smoothing. Let $Z \subset \mathbb{P}^3$ be a smooth hypersurface of degree ≥ 5, and let $D \subset Z$ be a smooth hyperplane section. Then $H^1(D, \mathcal{N}_{D/Z}) \neq 0$, so we can choose some $0 \neq \eta \in H^1(D, \mathcal{N}_{D/Z}) = \mathrm{Ext}^1(\mathcal{O}_D, \mathcal{N}_{D/Z})$. This defines a vector bundle $\mathcal{E}$ of rank 2, and we set $Y = \mathbb{P}_D(\mathcal{E})$. Now $V = Y \sqcup_D Z$ is a d-semistable normal crossing scheme. In particular, we find a semistable log smooth structure on V. However, by Persson and Pinkham [229, Thm. 3], every infinitesimal analytic deformation of V is locally trivial. Therefore, no semistable and hence no log smooth deformation exists over $(S_1, \mathcal{N}_1)$. $\diamondsuit$

Chan, Leung, and Ma observed fairly recently in [38] that this problem can be solved by replacing a dg Lie algebra with a $\mathbb{C}[\![Q]\!]$-linear *curved* Lie algebra.

Definition 1.93 (Curved Lie Algebras) Let Λ be a complete local Noetherian $\mathbb{C}$-algebra with residue field $\mathbb{C}$. Then a Λ-*linear curved Lie algebra* is a quadruple $(L^{\bullet}, [-, -], \bar{\partial}, \ell)$ where each L^i is a flat and complete Λ-module, $(L^{\bullet}, [-, -])$ is a graded Λ-linear Lie algebra, $\bar{\partial} \colon L^i \to L^{i+1}$ is a Λ-linear map with

$$\bar{\partial}[\theta, \xi] = [\bar{\partial}\theta, \xi] + (-1)^{|\theta|}[\theta, \bar{\partial}\xi]$$

for all homogeneous $\theta, \xi \in L^{\bullet}$, and $\ell \in \mathfrak{m}_\Lambda \cdot L^2$ is an element with $\bar{\partial}(\ell) = 0$ and $\bar{\partial}^2(\theta) = [\ell, \theta]$ for every $\theta \in L^{\bullet}$.

The operator $\bar{\partial}$ is called the *predifferential*, and $\ell \in L^2$ is called the *curvature*. We set $L_A^{\bullet} := L^{\bullet} \otimes_\Lambda A$ for $A \in \mathbf{Art}_\Lambda$. Every element $\phi \in \mathfrak{m}_A \cdot L_A^1$ gives rise to a modified predifferential $\bar{\partial}_\phi := \bar{\partial} + [\phi, -]$, which forms a new Λ-linear curved Lie algebra with the curvature

$$\ell_\phi := \bar{\partial}\phi + \frac{1}{2}[\phi, \phi] + \ell.$$

This is called the *twisting procedure*.[9] In order to obtain a deformation functor, we consider elements $\phi \in \mathfrak{m}_A \cdot L_A^1$ with $\ell_\phi = 0$, i.e., elements which satisfy the *extended* Maurer–Cartan equation

$$\bar{\partial}\phi + \frac{1}{2}[\phi, \phi] + \ell = 0.$$

These elements satisfy $\bar{\partial}_\phi^2 = [\ell_\phi, -] = 0$. The usual formulae, which can be found in Lemma 5.3, form a *gauge action* by elements in $\mathfrak{m}_A \cdot L_A^0$ on the Maurer–Cartan solutions, and gauge equivalence classes form the functor of Artin rings

$$\mathrm{Def}(L^\bullet, -)\colon \ \mathbf{Art}_\Lambda \to \mathbf{Set}.$$

Now $\phi = 0$ is not automatically a Maurer–Cartan solution for all $A \in \mathbf{Art}_\Lambda$, thus allowing for the existence of a $\mathbb{C}\llbracket t \rrbracket$-linear curved Lie algebra controlling the deformation functor in the examples of Persson–Pinkham.

1.8.2   Construction of the Curved Lie Algebra

Building upon ideas of [38], a curved Lie algebra which controls the deformation functor $\mathrm{LD}^{\mathcal{D}}_{X_0/S_0}$ has been first constructed by the author in [76].

Theorem 1.94 *Let $\Lambda = \mathbb{C}\llbracket Q \rrbracket$, and let $(S_0, N_0) = \mathrm{Spec}(Q \to \mathbb{C})$. Suppose that $f_0\colon (X_0, U_0, M_0) \to (S_0, N_0)$ is a separated generically log smooth family which is log Gorenstein, and let $\mathcal{D}$ be a system of deformations which has the base change property. Then there is a Λ-linear curved Lie algebra $L^\bullet_{X_0/\Lambda}$ and a natural isomorphism $\mathrm{LD}^{\mathcal{D}}_{X_0/S_0} \cong \mathrm{Def}(L^\bullet_{X_0/\Lambda}, -)$.*

The curved Lie algebra $L^\bullet_{X_0/\Lambda}$ is constructed via an intricate process, which we generalize (in comparison with [76]) in Chap. 13. Let us now sketch this process.

When $f\colon (X, U, M) \to (S, N)$ is a generically log smooth family of relative dimension $d \geq 1$, then we have the log de Rham complex $\mathcal{W}^\bullet_{X/S} = j_*\Omega^\bullet_{U/S}$ and the polyvector fields $\mathcal{V}^p_{X/S} = j_* \bigwedge^{-p} \Theta^1_{U/S}$. The pair

$$\mathcal{V}\mathcal{W}^\bullet_{X/S} := (\mathcal{V}^\bullet_{X/S}, \mathcal{W}^\bullet_{X/S})$$

---

[9] The twisting procedure is a more general phenomenon which is discussed in the nice new book [64] by Dotsenko, Shadrin, and Vallette. It applies classically to curved A_∞-algebras as well as curved L_∞-algebras, and moreover, there is an elaborate theory of the twisting procedure on the level of operads.

carries a number of operations which turn it into what we call a *two-sided Gerstenhaber calculus*, see Definition 3.22 and Proposition 8.5. In particular, $\mathcal{V}^{\bullet}_{X/S}$ is a Gerstenhaber algebra, and $\mathcal{V}^{-1}_{X/S}$ is a Lie algebra. They are the logarithmic analogs of the Gerstenhaber algebra and Lie algebra that we have seen in Sect. 1.4.

In a first step toward the construction of $L^{\bullet}_{X_0/\Lambda}$, we forget about the log structure on $f_0 \colon X_0 \to S_0$ and the local models $(V_{\alpha;A}, \mathcal{M}_{\alpha;A}) \to (S_A, \mathcal{N}_A)$ for its infinitesimal deformations, and we just retain the two-sided Gerstenhaber calculi. This yields a deformation problem for *geometric families of two-sided Gerstenhaber calculi* in the sense of Definition 12.1. Now the central fiber is a morphism $f_0 \colon X_0 \to S_0$ of schemes, carrying a two-sided Gerstenhaber calculus $\mathcal{V}\mathcal{W}^{\bullet}_0$, and the local models for deformations are morphism of schemes $V_{\alpha;A} \to S_A$, carrying a two-sided Gerstenhaber calculus $\mathcal{V}\mathcal{W}^{\bullet}_{\alpha;A}$. For morphisms $B \to A$ in $\mathbf{Art}_{\mathbb{C}[\![Q]\!]}$, they come with restriction maps, and on overlaps $V_\alpha \cap V_\beta$, we have isomorphisms

$$\mathcal{V}\mathcal{W}^{\bullet}_{\alpha;A}|_{\alpha\beta} \cong \mathcal{V}\mathcal{W}^{\bullet}_{\beta;A}|_{\alpha\beta}$$

which are induced from the isomorphisms of generically log smooth families. Now a deformation of $f_0 \colon X_0 \to S_0$ is a morphism $f_A \colon X_A \to S_A$ of schemes together with a two-sided Gerstenhaber calculus $\mathcal{V}\mathcal{W}^{\bullet}_A$ which is, on V_α, isomorphic to the local model $\mathcal{V}\mathcal{W}^{\bullet}_{\alpha;A}$. Unlike in the case of generically log smooth families, we now have to track the isomorphisms with the local models. The reason for this is that we need to distinguish between *inner* automorphisms of a geometric family of two-sided Gerstenhaber calculi, which can be constructed explicitly from the operations in the calculus, and *outer* automorphisms, which are simply structure-preserving invertible self-maps. The inner automorphisms are in one-to-one correspondence with automorphisms of the generically log smooth family, but there may be additional outer automorphisms.[10] From the one-to-one correspondence between automorphisms in the two cases, we obtain that

$$\mathrm{LD}^{\mathcal{D}}_{X_0/S_0} \cong \mathrm{GDef}^{\mathcal{D}}_{X_0/S_0}$$

as functors of Artin rings, where the latter deformation functor classifies equivalence classes of deformations of geometric families of two-sided Gerstenhaber calculi. Two deformations are equivalent if there is an (outer) isomorphism such that the induced automorphisms of the local models $V_{\alpha;A} \to S_A$ via the fixed comparison isomorphisms is an *inner* automorphism of the local model.

In Sect. 1.4, we have resolved the Gerstenhaber algebra $\Theta^{\bullet}_Y$ and the de Rham complex $\Omega^{\bullet}_Y$ by means of the Dolbeault resolution. In the algebraic case, and also in the singular holomorphic case, this resolution is not available. We may try to work with the Čech resolution for some open affine cover $\mathcal{U} = \{U_i\}_i$ instead. However, already in the case of a smooth algebraic variety Y, the Čech resolution $\check{C}^{\bullet}(\mathcal{U}, \mathcal{T}_Y)$ of the tangent sheaf $\mathcal{T}_Y$ does not carry the structure of a dg Lie algebra. The reason is

---

[10] The author does not know if non-inner automorphisms actually exist.

that the Čech resolution is not sufficiently compatible with operations defined on the resolved sheaf; we obtain the structure of a graded Lie algebra only in cohomology.

To circumvent this problem, we work with the *Thom–Whitney resolution*. This is a functorial resolution which associates a quasi-isomorphism

$$\mathcal{F} \to (\mathrm{TW}^\bullet(\mathcal{U}, \mathcal{F}), \bar\partial)$$

with every sheaf of $\mathbb{C}$-vector spaces $\mathcal{F}$. We retain the notation $\bar\partial$ for the differential from the Dolbeault resolution, but here, the construction of $\bar\partial$ is more closely related to the Čech differential. Unlike the Čech complex, the Thom–Whitney resolution is compatible with algebraic structures on $\mathcal{F}$. For example, $\mathrm{TW}^\bullet(\mathcal{U}, O_{X_0})$ is a sheaf of dg algebras, and $\mathrm{TW}^\bullet(\mathcal{U}, \mathcal{V}^{-1}_{X_0/S_0})$ is a sheaf of dg Lie algebras. If $\mathcal{F}$ is a quasi-coherent O_{X_0}-module, then each $\mathrm{TW}^j(\mathcal{U}, \mathcal{F})$ is a quasi-coherent O_{X_0}-module and acyclic for the global section functor. We review the Thom–Whitney resolution in Sect. 13.1.

Applying $\mathrm{TW}^\bullet(\mathcal{U}, -)$ to the two-sided Gerstenhaber calculus $\mathcal{V}\mathcal{W}_0^\bullet$ on the central fiber yields what we call a sheaf of *differential bigraded two-sided Gerstenhaber calculi*. The original sheaf of two-sided Gerstenhaber calculi can be recovered as the *cohomology* of the differential $\bar\partial$. We apply the same resolution construction also to the local models, obtaining the sheaves of differential bigraded two-sided Gerstenhaber calculi $\mathcal{V}\mathcal{W}_{\alpha;A}^{\bullet;\bullet}$. When we have a deformation $f_A \colon X_A \to S_A$ with sheaf of two-sided Gerstenhaber calculi $\mathcal{V}\mathcal{W}_A^\bullet$, then we can apply the resolution as well. The resulting sheaf of differential bigraded two-sided Gerstenhaber calculi $\mathcal{V}\mathcal{W}_A^{\bullet;\bullet}$ is locally isomorphic to the local models $\mathcal{V}\mathcal{W}_{\alpha;A}^{\bullet;\bullet}$. It is a curious but central fact of our theory that, once we forget the differential $\bar\partial$ and keep just the sheaf of *bigraded two-sided Gerstenhaber calculi*, all $\mathcal{V}\mathcal{W}_A^{\bullet;\bullet}$ are isomorphic, no matter what the original deformation was. This is essentially because isomorphism classes are classified by $H^1(X_0, \mathcal{V}_0^{-1,0}) = 0$, which vanishes due to acyclicity. In fact, there is a unique gluing

$$(\mathcal{P}\mathcal{V}_{X_0/A}^{\bullet;\bullet}, \mathcal{D}\mathcal{R}_{X_0/A}^{\bullet;\bullet}) \tag{1.1}$$

of the $\mathcal{V}\mathcal{W}_{\alpha;A}^{\bullet;\bullet}$, even if there is no deformation $f_A \colon X_A \to S_A$. This gluing carries all operations except the differential $\bar\partial$, which is not invariant under inner automorphisms of the local models and hence has no unique action on (1.1). We can, however, choose some differential operator $\bar\partial$ on (1.1) which is of the form $\bar\partial + [\varphi, -]$ on $\mathcal{V}\mathcal{W}_{\alpha;A}^{\bullet;\bullet}$ for $\varphi \in \mathcal{V}_{\alpha;A}^{-1,1}$ depending on the chosen isomorphism with $\mathcal{V}\mathcal{W}_{\alpha;A}^{\bullet;\bullet}$. This differential operator is called a *predifferential* because we may have $\bar\partial^2 \neq 0$.

A *differential* on (1.1) is now a differential operator $\bar\partial' := \bar\partial + [\varphi, -]$ with $(\bar\partial')^2 = 0$. When we have such a differential, then its cohomology is a deformation of geometric families of two-sided Gerstenhaber calculi, locally isomorphic to $\mathcal{V}\mathcal{W}_{\alpha;A}^\bullet$. For two differentials, the associated deformations are equivalent if and only if the two differentials are *gauge equivalent*, and any deformation of families of calculi is the cohomology of some differential. Thus, gauge equivalence classes of differentials form a functor of Artin rings isomorphic to $\mathrm{GDef}^{\mathscr{D}}_{X_0/S_0}$.

When taking the limit of the global sections of (1.1), we obtain a curved two-sided Gerstenhaber calculus

$$(PV^{\bullet,\bullet}_{X_0/\Lambda}, DR^{\bullet,\bullet}_{X_0/\Lambda})$$

over $\Lambda = \mathbb{C}[\![Q]\!]$. Now

$$L^{\bullet}_{X_0/\Lambda} := \Gamma(X_0, PV^{-1,\bullet}_{X_0/\Lambda})$$

is a Λ-linear curved Lie algebra, and gauge equivalence classes of Maurer–Cartan solutions are nothing but gauge equivalence classes of differentials as above. Thus, we have an isomorphism

$$\mathrm{LD}^{\mathscr{D}}_{X_0/S_0} \cong \mathrm{Def}(L^{\bullet}_{X_0/\Lambda}, -)\colon \quad \mathbf{Art}_{\mathbb{C}[\![Q]\!]} \to \mathbf{Set}$$

of functors of Artin rings.

Remark 1.95 In the earlier work [76] of the author, we have carried out the same construction for log smooth morphisms only, and with the Gerstenhaber *algebra* $\mathcal{V}^{\bullet}_{X/S}$ only, there denoted by $\mathcal{G}^{\bullet}_{X_0/S_0}$. For the construction of $L^{\bullet}_{X_0/\Lambda}$, this makes little difference, but we have indeed $\mathcal{W}^{\bullet}_{X_0/S_0}$ as well at our disposal, which greatly amplifies the possibilities to study properties of the deformations. For example, compare Theorem 1.98 with Theorem 1.101. The author's article [76] was in turn inspired by the work [38] of Chan, Leung, and Ma, where the Gerstenhaber calculus including $\mathcal{W}^{\bullet}_{X_0/S_0}$ is covered, albeit with one operation less (the left contraction $\vdash$ is missing), only in the analytical setting, and only for global sections, not on the sheaf level. We need the part of $\mathcal{W}^{\bullet}_{X_0/S_0}$ in order to study the logarithmic Bogomolov–Tian–Todorov theorem.   ◇

1.8.3 The Logarithmic Bogomolov–Tian–Todorov Theorem

Several proofs of the classical Bogomolov–Tian–Todorov Theorem 1.27 are known, including Ran's proof via the T^1-lifting criterion in [238] and Iacono's and Manetti's proof via the homotopy Abelianity (in the sense of Definition 1.41) of the dg Lie algebra $L^{\bullet}_Y$ in [152].

The $\mathbb{C}[\![Q]\!]$-linear curved Lie algebra $L^{\bullet}_{X_0/\Lambda}$ is the logarithmic analog of the dg Lie algebra $L^{\bullet}_X$. However, the proof of unobstructedness via homotopy Abelianity cannot be generalized to $L^{\bullet}_{X_0/\Lambda}$—it is not even clear what homotopy Abelianity of $L^\bullet_{X_0/\Lambda}$ should mean since it is not a complex. The argument of Iacono–Manetti shows in our situation that

$$L^\bullet_{X_0/\Lambda} \otimes_{\mathbb{C}[\![Q]\!]} \mathbb{C}$$

is homotopy Abelian, as we explained in our joint work [78, Thm. 3.3] with Andrea Petracci. This shows that LD_{X_0/S_0} is unobstructed on the full subcategory

$$\mathbf{Art} \cong \mathbf{Art}^\circ_{\mathbb{C}[\![Q]\!]} \subset \mathbf{Art}_{\mathbb{C}[\![Q]\!]}$$

of Artinian local $\mathbb{C}$-algebras where the Q-action is trivial, i.e., where $e_A \colon \mathbb{C}[\![Q]\!] \to A$ factors as $\mathbb{C}[\![Q]\!] \to \mathbb{C} \to A$. This captures only locally trivial logarithmic deformations of $f_0 \colon (X_0, \mathcal{M}_0) \to (S_0, \mathcal{N}_0)$ but no deformations in the smoothing direction.

At this point, the argument of Chan, Leung, and Ma in [38] takes advantage of the fact that we do not only have a $\mathbb{C}[\![Q]\!]$-linear curved Lie algebra but much more structure. The centerpiece of their argument is [38, Thm. 5.6], which they call the *abstract unobstructedness theorem*. They state the theorem solely for the structures as they arise in this article, but the abstract unobstructedness theorem is a purely algebraic fact once the correct definitions are collected. At the center of formulating the abstract unobstructedness theorem in purely algebraic terms is the notion of a Λ-*linear curved Batalin–Vilkovisky algebra*. Strictly speaking, the setup can be further weakened, but in practice, at least for our current purposes, we have that structure. Our definition has some minor improvements in that we stipulate necessary relations which are nowhere explicit in [38].

Definition 1.96 (Curved Batalin–Vilkovisky Algebras) Fix $d \geq 1$. A Λ-*linear curved Batalin–Vilkovisky algebra* of dimension d is a tuple

$$(G^{\bullet,\bullet}, \wedge, 1, [-, -], \Delta, \bar\partial, \ell, y)$$

where $G^{p,q}$ is a flat and complete Λ-module for $-d \leq p \leq 0$ and $q \geq 0$, the operation $\wedge$ is a Λ-bilinear product with unit $1 \in G^{0,0}$, the operation $[-, -]$ is a Λ-bilinear Lie bracket, $\Delta \colon G^{p,q} \to G^{p+1,q}$ is a Batalin–Vilkovisky operator satisfying $\Delta^2 = 0$, the operation $\bar\partial \colon G^{p,q} \to G^{p,q+1}$ is a predifferential, $\ell \in \mathfrak{m}_\Lambda \cdot G^{-1,2}$ is such that $\bar\partial^2(\theta) = [\ell, \theta]$, and $y \in \mathfrak{m}_\Lambda \cdot G^{0,1}$ is such that $\Delta\bar\partial(\theta) + \bar\partial\Delta(\theta) = [y, \theta]$. The precise relations can be found in Sect. 3.4.

When $G^{\bullet,\bullet}$ is a Λ-linear curved Batalin–Vilkovisky algebra, then $L^\bullet := G^{-1,\bullet}$ is a Λ-linear curved Lie algebra, giving rise to a deformation functor $\mathrm{Def}(L^\bullet, -)$. In order to show that this deformation functor is unobstructed, it is sufficient (and necessary) to show that the Maurer–Cartan functor $\mathrm{MC}(L^\bullet, -)$ is unobstructed. The subject of the abstract unobstructedness of [38] is however not the Maurer–Cartan functor $\mathrm{MC}(L^\bullet, -)$ but the *semi-classical Maurer–Cartan functor* $\mathrm{SMC}(G^{\bullet,\bullet}, -)$. This functor of Artin rings classifies solutions (ϕ, f) with $\phi \in \mathfrak{m}_A \cdot G_A^{-1,1}$ and

$f \in \mathfrak{m}_A \cdot G_A^{0,0}$ of the *semi-classical extended Maurer–Cartan equation*

$$\bar\partial\phi + \frac{1}{2}[\phi,\phi] + \ell = 0$$

$$\bar\partial f + [\phi, f] + y + \Delta\phi = 0.$$

Obviously, if (ϕ, f) is a semi-classical Maurer–Cartan solution, then ϕ is a classical Maurer–Cartan solution. These names come from the fact that we consider also a *quantum* Maurer–Cartan equation in proving the unobstructedness theorem. For $G^{\bullet,\bullet}$, the most natural condition which suffices for the unobstructedness theorem is the following:

Definition 1.97 (Quasi-perfectness)   A Λ-linear curved Batalin–Vilkovisky algebra is *quasi-perfect* if the following conditions hold:

(a) The first spectral sequence $'E_{0;r}$ associated with the double complex $(G_0^{\bullet,\bullet}, \Delta, \bar\partial)$ over $A_0 = \mathbb{C}$ degenerates at E_1.
(b) For every $A \in \mathbf{Art}_\Lambda$, the induced map $H^k(G_A^{\bullet}, \check d) \to H^k(G_0^{\bullet}, \check d)$ with

$$\check d = \Delta + \bar\partial + (\ell + y) \wedge (-)$$

  is surjective.

Over $\mathbb{C}$, we have $\bar\partial^2 = 0$ since $\ell \in \mathfrak{m}_\Lambda \cdot G^{-1,2}$; thus, $(G_0^{\bullet,\bullet}, \Delta, \bar\partial)$ is indeed a double complex. The map $\check d$ in the definition turns out to always satisfy $\check d^2 = 0$, and hence it is a differential on the total complex $G^{\bullet}$. The notion of *perfectness* incorporates an additional finiteness condition which is not necessary for unobstructedness.

Theorem 1.98 (First Abstract Unobstructedness Theorem)   *Let $G^{\bullet,\bullet}$ be a quasi-perfect Λ-linear curved Batalin–Vilkovisky algebra. Then* $\mathrm{SMC}(G^{\bullet,\bullet}, -)$ *is unobstructed.*

In the main text, this is Theorem 5.23. A weaker condition, called *semi-perfectness* and specified in Definition 6.7, is sufficient. Also note that Theorem 1.98 does *not* show unobstructedness of $\mathrm{Def}(L^{\bullet}, -)$—it only shows that *pairs* (ϕ, f) can be lifted along surjections $B \to A$ in $\mathbf{Art}_\Lambda$. We will come back to this point shortly.

Remark 1.99   In the case without log structures, the usage of Batalin–Vilkovisky algebras to show unobstructedness is by now classical—this is the proof of the classical analytic Bogomolov–Tian–Todorov theorem which we have explained in Sect. 1.4. Early uses in the mathematics literature include [19, 202]; a transition from the original use in physics to deformation theory seems to be made in [260]. The degeneracy of the spectral sequence as a condition for the unobstructedness of the deformation functor seems to be first mentioned in [171, §4.2.2]; in Proposition D.2, we check that the degeneration of the spectral sequence at E_1 is indeed equivalent to

the criterion as stated in [171, Defn. 4.13]. The degeneration of a somewhat different
spectral sequence is given as a criterion in [266]. ◇

While Theorem 1.98 is enough to show the *existence* of some deformation, it is
insufficient to conclude that $\mathrm{LD}^{\mathscr{D}}_{X_0/S_0}$ is unobstructed, i.e., that *every* infinitesimal
deformation over some A can be lifted to B. To show this stronger statement, we
take once again advantage of additional structure that we have in the geometric
situation. Instead of a Batalin–Vilkovisky algebra, we use a Batalin–Vilkovisky
calculus.

Definition 1.100 (Curved Batalin–Vilkovisky Calculi) A Λ-*linear curved
Batalin–Vilkovisky calculus* of dimension $d \geq 1$ is a tuple

$$(G^{\bullet,\bullet}, \wedge, 1_G, [-, -], \Delta, \bar{\partial}, \ell, y, A^{\bullet,\bullet}, \wedge, 1_A, \partial, \lrcorner, \omega, \bar{\partial}, \mathcal{L})$$

with a Λ-linear curved Batalin–Vilkovisky algebra, flat and complete Λ-modules
$A^{i,j}$ for $0 \leq i \leq d$ and $j \geq 0$, a product $\wedge$ on $A^{\bullet,\bullet}$ with unit 1_A, a de Rham
differential $\partial \colon A^{i,j} \to A^{i+1,j}$, a contraction map $\lrcorner \colon G^{p,q} \times A^{i,j} \to A^{p+i,q+j}$, a
volume form $\omega \in A^{d,0}$, a predifferential $\bar{\partial} \colon A^{i,j} \to A^{i,j+1}$ with $\bar{\partial}^2(\alpha) = \mathcal{L}_\ell(\alpha)$,
and a Lie derivative $\mathcal{L}$. These data must satisfy the relations given in Sect. 3.4; in
particular,

$$\kappa \colon \ G^{p,q} \to A^{p+d,q}, \quad \theta \mapsto (\theta \lrcorner \omega),$$

must be an isomorphism, and we have $\Delta(\theta) \lrcorner \omega = \partial(\theta \lrcorner \omega)$.

By Lemma 5.20, a Λ-linear curved Batalin–Vilkovisky calculus $(G^{\bullet,\bullet}, A^{\bullet,\bullet})$
is quasi-perfect (in the sense of Definition 5.18) if and only if the Batalin–
Vilkovisky algebra $G^{\bullet,\bullet}$ is quasi-perfect as defined above. When we exploit this
additional structure, we can show that any classical Maurer–Cartan solution ϕ
can be complemented to a semi-classical Maurer–Cartan solution (ϕ, f), see
Theorem 6.17. This yields a second abstract unobstructedness theorem, the actual
unobstructedness result which we have desired in the first place. This somewhat
stronger result is not contained in [38].

Theorem 1.101 (Second Abstract Unobstructedness Theorem) *Let* $(G^{\bullet,\bullet}, A^{\bullet,\bullet})$
be a quasi-perfect Λ-*linear curved Batalin–Vilkovisky calculus. Then* $\mathrm{MC}(L^\bullet, -)$
is unobstructed for the Λ-*linear curved Lie algebra* $L^\bullet = G^{-1,\bullet}$.

In the main text, this is Theorem 5.24.

Recall that, in the situation of a generically log smooth family $f_0 \colon (X_0, U_0, \mathcal{M}_0)$
$\to (S_0, \mathcal{N}_0)$ with a system of deformations $\mathscr{D}$, we have constructed a $\mathbb{C}[\![Q]\!]$-linear
curved *Gerstenhaber* calculus

$$(PV^{\bullet,\bullet}_{X_0/\Lambda}, DR^{\bullet,\bullet}_{X_0/\Lambda}). \tag{1.2}$$

In other words, we do not have the volume form ω, the Batalin–Vilkovisky operator Δ, and the element $y \in \mathfrak{m}_\Lambda \cdot G^{0,1}$ on this structure. However, when assuming that f_0 is log Calabi–Yau, then a volume form $\tilde{\omega}_0 \in \mathcal{W}^d_{X_0/S_0}$ gives rise to a volume form

$$\omega_0 \in DR^{d,0}_{X_0/\mathbb{C}} = DR^{d,0}_{X_0/\Lambda} \otimes_\Lambda \mathbb{C}.$$

This can be lifted order by order to obtain a (even after the initial choice of $\tilde{\omega}_0$ non-canonical) volume form $\omega \in DR^{d,0}_{X_0/\Lambda}$. Applying Proposition 3.28, we find a structure of Batalin–Vilkovisky calculus on our Gerstenhaber calculus after establishing the necessary properties of ω. This strategy is slightly different from the original approach in [38], which constructs ω and Δ directly along with the other structures of the Gerstenhaber calculus.

With Theorem 1.101, the final step to achieve the logarithmic Bogomolov–Tian–Todorov theorem for $f_0 \colon (X_0, U_0, \mathcal{M}_0) \to (S_0, \mathcal{N}_0)$ proper and log Calabi–Yau is to show that the curved Batalin–Vilkovisky calculus (1.2) is indeed quasi-perfect. This holds in the log smooth case and the more general log toroidal case.

Condition (a) in the Definition 1.97 of quasi-perfectness is equivalent to the degeneration of the log Hodge–de Rham spectral sequence

$$H^q(X_0, \mathcal{W}^p_{X_0/S_0}) \Rightarrow \mathbb{H}^{p+q}(X_0, \mathcal{W}^\bullet_{X_0/S_0})$$

at E_1. In case $f_0 \colon (X_0, \mathcal{M}_0) \to (S_0, \mathcal{N}_0)$ is log smooth, this has been known from the beginning of log geometry, Kato's seminal article [162, Thm. 4.12]. When f_0 is log toroidal, i.e., possibly non-accurately of class $\mathscr{C}^{\mathrm{tor}}_Q$ (see Example 1.87), the degeneration is a recent insight by Filip, Ruddat, and the author in [77]. The method of the proof is the same as the original algebraic approach by Deligne and Illusie in [59] via reduction to positive characteristic; where they use the infinitesimal lifting property of smooth morphisms, we rely heavily on the local models in $\mathscr{C}^{\mathrm{tor}}_Q$ to study the behavior of the method under the presence of log singularities.

Condition (b) in Definition 1.97 is somewhat subtle. For $A \in \mathbf{Art}_\Lambda$ with the property that there exists some deformation $f_A \colon (X_A, U_A, \mathcal{M}_A) \to (S_A, \mathcal{N}_A)$ of type $\mathscr{D}$ (our chosen system of deformations), the surjectivity in hypercohomology is equivalent to the following statement:

($*$)   The relative Hodge–de Rham spectral sequence

$$R^q(f_A)_* \mathcal{W}^p_{X_A/S_A} \Rightarrow R^{p+q}(f_A)_* \mathcal{W}^\bullet_{X_A/S_A}$$

degenerates at E_1, and the formation of $R^q(f_A)_* \mathcal{W}^p_{X_A/S_A}$ commutes with base change along $(S_0, \mathcal{N}_0) \to (S_A, \mathcal{N}_A)$.

However, the surjectivity in hypercohomology is also not a vacuous statement if no global deformation $f_A \colon (X_A, U_A, \mathcal{M}_A) \to (S_A, \mathcal{N}_A)$ exists for a given $A \in \mathbf{Art}_\Lambda$, and so Condition (b) is stronger than requiring ($*$) for every global deformation f_A of type $\mathscr{D}$. When we wrote [77], in particular the proof of Theorem 13.1

therein, we misunderstood [38] and wrongly believed that requiring (∗) for every global deformation would be a sufficient condition for the existence of a formal deformation. In particular, in [77], we showed (∗) for every global log toroidal deformation, but we did not show Condition (b). This is insufficient to conclude the existence of a formal deformation from Theorem 1.98 and thus a gap in [77].[11] This gap is bridged by a subtle variant of the argument which we also have used in [77] to show (∗) for log toroidal deformations, a variant already given by Chan, Leung, and Ma in [38, Lemma 4.18]. In Chap. 15, we show that the argument given by Chan, Leung, and Ma applies to our algebraically constructed Gerstenhaber calculus (1.2) as well, not only to the analytic version of [38]. Interestingly, we have to analytify the curved Gerstenhaber calculus nonetheless in order to conclude the quasi-perfectness of the algebraic Gerstenhaber calculus. In [77], we had to analytify as well to show (∗) for log toroidal deformations.

Theorem 1.102 (Quasi-perfectness) *Let* $f_0\colon (X_0, U_0, \mathcal{M}_0) \to (S_0, \mathcal{N}_0)$ *be a proper log Gorenstein generically log smooth family. Assume that* f_0 *is log toroidal, i.e., possibly non-accurately of class* $\mathscr{C}_Q^{\mathrm{tor}}$. *Let* $\mathscr{D}$ *be a system of deformations as a generically log smooth family, and assume that all local models* $(V_{\alpha;A}, U_{\alpha;A}, \mathcal{M}_{\alpha;A}) \to (S_A, \mathcal{N}_A)$ *in* $\mathscr{D}$ *are log toroidal. Then the* $\mathbb{C}[\![Q]\!]$-*linear curved (two-sided) Gerstenhaber calculus (1.2) is quasi-perfect.*

Corollary 1.103 (Log Toroidal Bogomolov–Tian–Todorov Theorem) *Consider a proper log toroidal family* $f_0\colon (X_0, U_0, \mathcal{M}_0) \to (S_0, \mathcal{N}_0)$ *which is log Calabi–Yau. Let* $\mathscr{D}$ *be a system of deformations such that all local models are log toroidal. Then* $\mathrm{LD}_{X_0/S_0}^{\mathscr{D}}$ *is unobstructed. In particular, this holds when* $f_0\colon (X_0, \mathcal{M}_0) \to (S_0, \mathcal{N}_0)$ *is saturated and log smooth, and* $\mathscr{D}$ *is a system of log smooth deformations.*

In the main text, both the theorem and its corollary are Theorem 15.2.

Example 1.104 The theorem holds for log smooth deformations of the log smooth morphism $g_0\colon (\widetilde{H}, \mathcal{M}_0) \to (S_0, \mathcal{N}_0)$ since g_0 is log Calabi–Yau, and proper. The theorem also holds for deformations accurately of class $\mathscr{C}^{\mathrm{anc}}$ (see Example 1.86) of the generically log smooth family $f_0\colon (H, U_0, \mathcal{R}_0) \to (S_0, \mathcal{N}_0)$ since deformations accurately of class $\mathscr{C}^{\mathrm{anc}}$ are log toroidal. $\Diamond$

Example 1.105 In the setting of the Gross–Siebert program, divisorial deformations of certain toric log Calabi–Yau spaces are unobstructed. After discussing the basics of the toric Gross–Siebert mirror construction, we give a precise statement in Theorem 1.149 below, whose content was known by the article [243] of Ruddat and Siebert before. Note that Corollary 1.103 has a far wider scope than any application to the toric Gross–Siebert construction. For example, we do not assume that the irreducible components of X_0 are toric varieties. $\Diamond$

---

[11] It is only *a posteriori* that we know the existence of a global deformation over every A and that thus (∗) implies Condition (b).

Remark 1.106 Both [38] (for the case of log toroidal families arising from the toric Gross–Siebert mirror construction) and [77] (for the case of more general log toroidal families without assumptions on the global geometry except for properness and the Calabi–Yau condition) consider only deformations over $\mathbb{C}[Q]/\mathfrak{m}_Q^{k+1}$, and both articles consider only the underlying flat deformation but not the deformed log structure.   ◊

Remark 1.107 For general bases but only in the log smooth case, we have stated Corollary 1.103 in [78, Thm. 3.7] in our joint work with Petracci. There, we deduce via [78, Prop. 2.8] that, in order for LD_{X_0/S_0} to be unobstructed, it is sufficient that the restriction map along the particular thickenings

$$A_{k+1} = \mathbb{C}[Q]/\mathfrak{m}_Q^{k+2} \to \mathbb{C}[Q]/\mathfrak{m}_Q^{k+1} = A_k$$

is surjective. Our description of this key step has, however, been very sketchy, deferring the details to a future article, and just mentioning the method of [38]. Here, we deliver the promised elaborate account of the theory in its algebraic version. The reduction to $A_{k+1} \to A_k$ is no longer necessary due to the more general form of Theorem 1.101.   ◊

In Sect. 1.12, we will see some examples of (enhanced) generically log smooth families which are not log toroidal but nonetheless very interesting from the perspective of deformation theory. It would be interesting to have a version of Corollary 1.103 in these situations. We establish some preliminary results in this direction (see Sect. 1.12), but the core result, the analog of Theorem 1.102, is still missing.

1.8.4 Smoothing Normal Crossing Schemes

We apply Corollary 1.103 to construct formal algebraic and analytic smoothings of normal crossing schemes. The classical result, which expands Friedman's work from the surface case to all dimensions, is [173, Thm. 4.2] due to Kawamata and Namikawa.

Theorem 1.108 (Kawamata–Namikawa 1994, Thm. 4.2) *Let V be a compact Kähler normal crossing complex space of dimension $d \geq 3$, and let $\tilde{V} \to V$ be the normalization. Assume that V is d-semistable, that $\omega_V^\circ \cong O_V$, that $H^{d-1}(V, O_V) = 0$, and that $H^{d-2}(\tilde{V}, O_{\tilde{V}}) = 0$. Then V admits a semistable analytic smoothing.*

With [77, Thm. 1.1], we have given a generalization of this result. Due to the generality of the Log Toroidal Bogomolov–Tian–Todorov Theorem 1.103, we do no longer need the two cohomological assumptions on V and its normalization $\tilde{V}$. More importantly, since we have a valid theory for generically log smooth families, we can drop the d-semistability assumption and replace it with the weaker assumption that $\mathcal{T}_V^1$ is globally generated. When V is not d-semistable, the smoothing will have singularities in the total space. The following smoothing result is a slightly more natural variant of [77, Thm. 1.1].

Let V be a normal crossing scheme. Recall from Sect. 1.2 that V carries a stratification into smooth locally closed subschemes; the set of strata of codimension m is denoted by $[\mathcal{S}_m V]$. Recall from Sect. 1.6 that V carries a sheaf $\mathcal{LS}_V$ whose sections correspond to log smooth structures on V. Finally, recall from Example 1.86 the class of almost normal crossing log singularities $\mathscr{C}^{\mathrm{anc}}$. The following result is a special case of Theorem 18.44 in the main text.

Theorem 1.109 (Smoothing Normal Crossing Schemes I) *Let $V/\mathbb{C}$ be a (proper) Calabi–Yau normal crossing scheme. Assume that every open stratum $S^\circ \in [\mathcal{S}_m V]$ is quasi-projective for $m \geq 1$, and assume that $\mathcal{T}_V^1$ is globally generated.*

(1) *There is a closed subset $Z \subset V$ of codimension ≥ 2, contained in $D = \mathrm{Sing}(V)$, such that $Z \cap S^\circ \subset S^\circ$ is a smooth effective Cartier divisor for every open stratum $S^\circ \in [\mathcal{S}_m V]$. The complement $U_0 = V \setminus Z$ carries a section $s \in \Gamma(U_0, \mathcal{LS}_V)$ such that the induced generically log smooth family $f_0 \colon (V, U_0, \mathcal{M}_0) \to (S_0, \mathcal{N}_0)$ over the standard log point is accurately of class $\mathscr{C}^{\mathrm{anc}}$ and log Calabi–Yau.*

(2) *Let $\mathscr{D}$ be a system of deformations which is accurately of class $\mathscr{C}^{\mathrm{anc}}$. Then $\mathrm{LD}_{V/S_0}^{\mathscr{D}}$ is unobstructed. In particular, V admits a formal algebraic one-parameter smoothing as well as an analytic one-parameter smoothing.*

Example 1.110 The four planes $H = \bigcup_{i=0}^{3} H_i$ together with the open subset $U_0 \subseteq H$ and the section $s \in \Gamma(U_0, \mathcal{LS}_H)$ constructed in Example 1.81 are as the data described by the theorem. $\diamond$

Example 1.111 This is an example which does not have an obvious explicit smoothing. It is a special case of [77, Ex. 1.3]. Let $P = \mathbb{P}^3$, and let $D \subset P$ be a smooth quartic surface, i.e., an anti-canonical hypersurface. We obtain a normal crossing threefold $N = P \sqcup_D P$ with two irreducible components by gluing two copies of $P = \mathbb{P}^3$ along one copy of the closed subset $D \subset P$. We denote the two copies of P by P_1 and P_2. In the notation of Sect. 1.2, we have $\omega_{(P_i, \partial P_i)} \cong O_P(-4) \otimes O_P(4) \cong O_P$ and $\omega_{(D, \partial D)} = \omega_D \cong O_D$. As in Example 1.33, this shows that $\omega_N^\circ \cong O_N$ so that N is a Calabi–Yau scheme. Since D is projective, and since $P \setminus D$ is quasi-projective, every open stratum (of any codimension) is quasi-projective.

We compute $\mathcal{T}_N^1$ as described in Sect. 1.5. We have $\mathcal{N}_{D/P} \cong O_D(4)$ so that $O_D(-N) \cong \mathcal{N}_{D/P_1}^\vee \otimes \mathcal{N}_{D/P_2}^\vee \cong O_D(-8)$ and $\mathcal{T}_N^1 \cong O_D(N) \cong O_D(8)$. In particular, $\mathcal{T}_N^1$ is globally generated.

By Theorem 1.109, we can find a smooth effective Cartier divisor $Z \subset D$ and a section $s \in \Gamma(N \setminus Z, \mathcal{LS}_N)$ such that the induced generically log smooth family $f_0 \colon (N, N \setminus Z, \mathcal{M}_0) \to (S_0, \mathcal{N}_0)$ is accurately of class $\mathscr{C}^{\mathrm{anc}}$. More concretely, we may take any section $\sigma \in \Gamma(D, \mathcal{T}_N^1)$ such that its associated divisor $Z = \mathrm{div}(\sigma) \subset$

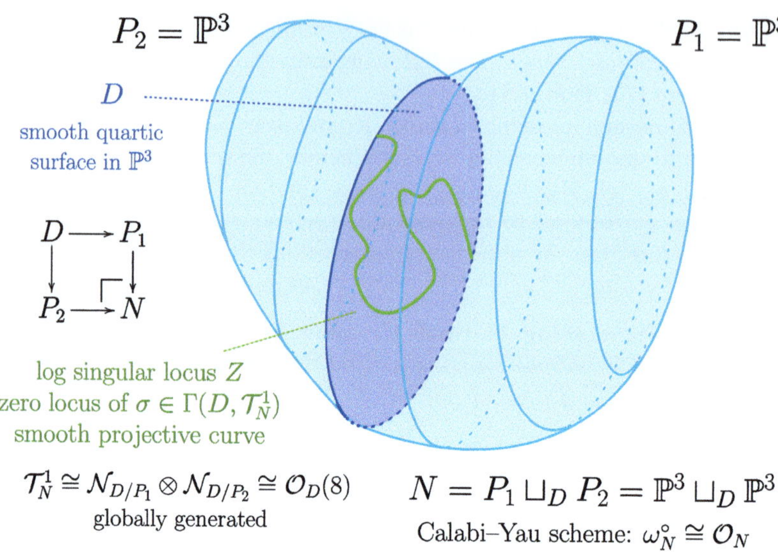

Fig. 1.15 Example 1.111. © Simon Felten 2025. All rights reserved

D is smooth (and in particular reduced). Then Proposition 18.54 shows that the associated generically log smooth family is étale locally around every point $v \in Z$ given by $\mu_{3;1}^{\mathrm{anc}}\colon \operatorname{Spec} \mathbb{C}[x, y, z, u, t]/(xy - tz) \to \mathbb{A}_t^1$ and hence accurately of class $\mathscr{C}^{\mathrm{anc}}$. Now Construction 1.91 gives us a log toroidal system of deformations $\mathscr{D}$, and we can apply Corollary 1.103. The total space of the resulting analytic one-parameter smoothing $X \to \Delta$ has singularities along $Z^{\mathrm{an}} \subset V^{\mathrm{an}} \subset X$. For an illustration, see Fig. 1.15.   ◊

Example 1.112 The construction of non-Kähler Calabi–Yau threefolds with arbitrarily large b_2 by Hashimoto and Sano in [140] is very similar to the previous example. We can simplify the construction because we do not need d-semistability. Then we start with $P = \mathbb{P}^1 \times \mathbb{P}^1 \times \mathbb{P}^1$ and a K3 surface $k\colon S \subset P$ of multidegree $(2, 2, 2)$ such that S has a certain automorphism $\iota\colon S \to S$ of infinite order. For any $a \geq 0$, we obtain a normal crossing scheme $V = P \sqcup_{S, \iota^a} P$ by pushout along $k\colon S \to P$ and $k \circ \iota^a\colon S \to P$. This V satisfies the assumptions of Theorem 1.109 so that we obtain an analytic one-parameter smoothing. It then remains to compute b_2 of the general fiber. This has not been carried out, but the author believes that this is possible without resolving the non-d-semistability because the log singularities of class $\mathscr{C}^{\mathrm{anc}}$ are closely related to relatively log smooth families, for which Nakayama and Ogus proved in [212, Thm. 3.7] that the associated morphism of Kato–Nakayama spaces is topologically a fiber bundle so that the topology of the nearby fiber is accessible via the log geometry of the central fiber.   ◊

1.8.5   Constancy of the Hodge Numbers

Let $f_0\colon (X_0, U_0, \mathcal{M}_0) \to (S_0, \mathcal{N}_0)$ be a proper log Gorenstein generically log smooth family with a system of deformations $\mathscr{D}$. Assume that (1.2) is quasi-perfect, as is the case for example if both f_0 and $\mathscr{D}$ are log toroidal. We do *not* assume that f_0 is log Calabi–Yau here. For any deformation $f_A\colon (X_A, U_A, \mathcal{M}_A) \to (S_A, \mathcal{N}_A)$ of type $\mathscr{D}$, we have an isomorphism

$$\mathbb{H}^m(X_A, \mathcal{W}^\bullet_{X_A/S_A}) \cong H^m(\mathrm{Tot}^\bullet(DR^{\bullet;\bullet}_{X_0/A}), d_\phi) \tag{1.3}$$

for the operator $d_\phi = \partial + \bar\partial_\phi$, where $\bar\partial_\phi$ is the differential corresponding to the Maurer–Cartan solution ϕ which in turn corresponds to $f_A\colon X_A \to S_A$. Namely,

$$(\mathcal{W}^\bullet_{X_A/S_A}, \partial) \to (\mathrm{Tot}^\bullet(\mathscr{D}\mathcal{R}^{\bullet;\bullet}_{X_0/A}), \partial + \bar\partial_\phi)$$

is an acyclic resolution. The right-hand side commutes with base change as part of the definition of quasi-perfectness. Thus, also the left-hand side commutes with base change. Then we can deduce from Proposition D.4 that the Hodge–de Rham spectral sequence

$$E^{pq}_1 = R^q(f_A)_* \mathcal{W}^p_{X_A/S_A} \Rightarrow R^{p+q}(f_A)_* \mathcal{W}^\bullet_{X_A/S_A}$$

degenerates at E_1, that E^{pq}_1 is a free A-module, and that its formation commutes with base change. Moreover, also the abutment $\mathbb{H}^m(X_A, \mathcal{W}^\bullet_{X_A/S_A})$ is a free A-module, and its formation commutes with base change. Since $f_A\colon X_A \to S_A$ is proper, $R^q(f_A)_* \mathcal{W}^p_{X_A/S_A}$ is finitely generated. We say that the Hodge numbers are *constant*. Namely, if we had a proper family $f\colon X \to S$ over a reduced base, then the constancy of the Hodge numbers would be equivalent to $R^q f_* \mathcal{W}^p$ being locally free.

In the case of a log toroidal family $f_A\colon (X_A, U_A, \mathcal{M}_A) \to (S_A, \mathcal{N}_A)$, at least for some bases A, we have known this before by Felten et al. [77, Thm. 1.10]. However, the above argument is slightly different in that we know that the right-hand side of (1.3) is free, and that its formation commutes with base change, even *without knowing or assuming the existence* of $f_A\colon (X_A, U_A, \mathcal{M}_A) \to (S_A, \mathcal{N}_A)$. Therefore, the hypercohomology is not so much associated with f_A as it is associated with the curved Gerstenhaber calculus (1.2), and thus with the *deformation problem*. Although we have chosen the specific total differential d_ϕ associated with a Maurer–Cartan solution ϕ above, the cohomology is independent of ϕ. Namely, if ϕ' is any other solution of the Maurer–Cartan equation, then we can transform d_ϕ into $d_{\phi'}$ via an automorphism of (1.2). Furthermore, even if ϕ does not solve the Maurer–Cartan equation, we can define a differential d_ϕ which gives rise to the same cohomology, see Lemma 5.28.

Remark 1.113 The reader may object that we have replaced the simpler problem—the constancy of the Hodge numbers—with the more difficult problem to show that the formation of the hypercohomology commutes with base change. However, the point is (1) that constancy of the Hodge numbers follows from the quasi-perfectness condition, and (2) that indeed our proof of the constancy of the Hodge numbers proceeds through analyzing the hypercohomology. (This is also true for Deligne's classical proof for families of smooth and proper varieties.)   ◊

1.8.6 *Comparison of the Local Moduli Spaces*

We illustrate the relation between logarithmic deformations and flat deformations on the level of local moduli spaces. To do so, we work with the log smooth morphism $g_0 \colon (\widetilde{H}, \mathcal{M}_0) \to (S_0, \mathcal{N}_0)$, the central fiber of the family in Example 1.49. We assume that $\widetilde{H}$ is in (-1)-form, which can be achieved by choosing the blowups in Example 1.49 appropriately.[12]

We have already discussed in Example 1.48 the base $(B \ni 0)$ of the miniversal analytic deformation of $\widetilde{H}$: It is a union

$$(B \ni 0) = (B_1 \ni 0) \cup (B_2 \ni 0)$$

of two complex space germs, both smooth. The component $(B_1 \ni 0)$ is a divisor in $(\mathbb{T}^1_{\widetilde{H}} \ni 0)$ and corresponds to locally trivial deformations. The component $(B_2 \ni 0)$ is always of dimension 20 and corresponds to semistable smoothings of $\widetilde{H}$. The intersection $(B_1 \cap B_2 \ni 0)$ corresponds to d-semistable locally trivial deformations.

The Log Smooth Bogomolov–Tian–Todorov Theorem 1.73 shows that $\mathrm{LD}_{\widetilde{H}/S_0}$ is unobstructed. Thus, its hull (in the sense of Schlessinger) is of the form $R = \mathbb{C}[\![t, s_1, \dots, s_r]\!]$ for

$$r = \dim H^1(\widetilde{H}, \Theta^1_{\widetilde{H}/S_0}) = \dim H^1(\widetilde{H}, \mathcal{W}^1_{\widetilde{H}/S_0}) = \dim H^1(K, \Omega^1_K) = 20,$$

where $K \subset \mathbb{P}^3$ is the Fermat quartic surface of Example 1.2, and where the second last equality is due to the constancy of the log Hodge numbers. Here, the parameter t corresponds to the log structure while the remaining parameters $s_1, \dots, s_r$ correspond to deformations over $(S_\varepsilon, \mathcal{N}_\varepsilon) = \mathrm{Spec}(\mathbb{N} \to \mathbb{C}[\varepsilon]/(\varepsilon^2), 1 \mapsto 0)$.

Given an infinitesimal log smooth deformation over $R_k = R/\mathrm{m}_R^{k+1}$, we can forget the log structure and analytify the underlying scheme. This gives rise to an analytic flat deformation and therefore to a map $\phi_k \colon O_{B,0} \to R_k$ by completeness of B. Versality of B shows that we can choose these maps compatible for all k so

---

[12] We use $\widetilde{H}$ being in (-1)-form to conclude $H^0(\widetilde{H}, \mathcal{T}_{\widetilde{H}/S_0}) = 0$ from [263, Lemma 3.3] or [97, Cor. 3.5, p. 292]. The author does not know if the condition is necessary.

that we obtain an induced map $\phi\colon O_{B,0} \to R$ since R is a complete local ring. Let us write $\varphi\colon \widehat{O}_{B,0} \to R$ for the induced map on completions.

Let $\mathfrak{p}_1, \mathfrak{p}_2 \subset O_{B,0}$ be the minimal prime ideals corresponding to B_1 respectively B_2. Since $O_{B,0}/\mathfrak{p}_i$ is regular, so is $\widehat{O_{B,0}/\mathfrak{p}_i} \cong \widehat{O}_{B,0}/\hat{\mathfrak{p}}_i$, and hence $\hat{\mathfrak{p}}_i \subset \widehat{O}_{B,0}$ is a prime ideal. Since $O_{B,0} \to \widehat{O}_{B,0}$ is flat and hence satisfies going down, we find $\mathfrak{p}_i = \hat{\mathfrak{p}}_i \cap O_{B,0}$. Furthermore, $\hat{\mathfrak{p}}_1$ and $\hat{\mathfrak{p}}_2$ are precisely the minimal prime ideals of $\widehat{O}_{B,0}$.

Let $\mathfrak{q} = \varphi^{-1}(0) \subset \widehat{O}_{B,0}$, which is a prime ideal. If $\hat{\mathfrak{p}}_1 \subseteq \mathfrak{q}$, then every logarithmic deformation would be locally trivial on underlying complex spaces, which is not the case. Thus, we have $\hat{\mathfrak{p}}_2 \subseteq \mathfrak{q}$. We may say that choosing a log smooth structure on $\widetilde{H}$ *selects the component* $(B_2 \ni 0)$ of $(B \ni 0)$. We now elaborate on this statement by further analyzing the map φ.

Let $\mathfrak{m} \subset \widehat{O}_{B,0}$ be the maximal ideal, and let $R^{\|} := R/\varphi(\mathfrak{m}) \cdot R$. In $R^{\|}$, we have $t = 0$ since all log smooth deformations obtained by base change along some $R^{\|} \to A$ are trivial on underlying schemes (at least after analytification). We compute the tangent space to $R^{\|}$. We have

$$T_{R^{\|}} = \mathrm{Hom}_{\mathbb{C}}(R^{\|}, \mathbb{C}[\varepsilon]/(\varepsilon^2)) = \mathrm{Hom}_{\mathbb{C}[\![t]\!]}(R^{\|}, \mathbb{C}[\varepsilon]/(\varepsilon^2))$$

$$= \{\chi \in \mathrm{Hom}_{\mathbb{C}[\![t]\!]}(R, \mathbb{C}[\varepsilon]/(\varepsilon^2)) \mid \chi(\varphi(\mathfrak{m})) = 0\}$$

$$= \ker\left(H^1(\widetilde{H}, \Theta^1_{\widetilde{H}/S_0}) \to \mathrm{Ext}^1(\Omega^1_{\widetilde{H}}, O_{\widetilde{H}})\right)$$

$$= \ker\left(H^1(\widetilde{H}, \Theta^1_{\widetilde{H}/S_0}) \to H^1(\widetilde{H}, \mathcal{T}_{\widetilde{H}/S_0})\right).$$

By forming the quotient of the injective map on the left, we obtain an exact sequence

$$0 \to \Theta^1_{\widetilde{H}/S_0} \to \mathcal{T}_{\widetilde{H}/S_0} \to \mathcal{B}_{\widetilde{H}/S_0} \to 0.$$

Thus, we have

$$T_{R^{\|}} \cong \mathrm{coker}\left(H^0(\widetilde{H}, \mathcal{T}_{\widetilde{H}/S_0}) \to H^0(\widetilde{H}, \mathcal{B}_{\widetilde{H}/S_0})\right).$$

In joint work in progress, Matthias Zach and the author compute the following result.

Lemma 1.114 *Let* $\pi\colon \widetilde{H} \to H$ *be the birational map from Example 1.48, and assume that* $\widetilde{H}$ *is in* (-1)-*form.*

(1) *We have a canonical homomorphism*

$$\pi_* \Theta^1_{\widetilde{H}/S_0} \to \Theta^1_{H/S_0}$$

due to reflexivity of the target. This map is an isomorphism.

(2) *We have a canonical homomorphism*

$$\pi_* \mathcal{T}_{\widetilde{H}/S_0} \to \mathcal{T}_{H/S_0}$$

due to reflexivity of the target. This map is injective, and its cokernel is isomorphic to the direct sum of skyscraper sheaves $\bigoplus_{z \in Z} \mathbb{C}_z$.

(3) *For $j \geq 1$, we have*

$$R^j \pi_* \Theta^1_{\widetilde{H}/S_0} = 0, \quad R^j \pi_* \mathcal{T}_{\widetilde{H}/S_0} = 0, \quad \text{and} \quad R^j \pi_* \mathcal{B}_{\widetilde{H}/S_0} = 0.$$

(4) *The cokernel $\mathcal{B}_{H/S_0}$ of $\Theta^1_{H/S_0} \to \mathcal{T}_{H/S_0}$ is a line bundle on the double locus $D = \operatorname{Sing}(H)$, and for each line $L_{ij} \subset D$, we have*

$$\mathcal{B}_{H/S_0}|_{L_{ij}} \cong \mathcal{O}_{L_{ij}}(4).$$

(5) *The direct image $\pi_* \mathcal{B}_{\widetilde{H}/S_0}$ is a line bundle on the double locus $D = \operatorname{Sing}(H)$, and for each line $L_{ij} \subset D$, we have*

$$(\pi_* \mathcal{B}_{\widetilde{H}/S_0})|_{L_{ij}} \cong \mathcal{O}_{L_{ij}}.$$

(6) *We have $H^0(H, \pi_* \mathcal{B}_{\widetilde{H}/S_0}) \neq 0$, and this implies $\pi_* \mathcal{B}_{\widetilde{H}/S_0} \cong \mathcal{O}_D$.*

(7) *The dimensions of the spaces constructed on H are as follows:*

$$h^0(H, \Theta^1_{H/S_0}) = 0, \qquad h^1(H, \Theta^1_{H/S_0}) = 20, \qquad h^2(H, \Theta^1_{H/S_0}) = 0,$$

$$h^0(H, \mathcal{T}_{H/S_0}) = 3, \qquad h^1(H, \mathcal{T}_{H/S_0}) = 1, \qquad h^2(H, \mathcal{T}_{H/S_0}) = 0,$$

$$h^0(H, \mathcal{B}_{H/S_0}) = 22, \qquad h^1(H, \mathcal{B}_{H/S_0}) = 0, \qquad h^2(H, \mathcal{B}_{H/S_0}) = 0.$$

(8) *The dimensions of the spaces constructed on $\widetilde{H}$ are as follows:*

$$h^0(H, \pi_* \Theta^1_{\widetilde{H}/S_0}) = 0, \quad h^1(H, \pi_* \Theta^1_{\widetilde{H}/S_0}) = 20, \quad h^2(H, \pi_* \Theta^1_{\widetilde{H}/S_0}) = 0,$$

$$h^0(H, \pi_* \mathcal{T}_{\widetilde{H}/S_0}) = 0, \quad h^1(H, \pi_* \mathcal{T}_{\widetilde{H}/S_0}) = 22, \quad h^2(H, \pi_* \mathcal{T}_{\widetilde{H}/S_0}) = 0,$$

$$h^0(H, \pi_* \mathcal{B}_{\widetilde{H}/S_0}) = 1, \quad h^1(H, \pi_* \mathcal{B}_{\widetilde{H}/S_0}) = 3, \quad h^2(H, \pi_* \mathcal{B}_{\widetilde{H}/S_0}) = 0.$$

Thus, $\dim T_{R^{\parallel}} = 1$, and we have $\dim(R^{\parallel}) \leq 1$ for the Krull dimension.

We consider the ring homomorphism $\widehat{\mathcal{O}}_{B,0}/\mathfrak{q} \to R$. By Authors [267, 00OM], we have

$$21 = \dim(R) \leq \dim(\widehat{\mathcal{O}}_{B,0}/\mathfrak{q}) + \dim(R^{\parallel}) \leq \dim(\widehat{\mathcal{O}}_{B,0}/\mathfrak{q}) + 1.$$

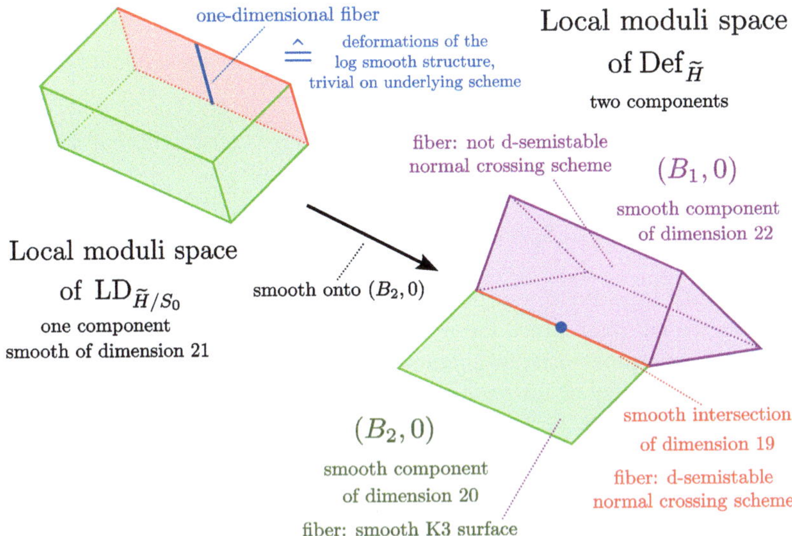

Fig. 1.16 The comparison map $\mathrm{LD}_{\widetilde{H}/S_0} \Rightarrow \mathrm{Def}_{\widetilde{H}}$.

Now $\dim(\widehat{O}_{B,0}/\mathfrak{q}) \leq \dim(\widehat{O}_{B_2,0}) = 20$ so that we have $\dim(\widehat{O}_{B,0}/\mathfrak{q}) = 20$ and $\dim(R^\parallel) = 1$. Since $\widehat{O}_{B_2,0}$ is regular and of dimension 20, this shows $\widehat{O}_{B,0}/\mathfrak{q} = \widehat{O}_{B_2,0}$. Furthermore, $R^\parallel$ is regular because the dimension of the tangent space is equal to the Krull dimension.

The conditions of the miracle flatness theorem [267, 00R4] are satisfied so that $\widehat{O}_{B_2,0} \to R$ is flat. Since the fiber $R^\parallel$ is (geometrically) regular, we find that the ring homomorphism $\widehat{O}_{B_2,0} \to R$ is formally smooth. In more geometric terms, the forgetful transformation $\mathrm{LD}_{\widetilde{H}/S_0} \Rightarrow \mathrm{Def}_{\widetilde{H}}$ surjects smoothly with one-dimensional fibers onto the irreducible component $(B_2 \ni 0)$ of $(B \ni 0)$. This situation is illustrated in Fig. 1.16.

The geometric meaning of the one-dimensional fiber is as follows. Let

$$g_\varepsilon \colon \ (\widetilde{H}_\varepsilon, \mathcal{M}_\varepsilon) \to (S_\varepsilon, \mathcal{N}_\varepsilon)$$

be the trivial log smooth deformation obtained by base change of g_0 along $(S_\varepsilon, \mathcal{N}_\varepsilon) \to (S_0, \mathcal{N}_0)$, and let $\tau = (1, 1) \in (\mathbb{C}^* \oplus \varepsilon \cdot \mathbb{C}) \oplus \mathbb{N} = \mathcal{N}_\varepsilon$. Then, for $\lambda \in \mathbb{C}$, we obtain a new log smooth deformation g_ε^λ by taking the same morphism of schemes and the same log structures but mapping $\tau \mapsto (1 + \lambda \cdot \varepsilon) \cdot g_\varepsilon^\flat(\tau)$. This new log morphism $g_\varepsilon^\lambda \colon (\widetilde{H}_\varepsilon, \mathcal{N}_\varepsilon) \to (S_\varepsilon, \mathcal{N}_\varepsilon)$ is not isomorphic to the original one unless $\lambda = 0$, but this is not seen on the level of underlying schemes (see also Remark 1.115 below).

The disappearance of the component B_1 in the local moduli space of $\mathrm{LD}_{\widetilde{H}/S_0}$ can be understood as a consequence of the fact that locally trivial log smooth deformations have fewer automorphisms than locally trivial flat deformations—

indeed, the natural map $\Theta^1_{\widetilde{H}/S_0} \to \mathcal{T}_{\widetilde{H}/S_0}$ is not surjective. This implies that there are fewer global gluings of trivial local log smooth deformations than of trivial local flat deformations.

Remark 1.115 It is not in general true that the above construction of g^λ_ε out of g_ε yields a new log smooth deformation. For example, take

$$f_0 \colon X_0 = \mathrm{Spec}(\mathbb{N}^2 \to \mathbf{k}[x, y]/(xy)) \to (S_0, \mathcal{N}_0).$$

This is an affine log smooth morphism, so any two log smooth deformations over $(S_\varepsilon, \mathcal{N}_\varepsilon)$ are isomorphic. Since the underlying flat deformations are the same, the comparison isomorphism acts as an automorphism on the underlying flat deformation, and the abundance of automorphisms of the underlying flat deformation in the affine case contributes to the local rigidity of the log smooth deformation.

In the proper case of $g_0 \colon \widetilde{H} \to S_0$, we have

$$H^0(\widetilde{H}, \mathcal{T}_{\widetilde{H}/S_0}) = 0$$

so that the identity is the only automorphism of the underlying flat deformation $\widetilde{H}_\varepsilon \to S_\varepsilon$. Using this, one can show that g^λ_ε is isomorphic to g_ε as a log smooth deformation only if $\lambda = 0$. $\diamondsuit$

1.9 Algebraic Smoothings over $\mathbb{C}[\![t]\!]$

Let $V/\mathbb{C}$ be a degenerate Calabi–Yau scheme. So far, we have a method to construct either a formal algebraic one-parameter smoothing or an analytic one-parameter smoothing. We now address the question of an algebraic smoothing over $\mathbb{C}[\![t]\!]$, i.e., of a proper algebraic deformation $f \colon (X \supset V) \to (\mathrm{Spec}\,\mathbb{C}[\![t]\!] \ni 0) =: (\Sigma \ni 0)$ such that the generic fiber X_η is smooth.

1.9.1 Projective Deformations

Let $V \subset \mathbb{P} := \mathbb{P}^s$ be a projective degenerate scheme. So far, we have considered deformations of V as a scheme or possibly log scheme, disregarding the embedding $V \subset \mathbb{P}$. We have captured those *abstract deformations* with the functor Def_V. We now turn to *embedded deformations*, which are deformations of V inside $\mathbb{P}$.

Definition 1.116 (Projective Deformations) Let $V \subset \mathbb{P} := \mathbb{P}^s$, and let $(S \ni 0)$ be a scheme germ. Then a *projective deformation* over $(S \ni 0)$ is a closed immersion of scheme germs

$$(X \supset V) \to (\mathbb{P} \times S \supset \mathbb{P} \times \{0\})$$

which extends $V \subset \mathbb{P}$ and gives rise to an algebraic deformation $(X \supset V) \to (S \ni 0)$ via composition with the projection $(\mathbb{P} \times S \supset \mathbb{P} \times \{0\}) \to (S \ni 0)$. Infinitesimal projective deformations (up to isomorphism of closed immersions) form the *local Hilbert functor*

$$\mathrm{H}_V^{\mathbb{P}}: \quad \mathbf{Art} \to \mathbf{Set}.$$

Forgetting the embedding $X \subset \mathbb{P} \times S$ yields the forgetful natural transformation $\mathrm{H}_V^{\mathbb{P}} \Rightarrow \mathrm{Def}_V$.

Example 1.117 Let us consider the smooth quartic K3 surface $K \subset \mathbb{P} := \mathbb{P}^3$ of Example 1.2. The classical Bogomolov–Tian–Todorov Theorem 1.27 shows that Def_K is unobstructed. The tangent space to this deformation functor is given by

$$H^1(K, \mathcal{T}_K) \cong H^1(K, \Omega_K^1) \cong \mathbb{C}^{20}$$

so that the hull of Def_K is of the form $R \cong \mathbb{C}[[s_1, \ldots, s_{20}]]$.

By Sernesi [254, Prop. 3.2.1], the tangent space to $\mathrm{H}_K^{\mathbb{P}}$ is given by $H^0(K, \mathcal{N}_{K/\mathbb{P}}) \cong \mathbb{C}^{34}$, and by Sernesi [254, Prop. 3.2.6], an obstruction space is given by $H^1(K, \mathcal{N}_{K/\mathbb{P}}) = 0$. Here, $\mathcal{N}_{K/\mathbb{P}}$ is the normal bundle of the embedding $K \subset \mathbb{P}$. In particular, the hull of $\mathrm{H}_K^{\mathbb{P}}$ is of the form $\mathbb{C}[[s_1, \ldots, s_{34}]]$. One may show that the 34-dimensional family of quartic surfaces in $\mathbb{P} = \mathbb{P}^3$ is a hull, cf. [254, Ex. 3.2.4].

On the level of tangent spaces, the forgetful natural transformation $\mathrm{H}_K^{\mathbb{P}} \Rightarrow \mathrm{Def}_K$ is given by the connection map $H^0(K, \mathcal{N}_{K/\mathbb{P}}) \to H^1(K, \mathcal{T}_K)$ induced by the short exact sequence

$$0 \to \mathcal{T}_K \to \mathcal{T}_{\mathbb{P}}|_K \to \mathcal{N}_{K/\mathbb{P}} \to 0.$$

We have $H^1(K, \mathcal{N}_{K/\mathbb{P}}) = 0$ and $H^1(K, \mathcal{T}_{\mathbb{P}}|_K) \cong \mathbb{C}^1$ so that the forgetful transformation $\mathrm{H}_K^{\mathbb{P}} \Rightarrow \mathrm{Def}_K$ is not surjective but has a 19-dimensional image on the level of tangent spaces. $\diamond$

The relevance of projective deformations is that they allow for an easy passage between algebraic deformations over $\mathbb{C}[[t]]$ and formal deformations. Let R be a complete local Noetherian $\mathbb{C}$-algebra with residue field $\mathbb{C}$, and let $R_k = R/\mathfrak{m}_R^{k+1}$. Let $S = \mathrm{Spec}\, R$ and $S_k = \mathrm{Spec}\, R_k$. Then every projective deformation $f: (X \supset V) \to (S \ni 0)$ yields a *formal projective deformation* $(f_k: (X_k \supset V) \to (S_k \ni 0))_{k \geq 0}$.

Conversely, let $(f_k)_{k \geq 0}$ be a formal projective deformation of $V \subset \mathbb{P}$. For each $k \geq 0$, we have a closed immersion $\iota_k: X_k \to \mathbb{P} \times S_k$. The sequence of bases S_k yields a Noetherian formal scheme $\mathfrak{S} = \mathrm{Spf}\, R$. Similarly, the sequence X_k yields a Noetherian formal scheme $\mathfrak{X}$ over $\mathfrak{S}$, and the sequence $\mathbb{P} \times S_k$ yields a Noetherian formal scheme $\mathfrak{P}$ over $\mathfrak{S}$. Furthermore, the ι_k induce a closed immersion $\iota_\infty: \mathfrak{X} \to \mathfrak{P}$ of formal schemes. The formal scheme $\mathfrak{P}$ is also equal to the completion of $\mathbb{P} \times S$

in the closed subscheme $\mathbb{P} \times \{0\} \subset \mathbb{P} \times S$. The *Grothendieck existence theorem* [130, Cor. 5.1.8] states that closed subschemes of $\mathbb{P} \times S$ are in one-to-one correspondence with formal closed subschemes of $\mathfrak{P}$. Thus, we can find a closed immersion $\iota : X \to \mathbb{P} \times S$ such that $\mathfrak{X}$ is the completion of X in $X \cap (\mathbb{P} \times \{0\})$. In particular, we have $X \cap (\mathbb{P} \times \{0\}) = V$ as well as $X_k = X \times_S S_k$. It follows from [267, 0523] that $f : X \to S$ is flat at all points in $V \subset X$. Since the flat locus is open and $f : X \to S$ is a closed morphism, this shows that $f : X \to S$ is flat, and therefore $(X \supset V) \to (\mathbb{P} \times S \supset \mathbb{P} \times \{0\})$ is a projective deformation.

In summary, in order to construct a proper algebraic deformation over $(\Sigma \ni 0) = (\mathrm{Spec}\, \mathbb{C}[\![t]\!] \ni 0)$, it is sufficient to construct a formal projective deformation. This can be achieved by analyzing the local Hilbert functor $\mathrm{H}_V^{\mathbb{P}}$.

1.9.2   Polarized Deformations

The embedding $\iota_0 : V \subset \mathbb{P}$ yields a very ample line bundle $\mathcal{L} = \iota_0^* \mathcal{O}_{\mathbb{P}}(1)$ together with a map $\mathbb{C}^{s+1} \to H^0(V, \mathcal{L})$ which defines a surjection $\mathcal{O}_V^{\oplus(s+1)} \to \mathcal{L}$. After replacing $\mathcal{L}$ with a sufficiently high tensor power, we may assume that $H^i(V, \mathcal{L}) = 0$ for all $i \geq 1$. Every projective deformation $f_A : (X_A \supset V) \to (S_A \ni 0)$ provides a line bundle

$$\mathcal{L}_A = \iota_A^* \mathcal{O}_{\mathbb{P} \times S_A}(1)$$

which deforms $\mathcal{L}$ and a homomorphism $A^{s+1} \to H^0(X_A, \mathcal{L}_A)$ which deforms $\mathbb{C}^{s+1} \to H^0(V, \mathcal{L})$. Conversely, given a deformation $\mathcal{L}_A$ of $\mathcal{L}$ and a deformation $A^{s+1} \to H^0(X_A, \mathcal{L}_A)$ of $\mathbb{C}^{s+1} \to H^0(V, \mathcal{L})$, we obtain a projective deformation $(X_A \supset V) \to (\mathbb{P} \times S_A \supset \mathbb{P} \times \{0\})$. Up to isomorphism, these two constructions are inverse to each other.

A *polarized deformation* of $V \subset \mathbb{P}$ is a deformation $f_A : (X_A \supset V) \to (S_A \ni 0)$ of V together with a deformation $\mathcal{L}_A$ of the line bundle $\mathcal{L} = \iota_0^* \mathcal{O}_{\mathbb{P}}(1)$. This allows us to factor the above forgetful natural transformation as

$$\mathrm{H}_V^{\mathbb{P}} \Rightarrow \mathrm{Def}_V(\mathcal{L}, -) \Rightarrow \mathrm{Def}_V,$$

where the middle term is the functor of isomorphism classes of polarized deformations.

Example 1.118 We continue Example 1.117. Polarized deformations are controlled by a sheaf of Lie algebras $\mathcal{T}_K(\mathcal{L})$ which fits into the *Atiyah extension*

$$0 \to \mathcal{O}_K \to \mathcal{T}_K(\mathcal{L}) \to \mathcal{T}_K \to 0$$

of coherent sheaves. This yields an exact sequence

$$0 \to H^1(K, \mathcal{T}_K(\mathcal{L})) \to H^1(K, \mathcal{T}_K) \to H^2(K, \mathcal{O}_K) \to H^2(\mathcal{T}_K(\mathcal{L})) \to 0,$$

which shows injectivity on the left. Since the image in $H^1(K, \mathcal{T}_K)$ is the same as the image of $H^0(K, \mathcal{N}_{K/\mathbb{P}^3}) \to H^1(K, \mathcal{T}_K)$, we find that

$$H^1(K, \mathcal{T}_K(\mathcal{L})) \cong \mathbb{C}^{19}$$

for the tangent space to $\mathrm{Def}_K(\mathcal{L}, -)$.

Let R be a hull for Def_K, and let R^+ be a hull for $\mathrm{Def}_K(\mathcal{L}, -)$. Then we obtain an induced map $R \to R^+$ which is compatible with $h_R \Rightarrow \mathrm{Def}_K$ and $h_{R^+} \Rightarrow \mathrm{Def}_K(\mathcal{L}, -)$. In particular, $h_{R^+}(\mathbb{C}[\varepsilon]/(\varepsilon^2)) \to h_R(\mathbb{C}[\varepsilon]/(\varepsilon^2))$ is injective so that $\mathfrak{m}_R/\mathfrak{m}_R^2 \to \mathfrak{m}_{R^+}/\mathfrak{m}_{R^+}^2$ is surjective. Then [264, Prop. B.2] shows that $R \to R^+$ is surjective. Thus, we can interpret $\mathrm{Def}_K(\mathcal{L}, -) \Rightarrow \mathrm{Def}_K$ as the embedding of a closed subscheme. $\diamond$

Let $B \to A$ be a surjection in **Art**, and let $(X_B \supset V; \mathcal{L}_B) \to (S_B \ni 0)$ and $(X_A \supset V; \mathcal{L}_A) \to (S_A \ni 0)$ be two polarized deformations with a morphism over $S_A \to S_B$ between them. Because $H^1(V, \mathcal{L}) = 0$, cohomology and base change (cf. [138, Thm. 12.11]) shows that $H^0(X_B, \mathcal{L}_B) \to H^0(X_A, \mathcal{L}_A)$ is surjective. Now, the natural map

$$\mathrm{H}_V^{\mathbb{P}}(B) \to \mathrm{H}_V^{\mathbb{P}}(A) \times_{\mathrm{Def}_V(\mathcal{L}, A)} \mathrm{Def}_V(\mathcal{L}, B)$$

is surjective so that $\mathrm{H}_V^{\mathbb{P}} \Rightarrow \mathrm{Def}_V(\mathcal{L}, -)$ is *smooth*. In particular, to construct a formal projective deformation of V, it is sufficient to construct a formal polarized deformation.

Example 1.119 We have seen in Example 1.117 that $\mathrm{H}_K^{\mathbb{P}}$ is unobstructed. Since $H^i(K, \mathcal{L}) = 0$ for all $i \geq 1$, we see that $\mathrm{Def}_K(\mathcal{L}, -)$ is unobstructed as well. Thus, its hull R^+ is of the form $\mathbb{C}[\![s_1, \ldots, s_{19}]\!]$, and $\mathrm{Def}_K(\mathcal{L}, -) \Rightarrow \mathrm{Def}_K$ can be interpreted as the embedding of a smooth hypersurface. $\diamond$

1.9.3 Unobstructedness of Pairs

Generalizing the notion of polarized deformations, a *deformation of the pair* $(V; \mathcal{L})$ is a deformation $f_A \colon (X_A \supset V) \to (S_A \ni 0)$ together with a line bundle $\mathcal{L}_A$ on X_A and an isomorphism $\mathcal{L}_A|_V \cong \mathcal{L}$. This gives rise to the deformation functor

$$\mathrm{Def}_V(\mathcal{L}, -) \colon \quad \mathbf{Art} \to \mathbf{Set}.$$

When $Y/\mathbb{C}$ is a smooth and proper Calabi–Yau variety, then $\mathrm{Def}_Y(\mathcal{L}, -)$ is unobstructed for every line bundle $\mathcal{L}$ on Y. As explained in [154, Rem. 2.6], this follows from Ran's T^1-lifting criterion. However, more is true: The main result of Iacono and Manetti's article [154] is that the dg Lie algebra controlling $\mathrm{Def}_Y(\mathcal{L}, -)$ is homotopy Abelian. Their proof is as follows: they construct a new log Calabi–Yau variety as the total space of the $\mathbb{P}^1$-bundle $p\colon P(\mathcal{L}) := \mathbb{P}(\mathcal{O}_Y \oplus \mathcal{L}) \to X$ endowed with the compactifying log structure from two sections $\Delta = \Delta_0 + \Delta_\infty$ of $p\colon P(\mathcal{L}) \to Y$, and then the dg Lie algebra controlling $\mathrm{Def}_Y(\mathcal{L}, -)$ is quasi-isomorphic to the dg Lie algebra controlling $\mathrm{LD}_{(P(\mathcal{L})|\Delta)/(\mathrm{Spec}\,\mathbb{C}|\varnothing)}$.

To construct formal projective smoothings, we would like to have a logarithmic analog. Given a generically log smooth family $f_0\colon (X_0, U_0, \mathcal{M}_0) \to (S_0, N_0)$ with a system of deformations $\mathcal{D}$, and given a line bundle $\mathcal{L}_0$ on X_0, we consider generically log smooth deformations of type $\mathcal{D}$ of the pair $f_0\colon (X_0, U_0, \mathcal{M}_0; \mathcal{L}_0) \to (S_0, N_0)$—they consist of a generically log smooth deformation $f_A\colon (X_A, U_A, \mathcal{M}_A) \to (S_A, N_A)$ of type $\mathcal{D}$ together with a line bundle $\mathcal{L}_A$ on X_A which deforms $\mathcal{L}_0$. Such deformations form a deformation functor

$$\mathrm{LD}^{\mathcal{D}}_{X_0/S_0}(\mathcal{L}_0, -)\colon \quad \mathbf{Art}_{\mathbb{C}[\![Q]\!]} \to \mathbf{Set}.$$

In this situation, we can form a similar $\mathbb{P}^1$-bundle $p_0\colon P_0(\mathcal{L}_0) \to X_0$, where we add the compactifying log structure of $\Delta = \Delta_0 + \Delta_\infty$ to the pull-back log structure from X_0 with the procedure described in Chap. 16. For $g_0\colon P_0(\mathcal{L}_0) \to S_0$, we can construct a new system of deformations $\mathcal{D}(\mathcal{L}_0)$ by applying the construction of $P_0(\mathcal{L}_0)$ to the system of deformations $\mathcal{D}$. Then we obtain an isomorphism

$$\mathrm{LD}^{\mathcal{D}}_{X_0/S_0}(\mathcal{L}_0, -) \cong \mathrm{LD}^{\mathcal{D}(\mathcal{L}_0)}_{P_0(\mathcal{L}_0)/S_0}$$

of functors of Artin rings. If $f_0\colon (X_0, U_0, \mathcal{M}_0) \to (S_0, N_0)$ is log Calabi–Yau, then $g_0\colon (P_0(\mathcal{L}_0), U_0', \mathcal{M}_0') \to (S_0, N_0)$ is log Calabi–Yau as well. Thus, for $\mathrm{LD}^{\mathcal{D}}_{X_0/S_0}(\mathcal{L}_0, -)$ to be unobstructed in the case where f_0 is log Calabi–Yau, it is sufficient that the characteristic curved (two-sided) Gerstenhaber calculus of g_0 is quasi-perfect. If $f_0\colon (X_0, U_0, \mathcal{M}_0) \to (S_0, N_0)$ is log toroidal, and the system of deformations $\mathcal{D}$ is log toroidal, then $g_0\colon (P_0(\mathcal{L}_0), U_0', \mathcal{M}_0') \to (S_0, N_0)$ and the system of deformations $\mathcal{D}(\mathcal{L}_0)$ are log toroidal as well. Thus, in this case, if $f_0\colon X_0 \to S_0$ is additionally proper, then $\mathrm{LD}^{\mathcal{D}}_{X_0/S_0}(\mathcal{L}_0, -)$ is unobstructed by Corollary 1.103. In other words, the pair $f_0\colon (X_0, U_0, \mathcal{M}_0; \mathcal{L}_0) \to (S_0, N_0)$ is unobstructed in this situation.

Theorem 1.120 (Unobstructedness of Pairs $(\mathbf{X_0}, \mathcal{L}_0)$) *Let Q be a sharp toric monoid with associated log point (S_0, N_0). Let $f_0\colon (X_0, U_0, \mathcal{M}_0) \to (S_0, N_0)$ be a proper log Calabi–Yau log toroidal family, and let $\mathcal{L}_0$ be a line bundle on X_0. Let $\mathcal{D}$ be a system of deformations which is log toroidal. Then $\mathrm{LD}^{\mathcal{D}}_{X_0/S_0}(\mathcal{L}_0, -)$ is unobstructed. In particular, this holds when $f_0\colon (X_0, \mathcal{M}_0) \to (S_0, N_0)$ is log smooth and $\mathcal{D}$ is a system of log smooth deformations.*

This is Theorem 16.2 in the main text.

Example 1.121 We consider the generically log smooth family $f_0\colon (H, U_0, \mathcal{R}_0) \to (S_0, \mathcal{N}_0)$ of Example 1.81, i.e., the central fiber of $f\colon (E|H) \to (\mathbb{P}^1|\{0\})$, together with a system of deformations $\mathscr{D}$ of class $\mathscr{C}^{\mathrm{anc}}$, which is defined in Example 1.86. In particular, both f_0 and $\mathscr{D}$ are log toroidal. For the line bundle, we take $\mathcal{L}_0 := \iota_0^* O_{\mathbb{P}^3}(1)$ for the embedding $\iota_0\colon H \to \mathbb{P}^3$. Now the theorem shows that $\mathrm{LD}^{\mathscr{D}}_{H/S_0}(\mathcal{L}_0, -)$ is unobstructed. In particular, we can find a formal deformation $(f_k\colon (X_k, U_k, \mathcal{M}_k) \to (S_k, \mathcal{N}_k))_{k\geq 0}$ together with compatible line bundles $\mathcal{L}_k$ on each X_k, for example $X_k = E_k$ and $\mathcal{L}_k = \iota_k^* O_{\mathbb{P}^3 \times S_k}(1)$ for the embedding $\iota_k\colon E_k \to \mathbb{P}^3 \times S_k$. When we forget the log structure, we obtain a formal polarized deformation. We have $H^i(H, \mathcal{L}_0) = 0$ for all $i \geq 1$ so that $\mathrm{H}^{\mathbb{P}^3}_V \Rightarrow \mathrm{Def}_V(\mathcal{L}_0, -)$ is smooth. Thus, we can extend the formal polarized deformation to a formal projective deformation. This, in turn, yields a projective deformation $f\colon X \to \Sigma$ over $\Sigma = \mathrm{Spec}\,\mathbb{C}[\![t]\!]$. Since each local model in $\mathscr{C}^{\mathrm{anc}}$ is a first-order smoothing, $(f_k\colon X_k \to S_k)_{k\geq 0}$ is a formal smoothing, and $f\colon X \to \Sigma$ is a smoothing by the algebraic variant of Corollary 1.22. An explicit example is given by $E \times_{\mathbb{P}^1} \Sigma \to \Sigma$.   ◇

Example 1.122 We consider again the deformation problem from the previous Example 1.121. The deformations of the pair are controlled by a sheaf of Lie algebras $\Theta^1_{H/S_0}(\mathcal{L}_0)$ which fits into the Atiyah extension

$$0 \to O_H \to \Theta^1_{H/S_0}(\mathcal{L}_0) \to \Theta^1_{H/S_0} \to 0.$$

The tangent space to $\mathrm{LD}_{H/S_0}(\mathcal{L}_0, -)$ is given by $H^1(H, \Theta^1_{H/S_0}(\mathcal{L}_0))$. From the short exact sequence, we obtain an exact sequence

$$0 \to H^1(H, \Theta^1_{H/S_0}(\mathcal{L}_0)) \to H^1(H, \Theta^1_{H/S_0}) \to H^2(H, O_H).$$

The dimensions of the middle and right terms are 20 and 1 because these numbers are constant in families. The deformation functor $\mathrm{LD}^{\mathscr{D}}_{H/S_0}(\mathcal{L}_0, -)$ is isomorphic to $\mathrm{LD}^{\mathscr{D}(\mathcal{L}_0)}_{P_0(\mathcal{L}_0)/S_0}$ so that the tangent space to the deformation functor is also isomorphic to $H^1(P_0(\mathcal{L}_0), \Theta^1_{P_0(\mathcal{L}_0)/S_0})$. In the family $f\colon (E|H) \to (\mathbb{P}^1|\{0\})$, we have a line bundle $\mathcal{L}$, which defines a new family $g\colon (P(\mathcal{L})|H(\mathcal{L}) \cup \Delta) \to (\mathbb{P}^1|\{0\})$ as above. This latter family is log toroidal and log Calabi–Yau as well. Thus, its log Hodge numbers are constant, and we find

$$H^1(P_0(\mathcal{L}_0), \Theta^1_{P_0(\mathcal{L}_0)/S_0}) \cong \mathbb{C}^{19}$$

by using Example 1.118. Thus, the hull of the deformation functor $\mathrm{LD}^{\mathscr{D}}_{H/S_0}(\mathcal{L}_0, -)$ is isomorphic to $\mathbb{C}[\![t, s_1, \ldots, s_{19}]\!]$, and $\mathrm{LD}^{\mathscr{D}}_{H/S_0}(\mathcal{L}_0, -) \Rightarrow \mathrm{LD}^{\mathscr{D}}_{H/S_0}$ can be interpreted as the embedding of a smooth hypersurface as in Example 1.118.   ◇

Example 1.123 In Theorem 1.120, assume that $\dim(X_0) = 3$ and that X_0 is strict, i.e., $H^1(X_0, O_{X_0}) = 0$ and $H^2(X_0, O_{X_0}) = 0$. Then

$$H^1(X_0, \Theta^1_{X_0/S_0}(\mathcal{L}_0)) = H^1(X_0, \Theta^1_{X_0/S_0}).$$

Since both hulls are power series rings over $\mathbb{C}[\![Q]\!]$, [264, Prop. B.5] shows that the hulls are isomorphic. Therefore, the forgetful map $\mathrm{LD}^{\mathscr{D}}_{X_0/S_0}(\mathcal{L}_0, -) \Rightarrow \mathrm{LD}^{\mathscr{D}}_{X_0/S_0}$ is smooth and an isomorphism on tangent spaces. $\diamond$

Remark 1.124 The forgetful natural transformation

$$\mathrm{LD}^{\mathscr{D}}_{X_0/S_0}(\mathcal{L}_0, -) \Rightarrow \mathrm{LD}^{\mathscr{D}}_{X_0/S_0}$$

is in general *not* smooth, even in the situation where we can show that each of the two functors is unobstructed. In other words, sometimes a line bundle $\mathcal{L}_A$ over $f_A\colon X_A \to S_A$ cannot be extended to a given thickening $f_B\colon X_B \to S_B$. We discuss an example of this situation in Example 11.13. $\diamond$

Remark 1.125 It is currently an open question if quasi-perfectness of the curved Gerstenhaber calculus (1.2) implies quasi-perfectness of

$$(PV^{\bullet,\bullet}_{P_0(\mathcal{L}_0)/\Lambda}, DR^{\bullet,\bullet}_{P_0(\mathcal{L}_0)/\Lambda}).$$

In practice, this does not seem to be too much of a problem as we may expect that most methods showing the quasi-perfectness of one of them also shows the quasi-perfectness of the other. $\diamond$

Remark 1.126 The analog of Theorem 1.120 is no longer true when $\mathcal{L}_0$ is a vector bundle of rank ≥ 2, see Example 11.14. $\diamond$

1.10 Modifications of the Log Structure

With Corollary 1.103, we have a powerful unobstructedness result for log toroidal log Calabi–Yau families $f_0\colon (X_0, U_0, \mathcal{M}_0) \to (S_0, \mathcal{N}_0)$. However, sometimes we want to deform a log scheme which is not log Calabi–Yau but for example log Fano. In some cases, we can *modify* the log structure and obtain a log Calabi–Yau scheme. In general, given a line bundle $\mathcal{L}_0$ on $(X_0, U_0, \mathcal{M}_0)$ and a section $s_0 \in \mathcal{L}_0$, we can define a modification $g_0\colon (X_0, U_0, \mathcal{M}_0(s_0)) \to (S_0, \mathcal{N}_0)$ with the property

$$\mathcal{W}^d_{X_0(s_0)/S_0} \cong \mathcal{W}^d_{X_0/S_0} \otimes \mathcal{L}_0.$$

Thus, when $\mathcal{L}_0 = (\mathcal{W}^d_{X_0/S_0})^\vee$ is the anti-canonical bundle, then the modified family $g_0\colon (X_0, U_0, \mathcal{M}_0(s_0)) \to (S_0, \mathcal{N}_0)$ is log Calabi–Yau. We have already used a variant of this construction in [77, §6]. In this monograph, we study a more general version systematically.

Construction 1.127 Let $f\colon (X, U, \mathcal{M}) \to (S, \mathcal{N})$ be a generically log smooth family, and let $\mathcal{L}$ be a line bundle on X, considered as a sheaf of $\mathcal{O}_X$-modules. We denote the total space by

$$L = \mathrm{Spec}_{\mathcal{O}_X} \bigoplus_{\ell \geq 0} (\mathcal{L}^\vee)^{\otimes \ell}.$$

We can add the log structure of the zero section $X \subset L$ to the pull-back log structure on L and obtain a log smooth morphism

$$p\colon \ L(X)^\dagger := (L, p^{-1}(U), \mathcal{M}_{L(X)}) \to (X, U, \mathcal{M}).$$

Then the *modification* of $f\colon (X, U, \mathcal{M}) \to (S, \mathcal{N})$ in $s \in \mathcal{L}$ is given by pulling back the log structure $\mathcal{M}_{L(X)}$ via the embedding $s\colon X \to L$ defined by $s \in \mathcal{L}$. We write $f \circ h\colon X(s)^\dagger := (X, U, \mathcal{M}(s)) \to (S, \mathcal{N})$ for this modified log structure.   ◇

The key difficulty is to find conditions on $s \in \mathcal{L}$ under which this construction is well-behaved. As such a condition, we propose *log regularity* in the sense of Definitions 18.7 and 18.9.[13] For us, the starting point in formulating this condition has been [222, IV, Thm. 3.2.2], which gives conditions for a strict closed subscheme in a log smooth scheme to be log smooth itself—in a sense, this is what we want for $X(s)^\dagger$ inside $f \circ p\colon L(X)^\dagger \to (S, \mathcal{N})$.

Now let $f_0\colon (X_0, U_0, \mathcal{M}_0) \to (S_0, \mathcal{N}_0)$ be a generically log smooth family with a system of deformations $\mathscr{D}$, let $\mathcal{L}_0$ be a line bundle on X_0, and let $s_0 \in \mathcal{L}_0$ be a log regular section. Then $g_0\colon X_0(s_0)^\dagger \to (S_0, \mathcal{N}_0)$ is generically log smooth as well. For any deformation $\mathcal{L}_A$ of $\mathcal{L}_0$, a section $s_A \in \mathcal{L}_A$ with $s_A|_0 = s_0$ is log regular again. After choosing a deformation of $\mathcal{L}_0$ over the local model $V_{\alpha;A} \to S_A$ together with a section $s_{\alpha;A}$ restricting to s_0, we obtain a generically log smooth deformation

$$(V_{\alpha;A}, U_{\alpha;A}, \widetilde{\mathcal{M}}_{\alpha;A}) \to (S_A, \mathcal{N}_A)$$

of $g_0\colon X_0(s_0)^\dagger \to (S_0, \mathcal{N}_0)$. Up to isomorphism, this deformation does not depend on the choices we made; thus, we obtain a system of deformations $\mathscr{D}(s_0)$ for $g_0\colon X_0(s_0)^\dagger \to (S_0, \mathcal{N}_0)$, and hence a deformation functor $\mathrm{LD}^{\mathscr{D}(s_0)}_{X_0(s_0)/S_0}$.

On the other hand, we can also consider deformations $f\colon (X_A, U_A, \mathcal{M}_A) \to (S_A, \mathcal{N}_A)$ of $f_0\colon (X_0, U_0, \mathcal{M}_0) \to (S_0, \mathcal{N}_0)$ together with a line bundle $\mathcal{L}_A$ and a section $s_A \in \mathcal{L}_A$ with $s_A|_0 = s_0$. This gives rise to a deformation functor $\mathrm{LD}^{\mathscr{D}}_{X_0/S_0}(\mathcal{L}_0, s_0, -)$. There is an obvious map

$$\mathrm{LD}^{\mathscr{D}}_{X_0/S_0}(\mathcal{L}_0, s_0, -) \Rightarrow \mathrm{LD}^{\mathscr{D}(s_0)}_{X_0(s_0)/S_0},$$

---

[13] Log quasi-regularity and log regularity coincide in the case of a plain generically log smooth family and are different only in the enhanced case.

which is in fact an isomorphism of functors of Artin rings. This ties the deformation theory of $f_0 \colon (X_0, U_0, \mathcal{M}_0) \to (S_0, \mathcal{N}_0)$ closely to the one of $g_0 \colon X_0(s_0)^\dagger \to (S_0, \mathcal{N}_0)$. In the case where $\mathcal{L}_0$ is the anti-canonical line bundle, unobstructedness of $\mathrm{LD}^{\mathscr{D}}_{X_0/S_0}$ is often equivalent to unobstructedness of $\mathrm{LD}^{\mathscr{D}(s_0)}_{X_0(s_0)/S_0}$, see Corollary 18.35.

We call log regular sections which satisfy the stronger condition in [77, §6] *log transversal*.[14] If $f \colon (X, U, \mathcal{M}) \to (S, \mathcal{N})$ is log toroidal, and $s \in \mathcal{L}$ is a log transversal section, then $g \colon X(s)^\dagger \to (S, \mathcal{N})$ is log toroidal as well. Thus, in the case of a proper log toroidal family family $f_0 \colon (X_0, U_0, \mathcal{M}_0) \to (S_0, \mathcal{N}_0)$ with a log transversal section $s_0 \in (\mathcal{W}^d_{X_0/S_0})^\vee$, the deformations of $g_0 \colon X_0(s_0)^\dagger \to (S_0, \mathcal{N}_0)$ of type $\mathscr{D}(s_0)$ are unobstructed.

With Theorem 1.109, we have already given a very strong smoothing result for proper Calabi–Yau normal crossing schemes. With the theory of modifications, we can state (and prove) the stronger variant where the Calabi–Yau hypothesis is relaxed. As a slight generalization of our original result [77, Thm. 1.1], we obtain the following theorem.

Theorem 1.128 (Smoothing Normal Crossing Spaces II) *Let $V/\mathbb{C}$ be a proper normal crossing scheme such that every open stratum $S^\circ \in [\mathcal{S}_m V]$ is quasi-projective for $m \geq 0$. Suppose that both $(\omega^\circ_V)^\vee$ and $\mathcal{T}^1_V$ are globally generated. Then V admits a formal algebraic one-parameter smoothing as well as an analytic one-parameter smoothing.*

In the main text, this is Theorem 18.44. Note that we require quasi-projectivity also for the open strata $S^\circ \in [\mathcal{S}_0 V]$ of maximal dimension, which we did not require in Theorem 1.109.

In [77, Thm. 1.1], the assumptions differ slightly. Due to the assumption of a given well-behaved section of $(\omega^\circ_V)^\vee$, the global generatedness of $(\omega^\circ_V)^\vee$ is dropped as well as the quasi-projectivity of open strata $S^\circ \in [\mathcal{S}_0 V]$, but then, concerning open strata in $[\mathcal{S}_m V]$ for $m \geq 1$, the stronger condition that the double locus D is projective is required. The argument is more or less the same, using a more refined version of Bertini's theorem to get the, in our opinion, more natural condition that the strata are quasi-projective.

If V is projective, then we can find a formal projective smoothing, which can be algebraized to a smoothing $f \colon (X \supset V) \to (\operatorname{Spec} \mathbb{C}[\![t]\!] \ni 0)$.

1.11 Mirror Pairs and Toroidal Crossing Schemes

We have already seen various examples of log toroidal families, but the only true application of the log toroidal Bogomolov–Tian–Todorov theorem Corollary 1.103 so far involved only the local models $\mu^{\mathrm{nc}}_{d;r}$ and $\mu^{\mathrm{anc}}_{d;r}$ in the class $\mathscr{C}^{\mathrm{anc}}$. We have used them to show smoothability of normal crossing schemes in Theorem 1.109; the

---

[14] It is actually an open question if the two conditions are equivalent.

proof of the more general application Theorem 1.128 involves only a slight variant of these local models. In this section, we give motivation and historical context for considering more general log toroidal families.

We explain how mirror symmetry motivated the construction of log smooth structures on more general degenerate schemes than normal crossing schemes. This led to the notion of a *toroidal crossing scheme*, which we discuss in Chap. 9.

The absence of global log smooth structures in the context of mirror symmetry led to the study of generically log smooth families. Also at log smooth points, we need more general local models than the local models $\mu_{d;r}^{\mathrm{nc}}$ of semistable degenerations. The log singularities need more general local models than $\mu_{d;r}^{\mathrm{anc}}$. Nonetheless, they are log toroidal, which originally motivated the study of log toroidal families. Corollary 1.103 yields Calabi–Yau varieties by smoothing (or normaling in dimension ≥ 4) explicitly constructed degenerate Calabi–Yau schemes.

1.11.1   Batyrev's Mirror Construction

Much of the interest in Calabi–Yau manifolds comes from *mirror symmetry*, which is the hypothesis that Calabi–Yau manifolds come in *mirror pairs* $(Y, \check{Y})$ that enjoy various duality properties. The simplest such property is the numerical duality

$$\dim H^q(Y, \Omega_Y^p) = \dim H^q(\check{Y}, \Omega_{\check{Y}}^{d-p}),$$

where $d = \dim(Y) = \dim(\check{Y})$. Other duality properties are defined in terms of Hodge structures, curve counts, coherent sheaves, and Lagrangian submanifolds (in the symplectic picture).

Much of the work in constructing Calabi–Yau manifolds focuses on constructing mirror pairs. We now discuss Batyrev's approach in [22]. His construction has two starting points. The first one is the simple observation that every anti-canonical hypersurface is a Calabi–Yau scheme. For later use, let us record this in good generality.

Lemma 1.129 *Let $P/\mathbb{C}$ be a connected Gorenstein normal variety of dimension $d + 1$, and let $\omega_P = j_*\Omega_{\mathrm{Reg}(P)}^{d+1}$ be the canonical bundle, where $j\colon \mathrm{Reg}(P) \to P$ is the inclusion of the regular locus. Let $\phi \in \Gamma(P, \omega_P^\vee)$ be a non-zero global section of the dual, and let $Y = \mathrm{div}(\phi) \subset P$ be the hypersurface defined by ϕ. Then Y is Gorenstein, and we have $\omega_Y^\circ := \mathcal{H}^{-d}(\omega_Y^\bullet) \cong \mathcal{O}_Y$.*

Proof Since P is Cohen–Macaulay, we have an isomorphism $\omega_P \cong \omega_P^\circ = \mathcal{H}^{-(d+1)}(\omega_P^\bullet)$ between the canonical bundle and the dualizing sheaf. Since P is Gorenstein, ω_P° is a line bundle. Since ϕ is a non-zero divisor when considered as an element of $\mathcal{O}_{P,x}$ via a local trivialization, [267, 0BJJ] shows that Y is Gorenstein.

Let $i\colon Y \to P$ be the inclusion, and let $p\colon P \to \operatorname{Spec}\mathbb{C}$ be the structure map. Then we have

$$\omega_Y^\bullet = (p \circ i)^! O_{\operatorname{Spec}\mathbb{C}} = i^! p^! O_{\operatorname{Spec}\mathbb{C}} = i^! \omega_P^\bullet = i^! \omega_P[d+1].$$

Because $i\colon Y \to P$ is a closed immersion, this can be computed as

$$i^! \omega_P[d+1] = \mathcal{RHom}(O_Y, \omega_P[d+1]).$$

Since $O_P(-Y)$ is a line bundle, we have a locally free resolution

$$0 \to O_P(-Y) \to O_P \to O_Y \to 0$$

so that

$$\begin{aligned}
\omega_Y^\bullet &= \mathcal{Hom}(O_P(-Y) \to O_P, \omega_P[d+1]) \\
&= (\omega_P \to O_P(Y) \otimes \omega_P) \\
&\cong (O_P(-Y) \to O_P) \cong O_Y[d].
\end{aligned}$$

Therefore, we have $\omega_Y^\circ = \mathcal{H}^{-d}(\omega_Y^\bullet) \cong O_Y$. $\qquad\square$

The second starting point is the polar duality of Gorenstein toric Fano varieties. Let $M \cong \mathbb{Z}^{d+1}$ be a lattice, and let $\Delta \subseteq M_\mathbb{R} = M \otimes_\mathbb{Z} \mathbb{R}$ be a polytope, i.e., the convex hull of finitely many rational points. We denote the dual lattice by $N = \operatorname{Hom}(M, \mathbb{Z})$. The *polar dual polytope* is

$$\Delta^\circ := \{n \in N_\mathbb{R} \mid \forall m \in \Delta\colon \langle m, n \rangle \geq -1\}.$$

The polytope Δ is *reflexive* if it is a lattice polytope and $(\Delta^\circ)^\circ = \Delta$. Reflexive polytopes are in one-to-one correspondence with Gorenstein toric Fano varieties via $\Delta \mapsto \mathbb{P}_\Delta$, the projective toric variety associated with Δ.

Example 1.130 Let

$$\Delta = \operatorname{Conv}((-1, -1, -1), (3, -1, -1), (-1, 3, -1), (-1, -1, 3)).$$

This is a reflexive polytope with $\mathbb{P}_\Delta \cong \mathbb{P}^3$. The dual polytope is

$$\Delta^\circ = \operatorname{Conv}((-1, -1, -1), (1, 0, 0), (0, 1, 0), (0, 0, 1)).$$

We have $\mathbb{P}_{\Delta^\circ} \cong \mathbb{P}^3$ as well. $\qquad\Diamond$

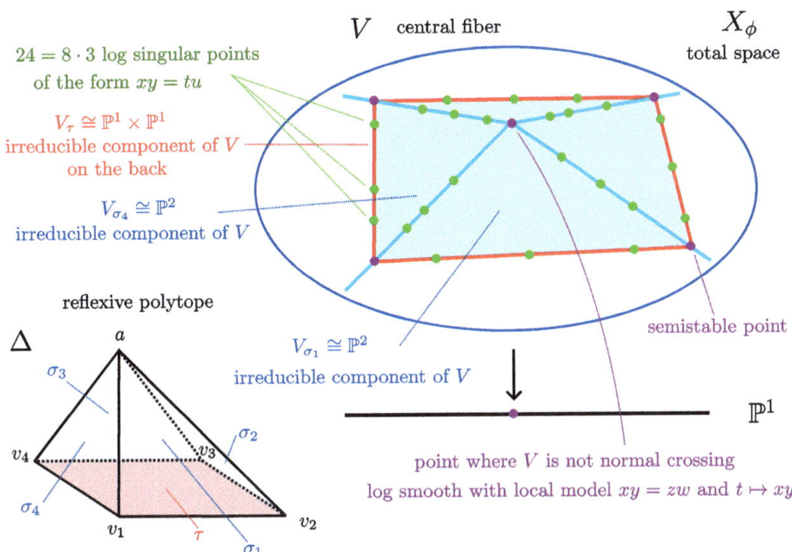

Fig. 1.17 Example 1.135. On the left, we have the reflexive polytope Δ from Example 1.131. On the right, we have the central fiber V of the toric degeneration $X_\phi \to \mathbb{P}^1$ from Example 1.135. It consists of five irreducible components, which correspond to the facets of Δ. We see four of them in the front in blue and one of them in the back in red. © Simon Felten 2025. All rights reserved

Example 1.131 This is an example of a reflexive polytope Δ such that $P = \mathbb{P}_\Delta$ is not smooth. We set $\Delta = \mathrm{Conv}(a, v_1, v_2, v_3, v_4) \subseteq \mathbb{R}^3$ with five vertices

$$a = (-1, -1, 2),$$

$$v_1 = (-1, -1, -1), \ v_2 = (2, -1, -1), \ v_3 = (2, 2, -1), \ v_4 = (-1, 2, -1).$$

This is a pyramid with apex a over the bottom facet $\tau = [v_1 v_2 v_3 v_4]$. We denote the four triangular facets on the sides of the pyramid by

$$\sigma_1 = [a v_1 v_2], \quad \sigma_2 = [a v_2 v_3], \quad \sigma_3 = [a v_3 v_4], \quad \sigma_4 = [a v_4 v_1].$$

For a depiction of Δ, see Fig. 1.17.

The polytope Δ defines a Gorenstein toric Fano variety $P = \mathbb{P}_\Delta$. When we denote the coordinates on the maximal torus by $X \,\hat{=}\, (1, 0, 0)$, $Y \,\hat{=}\, (0, 1, 0)$, $Z \,\hat{=}\, (0, 0, 1)$, then P is covered by five affine toric varieties:

$$U_0 = \mathrm{Spec}\, \mathbb{C}\left[\frac{X}{Z}, \frac{Y}{Z}, \frac{XY}{Z}, \frac{1}{Z}\right] \Big/ \left(\frac{X}{Z} \cdot \frac{Y}{Z} - \frac{XY}{Z} \cdot \frac{1}{Z}\right) =: \mathrm{Spec}\, \frac{\mathbb{C}[x_0, y_0, w_0, z_0]}{(x_0 y_0 - w_0 z_0)}$$

$$U_1 = \mathrm{Spec}\, \mathbb{C}\left[\frac{X}{1}, \frac{Y}{1}, \frac{Z}{1}\right] =: \mathrm{Spec}\, \mathbb{C}[x_1, y_1, z_1]$$

$$U_2 = \operatorname{Spec} \mathbb{C}\left[\frac{1}{X}, \frac{Y}{1}, \frac{Z}{X}\right] =: \operatorname{Spec} \mathbb{C}[x_2, y_2, z_2]$$

$$U_3 = \operatorname{Spec} \mathbb{C}\left[\frac{1}{X}, \frac{1}{Y}, \frac{Z}{XY}\right] =: \operatorname{Spec} \mathbb{C}[x_3, y_3, z_3]$$

$$U_4 = \operatorname{Spec} \mathbb{C}\left[\frac{X}{1}, \frac{1}{Y}, \frac{Z}{Y}\right] =: \operatorname{Spec} \mathbb{C}[x_4, y_4, z_4]$$

In particular, P is smooth outside a single isolated A_1-threefold singularity in U_0.
$\Diamond$

The key idea of Batyrev's mirror construction in [22] is to consider the duality $\Delta \leftrightarrow \Delta^\circ$ for reflexive polytopes as the source of mirror symmetry.

Let Δ be a reflexive polytope, and let $P = \mathbb{P}_\Delta$. For $0 \neq \phi \in \Gamma(P, \omega_P^\vee)$, we write $\overline{Y}_\phi := \operatorname{div}(\phi) \subset P$. We have seen above that this is a Calabi–Yau scheme. When ϕ is Δ-regular in the sense of [22, Defn. 3.1.1], then $\overline{Y}_\phi$ is normal and has toroidal singularities in the sense that it is étale locally isomorphic to toric varieties. In particular, $\overline{Y}_\phi$ is a Calabi–Yau variety in the sense of Definition 1.1. By Batyrev [22, Prop. 3.1.3], most global sections of $\omega_P^\vee$ are Δ-regular.

Let $P^\circ = \mathbb{P}_{\Delta^\circ}$, let $\psi \in \Gamma(P^\circ, \omega_{P^\circ}^\vee)$ be Δ°-regular, and let $\overline{Y}_\psi^\circ := \operatorname{div}(\psi)$. Then we can consider $(\overline{Y}_\phi, \overline{Y}_\psi^\circ)$ as a *naive Batyrev mirror pair*.

Example 1.132 Let Δ and Δ° be the polytopes of Example 1.130. Let $\phi = X_0^4 + X_1^4 + X_2^4 + X_3^4 \in \Gamma(P, \mathcal{O}_P(4))$ and $\psi = X_0^4 + X_1^4 + X_2^4 + X_3^4 \in \Gamma(P^\circ, \mathcal{O}_{P^\circ}(4))$. Then both $\overline{Y}_\phi$ and $\overline{Y}_\psi$ are smooth quartic surfaces. In other words, the Fermat quartic surface K is its own mirror under the naive Batyrev construction. $\Diamond$

Example 1.133 Let Δ be the reflexive polytope of Example 1.131. Let $\phi \in \Gamma(P, \omega_P^\vee)$ be the global section which is given by

$$\frac{1}{XYZ} + \frac{X^2}{YZ} + \frac{Y^2}{XZ} + \frac{Z^2}{XY} + \frac{X^2Y^2}{Z} + 1$$

on the maximal torus. In local coordinates, this becomes:

$$\overline{Y}_\phi \cap U_0 = \{x_0^3 + y_0^3 + z_0^3 + w_0^3 + 1 + x_0 y_0 = 0\}$$

$$\overline{Y}_\phi \cap U_i = \{x_i^3 + y_i^3 + z_i^3 + x_i^3 y_i^3 + 1 + x_i y_i z_i = 0\} \quad \text{for } i = 1, 2, 3, 4$$

One may check that $\overline{Y}_\phi$ is smooth so that it is an algebraic Calabi–Yau manifold. $\Diamond$

Batyrev's actual mirror construction is slightly more intricate. A morphism $\pi : \hat{Y} \to Y$ between connected Gorenstein normal varieties is a *maximal projective crepant partial resolution* if π is projective, birational, and crepant, and if $\hat{Y}$ is terminal and locally analytically $\mathbb{Q}$-factorial. By results of Batyrev in [22], every mpcp resolution of $P = \mathbb{P}_\Delta$ is a toric morphism. Furthermore, when $\overline{Y}_\phi \subset P$ is Δ-

regular, and when $\pi : \hat{P} \to P$ is an mpcp resolution, then $\hat{Y}_\phi := \pi^{-1}(\overline{Y}_\phi) \to \overline{Y}_\phi$ is also an mpcp resolution. In particular, it is crepant, and $\hat{Y}_\phi$ is a Calabi–Yau variety.

Let $P^\circ = \mathbb{P}_{\Delta^\circ}$, let $\psi \in \Gamma(P^\circ, \omega_{P^\circ}^\vee)$ be Δ°-regular, and let $\overline{Y}_\psi^\circ = \mathrm{div}(\psi)$. We also obtain an mpcp resolution $\hat{Y}_\psi^\circ \to \overline{Y}_\psi^\circ$. Then $(\hat{Y}_\phi, \hat{Y}_\psi^\circ)$ is a *Batyrev mirror pair*. For these, Batyrev proved the following mirror theorem: We have

$$h^{d-1,1}(\hat{Y}_\phi) = h^{1,1}(\hat{Y}_\psi^\circ)$$

for $d \geq 3$.

1.11.2 The Batyrev Construction from the Perspective of the Smoothing Strategy

We focus on the naive Batyrev construction for simplicity. So far, we have constructed naive Batyrev mirror pairs $(\overline{Y}_\phi, \overline{Y}_\psi^\circ)$ as subschemes of ambient spaces $\mathbb{P}_\Delta$ and $\mathbb{P}_{\Delta^\circ}$.

The ambient space $P = \mathbb{P}_\Delta$ has a distinguished anti-canonical hypersurface, which we have ignored so far: the full toric boundary $V \subset P$ with its reduced scheme structure. It is given by the section $\phi_0 \in \Gamma(P, \omega_P^\vee)$ which satisfies $\phi_0 = 1$ on the maximal torus. Lemma 1.129 shows that V is Gorenstein and Calabi–Yau.

Unlike a general $\overline{Y}_\phi$, the toric boundary V has a simple combinatorial structure: it is obtained as the colimit of projective toric varieties V_σ corresponding to the faces $\sigma \subset \Delta$, much like $H = \bigcup_{i=0}^3$, which we have described as a colimit in Example 1.10. We say that V is a *(reducible) toric scheme*.

The toric boundary V and a general $\overline{Y}_\phi$ are connected through the family

$$X_\phi = \mathrm{div}(T_0\phi - T_1\phi_0) \subset \mathbb{P}^1 \times P \to \mathbb{P}^1,$$

where we have $T_0\phi - T_1\phi_0 \in \omega_P^\vee \boxtimes O_{\mathbb{P}^1}(1)$, and where T_0, T_1 are the coordinates on the base $\mathbb{P}^1$. In other words, each $\overline{Y}_\phi$ degenerates to V. Thus, we can interpret the Batyrev mirror construction as providing a mirror pair (V, V°) of reducible toric schemes together with degenerations $X_\phi \to \mathbb{P}^1$ and $X_\psi^\circ \to \mathbb{P}^1$ with central fibers V and V°.

This perspective was taken by Gross and Siebert in a series of articles [119, 124–126]. The advantage is that we can reduce the amount (or better: strength) of the input data needed to construct a mirror pair, and that we therefore obtain a substantially more general construction. The relevant degenerations are then the *toric degenerations*, which may be defined as (vertical) generically log smooth families over $(\Sigma|\{0\}) = (\mathrm{Spec}\,\mathbb{C}[\![t]\!]|\{0\})$ such that the central fiber has the structure of a reducible toric scheme, and such that no toric stratum of the central fiber is contained in the log singular locus Z. For the precise definition used by Gross and

Siebert, see [124, Defn. 4.1]. Note that this notion is different from a degeneration to a normal toric variety, a concept which is also often called a "toric degeneration."

We call the mirror construction of the above-mentioned articles the *toric Gross–Siebert mirror construction* to distinguish it from the later *intrinsic Gross–Siebert mirror construction* of [128]. The toric Gross–Siebert mirror construction is the foundation of what is called the *Gross–Siebert program*. We say more about the construction in the following sections.

Let us now consider some examples of the family $X_\phi \to \mathbb{P}^1$ constructed above.

Example 1.134 Let Δ be the polytope from Example 1.130, and let $\phi = X_0^4 + X_1^4 + X_2^4 + X_3^4$. We have $\phi_0 = X_0 X_1 X_2 X_3$. Thus, the family $X_\phi \to \mathbb{P}^1$ is nothing but the smoothing $f\colon E \to \mathbb{P}^1$ from Example 1.11. $\diamond$

Example 1.135 Let Δ be the polytope from Example 1.131, i.e., the pyramid. Then we can describe $V \subset P = \mathbb{P}_\Delta$ in local coordinates as follows:

$$V \cap U_0 = \left\{ \frac{XY}{Z^2} = 0 \right\} = \{x_0 y_0 = 0\} = \operatorname{Spec} \mathbb{C}[x_0, y_0, z_0, w_0]/(x_0 y_0, z_0 w_0)$$

$$V \cap U_i = \{x_i y_i z_i = 0\} \quad \text{for } i = 1, 2, 3, 4$$

For the smoothing $f\colon X_\phi \to \mathbb{P}^1$, we obtain

$$X_\phi \cap (\mathbb{A}_t^1 \times U_0) = \{x_0 y_0 = z_0 w_0 = t(x_0^3 + y_0^3 + z_0^3 + w_0^3 + 1 + x_0 y_0)\}$$

and

$$X_\phi \cap (\mathbb{A}_t^1 \times U_i) = \{x_i y_i z_i = t(x_i^3 + y_i^3 + z_i^3 + x_i^3 y_i^3 + 1 + x_i y_i z_i)\}$$

for $1 \le i \le 4$.

Let us have a closer look at the log singularities $f\colon X_\phi \to \mathbb{P}^1$. In the point o given by $x_0 = y_0 = z_0 = w_0 = 0$ in $V \cap U_0$, the four irreducible components $V_{\sigma_1}, \ldots, V_{\sigma_4}$ of V meet, so V is not a normal crossing scheme in o.

The total space X_ϕ has singularities only over $t = 1$ and $t = 0$. Let Z be the locus of singularities lying over $t = 0$. Then we have $Z = Z' \cup \{o\}$, where Z' consists of 24 isolated singularities, three on each of the 8 lines in the double locus $D = \operatorname{Sing}(V)$. In particular, the deformation $(X \setminus Z \supset V \setminus Z) \to (\mathbb{P}^1 \ni 0)$ is semistable. In each $p \in Z'$, the family $f\colon X_\phi \to \mathbb{P}^1$ is étale locally isomorphic to $\mu_{2;1}^{\mathrm{anc}}\colon \{xy = tu\} \to \mathbb{A}_t^1$. Thus, f is of class $\mathscr{C}^{\mathrm{anc}}$ on $X_\phi \setminus \{o\}$, at least after restricting to some neighborhood of V in X_ϕ.

Since V is not a normal crossing scheme in o, we must leave the class $\mathscr{C}^{\mathrm{anc}}$ in o. Here, the family is étale locally isomorphic to

$$\left(\operatorname{Spec} \mathbb{C}[x, y, z, w]/(xy - zw) \middle| \{xy = 0\} \right) \to \left(\mathbb{A}_t^1 \middle| \{0\} \right), \quad t \mapsto xy.$$

This is nothing but the log smooth morphism associated with $\theta \colon \mathbb{N} \to P^{\mathrm{sze}}$, $1 \mapsto \bar{x} + \bar{y}$, where P^{sze} is the monoid from Example 1.79 (but the map is different). Therefore, $f \colon (X_\phi | V) \to (\mathbb{P}^1 | \{0\})$ is actually log smooth in $o \in V \subset X_\phi$. $\Diamond$

1.11.3 Toroidal Crossing Schemes

To see that Gross and Siebert's perspective on the Batyrev mirrors as smoothings or normalings of the toric boundary does not rely on the ambient space $\mathbb{P}_\Delta$ being given (and thus needs less or weaker input data than Batyrev's construction), we would like to construct a smoothing (or normaling) of the toric boundary $V \subset \mathbb{P}_\Delta$ without further reference to the ambient space $\mathbb{P}_\Delta$.

We have just seen in Example 1.135 that the Smoothing Theorem 1.109 for normal crossing schemes is not applicable because V is not always a normal crossing scheme. Nevertheless, the family considered in Example 1.135 is log smooth in the point where its central fiber V is not a normal crossing scheme. This suggests to adopt a strategy analogous to Theorem 1.109:

(A) construct a sheaf $\mathcal{LS}_V$ of log smooth structures on V;
(B) construct a closed subset $Z \subset V$ of codimension ≥ 2 and a section $s \in \Gamma(V \setminus Z, \mathcal{LS}_V)$ such that the induced generically log smooth family is log toroidal and log Calabi–Yau;
(C) construct a log toroidal system of deformations $\mathcal{D}$.

Once these steps are completed, Theorem 1.120 gives us an algebraic smoothing or normaling over $\mathbb{C}[\![t]\!]$.

On a normal crossing scheme V, we have defined $\mathcal{LS}_V$ as the sheaf of semistable log smooth structures in the sense of Definition 1.67. The first step is to establish the class of log smooth structures that we wish to allow for $V \subset P$. We use log smooth structures which arise locally in the étale topology from the following construction, which generalizes the local model $\mu^{\mathrm{nc}}_{d;r}$. They are precisely the local models of *saturated and vertical* log smooth morphisms to $(\mathbb{A}^1_t | \{0\})$.

Construction 1.136 We summarize Construction 9.2. We start with a lattice $M \cong \mathbb{Z}^r$ and a lattice polytope $\sigma \subseteq M_\mathbb{R} = M \otimes_\mathbb{Z} \mathbb{R}$. The cone $C(\sigma)$ over $\sigma \times \{1\} \subset M_\mathbb{R} \times \mathbb{R}$ defines a sharp toric monoid $P_\sigma = C(\sigma)^\vee \cap (N \oplus \mathbb{Z})$, where $N = M^\vee$ is the dual lattice. We have a distinguished element $\rho_\sigma = (0, 1) \in P_\sigma$; it is called the *Gorenstein degree* and satisfies $\mathrm{int}(P_\sigma) = \rho_\sigma + P_\sigma$. This in turn defines an affine toric variety $U(\sigma) = \operatorname{Spec} \mathbb{C}[P_\sigma]$ together with a map $\mu_\sigma \colon U(\sigma) \to \mathbb{A}^1_t$ such that $\mu_\sigma^{-1}(\{0\}) =: V(\sigma)$ is the full toric boundary. The induced morphism of log schemes $\mu_\sigma \colon (U(\sigma) | V(\sigma)) \to (\mathbb{A}^1_t | \{0\})$ is log smooth, and so also

$$(U(\sigma) | V(\sigma)) \times (\mathbb{A}^{d-r} | \varnothing) \to (\mathbb{A}^1_t | \{0\})$$

is log smooth. We denote the set of these local models by $\mathscr{C}^{\mathrm{ver}}$. $\Diamond$

In analogy with normal crossing schemes, a degenerate scheme V which is étale locally isomorphic to schemes of the form $V(\sigma) \times \mathbb{A}^{d-r}$ is called a *scheme with toroidal crossing singularities*.

Example 1.137 Let Δ be a reflexive polytope. Then the reduced full toric boundary $V \subset \mathbb{P}_\Delta$ is a scheme with toroidal crossing singularities.   ◊

One may now expect to define $\mathcal{LS}_V$ for a scheme V with toroidal crossing singularities simply as the sheaf of isomorphism classes of log smooth structures of the specified type. However, this would comprise too many log structures for an effective theory. The reason is as follows: For a normal crossing scheme V, the value $(d; r)$ for the local model $\mu_{d;r}^{\mathrm{nc}}$ is uniquely determined at every point $v \in V$. This is no longer true for schemes with toroidal crossing singularities. Namely, the choice $\sigma = [0, k]$ yields the log smooth morphisms

$$(\operatorname{Spec} \mathbb{C}[x, y, t]/(xy - t^k)|\{t = 0\}) \to (\mathbb{A}_t^1|\{0\})$$

from Example 1.63. They have the same central fiber $V = \{xy = 0\}$ for all $k \geq 1$, but the polytopes $[0, k]$ and hence the log structures are all different.

We would like to define $\mathcal{LS}_V$ in such a way that the local model μ_σ is uniquely determined at every point $v \in V$. This makes $\mathcal{LS}_V$ smaller so that it becomes easier to construct sections because the structure of $\mathcal{LS}_V$ is simpler. To this end, we endow V with additional structure.

Since μ_σ is a morphism of log schemes, $V(\sigma)$ carries the structure of a log smooth morphism $(V(\sigma), \mathcal{M}_{V(\sigma)}) \to (S_0, N_0)$ over the standard log point. In particular, we can form the ghost sheaf $\mathcal{P}_{V(\sigma)} := \mathcal{M}_{V(\sigma)}/\mathcal{O}_{V(\sigma)}^*$ and the section

$$\bar{\rho}_\sigma := [\mu_{\sigma;0}^\flat(\tau)] \in \Gamma(V(\sigma), \mathcal{P}_{V(\sigma)})$$

for $\tau = (1, 1) \in N_0 = \mathbb{N} \oplus \mathbb{C}^*$. The triple $(V(\sigma), \mathcal{P}_{V(\sigma)}, \bar{\rho}_\sigma)$ allows us to recover P_σ as the stalk $\mathcal{P}_{V(\sigma),0}$, and thus to recover μ_σ.

Definition 1.138 (Toroidal Crossing Schemes) A *toroidal crossing scheme* is a triple $(V, \mathcal{P}, \bar{\rho})$ consisting of a separated scheme $V/\mathbb{C}$ of finite type, a sheaf of monoids $\mathcal{P}$ in the étale topology, and a section $\bar{\rho} \in \Gamma(V, \mathcal{P})$ such that the triple is étale locally isomorphic to triples of the form

$$(V(\sigma), \mathcal{P}_{V(\sigma)}, \bar{\rho}_\sigma) \times \mathbb{A}^{d-r},$$

where the sheaf and the section are pulled back along the projection $V \times \mathbb{A}^{d-r} \to V$.

Remark 1.139 In its current form, we have briefly defined toroidal crossing schemes in [77], but we did not elaborate much on the concept. The concept itself goes back to [124], which also developed most of the theory we present in Chap. 9. The name "toroidal crossing" was coined in [253] by Siebert and Schröer in the early days of the Gross–Siebert program.   ◊

Example 1.140 Let Δ be a reflexive polytope, and let $i : V \to P = \mathbb{P}_\Delta$ be the inclusion of the full toric boundary. Let $\mathcal{P} := i^{-1}\overline{\mathcal{M}}_{(P|V)}$. Since every stalk $\mathcal{P}_{\bar{v}}$ is a Gorenstein monoid, there is a unique global section $\bar{\rho} \in \Gamma(V, \mathcal{P})$ which is the Gorenstein degree in every stalk. This turns $(V, \mathcal{P}, \bar{\rho})$ into a toroidal crossing scheme.

Note that it may not be possible to lift $\bar{\rho}$ to a global section of $i_{\log}^* \mathcal{M}_{(P|V)}$. To put this in context, note that there may not be a global toric morphism $P \to \mathbb{A}_t^1$ while such toric morphisms exist locally. $\Diamond$

Example 1.141 A normal crossing scheme V can be considered as a toroidal crossing scheme by setting $\mathcal{P} := \nu_* \underline{\mathbb{N}}_{\tilde{V}}$ and $\bar{\rho} = \nu_*(1)$, where $\nu : \tilde{V} \to V$ is the normalization, and where $\underline{\mathbb{N}}_{\tilde{V}}$ is the constant sheaf with stalk $\mathbb{N}$ on $\tilde{V}$. From the perspective of toroidal crossing schemes, this construction is the reason why we did not need to consider $(\mathcal{P}, \bar{\rho})$ when studying normal crossing schemes. $\Diamond$

Example 1.142 Unlike the situation for normal crossing schemes, the product of two toroidal crossing schemes $(V_1, \mathcal{P}_1, \bar{\rho}_1)$ and $(V_2, \mathcal{P}_2, \bar{\rho}_2)$ is a toroidal crossing scheme. For details, see Definition 9.47 and Lemma 9.48. $\Diamond$

For a toroidal crossing scheme $(V, \mathcal{P}, \bar{\rho})$, we now specialize our notion of a log smooth structure to reflect both the local models in $\mathscr{C}^{\mathrm{ver}}$ and the given ghost sheaf.

Definition 1.143 (Log Smooth Structures II)   Let $(V, \mathcal{P}, \bar{\rho})$ be a toroidal crossing scheme. A *log smooth structure* is a quadruple $(\mathcal{M}, \alpha, q, \rho)$, where $\alpha : \mathcal{M} \to O_V$ is a log structure, $q : \mathcal{M} \to \mathcal{P}$ is a homomorphism of sheaves of monoids which induces an isomorphism $\overline{\mathcal{M}} \cong \mathcal{P}$, and $\rho \in \Gamma(V, \mathcal{M})$ is a global section with $q(\rho) = \bar{\rho}$ such that the induced morphism $(V, \mathcal{M}) \to (S_0, \mathcal{N}_0)$ to the standard log point is log smooth.

When we have a log smooth structure in the sense of the definition, then it is automatically of class $\mathscr{C}^{\mathrm{ver}}$.

Just as in the case of a normal crossing scheme, we have a sheaf $\mathcal{LS}_V$ whose sections over $W \subseteq V$ are isomorphism classes of log smooth structures on W. We still have a map $\eta : \mathcal{LS}_V \to \mathcal{T}_V^1$; however, this map is not always injective, and thus not always suitable to construct sections of $\mathcal{LS}_V$.

For a log smooth structure on V, we have an isomorphism between the dualizing sheaf ω_V° and the log canonical bundle. Thus, $(V, \mathcal{M}) \to (S_0, \mathcal{N}_0)$ is log Calabi–Yau if and only if V is Calabi–Yau.

1.11.4   *Toward the Toric Gross–Siebert Mirror Construction*

For a reflexive polytope Δ, we now have two toroidal crossing schemes $(V, \mathcal{P}, \bar{\rho})$ and $(V^\circ, \mathcal{P}^\circ, \bar{\rho}^\circ)$ constructed as the full toric boundaries in $\mathbb{P}_\Delta$ and $\mathbb{P}_{\Delta^\circ}$. We leave the problem of constructing sections of $\mathcal{LS}_V$ and $\mathcal{LS}_{V^\circ}$ aside for a moment and focus instead on how Gross and Siebert generalized in [124] the construction of $(V, \mathcal{P}, \bar{\rho})$ out of Δ. This is the first step in the toric Gross–Siebert mirror construction.

Let Δ be a reflexive polytope. We consider the set of its proper faces as a poset $(\mathcal{P}, \preceq)$. Each $\tau \in \mathcal{P}$ is a polytope and hence gives rise to a projective toric variety $\mathbb{P}_\sigma$. For each inclusion $\tau \preceq \eta$, we have a closed immersion $\mathbb{P}_\tau \to \mathbb{P}_\eta$, and the full toric boundary $V \subset \mathbb{P}_\Delta$ is nothing but the colimit $V = \mathrm{colim}_{\tau \in \mathcal{P}}(\mathbb{P}_\tau)$. Thus, we can construct V from the *polyhedral complex* $\mathcal{P}$.

Actually, to construct V, less data are needed. Let $\tau \preceq \eta$ be an inclusion of cells in $\mathcal{P}$. When we move η inside the tangent space Λ_η of η to a position where τ passes through the origin and take the positive hull of η, then we obtain a toric monoid $E_\tau \eta$. For every $\tau \preceq \eta \preceq \rho$, we have an inclusion $E_\tau \eta \to E_\tau \rho$ and a localization $E_\tau \rho \to E_\eta \rho$. The inclusion can be interpreted as a closed immersion $\mathrm{Spec}\,\mathbb{C}[E_\tau \eta] \to \mathrm{Spec}\,\mathbb{C}[E_\tau \rho]$, and we obtain an affine scheme $V(\tau)$ as the colimit over the $\mathrm{Spec}\,\mathbb{C}[E_\tau \eta]$. Now V is obtained by open gluing along the open immersions induced by the localizations $E_\tau \rho \to E_\eta \rho$.

For a fixed $\tau \in \mathcal{P}$, the monoids of the form $E_\tau \eta$ for $\tau \preceq \eta$ form a *cone complex* E_τ. For a fixed η, the monoids of the form $E_\tau \eta$ for $\tau \preceq \eta$ are the information of the "lattice angles" of the polytope η. We may say that $E_\bullet \eta$ is a *polycone* because it is composed of many cones. Polycones are dual to fans in that $E_\bullet \eta$ retains the same information as the normal fan of η.

Let $B = \mathrm{colim}_{\tau \in \mathcal{P}}(\tau)$. This is a topological manifold and can be identified with the boundary $\partial \Delta$. Moreover, in the interior of each maximal cell $\sigma \in \mathcal{P}$, we have an *integral affine structure*. By definition, an integral affine structure is an atlas of the manifold with integral affine transition maps.

The integral affine structure on B can be extended around the vertices $v \in \mathcal{P}$. To do so, we choose an appropriate open neighborhood W_v of $v \in B = \partial \Delta$ and then define the integral affine structure by means of the composition $W_v \subset M_\mathbb{R} \to M_\mathbb{R}/(\mathbb{R} \cdot v)$. Under this construction, there remains a closed subset $D_B \subset B$ of codimension ≥ 2 on which we have not defined an integral affine structure. We say that B is an *integral affine manifold with singularities*.

Let $v \in \mathcal{P}$ be a vertex. We have already seen above the cone complex E_v. The integral affine structure around v turns this into a fan Σ_v. We denote the lattice in which this fan lives by Q_v.

Let $\tau \in \mathcal{P}$. The invertible elements in $E_\tau \eta$ are given by $E_\tau \tau = \Lambda_\tau$, the tangent lattice to τ. We set $K_\tau \eta := E_\tau \eta / \Lambda_\tau$. Now for our fixed τ, the collection of monoids $K_\tau \eta$ can be turned into a fan Σ_τ. We denote the corresponding monoid by Q_τ.

To turn V into a toroidal crossing scheme, we need the datum of a *polarization* φ on $(B, \mathcal{P})$. This amounts to the choice of a strictly convex piecewise linear function φ_v on each of the fans Σ_v, subject to some compatibility condition. To construct φ_v, we split

$$0 \to \mathbb{Z} \cdot v \to M \to M/(\mathbb{Z} \cdot v) \to 0,$$

move $\Delta \subset M_\mathbb{R}$ such that v lies in the origin, and take the (boundary of the) positive hull of $\Delta - v$ as the graph of φ_v.

To construct $(\mathcal{P}, \bar{\rho})$, we set

$$P_v := \{(m, h) \in Q_v \oplus \mathbb{Z} \mid h \geq \varphi_v(m)\}$$

and $\bar{\rho}_v := (0, 1)$. This defines a toric morphism $U(v) := \operatorname{Spec} \mathbb{C}[P_v] \to \mathbb{A}_t^1$ whose central fiber is the colimit $V(v)$ obtained from the spectrum of the cone complex E_v. Then we obtain $(\mathcal{P}, \bar{\rho})$ on $V(v)$ in the same way as we have obtained them above on the local models $(V(\sigma), \mathcal{P}_{V(\sigma)}, \bar{\rho}_\sigma)$ for toroidal crossing schemes. They are compatible and thus glue to a global structure $(\mathcal{P}, \bar{\rho})$ of toroidal crossing scheme on V.

The construction of Gross and Siebert starts with the data that we have just constructed out of Δ. In other words, they start with a topological manifold B, a closed subset $D_B \subset B$ of codimension ≥ 2, an integral affine structure on $B_0 = B \setminus D_B$, a polyhedral decomposition $\mathcal{P}$ of B, and a polarization φ. They give rise to a toroidal crossing scheme $\check{X}_0(B, \mathcal{P}, \varphi)$.

The language concerning the pair $(B, \mathcal{P})$ is not fully standardized. To make a precise statement in Theorem 1.149 below, we adopt the following definition, whose precise conditions are most interesting for the reader already familiar with the Gross–Siebert program.

Definition 1.144 (Polyhedral Affine Manifolds)   A *polyhedral affine manifold* is a polyhedral affine pseudo-manifold $(B, \mathcal{P})$ in the sense of [122, Constr. 1.1.1] such that:

(a)   there are only countably many cells;
(b)   each cell $\tau \in \mathcal{P}$ is bounded;
(c) $\partial B = \varnothing$; in other words, every cell $\rho \in \mathcal{P}$ of codimension one is contained in exactly two maximal cells;
(d)   for every vertex $v \in \mathcal{P}$, we have a fan structure S_v along v in the sense of [126, Defn. 1.1]; we assume this fan structure to be compatible with the given integral affine structure; in particular, the integral affine structure extends across the vertices; furthermore, B is a topological manifold since each fan must be complete due to our assumption that $\partial B = \varnothing$;
(e)   for every cell $\tau \in \mathcal{P}$ and any two vertices $v, w \in \tau$, the two fan structures along τ induced by the fan structures along v respectively w are equivalent as in [126, Defn. 1.2].

Remark 1.145   In the language of [126], this is an integral tropical manifold such that each cell $\tau \in \mathcal{P}$ is bounded and such that $\partial B = \varnothing$. In the language of [124], this is an integral affine manifold with singularities B and a toric polyhedral decomposition $\mathcal{P}$ such that no cell self-intersects, and such that we have only countably many cells. $\Diamond$

The construction of $\check{X}_0(B, \mathcal{P}, \varphi)$ that we have discussed is known as the *cone picture*. Note that the discussion of the cone picture in [124, § 2.1] does not use the polarization. Without the polarization, we can construct $\check{X}(B, \mathcal{P})$ only as a scheme, and for that, it even would have been sufficient to record just the polycones instead of the polyhedra. Constructing the toroidal crossing structure needs the polarization.

Originally, Gross and Siebert focused in [124] on a dual construction, the *fan picture*. The resulting toroidal crossing scheme is denoted by $X_0(B, \mathscr{P})$. Unlike the cone picture, we do not need the polarization φ to construct the toroidal crossing structure on $X_0(B, \mathscr{P})$. However, we do need the polyhedra here, not just the polycones.

We may now consider $X_0(B, \mathscr{P})$ and $\check{X}_0(B, \mathscr{P}, \varphi)$ as a mirror pair of toroidal crossing schemes. A more canonical way of doing so is obtained as follows. With any polarized compact orientable polyhedral affine manifold $(B, \mathscr{P}, \varphi)$, one can associate a dual polarized compact orientable polyhedral affine manifold $(\check{B}, \check{\mathscr{P}}, \check{\varphi})$, which is called the *discrete Legendre transform*. This construction is involutive, i.e., the discrete Legendre transform of $(\check{B}, \check{\mathscr{P}}, \check{\varphi})$ is $(B, \mathscr{P}, \varphi)$. Now we have

$$\check{X}_0(B, \mathscr{P}, \varphi) = X_0(\check{B}, \check{\mathscr{P}}) \quad \text{and} \quad X_0(B, \mathscr{P}) = \check{X}_0(\check{B}, \check{\mathscr{P}}, \check{\varphi})$$

as toroidal crossing schemes. Thus, it is sufficient to work either only in the cone picture or only in the fan picture.

Somewhat deviating from the original treatment in [124], the discrete Legendre transform is in the author's opinion best understood in three layers. For the first layer, we retain only the poset $\mathscr{P}$ and the quotients

$$K_\tau \eta := E_\tau \eta / E_\tau \eta^*,$$

where $E_\tau \eta^*$ is the group of invertibles in $E_\tau \eta$. Each $K_\tau \eta$ is a sharp toric monoid. We have already considered them above when briefly discussing the fans Σ_τ. The dual is given by the opposite poset $\check{\mathscr{P}}$ and the dual monoids

$$\check{K}_{\check{\eta}} \check{\tau} = \mathrm{Hom}(K_\tau \eta, \mathbb{N}).$$

For the first layer, involutivity is clear since the double dual of a sharp toric monoid P is canonically isomorphic to P.

The second layer of duality interchanges the polyconal data $E_\tau \eta$ (as opposed to only having the quotients $K_\tau \eta$) with the fan structure on Σ_τ (as opposed to only having cone complexes K_τ without their embedding into the lattice Q_τ). At this layer, we already have a duality of associated toric schemes. Namely, in the cone picture, we only need the poset $\mathscr{P}$ and the polyconal data for constructing the schemes.

The third layer of duality exchanges the polyhedral shape of τ (as opposed to only having $E_\omega \tau$ for the various cells $\omega \preceq \tau$) with the polarization φ_v on the fan Σ_v. We need the third layer for the duality of toroidal crossing structures. Namely, in the cone picture, the toroidal crossing structure is constructed from the polarization, and in the fan picture, the toroidal crossing structure is constructed from the polyhedral shapes.

The analysis shows that working with polyhedral affine manifolds is somewhat unnatural from the perspective of the discrete Legendre transform since retaining the polyhedra (instead of only the polycones) but discarding the polarization is not symmetric under the transform. The discrete Legendre transform of a non-polarized polyhedral affine manifold is polarized but not polyhedral.

Both the cone picture and the fan picture come in more flexible variants that allow to modify the gluing of the pieces by *gluing data*. In the (at least initially) better developed fan picture, Gross and Siebert distinguish three types of gluing data: *lifted gluing data*, *open gluing data*, and *closed gluing data*. While a closed gluing datum $\bar{s}$ only defines a scheme[15] $X_0(B, \mathscr{P}, \bar{s})$, an open gluing datum s gives rise to a toroidal crossing scheme $X_0(B, \mathscr{P}, s)$. In the cone picture, gluing data for $(B, \mathscr{P}, \varphi)$ are best considered as gluing data in the fan picture of $(\check{B}, \check{\mathscr{P}}, \check{\varphi})$. Note that this approach is more general than the gluing data for the cone picture defined in [124, § 2.1]. In particular, $\check{X}_0(B, \mathscr{P}, \varphi, \check{s}) := X_0(\check{B}, \check{\mathscr{P}}, \check{s})$ is not necessarily projective but only has projective irreducible components.

1.11.5 Constructing Log Smooth Structures on $\check{X}_0(B, \mathscr{P}, \varphi)$

We now have a quite general construction for mirror pairs of toroidal crossing schemes at hand, but we do not yet have the log smooth structures on them which we need to construct the desired toric degenerations by means of some logarithmic Bogomolov–Tian–Todorov theorem.

Let $(B, \mathscr{P}, \varphi)$ be a polarized compact orientable polyhedral affine manifold, and let $\check{X}_0 = X_0(\check{B}, \check{\mathscr{P}}, \check{s})$ for an open gluing datum $\check{s}$ on $(\check{B}, \check{\mathscr{P}})$. To construct a log smooth structure, Gross and Siebert computed $\mathcal{LS}_{\check{X}_0}$ in [124, Thm. 3.28]. It embeds into the direct sum

$$\mathcal{LS}^{+}_{\mathrm{pre}, \check{X}_0} := \bigoplus_{\rho \in \mathscr{P} : \dim(\rho) = d-1} \mathcal{N}_\rho,$$

where $\mathcal{N}_\rho$ is a certain line bundle on the codimension-1 stratum $\check{X}_\rho$ of $\check{X}_0$. In [124, Cor. 3.29], Gross and Siebert concluded from this that, if $\mathcal{LS}_{\check{X}_0}$ admits a global section, then the integral affine structure on $B \setminus D_B$ can be extended to the whole of B. This is usually not the case. In other words, $\check{X}_0$ rarely admits a global log smooth structure. We have seen this phenomenon more explicitly in Example 1.81, where we have seen that $H = \bigcup_{i=0}^{3} H_i$ does not admit a global section of $\mathcal{LS}_H$.

---

[15] For the expert: Recall that we assume $\mathscr{P}$ to have no self-intersections.

The explicit form of $\mathcal{N}_\rho$ allowed Gross and Siebert to define the concept of being *normalized* for a global section[16]

$$\check{f} \in \Gamma(\check{X}_0, \mathcal{LS}^+_{\mathrm{pre},\check{X}_0}).$$

If $(\check{B}, \check{\mathscr{P}})$ (or equivalently $(B, \mathscr{P})$) is *positive* and *simple*, and if $\check{s}$ underlies a lifted gluing datum, then $\check{X}_0$ admits a unique normalized section $\check{f}$. This $\check{f}$ lies in $\mathcal{LS}_{\check{X}_0}$ outside the double locus of $\check{X}_0$ and around each vertex v. Thus, $\check{f}$ turns $\check{X}_0$ into a generically log smooth family $f_0 \colon (\check{X}_0, \check{U}_0, \mathcal{M}_0) \to (S_0, \mathcal{N}_0)$ over the standard log point. Compactness of B ensures that f_0 is proper, and orientability of B ensures that

$$\omega^\circ_{\check{X}_0} \cong O_{\check{X}_0}$$

for the normalized dualizing sheaf. Since f_0 is vertical, this implies that f_0 is log Calabi–Yau. Gross and Siebert call f_0 a *toric log Calabi–Yau space.*[17]

The fact that the normalized section $\check{f}$ is usually not a global section of $\mathcal{LS}_{\check{X}_0}$ is the historical origin of the interest in generically log smooth families. When $(B, \mathscr{P})$ is positive and simple, then any pair of lifted gluing data defines a mirror pair of log Calabi–Yau generically log smooth families. Actual mirror statements for these pairs are beyond the scope of this introduction (and not the topic of this monograph); an example is given by Gross and Siebert [125, Cor. 3.24].

Remark 1.146 Both positivity and simplicity depend only on the polyconal structure of $(B, \mathscr{P})$ but not on the polyhedral shapes of $\tau \in \mathscr{P}$. Positivity of $(B, \mathscr{P})$ is a mild condition which is often satisfied. However, the same does not hold for simplicity. For example, the $(B, \mathscr{P})$ which we have constructed above out of a reflexive polytope Δ are rarely simple. Thus, we do not obtain a canonical log smooth structure on these V by means of normalized sections of $\mathcal{LS}^+_{\mathrm{pre},V}$. Moreover, to the author's knowledge, there is no direct construction of a section of $\mathcal{LS}^+_{\mathrm{pre},V}$ in the Batyrev situation in the literature—in other words, what we have given as motivation above has not yet been done, but it motivated the analogous result in the simple case. At least, Gross showed in [119] that, for every reflexive polytope Δ

---

[16] Note that this notion, at least a priori, depends on the actual open gluing datum $\check{s}$. It is not invariant under transitioning to a *cohomologous* open gluing datum but needs a finer notion of equivalence to become invariant.

[17] Note that a toric log Calabi–Yau space in the sense of [124] does not need to be log Calabi–Yau in our sense. This is because we require the Calabi–Yau condition on the level of the log canonical bundle (which is here equivalent to asking it for the normalized dualizing sheaf) while Gross and Siebert require the condition on the level of the irreducible components of X_0. For example, if B is compact but not orientable, then we have $\omega^{\otimes 2}_{X_0/S_0} \cong O_{X_0}$ but $\omega_{X_0/S_0} \not\cong O_{X_0}$ by Gross and Siebert [124, Thm. 2.39] (in the fan picture).

and every mpcp resolution $\hat{P} \to \mathbb{P}_\Delta$, there is some positive and simple $(\check{B}, \check{\mathscr{P}})$ such that the mpcp resolution $\hat{Y}_\phi$ of a general $\overline{Y}_\phi$ degenerates to $X_0(\check{B}, \check{\mathscr{P}})$. $\Diamond$

1.11.6 The Local Models of the Log Singularities

Let $(B, \mathscr{P}, \varphi)$ be positive and simple polarized compact orientable polyhedral affine manifold, and let $\check{s}$ be a lifted gluing datum of $(\check{B}, \check{\mathscr{P}})$. We denote $\check{X}_0(B, \mathscr{P}, \varphi, \check{s})$ with the log smooth structure defined by the unique normalized section by

$$\check{X}_0(B, \mathscr{P}, \varphi, \check{s})^\dagger \to (S_0, \mathcal{N}_0),$$

where $(-)^\dagger$ denotes the presence of a log structure. The log singularities of this generically log smooth family are not arbitrary—they admit very specific toric local models. This is the historical origin of the interest in log toroidal families. All relevant local models arise from the following construction, which we discuss in more detail in Sect. 14.1. It goes back to the sequel [125] to [124]; hence, we call them *local models of Gross–Siebert type*.

Construction 1.147 We start with a full-dimensional polytope $\tau \subseteq M'_\mathbb{R}$ for some lattice M', together with a sequence $\Delta_1, \ldots, \Delta_q \subseteq M'_\mathbb{R}$ of lattice polytopes which may not be of full dimension. We set $\Delta_0 := \tau$ and allow $q = 0$ so that we have only $\Delta_0 = \tau$. We assume that the normal fan $\check{\Sigma}_0$ of $\Delta_0 = \tau$ in the dual lattice N' is a subdivision of each normal fan $\check{\Sigma}_i$ of Δ_i. We say that $(M', N', \tau, \Delta_1, \ldots, \Delta_q)$ is a *Gross–Siebert local model datum*.

For each $0 \le i \le q$, we have a function $\check{\psi}_i(n) := -\inf\{\langle m, n \rangle \mid m \in \Delta_i\}$, and then we set

$$P := \left\{ n + a_0 e_0^* + \sum_{i=1}^q a_i e_i^* \in N' \oplus \mathbb{Z} \oplus \mathbb{Z}^q \;\middle|\; \forall i : a_i \ge \check{\psi}_i(n) \right\}.$$

This is a sharp toric monoid with a distinguished element $\rho = e_0^*$, which defines a toric morphism $f : \mathbb{A}_P = \operatorname{Spec} \mathbb{C}[P] \to \mathbb{A}_t^1$. The central fiber has the form $V(\tau) \times \mathbb{A}^q$, where $V(\tau)$ is as in Construction 1.136. With the appropriate log singular locus $Z = Z^\parallel \cup Z^*$, the datum defines a generically log smooth family

$$f : (\mathbb{A}_P, \mathbb{A}_P \setminus Z | V(\tau) \times \mathbb{A}^q) \to (\mathbb{A}_t^1 | \{0\}).$$

On the central fiber, the toroidal crossing scheme $V = V(\tau) \times \mathbb{A}^q$, this induces a section of $\mathcal{LS}_V$ outside $Z^\parallel$.

In Sect. 14.1, we will see a polytope $\Delta_+ \subseteq M'_\mathbb{R} \oplus \mathbb{R}^q$ which controls the singularities of the general fiber of $f : \mathbb{A}_P \to \mathbb{A}_t^1$. The general fiber is smooth if and only if Δ_+ is a standard simplex in the sense of Definition 14.6, and if Δ_+ is an elementary simplex in the sense of that definition, then the general fiber has very mild singularities.   ◇

Definition 1.148 ($\mathscr{C}^{\mathrm{GS}}$)   The class of singularities $\mathscr{C}^{\mathrm{GS}}$ is given by local models which arise from Construction 1.147. The class $\mathscr{C}^{\mathrm{eGS}} \subset \mathscr{C}^{\mathrm{GS}}$ consists of those models where Δ_+ is an elementary simplex, and the class $\mathscr{C}^{\mathrm{sGS}} \subset \mathscr{C}^{\mathrm{eGS}}$ consists of those local models where Δ_+ is a standard simplex.

At this point, two insights of Gross and Siebert in [125] are fundamental. First, if $(\check{B}, \check{\mathscr{P}})$ is positive and simple, then $\check{X}_0(B, \mathscr{P}, \varphi, \check{s})^\dagger \to (S_0, \mathcal{N}_0)$ is accurately of class $\mathscr{C}^{\mathrm{eGS}}$ (and in particular log toroidal). This is [125, Thm. 2.6]. Secondly, log deformations accurately of class $\mathscr{C}^{\mathrm{eGS}}$ are locally rigid. This is in [125, §2.2]. In particular, we can construct a log toroidal system of deformations $\mathscr{D}$ for $\check{X}_0(B, \mathscr{P}, \varphi, \check{s})^\dagger \to (S_0, \mathcal{N}_0)$ as in Construction 1.91. Generically log smooth deformations of type $\mathscr{D}$ are called *divisorial deformations* in the language of [125]. Now, the Log Toroidal Bogomolov–Tian–Todorov Theorem 1.103 applies.

Theorem 1.149 *Let $(B, \mathscr{P}, \varphi)$ be a polarized compact orientable polyhedral affine manifold which is positive and simple, and let $\check{s}$ be a lifted gluing datum for $(\check{B}, \check{\mathscr{P}})$. Then, divisorial deformations of $f_0\colon \check{X}_0(B, \mathscr{P}, \varphi, \check{s}) \to S_0$ are unobstructed.*

Remark 1.150   This result has been known before by work of Ruddat and Siebert in [243]. We will say more about this in a moment.   ◇

In dimension $d \leq 3$, any formal algebraic one-parameter deformation over $\mathrm{Spf}\,\mathbb{C}[\![t]\!]$ obtained from this theorem is a smoothing, but in dimension ≥ 4, this is often only a normaling. When the formal deformation can be algebraized to an algebraic deformation over $\mathbb{C}[\![t]\!]$, this is the desired toric degeneration.

In Chap. 14, we give a comprehensive discussion of the class of log singularities $\mathscr{C}^{\mathrm{eGS}}$. Among other things, we reproduce Gross and Siebert's proof of the local rigidity of $\mathscr{C}^{\mathrm{eGS}}$ for the reader's convenience.

1.11.7 The Scattering Algorithm

While Gross and Siebert originally hoped to apply a result like the Log Toroidal Bogomolov–Tian–Todorov Theorem 1.103 to obtain a smoothing of $\check{X}_0(B, \mathscr{P}, \varphi, \check{s})$ (see [125, Rem. 2.19]), such a result was not available until 2019 when Chan, Leung, and Ma made the preprint version of [38] available. Instead, Gross and Siebert devised the sophisticated *scattering algorithm* in [126], which constructs the formal algebraic one-parameter deformation explicitly. The name comes from a combinatorial device used in the construction, the *scattering diagram*, which is obtained by successive refinements of the polyhedral decomposition $\mathscr{P}$, both in the literal sense of a finer decomposition and in the broader sense of other additional data.

The explicit nature of the algorithm, although very complex, provides more insights into the smoothing than the Log Toroidal Bogomolov–Tian–Todorov Theorem 1.103, for example canonical coordinates on the local moduli space, which are further examined in [243]. On the other hand, the Bogomolov–Tian–Todorov is far more general since it does make few assumptions on the global geometry of $(X_0, U_0, \mathcal{M}_0) \to (S_0, \mathcal{N}_0)$.

Let us assume that $\check{X}_0(B, \mathscr{P}, \varphi, \check{s})$ is polarized by a line bundle $\mathcal{L}_0$. In [126, Rem. 1.32], Gross and Siebert announced that $\mathcal{L}_0$ admits a formal one-parameter deformation as well by applying the scattering algorithm to the total space of $\mathcal{L}_0$. Given our Theorem 1.120, the unobstructedness of pairs consisting of a log deformation of type $\mathscr{D}$ and a deformation of the line bundle, we see that indeed deformations of $f_0 \colon \check{X}_0(B, \mathscr{P}, \varphi, \check{s})^\dagger \to (S_0, \mathcal{N}_0)$ with any line bundle $\mathcal{L}_0$ are unobstructed, confirming their claim, yet with a totally different method.

The original version of the scattering algorithm produced a formal toric degeneration over $\mathrm{Spf}\,\mathbb{C}[\![t]\!]$. With [122, Thm. A.2.4], Gross, Hacking, and Siebert provided an expanded version that produces a canonical formal toric degeneration that also takes variations of the lifted gluing datum into account. Ruddat and Siebert proved with [243, Thm. 1.9] that this formal toric degeneration is miniversal for divisorial deformations of $\check{X}_0(B, \mathscr{P}, \varphi, \check{s}) \to S_0$. Since the base of the family is a formal power series ring over $\mathbb{C}[\![t]\!]$, this proves unobstructedness of divisorial deformations without relying on the Log Toroidal Bogomolov–Tian–Todorov Theorem 1.103. Note that Ruddat–Siebert does not assume $\partial B = \varnothing$ so that [243, Thm. 1.9] is somewhat more general than how we have stated Theorem 1.149.

Remark 1.151 We expect that this more general case also follows from the Log Toroidal Bogomolov–Tian–Todorov Theorem 1.103 once we have constructed the appropriate generically log smooth structure and the appropriate system of deformations. This needs a non-vertical version of toroidal crossing schemes and local rigidity for certain non-vertical log singularities, which we do not treat in this monograph.   ◊

1.12 Enhanced Generically Log Smooth Families

So far, we have focused on generically log smooth families $f \colon (X, U, \mathcal{M}) \to (S, \mathcal{N})$ which are log toroidal. When S has characteristic 0, they always have the base change property of Definition 1.78. More general generically log smooth families often do not have the base change property. For instance, the second family in Example 1.83 does not have the base change property.

The base change property is very important for studying deformations. To study deformations in situations where the base change property fails, we have developed the concept of an *enhanced generically log smooth family*. This is a generically log smooth family endowed with a *choice* of log de Rham complex $\mathscr{A}^\bullet_{X/S}$ and log polyvector fields $\mathcal{G}^{\bullet}_{X/S}$, and they may be chosen such that the base change property is satisfied.

Throughout the monograph, we have developed most of the theory for enhanced generically log smooth families rather than generically log smooth families. In this section, we give some motivation for doing so.

We have to admit that we have no general quasi-perfectness result so far for the curved two-sided Gerstenhaber calculus which occurs in the below examples. Therefore, applicability of Theorem 1.101 to conclude unobstructedness in these situations remains speculative for now. We included the theory as a preparation for future work.

1.12.1   *Enhanced Generically Log Smooth Deformations*

We explain the setup. Let $f_0 \colon (X_0, U_0, \mathcal{M}_0) \to (S_0, \mathcal{N}_0)$ be a generically log smooth family, say over the standard log point, and let $\mathscr{D}$ be a system of deformations in the sense of Definition 1.89. Suppose that we want to show that $\mathrm{LD}^{\mathscr{D}}_{X_0/S_0}$ is unobstructed by applying the Second Abstract Unobstructedness Theorem 1.101, as we have done in the log toroidal case. When we constructed the curved two-sided Gerstenhaber calculus

$$(PV^{\bullet,\bullet}_{X_0/\Lambda}, DR^{\bullet,\bullet}_{X_0/\Lambda}) \tag{1.4}$$

in Sect. 1.8, we assumed that all local models in $\mathscr{D}$ have the base change property. This assumption is deeply embedded into the construction and cannot easily be omitted.

Suppose now that some local models in $\mathscr{D}$ do not have the base change property. To obtain a curved two-sided Gerstenhaber calculus nonetheless, we would like to modify the definition of $\mathcal{W}^i$ and $\mathcal{V}^p$ so that their formation then commutes with base change. Namely, note that we have defined the log de Rham complex of a generically log smooth family f as $\mathcal{W}^i_{X/S} := j_* \Omega^i_{U/S}$ to obtain *some* coherent log differential forms, but that this is not necessarily the only possible choice.

The advantage of defining log differential forms via $j_*(-)$ is that this definition does not depend on further choices. When we want to force compatibility with base change, the log de Rham complex becomes a *choice* which must be recorded as part of the datum. This is what we call an *enhanced generically log smooth family*. More precisely, we record a log de Rham complex $\mathcal{A}^{\bullet}_{X/S}$ and a Gerstenhaber algebra of log polyvector fields $\mathcal{G}^{\bullet}_{X/S}$ as part of the structure. In Definition 10.8, we provide the details.

With the enhanced notion at hand, we can start with an enhanced generically log smooth family

$$f_0 \colon \ (X_0, U_0, \mathcal{M}_0; \mathcal{G}^{\bullet}_0, \mathcal{A}^{\bullet}_0) \to (S_0, \mathcal{N}_0)$$

and an *enhanced system of deformations* $\mathscr{D}$, whose local models are enhanced generically log smooth families such that the Gerstenhaber calculi are compatible with base change. We can then construct a curved two-sided Gerstenhaber calculus analogous to (1.4), whose curved Lie algebra controls the *enhanced generically log smooth deformation functor*

$$\mathrm{ELD}^{\mathscr{D}}_{X_0/S_0} \colon \quad \mathbf{Art}_{\mathbb{C}[\![t]\!]} \to \mathbf{Set}.$$

1.12.2 *Moving the Log Singular Locus into the Deepest Stratum*

This is a class of examples where we currently expect, with some preliminary computational evidence, that the curved two-sided Gerstenhaber calculus is quasi-perfect.

Let $H = \bigcup_{i=0}^{3} H_i$ be the normal crossing scheme from Example 1.10, and let $f_0 \colon (H, U_0, \mathcal{R}_0) \to (S_0, \mathcal{N}_0)$ be the structure of a generically log smooth family from Example 1.81. The log singularities of this family are very simple: They are described by $xy = tz$, or more precisely, by the central fiber of the generically log smooth family in Example 1.79. In a similar example, we now analyze what happens when we vary the equation of this family.

Example 1.152 As the central fiber, we use $F = \{X_1 X_2 X_3 = 0\} \subset \mathbb{P}^3$, the union of the three copies of $\mathbb{P}^2$ given by $F_i = \{X_i = 0\}$ for $i = 1, 2, 3$. This is a normal crossing scheme. The double locus $D = \mathrm{Sing}(F)$ has three irreducible components L_{12}, L_{23}, and L_{13}, defined as $L_{ij} := F_i \cap F_j$, each isomorphic to $\mathbb{P}^1$. They meet in a single point $T = \mathrm{Sing}(D) = \{o\}$.

The adjunction formula shows $\omega_F^\circ \cong \mathcal{O}_F(-1)$ so that F is a degenerate Fano scheme. In analogy with Example 1.47, we see that $\mathcal{T}_F^1|_{L_{ij}} = \mathcal{O}_{L_{ij}}(3)$ for every L_{ij}. In particular, F is not d-semistable.

First, we consider the family

$$f \colon \quad C = \{T_0(X_0^3 + X_1^3 + X_2^3 + X_3^3) - T_1 X_1 X_2 X_3 = 0\} \subset \mathbb{P}^1 \times \mathbb{P}^3 \to \mathbb{P}^1,$$

depicted in the upper half of Fig. 1.18. In a neighborhood of $F \subset C$, the singular locus of the total space is given by

$$Z = \mathrm{Sing}(F) \cap \{X_0^3 + X_1^3 + X_2^3 + X_3^3 = 0\} \subset F \subset C,$$

consisting of nine points in total, three on each $L_{ij} \setminus \{o\}$. Outside Z, the algebraic deformation $f \colon (C \supset F) \to (\mathbb{P}^1 \ni 0)$ is semistable. Around each $p \in Z$, a local model is given by $\mathrm{Spec}\,\mathbb{C}[x, y, z, t]/(xy - tz) \to \mathbb{A}_t^1$, which can be shown explicitly as in Example 1.80. Thus, $f \colon (C, C \setminus Z | F) \to (\mathbb{P}^1 | \{0\})$ is log toroidal, at least in a neighborhood of F.

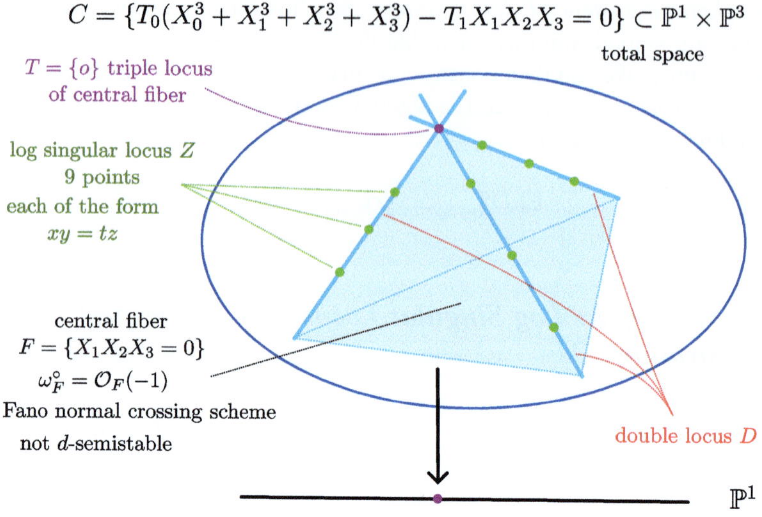

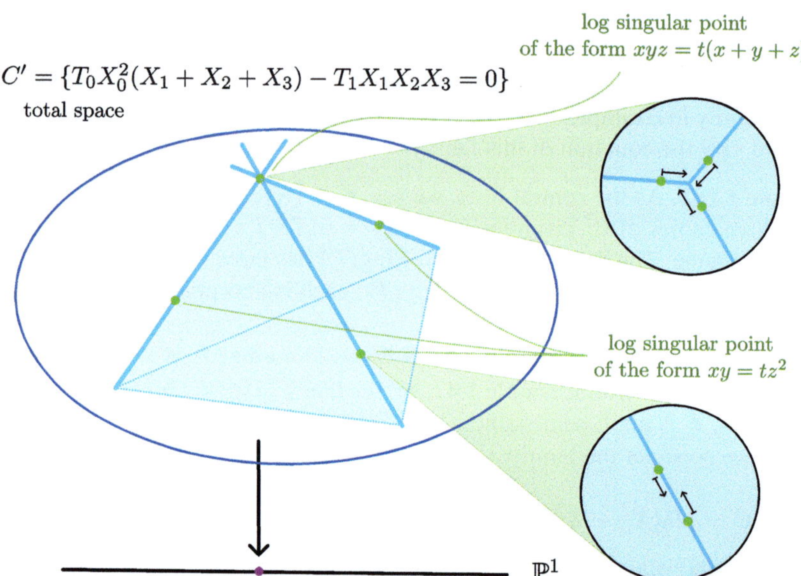

Fig. 1.18 Example 1.152. The upper picture shows the family $f : C \to \mathbb{P}^1$ with 9 log singular points of type $xy = tz$, three on each of the three irreducible components of $D = \mathrm{Sing}(F)$. The lower picture shows the family $f' : C' \to \mathbb{P}^1$. Here, on each irreducible component of D, two log singular points have come together to form a log singularity of type $xy = tz^2$. Furthermore, three log singular points have come together in the triple locus $T = \{o\}$ of F to form a log singular point of type $xyz = t(x + y + z)$. © Simon Felten 2025. All rights reserved

Now, we vary the defining equation of the family $f\colon C \to \mathbb{P}^1$ and consider

$$f'\colon \quad C' = \{T_0 X_0^2(X_1 + X_2 + X_3) - T_1 X_1 X_2 X_3 = 0\} \subset \mathbb{P}^1 \times \mathbb{P}^3 \to \mathbb{P}^1,$$

which is depicted in the lower half of Fig. 1.18. In a neighborhood of F, the family is locally given as follows:

(a) For $i = 1, 2, 3$, the open subset $\{X_i \neq 0\}$ is isomorphic to

$$\operatorname{Spec} \mathbb{C}[x, y, w, t]/(xy - tw^2(x + y + 1)).$$

The singular locus of this is given by $\{x = y = w = 0\}$, outside of which the family is semistable. This in turn is étale locally isomorphic to $\operatorname{Spec} \mathbb{C}[x, y, z, t]/(xy - tz^2) \to \mathbb{A}_t^1$.

(b) On the open subset $\{X_0 \neq 0\}$, the family is isomorphic to

$$M := \operatorname{Spec} \mathbb{C}[x, y, z, t]/(xyz - t(x + y + z)) \to \mathbb{A}_t^1.$$

The singular locus of the total space is given by the triple locus $T = \{o\}$, outside of which the family is semistable.

Let $Z = \operatorname{Sing}(C')$ and $U' = C' \setminus Z$ so that $f\colon (C', U'|F) \to (\mathbb{P}^1|\{0\})$ is generically log smooth, at least in a neighborhood of F.

First, we consider the local model in (a). Given by $xy = tz^2$, it differs from the local model $xy = tz$ of the log singularities of $f\colon (C|F) \to (\mathbb{P}^1|\{0\})$. We can think of it as two log singular points of type $xy = tz$ coming together, such as in the family of families

$$\operatorname{Spec} \mathbb{C}[x, y, z, t]/(xy - tz(z + s)) \to \mathbb{A}_t^1$$

parametrized by $s \in \mathbb{C}$. The local model given by $xy = tz^2$ is log toroidal so that $f\colon (C', U'|F) \to (\mathbb{P}^1|\{0\})$ is log toroidal in $\bigcup_{i=1}^3 \{X_i \neq 0\}$. In particular, there, the family has the base change property.

Next, we consider the chart $\{X_0 \neq 0\}$ in (b). We can think of this as three log singular points of type $xy = tz$ coming together in the triple locus, one from each of the meeting branches of the double locus. The relative log derivations can be computed as $\Theta^1_{M/\mathbb{A}_t^1} = \mathcal{T}_{M/\mathbb{A}_t^1}$, the relative classical derivations, so that $\mathcal{W}^1_{M/\mathbb{A}_t^1}$ becomes accessible as the dual of $\mathcal{T}_{M/\mathbb{A}_t^1}$. Then a computation in Macaulay2 shows that

$$\mathcal{W}^1_{M/\mathbb{A}_t^1} \otimes_{\mathbb{C}[t]} \mathbb{C}$$

is not reflexive so that $(M|\{t = 0\}) \to (\mathbb{A}_t^1|\{0\})$ does not have the base change property. In particular, $f'\colon (C', U'|F) \to (\mathbb{P}^1|\{0\})$ is not log toroidal.

To turn the central fiber into an enhanced generically log smooth family, we set

$$\mathcal{A}^i_{M_0/S_0} := \mathcal{W}^i_{M/\mathbb{A}^1_t} \otimes_{\mathbb{C}[t]} \mathbb{C} \quad \text{and} \quad \mathcal{G}^p_{M_0/S_0} := \mathcal{V}^p_{M/\mathbb{A}^1_t} \otimes_{\mathbb{C}[t]} \mathbb{C}.$$

An enhanced system of deformations can be defined similarly. $\Diamond$

In joint computational work with Matthias Zach that will be reported elsewhere, the author studies this and similar families. The results support the following hypothesis.

Hypothesis 1.153 *Let* $(V, \mathcal{P}, \bar{\rho})$ *be a proper toroidal crossing scheme of dimension two. Let* $Z \subset V$ *be a set of isolated points, and let* $s \in \Gamma(V \setminus Z, \mathcal{LS}_V)$ *be a section which induces a generically log smooth family* $f_0 \colon (V, U, M) \to (S_0, N_0)$. *Let* $(\mathcal{G}^{\bullet}_0, \mathcal{A}^{\bullet}_0)$ *be the structure of an enhanced generically log smooth family on* f_0, *and let* $\mathcal{D}$ *be an enhanced system of deformations such that, for every* α, *the local models (and in particular their Gerstenhaber calculi) arise from a generically log smooth family* $X_\alpha \to \mathrm{Spec}(\mathbb{N} \to \mathbb{C}[\![t]\!])$ *with its reflexive Gerstenhaber calculus. Then the curved two-sided Gerstenhaber calculus of* f_0 *is quasi-perfect.*

If the hypothesis is true, it gives us additional methods to smooth toroidal crossing Calabi–Yau surfaces. Furthermore, it enlarges the locus inside the moduli space (to be defined elsewhere) where the log Hodge sheaves are locally free.

Log singularities as in the hypothesis are closely connected with the mirror construction of Gross, Hacking, and Keel in [120]. An explicit example of such a mirror was computed by Gross, Hacking, Keel, and Siebert in [121].

1.12.3 Smoothing Fano Toroidal Crossing Schemes

This class of examples was our main motivation for developing enhanced generically log smooth families, but additional computations which we carried out in the meantime show that the curved two-sided Gerstenhaber calculus is not always quasi-perfect.

It has been known that the moduli space of Fano manifolds is bounded in any dimension since [177], but the classification of Fano manifolds is known only in dimension ≤ 3. On the other hand, Gorenstein toric Fano varieties are in one-to-one correspondence with reflexive polytopes, which are far easier to classify. One may therefore speculate that Fano manifolds can be classified as smoothings of (irreducible normal) Gorenstein toric Fano varieties $\mathbb{P}_\Delta$. Coates, Corti, and some others have considered this question from the perspective of mirror phenomena in [47–50].

It is well known that not every Gorenstein toric Fano variety $\mathbb{P}_\Delta$ admits a smoothing, see for example [231] and [232, Thm. 4.2]. Nonetheless, one may try to construct a smoothing under suitable restrictions on Δ. In dimension 3, such a restriction is being endowed with *admissible Minkowski decomposition data of*

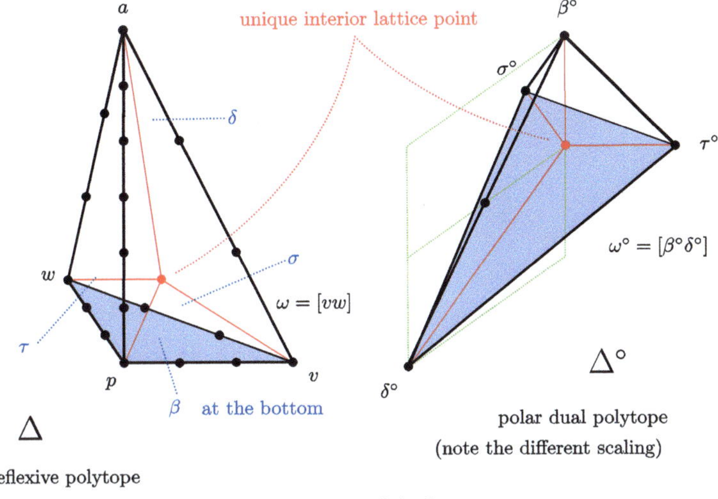

Fig. 1.19 Example 1.154. We see the reflexive polytope Δ from Example 1.154 together with its polar dual reflexive polytope $\Delta°$. Note that not all lattice points of Δ are marked, and that we see $\Delta°$ twice as large as it is in comparison with Δ. © Simon Felten 2025. All rights reserved

the dual $\Delta°$ as described in [49, 54].[18] It has been a conjecture for a while and is now a theorem of Corti, Hacking, and Petracci in [54] that admissible Minkowski decomposition data on $\Delta°$ give rise to a smoothing of $\mathbb{P}_\Delta$. The proof proceeds by analyzing the deformation theory of a certain partial resolution of singularities of $\mathbb{P}_\Delta$. In [49], Coates, Corti, and da Silva report computational work that shows that all 98 Fano threefolds with very ample anti-canonical bundle occur as smoothings of some $\mathbb{P}_\Delta$.

Example 1.154 We consider the reflexive polytope $\Delta \subseteq \mathbb{R}^3$ with vertex matrix

$$
\begin{pmatrix} p\, v\, w\, a \end{pmatrix} = \begin{pmatrix} -1 & 2 & -1 & -1 \\ -1 & -1 & 2 & -1 \\ -1 & -1 & -1 & 5 \end{pmatrix}.
$$

We denote the unique interior lattice point by $o = (0, 0, 0)$. We denote the facet at the bottom by $\beta = [pvw]$, the two vertical facets on the sides by $\sigma = [pva]$ and $\tau = [pwa]$, and the fourth facet by $\delta = [vwa]$. The polytope Δ is depicted on the left in Fig. 1.19.

Let $X = \mathbb{P}_\Delta$ be the toric variety associated with the normal fan of Δ. The open affine toric subvarieties $U(p)$ and $U(a)$ corresponding to the vertices p and a are

---

[18] They work with the toric variety associated with the *spanning fan* of Δ rather than the normal fan, which is just $\mathbb{P}_{\Delta°}$.

both isomorphic to $\mathbb{A}^3$ while the open affine toric subvarieties $U(v)$ and $U(w)$ are both isomorphic to $\operatorname{Spec}\mathbb{C}[x, y, z, w]/(xy - w^2)$. In particular, the singular locus of X is the toric stratum $V(\omega) \cong \mathbb{P}^1$ for the edge $\omega = [vw]$.

The dual reflexive polytope $\Delta^\circ \subseteq \mathbb{R}^3$ has vertices

$$\left(\tau^\circ\ \sigma^\circ\ \beta^\circ\ \delta^\circ\right) = \begin{pmatrix} 1 & 0 & 0 & -2 \\ 0 & 1 & 0 & -2 \\ 0 & 0 & 1 & -1 \end{pmatrix}.$$

This is the third entry in the Kreuzer–Skarke list of 4319 reflexive polytopes in dimension 3. The only lattice point which is neither in the interior nor a vertex is in the edge ω°. The two facets p° and a° are standard simplices while the two facets v° and w° are A_1-triangles in the sense of [54, Defn. 1.2]. Thus, they admit a trivial *admissible Minkowski decomposition* in the sense of [54, Defn. 1.4]. One may check that they can be completed to an admissible Minkowski decomposition datum in the sense of [54, Defn. 1.10]. Now [54, Thm. 1.17] guarantees the existence of a smoothing of $X = \mathbb{P}_\Delta$. $\Diamond$

The following alternative strategy has been communicated to the author by Corti and Ruddat. Instead of smoothing $X(\Delta) := \mathbb{P}_\Delta$ directly, one may degenerate $X(\Delta)$ into a far more singular space $X_0(\tilde{\Delta})$ and then try to apply logarithmic deformation theory to construct a smoothing of $X_0(\tilde{\Delta})$. Although we now have a more direct smoothing in dimension 3 by the result of Corti–Hacking–Petracci, this approach is still worthwhile with a view toward higher dimensions.

We construct $X_0(\tilde{\Delta})$. Let $\Delta \subseteq M_\mathbb{R}$ be a reflexive polytope. After taking the central subdivision $\mathscr{P}$, we have a strictly convex piecewise linear function $\varphi \colon \Delta \to \mathbb{R}$ with $\varphi(0) = 0$ and $\varphi(v) = 1$ for every vertex $v \in \Delta$. Let

$$\tilde{\Delta} := \{(m, h) \in M_\mathbb{R} \oplus \mathbb{R} \mid m \in \Delta, h \geq \varphi(m)\}.$$

This is an unbounded polyhedron. The toric variety $X(\tilde{\Delta})$ associated with the normal fan of $\tilde{\Delta}$ admits a toric morphism $f \colon X(\tilde{\Delta}) \to \mathbb{A}^1_t$ induced by the inclusion $\mathbb{Z} \to M \oplus \mathbb{Z}$ of the final summand. Over $\{t \neq 0\}$, the family is isomorphic to

$$X(\Delta) \times \{t \neq 0\} \to \{t \neq 0\}$$

while the central fiber is the colimit

$$X_0(\tilde{\Delta}) := \operatorname{colim}_{\sigma \in \mathscr{P}} V(\sigma)$$

over the toric varieties $V(\sigma)$ associated with the normal fans of the cells $\sigma \in \mathscr{P}$ of the central subdivision of Δ. The family $f \colon X(\tilde{\Delta}) \to \mathbb{A}^1_t$ is called the *Mumford degeneration*. By construction, $f \colon (X(\tilde{\Delta})|X_0(\tilde{\Delta})) \to (\mathbb{A}^1_t|\{0\})$ is log toroidal once we have turned it into a generically log smooth family by choosing the log singular locus.

Example 1.155 We continue Example 1.154. The polyhedron $\tilde{\Delta} \subseteq \mathbb{R}^3 \oplus \mathbb{R}$ has five vertices, which we denote by $\tilde{p}, \tilde{v}, \tilde{w}, \tilde{a}$, and $\tilde{o} = (0, 0, 0, 0)$. The toric variety $X(\tilde{\Delta})$ has an open cover by the five corresponding affine toric varieties. The open subset $U(\tilde{o})$ is the most complicated. It is given by

$$\operatorname{Spec} \mathbb{C}[C(\tilde{\Delta}) \cap \mathbb{Z}^4] = \operatorname{Spec} \mathbb{C}[C(\tilde{\Delta}^\circ)^\vee \cap \mathbb{Z}^4],$$

where $C(\tilde{\Delta})$ is the cone over $\tilde{\Delta} \times \{1\} \subset \mathbb{R}^4$ as in Construction 1.136. While $C(\tilde{\Delta}) \cap \mathbb{Z}^4$ has 30 generators and is thus not very amenable to study it by generators and relations, the given form of it shows that $(U(\tilde{o})|\{t = 0\}) \to (\mathbb{A}_t^1|\{0\})$ is log smooth, vertical, and saturated.

A Hilbert basis of the cone $E_{\tilde{p}} \tilde{\Delta} = \mathbb{R}_{\geq 0} \cdot (\tilde{\Delta} - \tilde{p})$ of $\tilde{\Delta}$ at $\tilde{p}$ is given by

$$x_p \,\hat{=}\, (1, 0, 0, 0), \quad y_p \,\hat{=}\, (0, 1, 0, 0), \quad z_p \,\hat{=}\, (0, 0, 1, 0),$$
$$t_p \,\hat{=}\, (0, 0, 0, 1), \quad u_p \,\hat{=}\, (1, 1, 1, -1),$$

which induces an isomorphism

$$U(\tilde{p}) \cong \operatorname{Spec} \mathbb{C}[x_p, y_p, z_p, u_p, t_p]/(x_p y_p z_p - t_p u_p).$$

A Hilbert basis of the cone $E_{\tilde{a}} \tilde{\Delta}$ of $\tilde{\Delta}$ at $\tilde{a}$ is given by

$$x_a \,\hat{=}\, (1, 0, -2, 0), \quad y_a \,\hat{=}\, (0, 1, -2, 0), \quad z_a \,\hat{=}\, (0, 0, -1, 0),$$
$$t_a \,\hat{=}\, (0, 0, 0, 1), \quad u_a \,\hat{=}\, (1, 1, -5, -1),$$

which induces an isomorphism

$$U(\tilde{a}) \cong \operatorname{Spec} \mathbb{C}[x_a, y_a, z_a, u_a, t_a]/(x_a y_a z_a - t_a u_a).$$

Thus, on the open subsets $U(\tilde{p})$ and $U(\tilde{a})$, the family is given by $xyz = tu$, a local model in the class $\mathscr{C}^{\mathrm{anc}}$.

A Hilbert basis of the cone $E_{\tilde{v}} \tilde{\Delta}$ of $\tilde{\Delta}$ at $\tilde{v}$ is given by

$$x_v \,\hat{=}\, (-1, 0, 0, 0), \quad y_v \,\hat{=}\, (-1, 0, 2, 0), \quad w_v \,\hat{=}\, (-1, 0, 1, 0),$$
$$z_v \,\hat{=}\, (-1, 1, 0, 0), \quad u_v \,\hat{=}\, (-2, 1, 1, -1), \quad t_v \,\hat{=}\, (0, 0, 0, 1),$$

which induces an isomorphism

$$U(\tilde{v}) \cong \operatorname{Spec} \mathbb{C}[x_v, y_v, z_v, w_v, u_v, t_v]/(x_v y_v - w_v^2, \; z_v w_w - t_v u_v).$$

A Hilbert basis of the cone $E_{\tilde{w}}\tilde{\Delta}$ of $\tilde{\Delta}$ at $\tilde{w}$ is given by

$$x_w \,\hat{=}\, (0, -1, 0, 0), \;\; y_w \,\hat{=}\, (0, -1, 2, 0), \;\; w_w \,\hat{=}\, (0, -1, 1, 0),$$
$$z_w \,\hat{=}\, (1, -1, 0, 0), \;\; u_w \,\hat{=}\, (1, -2, 1, -1), \;\; t_w \,\hat{=}\, (0, 0, 0, 1),$$

which induces an isomorphism

$$U(\tilde{w}) \cong \operatorname{Spec} \mathbb{C}[x_w, y_w, z_w, w_w, u_w, t_w]/(x_w y_w - w_w^2, \; z_w w_w - t_w u_w).$$

In particular, on $U(\tilde{v})$ and $U(\tilde{w})$, the family is isomorphic to the one given by

$$\operatorname{Spec} \mathbb{C}[x, y, z, w, u, t]/(xy - w^2, zw - tu) \to \mathbb{A}^1_t.$$

We have determined the explicit descriptions with a computation in Macaulay2, using the `Polyhedra` package.

For later use, we compute the coordinate transformation between $U(\tilde{v})$ and $U(\tilde{w})$. Let $\tilde{\omega} = [\tilde{v}\tilde{w}]$ be the edge in $\tilde{\Delta}$ connecting $\tilde{v}$ with $\tilde{w}$; we have $U(\tilde{v})\cap U(\tilde{w}) = U(\tilde{\omega})$. In coordinates, we have $U(\tilde{\omega}) = \{z_v \neq 0\} \subseteq U(\tilde{v})$ and $U(\tilde{\omega}) = \{z_w \neq 0\} \subseteq U(\tilde{w})$. The transformation is given by

$$\mathbb{C}[x_v, y_v, z_v, w_v, u_v, t_v]_{z_v}/(x_v y_v - w_v^2, \; z_v w_v - t_v u_v)$$
$$\to \mathbb{C}[x_w, y_w, z_w, w_w, u_w, t_w]_{z_w}/(x_w y_w - w_w^2, \; z_w w_w - t_w u_w),$$
$$x_v \mapsto x_w/z_w, \quad y_v \mapsto y_w/z_w, \quad w_v \mapsto w_w/z_w,$$
$$z_v \mapsto 1/z_w, \quad u_v \mapsto u_w/z_w^3, \quad t_v \mapsto t_w.$$

We describe the toric divisors in $X(\tilde{\Delta})$. For each facet β, δ, σ, τ, we obtain an unbounded facet

$$\beta^+ \cong \beta \times [0, \infty), \quad \delta^+ \cong \delta \times [0, \infty), \quad \sigma^+ \cong \sigma \times [0, \infty), \quad \tau^+ \cong \tau \times [0, \infty)$$

of $\tilde{\Delta}$. One may check that the *horizontal part* $H(\Delta) = V(\beta^+) \cup V(\delta^+) \cup V(\sigma^+) \cup V(\tau^+)$ of the toric boundary is given by $u = 0$ on $U(\tilde{p})$, $U(\tilde{a})$, $U(\tilde{v})$, and $U(\tilde{w})$.

Each face ρ of Δ gives rise to a bounded face ρ^Δ of $\tilde{\Delta}$ which can also be identified with the pyramid over ρ as a cell in the central subdivision $\mathscr{P}$ of Δ. Besides β^+, δ^+, σ^+, τ^+, the remaining facets of $\tilde{\Delta}$ are precisely β^Δ, δ^Δ, σ^Δ, and τ^Δ. The corresponding toric divisors are precisely the irreducible components of the central fiber $X_0(\tilde{\Delta})$.

We describe the central fiber $X_0(\tilde{\Delta})$ in more detail. The set $\mathscr{P}^{[3]}$ of maximal cells of the central subdivision $\mathscr{P}$ of Δ is given by the pyramids β^Δ, δ^Δ, σ^Δ, and τ^Δ. The set $\mathscr{P}^{[2]}$ of cells of dimension 2 is given by the six pyramids over the edges and the four facets β, δ, σ, τ of Δ. The set $\mathscr{P}^{[1]}$ of cells of dimension 1 is given by the four pyramids $p^\triangle$, $a^\triangle$, $v^\triangle$, $w^\triangle$ and by the six edges of $\triangle$. The set $\mathscr{P}^{[0]}$ of cells of dimension 0 consists of p, a, v, w, and the origin o.

Any two of the four irreducible components $V(\beta^\triangle)$, $V(\delta^\triangle)$, $V(\sigma^\triangle)$, $V(\tau^\triangle)$ of $X_0(\tilde{\triangle})$ meet in the toric stratum determined by the pyramid over their common edge. This is simultaneously the singular and the non-normal locus of $X_0(\tilde{\triangle})$. The four two-dimensional toric strata $V(\beta)$, $V(\delta)$, $V(\sigma)$, and $V(\tau)$ do not lie in the non-normal locus but are toric divisors in $V(\beta^\triangle)$, $V(\delta^\triangle)$, $V(\sigma^\triangle)$, respectively $V(\tau^\triangle)$. We write

$$\partial X_0(\tilde{\triangle}) := V(\beta) \cup V(\delta) \cup V(\sigma) \cup V(\tau)$$

for their union, and we have $\partial X_0(\tilde{\triangle}) = X_0(\tilde{\triangle}) \cap H(\triangle)$. In particular, in each of the four affine charts, $\partial X_0(\tilde{\triangle})$ is given by $u = 0$.

Any three irreducible components meet in a line $V(p^\triangle)$, $V(a^\triangle)$, $V(v^\triangle)$, respectively $V(w^\triangle)$. Each of them is isomorphic to $\mathbb{P}^1$. All four irreducible components meet in a single point $V(o)$. This is the only point of $X(\tilde{\triangle})$ which is not contained in any of the four simpler charts $U(\tilde{p})$, $U(\tilde{a})$, $U(\tilde{v})$, $U(\tilde{w})$. $\lozenge$

The central fiber $V = X_0(\tilde{\triangle})$ admits a natural structure of a toroidal crossing scheme. This is the Fano toroidal crossing scheme which we wish to smooth.

We first analyze the log geometry of the Mumford degeneration. Let us compute the log singular locus Z of the log morphism $f : (X(\tilde{\triangle})|X_0(\tilde{\triangle})) \to (\mathbb{A}^1_t|\{0\})$. Over the open subset $\{t \neq 0\}$, the family is trivial with fiber $X(\triangle)$, so we have

$$Z \cap \{t \neq 0\} = \mathrm{Sing}(X(\triangle)) \times \{t \neq 0\}.$$

We have already seen that the family is log smooth with smooth nearby fiber on $U(\tilde{o})$, so we have $Z \cap U(\tilde{o}) = \varnothing$. We have

$$X_0(\triangle) \cap U(\tilde{o}) = X_0(\triangle) \setminus \partial X_0(\triangle)$$

so that $Z \cap X_0(\tilde{\triangle}) \subset \partial X_0(\triangle)$. Let us now investigate the family on $U(\tilde{p})$ and $U(\tilde{a})$, where it is given by $xyz = tu$.

Example 1.156 (xyz = tu) We consider the log morphism

$$f : \quad (X|X_0) = \big(\mathrm{Spec}\,\mathbb{C}[x, y, z, u, t]/(xyz - tu)\big|\{t = 0\}\big) \to \big(\mathbb{A}^1_t\big|\{0\}\big).$$

It is illustrated in Fig. 1.20. The general fiber is smooth while the central fiber $X_0 = \mathrm{Spec}\,\mathbb{C}[x, y, z, u]/(xyz)$ is a normal crossing scheme. Let Z be the log singular locus; in particular, $Z \subset X_0$. On the regular locus $\mathrm{Reg}(X)$, the family is semistable so that we have $Z \subset \mathrm{Sing}(X)$. Now we have

$$\mathrm{Sing}(X) = \{t = 0, u = 0, xy = 0, yz = 0, xz = 0\} = \{u = 0\} \cap \mathrm{Sing}(X_0).$$

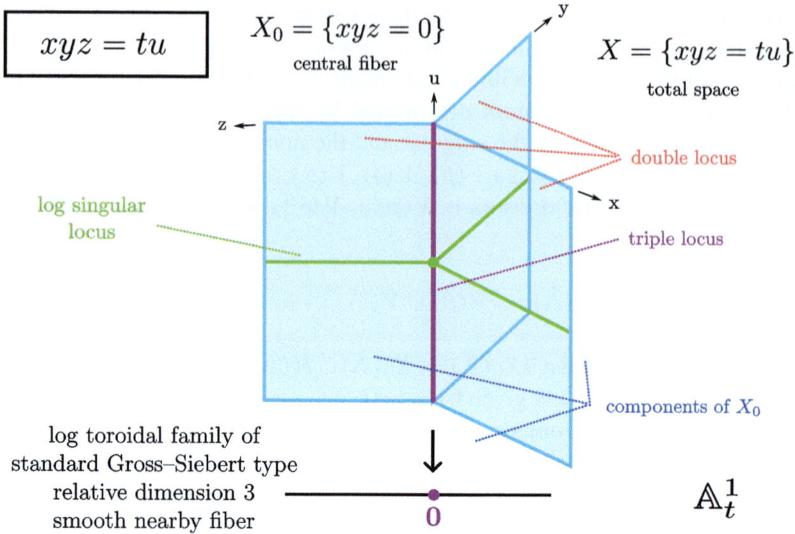

Fig. 1.20 Example 1.156. We see the double locus of the central fiber X_0 of the family given by $xyz = tu$ in blue. The three irreducible components of X_0 are omitted from the figure. The log singular locus has three branches, which meet in a single point inside the triple locus. © Simon Felten 2025. All rights reserved

This locus consists of three branches, one in each irreducible component of X_0, which meet in a single point $\{0\} = \{x = y = z = u = t = 0\}$ inside the triple locus $\mathbb{A}_u^1$. Around every $p \in \mathrm{Sing}(X) \setminus \{0\}$, the family is given by the local model $xy = tu$, i.e., by Example 1.79, so that it is actually log singular in these points. But then the family cannot be log smooth around $0 \in \mathrm{Sing}(X)$, and we find $Z = \mathrm{Sing}(X)$.

In Example 8.24, we will construct a monoid which exhibits the whole family as log toroidal, and in Sect. 18.8, we will see that the family is log toroidal of standard Gross–Siebert type.   ◇

Let $D_0(\tilde{\Delta}) \subset X_0(\tilde{\Delta})$ be the non-normal locus. The example shows that

$$Z \cap U(\tilde{p}) = \partial X_0(\tilde{\Delta}) \cap D_0(\tilde{\Delta}) \cap U(\tilde{p}),$$

and analogously for $U(\tilde{a})$. We now turn to the open subsets $U(\tilde{v})$ and $U(\tilde{w})$.

Example 1.157 ($xy = w^2$, $zw = tu$) We consider

$$f\colon \ (X|X_0) = \big(\mathrm{Spec}\,\mathbb{C}[x, y, z, w, u, t]/(xy - w^2, zw - tu)\big|\{t = 0\}\big) \to \big(\mathbb{A}_t^1\big|\{0\}\big),$$

which is illustrated in Fig. 1.21. The general fiber is a singular Gorenstein toric variety $\mathrm{Spec}\,\mathbb{C}[x, y, z, w]/(xy - w^2)$, i.e., a cA_1-singularity. The central fiber is

$$X_0 = \mathrm{Spec}\,\mathbb{C}[x, y, z, w, u]/(xy - w^2, zw),$$

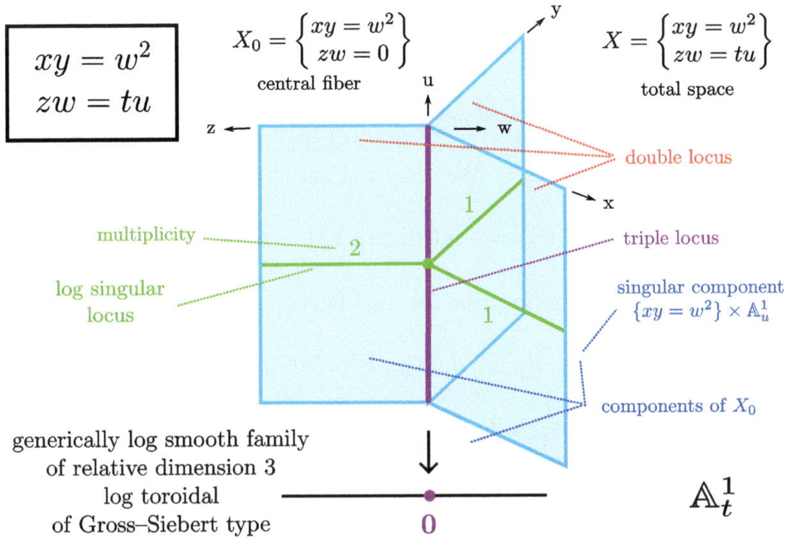

Fig. 1.21 Example 1.157. © Simon Felten 2025. All rights reserved

i.e., the product of the central fiber in Example 1.66 with $\mathbb{A}^1_u$. In particular, X_0 consists of three irreducible components of dimension 3, any two of them intersecting in a stratum of dimension 2, and all three intersecting in a stratum of dimension 1. The non-normal locus of X_0 coincides with $\mathrm{Sing}(X_0)$.

Let $Z_0 \subset X_0$ be the log singular locus inside the central fiber. On $\{u \neq 0\}$, the family is equivalent to the one given by $xy = w^2$ and $zw = t$. This is Example 1.66 and hence log smooth. In a neighborhood of the regular locus $\mathrm{Reg}(X_0)$, the family is smooth and hence log smooth since it is strict. In particular, we have

$$Z_0 \subset \{u = 0\} \cap \mathrm{Sing}(X_0).$$

The right-hand locus consists of three one-dimensional branches, one in each irreducible component of the non-normal locus of X_0. They meet in a single point $\{0\}$ inside the triple locus $\mathbb{A}^1_u$ of X_0.

Around a point $p \in \{u = 0\} \cap \mathrm{Sing}(X_0)$ inside one of the components $V_x = \mathbb{A}^1_{u,x}$ and $V_y = \mathbb{A}^1_{u,y}$ (but $p \neq 0$), the local model is given by $xy = tu$, i.e., by Example 1.79, so that p is indeed a log singular point in Z_0. Around a point $p \in \{u = 0\} \cap \mathrm{Sing}(X_0)$ which lies in $V_z = \mathbb{A}^1_{z,u}$ (but $p \neq 0$), the local model is given by $xy = t^2u^2$. In this local model, one may check explicitly that $\mathcal{W}^1$ is not locally free at $x = y = u = t = 0$ so that p is indeed a log singular point. Now f cannot be log smooth around $p = 0$, and we find that $Z_0 = \{u = 0\} \cap \mathrm{Sing}(X_0)$ is the log singular locus (once restricted to X_0).

To the two branches of Z_0 in V_x and V_y, we assign the multiplicity 1 since the local model is $xy = tu^1$. To the branch of Z_0 in V_z, we assign the multiplicity

2 because the local model is $xy = t^2 u^2$, referring to the exponent of u. The exponent 2 of t in this local model is related to the monoid P_2 of the ghost sheaf in Example 1.66.

This family is log toroidal of Gross–Siebert type. To see this, we follow the construction in Sect. 14.1 with $\tau \subseteq \mathbb{R}^2$ the so-called $(1, 1, 2)$-triangle given by the vertices $(0, 0)$, $(2, 0)$, and $(0, 1)$. We set $q = 1$ and $\Delta_1 = \tau$. Then we have

$$\check{\psi}_0(n_1, n_2) = -\inf\{0, 2n_1, n_2\} = \check{\psi}_1(n_1, n_2).$$

This function is piece-wise linear on the dual fan of τ. The monoid

$$P = \{(n_1, n_2, a_0, a_1) \mid a_i \geq \check{\psi}_i(n_1, n_2), \ i = 1, 2\}$$

gives rise to a family $f: \mathbb{A}_P = \operatorname{Spec} \mathbb{C}[P] \to \mathbb{A}^1$ by mapping $t \mapsto z^{(0,0,1,0)}$; this is a log toroidal family of Gross–Siebert type by definition. To see that $f: \mathbb{A}_P \to \mathbb{A}^1$ is isomorphic to $f: X \to \mathbb{A}^1$, note that

$$\bar{x} = (1, 0, 0, 0), \qquad \bar{y} = (-1, -2, 2, 2), \qquad \bar{z} = (0, 1, 0, 0),$$

$$\bar{w} = (0, -1, 1, 1), \qquad \bar{t} = (0, 0, 1, 0), \qquad \bar{u} = (0, 0, 0, 1)$$

are elements of P. They satisfy the relations $\bar{x} + \bar{y} = 2 \cdot \bar{w}$ and $\bar{z} + \bar{w} = \bar{t} + \bar{u}$; thus, we obtain a ring homomorphism

$$\mathbb{C}[x, y, z, w, u, t]/(xy - w^2, zw - tu) \to \mathbb{C}[P]$$

in the obvious way. A slightly cumbersome computation shows that the map is surjective; then $\mathbb{A}_P \subset X$ is a closed subscheme, and both are integral of the same dimension, i.e., $\mathbb{A}_P \cong X$. Both are endowed with the compactifying log structure coming from the central fiber, so they are isomorphic as log schemes as well. However, this family is not log toroidal of *elementary* Gross–Siebert type because the nearby fiber has singularities in codimension 2, which is not possible in a log toroidal family of elementary Gross–Siebert type by Lemma 14.9. $\Diamond$

In particular, we have

$$Z \cap X_0(\tilde{\Delta}) \cap U(\tilde{v}) = \partial X_0(\tilde{\Delta}) \cap D_0(\tilde{\Delta}) \cap U(\tilde{v}),$$

and analogously for $U(\tilde{w})$. In total, we find

$$Z = \left(\partial X_0(\tilde{\Delta}) \cap D_0(\tilde{\Delta})\right) \cup (\operatorname{Sing}(X(\Delta)) \times \{t \neq 0\}).$$

The strategy to find a smoothing rather than the already given normaling of $V = X_0(\tilde{\Delta})$ is to deform the log singular locus Z and the section of $\mathcal{LS}_V$, and to provide a system of deformations $\mathscr{D}$ whose local models are formal smoothings. Let us first

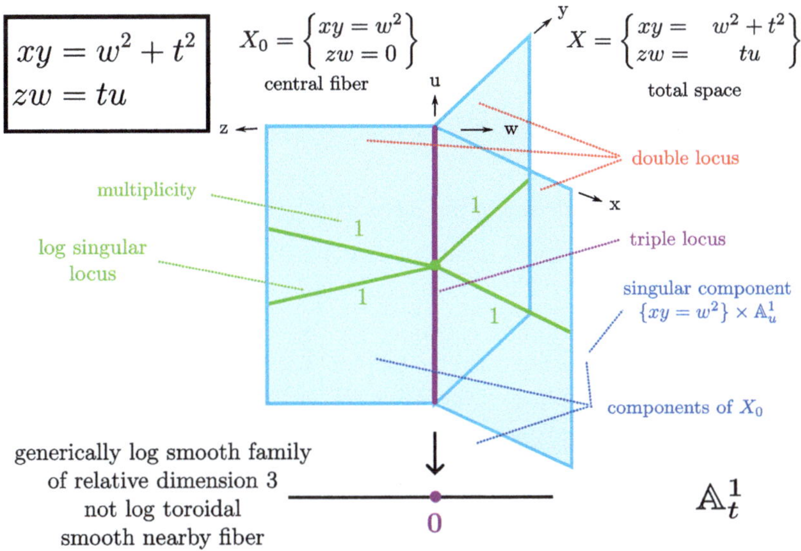

Fig. 1.22 Example 1.158. © Simon Felten 2025. All rights reserved

see that this is possible locally. The following log singularity does not have the base change property; hence it is necessary to work with enhanced generically log smooth families if we wish to allow it as a local model in our deformation problem.

Example 1.158 ($xy = w^2 + t^2$, $zw = tu$) We perturb the equations $xy = w^2$, $zw = tu$ of Example 1.157 and consider

$$f \colon (X|X_0) = \big(\operatorname{Spec} \mathbb{C}[x, y, z, w, u, t]/(xy - w^2 - t^2, \, zw - tu) \big| \{t = 0\} \big)$$
$$\to (\mathbb{A}^1_t | \{0\}),$$

which is depicted in Fig. 1.22. This is a generically log smooth family of relative dimension 3. The double line in the log singular locus of Example 1.157 has been separated into two single lines $u = iz$ and $u = -iz$ such that the log singular locus Z_0 now has four branches. The general fiber is smooth.

The family f is log toroidal with local models $xy = tz$ and $xy = t^2z$ outside $\{0\}$, but it is *not* log toroidal around the point $\{0\}$, and in particular not log toroidal of Gross–Siebert type. Namely, it does not satisfy the base change property of Definition 1.78, which is satisfied by all log toroidal families over $\mathbb{C}$. More precisely, when we denote the reflexive log differential forms of $f \colon (X|X_0) \to (\mathbb{A}^1|\{0\})$ by $\mathcal{W}^1_{X/\mathbb{A}^1}$, then $\mathcal{W}^1_{X/\mathbb{A}^1} \otimes \mathcal{O}_{X_0}$ is not reflexive on X_0. This can be verified with an explicit computation in Macaulay2. Namely, the sheaf of relative log derivations is isomorphic to the sheaf of relative classical derivations $\mathcal{T}_{X/\mathbb{A}^1}$, so $\mathcal{W}^1_{X/\mathbb{A}^1}$ can be computed as the dual of $\mathcal{T}_{X/\mathbb{A}^1}$. After pulling back to X_0, the canonical map

$$\mathcal{W}^1_{X/\mathbb{A}^1} \otimes \mathcal{O}_{X_0} \to (\mathcal{W}^1_{X/\mathbb{A}^1} \otimes \mathcal{O}_{X_0})^{\vee\vee}$$

to the bidual is not surjective. On the other hand, $\mathcal{W}^2_{X/\mathbb{A}^1}$ remains reflexive after pull-back to X_0. For the similar family in Example 10.5, also the pull-back of $\mathcal{W}^2$ is not reflexive. $\qquad\qquad\qquad\qquad\qquad\qquad\qquad\qquad\qquad\qquad\qquad\qquad\quad \Diamond$

When we take the perturbed equations $xy = w^2 + t^2$, $zw = tu$ on both $U(\tilde{v})$ and $U(\tilde{w})$, then the log singular loci on $X_0(\tilde{\Delta})$ are not compatible under the transition map that we have given explicitly in Example 1.155. However, they become compatible once we take yet another modification of the equations.

Example 1.159 ($xy = w^2 + t^{2z}$, $zw = tu$) We consider

$$f\colon \ (X|X_0) = \big(\operatorname{Spec} \mathbb{C}[x, y, z, w, u, t]/(xy - w^2 - t^2z, \ zw - tu)\big|\{t = 0\}\big)$$
$$\to \big(\mathbb{A}^1_t\big|\{0\}\big),$$

a generically log smooth family of relative dimension 3. For an illustration, see Fig. 1.23. The general fiber is smooth. The log singular locus Z_0 has three branches given by

$$Z_x = \{t = y = w = z = u = 0\} = \mathbb{A}^1_x,$$
$$Z_y = \{t = x = w = z = u = 0\} = \mathbb{A}^1_y,$$
$$Z_z = \{t = x = y = w = 0, \ u^2 + z^3 = 0\}.$$

They meet in a single point 0 inside the triple locus $\mathbb{A}^1_u$. Around points $p \in Z_x \setminus \{0\}$ and $p \in Z_y \setminus \{0\}$, the local model is $xy = tu$. The branch Z_z is a cuspidal cubic with cusp in 0. Around points $p \in Z_z \setminus \{0\}$, the local model is $xy = t^2u$.

With the same method as in Example 1.158, one may check that neither

$$\mathcal{W}^1_{X/\mathbb{A}^1_t} \otimes \mathcal{O}_{X_0} \quad \text{nor} \quad \mathcal{W}^2_{X/\mathbb{A}^1_t} \otimes \mathcal{O}_{X_0}$$

are reflexive so that f does not have the base change property. $\qquad\qquad\qquad \Diamond$

With this preparation, we can build a new structure of an enhanced generically log smooth family on $X_0(\tilde{\Delta})$ together with an enhanced system of deformations $\mathscr{D}$.

Example 1.160 First, we construct the structure of an enhanced generically log smooth family. It is depicted on the right-hand side of Fig. 1.24. The left-hand side shows the central fiber of the Mumford degeneration for comparison.

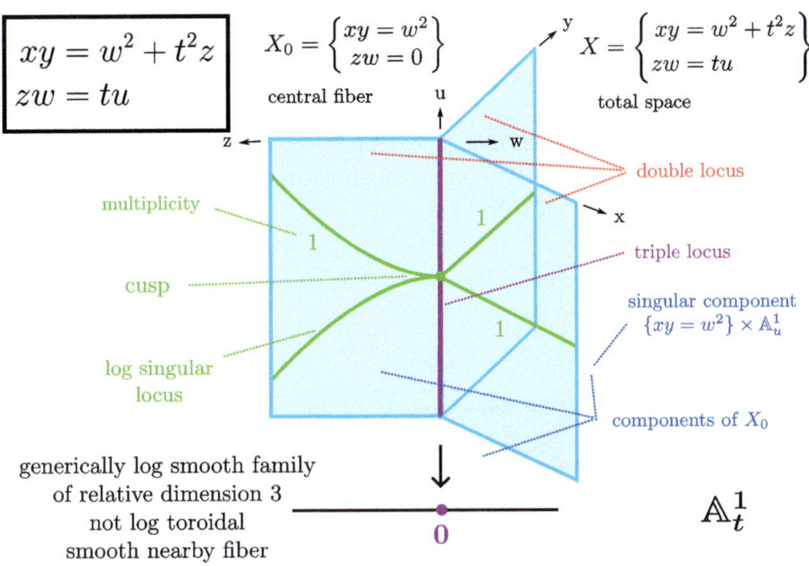

Fig. 1.23 Example 1.159. © Simon Felten 2025. All rights reserved

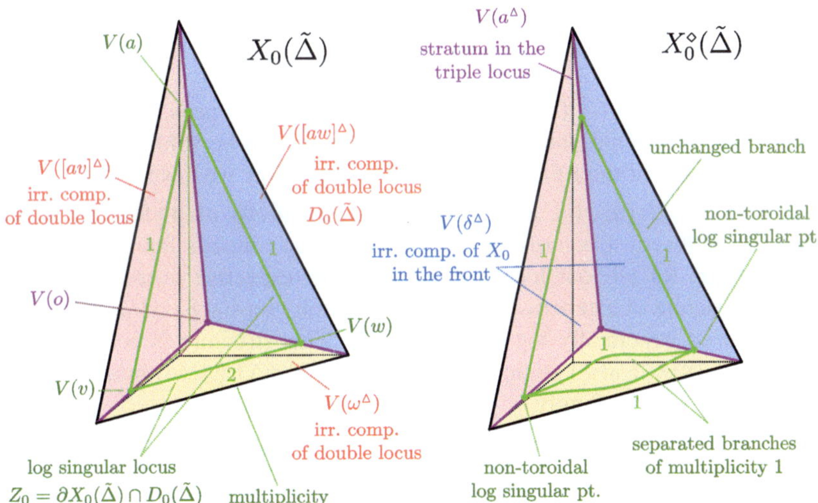

Fig. 1.24 The two log smooth structures on $X_0(\tilde{\Delta})$. Both sides show the scheme $X_0(\tilde{\Delta})$. We see in red, blue, and yellow three of the six irreducible components of the double locus $D_0(\tilde{\Delta})$. The irreducible component $V(\delta^\Delta)$ of $X_0(\tilde{\Delta})$ is in the front while the remaining three irreducible components are in the back. On the left-hand side, we have the generically log smooth structure from the Mumford degeneration. Its log singular locus is precisely the intersection of $D_0(\tilde{\Delta})$ with the toric boundary $\partial X_0(\tilde{\Delta})$. On the right-hand side, we have the enhanced generically log smooth structure constructed in Example 1.160. Here, the branch of Z_0 with multiplicity two (where the local model is $xy = tz^2$) has been separated into two branches of multiplicity one (with local model $xy = tz$). This introduces two non-toroidal log singular points where we need the enhanced structure because the local model does not have the base change property. © Simon Felten 2025. All rights reserved

Recall that

$$X_0(\tilde{\Delta}) = U_0(\tilde{o}) \cup U_0(\tilde{a}) \cup U_0(\tilde{p}) \cup U_0(\tilde{v}) \cup U_0(\tilde{w})$$

as a scheme, where $U_0(\tilde{u}) := U(\tilde{u}) \cap X_0(\tilde{\Delta})$. When we omit $U_0(\tilde{o})$, only a single point $V(o) = \{0\}$ is missing. We ignore this point for simplicity and construct the structures only on $X_0(\tilde{\Delta}) \setminus V(o)$.[19] Let

$$U^{\Diamond}(\tilde{v}) := \operatorname{Spec} \mathbb{C}[x_v, y_v, z_v, w_v, u_v, t_v]/(x_v y_v - w_v^2 - t_v^2 z_v, \; z_v w_v - t_v u_v)$$

and

$$U^{\Diamond}(\tilde{w}) := \operatorname{Spec} \mathbb{C}[x_w, y_w, z_w, w_w, u_w, t_w]/(x_w y_w - w_w^2 - t_w^2 z_w, \; z_w w_w - t_w u_w).$$

On the level of underlying schemes, the central fibers are $U_0^{\Diamond}(\tilde{v}) = U_0(\tilde{v})$ and $U_0^{\Diamond}(\tilde{w}) = U_0(\tilde{w})$. After localizing in z, both local models become isomorphic via

$$x_v \mapsto x_w/z_w, \quad y_v \mapsto y_w/z_w, \quad w_v \mapsto w_w/z_w,$$

$$z_v \mapsto 1/z_w, \quad u_v \mapsto u_w/z_w^3, \quad t_v \mapsto t_w,$$

as above. In particular, they define the same structure of an enhanced generically log smooth family on their overlap $U_0(\tilde{v}) \cap U_0(\tilde{w})$.

Next, let us set $U^{\Diamond}(\tilde{p}) := U(\tilde{p})$ and $U^{\Diamond}(\tilde{a}) := U(\tilde{a})$. Obviously, we have $U_0^{\Diamond}(\tilde{p}) = U_0(\tilde{p})$ and $U_0^{\Diamond}(\tilde{a}) = U_0(\tilde{a})$. Since, on these opens, $V = X_0(\tilde{\Delta})$ is a normal crossing scheme, the induced log structures can be studied as sections of $\mathcal{T}_V^1$, i.e., by means of their first-order deformations. Now we have $U_1^{\Diamond}(\tilde{v}) = U_1(\tilde{v})$ and $U_1^{\Diamond}(\tilde{w}) = U_1(\tilde{w})$ for the first-order deformations of underlying schemes so that the induced structures of a generically log smooth family coincide on $U_0(\tilde{p}) \cup U_0(\tilde{a})$ for all four local models on their respective overlaps. This gives rise to the structure of a generically log smooth family on $U_0(\tilde{p}) \cup U_0(\tilde{a})$, and hence of an enhanced generically log smooth family. This structure is compatible with the one on $U_0(\tilde{v}) \cup U_0(\tilde{w})$ so that we have turned $X_0(\tilde{\Delta}) \setminus V(o)$ into an enhanced generically log smooth family.

The system of deformations $\mathscr{D}$ is constructed as follows. As local models, we take $U_A^{\Diamond}(\tilde{u}) := U^{\Diamond}(\tilde{u}) \times_{\mathbb{A}_t^1} \operatorname{Spec} A$ for $u \in \{p, a, v, w\}$. Outside the two-point set $V(v) \cup V(w)$, the family and the local models are log toroidal of standard Gross–Siebert type. Thus, the local models are compatible. Now $V(v)$ is only contained in $U_0(\tilde{v})$, and $V(w)$ is only contained in $U_0(\tilde{w})$. Thus, we have no compatibility condition for the local models in these points. $\Diamond$

---

[19] The author believes that the log smooth structure can be extended log smoothly across $V(o)$ but did not do the necessary computations.

The example shows a second strategy to construct an (enhanced) system of deformations: We choose the open cover $\mathcal{V} = \{V_\alpha\}_\alpha$ of X_0 in such a way that every point around which we do not know local rigidity is contained only in a single patch V_α.

One may now try to modify the log structure on $V = X_0(\tilde{\Delta})$ with a log regular section s_0 of $\omega^\vee_{V/S_0}$ to obtain a log Calabi–Yau enhanced generically log smooth family. If the curved two-sided Gerstenhaber calculus associated with the modified system of deformations $\mathscr{D}(s_0)$ is quasi-perfect, then deformations are unobstructed.

Computational work of Matthias Zach and the author that will be reported elsewhere does not directly refute quasi-perfectness in this example, but the results also do not support the hypothesis. Namely, they disprove quasi-perfectness in related but less relevant examples.

1.13 Outlook

We conclude the introduction with some related problems of varying difficulty and importance that have remained unaddressed in this monograph.

Deformations of Vector Bundles We show that deformations of $f_0\colon (X_0, \mathcal{L}_0) \to S_0$ are unobstructed when $\mathcal{L}_0$ is a line bundle, and $f_0\colon X_0 \to S_0$ is log Calabi–Yau. By repeating the argument, we expect that simultaneous deformations of a finite collection $(\mathcal{L}_{0;1}, \ldots, \mathcal{L}_{0;m})$ of line bundles are unobstructed as well. It is also open to consider deformations of vector bundles of rank $r \geq 2$, or coherent sheaves, or simultaneous deformations of the category of coherent sheaves, including their morphisms.

Construction and Characterization of Log Regular Sections As mentioned in the main body, it is open if every log regular section $s_0 \in \mathcal{L}_0$ is also log transversal. If not, it is also open to show that the modification of a log toroidal family in a log regular section is always log toroidal. Furthermore, it is largely open to give criteria for the existence of global log regular sections, but cf. the results in [77] on the construction of log Calabi–Yau log structures on certain normal crossing spaces, a version of which we discuss in Theorem 18.44.

Propagation of Perfectness to Modifications and $\mathbb{P}^1$-Bundles Assume that $f_0\colon X_0 \to S_0$ is a proper enhanced generically log smooth family with perfect curved Gerstenhaber calculus. Then it is open if, given a line bundle $\mathcal{L}_0$, the family $P_0(\mathcal{L}_0) \to S_0$ constructed in Chap. 16 has perfect curved Gerstenhaber calculus. It is also open if, given a log regular section s_0 in some line bundle $\mathcal{L}_0$, the modification $X_0(s_0) \to S_0$ constructed in Chap. 18 has perfect curved Gerstenhaber calculus. Although it seems plausible to assume this, the perfectness of the new curved Gerstenhaber calculus does not appear to be an easy formal consequence in these settings.

Homotopy Theory of Gerstenhaber Calculi Although our construction of the characteristic curved Gerstenhaber calculus depends on the choice of two coverings, it is reasonable to expect that the resulting curved Gerstenhaber algebra is independent of these choices up to homotopy. Some results are already known in this direction, see [38, §3.3.2]. In the non-logarithmic case, where a homotopy theory is more straightforward due to the absence of true predifferentials, this seems to be classical, at least for dg Lie algebras/L_∞-algebras. Note also that there has been recent progress on the homotopy theory of curved Lie algebras [24, 194, 206].

Semi-Derived Deformations In practice, it might be difficult to construct a system of deformations $\mathscr{D}$ for a generically log smooth family $f_0\colon X_0 \to S_0$. One idea is to use local resolutions of singularities, and then take a (derived) direct image of the Gerstenhaber calculi on the resolutions. It is reasonable to expect that the resolutions do not glue, and that the derived direct images agree only up to homotopy. A variant of the theory should be developed that works in this setting. Here, we may start directly with curved Gerstenhaber calculi. While there may not be any well-defined actual Gerstenhaber calculus on such a deformation, it is reasonable to expect that there is a well-defined underlying deformation of schemes because many resolutions of log singularities satisfy $R\pi_*O_{\tilde{X}} = O_X$. It is then however unclear which invariants are preserved in such a deformation. It will probably be difficult to extract information about a smoothing from the general fiber in these settings. Using derived Gerstenhaber calculi (which are yet to be defined) may also rectify situations in which the curved Gerstenhaber calculus of a non-derived system of deformations is not perfect.

Integral Log Smooth Morphisms We assume throughout that our morphisms are saturated. There should be a similar theory for morphisms that are only integral. This situation is less rigid, so we have less information about the families. In the absence of non-coherent log singularities, one can take a base change which makes the map saturated, but this alters the underlying scheme, and then the behavior of our log singularities is not yet clear.

Local Deformation Theory In this monograph, we have studied extensively the passage from local deformation theory of log schemes to global deformation theory. For this, we assume that local deformations are fixed in a compatible way. However, some generically log smooth families admit many generically log smooth deformations while others admit none. It is an open problem to classify all generically log smooth log singularities over S_0 and their local deformations.

Chapter 2
Related Works

We discuss both historical and recent research developments related to the topic of this monograph.

Foundations of Logarithmic Geometry Logarithmic geometry was developed in the late 1980s by Fontaine and Illusie after structures with logarithmic poles had appeared in various contexts, most notably crystalline cohomology, the original context of logarithmic geometry. The first published account is by K. Kato in [162] in 1989.[1] A parallel theory, nowadays called Deligne–Faltings log structures, appears around the same time in [69] by Faltings. The first highlight of logarithmic geometry is Tsuji's celebrated proof of the C_{st}-conjecture in [271] in 1999, which contributed greatly to the popularity of the method.

In 2001, Olsson gives in [223] a new stack-theoretic interpretation of logarithmic geometry. Starting from there and from F. Kato's [161], log structures on stacks and stacks over the category of log schemes have been the interest of mostly recent research, see [1, 2, 43, 107, 127, 225, 258]. In 2018, Talpo and Vistoli developed in [265] yet another approach to logarithmic geometry via infinite root stacks. This approach is particularly suited to define a category of quasi-coherent sheaves on a log scheme which reflects the log geometry and not just the classical geometry of the underlying scheme.

Semistable Degenerations The simplest and best-known case of a log smooth family is a semistable family. Interest in semistable families seems to arise around

---

[1] According to private communication with Luc Illusie, he and Jean-Marc Fontaine conceived the concept of a log structure on the train from Paris to Oberwolfach in July 1988. There, Illusie quickly wrote four or five pages sketching the construction and handed them to Kazuya Kato who seemed interested. A few months later, Kato surprised them by presenting his preprint of [162], containing an elaborate theory of log structures with additional concepts like the chart of a log structure. While Illusie did not further pursue log structures in the meantime, Fontaine had started to work out the theory himself. He completely abandoned the project after he received Kato's preprint. Indeed, he never published on the subject of log geometry.

S. Felten, *Global Logarithmic Deformation Theory*, Lecture Notes
in Mathematics 2373, https://doi.org/10.1007/978-3-031-98751-9_2

1969 in the one-dimensional case with the Deligne–Mumford compactification $\overline{\mathcal{M}}_g$ of the stack of $\mathcal{M}_g$ of curves of genus g in [60]; this compactification is achieved precisely by adding curves with nodal singularities. In [174], it was proven in 1973 that any degeneration $f: X \to C$ of varieties of arbitrary dimension over a smooth curve C can be brought in semistable form after a finite base change; this is the well-known semistable reduction theorem.

For any (projective) holomorphic family $f: X^* \to \Delta^*$ over a small punctured disk, Schmid constructed in 1973 in [252] a limiting mixed Hodge structure on the cohomology $H^k(X_t, \mathbb{C})$ of the general fiber. If the family can be completed to a semistable degeneration $f: X \to \Delta$, the central fiber also carries a canonical mixed Hodge structure by work of Deligne. However, this mixed Hodge structure is not the same as Schmid's limiting mixed Hodge structure. In 1976, Steenbrink gave in [261] an interpretation of the limiting mixed Hodge structure in terms of the geometry of the semistable family $f: X \to \Delta$. This work already contains much of the later log differential calculus of semistable degenerations.

In [183], Kulikov classified in a 1977 article (in Russian) all possible central fibers of a semistable degeneration of K3 surfaces;[2] a more rigorous account was published by Persson and Pinkham in [228] in 1981. Two years later, Friedman showed in [87] that indeed all possible central fibers do occur in an analytic semistable degeneration of K3 surfaces. This is the first important smoothing result which turns out to be a special case of the unobstructedness of the logarithmic deformation functor in the Calabi–Yau case. Another important insight of [87] is that a normal crossing space V which occurs as the central fiber of a semistable degeneration must satisfy $\mathcal{T}_V^1 \cong O_{\mathrm{Sing}(V)}$, i.e., V is d-semistable. It was observed by Persson and Pinkham in the same year in [229] that d-semistability is not enough, i.e., that there are d-semistable normal crossing spaces which do not occur as the central fiber of a semistable degeneration.

In 1994, Friedman's smoothing result for normal crossing surfaces was generalized to arbitrary dimension by Kawamata and Namikawa in [173]. They use an early version of the language of logarithmic deformation theory; the argument essentially relies on flat deformation theory and the analysis of a specific mixed Hodge complex, which is related to Steenbrink's construction of the limiting mixed Hodge structure in a semistable degeneration. In the same year, in [262], Steenbrink himself, too, gives a construction of a mixed Hodge complex on what he calls a logarithmic deformation. By this, he means a normal crossing space together with the structure of a log smooth log morphism to the standard log point; it generalizes his earlier construction of the limiting mixed Hodge structure as well.

Theorem 1.1 in [77] by Filip, Ruddat, and the author, as well as Theorem 18.44 in this monograph finally give the most general existence result for semistable smoothings: it is sufficient to assume that V is d-semistable, that all strata of V are quasi-projective, and that $\omega_V^\vee$ is globally generated. This also removes the

---

[2] Kulikov's claim on the classification of degenerations of Enriques surfaces in the same article has been found to be false.

cohomological assumptions made in Kawamata–Namikawa (for the Calabi–Yau case considered there).

The smoothing result of Kawamata–Namikawa was applied and refined in a couple of instances. Already in 1990, Fujita classified the possible semistable degenerations of del Pezzo surfaces in [102], and in 2007, Kachi applied the result of Kawamata–Namikawa in [157] to show that all semistable del Pezzo surfaces occur indeed as the central fiber of a semistable degeneration. In 2010, Lee computed in [186] the topological invariants of many smoothings arising from Kawamata–Namikawa. In 2020, Lee constructs in [187] new examples of normal crossing Calabi–Yau threefolds which are smoothable by Kawamata–Namikawa, giving rise to smooth Calabi–Yau threefolds with topological invariants that no previously known Calabi–Yau threefold has. In 2021, Sano shows in [248] that, for every $N \geq 4$, there are infinitely many non-Kähler Calabi–Yau N-folds X with pairwise different second Betti number $b_2(X)$. In 2023, Hashimoto and Sano show in [140] that the same is true for $N = 3$, thus giving infinitely many different topological types of Calabi–Yau threefolds, albeit that none of them is projective. Infinitely many non-Kähler Calabi–Yau threefolds with $b_2(X) = 0$ and pairwise different third Betti number $b_3(X) = 2h$ for $h \geq 103$ have already been constructed by Friedman in [89] in 1991, attributing the construction to Clemens who used a similar construction in [46]. In 1994, Lu and Tian give a construction with $h = 25$ in [193].

In 2023, Doi and Yotsutani give in [62] an entirely different approach to smoothing normal crossing spaces in differential geometry.

Higher Methods in Deformation Theory Every dg Lie algebra $L^\bullet$ gives rise to an associated deformation functor $\mathrm{Def}_{L^\bullet}$, and many deformation problems admit a dg Lie algebra which controls the deformation problem in this way. For example, the holomorphic smooth deformation functor is governed by the Kodaira–Spencer dg Lie algebra $KS_X^\bullet$, i.e., the Dolbeault resolution of the holomorphic tangent sheaf Θ_X^1. Similarly, every L_∞-algebra $L^\bullet$ gives rise to a deformation functor, which depends only on the homotopy class of $L^\bullet$. Since every L_∞-algebra is homotopy equivalent to a dg Lie algebra, proven for example by Kriz and May in [185] in 1995, the deformation problems governed by some L_∞-algebra are the same as the deformation problems governed by some dg Lie algebra.

L_∞-algebras feature prominently in Kontsevich's celebrated proof of the formality conjecture in [179], published in 2003: The dg Lie algebras $T_{\mathrm{poly}}(X)$ and $D_{\mathrm{poly}}(X)$ are homotopy equivalent via an L_∞-morphism. An important technical result on L_∞-algebras is the homotopy transfer, which appears clearest and most general in the 2007 work [82] of Fiorenza and Manetti, but is much older, going back to a result of Kadeishvili in [158] on A_∞-algebras. This result essentially states that, when a complex is a homotopy retract of an L_∞-algebra, then there is an induced homotopy equivalent L_∞-structure on that complex.

It has long been a well-established philosophy that *every* deformation problem should be governed by dg Lie algebras. Indeed, for a large number of deformation problems, such a dg Lie algebra is known. This philosophy was established independently as a theorem by Pridham in [234] in 2010 and Lurie in [195] in

2011: They define an ∞-category of formal moduli problems as an ∞-category of certain functors of $\mathbb{E}_\infty$-algebras, and this ∞-category is equivalent to the ∞-category obtained from dg Lie algebras by localization of quasi-isomorphisms. A nice survey is given by Calaque and Grivaux in [33]. A new conceptual approach to the Lurie–Pridham theorem is given by Le Grignou and Roca i Lucio in [117].

Given a deformation problem, its deformation functor maps a local Artin ring A to the *set* of isomorphism classes. Similarly, given a dg Lie algebra $L^\bullet$, its deformation functor maps A to the *set* of equivalence classes. There is, however, an obvious 1-categorical enhancement: The deformation groupoid (over A) of a deformation problem has as objects deformations over A, and as morphisms isomorphisms between deformations. Similarly, the Deligne groupoid $C(L^\bullet)$ of a dg Lie algebra $L^\bullet$ has as objects Maurer–Cartan solutions, and as morphisms gauge transforms that send one solution to the other. Hinich proved in [144] in 1997 that the two groupoids are equivalent for certain deformation problems. In the course of doing so, he defines a Kan complex $\mathrm{MC}_\bullet(L^\bullet)$ out of the semisimplicial dga $(A_{\mathrm{PL}})_\bullet$ of polynomial differential forms on simplices. Then the Deligne groupoid $C(L^\bullet)$ is isomorphic to the Poincaré groupoid of $\mathrm{MC}_\bullet(L^\bullet)$, which is the category obtained from the left adjoint of the nerve functor $N \colon \mathrm{Cat} \to \mathrm{sSet}$. Since $\mathrm{MC}_\bullet(L^\bullet)$ is a Kan complex, we may consider it as an ∞-categorical realization of the deformation problem governed by $L^\bullet$.

The Kan complex $\mathrm{MC}_\bullet(L^\bullet)$ is not equivalent to the nerve $N_\bullet C(L^\bullet)$ of the Deligne groupoid. In 2009, Getzler constructed in [106] in the generality of L_∞-algebras a subcomplex $\gamma_\bullet(L^\bullet) \subseteq \mathrm{MC}_\bullet(L^\bullet)$ such that $\gamma_\bullet(L^\bullet)$ is isomorphic to $N_\bullet C(L^\bullet)$ if $L^\bullet$ is a nilpotent dg Lie algebra concentrated in degrees $[0, \infty)$, which is the relevant case in deformation theory. In 2020, Robert-Nicoud and Vallette gave a more explicit and more general interpretation of $\gamma_\bullet(L^\bullet)$ in [240].

The algebraic variant of the Kodaira–Spencer dg Lie algebra $KS_X^\bullet$, which is the classical counterpart of the global logarithmic deformation theory presented in this monograph, takes a Thom–Whitney resolution $L_X^\bullet$ of Θ_X^1 as the dg Lie algebra that controls Def_X. Then a canonical homotopy retract to the Čech complex $\check{C}(\mathcal{U}, \Theta_X^1)$ turns the latter into an L_∞-algebra controlling Def_X via the homotopy transfer. The idea to use the Thom–Whitney resolution of Θ_X^1 to control the smooth deformation functor goes at least back to the 1997 articles [145, 146] of Hinich and Schechtman. In these articles, the construction is used to describe the hull of Def_X in terms of Θ_X^1 in a case where Def_X is assumed to be unobstructed. The article [144] of Hinich discussed above in the context of $\mathrm{MC}_\bullet(L^\bullet)$ then also shows that the Thom–Whitney resolution of Θ_X^1 controls Def_X. Iacono and Manetti achieve the present form of the theory in [152] in 2010. Manetti and others further elaborated on various parts of the theory in [83, 84, 198, 200, 201]. The Thom–Whitney resolution has a long history; the variant which we use is the one used in Iacono–Manetti [152]; it was developed by Navarro Aznar in [217] in 1987. Hinich and Schechtman give their own work [147], also from 1987, as the source of the Thom–Whitney construction they use in [145, 146].

Non-logarithmic Bogomolov–Tian–Todorov Theorems The BTT theorem, asserting that infinitesimal deformations of a compact Calabi–Yau manifold are unobstructed, was first proven in a special case by Bogomolov in [28] in 1978. The general case was achieved independently by Tian in [269] in 1987 and by Todorov in [270] in 1989. Both proofs use differential geometry. The first algebraic proof was given by Ran in [238] in 1992 via the so-called T^1-lifting criterion, in the case of the complex numbers as the base field. In the same year, Kawamata generalized the algebraic proof to an arbitrary base field in [172]. In 1999, Fantechi and Manetti generalized in [72] the T^1-lifting criterion by weakening the necessary assumptions.

In the case of compact complex manifolds, the infinitesimal deformations are governed by the Kodaira–Spencer dg Lie algebra $KS_X^\bullet$, i.e., the Dolbeault resolution of the holomorphic tangent sheaf. In 1990, Goldman and Millson showed in [109] that this dg Lie algebra is homotopy Abelian in the Calabi–Yau case, a statement which implies the unobstructedness of the deformation functor. Their proof relies on the old trick of [58] to show that $KS_X^\bullet$ is quasi-isomorphic to the ∂-cohomology of the Dolbeault resolution $(A_X^{n-1,\bullet}, \bar\partial)$, showing formality, and then the (Bogomolov–)Tian–Todorov formula shows that this ∂-cohomology is an Abelian dg Lie algebra. The proof of unobstructedness of the extended deformation functor in Barannikov–Kontsevich [19] mentioned above relies on the same method.

In the algebraic case, Iacono and Manetti show the unobstructedness of the dg Lie algebra $L_X^\bullet$, constructed via the Thom–Whitney resolution, in [152] in 2010. Their proof relies on the theory of period maps and Cartan homotopies developed by Fiorenza and Manetti in [83]. Iacono generalizes the method to give an abstract unobstructedness theorem in [151] in 2017, giving a very explicit criterion for a dg Lie algebra to be homotopy Abelian. Iacono also uses this method in [150] in 2015 to show homotopy Abelianity for a pair (X, D) of a smooth variety X with a divisor $D \subset X$, and in [154] in 2021 (with Manetti) to show homotopy Abelianity for a pair (X, L) of a smooth variety X and a line bundle L. In [153] in 2019, they study deformations of pairs $(X, \mathcal{F})$ where $\mathcal{F}$ is a coherent sheaf.

The argument of Goldman–Millson, in particular before the background of extended deformations described above, suggests to study Batalin–Vilkovisky algebras instead of dg Lie algebras. In 2008, Terilla gives in [266] a smoothness criterion for the deformation functor associated with a BV algebra in terms of a quantization thereof. In the same year, Katzarkov, Kontsevich, and Pantev give in [171] a new perspective on the homotopy Abelianity of the dg Lie algebra inside a (dg) BV algebra; in particular, they define the degeneration property. These two articles lead up to the logarithmic variants pioneered in Chan–Leung–Ma [38], which we discuss below.

Formal Frobenius Manifolds and Batalin–Vilkovisky Algebras Given a Calabi–Yau manifold $\check{X}$, its genus-0 Gromov–Witten invariants can be encoded in the quantum cohomology ring, which in turn may be viewed as a structure of a formal Frobenius supermanifold on the formal stalk $\check{M}$ at 0 of the total cohomology ring

$H^\bullet = H^\bullet(\check{X}, \Omega^\bullet_{\check{X}})$.[3] After forgetting the metric/pairing, the latter is equivalent to a sequence of graded-symmetric products $m_n \colon (H^\bullet)^{\otimes n} \to H^\bullet$ of degree $2(2 - n)$ for $n \geq 2$, satisfying certain associativity relations. This is what Getzler calls a hypercommutative algebra in [105] in 1995.

In the article [178] of 1995, when expounding the foundational ideas of homological mirror symmetry, Kontsevich suggests that this structure of a formal Frobenius supermanifold on $\check{M}$ should correspond to a structure of a formal Frobenius supermanifold on the formal stalk M of the total Hochschild cohomology $H^\bullet(X, \bigwedge^\bullet \Theta^1_X)$ of the mirror X, and that M should be considered as an extended moduli space of X, extending the interpretation of $H^1(X, T_X)$ as the tangent space of the smooth deformation functor Def_X. Barannikov and Kontsevich give in [19] in 1998 a construction of such a formal Frobenius supermanifold on M by studying the differential BV algebra $\mathbf{t} = \bigoplus_{p,q} \Gamma(X, \bigwedge^q \overline{T}^*_X \otimes \bigwedge^p T_X)$ of polyvector fields; this formal Frobenius supermanifold conjecturally corresponds to the quantum cohomology of the mirror $\check{X}$. In [202], Manin gives, among many other things, an expanded exposition of the construction of the formal Frobenius manifold. The construction was further investigated, in particular from the perspective of Hodge theory, by Barannikov in [17, 18, 20]. Merkulov defines in [207] in 1999 a structure of an A_∞-algebra on the kernel of the BV operator on $\mathbf{t}$; he gives in [208] in 2000 an extension of the construction of the Frobenius manifold to the general case where X needs no longer be Calabi–Yau, with values in F_∞-manifolds. The construction of Barannikov–Kontsevich may be viewed as endowing the cohomology $H^\bullet(A^\bullet, d)$ of any dg Batalin–Vilkovisky algebra $A^\bullet$ with the structure of a hypercommutative algebra. This viewpoint has been further expanded by Losev and Shadrin in [192], and by Park in [226], both in 2007.

Commutative BV_∞-algebras have been first defined by Kravchenko in [182] in 2000, partially inspired by Barannikov–Kontsevich. A canonical notion of (not necessarily commutative) homotopy BV algebra was provided in 2012 by Gálvez-Carrillo, Tonks, and Vallette in [103] by taking the Koszul resolution of the BV-operad. In 2013, Drummond-Cole and Vallette gave in [67] an explicit description of the minimal model of this operad; an algebra over this operad is called a skeletal homotopy Batalin–Vilkovisky algebra. This allows them to extend the construction of Barannikov–Kontsevich: Every dg Batalin–Vilkovisky algebra $A^\bullet$ which is endowed with so-called non-commutative Hodge–de Rham degeneration data gives rise to a structure of *homotopy* hypercommutative algebra on the cohomology $H^\bullet(A^\bullet, d)$. Here, a homotopy hypercommutative algebra is defined via the Koszul resolution of the operad of hypercommutative algebras. Since $H^\bullet(A^\bullet, d)$ has trivial differential, this homotopy hypercommutative algebra has an underlying hypercommutative algebra, and this hypercommutative algebra is precisely the one constructed by Barannikov–Kontsevich. Unlike in the construction of Barannikov–Kontsevich, the dg Batalin–Vilkovisky algebra $A^\bullet$ can be reconstructed up to

---

[3] We are ignoring here that, in general, one has to work with coefficients in the Novikov ring.

homotopy from the homotopy hypercommutative algebra $H^\bullet(A^\bullet, d)$, again by [67]. The results were strengthened and extended by Dotsenko, Shadrin, and Vallette in [63]. As a consequence of results by Koroshkin, Markarian, and Shadrin in [175], there is also a hypercommutative structure on the dg BV algebra $A^\bullet$ (endowed with additional data), which is homotopy equivalent to the homotopy hypercommutative structure on its cohomology. In 2023, Cirici and Horel prove in [45] that, for a Calabi–Yau manifold, the Dolbeault resolution of polyvector fields is formal as a hypercommutative algebra, i.e., homotopy equivalent to its cohomology. Thus, in the case of Calabi–Yau manifolds, the hypercommutative algebra structure on the Hochschild cohomology looses no homotopical information compared to the dg BV algebra structure on polyvector fields. In the case of surfaces and threefolds, this was proven before by Drummond-Cole in [66]. This result is particularly interesting for us because it suggests that the Batalin–Vilkovisky calculi considered in this monograph may be interpreted as an incarnation of the B-model in Barannikov–Kontsevich mirror symmetry.

In 2013, Braun and Lazarev give in [32] a version of the theory in Katzarkov–Kontsevich–Pantev (discussed above) for commutative BV_∞-algebras. They define the degeneration property in this context, which implies the homotopy Abelianity of the associated L_∞-algebra. This degeneration property is, on the one hand, a generalization of Katzarkov–Kontsevich–Pantev's degeneration property, and on the other hand, a variant of the degeneration properties considered by Vallette and others in [63, 67, 103]. There has been some additional recent interest in commutative BV_∞-algebras, see [21, 203, 278].

A similar structure as the A_∞-algebra structure defined by Merkulov on the kernel of the BV operator also occurs in the recent work [196] of Mandel and Ruddat, where tropical quantum field theories are defined, and two new methods to compute the multiplicities of tropical curves are provided.

Logarithmic Deformation Theory In 1996, F. Kato gives in [160] the modern formulation of log smooth deformation theory as a functor from $\mathbf{Art}_Q$ to $\mathbf{Set}$. The notion of log smoothness itself was already introduced by K. Kato in the seminal paper [162] mentioned above, but we find a fully developed deformation theory for the first time in [160]. Two years later, in 1998, F. Kato developed a more general theory of functors of *log* Artin rings, which does not seem to have been in use since then. In 2000, F. Kato comes in [161] back to the original motivation of $\overline{\mathcal{M}}_g$, giving an interpretation of $\overline{\mathcal{M}}_g$ as the log stack of log smooth curves.

In 2015, Tziolas showed in [275] that a d-semistable Fano variety X_0 is smoothable in a semistable family by showing that the obstruction space $H^2(X_0, \Theta^1_{X_0/S_0})$ vanishes, applying F. Kato's log smooth deformation theory. In 2020, Shentu treated in [257] the more general case of semistable Fano varieties over the log point $\mathrm{Spec}(\mathbb{N}^r \to \mathbf{k})$; such families are also known as polystable families, following the terminology of Berkovich in [25].

Iacono's article [150] on the deformations of pairs (X, D) with a divisor $D \subset X$, and Iacono and Manetti's article [154] on deformations of pairs (X, L) with a line bundle also rely on logarithmic deformation theory over the trivial log point; namely, in both articles, the compactifying log structure of a divisor is considered.

In the case of an snc divisor $D \subset X$, logarithmic deformations need to be distinguished from deformations of the closed immersion $D \subset X$ where both X and D are allowed to vary. If (X, D) is a log CY pair, the first are always unobstructed, but the latter can be obstructed, as shown by Petracci, Robins, and the author in [79].

The Gross–Siebert Program The Gross–Siebert program is an explicit constructive approach to mirror symmetry via degenerations to reducible spaces which are glued from toric varieties. These degenerations are called toric degenerations and must be distinguished from degenerations to a normal toric variety, which are also called a toric degeneration in some other contexts. The central fiber of a toric degeneration is called a toric log Calabi–Yau space. Mirror pairs of toric log Calabi–Yau spaces admit a version of dual Lagrangian torus fibrations to an integral tropical manifold B. The original idea goes back to Batyrev's mirror symmetry of [22] from 1994, and the more general Batyrev–Borisov mirror symmetry of [23] from 1996. In the foundational article [124] from 2006, Gross and Siebert describe the construction of the degenerate Calabi–Yau spaces $X_0(B, \mathscr{P}, s)$ and $\check{X}_0(B, \mathscr{P}, s)$ as well as a log structure on these spaces and a mirror construction of $(B, \mathscr{P}, \varphi)$, the discrete Legendre transform $(\check{B}, \check{\mathscr{P}}, \check{\varphi})$. In the sequel [125] of 2010, they describe the deformation theory of these spaces. The first highlight of the Gross–Siebert program is [126] from 2011, where Gross and Siebert construct infinitesimal deformations of $\check{X}_0(B, \mathscr{P}, s)$ up to arbitrary order via an intricate algorithm. In [119], Gross explains how Batyrev–Borisov duality can be considered as a special case of their construction. There has been tremendous interest in toric degenerations since. In 2006, Nishinou and Siebert use the degeneration in [219] to count curves on toric varieties. In 2010, Ruddat generalizes in [241] the computation of certain Hodge numbers in the construction. In the same year, Nakayama and Ogus give with [212] a study on properties of relatively log smooth morphisms, the type of log singularity occurring in the Gross–Siebert construction. In 2014, Castaño-Bernard and Matessi study conifold transitions in this context in [26]. In 2015, Gross, Hacking, and Keel give a variant of toric degenerations for Looijenga pairs (Y, D) in [120]. In [235], Prince applies the Gross–Siebert algorithm to study Fano surfaces; in [236], he constructs new Calabi–Yau threefolds via the Gross–Siebert algorithm. In [12], Argüz studies the real as opposed to complex geometry of toric degenerations. In [243], Ruddat and Siebert show that the deformations constructed via the Gross–Siebert algorithm are formally versal as deformations of log schemes. In the same article, they construct canonical coordinates on the base of the formally miniversal deformation. In [122], Gross, Hacking, and Siebert construct canonical bases of ample line bundles for toric degenerations (as well as more general situations). There is also an extensive literature on questions related to Gromov–Witten invariants in the Gross–Siebert construction and other logarithmic setups, see [2, 4, 30, 31, 42, 81, 111, 123, 127] to give just a subjective sample.

Toroidal Crossing Spaces A toroidal crossing space $(V, \mathcal{P}, \bar{\rho})$ consists of a scheme V together with a sheaf of monoids $\mathcal{P}$ and a global section $\bar{\rho} \in \mathcal{P}$ which is locally the ghost sheaf of a vertical and saturated log smooth log morphism to the standard log point S_0. This notion is relevant in the construction of log singular log varieties, as one may construct $(V, \mathcal{P}, \bar{\rho})$ first, and then a section $s \in \mathcal{LS}_V$ in the sheaf of actual log smooth log structures with the specified ghost sheaf. The classical case is when V is a normal crossing space; in this situation, $(\mathcal{P}, \bar{\rho})$ can be canonically constructed from the geometry of V. The existence of a global section of $\mathcal{LS}_V$ is equivalent to d-semistability, introduced by Friedman in [87], and further investigated in [160, 173, 262]. In 2003, Olsson gives a more general treatment, still in the semistable case, in [224]. This, in turn, was further generalized by Li in [189] in 2007, where the case of weakly semistable families over a higher-dimensional base is treated. Also related to normal crossing spaces as toroidal crossing spaces, Tziolas provided in [274] methods to compute $\mathcal{T}_V^1$.

The case of toroidal crossings, where a semistable degeneration is replaced with a more general vertical and saturated log smooth morphism, was first investigated (and named) by Schröer and Siebert in [253] in 2006, under the name of Gorenstein toroidal crossings. In the same year, the variant of the theory that we consider now standard was developed by Gross and Siebert in [124] while expounding their mirror symmetry construction. In 2021, Filip, Ruddat, and the author established the name "toroidal crossing space" for a triple $(V, \mathcal{P}, \bar{\rho})$ as above in [77]. In Chap. 9, we give the first extended exposition on toroidal crossing spaces.

The Smoothing Method of Chan–Leung–Ma In 2019, Chan, Leung, and Ma achieve in [38] the major breakthrough needed to generalize the classical Bogomolov–Tian–Todorov theorem into a logarithmic one. Their key insight is that the dg Lie algebras of classical deformation theory must be replaced with curved Lie algebras in order to describe log deformations as solutions of a Maurer–Cartan equation. The pre-dg Lie algebra can be constructed by an adaptation of Iacono–Manetti's approach via the Thom–Whitney resolution. Then a pre-dg BV algebra is constructed, whose analysis in the spirit of Katzarkov–Kontsevich–Pantev [171] and Terilla [266] yields unobstructedness of pairs of a log deformation together with a volume form. The method can be applied whenever we have locally fixed an isomorphism type of log deformations, in particular for log smooth deformations and in the Gross–Siebert program. The article [38] also contains a treatment of semi-infinite variations of Hodge structures in that setting, generalizing the work of Barannikov.

In 2021, Filip, Ruddat, and the author applied in [77] the method of Chan–Leung–Ma to show a very general smoothing result for normal crossing spaces, relying on the local deformation theory for certain log singularities developed by Gross and Siebert in [125]. In 2022, the author modified in [76] slightly the method to prove explicitly that the log smooth deformation functor LD_{X_0/S_0} is controlled by a curved Lie algebra $L^\bullet_{X_0/\Lambda}$. Also in 2022, Petracci and the author announced and partially proved in [78] the logarithmic Bogomolov–Tian–Todorov theorem, i.e., the unobstructedness of LD_{X_0/S_0} if $f_0 \colon X_0 \to S_0$ is log Calabi–Yau; a full proof is contained in this monograph. If the base carries the trivial log structure, the log Bogomolov–Tian–Todorov theorem also follows from the method of Iacono–Manetti, as described in [78] in full generality, and in [150] in a special case.

In 2022, Chan and Ma apply in [40] their smoothing method to pairs of a space X and a bounded perfect complex C^*, very similar to the study of vector bundles and in particular line bundles in the present work. However, [40] misses the unobstructedness of line bundles on a log CY space, and hence needs to assume $H^2(X, O_X) = 0$ for the most interesting results. In the same year, Chan, Ma, and Suen apply in [41] the theory of [40] to deform certain vector bundles on some $X_0(B, \mathscr{P}, s)$ of dimension 2.

Chan, Leung, and Ma introduced as well methods of asymptotic analysis into the Gross–Siebert program in [36]. These methods were applied in [39] and [188] to curve counting, and in [37] to produce a variant of the above smoothing method where the Thom–Whitney resolution, applied on X_0, is replaced with a resolution constructed from asymptotic analysis, applied on B. This allows to tie solutions of the Maurer–Cartan equation explicitly to scattering diagrams, thus connecting the two methods to construct log deformations.

Global Smoothings of Normal Singularities Smoothings and deformations constructed via logarithmic deformation theory must be distinguished from smoothings constructed via the classical flat deformation functor Def_X. There is, in fact, a vast literature on the latter, both for normal and non-normal singularities, usually for either Calabi–Yau or Fano varieties. In 1986, Friedman gave in [88] a sufficient condition for a nodal Calabi–Yau threefold to be smoothable in terms of a small resolution of singularities. In 1994, Namikawa showed in [213] that Def_X of a projective 3-dimensional $\mathbb{Q}$-Calabi–Yau variety with terminal singularities is unobstructed, after a special case with Kleinian singularities had been treated by Ran in [237] in 1992. The deformations arising from these unobstructedness results need not always be smoothings, but under some special hypotheses they are. Related questions were further investigated in [118, 209, 210, 214–216] mostly by Namikawa and Minagawa, and more recently in [244–247] by Sano exclusively in the Fano case (and variants thereof). In particular, Namikawa gave in [216] a variant of Friedman's sufficient condition for smoothability of a nodal Calabi–Yau threefold which is a precise criterion. Other works focus on the relation between X and its smoothing; in 2011, Jahnke and Radloff show in [156] that the Picard numbers of a Gorenstein terminal Fano threefold X and its smoothing coincide.

New interest in this topic has arisen recently with a series of articles by Friedman and Laza on k-rational and k-du Bois singularities, see [90–95].

In a different direction, much work has been done to study the flat deformations of Gorenstein toric Fano varieties, starting with the work of Altmann [5–9] in the 1990s. An overview can be found in [232]. Conversely, different methods were developed to degenerate a smooth Fano variety to a Gorenstein toric Fano variety, for example the classical [110] for flag varieties, or Anderson's construction of degenerations to toric varieties from Okounkov bodies in [11] of 2013. The general philosophy is that moduli spaces of Fano varieties may be classified by the Gorenstein toric Fano varieties contained in them, as opposed to the more intricate philosophy of the Gross–Siebert program that moduli spaces can be classified by the degenerations to reducible toric varieties contained in them.

As a step toward understanding flat deformations of Gorenstein toric Fano varieties, for an affine Gorenstein toric threefold, Corti, Filip, and Petracci recently gave a conjectural characterization of the smoothing components of the miniversal base space (to be understood in an appropriate sense) in [53]. Even more recently, Corti, Hacking, and Petracci gave sufficient criteria for the existence of a global smoothing in [54]. We have discussed the latter article in Sect. 1.12.

A study of deformations in the non-normal case with classical methods was undertaken by Tziolas in [273] in 2010. In [70] and [71], Fantechi, Franciosi, and Pardini apply the result of Tziolas to semi-smooth varieties. Related to and motivated by this, Nobile reviews in [220] the relation between formal smoothings and geometric smoothings. Technically, also Friedman's classical [87] as well as Kawamata and Namikawa's [173] apply classical methods to construct global deformations of non-normal varieties, but we have discussed these works already above.

Log Hodge Structures Just as the curved Gerstenhaber calculi that we study here, log Hodge structures are another tool to study log spaces from a cohomological perspective. Log Hodge structures were developed by K. Kato, Nakayama, and Usui in a series of articles starting around 2000, see [165–170]. Somewhat surprisingly, the concern of these articles is not so much the construction of log Hodge structures from families of log schemes but rather the classification of log Hodge structures; in fact, the definition of a log Hodge structure is adapted to make their classification feasible, not their construction. An exception to that is the early work [204] of Matsubara in 1998, and possibly a preprint of K. Kato mentioned therein, as well as the work [164] of K. Kato, Matsubara, and Nakayama where they construct log Hodge structures if the base is log smooth over the trivial log point. The question of constructing a log Hodge structure geometrically over non-trivial log points (and more general bases) has been addressed quite recently by Fujisawa and Nakayama in [100, 101]; they obtain partial results which are already quite satisfactory. Usui's article [276] suggests that certain mirror phenomena can be captured by log Hodge structures, and this topic certainly deserves further investigation, as well as the relation with the log semi-infinite variations of Hodge structures that show up in [38].

Part I
Abstract Unobstructedness Theorems

Chapter 3
Algebraic Structures

The purpose of this chapter is to fix the definitions of several *algebraic structures* that will play an important role in our study of logarithmic deformation theory. We will study essentially seven different algebraic structures:

(1) Lie algebras;
(2) Lie–Rinehart algebras and pairs;
(3) Gerstenhaber algebras and calculi;
(4) Batalin–Vilkovisky algebras and calculi.[1]

Each of them comes in four variants. The singly graded original will usually appear as a collection of sheaves $(\mathcal{F}^i)_i$ on a space X together with a collection of operations between them; to each $\mathcal{F}^i$ will be assigned a degree $|\mathcal{F}^i| \in \mathbb{Z}$. While one may omit this degree in the case of Lie algebras and Lie–Rinehart algebras, it is essential for Gerstenhaber or Batalin–Vilkovisky algebras and calculi. The *differential bigraded* version shows up once we take an acyclic resolution $\mathcal{F}^i \to \mathcal{F}^{i,\bullet}$ of each $\mathcal{F}^i$ which is compatible with all operations. The (plainly) *bigraded* version is obtained by forgetting the differential $\bar{\partial}\colon \mathcal{F}^{i,j} \to \mathcal{F}^{i,j+1}$. Finally, the *curved* version arises when we relax the condition $\bar{\partial}^2 = 0$. In this case, $\bar{\partial}\colon \mathcal{F}^{i,j} \to \mathcal{F}^{i,j+1}$ is called a *predifferential*, and there will be a section ℓ of some sheaf $\mathcal{F}^i$, the *curvature*, which measures the defect of $\bar{\partial}$ to be a differential.

Many of our arguments in later chapters are equally valid for several of the above algebraic structures. To talk about all of them at once, we introduce an adhoc formalism of algebraic structures, which is essentially a simplified (and

---

[1] Gerstenhaber calculi and Batalin–Vilkovisky calculi will have two variants, a one-sided and a two-sided one, which differ by an additional operation whose relevance we discovered only toward the end of this project. This operation is relevant for some of our applications but not for all.

S. Felten, *Global Logarithmic Deformation Theory*, Lecture Notes in Mathematics 2373, https://doi.org/10.1007/978-3-031-98751-9_3

somewhat adapted) version of colored operads.[2] For the purpose of our discussion, an *algebraic structure* $\mathcal{P}$ comprises the following data:

(1) a non-empty set $D = D(\mathcal{P})$, called the set of *domains*;
(2) a *grading* $|\cdot|: D \to \mathbb{Z}$, assigning to every domain its degree;
(3) for every domain $P \in D$, a set $\mathcal{P}(P)$, the set of *constants*;
(4) for all domains $P_1, \ldots, P_n \in D$ ($n \geq 1$) and $Q \in D$, a set $\mathcal{P}(P_1, \ldots, P_n; Q)$, the set of *operations*;
(5) *axioms* between the constants and operations that should be satisfied.

Then in a *representation* of $\mathcal{P}$ (or $\mathcal{P}$-*algebra*), say in the category of Abelian sheaves on a topological space X, there is an Abelian sheaf $\mathcal{E}^P$ associated with every $P \in D$, a section $C^\gamma \in \mathcal{E}^P$ associated with every $\gamma \in \mathcal{P}(P)$, and a multilinear map

$$A^\mu: \ \mathcal{E}^{P_1} \times \ldots \times \mathcal{E}^{P_n} \to \mathcal{E}^Q$$

associated with every operation $\mu \in \mathcal{P}(P_1, \ldots, P_n; Q)$; they must satisfy the axioms, which are stated as equations for sums of compositions of operations and constants.

We will be concerned with representations of the seven algebraic structures in several contexts—as modules over a ring, as sheaves on a topological space etc. We unify these different setups as follows:

Definition 3.1 (Context) A *context* is a quadruple $(\mathcal{S}, C, O, \mathbf{C})$ where $\mathcal{S}$ is a site, C is a sheaf of rings on $\mathcal{S}$, the sheaf of *constants*, O is a sheaf of C-algebras, and $\mathbf{C}$ is a class of O-modules. In a representation of an algebraic structure $\mathcal{P}$ in the context $(\mathcal{S}, C, O, \mathbf{C})$, the domains are O-modules in the class $\mathbf{C}$, the constants are global sections of $\mathcal{E}^P$, and the operations are C-multilinear maps. They are not required to be O-multilinear.

Example 3.2 Typical examples of contexts include the following:

(1) Let $f: X \to S$ be a flat morphism of Noetherian schemes. Then we can take for $\mathcal{S}$ the small Zariski site of X and $C = f^{-1}(O_S)$, $O = O_X$. For $\mathbf{C}$, we take coherent O_X-modules which are flat over S. We denote this context by $\mathfrak{Coh}(X/S)$. When we take all quasi-coherent modules flat over S respectively all O_X-modules flat over S, we denote the context by $\mathfrak{QCoh}(X/S)$ respectively

---

[2] Since we work with constants, we have to allow operations of arity 0, which not all introductory texts on colored operads do. It is then often possible to treat (bi-)graded and differential (bi-)graded versions at once by changing the target of a representation but not the operad. Even a theory for the curved case has been recently worked out in [194]. We will do all this by hand, defining three algebraic structures $\mathcal{P}$, $\mathcal{P}^{\mathrm{bg}}$, and $\mathcal{P}^{\mathrm{crv}}$ in each case. The inclined reader familiar with operadic calculus may want to compare the result of the canonical procedure with our explicit definitions below.

by $\mathfrak{Mod}(X/S)$. When we wish to include also non-flat modules, we write $\mathfrak{Coh}'(X/S)$, $\mathfrak{QCoh}'(X/S)$, respectively $\mathfrak{Mod}'(X/S)$.

(2) Let $A \to R$ be a flat ring homomorphism. Then we can take S to be the site of a topological space with one point, $C = A$, and $O = R$. For **C**, we take finitely generated R-modules which are flat over A. We denote this context by $\mathfrak{Mod}^{\mathrm{fg}}(R/A)$.

(3) Let Λ be a complete local Noetherian **k**-algebra with residue field **k**. For S, we take the site of a topological space with one point; we take $C = O = \Lambda$. For **C**, we take flat and complete Λ-modules (which are not necessarily finitely generated). We denote this context by $\mathfrak{Comp}(\Lambda)$. $\Diamond$

3.1 Lie Algebras

Here are our three variants of Lie algebras.

Definition 3.3 (Lie Algebras) Let $(S, C, O, \mathbf{C})$ be a context.

(1) A *Lie algebra* $(\mathcal{L}, [-, -])$ consists of an O-module $\mathcal{L} \in \mathbf{C}$ and a C-bilinear map $[-, -] \colon \mathcal{L} \times \mathcal{L} \to \mathcal{L}$ such that $[\theta, \xi] = -[\xi, \theta]$ and the *Jacobi identity*

$$[\theta, [\xi, \eta]] = [[\theta, \xi], \eta] + [\xi, [\theta, \eta]]$$

holds. By convention, we assign to $\mathcal{L}$ the *degree* $|\mathcal{L}| = -1$; this is in order to fit later conventions.

(2) A *graded Lie algebra*[3] $(\mathcal{L}^\bullet, [-, -])$ consists of an O-module $\mathcal{L}^i \in \mathbf{C}$ of *bidegree* $(-1, i)$ and *total degree* $i - 1$ for every $i \geq 0$ and a C-bilinear map

$$[-, -] \colon \quad \mathcal{L}^i \times \mathcal{L}^j \to \mathcal{L}^{i+j}$$

for all indices $i, j \geq 0$ such that

$$[\theta, \xi] = -(-1)^{(|\theta|+1)(|\xi|+1)}[\xi, \theta]$$

(with $|\theta| = i - 1$ for $i \in \mathcal{L}^i$) and the *Jacobi identity*

$$[\theta, [\xi, \eta]] = [[\theta, \xi], \eta] + (-1)^{(|\theta|+1)(|\xi|+1)}[\xi, [\theta, \eta]]$$

holds.

---

[3] Or *bigraded* Lie algebra to put it more systematically, consistent with the fact that we have already assigned the degree $|\mathcal{L}| = -1$ to a Lie algebra.

(3) A *curved Lie algebra* $(\mathcal{L}^\bullet, [-,-], \bar{\partial}, \ell)$ consists of a graded Lie algebra $(\mathcal{L}^\bullet, [-,-])$, a C-linear map $\bar{\partial}\colon \mathcal{L}^i \to \mathcal{L}^{i+1}$ such that

$$\bar{\partial}[\theta, \xi] = [\bar{\partial}\theta, \xi] + (-1)^{|\theta|+1}[\theta, \bar{\partial}\xi],$$

and a global section $\ell \in \mathcal{L}^2$ such that $\bar{\partial}^2\theta = [\ell, \theta]$ and $\bar{\partial}\ell = 0$. The map $\bar{\partial}\colon \mathcal{L}^i \to \mathcal{L}^{i+1}$ is called the *predifferential*, and the global section $\ell \in \mathcal{L}^2$ is called the *curvature*.[4]

(4) A *differential graded Lie algebra* is a curved Lie algebra with $\ell = 0$. In this case, $\bar{\partial}^2 = 0$, but this condition is not sufficient in general (unless the curved Lie algebra is faithful as defined below).

(5) A Lie algebra is *faithful* or *centerless* if the map

$$\mathrm{ad}_L\colon \quad \mathcal{L} \to \mathcal{H}om_C(\mathcal{L}, \mathcal{L}), \quad \theta \mapsto (\xi \mapsto [\theta, \xi]),$$

is injective.

(6) A graded/dg/curved Lie algebra is *faithful* or *centerless* if the map

$$\mathrm{ad}_L^i\colon \quad \mathcal{L}^i \to \mathcal{H}om_C(\mathcal{L}^0, \mathcal{L}^i), \quad \theta \mapsto (\xi \mapsto [\theta, \xi]),$$

is injective for every $i \geq 0$. Note that the source of the Hom space in the target is always $\mathcal{L}^0$.

(7) A graded/dg/curved Lie algebra is *bounded* if $\mathcal{L}^i = 0$ for $i \gg 0$.

Remark 3.4 A couple of remarks are in order.

(1) What we call a (differential) graded Lie algebra is often called an *odd* differential graded Lie algebra to clarify the grading convention, which differs from the more obvious *even* grading convention with $[\theta, \xi] = -(-1)^{|\theta||\xi|}[\xi, \theta]$.

(2) The operator $\bar{\partial}\colon \mathcal{L}^i \to \mathcal{L}^{i+1}$ is called a *predifferential* to distinguish it from the *differentials*, which satisfy $\bar{\partial}^2 = 0$. We will later modify $\bar{\partial}$ to $\bar{\partial}_\phi = \bar{\partial} + [\phi, -]$ for $\phi \in \mathcal{L}^1$, and it will be useful to distinguish actual differentials from predifferentials also terminologically.

(3) The term *predifferential* seems to have arisen relatively recently in [24, 194, 206] as well as independently in the author's own previous article [76]. According to [24, 194], a predifferential graded object is a graded object $\mathcal{A}^\bullet$ together with a map $\bar{\partial}\colon \mathcal{A}^\bullet \to \mathcal{A}^{\bullet-1}$ (in homological grading convention) which may not satisfy $\bar{\partial}^2 = 0$. Thus, one may argue that a *predifferential graded Lie algebra* should be what we have defined as a curved Lie algebra without the conditions $\bar{\partial}^2\theta = [\ell, \theta]$ and $\bar{\partial}\ell = 0$.

---

[4] As already explained in the introduction, we always use the symbol $\bar{\partial}$ to denote the (pre-)differential of a curved Lie algebra. This notation is not intended to suggest any close relationship to the differential of the Dolbeault resolution in complex geometry.

(4) In [74, Defn. 10.4], we have defined a predifferential graded Lie algebra in such a way that we have an element $\ell \in \mathcal{L}^2$ with $\bar{\partial}^2\theta = [\ell, \theta]$ but without the condition $\bar{\partial}\ell = 0$, halfway between a predifferential graded Lie algebra as discussed in the previous remark and a curved Lie algebra as defined above. This definition has been modelled on the structures used by Chan, Leung, and Ma in [38], and there, the condition $\bar{\partial}\ell = 0$ is not explicit. Early in the course of writing this monograph, we have added the condition to the definition because we need it (and its variants) later in the proofs of Lemma 3.20 and Lemma 3.30. After making the manuscript available on the arXiv, the author has learned from Bruno Vallette that the condition $\bar{\partial}\ell = 0$ is (or should be) indeed part of the standard definition of a curved Lie algebra, for example as defined in [206, Defn. 4.1] or [194, Defn. 2.15], or in [233] for curved dg algebras[5] (although $\bar{\partial}\ell = 0$ is not required in [44], which also refers to predifferentials as differentials for lack of a better name as late as 2016). Thus, we have abandoned our original terminology "predifferential graded Lie algebra". This terminology seemed most natural to us, given that we consider $\bar{\partial}$ as a predifferential, even consistently with [24, 194], but it was chosen in ignorance of the concept of curved Lie algebras. We have replaced it with the already established standard name. This also prevents confusion with the concept of differential graded pre-Lie algebras, discussed for example in [191, §6.4.8].

(5) The name "faithful" refers to the fact that the adjoint representation is faithful. The name "centerless" is borrowed from group theory, where a group is called *centerless* if the identity is the only element which commutes with all other elements. For other algebraic structures, the two notions will differ, see for example Lie–Rinehart pairs.   ◇

Example 3.5 Let $f \colon X \to S$ be a smooth morphism of (Noetherian) schemes. Then the relative tangent sheaf $\mathcal{T}_{X/S}$ with its Lie bracket forms a Lie algebra in the context $\mathfrak{Coh}(X/S)$. The Lie bracket is $f^{-1}(O_S)$-linear but not O_X-linear.   ◇

3.2 Lie–Rinehart Algebras and Pairs

A *Lie–Rinehart algebra* consists of a Lie algebra acting on a ring of functions. The notion goes back to Rinehart's thesis [239], where it is called a (K, R)-*Lie algebra*. The name "Lie–Rinehart algebra" seems to arise first in [148] and is also mentioned in [180]; it seems firmly established in [149]. A Lie–Rinehart algebra can be considered as an algebraic analog of a Lie algebroid, but when we apply our definition of a Lie–Rinehart algebra in the context of a smooth variety $X/\mathbf{k}$, then we obtain a more general notion than a Lie algebroid.

---

[5] Positselski refers to curved algebras as curved dg algebras to emphasize their graded and predifferential nature. His definition requires $\bar{\partial}(h) = 0$ for the curvature h.

Definition 3.6 (Lie–Rinehart Algebras) Let $(\mathcal{S}, C, O, \mathbf{C})$ be a context.

(1) A *Lie–Rinehart algebra*

$$(\mathcal{F}, \mathcal{T}, *^F, *^T, \nabla^F, \nabla^T, 1_F)$$

consists of two O-modules $\mathcal{F} \in \mathbf{C}$ and $\mathcal{T} \in \mathbf{C}$ with degrees $|\mathcal{F}| = 0$ and $|\mathcal{T}| = -1$, a global section $1_F \in \mathcal{F}$, two O-bilinear maps

$$*^F : \ \mathcal{F} \times \mathcal{F} \to \mathcal{F}, \quad *^T : \ \mathcal{F} \times \mathcal{T} \to \mathcal{T},$$

and two C-bilinear maps

$$\nabla^F : \ \mathcal{T} \times \mathcal{F} \to \mathcal{F}, \quad \nabla^T : \ \mathcal{T} \times \mathcal{T} \to \mathcal{T}.$$

We write $* = *^F$, $\nabla = \nabla^F$, and $[-, -] = \nabla^T$ for short. They must satisfy the relations

$$(a * b) *^P p = a *^P (b *^P p), \quad a * b = b * a, \quad 1_F *^P p = p$$

for the two multiplications $*^F, *^T$, the anti-commutativity

$$[\theta, \xi] := \nabla_\theta^T(\xi) = -\nabla_\xi^T(\theta) = -[\xi, \theta],$$

the F-linearity condition $a *^T \nabla_\theta(b) = \nabla_{a *^T \theta}(b)$, the two Jacobi identities

$$\nabla_{[\theta, \xi]}^P(p) = \nabla_\theta^P \nabla_\xi^P(p) - \nabla_\xi^P \nabla_\theta^P(p),$$

and the two odd Poisson identities

$$\nabla_\theta^P(a *^P p) = \nabla_\theta(a) *^P p + a *^P \nabla_\theta^P(p).$$

(2) A *bigraded Lie–Rinehart algebra*

$$(\mathcal{F}^\bullet, \mathcal{T}^\bullet, *^F, *^T, \nabla^F, \nabla^T, 1_F)$$

consists of O-modules $\mathcal{F}^i \in \mathbf{C}$ (of bidegree $(0, i)$ and total degree i) and $\mathcal{T}^i \in \mathbf{C}$ (of bidegree $(-1, i)$ and total degree $i - 1$) for every $i \geq 0$, the O-bilinear maps

$$*^F : \ \mathcal{F}^i \times \mathcal{F}^j \to \mathcal{F}^{i+j}, \quad *^T : \ \mathcal{F}^i \times \mathcal{T}^j \to \mathcal{T}^{i+j},$$

the C-bilinear maps

$$\nabla^F : \ \mathcal{T}^i \times \mathcal{F}^j \to \mathcal{F}^{i+j}, \quad \nabla^T : \ \mathcal{T}^i \times \mathcal{T}^j \to \mathcal{T}^{i+j},$$

and the global section $1_F \in \mathcal{F}^0$. They must satisfy the relations

$$(a * b) *^P p = a *^P (b *^P p), \quad a * b = (-1)^{|a||b|} b * a, \quad 1_F *^P p = p$$

for the two multiplications $*^F$, $*^T$, the anti-commutativity

$$[\theta, \xi] = -(-1)^{(|\theta|+1)(|\xi|+1)}[\xi, \theta],$$

the F-linearity condition $a *^T \nabla_\theta(b) = \nabla_{a *^T \theta}(b)$, the two Jacobi identities

$$\nabla^P_{[\theta, \xi]}(p) = \nabla^P_\theta \nabla^P_\xi(p) - (-1)^{(|\theta|+1)(|\xi|+1)} \nabla^P_\xi \nabla^P_\theta(p),$$

and the two odd Poisson identities

$$\nabla^P_\theta(a *^P p) = \nabla_\theta(a) *^P p + (-1)^{(|\theta|+1)|a|} a *^P \nabla^P_\theta(p).$$

(3) A *curved Lie–Rinehart algebra*

$$(\mathcal{F}^\bullet, \mathcal{T}^\bullet, *^F, *^T, \nabla^F, \nabla^T, 1_F, \bar{\partial}^F, \bar{\partial}^T, \ell)$$

consists of a bigraded Lie–Rinehart algebra, the C-linear maps $\bar{\partial}^F : \mathcal{F}^i \to \mathcal{F}^{i+1}$ and $\bar{\partial}^T : \mathcal{T}^i \to \mathcal{T}^{i+1}$ satisfying the four derivation rules

$$\bar{\partial}^P \nabla^P_\theta(p) = \nabla^P_{\bar{\partial}^T \theta}(p) + (-1)^{|\theta|+1} \nabla^P_\theta(\bar{\partial}^P p)$$

and

$$\bar{\partial}^P(a *^P p) = \bar{\partial}^F(a) *^P p + (-1)^{|a|} a *^P \bar{\partial}^P p,$$

and a global section $\ell \in \mathcal{T}^2$ with $(\bar{\partial}^P)^2(p) = \nabla^P_\ell(p)$ and $\bar{\partial}^T(\ell) = 0$. We usually write $\bar{\partial} = \bar{\partial}^P$ for short; this map is called the *predifferential*. The section $\ell \in \mathcal{T}^2$ is called the *curvature*.

(4) A *differential bigraded Lie–Rinehart algebra* is a curved Lie–Rinehart algebra with $\ell = 0$.

(5) A Lie–Rinehart algebra is *centerless* if the map

$$\mathrm{ad}_T : \quad \mathcal{T} \to \mathcal{H}om_C(\mathcal{T}, \mathcal{T}), \quad \theta \mapsto (\xi \mapsto \nabla^F_\theta(\xi)),$$

is injective; it is *strictly faithful* if the map

$$\mathrm{ad}_F : \quad \mathcal{T} \to \mathcal{H}om_C(\mathcal{F}, \mathcal{F}), \quad \theta \mapsto (a \mapsto \nabla^F_\theta(a)),$$

is injective; it is *faithful* if the intersection of the two kernels is zero.

(6) A bigraded/dbg/curved Lie–Rinehart algebra is *centerless* if the map

$$\mathrm{ad}^i_T: \quad \mathcal{T}^i \to \mathcal{H}om_C(\mathcal{T}^0, \mathcal{T}^i), \quad \theta \mapsto (\xi \mapsto \nabla^F_\theta(\xi)),$$

is injective for all $i \geq 0$; it is *strictly faithful* if the map

$$\mathrm{ad}^i_F: \quad \mathcal{T}^i \to \mathcal{H}om_C(\mathcal{F}^0, \mathcal{F}^i), \quad \theta \mapsto (a \mapsto \nabla^F_\theta(a)),$$

is injective for all $i \geq 0$; it is *faithful* if the intersection of the two kernels is zero for all $i \geq 0$.

(7) A bigraded/dbg/curved Lie–Rinehart algebra is *bounded* if $\mathcal{F}^i = 0$ and $\mathcal{T}^i = 0$ for $i \gg 0$.

Example 3.7 Let $f: X \to S$ be a smooth morphism. Then we obtain a Lie–Rinehart algebra in the context $\mathfrak{Coh}(X/S)$ by setting $\mathcal{T} = \mathcal{T}_{X/S}$ and $\mathcal{F} = O_X$. The two products $*^F$ and $*^T$ are given by the O_X-module structure, ∇^T is the Lie bracket, and ∇^F is the action of $\mathcal{T}_{X/S}$ on O_X by derivations. The latter two operations are only $f^{-1}(O_S)$-linear but not O_X-linear. This Lie–Rinehart algebra is strictly faithful because a derivation in $\mathcal{T}_{X/S}$ is uniquely determined by its action on O_X. It is also centerless because a derivation which commutes with all other derivations must be trivial (look at $[D, fD](g) = D(f) \cdot D(g)$ for a derivation D). In particular, it is faithful. $\diamond$

A *Lie–Rinehart module* over a Lie–Rinehart algebra $(\mathcal{F}, \mathcal{T})$, as defined, for example, in [35], is an O-module $\mathcal{E}$ together with a C-bilinear operation

$$\nabla^E: \quad \mathcal{T} \times \mathcal{E} \to \mathcal{E}$$

which satisfies $\nabla^E_{a\theta}(e) = a\nabla^E_\theta(e)$, the Jacobi identity

$$\nabla^E_{[\theta,\xi]}(e) = \nabla^E_\theta \nabla^E_\xi(e) - \nabla^E_\xi \nabla^E_\theta(e),$$

and the derivation rule

$$\nabla^E_\theta(ae) = [\theta, a] \cdot e + a \cdot \nabla^E_\theta(e).$$

A *Lie–Rinehart pair* is then a Lie–Rinehart algebra together with a Lie–Rinehart module.

Definition 3.8 (Lie–Rinehart Pairs) Let $(S, C, O, \mathbf{C})$ be a context.

(1) A *Lie–Rinehart pair*

$$(\mathcal{F}, \mathcal{T}, \mathcal{E}, *^F, *^T, *^E, \nabla^F, \nabla^T, \nabla^E, 1_F)$$

consists of a Lie–Rinehart algebra, an O-module $\mathcal{E} \in \mathbf{C}$ of degree $|\mathcal{E}| = 0$, an O-bilinear map $*^E: \mathcal{F} \times \mathcal{E} \to \mathcal{E}$, and a C-bilinear map $\nabla^E: \mathcal{T} \times \mathcal{E} \to \mathcal{E}$ such that

$$(a *^F b) *^E e = a *^E (b *^E e), \quad 1_F *^E e = e,$$

the F-linearity

$$a *^E \nabla_\theta^E(e) = \nabla_{a*^T\theta}^E(e),$$

the Jacobi identity

$$\nabla_{[\theta,\xi]}^E(e) = \nabla_\theta^E \nabla_\xi^E(e) - \nabla_\xi^E \nabla_\theta^E(e),$$

and the derivation rule

$$\nabla_\theta^E(a *^E e) = \nabla_\theta^F(a) *^E e + a *^E \nabla_\theta^E(e)$$

hold.

(2) A *bigraded Lie–Rinehart pair*

$$(\mathcal{F}^\bullet, \mathcal{T}^\bullet, \mathcal{E}^\bullet, *^F, *^T, *^E, \nabla^F, \nabla^T, \nabla^E, 1_F)$$

consists of a bigraded Lie–Rinehart algebra, O-modules $\mathcal{E}^i \in \mathbf{C}$ (of bidegree $(0, i)$ and total degree i) for all $i \geq 0$, the O-bilinear maps $*^E : \mathcal{F}^i \times \mathcal{E}^j \to \mathcal{E}^{i+j}$, and the C-bilinear maps $\nabla^E : \mathcal{T}^i \times \mathcal{E}^j \to \mathcal{E}^{i+j}$ such that

$$(a *^F b) *^E e = a *^E (b *^E e), \quad 1_F *^E e = e,$$

the F-linearity

$$a *^E \nabla_\theta^E(e) = \nabla_{a*^T\theta}^E(e),$$

the Jacobi identity

$$\nabla_{[\theta,\xi]}^E(e) = \nabla_\theta^E \nabla_\xi^E(e) - (-1)^{(|\theta|+1)(|\xi|+1)} \nabla_\xi^E \nabla_\theta^E(e),$$

and the derivation rule

$$\nabla_\theta^E(a *^E e) = \nabla_\theta^F(a) *^E e + (-1)^{(|\theta|+1)|a|} a *^E \nabla_\theta^E(e)$$

hold.

(3) A *curved Lie–Rinehart pair*

$$(\mathcal{F}^\bullet, \mathcal{T}^\bullet, \mathcal{E}^\bullet, (*^P)_{P \in D}, (\nabla^P)_{P \in D}, (\bar{\partial}^P)_{P \in D}, 1_F, \ell)$$

consists of a curved Lie–Rinehart algebra and a C-linear map $\bar{\partial}^E : \mathcal{E}^i \to \mathcal{E}^{i+1}$ satisfying the derivation rules

$$\bar{\partial}^E \nabla_\theta^E(e) = \nabla_{\bar{\partial}^T \theta}^T(e) + (-1)^{|\theta|+1} \nabla_\theta^E(\bar{\partial}^E e)$$

and

$$\bar{\partial}^E(a *^E e) = \bar{\partial}^F(a) *^E e + (-1)^{|a|} a *^E \bar{\partial}^E(e)$$

as well as $(\bar{\partial}^E)^2(e) = \nabla_\ell^E(e)$.

(4) A *differential bigraded Lie–Rinehart pair* is a curved Lie–Rinehart pair with $\ell = 0$.

(5) A Lie–Rinehart pair is *centerless* respectively *strictly faithful* if the underlying Lie–Rinehart algebra is; it is *faithful* if the intersection of the kernels of the three maps

$$\mathrm{ad}_P : \ \mathcal{T} \to \mathcal{H}om_C(P, P), \quad \theta \mapsto (p \mapsto \nabla_\theta^P(p)),$$

for $P \in \{\mathcal{F}, \mathcal{T}, \mathcal{E}\}$ is zero.

(6) A bigraded/dbg/curved Lie–Rinehart pair is *centerless* respectively *strictly faithful* if the underlying (predifferential) bigraded Lie–Rinehart algebra is; it is *faithful* if, for each $i \geq 0$, the intersection of the kernels of the three maps

$$\mathrm{ad}^i : \ \mathcal{T}^i \to \mathcal{H}om_C(P^0, P^i), \quad \theta \mapsto (p \mapsto \nabla_\theta^P(p)),$$

is zero.

(7) A bigraded/dbg/curved Lie–Rinehart pair is *bounded* if $\mathcal{F}^i = 0$, $\mathcal{T}^i = 0$, and $\mathcal{E}^i = 0$ for $i \gg 0$.

Example 3.9 Let $X/\mathbf{k}$ be a smooth variety, let $\mathcal{F} = O_X$, $\mathcal{T} = \mathcal{T}_{X/\mathbf{k}}$, and let $\mathcal{E}$ be a vector bundle with an (algebraic) connection $\nabla : \mathcal{E} \to \mathcal{E} \otimes \Omega^1_{X/\mathbf{k}}$. Then we can consider $(\mathcal{F}, \mathcal{T}, \mathcal{E})$ as a Lie–Rinehart pair in the context $\mathfrak{Coh}(X/\mathbf{k})$ with the induced map $\nabla : \mathcal{T} \times \mathcal{E} \to \mathcal{E}$. $\Diamond$

3.3 Gerstenhaber Algebras and Calculi

When we endow the polyvector fields $\bigwedge^p \mathcal{T}_{X/S}$ with the Schouten–Nijenhuis bracket, the result is a *Gerstenhaber algebra*. The notion goes back to [104]. We fix an integer $d \geq 1$, which we call the *dimension* of the Gerstenhaber algebra. For the polyvector fields, it corresponds to the relative dimension of $f : X \to S$. We put Gerstenhaber algebras in negative degrees, i.e., they are in the range $[-d, 0]$. This makes the formulae for its modules more natural, and this convention has been used before in [38, 76].

Definition 3.10 (Gerstenhaber Algebras) Let $(\mathcal{S}, C, O, \mathbf{C})$ be a context.

(1) A *Gerstenhaber algebra* of dimension d

$$(\mathcal{G}^\bullet, \wedge, 1_G, [-,-])$$

consists of:

- an O-module $\mathcal{G}^p \in \mathbf{C}$ of degree $|\mathcal{G}^p| = p$ for every $-d \le p \le 0$; for p outside this range, we set $\mathcal{G}^p = 0$;
- an O-bilinear product $-\wedge-: \mathcal{G}^p \times \mathcal{G}^{p'} \to \mathcal{G}^{p+p'}$ and a global section $1_G \in \mathcal{G}^0$ such that

$$\theta \wedge (\xi \wedge \eta) = (\theta \wedge \xi) \wedge \eta; \quad \theta \wedge \xi = (-1)^{|\theta||\xi|}\xi \wedge \theta; \quad 1_G \wedge \theta = \theta;$$

- a C-bilinear product $[-,-]: \mathcal{G}^p \times \mathcal{G}^{p'} \to \mathcal{G}^{p+p'+1}$ such that

$$[\theta, \xi] = -(-1)^{(|\theta|+1)(|\xi|+1)}[\xi, \theta]$$

and both the Jacobi identity

$$[\theta, [\xi, \eta]] = [[\theta, \xi], \eta] + (-1)^{(|\theta|+1)(|\xi|+1)}[\xi, [\theta, \eta]]$$

and the odd Poisson identity

$$[\theta, \xi \wedge \eta] = [\theta, \xi] \wedge \eta + (-1)^{(|\theta|+1)|\xi|}\xi \wedge [\theta, \eta]$$

hold.

From these axioms, we also find $[\theta, 1_G] = 0$, the formula $[\varphi, [\varphi, \xi]] = [\frac{1}{2}[\varphi, \varphi], \xi]$ for $|\varphi|$ even, a second odd Poisson identity

$$[\theta \wedge \xi, \eta] = \theta \wedge [\xi, \eta] + (-1)^{(|\eta|+1)|\xi|}[\theta, \eta] \wedge \xi,$$

and $[\varphi^n, \theta] = n[\varphi, \theta] \wedge \varphi^{n-1}$ for $n \ge 1$ and $|\varphi|$ even, where we write $\varphi^n := \varphi \wedge \ldots \wedge \varphi$.

(2) A *bigraded Gerstenhaber algebra* of dimension d

$$(\mathcal{G}^{\bullet,\bullet}, \wedge, 1_G, [-,-])$$

consists of:

- the O-modules $\mathcal{G}^{p,q} \in \mathbf{C}$ of bidegree (p, q) and total degree $p + q$ for every $-d \le p \le 0$ and every $q \ge 0$; for (p, q) outside this range, we set $\mathcal{G}^{p,q} = 0$ whenever necessary;

- an O-bilinear product $-\wedge-\colon \mathcal{G}^{p,q} \times \mathcal{G}^{p',q'} \to \mathcal{G}^{p+p',q+q'}$ and a global section $1_G \in \mathcal{G}^{0,0}$ such that

$$\theta \wedge (\xi \wedge \eta) = (\theta \wedge \xi) \wedge \eta; \quad \theta \wedge \xi = (-1)^{|\theta||\xi|}\xi \wedge \theta; \quad 1_G \wedge \theta = \theta;$$

- a C-bilinear product $[-,-]\colon \mathcal{G}^{p,q} \times \mathcal{G}^{p',q'} \to \mathcal{G}^{p+p'+1,q+q'}$ such that

$$[\theta,\xi] = -(-1)^{(|\theta|+1)(|\xi|+1)}[\xi,\theta]$$

and both the Jacobi identity

$$[\theta,[\xi,\eta]] = [[\theta,\xi],\eta] + (-1)^{(|\theta|+1)(|\xi|+1)}[\xi,[\theta,\eta]]$$

and the odd Poisson identity

$$[\theta,\xi \wedge \eta] = [\theta,\xi] \wedge \eta + (-1)^{(|\theta|+1)|\xi|}\xi \wedge [\theta,\eta]$$

hold.

From these axioms, we find the same additional formulae as above, namely $[\theta,1_G] = 0$, the formula $[\varphi,[\varphi,\xi]] = [\frac{1}{2}[\varphi,\varphi],\xi]$ for $|\varphi|$ even, a second odd Poisson identity

$$[\theta \wedge \xi,\eta] = \theta \wedge [\xi,\eta] + (-1)^{(|\eta|+1)|\xi|}[\theta,\eta] \wedge \xi,$$

and $[\varphi^n,\theta] = n[\varphi,\theta] \wedge \varphi^{n-1}$ for $n \geq 1$ and $|\varphi|$ even, where we write $\varphi^n := \varphi \wedge \ldots \wedge \varphi$.

(3) A *curved Gerstenhaber algebra*[6] of dimension d

$$(\mathcal{G}^{\bullet,\bullet}, \wedge, 1, [-,-], \bar{\partial}, \ell)$$

consists of a bigraded Gerstenhaber algebra of dimension d, a C-linear map $\bar{\partial}\colon \mathcal{G}^{p,q} \to \mathcal{G}^{p,q+1}$ satisfying the two derivation rules

$$\bar{\partial}[\theta,\xi] = [\bar{\partial}\theta,\xi] + (-1)^{|\theta|+1}[\theta,\bar{\partial}\xi], \quad \bar{\partial}(\theta \wedge \xi) = \bar{\partial}\theta \wedge \xi + (-1)^{|\theta|}\theta \wedge \bar{\partial}\xi,$$

and a global section $\ell \in \mathcal{G}^{-1,2}$ with $\bar{\partial}^2(\theta) = [\ell,\theta]$ and $\bar{\partial}\ell = 0$. Then, it also satisfies $\bar{\partial}(1) = 0$ and $\ell \wedge \ell = 0$ as well as

$$\bar{\partial}(\varphi^n) = n \cdot \bar{\partial}\varphi \wedge \varphi^{n-1}$$

for $n \geq 1$ and $|\varphi|$ even, where we write $\varphi^n := \varphi \wedge \ldots \wedge \varphi$.

---

[6] Note that for us, *curved* always refers to the graded Lie algebra $\mathcal{G}^{-1,\bullet}$ in a bigraded Gerstenhaber algebra $\mathcal{G}^{\bullet,\bullet}$. We never use the term *curved Gerstenhaber algebra* to denote the structure of a curved Lie algebra on a singly graded Gerstenhaber algebra $\mathcal{G}^{\bullet}$, considered as a graded Lie algebra.

(4) A *differential bigraded Gerstenhaber algebra* of dimension d is a curved Gerstenhaber algebra of dimension d with $\ell = 0$.

(5) A Gerstenhaber algebra is *centerless* if the map

$$\mathrm{ad}_T\colon\ \mathcal{G}^{-1} \to \mathcal{H}om_C(\mathcal{G}^{-1}, \mathcal{G}^{-1}), \quad \theta \mapsto (\xi \mapsto [\theta, \xi]),$$

is injective; it is called *strictly faithful* if the map

$$\mathrm{ad}_F\colon\ \mathcal{G}^{-1} \to \mathcal{H}om_C(\mathcal{G}^0, \mathcal{G}^0), \quad \theta \mapsto (\xi \mapsto [\theta, \xi]),$$

is injective; it is *faithful* if the intersection of the kernels of all maps

$$\mathrm{ad}_{G^p}\colon\ \mathcal{G}^{-1} \to \mathcal{H}om(\mathcal{G}^p, \mathcal{G}^p), \quad \theta \mapsto (\xi \mapsto [\theta, \xi]),$$

is zero.

(6) A bigraded/dbg/curved Gerstenhaber algebra is *centerless* if the map

$$\mathrm{ad}_T^q\colon\ \mathcal{G}^{-1,q} \to \mathcal{H}om_C(\mathcal{G}^{-1,0}, \mathcal{G}^{-1,q}), \quad \theta \mapsto (\xi \mapsto [\theta, \xi]),$$

is injective for all $q \geq 0$; it is called *strictly faithful* if the map

$$\mathrm{ad}_F^q\colon\ \mathcal{G}^{-1,q} \to \mathcal{H}om_C(\mathcal{G}^{0,0}, \mathcal{G}^{0,q}), \quad \theta \mapsto (\xi \mapsto [\theta, \xi]),$$

is injective for all $q \geq 0$; it is *faithful* if the intersection of the kernels of all maps

$$\mathrm{ad}_{G^p}^q\colon\ \mathcal{G}^{-1,q} \to \mathcal{H}om(\mathcal{G}^{p,0}, \mathcal{G}^{p,q}), \quad \theta \mapsto (\xi \mapsto [\theta, \xi]),$$

is zero for each $q \geq 0$.

(7) A bigraded/dbg/curved Gerstenhaber algebra is *bounded* if $\mathcal{G}^{p,q} = 0$ for $q \gg 0$.

Remark 3.11 When we take $\mathcal{F} = \mathcal{G}^0$, $\mathcal{T} = \mathcal{G}^{-1}$, then we obtain a Lie–Rinehart algebra with $*^P = \wedge$ and $\nabla^P = [-, -]$. $\qquad\qquad \Diamond$

Example 3.12 Let $f\colon X \to S$ be a smooth morphism of relative dimension d. When we set $\mathcal{G}^p := \bigwedge^{-p} \mathcal{T}_{X/S}$ for $-d \leq p \leq 0$, then we obtain a Gerstenhaber algebra in the context $\mathfrak{Coh}(X/S)$ by taking for $[-, -]$ the Schouten–Nijenhuis bracket $[-, -]_{\mathrm{SN}}$. When we take instead $[-, -] = -[-, -]_{\mathrm{SN}}$—the *negative* of the Schouten–Nijenhuis bracket—then we obtain a Gerstenhaber algebra as well. We shall follow this latter convention. $\qquad\qquad \Diamond$

There are several notions of modules over a Gerstenhaber algebra possible, cf. for example [75]. The only notion interesting for us at this point is the one that fits to the de Rham complex as a module over the polyvector fields via contraction. We call a pair of a Gerstenhaber algebra and such a module (a *de Rham module*) a *Gerstenhaber calculus*.

Definition 3.13 (Gerstenhaber Calculi) Let $(S, C, O, \mathbf{C})$ be a context.

(1) A *Gerstenhaber calculus* of dimension d

$$\mathcal{GC}^\bullet = (\mathcal{G}^\bullet, \wedge, 1_G, [-, -], \mathcal{A}^\bullet, \wedge, 1_A, \partial, \lrcorner, \mathcal{L})$$

consists of a Gerstenhaber algebra of dimension d and:

- an O-module $\mathcal{A}^i \in \mathbf{C}$ of degree i for every $0 \le i \le d$; for i outside this range, we set $\mathcal{A}^i = 0$ whenever necessary;
- an O-bilinear product $-\wedge-: \mathcal{A}^i \times \mathcal{A}^{i'} \to \mathcal{A}^{i+i'}$ and a global section $1_A \in \mathcal{A}^0$ such that

$$(\alpha \wedge \beta) \wedge \gamma = \alpha \wedge (\beta \wedge \gamma); \quad \alpha \wedge \beta = (-1)^{|\alpha||\beta|} \beta \wedge \alpha; \quad 1_A \wedge \alpha = \alpha;$$

- a C-linear map $\partial: \mathcal{A}^i \to \mathcal{A}^{i+1}$ with $\partial^2 = 0$, satisfying the derivation rule

$$\partial(\alpha \wedge \beta) = \partial(\alpha) \wedge \beta + (-1)^{|\alpha|} \alpha \wedge \partial(\beta),$$

  the *de Rham differential*;
- an O-bilinear map $\lrcorner: \mathcal{G}^p \times \mathcal{A}^i \to \mathcal{A}^{p+i}$ such that

$$1_G \lrcorner \alpha = \alpha \quad \text{and} \quad (\theta \wedge \xi) \lrcorner \alpha = \theta \lrcorner (\xi \lrcorner \alpha);$$

  it is called the *contraction map*;
- a C-bilinear map $\mathcal{L}_-(-): \mathcal{G}^p \times \mathcal{A}^i \to \mathcal{A}^{p+i+1}$ such that

$$\mathcal{L}_{[\theta, \xi]}(\alpha) = \mathcal{L}_\theta(\mathcal{L}_\xi(\alpha)) - (-1)^{(|\theta|+1)(|\xi|+1)} \mathcal{L}_\xi(\mathcal{L}_\theta(\alpha));$$

  it is called the *Lie derivative*.

They must satisfy additionally:

- the mixed Leibniz rule

$$\theta \lrcorner \mathcal{L}_\xi(\alpha) = (-1)^{|\xi|+1}([\theta, \xi] \lrcorner \alpha) + (-1)^{|\theta|(|\xi|+1)} \mathcal{L}_\xi(\theta \lrcorner \alpha);$$

- the Lie–Rinehart homotopy formula

$$(-1)^{|\theta|} \mathcal{L}_\theta(\alpha) = \partial(\theta \lrcorner \alpha) - (-1)^{|\theta|}(\theta \lrcorner \partial\alpha);$$

consequently, we have

$$\mathcal{L}_{\theta \wedge \xi}(\alpha) = (-1)^{|\xi|} \mathcal{L}_\theta(\xi \lrcorner \alpha) + \theta \lrcorner \mathcal{L}_\xi(\alpha);$$

- for $\theta \in \mathcal{G}^0$, the identity $\theta \lrcorner \alpha = (\theta \lrcorner 1_A) \wedge \alpha$;

- for $\theta \in G^{-1}$, the identity

$$\theta \lrcorner (\alpha \wedge \beta) = (\theta \lrcorner \alpha) \wedge \beta + (-1)^{|\alpha||\theta|} \alpha \wedge (\theta \lrcorner \beta);$$

consequently, we have for $\theta \in G^{-1}$

$$\mathcal{L}_\theta(\alpha \wedge \beta) = \mathcal{L}_\theta(\alpha) \wedge \beta + (-1)^{|\alpha|(|\theta|+1)} \alpha \wedge \mathcal{L}_\theta(\beta);$$

- the map $\lambda \colon G^0 \to \mathcal{A}^0$, $\theta \mapsto \theta \lrcorner 1_A$, is an O-linear isomorphism with $\lambda(\theta \wedge \xi) = \lambda(\theta) \wedge \lambda(\xi)$.

Then we also have $\mathcal{L}_{1_G}(\alpha) = 0$ as well as $\mathcal{L}_\theta(1_A) = \partial(\theta \lrcorner 1_A)$ for $\theta \in G^0$ and $\mathcal{L}_\theta(1_A) = 0$ for $\theta \in G^p$ with $p \le -1$. For φ with $|\varphi|$ even and $n \ge 2$, we have

$$\mathcal{L}_{\varphi^n}(\alpha) = n\varphi^{n-1} \lrcorner \mathcal{L}_\varphi(\alpha) + \frac{n(n-1)}{2}([\varphi, \varphi] \wedge \varphi^{n-2}) \lrcorner \alpha.$$

The Lie–Rinehart homotopy formula yields $\partial \mathcal{L}_\theta(\alpha) = (-1)^{|\theta|+1} \mathcal{L}_\theta(\partial \alpha)$.

(2) A *bigraded Gerstenhaber calculus* of dimension d

$$GC^{\bullet,\bullet} = (G^{\bullet,\bullet}, \wedge, 1_G, [-,-], \mathcal{A}^{\bullet,\bullet}, \wedge, 1_A, \partial, \lrcorner, \mathcal{L})$$

consists of a bigraded Gerstenhaber algebra of dimension d and:

- an O-module $\mathcal{A}^{i,j} \in \mathbf{C}$ of bidegree (i, j) and total degree $i + j$ for every $0 \le i \le d$ and every $j \ge 0$; for (i, j) outside this range, we set $\mathcal{A}^{i,j} = 0$ whenever necessary;
- an O-bilinear product $-\wedge-\colon \mathcal{A}^{i,j} \times \mathcal{A}^{i',j'} \to \mathcal{A}^{i+i',j+j'}$ and a global section $1_A \in \mathcal{A}^{0,0}$ such that

$$(\alpha \wedge \beta) \wedge \gamma = \alpha \wedge (\beta \wedge \gamma); \quad \alpha \wedge \beta = (-1)^{|\alpha||\beta|} \beta \wedge \alpha; \quad 1_A \wedge \alpha = \alpha;$$

- a C-linear map $\partial \colon \mathcal{A}^{i,j} \to \mathcal{A}^{i+1,j}$ with $\partial^2 = 0$, satisfying the derivation rule

$$\partial(\alpha \wedge \beta) = \partial(\alpha) \wedge \beta + (-1)^{|\alpha|} \alpha \wedge \partial(\beta),$$

the *de Rham differential*;
- an O-bilinear map $\lrcorner \colon G^{p,q} \times \mathcal{A}^{i,j} \to \mathcal{A}^{p+i,q+j}$ such that

$$1_G \lrcorner \alpha = \alpha \quad \text{and} \quad (\theta \wedge \xi) \lrcorner \alpha = \theta \lrcorner (\xi \lrcorner \alpha);$$

it is called the *contraction map*;
- a C-bilinear map $\mathcal{L}_-(-)\colon G^{p,q} \times \mathcal{A}^{i,j} \to \mathcal{A}^{p+i+1,q+j}$ such that

$$\mathcal{L}_{[\theta,\xi]}(\alpha) = \mathcal{L}_\theta(\mathcal{L}_\xi(\alpha)) - (-1)^{(|\theta|+1)(|\xi|+1)} \mathcal{L}_\xi(\mathcal{L}_\theta(\alpha));$$

it is called the *Lie derivative*.

They must satisfy additionally:

- the mixed Leibniz rule

$$\theta \mathbin{\lrcorner} \mathcal{L}_\xi(\alpha) = (-1)^{|\xi|+1}([\theta, \xi] \mathbin{\lrcorner} \alpha) + (-1)^{|\theta|(|\xi|+1)}\mathcal{L}_\xi(\theta \mathbin{\lrcorner} \alpha);$$

- the Lie–Rinehart homotopy formula

$$(-1)^{|\theta|}\mathcal{L}_\theta(\alpha) = \partial(\theta \mathbin{\lrcorner} \alpha) - (-1)^{|\theta|}(\theta \mathbin{\lrcorner} \partial\alpha);$$

consequently, we have

$$\mathcal{L}_{\theta \wedge \xi}(\alpha) = (-1)^{|\xi|}\mathcal{L}_\theta(\xi \mathbin{\lrcorner} \alpha) + \theta \mathbin{\lrcorner} \mathcal{L}_\xi(\alpha);$$

- for $\theta \in \mathcal{G}^{0,q}$, the identity $\theta \mathbin{\lrcorner} \alpha = (\theta \mathbin{\lrcorner} 1_A) \wedge \alpha$;
- for $\theta \in \mathcal{G}^{-1,q}$, the identity

$$\theta \mathbin{\lrcorner} (\alpha \wedge \beta) = (\theta \mathbin{\lrcorner} \alpha) \wedge \beta + (-1)^{|\alpha||\theta|}\alpha \wedge (\theta \mathbin{\lrcorner} \beta);$$

consequently, we have for $\theta \in \mathcal{G}^{-1,q}$

$$\mathcal{L}_\theta(\alpha \wedge \beta) = \mathcal{L}_\theta(\alpha) \wedge \beta + (-1)^{|\alpha|(|\theta|+1)}\alpha \wedge \mathcal{L}_\theta(\beta);$$

- the map $\lambda \colon \mathcal{G}^{0,q} \to \mathcal{A}^{0,q}$, $\theta \mapsto \theta \mathbin{\lrcorner} 1_A$, is an O-linear isomorphism with $\lambda(\theta \wedge \xi) = \lambda(\theta) \wedge \lambda(\xi)$.

As above, we also have $\mathcal{L}_{1_G}(\alpha) = 0$ as well as $(-1)^{|\theta|}\mathcal{L}_\theta(1_A) = \partial(\theta \mathbin{\lrcorner} 1_A)$ for $\theta \in \mathcal{G}^{0,q}$ and $\mathcal{L}_\theta(1_A) = 0$ for $\theta \in \mathcal{G}^{p,q}$ with $p \le -1$. For φ with $|\varphi|$ even and $n \ge 2$, we have

$$\mathcal{L}_{\varphi^n}(\alpha) = n\varphi^{n-1} \mathbin{\lrcorner} \mathcal{L}_\varphi(\alpha) + \frac{n(n-1)}{2}([\varphi, \varphi] \wedge \varphi^{n-2}) \mathbin{\lrcorner} \alpha.$$

The Lie–Rinehart homotopy formula yields $\partial\mathcal{L}_\theta(\alpha) = (-1)^{|\theta|+1}\mathcal{L}_\theta(\partial\alpha)$.

(3) A *curved Gerstenhaber calculus* of dimension d

$$\mathcal{G}C^{\bullet,\bullet} = (\mathcal{G}^{\bullet,\bullet}, \wedge, 1_G, [-,-], \bar\partial, \ell, \mathcal{A}^{\bullet,\bullet}, \wedge, 1_A, \partial, \mathbin{\lrcorner}, \mathcal{L}, \bar\partial)$$

is simultaneously a curved Gerstenhaber algebra of dimension d and a bigraded Gerstenhaber calculus of dimension d together with a C-linear map $\bar\partial \colon \mathcal{A}^{i,j} \to \mathcal{A}^{i,j+1}$ satisfying the derivation rules

$$\bar\partial(\alpha \wedge \beta) = \bar\partial\alpha \wedge \beta + (-1)^{|\alpha|}\alpha \wedge \bar\partial\beta, \quad \bar\partial(\theta \mathbin{\lrcorner} \alpha) = (\bar\partial\theta) \mathbin{\lrcorner} \alpha + (-1)^{|\theta|}\theta \mathbin{\lrcorner} \bar\partial\alpha$$

as well as $\bar{\partial}^2(\alpha) = \mathcal{L}_\ell(\alpha)$ and $\partial\bar{\partial} + \bar{\partial}\partial = 0$. The Lie–Rinehart homotopy formula yields

$$\bar{\partial}\mathcal{L}_\theta(\alpha) = \mathcal{L}_{\bar{\partial}\theta}(\alpha) + (-1)^{|\theta|+1}\mathcal{L}_\theta(\bar{\partial}\alpha).$$

(4) A *differential bigraded Gerstenhaber calculus* of dimension d is a curved Gerstenhaber calculus of dimension d with $\ell = 0$.
(5) A Gerstenhaber calculus of dimension d is *centerless* respectively *strictly faithful* if the underlying Gerstenhaber algebra is; *faithfulness* is analogous to the previous definitions, taking into account all maps into $\mathcal{H}om_C(\mathcal{G}^{p,0}, \mathcal{G}^{p,q})$ and $\mathcal{H}om_C(\mathcal{A}^{i,0}, \mathcal{A}^{i,q})$.
(6) A bigraded/dbg/curved Gerstenhaber calculus of dimension d is *centerless* respectively *strictly faithful* if the underlying bigraded/dbg/curved Gerstenhaber algebra is; *faithfulness* is analogous to the previous definitions.
(7) A bigraded/dbg/curved Gerstenhaber calculus of dimension d is *bounded* if $\mathcal{G}^{p,q} = 0$ and $\mathcal{A}^{i,q} = 0$ for all $-d \leq p \leq 0, 0 \leq i \leq d$ for $q \gg 0$.

Remark 3.14 The symbol $\mathcal{G}C^\bullet$ which we use above as a short-hand notation for a Gerstenhaber calculus is in analogy with our notation for general algebraic structures; the bullet is meant to run over the index set $D = [-d, 0] \sqcup [0, d]$ although we will never use the notation $\mathcal{G}C^P$ for specific values $P \in D$. $\Diamond$

Example 3.15 Let $f \colon X \to S$ be a smooth morphism of relative dimension d. Continuing Example 3.12, we set $\mathcal{A}^i = \Omega^i_{X/S}$. Then we obtain a Gerstenhaber calculus of dimension d in the context $\mathfrak{Coh}(X/S)$ with the de Rham differential ∂, the contraction map $\lrcorner$ as defined, for example, in [75], and the Lie derivative $\mathcal{L}$ as defined by the Lie–Rinehart homotopy formula in the definition. $\Diamond$

3.3.1 The Local Batalin–Vilkovisky Condition

In a Gerstenhaber calculus $\mathcal{G}C^\bullet = (\mathcal{G}^\bullet, \mathcal{A}^\bullet)$, the $\wedge$-product turns $\mathcal{G}^0$ and $\mathcal{A}^0$ into sheaves of commutative rings, which are isomorphic via $\lambda \colon \mathcal{G}^0 \to \mathcal{A}^0$. We denote this sheaf of rings occasionally by $\mathcal{F}$, consistent with our notation for Lie–Rinehart pairs. Every $\mathcal{G}^p$ and every $\mathcal{A}^i$ comes with a structure of $\mathcal{F}$-module, and both $\wedge$-products as well as the contraction $\lrcorner$ are $\mathcal{F}$-bilinear.

Definition 3.16 (Gorenstein I) A Gerstenhaber calculus $(\mathcal{G}^\bullet, \mathcal{A}^\bullet)$ of dimension d is *Gorenstein*[7] if $\mathcal{A}^d$ is a locally free $\mathcal{F}$-module of rank 1.

---

[7] We have chosen this name because, in applications, $\mathcal{A}^d$ is usually some canonical bundle over a space.

When $\mathcal{G}C^\bullet$ is Gorenstein, then every local generator ω of $\mathcal{A}^d$ defines locally an $\mathcal{F}$-linear map

$$\kappa_\omega \colon \ \mathcal{G}^p \to \mathcal{A}^{p+d}, \quad \theta \mapsto \theta \lrcorner \omega.$$

In geometric applications, for example on a smooth or, more generally, log smooth morphism $f \colon X \to S$, this map is usually an isomorphism of $\mathcal{F}$-modules. Moreover, when $\mathcal{G}C^\bullet$ is in fact a (part of a) Batalin–Vilkovisky calculus as defined below, then $\mathcal{A}^d \cong \mathcal{A}^0$, and κ_ω is a global isomorphism. We consider the condition that κ_ω, defined locally, is an isomorphism as a property of Gerstenhaber calculi.

Definition 3.17 (Locally Batalin–Vilkovisky I) A Gerstenhaber calculus $(\mathcal{G}^\bullet, \mathcal{A}^\bullet)$ is *locally Batalin–Vilkovisky* if it is Gorenstein, and the map $\kappa_\omega \colon \mathcal{G}^p \to \mathcal{A}^{p+d}$ is an isomorphism of $\mathcal{F}$-modules for every local generator ω of $\mathcal{A}^d$ as an $\mathcal{F}$-module.

Note that it is sufficient to check this condition on some local generator in a neighborhood of each point.

We also introduce a bigraded variant of the two definitions. We write $\mathcal{F}^0 = \mathcal{G}^{0,0} = \mathcal{A}^{0,0}$.

Definition 3.18 (Gorenstein and Locally Batalin–Vilkovisky II) A bigraded Gerstenhaber calculus $(\mathcal{G}^{\bullet,\bullet}, \mathcal{A}^{\bullet,\bullet})$ is *Gorenstein* if we can find everywhere locally an element $\omega \in \mathcal{A}^{d,0}$ such that

$$v_\omega \colon \ \mathcal{A}^{0,j} \to \mathcal{A}^{d,j}, \quad \alpha \mapsto \alpha \wedge \omega,$$

is an isomorphism of $\mathcal{F}^0$-modules. It is *locally Batalin–Vilkovisky* if, for every such ω, the map $\kappa_\omega \colon \mathcal{G}^{p,q} \to \mathcal{A}^{p+d,q}$, $\theta \mapsto \theta \lrcorner \omega$, is an isomorphism of $\mathcal{F}^0$-modules.

Remark 3.19 We have to be careful with the notion of being locally Batalin–Vilkovisky for a curved Gerstenhaber calculus since κ_ω and v_ω are only compatible with $\bar{\partial}$ if $\bar{\partial}\omega = 0$. If we add this condition to the definition of being locally Batalin–Vilkovisky in the curved case, then we might end up with a Batalin–Vilkovisky calculus, as defined below, which is not locally Batalin–Vilkovisky. $\Diamond$

3.3.2 The Operator $d = \partial + \bar{\partial} + \ell \lrcorner (-)$ and the Hypercohomology

Let $(\mathcal{G}^{\bullet,\bullet}, \mathcal{A}^{\bullet,\bullet})$ be a curved Gerstenhaber calculus in the context $(S, C, O, \mathbf{C})$. Let us assume $\ell = 0$ for a moment. Then $(\mathcal{A}^{\bullet,\bullet}, \partial, \bar{\partial})$ is a double complex (with $\partial\bar{\partial} + \bar{\partial}\partial = 0$), and we can form its *total complex* $\mathrm{Tot}^\bullet(\mathcal{A}^{\bullet,\bullet}, \partial, \bar{\partial}) = (\mathcal{A}^\bullet, d)$ with

$$\mathcal{A}^m := \bigoplus_{i+j=m} \mathcal{A}^{i,j}, \quad d := \partial + \bar{\partial},$$

where $\mathcal{A}^m$ is a finite direct sum since both variables i and j are bounded below. However, when $\ell \neq 0$, then possibly $d^2 \neq 0$. Nonetheless, as was apparently first observed in [38, Defn. 4.10], we can define a differential on $\mathcal{A}^\bullet$ as the C-linear operator

$$d \colon \; \mathcal{A}^m \to \mathcal{A}^{m+1}, \quad \alpha \mapsto \partial \alpha + \bar{\partial} \alpha + (\ell \lrcorner \alpha),$$

called the *total de Rham differential*; it still preserves the total degree, albeit that $\mathcal{A}^{i,j}$ is mapped into the direct sum of three pieces $\mathcal{A}^{i+1,j}$, $\mathcal{A}^{i,j+1}$, and $\mathcal{A}^{i-1,j+2}$ instead of two as in the case $\bar{\partial}^2 = 0$.

Lemma and Definition 3.20 *The operator $d \colon \mathcal{A}^m \to \mathcal{A}^{m+1}$ is a differential, i.e., $d^2 = 0$, and the derivation rule*

$$d(\alpha \wedge \beta) = d\alpha \wedge \beta + (-1)^{|\alpha|} \alpha \wedge d\beta$$

holds. We denote its cohomology by $\mathbb{H}^m(\mathcal{A}^{\bullet,\bullet}) := H^m(\mathcal{A}^\bullet, d)$ *and call it the hypercohomology of* $(\mathcal{G}^{\bullet,\bullet}, \mathcal{A}^{\bullet,\bullet})$.

Proof The first assertion is a direct computation using $\partial(\ell \lrcorner \alpha) = -\mathcal{L}_\ell(\alpha) - (\ell \lrcorner \partial \alpha)$. The second assertion follows from the derivation rules for ∂, $\bar{\partial}$, and $\ell \lrcorner (-)$. $\qquad\square$

Remark 3.21 The name comes from the following situation: Let $f \colon X \to S$ be a (log) smooth morphism of relative dimension d, and consider the Gerstenhaber calculus of Example 3.15. Then there is a differential bigraded Gerstenhaber calculus in $\mathfrak{QCoh}(X/S)$ which forms a flasque resolution of the original singly graded Gerstenhaber calculus. Its hypercohomology in the sense of the above Definition is exactly the hypercohomology of the de Rham complex. $\qquad\diamond$

3.3.3 Two-sided Gerstenhaber Calculi

A Gerstenhaber calculus $\mathcal{G}C^\bullet$ as defined above comes with a contraction map $\lrcorner \colon \mathcal{G}^p \times \mathcal{A}^i \to \mathcal{A}^{p+i}$ which turns $\mathcal{A}^\bullet$ into a module over $\mathcal{G}^\bullet$. However, usually we can define a second analogous contraction map $\vdash \colon \mathcal{G}^p \times \mathcal{A}^i \to \mathcal{G}^{p+i}$, which turns $\mathcal{G}^\bullet$ into a module over $\mathcal{A}^\bullet$. This *left contraction*, as we shall call it to distinguish it from $\lrcorner$, which we then call the *right contraction*, has not been used in Chan, Leung, and Ma's article [38], and we have become aware of it only lately. Most of the theory does not need the left contraction, but it is occasionally convenient to have it at our disposal, especially in the setting of enhanced generically log smooth families, which carry a Gerstenhaber calculus which is not determined by the geometry of the underlying generically log smooth family. We have decided to include the left contraction not in our original definition of a Gerstenhaber calculus but in a separate definition, that of a *two-sided* Gerstenhaber calculus. In practice, many one-sided Gerstenhaber calculi carry a unique left contraction; we will see below that this is the case, for example, for Gerstenhaber calculi which are locally Batalin–Vilkovisky.[8] One may desire that $\lambda(\theta \vdash \alpha) = \theta \dashv \alpha$ for $|\theta| + |\alpha| = 0$ so that $\vdash$ would be an extension of $\dashv$ to the range $|\theta| + |\alpha| \leq 0$, but this is not compatible with the other natural relations that we want to impose, in particular (3.1) below. We will have $\lambda(\theta \vdash \alpha) = (-1)^{|\theta|}\theta \dashv \alpha$ for $|\theta| + |\alpha| = 0$ instead, and, given the other relations, this cannot be rectified by writing $\alpha \dashv \theta$ since this would also change the natural signs in the other relations, for example (3.1). At least, this convention yields the very satisfactory formula $[\theta, g] = \theta \vdash \partial(\lambda(g))$ for our conventions for the bracket $[-, -]$, which is the negative of the Schouten–Nijenhuis bracket in practice.

Definition 3.22 (Two-sided Gerstenhaber Calculi) Let $(S, C, O, \mathbf{C})$ be a context.

(1) A *two-sided Gerstenhaber calculus* of dimension d

$$\mathcal{GC}^{\bullet} = (\mathcal{G}^{\bullet}, \wedge, 1_G, [-, -], \mathcal{A}^{\bullet}, \wedge, 1_A, \partial, \dashv, \mathcal{L}, \vdash)$$

is a Gerstenhaber calculus of dimension d together with an O-bilinear operation

$$\vdash : \ \mathcal{G}^p \times \mathcal{A}^i \to \mathcal{G}^{p+i},$$

the *left contraction*, satisfying the following relations:

- $\theta \vdash 1_A = \theta$ and $\theta \vdash (\alpha \wedge \beta) = (\theta \vdash \alpha) \vdash \beta$;
- for $\alpha \in \mathcal{A}^0$, the identity $\theta \vdash \alpha = \theta \wedge (1_G \vdash \alpha)$;
- for $\alpha \in \mathcal{A}^1$, the identity

$$(\theta \wedge \xi) \vdash \alpha = (-1)^{|\xi||\alpha|}(\theta \vdash \alpha) \wedge \xi + \theta \wedge (\xi \vdash \alpha);$$

- for $\theta \in \mathcal{G}^{-i}$ and $\alpha \in \mathcal{A}^i$, the identity $\lambda(\theta \vdash \alpha) = (-1)^i \theta \dashv \alpha$;
- for $\omega \in \mathcal{A}^d$, the identity

$$(\theta \vdash \alpha) \dashv \omega = (-1)^{|\alpha||\omega|}(\theta \dashv \omega) \wedge \alpha; \tag{3.1}$$

- for $\omega^{\vee} \in \mathcal{G}^{-d}$, the identity

$$\omega^{\vee} \vdash (\theta \dashv \alpha) = (-1)^{|\omega^{\vee}||\theta|} \theta \wedge (\omega^{\vee} \vdash \alpha); \tag{3.2}$$

---

[8] The local Batalin–Vilkovisky condition is broken by taking global sections. Thus, while we can construct the left contraction on the sheaf level in this case, the global sections carry a left contraction as well which is not easily obtained from the right contraction on the level of global sections.

- for $\theta \in \mathcal{G}^{-1}$, the *special left mixed Leibniz rule*[9]

$$[\theta, \xi \vdash \alpha] = [\theta, \xi] \vdash \alpha + \xi \vdash \mathcal{L}_\theta(\alpha).$$

(2) A *bigraded two-sided Gerstenhaber calculus* of dimension d

$$\mathcal{GC}^{\bullet,\bullet} = (\mathcal{G}^{\bullet,\bullet}, \wedge, 1_G, [-,-], \mathcal{A}^{\bullet,\bullet}, \wedge, 1_A, \partial, \lrcorner, \mathcal{L}, \vdash)$$

is a bigraded Gerstenhaber calculus of dimension d together with an O-bilinear map

$$\vdash : \quad \mathcal{G}^{p,q} \times \mathcal{A}^{i,j} \to \mathcal{G}^{p+i,q+j},$$

the *left contraction*, satisfying the following relations:

- $\theta \vdash 1_A = \theta$ and $\theta \vdash (\alpha \wedge \beta) = (\theta \vdash \alpha) \vdash \beta$;
- for $\alpha \in \mathcal{A}^{0,j}$, the identity $\theta \vdash \alpha = \theta \wedge (1_G \vdash \alpha)$;
- for $\alpha \in \mathcal{A}^{1,j}$, the identity

$$(\theta \wedge \xi) \vdash \alpha = (-1)^{|\xi||\alpha|}(\theta \vdash \alpha) \wedge \xi + \theta \wedge (\xi \vdash \alpha);$$

- for $\theta \in \mathcal{G}^{-i,q}$ and $\alpha \in \mathcal{A}^{i,j}$, the identity $\lambda(\theta \vdash \alpha) = (-1)^i \theta \lrcorner \alpha$;[10]
- for $\omega \in \mathcal{A}^{d,j}$, the identity $(\theta \vdash \alpha) \lrcorner \omega = (-1)^{|\alpha||\omega|}(\theta \lrcorner \omega) \wedge \alpha$;
- for $\omega^\vee \in \mathcal{G}^{-d,q}$, the identity $\omega^\vee \vdash (\theta \lrcorner \alpha) = (-1)^{|\omega^\vee||\theta|} \theta \wedge (\omega^\vee \vdash \alpha)$;
- for $\theta \in \mathcal{G}^{-1,q}$, the *special left mixed Leibniz rule*

$$[\theta, \xi \vdash \alpha] = [\theta, \xi] \vdash \alpha + (-1)^{(|\theta|+1)|\xi|} \xi \vdash \mathcal{L}_\theta(\alpha).$$

(3) A *curved two-sided Gerstenhaber calculus* of dimension d

$$\mathcal{GC}^{\bullet,\bullet} = (\mathcal{G}^{\bullet,\bullet}, \wedge, 1_G, [-,-], \bar{\partial}, \ell, \mathcal{A}^{\bullet,\bullet}, \wedge, 1_A, \partial, \lrcorner, \mathcal{L}, \vdash, \bar{\partial})$$

is simultaneously a bigraded two-sided Gerstenhaber calculus of dimension d and a curved Gerstenhaber calculus of dimension d such that

$$\bar{\partial}(\theta \vdash \alpha) = \bar{\partial}\theta \vdash \alpha + (-1)^{|\theta|} \theta \vdash \bar{\partial}\alpha.$$

---

[9] This relation is dictated by the compatibility of $\vdash$ with gauge transforms, cf. also Definition 3.37. As stated, the relation does not hold for θ of other degrees than -1 in those Gerstenhaber calculi in which we are interested. Unfortunately, we could not find a general relation of which this is a special case.

[10] This relation does not acquire the usual sign from the total degree when we obtain a bigraded two-sided Gerstenhaber calculus as the Thom–Whitney resolution of a singly graded one.

(4) A *differential bigraded two-sided Gerstenhaber calculus* of dimension d is a curved one with $\ell = 0$.

(5) The notions of faithfulness, strict faithfulness, centerlessness, and boundedness apply to the underlying one-sided Gerstenhaber calculus.

When $\mathcal{GC}^\bullet$ is a one-sided Gerstenhaber calculus which is locally Batalin–Vilkovisky, then we can use the relation

$$(\theta \vdash \alpha) \dashv \omega = (-1)^{|\alpha||\omega|}(\theta\omega) \wedge \alpha$$

to *define* the left contraction. Since this relation is $\mathcal{A}^0$-linear in ω on both sides, this definition is independent of the choice of local generator ω and thus globally consistent. Similarly, we can define the left contraction on a bigraded one-sided Gerstenhaber calculus which is locally Batalin–Vilkovisky. With this definition, most relations for $\vdash$ follow from the relations in a one-sided Gerstenhaber calculus, but not all.

Lemma 3.23 *Let $(\mathcal{S}, C, O, \mathbf{C})$ be a context.*

(1) *Let $\mathcal{GC}^\bullet$ be a one-sided Gerstenhaber calculus of dimension d which is locally Batalin–Vilkovisky. Define $\vdash$ as above. Then $\vdash : \mathcal{G}^p \times \mathcal{A}^i \to \mathcal{G}^{p+i}$ is O-bilinear and satisfies the following relations:*

- *$\theta \vdash 1_A = \theta$ and $\theta \vdash (\alpha \wedge \beta) = (\theta \vdash \alpha) \vdash \beta$;*
- *for $\alpha \in \mathcal{A}^0$, the identity $\theta \vdash \alpha = \theta \wedge (1_G \vdash \alpha)$;*
- *for $\theta \in \mathcal{G}^0$ and $\alpha \in \mathcal{A}^0$, the identity $\lambda(\theta \vdash \alpha) = \theta \dashv \alpha$;*
- *for $\theta \in \mathcal{G}^{-1}$ and $\alpha \in \mathcal{A}^1$, the identity $\lambda(\theta \vdash \alpha) = -\theta \dashv \alpha$;*
- *for $\omega^\vee \in \mathcal{G}^{-d}$, the identity $\omega^\vee \vdash (\theta \dashv \alpha) = (-1)^{|\omega^\vee||\theta|}\theta \wedge (\omega^\vee \vdash \alpha)$;*
- *for $\theta \in \mathcal{G}^{-1}$, the special left mixed Leibniz rule*

$$[\theta, \xi \vdash \alpha] = [\theta, \xi] \vdash \alpha + \xi \vdash \mathcal{L}_\theta(\alpha).$$

In other words, only the identity

$$(\theta \wedge \xi) \vdash \alpha = (-1)^{|\xi||\alpha|}(\theta \vdash \alpha) \wedge \xi + \theta \wedge (\xi \vdash \alpha),$$

for $\alpha \in \mathcal{A}^1$ as well as $\lambda(\theta \vdash \alpha) = (-1)^i \theta \dashv \alpha$ for $\theta \in \mathcal{G}^{-i}$ and $\alpha \in \mathcal{A}^i$ with $i \geq 2$ are missing.

(2) *Let $\mathcal{GC}^{\bullet,\bullet}$ be a bigraded one-sided Gerstenhaber calculus of dimension d which is locally Batalin–Vilkovisky. Define $\vdash$ as above. Then $\vdash : \mathcal{G}^{p,q} \times \mathcal{A}^{i,j} \to \mathcal{G}^{p+i,q+j}$ is an O-bilinear map satisfying all relations from the definition of a bigraded two-sided Gerstenhaber calculus except for possibly the relation for $(\theta \wedge \xi) \vdash \alpha$ and $\lambda(\theta \vdash \alpha) = (-1)^i \theta \dashv \alpha$ for $i \geq 2$.*[11]

---

[11] As above, in the curved case, if $\bar{\partial}\omega \neq 0$, we cannot conclude that the relation for $\bar{\partial}(\theta \vdash \alpha)$ holds automatically.

Proof We start with the singly graded case. That $\vdash$ respects the indicated degrees follows from the degrees in the defining relation of $\vdash$. Similarly, since both $\wedge$ and $\lrcorner$ are O-bilinear, we find that $\vdash$ is O-bilinear. The claimed relations now all follow from applying the local isomorphism $\kappa_\omega(-) = (-)\lrcorner\omega$ to the relation. To obtain the special left mixed Leibniz rule, one uses the usual mixed Leibniz rule. The proof of $\lambda(\theta\vdash\alpha) = -\theta\lrcorner\alpha$ for $i = 1$ uses the relation for $\theta\lrcorner(\alpha\wedge\beta)$, and that we have no such relation for $|\theta| \neq -1$ is the reason why we cannot show $\lambda(\theta\vdash\alpha) = (-1)^i\,\theta\lrcorner\alpha$ for $i \geq 2$ at this point. The proof in the bigraded case is similar. The computation for the bigraded special left mixed Leibniz rule is very tedious due to many signs that need to be tracked. $\qquad\square$

3.4 Batalin–Vilkovisky Algebras and Calculi

A *Batalin–Vilkovisky algebra* is a Gerstenhaber algebra with an additional operator $\Delta\colon G^p \to G^{p+1}$. This operator is usually obtained inside a Gerstenhaber *calculus* from the de Rham differential ∂ by means of an isomorphism $G^p \cong \mathcal{A}^{p+d}$ induced by a chosen global volume form $\omega \in \mathcal{A}^d$, which exists on Calabi–Yau spaces. For theoretical purposes, it is also worthwhile to study Batalin–Vilkovisky algebras without a de Rham module. We fix an integer $d \geq 1$.

Definition 3.24 (Batalin–Vilkovisky Algebras) Let $(\mathcal{S}, C, O, \mathbf{C})$ be a context.

(1) A *Batalin–Vilkovisky algebra* of dimension d

$$(G^\bullet, \wedge, 1_G, [-, -], \Delta)$$

is a Gerstenhaber algebra of dimension d together with a C-linear *Batalin–Vilkovisky operator* $\Delta\colon G^p \to G^{p+1}$ which satisfies

$$\Delta(1_G) = 0, \quad \Delta^2 = 0, \quad \Delta[\theta, \xi] = [\Delta(\theta), \xi] + (-1)^{|\theta|+1}[\theta, \Delta(\xi)],$$

and the *Bogomolov–Tian–Todorov formula*

$$(-1)^{|\theta|}[\theta, \xi] = \Delta(\theta \wedge \xi) - \Delta(\theta) \wedge \xi - (-1)^{|\theta|}\theta \wedge \Delta(\xi).$$

(2) A *bigraded Batalin–Vilkovisky algebra* of dimension d

$$(G^{\bullet,\bullet}, \wedge, 1_G, [-, -], \Delta)$$

is a bigraded Gerstenhaber algebra of dimension d together with a C-linear *Batalin–Vilkovisky operator* $\Delta\colon G^{p,q} \to G^{p+1,q}$ which satisfies

$$\Delta(1_G) = 0, \quad \Delta^2 = 0, \quad \Delta[\theta, \xi] = [\Delta(\theta), \xi] + (-1)^{|\theta|+1}[\theta, \Delta(\xi)],$$

and the *Bogomolov–Tian–Todorov formula*

$$(-1)^{|\theta|}[\theta, \xi] = \Delta(\theta \wedge \xi) - \Delta(\theta) \wedge \xi - (-1)^{|\theta|}\theta \wedge \Delta(\xi).$$

(3) A *curved Batalin–Vilkovisky algebra* of dimension d

$$(\mathcal{G}^{\bullet,\bullet}, \wedge, 1_G, [-,-], \Delta, \bar{\partial}, \ell, y)$$

is a bigraded Batalin–Vilkovisky algebra of dimension d together with a C-linear operator $\bar{\partial} \colon \mathcal{G}^{p,q} \to \mathcal{G}^{p,q+1}$ satisfying the two derivation rules

$$\bar{\partial}[\theta, \xi] = [\bar{\partial}\theta, \xi] + (-1)^{|\theta|+1}[\theta, \bar{\partial}\xi], \quad \bar{\partial}(\theta \wedge \xi) = \bar{\partial}\theta \wedge \xi + (-1)^{|\theta|}\theta \wedge \bar{\partial}\xi,$$

and global sections $\ell \in \mathcal{G}^{-1,2}$ and $y \in \mathcal{G}^{0,1}$ which satisfy

$$\bar{\partial}^2(\theta) = [\ell, \theta], \quad \bar{\partial}(\ell) = 0, \quad \bar{\partial}\Delta(\theta) + \Delta\bar{\partial}(\theta) = [y, \theta], \quad \bar{\partial}(y) + \Delta(\ell) = 0.$$

We then have

$$\bar{\partial}(1_G) = 0, \quad \ell \wedge \ell = 0, \quad y \wedge y = 0, \quad y \wedge \ell + \ell \wedge y = 0, \quad [\ell, y] + [y, \ell] = 0.$$

(4) A *differential bigraded Batalin–Vilkovisky algebra* of dimension d is a curved Batalin–Vilkovisky algebra of dimension d with $\ell = 0$ and $y = 0$.
(5) The notions of *centerless*, *strictly faithful*, *faithful*, and *bounded* apply to the underlying bigraded/dbg/curved Gerstenhaber algebra.

Remark 3.25 The condition $\bar{\partial}(\ell) + \Delta(y) = 0$ is not explicitly discussed in [38]; however, the condition is satisfied in their context, and we need it to show Lemma 3.30. Moreover, the proof of Proposition 3.28 suggests that the condition is natural. $\Diamond$

A *Batalin–Vilkovisky calculus* is the analog of a Batalin–Vilkovisky algebra for a Gerstenhaber calculus.

Definition 3.26 (Batalin–Vilkovisky Calculi) Let $(S, C, O, \mathbf{C})$ be a context.

(1) A *Batalin–Vilkovisky calculus* of dimension d

$$(\mathcal{G}^{\bullet}, \wedge, 1_G, [-,-], \Delta, \mathcal{A}^{\bullet}, \wedge, 1_A, \partial, \lrcorner, \mathcal{L}, \omega)$$

is simultaneously a bigraded Gerstenhaber calculus of dimension d and a bigraded Batalin–Vilkovisky algebra of dimension d together with a global section $\omega \in \mathcal{A}^d$ such that

$$\kappa \colon \mathcal{G}^p \to \mathcal{A}^{p+d}, \quad \theta \mapsto \theta \lrcorner \omega,$$

is an isomorphism of O-modules,

$$v\colon \; \mathcal{A}^0 \to \mathcal{A}^d, \quad \alpha \mapsto \alpha \wedge \omega,$$

is an isomorphism of O-modules, and

$$\Delta(\theta) \lrcorner\, \omega = \partial(\theta \lrcorner\, \omega)$$

holds.

(2) A *bigraded Batalin–Vilkovisky calculus* of dimension d

$$(\mathcal{G}^{\bullet,\bullet}, \wedge, 1_G, [-,-], \Delta, \mathcal{A}^{\bullet,\bullet}, \wedge, 1_A, \partial, \lrcorner, \mathcal{L}, \omega)$$

is simultaneously a Gerstenhaber calculus of dimension d and a Batalin–Vilkovisky algebra of dimension d together with a global section $\omega \in \mathcal{A}^{d,0}$ such that

$$\kappa\colon \; \mathcal{G}^{p,q} \to \mathcal{A}^{p+d,q}, \quad \theta \mapsto \theta \lrcorner\, \omega,$$

is an isomorphism of O-modules,

$$v\colon \; \mathcal{A}^{0,q} \to \mathcal{A}^{d,q}, \quad \alpha \mapsto \alpha \wedge \omega,$$

is an isomorphism of O-modules, and

$$\Delta(\theta) \lrcorner\, \omega = \partial(\theta \lrcorner\, \omega)$$

holds.

(3) A *curved Batalin–Vilkovisky calculus* of dimension d

$$(\mathcal{G}^{\bullet,\bullet}, \wedge, 1_G, [-,-], \Delta, \bar{\partial}, \ell, y, \mathcal{A}^{\bullet,\bullet}, \wedge, 1_A, \partial, \lrcorner, \mathcal{L}, \bar{\partial}, \omega)$$

is simultaneously a curved Gerstenhaber calculus of dimension d and a bigraded Batalin–Vilkovisky calculus of dimension d such that $y \lrcorner\, \omega = \bar{\partial}\omega$.

(4) A *differential bigraded Batalin–Vilkovisky calculus* of dimension d is a curved Batalin–Vilkovisky calculus of dimension d with $\ell = 0$ and $y = 0$.

(5) The notions of *centerless*, *strictly faithful*, *faithful*, and *bounded* apply to the underlying bigraded/dbg/curved Gerstenhaber calculus.

Example 3.27 Let $f\colon X \to S$ be a smooth morphism of Noetherian schemes of relative dimension d. Assume that $\omega_{X/S} \cong O_X$. Then the polyvector fields together with the de Rham complex and a chosen volume form $\omega \in \omega_{X/S}$ form a Batalin–Vilkovisky calculus of dimension d in the context $\mathfrak{Coh}(X/S)$. Namely, we use the Gerstenhaber calculus of Example 3.15, and apply Proposition 3.28 below.   ◇

3.4.1 Batalin–Vilkovisky Calculi from Gerstenhaber Calculi

Batalin–Vilkovisky calculi arise naturally from Gerstenhaber calculi in the Calabi–Yau setting.

Proposition 3.28 *Let $(S, C, O, \mathbf{C})$ be a context.*

(1) *Let $(\mathcal{G}^\bullet, \mathcal{A}^\bullet)$ be a Gerstenhaber calculus of dimension d, and let $\omega \in \mathcal{A}^d$ be a global section such that*

$$\kappa \colon \quad \mathcal{G}^p \to \mathcal{A}^{p+d}, \quad \theta \mapsto (\theta \lrcorner \omega),$$

is an isomorphism of O-modules for every $-d \le p \le 0$. Then $(\mathcal{G}^\bullet, \mathcal{A}^\bullet)$ is a Batalin–Vilkovisky calculus with $\Delta := \kappa^{-1} \circ \partial \circ \kappa$.

(2) *Let $(\mathcal{G}^{\bullet,\bullet}, \mathcal{A}^{\bullet,\bullet})$ be a bigraded Gerstenhaber calculus of dimension d, and let $\omega \in \mathcal{A}^{d,0}$ be a global section such that*

$$\kappa \colon \quad \mathcal{G}^{p,q} \to \mathcal{A}^{p+d,q}, \quad \theta \mapsto (\theta \lrcorner \omega),$$

is an isomorphism of O-modules for all $-d \le p \le 0$ and $q \ge 0$. Then $(\mathcal{G}^{\bullet,\bullet}, \mathcal{A}^{\bullet,\bullet})$ is a bigraded Batalin–Vilkovisky calculus with $\Delta := \kappa^{-1} \circ \partial \circ \kappa$.

(3) *Let $(\mathcal{G}^{\bullet,\bullet}, \mathcal{A}^{\bullet,\bullet})$ be a curved Gerstenhaber calculus of dimension d, and let $\omega \in \mathcal{A}^{d,0}$ be a global section such that*

$$\kappa \colon \quad \mathcal{G}^{p,q} \to \mathcal{A}^{p+d,q}, \quad \theta \mapsto (\theta \lrcorner \omega),$$

is an isomorphism of O-modules for all $-d \le p \le 0$ and $q \ge 0$. Then $(\mathcal{G}^{\bullet,\bullet}, \mathcal{A}^{\bullet,\bullet})$ is a curved Batalin–Vilkovisky calculus with $\Delta := \kappa^{-1} \circ \partial \circ \kappa$ and $y := \kappa^{-1}(\bar\partial \omega) \in \mathcal{G}^{0,1}$.

Proof Since $v \circ \lambda = \kappa$, the map $v \colon \mathcal{A}^0 \to \mathcal{A}^d$, $\alpha \mapsto \alpha \wedge \omega$, must be an O-linear isomorphism. With our definition, Δ is a C-linear operator with $\Delta(1) = 0$ (for degree reasons) and $\Delta^2 = 0$. The Bogomolov–Tian–Todorov formula can be shown by evaluating

$$\left(\Delta(\theta \wedge \xi) - \Delta(\theta) \wedge \xi - (-1)^{|\theta|} \theta \wedge \Delta(\xi) \right) \lrcorner \omega$$

with the Lie–Rinehart homotopy formula and then applying the mixed Leibniz rule. The derivation rule for $\Delta[\theta, \xi]$ then follows by applying Δ to the Bogomolov–Tian–Todorov formula. This concludes the proof in the singly graded case. The bigraded case is similar.

To show $\bar\partial \Delta + \Delta \bar\partial = [y, -]$ in the curved case, note that

$$(\bar\partial \Delta(\theta) + \Delta \bar\partial(\theta)) \lrcorner \omega = \left((-1)^{|\theta|} \Delta\theta \lrcorner \bar\partial \omega \right) - \mathcal{L}_\theta(\bar\partial \omega).$$

The Bogomolov–Tian–Todorov formula yields

$$- [y, \theta] \lrcorner\, \omega = (-1)^{|\theta|} \partial(\theta \lrcorner\, \bar{\partial}\omega) + (-1)^{|\theta|+1} \Delta\theta \lrcorner\, \bar{\partial}\omega.$$

Then the Lie–Rinehart homotopy formula applied to the left summand on the right yields

$$(\bar{\partial}\Delta(\theta) + \Delta\bar{\partial}(\theta)) \lrcorner\, \omega = [y, \theta] \lrcorner\, \omega.$$

Next, by a direct computation

$$(\bar{\partial}y + \Delta\ell) \lrcorner\, \omega = \bar{\partial}^2(\omega) - \mathcal{L}_\ell(\omega) - (y \wedge y) \lrcorner\, \omega = 0$$

since $y \wedge y = 0$ (because $y \in \mathcal{G}^{0,1}$), so $\bar{\partial}(y) + \Delta(\ell) = 0$. The remaining condition $y \lrcorner\, \omega = \bar{\partial}\omega$ is exactly the construction of y. $\qquad\qquad\qquad\square$

3.4.2 Two-sided Batalin–Vilkovisky Calculi

Similar to the case of Gerstenhaber calculi, we define a two-sided Batalin–Vilkovisky calculus as one endowed with a left contraction $\vdash$. Since every Batalin–Vilkovisky calculus $\mathcal{G}C^\bullet$ comes already equipped with a volume form $\omega \in \mathcal{A}^d$ which induces isomorphisms $\kappa \colon \mathcal{G}^p \cong \mathcal{A}^{p+d}$, there is only one possible $\vdash$ compatible with (3.1). By Lemma 3.23, it satisfies already most of the relations imposed in a two-sided Gerstenhaber calculus.

Definition 3.29 (Two-sided Batalin–Vilkovisky Calculi) A Batalin–Vilkovisky calculus $\mathcal{G}C^\bullet$ of dimension d is *two-sided* if the left contraction $\vdash \colon \mathcal{G}^p \times \mathcal{A}^i \to \mathcal{G}^{p+i}$ turns it into a two-sided Gerstenhaber calculus. Bigraded two-sided Batalin–Vilkovisky calculi and curved two-sided Batalin–Vilkovisky calculi are defined analogously.

We do not impose any additional relations. In particular, the obvious analog of Proposition 3.28 holds.

3.4.3 The Operator $\check{d} = \Delta + \bar{\partial} + (\ell + y) \wedge (-)$

Let $\mathcal{G}^{\bullet,\bullet}$ be a curved Batalin–Vilkovisky *algebra*. Assume $\ell = 0$ and $y = 0$ for a moment. Then $(\mathcal{G}^{\bullet,\bullet}, \Delta, \bar{\partial})$ is a double complex, and we can form its *total complex* $(\mathcal{G}^\bullet, \check{d})$ as

$$\mathcal{G}^m := \bigoplus_{p+q=m} \mathcal{G}^{p,q}, \quad \check{d} := \Delta + \bar{\partial};$$

here, $\mathcal{G}^m$ is a finite direct sum since both p and q are bounded below. If $\ell \neq 0$ or $y \neq 0$, then we may have $\check{d}^2 \neq 0$. To correct this, we set

$$\check{d}\colon \ \mathcal{G}^m \to \mathcal{G}^{m+1}, \quad \theta \mapsto \Delta\theta + \bar{\partial}\theta + (\ell + y) \wedge \theta;$$

this is a C-linear operator compatible with the total degree but not the bidegree.

Lemma 3.30 *The operator $\check{d}$ is a C-linear differential, i.e., we have $\check{d}^2 = 0$. We denote the cohomology of $(\mathcal{G}^\bullet, \check{d})$ by $\mathbb{H}^\bullet(\mathcal{G}^{\bullet,\bullet})$.*

Proof A direct computation yields

$$\check{d}^2(\theta) = \Delta(y) \wedge \theta + \bar{\partial}(\ell) \wedge \theta + \Delta(\ell) \wedge \theta + \bar{\partial}(y) \wedge \theta;$$

it is here that we use our assumptions $\bar{\partial}(\ell) = 0$ and $\bar{\partial}(y) + \Delta(\ell) = 0$ in order to conclude $\check{d}^2(\theta) = 0$ (note that $\Delta y = 0$ by degree reasons). $\qquad\square$

Remark 3.31 The derivation rules for $\check{d}(\theta \wedge \xi)$ and $\check{d}[\theta, \xi]$ which one may naively expect do not hold. $\qquad\Diamond$

In a curved Batalin–Vilkovisky *calculus* $(\mathcal{G}^{\bullet,\bullet}, \mathcal{A}^{\bullet,\bullet})$, the operator $\check{d}$ is nothing but what is obtained when transferring d to $\mathcal{G}^{\bullet,\bullet}$ with the isomorphism κ.

Lemma 3.32 *Let $(\mathcal{G}^{\bullet,\bullet}, \mathcal{A}^{\bullet,\bullet})$ be a curved Batalin–Vilkovisky calculus. Then we have $\kappa \circ \check{d} = d \circ \kappa$.*

Proof This is a short and direct computation using $(-1)^{|\theta|}\theta \lrcorner \bar{\partial}\omega = (y \wedge \theta) \lrcorner \omega$. $\qquad\square$

3.5 Semi-Cartan Structures and Cartan Structures

We introduce two notions of additional structures on an algebraic structure $\mathcal{P}$ to capture the similarities between our above definitions: *semi-Cartan structures* and *Cartan structures*. The name is meant to suggest that an algebraic structure with a (semi-)Cartan structure on it is similar to the Cartan calculus in differential geometry: Given a smooth manifold M, *Cartan calculus* refers to the exterior derivative, the Lie derivative, and other natural operations on differential forms and polyvector fields.

When discussing (semi-)Cartan structures on an algebraic structure $\mathcal{P}$ abstractly, the axioms of $\mathcal{P}$ play no role. If all the data of an algebraic structure except axioms are given, we call this a *pre-algebraic structure*. If $\mathcal{P}$ is a pre-algebraic structure, then we call a representation of $\mathcal{P}$ in a context a *$\mathcal{P}$-pre-algebra*.

Our notion of a context $(\mathcal{S}, C, O, \mathbf{C})$ allows us to distinguish between O-linear maps and C-linear maps. The notion of a *multilinear differential operator of finite order* interpolates between the two. The classical notion of a linear differential operator can be found e.g. in [267, 0G3Q]. Here is our multilinear variant.

Definition 3.33 (Multilinear Differential Operators) A C-multilinear map

$$\mu \colon \quad P_1 \times \ldots \times P_n \to Q$$

is a *differential operator of order* 0 if it is O-multilinear. Then, inductively, for $n \geq 1$, it is a *differential operator of order* $N + 1$ if, for all indices $1 \leq \ell \leq n$ and all local sections $a \in \Gamma(U, O_X)$, the induced map

$$D_{\ell;a}(\mu) \colon \quad P_1|_U \times \ldots \times P_n|_U \to Q|_U,$$

$$(p_1, \ldots, p_n) \mapsto \mu(p_1, \ldots, a \cdot p_\ell, \ldots, p_n) - a \cdot \mu(p_1, \ldots, p_n),$$

is a differential operator of order N. The map μ is a *differential operator of finite order* if it is a differential operator of order N for some $N \geq 0$. Differential operators of order N form a sheaf $\mathcal{D}\!i\!f\!f^N_{O/C}(P_1, \ldots, P_n; Q)$.

Remark 3.34

(1) If a map is a differential operator of order N, it is also one of any order $M \geq N$. If μ is a differential operator of order N, and $\nu_1, \ldots, \nu_n$ are differential operators of order M, then the composition $\mu(\nu_1, \ldots, \nu_n)$ is a differential operator of order $M + N$.
(2) Differential operators of finite order play an important role in the context $\mathfrak{Coh}(X/S)$ since they admit base change along morphisms $T \to S$, which we do not have for general C-multilinear maps.
(3) If $\mathcal{F}$ is a sheaf of O-algebras, and all sheaves $P_1, \ldots, P_n, Q$ are sheaves of $\mathcal{F}$-modules, then the same definition of multilinear differential operators makes sense with O replaced by $\mathcal{F}$. $\Diamond$

A (pre-)algebraic structure $\mathcal{P}$ itself does not keep track of the O-module structure and possible differential operator properties of the maps. We use *semi-Cartan structures* for this purpose.

Definition 3.35 (Semi-Cartan Structures) Let $\mathcal{P}$ be a pre-algebraic structure. Then a *semi-Cartan structure*

$$(F, 1_F, (*^P)_{P \in D}, \mathcal{P}_\bullet)$$

on $\mathcal{P}$ consists of a domain $F \in D(\mathcal{P})$ of degree $|F| = 0$, a constant $1_F \in \mathcal{P}(F)$, an operation $*^P \in \mathcal{P}(F, P; P)$ for every $P \in D(\mathcal{P})$, and an exhaustive increasing filtration

$$\mathcal{P}_0(P_1, \ldots, P_n; Q) \subseteq \mathcal{P}_1(P_1, \ldots, P_n; Q) \subseteq \ldots \subseteq \mathcal{P}(P_1, \ldots, P_n; Q)$$

with $*^P \in \mathcal{P}_0(F, P; P)$. A $\mathcal{P}$-pre-algebra $\mathcal{E}^\bullet$ is *semi-Cartanian* if $*^P : \mathcal{E}^F \times \mathcal{E}^P \to \mathcal{E}^P$ is O-bilinear, we have

$$a *^F b = b *^F a, \quad a *^P (b *^P p) = (a *^F b) *^P p, \quad 1_F *^P p = p,$$

i.e., $\mathcal{F} := \mathcal{E}^F$ is an O-algebra with unit 1_F and $\mathcal{E}^P$ is an $\mathcal{F}$-module via $*^P$, and, for every $\mu \in \mathcal{P}_N(P_1, \ldots, P_n; Q)$, the map

$$\mu : \ \mathcal{E}^{P_1} \times \ldots \times \mathcal{E}^{P_n} \to \mathcal{E}^Q$$

is a multilinear differential operator of order N with respect to $\mathcal{F}/C$. In particular, it is a multilinear differential operator of order N with respect to O/C.

In the context $\mathfrak{Coh}(X/S)$, we will usually have $\mathcal{E}^F = O_X$, but in the context $\mathfrak{Comp}(\Lambda)$, we have $O = C$, and $\mathcal{E}^F$ will be often a non-trivial O-algebra.

Example 3.36 The following algebraic structures carry a natural semi-Cartan structure in their singly graded version:

(1) Lie–Rinehart algebras; we have $*^P \in \mathcal{P}_0(F, P; P)$ and $\nabla^P \in \mathcal{P}_1(T, P; P)$;
(2) Lie–Rinehart pairs;
(3) Gerstenhaber algebras of dimension d; we take $F = \mathcal{G}^0$; the operation $*^P$ is given by the $\wedge$-product with F; for the filtration, we have $\wedge \in \mathcal{P}_0$ and $[-, -] \in \mathcal{P}_1$;
(4) Gerstenhaber calculi of dimension d; we take $F = \mathcal{G}^0$; for $\mathcal{A}^i$, the operation $*^P$ is given by the $\wedge$-product with $\mathcal{A}^0$ via the isomorphism λ; for the filtration, we have $\lrcorner \in \mathcal{P}_0$, $\partial \in \mathcal{P}_1$, $\mathcal{L} \in \mathcal{P}_1$;
(5) Batalin–Vilkovisky algebras and calculi of dimension d with the same semi-Cartan structure as the underlying Gerstenhaber algebras and calculi; for the filtration, we have $\Delta \in \mathcal{P}_1$.

They carry a natural semi-Cartan structure in their bigraded and curved versions as well, but for these versions, we will introduce separate conventions below. For each of these algebraic structures, a $\mathcal{P}$-algebra as defined above is automatically semi-Cartanian. Note that the condition that a map should be not only C-multilinear but O-multilinear, which recurs often in our above definitions, fits nicely into the framework of semi-Cartan structures, while an algebraic structure itself has no direct means of distinguishing C-multilinear from O-multilinear maps. Also note that the algebraic structure of Lie algebras does not naturally carry a semi-Cartan structure since it has no object F. $\qquad\qquad\qquad\qquad\qquad\qquad\qquad\qquad\qquad\qquad\qquad \diamond$

A *Cartan structure* also carries information about the tangent sheaf and the action by derivations.

Definition 3.37 (Cartan Structures) Let $\mathcal{P}$ be a pre-algebraic structure. Then a *Cartan structure*

$$(F, \ T, \ Z, \ 1_F, \ (*^P)_{P \in D}, \ (\nabla^P)_{P \in D}, \ \mathcal{P}_\bullet)$$

on $\mathcal{P}$ is a semi-Cartan structure together with a domain $T \in D(\mathcal{P})$ of degree $|T| = -1$ (in particular, $F \neq T$), a subset $Z \subseteq D(\mathcal{P})$ with $F \in Z$,[12] and an operation $\nabla^P \in \mathcal{P}_1(T, P; P)$ for every domain $P \in D(\mathcal{P})$. We assume furthermore that

$$\mathcal{P}(P_1, \ldots, P_n; Q) = \emptyset \ \text{ if } n \geq 3.$$

A $\mathcal{P}$-pre-algebra is *Cartanian* if it is semi-Cartanian, satisfies $\nabla_\theta^T(\xi) = -\nabla_\xi^T(\theta)$ for $\theta, \xi \in \mathcal{T}$ and

$$a *^M \nabla_\theta^M(m) = \nabla_{a*^T\theta}^M(m)$$

for every $M \in Z$ (in particular for $M = F$), has $\nabla_\theta^P(\gamma) = 0$ for all constants $\gamma \in \mathcal{P}(P)$, and for all $\mu \in \mathcal{P}(P_1, \ldots, P_n; Q)$ (for $n \geq 1$), the relation

$$\nabla_\theta^Q(\mu(p_1, \ldots, p_n)) = \sum_{s=1}^{n} \mu(p_1, \ldots, \nabla_\theta^{P_s}(p_s), \ldots, p_n) \tag{3.3}$$

holds.

Lie–Rinehart algebras, Gerstenhaber algebras, and (one- and two-sided) Gerstenhaber calculi (and, technically, also their bigraded counterparts) carry a natural Cartan structure with $Z = \{F\}$. We have $\nabla_\theta^P(\xi) = [\theta, \xi]$ respectively $\nabla_\theta^P(\alpha) = \mathcal{L}_\theta(\alpha)$. Lie–Rinehart *pairs* carry a natural Cartan structure with $Z = \{F, E\}$. In all these cases, the $\mathcal{P}$-algebras are automatically Cartanian. Batalin–Vilkovisky algebras and calculi also carry a natural Cartan structure—the same as the underlying Gerstenhaber algebras or calculi—but the $\mathcal{P}$-algebras are not Cartanian since we have $\nabla_\theta(\Delta(\xi)) = [\theta, \Delta(\xi)] \neq \Delta[\theta, \xi] = \Delta(\nabla_\theta(\xi))$. (Similarly, the curved algebras are not Cartanian since ∇^P is not compatible with the predifferential $\bar{\partial}$, and since $\nabla_\theta(\ell) \neq 0$ in general; however, we will introduce a modified version of Cartanianity below to account for this.)

Remark 3.38 The relation (3.3) can be considered as a partial unification of many relations that we have required in the definitions. For example, it yields the Poisson identities, the definition of the Lie bracket, and the Jacobi identity in a Lie–Rinehart algebra. It yields with the Jacobi identity and the derivation rule two of the three relations of a Lie–Rinehart module. Ultimately, a Cartan structure on an algebraic structure determines a Lie–Rinehart algebra inside it and exhibits every domain in $D(\mathcal{P})$ as a sort of weak Lie–Rinehart module over it (not necessarily satisfying the third condition of a Lie–Rinehart module unless $P \in Z$). $\Diamond$

Remark 3.39 The condition $\mathcal{P}(P_1, \ldots, P_n; Q) = \emptyset$ for $n \geq 3$ plays an important role when we construct the Thom–Whitney resolution of a Cartanian $\mathcal{P}$-algebra in $\mathfrak{Cob}(X/S)$. In the general case, it is not clear (to the author) how to define the signs of

---

[12] See the end of the section for the purpose of Z.

the operations in the resolution correctly. Also, by giving the definitions of bigraded and curved versions explicitly for all algebraic structures we are interested in, we have circumvented the problem of defining the correct relations of the operations in the Thom–Whitney resolution in general.   ◊

The purpose of the subset $Z \subseteq D(\mathcal{P})$ is to give a modified variant of faithfulness which takes only $P \in Z$ into account. For $Z = \{F\}$, we recover strict faithfulness, but Lie–Rinehart pairs are often only faithful when we take $\mathcal{E}^E$ into account as well.

Definition 3.40 Let $\mathcal{P}$ be a pre-algebraic structure carrying a Cartan structure.

(1) A Cartanian $\mathcal{P}$-pre-algebra is *strictly faithful* if the map

$$\mathrm{ad}_F: \quad \mathcal{T} \to \mathcal{H}om_C(\mathcal{F}, \mathcal{F}), \quad \theta \mapsto (a \mapsto \nabla_\theta^F(a)),$$

   is injective.
(2) A Cartanian $\mathcal{P}$-pre-algebra is *Z-faithful* if the intersection of the kernels of the maps

$$\mathrm{ad}_M: \quad \mathcal{T} \to \mathcal{H}om_C(\mathcal{E}^M, \mathcal{E}^M), \quad \theta \mapsto (m \mapsto \nabla_\theta^M(m)),$$

   for $M \in Z$ is zero.

Due to the condition $a *^M \nabla_\theta^M(m) = \nabla_{a*^T\theta}^M(m)$, the notion of Z-faithfulness is easier to study, but it is sufficient for our purposes.

3.6 (Semi-)Cartan Structures in the Bigraded/Curved Case

Given an algebraic structure $\mathcal{P}$ such as Lie–Rinehart algebras, it is difficult to give a general rule to construct the relations of the bigraded version $\mathcal{P}^{\mathrm{bg}}$ out of the relations of $\mathcal{P}$ such that we get back our explicit definitions of the bigraded versions above. However, on the level of the pre-algebraic structures, i.e., omitting the relations, the construction of $\mathcal{P}^{\mathrm{bg}}$ out of $\mathcal{P}$ is straightforward.

Definition 3.41 Let $\mathcal{P}$ be a pre-algebraic structure. Then we define a new pre-algebraic structure $\mathcal{P}^{\mathrm{bg}}$ as follows: The set of domains is $D(\mathcal{P}^{\mathrm{bg}}) := \{P^j \mid P \in D(\mathcal{P}), j \geq 0\}$, and the grading is $|P^j| := |P| + j$. The set of constants is $\mathcal{P}^{\mathrm{bg}}(P^0) = \mathcal{P}(P)$ and $\mathcal{P}^{\mathrm{bg}}(P^j) = \emptyset$ for $j \geq 1$. For the operations, we have

$$\mathcal{P}^{\mathrm{bg}}(P_1^{j_1}, \ldots, P_n^{j_n}; Q^j) = \mathcal{P}(P_1, \ldots, P_n; Q)$$

if $j_1 + \ldots + j_n = j$, and

$$\mathcal{P}^{\mathrm{bg}}(P_1^{j_1}, \ldots, P_n^{j_n}; Q^j) = \emptyset$$

otherwise. If $\gamma \in \mathcal{P}(P)$ or $\mu \in \mathcal{P}(P_1, \ldots, P_n; Q)$, then we keep the same notation for the constant in $\mathcal{P}^{\mathrm{bg}}(P^0)$ and the operation in $\mathcal{P}^{\mathrm{bg}}(P^{j_1}, \ldots, P^{j_n}; Q^j)$ if confusion is unlikely.

If $\mathcal{P}$ carries a semi-Cartan structure or a Cartan structure, then $\mathcal{P}^{\mathrm{bg}}$ carries a variant of the respective structure as well. However, for a $\mathcal{P}^{\mathrm{bg}}$-pre-algebra to be semi-Cartanian respectively Cartanian, we wish to impose additional conditions that do not come from the general definition of (semi-)Cartanian pre-algebras, and which reflect the bigraded nature of $\mathcal{P}^{\mathrm{bg}}$.

Definition 3.42 If $\mathcal{P}$ carries a semi-Cartan structure, then we say that a $\mathcal{P}^{\mathrm{bg}}$-pre-algebra $\mathcal{E}^{\bullet,\bullet}$ is *semi-Cartanian* if

- $*^P : \mathcal{F}^{j_1} \times \mathcal{E}^{P,j_2} \to \mathcal{E}^{P,j_1+j_2}$ is O-bilinear;
- $a *^F b = (-1)^{|a||b|} b *^F a$ for $a \in \mathcal{F}^{j_1}, b \in \mathcal{F}^{j_2}$ and the total degree $|a| = 0 + j_1$, $|b| = 0 + j_2$;
- $a *^P (b *^P p) = (a *^F b) *^P p$ for $a \in \mathcal{F}^{j_1}, b \in \mathcal{F}^{j_2}, p \in \mathcal{E}^{P,j_3}$;
- $1_F *^P p = p$ for $p \in \mathcal{E}^{P,j}$;
- every operation $\mu \in \mathcal{P}^{\mathrm{bg}}_N(P_1^{j_1}, \ldots, P_n^{j_n}; Q^j)$ (for the induced filtration from $\mathcal{P}$) is a multilinear differential operator of order N with respect to $\mathcal{F}^0/\mathcal{C}$.

If $\mathcal{P}$ carries a Cartan structure, then we say a $\mathcal{P}^{\mathrm{bg}}$-pre-algebra $\mathcal{E}^{\bullet,\bullet}$ is *Cartanian* if it is semi-Cartanian and:

- $\nabla_\theta^T(\xi) = -(-1)^{(|\theta|+1)(|\xi|+1)} \nabla_\xi^T(\theta)$ for $\theta \in \mathcal{T}^{j_1}, \xi \in \mathcal{T}^{j_2}$;
- $a *^M \nabla_\theta^M(m) = \nabla_{a*^T\theta}^M(m)$ for $M \in Z$ and $a \in \mathcal{F}^{j_1}, \theta \in \mathcal{T}^{j_2}, m \in \mathcal{E}^{M,j_3}$;
- $\nabla_\theta^P(\gamma) = 0$ for $\theta \in \mathcal{T}^j$ and constants $\gamma \in \mathcal{E}^{P,0}$;
- $\nabla_\theta^Q(\mu(p)) = (-1)^{|\mu|(|\theta|+1)} \mu(\nabla_\theta^P(p))$ for $\theta \in \mathcal{T}^{j_1}$, $p \in \mathcal{E}^{P,j_2}$ for unary operations $\mu \in \mathcal{P}(P; Q)$, where $|\mu| := |Q| - |P|$ is the degree of μ;
- $\nabla_\theta^R \mu(p, q) = \mu(\nabla_\theta^P(p), q) + (-1)^{(|\theta|+1)(|\mu|+|p|)} \mu(p, \nabla_\theta^Q(q))$ for $p \in \mathcal{E}^{P,j_1}, q \in \mathcal{E}^{Q,j_2}, \theta \in \mathcal{T}^j$ for binary operations $\mu \in \mathcal{P}(P, Q; R)$, where $|\mu| := |R| - |P| - |Q|$ is the degree of μ.

Remark 3.43 These sign conventions are compatible with the explicit definitions of bigraded structures that we have made. We do not know the correct signs in the formulae if μ is a ternary or higher operation. Calculations with various compositions of binary operations in the Thom–Whitney resolution (see Chap. 13.1) suggest that there may not be a universal sign correct for all ternary operations. This is essentially the reason why we exclude higher operations from the definition of a Cartan structure. All sign conventions that we make are chosen in a way to fit our definition of Thom–Whitney resolutions in Chap. 13.1.   ◇

Similarly, we can define the curved version on the level of pre-algebraic structures.

Definition 3.44 Let $\mathcal{P}$ be a pre-algebraic structure which carries a Cartan structure. Then we define a new pre-algebraic structure $\mathcal{P}^{\mathrm{crv}}$ as follows: First, we take the pre-algebraic structure $\mathcal{P}^{\mathrm{bg}}$. Then we add a constant $\ell \in \mathcal{P}(T^2)$ and an operation

$\bar{\partial} \in \mathcal{P}_1(P^j; P^{j+1})$ for every $P^j \in D(\mathcal{P}^{\mathrm{bg}})$. We say that a $\mathcal{P}^{\mathrm{crv}}$-pre-algebra is *Cartanian* if it is Cartanian for $\mathcal{P}^{\mathrm{bg}}$ and:

- $\bar{\partial}(\gamma) = 0$ for constants $\gamma \in \mathcal{P}(P)$;
- $\bar{\partial}(\mu(p)) = (-1)^{|\mu|} \mu(\bar{\partial}(p))$ for unary operations $\mu \in \mathcal{P}(P; Q)$ and $p \in \mathcal{E}^{P,j}$;
- $\bar{\partial}\mu(p, q) = \mu(\bar{\partial}(p), q) + (-1)^{|\mu|+|p|}\mu(p, \bar{\partial}(q))$ for binary operations $\mu \in \mathcal{P}(P, Q; R)$ and $p \in \mathcal{E}^{P,j_1}$, $q \in \mathcal{E}^{Q,j_2}$;
- $\bar{\partial}^2(p) = \nabla_\ell(p)$ for all $P^j \in D(\mathcal{P}^{\mathrm{bg}})$;
- $\bar{\partial}(\ell) = 0$.

Definition 3.45 Let $\mathcal{P}$ be a pre-algebraic structure carrying a Cartan structure.

(a) A Cartanian $\mathcal{P}^{\mathrm{bg}}$-pre-algebra is *strictly faithful* if, for every $j \geq 0$, the map

$$\mathrm{ad}_F^j : \quad \mathcal{T}^j \to \mathcal{H}om_C(\mathcal{F}^0, \mathcal{F}^j), \quad \theta \mapsto (a \mapsto \nabla_\theta^{F^0}(a)),$$

   is injective.

(b) A Cartanian $\mathcal{P}^{\mathrm{bg}}$-pre-algebra is *Z-faithful* if, for every $j \geq 0$, the intersection of the kernels of the maps

$$\mathrm{ad}_M^j : \quad \mathcal{T}^j \to \mathcal{H}om_C(\mathcal{E}^{M,0}, \mathcal{E}^{M,j}), \quad \theta \mapsto (m \mapsto \nabla_\theta^{M^0}(m)),$$

   for $M \in Z$ is zero.

A Cartanian $\mathcal{P}^{\mathrm{crv}}$-pre-algebra is strictly faithful respectively Z-faithful if the underlying $\mathcal{P}^{\mathrm{bg}}$-pre-algebra is.

Chapter 4
Gauge Transforms

We review the theory of gauge transforms in the algebraic structures defined in the previous chapter. Gauge transforms are a type of inner automorphisms which can be computed by means of the operations of the algebraic structure. They are relevant for logarithmic deformation theory because infinitesimal automorphisms of logarithmic deformations can be computed as gauge transforms of the associated Gerstenhaber calculus. Let us work in the following setup:

Situation 4.1 Let Λ be a complete local Noetherian $\mathbf{k}$-algebra with residue field $\mathbf{k}$, let X be a topological space (with associated site $\mathcal{S}$), and let C be the constant sheaf of rings with stalk Λ. We form a context $\mathfrak{Flat}(X; \Lambda) := (\mathcal{S}, C, O, \mathbf{C})$ by setting $O = C$ and by taking for $\mathbf{C}$ the class of flat sheaves of Λ-modules, i.e., sheaves of Λ-modules $\mathcal{E}$ with flat stalks $\mathcal{E}_x$.

For every $A \in \mathbf{Art}_\Lambda$, we can form the tensor product $\mathcal{E}_A := \mathcal{E} \otimes_\Lambda A$ in the category of sheaves of Λ-modules; it is a flat sheaf of A-modules. We consider it as an infinitesimal thickening of $\mathcal{E}_0 := \mathcal{E} \otimes_\Lambda \mathbf{k}$. If $B' \to B$ is a surjection in $\mathbf{Art}_\Lambda$ with kernel $I \subset B'$, then we have an exact sequence

$$0 \to I \otimes_\Lambda \mathcal{E}_x \to \mathcal{E}_{B',x} \to \mathcal{E}_{B,x} \to 0$$

of stalks at $x \in X$ due to flatness. If $B' \to B$ is a first-order extension, then $I \otimes_\Lambda \mathcal{E}_x \cong I \otimes_\mathbf{k} \mathcal{E}_{0,x}$.

Let $\mathcal{P}$ be a pre-algebraic structure—we shall be primarily interested in our six algebraic structures carrying a Cartan structure (which are listed in Example 3.36) and in their bigraded counterparts—and let $\mathcal{E}^\bullet$ be a $\mathcal{P}$-pre-algebra in $\mathfrak{Flat}(X; \Lambda)$. For every $A \in \mathbf{Art}_\Lambda$, the tensor product $\mathcal{E}_A^\bullet$ is a $\mathcal{P}$-pre-algebra in $\mathfrak{Flat}(X; A)$; indeed, all operations allow a base change along $\Lambda \to A$ because they are Λ-multilinear, and the underlying topological space remains the same.

For a surjection $B' \to B$ in $\mathbf{Art}_\Lambda$, an *infinitesimal automorphism* of $\mathcal{E}_{B'}^\bullet$ over $\mathcal{E}_B^\bullet$ consists of an B'-linear automorphism $\phi^P : \mathcal{E}_{B'}^P \to \mathcal{E}_{B'}^P$ for every domain $P \in D(\mathcal{P})$

S. Felten, *Global Logarithmic Deformation Theory*, Lecture Notes in Mathematics 2373, https://doi.org/10.1007/978-3-031-98751-9_4

which restricts to the identity on $\mathcal{E}_B^P$ along the canonical map $\mathcal{E}_{B'}^P \to \mathcal{E}_B^P$, preserves all constants in $\mathcal{P}(P)$, and commutes with all operations

$$\mu\colon \ \mathcal{E}_{B'}^{P_1} \times \ldots \times \mathcal{E}_{B'}^{P_n} \to \mathcal{E}_{B'}^Q.$$

Infinitesimal automorphisms form a sheaf of groups $\mathcal{A}ut_{B'/B}(\mathcal{E}_{B'}^\bullet)$ on X. In general, it is not so easy to determine this sheaf, but if $\mathcal{P}$ carries a Cartan structure and $\mathcal{E}^\bullet$ is Cartanian, then we can easily describe specific infinitesimal automorphisms, the *gauge transforms*. They are controlled by elements $\theta \in \mathcal{T}_{B'} = \mathcal{E}_{B'}^T$ and constructed from the operations ∇^P—which is the reason why we require these operations as part of a Cartan structure. In good situations, θ can be recovered from its gauge transform, and the gauge transforms are precisely those infinitesimal automorphisms which "arise from geometry". Before we define gauge transforms, we need an easy lemma.

Lemma 4.2 *Let $\mathcal{P}$ be a pre-algebraic structure carrying a Cartan structure, and let $\mathcal{E}^\bullet$ be a Cartanian $\mathcal{P}$-pre-algebra in $\mathfrak{Flat}(X;\Lambda)$. Then $\mathcal{E}_A^\bullet$ is a Cartanian $\mathcal{P}$-pre-algebra in $\mathfrak{Flat}(X;A)$.*

Proof All requirements can be expressed as relations of (compositions of) operations (possibly with inserted constants). This holds in particular for the differential operator property, which can be expressed as relations for maps

$$\mathcal{F} \times \ldots \times \mathcal{F} \times \mathcal{E}^{P_1} \times \ldots \times \mathcal{E}^{P_n} \to \mathcal{E}^Q.$$

The relations can be checked on the stalks, and there we have $\mathcal{E}_{A,x}^P = \mathcal{E}^P \otimes_\Lambda A$. By A-multilinearity of all relations, it is sufficient to check them on entries which are in the image of the natural maps $\mathcal{E}_x^P \to \mathcal{E}_{A,x}^P$, and there, the relations hold by assumption. All O-linearity requirements are automatic since $C = O$. $\square$

Definition 4.3 Let $B' \to B$ be a surjection in $\mathbf{Art}_\Lambda$ with kernel $I \subset B'$, and let $\mathcal{E}^\bullet$ be a Cartanian $\mathcal{P}$-pre-algebra. Let $\theta \in \Gamma(U, I \cdot \mathcal{T}_{B'})$ for some open $U \subseteq X$. Then the *gauge transform* is the map

$$\exp_\theta\colon \ \mathcal{E}_{B'}^P|_U \to \mathcal{E}_{B'}^P|_U, \quad p \mapsto \sum_{n=0}^\infty \frac{(\nabla_\theta^P)^n(p)}{n!} = p + \nabla_\theta^P(p) + \frac{1}{2}\nabla_\theta^P \nabla_\theta^P(p) + \ldots.$$

This definition is one of the key reasons why we assume $\mathrm{char}(\mathbf{k}) = 0$. Let us write $[\theta, \xi] := \nabla_\theta^T(\xi)$ for $\theta, \xi \in \mathcal{T}$ as usual. This turns $\mathcal{T}$ into a sheaf of Lie algebras.

Lemma 4.4 *The gauge transform has the following properties:*

(1) $\exp_\theta\colon \mathcal{E}_{B'}^P \to \mathcal{E}_{B'}^P$ *is a well-defined B'-linear sheaf homomorphism.*
(2) $\exp_\theta\colon \mathcal{E}_{B'}^P \to \mathcal{E}_{B'}^P$ *restricts to the identity on $\mathcal{E}_B^P$.*
(3) *For a constant $\gamma \in \mathcal{P}(P)$, we have $\exp_\theta(\gamma) = \gamma$.*

(4) *For an operation $\mu \in \mathcal{P}(P_1, \ldots, P_n; Q)$, we have*

$$\mu(\exp_\theta(p_1), \ldots, \exp_\theta(p_n)) = \exp_\theta(\mu(p_1, \ldots, p_n)).$$

(5) *For $\theta = 0 \in I \cdot \mathcal{T}_{B'}$, we have $\exp_0(p) = p$.*

(6) *For $\theta, \xi \in I \cdot \mathcal{T}_{B'}$, the composition is given by $\exp_\theta \circ \exp_\xi(p) = \exp_{\theta \odot \xi}(p)$, where $\theta \odot \xi$ is defined by the* Baker–Campbell–Hausdorff *formula*

$$\theta \odot \xi := \sum_{n=1}^{\infty} \frac{(-1)^{n+1}}{n} \sum_{r_1+s_1>0} \cdots \sum_{r_n+s_n>0}$$

$$\times \frac{1}{\sum_{i=1}^{n}(r_i + s_i)} \frac{[\theta^{(r_1)}, \xi^{(s_1)}, \ldots, \theta^{(r_n)}, \xi^{(s_n)}]}{\prod_{i=1}^{n} r_i! s_i!}$$

$$= \theta + \xi + \frac{1}{2}[\theta, \xi] + \frac{1}{12}[\theta, [\theta, \xi]] - \frac{1}{12}[\xi, [\theta, \xi]] + \ldots$$

where $[\theta] := \theta$ and $[\theta_1, \theta_2, \ldots, \theta_{n+1}] := [\theta_1, [\theta_2, \ldots, \theta_n]]$ inductively for $n \geq 2$, and where $\theta^{(r)}$ denotes the repetition of r entries of θ.

(7) *The inverse of $\exp_\theta$ is given by $\exp_{-\theta}$.*

This shows that $\exp_\theta$ is an infinitesimal automorphism of the $\mathcal{P}$-pre-algebra $\mathcal{E}_{B'}^{\bullet}$ over $\mathcal{E}_B^{\bullet}$. When we endow $I \cdot \mathcal{T}_{B'}$ with the Baker–Campbell–Hausdorff product $\odot$ defined above, it becomes a sheaf of groups with neutral element 0. Then

$$\exp: \quad (I \cdot \mathcal{T}_{B'}, \odot) \to (\mathcal{A}ut_{B'/B}(\mathcal{E}_{B'}^{\bullet}), \circ) \tag{4.1}$$

is a homomorphism of sheaves of groups. If $B' \to B$ is a first-order extension, then $\theta \odot \xi = \theta + \xi$, and $(I \cdot \mathcal{T}_{B'}, \odot)$ is Abelian.

Proof Since ∇_θ^P is B'-linear, it follows from the nilpotency of I that $\exp_\theta$ is a well-defined B'-linear operator. The induced map on $\mathcal{E}_B^P$ is the identity since $\theta \in I \cdot \mathcal{T}_{B'}$. For the constants $\gamma \in \mathcal{P}(P)$, we find $\exp_\theta(\gamma) = \gamma$ since $\nabla_\theta^P(\gamma) = 0$. For unary operations $\mu \in \mathcal{P}(P; Q)$, we have

$$\exp_\theta(\mu(p)) = \sum_{n=0}^{\infty} \frac{\nabla_\theta^n(\mu(p))}{n!} = \sum_{n=0}^{\infty} \frac{\mu(\nabla_\theta^n(p))}{n!} = \mu(\exp_\theta(p)).$$

For binary operations $\mu \in \mathcal{P}(P_1, P_2; Q)$, we find

$$\nabla_\theta^n(\mu(p_1, p_2)) = \sum_{k=0}^{n} \binom{n}{k} \mu(\nabla_\theta^k(p_1), \nabla_\theta^{n-k}(p_2))$$

by induction on n, and then $\exp_\theta(\mu(p_1, p_2)) = \mu(\exp_\theta(p_1), \exp_\theta(p_2))$. By assumption on a Cartan structure, there are no higher operations, but a similar proof would show the corresponding formula. If $\theta = 0$, then $\nabla_0(p) = 0$, so $\exp_\theta(p) = p$.

We can reduce the Baker–Campbell–Hausdorff formula in our case to a general Baker–Campbell–Hausdorff formula for operators on $\mathbf{k}$-vector spaces. Note that

$$\exp_\theta = \exp(\nabla_\theta) := \sum_{n=0}^{\infty} \frac{\nabla_\theta^n}{n!}$$

as operators. Now $\exp(\nabla_\theta) \circ \exp(\nabla_\xi) - \mathrm{Id}$ is a nilpotent B'-linear operator; when we apply the formula for the logarithm, then we obtain an operator

$$C := \sum_{n=1}^{\infty} \frac{(-1)^{n+1}}{n} (\exp(B_\theta) \circ \exp(B_\xi) - \mathrm{Id})^n : \ \mathcal{E}_{B'}^P \to \mathcal{E}_{B'}^P$$

with the property that $\exp(C) = \exp_\theta \circ \exp_\xi$. When we expand this formula, we get

$$C = \sum_{n=1}^{\infty} \frac{(-1)^{n+1}}{n} \sum_{r_1+s_1>0} \cdots \sum_{r_n+s_n>0} \frac{\nabla_\theta^{r_1} \nabla_\xi^{s_1} \cdots \nabla_\theta^{r_n} \nabla_\xi^{s_n}}{\prod_{i=1}^{n} r_i! s_i!}.$$

For general operators $P_1, P_2, \ldots$, we set $[P_1] := P_1$, $[P_1, P_2] := P_1 P_2 - P_2 P_1$, and $[P_1, \ldots, P_{n+1}] := [P_1, [P_2, \ldots, P_{n+1}]]$ inductively, i.e., they are the commutators. Let us introduce notation

$$I(n; e) := \left\{ (r_i, s_i)_i \in \prod_{i=1}^{n} (\mathbb{N} \times \mathbb{N}) \ \middle| \ r_i + s_i > 0, \ \sum_{i=1}^{n} (r_i + s_i) = e \right\}$$

for the part of the index set where the sum of all indices is e. Then the classical Baker–Campbell–Hausdorff formula for operators yields

$$\sum_{n=1}^{e} \frac{(-1)^{n+1}}{n} \sum_{(r_i,s_i)_i \in I(n;e)} \frac{\nabla_\theta^{r_1} \nabla_\xi^{s_1} \cdots \nabla_\theta^{r_n} \nabla_\xi^{s_n}}{\prod_{i=1}^{n} r_i! s_i!}$$

$$= \frac{1}{e} \cdot \sum_{n=1}^{e} \frac{(-1)^{n+1}}{n} \sum_{(r_i,s_i)_i \in I(n;e)} \frac{[\nabla_\theta^{(r_1)}, \nabla_\xi^{(s_1)}, \ldots, \nabla_\theta^{(r_n)}, \nabla_\xi^{(s_n)}]}{\prod_{i=1}^{n} r_i! s_i!}$$

where the sum over n stops at e because $I(n; e) = \emptyset$ for $n > e$. The formula does only hold for the sum but not for each individual value of n. The Jacobi identity yields via induction over m the formula

$$[\nabla_{\theta_1}, \ldots, \nabla_{\theta_m}](p) = \nabla_{[\theta_1, \ldots, \theta_m]}(p),$$

where we have on the left the iterated commutator of operators, and on the right the iterated Lie bracket in $\mathcal{T}_{B'}$. The formula shows that $C = \nabla_{\theta \odot \xi}$ with the definition of $\theta \odot \xi$ in the statement; thus $\exp_\theta \circ \exp_\xi = \exp_{\theta \odot \xi}$.

Evaluating the Baker–Campbell–Hausdorff formula, we find $\theta \odot (-\theta) = 0$ since $[\theta, \theta] = -[\theta, \theta]$ and hence $[\theta, \theta] = 0$. Thus, $\exp_\theta$ and $\exp_{-\theta}$ are inverse to each other.

The sheaf of magmas (sets with a binary operation) $(I \cdot \mathcal{T}_{B'}, \odot)$ is a sheaf of groups because $\odot$ defines a group structure on every nilpotent $\mathbf{k}$-Lie algebra, see e.g. [197, V, §§1-3]. That exp is a group homomorphism is now clear. If $B' \to B$ is a first-order extension, we have $I \cdot \mathfrak{m}_{B'} = 0$ and hence $I^2 = 0$, so $[I \cdot \mathcal{T}_{B'}, I \cdot \mathcal{T}_{B'}] = 0$, and hence $\theta \odot \xi = \theta + \xi$. $\qquad\square$

Example 4.5 Let $f_A \colon X_A \to S_A$ be a separated, log smooth, and saturated morphism to the log point $S_A = \mathrm{Spec}(Q \to A)$. Let $\mathcal{G}^\bullet_{X_A/S_A}$ be its Gerstenhaber algebra of polyvector fields (with the conventions of [76]). Then the map

$$(I \cdot \mathcal{G}^{-1}_{X_A/S_A}, \odot) \to (\mathcal{A}ut_{A/\mathbf{k}}(\mathcal{G}^\bullet_{X_A/S_A}), \circ)$$

is injective. By [76, §§2-3], the gauge transforms of the Gerstenhaber algebra $\mathcal{G}^\bullet_{X_A/S_A}$ are precisely those automorphisms which are induced from infinitesimal automorphisms of the log smooth deformation $f_A \colon X_A \to S_A$ of its central fiber $f_0 \colon X_0 \to S_0$. $\qquad\Diamond$

A priori, the exponential map in (4.1) is not injective in general. However, under mild conditions, it is. For our purposes, this condition is (virtually) always satisfied.

Lemma 4.6 *Let $\mathcal{E}^\bullet$ be a Cartanian $\mathcal{P}$-pre-algebra in the context $\mathfrak{Flat}(X; \Lambda)$. Assume that $\mathcal{E}^\bullet_0$ is faithful, i.e., the intersection of the kernels of the maps*

$$\mathrm{ad}_{P,0} \colon \ \mathcal{T}_0 \to \mathcal{H}om_{\mathbf{k}}(\mathcal{E}^P_0, \mathcal{E}^P_0), \quad \theta \mapsto (p \mapsto \nabla^P_\theta(p)),$$

is zero. Then, for all $A \in \mathbf{Art}_\Lambda$, the intersection of the kernels of the maps $\mathrm{ad}_{P,A} \colon \mathcal{T}_A \to \mathcal{H}om_A(\mathcal{E}^P_A, \mathcal{E}^P_A)$ *is zero. Moreover, for a surjection $B' \to B$ in $\mathbf{Art}_\Lambda$ with kernel $I \subset B'$, the exponential map*

$$\exp \colon \ (I \cdot \mathcal{T}_{B'}, \odot) \to (\mathcal{A}ut_{B'/B}(\mathcal{E}^\bullet_{B'}), \circ)$$

is injective. In particular, it is injective if $\mathcal{E}^\bullet_0$ is centerless or strictly faithful.

Proof We prove the first statement by induction over small extensions $B' \to B$ with kernel I. Let $\theta \in \mathcal{T}_{B'}$ with $\nabla^P_\theta = 0$ for all $P \in D(\mathcal{P})$. Then, in particular, $\nabla^P_\theta|_B = 0$. By the induction hypothesis, we find $\theta|_B = 0$, so $\theta \in I \cdot \mathcal{T}_{B'}$. Since $B' \to B$ is a small extension, I is a one-dimensional $\mathbf{k}$-vector space, so we can choose a generator $i \in I$ with $I = i \cdot \mathbf{k}$. Then, at least locally, $\theta = i \cdot \xi$ for some

$\xi \in \mathcal{T}_{B'}$. Now the kernel of $\mu_i : B' \to B'$, $b \mapsto ib$, is $\mathfrak{m}_{B'}$; thus, since $\mathcal{E}_{B'}^P$ is flat over B', the kernel of $\mu_i : \mathcal{E}_{B'}^P \to \mathcal{E}_{B'}^P$, $p \mapsto ip$, is $\mathfrak{m}_{B'} \cdot \mathcal{E}_{B'}^P$. We have

$$0 = \nabla_\theta^P (p) = i \cdot \nabla_\xi^P (p),$$

hence $\nabla_\xi^P (p) \in \mathfrak{m}_{B'} \cdot \mathcal{E}_{B'}^P$. This shows that $\nabla_{\xi_0}^P : \mathcal{E}_0^P \to \mathcal{E}_0^P$ is the zero map, where $\xi_0 := \xi|_0$. Since this holds for all $P \in D(\mathcal{P})$, the faithfulness assumption on $\mathcal{E}_0^\bullet$ shows $\xi_0 = 0$. However, in this case, $\theta = i \cdot \xi = 0$.

By decomposing $B' \to B$ into a sequence of small extensions, we see that it is sufficient to prove the second statement in the case of a small extension $B' \to B$. In this case, $\exp_\theta(p) = p + \nabla_\theta^P(p)$, so $\exp_\theta = \mathrm{id}$ implies $\nabla_\theta^P(p) = 0$ for all $P \in D(\mathcal{P})$ and $p \in \mathcal{E}_{B'}^P$. Then the claim follows from our first statement. $\square$

Remark 4.7 If $\mathcal{E}^P$ is *complete*, i.e., $\mathcal{E}^P$ is the limit of the system $\mathcal{E}_k^p := \mathcal{E}^P \otimes_\Lambda \Lambda/\mathfrak{m}_\Lambda^{k+1}$, and $\mathcal{E}_0^\bullet$ is faithful, then $\mathcal{E}^\bullet$ is faithful as well. More generally, it suffices that the map from $\mathcal{E}^P$ to the limit is injective to conclude faithfulness of $\mathcal{E}^\bullet$. The converse is false: Let $\Lambda = \mathbf{k}[\![t]\!]$, and let $X = \{*\}$ be a point. Consider $F = \Lambda[s]$ and $T = \Lambda[s] \cdot D$. These flat modules are not complete, but they inject into their t-adic completions. We turn them into a Lie–Rinehart algebra by setting $D = t \cdot \partial_s$, i.e., $D(s) = t\partial_s(s) = t$ etc., and

$$[q(t,s) \cdot D, r(t,s) \cdot D] := \big(q(t,s) \cdot D(r(t,s)) - r(t,s) \cdot D(q(t,s))\big) \cdot D.$$

This Lie–Rinehart algebra is both strictly faithful and centerless. However, in the base change (F_0, T_0) to $\mathbf{k}$, the generator D of T_0 acts as the zero operator, and the Lie bracket vanishes as well. Thus, (F_0, T_0) is neither strictly faithful nor centerless nor faithful. $\lozenge$

Remark 4.8 In [76, §10], a curved Lie algebra $L^\bullet$ (in the context $\mathfrak{Comp}(\Lambda)$ of flat and complete Λ-modules) is defined to be *faithful* if, in our language, $L_A^\bullet$ is centerless for all $A \in \mathbf{Art}_\Lambda$. The proof of the lemma shows that it would have been enough to require that $L_0^\bullet$ is centerless. $\lozenge$

Chapter 5
The Extended Maurer–Cartan Equations

We discuss homological aspects of logarithmic deformation theory. Every differential graded Lie algebra $L^\bullet$ (in the context $\mathfrak{Comp}(\Lambda)$) gives rise to a deformation functor $\mathrm{Def}(L^\bullet, -)$ by taking gauge equivalence classes of Maurer–Cartan solutions. This is the reason why dg Lie algebras are ubiquitous in deformation theory: For many deformation problems, one can find a dg Lie algebra $L^\bullet$ such that $\mathrm{Def}(L^\bullet, -)$ classifies isomorphism classes of deformations.

Less well known is that every *curved* Lie algebra gives rise to a deformation functor $\mathrm{Def}(L^\bullet, -)$ as well, by considering solutions of the *extended* Maurer–Cartan equation. This is absolutely necessary in logarithmic deformation theory because these deformation problems are not controlled by any dg Lie algebra.

We work in the context $\mathfrak{Comp}(\Lambda)$ throughout. Recall that objects in $\mathbf{C}$ are flat and complete Λ-modules. We work with the following structures in this chapter.

Definition 5.1 (Λ-Linear Curved Algebras)

(1) A Λ-*linear curved Lie algebra* is a curved Lie algebra $L^\bullet$ in the context $\mathfrak{Comp}(\Lambda)$ with $\ell \in \mathfrak{m}_\Lambda \cdot L^2$.

(2) Let $\mathcal{P}$ be a pre-algebraic structure carrying a Cartan structure. Then a Λ-*linear curved $\mathcal{P}$-algebra* is a Cartanian $\mathcal{P}^{\mathrm{crv}}$-pre-algebra in $\mathfrak{Comp}(\Lambda)$ in the sense of Definition 3.44 with $\ell \in \mathfrak{m}_\Lambda \cdot \mathcal{T}^2$.

(3) A Λ-*linear curved Lie–Rinehart algebra* respectively *Lie–Rinehart pair* respectively *Gerstenhaber algebra* respectively *Gerstenhaber calculus* is a curved Lie–Rinehart algebra (respectively any of the other notions) in the context $\mathfrak{Comp}(\Lambda)$ with $\ell \in \mathfrak{m}_\Lambda \cdot \mathcal{T}^2$.

(4) A Λ-*linear Batalin–Vilkovisky curved algebra* respectively *Batalin–Vilkovisky calculus* is a curved Batalin–Vilkovisky algebra respectively calculus in $\mathfrak{Comp}(\Lambda)$ with $\ell \in \mathfrak{m}_\Lambda \cdot \mathcal{G}^{-1,2}$ and $y \in \mathfrak{m}_\Lambda \cdot \mathcal{G}^{0,1}$.

Note that, for example, if $\mathcal{G}_d$ is the algebraic structure of Gerstenhaber algebras of dimension d, then every Λ-linear curved Gerstenhaber algebra is a Λ-linear curved $\mathcal{G}_d$-algebra but not conversely since we assume fewer relations for the latter notion.

© The Author(s), under exclusive license to Springer Nature Switzerland AG 2025

S. Felten, *Global Logarithmic Deformation Theory*, Lecture Notes in Mathematics 2373, https://doi.org/10.1007/978-3-031-98751-9_5

5.1   The Classical Extended Maurer–Cartan Equation

We introduce the classical extended Maurer–Cartan equation. After reviewing the most basic case of curved Lie algebras, we briefly discuss the case of curved Lie–Rinehart algebras which is important for deformations of pairs, and then go to the main case of a curved Gerstenhaber algebra.

5.1.1   Curved Lie Algebras

Let $L^\bullet$ be a Λ-linear curved Lie algebra. Here, we recall how it gives rise to a deformation functor. By definition, it comes with a predifferential $\bar{\partial}$, inducing a predifferential $\bar{\partial}\colon L_A^q \to L_A^{q+1}$ for every $A \in \mathbf{Art}_\Lambda$. However, for every $\phi \in \mathfrak{m}_A \cdot L_A^1$, we can define a map

$$\bar{\partial}_\phi\colon \; L_A^q \to L_A^{q+1}, \quad \xi \mapsto \bar{\partial}\xi + [\phi, \xi].$$

It is an A-linear predifferential as well; in particular, we have

$$\bar{\partial}_\phi^2(\xi) = [\bar{\partial}(\phi) + \frac{1}{2}[\phi, \phi] + \ell, \xi]$$

so that $(L_A^\bullet, [-, -], \bar{\partial}_\phi, \ell_\phi)$ with

$$\ell_\phi := \bar{\partial}(\phi) + \frac{1}{2}[\phi, \phi] + \ell$$

is another curved Lie algebra. In particular, we have $\bar{\partial}_\phi(\ell_\phi) = 0$ since $[\phi, [\phi, \phi]] = 0$. We can consider $\mathrm{PDiff}_A(L^\bullet) := \mathfrak{m}_A \cdot L_A^1$ as the space of predifferentials. Since ϕ can always be lifted along a surjection $B' \to B$ in $\mathbf{Art}_\Lambda$, there is not really a distinguished predifferential; we may consider a curved Lie algebra as a graded Lie algebra which comes with an affine space of predifferentials, none of which is intrinsically distinguished.

If the central fiber $L_0^\bullet$ is centerless, then $\bar{\partial}_\phi$ is a differential if and only if $\ell_\phi = 0$ (by the argument in the proof of Lemma 4.6); in general, this is a sufficient condition. If $\ell = 0$, then this becomes the classical *Maurer–Cartan equation*

$$\bar{\partial}(\phi) + \frac{1}{2}[\phi, \phi] = 0.$$

In our case, we have an additional term ℓ so that we talk about an *extended* Maurer–Cartan equation.

Definition 5.2 (Classical Extended Maurer–Cartan Equation) Let $L^\bullet$ be a Λ-linear curved Lie algebra, and let $A \in \mathbf{Art}_\Lambda$. Then the *classical extended Maurer–Cartan equation* is the equation

$$\ell_\phi = \bar{\partial}(\phi) + \frac{1}{2}[\phi, \phi] + \ell = 0$$

which is to be satisfied by elements $\phi \in \mathfrak{m}_A \cdot L_A^1$. Its solutions form a homogeneous[1] functor

$$\mathrm{MC}(L^\bullet, -) : \quad \mathbf{Art}_\Lambda \to \mathbf{Set}$$

of Artin rings in the sense of Definition A.8.

For a fixed predifferential $\bar{\partial}$, the solutions of the classical extended Maurer–Cartan equation form the *differentials* on $L_A^\bullet$ in the sense that $(L_A^\bullet, [-, -], \bar{\partial}_\phi, \ell_\phi)$ becomes a differential graded Lie algebra in our sense.

Observe that the Definition 4.3 of gauge transforms makes sense also for curved Lie algebras, although they do not come with a Cartan structure. The predifferential $\bar{\partial}_\phi$ is not compatible with gauge transforms. Nonetheless, we can formulate precisely how it behaves under gauge transforms. In order to do this, we need the power series expansion

$$T(x) := \frac{e^x - 1}{x} = \sum_{n=0}^{\infty} \frac{x^n}{(n+1)!} = 1 + \frac{x}{2} + \frac{x^2}{6} + \frac{x^3}{24} + \dots .$$

Lemma and Definition 5.3 (Gauge Action I) *Let $L^\bullet$ be a Λ-linear curved Lie algebra, and let $B' \to B$ be a surjection in $\mathbf{Art}_\Lambda$ with kernel $I \subset B'$. Let $\theta \in I \cdot L_{B'}^0$ and $\phi \in \mathfrak{m}_{B'} \cdot L_{B'}^1$. Then we have*

$$\exp_\theta(\bar{\partial}_\phi(\xi)) = \exp_\theta(\bar{\partial}\xi + [\phi, \xi])$$

$$= \left(\bar{\partial}(-) + [\exp_\theta(\phi) - T(\nabla_\theta)(\bar{\partial}\theta), -]\right)(\exp_\theta(\xi))$$

with $\nabla_\theta = [\theta, -]$ and

$$T(\nabla_\theta)(\bar{\partial}\theta) = \bar{\partial}\theta + \frac{1}{2}[\theta, \bar{\partial}\theta] + \frac{1}{6}[\theta, [\theta, \bar{\partial}\theta]] + \dots .$$

This gives rise to the gauge action

$$\exp_\theta * \phi := \exp_\theta(\phi) - T(\nabla_\theta)(\bar{\partial}\theta) = \phi + T(\nabla_\theta)([\theta, \phi] - \bar{\partial}\theta),$$

---

[1] This is standard. See e.g. [76, 10.5] for some hints how to show this.

which satisfies $\exp_\theta * (\exp_\xi * \phi) = \exp_{\theta \odot \xi} * \phi$ *with the Baker–Campbell–Hausdorff product* $\theta \odot \xi$. *We say that* $\phi, \phi' \in \mathfrak{m}_{B'} \cdot L_{B'}^1$ *are gauge equivalent relative to* B *if there is some* $\theta \in I \cdot L_{B'}^0$ *with* $\exp_\theta * \phi = \phi'$. *This is an equivalence relation.*

Proof See [199, §6.3] for a general discussion about the gauge action. The formula for the behavior of $\bar{\partial}_\phi$ under gauge transforms is taken from [38, Lemma 2.5]. The statements remain true for a curved Lie algebra since the square $\bar{\partial}^2$ does not show up anywhere. To see the first statement explicitly, note that

$$\bar{\partial}\nabla_\theta^n(\xi) = \nabla_\theta^n(\bar{\partial}\xi) + \sum_{s=1}^{n} \binom{n}{s} \nabla_{\nabla_\theta^{s-1}(\bar{\partial}\theta)} \nabla_\theta^{n-s}(\xi)$$

for $n \geq 1$. Then we obtain

$$\bar{\partial}\exp_\theta(\xi) = \exp_\theta(\bar{\partial}\xi) + [T(\nabla_\theta)(\bar{\partial}\theta), \exp_\theta(\xi)]$$

by a direct computation, from which the version for $\bar{\partial}_\phi$ follows easily. $\square$

We are mostly interested in gauge equivalence relative to $A_0 = \mathbf{k}$. If $L_0^\bullet$ is centerless, then $\phi, \phi' \in \mathfrak{m}_A \cdot L_A^1$ are gauge equivalent if and only if there is some $\theta \in \mathfrak{m}_A \cdot L_A^0$ with

$$\exp_\theta \circ \bar{\partial}_\phi = \bar{\partial}_{\phi'} \circ \exp_\theta.$$

Remark 5.4 It is tempting to use this as a definition of gauge equivalence. In fact, we stated this wrongly in [76, §10.2], where gauge equivalence has been defined by this equation without assuming that $L_0^\bullet$ is centerless. The problem with the latter definition is that it can depend on which space $\bar{\partial}_\phi$ acts on, i.e., when we require the above equation in a bigraded Gerstenhaber calculus then it might give a different notion of gauge equivalence than when we consider it only in the graded Lie algebra which is the (-1)-part of the bigraded Gerstenhaber algebra. However, the definition of gauge equivalence in [76, Defn. 7.1] is still correct because of strict faithfulness (i.e., (-1)-injectivity)—there, $\exp_\theta \circ \bar{\partial}_\phi = \bar{\partial}_{\phi'} \circ \exp_\theta$ is required on the level of the bigraded Gerstenhaber algebra. $\Diamond$

If $\phi, \phi' \in \mathfrak{m}_A \cdot L_A^1$ are gauge equivalent, then ϕ is a Maurer–Cartan solution if and only if ϕ' is a Maurer–Cartan solution. If $L_0^\bullet$ is centerless, then this is easy to see, but it is true in general. For the general case, first observe that

$$\bar{\partial}T(\nabla_\theta)(\bar{\partial}\theta) = T(\nabla_\theta)(\bar{\partial}^2\theta) + \frac{1}{2}[T(\nabla_\theta)(\bar{\partial}\theta), T(\nabla_\theta)(\bar{\partial}\theta)],$$

hence we have

$$\exp_\theta(\ell) - \ell = \frac{1}{2}[T(\nabla_\theta)(\bar{\partial}\theta), T(\nabla_\theta)(\bar{\partial}\theta)] - \bar{\partial}T(\nabla_\theta)(\bar{\partial}\theta).$$

With this, we find

$$\bar{\partial}\phi' + \frac{1}{2}[\phi', \phi'] + \ell = \exp_\theta(\bar{\partial}\phi + \frac{1}{2}[\phi, \phi] + \ell)$$

by using Lemma 5.3 in a straightforward computation of the left-hand side since $\exp_\theta$ commutes with $[-, -]$. Thus ϕ is a Maurer–Cartan solution if and only if $\phi' = \exp_\theta * \phi$ is a Maurer–Cartan solution.

We assemble gauge equivalence classes of Maurer–Cartan solutions in the functor of Artin rings

$$\mathrm{Def}(L^\bullet, -)\colon \quad \mathbf{Art}_\Lambda \to \mathbf{Set}. \tag{5.1}$$

It is in general not homogeneous, but it is a deformation functor.

Lemma 5.5 *The functor (5.1) is a neat deformation functor in the sense of Definition A.9, i.e., it satisfies (H_0), (H_1), (H_2), and (H_2^+).*

Proof Except for (H_2^+), we gave already a proof in [76, Prop. 10.5]. The Maurer–Cartan functor $\mathrm{MC}(L^\bullet, -)$ is homogeneous, and we have a smooth map of functors of Artin rings $\mathrm{MC}(L^\bullet, -) \Rightarrow \mathrm{Def}(L^\bullet, -)$. This shows that $\mathrm{Def}(L^\bullet, -)$ satisfies (H_0) and (H_1). For (H_2) and (H_2^+), we use that $A \mapsto \mathfrak{m}_A \cdot L_A^0$ is a homogeneous functor of Artin rings, which follows e.g. from considering the diagram in the proof of [76, Prop. 10.5]. $\qquad\square$

The key insight of [38] and, subsequently, of [76], is that deformation functors in log geometry are isomorphic to this deformation functor for an appropriate choice of $L^\bullet$.

Example 5.6 Let $f_0\colon X_0 \to S_0$ be proper, log smooth, and saturated over the punctual log scheme $S_0 = \mathrm{Spec}(Q \to \mathbf{k})$. Then for the curved Lie algebra $L^\bullet_{X_0/\Lambda}$ constructed in [76, §8], the log smooth deformation functor LD_{X_0/S_0} is isomorphic to $\mathrm{Def}(L^\bullet_{X_0/\Lambda}, -)$. $\qquad\Diamond$

5.1.2   Curved $\mathcal{P}$-Algebras

Let $\mathcal{P}$ be an algebraic structure carrying a Cartan structure, and let $E^{\bullet,\bullet}$ be a Λ-linear curved $\mathcal{P}$-algebra, i.e., a Cartanian $\mathcal{P}^{\mathrm{crv}}$-*pre*-algebra in $\mathfrak{Comp}(\Lambda)$ with $\ell \in \mathfrak{m}_\Lambda \cdot T^2$. Then $T^\bullet$ is a Λ-linear curved Lie algebra, so we obtain a deformation functor

$$\mathrm{Def}(E^{\bullet,\bullet}, -) := \mathrm{Def}(T^\bullet, -)\colon \quad \mathbf{Art}_\Lambda \to \mathbf{Set}.$$

A general $\phi \in \mathfrak{m}_A \cdot T_A^1$ gives a new predifferential

$$\bar{\partial}_\phi\colon \quad \mathcal{E}^{P,j} \to \mathcal{E}^{P,j+1}, \quad p \mapsto \bar{\partial}p + \nabla_\phi(p),$$

in the sense that $(E_A^{\bullet,\bullet}, \bar{\partial}_\phi, \ell_\phi)$ is again a Cartanian curved $\mathcal{P}$-algebra with

$$\ell_\phi := \bar{\partial}\phi + \frac{1}{2}[\phi, \phi] + \ell \in T_A^2.$$

Indeed, if $\bar{\partial}$ satisfies the five conditions in Definition 3.44, then $\bar{\partial}_\phi$ satisfies them as well.

Remark 5.7 If $E^{\bullet,\bullet}$ is in fact a Lie–Rinehart algebra or pair or a Gerstenhaber algebra or calculus, then it still is with the new predifferential $\bar{\partial}_\phi$ and the new ℓ_ϕ. This is because the axioms we are missing with a Cartanian curved $\mathcal{P}$-algebra do not involve $\bar{\partial}$. On the other side, this discussion does not apply to Batalin–Vilkovisky algebras or calculi since already their plain, singly graded version does not admit a Cartan structure. $\diamond$

If ϕ is a solution of the classical extended Maurer–Cartan equation, then $\ell_\phi = 0$ and $\bar{\partial}_\phi$ is a differential on $E_A^{\bullet,\bullet}$. If two solutions ϕ, ϕ' of the classical extended Maurer–Cartan equation are gauge equivalent, $\phi' = \exp_\theta * \phi$, then we have

$$\exp_\theta \circ \bar{\partial}_\phi = \bar{\partial}_{\phi'} \circ \exp_\theta$$

on all $E_A^{P,j}$ by the proof of Lemma 5.3. If $E_0^{\bullet,\bullet}$ is faithful, e.g. centerless or strictly faithful, then we can identify the set of Maurer–Cartan solutions with those $\phi \in \mathfrak{m}_A \cdot T_A^1$ such that $\bar{\partial}_\phi^2 = 0$, and two differentials $\bar{\partial}_\phi$ and $\bar{\partial}_{\phi'}$ are gauge equivalent if and only if they are transformed into each other by a gauge transform. Here, it is crucial that, in our definition of faithfulness in the bigraded case, we have required the condition not only for T^0 but for all T^j.

Example 5.8 Let $f_0 \colon X_0 \to S_0$ be a separated, log smooth, and saturated morphism to a log point S_0. Then the construction in Chap. 13 yields a curved two-sided Gerstenhaber calculus. Since the underlying bigraded two-sided Gerstenhaber calculus is strictly faithful, we can interpret Maurer–Cartan solutions as differentials on a Gerstenhaber calculus, not only on a Lie algebra. $\diamond$

Example 5.9 When we study deformations of pairs $f_0 \colon X_0 \to S_0$ and a line bundle $\mathcal{L}_0$ on X_0, then we apply this discussion to Λ-linear curved Lie–Rinehart pairs. $\diamond$

Example 5.10 In a Λ-linear curved Batalin–Vilkovisky algebra or calculus $\mathcal{E}^{\bullet,\bullet}$, we consider the classical extended Maurer–Cartan equation and deformation functor of the underlying Λ-linear Gerstenhaber algebra or calculus as the ones of $\mathcal{E}^{\bullet,\bullet}$. $\diamond$

When showing unobstructedness, we can work equivalently with the Maurer–Cartan functor $\mathrm{MC}(E^{\bullet,\bullet}, -)$.

Lemma 5.11 *Let $E^{\bullet,\bullet}$ be a Λ-linear curved $\mathcal{P}$-algebra. Then $\mathrm{Def}(E^{\bullet,\bullet}, -)$ is unobstructed if and only if $\mathrm{MC}(E^{\bullet,\bullet}, -)$ is unobstructed.*

Proof If $\mathrm{MC}(G^{\bullet,\bullet}, -)$ is unobstructed, then $\mathrm{Def}(G^{\bullet,\bullet}, -)$ is unobstructed as well because the natural transformation is surjective. Conversely, let $B' \to B$ be a small extension with kernel I. Let ϕ be a Maurer–Cartan solution over B, and let ψ' be a Maurer–Cartan solution over B' such that $\psi := \psi'|_B$ is gauge equivalent to ϕ. Let $\theta \in \mathfrak{m}_B \cdot T_B^0$ be such that $\exp_\theta * \psi = \phi$, and let $\theta' \in \mathfrak{m}_{B'} \cdot T_{B'}^0$ be a lift of θ. Then define $\phi' := \exp_{\theta'} * \psi'$. This is a Maurer–Cartan solution as well. Since the gauge action commutes with base change along $B' \to B$, we find $\phi'|_B = \phi$. Thus $\mathrm{MC}(E^{\bullet,\bullet}, -)$ is unobstructed. $\square$

5.2   The Semi-classical Extended Maurer–Cartan Equation

Now let $G^{\bullet,\bullet}$ be a Λ-linear curved *Batalin–Vilkovisky* algebra. Let us assume (for simplicity) that $G_0^{\bullet,\bullet}$ is strictly faithful. We have seen above that a classical Maurer–Cartan solution ϕ gives rise to a differential $\bar{\partial}_\phi$ on $G^{\bullet,\bullet}$ by modifying the predifferential $\bar{\partial}$ with $[\phi, -]$. However, on a curved Batalin–Vilkovisky algebra, we also have the Batalin–Vilkovisky operator Δ. The anti-commutation rule becomes

$$\bar{\partial}_\phi \Delta + \Delta \bar{\partial}_\phi = [y + \Delta\phi, -],$$

which is in general not zero. To achieve that this is also zero, i.e., that we do not only have $\bar{\partial}_\phi^2 = 0$ but also an anti-commuting Batalin–Vilkovisky operator, we also want to modify Δ with an element $f \in \mathfrak{m}_A \cdot G_A^{0,0}$, i.e., $\Delta_f := \Delta + [f, -]$. Then $\Delta_f^2 = 0$ is automatic; moreover, we have

$$\bar{\partial}_\phi \Delta_f + \Delta_f \bar{\partial}_\phi = [\bar{\partial} f + [\phi, f] + y + \Delta\phi, -].$$

The term in the bracket on the right-hand side is in $G_A^{0,2}$, so we cannot conclude from strict faithfulness[2] that it is necessary to have

$$\bar{\partial} f + [\phi, f] + y + \Delta\phi = 0$$

in order to have $\bar{\partial}_\phi \Delta_f + \Delta_f \bar{\partial}_\phi = 0$, but it is certainly sufficient.

---

[2] The fact that $[1, -] = 0$ suggests that we cannot expect an equivalent of faithfulness for elements that are in $G_A^{0,\bullet}$ rather than $G_A^{-1,\bullet}$.

Definition 5.12 (Semi-classical Extended Maurer–Cartan Equation) The *semi-classical*[3] *extended Maurer–Cartan equation* is the system of equations

$$\bar\partial\phi + \frac{1}{2}[\phi, \phi] + \ell = 0$$

$$\bar\partial f + [\phi, f] + y + \Delta\phi = 0$$

which is to be satisfied by pairs (ϕ, f) with $\phi \in \mathfrak{m}_A \cdot G_A^{-1,1}$ and $f \in \mathfrak{m}_A \cdot G_A^{0,0}$. Its solutions form a homogeneous[4] functor

$$\mathrm{SMC}(G^{\bullet,\bullet}, -): \quad \mathbf{Art}_\Lambda \to \mathbf{Set}$$

of Artin rings.

Lemma 5.13 *If $(\phi, f) \in \mathrm{SMC}(G^{\bullet,\bullet}, A)$, then $G_A^{\bullet,\bullet}$ forms an A-linear differential bg Batalin–Vilkovisky algebra with the differentials Δ_f and $\bar\partial_\phi$. More precisely:*

(1) $\bar\partial_\phi^2(\theta) = 0$ *and* $\bar\partial_\phi\Delta_f(\theta) + \Delta_f\bar\partial_\phi(\theta) = 0$;
(2) *the two derivation rules for $\bar\partial_\phi(\theta \wedge \xi)$ and $\bar\partial_\phi[\theta, \xi]$ hold;*
(3) $\Delta_f^2(\theta) = 0$, $\Delta_f(1) = 0$, *and the derivation rule for $\Delta_f[\theta, \xi]$ holds;*
(4) *the Bogomolov–Tian–Todorov formula for $\Delta_f(\theta \wedge \xi)$ holds.*

Proof All statements follow from easy and straightforward computations.   □

For a Λ-linear curved Gerstenhaber algebra, or, more generally, a Λ-linear curved $\mathcal{P}$-algebra as in Definition 5.1, the gauge transform $\exp_\theta$ for an element $\theta \in \mathfrak{m}_A \cdot T_A^0$ gives an isomorphism between $(\mathcal{E}_A^{\bullet,\bullet}, \bar\partial_\phi)$ and $(\mathcal{E}_A^{\bullet,\bullet}, \bar\partial_{\phi'})$ for $\phi' = \exp_\theta * \phi$. The same is true for Λ-linear curved Batalin–Vilkovisky algebras as soon as we extend the gauge action to $f \in \mathfrak{m}_A \cdot G_A^{0,0}$ appropriately.

Lemma and Definition 5.14 (Gauge Action II) *Let $G^{\bullet,\bullet}$ be a Λ-linear curved Batalin–Vilkovisky algebra, and let $B' \to B$ be a surjection in $\mathbf{Art}_\Lambda$ with kernel $I \subset B'$. Let $\theta \in I \cdot G_{B'}^{-1,0}$ and $f \in \mathfrak{m}_{B'} \cdot G_{B'}^{0,0}$. Then we have*

$$\exp_\theta \circ \Delta_f(\xi) = \Delta \circ \exp_\theta(\xi) + [\exp_\theta(f) - T(\nabla_\theta)(\Delta\theta), \ \exp_\theta(\xi)].$$

This gives rise to the gauge action

$$\exp_\theta * f := \exp_\theta(f) - T(\nabla_\theta)(\Delta\theta) = f + T(\nabla_\theta)([\theta, f] - \Delta\theta),$$

<hr>

[3] This notion is intermediate between the *classical* extended Maurer–Cartan equation and the *quantum* extended Maurer–Cartan equation that we consider below. The latter amounts to an infinite sequence of equations, and the second equation of the semi-classical extended Maurer–Cartan equation is but one of them.

[4] The proof is the same as for $\mathrm{MC}(L^\bullet, -)$.

which satisfies $\exp_\theta * (\exp_\xi * f) = \exp_{\theta \odot \xi} * f$ *with the Baker–Campbell–Hausdorff product* $\theta \odot \xi$.

Proof The proof of the first statement is the same as the one of Lemma 5.3 with $\bar\partial$ replaced by Δ and ϕ replaced by f. The formula $\exp_\theta * (\exp_\xi * f) = \exp_{\theta \odot \xi} * f$ comes from the fact that $\exp_\theta * f$ is the gauge action in the differential graded Lie algebra $M^\bullet := (G_{B'}^{\bullet+1,0}, [-,-], \Delta)$. Note that the grading is such that $M^0 = G_{B'}^{-1,0}$ and $M^1 = G_{B'}^{0,0}$. $\qquad\square$

We say that two pairs (ϕ, f) and (ϕ', f') are *gauge equivalent* if there is some $\theta \in \mathfrak{m}_A \cdot G_A^{-1,0}$ with $\exp_\theta * \phi = \phi'$ and $\exp_\theta * f = f'$. In this case, $\exp_\theta$ induces an isomorphism between the Λ-linear curved Batalin–Vilkovisky algebras $(G_A^{\bullet,\bullet}, \Delta_f, \bar\partial_\phi)$ and $(G_A^{\bullet,\bullet}, \Delta_{f'}, \bar\partial_{\phi'})$. Just as in the case of ϕ, the pair (ϕ, f) is a solution of the semi-classical Maurer–Cartan equation if and only if (ϕ', f') is. To see the second part, observe that

$$\bar\partial T(\nabla_\theta)(\Delta\theta) + \Delta T(\nabla_\theta)(\bar\partial\theta) = T(\nabla_\theta)(\bar\partial\Delta\theta) + T(\nabla_\theta)(\Delta\bar\partial\theta)$$
$$+ [T(\nabla_\theta)(\Delta\theta), T(\nabla_\theta)(\bar\partial\theta)].$$

This implies

$$\exp_\theta(y) - y = [T(\nabla_\theta)(\Delta\theta), T(\nabla_\theta)(\bar\partial\theta)] - \bar\partial T(\nabla_\theta)(\Delta\theta) - \Delta T(\nabla_\theta)(\bar\partial\theta).$$

From here, we obtain

$$\bar\partial f' + [\phi', f'] + y + \Delta\phi' = \exp_\theta(\bar\partial f + [\phi, f] + y + \Delta\phi)$$

by a straightforward computation using Lemmas 5.3 and 5.14. Thus, (ϕ, f) is a semi-classical Maurer–Cartan solution if and only if (ϕ', f') is. Gauge equivalence classes of elements $(\phi, f) \in \mathrm{SMC}(G^{\bullet,\bullet}, A)$ form a functor of Artin rings

$$\mathrm{SDef}(G^{\bullet,\bullet}, -): \quad \mathbf{Art}_\Lambda \to \mathbf{Set}. \tag{5.2}$$

Lemma 5.15 *The functor* (5.2) *is a neat deformation functor in the sense of Definition A.9, i.e., it satisfies* (H_0), (H_1), (H_2), *and* (H_2^+).

Proof Condition (H_1) holds since $\mathrm{SMC}(G^{\bullet,\bullet}, -)$ is homogeneous, and (H_2) and (H_2^+) follow (as in the classical case) from a careful analysis of lifts of gauge transforms in the diagram in [76, Prop. 10.5]. $\qquad\square$

As in the classical case, $\mathrm{SDef}(G^{\bullet,\bullet}, -)$ is unobstructed if and only if $\mathrm{SMC}(G^{\bullet,\bullet}, -)$ is unobstructed.

5.2.1 The Volume Form Associated with a Semi-classical Maurer–Cartan Solution

Suppose that $G^{\bullet,\bullet}$ is part of a curved Batalin–Vilkovisky calculus $(G^{\bullet,\bullet}, A^{\bullet,\bullet})$. Recall that, for $\phi \in \mathfrak{m}_A \cdot G_A^{-1,1}$, we have on $A_A^{\bullet,\bullet}$ the new predifferential

$$\bar{\partial}_\phi(\alpha) := \bar{\partial}\alpha + \mathcal{L}_\phi(\alpha)$$

since $\nabla_\theta = \mathcal{L}_\theta$ on $A^{\bullet,\bullet}$. We wish to find a modification $\omega_{(\phi,f)}$ of ω such that:

(a) $(G_A^{\bullet,\bullet}, A_A^{\bullet,\bullet}, \Delta_f, \bar{\partial}_\phi, \omega_{(\phi,f)})$ is a Λ-linear curved Batalin–Vilkovisky calculus for all pairs (ϕ, f) with ℓ_ϕ and $y_{(\phi,f)} := \bar{\partial}f + [\phi, f] + y + \Delta\phi$;
(b) if (ϕ, f) and (ϕ', f') are gauge equivalent via θ, then $(G_A^{\bullet,\bullet}, A_A^{\bullet,\bullet}, \Delta_f, \bar{\partial}_\phi, \omega_{(\phi,f)})$ and $(G_A^{\bullet,\bullet}, A_A^{\bullet,\bullet}, \Delta_{f'}, \bar{\partial}_{\phi'}, \omega_{(\phi',f')})$ are isomorphic via $\exp_\theta$.

For a semi-classical Maurer–Cartan solution (ϕ, f), $(G_A^{\bullet,\bullet}, A_A^{\bullet,\bullet}, \Delta_f, \bar{\partial}_\phi, \omega_{(\phi,f)})$ is automatically a *differential* bg Batalin–Vilkovisky calculus.

Since $\omega_{(\phi,f)} \in A_A^{d,0}$ would be a constant, we expect that

$$\omega_{(\phi',f')} = \exp_\theta(\omega_{(\phi,f)}). \tag{5.3}$$

Lemma 5.16 *The choice*

$$\omega_{(\phi,f)} := \omega_f := e^f \lrcorner \omega := \sum_{n=0}^{\infty} \frac{(f\lrcorner)^n(\omega)}{n!} = \omega + (f \lrcorner \omega) + \frac{f \lrcorner (f \lrcorner \omega)}{2} + \dots$$

$$= \left(\sum_{n=0}^{\infty} \frac{f^n}{n!} \right) \lrcorner \omega := (e^f) \lrcorner \omega,$$

satisfies (5.3).

Proof We have to show $\exp_\theta(e^f \lrcorner \omega) = e^{f'} \lrcorner \omega$ with $f' = \exp_\theta(f) - T(\nabla_\theta)(\Delta\theta)$. It is straightforward to see that it is sufficient to show

$$\exp(-T(\nabla_\theta)(\Delta\theta)) \lrcorner \omega = \exp_\theta(\omega)$$

where the exp on the left-hand side is the exponential of the nilpotent element $-T(\nabla_\theta)(\Delta\theta)$ in the ring $G_A^{0,0}$. We can expand the left-hand side to

$$\left(1 + \sum_{t=1}^{\infty} \sum_{n=1}^{t} \frac{(-1)^n}{n!} \sum_{\substack{k_1+\dots+k_n=t \\ k_i \geq 1}} \prod_{i=1}^{n} \frac{1}{k_i!} \nabla_\theta^{k_i-1} \Delta(\theta) \right) \lrcorner \omega.$$

Let us denote the summand for $t \geq 1$ in this formula by $S(t)$. The right-hand side of the equation is

$$\exp_\theta(\omega) = \omega + \sum_{t=1}^{\infty} \frac{1}{t!} \mathcal{L}_\theta^t(\omega).$$

To conclude the proof, we show

$$(S(t) \lrcorner 1) \wedge \omega = \frac{1}{t!} \mathcal{L}_\theta^t(\omega)$$

by induction on $t \geq 1$. We have $S(1) = -\Delta\theta$ and $\mathcal{L}_\theta(\omega) = -\Delta\theta \lrcorner \omega$ for the base case. On the right-hand side, the induction step yields

$$\frac{1}{(t+1)!} \mathcal{L}_\theta^{t+1}(\omega) = \frac{1}{t+1} \Big(\big([\theta, S(t)] - S(t) \wedge \Delta\theta\big) \lrcorner 1 \Big) \wedge \omega$$

so that it remains to show

$$S(t+1) = \frac{1}{t+1} \big([\theta, S(t)] - S(t) \wedge \Delta\theta\big).$$

This is slightly tricky combinatorics. Let $I(t)$ be the set of tuples $(k_1, \ldots, k_n)$ of natural numbers ≥ 1 of variable length ≥ 1 with $k_1 + \ldots + k_n = t$, and let

$$s(k_1, \ldots, k_n) = \frac{(-1)^n}{n!} \prod_{i=1}^{n} \frac{1}{k_i!} \nabla_\theta^{k_i - 1} \Delta\theta$$

so that $S(t) = \sum_{\underline{k} \in I(t)} s(\underline{k})$. Then

$$[\theta, s(k_1, \ldots, k_n)] - s(k_1, \ldots, k_n) \wedge \Delta\theta$$

$$= \sum_{j=1}^{n} (k_j + 1) s(k_1, \ldots, k_j + 1, \ldots, k_n)$$

$$+ \sum_{j=0}^{n} s(k_1, \ldots, k_j, 1, k_{j+1}, \ldots, k_n).$$

Collecting all summands $s(\ell_1, \ldots, \ell_n)$ for $\ell_1 + \ldots + \ell_n = t + 1$, we find the assertion. $\square$

The volume form ω_f does what we want.

Lemma 5.17 *Let $\phi \in \mathfrak{m}_A \cdot G_A^{-1,1}$ and $f \in \mathfrak{m}_A \cdot G_A^{0,0}$. Then $(G_A^{\bullet,\bullet}, A_A^{\bullet,\bullet}, \Delta_f, \bar{\partial}_\phi, \omega_f)$ is a Λ-linear curved Batalin–Vilkovisky calculus with ℓ_ϕ and $y_{(\phi,f)}$. If $(\phi, f) \in \mathrm{SMC}(G^{\bullet,\bullet}, A)$, then it is a differential bg Batalin–Vilkovisky calculus. Concretely, we have:*

(1) *$\bar{\partial}_\phi^2(\alpha) = \mathcal{L}_{\ell_\phi}(\alpha)$ and $\bar{\partial}_\phi \partial(\alpha) + \partial \bar{\partial}_\phi(\alpha) = 0$;*

(2) *the two derivation rules hold for $\bar{\partial}_\phi(\alpha \wedge \beta)$ and $\bar{\partial}_\phi(\theta \neg \alpha)$;*

(3) *$\bar{\partial}_\phi \omega_f = y_{(\phi,f)} \neg \omega_f$; in particular, if (ϕ, f) is a semi-classical Maurer–Cartan solution, then*

$$\bar{\partial}_\phi \omega_f = 0$$

and $\bar{\partial}_\phi(\theta) \neg \omega_f = \bar{\partial}_\phi(\theta \neg \omega_f)$;

(4) *$\kappa_f : G_A^{p,q} \to G_A^{d+p,q}$, $\theta \mapsto \theta \neg \omega_f$, is an isomorphism of A-modules;*

(5) *$\upsilon_f : A_A^{0,q} \to A_A^{d,q}$, $\alpha \mapsto \alpha \wedge \omega_f$, is an isomorphism of A-modules;*

(6) *$\Delta_f(\theta) \neg \omega_f = \partial(\theta \neg \omega_f)$, i.e., the modified Batalin–Vilkovisky operator Δ_f arises from the de Rham differential ∂ via the new isomorphism κ_f.*

Proof The first two statements follow from the general theory of Λ-linear curved $\mathcal{P}$-algebras. For $\bar{\partial}_\phi \omega_f = y_{(\phi,f)} \neg \omega_f$, first note that $\bar{\partial}_\phi(e^f) = \bar{\partial}_\phi(f) \wedge e^f$. Then we have

$$\bar{\partial}_\phi(\omega_f) = \bar{\partial}_\phi(e^f) \neg \omega + e^f \neg \bar{\partial}_\phi \omega = e^f \neg (\bar{\partial}_\phi f \neg \omega + \bar{\partial}_\phi \omega)$$

$$= e^f \neg (\bar{\partial}_\phi f \neg \omega + y \neg \omega + \partial(\phi \neg \omega))$$

$$= e^f \neg (\bar{\partial} f + [\phi, f] + y + \Delta\phi) \neg \omega = y_{(\phi,f)} \neg \omega_f$$

since $y \neg \omega = \bar{\partial}\omega$ and $\mathcal{L}_\phi(\omega) = \partial(\phi \neg \omega) = \Delta\phi \neg \omega$. Next, let $\mu_f(\theta) := e^f \wedge \theta$; it is an isomorphism of A-modules with inverse μ_{-f}. Since $\kappa_f = \kappa \circ \mu_f$ and κ is an isomorphism, κ_f is an isomorphism as well. The proof that υ_f is an isomorphism is similar. The last statement follows from an application of the Bogomolov–Tian–Todorov formula when we also use $[e^f, \theta] = [f, \theta] \wedge e^f$. $\square$

If (ϕ, f) and (ϕ', f') are gauge equivalent via θ, then the two Λ-linear curved Batalin–Vilkovisky calculi $(G_A^{\bullet,\bullet}, A_A^{\bullet,\bullet}, \Delta_f, \bar{\partial}_\phi, \omega_f)$ and $(G_A^{\bullet,\bullet}, A_A^{\bullet,\bullet}, \Delta_{f'}, \bar{\partial}_{\phi'}, \omega_{f'})$ are isomorphic via $\exp_\theta$. Indeed, the compatibilities with most operations follow from generalities on (Cartanian) Λ-linear curved $\mathcal{P}$-algebras, the compatibility with Δ is Lemma 5.14, the compatibility with ω is Lemma 5.16, and we have checked the compatibility with ℓ and y when we have shown that solving the semi-classical Maurer–Cartan equation is a gauge invariant property.

5.3 Perfect Curved Batalin–Vilkovisky Algebras and Calculi

In this section, we present conditions under which we can show in Chap. 6 unobstructedness of our Maurer–Cartan functors using the argument of Chan–Leung–Ma in [38].

If a Λ-linear curved *Gerstenhaber* calculus $(G^{\bullet,\bullet}, A^{\bullet,\bullet})$ arises from a *proper*, saturated, and log smooth morphism, log Calabi–Yau or not, then it is perfect in the sense of the following definition. When $A^{\bullet,\bullet}$ is a double complex (with $\partial\bar\partial + \bar\partial\partial = 0$), then its *first* spectral sequence is the one with E_1-page ${}'E_1^{p,q} = H^q(A^{p,\bullet}, \bar\partial)$ and boundary maps

$$d_r^{p,q} : \quad {}'E_r^{p,q} \to {}'E_r^{p+r,q+1-r}.$$

Definition 5.18 (Quasi-perfectness and Perfectness I) Let $(G^{\bullet,\bullet}, A^{\bullet,\bullet})$ be a Λ-linear curved Gerstenhaber calculus. We say that it is *quasi-perfect* if it is strictly faithful and the following conditions hold:

(a) The first spectral sequence ${}'E_{0;r}$ associated with the double complex $(A_0^{\bullet,\bullet}, \partial, \bar\partial)$ degenerates at E_1.

(b) For every $A \in \mathbf{Art}_\Lambda$ and every $k \geq 0$, the induced map

$$H^k(A_A^\bullet, d) \to H^k(A_0^\bullet, d)$$

   is surjective, where $d = \partial + \bar\partial + \ell \dashv (-)$.

We say that it is *perfect* if, furthermore, for every $A \in \mathbf{Art}_\Lambda$, the A-module $H^k(A_A^\bullet, d)$ is finitely generated.

Remark 5.19 Several remarks are in order.

(1) If $(G^{\bullet,\bullet}, A^{\bullet,\bullet})$ is perfect, and if $\phi \in \mathfrak{m}_A \cdot G_A^{-1,1}$ is a Maurer–Cartan solution, then the first spectral sequence of the double complex $(A_A^{\bullet,\bullet}, \partial, \bar\partial_\phi)$ degenerates at E_1 by Proposition D.4. The cohomology $H^q(A_A^{p,\bullet}, \bar\partial_\phi)$ is a flat A-module whose formation commutes with base change, as is the total cohomology $H^k(A_A^\bullet, \partial + \bar\partial_\phi)$. This is possible without any finiteness conditions by the very general nature of Proposition D.4, and will be discussed in more detail toward the end of this section, especially Lemma 5.28.

(2) Geometrically, the E_1-degeneration of ${}'E_{0;r}$ corresponds to the Hodge–de Rham degeneration over the point, the surjectivity of $H^k(A_A^\bullet, d) \to H^k(A_0^\bullet, d)$ corresponds to the freeness of the relative algebraic de Rham cohomology $R^k f_* \Omega_{X_A/S_A}^\bullet$, and the finiteness condition corresponds to properness of $f : X_A \to S_A$. However, especially the surjectivity condition is stronger than its geometric interpretation because it even has a content if there is no Maurer–Cartan element which would correspond to a geometric map $f : X_A \to S_A$.

(3) For a discussion of Condition (b), see also Remark 5.22 below.

(4) We will see in Theorem 15.2 that the Λ-linear curved Gerstenhaber calculus associated with a log toroidal family $f_0 \colon X_0 \to S_0$ and a log toroidal system of deformations $\mathscr{D}$ is (quasi-)perfect. This is the most general (quasi-)perfectness theorem that we have so far. $\Diamond$

If we have a Λ-linear curved *Batalin–Vilkovisky* calculus, then we can use the isomorphisms κ to state (quasi-)perfectness in terms of the Batalin–Vilkovisky *algebra*.

Lemma 5.20 *Let $(G^{\bullet,\bullet}, A^{\bullet,\bullet})$ be a Λ-linear curved Batalin–Vilkovisky calculus. Assume that $G^{\bullet,\bullet}$ is strictly faithful. Then the calculus is quasi-perfect if and only if the following conditions hold:*

(a) *The first spectral sequence $'E_{0;r}$ associated with the double complex $(G_0^{\bullet,\bullet}, \Delta, \bar{\partial})$ over $A_0 = \mathbf{k}$ degenerates at E_1.*
(b) *For every $A \in \mathbf{Art}_\Lambda$, the induced map $H^k(G_A^\bullet, \check{d}) \to H^k(G_0^\bullet, \check{d})$ with*

$$\check{d} = \Delta + \bar{\partial} + (\ell + y) \wedge (-)$$

is surjective.

In this case, $(G^{\bullet,\bullet}, A^{\bullet,\bullet})$ is perfect if and only if, for every $A \in \mathbf{Art}_\Lambda$, the A-module $H^k(G_A^\bullet, \check{d})$ is finitely generated.

Proof Over $A_0 = \mathbf{k}$, the two double complexes $(G_0^{\bullet,\bullet}, \Delta, \bar{\partial})$ and $(A_0^{\bullet,\bullet}, \partial, \bar{\partial})$ are isomorphic via $\kappa = (-) \lrcorner \, \omega$. This is because $\bar{\partial}\omega \in \mathfrak{m}_\Lambda \cdot A^{d,1}$. In particular, their spectral sequences are isomorphic. Since $\kappa \circ \check{d} = d \circ \kappa$, the isomorphism κ also induces an isomorphism $H^k(G_A^\bullet, \check{d}) \cong H^k(A_A^\bullet, d)$, which is compatible with base change. $\square$

For a Λ-linear curved Batalin–Vilkovisky algebra, we turn this into a definition:

Definition 5.21 (Quasi-perfectness and Perfectness II) Let $G^{\bullet,\bullet}$ be a Λ-linear curved Batalin–Vilkovisky algebra. We say that $G^{\bullet,\bullet}$ is *quasi-perfect* respectively *perfect* if it is strictly faithful and satisfies the corresponding conditions in Lemma 5.20.

Remark 5.22 Condition 5.20 in Lemma 5.20 might seem a bit odd. In view of the T^1-lifting criterion of Z. Ran in [238], one might expect a condition about the freeness of, for example, $H^j(G_A^{-1,\bullet}, \bar{\partial}_\phi)$ for every Maurer–Cartan solution ϕ. We will see in Sect. 5.4 that, what is required here, is in fact stronger, and, in a sense, more elegant. We do not know if the weaker condition would be sufficient to prove unobstructedness theorems, but if the weaker condition (for all $H^j(G_A^{i,\bullet}, \bar{\partial}_\phi)$) holds and the functor is unobstructed, then also the stronger condition as stated above holds. $\Diamond$

The following is the abstract unobstructedness theorem of [38, Thm. 5.6], stated in its proper clarity and generality.[5] We will prove this theorem in Sect. 6.4.

Theorem 5.23 (First Abstract Unobstructedness Theorem) *Let $G^{\bullet,\bullet}$ be a quasi-perfect Λ-linear curved Batalin–Vilkovisky algebra. Then $\mathrm{SMC}(G^{\bullet,\bullet}, -)$ is unobstructed, i.e., for every surjection $B' \to B$ in $\mathbf{Art}_\Lambda$, the induced map*

$$\mathrm{SMC}(G^{\bullet,\bullet}, B') \to \mathrm{SMC}(G^{\bullet,\bullet}, B)$$

is surjective.

This is sufficient as long as we only want to know that a solution of the (classical extended) Maurer–Cartan equation exists. Namely, we can start with the pair $(0, 0)$ over A_0, and then we can lift it from order to order and obtain a solution (ϕ_k, f_k) of the semi-classical extended Maurer–Cartan equation. Then ϕ_k is also a solution of the classical extended Maurer–Cartan equation. This is sufficient e.g. for the situation in our previous article [77] (with Filip and Ruddat), where we were interested in the existence of a deformation in the geometric setup. When we want to show that *every* classical Maurer–Cartan solution can be lifted independently of an f that makes (ϕ, f) into a semi-classical Maurer–Cartan solution, i.e., when we want to show that the deformation functor in a geometric setup is unobstructed, we have to work again with Λ-linear curved Batalin–Vilkovisky *calculi*. We will prove the following theorem in Sect. 6.5.

Theorem 5.24 (Second Abstract Unobstructedness Theorem) *Let $(G^{\bullet,\bullet}, A^{\bullet,\bullet})$ be a quasi-perfect Λ-linear curved Batalin–Vilkovisky calculus. Then $\mathrm{MC}(G^{\bullet,\bullet}, -)$ is unobstructed.*

Note that we require a perfect curved Batalin–Vilkovisky calculus. It is not sufficient to have a perfect curved Gerstenhaber calculus, as they arise from any geometry, i.e., not necessarily Calabi–Yau. On the other hand, the finiteness condition is usually satisfied in geometry, but it is not necessary for the abstract theorem. This second abstract unobstructedness theorem has, to the author's knowledge, not been addressed in [38].

5.4 Freeness of $H^j(G_A^{i;\bullet}, \bar{\partial}_\phi)$ and $H^j(A_A^{i;\bullet}, \bar{\partial}_\phi)$

Let, for a moment, Q be a sharp toric monoid, and let $f_A \colon X_A \to S_A$ be a proper, log smooth, and saturated morphism of fs log schemes, where $S_A = \mathrm{Spec}(Q \to A)$ for an Artinian local $\mathbf{k}[\![Q]\!]$-algebra A with residue field $\mathbf{k}$. Then, by the Hodge–de

---

[5] In fact, quasi-perfectness can further be weakened to semi-perfectness, which we discuss below when we prove the theorem.

Rham degeneration theorem,

$$R^q (f_A)_* \Omega^p_{X_A/S_A}$$

is a finitely generated free A-module whose formation commutes with base change.
If $f_A \colon X_A \to S_A$ is log Calabi–Yau, i.e., we have $\omega_{X_A/S_A} \cong O_{X_A}$, then also

$$R^q (f_A)_* \Theta^p_{X_A/S_A}$$

is a finitely generated free A-module whose formation commutes with base change.
This suggests that, for a Maurer–Cartan solution ϕ, the A-modules $H^j(A_A^{i,\bullet}, \bar{\partial}_\phi)$
and $H^j(G_A^{i,\bullet}, \bar{\partial}_\phi)$ might be free and commute with base change. This is in fact true
under the quasi-perfectness assumption, not even assuming finite generation.

5.4.1   Freeness of $H^j(G_A^{i,\bullet}, \bar{\partial}_\phi)$ and the Relevance of $\check{d} = \bar{\partial} + \Delta + (\ell + y) \wedge (-)$

Let $G^{\bullet,\bullet}$ be a quasi-perfect Λ-linear curved Batalin–Vilkovisky algebra. Suppose
we have a semi-classical Maurer–Cartan solution (ϕ, f) over $A \in \mathbf{Art}_\Lambda$. We have
seen above that then

$$(C^{\bullet,\bullet}, d, d') := (G_A^{\bullet,\bullet}, \Delta_f, \bar{\partial}_\phi)$$

is a double complex. For the moment, we are interested in the cohomology
$H^k(G_A^\bullet, \bar{\partial}_\phi + \Delta_f)$ of its total complex. By our quasi-perfectness assumption, we
know that

$$H^k(G_A^\bullet, \check{d}) \to H^k(G_0^\bullet, \check{d})$$

is surjective, where $\check{d}\theta = \bar{\partial}\theta + \Delta\theta + (\ell + y) \wedge \theta$. Since ϕ and f are nilpotent for
the $\wedge$-product, we can define $E := \sum_{n=0}^\infty \frac{(f+\phi)^n}{n!}$; then we have a map

$$\Phi_E \colon \ G_A^m \to G_A^m, \quad \theta \mapsto E \wedge \theta.$$

This map is an isomorphism of A-modules, for its inverse is

$$\Psi_E(\theta) = \sum_{n=0}^\infty \frac{(-1)^n (f+\phi)^n}{n!} \wedge \theta.$$

Lemma 5.25 *We have*

$$\Phi_E \circ (\bar{\partial}_\phi + \Delta_f) = \check{d} \circ \Phi_E.$$

In particular, we have

$$H^k(G_A^\bullet, \bar{\partial}_\phi + \Delta_f) \cong H^k(G_A^\bullet, \check{d}),$$

so that $H^k(G_A^\bullet, \Delta_f + \bar{\partial}_\phi) \to H^k(G_0^\bullet, \Delta + \bar{\partial})$ is surjective.

Proof Recall the formulae

$$\bar{\partial}(\theta^n) = n\bar{\partial}(\theta) \wedge \theta^{n-1} \quad \text{and} \quad \Delta(\theta^n) = n\Delta(\theta) \wedge \theta^{n-1} + \frac{n(n-1)}{2}\theta^{n-2} \wedge [\theta, \theta]$$

for $\theta \in G_A^\bullet$ of even degree and $n \geq 2$. By computation, the assertion

$$\Phi_E \circ (\bar{\partial}_\phi + \Delta_f)(\xi) = \check{d} \circ \Phi_E(\xi)$$

is equivalent to

$$E \wedge [f + \phi, \xi] = [E, \xi] + \Big(\bar{\partial}(f + \phi) + \Delta(f + \phi)$$
$$+ \frac{1}{2}[f + \phi, f + \phi] + \ell + y\Big) \wedge E \wedge \xi.$$

Since $E \wedge [f + \phi, \xi] = [E, \xi]$ and $\check{d}(f + \phi) = 0$, this holds. $\square$

Proposition 5.26 *Let $G^{\bullet,\bullet}$ be a quasi-perfect Λ-linear curved Batalin–Vilkovisky algebra. Let (ϕ, f) be a semi-classical Maurer–Cartan solution over $A \in \mathbf{Art}_\Lambda$. Then*

$$H^j(G_A^{i,\bullet}, \bar{\partial}_\phi)$$

is a free A-module, and its formation commutes with base change in $\mathbf{Art}_A$. If $G^{\bullet,\bullet}$ is perfect, then this cohomology module is finitely generated.

Proof We apply Proposition D.4 and Corollary D.5 to the double complex

$$(C^{\bullet,\bullet}, d, d') = (G_A^{\bullet,\bullet}, \Delta_f, \bar{\partial}_\phi).$$

If $H^k(G_A^\bullet, \Delta_f + \bar{\partial}_\phi)$ is finitely generated, so are its subquotients. $\square$

Remark 5.27 Whenever $G^{\bullet,\bullet}$ is part of a perfect Λ-linear curved Batalin–Vilkovisky calculus $(G^{\bullet,\bullet}, A^{\bullet,\bullet})$, then, by Theorem 6.17 below, a classical Maurer–Cartan solution ϕ can be extended to a semi-classical Maurer–Cartan solution (ϕ, f); thus, in this case, the proposition also holds for classical Maurer–Cartan solutions ϕ. $\Diamond$

5.4.2 Freeness of $H^j(A_A^{i;\bullet}, \bar{\partial}_\phi)$ and the Relevance of $d = \partial + \bar{\partial} + \ell \lrcorner (-)$

Let $(G^{\bullet,\bullet}, A^{\bullet,\bullet})$ be a quasi-perfect Λ-linear curved Gerstenhaber calculus. It may be actually a Batalin–Vilkovisky calculus, but we do not require this extra structure here. Suppose we have a classical Maurer–Cartan solution ϕ over $A \in \mathbf{Art}_\Lambda$, not necessarily part of a semi-classical Maurer–Cartan solution (ϕ, f). Then we can form a double complex

$$(C^{\bullet,\bullet}, d, d') := (A_A^{\bullet,\bullet}, \partial, \bar{\partial}_\phi).$$

Here, we know that $H^k(A_A^\bullet, d) \to H^k(A_0^\bullet, d)$ is surjective with

$$d\alpha := \partial\alpha + \bar{\partial}\alpha + (\ell \lrcorner \alpha).$$

After setting $E := e^\phi = \sum_{n=0}^\infty \frac{\phi^n}{n!}$, we have an isomorphism

$$\Phi_E: \quad A_A^m \to A_A^m, \quad \alpha \mapsto E \lrcorner \alpha,$$

of A-modules.

Lemma 5.28 *We have*

$$\Phi_E \circ (\partial + \bar{\partial}_\phi)(\alpha) = d \circ \Phi_E(\alpha).$$

In particular, we have

$$H^k(A_A^\bullet, \partial + \bar{\partial}_\phi) \cong H^k(A_A^\bullet, d),$$

so that $H^k(A_A^\bullet, \partial + \bar{\partial}_\phi) \to H^k(A_0^\bullet, d)$ is surjective.

Proof First, we have $\mathcal{L}_{e^\phi}(\alpha) = e^\phi \lrcorner \mathcal{L}_\phi(\alpha) + \frac{1}{2}[\phi, \phi] \lrcorner e^\phi \lrcorner \alpha$. Then we compute

$$d\circ\Phi_E(\alpha) - \Phi_E \circ (\partial + \bar{\partial}_\phi)(\alpha)$$

$$= \partial(e^\phi \lrcorner \alpha) + \bar{\partial}(e^\phi \lrcorner \alpha) + (\ell \lrcorner e^\phi \lrcorner \alpha)$$

$$\quad - (e^\phi \lrcorner \partial\alpha) - (e^\phi \lrcorner \bar{\partial}\alpha) - (e^\phi \lrcorner \mathcal{L}_\phi(\alpha))$$

$$= \mathcal{L}_{e^\phi}(\alpha) + \bar{\partial}(e^\phi) \lrcorner \alpha + (\ell \lrcorner e^\phi \lrcorner \alpha) - (e^\phi \lrcorner \mathcal{L}_\phi(\alpha))$$

$$= (\bar{\partial}\phi \lrcorner e^\phi \lrcorner \alpha) + \left(\frac{1}{2}[\phi, \phi] \lrcorner e^\phi \lrcorner \alpha\right) + (\ell \lrcorner e^\phi \lrcorner \alpha) = 0$$

because ϕ is a classical Maurer–Cartan solution. $\square$

Then we have:

Proposition 5.29 *Let $(G^{\bullet,\bullet}, A^{\bullet,\bullet})$ be a quasi-perfect Λ-linear curved Gerstenhaber calculus. Let ϕ be a classical Maurer–Cartan solution over $A \in \mathbf{Art}_\Lambda$. Then*

$$H^j(A_A^{i,\bullet}, \bar{\partial}_\phi)$$

is a free A-module, and its formation commutes with base change in $\mathbf{Art}_A$. If $(G^{\bullet,\bullet}, A^{\bullet,\bullet})$ is perfect, then the cohomology module is finitely generated.

Proof We apply Proposition D.4 and Corollary D.5 to the double complex $(A_A^{\bullet,\bullet}, \partial, \bar{\partial}_\phi)$. $\qquad\square$

Chapter 6
The Two Abstract Unobstructedness Theorems

We prove the two abstract unobstructedness theorems announced in the previous chapter. We achieve this by solving a third extended Maurer–Cartan equation, the *quantum extended Maurer–Cartan equation*. The name "quantum" comes from Terilla's article [266]; it is the motivation for the names "classical" and "semiclassical" for the other two extended Maurer–Cartan equations. The proofs follow the article [38] of Chan, Leung, and Ma, who in turn followed the ideas of [171, 266].

6.1 Formal Power Series and Formal Laurent Series in $G_A^{p,q}$

Let $G^{\bullet,\bullet}$ be a Λ-linear curved Batalin–Vilkovisky algebra. Let us write $G_\Lambda^{p,q} := G^{p,q}$. Then we set

$$G_A^{p,q}[\![\hbar]\!] := \left\{ \sum_{i=0}^{\infty} g_i \hbar^i \;\middle|\; g_i \in G_A^{p,q} \right\}$$

for either $A \in \mathbf{Art}_\Lambda$ or $A = \Lambda$ itself, a convention which we follow in this section. Compared with [38], we have changed the notation from t to $\hbar$ in order to avoid confusion with the frequent use of t as a variable in Λ. We also set

$$G_A^{p,q}[\![u]\!] := \left\{ \sum_{i=0}^{\infty} g_i u^i \;\middle|\; g_i \in G_A^{p,q} \right\},$$

S. Felten, *Global Logarithmic Deformation Theory*, Lecture Notes in Mathematics 2373, https://doi.org/10.1007/978-3-031-98751-9_6

where we think of u as a variable with $u^2 = \hbar$, i.e., $G_A^{p,q}[\![\hbar]\!] \subset G_A^{p,q}[\![u]\!]$ is the subspace of formal power series with entries in even degrees.[1] Note that, as an A-module, we have

$$G_A^{p,q}[\![u]\!] = \varprojlim G_A^{p,q}[u]/(u^m).$$

Lemma 6.1 *Let $A \in \mathbf{Art}_\Lambda$ or $A = \Lambda$. Also, let $B' = \Lambda$ or $B' \in \mathbf{Art}_\Lambda$, let $B \in \mathbf{Art}_\Lambda$, and let $B' \to B$ be a homomorphism in $\mathbf{Comp}_\Lambda$.*

(1) We have isomorphisms

$$G_{B'}^{p,q}[\![\hbar]\!] \otimes_{B'} B \xrightarrow{\cong} G_B^{p,q}[\![\hbar]\!] \quad and \quad G_{B'}^{p,q}[\![u]\!] \otimes_{B'} B \xrightarrow{\cong} G_B^{p,q}[\![u]\!].$$

(2) The A-modules $G_A^{p,q}[\![\hbar]\!]$ and $G_A^{p,q}[\![u]\!]$ are flat.
(3) The Λ-modules $G^{p,q}[\![\hbar]\!]$ and $G^{p,q}[\![u]\!]$ are complete with respect to $\mathfrak{m}_\Lambda$.

Before we embark into the proof, we need the following lemma.

Lemma 6.2 *Let R be a (commutative and unital) ring. Let $(M_i)_{i \in \mathbb{N}}$ be an inverse system of flat R-modules with surjective restriction maps. Let N be a finitely presented R-module. Then we have an isomorphism*

$$\Phi_N : \left(\varprojlim M_i \right) \otimes_R N \to \varprojlim (M_i \otimes_R N).$$

Proof First note that the statement holds for $N = R^n$. If N is finitely *generated*, we have an exact sequence

$$0 \to K \to R^n \to N \to 0.$$

The system $(M_i \otimes_R K)_i$ has surjective restriction maps, so, together with the flatness of M_i, the upper line in the following diagram is exact:

$$
\begin{array}{ccccccccc}
0 & \longrightarrow & \varprojlim(M_i \otimes_R K) & \longrightarrow & \varprojlim(M_i \otimes_R R^n) & \longrightarrow & \varprojlim(M_i \otimes_R N) & \longrightarrow & 0 \\
 & & \Big\uparrow{\scriptstyle \Phi_K} & & \Big\uparrow{\scriptstyle \cong} & & \Big\uparrow{\scriptstyle \Phi_N} & & \\
 & & \left(\varprojlim M_i \right) \otimes_R K & \longrightarrow & \left(\varprojlim M_i \right) \otimes_R R^n & \longrightarrow & \left(\varprojlim M_i \right) \otimes_R N & \longrightarrow & 0
\end{array}
$$

The lower row is exact since taking the tensor product is right exact. From the diagram, we find that Φ_N is surjective. Since our N is assumed to be finitely *presented*, the kernel K is actually finitely *generated*, so Φ_K is surjective. Now a diagram chase shows that Φ_N is also injective. $\qquad\square$

---

[1] In [38], the notation $t^{\frac{1}{2}}$ is used instead of u, which the author finds somewhat cumbersome.

Proof of Lemma 6.1 It suffices to show the Lemma for u; then the proof for $\hbar$ is the same.

Claim (1):   Note that B is finitely presented as an B'-module because it is actually a **k**-vector space of finite dimension. Note also that $G_{B'}^{p,q}[u]/(u^m)$ is a flat B'-module, and that $G_{B'}^{p,q}[u]/(u^m) \otimes_{B'} B \cong G_B^{p,q}[u]/(u^m)$.

Claim (2):   Note that $G_A^{p,q}[u]/(u^m)$ is a flat $A[u]/(u^m)$-module because a tensor product preserves flatness. Thus, by [267, 0912], the limit $G_A^{p,q}[\![u]\!]$ is a flat $A[u]$-module, hence a flat A-module.

Claim (3):   We have

$$G^{p,q}[\![u]\!] = \varprojlim_m \varprojlim_k G_k^{p,q}[u]/(u^m) = \varprojlim_k \varprojlim_m G_k^{p,q}[u]/(u^m) = \varprojlim_k G_k^{p,q}[\![u]\!]$$

since inverse limits commute with each other.

$\square$

Next, we set, for either $A \in \mathbf{Art}_\Lambda$ or $A = \Lambda$,

$$G_A^{p,q}(\!(u)\!) := \left\{ \sum_{i=-N}^{\infty} g_i u^i \;\middle|\; N \geq 0,\, g_i \in G_A^{p,q} \right\}$$

and also

$$G_A^{p,q}(\!(\hbar)\!) := \left\{ \sum_{i=-N}^{\infty} g_i \hbar^i \;\middle|\; N \geq 0,\, g_i \in G_A^{p,q} \right\}.$$

We have

$$G_A^{p,q}(\!(u)\!) = G_A^{p,q}[\![u]\!]_u,$$

i.e., it is the localization of the $A[u]$-module $G_A^{p,q}[\![u]\!]$ in the non-zero divisor u. Thus, we also have

$$G_A^{p,q}(\!(u)\!) = G_A^{p,q}[\![u]\!] \otimes_{A[u]} A[u, u^{-1}].$$

In particular, $G_A^{p,q}(\!(u)\!)$ is a flat $A[u, u^{-1}]$-module. Similarly, we have

$$G_A^{p,q}(\!(u)\!) = G_A^{p,q}[\![u]\!] \otimes_{\mathbf{k}[u]} \mathbf{k}[u, u^{-1}].$$

Note also that

$$G_{B'}^{p,q}(\!(u)\!) \otimes_{B'} B = G_{B'}^{p,q}[\![u]\!] \otimes_{B'[u]} B[u, u^{-1}]$$

$$= (G_{B'}^{p,q}[\![u]\!] \otimes_{B'} B) \otimes_{B[u]} B[u, u^{-1}] = G_B^{p,q}(\!(u)\!)$$

for $B' = \Lambda$ or $B' \in \mathbf{Art}_\Lambda$ and $B \in \mathbf{Art}_\Lambda$, along a map $B' \to B$.

Remark 6.3 Unlike the case of $G^{p,q}[\![u]\!]$, the natural map

$$G^{p,q}(\!(u)\!) \to \varprojlim_k G_k^{p,q}(\!(u)\!)$$

is *not* an isomorphism. Namely, an element in the limit on the right might have components in arbitrarily negative degree while the degree of the non-zero components of an element on the left is bounded below. The map is, however, injective. $\qquad\qquad\qquad\qquad\qquad\qquad\qquad\qquad\qquad\qquad\qquad\qquad\quad \diamond$

6.2 The Quantum Extended Maurer–Cartan Equation

Instead of the original extended Maurer–Cartan equations, we study the *quantum extended Maurer–Cartan equation*

$$(\bar\partial + \hbar\Delta)\varphi_A + \frac{1}{2}[\varphi_A, \varphi_A] + (\ell + \hbar y) = 0 \in G_A^1[\![\hbar]\!]$$

which is to be satisfied by elements

$$\varphi_A = \sum_{i=0}^\infty C_i^{0,0}(\varphi_A)\hbar^i + \sum_{i=0}^\infty C_i^{-1,1}(\varphi_A)\hbar^i + \ldots + \sum_{i=0}^\infty C_i^{-d,d}(\varphi_A)\hbar^i \in \mathfrak{m}_A \cdot G_A^0[\![\hbar]\!].$$

Here, we write

$$G_A^m[\![\hbar]\!] := \bigoplus_{p+q=m} G_A^{p,q}[\![\hbar]\!];$$

since p is bounded below by $-d$, and q is bounded below by 0, this direct sum is actually finite. We denote the set of solutions by $\mathrm{QMC}(G^{\bullet,\bullet}, A)$. We chose the term "quantum" for this equation in analogy with Terilla's article [266].[2]

Lemma 6.4 *Assume that $\varphi_A \in \mathfrak{m}_A \cdot G_A^0[\![\hbar]\!]$ satisfies the quantum extended Maurer–Cartan equation. Then $(\bar\partial + \hbar\Delta + [\varphi_A, -])^2 = 0$ as an operator $G_A^m[\![\hbar]\!] \to G_A^{m+1}[\![\hbar]\!]$.*

---

[2] In [266, §3.3], Terilla gives some physical motivation for this terminology. The author (of this monograph) does not know if there is also a relation with Kontsevich's deformation quantization of [179], and if yes, what the precise relation is.

Proof For $\theta \in G_A^m[\![\hbar]\!]$, we have

$$(\bar{\partial} + \hbar \Delta + [\varphi_A, -])^2(\theta) = [\bar{\partial}\varphi_A + \hbar \Delta(\varphi_A) + \frac{1}{2}[\varphi_A, \varphi_A] + \ell + \hbar y, \theta] = 0$$

by direct computation. $\square$

Remark 6.5 In [38], Chan, Leung, and Ma claim that the two conditions would be equivalent. While this is the case for the classical extended Maurer–Cartan equation, it is unclear why it should be true here. Namely, not for every (p, q) it is true that, for $\theta \in G_A^{p,q}$, the condition $[\theta, \xi] = 0$ for all ξ implies $\theta = 0$. For example, we have $[1, \xi] = 0$ for all ξ (proven by using the odd Poisson identity) but $1 \neq 0$. $\Diamond$

We define a *scaling operator*

$$\sigma: \ G_A^{p,q}(\!(u)\!) \to G_A^{p,q}(\!(u)\!), \quad \theta \mapsto u^{q-p-2}\theta;$$

it is an automorphism. Since $\varphi_A \in \mathfrak{m}_A \cdot G_A^0[\![\hbar]\!]$, the sum

$$e^{\sigma(\varphi_A)} = \sum_{n=0}^{\infty} \frac{\sigma(\varphi_A)^n}{n!} \in G_A^0(\!(u)\!)$$

is actually finite and hence well-defined. Recall that we have $\check{d} = \bar{\partial} + \Delta + (\ell + y) \wedge (-)$.

Lemma 6.6 *An element $\varphi_A \in \mathfrak{m}_A \cdot G_A^0[\![\hbar]\!]$ satisfies the quantum extended Maurer–Cartan equation if and only if*

$$\check{d}(e^{\sigma(\varphi_A)}) = 0.$$

Proof We write $\varphi = \varphi_A$ for short. Let $\varphi = \varphi^{0,0} + \varphi^{-1,1} + \ldots + \varphi^{-d,d}$ be the bidegree decomposition with $\varphi^{-i,i} \in G_k^{-i,i}[\![\hbar]\!]$. Then the quantum extended Maurer–Cartan equation is equivalent to the following system of equations:

$$(0, 1): \quad \bar{\partial}\varphi^{0,0} + \hbar \Delta \varphi^{-1,1} + \frac{1}{2}[\varphi^{0,0}, \varphi^{-1,1}]$$

$$+ \frac{1}{2}[\varphi^{-1,1}, \varphi^{0,0}] + \hbar y = 0 \tag{6.1}$$

$$\bar{\partial}\varphi^{-1,1} + \hbar \Delta \varphi^{-2,2} + \frac{1}{2}[\varphi^{0,0}, \varphi^{-2,2}]$$

$$(-1, 2): \tag{6.2}$$

$$+ \frac{1}{2}[\varphi^{-1,1}, \varphi^{-1,1}] + \frac{1}{2}[\varphi^{-2,2}, \varphi^{0,0}] + \ell = 0$$

$$(-j, j+1): \quad \bar{\partial}\varphi^{-j,j} + \hbar \Delta \varphi^{-j-1,j+1} + \sum_{i=0}^{j+1} \frac{1}{2}[\varphi^{-i,i}, \varphi^{-(j+1-i),j+1-i}] = 0$$

$$\tag{6.3}$$

where we have one equation for every $2 \le j \le d$ (or only two in total if $d = 1$). By multiplication with powers of $\hbar$, this system is equivalent to the system

$$(0, 1): \qquad \begin{aligned} &\hbar^{-1}\bar{\partial}\varphi^{0,0} + \Delta\varphi^{-1,1} \\ &\quad + \frac{1}{2}\hbar^{-1}[\varphi^{0,0}, \varphi^{-1,1}] + \frac{1}{2}\hbar^{-1}[\varphi^{-1,1}, \varphi^{0,0}] \\ &\qquad\qquad\qquad + y = 0 \end{aligned} \tag{6.4}$$

$$(-1, 2): \qquad \begin{aligned} &\bar{\partial}\varphi^{-1,1} + \hbar\Delta\varphi^{-2,2} + \frac{1}{2}[\varphi^{0,0}, \varphi^{-2,2}] \\ &\quad + \frac{1}{2}[\varphi^{-1,1}, \varphi^{-1,1}] + \frac{1}{2}[\varphi^{-2,2}, \varphi^{0,0}] + \ell = 0 \end{aligned} \tag{6.5}$$

$$(-j, j+1): \qquad \begin{aligned} &\hbar^{j-1}\bar{\partial}\varphi^{-j,j} + \hbar^j \Delta\varphi^{-j-1,j+1} \\ &\quad + \sum_{i=0}^{j+1} \frac{1}{2}\hbar^{j-1}[\varphi^{-i,i}, \varphi^{-(j+1-i),j+1-i}] = 0. \end{aligned} \tag{6.6}$$

Since $\sigma(\varphi) = \hbar^{-1}\varphi^{0,0} + \varphi^{-1,1} + \hbar\varphi^{-2,2} + \dots$, putting everything together again, this system is equivalent to

$$\bar{\partial}(\sigma(\varphi)) + \Delta(\sigma(\varphi)) + \frac{1}{2}[\sigma(\varphi), \sigma(\varphi)] + \ell + y = 0 \ \in \ G_A^1[\![\hbar]\!].$$

A direct computation yields

$$\check{d}(e^{\sigma(\varphi)}) = \left(\bar{\partial}(\sigma(\varphi)) + \Delta(\sigma(\varphi)) + \frac{1}{2}[\sigma(\varphi), \sigma(\varphi)] + \ell + y\right) \wedge e^{\sigma(\varphi)}.$$

Thus, if φ satisfies the quantum extended Maurer–Cartan equation, then $\check{d}(e^{\sigma(\varphi)}) = 0$. Conversely, assume that $\check{d}(e^{\sigma(\varphi)}) = 0$. Let μ be the above term such that $\check{d}(e^{\sigma(\varphi)}) = \mu \wedge e^{\sigma(\varphi)} = 0$. Assume that $\mu \ne 0$. Let m be the number such that $\mu \in \mathfrak{m}_A^m \cdot G_A^0(\!(u)\!)$ but $\mu \notin \mathfrak{m}_A^{m+1} \cdot G_A^0(\!(u)\!)$. Now

$$\mu \wedge e^{\sigma(\varphi)} = \mu + \mu \wedge \sigma(\varphi) + \dots = 0.$$

Since $\mu \wedge \sigma(\varphi) + \dots \in \mathfrak{m}_A^{m+1} \cdot G_A^0(\!(u)\!)$, we find that $\mu \in \mathfrak{m}_A^{m+1} \cdot G_A^0(\!(u)\!)$, a contradiction. Thus, $\mu = 0$, and φ satisfies the quantum extended Maurer–Cartan equation. $\qquad\square$

6.3 Solving the Quantum Extended Maurer–Cartan Equation

Here, we show that the quantum extended Maurer–Cartan functor $\mathrm{QMC}(G^{\bullet,\bullet}, -)$ is unobstructed if $G^{\bullet,\bullet}$ is quasi-perfect. In fact, an even weaker assumption is sufficient.

Definition 6.7 (Semi-perfectness) Let $G^{\bullet,\bullet}$ be a strictly faithful Λ-linear curved Batalin–Vilkovisky algebra of dimension d. Then $G^{\bullet,\bullet}$ is *semi-perfect* if:

1. the first spectral sequence $'E_{0;r}$ associated with the double complex $(G_0^{\bullet,\bullet}, \Delta, \bar{\partial})$ over $A_0 = \mathbf{k}$ degenerates at E_1 in $[0, 1]$ (see Definition D.1);
2. for every $A \in \mathbf{Art}_\Lambda$, the induced map $H^k(G_A^\bullet, \check{d}) \to H^k(G_0^\bullet, \check{d})$ is surjective for $k \in \{-1, 0\}$.

Clearly, every quasi-perfect $G^{\bullet,\bullet}$ is also semi-perfect. If the cohomology groups over $A_0 = \mathbf{k}$ are finitely generated as $\mathbf{k}$-vector spaces, in order to satisfy the first condition, it is sufficient to have that the dimensions of $'E_{0;1}^{p;q}$ sum up to the dimension of $H^k(G_0^\bullet, \check{d})$ for either $k = 0$ or $k = 1$. The second condition for both $k = -1$ and $k = 0$ ensures that $H^0(G_A^\bullet, \check{d})$ is a flat A-module.

Theorem 6.8 *Let $G^{\bullet,\bullet}$ be a Λ-linear curved Batalin–Vilkovisky algebra of dimension d. Assume that $G^{\bullet,\bullet}$ is semi-perfect. Then the quantum extended Maurer–Cartan functor*

$$\mathrm{QMC}(G^{\bullet,\bullet}, -)\colon \quad \mathbf{Art}_\Lambda \to \mathbf{Set}$$

is unobstructed. In particular, if $G^{\bullet,\bullet}$ is quasi-perfect, then the functor is unobstructed.

Remark 6.9 This is the most fundamental unobstructedness theorem which we consider here. We have learned the proof from Chan–Leung–Ma in [38, Thm. 5.6], where a more general statement in a more restrictive context is given. Here, we give the least restrictive conditions possible for the argument to work. The proof in [38, Thm. 5.6] in turn has been inspired by [18, 171, 266], which all deal with the classical, not logarithmic, case. In particular, [266] contains a very general unobstructedness theorem, which is proven by studying $V[\![\hbar]\!]$ with a differential $Q + \hbar\Delta$ instead of the original BV algebra V with differential Q and BV operator Δ. However, Theorem 6.8 is not a mere consequence of this unobstructedness theorem due to the predifferential nature of our situation. $\Diamond$

Let $\varphi_B \in \mathfrak{m}_B \cdot G_B^0[\![\hbar]\!]$ be a solution of the quantum extended Maurer–Cartan equation over $B \in \mathbf{Art}_\Lambda$, and let $B' \to B$ be a first-order extension in $\mathbf{Art}_\Lambda$ with kernel $I \subset B'$. We have an exact sequence

$$0 \to I \cdot G_{B'}^{p;q}[\![\hbar]\!] \to G_{B'}^{p;q}[\![\hbar]\!] \to G_B^{p;q}[\![\hbar]\!] \to 0$$

where we have additionally $I \cdot G_{B'}^{p,q}[\![\hbar]\!] = G_0^{p,q}[\![\hbar]\!] \otimes_{\mathbf{k}} I$ when we consider I as a $\mathbf{k}$-vector space.

Let us write $\varphi := \varphi_B$. In the rest of the section, we show the existence of a solution $\varphi' := \varphi_{B'}$ of the quantum extended Maurer–Cartan equation over B' with $\varphi'|_B = \varphi_B = \varphi$, where $(-)|_B$ is the restriction map in the above exact sequence.

As the first step, we show another variant of Lemma 6.6.

Lemma 6.10 *Let $\varphi_A \in \mathfrak{m}_A \cdot G_A^0[\![\hbar]\!]$. Then φ_A is a solution of the quantum extended Maurer–Cartan equation if and only if*

$$\check{d}_\hbar(e^{\varphi_A/\hbar}) = 0$$

in $G_A^\bullet(\!(\hbar)\!)$ where $\check{d}_\hbar := \bar{\partial} + \hbar\Delta + \hbar^{-1}(\ell + \hbar y) \wedge (-)$. Moreover, this map satisfies $\check{d}_\hbar^2 = 0$.

Proof The element $e^{\varphi_A/\hbar} \in G_A^0(\!(\hbar)\!)$ is well-defined because $\mathfrak{m}_A$ is nilpotent, so exponents of $\hbar$ are bounded below when computing the power series.

By direct computation, we have $\check{d}_\hbar(\theta) = u \cdot \sigma^{-1} \circ \check{d} \circ \sigma(\theta)$ in $G_A^\bullet(\!(u)\!)$, so $\check{d}_\hbar(e^{\varphi_A/\hbar}) = 0$ if and only if $\check{d} \circ \sigma(e^{\varphi_A/\hbar}) = 0$.

Since $\sigma(\theta \wedge \xi) = \hbar \cdot \sigma(\theta) \wedge \sigma(\xi)$, we find $\sigma((\varphi_A/\hbar)^n) = \sigma(\varphi_A)^n \cdot \hbar^{-1}$ and thus

$$\sigma(e^{\varphi_A/\hbar}) = \hbar^{-1} \cdot e^{\sigma(\varphi_A)}.$$

Consequently, $\check{d}_\hbar(e^{\varphi_A/\hbar}) = 0$ is equivalent to $\check{d}(e^{\sigma(\varphi_A)}) = 0$, the statement of Lemma 6.6.   □

Since we assume φ to be a solution of the quantum extended Maurer–Cartan equation, we obtain $\check{d}_\hbar(e^{\varphi/\hbar}) = 0$. By a direct computation, we have

$$\check{d}_\hbar(e^{\varphi/\hbar}) = \left(\bar{\partial}(\varphi/\hbar) + \hbar\Delta(\varphi/\hbar) + \frac{1}{2}\hbar[\varphi/\hbar, \varphi/\hbar] + \hbar^{-1}(\ell + \hbar y)\right) \wedge e^{\varphi/\hbar}.$$

Let μ be the term in the left bracket on the right-hand side. As in the proof of Lemma 6.6, we have $\mu = 0$.

Now we choose an arbitrary lift $\tilde{\varphi}$ of φ in $\mathfrak{m}_{B'} \cdot G_{B'}^0[\![\hbar]\!]$, i.e., $\tilde{\varphi}|_B = \varphi$. Let

$$\tilde{\mu} := \bar{\partial}(\tilde{\varphi}/\hbar) + \hbar\Delta(\tilde{\varphi}/\hbar) + \frac{1}{2}\hbar[\tilde{\varphi}/\hbar, \tilde{\varphi}/\hbar] + \hbar^{-1}(\ell + \hbar y);$$

then $\tilde{\mu} \in \hbar^{-1} \cdot \mathfrak{m}_{B'} \cdot G_{B'}^1[\![\hbar]\!]$, i.e., the lowest $\hbar$-term is at $\hbar^{-1}$. Since $\tilde{\mu}|_B = \mu = 0$, we have $\tilde{\mu} \in I \cdot G_{B'}^1(\!(\hbar)\!)$. Because $I \cdot \mathfrak{m}_{B'} = 0$, we have moreover

$$\check{d}_\hbar(e^{\tilde{\varphi}/\hbar}) = \tilde{\mu} \wedge e^{\tilde{\varphi}/\hbar} = \tilde{\mu}.$$

Lemma 6.11 *Let $G^{\bullet,\bullet}$ be as above. Then we have an exact sequence*

$$0 \to (G_0^\bullet(\!(\hbar)\!), \check{d}_\hbar) \otimes_{\mathbf{k}} I \to (G_{B'}^\bullet(\!(\hbar)\!), \check{d}_\hbar) \to (G_B^\bullet(\!(\hbar)\!), \check{d}_\hbar) \to 0$$

of complexes; it induces an exact sequence

$$0 \to H^1(G_0^\bullet(\!(\hbar)\!), \check{d}_\hbar) \otimes_{\mathbf{k}} I \to H^1(G_{B'}^\bullet(\!(\hbar)\!), \check{d}_\hbar) \to H^1(G_B^\bullet(\!(\hbar)\!), \check{d}_\hbar)$$

in cohomology.[3]

Proof The first sequence is exact on the level of graded pieces, and the horizontal maps are compatible with the differentials, so the first sequence of complexes is exact.

For $A \in \mathbf{Art}_\Lambda$, we have a series of functorial isomorphisms. Denote by $SG_A^\bullet(\!(\hbar)\!)$ the complex with graded pieces $SG_A^m(\!(\hbar)\!) = u^m G_A^m(\!(\hbar)\!) \subseteq G_A^m(\!(u)\!)$. Then we have an isomorphism

$$\sigma: \ (G_A^\bullet(\!(\hbar)\!), \check{d}_\hbar) \xrightarrow{\cong} (SG_A^\bullet(\!(\hbar)\!), u\check{d})$$

of complexes since $\check{d}_\hbar = u \cdot \sigma^{-1} \circ \check{d} \circ \sigma$. Next, multiplication with u^m in degree m gives an isomorphism

$$u^\bullet: \ (G_A^\bullet(\!(\hbar)\!), \check{d}) \xrightarrow{\cong} (SG_A^\bullet(\!(\hbar)\!), u\check{d})$$

of complexes. They induce an isomorphism $H^k(G_A^\bullet(\!(\hbar)\!), \check{d}_\hbar) \cong H^k(G_A^\bullet(\!(\hbar)\!), \check{d})$ of A-modules, functorial in $A \in \mathbf{Art}_\Lambda$. Further, we have a functorial isomorphism

$$H^k(G_A^\bullet(\!(\hbar)\!), \check{d}) \xrightarrow{\cong} H^k(G_A^\bullet, \check{d})(\!(\hbar)\!).$$

Since $H^k(G_B^\bullet, \check{d}) \to H^k(G_0^\bullet, \check{d})$ is surjective for both $k = -1$ and $k = 0$, we find that $H^0(G_B^\bullet, \check{d})$ is flat over B by Proposition D.3. Then the proof of Corollary D.5 shows that $H^0(G_{B'}^\bullet, \check{d}) \to H^0(G_B^\bullet, \check{d})$ is surjective so that

$$H^0(G_{B'}^\bullet(\!(\hbar)\!), \check{d}_\hbar) \to H^0(G_B^\bullet(\!(\hbar)\!), \check{d}_\hbar)$$

is surjective as well. This proves the second part of the assertion. $\square$

Since $\check{d}_\hbar(e^{\tilde{\varphi}/\hbar}) = \tilde{\mu}$, we have $\check{d}_\hbar(\tilde{\mu}) = 0$. Thus, we can find an element $a \in G_0^1(\!(\hbar)\!) \otimes_{\mathbf{k}} I$ with $a \mapsto \tilde{\mu}$ and $\check{d}_\hbar(a) = 0$; in particular, we have $[a] \mapsto [\tilde{\mu}] = 0$ on cohomology classes of $\check{d}_\hbar$. Since the map is injective on cohomology classes by the Lemma, we have $[a] = 0$; thus, there is $b \in G_0^0(\!(\hbar)\!) \otimes_{\mathbf{k}} I$ with $\check{d}_\hbar(b) = a$. Expanding

---

[3] Here we use the second condition of Theorem 6.8.

the definition of $\check{d}_\hbar$, we find $(\bar{\partial} + \hbar\Delta)(b) = a$. Note that in fact $a \in \hbar^{-1}G_0^1[\![\hbar]\!] \otimes_{\mathbf{k}} I$, i.e., a has no $\hbar$-terms below $\hbar^{-1}$, because this is true for $\tilde{\mu}$. We want to choose b in such a way that $b \in \hbar^{-1}G_0^0[\![\hbar]\!] \otimes_{\mathbf{k}} I$ as well.

Lemma 6.12 *In the above situation, let* $\alpha \in G_0^1[\![\hbar]\!] \subset G_0^1(\!(\hbar)\!)$, *and assume that we have* $\beta \in G_0^0(\!(\hbar)\!)$ *with* $(\bar{\partial} + \hbar\Delta)(\beta) = \alpha$. *Then we can find* $\gamma \in G_0^0[\![\hbar]\!]$ *with* $(\bar{\partial} + \hbar\Delta)(\gamma) = \alpha.$[4]

Proof We have $(G_0^\bullet(\!(\hbar)\!), \bar{\partial} + \hbar\Delta) = (G_0^\bullet[\![\hbar]\!], \bar{\partial} + \hbar\Delta) \otimes_{\mathbf{k}[\![\hbar]\!]} \mathbf{k}(\!(\hbar)\!)$. Since $\mathbf{k}[\![\hbar]\!] \to \mathbf{k}[\![\hbar]\!]_\hbar = \mathbf{k}(\!(\hbar)\!)$ is flat, we find

$$H^k(G_0^\bullet(\!(\hbar)\!), \bar{\partial} + \hbar\Delta) = H^k(G_0^\bullet[\![\hbar]\!], \bar{\partial} + \hbar\Delta) \otimes_{\mathbf{k}[\![\hbar]\!]} \mathbf{k}(\!(\hbar)\!).$$

Since $G_0^\bullet[\![\hbar]\!] \to G_0^\bullet(\!(\hbar)\!)$ is injective, α defines a class $[\alpha] \in H^1(G_0^\bullet[\![\hbar]\!], \bar{\partial} + \hbar\Delta)$; by assumption, its image in $H^1(G_0^\bullet(\!(\hbar)\!), \bar{\partial} + \hbar\Delta)$ is 0. The kernel of

$$H^1(G_0^\bullet[\![\hbar]\!], \bar{\partial} + \hbar\Delta) \to H^1(G_0^\bullet[\![\hbar]\!], \bar{\partial} + \hbar\Delta) \otimes_{\mathbf{k}[\![\hbar]\!]} \mathbf{k}(\!(\hbar)\!)$$

consists of those classes δ on the left with $\hbar^m\delta = 0$ for some $m > 0$. By assumption, the first spectral sequence degenerates at E_1 in $[0, 1]$ so that $H^1(G_0^\bullet[\![\hbar]\!], \bar{\partial} + \hbar\Delta)$ is flat, hence torsion-free, by Proposition D.2. Thus $[\alpha] = 0$ as a cohomology class in $H^1(G_0^\bullet[\![\hbar]\!], \bar{\partial} + \hbar\Delta)$, and the assertion follows. $\square$

The lemma holds as well for $G_0^\bullet[\![\hbar]\!] \otimes_{\mathbf{k}} I$ because I is a $\mathbf{k}$-vector space of dimension 1. Replacing a with $\hbar a$ for a moment, we can find $c \in G_0^0[\![\hbar]\!] \otimes_{\mathbf{k}} I$ with $(\bar{\partial} + \hbar\Delta)(c) = \hbar a$; then $b = \hbar^{-1}c$ is our desired b. With this b, we set

$$\varphi' := \tilde{\varphi} - \hbar b \in \mathfrak{m}_{B'} \cdot G_{B'}^0[\![\hbar]\!].$$

Then we have $\varphi'|_B = \varphi$ and

$$\check{d}_\hbar(e^{\varphi'/\hbar}) = \check{d}_\hbar(e^{\tilde{\varphi}/\hbar - b}) = \check{d}_\hbar(1 + (\tilde{\varphi}/\hbar - b) + \ldots) = \check{d}_\hbar(e^{\tilde{\varphi}/\hbar} - b) = \tilde{\mu} - a = 0$$

since $b \wedge \tilde{\varphi} = 0$ and $b \wedge b = 0$. Thus φ' is a solution of the quantum extended Maurer–Cartan equation, which is a lift of $\varphi = \varphi_B$, concluding the proof of Theorem 6.8.

---

[4] Here we use the first assumption of Theorem 6.8.

6.4   The First Abstract Unobstructedness Theorem

In this section, we show that the semi-classical Maurer–Cartan functor $\mathrm{SMC}(G^{\bullet,\bullet}, -)$ is unobstructed if $G^{\bullet,\bullet}$ is quasi-perfect or, more generally, semi-perfect. To do so, we introduce yet another variant $\mathrm{QMC}_0(G^{\bullet,\bullet}, -)$.

Definition 6.13 (Obtusity) Let $G^{\bullet,\bullet}$ be a Λ-linear curved Batalin–Vilkovisky algebra. A solution

$$\varphi_A = \sum_{i=0}^{\infty} \left(\sum_{p=0}^{d} C_i^{-p,p}(\varphi_A) \right) \hbar^i \in \mathfrak{m}_A \cdot G_A^0[\![\hbar]\!]$$

with $C_i^{-p,p}(\varphi_A) \in \mathfrak{m}_A \cdot G_A^{-p,p}$ of the quantum extended Maurer–Cartan equation

$$(\bar{\partial} + \hbar\Delta)\varphi_A + \frac{1}{2}[\varphi_A, \varphi_A] + (\ell + \hbar y) = 0$$

is called *obtuse* if $C_0^{0,0}(\varphi_A) = 0$. We denote the set of obtuse solutions of the quantum extended Maurer–Cartan equation by $\mathrm{QMC}_0(G^{\bullet,\bullet}, A)$.

Remark 6.14 The name "obtuse," usually used to refer to an angle of more than $90°$, refers to the missing apex in $(0, 0)$ when we plot the non-zero coefficients $C_i^{-p,p}$ at (i, p) into the $\mathbb{N}^2$-plane. $\qquad\qquad\Diamond$

Proposition 6.15 (cf. [38, Lemma 5.12]) *Let $G^{\bullet,\bullet}$ be a semi-perfect Λ-linear curved Batalin–Vilkovisky algebra. Then the functor $\mathrm{QMC}_0(G^{\bullet,\bullet}, -)$ is unobstructed.*

Proof Let $B' \to B$ be a first-order extension in $\mathbf{Art}_\Lambda$ with kernel $I \subset B'$. Let $\varphi \in \mathrm{QMC}_0(G^{\bullet,\bullet}, B)$ be an obtuse solution of the quantum extended Maurer–Cartan equation, and let φ' be a (possibly not obtuse) lift to B', which exists by Theorem 6.8. By assumption, we have

$$(\bar{\partial} + \hbar\Delta)\varphi' + \frac{1}{2}[\varphi', \varphi'] + (\ell + \hbar y) = 0;$$

considering only the coefficient of $\hbar^0$, we then find

$$\bar{\partial}\left(\sum_{p=0}^{d} C_0^{-p,p}(\varphi') \right) + \frac{1}{2}\left[\sum_{p=0}^{d} C_0^{-p,p}(\varphi'), \sum_{p=0}^{d} C_0^{-p,p}(\varphi') \right] + \ell = 0$$

$$\in G_{B'}^1 = \bigoplus_{p=0}^{d} G_{B'}^{-p,p+1}.$$

Considering the pieces in the various bidegrees separately, we find, for $G_{B'}^{0,0}$, that

$$\bar{\partial} C_0^{0,0}(\varphi') + [C_0^{0,0}(\varphi'), C_0^{-1,1}(\varphi')] = 0.$$

Since φ is obtuse, we have $C_0^{0,0}(\varphi') \in I \cdot G_{B'}^{0,0}$; because $I \cdot \mathfrak{m}_{B'} = 0$, we find $\bar{\partial} C_0^{0,0}(\varphi') = 0$. We set $\hat{\varphi}' := \varphi' - C_0^{0,0}(\varphi')$. Again using $I \cdot \mathfrak{m}_{B'} = 0$, we find that $\hat{\varphi}'$ is a solution of the quantum extended Maurer–Cartan equation; it is obtuse by construction. $\square$

With this preparation, it is now easy to prove the first abstract unobstructedness theorem.

Theorem 6.16 (First Abstract Unobstructedness Theorem) *Let $G^{\bullet,\bullet}$ be a quasi-perfect or just semi-perfect Λ-linear curved Batalin–Vilkovisky algebra. Then the semi-classical Maurer–Cartan functor* $\mathrm{SMC}(G^{\bullet,\bullet}, -)$ *is unobstructed.*

Proof Let $B' \to B$ be a first-order extension in $\mathbf{Art}_\Lambda$ with kernel $I \subset B'$. Let (ϕ, f) be a semi-classical Maurer–Cartan solution over B. Then $\varphi := \phi + \hbar f$ is an obtuse solution of the quantum extended Maurer–Cartan equation. By Proposition 6.15, we find an obtuse lift φ'. We set $\phi' := C_0^{-1,1}(\varphi')$ and $f' := C_1^{0,0}(\varphi')$. Then the $(-1, 2)$-part of the $\hbar^0$-piece of the quantum extended Maurer–Cartan equation for φ' yields

$$\bar{\partial} \phi' + \frac{1}{2}[\phi', \phi'] + \ell = 0,$$

and the $(0, 1)$-part of the $\hbar^1$-piece yields

$$\bar{\partial} f' + [\phi', f'] + y + \Delta\phi' = 0.$$

Thus, (ϕ', f') is a semi-classical Maurer–Cartan solution lifting (ϕ, f) to B'. $\square$

6.5 The Second Abstract Unobstructedness Theorem

We deduce the Second Abstract Unobstructedness Theorem 5.24 from the first one. In fact, it is sufficient to show that the forgetful transformation

$$\mathrm{SMC}(G^{\bullet,\bullet}, -) \Rightarrow \mathrm{MC}(G^{\bullet,\bullet}, -)$$

is a smooth morphism of functors of Artin rings. Namely, if it is smooth, then it is also surjective, i.e., surjective on every $A \in \mathbf{Art}_\Lambda$; then smoothness of $\mathrm{MC}(G^{\bullet,\bullet}, -)$ follows from smoothness of $\mathrm{SMC}(G^{\bullet,\bullet}, -)$ by [251, 2.5(iii)]. For the second abstract unobstructedness theorem, semi-perfectness alone is not sufficient, and we assume quasi-perfectness for simplicity.

Theorem 6.17 *Let $(G^{\bullet,\bullet}, A^{\bullet,\bullet})$ be a quasi-perfect Λ-linear curved Batalin–Vilkovisky calculus. Then*

$$\mathrm{SMC}(G^{\bullet,\bullet}, -) \Rightarrow \mathrm{MC}(G^{\bullet,\bullet}, -)$$

is smooth.

Proof Let $B' \to B$ be a first-order extension in $\mathbf{Art}_\Lambda$ with kernel $I \subset B'$. Let (ϕ, f) be a semi-classical Maurer–Cartan solution over B, and let ϕ' be a classical Maurer–Cartan solution over B' with $\phi'|_B = \phi$. We have to show that it can be extended to a semi-classical Maurer–Cartan solution (ϕ', f') over B' with $f'|_B = f$.

We write $\omega_f = e^f \lrcorner\, \omega$; then we have $\bar{\partial}_\phi \omega_f = 0$ by Lemma 5.17. Thus, it defines a class $[\omega_f] \in H^0(A_B^{0,\bullet}, \bar{\partial}_\phi)$. By Proposition 5.29, the formation of $H^0(A_{B'}^{0,\bullet}, \bar{\partial}_{\phi'})$ commutes with base change; in particular, $H^0(A_{B'}^{0,\bullet}, \bar{\partial}_{\phi'}) \to H^0(A_B^{0,\bullet}, \bar{\partial}_\phi)$ is surjective, and we find $\omega' \in A_{B'}^{0,0}$ representing a class $[\omega']$—i.e., $\bar{\partial}_{\phi'} \omega' = 0$—with $[\omega']|_B = [\omega_f]$. Since

$$H^0(A_{B'}^{0,\bullet}, \bar{\partial}_{\phi'}) = \ker\left(\bar{\partial}_{\phi'} \colon\ A_{B'}^{0,0} \to A_{B'}^{0,1}\right)$$

and analogously for B, we have in fact $\omega'|_B = \omega_f$.

Let $\tilde{f} \in \mathfrak{m}_{B'} \cdot G_{B'}^{0,0}$ be an arbitrary lift of f, and set $\omega_{\tilde{f}} := e^{\tilde{f}} \lrcorner\, \omega$. Then $\omega_{\tilde{f}}|_B = \omega_f$. Let $g \in G_{B'}^{0,0}$ be the unique element with $g \lrcorner\, \omega = \omega'$, and let $h := g - e^{\tilde{f}}$. Then we have

$$h \lrcorner\, \omega = \omega' - \omega_{\tilde{f}} \in I \cdot A_{B'}^{d,0},$$

so $h \in I \cdot G_{B'}^{0,0}$. Set $f' := \tilde{f} + h$. Then we have $f'|_B = f$ and

$$e^{f'} \lrcorner\, \omega = e^{\tilde{f}+h} \lrcorner\, \omega = (e^{\tilde{f}} + h) \lrcorner\, \omega = \omega'$$

since $I \cdot \mathfrak{m}_{B'} = 0$, so $e^{\tilde{f}+h} = e^{\tilde{f}} + h$.

We have $e^{f'} \wedge e^{-f'} = 1$, so multiplication with $e^{f'}$ is an isomorphism of B'-modules. Consequently, the contraction map

$$\kappa' \colon\ G_{B'}^{p,q} \to A_{B'}^{d+p,q}, \quad \theta \mapsto \theta \lrcorner\, \omega' = (\theta \wedge e^{f'}) \lrcorner\, \omega,$$

is an isomorphism of B'-modules.

Although we do not yet know if (ϕ', f') is a semi-classical Maurer–Cartan solution, we define $\Delta_{f'}(\theta) := \Delta(\theta) + [f', \theta]$ as in that case. Then, using

$e^{f'} \wedge [f', \theta] = [e^{f'}, \theta]$ and the Bogomolov–Tian–Todorov formula for $\Delta(e^{f'} \wedge \theta)$, we find

$$\partial(\theta \lrcorner \omega') = \Delta_{f'}(\theta) \lrcorner \omega'.$$

Now we have

$$(\bar{\partial} f' + [\phi', f'] + y + \Delta\phi') \lrcorner \omega' = \bar{\partial} f' \lrcorner \omega' + \Delta_{f'}(\phi') \lrcorner \omega' + y \lrcorner \omega'$$

$$= (\bar{\partial} e^{f'}) \lrcorner \omega + \partial(\phi' \lrcorner \omega') + e^{f'} \lrcorner \bar{\partial}\omega$$

$$= \bar{\partial}(e^{f'} \lrcorner \omega) + \mathcal{L}_{\phi'}(\omega') = \bar{\partial}_{\phi'}(\omega') = 0$$

where we have $y \lrcorner \omega' = e^{f'} \lrcorner \bar{\partial}\omega$ due to $y \lrcorner \omega = \bar{\partial}\omega$, and $\mathcal{L}_{\phi'}(\omega') = \partial(\phi' \lrcorner \omega')$ by the Lie–Rinehart homotopy formula. Since contraction with ω' is an isomorphism, (ϕ', f') is a semi-classical Maurer–Cartan solution. $\qquad\square$

Part II
Logarithmic Geometry

Chapter 7
Logarithmic Geometry

In this chapter, we give an overview over the logarithmic geometry used in this monograph. The main reference is Ogus' book [222] as well as F. Kato's article [160]. The reader already acquainted with logarithmic geometry may skip this chapter and recur to it as needed. For the reader unacquainted with log geometry, this can only be a quick introduction which does not replace a more in-depth study of the more comprehensive sources. However, it should be sufficient to read most of this monograph. This chapter follows closely certain chapters and sections of Ogus' book [222] and can serve as the starting point to read that book. Section 7.6 follows closely F. Kato's article [160], and Sect. 7.7 revisits a result of Tsuji in [272].

7.1 Advice for Reading Ogus' Lecture Notes

The reader who wishes to study logarithmic geometry in more depth may want to study [222] in the following order: First, read chapter I.1.1 to get acquainted with the basic definitions concerning monoids. Then jump to part III and read chapters III.1.1 and III.1.2 on log structures and log schemes. After this, deepen your understanding of basic properties of monoids by reading I.1.3 and I.1.4 as well as I.2.1. You can skip I.1.2 at the beginning of your studies. Now go back to log schemes and read III.1.4 to become aware of the two conventions of defining $\mathcal{M}_X$ either in the Zariski or étale topology, and then go to III.1.6 to see why this matters, in addition to finding a very important construction of log schemes. Then go to III.2.1 and III.2.2 and read in addition what you need to read to understand these chapters. After that, you may want to acquaint yourself more with the relation between toric varieties and log geometry. To do this, read a selection of I.3 of your choice. Then read I.4.6 and I.4.8 to study the important notions of integral and saturated homomorphisms. The geometric version is then discussed in III.2.5. After this, we recommend IV.1.1 until IV.3.3 to learn about logarithmic derivations and differentials as well as log smooth

© The Author(s), under exclusive license to Springer Nature Switzerland AG 2025
S. Felten, *Global Logarithmic Deformation Theory*, Lecture Notes
in Mathematics 2373, https://doi.org/10.1007/978-3-031-98751-9_7

morphisms. In the beginning, the whole part II of the book should only be read as needed. The last part V essentially concerns de Rham cohomology and can be read after a solid understanding of the basics of log geometry is achieved. Interesting articles for further study are K. Kato's seminal paper [162], F. Kato's log smooth deformation theory [160], and the work [155] of Illusie, K. Kato, and Nakayama on the logarithmic Riemann–Hilbert correspondence. The arithmetically inclined reader may also want to look into Tsuji's proof of the C_{st}-conjecture in [271].

7.2 Monoids

A *monoid* is a set P together with a binary operation $+\colon P \times P \to P$ which is associative, commutative, and has an identity element $0_P \in P$, i.e., $0_P + p = p = p + 0_P$ for all $p \in P$. As in the case of groups, the identity element is unique. A *homomorphism* of monoids is a map $\theta\colon Q \to P$ of sets such that $\theta(0_Q) = 0_P$ and $\theta(q + q') = \theta(q) + \theta(q')$. Monoids and their homomorphisms form a category **Mon**. A *submonoid* is a subset $Q \subseteq P$ with $0_P \in Q$ and $p + p' \in Q$ if $p, p' \in Q$.

Example 7.1 $\{0\}$ is a monoid in the obvious way. It is the zero object of **Mon** in the sense that it is both initial and terminal. The natural numbers $\mathbb{N} = \{0, 1, 2, \ldots\}$ form a monoid. Any Abelian group G is a monoid upon forgetting the existence of inverses. $\diamond$

Example 7.2 Let $M \cong \mathbb{Z}^r$ be a free Abelian group of finite rank, and let $C \subseteq M_{\mathbb{R}}$ be a convex rational polyhedral cone. Then $P := C \cap M \subseteq M$ is a monoid. $\diamond$

An element $p \in P$ is *invertible* if there is some $p' \in P$ with $p + p' = 0$. Such an inverse is unique if it exists. The invertible elements form a subgroup $P^* \subseteq P$. A monoid is called *dull* if $P^* = P$, i.e., if it is a group.

At the other extreme, a monoid is called *sharp* if $P^* = \{0\}$. For any monoid P, we can form the quotient $\overline{P} = P/P^*$, which is a sharp monoid.[1] Any homomorphism $\theta\colon P \to R$ to a sharp monoid R factors uniquely through $P \to \overline{P}$.

Given a monoid P, we can form a group P^{gp}—its *Grothendieck group* or *groupification*—as follows: $P^{\mathrm{gp}} = \{(p, q) \in P \times P\}/\sim$ where $(p_1, q_1) \sim (p_2, q_2)$ if there is some $r \in P$ with $p_1 + q_2 + r = p_2 + q_1 + r$. This is an equivalence relation, and a group structure is given by $(p_1, q_1) + (p_2, q_2) := (p_1 + p_2, q_1 + q_2)$. The canonical homomorphism $P \to P^{\mathrm{gp}}$, $p \mapsto (p, 0)$, has the universal property that any homomorphism $\theta\colon P \to G$ to an Abelian group G factors uniquely through $P \to P^{\mathrm{gp}}$.

Example 7.3 We have $\mathbb{N}^{\mathrm{gp}} = \mathbb{Z}$. $\diamond$

---

[1] The quotient is given by equivalence classes of elements $p \in P$ where $p \sim p'$ if there are elements $q, q' \in P^*$ with $p + q = p' + q'$.

The category **Mon** of monoids admits all small limits and all small colimits. The construction of limits is straightforward: The limit of underlying sets carries a canonical monoid structure. The construction of colimits is more involved. The easiest example are direct sums $\bigoplus_i P_i$, which are constructed in the same way as direct sums of Abelian groups. Another situation that admits a relatively simple construction is the *amalgamated sum* or *push-out* of a diagram

$$
\begin{array}{ccc}
Q & \xrightarrow{\ u_1\ } & P_1 \\
{\scriptstyle u_2}\big\downarrow & & \big\downarrow \\
P_2 & \longrightarrow & P_1 \oplus_Q P_2
\end{array}
$$

in case at least one of Q, P_1, P_2 is a group. In this case,

$$
P_1 \oplus_Q P_2 = \{(p_1, p_2) \in P_1 \times P_2\}/\sim
$$

with $(p_1, p_2) \sim (p_1', p_2')$ if there are $q, q' \in Q$ with $p_1 + u_1(q') = p_1' + u_1(q)$ and $p_2 + u_2(q) = p_2' + u_2(q')$. The general construction of colimits—for example of the push-out when none of Q, P_1, P_2 is a group—is more involved and can be found in [222, I, 1.1].

Example 7.4 $\mathbb{N}^r = \{(n_1, \ldots, n_r) \mid n_i \in \mathbb{N}\}$ is the direct sum of r copies of $\mathbb{N}$. We thus also write $\mathbb{N}^{\oplus r}$. $\qquad\qquad\Diamond$

Example 7.5 Arbitrary intersections of submonoids form a submonoid. We say that a submonoid $Q \subseteq P$ is *generated* by a set S of elements of P if Q is the intersection of all submonoids of P which contain S. In this case, we write $\langle S \rangle = Q$. $\qquad\Diamond$

7.2.1 Integral and Saturated Monoids

A monoid P is *finitely generated* if there is a surjective homomorphism $\mathbb{N}^{\oplus r} \to P$ for some $r \geq 0$. Equivalently, there is a finite subset $S \subseteq P$ such that $P = \langle S \rangle$.

A monoid P is *u-integral* if, for any $p \in P$ and $v \in P^*$, the equality $p = p + v$ implies $v = 0$. It is *integral* if the cancellation law does not only hold for units v but in general: $p + q = p + r$ implies $q = r$ for $p, q, r \in P$. A monoid is integral if and only if the canonical map $P \to P^{\mathrm{gp}}$ is injective, and it is u-integral if and only if $P^* \to P^{\mathrm{gp}}$ is injective. A monoid is integral if and only if it is u-integral, and $\overline{P}$ is integral.

A monoid is *fine* if it is finitely generated and integral.

Example 7.6 The monoid $P := (\mathbb{Z} \sqcup \mathbb{N}_{\geq 1}, \bullet)$ with $z \bullet z' := z + z'$, $z \bullet n := n$, $n \bullet n' := n + n'$ for $z, z' \in \mathbb{Z}$, $n, n' \in \mathbb{N}_{\geq 1}$ is not u-integral but finitely generated. Its submonoid $\mathbb{N} \sqcup \mathbb{N}_{\geq 1}$ is finitely generated as well, and it is u-integral because it is sharp, but it is not integral. The submonoid

$$Q = \{(0,0)\} \cup \{(m,n) \mid m,n \geq 1\} \subseteq \mathbb{N}^2$$

is integral but not finitely generated. The submonoid $\mathbb{N} \setminus \{1,2\} = \{0,3,4,5,\ldots\} \subseteq$ $\mathbb{N}$ is integral and finitely generated, so it is fine. Another monoid which is not finitely generated is $(\mathbb{Q},+)$. $\Diamond$

A monoid is *saturated* if it is integral and the following holds: Let $p \in P^{\mathrm{gp}}$, and assume there is some $n \in \mathbb{Z}_{>0}$ with $np := p + \ldots + p \in P \subseteq P^{\mathrm{gp}}$. Then $p \in P$. An integral monoid P is saturated if and only if $\overline{P}$ is saturated.

A monoid is *toric* if it is fine and saturated, and if there is additionally an isomorphism $P^{\mathrm{gp}} \cong \mathbb{Z}^r$ for some $r \geq 0$. Toric monoids are precisely the monoids which arise as intersections of lattices $M \cong \mathbb{Z}^r$ with rational polyhedral cones $C \subseteq M_{\mathbb{R}}$.

Example 7.7 The monoid $\mathbb{N} \setminus \{1,2\} = \{0,3,4,\ldots\}$ is fine but not saturated. The monoid

$$\{(0,0)\} \cup \{(m,n) \mid m,n \geq 1\}$$

is integral and saturated but not finitely generated. The monoid $\mathbb{N} \oplus \mathbb{Z}/2\mathbb{Z}$ is fine and saturated but not toric. For $k \geq 1$, the monoid

$$P_k := \{(m,n) \in \mathbb{Z}^2 \mid n \geq 0, n \geq k \cdot m\} \subseteq \mathbb{Z}^2$$

is toric. We have $P_1 \cong \mathbb{N}^2$, and for $k \geq 2$, it needs three generators $(1,0)$, $(0,1)$, $(-1,k)$. $\Diamond$

For an arbitrary monoid P, we have an associated integral monoid P^{int} given by the image of the canonical map $P \to P^{\mathrm{gp}}$. For an integral monoid P, we have an associated saturated monoid P^{sat} given by the set of all $p \in P^{\mathrm{gp}}$ such that $np \in P \subset P^{\mathrm{gp}}$ for some integer $n \geq 1$. For a general monoid P, we write $P^{\mathrm{sat}} := (P^{\mathrm{int}})^{\mathrm{sat}}$. These functors are left adjoint to the inclusion $\mathbf{Mon}^{\mathrm{int}} \to \mathbf{Mon}$ of the category of integral monoids into all monoids respectively to the inclusion $\mathbf{Mon}^{\mathrm{sat}} \to \mathbf{Mon}^{\mathrm{int}}$ of the category of saturated monoids into the category of integral monoids. Arbitrary small limits of integral monoids are integral again, and arbitrary small limits of saturated monoids are saturated again. Small colimits in $\mathbf{Mon}^{\mathrm{int}}$ and $\mathbf{Mon}^{\mathrm{sat}}$ are given by applying $(-)^{\mathrm{int}}$ and $(-)^{\mathrm{sat}}$.

7.2.2   The Monoid Ring

Let P be a monoid and R a ring. Then we write $R[P] := \bigoplus_{p \in P} R \cdot z^p$. This is a ring with $z^p \cdot z^q := z^{p+q}$, called a *monoid ring*. If $\theta \colon Q \to P$ is a monoid homomorphism, then we have an induced ring homomorphism $R[\theta]\colon R[Q] \to R[P]$. We write $A_\theta\colon A_P \to A_Q$ for the corresponding morphism of affine schemes. For a field $\mathbf{k}$ and a toric monoid P, we obtain an affine toric variety A_P.

7.2.3 Ideals and Faces

An *ideal* $I \subseteq P$ in a monoid is a subset such that $k \in I$ and $p \in P$ implies $p + k \in I$. An ideal $I \subseteq P$ is called *prime* if $I \neq P$ and $p + q \in I$ implies that either $p \in I$ or $q \in I$ for elements $p, q \in P$. The motivation for this notion is that $R[I] \subseteq R[P]$ is an ideal, where R is a ring and $R[P]$ is the associated monoid ring. When R is an integral ring, P is an integral monoid, and $I \subseteq P$ is a prime ideal, then $R[I] \subseteq R[P]$ is a prime ideal.

The empty set $\varnothing \subseteq P$ is the unique minimal prime ideal of P, and the set $P^+ := P \setminus P^*$ of non-units is the unique maximal prime ideal of P.

A *face* is a submonoid $F \subseteq P$ such that, for $p, q \in P$, we have that $p + q \in F$ implies $p \in F$ and $q \in F$. A face is nothing but a submonoid whose complement is an ideal, and a prime ideal is nothing but an ideal whose complement is a submonoid. Faces are much like faces of a convex polyhedral cone.

Example 7.8 The faces of $\mathbb{N}^2$ are precisely $\{(0, 0)\}$, $\mathbb{N} \times \{0\}$, $\{0\} \times \mathbb{N}$, and $\mathbb{N} \times \mathbb{N}$. Examples of ideals of $\mathbb{N}^2$ are $\mathbb{N} \times \{n \geq 3\}$ and $\{m \geq 5\} \times \mathbb{N}$. There are infinitely many different ideals in $\mathbb{N}^2$, but all of them are finitely generated, i.e., of the form $\bigcup_{s \in S}(s + \mathbb{N}^2)$ for a finite subset $S \subseteq \mathbb{N}^2$. There are precisely four prime ideals, namely the complements of the four faces. $\qquad\qquad\diamond$

7.2.4 Basic Properties of Homomorphisms

A homomorphism $\theta\colon Q \to P$ of monoids is *local* if $\theta^{-1}(P^*) = Q^*$, or, equivalently, $\theta^{-1}(P^+) = Q^+$. It is *sharp* if the induced homomorphism $Q^* \to P^*$ is an isomorphism, and it is *logarithmic* if $\theta^{-1}(P^*) \to P^*$ is an isomorphism. A homomorphism $\theta\colon Q \to P$ is logarithmic if and only if it is sharp and local. The relevance of logarithmic homomorphisms comes mainly from the fact that a prelog structure $\alpha_X\colon \mathcal{M}_X \to (\mathcal{O}_X, \cdot)$ on a scheme X is, by definition, a log structure if and only if it is a logarithmic homomorphism of sheaves of monoids, see below.

A homomorphism $\theta\colon Q \to P$ is *strict* if $\bar\theta\colon \overline{Q} \to \overline{P}$ is an isomorphism, and it is *s-injective* if $\bar\theta\colon \overline{Q} \to \overline{P}$ is injective. The s refers to the fact that the homomorphism becomes injective after sharpening.

The most fundamental property that a homomorphism $\theta : Q \to P$ can have is *exactness*; θ is called *exact* if the diagram

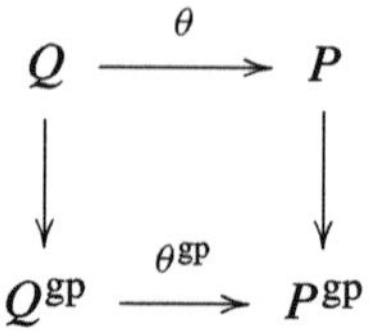

is Cartesian in **Mon**. A monoid P is integral if and only if the diagonal $\Delta : P \to P \times P$ is exact, and every exact homomorphism $\theta : Q \to P$ is local. An *exact submonoid* is a submonoid $Q \subseteq P$ such that the inclusion is exact. Unlike general submonoids, exact submonoids of fine, fine and saturated, or toric monoids have the same property again. A face $F \subseteq P$ in an integral monoid is an exact submonoid. If P is fine, then F is finitely generated as a monoid, and *monogenic* as a face, i.e., there is a single element $p \in F$ such that F is the smallest face containing p. More information on exactness can be found in [222, I, 2.1].

7.2.5 *Integral and Saturated Homomorphisms*

In general, the push-out $P \oplus_Q Q'$ of integral monoids does not need to be integral when formed in the category **Mon** of all monoids. Similarly, the push-out $P \oplus_Q Q'$ of saturated monoids does not need to be saturated when formed in **Mon**. In general, this situation can be remedied by applying $(-)^{\mathrm{int}}$ and $(-)^{\mathrm{sat}}$, but for reasons that will become clear below when we discuss fiber products of log schemes, we are—within the scope of this monograph—mostly interested in homomorphisms $\theta : Q \to P$ such that $P \oplus_Q Q'$ is automatically integral respectively saturated.

Definition 7.9 (Integral and Saturated Homomorphisms) A homomorphism $\theta : Q \to P$ between integral monoids is *integral* if, for every homomorphism $\phi : Q \to Q'$ of integral monoids, the push-out $P \oplus_Q Q'$ in **Mon** is an integral monoid. A homomorphism $\theta : Q \to P$ of saturated monoids is *saturated* if it is integral, and for every homomorphism $\phi : Q \to Q'$ of saturated monoids, the push-out $P \oplus_Q Q'$ in **Mon** is saturated.

A homomorphism $\theta : Q \to P$ is integral respectively saturated if and only if $\bar{\theta} : \overline{Q} \to \overline{P}$ has the same property.

We have an easy criterion to check integrality of a map $\theta : Q \to P$.

Lemma 7.10 *Let $\theta : Q \to P$ be a homomorphism of integral monoids. Then θ is integral if and only if the following condition holds: Let $q_1, q_2 \in Q$ and $p_1, p_2 \in P$ be such that $\theta(q_1) + p_1 = \theta(q_2) + p_2$. Then there exist $p' \in P$ and $q'_1, q'_2 \in Q$ such that $p_i = p' + \theta(q'_i)$ and $q_1 + q'_1 = q_2 + q'_2$.*

Proof See [222, I, Defn. 4.6.2]. □

Example 7.11 The map $\theta : \mathbb{N} \to \mathbb{N}^2$, $1 \mapsto p$, is integral for any $p \in \mathbb{N}^2$. ◇

There is no similar criterion for saturated homomorphisms in general. However, in practice, we can often apply the following criterion. A homomorphism $\theta : Q \to P$ of integral monoids is called *locally exact* if, for every face $G \subseteq P$, the induced map $Q_{\theta^{-1}(G)} \to P_G$ of localizations is exact, see [222, I, Defn. 4.2.12]. Here, $P_G \subseteq P^{\mathrm{gp}}$ is the monoid obtained from P by inverting only the elements $g \in G$.

Theorem 7.12 *Let* $\theta : Q \to P$ *be a locally exact homomorphism of toric monoids. Then the following conditions are equivalent:*

(a) θ *is saturated.*
(b) *For every face* $G \subseteq P$, *the cokernel of* $Q^{\mathrm{gp}} \to P^{\mathrm{gp}}/G^{\mathrm{gp}}$ *is torsion-free.*
(c) *The ideal* $K_\theta := \theta(\theta^{-1}(P^+)) + P$ *is a radical ideal, i.e.,* $p + \ldots + p \in K_\theta$ *implies* $p \in K_\theta$ *for* $p \in P$.

Proof See [222, I, Thm. 4.8.14]. □

Example 7.13 The homomorphism $\theta : \mathbb{N} \to \mathbb{N}^2$, $1 \mapsto (1, 1)$, is saturated. The homomorphism $\phi : \mathbb{N} \to \mathbb{N}^2$, $1 \mapsto (2, 2)$, is integral but not saturated. ◇

7.2.6 Vertical Homomorphisms

Definition 7.14 (Vertical Homomorphisms) A homomorphism $\theta : Q \to P$ of integral monoids is *vertical* if the image of P in $\mathrm{Coker}(\theta^{\mathrm{gp}})$ is a dull monoid, i.e., a subgroup.

This is equivalent to the condition that the image of θ is not contained in any proper face $F \subsetneq P$. A homomorphism $\theta : Q \to P$ is vertical if and only if $\bar{\theta} : \overline{Q} \to \overline{P}$ is vertical.

The geometric meaning of this notion is as follows: Let $\theta : Q \to P$ be a homomorphism of sharp toric monoids, giving rise to a toric morphism $A_\theta : A_P \to A_Q$ of affine toric varieties. Then $\theta : Q \to P$ is vertical if and only if the preimage $A_\theta^{-1}(D_Q)$ of the toric boundary divisor $D_Q \subset A_Q$ is the toric boundary divisor $D_P \subset A_P$. In other words, there is no horizontal part of D_P which would (partially) map to the big torus orbit A_Q^*.

Example 7.15 The homomorphism $\theta : \mathbb{N} \to \mathbb{N}^2$, $1 \mapsto (1, 1)$, is vertical. The homomorphism $\phi : \mathbb{N} \to \mathbb{N}^2$, $1 \mapsto (1, 0)$, is not vertical. ◇

7.3 Logarithmic Schemes

7.3.1 Log Schemes

Here is our central definition:

Definition 7.16 (Log Schemes) Let X be a scheme. A *log structure* is a pair $(\mathcal{M}_X, \alpha_X)$ where $\mathcal{M}_X$ is a sheaf of monoids and $\alpha_X \colon (\mathcal{M}_X, \cdot) \to (\mathcal{O}_X, \cdot)$ is a homomorphism of sheaves of monoids such that $\alpha_X \colon \alpha_X^{-1}(\mathcal{O}_X^*) \to \mathcal{O}_X^*$ is an isomorphism. The triple $(X, \mathcal{M}_X, \alpha_X)$ is called a *log scheme*. A *homomorphism of log structures* is a homomorphism $\gamma \colon \mathcal{M}_X \to \mathcal{N}_X$ of sheaves of monoids which is compatible with the two maps $\alpha_X \colon \mathcal{M}_X \to \mathcal{O}_X$ and $\beta_X \colon \mathcal{N}_X \to \mathcal{O}_X$.

We say that $(\mathcal{M}_X, \alpha_X)$ is a log structure in the *Zariski* respectively *étale topology* if $\mathcal{M}_X$ and α_X are given in the Zariski respectively étale topology.

Unlike the situation for coherent sheaves, the two notions of a log structure given in the Zariski respectively étale topology are different. We will discuss their relation in more detail below once we have more theory at our disposal. For now, we consider them as variants which we develop in parallel.

Example 7.17 For any scheme X, the inclusion $\mathcal{O}_X^* \to \mathcal{O}_X$ is a log structure, called the *trivial log structure*. On the point $* = \operatorname{Spec} \mathbb{C}$, we can define a log structure by setting $\mathcal{M}(*) = \mathbb{C}^* \oplus \mathbb{N}$ with $(\lambda, n) \mapsto \lambda \cdot 0^n$. This is called the *standard log point*. $\diamond$

A homomorphism $\alpha \colon \mathcal{M}_X \to (\mathcal{O}_X, \cdot)$ of sheaves of monoids without the condition on the invertible functions is called a *prelog structure*. Given any prelog structure $\beta \colon \mathcal{P} \to \mathcal{O}_X$, there is always a factorization

$$\mathcal{P} \xrightarrow{\gamma} \mathcal{P}^{\log} \xrightarrow{\beta^{\log}} \mathcal{O}_X$$

such that $\beta^{\log} \colon \mathcal{P}^{\log} \to \mathcal{O}_X$ is a log structure, the *associated log structure* of $\beta \colon \mathcal{P} \to \mathcal{O}_X$. It has the universal property that, for any factorization $\mathcal{P} \to \mathcal{M} \to \mathcal{O}_X$ of β with a log structure $\alpha \colon \mathcal{M} \to \mathcal{O}_X$, there is a unique homomorphism $\mathcal{P}^{\log} \to \mathcal{M}$ of sheaves of monoids which fits both in a factorization $\mathcal{P} \to \mathcal{P}^{\log} \to \mathcal{M}$ and a factorization $\mathcal{P}^{\log} \to \mathcal{M} \to \mathcal{O}_X$. Explicitly, the associated log structure is constructed as follows: Let $\mathcal{P}^\circ \subset \mathcal{P}$ be the sheaf of submonoids of elements $p \in \mathcal{P}$ such that $\beta(p) \in \mathcal{O}_X^*$. Then we can form a diagram

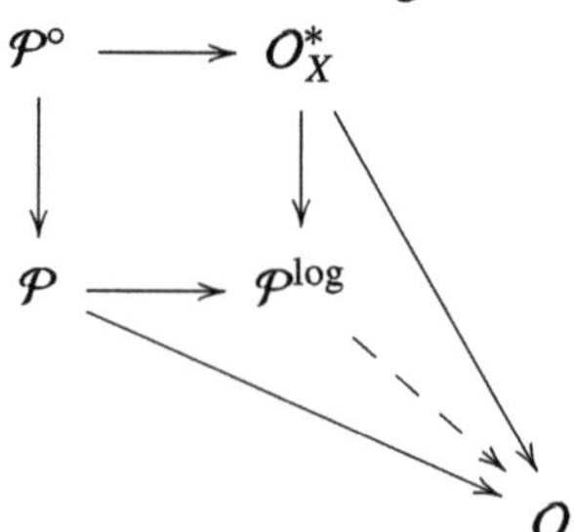

where $\mathcal{P}^{\log}$ is the push-out of the square in the category of sheaves of monoids.

The difference between $\mathcal{M}_X$ and $\mathcal{O}_X^*$ is captured in the *ghost sheaf*, sometimes called the *characteristic sheaf*.

Definition 7.18 (Ghost Sheaf) Let $(X, \mathcal{M}_X, \alpha_X)$ be a log scheme. Then the *ghost sheaf* is the quotient

$$\overline{\mathcal{M}}_X := \mathcal{M}_X / \mathcal{O}_X^*.$$

We denote the binary operation on $\overline{\mathcal{M}}_X$ by $+$.

Here, we write $\mathcal{O}_X^*$ for the subsheaf of units $\mathcal{M}_X^* \subset \mathcal{M}_X$ since we have an isomorphism $\alpha_X \colon \mathcal{M}_X^* \cong \mathcal{O}_X^*$. Also note that $\mathcal{M}_X^*(U) = \mathcal{M}_X(U)^*$, i.e., the *sheaf of units* consists precisely of the units in each $\mathcal{M}_X(U)$. The ghost sheaf $\overline{\mathcal{M}}_X$ is computed by sheafification of $\mathcal{M}_X(U)/\mathcal{M}_X(U)^*$. Unlike $\mathcal{M}_X$ itself, the quotient $\overline{\mathcal{M}}_X$ can often be described explicitly.

Example 7.19 When $\mathcal{M}_X = \mathcal{O}_X^*$, then $\overline{\mathcal{M}}_X = 0$. For the standard log point, we have $\overline{\mathcal{M}}_X(*) = \mathbb{N}$. $\Diamond$

Let us now define morphisms of log schemes.

Definition 7.20 (Morphisms of Log Schemes) Let $(X, \mathcal{M}_X, \alpha_X)$ and $(Y, \mathcal{M}_Y, \alpha_Y)$ be log schemes, defined either in the Zariski or étale topology. Then a *morphism* $f \colon (X, \mathcal{M}_X, \alpha_X) \to (Y, \mathcal{M}_Y, \alpha_Y)$ consists of a morphism of schemes $(f, f^\sharp) \colon (X, \mathcal{O}_X) \to (Y, \mathcal{O}_Y)$ together with a homomorphism of sheaves of monoids $f^\flat \colon \mathcal{M}_Y \to f_* \mathcal{M}_X$ which fits into a commutative diagram

$$
\begin{array}{ccc}
\mathcal{M}_Y & \xrightarrow{\ f^\flat\ } & f_* \mathcal{M}_X \\
{\scriptstyle \alpha_Y}\downarrow & {\scriptstyle f_*\alpha_X} & \downarrow \\
& {\scriptstyle f^\sharp} & \\
\mathcal{O}_Y & \xrightarrow{\quad} & f_* \mathcal{O}_X
\end{array}
$$

of sheaves of monoids.

Log schemes thus form a category, which we denote by **LSch**. There is a forgetful functor

$$\mathbf{LSch} \to \mathbf{Sch}, \quad (X, \mathcal{M}_X, \alpha_X) \mapsto X.$$

It has a right adjoint given by $X \mapsto (X, \mathcal{O}_X^*)$, the trivial log structure already introduced in the example above.

We usually suppress the homomorphism α in the notation and write $(X, \mathcal{M}_X)$ for a log scheme. Often, we will even drop the sheaf of monoids $\mathcal{M}_X$. In this situation, we sometimes use the notation $\underline{X}$ to denote the underlying scheme without log structure.

7.3.2 The Compactifying Log Structure

A large and important class of log schemes arises from the following construction.

Definition 7.21 (Compactifying Log Structures) Let X be a scheme, and let $D \subset X$ be a closed subset. Then

$$\mathcal{M}_{(X|D)}(U) := \{ f \in \mathcal{O}_X(U) \mid f|_{U \setminus D} \in \mathcal{O}_X^*(U \setminus D) \}$$

for opens $U \to X$ together with the obvious map $\mathcal{M}_{(X|D)} \to \mathcal{O}_X$ is a log structure, called the *compactifying* log structure of $D \subset X$. We denote the log scheme $(X, \mathcal{M}_{(X|D)})$ by $(X|D)$ for short.[2]

Especially when D is a divisor, we also say that $\mathcal{M}_{(X|D)}$ is the *divisorial* log structure of $D \subset X$. It is easy to see that $\mathcal{M}_{(X|D)}$ is indeed a log structure. The construction makes sense both in the Zariski and the étale topology, and in many examples, the two versions of $\mathcal{M}_{(X|D)}$ in the two topologies are fundamentally different. The name is derived from situations where X is a compactification of $X \setminus D$, such as in the construction of Deligne's mixed Hodge structure on a non-compact smooth variety.

Example 7.22 Here are some important examples of compactifying log structures.

(1) $(X|\varnothing)$ is X endowed with the trivial log structure $\mathcal{O}_X^*$.
(2) $(X|D) = (\mathbb{A}^1|\{t = 0\})$: The log structure is trivial everywhere on the line $\mathbb{A}^1$ except in $\{0\}$, where the stalk of the ghost sheaf is $\mathbb{N}$.
(3) $(X|D) = (\mathbb{A}^2|\{xy = 0\})$: Outside the two coordinate axes, the log structure is trivial. Inside each of the coordinate axes, the stalk of the ghost sheaf is $\mathbb{N}$. In $\{0\}$, the stalk of the ghost sheaf is $\mathbb{N}^2$.
(4) When $f \colon X \to Y$ is a morphism of schemes and $D \subset Y$, then we obtain a morphism of log schemes $(X|f^{-1}(D)) \to (Y|D)$. An example of this situation is given by $\mathbf{k}[t] \to \mathbf{k}[x, y]$, $t \mapsto xy$, with $D = \{t = 0\}$. $\qquad\qquad \Diamond$

7.3.3 First Properties of Log Schemes

We say that a log scheme $(X, \mathcal{M}_X)$ is *u-integral* if every stalk $\mathcal{M}_{X,x}$ is u-integral. This is equivalent to asking that $\mathcal{M}_X(U)$ is u-integral for every open $U \to X$. This is a very basic well-behavedness condition; log schemes which are not u-integral are

---

[2] The author has learned the notation $(X|D)$ (instead of the more common (X, D)) from the article [205] of Maulik and Ranganathan.

rarely studied in log geometry, and the intuition derived from working with more commonly studied log schemes may be misleading for them.

We say that a log scheme $(X, \mathcal{M}_X)$ is *integral* if each stalk $\mathcal{M}_{X,x}$ is an integral monoid, i.e., the cancellation law holds in $\mathcal{M}_{X,x}$. This is equivalent to asking that every $\mathcal{M}_X(U)$ is integral for opens $U \to X$. Obviously, every integral log scheme is u-integral. A log scheme $(X, \mathcal{M}_X)$ is integral if and only if it is u-integral and the ghost sheaf $\overline{\mathcal{M}}_X$ is integral (in the sense that every stalk $\overline{\mathcal{M}}_{X,x}$ is integral, which is equivalent to that every $\overline{\mathcal{M}}_X(U)$ is integral). When $X \setminus D \subseteq X$ is scheme-theoretically dense, then the compactifying log structure $\mathcal{M}_{(X|D)}$ is integral.

A log scheme $(X, \mathcal{M}_X)$ is called *saturated* if every stalk $\mathcal{M}_{X,x}$ is (integral and) saturated. Again, this is equivalent to asking that every $\mathcal{M}_X(U)$ is saturated. A log scheme $(X, \mathcal{M}_X)$ is saturated if and only if it is u-integral, and the ghost sheaf $\overline{\mathcal{M}}_X$ is saturated. Most log structures studied in the literature are saturated or at least integral. This will also be the case for log structures studied in this monograph.

Example 7.23 We have already considered above the log structure on $* = \operatorname{Spec} \mathbb{C}$ given by $\mathcal{M}(*) = \mathbb{C}^* \oplus \mathbb{N}$ with $(\lambda, n) \mapsto \lambda \cdot 0^n \in \mathbb{C}$. This log scheme is saturated. Now let $P = \mathbb{C}^* \sqcup \mathbb{N}_{\geq 1}$ with $\lambda + \mu := \lambda\mu, \lambda + n := n, n + n' := n + n'$. This is a monoid, and we can define a log structure on $*$ with $\mathcal{N}(*) = P$ by setting $\lambda \mapsto \lambda$ and $n \mapsto 0$. Then the log scheme $(*, \mathcal{N})$ is not u-integral, but its ghost sheaf is nonetheless integral and even saturated, namely $\overline{\mathcal{N}}(*) = \mathbb{N}$. We have a map of log structures $\mathcal{M} \to \mathcal{N}$ given by $(\lambda, n) \mapsto \lambda$ if $n = 0$ and $(\lambda, n) \mapsto n$ if $n \geq 1$. This map induces an isomorphism on ghost sheaves, but it is not an isomorphism of log structures—a phenomenon that cannot occur for morphisms between u-integral log structures. $\Diamond$

7.3.4 Charts and Coherence

Let $(X, \mathcal{M}_X)$ be a log scheme. Assume that we have a monoid P and a morphism $\beta\colon \underline{P} \to \mathcal{M}_X|_U -: \mathcal{M}_U$ of sheaves of monoids for some (Zariski or étale) open subset $U \to X$, where $\underline{P}$ is the constant sheaf of monoids with stalk P. Then the composition $\underline{P} \to \mathcal{M}_U \to \mathcal{O}_U$ is a prelog structure. By the universal property of the associated log structure, we obtain a homomorphism $(\underline{P})^{\log} \to \mathcal{M}_U$ of sheaves of monoids. We say that $\underline{P} \to \mathcal{M}_U$ (or $P \to \Gamma(U, \mathcal{M}_X)$), is a *chart* for the log structure $\mathcal{M}_U \to \mathcal{O}_U$ if this map is an isomorphism. Most log structures that are studied in the literature admit locally a chart. This is very convenient from a technical perspective, but unfortunately, this condition is not always satisfied in the setup of this monograph.

Definition 7.24 (Coherent and Fine Log Structures) Let $(X, \mathcal{M}_X)$ be a log scheme in the Zariski or étale topology. We say that:

(1) $\mathcal{M}_X$ is *quasi-coherent* if there is a cover $\{U_i \to X\}_i$ in the respective topology such that there are charts $P_i \to \mathcal{M}_{U_i}$ for the log structure on U_i;[3]
(2) $\mathcal{M}_X$ is *coherent* if each P_i can be chosen to be finitely generated;
(3) $\mathcal{M}_X$ is *fine* if each P_i can be chosen to be finitely generated and integral;
(4) $\mathcal{M}_X$ is *fine and saturated* (or simply *fs*) if each P_i can be chosen to be fine and saturated.

If P is integral or saturated, then $(\underline{P})^{\log}$ is integral respectively saturated as well. In particular, a fine log structure is integral, and an fs log structure is integral and saturated. It is in general *not* true that $(\underline{P})^{\log}$ is u-integral as soon as P is u-integral.

Charts are very convenient to *define* log structures. Besides compactifying log structures, this is the second big construction of log schemes. A *log ring* is a map $\beta \colon P \to R$ from a monoid P into a ring R such that $\beta(p + q) = \beta(p) \cdot \beta(q)$ and $\beta(0) = 1$. When $X = \operatorname{Spec} R$, then we obtain a map $\beta \colon \underline{P} \to O_X$, and its associated log structure $\mathcal{M}_X$ is quasi-coherent with chart $\underline{P} \to \mathcal{M}_X$. We denote this log scheme by $\operatorname{Spec}(P \to R)$ and say it is the *spectrum* of the log ring $P \to R$. More generally, we obtain a quasi-coherent log structure from a homomorphism $\beta \colon P \to \Gamma(X, O_X)$ for a general scheme X.

Example 7.25 The standard log point introduced above is isomorphic to the spectrum $\operatorname{Spec}(\mathbb{N} \to \mathbb{C})$ with $0 \mapsto 1$ and $n \mapsto 0$ for $n \geq 1$. $\Diamond$

Example 7.26 Fix a base ring R. For a monoid P, we write

$$A_P := \operatorname{Spec}(P \to R[P]),$$

where $R[P] := \bigoplus_{p \in P} R \cdot z^p$ is the associated *monoid ring*, and the map $P \to R[P]$ is given by $p \mapsto z^p$. If $\theta \colon Q \to P$ is a homomorphism of monoids, then we have a morphism $A_\theta \colon A_P \to A_Q$ of log schemes. $\Diamond$

Example 7.27 Let $\{0\} \subset \mathbb{A}^1$. Then the compactifying log structure $\mathcal{M}_{(\mathbb{A}^1|0)}$ is isomorphic to the log structure of $A_{\mathbb{N}}$. The isomorphism arises from $\mathbb{N} \ni 1 \mapsto t \in \mathcal{M}_{(\mathbb{A}^1|0)}$ through the universal property of the log structure associated with the prelog structure $\underline{\mathbb{N}} \to O_{\mathbb{A}^1}$. Thus, $\mathcal{M}_{(\mathbb{A}^1|0)}$ is fine and saturated, in particular (quasi-)coherent. Similarly, the log structure of $A_{\mathbb{N}^2}$ is isomorphic to the compactifying log structure defined by $D = \{xy = 0\} \subset \mathbb{A}^2$. $\Diamond$

---

[3] Note that we actually could define four notions, namely in addition that a Zariski log structure has étale local charts, or that an étale log structure admits charts defined in the Zariski topology. In the latter case, assuming the log structure to be u-integral (a weak condition which is virtually always satisfied), the log structure is actually defined in the Zariski topology and canonically extended to the étale topology. In practice, it usually suffices to consider the two notions of the definition.

Remark 7.28 Not every quasi-coherent log structure is a compactifying log structure: There is no closed subset $D \subset S_0$ of the standard log point which would define its log structure. Conversely, not every compactifying log structure is quasi-coherent. An example is given by $\mathrm{Spec}\,\mathbf{k}[x, y, t, z]/(xy - tz)$ with the compactifying log structure defined by $\{t = 0\}$, but we cannot yet prove this with the material presented so far in this chapter. $\diamond$

7.3.5 The Pull-Back of Log Structures and Strict Morphisms

Given a morphism of schemes $f : X \to Y$ and a log structure on Y, there is a minimal way of putting a log structure on X:

Proposition 7.29 (Pull-Back of Log Structures) *Let $f : X \to Y$ be a morphism of schemes, and let $(\mathcal{M}_Y, \alpha_Y)$ be a log structure on Y, defined either in the Zariski or étale topology. Then there exists a log structure $f_{\log}^* \mathcal{M}_Y \to \mathcal{O}_X$ on X together with a morphism*

$$f : \ (X, f_{\log}^* \mathcal{M}_Y) \to (Y, \mathcal{M}_Y)$$

of log schemes with the following universal property:

Let $(Z, \mathcal{M}_Z)$ be a log scheme, let $h : (Z, \mathcal{M}_Z) \to (Y, \mathcal{M}_Y)$ be a morphism of log schemes, and suppose that we have a morphism $g : Z \to X$ of schemes such that $f \circ g = h$ as morphisms of schemes. Then there is a unique enhancement of g to a morphism of log schemes $(Z, \mathcal{M}_Z) \to (X, f_{\log}^* \mathcal{M}_Y)$ such that $f \circ g = h$ as morphisms of log schemes.

Definition 7.30 (Strict Morphisms) A morphism $f : (X, \mathcal{M}_X) \to (Y, \mathcal{M}_Y)$ is called *strict* if

$$\mathcal{M}_X = f_{\log}^* \mathcal{M}_Y$$

with the identification from the universal property.

If $f : (X, \mathcal{M}_X) \to (Y, \mathcal{M}_Y)$ is strict, then $\overline{\mathcal{M}}_X = f^{-1}(\overline{\mathcal{M}}_Y)$. If $\mathcal{M}_Y$ is u-integral, integral, or saturated, then $f_{\log}^* \mathcal{M}_Y$ has the same property. The pull-back of log structures is compatible with charts in the following sense:

Lemma 7.31 *Let $\beta : P \to \Gamma(Y, \mathcal{O}_Y)$ be a homomorphism of monoids, and let $f : X \to Y$ be a morphism of schemes. Then the log structure $f_{\log}^* \underline{P}_Y^{\log}$ obtained via pull-back of the log structure $\underline{P}_Y^{\log}$ associated with $\beta : P \to \Gamma(Y, \mathcal{O}_Y)$ coincides with the log structure associated with $P \to \Gamma(Y, \mathcal{O}_Y) \to \Gamma(X, \mathcal{O}_X)$.*

Thus, if $\mathcal{M}_Y$ is quasi-coherent, coherent, fine, or fs, then $f_{\log}^* \mathcal{M}_Y$ has the same property. Furthermore, we find the following reformulation of quasi-coherence: A log scheme $(X, \mathcal{M}_X)$ is quasi-coherent if and only if we can find a cover $\{U_i \to X\}_i$ and monoids P_i such that there are strict morphisms $b_i : U_i \to A_{P_i}$, where A_{P_i} is formed over the base ring $\mathbb{Z}$.

Example 7.32 We have a strict morphism $\mathrm{Spec}(\mathbb{N} \to \mathbb{C}) \to A_{\mathbb{N}}$ which is given by $z^1 \mapsto 0$. $\diamond$

7.3.6 Log Structures in the Zariski and in the Étale Topology

We are now ready to discuss the difference between log structures defined in the Zariski respectively étale topology. When $\mathcal{M}_X$ is a *u-integral* log structure in the Zariski topology of a scheme X, then we can define its associated étale log structure $(\mathcal{M}_X)^{\text{ét}}$ by pull-back of the Zariski log structure along étale morphisms $h : U \to X$:

$$\Gamma(U, \mathcal{M}_X^{\text{ét}}) = \Gamma(U, h^*_{\log}\mathcal{M}_X).$$

Conversely, when $\mathcal{M}_X$ is a log structure in the étale topology, then we can just restrict it to Zariski open subsets $U \to X$ and obtain a Zariski log structure $(\mathcal{M}_X)^{\text{Zar}}$. We see easily from the construction that we have

$$((\mathcal{M}_X)^{\text{ét}})^{\text{Zar}} = \mathcal{M}_X$$

because the pull-back along a Zariski open immersion $U \to X$ does not change $\mathcal{M}_X$. However, the converse is not true.

Example 7.33 Let $X = \mathbb{A}^2$ and $D = \{y^2 = x^2 - x^3\}$, i.e., a nodal cubic. Then a chart for $\mathcal{M} := \mathcal{M}^{\text{Zar}}_{(X|D)}$, the compactifying log structure in the Zariski topology, is given by

$$\mathbb{N} \to \mathbb{C}[x, y], \quad 1 \mapsto y^2 - x^2 + x^3.$$

The stalk of the ghost sheaf of this log structure at $0 \in \mathbb{A}^2$ is $\overline{\mathcal{M}}_0 = \mathbb{N}$. In particular, for any étale map $h : U \to X$ and point $u \in U$ with $h(u) = 0$, we have $\overline{h^*_{\log}\mathcal{M}}_u = \mathbb{N}$. Now let specifically $h : U \to X$ be an étale map which separates the two local branches of the node at $0 \in \mathbb{A}^2$, and fix a point $u \in U$ with $h(u) = 0$. Let $\mathcal{N} = \mathcal{M}^{\text{Zar}}_{(U|h^{-1}(D))}$ be again the compactifying log structure in the Zariski topology. Now the two branches of $h^{-1}(D)$ contribute independently to the ghost sheaf, and we have $\overline{\mathcal{N}}_u = \mathbb{N}^2$. This shows $(\mathcal{M}^{\text{Zar}}_{(X|D)})^{\text{ét}} \neq \mathcal{M}^{\text{ét}}_{(X|D)}$ for the compactifying log structure defined in the étale topology. Since we obviously have $\mathcal{M}^{\text{Zar}}_{(X|D)} = (\mathcal{M}^{\text{ét}}_{(X|D)})^{\text{Zar}}$, this shows that the converse of the above statement is not true. $\diamond$

Thus, as long as we are only interested in *u*-integral log structures, log structures in the étale topology are a strict generalization of log structures defined in the Zariski topology. If a log structure $\mathcal{M}_X$ in the Zariski topology is u-integral, integral, or saturated, then $(\mathcal{M}_X)^{\text{ét}}$ has the same property.

Besides that (u-integral) log structures in the étale topology are strictly more general than log structures in the Zariski topology, we have a second and more important reason to prefer log structures in the étale topology: A compactifying log structure defined in the étale topology represents more adequately the local geometry of the embedding $D \subset X$. In the example of the nodal cubic D, only the compactifying log structure in the étale topology detects the node of D through a change in the ghost sheaf, while the ghost sheaf of the one defined in the Zariski topology is $\mathbb{N}$ everywhere on the cubic D. Furthermore, the formation of the compactifying log structure in the Zariski topology does not commute with étale morphisms, a very inconvenient fact in using the construction in geometry, as for example two singularities are usually considered equivalent if they are étale locally isomorphic. This is remedied by working with log structures defined in the étale topology.

Let us now consider the behavior of quasi-coherence with respect to the two topologies. If $\mathcal{M}_X$ is the log structure in the Zariski topology obtained from $\beta \colon \underline{P} \to O_X$, then $(\mathcal{M}_X)^{\text{ét}}$ is the log structure in the étale topology obtained from $\beta \colon \underline{P} \to O_X$, considered as a homomorphism of sheaves in the étale topology. In particular, if $\mathcal{M}_X$ is a quasi-coherent Zariski log structure, then $(\mathcal{M}_X)^{\text{ét}}$ is also quasi-coherent. Only a partial converse is true as well, see [222, III, Cor. 1.4.4]: If $\mathcal{M}_X$ is an integral and (quasi-)coherent log structure in the étale topology, then there is some étale cover $h \colon X' \to X$ and a (quasi-)coherent Zariski log structure $\mathcal{M}'$ on X' such that $\mathcal{M}_X|_{X'} = (\mathcal{M}')^{\text{ét}}$.[4]

Finally, let us provide the following comparison result for compactifying log structures defined in the Zariski respectively étale topology, which is [222, III, Prop. 1.6.5].

Proposition 7.34 *Assume that X is a locally Noetherian and normal scheme, and let $D \subset X$ be such that $X \setminus D$ contains all the associated points of X. Assume that D is pure of codimension 1, and that each irreducible component of D is geometrically unibranch.[5] Then $(\mathcal{M}^{\text{Zar}}_{(X|D)})^{\text{ét}} = \mathcal{M}^{\text{ét}}_{(X|D)}$.*

An important example is given by a toric variety X and a union of (some or all) toric boundary divisors D. Note that the proposition does not assume quasi-coherence of $\mathcal{M}_{(X|D)}$.

---

[4] Even when $\mathcal{M}_X$ is the étale extension of a log structure in the Zariski topology, it can happen, at least a priori, that it admits charts only étale locally so that $\mathcal{M}_X$ is quasi-coherent, but its Zariski version is not. The author does not currently know an example of this.

[5] Normal schemes are geometrically unibranch, so it is sufficient when every irreducible component of D is normal.

7.3.7 Fiber Products and Integral/Saturated Morphisms

Let us now switch to the convention that X denotes a log scheme and $\underline{X}$ denotes its underlying scheme. The category of log schemes admits fiber products, which are constructed as follows, cf. [222, III, Prop. 2.1.2]:

Construction 7.35 Let $f\colon X \to Z$ and $g\colon Y \to Z$ be two morphisms of log schemes. Let $p\colon \underline{X} \times_{\underline{Z}} \underline{Y} \to \underline{X}$ and $q\colon \underline{X} \times_{\underline{Z}} \underline{Y} \to \underline{Y}$ be the two projections from the fiber product of underlying schemes, and let $h\colon \underline{X} \times_{\underline{Z}} \underline{Y} \to \underline{Z}$ be the composition $h = f \circ p = g \circ q$. Then we have morphisms $h_{\log}^{*}\mathcal{M}_Z \to p_{\log}^{*}\mathcal{M}_X$ and $h_{\log}^{*}\mathcal{M}_Z \to q_{\log}^{*}\mathcal{M}_Y$ of log structures on $\underline{X} \times_{\underline{Z}} \underline{Y}$. The push-out of these morphisms in the category of log structures on $\underline{X} \times_{\underline{Z}} \underline{Y}$, i.e., the associated log structure of the push-out in the category of sheaves of monoids, gives the fiber product $X \times_Z Y$ in the category of log schemes. $\diamondsuit$

For lack of reference, we record the following property of the fiber product. We will use it in Lemma 9.48 to show that products of toroidal crossing spaces are toroidal crossing spaces.

Lemma 7.36 *Let $f\colon X \to Z$ and $g\colon Y \to Z$ be two morphisms of log schemes. Let $W = X \times_Z Y$ be the fiber product, and let $p\colon W \to X$ and $q\colon W \to Y$ be the two projections. Let $h = f \circ p = g \circ q$. Then the diagram*

$$
\begin{array}{ccc}
h^{-1}\overline{\mathcal{M}}_Z & \longrightarrow & p^{-1}\overline{\mathcal{M}}_X \\
\downarrow & & \downarrow \\
q^{-1}\overline{\mathcal{M}}_Y & \longrightarrow & \overline{\mathcal{M}}_W
\end{array}
$$

is a push-out square in the category of sheaves of sharp monoids on W.

Proof Let

$$
\mathcal{N} = p_{\log}^{*}\mathcal{M}_X \oplus_{h_{\log}^{*}\mathcal{M}_Z} q_{\log}^{*}\mathcal{M}_Y
$$

be the push-out in the category of sheaves of monoids. Since the functor $\mathcal{M} \mapsto \overline{\mathcal{M}}$ is left adjoint to the inclusion of sheaves of sharp monoids, $\overline{\mathcal{N}}$ is the push-out of $h^{-1}\overline{\mathcal{M}}_Z \to p^{-1}\overline{\mathcal{M}}_X$ and $h^{-1}\overline{\mathcal{M}}_Z \to q^{-1}\overline{\mathcal{M}}_Y$ in the category of sheaves of sharp monoids.

Let $\beta\colon \mathcal{N} \to \mathcal{O}_W$ be the induced homomorphism of sheaves of monoids. Surjectivity of $p_{\log}^{*}\mathcal{M}_X \oplus q_{\log}^{*}\mathcal{M}_Y \to \mathcal{N}$ together with the fact that $p_{\log}^{*}\mathcal{M}_X$ and $q_{\log}^{*}\mathcal{M}_Y$ are log structures implies that $\beta^{-1}(\mathcal{O}_W^{*}) = \mathcal{N}^{*}$. Now $\mathcal{M}_W = \mathcal{N} \oplus_{\mathcal{N}^{*}} \mathcal{O}_W^{*}$ in the category of sheaves of monoids, so we have $\overline{\mathcal{M}}_W = \overline{\mathcal{N}} \oplus_{\{1\}} \{1\} = \overline{\mathcal{N}}$ in the category of sheaves of sharp monoids. $\qquad\square$

If X, Y, and Z are quasi-coherent or coherent, then their fiber product $X \times_Z Y$ is quasi-coherent respectively coherent as well. This is no longer true for fine or fs log schemes, just as push-outs of integral or saturated monoids in **Mon** do not need to be integral or saturated again: Even if X, Y, and Z are all fine, the fiber product $X \times_Z Y$ may not be fine, and if they are all fine and saturated, the fiber product may not be fine and saturated. This situation can be remedied by the following adjunction, which comes from the functors $(\cdot)^{\mathrm{int}}$ and $(\cdot)^{\mathrm{sat}}$ of monoids:

Proposition 7.37 *Let* $\mathbf{LSch}^{\mathrm{coh}}$ *be the category of coherent log schemes,* $\mathbf{LSch}^{\mathrm{fin}}$ *be the category of fine log schemes, and* $\mathbf{LSch}^{\mathrm{fs}}$ *be the category of fine and saturated log schemes.*

(1) *The inclusion* $\mathbf{LSch}^{\mathrm{fin}} \to \mathbf{LSch}^{\mathrm{coh}}$ *has a right adjoint* $X \mapsto X^{\mathrm{int}}$ *such that the canonical map* $X^{\mathrm{int}} \to X$ *is a closed immersion on underlying schemes.*
(2) *The inclusion* $\mathbf{LSch}^{\mathrm{fs}} \to \mathbf{LSch}^{\mathrm{fin}}$ *has a right adjoint* $X \mapsto X^{\mathrm{sat}}$ *such that the canonical map* $X^{\mathrm{sat}} \to X$ *is finite and surjective on underlying schemes.*

Proof See [222, III, Prop. 2.1.5]. $\qquad\qquad\qquad\qquad\qquad\qquad\qquad\qquad\qquad\square$

Note that, in this proposition, the adjoints are only defined for *coherent* log schemes, not for quasi-coherent or more general log schemes. Now $(X \times_Z Y)^{\mathrm{int}}$ is the fiber product in the category of fine log schemes $\mathbf{LSch}^{\mathrm{fin}}$, and $(X \times_Z Y)^{\mathrm{sat}}$ is the fiber product in the category of fs log schemes $\mathbf{LSch}^{\mathrm{fs}}$. Here, for a coherent log schemes X, we write $X^{\mathrm{sat}} := (X^{\mathrm{int}})^{\mathrm{sat}}$.

Unfortunately, these adjunctions change the underlying scheme. In particular, the underlying scheme of $(X \times_Z Y)^{\mathrm{int}}$ or $(X \times_Z Y)^{\mathrm{sat}}$ may not be the fiber product $\underline{X} \times_{\underline{Z}} \underline{Y}$. A partial remedy is achieved with the following definitions:

Definition 7.38 (Integral and Saturated Morphisms) A morphism $f : X \to Y$ of integral log schemes is *integral* if the induced morphism $f^{\flat} : \mathcal{M}_{Y,f(x)} \to \mathcal{M}_{X,x}$ is integral for all points $x \in X$. A morphism $f : X \to Y$ of saturated log schemes is *saturated* if $f^{\flat} : \mathcal{M}_{Y,f(x)} \to \mathcal{M}_{X,x}$ is saturated for all points $x \in X$.

Since a homomorphism $\theta : Q \to P$ of monoids is integral or saturated if and only if the induced homomorphism $\bar{\theta} : \overline{Q} \to \overline{P}$ has the respective property, integrality and saturatedness of a morphism of log schemes can be checked on the ghost sheaves.

Lemma 7.39 *Let* $f : X \to S$ *be an integral morphism of fine log schemes, and let* $b : T \to S$ *be a morphism from a fine log scheme* T. *Then the fiber product* $X \times_S T$, *formed in* **LSch**, *is a fine log scheme, i.e.,* $(X \times_S T)^{\mathrm{int}} = X \times_S T$. *If* $f : X \to S$ *is a saturated morphism of fs log schemes, and if* T *is fine and saturated, then* $X \times_S T$ *is fine and saturated so that* $(X \times_S T)^{\mathrm{sat}} = X \times_S T$.

This allows us to bypass the integralizations and saturations of the fiber product in certain situations. In this monograph, when we consider a morphism of log schemes $f : X \to S$, we will usually assume that it is saturated.

Example 7.40 Let $\theta\colon Q \to P$ be an integral or saturated homomorphism of sharp toric monoids. Then $A_\theta\colon A_P \to A_Q$ is integral respectively saturated. $\qquad\Diamond$

Example 7.41 The diagonal map $\mathbb{N} \to \mathbb{N}^2$, $1 \mapsto (1, 1)$, is integral. The associated map $A_{\mathbb{N}^2} \to A_{\mathbb{N}}$ is the semistable degeneration given by $t \mapsto xy$. The map

$$\mathbb{N}^2 \to \mathbb{N}^2, \quad (1, 0) \mapsto (1, 0), \quad (0, 1) \mapsto (1, 1),$$

is not integral. The associated map $A_{\mathbb{N}^2} \to A_{\mathbb{N}^2}$ is the blow-up of $\mathbb{A}^2$ in a point, restricted to an affine patch of the source. $\qquad\Diamond$

7.4　Logarithmic Derivations and Differential Forms

Consider a smooth and proper algebraic variety $X/\mathbb{C}$ together with a simple normal crossing divisor $D \subset X$. Classically, we can form the sheaf of *differential forms with log poles* $\Omega^1_X(\log D)$: When the divisor is given by $x_1 \cdot \ldots \cdot x_r = 0$ inside $\mathbb{A}^n$, then they are given by

$$O_{\mathbb{A}^n} \cdot \frac{dx_1}{x_1} \oplus \ldots \oplus O_{\mathbb{A}^n} \cdot \frac{dx_r}{x_r} \oplus O_{\mathbb{A}^n} \cdot dx_{r+1} \oplus \ldots \oplus O_{\mathbb{A}^n} \cdot dx_n.$$

They can be extended to a whole de Rham complex $\Omega^\bullet_X(\log D)$ of differential forms with log poles, whose hypercohomology computes the cohomology of $H^k(U^{\mathrm{an}}, \mathbb{C})$, where $U = X \setminus D$. This plays an important role in Deligne's construction of the mixed Hodge structure on $H^k(U^{\mathrm{an}}, \mathbb{C})$. In this section, we discuss the interpretation of $\Omega^1_X(\log D)$ from the perspective of logarithmic geometry.

7.4.1　Logarithmic Derivations

Let X be a log scheme, and let $\mathcal{E}$ be a sheaf of O_X-modules. The notion of a log derivation with values in $\mathcal{E}$ extends the classical notion of derivations. We define directly the relative version:

Definition 7.42 (Log Derivations) Let $f\colon X \to S$ be a morphism of log schemes, and let $\mathcal{E}$ be a sheaf of O_X-modules, not necessarily quasi-coherent. Then a relative *log derivation* with values in $\mathcal{E}$ is a pair (D, Δ) where $D\colon O_X \to \mathcal{E}$ is a derivation, and $\Delta\colon (\mathcal{M}_X, \cdot) \to (\mathcal{E}, +)$ is a homomorphism of sheaves of monoids such that

$$\alpha_X(m) \cdot \Delta(m) = D(\alpha_X(m)),$$

and such that $D(f^\sharp(g)) = 0$ for $g \in f^{-1}O_S$ and $\Delta(f^\flat(m)) = 0$ for $m \in f^{-1}\mathcal{M}_S$. We denote the sheaf of relative log derivations with values in $\mathcal{E}$ by $\mathcal{D}er_{X/S}(\mathcal{E})$. The sheaf of relative log derivations with values in O_X is called the *log tangent sheaf* and is denoted by $\Theta^1_{X/S}$.

When X carries the trivial log structure $M_X = O_X^*$, then $\Delta(m) = \alpha_X(m)^{-1} \cdot D(\alpha_X(m))$, and a relative log derivation is nothing but a relative classical derivation with values in $\mathcal{E}$.

The sheaf $\mathcal{D}er_{X/S}(\mathcal{E})$ of relative derivations is an O_X-module under

$$g \cdot (D, \Delta) := (g \cdot D, g \cdot \Delta).$$

If $\phi \colon \mathcal{E} \to \mathcal{F}$ is a homomorphism of O_X-modules, and if $(D, \Delta) \colon (O_X, M_X) \to \mathcal{E}$ is a relative log derivation, then $(\phi \circ D, \phi \circ \Delta) \colon (O_X, M_X) \to \mathcal{F}$ is a relative log derivation as well.

We always have the forgetful map

$$\Theta^1_{X/S} \to \mathcal{T}_{X/S}$$

into the sheaf of relative classical derivations $\mathcal{T}_{X/S}$. If every $\alpha_X(m)$ is a non-zero divisor, then this map is injective. In the case of compactifying log structures, $\Theta^1_{X/S}$ is often easy to compute:

Lemma 7.43 *Let $X/\mathbf{k}$ be a normal integral scheme over a field $\mathbf{k}$, and let $C \subset X$ be a reduced Weil divisor. Endow X with the compactifying log structure defined by C. Then $\Theta^1_{X/\mathbf{k}} \to \mathcal{T}_{X/\mathbf{k}}$ is injective, and $D \in \mathcal{T}_{X/\mathbf{k}}$ is in the image of the inclusion if and only if $D(\mathcal{I}_C) \subseteq \mathcal{I}_C$ for the ideal $\mathcal{I}_C$ defining $C \subset X$.*

Proof See [125, Ex. 1.4]. $\square$

Example 7.44 Consider $X = (\mathbb{A}^1 | 0)$ with the compactifying log structure. Then $\Theta^1_{X/\mathbf{k}}$ is freely generated by $t \partial_t$. $\diamondsuit$

7.4.2   Logarithmic Differential Forms

Just as in the case of classical derivations, there is a universal logarithmic derivation. This defines the sheaf of *log differential forms*.

Proposition 7.45 *Let $f \colon X \to S$ be a morphism of log schemes. Then there is a sheaf of O_X-modules $\Omega^1_{X/S}$ together with a log derivation*

$$(d, \delta) \colon \ (O_X, M_X) \to \Omega^1_{X/S}$$

with values in $\Omega^1_{X/S}$ such that the map

$$\mathcal{H}om(\Omega^1_{X/S}, \mathcal{E}) \to \mathcal{D}er_{X/S}(\mathcal{E})$$

induced by composition is an isomorphism for all O_X-modules $\mathcal{E}$.

Proof Explicitly, we have

$$\Omega^1_{X/S} = \left(\Omega^1_{\underline{X}/\underline{S}} \oplus ((O_X, +) \otimes_{\mathbb{Z}} \mathcal{M}^{\mathrm{gp}}_X) \right) / \mathcal{R}$$

where $\mathcal{R}$ the O_X-submodule generated by

$$(d\alpha_X(m), -\alpha_X(m) \otimes m) \ \text{ for } m \in \mathcal{M}_X, \quad (0, 1 \otimes f^*m) \ \text{ for } m \in f^{-1}\mathcal{M}_S.$$

Here, $\Omega^1_{\underline{X}/\underline{S}}$ is the sheaf of relative classical differential forms. See [222, IV, Thm. 1.2.4] for more discussion. □

Unfortunately, $\Omega^1_{X/S}$ is not always (quasi-)coherent. Our standard example of this situation is the following.

Example 7.46 Let $X = \mathrm{Spec}\,\mathbf{k}[x, y, z, t]/(xy - tz)$ endowed with the compactifying log structure defined by $\{t = 0\}$. Let $S = \mathrm{Spec}\,\mathbf{k}[t]$ with the compactifying log structure defined by $\{t = 0\}$, and let $f : X \to S$ be the obvious map. Then $\Omega^1_{X/S}$ is not quasi-coherent in 0. Cf. [125, Ex. 1.11].[6]   ◇

We will see later how we deal with this situation. One of the reasons why coherent log schemes are so convenient is that, for coherent log schemes, $\Omega^1_{X/S}$ is always quasi-coherent. This also completes the proof that the example is an example of an non-coherent log scheme.

Proposition 7.47 *Let $f : X \to S$ be a morphism of coherent log schemes. Then $\Omega^1_{X/S}$ is quasi-coherent. If $\underline{f} : \underline{X} \to \underline{S}$ is of finite type, then $\Omega^1_{X/S}$ is of finite type. If $\underline{f}$ is of finite presentation, then $\Omega^1_{X/S}$ is of finite presentation.*

Proof See [222, IV, Cor. 1.2.8]. □

7.4.3 *Functoriality and Fundamental Exact Sequences*

Let $f : X \to Y$ and $g : Y \to S$ be morphisms of log schemes. Then we have canonical maps $f^*\Omega^1_{Y/S} \to \Omega^1_{X/S}$ and $\Omega^1_{Y/S} \to f_*\Omega^1_{Y/S}$, as well as $\Omega^1_{X/S} \to \Omega^1_{X/Y}$. They fit into the usual exact sequence

$$f^*\Omega^1_{Y/S} \to \Omega^1_{X/S} \to \Omega^1_{X/Y} \to 0.$$

If $i : X \to Y$ is a strict closed immersion defined by an ideal $\mathcal{I} \subseteq O_Y$, then we have an exact sequence

$$\mathcal{I}/\mathcal{I}^2 \to i^*\Omega^1_{Y/S} \to \Omega^1_{X/S} \to 0.$$

---

[6] A detailed proof of this fact can be found in the author's master thesis.

Given a commutative diagram

$$
\begin{array}{ccc}
Y & \xrightarrow{\;\;c\;\;} & X \\
\downarrow{\scriptstyle g} & & \downarrow{\scriptstyle f} \\
T & \xrightarrow{\;\;b\;\;} & S
\end{array}
$$

of log schemes, we have a canonical map

$$
c^*\Omega^1_{Y/T} \to \Omega^1_{X/S}.
$$

If the diagram is Cartesian in either **LSch**, **LSch**^{fin}, or **LSch**^{fs}, then this map is an isomorphism.

7.4.4 The Logarithmic de Rham Complex

Much like in the classical case, we can construct a de Rham complex for a morphism $f : X \to S$ of coherent log schemes.

Proposition 7.48 *Let $f : X \to S$ be a morphism of coherent log schemes, and let $\Omega^i_{X/S}$ be the i-th exterior power of $\Omega^1_{X/S}$. Then there is a unique collection of homomorphisms of sheaves of Abelian groups*

$$
\partial^i : \quad \Omega^i_{X/S} \to \Omega^{i+1}_{X/S}
$$

for $i \geq 0$ satisfying the following conditions:

(a) *$\partial^0 = d : O_X \to \Omega^1_{X/S}$ is the classical part of the universal derivation;*
(b) *$\partial^1 \delta(m) = 0$ for $m \in M_X$, where $\delta : M_X \to \Omega^1_{X/S}$ is the log part of the universal derivation;*
(c) *$\partial^{i+1}\partial^i(\alpha) = 0$ for $\alpha \in \Omega^i_{X/S}$;*
(d) *$\partial^{i+j}(\alpha \wedge \beta) = \partial^i(\alpha) \wedge \beta + (-1)^i \alpha \wedge \partial^j(\beta)$ for $\alpha \in \Omega^i_{X/S}$ and $\beta \in \Omega^j_{X/S}$.*

Proof See [222, V, Prop. 2.1.1]. $\qquad\qquad\qquad\qquad\qquad\qquad\qquad\qquad\quad\square$

We are usually interested in the case where S is Noetherian and $f : X \to S$ of finite type so that the log de Rham complex $\Omega^\bullet_{X/S}$ is actually a bounded complex.

When we have a commutative diagram of log schemes as above, then we have a morphism

$$
\Omega^\bullet_{X/S} \to c_*\Omega^\bullet_{Y/T}
$$

of complexes. If the diagram is Cartesian, the adjoint map $c^*\Omega^i_{X/S} \to \Omega^i_{Y/T}$ is an isomorphism for each $i \geq 0$, but note that we cannot form a differential $c^*\partial$ on $c^*\Omega^\bullet_{X/S}$ from mere abstract non-sense.

7.5   Log Smooth Morphisms

Log smooth morphisms are defined in analogy with Grothendieck's geometric-functorial characterization of smoothness, also known as the infinitesimal lifting criterion. Thus, we have to define thickenings of log schemes first.

Definition 7.49 (Log Thickenings) A *log thickening* of (finite) order $n \geq 1$ is a strict closed immersion $i \colon T_0 \to T$ of log schemes such that the ideal $\mathcal{I}$ of T_0 in T is nilpotent with $\mathcal{I}^{n+1} = 0$, and the subgroup $1 + \mathcal{I}$ of $O_T^* \cong \mathcal{M}_T^*$ operates freely on $\mathcal{M}_T$.

A log thickening induces an isomorphism of underlying topological spaces $|T_0| \cong |T|$. The log scheme T is coherent, integral, fine, or fine and saturated if and only if T_0 is, see [222, IV, Prop. 2.1.3].

Definition 7.50 (Deformations) Consider a solid diagram

$$
\begin{array}{ccc}
T_0 & \xrightarrow{\ g\ } & X \\[2pt]
{\scriptstyle i}\big\downarrow & {\scriptstyle k}\ \ \nearrow & \big\downarrow {\scriptstyle f} \\[2pt]
T & \xrightarrow[\ h\ ]{} & S
\end{array}
$$

of log schemes where $i \colon T_0 \to T$ is a log thickening of some finite order n. A *deformation of g to T* is a morphism of schemes $k \colon T \to X$ which makes the above diagram commutative. Deformations of g to T form a sheaf $\mathrm{Def}_{X/S}(g, T)$ in the étale topology of T.

If $i \colon T_0 \to T$ is a log thickening of first order, then $\mathrm{Def}_{X/S}(g, T)$ has a particularly easy structure:

Theorem 7.51 *Let $i \colon T_0 \to T$ be a first-order log thickening, and assume that we have a diagram as in the above definition. Let $\mathcal{I}_T$ be the ideal sheaf defining the inclusion $i \colon T_0 \to T$, considered as an O_{T_0}-module. Then there is an action of $\mathrm{Der}_{X/S}(g_*\mathcal{I}_T)$ on $g_*\mathrm{Def}_{X/S}(g, T)$ which turns the latter into a pseudo-torsor. Similarly, there is an action of $\mathcal{H}om(g^*\Omega^1_{X/S}, \mathcal{I}_T)$ on $\mathrm{Def}_{X/S}(g, T)$ which turns the latter into a pseudo-torsor.*

Proof See [222, IV, Thm. 2.2.2] for the first statement. The proof relies in particular on the above assumption that $1 + \mathcal{I}_T$ operates freely on $\mathcal{M}_T$. For the second statement, use additionally the following: A prelog ringed site is a site $\mathcal{S}$ together with a sheaf of rings O and a sheaf of monoids $\mathcal{M}$ together with a map $\alpha \colon \mathcal{M} \to (O, \cdot)$. For a map of prelog ringed sites $(\mathcal{M}_1, O_1) \to (\mathcal{M}_2, O_2)$, there is a relative module of differentials $\Omega^1_{(O_2, \mathcal{M}_2)/(O_1, \mathcal{M}_1)}$. Then if $f \colon \mathcal{T} \to \mathcal{S}$ is a morphism of sites, we have

$$f^{-1}\Omega^1_{(O_2, \mathcal{M}_2)/(O_1, \mathcal{M}_1)} = \Omega^1_{(f^{-1}O_2, f^{-1}\mathcal{M}_2)/(f^{-1}O_1, f^{-1}\mathcal{M}_1)}.$$

The comparison isomorphism is induced by the universal property.   □

Next, we define logarithmic smoothness. In the definition, we have to make a number of choices, which do not affect the notion of smoothness in the end, at least for morphisms between coherent log schemes. Here is our variant.[7]

---

[7] We could strengthen this definition by requiring the existence of $k \in \mathrm{Def}_{X/S}(g, T)$ Zariski locally or on every affine open subset, or we could weaken this definition by requiring the existence only for first-order thickenings $i \colon T_0 \to T$ where T_0 and T are coherent. If $f \colon X \to S$ is a morphism between log schemes defined in the Zariski topology, we also have a choice of considering only $i \colon T_0 \to T$ defined in the Zariski topology, or allowing $i \colon T_0 \to T$ to carry étale log structures which cannot be defined in the Zariski topology. The following lemma essentially explains why these choices do not matter, at least for coherent log schemes.

Lemma 7.52 *Let $f \colon X \to S$ be a morphism between coherent log schemes. Then $f \colon X \to S$ is formally log smooth if and only if the following condition holds: Let $g \colon U \subseteq X$ be an affine Zariski open subset, and let $i \colon U \to V$ be a first-order log thickening over S via the map $h \colon V \to S$. Then there is a map $r \colon V \to U$ with $r \circ i = \mathrm{id}_U$ and $h \circ i \circ r = h$.*

Proof First assume that $f \colon X \to S$ is formally log smooth in the sense of the above definition. Then $r \colon V \to U$ exists étale locally on V. By Theorem 7.51, we know that such $r \colon V \to U$ form a torsor in the étale topology under the Abelian sheaf $\mathcal{G} = \mathcal{H}om(\Omega^1_{X/S}|_U, \mathcal{I}_V)$. Since X and S are coherent log schemes, $\Omega^1_{X/S}$ is quasi-coherent, so $H^1(U, \mathcal{G}) = 0$, and $r \colon V \to U$ exists globally on V.

To illustrate the converse, let us assume that $f \colon X \to S$ is defined in the Zariski topology, and that the condition holds for Zariski log schemes. Assume that we have a diagram as in Definition 7.50 with $i \colon T_0 \to T$ defined only in the étale topology, and without any coherence assumption. By shrinking X and T_0 in the Zariski topology, we can assume that $g \colon T_0 \to X$ is an affine morphism between affine schemes. Then we can apply the construction in [222, IV, Ex. 2.1.7] to obtain a diagram

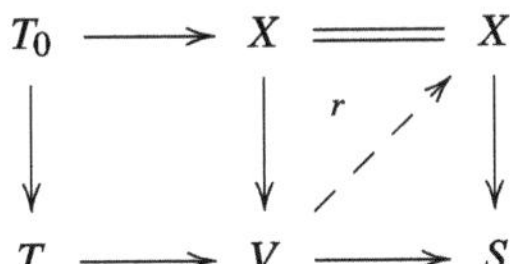

where $X \to V$ is a first-order log thickening. Since the log structure of X can be defined in the Zariski topology, the same is true for the thickening V although its log structure is a priori constructed in the étale topology. Then we can apply the condition, interpreted purely in the Zariski topology, and obtain $r \colon V \to X$. Composition with $T \to V$ yields the desired deformation. Thus, deformations exist locally in the Zariski topology of T. As in the first direction, we see that they exist on every Zariski open subset of T from coherence of $\Omega^1_{X/S}$.   □

Definition 7.53 (Formal Log Smoothness) Let $f \colon X \to S$ be a morphism of log schemes in the étale topology. Consider a solid diagram as in Definition 7.50. Then we say that:

(1) $f \colon X \to S$ is *formally log smooth* if, for all such diagrams with a first-order thickening $i \colon T_0 \to T$, a deformation $k \in \mathrm{Def}_{X/S}(g, T)$ exists étale locally on T;

(2) $f \colon X \to S$ is *formally log unramified* if any two such deformations $k \in \mathrm{Def}_{X/S}(g, T)$ are identical;

(3) $f \colon X \to S$ is *formally log étale* if it is formally log smooth and formally log unramified.

We say that $f \colon X \to S$ is *log unramified* if it is formally log unramified and $\underline{f} \colon \underline{X} \to \underline{S}$ is of finite type. We say that $f \colon X \to S$ is *log smooth* respectively *log étale* if it is formally log smooth respectively formally log étale, $\underline{f}$ is of finite presentation, and X and S are coherent log schemes.

If X and S are coherent, then $\Omega^1_{X/S}$ is quasi-coherent, and, in the formally log smooth case, the deformation $k \in \mathrm{Def}_{X/S}(g, T)$ exists whenever T_0 (or, equivalently, T) is affine by Theorem 7.51.

Any composition of log smooth, log étale, or log unramified morphisms has the same property again. All three notions are stable under base change in the category **LSch** of all log schemes. If $f \colon X \to S$ is a strict morphism of coherent log schemes and S is u-integral, then it is log smooth, log étale, respectively log unramified if and only if $\underline{f} \colon \underline{X} \to \underline{S}$ is smooth, étale, respectively unramified, cf. [222, IV, Prop. 3.1.6].

7.5.1   The Local Structure of Log Smooth Morphisms

Let $\theta \colon Q \to P$ be a homomorphism of finitely generated monoids. Then we have a very convenient criterion for smoothness, étaleness, and unramifiedness of $A_\theta \colon A_P \to A_Q$.

Theorem 7.54 *Let* $\theta \colon Q \to P$ *be a homomorphism between finitely generated monoids. Let* $A_\theta \colon A_P \to A_Q$ *be the corresponding morphism of coherent log schemes, defined over the base ring* R.

(1) *The morphism* $A_\theta \colon A_P \to A_Q$ *is log smooth if and only if the kernel and the torsion part of the cokernel of* $\theta^{\mathrm{gp}} \colon Q^{\mathrm{gp}} \to P^{\mathrm{gp}}$ *are finite groups whose order is invertible in* R.

(2) *The morphism* $A_\theta \colon A_P \to A_Q$ *is log étale if and only if the kernel and the cokernel of* $\theta^{\mathrm{gp}} \colon Q^{\mathrm{gp}} \to P^{\mathrm{gp}}$ *are finite groups whose order is invertible in* R.

(3) *The morphism* $A_\theta \colon A_P \to A_Q$ *is log unramified if and only if the cokernel of* $\theta^{\mathrm{gp}} \colon Q^{\mathrm{gp}} \to P^{\mathrm{gp}}$ *is a finite group whose order is invertible in* R.

Proof See [222, IV, Thm. 3.1.7, Thm. 3.1.8]. $\qquad\qquad\qquad\qquad\qquad\qquad\qquad\square$

Example 7.55 Let $R = \mathbf{k}$ be a field of characteristic zero. Let $0 \neq \rho \in \mathbb{N}^r$. Then the map $A_\theta \colon A_{\mathbb{N}^r} \to A_{\mathbb{N}}$ corresponding to $\theta \colon \mathbb{N} \to \mathbb{N}^r$ with $\theta(1) = \rho$ is log smooth. In other words, the map $f \colon \mathbb{A}^r \to \mathbb{A}^1$, $t \mapsto x_1^{m_1} \cdot \ldots \cdot x_r^{m_r}$, is log smooth as soon as we have $m_i \geq 1$ for some $1 \leq i \leq r$. $\Diamond$

Example 7.56 The map $A_\theta \colon A_{\mathbb{N}^2} \to A_{\mathbb{N}^2}$ for

$$\theta \colon \ \mathbb{N}^2 \to \mathbb{N}^2, \quad (1,0) \mapsto (1,0), \quad (0,1) \mapsto (1,1),$$

is log étale for any base ring R. In particular, a log étale morphisms does not need to be either flat or unramified in the classical sense. $\Diamond$

Example 7.57 Let X be a coherent log scheme. Then the canonical maps $X^{\mathrm{sat}} \to X^{\mathrm{int}}$ and $X^{\mathrm{int}} \to X$ are log étale, see [222, IV, Cor. 3.1.11]. $\Diamond$

The somewhat surprising fact is that these are essentially all log smooth morphisms. We say that $A_\theta \colon A_P \to A_Q$ is a *local model* for log smooth morphisms. More precisely, we have the following result which goes back to K. Kato's seminal article [162] and is thus also called *Kato's toroidal characterization of log smoothness*. It has several versions; the basic version is as follows.

Theorem 7.58 (Toroidal Characterization of Log Smoothness I) *Let $f \colon X \to S$ be a log smooth morphism of fine log schemes over a base ring R. Let $a \colon S \to A_Q$ be a strict morphism with fine monoid Q, and let $\bar{x} \in X$ be a geometric point. Then, after replacing X and S with appropriate étale neighborhoods of $\bar{x}$ and $f(\bar{x})$, we can find an injective homomorphism $\theta \colon Q \to P$ of fine monoids such that the torsion part of the cokernel of $\theta^{\mathrm{gp}} \colon Q^{\mathrm{gp}} \to P^{\mathrm{gp}}$ is a finite group whose order is invertible in O_X, and a commutative diagram*

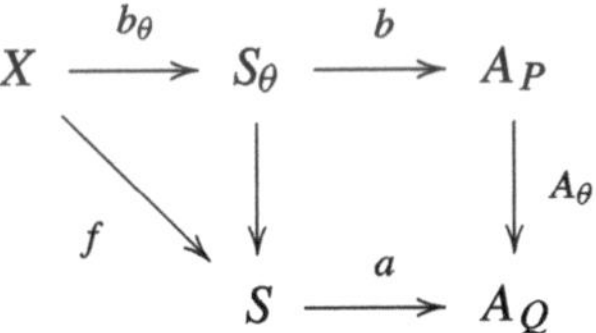

where the square is Cartesian in the category of log schemes, and $b_\theta \colon X \to S_\theta$ is strict and étale.

Proof See [222, IV, Thm. 3.3.1]. $\Box$

For our purposes, the following version is the most important one.

Theorem 7.59 (Toroidal Characterization of Log Smoothness II) *Assume that $f \colon X \to S$ is a log smooth and saturated morphism of fs log schemes, and let $\bar{x} \to X$ be a geometric point with image $\bar{s} = f(\bar{x})$. Let $q \colon (S', \bar{s}') \to (S, \bar{s})$ be a strict étale neighborhood of $\bar{s}$, let Q be a sharp toric monoid, and let*

$$a \colon \ S' \to A_Q = \mathrm{Spec}(Q \to \mathbb{Z}[Q])$$

be a strict morphism such that $Q \to \overline{\mathcal{M}}_{S,\bar{s}}$ is an isomorphism. Then, there is a commutative diagram of log schemes

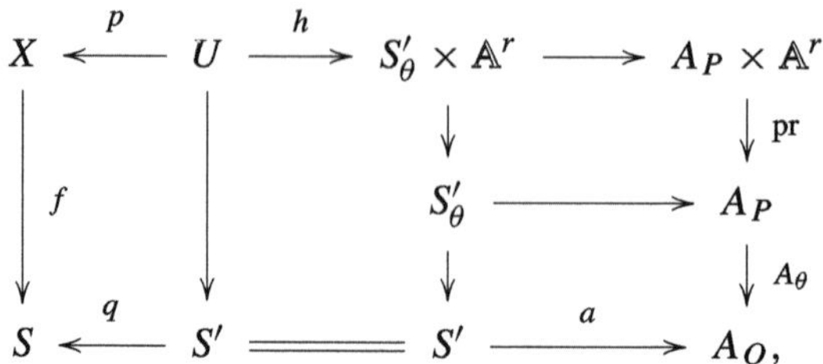

where:

(a) $\theta\colon Q \to P$ *is an injective and saturated homomorphism of sharp toric monoids;* $A_\theta\colon A_P \to A_Q$ *is the associated map of log schemes;*
(b) $\mathrm{pr}\colon A_P \times \mathbb{A}^r \to A_P$ *is the strict and smooth projection for some* $r \geq 0$;
(c) $p\colon (U, \bar{u}) \to (X, \bar{x})$ *is a strict étale neighborhood of* $\bar{x}$;
(d) $h\colon U \to S'_\theta \times \mathbb{A}^r$ *is strict étale;*
(e) *the two right squares are Cartesian in the category of log schemes;*
(f) *the induced map* $P \to \overline{\mathcal{M}}_{X,\bar{x}}$ *is an isomorphism.*

In particular, $\theta\colon Q \to P$ *can be identified with* $f^\flat\colon \overline{\mathcal{M}}_{S,\bar{s}} \to \overline{\mathcal{M}}_{X,\bar{x}}$.

Proof See [222, IV, Thm. 3.3.3] and note that f is s-injective because it is saturated and local, hence exact. Furthermore, saturatedness guarantees that $\mathcal{M}^{\mathrm{gp}}_{X/S,\bar{x}}$ is torsion-free. $\qquad\square$

7.5.2 Differential Forms and Log Smoothness

If $f\colon X \to S$ is a log smooth morphism of coherent log schemes, then $\Omega^1_{X/S}$ is locally free of finite type, see [222, IV, Prop. 3.2.1]. However, the converse is not true. For example, $\Omega^1_{X/S}$ is locally free whenever we can enhance $f\colon X \to S$ to what is called an *ideally log smooth* morphism of *idealized log schemes*. Essentially, this means that we have chosen sheaves of ideals $\mathcal{K}_X \subseteq \mathcal{M}_X$ and $\mathcal{K}_S \subseteq \mathcal{M}_S$ such that $f^\flat\colon f^{-1}\mathcal{M}_S \to \mathcal{M}_X$ is compatible with these sheaves of ideals, see [222, III, 1.3] for more discussion. A concrete example is given by

$$S_0 = \mathrm{Spec}(\mathbb{N} \to \mathbb{C}) \to \mathrm{Spec}(0 \to \mathbb{C}) = *,$$

the map from the standard log point to the trivial log point. Here, we have $\Omega^1_{S_0/*} \cong \mathcal{O}_{S_0}$. In particular, the rank of $\Omega^1_{X/S}$ does not need to be the relative dimension of $\underline{f}\colon \underline{X} \to \underline{S}$. The situation is better when $f\colon X \to S$ is a log smooth and *saturated* morphism of fs log schemes. In this case, there is an open subset $W \subseteq X$, dense in every fiber, such that $f : W \to S$ is strict and log smooth, hence smooth.

The following result is another nice illustration of the behavior of log smooth morphisms.

Proposition 7.60 *Let $f : X \to Y$ and $g : Y \to S$ be morphisms of coherent log schemes. Consider the exact sequence*

$$f^*\Omega^1_{Y/S} \xrightarrow{s} \Omega^1_{X/S} \xrightarrow{t} \Omega^1_{X/Y} \to 0.$$

(1) *If $f : X \to Y$ is log smooth, then s is injective and locally split.*
(2) *If $g \circ f : X \to S$ is smooth, and s is injective and locally split, then f is log smooth.*

Proof See [222, IV, Prop. 3.2.3]. $\qquad\qquad\qquad\qquad\qquad\qquad\qquad\qquad\square$

7.5.3 Smoothness, Integrality, and Saturatedness

Let $f : X \to S$ be a morphism of finite type between locally Noetherian fine log schemes. Here, of course, a log scheme is called (locally) Noetherian if its underlying scheme has this property. Assume that we already know that $f : X \to S$ is log smooth. In general, the underlying morphism $\underline{f} : \underline{X} \to \underline{S}$ does not need to be flat. An example is given by $A_{\mathbb{N}^2} \to A_{\mathbb{N}^2}$ defined via $(1, 0) \mapsto (1, 0)$ and $(0, 1) \mapsto (1, 1)$—this map is even log étale but not flat. However, when we know that $f : X \to S$ is integral, then it is proven in, for example, [222, IV, Thm. 4.3.5] that $\underline{f} : \underline{X} \to \underline{S}$ is flat. Moreover, when X and Y are not only fine but fs log schemes, then the fibers of f are Cohen–Macaulay.[8]

Now suppose that $f : X \to S$ is a log smooth and integral morphism of fine and saturated log schemes. Then $f : X \to S$ is saturated if and only if the fibers of $\underline{f} : \underline{X} \to \underline{S}$ are reduced, and this, in turn, is the case if and only if the fibers of $\underline{f} : \underline{X} \to \underline{S}$ are generically reduced, i.e., satisfy the condition (R_0). See [222, IV, Thm. 4.3.6] for a proof of this result which was first proven by Tsuji.

Example 7.61 The log smooth and integral morphism $A_{\mathbb{N}^2} \to A_{\mathbb{N}}$ given by $1 \mapsto (1, 1)$ has reduced fibers, so it is saturated. The log smooth and integral morphism $A_{\mathbb{N}^2} \to A_{\mathbb{N}}$ given by $1 \mapsto (1, 2)$ has a non-reduced fiber, so it is not saturated. $\qquad \Diamond$

---

[8] The same theorem also states that $f : X \to S$ is *log flat*. Logarithmic flatness is the logarithmic analog of flatness, defined by imitating Kato's toroidal characterization of smoothness with étaleness of $X \to S_\theta$ replaced with flatness, and log smoothness of $A_\theta : A_P \to A_Q$ replaced with the condition that $\theta : Q \to P$ is injective. The local models are required to exist fppf locally. For the purpose of the theory in this monograph, log flatness does not play an independent role from log smoothness, so we do not discuss it here.

7.6 Log Smooth Deformation Theory

In this section, we give a slightly more expanded review of F. Kato's log smooth deformation theory than in the introduction.

Let $\mathbf{k}$ be a field of characteristic 0. We fix a sharp toric monoid Q. This gives rise to a log point $S_0 = \mathrm{Spec}(Q \to \mathbf{k})$. Here, the log ring $Q \to \mathbf{k}$ is given by $0 \mapsto 1$ and $q \mapsto 0$ for $q \neq 0$. Explicitly, we have $\mathcal{M}_{S_0}(*) = Q \oplus \mathbf{k}^*$; the ghost sheaf is given by $\overline{\mathcal{M}}_{S_0}(*) = Q$.

The object that we wish to deform is an integral and log smooth morphism $f_0 \colon X_0 \to S_0$ of fs log schemes which we usually assume to be separated. The underlying scheme of X_0 may or may not be reduced, and we have seen in Chap. 7 that it is reduced if and only if $f_0 \colon X_0 \to S_0$ is saturated.

Just as in classical deformation theory, our goal is to classify *infinitesimal* log smooth deformations of $f_0 \colon X_0 \to S_0$, i.e., Cartesian diagrams

$$
\begin{array}{ccc}
X_0 & \xrightarrow{\;i\;} & X \\
\Big\downarrow{\scriptstyle f_0} & & \Big\downarrow{\scriptstyle f} \\
S_0 & \xrightarrow{\;i\;} & S
\end{array}
$$

of log schemes where $f \colon X \to S$ is log smooth and $i \colon S_0 \to S$ is a log thickening as in Definition 7.49. Since $S_0 \to S$ is a log thickening, the fact that S_0 is fine and saturated implies that S is fine and saturated as well. Similarly, X is fine and saturated. By Cartesian, we mean that the diagram should be Cartesian in the category of all log schemes. However, since $i \colon S_0 \to S$ is strict, this is equivalent to asking for Cartesianity in $\mathbf{LSch}^{\mathrm{fs}}$.

We assume that all our schemes are defined over $\mathbf{k}$. Since $i \colon S_0 \to S$ is a thickening on underlying schemes, we find $\underline{S} = \mathrm{Spec}\,A$ for a local Artinian $\mathbf{k}$-algebra A with residue field $\mathbf{k}$. One can then show that there is a map $Q \to (A, \cdot)$ of monoids whose composition with $A \to \mathbf{k}$ is the given map $Q \to \mathbf{k}$, and such that $S = \mathrm{Spec}(Q \to A)$.

From the monoid Q, we obtain the monoid algebra $\mathbf{k}[Q] := \bigoplus_{q \in Q} \mathbf{k} \cdot z^q$ with multiplication $z^q \cdot z^{q'} := z^{q+q'}$. Since Q is sharp, $\mathfrak{m}_Q = \mathbf{k}[Q^+] := \bigoplus_{q \in Q^+} \mathbf{k} \cdot z^q$ is a maximal ideal, where $Q^+ = Q \setminus Q^* = Q \setminus \{0\}$. The completion of $\mathbf{k}[Q]$ in $\mathbf{k}[Q^+]$ is

$$
\Lambda := \mathbf{k}[\![Q]\!] = \prod_{q \in Q} \mathbf{k} \cdot z^q.
$$

Then the monoid homomorphism $Q \to (A, \cdot)$ is equivalent to a ring homomorphism $\mathbf{k}[\![Q]\!] \to A$ over $\mathbf{k}$. Thus, a log thickening $i \colon S_0 \to S$ is equivalent to an Artinian local $\mathbf{k}[\![Q]\!]$-algebra with residue field $\mathbf{k}$. From this datum, we obtain S

back as $S = S_A := \mathrm{Spec}(Q \to \mathbf{k}[\![Q]\!] \to A)$. We denote the category of Artinian local $\mathbf{k}[\![Q]\!]$-algebras with residue field $\mathbf{k}$ by

$$\mathbf{Art}_Q \quad \text{or} \quad \mathbf{Art}_{\mathbf{k}[\![Q]\!]}.$$

Special objects in this category are

$$A_0 := \mathbf{k}, \quad A_k := \mathbf{k}[\![Q]\!]/\mathfrak{m}_Q^{k+1}, \quad \text{and} \quad A_\varepsilon := \mathbf{k}[\varepsilon]/(\varepsilon^2)$$

with $0 \mapsto 1$ and $q \mapsto 0$ for $q \neq 0$. Note that $A_1 \neq A_\varepsilon$ unless $Q = 0$. We denote the corresponding log schemes by

$$S_0 = \mathrm{Spec}(Q \to A_0), \quad S_k = \mathrm{Spec}(Q \to A_k), \quad \text{and} \quad S_\varepsilon = \mathrm{Spec}(Q \to A_\varepsilon).$$

For a general $A \in \mathbf{Art}_Q$, we write $S_A := \mathrm{Spec}(Q \to A)$.

We encode the log smooth deformations of $f_0 \colon X_0 \to S_0$ in the *log smooth deformation functor*

$$\mathrm{LD}_{X_0/S_0} \colon \quad \mathbf{Art}_Q \to \mathbf{Set}.$$

This is a covariant functor which sends $A \in \mathbf{Art}_Q$ to the set of isomorphism classes of Cartesian diagrams as above, where two log smooth deformations $f_1 \colon X_1 \to S$ and $f_2 \colon X_2 \to S$ are isomorphic if there is an isomorphism of log schemes $\phi \colon X_1 \to X_2$ with $\phi \circ i_1 = i_2$ and $f_2 \circ \phi = f_1$. On ring homomorphisms in $\mathbf{Art}_Q$, this functor is defined by fiber products, which preserve log smoothness.

Example 7.62 Consider the saturated and log smooth morphism

$$f \colon \quad X = (\mathbb{A}^2 | \{xy = 0\}) \to (\mathbb{A}^1 | \{t = 0\}), \quad t \mapsto xy.$$

Then the central fiber $f_0 \colon X_0 \to S_0$, obtained via base change along $S_0 \to \mathbb{A}^1$, is saturated and log smooth over the standard log point S_0. For any Artinian local $\mathbf{k}[\![t]\!]$-algebra A with residue field $\mathbf{k}$, the base change along $S_A \to \mathbb{A}^1$ is a log smooth deformation of $f_0 \colon X_0 \to S_0$. This example also illustrates the difference between A_1 and A_ε: Namely, the underlying scheme of the corresponding log smooth deformation is $\underline{X}_1 = \mathrm{Spec}\,\mathbf{k}[x, y]/(x^2 y^2)$ respectively $\underline{X}_\varepsilon = \mathrm{Spec}\,\mathbf{k}[x, y, t]/(t^2, xy)$. Thus, different flat deformations can become log smooth over different bases.   ◇

Example 7.63 Let $\theta \colon Q \to P$ be an injective and saturated homomorphism of sharp toric monoids. Then $A_\theta \colon A_P \to A_Q$ is log smooth and saturated, in particular integral. For any $A \in \mathbf{Art}_Q$, we have a strict morphism $S_A \to A_Q$. Base change along $S_0 \to A_Q$ yields a log smooth family $f_0 \colon L_0(P/Q) \to S_0$, and the base change $f \colon L_A(P/Q) \to S_A$ is a log smooth deformation thereof.   ◇

Since both $i \colon S_0 \to S$ and $i \colon X_0 \to X$ are strict, integrality of $f_0 \colon X_0 \to S_0$ implies integrality of $f \colon X \to S$. Since we have assumed that $f \colon X \to S$ is log smooth, this implies that $\underline{f} \colon \underline{X} \to \underline{S}$ is a flat morphism of schemes. This is not in general true when we drop the integrality assumption.

In [160, Rem. 8.2], F. Kato makes a point that log smoothness of $f \colon X \to S$ is not automatic, i.e., there are Cartesian diagrams as above where $f \colon X \to S$ is not log smooth. However, we have the following nice criterion:

Lemma 7.64 *Let $f_0 \colon X_0 \to S_0$ be log smooth and integral, and let $f \colon X \to S$ be a morphism of log schemes fitting into a diagram as above with log thickenings $S_0 \to S$ and $X_0 \to X$. Assume that $f \times_S S_0 = f_0$. Then $f \colon X \to S$ is log smooth if and only if $\underline{f} \colon \underline{X} \to \underline{S}$ is a flat morphism of schemes.*

Proof We have already seen that log smoothness of $f \colon X \to S$ together with integrality implies flatness. For the converse, it suffices to ask that $f \colon X \to S$ is log flat by [222, IV, Cor. 4.2.3], and that $\underline{f} \colon \underline{X} \to \underline{S}$ is flat by [222, IV, Prop. 4.2.4]. $\qquad\square$

The most fundamental property of log smooth deformations is that they exist locally and are unique locally, up to non-unique isomorphism. This is in parallel with the classical infinitesimal flat deformation theory of smooth morphisms. We will later mimick this behavior in much greater generality with the definition of a system of deformations in Definition 10.15.

Proposition 7.65 *Let $f_0 \colon X_0 \to S_0$ be log smooth and integral. Let $\bar{x} \in X_0$ be a geometric point.*

(1) *The point $\bar{x}$ has an étale neighborhood $V_0 \to X_0$ such that there is a log smooth deformation $V_A \to S_A$ for all $A \in \mathbf{Art}_Q$.*
(2) *Let $A \in \mathbf{Art}_Q$, and let $f \colon X_A \to S_A$ and $g \colon X'_A \to S_A$ be two log smooth deformations. Then $\bar{x}$ has an étale neighborhood $V_0 \to X_0$ such that the induced log smooth deformations $V_A \to S_A$ and $V'_A \to S_A$ are isomorphic.*

Proof The existence follows from the toroidal characterization of log smoothness. Namely, we have a homomorphism $\theta \colon Q \to P$ of toric monoids (with the given Q) such that $A_\theta \colon A_P \to A_Q$ is log smooth, and such that we have a factorization $V_0 \to A_P \times_{A_Q} S_0 \to S_0$ of $V_0 \to X_0 \to S_0$ with $V_0 \to A_P \times_{A_Q} S_0$ strict and étale. Then lifting $A_P \times_{A_Q} S_A$ along the étale morphism $V_0 \to A_P \times_{A_Q} S_0$ yields the log smooth deformation $V_A \to S_A$. It is clear from the construction that we can choose one V_0 for all A. The uniqueness is proven similarly via [160, Lemma 8.3]. $\qquad\square$

Recall that a *first-order extension* is a surjection $B' \to B$ in $\mathbf{Art}_Q$ such that the kernel $I \subset B'$ satisfies $I \cdot \mathfrak{m}_{B'} = 0$. Every surjection in $\mathbf{Art}_Q$ can be factored into a composition of first-order extensions (non-uniquely).

Now let $f \colon X_B \to S_B$ be a log smooth deformation, and assume that we have a lift $f' \colon X_{B'} \to S_{B'}$ for a first-order extension $B' \to B$ with kernel $I \subset B'$. The sheaf $\Theta^1_{X_0/S_0} \otimes_{\mathbf{k}} I$ of log derivations acts as infinitesimal automorphisms on X' which fix X. Given a log derivation $\theta = (D, \Delta)$, the action is given by

$$\mathrm{Exp}_{\theta \otimes i}(g) = g + i D(g), \quad \mathrm{Exp}_{\theta \otimes i}(m) = m + \alpha^{-1}(1 + i \Delta(m)),$$

for $g \in O_{X_{B'}}$, $m \in \mathcal{M}_{X_{B'}}$, and $i \in I$. This defines an isomorphism

$$\mathrm{Exp}: \quad \Theta^1_{X_0/S_0} \otimes_{\mathbf{k}} I \to \mathcal{A}ut_{X_{B'}/X_B}$$

of sheaves of groups, where the left-hand side carries the addition from its Abelian sheaf structure. The right-hand side is the sheaf of automorphisms of $X_{B'}$ over $S_{B'}$ which induce the identity on X_B. Note that we consider here both sheaves as sheaves in the étale topology of $X_{B'}$, which coincides, as a site, with the étale topology of X_0. We then obtain the following standard result:

Proposition 7.66 *Let $f_0 \colon X_0 \to S_0$ be log smooth and integral, and let $f \colon X_B \to S_B$ be a log smooth deformation over $B \in \mathbf{Art}_Q$. Let $B' \to B$ be a first-order extension with kernel $I \subset B'$.*

(1) *If $f' \colon X_{B'} \to S_{B'}$ is a log smooth lifting over $S_{B'}$, i.e., a log smooth deformation inducing f after base change along $S_B \to S_{B'}$, then the group of infinitesimal automorphisms is*

$$H^0(X_0, \Theta^1_{X_0/S_0} \otimes_{\mathbf{k}} I).$$

(2) *The set of all log smooth liftings $f' \colon X_{B'} \to S_{B'}$ is a (pseudo-)torsor under*

$$H^1(X_0, \Theta^1_{X_0/S_0} \otimes_{\mathbf{k}} I).$$

(3) *The obstruction class for the existence of a log smooth lifting $f' \colon X_{B'} \to S_{B'}$ is in*

$$H^2(X_0, \Theta^1_{X_0/S_0} \otimes_{\mathbf{k}} I).$$

Proof This is [160, Prop. 8.6]. $\qquad\qquad\qquad\qquad\qquad\qquad\qquad\qquad\qquad\square$

Corollary 7.67 *Let $f_0 \colon X_0 \to S_0$ be log smooth and integral, and assume that X_0 is an affine scheme. Then, for every $A \in \mathbf{Art}_Q$, there is a unique log smooth deformation $f \colon X_A \to S_A$ of f_0. It is unique up to non-unique isomorphism.*

Finally, the obligatory result proven by F. Kato is that LD_{X_0/S_0} is a deformation functor in the sense of Schlessinger. For the notion of a deformation functor, cf. our discussion in Appendix A, in particular for the definition of Condition (H_2^+) in Definition A.9.

Proposition 7.68 *Let $S_0 = \mathrm{Spec}(Q \to \mathbf{k})$, and let $f_0 \colon X_0 \to S_0$ be integral and log smooth. Let $F = \mathrm{LD}_{X_0/S_0}$.*

(H_0) *We have $F(A_0) = \{*\}$.*

(H_1) *The natural map*

$$\eta: \quad F(A' \times_A B) \to F(A') \times_{F(A)} F(B)$$

is surjective for every surjection $B \to A$. Here, $A' \times_A B$ is the fiber product of rings.

(H_2) *The natural map η is bijective for $B = A_\varepsilon$ and $A = A_0 = \mathbf{k}$.*

(H_2^+) *The natural map η is bijective for $A = A_0 = \mathbf{k}$.*

Thus, LD_{X_0/S_0} is a deformation functor in the sense of Schlessinger, and it is neat in the sense of Definition A.9.

(H_3) *If $f_0 \colon X_0 \to S_0$ is proper (as a morphism of schemes), then the $\mathbf{k}$-vector space*

$$T(X_0/S_0) := \mathrm{T}_F := F(A_\varepsilon)$$

is finite-dimensional. In particular, F has a hull.

Proof This is [160, Thm. 8.7]. The proof of (H_2^+), which is not in [160], is analogous to the proof of (H_2). $\qquad\qquad\qquad\qquad\qquad\qquad\qquad\qquad\qquad\qquad\square$

Unlike for A_1, there is a map $A_0 \to A_\varepsilon$ in $\mathbf{Art}_Q$, inducing a map $S_\varepsilon \to S_0$. Thus, $f_0 \times_{S_0} S_\varepsilon$ is a canonical log smooth deformation over S_ε. Now Proposition 7.66 yields

$$T(X_0/S_0) = \mathrm{LD}_{X_0/S_0}(A_\varepsilon) = H^1(X_0, \Theta^1_{X_0/S_0} \otimes_{\mathbf{k}} (\varepsilon))$$

as sets, but this actually holds as $\mathbf{k}$-vector spaces. In particular, it is easy to see how properness of $f_0 \colon X_0 \to S_0$ implies that $T(X_0/S_0)$ is finite-dimensional. On the other hand, there is no canonical log smooth deformation over S_1. In fact, Example 1.92 shows that, in some cases, there is no global log smooth deformation over S_1.

7.7   The Normalized Dualizing Sheaf

Let $S_0 = \mathrm{Spec}(\mathbb{N} \to \mathbf{k})$ be the standard log point, and let $f_0 \colon X_0 \to S_0$ be log smooth, saturated, and vertical. In this section, we give a more careful proof of Tsuji's result [272, Thm. 2.21] that $\Omega^d_{X_0/S_0}$ can be identified with the normalized dualizing sheaf

$$\omega^\circ_{X_0} = \mathcal{H}^{-d}(f_0^! O_{S_0}).$$

Tsuji's result is more general, but we restrict our attention to the vertical case here.

Remark 7.69 Judging from [51, 190, 218, 249], choosing the correct morphisms and establishing their naturality seems to be a notoriously difficult problem in Grothendieck duality. In the proof of [272, Thm. 2.21], Tsuji uses some of this naturality without much discussion while the references that we found appeared several years after Tsuji's article. Therefore, it seems useful to us to revisit the proof with these newer references at hand. Note however the footnote in our proof, showing that we did not fully succeed in finding references for all necessary statements. $\Diamond$

For a morphism of log schemes $f\colon X \to S$, in the following, we write $\Omega^p_{X/S}$ for the sheaf of logarithmic differential forms and $\Omega^p_{\underline{X}/\underline{S}}$ for the sheaf of classical differential forms. For f a smooth morphism of schemes of relative dimension d, Verdier established in [277] a concrete isomorphism

$$\Theta_{X/S}\colon \quad \Omega^d_{\underline{X}/\underline{S}} \to \mathcal{H}^{-d}(f^! O_S).$$

Following the article [249] of Sastry, we call this isomorphism with Sastry's sign choices the *Verdier isomorphism*. When f is strict and smooth, we can consider $\Omega^d_{X/S}$ rather than $\Omega^d_{\underline{X}/\underline{S}}$ as the source of the Verdier isomorphism.

Proposition 7.70 ([272, Thm. 2.21]) *Let $f_0\colon X_0 \to S_0$ be log smooth, saturated, and vertical. Assume furthermore that f_0 is separated and of finite type, and that X_0 is equidimensional of dimension d. Let $V_0 \subseteq X_0$ be the open subset where f_0 is strict. Then, there is a unique isomorphism*

$$\vartheta_{X_0/S_0}\colon \quad \Omega^d_{X_0/S_0} \xrightarrow{\ \cong\ } \omega^\circ_{X_0}$$

which coincides with the Verdier isomorphism $\Omega^d_{V_0/S_0} \cong \omega^\circ_{V_0}$ on the strict and smooth locus $V_0 \subseteq X_0$.

Proof The proof proceeds in several steps. First, the left-hand side is a line bundle because f_0 is saturated and log smooth of relative dimension d. Since f_0 is vertical, X_0 is Gorenstein, and the right-hand side is also a line bundle.

Since f_0 is saturated, the strict and smooth locus $V_0 \subseteq X_0$ is scheme-theoretically dense. We will review this fact in Lemma 8.34. We now see that there can be *at most* one isomorphism ϑ_{X_0/S_0} which restricts to the Verdier isomorphism on V_0.

Let $\xi\colon \mathbb{N} \to P$ be an injective, saturated, and vertical homomorphism between sharp toric monoids. The reader can find an explicit description of such a homomorphism in Lemma 9.10 together with Lemma 9.12. Let $L = A_P$, $S = A_{\mathbb{N}}$, and consider $p = A_\xi\colon L \to S$. This map is log smooth, saturated, and vertical. Let d be its relative dimension. Let $W \subseteq L$ be the open subset where p is strict and smooth. The complement $L \setminus W$ is of codimension ≥ 2 since $p|_{t \neq 0}$ is smooth, and since $L_0 = p^{-1}(0)$ is reduced. Let $j\colon W \to L$ be the inclusion, and let $q = p \circ j$.

Since $q\colon W \to S$ is smooth, we have the Verdier isomorphism

$$\Theta_{W/S}\colon \quad \Omega^d_{\underline{W}/\underline{S}} \to \omega^\circ_{W/S} = \mathcal{H}^{-d}(q^! O_S).$$

Let $b\colon S_0 \to S$, $e\colon W_0 \to W$, and $c\colon L_0 \to L$ be the inclusions. Then [249, Thm. 2.3.5] provides us with a canonical comparison isomorphism

$$\theta_b^q\colon \quad e^*\omega_{W/S}^\circ \to \omega_{W_0/S_0}^\circ$$

such that

$$
\begin{array}{ccc}
e^*\Omega_{\underline{W}/\underline{S}}^d & \xrightarrow{\ e^*\Theta_{W/S}\ } & e^*\omega_{W/S}^\circ \\
\big\downarrow & & \big\downarrow{\scriptstyle\theta_b^q} \\
\Omega_{\underline{W}_0/\underline{S}_0}^d & \xrightarrow{\ \Theta_{W_0/S_0}\ } & \omega_{W_0/S_0}^\circ
\end{array}
$$

commutes. The existence of θ_b^q in [249] does not rely on smoothness of $q\colon W \to S$ but only on the Cohen–Macaulay property. Therefore, it extends to a comparison isomorphism $\theta_b^p\colon c^*\omega_{L/S}^\circ \to \omega_{L_0/S_0}^\circ$.

The log canonical bundle $\Omega_{L/S}^d$ is a line bundle because $p\colon L \to S$ is log smooth, and $\omega_{L/S}^\circ$ is a line bundle because $p\colon L \to S$ is a Gorenstein morphism. Since we already have an isomorphism on $W \subseteq L$ whose complement has codimension ≥ 2, this extends to a unique isomorphism

$$\vartheta_{L/S}\colon \quad \Omega_{L/S}^d \to \omega_{L/S}^\circ$$

whose restriction to W is $\Theta_{W/S}$ (when using the identification of log differentials with classical differentials on W). Now we *define* ϑ_{L_0/S_0} as the unique isomorphism which makes the diagram

$$
\begin{array}{ccc}
c^*\Omega_{L/S}^d & \xrightarrow{\ c^*\vartheta_{L/S}\ } & c^*\omega_{L/S}^\circ \\
\big\downarrow & & \big\downarrow{\scriptstyle\theta_b^p} \\
\Omega_{L_0/S_0}^d & \xrightarrow{\ \vartheta_{L_0/S_0}\ } & \omega_{L_0/S_0}^\circ
\end{array}
$$

commutative. When restricting to $W_0 \subseteq L_0$, we obtain the above diagram. Therefore, we have $\vartheta_{L_0/S_0}|_{W_0} = \Theta_{W_0/S_0}$. This proves the claim for $p_0\colon L_0 \to S_0$. For the morphism $p_0\colon L_0 \times \mathbb{A}^r \to S_0$, the claim can be proven analogously.

Let now $f_0\colon X_0 \to S_0$ be a general morphism as in the statement, and let d be the relative dimension. Let $x \in X_0$ be a point. By the toroidal characterization of log smoothness Theorem 7.59, we can find strict étale maps $\pi_0\colon U_0 \to X_0$ and $\tau_0\colon U_0 \to L_0 \times \mathbb{A}^r$, where $x \in \pi_0(U_0)$ and $L_0 \to S_0$ is of the form discussed above. Pulling back ϑ_{L_0/S_0} along τ_0 yields an isomorphism $\vartheta_{U_0/S_0}\colon \Omega_{U_0/S_0}^d \to \omega_{U_0}^\circ$ which

coincides with the Verdier isomorphism on the strict open subset $\tau_0^{-1}(W_0 \times \mathbb{A}^r)$.[9]
When considering an étale cover of X_0 consisting of étale opens of the preceding
form, then the locally defined isomorphisms ϑ_{U_0/S_0} descend to an isomorphism
$\vartheta_{X_0/S_0} : \Omega^d_{X_0/S_0} \to \omega^\circ_{X_0}$ because they coincide over the scheme-theoretically (and
topologically) dense open subset $V_0 \subseteq X_0$. Now ϑ_{V_0/S_0} is the Verdier isomorphism
étale locally, and again since the formation of the Verdier isomorphism commutes
with étale maps, ϑ_{V_0/S_0} coincides with the Verdier isomorphism. $\square$

---

[9] The author, who is not an expert in Grothendieck duality, could not find a reference for the
commutation of the Verdier isomorphism with étale pullback. Although a very elementary and
straightforward statement, the proof seems to be surprisingly intricate: In [190, pp. 9–10], the
introduction of a comprehensive monograph on Grothendieck duality, Lipman states the result
informally and writes "The proofs of these down-to-earth statements are not easy, and will not
appear in these notes."

Chapter 8
Families of Singular Log Schemes

Often, we can improve the technical properties of a degenerating family of varieties by introducing a log structure on the base and the total space. For example, when $f\colon X \to C$ is a semistable degeneration over a smooth curve C, and $\Delta \subset C$ is a finite set of points such that $f\colon X \setminus f^{-1}(\Delta) \to C \setminus \Delta$ is smooth, then we obtain a log smooth morphism $f\colon (X|D) \to (C|\Delta)$ once we endow the total space with the divisorial log structure defined by $D = f^{-1}(\Delta)$, and the base with the divisorial log structure defined by Δ. This remains true when we allow certain more general local models for the degeneration $f\colon X \to C$, such as the ones defined in Construction 9.2, which give rise to *toroidal crossing degenerations* in the sense of Definition 9.1 (with the appropriate modifications).

This suggests to consider log smooth families $f\colon X \to S$ as generalizations of smooth families of varieties. However, the notion of log smoothness can be quite badly behaved from the perspective of moduli theory. For example, the blow-up $\pi\colon \mathrm{Bl}_0\mathbb{A}^2 \to \mathbb{A}^2$ becomes log smooth once we endow the target with the divisorial log structure given by $xy = 0$ and the source with the divisorial log structure given by $\pi^*(xy) = 0$. In other words, this blow-up is not even flat as a morphism of schemes. To ensure flatness, it is convenient to require that $f\colon X \to S$ is an integral morphism of log schemes. Then the underlying morphism of schemes $\underline{f}\colon \underline{X} \to \underline{S}$ is not only flat but also has Cohen–Macaulay fibers.

In addition to integrality, we also impose the condition that $f\colon X \to S$ should be a saturated morphism of log schemes. This has several advantages over asking only for integrality. First, we wish to work in the category of fs log schemes instead of the category of fine log schemes. If $f\colon X \to S$ is only integral, it can happen that, for a base change $b\colon T \to S$ of fs log schemes, the fiber product $Y = X \times_S T$ is only fine but not fs, integrality of Y of course being ensured by integrality of $f\colon X \to S$. Then, in order to force the fiber product to be in the category of fs log schemes, we have to pass to its saturation $Y^{\mathrm{sat}} \to Y$. However, this map is not an isomorphism of underlying schemes, so our base change would no longer be a base change of underlying families of schemes. We certainly do not want to have this

© The Author(s), under exclusive license to Springer Nature Switzerland AG 2025
S. Felten, *Global Logarithmic Deformation Theory*, Lecture Notes
in Mathematics 2373, https://doi.org/10.1007/978-3-031-98751-9_8

situation when we use log geometry as a tool to analyze degenerating families of varieties.

Secondly, a log smooth and integral morphism $f\colon X \to S$ of fs log schemes is saturated if and only if its fibers are reduced, and this holds if and only if its fibers are geometrically reduced. In this situation, there is an open subset $X^\circ \subseteq X$ which is dense in every fiber and has the property that $X^\circ \to S$ is not only log smooth but strict and smooth. Besides that families with reduced fibers seem to be more geometric, this also has some technical advantages.

Thirdly, saturated log smooth morphisms have better Hodge-theoretic properties than integral log smooth morphisms. While the Hodge–de Rham spectral sequence degenerates at E_1 also for integral log smooth morphisms in characteristic 0, we have local freeness of $R^q f_* \Omega^p_{X/S}$ in the Zariski or étale topology only for saturated $f\colon X \to S$, see [155, Thm. 7.1], its proof, and its corollary. Due to our preference for saturated over integral log smooth morphisms, we define:

Definition 8.1 (Log Smooth Families) A *log smooth family* is a saturated and log smooth morphism $f\colon X \to S$ of locally Noetherian fs log schemes such that $\underline{f}\colon \underline{X} \to \underline{S}$ is separated and of finite type. Such a family is automatically flat and has geometrically reduced Cohen–Macaulay fibers.

The examples at the beginning of this section suggest the following recipe for improving the technical properties of flat and proper families $f\colon X \to S$. First, choose some divisor $\Delta \subset S$ such that $f\colon X \setminus f^{-1}(\Delta) \to S \setminus \Delta$ is smooth. Then set $D = f^{-1}(\Delta)$ and consider the morphism of log schemes $f\colon (X|D) \to (S|\Delta)$. As we have already discussed in the Introduction, it is rather the exception than the rule that this procedure yields a log smooth family. In general, the result $f\colon (X|D) \to (S|\Delta)$ will have *log singularities*. Our standard Examples 1.11 and 1.79 are examples of this situation.

While $f\colon (X|D) \to (S|\Delta)$ carries a log structure which is defined everywhere, we have adopted the convention that we consider the log structure to be defined only on a large open subset $j\colon U \subseteq X$ where $f\colon (X|D) \to (S|\Delta)$ is actually log smooth. The choice of this open subset is part of the datum of the families we consider. We have adopted this convention essentially because it is difficult to control the behavior of the log structure in the log singularities under standard constructions of log geometry, in particular fiber products.

With the log structure defined only on an open subset $U \subseteq X$, it becomes more urgent to restrict to saturated morphisms (or integral morphisms if one wishes to work with fine log schemes instead of fs ones). Namely, consider a morphism $b\colon T \to S$ of locally Noetherian fs log schemes and the base change $g\colon Y = X \times_S T \to T$. Then the log structure is defined on $V = U \times_S T$, but if $f\colon U \to S$ is not saturated, V may not be a saturated log scheme. When passing to the saturation $V^{\mathrm{sat}} \to V$, the log structure would no longer be defined on an open subset $V \subseteq Y$, but on the source of a morphism $V^{\mathrm{sat}} \to Y$ which factors as an open immersion composed with a finite morphism. It might be possible to build a theory strong enough to incorporate this phenomenon, but we do not wish to do so in this monograph.

In this chapter, we present the foundations of our theory of families of singular log schemes, i.e., of *generically log smooth families*. In particular, we discuss generically log smooth families whose singularities are controlled by certain toric local models, the *log toroidal families*. Except for Sects. 8.7 and 8.8, this chapter contains no new material. We follow closely our prior publications [77] and [75].

8.1 Generically Log Smooth Families

We introduced generically log smooth families in [77], and a further discussion is contained in the author's doctoral thesis [75]. In this monograph, we wish to work with a slight generalization of this notion, which we define in a moment.

A generically log smooth family is a family of schemes $f : X \to S$ together with an open subset $j : U \to X$ on which we have the structure of a log smooth morphism. The complement $Z := X \setminus U$, on which no log structure is given, should be rather small: We require that $U \subseteq X$ satisfies the *codimension condition*, by which we mean that

$$\mathrm{codim}(X_s \setminus U_s, X_s) \geq 2$$

for all points $s \in S$. Slightly more elegantly, we may say that $\dim(O_{X_s,z}) \geq 2$ for every $z \in X_s \setminus U_s$ and every $s \in S$.

Recall the notion of *partial log schemes* from Definition 1.75, which we have introduced to formulate generically log smooth families. Here is our more precise definition, which we announced in the introduction.

Definition 8.2 (Generically Log Smooth Families) A *generically log smooth family* is a morphism $f : (X, U) \to S$ of partial log schemes such that:

(a) (S, O_S) is a locally Noetherian scheme, and $\alpha_S : \mathcal{M}_S \to O_S$ is a fine and saturated log structure;
(b) (X, O_X) is a locally Noetherian scheme, and $(f, f^{\sharp}) : (X, O_X) \to (S, O_S)$ is a flat and separated morphism of finite type; its fibers are geometrically reduced, satisfy Serre's condition (S_2), and are equidimensional of some fixed dimension $d \geq 1$;
(c) the open immersion $j : U \to X$ satisfies the codimension condition;
(d) $\alpha_U : \mathcal{M}_U \to O_U$ is a fine and saturated log structure on U;
(e) the morphism of log schemes $f : U \to S$ is log smooth and saturated.

We say that $U \subseteq X$ is the *log smooth locus* of $f : (X, U) \to S$, and that $Z := X \setminus U$ is the *log singular locus* of $f : (X, U) \to S$.

For two generically log smooth families $f\colon (X, U) \to S$ and $g\colon (Y, V) \to T$, a *morphism* consists of a morphism $b\colon T \to S$ of log schemes and a morphism $c\colon (Y, V) \to (X, U)$ of partial log schemes such that $f \circ c = b \circ g$ as morphisms of partial log schemes. A morphism between generically log smooth families is *accurate* if $c\colon (Y, V) \to (X, U)$ is accurate.

Given a generically log smooth family $f\colon (X, U) \to S$ and a morphism $b\colon T \to S$ of locally Noetherian fs log schemes, we define the *base change* $g\colon (Y, V) \to T$ as the fiber product $(Y, V) = (X, U) \times_S T \to T$ in the category of partial log schemes.

We discuss this definition.

(1) Generically log smooth families can be considered as the logarithmic analog of a family of normal varieties. Namely, we replace regularity in (relative) codimension 1 with log smoothness in relative dimension 1 while we retain Serre's condition (S_2) on the fibers.

(2) Although most examples of generically log smooth families come with a log structure defined on the whole of X, we have chosen to consider only the log structure on the log smooth locus U as part of the datum. The reason for this is that the log structure may be very poorly behaved in Z. Our standard example is

$$X^{\mathrm{sze}} = \mathrm{Spec}\, \mathbf{k}[x, y, t, z]/(xy - tz) \to \mathbb{A}^1_t -\colon S$$

with the divisorial log structures given by $t = 0$. As shown in [125, Ex. 1.11], the sheaf of log differential forms $\Omega^1_{X^{\mathrm{sze}}/S}$ is not a quasi-coherent sheaf of $\mathcal{O}_X$-modules around $0 \in X$. This implies that the log structure is not coherent in 0. Since we have little control over such log structures, we drop it from the definition of a generically log smooth family.

(3) Given a generically log smooth family $f\colon (X, U) \to S$, we always have a log structure $j_*\mathcal{M}_U \to \mathcal{O}_X$ on the whole of X since $j_*\mathcal{O}_U = \mathcal{O}_X$, as we will discuss below. However, there are two reasons why we do not consider this log structure here:

 (i) We do not know if the formation of $j_*\mathcal{M}_U$ commutes with base change along $b\colon T \to S$.
 (ii) We do not know if the formation of $j_*\mathcal{M}_U$ is invariant under shrinking U.[1]

(4) There is a theory of *relatively log smooth* morphisms, which goes back to the work [212] of Nakayama and Ogus and was further developed by Ogus in

---

[1] We believe that this construction is stable under shrinking U, at least under additional hypotheses. The problem is somewhat subtle. For example, the ghost sheaves of two different log smooth families can become isomorphic after shrinking U, but in the examples that we have, the log morphisms themselves are not isomorphic.

[222]. A relatively log smooth morphism $f: (X, \mathcal{F}) \to (S, \mathcal{N})$ is such that the possibly non-coherent log structure $\mathcal{F}$ can be locally embedded as a sheaf of faces in a (locally defined) coherent log structure $\mathcal{M}$ such that $(X, \mathcal{M}) \to (S, \mathcal{N})$ is log smooth. Every relatively log smooth family $f: (X, \mathcal{F}) \to (S, \mathcal{N})$ can be turned into a generically log smooth family by selecting the log smooth locus $U \subseteq X$ and restricting $\mathcal{F}$ to U. The author does not know if replacing a relatively log smooth family with its associated generically log smooth family is faithful, in particular when we also allow to shrink U further than necessary. Relatively log smooth families are closely related with but less general than *log toroidal families*, which we discuss in Sects. 8.3 and 8.5. In particular, many explicit examples that we consider in this monograph (though not the most important ones) cannot be interpreted as relatively log smooth families, for instance Example 1.159. For some further discussion, see the author's doctoral thesis [75]. We do not use the theory of relatively log smooth families in this monograph.

(5) We always call U the "log smooth locus" and Z the "log singular locus," even when our generically log smooth family arises from deleting points in a globally log smooth family. The choice of an open $U \subseteq X$ is an integral part of the data of a generically log smooth family, simplifying the theory considerably in comparison to a scenario where we wish to consider all possible choices of U at once (as the author did in his Bachelor thesis). Nonetheless, many properties of generically log smooth families are invariant under shrinking U.

(6) Serre's condition (S_2) is assumed to hold for the fibers over the residue fields $\kappa(s)$ of points $s \in S$. Then also the fiber over any field extension $\kappa(s) \to K$ has Serre's property (S_2). Namely, let $x \in X_s$, and let $y \in X_K$ be a point lying over x. Then we have

$$\dim(O_{X_K, y}) = \dim(O_{X, x}) + \dim(O_{X_K, y} / \mathfrak{m}_x O_{X_K, y})$$

and the analogous formula for the depth since $O_{X_s, x} \to O_{X_K, y}$ is flat and local. Since $K \otimes_{\kappa(s)} \kappa(x)$ is Cohen–Macaulay (as a base change of the Cohen–Macaulay morphism $\kappa(s) \to K$), one can show that Serre's condition (S_2) for $O_{X_s, x}$ implies Serre's condition (S_2) for $O_{X_K, y}$. If $\dim(K \otimes_{\kappa(s)} \kappa(x)) = 0$, then the converse holds. This is the case, for example, if $\kappa(s) \to K$ is an algebraic field extension. In particular, if the geometric fibers (over the algebraic closures of the residue field) satisfy Serre's condition (S_2), then also the fibers over $\kappa(s)$ satisfy Serre's condition (S_2).

(7) Serre's condition (S_2) implies that every connected component of X_s is equidimensional, i.e., every irreducible component of the same connected component has the same dimension, and that every connected component of X_s is connected in codimension 1, i.e., remains connected after removing any closed subset of codimension ≥ 2. See [136].

(8) Flatness of $f: X \to S$ together with the (S_2)-assumption on the fibers and the codimension condition on $U \subseteq X$ implies that $j_* O_U = O_X$, see Chap. B.

Thus, if $\mathcal{F}$ is a reflexive coherent sheaf on U, then $j_*\mathcal{F} = (j_*\mathcal{F})^{\vee\vee}$ is not only a quasi-coherent but a coherent sheaf on X. Namely, $\mathcal{F}$ admits *some* coherent extension $\mathcal{G}$ on X with $\mathcal{G}|_U = \mathcal{F}$, and then $j_*\mathcal{F} = \mathcal{G}^{\vee\vee}$. In particular, in the example mentioned above, $j_*\Omega^1_{U^{\mathrm{sze}}/S}$ is a coherent sheaf. Furthermore, the formation of $j_*\mathcal{F}$ is invariant under shrinking U.

(9) In [77], we have required that the fibers of $f: X \to S$ should be Cohen–Macaulay instead of only satisfying (S_2). This stronger assumption, inspired by the fact that every saturated log smooth family has Cohen–Macaulay fibers, serves the same purpose of ensuring $j_*\mathcal{O}_U = \mathcal{O}_X$, but the (S_2)-assumption is more natural, allows for all normal varieties to be considered as generically log smooth, and is compatible with the conventions in the moduli theory of varieties of general type.

(10) The saturatedness assumption on $f: U \to S$ is important in the construction of the base change $g: (Y, V) \to T$. Namely, it ensures that $V = U \times_S T$, formed in the category of all log schemes **LSch**, is already a fine and saturated log scheme. When $f: U \to S$ were not saturated, we would need to take the saturation $V^{\mathrm{sat}} \to V$ in order to stay in the category of fs log schemes. However, $V^{\mathrm{sat}} \to Y$ is no longer an open immersion.

(11) If $f: U \to S$ is integral, then it is automatically saturated because we assumed the fibers of $f: X \to S$ to be geometrically reduced.

(12) In [77], we assumed that the base S of a generically log smooth family should be Noetherian. We have relaxed this to being locally Noetherian since our theory is local on the base.

(13) Since $f: U \to S$ is saturated and log smooth, there is an open subset $U^\circ \subseteq U$ which is scheme-theoretically dense in every fiber, such that $f: U^\circ \to S$ is strict and smooth, cf. Lemma 8.34 below. Thus, $\Omega^1_{U/S}$, which is a locally free sheaf since $f: U \to S$ is log smooth, is of constant rank d, see also Corollary 8.36.

(14) The fibers of $f: X \to S$ may not be geometrically connected, even if they are connected. For an example, just take the smooth morphism $\mathbb{Q} \to \mathbb{Q}[i][x]$.

8.2 The Reflexive Log de Rham Complex and Polyvector Fields

Let $f: (X, U) \to S$ be a generically log smooth family. Then we have the log de Rham complex $\Omega^\bullet_{U/S}$ of $f: U \to S$. As in [77], we extend it to the whole of X as follows.

Definition 8.3 (Reflexive Log de Rham Complex) Let $f: (X, U) \to S$ be a generically log smooth family, and let $j: U \to X$ be the inclusion. The *reflexive* *log de Rham complex* is given by

$$\mathcal{W}^{\bullet}_{X/S} := j_* \Omega^{\bullet}_{U/S}.$$

The pieces $\mathcal{W}^i_{X/S}$ are a coherent and reflexive replacement for $\Omega^i_{X/S}$, which is, of course, not defined in this setup, but which is not a quasi-coherent O_X-module in our standard Example 1.79 given by the equation $xy = tz$.[2] The reflexive log de Rham complex $\mathcal{W}^{\bullet}_{X/S}$ is invariant under shrinking U.

Since we have assumed $f \colon (X, U) \to S$ to have some fixed relative dimension $d \geq 1$,

$$\omega_{X/S} := \mathcal{W}^d_{X/S}$$

is a reflexive sheaf of rank one. In many applications, $\omega_{X/S}$ will actually be a line bundle, even if the other pieces $\mathcal{W}^i_{X/S}$ are not locally free.

Definition 8.4 (Reflexive Log Canonical Sheaf) Let $f \colon (X, U) \to S$ be a generically log smooth family of relative dimension d. Then $\omega_{X/S} := \mathcal{W}^d_{X/S}$ is the *reflexive log canonical sheaf*. The family f is *log Gorenstein* if $\omega_{X/S}$ is a line bundle, and the family f is *log Calabi–Yau* if $\omega_{X/S} \cong O_X$.[3]

Similarly to $\Omega^1_{U/S}$, we have the *log tangent sheaf* $\Theta^1_{U/S} = \mathcal{H}om(\Omega^1_{U/S}, O_U)$, which is locally free on U and classifies relative log derivations (D, Δ) with values in O_X. Since, in general, the sheaf of log derivations is naturally reflexive, we also write

$$\Theta^1_{X/S} := j_* \Theta^1_{U/S};$$

we consider this as a sheaf of Lie algebras with the natural Lie bracket. Its sections (D, Δ) on an open subset $W \subseteq X$ act as derivations $D \colon O_W \to O_W$ with log part $\Delta \colon \mathcal{M}_{U \cap W} \to O_{U \cap W}$. We denote the exterior powers of $\Theta^1_{U/S}$ by $\Theta^p_{U/S} := \Lambda^p \Theta^1_{U/S}$. Then we define

$$\mathcal{V}^{-1}_{X/S} := j_* \Theta^1_{U/S} \quad \text{and} \quad \mathcal{V}^{-p}_{X/S} := j_* \Theta^p_{U/S}$$

as well as $\mathcal{V}^0_{X/S} := O_X$. We consider $\mathcal{V}^{\bullet}_{X/S}$ as a Gerstenhaber algebra endowed with the *negative* of the usual Schouten–Nijenhuis bracket, see also Proposition 8.5

---

[2] The idea of using a direct image or, equivalently, the reflexive hull in forming the de Rham complex goes back at least to Danilov's seminal article [57] and has been employed by Gross and Siebert in [125] in the context of degenerating families of varieties. In the classical setup of Danilov, the original de Rham complex has coherent pieces, but the direct image from the smooth locus has better properties.

[3] When studying moduli, one may want to relax this condition to $\omega_{X/S} \cong f^* N$ for some line bundle N on S.

below. In particular, the difference between $\mathcal{V}_{X/S}^{-1}$ and $\Theta_{X/S}^1$ is the sign of the Lie bracket as well as the sign of the action on O_X. We call $\mathcal{V}_{X/S}^{\bullet}$ the *reflexive Gerstenhaber algebra of log polyvector fields*. Each $\mathcal{V}_{X/S}^p$ is a reflexive coherent sheaf.[4]

8.2.1 The Gerstenhaber Calculus

Polyvector fields and the de Rham complex carry more structure which is very relevant to our deformation theory. The obvious $\wedge$-product, the unit element $1 \in \mathcal{V}_{X/S}^0$, and the *negative* Schouten–Nijenhuis bracket $[-, -] = -[-, -]_{\text{sn}}$ (already mentioned above) turn $\mathcal{V}_{X/S}^{\bullet}$ into a Gerstenhaber algebra in the context $\mathfrak{Coh}'(X/S)$.[5,6] Similarly, $\mathcal{W}_{X/S}^{\bullet}$ naturally comes with a $\wedge$-product, $1 \in \mathcal{W}_{X/S}^0$, and a de Rham differential $\partial \colon \mathcal{W}_{X/S}^i \to \mathcal{W}_{X/S}^{i+1}$ which is only $f^{-1}(O_S)$-linear. As is shown e.g. in [75, Lemma 2.32] in the author's thesis, we can construct a natural (right) contraction map $\neg$. Then we can define the Lie derivative $\mathcal{L}$ via the Lie–Rinehart homotopy formula. As is checked in [75, Lemma 2.36], this turns $(\mathcal{V}_{X/S}^{\bullet}, \mathcal{W}_{X/S}^{\bullet})$ into a (one-sided) Gerstenhaber calculus in the context $\mathfrak{Coh}'(X/S)$. We can also endow it with a left contraction. Summarizing the discussion, we have:

Proposition 8.5 *Let $f \colon (X, U) \to S$ be a generically log smooth family of relative dimension d.*

(1) *We have a two-sided Gerstenhaber calculus*

$$\mathcal{V}\mathcal{W}_{X/S}^{\bullet} := (\mathcal{V}_{X/S}^{\bullet}, \wedge, 1, [-, -], \mathcal{W}_{X/S}^{\bullet}, \wedge, 1, \partial, \neg, \mathcal{L}, \vdash)$$

of dimension d in the context $\mathfrak{Coh}'(X/S)$. The graded pieces are reflexive coherent sheaves. The $\wedge$-products and the two contractions $\vdash$ and $\neg$ are O_X-linear whereas $[-, -]$, ∂, and $\mathcal{L}$ are only $f^{-1}(O_S)$-linear. The Gerstenhaber calculus is strictly faithful.

(2) *If $f \colon (X, U) \to S$ is log Gorenstein, then $\mathcal{V}\mathcal{W}_{X/S}^{\bullet}$ is locally Batalin–Vilkovisky.*

---

[4] We put $\mathcal{V}_{X/S}^{\bullet}$ in negative degrees since this convention is more natural from the perspective of Gerstenhaber calculi, in particular the two contraction maps $\vdash$ and $\neg$, consistent with our definition of a Gerstenhaber calculus. The notation $\mathcal{W}_{X/S}^{\bullet}$ originally grew out of denoting Ω as W in emails, for its vague similarity of shapes. The notation $\mathcal{V}_{X/S}^{\bullet}$ is intended to suggest *vector fields*. The symbol $\Theta^{\bullet}$ is reserved for actual log morphisms, and the symbol $\mathcal{T}$ is used for the sheaves computed from the classical, non-logarithmic cotangent complex.

[5] The context $\mathfrak{Coh}'(X/S)$, as opposed to $\mathfrak{Coh}(X/S)$, allows also modules which are not flat over S.

[6] The sign convention for the bracket is used in various places, including [38], [76], and [75].

(3) *If $f \colon (X, U) \to S$ is log Calabi–Yau, then $\mathcal{V}\mathcal{W}^\bullet_{X/S}$ carries the structure of a two-sided Batalin–Vilkovisky calculus, depending on the choice of a global trivializing section $\omega \in \mathcal{W}^d_{X/S}$.*

(4) *If $f \colon (X, U) \to S$ has the base change property (defined below in Definition 8.47), then $\mathcal{W}^i_{X/S}$ is flat over S. If $f \colon (X, U) \to S$ is moreover log Gorenstein, then $\mathcal{V}^p_{X/S}$ is flat over S as well. In this case, we have a two-sided Gerstenhaber calculus in the context $\mathfrak{Coh}(X/S)$.*

Proof We have a one-sided Gerstenhaber calculus by [75, Lemma 2.32]. On U, we have that $\mathcal{W}^d_{U/S}$ is a line bundle, so we can construct the left contraction $\vdash$ on U as discussed before Lemma 3.23. Then the left contraction on X is obtained as the direct image from U. Most relations then follow from Lemma 3.23, first on U, and then on X. By induction on r, we see that

$$
0 = (\theta_r \wedge \ldots \wedge \theta_1) \lrcorner (\omega \wedge \alpha)
$$

$$
= ((\theta_r \wedge \ldots \wedge \theta_1) \lrcorner \omega) \wedge \alpha
$$

$$
+ \sum_{i=1}^{r} (-1)^{d+1-i} ((\theta_r \wedge \ldots \wedge \hat{\theta}_i \wedge \ldots \wedge \theta_1) \lrcorner \omega) \wedge (\theta_i \lrcorner \alpha)
$$

for $r \geq 1$ and $\theta_i \in \mathcal{V}^{-1}_{X/S}$ and $\alpha \in \mathcal{W}^1_{X/S}$. Thus, we have

$$
(\theta_r \wedge \ldots \wedge \theta_1) \vdash \alpha = \sum_{i=1}^{r} (-1)^{i+1} (\theta_r \wedge \ldots \wedge \hat{\theta}_i \wedge \ldots \wedge \theta_1) \wedge (\theta_i \vdash \alpha).
$$

From this, a direct computation yields the relation for $(\theta \wedge \xi) \vdash \alpha$ when $\theta = \theta_r \wedge \ldots \wedge \theta_1$ and $\xi = \xi_s \wedge \ldots \wedge \xi_1$. However, on U, every section of $\mathcal{V}^p_{X/S}$ is locally of this form, so the relation follows. As usual, the above formula yields

$$
(\theta_r \wedge \ldots \wedge \theta_1) \vdash (\alpha_1 \wedge \ldots \wedge \alpha_r) = \sum_{\tau \in S_r} (-1)^\tau \prod_{i=1}^{r} \theta_i \vdash \alpha_{\tau(i)},
$$

and since we also have

$$
(\theta_r \wedge \ldots \wedge \theta_1) \lrcorner (\alpha_1 \wedge \ldots \wedge \alpha_r) = \sum_{\tau \in S_r} (-1)^\tau \prod_{i=1}^{r} \theta_i \lrcorner \alpha_{\tau(i)},
$$

we obtain $\lambda(\theta \vdash \alpha) = (-1)^r \theta \lrcorner \alpha$ for $r \geq 2$ from the case $r = 1$, using that, on U, every element of $\mathcal{V}^{-r}_{X/S}$ respectively $\mathcal{W}^r_{X/S}$ is locally of the product form.

On the strict locus U°, strict faithfulness is clear from the construction. Since the strict locus is scheme-theoretically dense, the restriction $\mathcal{V}^{-1}_{X/S} \to j_* \mathcal{V}^{-1}_{U^\circ/S}$ is injective; hence, $\mathcal{V}^\bullet_{X/S}$ is strictly faithful.

If $f\colon (X, U) \to S$ is log Gorenstein, then a local generator $\omega \in \mathcal{W}^d_{X/S}$ on an open subset $W \subseteq X$ induces an isomorphism $\mathcal{V}^p_{(U \cap W)/S} \cong \mathcal{W}^{p+d}_{(U \cap W)/S}$. Since both sides are reflexive, this is an isomorphism on W. In the log Calabi–Yau case, we use Proposition 3.28. For flatness in the case where $f\colon (X, U) \to S$ has the base change property, we use Corollary 8.49. The remaining statements are clear. $\qquad\square$

When we keep only the data of a Lie–Rinehart algebra in the context $\mathfrak{Cob}'(X/S)$, then we denote it by $\mathcal{LR}^\bullet_{X/S}$; it has two O_X-modules

$$\mathcal{LR}^F_{X/S} = O_X, \quad \mathcal{LR}^T_{X/S} = \Theta^1_{X/S}.$$

As in the case of the Gerstenhaber algebra, we consider $\mathcal{LR}^T_{X/S}$ endowed with the negative of the usual Lie bracket on the log tangent sheaf, and it acts on $\mathcal{LR}^F_{X/S}$ via the negative of the evaluation of the log derivation on functions. However, when we consider $\Theta^1_{X/S}$ as a Lie algebra, then we use the usual Lie bracket, as explained above.

8.2.2 Functoriality of the Gerstenhaber Calculus

Let $f\colon (X, U) \to S$ be a generically log smooth family, and let $b\colon T \to S$ be a morphism of locally Noetherian fs log schemes. Let $g\colon (Y, V) = (X, U) \times_S T \to T$ be the base change with induced map $c\colon (Y, V) \to (X, U)$. Then we have an induced isomorphism $c^*\Omega^1_{U/S} \xrightarrow{\cong} \Omega^1_{V/T}$, and hence, since $\mathcal{W}^i_{Y/T} = j_*\mathcal{W}^i_{V/T}$, an induced homomorphism

$$c^*\mathcal{W}^i_{X/S} \to \mathcal{W}^i_{Y/T}.$$

Since $\Omega^i_{U/S}$ is locally free, we also have an induced isomorphism

$$c^*\Theta^1_{U/S} = c^*\mathcal{H}om(\Omega^1_{U/S}, O_U) \cong \mathcal{H}om(\Omega^1_{V/T}, O_V) = \Theta^1_{V/T}.$$

Via the exterior power, this yields $c^*\Theta^p_{U/S} \cong \Theta^p_{V/T}$, and then we obtain a homomorphism $c^*\mathcal{V}^p_{X/S} \to \mathcal{V}^p_{Y/T}$. The adjoint maps on X are compatible with all structures.

Proposition 8.6 *In the above situation, let*

$$e_i\colon \ \mathcal{W}^i_{X/S} \to c_*\mathcal{W}^i_{Y/T} \quad \text{and} \quad e_p\colon \ \mathcal{V}^p_{X/S} \to c_*\mathcal{V}^p_{Y/T}$$

be the adjoint homomorphisms of O_X-modules. Then e is a map of two-sided Gerstenhaber calculi.

Proof We have to show the compatibility with all operations and constants. We start with the O-linear part. This can be checked on Y. The compatibility with $\wedge$ and the two units follows directly from the construction. Similarly, we have the compatibility with the contraction $\lrcorner : \mathcal{V}^{-1} \times \mathcal{W}^1 \to O$. The compatibility with all left and right contractions is then a formal consequence from the relations, given that, on U respectively V, $\mathcal{V}^\bullet$ is generated in degree -1, and $\mathcal{W}^\bullet$ is generated in degree 1. The compatibility with the de Rham differential ∂ is a well-known fact. Then the compatibility with $\mathcal{L}$ follows from the Lie–Rinehart homotopy formula. Now let ω be a local volume form on U/S. Then the mixed Leibniz rule shows

$$e([\theta, \xi]) \lrcorner e(\omega) = [e(\theta), e(\xi)] \lrcorner e(\omega).$$

Since $e(\omega)$ is a local volume form of V/T, this shows compatibility with $[-, -]$.
$\square$

8.3 Elementary Log Toroidal Families

We introduce *elementary log toroidal families*, which are the local models for *log toroidal families*, defined in the next Sect. 8.5.

8.3.1 Overview

The notion of a generically log smooth family $f : (X, U) \to S$ is deliberately very general, and its log singularities might be quite poorly behaved. In practice, we often want to restrict our attention to situations where we have additional information about the log singularities of $f : (X, U) \to S$. The most important approach is to ask that $f : (X, U) \to S$ should be locally isomorphic to some generically log smooth family in an explicitly given class of generically log smooth families. *Elementary log toroidal* families are one such class.

Let $\theta : Q \to P$ be an injective and saturated homomorphism of sharp toric monoids. Then we have an associated morphism

$$A_\theta : \quad A_P = \mathrm{Spec}(P \to \mathbb{Z}[P]) \to \mathrm{Spec}(Q \to \mathbb{Z}[Q]) = A_Q$$

of log schemes. This morphism is saturated and log smooth. Conversely, every saturated and log smooth morphism $f : X \to S$ is étale locally a base change of $A_P \times \mathbb{A}^r \to A_Q$ for some injective saturated $\theta : Q \to P$ and $r \geq 0$—this is a variant of K. Kato's toroidal characterization of log smoothness. Here, $\mathbb{A}^r$ carries the trivial log structure, not the one coming from $\mathbb{N}^r$. Otherwise, we would have the local model only locally in the smooth topology.

Before we continue, let us briefly discuss the geometry of the affine toric variety $\mathbb{A}_P = \operatorname{Spec} \mathbb{Z}[P]$.[7]

Construction 8.7 Let P be a toric monoid. For every face $F \subseteq P$, we have a closed embedding

$$V_F := \operatorname{Spec} \mathbb{Z}[F] \subset \operatorname{Spec} \mathbb{Z}[P] = \mathbb{A}_P$$

defined by $z^p \mapsto z^p$ for $p \in F$ and $z^p \mapsto 0$ for $p \notin F$. We say that $V_F \subset \mathbb{A}_P$ is a *closed toric stratum*. If $F \subseteq P$ is a facet (= a face of codimension 1), then $V_F \subset \mathbb{A}_P$ is a Weil divisor, called a *toric divisor*. The union of all toric divisors is called the *(full) toric boundary* and is denoted by D_P.

For a face $F \subseteq P$, we also have the localization P_F of P in F, which equals the submonoid in P^{gp} generated by P and $-F$. It gives an open embedding

$$U_F := \operatorname{Spec} \mathbb{Z}[P_F] \subseteq \operatorname{Spec} \mathbb{Z}[P] = \mathbb{A}_P$$

via the inclusion $P \subseteq P_F$ in P^{gp}. We denote their intersection by

$$V_F^{\circ} := V_F \cap U_F \subseteq \mathbb{A}_P,$$

and we call it an *open toric stratum*, although it is not open but locally closed in $\mathbb{A}_P$. We have $V_F^{\circ} \cong \operatorname{Spec} \mathbb{Z}[F^{\mathrm{gp}}]$ so that V_F° is smooth.

Set-theoretically, we have a disjoint decomposition

$$\mathbb{A}_P = \bigsqcup_{F \subseteq P} V_F^{\circ},$$

where the union is over faces $F \subseteq P$. In particular, we have the formula

$$\mathbb{A}_P \setminus U_F = \bigcup_{F \nsubseteq K} V_K,$$

where the union is over faces $K \subseteq P$ such that $F \nsubseteq K$, i.e., such that there is some $f \in F$ with $f \notin K$. $\Diamond$

It is proven in [222, III, Thm. 1.9.4] that the log structure on $\mathbb{A}_P$, defined via the chart $P \to \mathbb{Z}[P]$, is isomorphic to the compactifying log structure defined by the full toric boundary $D_P \subset \mathbb{A}_P$. To obtain elementary log toroidal families as a generalization of $A_\theta \colon A_P \to A_Q$, we now allow for some of the toric boundary divisors in D_P to be removed from the log structure. To explain this more precisely,

---

[7] Here, our convention is that A_P carries the log structure given by $P \to \mathbb{Z}[P]$, and $\mathbb{A}_P$ is its underlying scheme $\operatorname{Spec} \mathbb{Z}[P]$. We may also use this notation to denote $\operatorname{Spec} \mathbb{Z}[P]$ endowed with the trivial log structure.

let us denote by $\mathcal{F}_{\max}$ the set of all facets $F \subset P$, corresponding to the toric boundary divisors in D_P. For $F \in \mathcal{F}_{\max}$, the image $A_\theta(V_F)$ is contained in the toric boundary D_Q of A_Q if and only if $Q \not\subset F$. Let us denote the set of these facets by $\mathcal{F}_{\min}$. With this notation, we additionally have

$$A_\theta^{-1}(D_Q) = \bigcup_{F \in \mathcal{F}_{\min}} V_F \subset A_P.$$

Thus, for every set of facets $\mathcal{F}_{\min} \subseteq \mathcal{F} \subseteq \mathcal{F}_{\max}$, we have a closed subset $A_\theta^{-1}(D_Q) \subseteq D_{\mathcal{F}} \subseteq D_P$ of A_P, and hence we can define a family of log schemes

$$f \colon A_{P,\mathcal{F}} \to A_Q$$

where $A_{P,\mathcal{F}}$ carries the compactifying log structure defined by $D_{\mathcal{F}} = \bigcup_{F \in \mathcal{F}} V_F$. This is an *elementary log toroidal family*. We will exhibit an open subset $U_{P/Q} \subseteq A_{P,\mathcal{F}}$ which turns it into a generically log smooth family below, see (8.3). Note that $\mathcal{F}_{\min} \subseteq \mathcal{F}$ is a necessary condition in this construction since we need to have $A_\theta^{-1}(D_Q) \subseteq D_{\mathcal{F}}$ in order to obtain a morphism of log schemes.

A *log toroidal family* will be defined below as a generically log smooth family which is locally isomorphic to the base change of some elementary log toroidal family $f \colon A_{P,\mathcal{F}} \to A_Q$.

We studied elementary log toroidal families extensively in [77] and in the author's doctoral thesis [75]. They were our motivating example for the definition of a generically log smooth family, and they are still the best-understood class of generically log smooth families beyond log smooth families. The usefulness of this class of log singularities has been discovered by Gross and Siebert in [124, 125] when attempting to construct the structure of a log smooth morphism to the standard log point S_0 on a prescribed degenerate variety $X_0(B, \mathscr{P}, s)$. Namely, most of these degenerate varieties do not carry a global log structure with a log smooth morphism to S_0, but they all admit a canonical structure of a generically log smooth family with log toroidal singularities modeled by $f \colon A_{P,\mathcal{F}} \to A_Q$. A closely related example of the usefulness of the class of log toroidal singularities is our main result [77, Thm. 1.1], whose proof crucially relies on our good understanding of this class of log singularities.

In Chap. 14.1, we discuss in detail *local models of Gross–Siebert type*, which are a special class of elementary log toroidal families—except for that they come with a different log smooth locus.

In the rest of this section, we follow closely the exposition in [77] and [75].

8.3.2   Details

Definition 8.8 (Elementary Log Toroidal Data) An *elementary log toroidal datum* $(Q \subset P, \mathcal{F})$ (elt datum or ETD for short) consists of a saturated injective homomorphism $\theta \colon Q \to P$ of sharp toric monoids and a set $\mathcal{F}$ of facets of P containing all facets that do not contain Q. After setting

$$\mathcal{F}_{\min} := \underbrace{\{F \subset P \text{ a facet} \mid Q \not\subset F\}}_{\text{vertical facets}},$$

this means $\mathcal{F}_{\min} \subseteq \mathcal{F} \subseteq \mathcal{F}_{\max}$, where $\mathcal{F}_{\max}$ is the set of all facets of P. We set $d := \mathrm{rk}(P^{\mathrm{gp}}/Q^{\mathrm{gp}})$ and assume $d \geq 1$.

We interpret an elt datum $(Q \subset P, \mathcal{F})$ geometrically as the morphism

$$\underline{f} \colon \quad \mathbb{A}_P = \mathrm{Spec}\,\mathbb{Z}[P] \to \mathrm{Spec}\,\mathbb{Z}[Q] = \mathbb{A}_Q$$

of schemes. By [77, Lemma 3.4], this is a flat morphism whose fibers are Cohen–Macaulay and of dimension $d = \mathrm{rk}(P^{\mathrm{gp}}/Q^{\mathrm{gp}})$. Furthermore, its fibers are geometrically reduced. We have

$$\underline{f}^{-1}(D_Q) = \bigcup_{F \in \mathcal{F}_{\min}} V_F \subset \mathbb{A}_P$$

whereas, for a facet $F \notin \mathcal{F}_{\min}$, the composition $V_F \subset \mathbb{A}_P \to \mathbb{A}_Q$ is flat and surjective. Thus, we call the divisor V_F *vertical* if $F \in \mathcal{F}_{\min}$, and *horizontal* if $F \notin \mathcal{F}_{\min}$. Endowing $\mathbb{A}_Q$ with the divisorial log structure from D_Q and $\mathbb{A}_P$ with the divisorial log structure from $D_{\mathcal{F}} := \bigcup_{F \in \mathcal{F}} V_F$, we obtain a log morphism $f \colon A_{P,\mathcal{F}} \to A_Q$ which we call an *elementary log toroidal family*. We work here with divisorial log structures defined in the étale topology; however, by [222, III, Prop. 1.6.5], for $A_{P,\mathcal{F}}$ and A_Q, the canonical homomorphism from the étale log structure associated with the divisorial log structure defined in the Zariski topology is an isomorphism.

The composition

$$A_\theta \colon \quad A_P = A_{P,\mathcal{F}_{\max}} \to A_{P,\mathcal{F}} \xrightarrow{f} A_Q$$

is saturated since $\theta \colon Q \to P$ is saturated. Because $A_P \to A_{P,\mathcal{F}}$ is given by embedding one log structure as a sheaf of faces into another, it is exact. Then $f \colon A_{P,\mathcal{F}} \to A_Q$ is saturated by [222, I, Prop. 4.8.5(2)].

In the rest of this section, we show that we can consider $f \colon A_{P,\mathcal{F}} \to A_Q$ as a generically log smooth family, i.e., that we can find an appropriate open subset $U \subseteq \mathbb{A}_P$ such that $f \colon U \to A_Q$ is log smooth.

Example 8.9 Let P be a sharp toric monoid. Then $(0 \subset P, \mathcal{F}_{\max})$ is an elt datum. Its elementary log toroidal family is the affine toric variety A_P with its standard log structure given by the full toric boundary. $\Diamond$

Example 8.10 Consider the map $\theta \colon \mathbb{N} \to P$ of Example 1.79, and let

$$\mathcal{F} = \{\langle \bar{x}, \bar{w} \rangle, \langle \bar{y}, \bar{w} \rangle\}$$

where $\langle \cdot \rangle$ denotes the generated face. Then $\mathcal{F}$ equals $\mathcal{F}_{\min}$ and corresponds to the orange faces in Fig. 8.1. We obtain an elt datum $(Q \subset P, \mathcal{F})$ whose geometric realization is the family $f \colon X^{\mathrm{sze}} = \operatorname{Spec} \mathbf{k}[x, y, t, z]/(xy - tz) \to \mathbb{A}^1$ with the divisorial log structure defined by $t = 0$ on source and target. $\Diamond$

Example 8.11 If $(Q \subset P, \mathcal{F})$ is an elt datum and $r \geq 0$, then we obtain another elt datum

$$(Q \times \{0\} \subset P \times \mathbb{N}^r, \mathcal{F}')$$

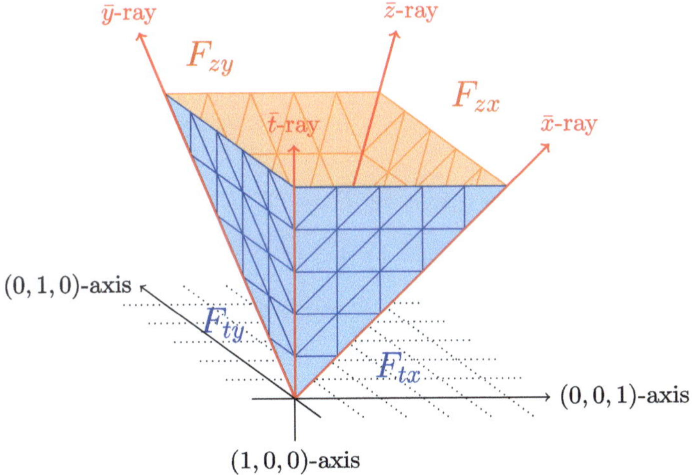

Fig. 8.1 Example 8.10. We see the four facets of the monoid P in Example 8.10. We have in blue the two facets $F_{tx} = \langle \bar{t}, \bar{x} \rangle$ and $F_{ty} = \langle \bar{t}, \bar{y} \rangle$ containing Q, and in orange the two maximal essential faces $F_{zx} = \langle \bar{x}$ and $\bar{z} \rangle$, $F_{zy} = \langle \bar{y}, \bar{z} \rangle$. The elements of the facets are the lattice points on the intersection of the lines of the grid. We also see $E = F_{zx} \cup F_{zx}$ in orange. This graphic is a slight modification of a graphic which has been created by the author for his master thesis and which is also contained in his doctoral thesis [75]. © Simon Felten 2018. All rights reserved

where $\mathcal{F}' = \{F \times \mathbb{N}^r \mid F \in \mathcal{F}\}$. The canonical morphism $A_{P \times \mathbb{N}^r, \mathcal{F}'} \to A_{P, \mathcal{F}}$ is smooth (on underlying schemes) and strict on the log smooth locus $U_{P/Q} \times \mathbb{A}^r$ (defined below). $\Diamond$

We analyze the combinatorics of $Q \subset P$. The homomorphism $\theta\colon Q \to P$ being saturated is equivalent to P being a free Q-set whose canonical basis is a union of faces of P, cf. [222, I, Cor. 4.6.11, Thm. 4.8.14, Cor. 1.4.3]. We denote this canonical basis by E. In other words, we have a subset $E \subseteq P$ such that every $p \in P$ has a unique decomposition $p = e + q$ with $e \in E$ and $q \in Q$, i.e.,

$$E \times Q \to P, \quad (e, q) \mapsto e + q, \tag{8.1}$$

is bijective, and this subset E is a union of faces of P. We say that a face F of P contained in E is an *essential face*.

We see that $E = P \setminus (Q^+ + P)$ where $Q^+ = Q \setminus 0$ is the maximal ideal. Moreover, projecting E to $P^{\mathrm{gp}}/Q^{\mathrm{gp}}$ is injective, and its image $\bar{P} \subset P^{\mathrm{gp}}/Q^{\mathrm{gp}}$ is a monoid since it is also the image of P. Note that $\bar{P}^{\mathrm{gp}} = P^{\mathrm{gp}}/Q^{\mathrm{gp}}$. A choice of splitting $P^{\mathrm{gp}} \cong \bar{P}^{\mathrm{gp}} \oplus Q^{\mathrm{gp}}$ yields a unique map of sets $\varphi\colon \bar{P} \to Q^{\mathrm{gp}}$ so that $\mathrm{id} \times \varphi\colon \bar{P} \to \bar{P} \oplus Q^{\mathrm{gp}}$ is a section of the projection $P \to \bar{P}$ with the property that its image is E, so

$$P = \{(\bar{p}, q) \in \bar{P} \oplus Q^{\mathrm{gp}} \mid \exists \tilde{q} \in Q\colon q = \varphi(\bar{p}) + \tilde{q}\}. \tag{8.2}$$

Next, we define the open subset $U = U_{P/Q} \subseteq \mathbb{A}_P$ that satisfies the codimension condition and has the property that $f\colon A_{P,\mathcal{F}} \to A_Q$ is log smooth on U. We actually define the complement $Z_{P/Q}$ of $U_{P/Q}$, for which we use the following description of the faces of P.

Lemma 8.12 *Let $F \subseteq P$ be a face. Set $\bar{F} := F \cap E$, $Q' \cap F$, then*

$$F = \bar{F} + Q' := \{\bar{f} + q' \mid \bar{f} \in \bar{F}, q' \in Q'\}.$$

Since E is a union of faces of P, so is $\bar{F}$. Note also that Q' is a face of Q.

Proof This is [77, Lemma 3.5]. $\square$

With this description at hand, we define the set of *bad faces* of P as

$$\mathcal{B} = \left\{ \bar{F} + Q' \;\middle|\; \begin{array}{l} \bar{F} \text{ is a union of essential faces of rank at most } d - 2 \\ Q' \text{ is a face of } Q,\ \bar{F} + Q' \text{ is a face of } P \end{array} \right\}.$$

Let $\bar{F} + Q' \in \mathcal{B}$. Then, by [77, Lemma 3.6], the closed subset $V_{\bar{F}+Q'} \subset \mathbb{A}_P$ is flat over $V_{Q'} \subset \mathbb{A}_Q$, and $V_{\bar{F}+Q'}$ has codimension ≥ 2 in every fiber of $\underline{f}\colon \mathbb{A}_P \to \mathbb{A}_Q$. In other words,

$$U := U_{P/Q} := \mathbb{A}_P \setminus \left(\bigcup_{B \in \mathcal{B}} V_B \right) \tag{8.3}$$

satisfies the codimension condition. By [77, Lemma 3.7], we also have

$$U_{P/Q} = \bigcup_{F \in \mathcal{E}} U_F$$

where $\mathcal{E}$ is the set of essential faces of P which have rank $d - 1$. In this formula, $U_F \subseteq \mathbb{A}_P$ is the open subset defined by localization in the face $F \subseteq P$.

Proposition 8.13 *The morphism of partial log schemes $f \colon (\mathbb{A}_{P,\mathcal{F}}, U_{P/Q}) \to \mathbb{A}_Q$ is a generically log smooth family.*

Definition 8.14 (Elementary Log Toroidal Families) Let $(Q \subset P, \mathcal{F})$ be an elt datum. Then the associated *elementary log toroidal family* is the generically log smooth family $f \colon (\mathbb{A}_{P,\mathcal{F}}, U_{P/Q}) \to \mathbb{A}_Q$.

Proof of Proposition 8.13 The essential point is to show that $f \colon U_{P/Q} \to \mathbb{A}_Q$ is log smooth. Let $F \subseteq P$ be an essential face of rank $d - 1$. We show that $U_F \to \mathbb{A}_Q$ is log smooth. Set $\bar{P}_F := P_F / F^{\mathrm{gp}}$ and note that the projection of Q to $\bar{P}_F$ is injective because $F^{\mathrm{gp}} \cap Q = \{0\}$. There is an isomorphism $P_F \cong F^{\mathrm{gp}} \times \bar{P}_F$ commuting with the injection of Q that is $\{0\} \times Q$ on the right. The log structure on U_F is a divisorial log structure given by a set of divisors each of which pulls back from $\operatorname{Spec} \mathbb{Z}[\bar{P}_F]$, so we may consider the corresponding divisorial log structure on $\operatorname{Spec} \mathbb{Z}[\bar{P}_F]$ to upgrade this to a log scheme $\bar{U}_F$. We have a factorization $U_F \to \bar{U}_F \to \mathbb{A}_Q$ with the first map a smooth projection from a product that is therefore strict, hence log smooth. It thus suffices to show that $\bar{U}_F \to \mathbb{A}_Q$ is log smooth. Note that $\bar{U}_F \to \mathbb{A}_Q$ is the log morphism of an elt datum with $d = 1$. The following lemma finishes the proof. $\qquad\qquad\square$

Lemma 8.15 *Assume that $f \colon \mathbb{A}_{P,\mathcal{F}} \to \mathbb{A}_Q$ has one-dimensional fibers, i.e., $d = 1$. Then f is log smooth.*

Proof If $\mathcal{F} = \mathcal{F}_{\max}$, then $\mathbb{A}_{P,\mathcal{F}} = \mathbb{A}_P$ and $f \colon \mathbb{A}_P \to \mathbb{A}_Q$ is log smooth. This always holds if Q meets the interior of P, so assume Q is contained in a proper face of P. Then, by Lemma 8.12, it is in fact a facet of P, and then $\bar{P} = \mathbb{N}$ so that $P = \mathbb{N} \times Q$. A facet of P that is not Q is in $\mathcal{F}_{\min} = \{\mathbb{N} \times F \mid F \text{ is a facet of } Q\}$. Hence $\mathcal{F} \subsetneq \mathcal{F}_{\max}$ implies $\mathcal{F} = \mathcal{F}_{\min}$, and thus f is strict. Since $\underline{f}$ is smooth, we find that f is log smooth. $\qquad\qquad\square$

Remark 8.16 The third situation of Fig. 8.2 is an example of this situation. $\qquad\quad\Diamond$

Corollary 8.17 *It is possible to find a decomposition $U_{P/Q} = U_1 \cup U_2$ into open subsets such that $\mathbb{A}_P|_{U_1} = \mathbb{A}_{P,\mathcal{F}}|_{U_1}$, and $f \colon U_2 \subseteq \mathbb{A}_{P,\mathcal{F}} \to \mathbb{A}_Q$ is strict and smooth.*

Proof Let $\mathcal{E}_1$ be the set of essential faces of rank $d - 1$ such that when applying the proof of Lemma 8.15 to $\bar{U}_F \to \mathbb{A}_Q$ from the proof of Proposition 8.13, we are in the case $\mathcal{F} = \mathcal{F}_{\max}$, and let $\mathcal{E}_2$ be the set of faces where we are in case $\mathcal{F} = \mathcal{F}_{\min}$. Then

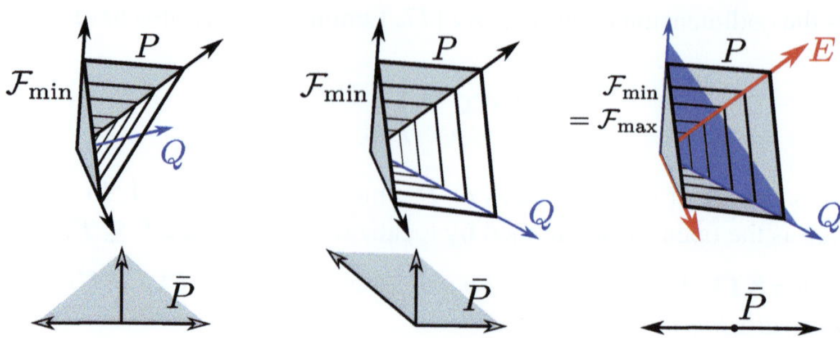

Fig. 8.2 Examples of $Q \subset P$. These are three examples of a saturated injection $\theta : Q \subset P$ together with the image $\bar{P} \subseteq P^{\mathrm{gp}}/Q^{\mathrm{gp}}$ of P. The outer two examples are log smooth, and the middle one gives Example 1.79. This graphic is a slight modification of a graphic which has been created by Helge Ruddat for the article [77] and which is also contained in the author's doctoral thesis [75]. A CC BY 4.0 license applies to it

for $F \in \mathcal{E}_1$ we have $A_P|_{U_F} = A_{P,\mathcal{F}}|_{U_F}$, and for $F \in \mathcal{E}_2$, the morphism $U_F \to A_Q$ is strict and smooth. Now we define $U_1 = \bigcup_{F \in \mathcal{E}_1} U_F$ and $U_2 = \bigcup_{F \in \mathcal{E}_2} U_F$. $\square$

Remark 8.18 If $F \subseteq P$ is a face, we can easily determine whether $A_P|_{U_F} = A_{P,\mathcal{F}}|_{U_F}$ or not: Equality holds if and only if, for every facet $H \subseteq P$ with $F \subseteq H$, we have $H \in \mathcal{F}$. Indeed, the facets with $F \subseteq H$ correspond to the toric divisors intersecting U_F. Moreover, we need to test only facets with $Q \subseteq H$, too. In case F is an essential face of rank $d - 1$, the generated face $\langle F, Q \rangle$ is at least a facet by a dimension argument, so we need to test at most one facet: Thus $A_P|_{U_F} = A_{P,\mathcal{F}}|_{U_F}$ for $\langle F, Q \rangle = P$ or $\langle F, Q \rangle \in \mathcal{F}$, and $U_F \to A_Q$ is strict and smooth otherwise. $\Diamond$

Remark 8.19 Every elementary log toroidal family $f : A_{P,\mathcal{F}} \to A_Q$ is *generically strict*, by which we mean that there is an open $U_\circ \subseteq U$, dense in every fiber, on which f is strict. Indeed, for an essential face F of rank d, i.e., a maximal essential face, we have $\langle F, Q \rangle = P$, so $A_{P,\mathcal{F}}|_{U_F} = A_P|_{U_F}$. Moreover, the map $U_F \to A_Q$ is the projection from the product $A_Q \times A_{P_F^{\mathrm{gp}}}$, so it is strict and smooth. Now the complement of $\bigcup_{F \in \mathcal{E}(d)} U_F$ has dimension $\leq d - 1$ in every fiber where $\mathcal{E}(d)$ is the set of essential faces of rank d. $\Diamond$

Remark 8.20 Let R be a regular ring. Then base change along $\operatorname{Spec} R \to \operatorname{Spec} \mathbb{Z}$ gives a generically log smooth family $A_{P,\mathcal{F}} \times R \to A_Q \times R$. Since R is regular, $A_Q \times R$ is log regular and hence carries a compactifying log structure by [163, Thm. 11.6]. We see that it is the divisorial log structure induced by $D_Q \times R$. Similarly, $U_{P/Q} \times R$ is log regular, so it carries the divisorial log structure induced by $\bigcup_{F \in \mathcal{F}} \operatorname{Spec} R[F]$. This means that the construction of $f : A_{P,\mathcal{F}} \to A_Q$ can be carried out with any regular domain R instead of $\mathbb{Z}$, yielding a generically log smooth family $A_{P,\mathcal{F}} \times R \to A_Q \times R$. We do not know (and do not care) if the two log structures (from base change and from the direct divisor construction) also coincide outside $U_{P/Q} \times R$. $\diamond$

Remark 8.21 Let $(Q \subset P, \mathcal{F})$ be an elt datum, and let $\beta : Q \to Q'$ be a homomorphism of sharp toric monoids. In general, the push-out $Q' \oplus_Q P$ is not a sharp monoid, so we cannot find an elt datum $(Q' \subset Q' \oplus_Q P, \mathcal{F}')$ in the sense of Definition 8.8 as base change along $\beta : Q \to Q'$. However, we get very close in the sense that there is some elt datum $(Q' \subset P'', \mathcal{F}'')$ such that the base change $A_{P,\mathcal{F}} \times_{A_Q} A_{Q'} \to A_{Q'}$ is an open subset of $A_{P'',\mathcal{F}''} \to A_{Q'}$. This is explained in [75, Prop. 3.25] and the discussion preceding it. $\diamond$

8.4 Examples of ELT Data

Figure 8.2 shows some more abstract examples of saturated injections $Q \subset P$. In this section, we give some concrete examples of elt data. Moreover, many examples in the introduction are elementary log toroidal families, e.g. Example 1.83.

Example 8.22 The saturated injection $\mathbb{N} \to \mathbb{N}^2$, $1 \mapsto (1, 1)$, satisfies $\mathcal{F}_{\min} = \mathcal{F}_{\max}$ and gives an elt datum $(\mathbb{N} \subset \mathbb{N}^2, \mathcal{F})$. Geometrically, this is

$$f : \quad \operatorname{Spec} \mathbb{Z}[x, y] \to \operatorname{Spec} \mathbb{Z}[t], \quad t \mapsto xy,$$

with the divisorial log structure defined by $\{t = 0\}$ respectively $\{xy = 0\}$. Applying the construction from Example 8.11 with $r = 1$ to it, we obtain a log smooth and saturated morphism

$$f : \quad (\mathbb{A}^3 | \{xy = 0\}) \to (\mathbb{A}^1 | \{0\})$$

which is associated with the elt datum with homomorphism $\mathbb{N} \to \mathbb{N}^3$, $1 \mapsto (1, 1, 0)$, and set of facets $\mathcal{F}_{\min} = \{\mathbb{N} \times 0 \times \mathbb{N}, 0 \times \mathbb{N} \times \mathbb{N}\}$. This is the same example as Example 1.62. $\diamond$

Non-Example 8.23 The map $\theta : \mathbb{N} \to \mathbb{N}^r$, $1 \mapsto (a_1, \ldots, a_r)$, is saturated if and only if $0 \le a_i \le 1$ for all i. Thus, if $a_i \ge 2$ for some i, then $(\mathbb{N} \subset \mathbb{N}^r, \mathcal{F})$ is *not* an elt datum. Indeed, e.g.

$$f : \quad \mathbb{A}^2 \to \mathbb{A}^1, \quad (x, y) \mapsto x^2 y,$$

has a non-reduced fiber $f^{-1}(0) \cong \operatorname{Spec} k[x, y]/(x^2 y)$, which is not possible for an elementary log toroidal family. Nonetheless, f is log smooth and integral. $\diamond$

Example 8.24 Generalizing Examples 1.79, 8.10, and 1.156, the family

$$\{tu - x_1 \cdot \ldots \cdot x_k = 0\} \to \mathbb{A}^1_t$$

has the structure of an elementary log toroidal family. For an illustration in the case $k = 2$, see Fig. 1.12 on page 55, and for an illustration in the case $k = 3$, see Fig. 1.20 on page 118.

Let $k \geq 1$, and let $P_k^{\mathrm{anc}} \subset \mathbb{N}^{k+1}$ be the submonoid spanned by

$$\bar{t} = e_0, \quad \bar{w} = (k - 1) \cdot e_0 + e_1 + \ldots + e_k, \quad \bar{z}_i = e_0 + e_i \quad (1 \leq i \leq k)$$

where $e_0, \ldots, e_k$ is the standard basis of $\mathbb{N}^{k+1}$. Since

$$P_k^{\mathrm{anc}} = \left\{ p \in \mathbb{Z}^{k+1} \,\middle|\, \forall 1 \leq i \leq k \colon \langle e_i, p \rangle \geq 0, \left\langle e_0 - \sum_{j \neq 0, i} e_j, p \right\rangle \geq 0 \right\},$$

it is a sharp toric monoid. The homomorphism $\theta \colon \mathbb{N} \to P_k^{\mathrm{anc}}, 1 \mapsto \bar{t}$, is locally exact by the exactness criterion [222, I, Prop. 2.1.16(5)]. Because the complement of the ideal $J = T + P_k^{\mathrm{anc}}$ is

$$E_k^{\mathrm{anc}} = \bigcup_{i=1}^{k} \left\{ p \in P_k^{\mathrm{anc}} \,\middle|\, \left\langle e_0 - \sum_{j \neq 0, i} e_j, p \right\rangle = 0 \right\} =: \bigcup_{i=1}^{k} F_i^{\mathrm{ver}},$$

it is a radical ideal, and $\theta \colon \mathbb{N} \to P_k^{\mathrm{anc}}$ is saturated by [222, I, Thm. 4.8.14(5)]. Any facet of P_k^{anc} is $h^{-1}(0)$ for a function $h \colon P_k^{\mathrm{anc}} \to \mathbb{N}$ that evaluates 0 on at least k elements of $\{\bar{t}, \bar{w}, \bar{z}_1, \ldots, \bar{z}_k\}$, so the facets of P_k^{anc} are F_i^{ver} defined by the formula above, and $F_i^{\mathrm{hor}} = \{p \in P_k^{\mathrm{anc}} \mid \langle e_i, p \rangle = 0\}$. Thus, after setting

$$\mathcal{F}_{\min} := \{F_1^{\mathrm{ver}}, \ldots, F_k^{\mathrm{ver}}\},$$

we obtain an elt datum $(\mathbb{N} \subset P_k^{\mathrm{anc}}, \mathcal{F}_{\min})$. Geometrically, we have a log morphism

$$X_k^{\mathrm{anc}} = \mathrm{Spec}\, \mathbb{Z}[t, u, x_1, \ldots, x_k]/(tu - x_1 \cdot \ldots \cdot x_k) \to \mathbb{A}^1_t,$$

where both schemes carry the divisorial log structure defined by $\{t = 0\}$.

"anc" is short for "almost normal crossing" as in Example 1.86, where we defined the class $\mathscr{C}^{\mathrm{anc}}$ of almost normal crossing log singularities. Below in Example 8.44, we use the monoids P_k^{anc} to show that a pencil of normal crossing divisors (with only finitely many of them not smooth) is log toroidal. More importantly, local models of this type feature prominently in the proof of Theorem 18.44, which provides us with smoothings of many normal crossing spaces. We will see in Construction 18.46 that $X_k^{\mathrm{anc}} \to \mathbb{A}^1_t$ is a local model of standard Gross–Siebert type. $\Diamond$

Example 8.25 Any local model of weak Gross–Siebert type is an elementary log toroidal family when we use $U_{P/Q}$ as the log smooth locus. See Sect. 14.1 for more information.   ◊

8.5 Log Toroidal Families

After studying elementary log toroidal families, in this section, we introduce log toroidal families, i.e., generically log smooth families whose log singularities are given by elementary log toroidal families. This section follows closely the corresponding section in the author's doctoral thesis [75].

8.5.1 Definition and Basic Properties

Our precise definition of a log toroidal family is slightly technical and intricate. We start with the definition of a local model. This definition should be compared with the toroidal characterization of log smoothness in the form of Theorem 7.59.

Definition 8.26 (Log Toroidal Local Models) Let $f\colon (X, U) \to S$ be a generically log smooth family.

(1) For a geometric point $\bar{s} \to S$, a *base chart* at $\bar{s}$ is a strict étale morphism $q\colon (S', \bar{s}') \to (S, \bar{s})$ and a map $a\colon S' \to A_Q$ given by a chart $Q \to \mathcal{M}_{\bar{s}}$ of the log structure which is neat at $\bar{s}$, i.e., $Q \to \overline{\mathcal{M}}_{S,\bar{s}}$ is an isomorphism. In this case, Q is a sharp toric monoid, and $q(\bar{s}')$ is in the closed subset of A_Q which is cut out by the ideal $\mathbb{Z}[Q \setminus \{0\}]$.[8]
(2) A *base chart* (without reference to a point) is a strict étale morphism $q\colon S' \to S$ together with a strict morphism $S' \to A_Q$ for some sharp toric monoid Q.
(3) For a geometric point $\bar{x} \in X$ and a base chart at $f(\bar{x})$, a *log toroidal local model* at $\bar{x}$ is a commutative diagram

$$
\begin{array}{ccccccc}
(X, U) & \xleftarrow{\ p\ } & (V, U_V) & \xrightarrow{\ h\ } & (L, U_L) & \xrightarrow{\ c\ } & (A_{P,\mathcal{F}}, U_{P/Q}) \\
\ \downarrow{\scriptstyle f} & & \downarrow & & \downarrow & & \downarrow \\
S & \xleftarrow{\ q\ } & S' & = & S' & \xrightarrow{\ a\ } & A_Q
\end{array}
\qquad\text{(LM)}
$$

---

[8] This closed subset is isomorphic to $\operatorname{Spec}\mathbb{Z}$, and the image depends only on the characteristic of the residue field $\kappa(\bar{s})$.

of partial log schemes where:

(a) the right vertical map is the elementary log toroidal family associated with an elt datum $(Q \subset P, \mathcal{F})$;

(b) the right square is Cartesian in the category of partial log schemes; in particular, $U_L = c^{-1}(U_{P/Q})$, and $(L, U_L) \to S'$ is generically log smooth;

(c) $h \colon (V, U_V) \to (L, U_L)$ is strict and étale but not necessarily accurate; $U_V \subseteq V$ satisfies the codimension condition so that $(V, U_V) \to S'$ is generically log smooth;

(d) $p \colon (V, U_V) \to (X, U)$ is a strict étale neighborhood of $\bar{x}$ but not necessarily accurate;

(e) $c \circ h(\bar{x})$ is in the closed subset of $A_{P,\mathcal{F}}$ which is cut out by the ideal $\mathbb{Z}[P \setminus \{0\}]$.

(4) For a base chart $S \leftarrow S' \to A_Q$ and a Zariski open $W \subseteq X$, a *log toroidal local model* is a diagram (LM) as above with $p(V) = W$ (and no requirement on some $\bar{x} \in X$).

Definition 8.27 (Log Toroidal Families)

(1) A *log toroidal family* is a generically log smooth family $f \colon (X, U) \to S$ such that we can find finitely many base charts $S \leftarrow S'_i \to A_{Q_i}$, Zariski open subsets $W_i \subseteq X$ with $\bigcup_i W_i = X$, and a log toroidal local model for each W_i.

(2) Let $S \cong \mathrm{Spec}(Q \to B)$, and consider the base chart given by $S' = S$ and $a \colon S \to A_Q$ given by the chart $Q \to B$. Then $f \colon (X, U) \to S$ is *log toroidal with respect to* $a \colon S \to A_Q$ if we can choose the log toroidal local models over this base chart.

Example 8.28 Every elementary log toroidal family $f \colon (A_{P,\mathcal{F}}, U_{P/Q}) \to A_Q$ is log toroidal. $\Diamond$

Example 8.29 Toric varieties X are log toroidal families over a trivial base. Indeed, let $\underline{X}$ be a toric variety over $S = \mathrm{Spec}\,\mathbb{Z}$ and $D \subset X$ a reduced toric divisor. Let X be the log space $\underline{X}$ with divisorial log structure defined by D and let S have the trivial log structure, then $X \to S$ is a log toroidal family since it is Zariski locally Example 8.28 with $Q = 0$. $\Diamond$

We give more examples below in Sect. 8.6. We recommend to browse the examples before reading on. In particular, we will briefly comment on the fact that saturated log smooth morphisms are log toroidal in Example 8.41.

Remark 8.30 We do not require $p \colon (V, U_V) \to (X, U)$ and $h \colon (V, U_V) \to (L, U_L)$ to be accurate morphisms of partial log schemes since we do not care about the precise chosen log smooth locus $U \subseteq X$. In this way, e.g. $f \colon A_P \to A_Q$ with $U = A_P$ instead of $U = U_{P/Q}$ can be considered as a log toroidal family. However, this definition means we cannot easily understand the local structure of constructions that do depend on the precise U just by considerations in the local models. This applies in particular to the deformation theory of log toroidal families. $\Diamond$

Remark 8.31 There are three natural options to define a log toroidal family: First, we could require a local model *at* every point $\bar{x} \in X$, i.e., we require $(coh)(\bar{x}) = 0 \in A_{P,\mathcal{F}}$. Secondly, we could require finitely many points $\bar{x}_i \in X$ and local models at them such that $X = \bigcup_i p_i(V_i)$. For the definition above, we decided to use the third option employing local models for Zariski opens $W \subseteq X$. This gives us (compared to the second option) more flexibility since we do not need to give a point mapping to $0 \in A_{P,\mathcal{F}}$. With our definition, the family $f : A_{P,\mathcal{F}} \setminus \{0\} \to A_Q$ is obviously log toroidal, and this flexibility is also crucial for the proof of Proposition 8.32 below. However, we prove in Proposition 8.38 that, over an algebraically closed base field **k**, there is no difference between the three options. In other words, in this situation, we actually have local models controlling the local geometry around every point (which is the first option). $\diamond$

8.5.2   *Base Change*

Due to our assumption that $a : \tilde{S} \to A_Q$ must be a chart, a priori, the notion of a log toroidal family is stable only under strict base change (along a strict morphism $b : T \to S$) but not under general base change. Unfortunately, we have never proven the stability in full generality, but we briefly mention the following result proven in the author's doctoral thesis [75].

Proposition 8.32 *Let $f : (X, U) \to S$ be a log toroidal family, and let $b : T \to S$ be a morphism from a locally Noetherian fs log scheme T. Assume that, for every geometric point $\bar{t} \in T$, the order of the torsion subgroup of $\mathcal{M}^{\mathrm{gp}}_{T/S,\bar{t}}$ is invertible in the residue field $\kappa(\bar{t})$. Then the fiber product $g : (Y, V) = (X, U) \times_S T \to T$ is a log toroidal family.*

Proof This is [75, Prop. 4.7]. See also Remark 8.21 above. $\square$

Remark 8.33 The order of the torsion subgroup of $\mathcal{M}^{\mathrm{gp}}_{T/S,\bar{t}}$ being invertible in $\kappa(\bar{t})$ is a very technical condition. We need it here to ensure the existence of a chart for $T \to S$ which is neat, not only exact, at $\bar{t}$. We expect that a stronger version of [75, Prop. 3.25] would render the assumption superfluous. However, the condition is virtually always satisfied, e.g. when $T \to S$ is strict or more generally saturated, or if $S/\mathbb{Q}$. $\diamond$

8.5.3   *Generic Strictness*

As we have seen in Remark 8.19, every elementary log toroidal family is generically strict. This easily generalizes to log toroidal families and even generically log smooth families.

Lemma 8.34 *Let $f : (X, U) \to S$ be a log toroidal family. Then there is an open subset $U^\circ \subseteq U$, dense in every fiber, on which f is strict. The same holds for generically log smooth families.*

Proof In a log toroidal local model (LM), denote the strict locus of Remark 8.19 by $U^\circ_{P/Q} \subseteq U_{P/Q}$, and take the union over all $g((c \circ h)^{-1}(U^\circ_{P/Q}) \cap \tilde{U}) \subseteq X$ for all log toroidal local models. For generically log smooth families, note that the saturated log smooth morphism $f : U \to S$ is a log toroidal family by Example 8.41 below.
□

Corollary 8.35 *Let $f : (X, U) \to S$ be a generically log smooth family. Then the forgetful map $\Theta^1_{X/S} \to \Theta^1_{\underline{X}/\underline{S}}$ forgetting the log part of the derivation is injective.*

Proof Since $O_X \to j_* O_{U^\circ}$ is injective, this follows from [125, Prop. 1.3].
□

Corollary 8.36 *Let $f : (X, U) \to S$ be a generically log smooth family of relative dimension d. Then rk $\Omega^1_{U/S} = d$.*

Remark 8.37 In general, if $f : X \to S$ is a flat finite-type morphism of locally Noetherian schemes with reduced fibers, and $j : U \to X$ is an open subset which is dense in every fiber, then $O_X \to j_* O_U$ is injective. For lack of reference, we briefly indicate the proof: Injectivity is equivalent to $U \subseteq X$ being scheme-theoretically dense by [267, 01RE], which, since $U \to X$ is a quasi-compact morphism, holds if and only if the scheme-theoretic closure $\overline{U} \subseteq X$ is X. If $S = \operatorname{Spec} k$ for a field k, then this holds because X is reduced. If $S = \operatorname{Spec} A$ for an Artinian ring A, then the base change $(\overline{U})_0$ to the reduction is the scheme-theoretic image of $U_0 \subseteq X_0$. Because $A \to A/\mathfrak{m}_A$ is a finite-order thickening, the closed immersion $\overline{U} \subseteq X$ is an isomorphism by [267, 06AG], so $O_X \to j_* O_U$ is injective. From here, we easily generalize to complete local Noetherian, local Noetherian, and finally arbitrary locally Noetherian bases.
◇

8.5.4 Local Models at Points

As we have seen in Remark 8.31, a priori there is a difference between log toroidal local models for open subsets $W \subseteq X$ and log toroidal local models at points $\bar{x}$. The first one is easier to check, the second one is more convenient to study the local structure of log toroidal families. In this section, we show that, given a log toroidal family $f : (X, U) \to S$ over an algebraically closed field $\mathbf{k}$ (which has log toroidal local models for opens $W_i \subseteq X$ by definition), there exist log toroidal local models at points $\bar{x} \in X$. This is important for the proof of the relative degeneration Theorem [77, Thm. 1.10] (Theorem 8.65 below) since it depends on a local computation that we carry out on $0 \in A_{P,\mathcal{F}}$ of a local model.

Proposition 8.38 *Let* **k** *be an algebraically closed field, let* $f : (X, U) \to S$ *be a log toroidal family where* $S/\mathbf{k}$*, and let* $\bar{x} \in X$ *be a* **k***-valued point. Then there is a base chart* $S \leftarrow S' \to A_Q$ *at* $f(\bar{x})$ *and a log toroidal local model at* $\bar{x}$*.*

Proof This is [75, Prop. 4.15]. The result is not included in [77], but the method of the proof is a standard method. $\square$

Corollary 8.39 *Let* **k** *be an algebraically closed field, let* A *be an Artinian* $\mathbf{k}[\![Q]\!]$*-algebra, and let* $S = \mathrm{Spec}(Q \to A)$*. Let* $f : (X, U) \to S$ *be a log toroidal family with respect to* $S \to A_Q$ *and let* $\bar{x} \in X$ *be a* **k***-valued point. Then there is a log toroidal local model at* $\bar{x}$ *with base chart* $S \to A_Q$*.*

Proof This is [75, Cor. 4.16]. $\square$

Remark 8.40 For a log toroidal family $f : (X, U) \to S$ with respect to $S \to A_Q$ where S is not a punctual scheme, there is no analog of Corollary 8.39: For some points $\bar{s} \to S$, the induced map $Q \to \overline{\mathcal{M}}_{S,\bar{s}}$ might not be an isomorphism, so we need to change the base chart. $\Diamond$

8.6 Examples of Log Toroidal Families

We illustrate the concept of a log toroidal family with several examples.

8.6.1 Saturated Log Smooth Morphisms

Example 8.41 Let $f : X \to S$ be a saturated log smooth morphism. Then it is a log toroidal family with $U = X$.

Proof This is a consequence of Kato's toroidal characterization of log smoothness in the version of Theorem 7.59, see also [222, IV, Thm. 3.3.3]. Let $\bar{x} \to X$ be a geometric point, and let $S \leftarrow S' \to A_Q$ be a base chart at $f(\bar{x})$ in the sense of Definition 8.26. Then Theorem 7.59 yields a chart as in that theorem. It remains to see that $A_P \times \mathbb{A}^r \to A_Q$ arises from an elt datum. We have already seen in Theorem 7.59 that $Q \to P$ is injective and saturated. Now the desired elt datum is $(Q \subset P \oplus \mathbb{N}^r, \mathcal{F}')$ obtained by applying the construction in Example 8.11 to $(Q \subset P, \mathcal{F}_{\max})$. $\square$

8.6.2 Toroidal Embeddings

Let us work over an algebraically closed field $\mathbf{k} \supset \mathbb{Q}$. Toroidal embeddings $U \subseteq X$ which have been originally defined in [174] give rise to log smooth and saturated log schemes over $\operatorname{Spec} \mathbf{k}$ by endowing X with the divisorial log structure defined by $X \setminus U$. As observed e.g. in [10] the definition is equivalent to the existence of an étale roof

$$(X, x) \overset{u}{\leftarrow} (X', x') \overset{v}{\to} (A_P, p)$$

for every closed point $x \in X$ where A_P is an affine toric variety, $p \in A_P$ is a point and we have $u^{-1}(U) = v^{-1}(A_P^*)$. Toroidal morphisms $f : (X, U_X) \to (S, U_S)$ of toroidal embeddings have been defined e.g. in [3]. Using [15, Cor. 2.6] we can find étale roofs as above, a toric morphism $A_P \to A_Q$ and the middle map such that we have a commutative diagram

$$
\begin{array}{ccccc}
(X, x) & \longleftarrow & (X', x') & \longrightarrow & (A_P, p) \\
\downarrow & & \downarrow & & \downarrow \\
(S, f(x)) & \longleftarrow & (S', s') & \longrightarrow & (A_Q, q)
\end{array}
$$

Since f is dominant and $\hat{O}_{S, f(x)}$ is a domain the induced map $\hat{O}_{S, f(x)} \to \hat{O}_{X, x}$ is injective by [135, I. Cor. 3.9.8]. We find that $\mathbf{k}[Q] \to \mathbf{k}[P]$ is injective, so the kernel and the torsion part of the cokernel of $Q^{\mathrm{gp}} \to P^{\mathrm{gp}}$ are finite groups of order invertible in $\mathbf{k}$. This proves $f : X \to S$ log smooth, a fact that has been also observed e.g. in [61] with a slightly different definition. On the other hand, not every toroidal morphism is saturated, e.g. $\mathbf{k}[t] \to \mathbf{k}[t], t \mapsto t^3$. However, weakly semistable families in the sense of Abramovich–Karu as defined in [3] are. They have been introduced to generalize semistable reduction to higher dimensional bases. To us, they are just an example illustrating the wide range where log toroidal families occur. We recall the definition for convenience of the reader.

Definition 8.42 ([3]) A *weakly semistable family* is a flat morphism $f : X \to S$ of projective varieties with connected fibers together with structures of toroidal embeddings $U_S \subseteq S, U_X \subseteq X$ such that $f^{-1}(U_S) = U_X$ and f is toroidal, equidimensional and has reduced fibers. Moreover S is nonsingular. The family $f : X \to S$ is *semistable* if additionally X is smooth.

Note that this definition of semistability is more general than the classical one as we do not assume S to be one-dimensional.

Example 8.43 Weakly semistable families are log smooth and saturated.

Proof The log smooth map $A_P \to A_Q$ is log flat. Since $X' \to S'$ is flat and S' is log regular the result [222, IV, Thm. 4.3.5] yields $X' \to S'$ integral. It has reduced fibers, so it is saturated by [222, IV, Thm. 4.3.6].   □

8.6.3 Degenerations

The following construction is a generalization of the degeneration of the smooth quartic surface in Example 1.11.

Example 8.44 Pencils of normal crossing divisors are log toroidal families. More precisely, let $\mathbf{k} = \bar{\mathbf{k}}$ be an algebraically closed field, let $H = \{F = 0\} \subset \mathbb{P}^n$ be a normal crossing divisor of degree d, let $K = \{G = 0\} \subset \mathbb{P}^n$ be a smooth hypersurface of degree d and assume that $H \cup K = \{FG = 0\}$ is a normal crossing divisor. After defining

$$Y := \{TF - SG = 0\} \subset \mathbb{P}^1 \times \mathbb{P}^n$$

where $[S : T]$ are the variables of $\mathbb{P}^1$, we obtain a morphism $\mathbf{f} \colon Y \to \mathbb{P}^1$ via projection. Then there is a neighborhood $S \subseteq \mathbb{P}^1$ of $0 = [0 : 1]$ and an open subset $U \subseteq X := \mathbf{f}^{-1}(S)$ such that $f \colon (X, U) \to S$ with the divisorial log structures defined by $f^{-1}(0)$ and 0 is a proper log toroidal family.

Proof For a $\mathbf{k}$-valued point $\bar{x} \in \mathbb{P}^n$, let $\Phi = \Phi_{\bar{x}}$ be a homogeneous polynomial of degree d such that $\bar{x} \in \{\Phi \neq 0\}$. Since $H \cup K$ is normal crossing, we can find étale morphisms

$$\mathbb{P}^n \xleftarrow{\gamma} V_{\bar{x}} \xrightarrow{\eta} \mathbb{A}^n$$

such that $\gamma^*(F/\Phi \cdot G/\Phi) = \eta^*(z_1 \cdot \ldots \cdot z_k)$ for some $0 \leq k \leq n$ (where the empty product is 1). Because z_i is a prime in the étale local ring $(\mathbb{A}^n)_{\bar{0}}$, we can choose $V_{\bar{x}}, \gamma, \eta$ such that $\gamma^*(F/\Phi) = \eta^*(z_1 \cdot \ldots \cdot z_\ell)$ and $\gamma^*(G/\Phi) = \eta^*(z_{\ell+1} \cdot \ldots \cdot z_k)$. In particular, using [68, Prop. IV-25], the second projection $\pi \colon Y \to \mathbb{P}^n$ can be identified with the blow-up of $\mathbb{P}^n$ in $H \cap K$. Therefore, the scheme Y is integral, and $f \colon Y \to \mathbb{P}^1$ is flat because Y dominates the smooth curve $\mathbb{P}^1$.

Because $\mathbf{f} \colon Y \to \mathbb{P}^1$ is proper and has a non-singular fiber $K = \mathbf{f}^{-1}(\infty)$, there is an open neighborhood $S \ni 0$ such that $\mathbf{f}^{-1}(0) = H$ is the only singular fiber over S. We take this S as the base of $f \colon X \to S$. After writing $D = \mathrm{Sing}(H) \subset H$ for the double locus of the normal crossing divisor H, we set $U = X \setminus (D \cap K)$ under the identification $H = f^{-1}(0)$. The scheme $\pi^{-1}(\mathbb{P}^n \setminus (D \cap K))$ is étale locally the blow-up of $\mathbb{A}^{n-2} \subset \mathbb{A}^n$, so it is a regular scheme, showing that U is a regular scheme. Thus, $f \colon U \to S$ is a semistable degeneration and hence saturated and log smooth with the given log structures. Around a point $\bar{x} \in X \setminus U$, on $\mathbb{A}^1 \times V_{\bar{x}}$, the scheme X is locally given by the equation

$$\gamma^*(F/\Phi) - s \cdot \gamma^*(G/\Phi) = \eta^*(z_1 \cdot \ldots \cdot z_{k-1}) - s\eta^*(z_k) = 0$$

where $\gamma^*(G/\Phi) = \eta^*(z_k)$, i.e., $\ell = k - 1$ because K is a smooth hypersurface. Thus, a log toroidal local model is given by the elt datum $(\mathbb{N} \subset P_{k-1}^{\mathrm{anc}}, \mathscr{F}_{\mathrm{min}})$ from Example 8.24. $\qquad\square$

Example 8.45 By setting $H = \{XYZ = 0\}$ and $K = \{X^3 + Y^3 + Z^3 = 0\}$, we obtain the *Hesse pencil*

$$\lambda(X^3 + Y^3 + Z^3) + \mu XYZ = 0$$

of cubic curves, cf. [13]. Since it is a log toroidal family with fiber dimension 1, it is actually log smooth.   ◊

Example 8.46 By setting $H = \{X_0 X_1 X_2 X_3 = 0\} \subset \mathbb{P}^3$ the union of the four coordinate planes and $K = \{X_0^4 + X_1^4 + X_2^4 + X_3^4 = 0\} \subset \mathbb{P}^3$ the Fermat quartic, we obtain the degeneration of quartic surfaces from Example 1.11. For

$$H = \{X_0 X_1 X_2 X_3 X_4 = 0\} \subset \mathbb{P}^4 \quad \text{and} \quad K = \{X_0^5 + X_1^5 + X_2^5 + X_3^5 + X_4^5 = 0\} \subset \mathbb{P}^4,$$

we obtain the famous degeneration of quintic threefolds which is often studied in mirror symmetry, e.g. in [34, 211].   ◊

8.7 The Base Change Property

Let $f \colon (X, U) \to S$ be a generically log smooth family, and let $b \colon T \to S$ be a strict morphism. Let $g \colon (Y, V) \to T$ be the generically log smooth family obtained by base change, and let $c \colon (Y, V) \to (X, U)$ be the induced map. On $V = c^{-1}(U)$, we have an isomorphism $c^* \Omega^i_{U/S} \cong \Omega^i_{V/T}$, which can be extended to a map $c^* \mathcal{W}^i_{X/S} \to \mathcal{W}^i_{Y/T}$ since the target is reflexive. In general, this map is not an isomorphism, so in [77], we have attached a name to the situation where it is an isomorphism:

Definition 8.47 (Base Change Property)   Let $f \colon (X, U) \to S$ be a generically log smooth family. Then $f \colon (X, U) \to S$ has the *base change property* if, for every strict morphism $b \colon T \to S$ of Noetherian fs log schemes with base change $g \colon (Y, V) \to T$ and map $c \colon (Y, V) \to (X, U)$, the map $c^* \mathcal{W}^i_{X/S} \to \mathcal{W}^i_{Y/T}$ is an isomorphism.

Remark 8.48 If $f \colon (X, U) \to S$ is log Gorenstein, then we have isomorphisms $c^* \mathcal{V}^p_{X/S} \cong \mathcal{V}^p_{Y/T}$ as well. For a more comprehensive discussion of this fact and an interpretation of the map, see [75, §6.3] in the author's thesis.   ◊

One of the reasons why log toroidal families are a well-behaved class of generically log smooth families is that many of them have the base change property. We have included the two main results in this direction as Theorems 8.63 and 8.64 below. However, not every log toroidal family has the base change property. A counterexample is [77, Ex. 7.5].

We investigate the behavior of the base change property for general generically log smooth families. For such families $f_A \colon X_A \to S_A$ over a punctual log scheme S_A, it has a particularly simple behavior:

Lemma 8.49 *Let Q be a sharp toric monoid, and let $A \in \mathbf{Art}_Q$. Let $S_A = \mathrm{Spec}(Q \to A)$, and let $f_A \colon X_A \to S_A$ be a generically log smooth family with central fiber $f_0 \colon X_0 \to S_0$. Let $i \colon X_0 \to X_A$ be the inclusion. Then f_A has the base change property if and only if $i^* \mathcal{W}^i_{X_A/S_A} \cong \mathcal{W}^i_{X_0/S_0}$ for all $0 \le i \le d$; this is the case if and only if $\mathcal{W}^i_{X_A/S_A}$ is flat over S_A for all $0 \le i \le d$.*

Proof The equivalence of the latter two statements follows directly from [279, Thm. 0.4]. If f_A has the base change property, we obviously have the isomorphism for $S_0 \to S_A$. For the converse, it is sufficient to assume that $X_A = \mathrm{Spec}\, S$ is affine, and that $T = \mathrm{Spec}\, R$ is affine as well. If $R = \ell$ is a field, then $A \to \ell$ factors through a flat map $\mathbf{k} \to \ell$, so the isomorphism holds. Then, in the general case, we already know that we have an isomorphism $H^0(Y_t, (c^* \mathcal{W}^i_{X_A/S_A})_t) \to H^0(V_t, (c^* \mathcal{W}^i_{X_A/S_A})_t)$ in the fibers of $Y \to T$, so the claim follows from Lemma B.5. $\qquad\square$

This allows us to give a characterization of the base change property in general.

Corollary 8.50 *Let $f \colon (X, U) \to S$ be a generically log smooth family of relative dimension d. Then f has the base change property if and only if the following two conditions hold for each $0 \le i \le d$:*

(a) *$\mathcal{W}^i_{X/S}$ is flat over S;*
(b) *for every point $s \in S$, the map $\mathcal{W}^i_{X/S}|_{X_s} \to \mathcal{W}^i_{X_s/\kappa(s)}$ is an isomorphism.*

Proof If f has the base change property, then (ii) holds by assumption. To show that $\mathcal{W}^i_{X/S}$ is flat, it is sufficient to assume that S is the spectrum of a local Noetherian ring, and that X is affine. By Lemma 8.49, the base change of $\mathcal{W}^i_{X/S}$ to every Artinian ring is flat. Then a combination of [267, 00HP, 00MC, 0584] respectively their methods allows to conclude that $\mathcal{W}^i_{X/S}$ is flat over S. For the converse, first observe that the base change property holds for a base change to a field ℓ mapping to S since it factors through a field extension $\kappa(s) \to \ell$ for some $s \in S$. Then the assertion follows from Lemma B.5. $\qquad\square$

With this criterion, we can give examples of generically log smooth families which are log Gorenstein and have the base change property, but which are not log toroidal.

Example 8.51 Let

$$X = \mathrm{Spec}\, \mathbf{k}[x, y, z, w, t]/(xy - tzw(z + w)),$$

and let $f \colon X \to S = \mathbb{A}^1_t$ be the obvious map. Endow both the source and target with the divisorial log structure defined by $t = 0$. This family is illustrated in Fig. 8.3. The nearby fiber $X_s = f^{-1}(s)$ is a three-dimensional normal variety with an isolated

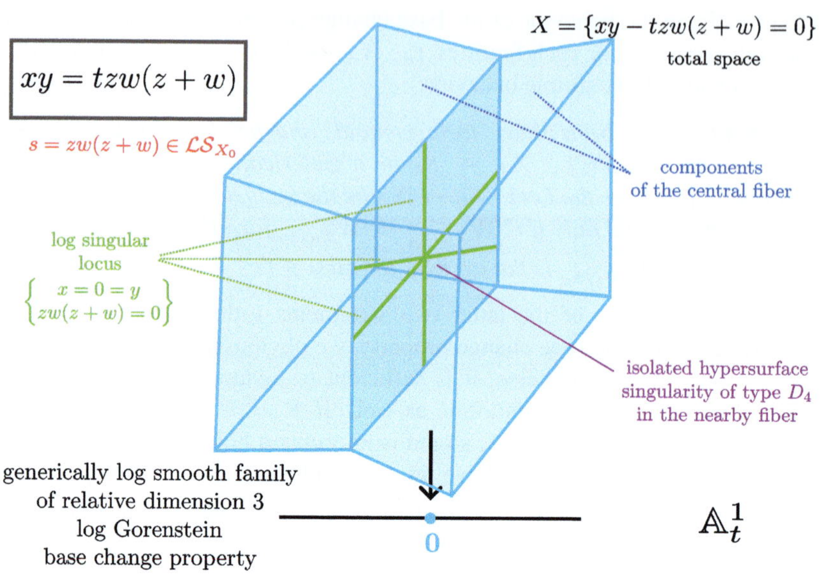

Fig. 8.3 Example 8.51. © Simon Felten 2025. All rights reserved

hypersurface singularity at 0. In fact, $xy = zw(z+w)$ defines a threefold singularity of type D_4, in particular, a terminal Gorenstein singularity. Since $xy = zw$ is the only toric terminal Gorenstein threefold singularity, X_s is not toroidal, and $f : X \to \mathbb{A}^1_t$ is not log toroidal. The central fiber $V = \operatorname{Spec} \mathbf{k}[x, y, z, w]/(xy)$ is a normal crossing space with two components intersecting in $\mathbb{A}^2_{z,w}$. Inside V, the log singular locus is $Z = \{zw(z + w) = 0\}$, i.e., it consists of three lines intersecting in 0. The induced log structure is given by $s = zw(z + w) \in \mathcal{LS}_V \subseteq \mathcal{T}^1_V$ outside Z. Outside the very singular point 0, the local model for the log singularity is $xy = tz$, so the family is log toroidal outside 0. To show that $f : X \to S$ has the base change property and is log Gorenstein, we compute first (in Macaulay2) the relative classical tangent sheaf $\mathcal{T}^0_{X/S}$ as the kernel of the map

$$\left(\frac{\partial f}{\partial x} \; \frac{\partial f}{\partial y} \; \frac{\partial f}{\partial z} \; \frac{\partial f}{\partial w} \right): \quad \mathcal{O}^{\oplus 4}_X \to \mathcal{O}_X,$$

where $f = xy - tzw(z + w)$. Then $\mathcal{V}^{-1}_{X/S} = \mathcal{T}^0_{X/S}$, and we can compute $\mathcal{W}^1_{X/S}$ as the dual of $\mathcal{V}^{-1}_{X/S}$. It turns out that $\mathcal{W}^3_{X/S} \cong \mathcal{O}_X$, so $f : X \to S$ is log Gorenstein. Flatness of $\mathcal{W}^1_{X/S}$ and $\mathcal{W}^2_{X/S}$ follows from the fact that they have no $\mathbf{k}[t]$-torsion. Finally, a direct computation shows that both $\mathcal{W}^1_{X/S}|_{X_s}$ and $\mathcal{W}^2_{X/S}|_{X_s}$ are reflexive for all $s \in \mathbf{k}$. For this, it suffices to check reflexivity for $s = 0$ and $s = 1$. Thus, $f : X \to S$ has the base change property. $\diamond$

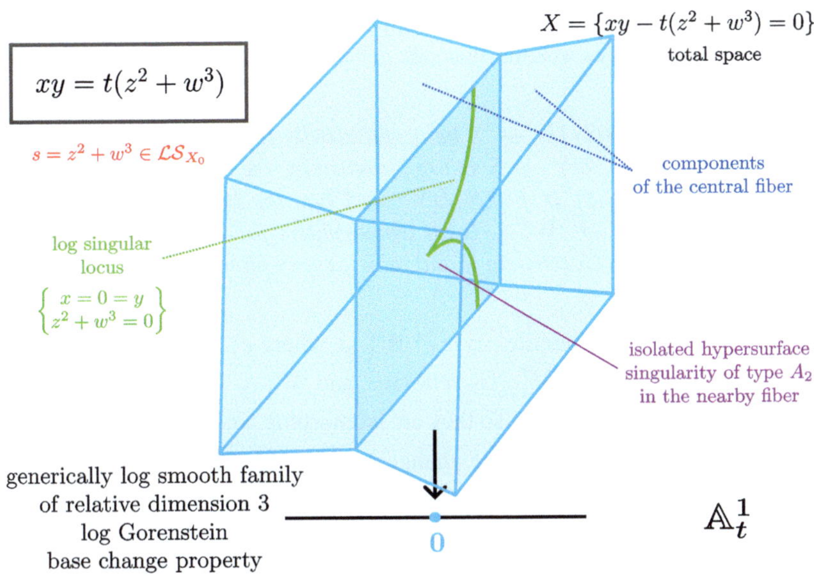

Fig. 8.4 Example 8.52. © Simon Felten 2025. All rights reserved

Example 8.52 Let

$$X = \mathrm{Spec}\, \mathbf{k}[x, y, z, w, t]/(xy - t(z^2 + w^3)),$$

and let $f \colon X \to S = \mathbb{A}^1_t$ be the obvious map. Endow both the source and target
with the divisorial log structure defined by $t = 0$. This family is depicted in Fig. 8.4.
Computations in Macaulay2 analogous to Example 8.51 show that $f \colon X \to S$ is a
log Gorenstein generically log smooth family with the base change property as well.
Now the log singular locus inside the central fiber $V = \mathrm{Spec}\, \mathbf{k}[x, y, z, w]/(xy)$ is
a cuspidal cubic with cusp at 0. The local model for the log singularity outside
the very singular point 0 is again $xy = tz$. The nearby fiber $X_s = f^{-1}(s)$ is a
normal threefold with isolated hypersurface singularity at 0. More precisely, X_s has
a Kleinian singularity of type A_2, i.e., again a terminal Gorenstein singularity which
is not toroidal. Thus, also this $f \colon X \to \mathbb{A}^1_t$ is not log toroidal.   ◇

8.8 The Log Gorenstein Property

Here, we analyze the log Gorenstein property, which we have already defined in
Sect. 8.2. On the one hand, every log Calabi–Yau generically log smooth family is
log Gorenstein; on the other hand, if a generically log smooth family $f \colon (X, U) \to$
S has the base change property, and we know that it is log Gorenstein, then also

$\mathcal{V}^p_{X/S}$ is flat over S, and its formation commutes with base change. For these reasons, we will virtually always assume that our generically log smooth families are log Gorenstein.

Lemma 8.53 *Let $f\colon (X, U) \to S$ be a generically log smooth family of relative dimension d. If $f\colon (X, U) \to S$ is log Gorenstein, the same holds for any base change $g\colon (Y, V) \to T$. If $f\colon (X, U) \to S$ has the base change property for $\mathcal{W}^d_{X/S}$, i.e., $c^*\mathcal{W}^d_{X/S} \to \mathcal{W}^d_{Y/T}$ is an isomorphism for every base change, then $f\colon (X, U) \to S$ is log Gorenstein if and only if every fiber $(X_s, U_s) \to \kappa(s)$ is log Gorenstein.*

Proof If $\mathcal{W}^d_{X/S}$ is a line bundle, so is $c^*\mathcal{W}^d_{X/S}$, where $c\colon (Y, V) \to (X, U)$ is the base change map. Then $c^*\mathcal{W}^d_{X/S}$ is reflexive, and $\mathcal{W}^d_{Y/T}$ is reflexive as well; both are isomorphic on $V = c^{-1}(U)$, so they are isomorphic, and $\mathcal{W}^d_{Y/T}$ is a line bundle.

If $f\colon (X, U) \to S$ has the base change property for $\mathcal{W}^d_{X/S}$, then $\mathcal{W}^d_{X/S}|_{X_s} \cong \mathcal{W}^d_{X_s/\kappa(s)}$. The latter is a line bundle by assumption. In particular, we have $\dim_{\kappa(x)}(\mathcal{W}^d_{X/S} \otimes \kappa(x)) = 1$ for all points $x \in X$. Since $\mathcal{W}^d_{X/S}$ is a line bundle on U, this implies that it is a line bundle on X. Namely, we find locally a surjection $\mathcal{O}_X \to \mathcal{W}^d_{X/S}$, and it must be injective because it is injective on U and $\mathcal{O}_X$ is Z-closed. $\qquad\square$

Recall from Definition 7.14 that a homomorphism $\theta\colon Q \to P$ of integral monoids is *vertical* if its image is not contained in any proper face of P. If $\theta\colon \mathbb{N} \to P$ is a vertical saturated injection of sharp toric monoids, then P is a Gorenstein monoid, as we will see in Construction 9.2 in conjunction with Lemma 9.12. In particular, $A_\theta\colon A_P \to A_{\mathbb{N}}$ is a flat morphism with Gorenstein fibers. Since every vertical saturated log smooth morphism $f\colon U_0 \to S_0$ over $S_0 = \mathrm{Spec}(\mathbb{N} \to \mathbf{k})$ admits a local model by some vertical saturated injection $\theta\colon \mathbb{N} \to P$, we find that U_0 is a Gorenstein scheme. With this preparation, we prove:

Lemma 8.54 *Let $f\colon (X, U) \to S$ be a vertical generically log smooth family, i.e., for every geometric point $\bar{x} \to\in U$, the map $\overline{\mathcal{M}}_{S, f(\bar{x})} \to \overline{\mathcal{M}}_{U, \bar{x}}$ is vertical. Assume that every fiber of $f\colon X \to S$ is Cohen–Macaulay, and that $f\colon (X, U) \to S$ is log Gorenstein. Then every fiber is a Gorenstein scheme. Conversely, if every fiber is a Gorenstein scheme, then $f\colon (X, U) \to S$ is log Gorenstein provided that either S is a log point or that $f\colon (X, U) \to S$ has the base change property for $\mathcal{W}^d_{X/S}$.*[9]

Proof First, we show that every fiber of $f\colon U \to S$ is a Gorenstein scheme. We may assume that $S = \mathrm{Spec}(Q \to \mathbf{k})$ for some sharp toric monoid Q and field $\mathbf{k}$. If $Q = 0$, then verticality implies that $f\colon U \to S$ is strict, hence smooth, and U is a

---

[9] Presumably, the converse holds in general. For example, when we know that, for a vertical saturated log smooth morphism $f\colon X \to S$, $\Omega^d_{X/S}$ is isomorphic to the relative dualizing sheaf $\omega^\circ_{X/S} = \mathcal{H}^{-d}(f^!\mathcal{O}_S)$, then the converse holds in general. Unfortunately, this has been established only for $S = \mathrm{Spec}(\mathbb{N} \to \mathbf{k})$, see [272, Thm. 2.21] and our discussion thereof in Proposition 7.70.

Gorenstein scheme. If $Q = \mathbb{N}$, then our claim follows from the above discussion. In the general case, we find a local homomorphism $h\colon Q \to \mathbb{N}$, giving rise to a map of log schemes $b\colon T = \mathrm{Spec}(\mathbb{N} \to \mathbf{k}) \to \mathrm{Spec}(Q \to \mathbf{k})$. After base change along $b\colon T \to S$, the underlying scheme of U remains unchanged, and we are in the case $Q = \mathbb{N}$. Thus, here U is a Gorenstein scheme as well.

Now assume that $f\colon (X, U) \to S$ is log Gorenstein and has Cohen–Macaulay fibers. Then every fiber is log Gorenstein, and after base change induced by $h\colon Q \to \mathbb{N}$ as above, we can assume that the base is $S = \mathrm{Spec}(\mathbb{N} \to \mathbf{k})$. Let ω_X° be the dualizing sheaf of X. From Proposition 7.70, we have an isomorphism $\omega_X^\circ|_U \cong \mathcal{W}_{U/S}^d$. By [267, 0AWN], the dualizing sheaf ω_X° has the property (S_2), so we have $\omega_X^\circ = j_* \omega_X^\circ|_U$. Since $\mathcal{W}_{X/S}^d$ is a line bundle by assumption, this implies $\omega_X^\circ \cong \mathcal{W}_{X/S}^d$. But then the dualizing sheaf is a line bundle, so X is Gorenstein.

Conversely, if each fiber of $f\colon X \to S$ is Gorenstein, then $\omega_{X_s}^\circ$ is a line bundle on each fiber X_s, and after base change along $h\colon Q \to \mathbb{N}$, we obtain that $\mathcal{W}_{X_s/\kappa(s)}^d$ is a line bundle from Proposition 7.70. If S is a log point itself, then we are finished. If $f\colon (X, U) \to S$ has the base change property for $\mathcal{W}_{X/S}^d$, then the result follows from Lemma 8.53. $\qquad\square$

If $f\colon (X, U) \to S$ is not vertical, then it may be log Gorenstein even if the fibers are not Gorenstein schemes.

Example 8.55 Let $P \subseteq \mathbb{Z}^2$ be the sharp toric monoid generated by ray generators $(1, 0)$ and $(-2, 3)$ as well as interior generators $(0, 1)$ and $(-1, 2)$. Then P is not a Gorenstein monoid, hence A_P is not a Gorenstein scheme. Nonetheless, $A_P \to A_0$ is of course log Gorenstein because it is log smooth. Next, let $\tilde{P} \subseteq \mathbb{Z}^3$ be the sharp toric monoid consisting of elements (x, y, z) with either $x < 0, z \geq 0$, and $(x, y) \in P$ or $x \geq 0, z \geq x$, and $(x, y) \in P$. Let $\rho = (0, 0, 1)$. Then $\theta\colon \mathbb{N} \to \tilde{P}$, $1 \mapsto \rho$, is a saturated injection, so $A_\theta\colon A_P \to A_\mathbb{N}$ is saturated and log smooth, in particular log Gorenstein, but neither the special fiber nor the generic fiber of A_θ is a Gorenstein scheme. $\qquad\diamond$

Not every log toroidal family is log Gorenstein. Simply take an affine toric variety $X = A_P$ which is not Gorenstein, and consider it as a trivial log scheme. Then $\mathcal{W}_{X/\mathbf{k}}^d$ is the dualizing sheaf, which is not a line bundle in this case.

8.9   The de Rham Complex of Log Toroidal Families

The explicit nature of elementary log toroidal families allows us to prove properties of the de Rham complex or the polyvector fields of log toroidal families by explicit computation. In this section, we give an overview over the results in this direction contained in [77] and [75].

Fix a principal ideal domain R such as $\mathbb{Z}$ or $\mathbb{C}$ as the base ring of an elementary log toroidal family. In the absolute case, i.e., $Q = 0$, the pieces of the de Rham complex can be computed as follows:

Proposition 8.56 *Let $(0 \subset P, \mathcal{F})$ be an elt datum, and let $\mathcal{W}^{\bullet}$ be the reflexive log de Rham complex of $f : \mathbb{A}_{P,\mathcal{F}} \to \mathbb{A}_0$. Then, we have $\Gamma(\mathbb{A}_P, \mathcal{W}^m) = \bigoplus_{p \in P} (\mathcal{W}^m)_p$ with*

$$(\mathcal{W}^m)_p = \bigwedge_R^m \left(\bigcap_{\substack{F \in \mathcal{F}_{\max} \setminus \mathcal{F} \\ p \in F}} F^{\mathrm{gp}} \otimes_{\mathbb{Z}} R \right)$$

where the intersection is $P^{\mathrm{gp}} \otimes_{\mathbb{Z}} R$ if the index set is empty.

Proof See [77, Prop. 7.2]. $\square$

In the relative case, this generalizes to:

Proposition 8.57 *Let $(Q \subset P, \mathcal{F})$ be an elt datum, and let $\mathcal{W}^{\bullet}_f$ be the reflexive log de Rham complex of $f : \mathbb{A}_{P,\mathcal{F}} \to \mathbb{A}_Q$. Then, we have $\Gamma(\mathbb{A}_P, \mathcal{W}^m_f) = \bigoplus_{p \in P} (\mathcal{W}^m_f)_p$ with*

$$(\mathcal{W}^m_f)_p = \bigwedge_R^m \left(\left(\bigcap_{\substack{F \in \mathcal{F}_{\max} \setminus \mathcal{F} \\ p \in F}} F^{\mathrm{gp}} \otimes_{\mathbb{Z}} R \right) \Big/ (Q^{\mathrm{gp}} \otimes_{\mathbb{Z}} R) \right)$$

where the intersection is $P^{\mathrm{gp}} \otimes_{\mathbb{Z}} R$ if the index set is empty. Since $Q^{\mathrm{gp}} \subset P^{\mathrm{gp}}$ splits, we can equivalently take the quotient before the intersection.

Proof See [77, Prop. 7.3]. $\square$

Let us explain the meaning of the description of $\mathcal{W}^1_f$. The open subscheme $A^*_P = \mathrm{Spec}\, R[P^{\mathrm{gp}}] \subseteq A_{P,\mathcal{F}}$ is log trivial. The R-linear derivations on it are given by

$$\mathrm{Der}(A^*_P) = \bigoplus_{p \in P^{\mathrm{gp}}} z^p \cdot \mathrm{Hom}(P^{\mathrm{gp}}, R)$$

acting as $(z^p \phi)(z^q) = \phi(q) z^{p+q} \in R[P^{\mathrm{gp}}]$. The $R[Q]$-linear derivations are those with $\phi(q) = 0$ for $q \in Q$, so they are given by $\bigoplus_{p \in P^{\mathrm{gp}}} z^p \cdot \mathrm{Hom}(P^{\mathrm{gp}}/Q^{\mathrm{gp}}, R)$. The restriction map

$$\Gamma(A_{P,\mathcal{F}}, \Theta^1_f) \to \bigoplus_{p \in P^{\mathrm{gp}}} z^p \cdot \mathrm{Hom}(P^{\mathrm{gp}}/Q^{\mathrm{gp}}, R)$$

is injective and thus identifies Θ^1_f with some submodule. The differentials $\mathcal{W}^1_f$ are the dual of Θ^1_f and are thus identified with a submodule of the module $\bigoplus_{p \in P^{\mathrm{gp}}} z^p \cdot (P^{\mathrm{gp}}/Q^{\mathrm{gp}}) \otimes R$ via the pairing

$$\bigoplus_{p \in P^{\mathrm{gp}}} z^p \cdot \mathrm{Hom}(P^{\mathrm{gp}}/Q^{\mathrm{gp}}, R) \times \bigoplus_{p \in P^{\mathrm{gp}}} z^p \cdot (P^{\mathrm{gp}}/Q^{\mathrm{gp}}) \otimes R \to R[P^{\mathrm{gp}}]$$

given by $(z^p \phi, z^q r) \mapsto \phi(r) z^{p+q}$. Indeed, it is possible but very tedious to compute $\mathcal{W}^1_f$ along these lines. Within this framework, one can compute the differential of the de Rham complex: We find first $d(z^p) = z^p \cdot [p]$ for functions $z^p \in \mathcal{W}^0_f$ and then $d(z^p \cdot n) = z^p \cdot [p] \wedge n$ for $z^p \cdot n \in \mathcal{W}^m_f$.

Example 8.58 For the elementary log toroidal family $f \colon X^{\mathrm{sze}} \to \mathbb{A}^1$ of Example 1.79, Proposition 8.57 yields the following: The monoid P is divided into four regions

$$P_0 := P \setminus (F_{tx} \cup F_{ty}), \quad P_x := F_{tx} \setminus Q, \quad P_y := F_{ty} \setminus Q, \quad P_{xy} := Q$$

on which the graded piece $(\mathcal{W}^1_f)_p$ is constant. Here, F_{tx} and F_{ty} denote facets of P as in Fig. 8.1 at Example 8.10. Using the isomorphism $P^{\mathrm{gp}} \otimes \mathbf{k}/Q^{\mathrm{gp}} \otimes \mathbf{k} \cong \mathbf{k} \oplus \mathbf{k}$, we get:

$$\begin{aligned}
(\mathcal{W}^1_f)_p &= (P^{\mathrm{gp}} \otimes \mathbf{k})/(Q^{\mathrm{gp}} \otimes \mathbf{k}) = \mathbf{k} \oplus \mathbf{k} \ \ \text{for } p \in P_0 \\
(\mathcal{W}^1_f)_p &= (F^{\mathrm{gp}}_{tx} \otimes \mathbf{k})/(Q^{\mathrm{gp}} \otimes \mathbf{k}) = 0 \oplus \mathbf{k} \ \ \text{for } p \in P_x \\
(\mathcal{W}^1_f)_p &= (F^{\mathrm{gp}}_{ty} \otimes \mathbf{k})/(Q^{\mathrm{gp}} \otimes \mathbf{k}) = \mathbf{k} \oplus 0 \ \ \text{for } p \in P_y \\
(\mathcal{W}^1_f)_p &= (Q^{\mathrm{gp}} \otimes \mathbf{k})/(Q^{\mathrm{gp}} \otimes \mathbf{k}) = 0 \oplus 0 \ \ \text{for } p \in P_{xy}
\end{aligned}$$

This is visualized in Fig. 8.5. $\Diamond$

Corollary 8.59 *For all m, the module $\mathcal{W}^m_f$ is flat over A_Q.*

8.9.1 Base Change for Elementary Log Toroidal Families

The explicit description of $\mathcal{W}^\bullet_f$ allows us to give criteria for when the formation of $\mathcal{W}^\bullet_f$ commutes with base change along morphisms $b \colon T = \mathrm{Spec}\,\mathcal{T} \to \mathrm{Spec}\,R[Q]$. Let $Y = A_{P,\mathcal{F}} \times_{A_Q} T$, and let $c \colon Y \to A_{P,\mathcal{F}}$ be the fiber product. We denote the open subset of log smoothness by $V = c^{-1}(U_{P/Q})$. We always have a canonical map $c^* \mathcal{W}^m_f \to \mathcal{W}^m_{Y/T}$; that the formation of $\mathcal{W}^\bullet$ commutes with base change just means that this is an isomorphism, which in turn is equivalent to that $c^* \mathcal{W}^m_f$ is reflexive. [77, Ex. 7.5] shows that this is not always the case.

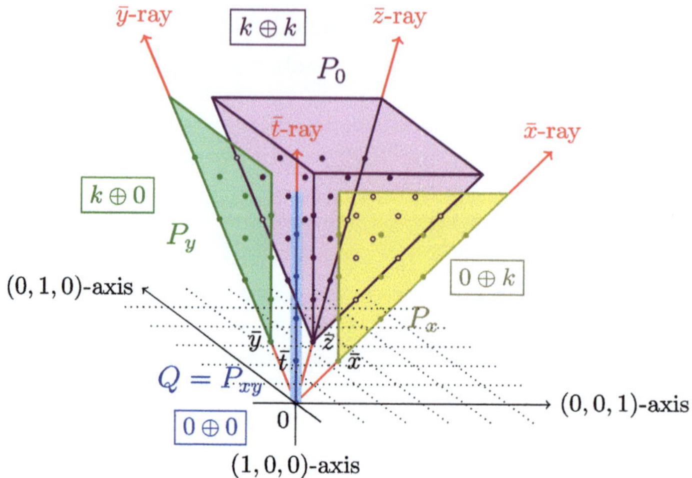

Fig. 8.5 Example 8.58. We see the four regions P_{xy}, P_x, P_y, P_0 of the monoid P from Example 8.58 with the respective graded pieces $(\mathcal{W}_f^1)_p$. This graphic is a slight modification of a graphic which has been created by the author for his master thesis and which is also contained in his doctoral thesis [75].

We compute O_Y as $\Gamma(Y, O_Y) = \bigoplus_{e \in E} z^e \cdot \mathcal{T}$ with multiplication

$$z^{e_1} \cdot z^{e_2} = z^e \cdot \sigma(q) \quad \text{whenever} \quad e_1 + e_2 = e + q$$

with $e \in E, q \in Q$ under the canonical decomposition from (8.1). Here, $\sigma \colon Q \to R[Q] \to \mathcal{T}$ is the obvious map. Then, we can compute $c^*\mathcal{W}_f^m$ explicitly as

$$\Gamma(Y, c^*\mathcal{W}_f^m) = \bigoplus_{e \in E} z^e \cdot ((\mathcal{W}_f^m)_e \otimes_R \mathcal{T}). \tag{8.4}$$

We have shown in [77, Lemma 7.7] that the reflexivity of $c^*\mathcal{W}_f^m$ is equivalent to the surjectivity of the restriction map $\rho \colon \Gamma(Y, c^*\mathcal{W}_f^m) \to \Gamma(V, c^*\mathcal{W}_f^m)$. For the surjectivity of this map, we have the following criterion. Let $\mathcal{E}$ be the set of essential faces of P of rank $d - 1$. Then U is covered by $\{U_F | F \in \mathcal{E}\}$. Set $V_F = c^{-1}(U_F)$, so these cover V. For each $F \in \mathcal{E}$, choose $e_F \in F$ in the relative interior, i.e., $\langle e_F \rangle = F$ is the smallest face containing e_F.

Theorem 8.60 *Write $M_p := (\mathcal{W}_f^m)_p$ for short, and assume that, for every subset $\mathcal{E}' \subset \mathcal{E}$ and every $e \in E$, the natural map*

$$\left(\bigcap_{F \in \mathcal{E}'} M_{e+e_F} \right) \otimes_R \mathcal{T} \to \bigcap_{F \in \mathcal{E}'} (M_{e+e_F} \otimes_R \mathcal{T})$$

is an isomorphism. Then ρ is surjective.

Proof See [77, Thm. 7.8]. □

From this, we deduce two useful criteria for the reflexivity of $c^* \mathcal{W}_f^m$.

Corollary 8.61 *Let $(Q \subset P, \mathcal{F})$ be an elt datum, let $\mathcal{T}$ be a Noetherian ring, and $T = \operatorname{Spec} \mathcal{T} \to A_Q$ a strict morphism of log schemes. Then $c^* \mathcal{W}_f^m$ is reflexive and $c^* \mathcal{W}_f^m \to \mathcal{W}_{Y/T}^m$ is an isomorphism provided that the composition*

$$R \to R[Q] \to \mathcal{T}$$

is flat, e.g. when R is a field.

We say that the ring $\mathcal{T}$ has *characteristic* $\geq p$ if, for every prime $\mathfrak{p} \subset \mathcal{T}$, we have either $\operatorname{char}(\kappa(\mathfrak{p})) = 0$ or $\operatorname{char}(\kappa(\mathfrak{p})) \geq p$.

Corollary 8.62 *Let $(Q \subset P, \mathcal{F})$ be an elt datum, and assume $f \colon A_{P,\mathcal{F}} \to A_Q$ to be defined over $R = \mathbb{Z}$. Then there is a $p_0 = p_0(Q \subset P, \mathcal{F})$ such that for $T = \operatorname{Spec} \mathcal{T} \to A_Q$ with a Noetherian ring $\mathcal{T}$ of characteristic $\geq p_0$, the sheaf $c^* \mathcal{W}_f^m$ is reflexive, and $c^* \mathcal{W}_f^m \to \mathcal{W}_{Y/T}^m$ is an isomorphism.*

8.9.2   Base Change for Log Toroidal Families

The local results of the previous section can easily be generalized to the global case.

Theorem 8.63 (Base Change over Fields) *Let $f \colon (X, U) \to S$ be a log toroidal family such that S is defined over a field $\mathbf{k}$. Then f has the base change property.*

Proof This is [77, Thm. 8.2]. □

Theorem 8.64 (Generic Base Change) *Let $f \colon (X, U) \to S$ be a log toroidal family. Then there is a finite set of prime numbers $p_1, \ldots, p_N \in \mathbb{Z}$ such that if $f^\circ \colon X^\circ \to S^\circ$ is obtained from f by inverting $p_1, \ldots, p_N$ (i.e., base change to $\operatorname{Spec} \mathbb{Z}_{p_1 \cdots p_N}$), then f° has the base change property.*

Proof This is [77, Thm. 8.3]. □

8.9.3   Hodge–de Rham Degeneration

Let $f : (X, U) \to S$ be a proper generically log smooth family. Then we have the Hodge–de Rham spectral sequence

$$E_1^{pq} = R^q f_* \mathcal{W}^p_{X/S} \Rightarrow R^{p+q} f_* \mathcal{W}^\bullet_{X/S}.$$

From a technical perspective, the main result of [77] is that this spectral sequence degenerates at E_1 for many log toroidal families. Moreover, $R^q f_* \mathcal{W}^p_{X/S}$ is locally free of finite rank, and its formation commutes with base change. This is then a crucial ingredient in our proof of the main geometric Unobstructedness Theorem 15.2. More precisely, we have the following:

Theorem 8.65 *Let Q be a sharp toric monoid, let $\mathbf{k}$ be a field of characteristic 0, and let S be one of the following:*

(1) $S = \mathrm{Spec}(Q \to \mathbf{k})$;
(2) $S = \mathrm{Spec}(Q \to A_k)$ with $A_k = \mathbf{k}[\![Q]\!]/\mathfrak{m}_Q^{k+1}$;
(3) $S = \mathrm{Spec}(Q \to \mathbf{k}[\![Q]\!])$.

Let $a : S \to A_Q$ be the obvious chart, and let $f : (X, U) \to S$ be a proper log toroidal family with respect to $a : S \to A_Q$. Then the Hodge–de Rham spectral sequence degenerates at E_1, the Hodge sheaves $R^q f_* \mathcal{W}^p_{X/S}$ are locally free of finite rank, and their formation commutes with base change.

In the statement of the theorem, note that we have assumed log toroidality with respect to the specified chart $a : S \to A_Q$. We do not allow more general log toroidality over other charts.

Proof Case (a) is [77, Thm. 1.9] and Case (b) with $Q = \mathbb{N}$ is [77, Thm. 1.10]. Case (b) for general Q is [75, Thm. 8.12], and Case (c) is [75, Cor. 8.13]. $\square$

Chapter 9
Toroidal Crossing Spaces

In this chapter, we give a synopsis of the basic theory of toroidal crossing spaces. The first appearance of the name and the essential idea of the concept is by Schröer and Siebert in [253], which has a slightly different focus than we have. Most theory was developed by Gross and Siebert in [124], but toroidal crossing spaces do not show up as a concept and a definition there. Finally, in [77], we have used toroidal crossing spaces as a concept, but the treatment is rather sloppy. As of now, toroidal crossing spaces are the single most important source of generically log smooth families.

A toroidal crossing space is a generalization of a *normal* crossing space which is particularly suitable to the study of degenerations and more natural from the perspective of logarithmic geometry.

Let R be a discrete valuation **k**-algebra with residue field **k**, choose a uniformizer $t \in \mathfrak{m}_R$, and write $S = \operatorname{Spec} R$. A normal crossing space V has local models $\{z_1 \cdot \ldots \cdot z_r = 0\}$. Every semistable degeneration $f : X \to S$ (with central fiber V) becomes a log smooth and saturated as well as vertical[1] morphism once we endow both the base and the total space with the compactifying log structure defined by $t = 0$, as usual in the étale topology, as also the local models are in the étale topology. This induces a log structure $\mathcal{M}_{X_0}$ on V, turning it into a log scheme $X_0 = (V, \mathcal{M}_{X_0})$, and a log smooth morphism to $S_0 = \operatorname{Spec}(\mathbb{N} \to \mathbf{k})$. Different semistable degenerations to V may induce different log structures on V, but they all have the same ghost sheaf, namely $\mathcal{P} := \nu_* \underline{\mathbb{N}}_{\tilde{V}}$ where $\nu : \tilde{V} \to V$ is the normalization. The image of the generator $1_{S_0} \in \mathcal{M}_{S_0}$ in $\mathcal{P}$ is always $\nu_*(1_{\tilde{V}})$, where $1_{\tilde{V}} \in \underline{\mathbb{N}}_{\tilde{V}}$ is the global section which is equal to 1 in every stalk. The isomorphism classes of log structures on V together with the structure of a log morphism to S_0 which can arise from local semistable degenerations to V form not only a presheaf but indeed a sheaf $\mathcal{LS}_V$ on V, which turns out to be isomorphic to the subsheaf of local generators inside the line bundle $\mathcal{T}_V^1 = \mathcal{E}xt^1(\Omega_V^1, \mathcal{O}_V)$ on the double locus $D = \operatorname{Sing}(V)$ of

---

[1] In the sense of [222, I, Defn. 4.3.1], cf. Definition 7.14.

S. Felten, *Global Logarithmic Deformation Theory*, Lecture Notes in Mathematics 2373, https://doi.org/10.1007/978-3-031-98751-9_9

V. d-semistability of V, a well-known necessary condition for the existence of a *global* semistable degeneration to V, then becomes equivalent to the existence of a global section in $\mathcal{LS}_V$. Generically log smooth families over S_0 now arise from sections of $\mathcal{LS}_V$ on $V \setminus Z$ which cannot be extended across the log singular locus $Z \subset X$.

Toroidal crossing spaces arise when we replace semistable degenerations with the more general degenerations $f : X \to S$ which become log smooth, saturated, and vertical once we endow source and target with the compactifying log structure defined by $t = 0$.

Definition 9.1 (Toroidal Crossing Degenerations I)   A *toroidal crossing degeneration* is a separated morphism of finite type $f : X \to S$ of some constant relative dimension d which becomes log smooth, saturated, and vertical once we endow X and S with the compactifying log structure defined by $t = 0$. In particular, the log structure $\mathcal{M}_X$ is fine and saturated, the morphism $f : X \to S$ is flat, the generic fiber X_η is smooth, the central fiber X_0 is Gorenstein, and the total space X is normal and Cohen–Macaulay.

The following construction yields such degenerations.

Construction 9.2   Let $M \cong \mathbb{Z}^r$ be a lattice with dual $N = \mathrm{Hom}_{\mathbb{Z}}(M, \mathbb{Z})$, and set $M_{\mathbb{R}} = M \otimes_{\mathbb{Z}} \mathbb{R}$. Let $\sigma \subseteq M_{\mathbb{R}}$ be a lattice polytope of full dimension r; in particular, it is bounded.[2] The cone over $\sigma \times \{1\}$ is

$$C(\sigma) = \{(sm, s) \mid m \in \sigma, s \geq 0\} \subseteq M_{\mathbb{R}} \times \mathbb{R};$$

it gives rise to a sharp toric monoid

$$P_\sigma := C(\sigma)^{\vee} \cap (N \times \mathbb{Z}),$$

which consists of the lattice points in the dual cone $C(\sigma)^{\vee}$. The monoid P_σ defines the affine toric variety

$$U(\sigma) = \mathrm{Spec}\, \mathbf{k}[P_\sigma].$$

It is Gorenstein according to the criterion in [221, p. 126] because we have

$$\mathrm{int}(P_\sigma) = \rho_\sigma + P_\sigma.$$

---

[2] Our convention is that lattice *polytopes* are bounded (= compact) lattice *polyhedra*. The latter are intersections of integral affine half-spaces.

Here, $\text{int}(P_\sigma)$ denotes the *interior*—the elements which are not in any proper face of P_σ—and $\rho_\sigma = (0, 1) \in P_\sigma \subseteq N \times \mathbb{Z}$ is the so-called *Gorenstein degree*. By Lemma 9.12, $\theta_\sigma : \mathbb{N} \to P_\sigma$, $1 \mapsto \rho_\sigma$, is saturated and vertical.

The toric boundary of $U(\sigma)$ is

$$V(\sigma) := \text{Spec } \mathbf{k}[P_\sigma]/(z^{\rho_\sigma}) = \{z^{\rho_\sigma} = 0\} \subset U(\sigma)$$

since the ideal $\mathbf{k}[\text{int}(P_\sigma)] \subset \mathbf{k}[P_\sigma]$ cutting out the toric boundary is generated by z^{ρ_σ}. We obtain a family

$$U(\sigma) \to \mathbb{A}^1_t, \quad t \mapsto z^{\rho_\sigma},$$

with central fiber $V(\sigma)$ and smooth generic fiber. We write $\mathcal{M}_{U(\sigma)}$ for the divisorial log structure induced by the inclusion $V(\sigma) \subset U(\sigma)$. Then $(U(\sigma), \mathcal{M}_{U(\sigma)}) \to A_{\mathbb{N}}$ is a log smooth, saturated, and vertical family, and $(V(\sigma), \mathcal{M}_{V(\sigma)}) \to S_0$ is log smooth, saturated, and vertical as well. We denote the base change of $U(\sigma) \to \mathbb{A}^1_t$ to $S = \text{Spec } R$ along $\mathbf{k}[t] \ni t \mapsto t \in R$ by

$$U_S(\sigma) \to S.$$

On $U_S(\sigma)$, the induced log structure from $U(\sigma)$ coincides with the compactifying log structure defined by $t = 0,$[3] as is the case on S. $\Diamond$

When considering the following examples, the reader may also want to look at Sect. 9.1, which contains additional discussion of the construction.

Example 9.3 Let $M = \mathbb{Z}$ and $\sigma = [0, k] \subseteq \mathbb{R} = \mathbb{Z} \otimes_{\mathbb{Z}} \mathbb{R}$ for $k \geq 1$. Then $P_\sigma \subseteq N \oplus \mathbb{Z}$ is generated by the three vectors $(-1, k), (0, 1), (1, 0)$. Thus,

$$U(\sigma) = \text{Spec } \mathbf{k}[x, y, t]/(xy - t^k)$$

is the two-dimensional A_{k-1}-singularity. The central fiber is

$$V(\sigma) \cong \text{Spec } \mathbf{k}[x, y]/(xy);$$

in particular, it does not depend on k. In the language of Definition 9.50 below, we say that k is the *kink* along the double locus of $V(\sigma)$. $\Diamond$

Example 9.4 Let $M = \mathbb{Z}^2$, let $m \geq 1$, and set

$$\sigma = \Delta(1, 1, m) = \text{Conv}((0, 0), (m, 0), (0, 1)).$$

---

[3] To see this, use [222, I, Prop. 1.6.3] together with the proof of [222, I, Lemma 1.6.7]. Use that a reflexive sheaf of rank 1 on $U(\sigma)$ which is locally free outside $V(\sigma)$ is a line bundle if and only if its pull-back to $U_S(\sigma)$ is a line bundle.

Then, P_σ is generated by

$$\bar{x} = (1, 0, 0), \quad \bar{y} = (-1, -m, m), \quad \bar{z} = (0, 1, 0), \quad \bar{w} = (0, -1, 1), \quad \bar{t} = (0, 0, 1)$$

so that $U(\sigma) = \operatorname{Spec} \mathbf{k}[x, y, z, w, t]/(xy - w^m, zw - t)$ and

$$V(\sigma) = \operatorname{Spec} \mathbf{k}[x, y, z, w]/(xy - w^m, zw).$$

This central fiber has three irreducible components, any two of which meet in a line. In the language of Definition 9.50 below, the kink along two of these double lines is one while the kink along the third double line is m. Note that $U(\sigma) \to \mathbb{A}_t^1$ coincides with the family considered in Example 1.66. There, the reader can also find an illustration. $\qquad\qquad\Diamond$

Example 9.5 Let $M = \mathbb{Z}^2$ and $\sigma = \operatorname{Conv}((0, 0), (3, 0), (0, 2))$. This is similar to the previous Example 9.4, but now the kinks are 1, 2, and 3. $\qquad\qquad\Diamond$

The name "toroidal crossing degeneration" is justified by the analogy with normal crossing degenerations through the following result. In other words, a toroidal crossing degeneration has *toric* crossings étale locally.

Lemma 9.6 *Let $f : X \to S$ be a separated morphism of finite type. Then $f : X \to S$ is a toroidal crossing degeneration if and only if it is, locally in the étale topology on X, isomorphic to families of the form $U_S(\sigma) \times_S \mathbb{A}_S^{d-r} \to S$.*[4]

Proof We have already seen that every family of the specified form is a toroidal crossing degeneration. For the converse, the local model is clear around points $x \in X_\eta$ since X_η is smooth. If $x \in X_0$, let $h : \mathbb{N} \to P$ be the map on ghost stalks. By assumption, P is a sharp toric monoid. By [222, IV, Thm. 3.3.1] (see also Theorem 7.59), $f : X \to S$ has a local model by $A_h \times \mathbb{A}^{d-r} : A_P \times \mathbb{A}^{d-r} \to A_\mathbb{N}$. It thus suffices to show that $A_h : A_P \to A_\mathbb{N}$ is of the form $U(\sigma) \to \mathbb{A}_t^1$. The map $h : \mathbb{N} \to P$ is injective by [222, IV, Thm. 3.3.1]. Verticality implies $h(1) \in \operatorname{int}(P)$, and saturatedness implies that $K_h = h(1) + P \subseteq \operatorname{int}(P)$ is a radical ideal. Then K_h is the intersection of the prime ideals containing it, hence $K_h = \operatorname{int}(P)$. Thus, P is a Gorenstein toric monoid, which are all of the form P_σ. $\qquad\qquad\square$

As in the case of normal crossings, the local structure of the central fiber of a toroidal crossing degeneration $f : X \to S$ is dictated by the local models $U_S(\sigma) \to S$. Thus, they are all of the following form:

Definition 9.7 (Toroidal Crossing Singularities) A *space with toroidal crossing singularities* is a separated scheme $V/\mathbf{k}$ of finite type, pure of some dimension

---

[4] Note that this is independent of the choice uniformizer $t \in \mathfrak{m}_R$ as a different choice differs by an automorphism of the base S.

d, which is, locally in the étale topology, isomorphic to local models of the form $V(\sigma) \to \mathbf{k}$. In particular, V is reduced and Gorenstein.

Our goal is now to understand toroidal crossing degenerations to a given space V with toroidal crossing singularities. When studying semistable degenerations $f \colon X \to S$ to a normal crossing space V, then the local model $z_1 \cdot \ldots \cdot z_r = t$ of $f \colon X \to S$ around a point $x \in V$ is dictated by the geometry of V: the number r is precisely the number of étale local components of V which meet at x. Thus, in a way, all semistable degenerations to V are the same around $x \in V$ even though they do not induce the same log structure on V. Example 9.3 shows that this is no longer true for a space with toroidal crossing singularities: each toroidal crossing degeneration $\{xy = t^k\}$ has the same central fiber $\{xy = 0\}$, but the total spaces have non-isomorphic singularities. However, to have a well-behaved theory analogous to the case of normal crossing spaces, we wish to prescribe the toroidal crossing degeneration up to étale local isomorphy, i.e., fix the local model at every point. Now a local model is always of the form $A_P \times \mathbb{A}^{d-r} \to A_{\mathbb{N}}$ so that it can be recovered from the induced map on ghost sheaves of the log structure (plus the dimension to recover $d - r$ where r is the rank of P). Thus, we endow V with a sheaf of monoids $\mathcal{P}$, which is supposed to be the ghost sheaf of the log structure induced on the central fiber, and with a global section $\bar{\rho} \in \mathcal{P}$, which is supposed to be the image of the generator $1_{S_0} \in \mathcal{M}_{S_0}$ under the induced map of log schemes. Then [222, IV, Thm. 3.3.1] (see also Theorem 7.59) implies that any toroidal crossing degeneration whose induced map in the central fiber on the level of ghost sheaves is $\mathbb{N} \to \mathcal{P},\ 1 \mapsto \bar{\rho}$, has, around each point $\bar{v} \in V$, the same local model $A_P \times \mathbb{A}^{d-r} \to A_{\mathbb{N}}$ with $P = \mathcal{P}_{\bar{v}}$. A *toroidal crossing space* is such a triple $(V, \mathcal{P}, \bar{\rho})$. In the following sections, we compile the basic theory of $(V, \mathcal{P}, \bar{\rho})$.

The following result is sometimes useful.

Lemma 9.8 *Let V be a space with toroidal crossing singularities. Then V has semi–log canonical singularities.*

Proof Our reference for semi–log canonical singularities is [176]. Since V is deminormal and Gorenstein, what we have to show is that $(\bar{V}, \bar{D})$ is log canonical, where $v \colon \bar{V} \to V$ is the normalization of V, and $\bar{D} \subset \bar{V}$ is the conductor locus. The construction of $(\bar{V}, \bar{D})$ commutes with étale morphisms $f \colon Y \to V$, and the condition of having log canonical singularities is local in the étale topology. Thus, it is sufficient to show that $V(\sigma)$ has semi–log canonical singularities. The normalization $\bar{V}(\sigma)$ of $V(\sigma)$ is the disjoint union of the irreducible components of $V(\sigma)$ because they are normal. The conductor locus $D(\sigma) \subseteq V(\sigma)$ is the locus where at least two components meet with its reduced scheme structure, and $\bar{D}(\sigma) \subseteq \bar{V}(\sigma)$ is the toric boundary of each irreducible component with its reduced scheme structure. Thus, $(\bar{V}(\sigma), \bar{D}(\sigma))$ is log canonical by [56, Cor. 11.4.25], and $V(\sigma)$ is semi–log canonical. $\qquad\square$

Corollary 9.9 *Let V be a space with toroidal crossing singularities which is projective and of dimension d. Let $\mathcal{L}$ be an ample line bundle on V. Then $H^k(V, \omega_V^\circ \otimes \mathcal{L}) = 0$ for $k \geq 1$ and $H^k(V, \mathcal{L}^\vee) = 0$ for $0 \leq k \leq d - 1$. Here, ω_V° is the normalized dualizing sheaf of the Gorenstein scheme V.*

Proof The first statement is explicit in [98, Thm. 1.8] (for slc singularities which may not be Cohen–Macaulay), and the second statement is explicit in [181, Cor. 6.6] (over $\mathbf{k} = \mathbb{C}$, for weakly slc singularities which are required to be Cohen–Macaulay). Both statements are equivalent in our case by [138, III, Thm. 7.6]. $\qquad\square$

9.1 The Monoid P_σ

Let $\sigma \subseteq M_\mathbb{R}$ be a full-dimensional lattice polytope as in Construction 9.2. Here, we collect some additional basic facts about P_σ. We start with a more explicit description.

Lemma 9.10 *Let $\sigma \subseteq M_\mathbb{R}$ be a full-dimensional lattice polytope, and let*

$$\check{\psi}_\sigma(n) := -\inf\{\langle m, n \rangle \mid m \in \sigma\}.$$

Then, we have

$$P_\sigma = \{n + a_0 e_0^* \in N \oplus \mathbb{Z} \cdot e_0^* \mid a_0 \geq \check{\psi}_\sigma(n)\}.$$

This is a sharp toric monoid.

Proof Note $C(\sigma)^\vee = \{n + a_0 e_0^* \in N_\mathbb{R} \oplus \mathbb{R} \cdot e_0^* \mid \forall m \in \sigma : \langle m, n \rangle \geq -a_0\}$ to deduce the given explicit description. This description also shows that P_σ is toric. For sharpness, assume that $n + a_0 e_0^* \in P_\sigma$ and $-n - a_0 e_0^* \in P_\sigma$. Then, we have $a_0 \geq \check{\psi}_\sigma(n)$ and $-a_0 \geq \check{\psi}_\sigma(-n)$, so $\check{\psi}_\sigma(n) + \check{\psi}_\sigma(-n) \leq 0$. This implies that $\langle m, n \rangle$ is constant for $m \in \sigma$. Since σ is full-dimensional, we find $n = 0$. Now $-e_0^* \notin P_\sigma$, so $a_0 = 0$. $\qquad\square$

When we move σ, the monoid P_σ changes by a canonical isomorphism. Namely, let $p \in M$, and consider the lattice polytope $p + \sigma$. Then, we have $\check{\psi}_{p+\sigma}(n) = \check{\psi}_\sigma(n) + \langle p, n \rangle$ and therefore an isomorphism

$$\phi \colon P_\sigma \to P_{p+\sigma}, \quad n + ae^* \mapsto n + (\langle p, n \rangle + a)e^*,$$

of monoids with $\phi(\rho_\sigma) = \rho_{p+\sigma}$.

Next, we describe the faces of P_σ.

Lemma 9.11 *Let $\varnothing \neq \tau \subseteq \sigma$ be a non-empty face of the polytope. Then*

$$\check{\tau} = \{n \in N_\mathbb{R} \mid \forall m \in \sigma, m_0 \in \tau : \langle m, n \rangle \geq \langle m_0, n \rangle\}$$

is a rational polyhedral cone in $N_\mathbb{R}$. The function $\check{\psi}_\sigma : N_\mathbb{R} \to \mathbb{R}$ is linear on $\check{\tau}$, and

$$F_{\tau/\sigma} := \{n + \check{\psi}_\sigma(n) \cdot e_0^* \mid n \in \check{\tau} \cap N\} \subseteq P_\sigma$$

is a proper face of P_σ. This induces an order-reversing one-to-one correspondence between non-empty faces $\tau \subseteq \sigma$ and proper faces $F \subsetneq P_\sigma$.

Proof Let $\sigma^{[0]}$ be the set of vertices of σ, and let $\tau^{[0]}$ be the set of vertices of τ. Let $m_0 \in \tau^{[0]}$. Then, we have

$$\check{\tau} = \{n \in N_\mathbb{R} \mid \forall m \in \sigma^{[0]} : \langle m - m_0, n \rangle \geq 0, \ \forall m \in \tau^{[0]} : \langle m - m_0, n \rangle = 0\}$$

so that $\check{\tau}$ is a rational polyhedral cone. When $n \in \check{\tau}$, then $\langle m, n \rangle$ achieves an infimum for $m \in \sigma$ at m_0, so we have $\check{\psi}_\sigma(n) = -\langle m_0, n \rangle$ for $n \in \check{\tau}$, and $\check{\psi}_\sigma$ is linear on $\check{\tau}$. This shows that $F_{\tau/\sigma} \subseteq P_\sigma$ is a submonoid.

Next, we show that $F_{\tau/\sigma}$ is a face. Let $n' + a_0' e_0^*, n'' + a_0'' e_0^* \in P_\sigma$ be such that their sum $n + a_0 e_0^*$ is in $F_{\tau/\sigma}$. In particular, $a_0' + a_0'' = \check{\psi}_\sigma(n)$. A computation exploiting the inequalities which characterize P_σ now shows that $\check{\psi}_\sigma(n') + \check{\psi}_\sigma(n'') = \check{\psi}_\sigma(n)$ and thus $a_0' = \check{\psi}_\sigma(n')$ and $a_0'' = \check{\psi}_\sigma(n'')$. A similar computation shows that $\check{\psi}_\sigma(n') + \check{\psi}_\sigma(n'') = \check{\psi}_\sigma(n)$ implies $n', n'' \in \check{\tau}$. Therefore, $n' + a_0' e_0^*, n'' + a_0'' e_0^* \in F_{\tau/\sigma}$, which is hence a face. We have $\rho_\sigma = e_0^* \notin F_{\tau/\sigma}$ so that $F_{\tau/\sigma}$ is a proper face.

When $\omega \subseteq \tau$, then we have $\check{\omega} \supseteq \check{\tau}$ so that the construction reverses the order given by inclusion.

For every non-empty face $\tau \subseteq \sigma$ and any vertex $m_0 \in \tau^{[0]}$, there is some element $n_\tau \in N$ with $\langle m, n_\tau \rangle = \langle m_0, n_\tau \rangle$ for $m \in \tau^{[0]}$ and $\langle m, n_\tau \rangle > \langle m_0, n_\tau \rangle$ for $m \in \sigma^{[0]} \setminus \tau^{[0]}$. Therefore, $\check{\tau}_1 = \check{\tau}_2$ implies $\tau_1 = \tau_2$, and the assignment $\tau \mapsto F_{\tau/\sigma}$ is injective.

It remains to show that every proper face of P_σ is of the form $F_{\tau/\sigma}$. Let $F \subsetneq P_\sigma$ be a proper face. Assume that we have $\rho_\sigma \in F$. Let $n \in N$. Then, we have $n + \check{\psi}_\sigma(n) e_0^* \in P_\sigma$ and $-n + \check{\psi}_\sigma(-n) e_0^* \in P_\sigma$. Now their sum is $\check{\psi}_\sigma(n) e_0^* + \check{\psi}_\sigma(-n) e_0^*$. This is a positive multiple of $e_0^* = \rho_\sigma$ because P_σ is sharp, so the face property implies that $n + \check{\psi}_\sigma(n) e_0^* \in F$. But then we have $F = P_\sigma$, contradicting our assumption that F is a proper face. In other words, we have $\rho_\sigma \notin F$. This implies that every element of F is of the form $n + \check{\psi}_\sigma(n) e_0^*$ for some $n \in N$.

Every face of P_σ is monogenic, i.e., there is some element $f \in F$ such that F is the smallest face containing f. Let $n_F \in N$ be such that $n_F + \check{\psi}_\sigma(n_F) e_0^* = f$. As a function on $\sigma \subseteq M_\mathbb{R}$, n_F assumes its minimum on a non-empty face $\tau \subseteq \sigma$. We have $n_F \in \check{\tau}$ so that $f \in F_{\tau/\sigma}$ and hence $F \subseteq F_{\tau/\sigma}$ by the characterization of F as the smallest face containing f.

Let $H = \{n \in N \mid n + \check{\psi}_\sigma(n) e_0^* \in F\}$. Then, $H \subseteq \check{\tau} \cap N$ is a face. In particular, $\check{\tau}$ admits a face γ in the sense of polyhedral geometry such that $H = \gamma \cap N$. However, the faces of $\check{\tau}$ are precisely the cones $\check{\omega}$ for $\omega \subseteq \tau$, cf. [56, Prop. 2.3.7]. But then $F = F_{\omega/\sigma}$.[5] $\square$

At this point, it is also clear that $\partial P_\sigma = \{n + \check{\psi}_\sigma(n)e_0^* \mid n \in N\}$ and that $\mathrm{int}(P_\sigma) = \rho_\sigma + P_\sigma$, as we claimed in Construction 9.2.

9.1.1   Verticality

We have reviewed vertical monoid homomorphisms in Definition 7.14. In particular, for a homomorphism $\theta: Q \to P$ between integral monoids, θ is vertical if and only if its image is not contained in any proper face of P. Verticality characterizes the monoids of the form P_σ.

Lemma 9.12 *Let $\theta: \mathbb{N} \to P$ be a homomorphism between monoids. Then θ is of the form $\theta_\sigma: \mathbb{N} \to P_\sigma$, $1 \mapsto \rho_\sigma$, for a full-dimensional lattice polytope $\sigma \subseteq M_\mathbb{R}$ if and only if P is a sharp toric monoid, and θ is injective, saturated, and vertical.*

Proof First, let $\sigma \subseteq M_\mathbb{R}$ be a full-dimensional lattice polytope. We have already seen in Lemma 9.10 that P_σ is a sharp toric monoid. The map θ_σ is clearly injective. Since $\mathbb{N}$ is valuative, [222, Prop. 4.6.3] shows that θ_σ is integral and hence locally exact. Now the assumptions of [222, Thm. 4.8.14] are satisfied. We have $K_{\theta_\sigma} = \theta_\sigma(\theta_\sigma^{-1}(P_\sigma^+)) + P_\sigma = \rho_\sigma + P_\sigma$, which is a radical monoid ideal. Therefore, [222, Thm. 4.8.14] implies that θ_σ is saturated. Verticality is clear since ρ_σ is not contained in any proper face.

Conversely, let P be a sharp toric monoid, and let $\theta: \mathbb{N} \to P$ be a homomorphism which is injective, saturated, and vertical. Since θ is saturated, the cokernel $N = \mathrm{coker}(\theta^{\,\mathrm{gp}}: \mathbb{Z} \to P^{\,\mathrm{gp}})$ is a lattice, and since θ is injective, we have a short exact sequence

$$0 \to \mathbb{Z} \to P^{\,\mathrm{gp}} \to N \to 0.$$

Let $\xi: N \to P^{\,\mathrm{gp}}$ be a splitting of $\varpi: P^{\,\mathrm{gp}} \to N$. Then, we have an isomorphism $P^{\,\mathrm{gp}} \cong N \oplus \mathbb{Z}$, and hence we can consider $P \subseteq N \oplus \mathbb{Z}$.

Since θ is vertical, $\rho := \theta(1)$ is not contained in any proper face. Therefore, any element of $P^{\,\mathrm{gp}}$ has the form $p + k\rho$ for $p \in P$ and $k \in \mathbb{Z}$. In particular, the map $P \to N$ is surjective. Furthermore, for any $n \in N$, the set $L(n) = \{k \in \mathbb{Z} \mid \xi(n) + k\rho \in P\}$ is non-empty, and $k \in L(n)$ implies $k + 1 \in L(n)$.

Assume that $L(n) = \mathbb{Z}$. Since $L(-n) \neq \varnothing$, we can find $\ell \in L(-n)$. Then, we have $\xi(n) + k\rho + \xi(-n) + \ell\rho = (k + \ell)\rho \in P$ for any $k \in \mathbb{Z}$, contradicting

---

[5] We actually have $\omega = \tau$ in our construction, but we do not need to prove this in order to prove the lemma.

$-\rho \notin P$, which holds since $\rho \neq 0$ and P is sharp. Thus, there is some $\check{\psi}(n) \in \mathbb{Z}$ with $L(n) = \check{\psi}(n) + \mathbb{N}$. Now, we have

$$P = \{\xi(n) + a_0\rho \mid a_0 \geq \check{\psi}(n)\}.$$

Let us write $E = \{\xi(n) + \check{\psi}(n)\rho \mid n \in N\} \subseteq P$.

Let $F \subsetneq P$ be a proper face. Then $\rho \notin F$ since ρ is not contained in any proper face. Now, if $\xi(n) + a_0\rho \in F$ for some $a_0 > \check{\psi}(n)$, then we would also have $\rho \in F$. Therefore, we have $F \subseteq E$.

Conversely, we would like to show that every element of E is contained in a proper face. Since θ is integral and hence locally exact, [222, Thm. 4.8.14] applies, and $K_\theta = \theta(\theta^{-1}(P^+)) + P = \rho + P$ is a radical monoid ideal.

Let $p \in \text{int}(P)$ be not contained in any proper face. Then, the face generated by p is equal to P, so we can find $q \in P$ and $n \geq 1$ with $\rho + q = np$. Thus, $np \in K_\theta$ and so $p \in K_\theta$ because K_θ is radical. Conversely, if $p = \rho + q$, then $p \in \text{int}(P)$ because the face generated by p must contain ρ. Therefore, we have $\rho + P = K_\theta = \text{int}(P)$. Now, $\xi(n) + \check{\psi}(n)\rho$ is not of the form $\rho + q$ for any $q \in P$, so $\xi(n) + \check{\psi}(n)\rho \notin \text{int}(P)$, and $\xi(n) + \check{\psi}(n)\rho$ is contained in a proper face $F \subsetneq P$.

On the level of lattices, a fan in N is nothing but a finite collection of subsets $\check{\Sigma} = \{\check{\tau}\}$ such that each $\check{\tau} \subseteq N$ is a finitely generated sharp saturated submonoid, such that $\check{\tau}_1 \cap \check{\tau}_2 \in \check{\Sigma}$ for any $\check{\tau}_1, \check{\tau}_2 \in \check{\Sigma}$, and such that any face of some $\check{\tau} \in \check{\Sigma}$ is again contained in $\check{\Sigma}$.

The collection $\check{\Sigma}$ of subsets of the form $\varpi(F)$ for a proper face $F \subsetneq P$ is a fan in this sense. That $\varpi(F)$ is a finitely generated sharp submonoid follows from linearity of ϖ and the fact that F is sharp. Now assume that $k \cdot n \in \varpi(F)$ for $k > 1$. Since every element of E is contained in a proper face, there is some proper face $G \subsetneq P$ with $\xi(n) + \check{\psi}(n)\rho \in G$. Then, we also have $k\xi(n) + k\check{\psi}(n)\rho \in G \subseteq E$. By definition, we have $\xi(kn) + \check{\psi}(kn)\rho \in F$. Thus, if $k\check{\psi}(n) > \check{\psi}(kn)$, then we would have $k\xi(n) + k\check{\psi}(n)\rho \in \text{int}(P)$, a contradiction. In other words, we have $k\check{\psi}(n) = \check{\psi}(kn)$, and therefore $k(\xi(n) + \check{\psi}(n)\rho) \in F$. Since F is a face, this implies $\xi(n) + \check{\psi}(n)\rho \in F$, and therefore $n \in \varpi(F)$. In other words, $\varpi(F) \subseteq N$ is saturated. The condition on intersections follows from bijectivity of $\varpi : E \to N$, and the last condition is clear. Furthermore, the fan is complete.

The function $\check{\psi} : N \to \mathbb{Z}$ is clearly linear on each $\varpi(F)$, and it is convex by construction. It is strictly convex in the sense that $\check{\psi}(n_1) + \check{\psi}(n_2) > \check{\psi}(n_1 + n_2)$ whenever there is no $\varpi(F)$ which contains both n_1 and n_2. Now, there is a unique[6] full-dimensional lattice polytope $\sigma \subseteq M_\mathbb{R}$ for $M = \text{Hom}_\mathbb{Z}(N, \mathbb{Z})$ with $\check{\psi} = \check{\psi}_\sigma$. Thus, $\theta : \mathbb{N} \to P$ is of the desired form. $\qquad\square$

---

[6] Note however that σ depends on our choice of a splitting $\xi : N \to P^{\text{gp}}$.

9.1.2 The Toric Stratification of $V(\sigma)$

Let $P = P_\sigma$. In Construction 8.7, we introduced, for any face $F \subseteq P$, the open subset $U_F = \operatorname{Spec} \mathbf{k}[P_F] \subseteq U(\sigma)$ and the closed subset $V_F = \operatorname{Spec} \mathbf{k}[F] \subset U(\sigma)$ as well as their intersection $V_F^\circ = V_F \cap U_F = \operatorname{Spec} \mathbf{k}[F^{\mathrm{gp}}]$. Then we have disjoint decompositions

$$U(\sigma) = \bigsqcup_{F \subseteq P} V_F^\circ \quad \text{and} \quad V(\sigma) = \bigsqcup_{F \subsetneq P} V_F^\circ,$$

where the first union runs over all faces $F \subseteq P$ while the second union only runs over proper faces $F \subsetneq P$. We write 0 for the unique point of $V_{\{0\}}^\circ$, where $\{0\} \subseteq P$ is the trivial face.

9.1.3 $\mathcal{O}_{V(\sigma)}$ as a Limit

Let $P = P_\sigma$ and $\rho = \rho_\sigma$. Then, we can describe the ring $\mathbf{k}[P]/(z^\rho)$ defining $V(\sigma)$ as follows. Let $E = P \setminus (\rho + P)$, and consider $\mathbf{k}[E] = \bigoplus_{e \in E} \mathbf{k} \cdot z^e$ as a vector space. Then, we have an isomorphism of vector spaces $\mathbf{k}[E] \to \mathbf{k}[P]/(z^\rho)$.

We have seen in the proof of Lemma 9.12 that $E = \bigcup_{F \subsetneq P} F$ is the union of the proper faces. We turn $\mathbf{k}[E]$ into a ring as follows: If there is a proper face $F \subsetneq P$ with $e_1, e_2 \in F$, then we set $z^{e_1} \cdot z^{e_2} = z^{e_1 + e_2}$. Otherwise, we set $z^{e_1} \cdot z^{e_2} = 0$. This turns $\mathbf{k}[E]$ into a ring, and $\mathbf{k}[E] \to \mathbf{k}[P]/(z^\rho)$ is an isomorphism of $\mathbf{k}$-algebras.

For every proper face $F \subsetneq P$, we consider $\mathbf{k}[F]$ as a $\mathbf{k}[E]$-module via $z^e \cdot z^f = z^{e+f}$ if $e \in F$ and $z^e \cdot z^f = 0$ otherwise. For an inclusion $G \subseteq F$, the map $\mathbf{k}[F] \to \mathbf{k}[G]$ (given by $z^f \mapsto z^f$ if $f \in G$ and $z^f \mapsto 0$ otherwise) is a homomorphism of $\mathbf{k}[E]$-modules. When $\mathcal{F}$ is the set of proper faces of P ordered by inclusion, considered as a category, then we obtain a diagram $\mathcal{F}^{\mathrm{op}} \to \mathbf{mod}(\mathbf{k}[E])$ of finitely generated $\mathbf{k}[E]$-modules.

Lemma 9.13 *The canonical map $\phi \colon \mathbf{k}[E] \to \lim_{F \subsetneq P}(\mathbf{k}[F])$ is an isomorphism of $\mathbf{k}[E]$-modules. In particular, the canonical map*

$$\mathcal{O}_{V(\sigma)} \to \lim_{F \subsetneq P}(\mathcal{O}_{V_F})$$

is an isomorphism of $\mathcal{O}_{V(\sigma)}$-modules.

Proof Let $\phi(\sum_i \lambda_i z^{e_i}) = 0$. For every e_i, there is some $F_i \subsetneq P$ with $e_i \in F_i$. Then, we have $\sum_{j : j \in F_i} \lambda_j z^{e_j} = 0$ and thus $\lambda_i = 0$. Therefore, ϕ is injective. Conversely, an element of the limit is given by an element $\sum_{e \in F} \lambda_{F;e} z^e$ for every proper face $F \subsetneq P$, where the sum is finite for every F. If $e \in G \subseteq F$, then we must have $\lambda_{G;e} = \lambda_{F;e}$. In particular, $\lambda_e := \lambda_{F;e}$ is well-defined. Furthermore, only finitely many λ_e are non-zero so that $\sum_{e \in E} \lambda_e z^e \in \mathbf{k}[E]$ is a well-defined element. Its image under ϕ is the original element of the limit. Thus, ϕ is surjective. For the second claim, note that the functor $M \mapsto \widetilde{M}$ mapping a $\mathbf{k}[E]$-module to its associated quasi-coherent sheaf is exact and thus preserves finite limits. $\qquad\square$

9.1.4 The Local Structure of $V(\sigma)$

Let us denote the ghost sheaf of $\mathcal{M}_{V(\sigma)}$ by $\mathcal{P}_\sigma$. On each V_F° for a proper face $F \subsetneq P$, the stalk of $\mathcal{P}_\sigma$ is P/F. Let $\varpi \colon P \to P/F$ be the quotient map. For every proper face $\widetilde{G} \subsetneq P/F$, the preimage $G = \varpi^{-1}(\widetilde{G})$ is a proper face of P. Thus, $\tilde{\rho}_\sigma = \varpi(\rho_\sigma) \in \mathrm{int}(P/F)$. Let now $h \in P$ be such that $\varpi(h) \in \mathrm{int}(P/F)$. Then $h \notin G$ for every proper face $G \subsetneq P$ with $F \subseteq G$. Let $f \in F$ be such that $F = \langle f \rangle$ for the smallest face containing f. Since no proper face contains both f and h, we have $f + h \in \mathrm{int}(P)$. But then, we can find $g \in P$ with $f + h = \rho_\sigma + g$, i.e., $\varpi(h) \in \tilde{\rho}_\sigma + P/F$. Therefore, P/F is a Gorenstein monoid, and $\tilde{\rho}_\sigma = \varpi(\rho_\sigma)$ is the Gorenstein degree. We can also compute P/F in terms of σ.

Lemma 9.14 *Let $\sigma \subseteq M_\mathbb{R}$ be a full-dimensional lattice polytope, and let $P = P_\sigma$. Let $\tau \subseteq \sigma$ be a non-empty face, and let $F = F_{\tau/\sigma}$. Embed τ into some $M'_\mathbb{R}$ such that τ becomes a full-dimensional lattice polytope. Then, we have an isomorphism $P/F \cong P'_\tau$, where P'_τ is constructed in analogy with P_σ.*

Proof Since P_σ is, up to canonical isomorphism, invariant under translation of σ, we can assume that $0 \in \tau$. Let $M' \subseteq M$ be the integral affine tangent space to τ so that $\tau \subseteq M'_\mathbb{R}$ is a full-dimensional lattice polytope. We set

$$\check{\psi}'_\tau(n') := -\inf\{\langle m', n' \rangle \mid m' \in \tau\}$$

as a function $N' = \mathrm{Hom}_\mathbb{Z}(M', \mathbb{Z}) \to \mathbb{Z}$ so that

$$P'_\tau = \{n' + a_0 e_0^* \mid a_0 \geq \check{\psi}'_\tau(n')\}.$$

We also have a function $\check{\psi}_\tau \colon N \to \mathbb{Z}$ given by

$$\check{\psi}_\tau(n) = -\inf\{\langle m, n \rangle \mid m \in \tau\}.$$

When $\varrho \colon N \to N'$ is the projection, we have $\check{\psi}_\tau = \check{\psi}'_\tau \circ \varrho$. The function $\check{\psi}_\tau$ defines a toric monoid $P_\tau \subseteq N \oplus \mathbb{Z}$ in analogy with P_σ. The projection $\varrho \colon N \to N'$ then induces a projection $\rho \colon P_\tau \to P'_\tau$, which is clearly surjective. Since P'_τ is sharp, any invertible element in $P_\tau^* \subseteq P_\tau$ must map to zero in P'_τ, so we have $P_\tau^* \subseteq (M')^\perp = \ker(N \oplus \mathbb{Z} \to N' \oplus \mathbb{Z})$. Conversely, the elements in $(M')^\perp$ are clearly invertible in P_τ. It is now clear that $P_\tau \to P'_\tau$ is the quotient map by the face $P_\tau^* \subseteq P_\tau$.

Since $\tau \subseteq \sigma$, we have $\check{\psi}_\tau(n) \le \check{\psi}_\sigma(n)$ for all $n \in N$ so that $P_\sigma \subseteq P_\tau$. This is the localization map in $F_{\tau/\sigma}$ so that we obtain an isomorphism $P_\sigma/F_{\tau/\sigma} \cong P'_\tau$. $\qquad\square$

Corollary 9.15 *Let $\sigma \subseteq M_\mathbb{R}$ be a full-dimensional lattice polytope, and let $v \in V(\sigma)$ be a point. Then, there is an open neighborhood $v \in U \subseteq V(\sigma)$, a full-dimensional lattice polytope $\tau \subseteq M'_\mathbb{R}$ in some other lattice M', and a strict and smooth morphism $h\colon U \to V(\tau)$ of log schemes with $h(v) = 0$.*

Proof Let $F \subsetneq P_\sigma$ be the proper face with $v \in V_F^\circ$. Let $\tau \subseteq \sigma$ be the non-empty face with $F = F_{\tau/\sigma}$, and move σ in M such that $0 \in \tau$. Now

$$U := \bigsqcup_{F \subseteq G \subsetneq P} V_G^\circ = U_F \cap V(\sigma)$$

is an open neighborhood of V_F°. We have a non-canonical isomorphism $P_F \cong F^{\mathrm{gp}} \oplus P/F$ which can be chosen to respect ρ_σ. Therefore, we have a strict and smooth map $U_F \to A_{P/F}$ over $A_\mathbb{N}$, and hence a strict and smooth map $U_F \cap V(\sigma) \to V(\tau)$ since $P/F \cong P_\tau$. $\qquad\square$

9.2 Spaces with a Toric Ghost Sheaf

This is the first device we define on our way toward toroidal crossing spaces. It is a space V together with a sheaf of monoids $\mathcal{P}$ that might occur as the ghost sheaf of a log structure. First, we need to specify which sheaves $\mathcal{P}$ can occur as the ghost sheaves of a fine and saturated log structure.

Definition 9.16 (Sharp and Toric Sheaves of Monoids) A sheaf of monoids $\mathcal{P}$ in the étale topology of V is *sharp and toric* if the following two conditions hold:

(a) for every geometric point $\bar{v} \to V$, the stalk $\mathcal{P}_{\bar{v}}$ is a sharp toric monoid—i.e., fine, saturated, $\mathcal{P}_{\bar{v}}^{\mathrm{gp}}$ is torsion-free, and $\mathcal{P}_{\bar{v}}^* = \{0\}$ for the subgroup of invertibles;

(b) every geometric point $\bar{v} \to V$ admits an étale neighborhood U and a map $\beta\colon \underline{P}_U \to \mathcal{P}|_U$ of sheaves of monoids from a constant sheaf $\underline{P}_U$ such that $P \to \mathcal{P}_{\bar{v}}$ is an isomorphism, and such that in the logification factorization $\underline{P}_U \to \underline{P}_U^{\mathrm{log}} \to \mathcal{P}|_U$ in the sense of [222, II, Prop. 1.1.5], we have $\underline{P}_U^{\mathrm{log}} = \mathcal{P}|_U$; under all prior conditions, the last condition is equivalent to requiring that $P \to \mathcal{P}_{\bar{u}}$ is the quotient map by a face of P for every geometric point $\bar{u} \to U$.

With this definition, a u-integral log scheme $(V, \mathcal{M})$ is fine and saturated if and only if $\overline{\mathcal{M}}$ is a sharp and toric sheaf of monoids. This follows from the argument given in [124, Prop. 3.7]. Unfortunately, there in the original setting, the argument is wrong in several respects. First, without the u-integrality assumption, the given sequence

$$1 \to O^*_{X,\bar{x}} \to M^{\mathrm{gp}}_{X,\bar{x}} \to \overline{M}^{\mathrm{gp}}_{X,\bar{x}} \to 0$$

at the beginning of the proof is not injective on the left. Secondly, if we assume $\mathcal{M}$ only to be fine—as Gross–Siebert does—the exact sequence in the proof of [124, Prop. 3.7] does not necessarily split—at least the argument is wrong. This is because there are integral and sharp monoids P such that P^{gp} has torsion. An example can be found before [222, I, Prop. 1.3.5]. Adding the saturatedness hypothesis remedies this situation. Cf. also [222, III, Thm. 1.2.7]. Thirdly, the condition on $\beta \colon \underline{P}_U \to \mathcal{P}|_U$ is wrongly stated as being merely surjective and an isomorphism at $\bar{v}$ instead of asking that every $P \to \mathcal{P}_{\bar{u}}$ is the quotient map by a face. This issue becomes relevant at the end of the proof of [124, Prop. 3.7], where it is wrongly assumed that a surjective monoid homomorphism with trivial kernel would be an isomorphism.

When $\mathcal{P}$ is a sharp and toric sheaf of monoids, note that $\mathcal{P}_{\bar{v}}$ is, up to isomorphism, independent of the geometric point $\bar{v}$ lying over $v \in V$.

Definition 9.17 (Spaces with a Toric Ghost Sheaf)   A *space with a toric ghost sheaf* is a pair $(V, \mathcal{P})$ where V is a reduced scheme and $\mathcal{P}$ is a sharp and toric sheaf of monoids. A *morphism* between spaces with a toric ghost sheaf $(Y, \mathcal{P}_Y)$ and $(V, \mathcal{P}_V)$ is a morphism of schemes $f \colon Y \to V$ together with a homomorphism of sheaves of monoids $f^\flat \colon \mathcal{P}_V \to f_* \mathcal{P}_Y$. A morphism is *strict* if the induced map $f^\flat \colon f^{-1} \mathcal{P}_V \to \mathcal{P}_Y$ is an isomorphism.

Example 9.18   Let $V(\sigma)$ be as in Construction 9.2. Then with $\mathcal{P}_\sigma := i^{-1} \overline{\mathcal{M}}_{U(\sigma)}$, we obtain a space with a toric ghost sheaf $(V(\sigma), \mathcal{P}_\sigma)$. Here, $i \colon V(\sigma) \to U(\sigma)$ denotes the inclusion. $\Diamond$

Our goal is to classify all u-integral log structures on V that have the given sheaf $\mathcal{P}$ as the ghost sheaf. We consider the log structure as a triple $(\mathcal{M}, \alpha, q)$ where $\mathcal{M}$ is a sheaf of monoids, $\alpha \colon \mathcal{M} \to O_V$ is a homomorphism turning $\mathcal{M}$ into a log structure, and $q \colon \mathcal{M} \to \mathcal{P}$ is a homomorphism establishing $\mathcal{P}$ as the ghost sheaf; an isomorphism between $(\mathcal{M}, \alpha, q)$ and $(\mathcal{M}', \alpha', q')$ is an isomorphism $\varphi \colon \mathcal{M} \cong \mathcal{M}'$ which is compatible with all data. Isomorphism classes of log structures do *not* form a sheaf, so we define:

Definition 9.19 $(\widehat{\mathcal{L}}_V)$ Let $(V, \mathcal{P})$ be a space with a toric ghost sheaf. Then $\widehat{\mathcal{L}}_V$ is the sheaf of equivalence classes of u-integral log structures $(\mathcal{M}, \alpha, q)$ where two log structures are *equivalent* if they are locally isomorphic.[7]

---

[7] The sections of $\widehat{\mathcal{L}}_V$ are not in general log structures on their locus of definition; rather, they are log structures on the open subsets of a cover which are equivalent on overlaps.

To compute the stalk $\widehat{\mathcal{L}}_{V,\bar{v}}$ at a geometric point $\bar{v} \in V$, we construct a canonical injection

$$\iota_{V,\bar{v}}\colon \; \widehat{\mathcal{L}}_{V,\bar{v}} \to \mathcal{E}xt^1(\mathcal{P}^{\mathrm{gp}}, O_V^*)_{\bar{v}}$$

and determine its image. Here, $\mathcal{P}^{\mathrm{gp}}$ is the sheaf of groups associated to $\mathcal{P}$. Note that the target is *not* the extension group $\mathrm{Ext}^1(\mathcal{P}_{\bar{v}}^{\mathrm{gp}}, O_{V,\bar{v}}^*)$.

For $\iota_{V,\bar{v}}$, we actually have two constructions, which yield the same map. Firstly, a germ $[(\mathcal{M}, \alpha, q)] \in \widehat{\mathcal{L}}_{V,\bar{v}}$—represented by a log structure defined in a neighborhood of $\bar{v}$—induces a short exact sequence

$$1 \to O_V^* \to \mathcal{M}^{\mathrm{gp}} \xrightarrow{q^{\,\mathrm{gp}}} \mathcal{P}^{\mathrm{gp}} \to 0$$

of Abelian groups and thus an extension class in $\mathcal{E}xt^1(\mathcal{P}^{\mathrm{gp}}, O_V^*)_{\bar{v}}$. Injectivity follows from the argument in [124, Prop. 3.11, Cor. 3.12]. The second construction is more involved.

Construction 9.20 We set $P := \mathcal{P}_{\bar{v}}$, which is a toric monoid. Because $\mathcal{P}$ is a sharp and toric sheaf of monoids, there is—possibly after shrinking V to an appropriate étale neighborhood—a surjection $P_V \to \mathcal{P}$ which is an isomorphism at $\bar{v}$. Here, P_V denotes the constant sheaf with stalk P. Groupification yields the surjection $P_V^{\mathrm{gp}} \to \mathcal{P}^{\mathrm{gp}}$ of Abelian sheaves, whose kernel $\mathcal{R}$ is called the *relation sheaf*. Since $\mathcal{E}xt^1(P_V^{\mathrm{gp}}, O_V^*) = 0$, we get a presentation

$$\mathcal{H}om(P_V^{\mathrm{gp}}, O_V^*) \to \mathcal{H}om(\mathcal{R}, O_V^*) \to \mathcal{E}xt^1(\mathcal{P}^{\mathrm{gp}}, O_V^*) \to 0. \tag{9.1}$$

The stalk $\mathcal{H}om(\mathcal{R}, O_V^*)_{\bar{v}}$ has an explicit description: An element $p \in P^{\mathrm{gp}}$ induces an element in $\Gamma(V, \mathcal{P}^{\mathrm{gp}})$; its *support* is

$$\mathrm{supp}(p) = \{\bar{x} \in V \mid 0 \neq p_{\bar{x}} \in \mathcal{P}_{\bar{x}}^{\mathrm{gp}}\} \subseteq V,$$

which is already a closed subset of V, so we do not need to take its closure to obtain the support. For $n \in \mathbb{Z} \setminus \{0\}$, we have $\mathrm{supp}(np) = \mathrm{supp}(p)$, and for $p, q \in P^{\mathrm{gp}}$ we have

$$(V \setminus \mathrm{supp}(p)) \cap (V \setminus \mathrm{supp}(q)) \subseteq (V \setminus \mathrm{supp}(p + q)).$$

We say p and q are *equifacial* if, for every face $F \subseteq P$, it holds that $p + q \in F^{\mathrm{gp}}$ implies $p \in F^{\mathrm{gp}}$ and $q \in F^{\mathrm{gp}}$. In this case, the above inequality is an equality. Let us write

$$V(p)^\circ := V \setminus \mathrm{supp}(p)$$

for short. After writing $j_p \colon V(p)^\circ \subseteq V$ for the inclusion, the stalk $\mathcal{H}om(\mathcal{R}, O_V^*)_{\bar{v}}$ is

$$\left\{ (h_p)_{p \in P^{\mathrm{gp}}} \;\middle|\; h_p \in (j_{p*} O_{V(p)^\circ}^*)_{\bar{v}}, \;\; h_p \cdot h_q = h_{p+q} \text{ on } V(p)^\circ \cap V(q)^\circ \right\}.$$

At this point, there is a minor mistake in the statement of [124, Prop. 3.14] where they write $V \setminus \mathrm{supp}(p+q)$ instead of $(V \setminus \mathrm{supp}(p)) \cap (V \setminus \mathrm{supp}(q))$. To obtain the map between the stalk and its explicit description, let $\theta \colon \mathcal{R} \to O_V^*$ be a homomorphism defined on a neighborhood $\bar{v} \in W \subseteq V$. For $p \in P^{\mathrm{gp}}$, we have $p \in \Gamma(W \cap V(p)^\circ, \mathcal{R})$, and hence an element $\theta(p) \in \Gamma(W \cap V(p)^\circ, O_V^*)$. We then set $h_p = \theta(p)|_{\bar{v}}$ as the restriction to the stalk at $\bar{v}$. We obtain an equivalent description when we drop the h_p for $p \notin P$; this yields

$$\mathcal{H}om(\mathcal{R}, O_V^*)_{\bar{v}}$$
$$= \left\{ (h_p)_{p \in P} \;\middle|\; h_p \in (j_{p*} O_{V(p)^\circ}^*)_{\bar{v}}, \;\; h_p \cdot h_q = h_{p+q} \text{ on } V(p+q)^\circ \right\}$$

for the stalk. Now let $[(\mathcal{M}, \alpha, q)] \in \widehat{\mathcal{L}}_{V,\bar{v}}$. The surjection $q^{\mathrm{gp}} \colon \mathcal{M}_{\bar{v}}^{\mathrm{gp}} \to \mathcal{P}_{\bar{v}}^{\mathrm{gp}}$ of stalks splits because $\mathcal{P}_{\bar{v}}$ is toric; the splitting then induces locally around $\bar{v}$ a splitting $\sigma \colon \mathcal{P} \to \mathcal{M}$ of the sheaf homomorphism $\mathcal{M} \to \mathcal{P}$. The composition as in the diagram

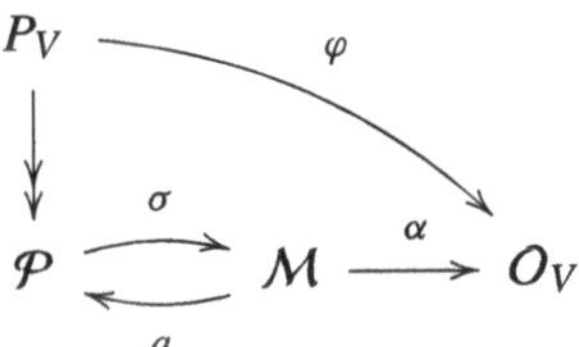

yields a map $\varphi \colon P \to \Gamma(V, O_V)$; then $\varphi(p)|_{V \setminus \mathrm{supp}(p)}$ is an invertible function. Indeed, we have $p_{\bar{x}} = 0 \in \mathcal{P}_{\bar{x}}^{\mathrm{gp}}$ if and only if $\varphi(p)_{\bar{x}} \in O_{V,\bar{x}}^*$. Thus, $\varphi(p)$ defines an element h_p of $(j_{p*} O_{V \setminus \mathrm{supp}(p)}^*)_{\bar{v}}$, and we get $(h_p)_{p \in P} \in \mathcal{H}om(\mathcal{R}, O_V^*)_{\bar{v}}$. The image in $\mathcal{E}xt^1(\mathcal{P}^{\mathrm{gp}}, O_V^*)_{\bar{v}}$ is independent of the choice of splitting σ; this yields our second construction of $\iota_{V,\bar{v}}$, which coincides with the first one. $\Diamond$

We determine the image of $\iota_{V,\bar{v}}$. Assume first that $\xi = [(h_p)_p]$ comes from a log structure $(\mathcal{M}, \alpha, q)$. Then each h_p extends by 0 to V—this means that there are $\tilde{h}_p := \varphi(p) \in O_{V,\bar{v}}$ such that $\tilde{h}_p \mapsto h_p$ under $O_{V,\bar{v}} \to (j_{p*} O_{V \setminus \mathrm{supp}(p)})_{\bar{v}}$ and $\tilde{h}_p \mapsto 0$ under $O_{V,\bar{v}} \to O_{\mathrm{supp}(p),\bar{v}}$. Such an extension $\tilde{h}_p$ of h_p is unique if it exists; in fact, the map

$$O_{V,\bar{v}} \to O_{\mathrm{supp}(p),\bar{v}} \times (j_{p*} O_{V \setminus \mathrm{supp}(p)})_{\bar{v}} \tag{9.2}$$

is injective because V is reduced. Conversely, if each h_p admits an extension $\tilde{h}_p$ by 0, then $P \to O_V$, $p \mapsto \tilde{h}_p$ is a chart of a log structure (in a neighborhood of $\bar{v}$) because it is a homomorphism due to uniqueness of $\tilde{h}_p$. Then $\xi = [(h_p)_p]$ is the image of this log structure under $\iota_{V,\bar{v}}$.

Lemma 9.21 (cf. [124, Prop. 3.14]) *A germ $\xi = [(h_p)_p] \in \mathcal{E}xt^1(\mathcal{P}^{\mathrm{gp}}, O_V^*)_{\bar{v}}$ is in the image of*

$$\iota_{V,\bar{v}} \colon \widehat{\mathcal{L}}_{V,\bar{v}} \to \mathcal{E}xt^1(\mathcal{P}^{\mathrm{gp}}, O_V^*)_{\bar{v}}$$

if and only if every h_p extends to V by 0. This property does not depend on the choice of representative $(h_p)_p \in \mathcal{H}om(\mathcal{R}, O_V^*)_{\bar{v}}$.

We introduce an equivalence relation on the germs of log structures—the $\xi = [(h_p)_p]$ that extend by 0—called the *type*. Let

$$V(p) := \overline{V \setminus \mathrm{supp}(p)}.$$

Definition 9.22 (Being of the Same Type) Two germs $\xi = [(h_p)_p]$ and $\xi' = [(h'_p)_p]$ that extend by 0 are *of the same type* if, for every $p \in P$, we can find $e_p \in O_{V(p),\bar{v}}^*$ such that

$$h'_p = e_p h_p \in (j_{p*} O_{V \setminus \mathrm{supp}(p)}^*)_{\bar{v}}.$$

Note the canonical restriction map

$$O_{V(p),\bar{v}}^* \to (j_{p*} O_{V \setminus \mathrm{supp}(p)}^*)_{\bar{v}},$$

which we apply to e_p. There is no condition relating e_p and e_q; moreover, there is no condition for $p \in P^{\mathrm{gp}} \setminus P$. Of course, being of the same type is an equivalence relation. We have the following more intrinsic characterization, which is not explicit in [124].

Lemma 9.23 *Let $(V, \mathcal{P})$ be a space with a toric ghost sheaf, and let $\bar{v} \to V$ be a geometric point. Let $\xi, \xi' \in \widehat{\mathcal{L}}_{V,\bar{v}}$ be represented by two (locally defined) log structures $\alpha \colon M \to O_V$ and $\alpha' \colon M' \to O_V$. Then, ξ and ξ' are of the same type if and only if the induced maps $\bar{\alpha} \colon \mathcal{P}_{\bar{v}} \to O_{V,\bar{v}}/O_{V,\bar{v}}^*$ and $\bar{\alpha}' \colon \mathcal{P}_{\bar{v}} \to O_{V,\bar{v}}/O_{V,\bar{v}}^*$ coincide.*

Proof First, assume that the two germs are of the same type. Let $(\tilde{h}_p)_p$ and $(\tilde{h}'_p)_p$ be the two extensions by zero, and let $\tilde{e}_p$ be a lift of e_p under $O_{V,\bar{v}} \to O_{V(p),\bar{v}}$. Then, we have $\tilde{e}_p \in O_{V,\bar{v}}^*$. Now $\tilde{h}'_p$ and $\tilde{e}_p \cdot \tilde{h}_p$ have the same image under the map (9.2) above so that we have $\tilde{h}'_p = \tilde{e}_p \cdot \tilde{h}_p$. Thus, we have $\bar{\alpha}(p) = \bar{\alpha}'(p)$.

Conversely, assume that we have $\bar{\alpha} = \bar{\alpha}'$ for the two germs ξ and ξ'. Then, we have $\tilde{h}_p = \tilde{h}'_p$ in $O_{V,\bar{v}}/O_{V,\bar{v}}^*$ so that we can find $\tilde{e}_p \in O_{V,\bar{v}}^*$ with $\tilde{h}'_p = \tilde{e}_p \cdot \tilde{h}_p$. In particular, we have $h'_p = e_p \cdot h_p$ in $(j_{p*}O^*_{V \setminus \mathrm{supp}(p)})_{\bar{v}}$, where e_p is the image of $\tilde{e}_p$ in $O_{V(p),\bar{v}}$. $\qquad \square$

The following is an important source of log structures which are of the same type. This result is implicit in [124].

Lemma 9.24 *Let* $\sigma \subseteq M_{\mathbb{R}}$ *be a full-dimensional lattice polytope, and let* $(V(\sigma), \mathcal{P}_\sigma)$ *be the space with a toric ghost sheaf from Example 9.18. Let* $(U, \mathcal{P}_U)$ *be another space with a toric ghost sheaf, and let* $\tau_1, \tau_2 \colon U \rightrightarrows V(\sigma)$ *be two strict and smooth morphisms of spaces with a toric ghost sheaf. Let* $\bar{u} \to U$ *be a geometric point lying over* $u \in U$, *and assume that* $\tau_1(u) = \tau_2(u) = 0 \in V(\sigma)$. *Assume that*

$$\tau_1^{\flat} = \tau_2^{\flat} \colon \quad P_\sigma \xrightarrow{\cong} \mathcal{P}_{\sigma, \tau_i(\bar{u})} \xrightarrow{\cong} \mathcal{P}_{U, \bar{u}}.$$

Then, the two log structures $(\tau_1)^*_{\log} \mathcal{M}_{V(\sigma)}$ *and* $(\tau_2)^*_{\log} \mathcal{M}_{V(\sigma)}$ *are of the same type in* $\bar{u}$.

Proof In Sect. 9.6 below, we will see a stratification which we can also apply to U and $V(\sigma)$. After shrinking U in the Zariski topology, we can assume that all closed strata of U contain u. On $V(\sigma)$, the stratification coincides with $V(\sigma) = \bigsqcup_{F \subseteq P_\sigma} V_F^{\circ}$. When arguing as in the proof of Lemma 9.54, we see that both τ_1 and τ_2 induce a bijection between the open strata of $V(\sigma)$ and the open strata of U given by $V_F^{\circ} \mapsto \tau_1^{-1}(V_F^{\circ})$ respectively $V_F^{\circ} \mapsto \tau_2^{-1}(V_F^{\circ})$.

Let η be the generic point of $\tau_1^{-1}(V_F^{\circ})$, and let η' be the generic point of $\tau_2^{-1}(V_F^{\circ})$. Then, we have induced maps $\mathcal{P}_{U, \bar{u}} \to \mathcal{P}_{U, \bar{\eta}}$ and $\mathcal{P}_{U, \bar{u}} \to \mathcal{P}_{U, \bar{\eta}'}$. Under $\tau_1^{\flat}$ respectively $\tau_2^{\flat}$, they are both identified with $P_\sigma \to P_\sigma / F$. By assumption, $\tau_1^{\flat} = \tau_2^{\flat}$ at $u \in U$, so $\mathcal{P}_{U, \bar{u}} \to \mathcal{P}_{U, \bar{\eta}}$ and $\mathcal{P}_{U, \bar{u}} \to \mathcal{P}_{U, \bar{\eta}'}$ is the quotient by the same face. Now there is also a face $G \subsetneq P_\sigma$ with $\tau_1^{-1}(V_G^{\circ}) = \tau_2^{-1}(V_F^{\circ})$. Then $\mathcal{P}_{U, \bar{u}} \to \mathcal{P}_{U, \bar{\eta}'}$ is also the quotient by G under the identification $\tau_1^{\flat}$. Thus, we have $F = G$. In other words, we have $S_F^{\circ} := \tau_1^{-1}(V_F^{\circ}) = \tau_2^{-1}(V_F^{\circ})$ for all proper faces $F \subsetneq P_\sigma$. Then, we also have $S_F := \tau_1^{-1}(V_F) = \tau_2^{-1}(V_F)$. These equalities hold a priori set-theoretically, but because they are reduced locally closed subschemes, they are also equal as schemes.

To show that $(\tau_1)^*_{\log} \mathcal{M}_{V(\sigma)}$ and $(\tau_2)^*_{\log} \mathcal{M}_{V(\sigma)}$ are of the same type, we have to show that, for every $p \in P_\sigma$, the two functions $\tau_1^*(z^p)$ and $\tau_2^*(z^p)$ differ by an invertible function.

First, we show that they differ by an invertible function on each S_F. When $p \notin F$, then $z^p|_{V_F} = 0$, and we are done. So we assume $p \in F$. Since S_F is an irreducible normal scheme, it is sufficient to show that $\mathrm{div}(\tau_1^*(z^p)) = \mathrm{div}(\tau_2^*(z^p))$ as Weil divisors. Set-theoretically, the zero locus of z^p on V_F is $\bigcup_{p \notin G \subseteq F} V_G$. Therefore, the set-theoretic zero locus of both $\tau_1^*(z^p)$ and $\tau_2^*(z^p)$ on S_F is $\bigcup_{p \notin G \subseteq F} S_G$. Thus, it is sufficient to show $\mathrm{val}_{S_G}(\tau_1^*(z^p)) = \mathrm{val}_{S_G}(\tau_2^*(z^p))$ for every $G \subsetneq F$ of codimension 1 with $p \notin G$.

Let η be the generic point of S_G. Then, we have $\tau_1(\eta) = \tau_2(\eta) =: \eta_G$ and thus two maps $\tau_1^*, \tau_2^* : O_{V_F, \eta_G} \to O_{S_F, \eta}$ between discrete valuation rings. These two maps have reduced fibers because $\tau_1, \tau_2 : U \to V(\sigma)$ are smooth. When π is a uniformizer of O_{V_F, η_G}, this implies that $\tau_1^*(\pi) = u_1 \cdot \pi'$ and $\tau_2^*(\pi) = u_2 \cdot \pi'$ for a uniformizer π' of $O_{S_F, \eta}$ and two invertibles u_1 and u_2. In other words, we have $\mathrm{val}_{S_G}(\tau_1^*(z^p)) = \mathrm{val}_{V_G}(z^p) = \mathrm{val}_{S_G}(\tau_2^*(z^p))$. This shows that, possibly after shrinking U, we have some $e_{p;F} \in O_{S_F}^*$ with $\tau_2^*(z^p)|_{S_F} = e_{p;F} \cdot \tau_1^*(z^p)|_{S_F}$.

Let now $p \in G \subseteq F$ be two faces containing p. Since S_G is integral, we have $e_{p;G} = e_{p;F}|_{S_G}$. Since $O_{U(p)}$ is the limit of the O_{S_F} with $p \in F$, this shows that the $e_{p;F}$ glue to an element of $O_{U(p)}^*$, which then can be lifted to an element e_p of O_U^*. It satisfies $\tau_2^*(z^p) = e_p \cdot \tau_1^*(z^p)$. $\qquad\square$

Remark 9.25 Note that the condition $\tau_1^\flat = \tau_2^\flat$ at $\bar{u}$ is necessary. For an example where both this condition and the statement of the lemma fail, consider $U(\sigma) = \mathrm{Spec}\,\mathbf{k}[x, y, t]/(xy - t)$, $\tau_1 = \mathrm{id}$, and $\tau_2^*(x) = y$, $\tau_2^*(y) = x$. $\qquad\Diamond$

9.3 Pre–Ghost Structures

We turn from the *absolute* problem of classifying log structures to the *relative* problem of classifying log morphisms to $S_0 = \mathrm{Spec}(\mathbb{N} \to \mathbf{k})$. Once we fix a log structure $\mathcal{M}$ on V, the datum of a log morphism to S_0 is equivalent to the datum of a global section $\rho \in \Gamma(V, \mathcal{M})$ with $\alpha(\rho) = 0$. At the level of ghost sheaves, this yields $\bar{\rho} := [\rho] \in \Gamma(V, \overline{\mathcal{M}})$. When classifying log morphisms, we assume $\bar{\rho} \in \Gamma(V, \mathcal{P})$ to be fixed.

Definition 9.26 (Spaces with a Pre–Ghost Structure) A *space with a pre–ghost structure* is a triple $(V, \mathcal{P}, \bar{\rho})$ where $(V, \mathcal{P})$ is a space with a toric ghost sheaf, and $\bar{\rho} \in \Gamma(V, \mathcal{P})$ is a global section such that, for every geometric point $\bar{v} \to V$, the induced map $\mathbb{N} \to \mathcal{P}_{\bar{v}}$, $1 \mapsto \bar{\rho}_{\bar{v}}$, is injective and saturated. A *morphism* between spaces with a pre–ghost structure $(Y, \mathcal{P}_Y, \bar{\rho}_Y)$ and $(V, \mathcal{P}_V, \bar{\rho}_V)$ is a morphism of spaces with a toric ghost sheaf $f : (Y, \mathcal{P}_Y) \to (V, \mathcal{P}_V)$ such that $f^\flat(\bar{\rho}_V) = \bar{\rho}_Y$.

Remark 9.27 The notion of a pre–ghost structure is not explicit in [124]. Gross and Siebert assume that $\bar{\rho}_{\bar{v}}$ is never nilpotent, but in our setting, this is automatic when $\bar{\rho}_{\bar{v}} \neq 0$ because $\mathcal{P}_{\bar{v}}$ is a sharp toric monoid. Gross and Siebert do not (explicitly?) assume $\mathbb{N} \to \mathcal{P}_{\bar{v}}$, $1 \mapsto \bar{\rho}_{\bar{v}}$, to be saturated. The first effect of this is that $\bar{\rho}_{\bar{v}}$ is primitive, which implies that $\mathcal{P}_{\bar{v}}^{\mathrm{gp}}/\mathbb{Z}\bar{\rho}_{\bar{v}}$ is a free Abelian group; this will simplify our arguments below. Primitivity is weaker than saturatedness, but saturatedness ensures that primitivity is preserved when passing to a quotient of $\mathcal{P}_{\bar{v}}$ by a face which does not contain $\bar{\rho}_{\bar{v}}$. In the cases of interest for us, saturatedness is always satisfied. $\qquad\Diamond$

Example 9.28 Let $V(\sigma)$ be as in Construction 9.2. According to Example 9.18, it can be enhanced to a space with a toric ghost sheaf $(V(\sigma), \mathcal{P}_\sigma)$; the element

$\rho_\sigma \in P_\sigma$ defines a global section $\bar\rho_\sigma \in \Gamma(V(\sigma), \mathcal{P}_\sigma)$, which yields a space with a pre–ghost structure $(V(\sigma), \mathcal{P}_\sigma, \bar\rho_\sigma)$. $\qquad\qquad\qquad\qquad\qquad\qquad\qquad\quad \diamond$

A log morphism from V to S_0 is a quadruple $(\mathcal{M}, \alpha, q, \rho)$ where $\alpha \colon \mathcal{M} \to O_V$ is a u-integral log structure, $q \colon \mathcal{M} \to \mathcal{P}$ exhibits $\mathcal{P}$ as the ghost sheaf, and $\rho \in \Gamma(V, \mathcal{M})$ satisfies $q(\rho) = \bar\rho$ and $\alpha(\rho) = 0$.

Definition 9.29 ($\widehat{\mathcal{LS}}_V$) Let $(V, \mathcal{P}, \bar\rho)$ be a space with a pre–ghost structure. Then $\widehat{\mathcal{LS}}_V$ is the sheaf of equivalence classes of log morphisms $(\mathcal{M}, \alpha, q, \rho)$ where two log morphisms are *equivalent* if they are locally isomorphic on V.

Let $\xi \in \widehat{\mathcal{LS}}_{V,\bar v}$ be a germ, given by a log morphism $(\mathcal{M}, \alpha, q, \rho)$ defined in a neighborhood of $\bar v \in V$. It gives rise to an extension

$$0 \to O_V^* \to \mathcal{M}^{\mathrm{gp}}/\rho \to \mathcal{P}^{\mathrm{gp}}/\bar\rho \to 0;$$

thus, we obtain a canonical map

$$\bar\iota_{V,\bar v}\colon \ \widehat{\mathcal{LS}}_{V,\bar v} \to \mathcal{E}xt^1(\mathcal{P}^{\mathrm{gp}}/\bar\rho, O_V^*)_{\bar v},$$

which is injective. We can reconstruct the log structure $(\mathcal{M}, \alpha, q)$ from the extension by pull-back along $\mathcal{P}^{\mathrm{gp}} \to \mathcal{P}^{\mathrm{gp}}/\bar\rho$. We can reconstruct ρ as well; namely, we get the subsheaf $\mathbb{Z}\rho \subseteq \mathcal{M}^{\mathrm{gp}}$ as the kernel of the map to $\mathcal{M}^{\mathrm{gp}}/\rho$, and we choose the "correct" generator ρ (instead of the wrong one $-\rho$) because only one of them maps to $\mathcal{P} \subseteq \mathcal{P}^{\mathrm{gp}}$. We have a commutative diagram

$$
\begin{array}{ccc}
\widehat{\mathcal{LS}}_{V,\bar v} & \xrightarrow{\ \bar\iota_{V,\bar v}\ } & \mathcal{E}xt^1(\mathcal{P}^{\mathrm{gp}}/\bar\rho, O_V^*)_{\bar v} \\[2mm]
\big\downarrow & & \big\downarrow \\[2mm]
\widehat{\mathcal{L}}_{V,\bar v} & \xrightarrow{\ \iota_{V,\bar v}\ } & \mathcal{E}xt^1(\mathcal{P}^{\mathrm{gp}}, O_V^*)_{\bar v}
\end{array}
\tag{9.3}
$$

where the left vertical arrow is the forgetful map $(\mathcal{M}, \alpha, q, \rho) \mapsto (\mathcal{M}, \alpha, q)$.

The explicit description of $\mathcal{E}xt^1(\mathcal{P}^{\mathrm{gp}}, O_V^*)_{\bar v}$ in the absolute case has an analog for the sheaf $\mathcal{E}xt^1(\mathcal{P}^{\mathrm{gp}}/\bar\rho, O_V^*)_{\bar v}$ in the relative case. Because the surjection $P_V^{\mathrm{gp}} \to \mathcal{P}^{\mathrm{gp}}$ is an isomorphism on the stalk at $\bar v$, there is—for V small enough and connected—a unique $\tilde\rho \in P^{\mathrm{gp}} = \Gamma(V, P_V^{\mathrm{gp}})$ with $\tilde\rho \mapsto \bar\rho \in \Gamma(V, \mathcal{P}^{\mathrm{gp}})$. This gives a short exact sequence

$$0 \to \mathcal{R} \to P_V^{\mathrm{gp}}/\tilde\rho \to \mathcal{P}^{\mathrm{gp}}/\bar\rho \to 0$$

with the same sheaf of relations $\mathcal{R}$ as above. Using $\mathcal{E}xt^1(P_V^{\mathrm{gp}}/\tilde{\rho}, O_V^*) = 0$, we find a morphism

$$\mathcal{H}om(P_V^{\mathrm{gp}}/\tilde{\rho}, O_V^*) \longrightarrow \mathcal{H}om(\mathcal{R}, O_V^*) \longrightarrow \mathcal{E}xt^1(\mathcal{P}^{\mathrm{gp}}/\tilde{\rho}, O_V^*) \longrightarrow 0$$
$$\downarrow \qquad\qquad\qquad \| \qquad\qquad\qquad \downarrow$$
$$\mathcal{H}om(P_V^{\mathrm{gp}}, O_V^*) \longrightarrow \mathcal{H}om(\mathcal{R}, O_V^*) \longrightarrow \mathcal{E}xt^1(\mathcal{P}^{\mathrm{gp}}, O_V^*) \longrightarrow 0$$

of presentations. Thus, a germ of a log morphism $(\mathcal{M}, \alpha, q, \rho)$ can be represented by functions $(h_p)_p \in \mathcal{H}om(\mathcal{R}, O_V^*)_{\bar{v}}$ as above—we just divide by a smaller subgroup. The result is:

Lemma 9.30 *A germ $\xi = [(h_p)_p] \in \mathcal{E}xt^1(\mathcal{P}^{\mathrm{gp}}/\bar{\rho}, O_V^*)_{\bar{v}}$ is in the image of*

$$\bar{\iota}_{V,\bar{v}}\colon \widehat{\mathcal{LS}}_{V,\bar{v}} \to \mathcal{E}xt^1(\mathcal{P}^{\mathrm{gp}}/\bar{\rho}, O_V^*)_{\bar{v}}$$
$$= \mathrm{coker}\left(\mathcal{H}om(P_V^{\mathrm{gp}}/\tilde{\rho}, O_V^*)_{\bar{v}} \to \mathcal{H}om(\mathcal{R}, O_V^*)_{\bar{v}}\right)$$

if and only if every h_p extends to V by 0. This property does not depend on the choice of representative $(h_p)_p \in \mathcal{H}om(\mathcal{R}, O_V^*)_{\bar{v}}$.

Corollary 9.31 *The square (9.3) is Cartesian, and the forgetful map $\widehat{\mathcal{LS}}_{V,\bar{v}} \to \widehat{\mathcal{L}}_{V,\bar{v}}$ is surjective.*

Remark 9.32 Intuitively, the reader might feel that the surjectivity would conflict with the additional condition $\alpha(\rho) = 0$ in our definition of $\widehat{\mathcal{LS}}_V$. But because $\mathrm{supp}(\bar{\rho}) = V$, we have $h_{\bar{\rho}} = 0 \in (j_{\bar{\rho}*}O_{\emptyset}^*)_{\bar{v}}$; in particular, $h_{\bar{\rho}} = 0$ as a function after extending by 0. $\Diamond$

9.4 Ghost Structures

Ghost structures are not just pre–ghost structures that enjoy a special property; they encode more information. Once we have a ghost structure, we can specify a subsheaf $\mathcal{LS}_V \subseteq \widehat{\mathcal{LS}}_V$ whose sections (1) correspond to actual log morphisms, not just equivalence classes,[8] and (2) correspond to log morphisms which are log smooth, not just some morphisms.

In the definition, we use the following construction. Let $\theta\colon \mathbb{N} \to P$ be an injective and saturated (but not necessarily vertical) homomorphism between sharp

---

[8] When $s \in \Gamma(U, \mathcal{LS}_V)$, there is one (globally) defined morphism of log schemes on U, not just morphisms of log schemes which are defined on U_i for an open cover $U = \bigcup_i U_i$, and which would only be equivalent on overlaps $U_i \cap U_j$.

toric monoids. Then, we have an induced morphism of log schemes $A_\theta : A_P \to A_\mathbb{N}$. Let $(L(\theta), \mathcal{M}_\theta) \to S_0$ be the central fiber. By forgetting the log structure, we obtain a space with a pre–ghost structure $(L(\theta), \mathcal{P}_\theta, \bar{\rho}_\theta)$. Namely, for every $x \in L(\theta)$, the ghost stalk is of the form P/G for a face $G \subseteq P$ which does not contain $\theta(1)$; hence, $\mathbb{N} \to P/G$ is injective. Unlike in the case of $V(\sigma) \subset U(\sigma)$, the central fiber $L(\theta) \subset A_P$ may not be the full toric boundary.

The following definition is an adaptation of [124, Defn. 3.16].

Definition 9.33 (Spaces with a Ghost Structure) Let $(V, \mathcal{P}, \bar{\rho})$ be a space with a pre–ghost structure. Then a *ghost chart* is a tuple $(v, U, \pi, u, \theta, \tau)$ where:

(a) $v \in V$ is a point, $\pi : U \to V$ is an étale neighborhood, and $u \in U$ is a point with $\pi(u) = v$;
(b) $\theta : \mathbb{N} \to P$ is injective and saturated, and P is a sharp toric monoid;
(c) $\tau : (U, \mathcal{P}|_U, \bar{\rho}) \to (L(\theta), \mathcal{P}_\theta, \bar{\rho}_\theta)$ is a strict and smooth morphism of spaces with a pre–ghost structure such that $\tau(u) = 0$, where $0 \in A_P$ is the point defined by $\mathbf{k}[P \setminus \{0\}]$.

A *space with a ghost structure* V^g is a space with a pre–ghost structure $(V, \mathcal{P}, \bar{\rho})$ together with finitely many ghost charts $(v_i, U_i, \pi_i, u_i, \theta_i, \tau_i)$ such that:

(a) $V = \bigcup_{i=1}^m \pi_i(U_i)$;
(b) the two log structures $(\tau_i)^*_{\log} \mathcal{M}_{\theta_i}$ (on U_i) and $(\tau_j)^*_{\log} \mathcal{M}_{\theta_j}$ (on U_j) are of the same type on overlaps $U_i \times_V U_j$; this means that the log structures are of the same type in every point and is equivalent to asking for the induced maps $\mathcal{P}|_{U_i} \to O_{U_i}/O^*_{U_i}$ to be compatible; in particular, we have an induced global map

$$\bar{\alpha} : \quad \mathcal{P} \to O_V/O^*_V$$

of sheaves in the étale topology.

Definition 9.34 ($\mathcal{LS}_V$) Let V^g be a space with a ghost structure. The subsheaf $\mathcal{LS}_V \subseteq \widehat{\mathcal{LS}}_V$ consists of equivalence classes of log morphisms $(\mathcal{M}, \alpha, q, \rho)$ such that in every point, the log structure $(\mathcal{M}, \alpha, q)$ is of the same type as prescribed by the ghost structure, i.e., of the same type as $(\tau_i)^*_{\log} \mathcal{M}_{\theta_i}$. Equivalently, we may ask that $\alpha : \mathcal{M} \to O_V$ induces $\bar{\alpha} : \mathcal{P} \to O_V/O^*_V$ via $q : \mathcal{M} \to \mathcal{P}$.

Note that $\mathcal{LS}_{V,\bar{v}} \neq \varnothing$ for every geometric point $\bar{v} \to V$.

A *log smooth structure* of *ghost type* V^g is a log morphism $(\mathcal{M}, \alpha, q, \rho) \in \mathcal{LS}_V$ such that the corresponding map $(V, \mathcal{M}) \to S_0$ is log smooth. The key point about ghost structures is that—under one more technical hypothesis—they correspond precisely with sections of $\mathcal{LS}_V$. The purpose of all the above structures is to ensure that $(V, \mathcal{M}) \to S_0$ is log smooth.

Proposition 9.35 ([124, Prop. 3.20]) *Let V^g be a space with a ghost structure, and assume that $\mathcal{P}^{gp}_{\bar{\eta}}$ is generated by $\bar{\rho}_{\bar{\eta}}$ at geometric generic points $\bar{\eta}$ of irreducible components of V. Then, for every étale open $\pi : U \to V$, sections in $\Gamma(U, \mathcal{LS}_V)$ correspond to actual log structures $(\mathcal{M}, \alpha, q, \rho)$ on U (rather than equivalence* *classes of locally defined log structures), and the corresponding morphism of log schemes $(U, \mathcal{M}) \to S_0$ is log smooth.*

9.5 Toroidal Crossing Spaces

Toroidal crossing spaces are spaces with a pre–ghost structure locally isomorphic to the spaces $(V(\sigma), \mathcal{P}_\sigma, \bar{\rho}_\sigma)$ which we introduced in Example 9.28. The following definition is adapted from [77, Defn. 1.5].

Definition 9.36 (Toroidal Crossing Spaces) A *toroidal crossing space* is a space with a pre–ghost structure $(V, \mathcal{P}, \bar{\rho})$ such that, for every point $v \in V$, there is a full-dimensional lattice polytope $\sigma \subseteq M_{\mathbb{R}}$, a strict and *smooth* morphism of spaces with a pre–ghost structure

$$\tau \colon \ (U, \mathcal{P}', \bar{\rho}') \to (V(\sigma), \mathcal{P}_\sigma, \bar{\rho}_\sigma),$$

a strict and étale morphism of spaces with a pre–ghost structure

$$\pi \colon \ (U, \mathcal{P}', \bar{\rho}') \to (V, \mathcal{P}, \bar{\rho}),$$

and a point $u \in U$ with $\pi(u) = v$ and $\tau(u) = 0$, where $0 \in V(\sigma) \subset U(\sigma)$ is the point defined by the ideal $\mathbf{k}[P_\sigma \setminus \{0\}]$.

Example 9.37 Let $\sigma \subseteq M_{\mathbb{R}}$ be a full-dimensional lattice polytope. Then, the space with a pre–ghost structure $(V(\sigma), \mathcal{P}_\sigma, \bar{\rho}_\sigma)$ from Example 9.28 is a toroidal crossing space. This follows from Corollary 9.15. $\Diamond$

Example 9.38 In a two-dimensional normal crossing space, only three irreducible components can meet in a point. For a two-dimensional toroidal crossing space, this number can be arbitrarily large.

(1) Take $M = \mathbb{Z}^2$ and $\sigma = \mathrm{Conv}((0, 0), (1, 0), (1, 1), (0, 1))$, a unit square. Then $V(\sigma)$ consists of four copies of $\mathbb{A}^2$ which all meet in a single point. This $U(\sigma) \to \mathbb{A}^1_t$ coincides with Example 14.13, where the reader can also find an illustration.
(2) Now take $\sigma' = \mathrm{Conv}((0, 0), (2, 0), (1, 1), (0, 1))$. Then, $V(\sigma)$ also consists of four copies of $\mathbb{A}^2$ which all meet in a single point. However, P_σ and $P_{\sigma'}$ are not isomorphic, so also $V(\sigma) \not\cong V(\sigma')$ as toroidal crossing spaces.
(3) When we take $\sigma'' = \mathrm{Conv}((0, 0), (2, 0), (2, 1), (1, 2), (0, 2))$, then $V(\sigma'')$ consists of five copies of $\mathbb{A}^2$ which all meet in a single point. $\Diamond$

For any geometric point $\bar{v}$ lying over a point $u \in U$ in a chart as in the definition, note that the data of the definition induce an isomorphism $\mathcal{P}_{\bar{v}} \cong P_\sigma$ respecting $\bar{\rho}$ and $\bar{\rho}_\sigma$. Since $\mathcal{P}_{\bar{v}}$ is, up to isomorphism, independent of the geometric point lying over $v \in V$, we have $\mathcal{P}_{\bar{v}} \cong P_\sigma$ for any geometric point $\bar{v}$ lying over $v \in V$.

To turn a toroidal crossing space $(V, \mathcal{P}, \bar{\rho})$ into a space with a ghost structure, we choose points $v_1, \ldots, v_m$ of V such that the corresponding étale neighborhoods $\pi_i : U_i \to V$ of Definition 9.36 form a cover. After writing σ_i for the polytope associated to v_i, we obtain smooth morphisms $\tau_i : U_i \to V(\sigma_i)$ with $\tau_i(u_i) = 0$ as required. Let us show that the two induced log morphisms are of the same type on $U_i \times_V U_j$ so that we obtain a ghost structure in the sense of Definition 9.33.

Lemma 9.39 *In the above situation, the two log structures on $U_i \times_V U_j$ are of the same type.*

Proof Let $U = U_i \times_V U_j$, let $\varrho_i : U \to U_i$ and $\varrho_j : U \to U_j$ be the two projections, and let $\phi_i = \tau_i \circ \varrho_i : U \to V(\sigma_i)$ and $\phi_j = \tau_j \circ \varrho_j : U \to V(\sigma_j)$. Let $\mathcal{M}_i = (\phi_i)^*_{\log} \mathcal{M}_{V(\sigma_i)}$ and $\mathcal{M}_j = (\phi_j)^*_{\log} \mathcal{M}_{V(\sigma_j)}$ be the two log structures. Let $u \in U$ be a point. Possibly after shrinking U around u in the Zariski topology, Corollary 9.15 gives us full-dimensional lattice polytopes σ_i' and σ_j' together with strict and smooth morphisms $\psi_i : (U, \mathcal{M}_i) \to V(\sigma_i')$ and $\psi_j : (U, \mathcal{M}_j) \to V(\sigma_j')$ with $\psi_i(u) = 0$ and $\psi_j(u) = 0$.

Choose a geometric point $\bar{u}$ lying over u. Then, we have isomorphisms $P_{\sigma_i'} = \mathcal{P}_{\bar{u}}$ and $P_{\sigma_j'} \cong \mathcal{P}_{\bar{u}}$. This yields an identification $V(\sigma_i') \cong V(\sigma_j')$. Let us write $V(\sigma')$ for this log scheme. Note that we now have $\psi_i^\flat = \psi_j^\flat$ at u for the induced map of spaces with a toric ghost sheaf. Therefore, we are in the situation of Lemma 9.24, which shows that $\mathcal{M}_i$ and $\mathcal{M}_j$ are of the same type. $\qquad\qquad\square$

A priori, the subsheaf $\mathcal{LS}_V \subseteq \widehat{\mathcal{LS}}_V$ might depend on the choices we made to turn V into a space with a ghost structure. However, it is in fact independent of the choices; namely, when we add further ghost charts $\pi : U \to V$, $\tau : U \to V(\sigma)$ to the ghost structure, then they are still mutually of the same type, so this larger ghost structure defines the same $\mathcal{LS}_V$. When we have two ghost structures, we can compare both with their union.

Corollary 9.40 *Let $(V, \mathcal{P}, \bar{\rho})$ be a toroidal crossing space. Then the subsheaf*

$$\mathcal{LS}_V \subseteq \widehat{\mathcal{LS}}_V$$

is well-defined. The log morphism corresponding to a section $s \in \Gamma(U, \mathcal{LS}_V)$ is log smooth.

Definition 9.41 (Log Smooth Structures)   Let $(V, \mathcal{P}, \bar{\rho})$ be a toroidal crossing space. A *log smooth structure* on an étale open $\pi : U \to V$ is a section $s \in \Gamma(U, \mathcal{LS}_V)$.

Remark 9.42   The definition and theory of a toroidal crossing space by Schröer–Siebert in [253] is slightly different. They are not so much interested in log morphisms, but in log structures that we classify with $\widehat{\mathcal{L}}_V$. Moreover, they restrict to *tame* toroidal crossing spaces in the sense of our definition below. $\qquad\qquad\Diamond$

9.5.1 Normal Crossing Spaces as Toroidal Crossing Spaces

Every normal crossing space carries a canonical structure of a toroidal crossing space as follows.

Example 9.43 (Normal Crossing Spaces) Let V be a normal crossing space, let $\nu : \tilde{V} \to V$ be the normalization map, and let $\underline{\mathbb{N}}_{\tilde{V}}$ be the constant sheaf on $\tilde{V}$ with stalk $\mathbb{N}$, formed in the étale topology; it has a distinguished global section $1_{\tilde{V}} \in \Gamma(\tilde{V}, \underline{\mathbb{N}}_{\tilde{V}})$. Now the space with a pre–ghost structure

$$(V, \nu_* \underline{\mathbb{N}}_{\tilde{V}}, \bar{\rho}_V := \nu_* 1_{\tilde{V}})$$

is a toroidal crossing space. All ghost stalks are of the form $\mathbb{N}^r$ with $\bar{\rho}_{\bar{v}} = (1, \dots, 1)$ and $r \le d = \dim V$. $\Diamond$

Remark 9.44 A normal crossing space V might carry other structures of a toroidal crossing space besides the canonical one. For example, taking $\sigma = [0, k] \subseteq \mathbb{R}$, we get a toroidal crossing space $V(\sigma)$ with underlying space $\{xy = 0\} \subseteq \mathbb{A}^2_{x,y}$; for $k = 1$, this is the canonical structure of a toroidal crossing space, but for $k \ge 2$, it is another one. $\Diamond$

9.5.2 A Coordinate-Free Characterization of Toroidal Crossing Spaces

Above, we have given the definition of toroidal crossing spaces closely following [124] and [77]. This definition has the advantage of being very explicit about the local structure. The following characterization is often more convenient from a theoretical perspective.

Lemma 9.45 *Let $V/\mathbf{k}$ be separated and of finite type, let $\mathcal{P}$ be a sheaf of monoids in the étale topology, and let $\bar{\rho} \in \Gamma(V, \mathcal{P})$ be a global section. Then $(V, \mathcal{P}, \bar{\rho})$ is a toroidal crossing space if and only if, locally in the étale topology, we can find a fine and saturated log structure $\alpha : \mathcal{M} \to \mathcal{O}_V$, a homomorphism $q : \mathcal{M} \to \mathcal{P}$ which induces an isomorphism $\overline{\mathcal{M}} \cong \mathcal{P}$, and a section $\rho \in \mathcal{M}$ with $q(\rho) = \bar{\rho}$ and $\alpha(\rho) = 0$ such that the induced morphism of log schemes $(V, \mathcal{M}) \to S_0$ is log smooth, saturated, and vertical.*

Proof First assume that $(V, \mathcal{P}, \bar{\rho})$ is a toroidal crossing space. Let $\pi : U \to V$ and $\tau : U \to V(\sigma)$ be morphisms as in the definition. The morphism $V(\sigma) \to S_0$ is log smooth, saturated, and vertical. In particular, also $(U, \tau^*_{\log} \mathcal{M}_{V(\sigma)}) \to S_0$ is log smooth, saturated, and vertical. This is the desired étale local structure of a log smooth, saturated, and vertical morphism.

Conversely, assume that we have the structure of a log smooth, saturated, and vertical morphism on an étale open $W \to V$. Let $S_0 \to A_{\mathbb{N}}$ be the standard chart, which is neat at 0. This means that the induced map $\mathbb{N} \to \overline{\mathcal{M}}_{S_0,0}$ is an isomorphism. Fix a geometric point $\bar{v} \in W$. By Kato's toroidal characterization of log smoothness in the form of Theorem 7.59, we can find, after shrinking W if necessary, a chart $W \to A_P$ which is neat at $\bar{v}$ (i.e., $P \to \overline{\mathcal{M}}_{\bar{v}}$ is an isomorphism, where $\mathcal{M}$ is the log structure on W), such that we have a map $\theta : \mathbb{N} \to P$ which turns it into a chart for the log morphism, and such that the induced map $\tau : W \to A_P \times_{A_{\mathbb{N}}} S_0$ is strict and smooth. Since θ is injective (because it is integral and local), saturated, and vertical, Lemma 9.12 shows that there is a full-dimensional lattice polytope $\sigma \subseteq M_{\mathbb{R}}$ with $P \cong P_\sigma$ and $\theta(1) = \rho_\sigma$ under this identification. Consequently, we have $A_P \times_{A_{\mathbb{N}}} S_0 = V(\sigma)$ as a log scheme. Therefore, $(V, \mathcal{P}, \bar{\rho})$ has the required local form. Now $(V, \mathcal{P}, \bar{\rho})$ is automatically a space with a pre–ghost structure and hence a toroidal crossing space. $\qquad\square$

Then, we also obtain a characterization of the sections of $\mathcal{LS}_V$ without reliance on the notion of being of the same type.

Corollary 9.46 *Let $(V, \mathcal{P}, \bar{\rho})$ be a toroidal crossing space. For a section $s = (\mathcal{M}, \alpha, q, \rho)$ of $\widehat{\mathcal{LS}}_V$, we have:*

(1) $s \in \mathcal{LS}_V$ if and only if the induced morphism of log schemes $(V, \mathcal{M}) \to S_0$ is log smooth;

(2) $(V, \mathcal{M}) \to S_0$ is saturated and vertical.

Proof First, let $s \in \mathcal{LS}_V$. We have already seen in Proposition 9.35 that $(V, \mathcal{M}) \to S_0$ is log smooth in this case. Conversely, if $s \in \widehat{\mathcal{LS}}_V$ is such that the induced map is log smooth, then by Theorem 7.59, it is étale locally pulled back from $V(\sigma) \to S_0$ along a smooth morphism as in the proof of the lemma. Now Lemma 9.24 shows that the log structure is of the type prescribed by the ghost structure so that we have $s \in \mathcal{LS}_V$.

For a section $s \in \widehat{\mathcal{LS}}_V$, the corresponding log morphism is saturated and vertical since these properties depend only on the map on the level of ghost sheaves. $\qquad\square$

9.5.3   Products of Toroidal Crossing Spaces

The product of two normal crossing spaces is rarely a normal crossing space. For an example, just consider $\operatorname{Spec} \mathbf{k}[x, y]/(xy) \times \operatorname{Spec} \mathbf{k}[z, w]/(zw)$. However, the product of two toroidal crossing spaces is always a toroidal crossing space.

Definition 9.47 (Products)   Let $(V_1, \mathcal{P}_1, \bar{\rho}_1)$ and $(V_2, \mathcal{P}_2, \bar{\rho}_2)$ be two toroidal crossing spaces. Then their *product* $(V, \mathcal{P}, \bar{\rho})$ is given as follows: $V = V_1 \times V_2$ is the product of schemes; $\mathcal{P}$ is the push-out of $\mathbb{N} \to p^{-1}\mathcal{P}_1$, $1 \mapsto p^*\bar{\rho}_1$, and $\mathbb{N} \to q^{-1}\mathcal{P}_2$, $1 \mapsto q^*\bar{\rho}_2$, in the category of sheaves of sharp monoids, where $p : V \to V_1$ and $q : V \to V_2$ are the projections; $\bar{\rho} = p^*\bar{\rho}_1 + q^*\bar{\rho}_2$.

Lemma 9.48 *The product* $(V, \mathcal{P}, \bar{\rho}) = (V_1, \mathcal{P}_1, \bar{\rho}_1) \times (V_2, \mathcal{P}_2, \bar{\rho}_2)$ *is a toroidal crossing space.*

Proof By Lemma 9.45, we have, locally in the étale topology, structures of log smooth, saturated, and vertical morphisms $(V_1, \mathcal{M}_1) \to S_0$ and $(V_2, \mathcal{M}_2) \to S_0$. Their product $(V_1 \times V_2, \mathcal{M}_1 \times \mathcal{M}_2) \to S_0$ is then log smooth, saturated, and vertical as well. Lemma 7.36 shows that the ghost sheaf of $\mathcal{M}_1 \times \mathcal{M}_2$ is precisely the push-out in the definition of the product of toroidal crossing spaces. $\square$

9.6 The Stratification of a Toroidal Crossing Space

Let $V = (V, \mathcal{P}, \bar{\rho})$ be a toroidal crossing space. If $\bar{v}$ is a geometric point over $v \in V$, then $\mathcal{P}_{\bar{v}}$ depends only on v and not on the choice of $\bar{v}$. This is because, on a roof $V \leftarrow U \to V(\sigma)$ as in the definition of a toroidal crossing space, we have $\mathcal{P}|_U = \tau^{-1}(\mathcal{P}_\sigma)$, i.e., the stalks of the restriction of $\mathcal{P}$ to the Zariski topology on U do not change under a further étale open $\pi' \colon U' \to U$. It is, however, not always true that $\mathcal{P}_{\bar{v}} = \mathcal{P}_v^{\mathrm{Zar}}$ when $\mathcal{P}^{\mathrm{Zar}}$ is the restriction of $\mathcal{P}$ to the Zariski topology of V, as the example of the nodal cubic with $\mathcal{P}$ as constructed in Example 9.43 shows. Since $\mathcal{P}_{\bar{v}}$ is independent of the geometric point $\bar{v}$ over v, the following loci are well-defined for every $m \geq 0$:

$$\mathcal{U}_m V := \{ v \in V \mid \mathrm{rk}(\mathcal{P}_{\bar{v}}) \leq m + 1 \}$$

$$C_m V := \{ v \in V \mid \mathrm{rk}(\mathcal{P}_{\bar{v}}) \geq m + 1 \}$$

$$S_m V := \{ v \in V \mid \mathrm{rk}(\mathcal{P}_{\bar{v}}) = m + 1 \} \;=\; \mathcal{U}_m V \cap C_m V$$

It is easy to see from the local models that $\mathcal{U}_m V \subseteq V$ is open, and that $C_m V \subseteq V$ is closed. Hence, $S_m V \subseteq V$ is locally closed. We also see from the local models that $\mathcal{U}_0 V = S_0 V$ is smooth, and that its complement $C_1 V$ is the locus where at least two étale local components meet, which is also equal to the non-normal locus of V. On $S_1 V$, we have only double normal crossing singularities given by $\{xy = 0\}$; the locus $C_2 V$ contains more complicated singularities, where always at least three étale local components meet.

We derive from the local models as well that $S_m V$ is smooth of dimension $d - m$, where $d = \dim(V)$. This provides us with a stratification of V by irreducible components of $S_m V$.

Definition 9.49 (Stratification) Let $(V, \mathcal{P}, \bar{\rho})$ be a toroidal crossing space. An *open stratum* is an irreducible component $S^\circ \subseteq S_m V$ for any $m \geq 0$. We denote the set of irreducible components of $S_m V$ by $[S_m V]$. A *closed stratum* is the closure $S = \overline{S^\circ}$ of any open stratum S°. We denote the set of closures of $S^\circ \in [S_m V]$ by $[C_m V]$.

Every open stratum $S^\circ \in [\mathcal{S}_m V]$ is not only an irreducible but also a connected component of $\mathcal{S}_m$. It is open in $\mathcal{C}_m V$ and locally closed in V. A closed stratum S is closed in V. The closed strata in $[\mathcal{C}_0 V]$ are precisely the irreducible components of V; we also denote them by $V_1, \ldots, V_c$.

9.6.1 Tame Toroidal Crossing Spaces and the Kink

If $v \in \mathcal{S}_1 V$, then we must have

$$\mathcal{P}_{\bar{v}} \cong \langle (-1, k), (0, 1), (1, 0) \rangle \subseteq \mathbb{Z}^2$$

with $\bar{\rho}_{\bar{v}} = (0, 1)$ for some $k \geq 1$. The corresponding polytope is $\sigma = [0, k] \subseteq \mathbb{R}$. We get a function

$$\kappa \colon \ \mathcal{S}_1 V \to \mathbb{N}, \quad v \mapsto k,$$

which is locally constant because the structure of toroidal crossing space is locally around $\bar{v}$ pulled back from $V(\sigma)$—no change of k is possible. Thus, we consider it as a function

$$\kappa \colon \ [\mathcal{S}_1 V] \to \mathbb{N}.$$

Definition 9.50 (Kink and Tameness) Let $(V, \mathcal{P}, \bar{\rho})$ be a toroidal crossing space. Then $\kappa \colon [\mathcal{S}_1 V] \to \mathbb{N}$ is the *kink function*. For $D^\circ \in [\mathcal{S}_1 V]$ an open stratum of codimension 1, its value $k = \kappa(D^\circ)$ is the *kink* of D°. The toroidal crossing space $(V, \mathcal{P}, \bar{\rho})$ is *tame* if $\kappa(D^\circ) = 1$ for all $D^\circ \in [\mathcal{S}_1 V]$.

When we consider a piecewise linear function $h \colon \mathbb{Z} \to \mathbb{Z}$ with

$$\mathcal{P}_{\bar{v}} \cong \{ (m_1, m_2) \in \mathbb{Z}^2 \mid m_2 \geq h(m_1) \},$$

i.e., $h(m) = -k \cdot m$ for $m \leq 0$ and $h(m) = 0$ for $m \geq 0$, then k becomes the change of slope of h—hence the name "kink." The relevance of being tame will become clear in Sect. 9.12.

9.7 Simple Toroidal Crossing Spaces

In analogy with normal crossing spaces, a toroidal crossing space is simple if its components do not have self-intersections. More precisely:

Definition 9.51 (Simple Toroidal Crossing Spaces) Let $(V, \mathcal{P}, \bar{\rho})$ be a toroidal crossing space. Then $(V, \mathcal{P}, \bar{\rho})$ is *simple* if each top-dimensional closed stratum $V_i \in [\mathcal{C}_0 V]$ is normal.

Example 9.52 Every simple normal crossing space is a simple toroidal crossing space. For some concrete examples, see Example 1.32.   ◊

Example 9.53 Let V be a nodal cubic with node $v \in V$, considered as a toroidal crossing space by Example 9.43. Then $\mathcal{U}_0 V = \mathcal{S}_0 V = V \setminus \{v\}$ is irreducible although it has, étale locally around v, two strata of codimension 0. Since $V = \overline{V \setminus \{v\}}$ is not normal, V is not simple. We have $\mathcal{C}_1 V = \mathcal{S}_1 V = \{v\}$.   ◊

9.7.1 Normality of the Closed Strata

We prove that every closed stratum of a simple toroidal crossing space is normal. As a preparation, we construct special charts. Recall the stratification of $V(\sigma)$ into subsets of the form $V_F^\circ = \operatorname{Spec} \mathbf{k}[F^{\mathrm{gp}}]$ from Sect. 9.1.

Lemma 9.54 *Let $(V, \mathcal{P}, \bar{\rho})$ be a simple toroidal crossing space, and let $v \in V$ be a point. Then there is a full-dimensional lattice polytope $\sigma \subseteq M_\mathbb{R}$, an étale neighborhood $\pi : U \to V$ of v, and a smooth morphism $\tau : U \to V(\sigma)$ as in the definition of a toroidal crossing space such that the following conditions are satisfied:*

(1) *for each open stratum V_F° of $V(\sigma)$, the preimage $\tau^{-1}(V_F^\circ)$ is an open stratum of U, and the assignment $V_F^\circ \mapsto \tau^{-1}(V_F^\circ)$ induces a bijection between open strata of $V(\sigma)$ and open strata of U;*
(2) *we have $\tau^{-1}(V_F) = \overline{\tau^{-1}(V_F^\circ)}$;*
(3) *for each open stratum $T^\circ \in [\mathcal{S}_m V]$ such that $v \in T$, the preimage $\pi^{-1}(T^\circ)$ is an open stratum of U, and the assignment $T^\circ \mapsto \pi^{-1}(T^\circ)$ induces a bijection between open strata $T^\circ \in [\mathcal{S}_m V]$ with $v \in T$ and open strata of U;*
(4) *for each open stratum $T^\circ \in [\mathcal{S}_m V]$ with $v \in T$, we have $\pi^{-1}(T) = \overline{\pi^{-1}(T^\circ)}$.*

Proof Since V is quasi-compact, it has only finitely many open strata. Therefore, the union of all $T^\circ \in [\mathcal{S}_m V]$ with $v \notin T$ is closed, and their complement $V' \subseteq X$ is open. Now we can choose a polytope σ, $\pi : U' \to V'$ étale, $\tau : U' \to V(\sigma)$ smooth, and a point $u \in U'$ with $\pi(u) = v$ and $\tau(u) = 0$ such as in the definition of a toroidal crossing space. Also U' has only finitely many open strata, so we can remove all closed strata which do not contain u and obtain an open subset $u \in U \subseteq U'$. It remains to show that this satisfies the claimed conditions.

Let $d = \dim(V)$, let r be the relative dimension of τ, and let $n = d - r$ be the dimension of $V(\sigma)$. Let $F \subset P_\sigma$ be a proper face of rank $n - m$, and assume that $\tau^{-1}(V_F^\circ) = \varnothing$. Since $0 \in V_F$, we have $\tau^{-1}(V_F) \neq \varnothing$, so $\dim(\tau^{-1}(V_F)) = n - m + r$. Now $\dim(V_F \setminus V_F^\circ) = n - m - 1$, so $\dim(\tau^{-1}(V_F \setminus V_F^\circ)) = n - m + r - 1$. Since $\tau^{-1}(V_F) = \tau^{-1}(V_F \setminus V_F^\circ)$ set-theoretically, this is a contradiction. Therefore, we have $\tau^{-1}(V_F^\circ) \neq \varnothing$.

When $\mathcal{F}(m)$ is the set of faces of P_σ of rank $n - m$, then we have $\mathcal{S}_m V(\sigma) = \bigsqcup_{F \in \mathcal{F}(m)} V_F^\circ$ and hence $\mathcal{S}_m U = \bigsqcup_{F \in \mathcal{F}(m)} \tau^{-1}(V_F^\circ)$. Therefore, open strata of U are in one-to-one correspondence with connected components of $\tau^{-1}(V_F^\circ)$.

Next, we show that $\tau^{-1}(V_F^\circ)$ is irreducible. Assume the contrary. Since $\tau^{-1}(V_F^\circ)$ is normal, we can find two connected components S_1°, S_2° of $\tau^{-1}(V_F^\circ)$. They must also be open strata of U. Therefore, both closures S_1 and S_2 contain u. However, $\tau^{-1}(V_F)$ is normal and of the same dimension as $\tau^{-1}(V_F^\circ)$, so S_1 and S_2 are connected components of $\tau^{-1}(V_F)$. Since they both contain u, we have $S_1 = S_2$. But then we also have $S_1^\circ \cap S_2^\circ \neq \varnothing$ because they are full-dimensional open subsets of the same irreducible scheme, so $S_1^\circ = S_2^\circ$ because they are connected components. This shows that $\tau^{-1}(V_F^\circ)$ is irreducible. From the above decomposition of $\mathcal{S}_m U$, we find that $V_F^\circ \mapsto \tau^{-1}(V_F^\circ)$ induces a bijection between open strata of $V(\sigma)$ and open strata of U.

We have already seen that $\overline{\tau^{-1}(V_F^\circ)} \subset \tau^{-1}(V_F)$ is a connected component because $\tau^{-1}(V_F)$ is normal and both have the same dimension. If $\tau^{-1}(V_F)$ had another connected component, then it would map to $V_F \setminus V_F^\circ$. However, the map would have the same relative dimension as τ, and the total space would have the same dimension as $\tau^{-1}(V_F)$. This is a contradiction, and we find $\overline{\tau^{-1}(V_F^\circ)} = \tau^{-1}(V_F)$.

We now turn to the map $\pi : U \to V$. We start with $m = 0$, so let $H^\circ \in [\mathcal{S}_0 V]$ be an open stratum with $v \in H$. The same argument as above shows that $\pi^{-1}(H^\circ) \neq \varnothing$.

By the assumption that $(V, \mathcal{P}, \bar\rho)$ is simple, H is normal. Then $\pi^{-1}(H)$ is normal, and the irreducible components of $\pi^{-1}(H)$ are also its connected components. When arguing as in the case of τ, we find that $\pi^{-1}(H^\circ)$ is irreducible and an open stratum of U, and that $\pi^{-1}(H)$ is irreducible. Then, we also have $\overline{\pi^{-1}(H^\circ)} = \pi^{-1}(H)$.

Now let $m \geq 1$, and let $T^\circ \in [\mathcal{S}_m V]$ be an open stratum with $v \in T$. Again, we have $\pi^{-1}(T^\circ) \neq \varnothing$ because $v \in T$. Also, arguing as in the case of τ, we see that the connected components of $\pi^{-1}(T^\circ)$ are open strata of U.

Let $S^\circ \subseteq \pi^{-1}(T^\circ)$ be one of the connected components. Then $S^\circ = \tau^{-1}(V_F^\circ)$ for some proper face $F \subset P_\sigma$. Since every face P_σ is the intersection of the facets containing it, we find that S is the intersection of all top-dimensional closed strata in U containing it: $S = \bigcap_{E \in [\mathcal{C}_0 U] : S \subseteq E} E$. Now each E is of the form $\pi^{-1}(H)$ for a unique top-dimensional closed stratum H of V with $v \in H$. Therefore, there is some intersection C of top-dimensional closed strata of V such that $S = \pi^{-1}(C)$. Furthermore, if η_S is the generic point of S, then $\pi(\eta_S)$ is the generic point of T, so we have $T \subset H$ for each H that yields some $E = \pi^{-1}(E)$. Thus, we have $T \subset C$. Now $S \subset \pi^{-1}(T) \subset \pi^{-1}(C) = S$, so we have $S = \pi^{-1}(T)$. In particular, $\pi^{-1}(T)$, and hence its open subset $\pi^{-1}(T^\circ)$, is irreducible. This also shows $S^\circ = \pi^{-1}(T^\circ)$. Since $\mathcal{S}_m U = \bigsqcup_{T^\circ \in [\mathcal{S}_m V] : v \in T} \pi^{-1}(T^\circ)$, we find that the assignment $T^\circ \mapsto \pi^{-1}(T^\circ)$ is bijective. $\square$

Corollary 9.55 *Let $(V, \mathcal{P}, \bar{\rho})$ be a simple toroidal crossing space. Then every closed stratum $S \in [C_m V]$ is normal.*

Proof Let $v \in S$, and choose a chart as in Lemma 9.54. Then $\pi^{-1}(S)$ is a closed stratum of U and hence of the form $\tau^{-1}(V_F)$ for some proper face $F \subset P_\sigma$. In particular, $\pi^{-1}(S)$ is normal, and since $\pi^{-1}(S) \to S$ is étale, we find that S is normal at v. $\square$

Let us now assume that we have a section $s \in \Gamma(V, \mathcal{LS}_V)$ which yields a log smooth structure given by a log structure $\alpha \colon \mathcal{M} \to O_V$, a homomorphism $q \colon \mathcal{M} \to \mathcal{P}$, and a section $\rho \in \Gamma(V, \mathcal{M})$. In a special chart as constructed in Lemma 9.54, we then would like to have $\pi_{\log}^* \mathcal{M} = \tau_{\log}^* \mathcal{M}_{V(\sigma)}$ compatibly with the morphism to S_0. This is indeed possible. We use the following result in Sect. 9.9 in the proof of Lemma 9.75.

Corollary 9.56 *Let $(V, \mathcal{P}, \bar{\rho})$ be a simple toroidal crossing space, and let $s \in \Gamma(V, \mathcal{LS}_V)$ be a log smooth structure. Let $f_0 \colon (V, \mathcal{M}) \to S_0$ be the induced morphism of log schemes. Let $v \in V$ be a point. Then there is a full-dimensional lattice polytope $\sigma \subseteq M_{\mathbb{R}}$, a strict étale morphism $\pi \colon U \to V$ of log schemes, a strict and smooth morphism of log schemes $\tau \colon U \to V(\sigma)$ such that $f_0 \circ \pi = p_0 \circ \tau$ for $p_0 \colon V(\sigma) \to S_0$, and a point $u \in U$ with $\pi(u) = v$ and $\tau(u) = 0$, such that the combinatorial conditions listed in Lemma 9.54 are satisfied.*

Proof The proof is the same as the one of Lemma 9.54 once we start with a chart for the log smooth morphism $f_0 \colon (V, \mathcal{M}) \to S_0$ as in the proof of Lemma 9.45. $\square$

9.7.2 O_V as a Limit

Let $[C_\bullet V]$ be the set of closed strata, ordered by inclusion and considered as a category. Then, we obtain a diagram $O_\bullet \colon [C_\bullet V]^{\mathrm{op}} \to \mathrm{Coh}(V)$, $S \mapsto O_S$, together with a homomorphism $\phi \colon O_V \to \lim_S(O_S)$ of O_V-modules. The following lemma confirms that we can consider V as a closed gluing of its closed strata. Furthermore, we will use it in the proof of Lemma 9.78.

Lemma 9.57 *Let $(V, \mathcal{P}, \bar{\rho})$ be a simple toroidal crossing space. Then the map $\phi \colon O_V \to \lim_S(O_S)$ is an isomorphism of O_V-modules.*

Proof Since the category of coherent sheaves has finite limits which are computed in the category of presheaves of O_V-modules, the limit exists and is coherent.

It is sufficient to prove isomorphy locally around a point $v \in V$. Let $[C_\bullet V]_v \subseteq [C_\bullet V]$ be the subcategory of those strata where $v \in S$. Then, we have a canonical map $\psi \colon \lim_S(O_S) \to \lim_{S \ni v}(O_S)$, which is an isomorphism around v. Let $\pi \colon U \to V$ and $\tau \colon U \to V(\sigma)$ be a special chart around v as in Lemma 9.54. Since π^* is an exact functor and thus preserves finite limits, it is sufficient to show that $O_U \to \lim_{S \ni v} O_{\pi^{-1}(S)}$ is an isomorphism. This follows from the combinatorics of a special chart and Lemma 9.13 because τ^* preserves finite limits. $\square$

9.7.3 The Ghost Sheaf in the Zariski Topology

For later use in computing the dualizing sheaf, we prove that the ghost sheaf $\mathcal{P}$ of a simple toroidal crossing space $(V, \mathcal{P}, \bar{\rho})$ is defined in the Zariski topology.

Let $\mathcal{S}$ be a sheaf of sets in the étale topology of V, and let $\mathcal{S}^{\mathrm{Zar}}$ be its restriction to the Zariski topology. For any étale map $p\colon W \to V$, we can form the Zariski sheaf $p^{-1}\mathcal{S}^{\mathrm{Zar}}$ on W. Then the presheaf $W \mapsto \Gamma(W, p^{-1}\mathcal{S}^{\mathrm{Zar}})$ in the étale topology is actually a sheaf by [222, Claim 1.4.3], which we denote by $(\mathcal{S}^{\mathrm{Zar}})^{\mathrm{\acute{e}t}}$. It comes with a natural comparison map $(\mathcal{S}^{\mathrm{Zar}})^{\mathrm{\acute{e}t}} \to \mathcal{S}$, which is an isomorphism if and only if the comparison map at the stalks $\mathcal{S}^{\mathrm{Zar}}_v \to \mathcal{S}_{\bar{v}}$ is an isomorphism for all geometric points $\bar{v}$ lying over any $v \in V$.

Proposition 9.58 *Let $(V, \mathcal{P}, \bar{\rho})$ be a simple toroidal crossing space, and let $\mathcal{P}^{\mathrm{Zar}}$ be the restriction of $\mathcal{P}$ to the Zariski topology. Then $(\mathcal{P}^{\mathrm{Zar}})^{\mathrm{\acute{e}t}} \to \mathcal{P}$ is an isomorphism.*

Proof Let $\bar{v}$ be a geometric point lying over $v \in V$. Choose a special chart $\pi\colon U \to V$, $\tau\colon U \to V(\sigma)$ as in Lemma 9.54. Then, we have a monoid homomorphism $\beta\colon P_\sigma \to \Gamma(U, \mathcal{P})$ which induces a chart of the sheaf of monoids $\mathcal{P}|_U$.

Let $V' = \pi(U) \subseteq V$, which is open in the Zariski topology. Our first goal is to show that we have a factorization $P_\sigma \to \Gamma(V', \mathcal{P}) \to \Gamma(U, \mathcal{P})$ of β. Since $\mathcal{P}$ is a sheaf, $\Gamma(V', \mathcal{P})$ is the equalizer of

$$\Gamma(U, \mathcal{P}) \rightrightarrows \Gamma(U \times_{V'} U, \mathcal{P}).$$

Let $\varpi_1, \varpi_2\colon U \times_{V'} U \rightrightarrows V'$ be the two maps. We have to show $\varpi_1^* \beta(p) = \varpi_2^* \beta(p)$ for every $p \in P_\sigma$. Note that we do not know if $\tau \circ \varpi_1 = \tau \circ \varpi_2$.

The map $\pi\colon U \to V'$ decomposes disjointly into maps between open strata $\pi^{-1}(S^\circ) \to S^\circ$. Therefore, we also have a disjoint decomposition

$$U \times_{V'} U = \bigsqcup_{S^\circ \in [S_m V']} \pi^{-1}(S^\circ) \times_{S^\circ} \pi^{-1}(S^\circ),$$

and ϖ_i decomposes into $\pi^{-1}(S^\circ) \times_{S^\circ} \pi^{-1}(S^\circ) \to \pi^{-1}(S^\circ)$.

On $\pi^{-1}(S^\circ)$ and $\pi^{-1}(S^\circ) \times_{S^\circ} \pi^{-1}(S^\circ)$, the sheaf $\mathcal{P}$ is constant. Let $\bar{w} \to \pi^{-1}(S^\circ) \times_{S^\circ} \pi^{-1}(S^\circ)$ be a geometric point, and let $\bar{\eta} \to \pi^{-1}(S^\circ) \times_{S^\circ} \pi^{-1}(S^\circ)$ be the geometric generic point of the component which contains $\bar{w}$. Then, we have isomorphisms $\mathcal{P}^{\mathrm{Zar}}_{\varpi_i(\eta)} \to \mathcal{P}_{\bar{\eta}}$ for $i = 1, 2$. Here, $\eta \in \pi^{-1}(S^\circ) \times_{S^\circ} \pi^{-1}(S^\circ)$ is the underlying generic point, and $\varpi_i(\eta)$ is its image in $\pi^{-1}(S^\circ)$. The latter must be the generic point of the irreducible scheme $\pi^{-1}(S^\circ)$, and the first map is an isomorphism because $\mathcal{P}|_{\pi^{-1}(S^\circ)}$ is defined in the Zariski topology. Next, note that $\varpi_1(\eta) = \varpi_2(\eta)$. Although the two geometric points $\bar{\eta} \to \pi^{-1}(S^\circ) \times_{S^\circ} \pi^{-1}(S^\circ) \xrightarrow{\varpi_i} \pi^{-1}(S^\circ)$ are a priori different, an easy diagram chase shows that the two induced maps $\mathcal{P}^{\mathrm{Zar}}_{\varpi_i(\eta)} \to \mathcal{P}_{\bar{\eta}}$ are the same. Now, for $i = 1, 2$, we have a commutative diagram

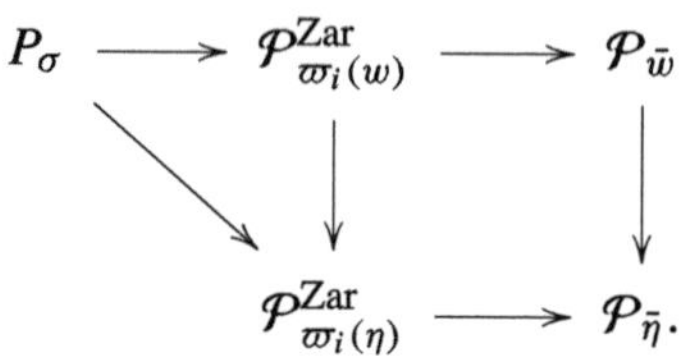

Note that we may have $\varpi_1(w) \neq \varpi_2(w)$. The vertical maps are isomorphisms due to the constancy of $\mathcal{P}$ on the respective strata. Therefore, we find $\varpi_1^*\beta(p)_{\bar{w}} = \varpi_2^*\beta(p)_{\bar{w}}$ on the stalks, as desired.

We now have a factorization $P_\sigma \to \mathcal{P}^{\mathrm{Zar}}_v \to \mathcal{P}_{\bar{v}}$ of the isomorphism $P_\sigma \to \mathcal{P}_{\bar{v}}$. The second map is injective by general considerations, so we find that $\mathcal{P}^{\mathrm{Zar}}_v \to \mathcal{P}_{\bar{v}}$ is bijective. $\qquad\square$

9.8　Toroidal Embeddings

Due to their close connection with toroidal crossing spaces and in particular their relevance to compute the dualizing sheaf of a toroidal crossing space, we review toroidal embeddings in this section.

A *toroidal embedding* may be defined as a normal scheme X together with a smooth Zariski open subset $U \subseteq X$ such that every point $x \in X$ admits an étale neighborhood $p\colon W \to X$ and an étale morphism $h\colon W \to \mathbb{A}_P$ to an affine normal toric variety $\mathbb{A}_P$ such that $p^{-1}(U) = h^{-1}(\mathbb{A}_P^*)$ for the big torus orbit $\mathbb{A}_P^* \subseteq \mathbb{A}_P$. When we endow X with the divisorial log structure defined by $X \setminus U$, then we obtain a log smooth and saturated morphism to the trivial log point $\mathbf{0} = \mathrm{Spec}(0 \to \mathbf{k})$. Conversely, when we have a log smooth and saturated morphism $X \to \mathbf{0}$, then the embedding of the log trivial locus is a toroidal embedding in the above sense. Therefore, we adopt the following definition.

Definition 9.59 (Toroidal Embeddings)　A *toroidal embedding* is a morphism of log schemes $f\colon X \to \mathbf{0}$ which is log smooth, saturated, separated, and of finite type, where X is equidimensional of dimension d.

9.8.1   The Stratification of a Toroidal Embedding

Just like a toroidal crossing space, a toroidal embedding is stratified according to the rank of $\overline{\mathcal{M}}_X$. We set

$$\mathcal{U}_m X = \{x \in X \mid \mathrm{rk}(\overline{\mathcal{M}}_{X,\bar{x}}) \leq m\},$$

$$C_m X = \{x \in X \mid \mathrm{rk}(\overline{\mathcal{M}}_{X,\bar{x}}) \geq m\}, \text{ and}$$

$$\mathcal{S}_m X = \{x \in X \mid \mathrm{rk}(\overline{\mathcal{M}}_{X,\bar{x}}) = m\}$$

as in the case of a toroidal crossing space, but note the different bounds.

For a face $F \subseteq P$ in a toric monoid P, we have a locally closed subset

$$V_F^\circ := \mathrm{Spec}\,\mathbf{k}[F^{\mathrm{gp}}] \subseteq \mathrm{Spec}\,\mathbf{k}[F] \subset A_P.$$

The toric variety A_P then decomposes disjointly as $A_P = \bigsqcup_{F \subseteq P} V_F^\circ$. For $d = \mathrm{rk}(P^{\mathrm{gp}})$, we have $\mathrm{rk}(\overline{\mathcal{M}}_{A_P,x}^{\mathrm{gp}}) = d - e$ for $e = \mathrm{rk}(F^{\mathrm{gp}})$ and every $x \in V^\circ(F)$. This shows that $\mathcal{U}_m A_P$ is open, that $C_m A_P$ is closed, and that $\mathcal{S}_m A_P$ is smooth of dimension $d - m$.

The formation of $\mathcal{U}_m X$, $C_m X$, and $\mathcal{S}_m X$ commutes with strict étale morphisms $p \colon W \to X$. Therefore, $\mathcal{U}_m X \subseteq X$ is open, $C_m X \subset X$ is closed, and $\mathcal{S}_m X$ is smooth. An *open stratum* is then a connected (or equivalently irreducible) component of $\mathcal{S}_m X$. We denote the set of open strata of codimension m by $[\mathcal{S}_m X]$.

A *closed stratum* of codimension m is the closure of any $S^\circ \in [\mathcal{S}_m X]$. We denote the set of closed strata of codimension m by $[C_m X]$. For $S \in [C_m X]$, we write S° for the corresponding open stratum, and $\partial S = S \setminus S^\circ$.

Example 9.60 The open strata of A_P are given by V_F° for the faces $F \subseteq P$. The closed strata are given by $V_F = \mathrm{Spec}\,\mathbf{k}[F]$ for the faces $F \subseteq P$. $\Diamond$

Let us now turn to *simple* toroidal embeddings, which are also called toroidal embeddings *without self-intersections*.

Definition 9.61 (Simple Toroidal Embeddings) Let $X \to \mathbf{0}$ be a toroidal embedding. Then it is *simple* if each closed stratum $D \in [C_1 X]$ is normal.

Just like simple toroidal crossing spaces, simple toroidal embeddings admit special charts. They allow us to show that every closed stratum of a simple toroidal embedding is normal, not just the closed strata of codimension 1.

Lemma 9.62 *Let $X \to \mathbf{0}$ be a simple toroidal embedding, and let $x \in X$ be a point. Then there is a toric monoid P, a strict étale neighborhood $p \colon W \to X$ of x, and a strict étale morphism $h \colon W \to A_P$ with the following properties:*

(a) *for each open stratum V_F° of A_P, the preimage $h^{-1}(V_F^\circ)$ is an open stratum of W, and the assignment $V_F^\circ \mapsto h^{-1}(V_F^\circ)$ induces a one-to-one correspondence between open strata of A_P and open strata of W;*

(b) *we have $h^{-1}(V_F) = \overline{h^{-1}(V_F^\circ)}$ for the closed strata;*
(c) *for each open stratum T° of X whose closure contains x, the preimage $p^{-1}(T^\circ)$ is an open stratum of W, and the assignment $T^\circ \mapsto p^{-1}(T^\circ)$ induces a one-to-one correspondence between open strata of X whose closure contains x and open strata of W;*
(d) *for each open stratum T° of X whose closure contains x, we have $p^{-1}(T) = \overline{p^{-1}(T^\circ)}$.*

Proof The proof is completely analogous to the proof of Lemma 9.54.   □

Corollary 9.63 *Let $X \to \mathbf{0}$ be a simple toroidal embedding. Let $S \in [C_m X]$ be a closed stratum. Then $(S|\partial S)$ (with its compactifying log structure in the étale topology) is a simple toroidal embedding. In particular, S is normal.*

Proof For a point $x \in S$, choose a chart as in Lemma 9.62. Then $(S|\partial S)$ is étale locally isomorphic to $(V_F|\partial V_F)$ for some face $F \subseteq P$ and $\partial V_F := V_F \setminus V_F^\circ$. Separatedness, finite type, and the dimension are clear.   □

9.8.2 *About the Closed Strata of a Simple Toroidal Crossing Space*

The closed strata of a simple toroidal crossing space are simple toroidal embeddings.

Lemma 9.64 *Let $(V, \mathcal{P}, \bar{\rho})$ be a simple toroidal crossing space, and let $S \in [C_m V]$ be a closed stratum. Then $(S|\partial S) \to \mathbf{0}$ is a simple toroidal embedding. The stratification obtained by restricting the stratification of V coincides with the stratification of S as a toroidal embedding.*

Proof The log scheme $(S|\partial S)$ is locally in the smooth topology of the form $(V_F|\partial V_F)$ for a proper face $F \subset P_\sigma$ and $\partial V_F := V_F \setminus V_F^\circ$. Therefore, $(S|\partial S)$ is a toroidal embedding.

In the local model $(V_F|\partial V_F)$, the open strata from the structure of a toroidal embedding are precisely the open strata of the toroidal crossing space $V(\sigma)$ which are contained in V_F. Namely, both are precisely the subsets of the form V_G° for faces $G \subseteq F$. Therefore, the same is true on V, and the two stratifications of S coincide. Since every closed stratum of V is normal, S is a simple toroidal embedding.   □

9.8.3 *Obtaining a Toroidal Crossing Space from a Toroidal Embedding*

Gorenstein toroidal embeddings are a rich source of toroidal crossing spaces. In particular, every Gorenstein toric variety gives rise to a toroidal crossing space by means of the following construction. We have seen a special case in Example 1.140.

Construction 9.65 Let $U \subseteq X$ be a simple toroidal embedding, and write $V := X \setminus U = \bigcup_i E_i$ for the decomposition of the boundary in irreducible components. By assumption, each E_i is normal. We endow X with the divisorial log structure $\mathcal{M}_X$ induced from $V \subseteq X$; a priori, we should carry out this construction in the étale topology, but because $U \subseteq X$ is simple, we get the same answer in the Zariski topology, cf. [222, III. Prop. 1.6.5]. We have an isomorphism $\overline{\mathcal{M}}_X \cong \underline{\Gamma}_V(\mathrm{Div}_X^+)$, the sheaf of effective Cartier divisors with support in V, formed in the Zariski topology. When we set

$$\mathcal{P} := \overline{\mathcal{M}}_X|_V = \Gamma_V(\mathrm{Div}_X^+)|_V,$$

then $(V, \mathcal{P})$ is a space with a toric ghost sheaf. If X is *Gorenstein*, then $V \subseteq X$ itself is an effective Cartier divisor. It gives a global section $\bar{\rho} \in \mathcal{P}$; by definition of a toroidal embedding, now $(V, \mathcal{P}, \bar{\rho})$ is a toroidal crossing space in our sense. In [174], a *conical polyhedral complex* is associated to the toroidal embedding $U \subseteq X$; its cones are dual to the monoids of effective Cartier divisors on an open neighborhood $\mathrm{Star}(Y)$ of a stratum of $\bigcup_i E_i$, so we may consider this conical polyhedral complex a precursor to the structure of a toroidal crossing space.　　　$\Diamond$

9.9　The Dualizing Sheaf of a Simple Toroidal Crossing Space

In this section, we compute the dualizing sheaf of a simple toroidal crossing space $(V, \mathcal{P}, \bar{\rho})$. Let us first formulate the result that we prove here. A closed stratum $X \subset V$ is a simple toroidal embedding and thus has a log canonical sheaf $\omega_{X/0} = \Omega_{X/0}^{d-m}$, where $m = \mathrm{codim}(X, V)$ and $d = \dim(V)$.

Theorem 9.66 *Let $(V, \mathcal{P}, \bar{\rho})$ be a simple toroidal crossing space. Then, for every closed stratum $X \subset V$, we have a constant sheaf $\mho_{X/V}$ on X with stalk $\mathbb{Z}$, and for every inclusion $D \subset X$ of closed strata, we have a restriction map*

$$\rho_{D/X} \colon \quad \omega_{X/0} \otimes_{\mathbb{Z}} \mho_{X/V} \to \omega_{D/0} \otimes_{\mathbb{Z}} \mho_{D/V}$$

between coherent sheaves on V (via the closed embeddings into V) such that

$$\rho_{S/D} \circ \rho_{D/X} = \rho_{S/X}$$

for inclusions $S \subset D \subset X$ of closed strata. The limit of the diagram of coherent sheaves on V is a coherent sheaf ω_V. When $\pi \colon U \to V$ is étale, and $s \in \Gamma(U, \mathcal{LS}_V)$ is a log smooth structure, then we have a canonical isomorphism $\omega_V \cong \omega_{U_s/S_0}$ with the relative log canonical sheaf, where U_s is U together with the log structure from s. Furthermore, we have a canonical isomorphism $\omega_V \cong \omega_V^\circ$ with the normalized dualizing sheaf.

The constant sheaf $\mho_{X/V}$ with stalk $\mathbb{Z}$ does not in general have a canonical generator.

9.9.1 Comparing the Log Canonical Sheaves of Simple Toroidal Embeddings

We start by analyzing the log canonical sheaf of simple toroidal embeddings.

Let X be a simple toroidal embedding, let X° be the log trivial locus, and let $\partial X = X \setminus X^\circ$. Then X carries the divisorial log structure $\mathcal{M}_{(X|\partial X)}$ defined in the étale topology. However, since all irreducible components of ∂X are normal, [222, III, Prop. 1.6.5] shows that this log structure can be defined in the Zariski topology. In particular, we can use the log structure in the Zariski topology to compute the log canonical sheaf

$$\omega_{X/\mathbf{0}} = \Omega^d_{X/\mathbf{0}},$$

where $d = \dim(X)$. We now write $\mathcal{M}_X$ for the log structure in the Zariski topology and $\mathcal{M}_X^{\text{ét}}$ for the log structure in the étale topology.

Let $D \subset X$ be a closed stratum. Our goal is to compare $\omega_{X/\mathbf{0}}|_D$ with $\omega_{D/\mathbf{0}}$. Let $\mathcal{I}_{D/X} \subseteq \mathcal{O}_X$ be the ideal sheaf which cuts out the closed stratum $D \subset X$. For a face $F \subseteq P$ and the closed stratum $V_F \subset A_P$, this ideal sheaf is given by $\mathbf{k}[K]$ for the monoid ideal $K = P \setminus F$.

Next, we set $\widetilde{\mathcal{K}}^{\text{ét}}_{D/X} := \alpha^{-1}(\mathcal{I}^{\text{ét}}_{D/X}) \subseteq \mathcal{M}_X^{\text{ét}}$. This is a sheaf of monoid ideals. In the local model $V_F \subset A_P$, this contains in particular the images of $K \subseteq P$ under $\beta\colon P \to \mathcal{M}_{A_P}^{\text{ét}}$, so we have

$$\mathcal{I}^{\text{ét}}_{V_F/A_P} = \alpha(\widetilde{\mathcal{K}}^{\text{ét}}_{V_F/A_P}) \cdot \mathcal{O}^{\text{ét}}_{A_P}.$$

But then we also have $\mathcal{I}^{\text{ét}}_{D/X} = \alpha(\widetilde{\mathcal{K}}^{\text{ét}}_{X/D}) \cdot \mathcal{O}^{\text{ét}}_X$ globally. Since $\widetilde{\mathcal{K}}^{\text{ét}}_{V_F/A_P}$ is also generated by the images of $K \subseteq P$ under β, we find that $\widetilde{\mathcal{K}}^{\text{ét}}_{D/X} \subseteq \mathcal{M}_X^{\text{ét}}$ is a coherent sheaf of monoid ideals.

By [222, III, Prop. 1.3.4], we obtain an idealized log structure on the vanishing locus $D \subset X$ of $\mathcal{I}^{\text{ét}}_{D/X}$. Let us denote the log structure by $\mathcal{M}_{D/X}^{\text{ét}}$ and its sheaf of ideals by $\mathcal{K}_{D/X}^{\text{ét}}$. Then, we have a morphism of idealized log schemes

$$i\colon \ [D/X] := (D, \mathcal{M}_{D/X}^{\text{ét}}, \mathcal{K}_{D/X}^{\text{ét}}) \to (X, \mathcal{M}_X^{\text{ét}}, \varnothing).$$

On underlying morphisms of log schemes, this is strict. In particular, $\mathcal{M}_{D/X}^{\text{ét}}$ is the étale log structure associated with the Zariski log structure $\mathcal{M}_{D/X} := i^*_{\log} \mathcal{M}_X$ for the inclusion $i \colon D \to X$.[9] By [222, IV, Var. 3.1.21], the morphism $i \colon [D/X] \to X$ is ideally log étale.

Lemma 9.67 *The canonical map* $i^*\Omega^1_{X/0} \to \Omega^1_{[D/X]/0}$ *is an isomorphism. In particular,* $\Omega^1_{[D/X]/0}$ *is locally free of rank* $d = \dim(X)$.

Proof This should be true for any ideally log étale morphism, but we did not find a direct reference, so let us briefly explain why it is true here. On underlying log schemes, $i \colon [D/X] \to X$ is a strict closed immersion, so we have an exact sequence

$$\mathcal{I}_{D/X}/\mathcal{I}^2_{D/X} \to i^*\Omega^1_{X/0} \to \Omega^1_{[D/X]/0} \to 0.$$

It is then sufficient to show that the left-hand map is zero, which can be done étale locally. In the local model given by $F \subseteq P$, we find $z^k \mapsto z^k \cdot \frac{dz^k}{z^k}$ for any generator of $\mathbf{k}[K]$, so this is zero after pulling back to A_F. $\qquad\qquad\square$

Since D is reduced an irreducible, $\mathcal{I}_{D/X,\bar{x}} \subset O_{X,\bar{x}}$ is a prime ideal for every $\bar{x} \to D$. This implies that $\widetilde{\mathcal{K}}^{\text{ét}}_{D/X,\bar{x}} \subset M^{\text{ét}}_{X,\bar{x}}$ is a monoid prime ideal, and therefore also $\mathcal{K}^{\text{ét}}_{D/X,\bar{x}} \subset M^{\text{ét}}_{D/X,\bar{x}}$ is a monoid prime ideal. In particular, its complement

$$\mathcal{F}^{\text{ét}}_{D/X,\bar{x}} := M^{\text{ét}}_{D/X,\bar{x}} \setminus \mathcal{K}^{\text{ét}}_{D/X,\bar{x}}$$

is a face. But then $\mathcal{F}^{\text{ét}}_{D/X} \to O_D$ is a log structure, and

$$\ell \colon \quad (D, M^{\text{ét}}_{D/X}, \mathcal{K}^{\text{ét}}_{D/X}) \to (D, M^{\text{ét}}_{D/X}, \varnothing) \to (D, \mathcal{F}^{\text{ét}}_{D/X}, \varnothing)$$

is a morphism of idealized log schemes.

Lemma 9.68 *We have a canonical isomorphism* $\mathcal{F}^{\text{ét}}_{D/X} \cong M^{\text{ét}}_D$ *of log structures, where* $M^{\text{ét}}_D$ *is the divisorial log structure defined by* $\partial D \subset D$.

Proof First, we consider the local model given by a face $F \subseteq P$. As a log scheme, we then have $[D/X] = \mathrm{Spec}(P \to \mathbf{k}[F]) =: A_{P,K}$ with the map $p \mapsto p$ if $p \in F$ and $p \mapsto 0$ otherwise. The sheaf of ideals is generated by the image of $\beta \colon K \to M_{D/X}$. At any point, the ghost stalk is P/G for a face $G \subseteq F$. Therefore, the sheaf of ideals generated by the image of K consists precisely of elements of the form $u \cdot \beta(k)$ for invertibles u. The complement then consists precisely of elements of the form $u \cdot \beta(f)$ for $f \in F$ and u invertible. This is nothing but the log structure associated with $F \to \mathbf{k}[F]$. In particular, $\mathcal{F}_{D/X}$ is trivial on $D^\circ = V^\circ_F$. Then the universal property of the divisorial log structure yields a map $\mathcal{F}_{D/X} \to M_{(D|\partial D)}$ which is an isomorphism here.

---

[9] Note that it is not clear if also $\mathcal{K}^{\text{ét}}_{D/X}$ is defined in the Zariski topology. In any case, for our purposes, it is sufficient to have $\mathcal{K}^{\text{ét}}_{D/X}$ in the étale topology.

In the general case, for a point $x \in D^\circ$, we choose a special chart as in Lemma 9.62. Then x corresponds to a point in V_F° of the local model, and we find that $\mathcal{F}_{D/X}^{\text{ét}}|_{D^\circ}$ is a trivial log structure. Now the universal property of the divisorial log structure yields a map $\mathcal{F}_{D/X}^{\text{ét}} \to \mathcal{M}_D^{\text{ét}}$, and we have already seen that this map is locally an isomorphism. $\qquad\square$

At this point, we see that also $\mathcal{F}_{D/X}^{\text{ét}}$ can be defined in the Zariski topology.

Lemma 9.69 *The morphism $\ell\colon [D/X] \to D$ is ideally log smooth.*

Proof It is sufficient to prove this in the local model, so let $F \subseteq P$ be a face. As a log scheme, we have $[D/X] = \mathrm{Spec}(P \to \mathbf{k}[F])$, as we have already seen, and we also know that $[D/X] \to X = A_P$ is ideally log étale. The inclusion $F \subseteq P$ is saturated and induces a log smooth morphism $A_P \to A_F$. The composition is given by $(F \to \mathbf{k}[F]) \to (P \to \mathbf{k}[F])$ and hence the same map as $\ell\colon [D/X] \to D$. Thus, $[D/X] \to D$ is ideally log smooth. $\qquad\square$

Corollary 9.70 *We have a locally split exact sequence*

$$0 \to \Omega^1_{D/0} \xrightarrow{d\ell^*} \Omega^1_{[D/X]/0} \xrightarrow{\varrho} \Omega^1_{[D/X]/D} \to 0$$

of locally free sheaves. When $m = \mathrm{codim}(D, X)$, then the right-hand sheaf is locally free of rank m, and the left-hand sheaf is locally free of rank $d - m$, where $d = \dim(X)$ as above.

Proof By [222, IV, Prop. 3.2.1], ideal log smoothness of ℓ implies that $\Omega^1_{[D/X]/D}$ is locally free. To determine its rank, we use the local model. There, we have a factorization of $[A_{P,K}/A_P] \to A_F$ into $[A_{P,K}/A_P] \xrightarrow{i} A_P \to A_F$, which yields an exact sequence

$$I/I^2 \to i^*\Omega^1_{A_P/A_F} \to \Omega^1_{[A_{P,K}/A_F]/A_F} \to 0$$

since i is a strict closed immersion. As in the proof of Lemma 9.67, we see that the left-hand map is zero so that the right-hand map is an isomorphism. Now the rank of $\Omega^1_{A_P/A_F}$ is m.

It remains to show that $d\ell^*$ is injective. From the previous paragraph, we see that the kernel of ϱ is locally free of rank $d - m$. But also $\Omega^1_{D/0}$ is locally free of rank $d - m$, and it surjects onto the kernel of ϱ, so $d\ell^*$ is injective. $\qquad\square$

From the corollary, we obtain an isomorphism

$$\psi_{D/X}\colon \ \omega_{D/0} \otimes \omega_{[D/X]/D} \to \omega_{[D/X]/0}, \quad \alpha \otimes \beta \mapsto d\ell^*\alpha \wedge \tilde{\beta},$$

where $\tilde{\beta}$ is any section with $(\bigwedge^m \varrho)(\tilde{\beta}) = \beta$. It is important to keep track of the exact isomorphism since, for example, the isomorphism involves a choice of order and thus does not necessarily commute with changing the order of the factors on the left. Together with the isomorphism $di^* \colon i^*\omega_{X/0} \to \omega_{[D/X]/0}$, we obtain an isomorphism

$$\tilde{\rho}_{D/X} \colon \ i^*\omega_{X/0} \to \omega_{D/0} \otimes \omega_{[D/X]/D}, \quad \alpha \mapsto \psi_{D/X}^{-1} \circ di^*(\alpha).$$

9.9.2 A Description of $\omega_{[D/X]/D}$

Recall that all three log structures on X, D, and $[D/X]$ can be defined in the Zariski topology, and note that then also the morphisms arise from morphisms of log structures in the Zariski topology. We now consider the two exact sequences

$$
\begin{array}{ccccccccc}
0 & \longrightarrow & \mathcal{M}_D^{\mathrm{gp}} & \longrightarrow & \mathcal{M}_{[D/X]}^{\mathrm{gp}} & \longrightarrow & \mathcal{M}_{[D/X]/D}^{\mathrm{gp}} & \longrightarrow & 0 \\
& & \downarrow & & \downarrow & & \| & & \\
0 & \longrightarrow & \overline{\mathcal{M}}_D^{\mathrm{gp}} & \longrightarrow & \overline{\mathcal{M}}_{[D/X]}^{\mathrm{gp}} & \longrightarrow & \overline{\mathcal{M}}_{[D/X]/D}^{\mathrm{gp}} & \longrightarrow & 0
\end{array}
$$

in the Zariski topology.

Lemma 9.71 *The sheaf of Abelian groups $\mathcal{M}_{[D/X]/D}^{\mathrm{gp}}$ is constant with stalk $\mathbb{Z}^m$, where $m = \mathrm{codim}(D, X)$.*

Proof Let $p \colon W \to X$ and $h \colon W \to A_P$ be a chart. Note that

$$p^{-1}\mathcal{M}_{[D/X]/D}^{\mathrm{gp}} \cong h^{-1}\mathcal{M}_{[A_F/A_P]/A_F}^{\mathrm{gp}},$$

so we can compute the étale local form of $\mathcal{M}_{[D/X]/D}^{\mathrm{gp}}$ on the local models. Let us now write $X = A_P$ and $D = A_F$. For any point $x \in D$, we have $\overline{\mathcal{M}}_{[D/X],x} = P/G$ and $\overline{\mathcal{M}}_{D,x} = F/G$ for the same face $G \subseteq F$. Therefore, we have $\overline{\mathcal{M}}_{[D/X]/D}^{\mathrm{gp}} = P^{\mathrm{gp}}/F^{\mathrm{gp}}$, and the isomorphism is indeed global on $D = A_F$. For a general X, this shows that $\mathcal{M}_{[D/X]/D}^{\mathrm{gp}}$ is locally constant in the étale topology. Using finite generatedness, one may now show that $\mathcal{M}_{[D/X]/D}^{\mathrm{gp}}$ is locally constant already in the Zariski topology. Since D is integral, this shows that $\mathcal{M}_{[D/X]/D}^{\mathrm{gp}}$ is constant. $\qquad\square$

The log part of the universal derivation gives a group homomorphism

$$\delta \colon \ \mathcal{M}_{[D/X]}^{\mathrm{gp}} \to \Omega^1_{[D/X]/D}.$$

By construction, this vanishes on $\mathcal{M}_D^{\mathrm{gp}} \subseteq \mathcal{M}_{[D/X]}^{\mathrm{gp}}$. Therefore, we obtain an induced group homomorphism

$$\bar{\delta}_{D/X} \colon \ \mathcal{M}_{[D/X]/D}^{\mathrm{gp}} \to \Omega^1_{[D/X]/D}.$$

Let us now write $\Phi_{D/X} := \mathcal{M}_{[D/X]/D}^{\mathrm{gp}}$ for short.

Lemma 9.72 *The map*

$$O_D \otimes_{\mathbb{Z}} \Phi_{D/X} \to \Omega^1_{[D/X]/D}$$

is an isomorphism of O_D-modules.

Proof It is sufficient to prove this in the local model. It follows from the results in Sect. 8.9 that $O_{A_P} \otimes_{\mathbb{Z}} P^{\mathrm{gp}}/F^{\mathrm{gp}} \to \Omega^1_{A_P/A_F}$ is an isomorphism. Pulling this back to $A_{P,K} \subset A_P$ yields precisely the map in the statement of the lemma and hence proves that it is an isomorphism. $\qquad\square$

Corollary 9.73 *When setting*

$$\mho_{D/X} := \bigwedge_{\mathbb{Z}}^m \mathcal{M}^{\mathrm{gp}}_{[D/X]/D},$$

we obtain an isomorphism $O_D \otimes_{\mathbb{Z}} \mho_{D/X} \to \omega_{[D/X]/D}$ and thus an isomorphism

$$\hat{\rho}_{D/X} \colon \ i^* \omega_{X/0} \to \omega_{D/0} \otimes_{\mathbb{Z}} \mho_{D/X}.$$

9.9.3 Functoriality with Respect to Two Inclusions

We analyze what happens when we have two inclusions $S \subset D$ and $D \subset X$ of closed strata. On the one hand, we have the map

$$\tilde{\rho}_{S/X} \colon \ i^*_{S/X} \omega_{X/0} \to \omega_{S/0} \otimes \omega_{[S/X]/S}$$

corresponding to $S \subset X$. On the other hand, we have the composition

$$i^*_{S/D} i^*_{D/X} \omega_{X/0} \xrightarrow{\ i^*_{S/D} \tilde{\rho}_{D/X}\ } i^*_{S/D}(\omega_{D/0} \otimes \omega_{[D/X]/D})$$

$$\xrightarrow{\ \tilde{\rho}_{S/D} \otimes \mathrm{id}\ } \omega_{S/0} \otimes \omega_{[S/D]/S} \otimes i^*_{S/D} \omega_{[D/X]/D}.$$

At least on underlying log schemes, the diagram

$$
\begin{array}{ccc}
[S/X] & \longrightarrow & [D/X] \\
\downarrow & & \downarrow \\
[S/D] & \longrightarrow & D
\end{array}
$$

is Cartesian so that we have $i_{S/D}^* \Omega^1_{[D/X]/D} \cong \Omega^1_{[S/X]/[S/D]}$. The composition $[S/X] \to [S/D] \to S$ yields a short exact sequence

$$0 \to \Omega^1_{[S/D]/S} \to \Omega^1_{[S/X]/S} \to \Omega^1_{[S/X]/[S/D]} \to 0$$

whose exactness on the left is seen by a rank argument as in the proof of Corollary 9.70. This induces an isomorphism

$$\chi_{S/D/X}: \quad \omega_{[S/D]/S} \otimes i_{S/D}^* \omega_{[D/X]/D} \cong \omega_{[S/X]/S}.$$

Similarly, we have an exact sequence

$$0 \to \mathcal{M}^{\mathrm{gp}}_{[S/D]/S} \to \mathcal{M}^{\mathrm{gp}}_{[S/X]/S} \to i^{-1} \mathcal{M}^{\mathrm{gp}}_{[D/X]/D} \to 0$$

which induces an isomorphism

$$\hat{\chi}_{S/D/X}: \quad \mho_{S/D} \otimes_{\mathbb{Z}} i^{-1} \mho_{D/X} \cong \mho_{S/X}.$$

Lemma 9.74 *The diagram*

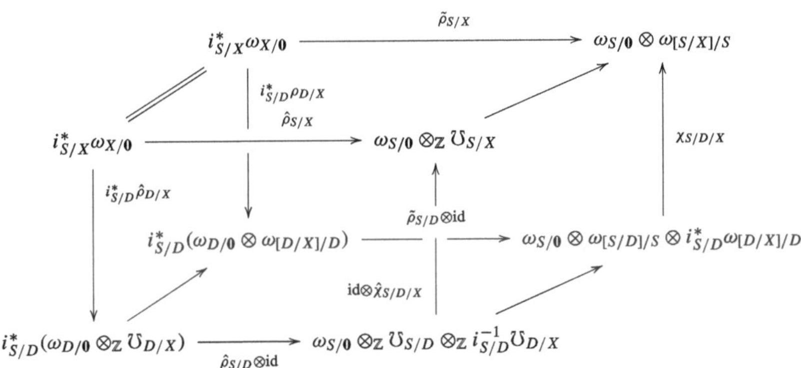

commutes.

Proof Let $m = \mathrm{codim}(D, X)$ and $c = \mathrm{codim}(S, X)$. The top square commutes by definition of $\hat{\rho}_{S/X}$, and the left square and bottom square commute for similar reasons. To see that the right square commutes, it is sufficient to show that

$$
\begin{array}{ccc}
\mho_{S/D} \otimes_{\mathbb{Z}} \mho_{D/X}|_S & \xrightarrow{\hat{\chi}_{S/D/X}} & \mho_{S/X} \\
\downarrow & & \downarrow \\
\omega_{[S/D]/S} \otimes i_{S/D}^* \omega_{[D/X]/D} & \xrightarrow{\chi_{S/D/X}} & \omega_{[S/X]/S}
\end{array}
$$

commutes. When $\alpha \in \bigwedge^{c-m}$ and $\beta \in \bigwedge^{m} \Phi_{D/X}|_S$, then the upper right composition is $\delta(\alpha \wedge \beta')$ for some $\beta' \in \bigwedge^{m} \Phi_{S/X}$ which lifts β under $\Phi_{S/X} \to \Phi_{D/X}|_S$. The lower left composition is $\delta(\alpha) \wedge \delta(\beta)'$, where $\delta(\beta)'$ lifts $\delta(\beta)$ under $\bigwedge^{m} \Omega^1_{[S/X]/S} \to \bigwedge^{m} \Omega^1_{[D/X]/D}|_S$. This lift can be chosen to be $\delta(\beta')$, so the right square commutes.

All arrows are isomorphisms. Thus, it is now sufficient to show that the back square commutes. Let

$$\alpha \in \Omega^{d-c}_{S/0}, \quad \beta \in \Omega^{c-m}_{[S/D]/S}, \quad \text{and} \quad \gamma \in i^*_{S/D}\Omega^m_{[D/X]/D} = \Omega^m_{[S/X]/[S/D]}.$$

Let $\gamma' \in \Omega^m_{[S/X]/S}$ be a lift of γ under $\Omega^m_{[S/X]/S} \to \Omega^m_{[S/X]/[S/D]}$. Then, we have

$$\chi_{S/D/X}(\alpha \otimes \beta \otimes \gamma) = \alpha \otimes (\beta \wedge \gamma').$$

Let now γ'' be a lift of γ' under $\Omega^m_{[S/X]/0} \to \Omega^m_{[S/X]/S}$, and let β' be a lift of β under $\Omega^{c-m}_{[S/D]/0} \to \Omega^{c-m}_{[S/D]/S}$, which can be considered as an element of $\Omega^{c-m}_{[S/X]/0}$ via $\Omega^{c-m}_{[S/D]/0} \to \Omega^{c-m}_{[S/X]/0}$. Then, we have

$$\tilde{\rho}^{-1}_{S/X}(\alpha \otimes (\beta \wedge \gamma')) = \alpha \wedge \beta' \wedge \gamma''.$$

Similarly, we have $\tilde{\rho}^{-1}_{S/D}(\alpha \otimes \beta) = \alpha \wedge \beta'$, where we use the same lift β' as above. Then $\tilde{\rho}^{-1}_{S/X}((\alpha \wedge \beta') \otimes \gamma) = \alpha \wedge \beta' \wedge \gamma''$ with the lift γ'' from above. Therefore, the back square commutes. $\qquad\square$

9.9.4 The Embedding of the Closed Strata into a Simple Toroidal Crossing Space

So far, we have focused on simple toroidal embeddings. Let now $(V, \mathcal{P}, \bar{\rho})$ be a simple toroidal crossing space. We have seen in Proposition 9.58 that $\mathcal{P}$ is defined in the Zariski topology of V. Let $X \in [C_m V]$ be a closed stratum, and let η_X be its generic point. We define a sheaf of ideals $\widehat{\mathcal{K}}_{X/V} \subseteq \mathcal{P}$ as follows. On an open subset $U \subseteq V$ with $\eta_X \in U$, we have a restriction map $\varrho \colon \Gamma(U, \mathcal{P}) \to \mathcal{P}_{\eta_X}$. Then, we set $\Gamma(U, \widehat{\mathcal{K}}_{X/V}) := \{p \in \Gamma(U, \mathcal{P}) \mid \varrho(p) \neq 0\}$. On an open subset $U \subseteq V$ with $\eta_X \notin U$, we set $\Gamma(U, \widehat{\mathcal{K}}_{X/V}) := \Gamma(U, \mathcal{P})$. It is easy to see that this is in fact a sheaf of ideals. We then set

$$\overline{\mathcal{K}}_{X/V} := i^{-1}\widehat{\mathcal{K}}_{X/V} \subseteq \mathcal{P}|_X$$

for the inclusion $i\colon X \to V$. This is clearly a sheaf of proper ideals because $\eta_X \in U$ for any neighborhood U of some point in X. Since $\mathcal{P}_{\eta_X}$ is sharp, this is furthermore a sheaf of prime ideals so that the complement

$$\overline{\mathcal{F}}_{X/V} := \mathcal{P}|_X \setminus \overline{\mathcal{K}}_{X/V}$$

is a sheaf of faces. We interpret this sheaf as follows.

Lemma 9.75 *Let $(V, \mathcal{P}, \bar{\rho})$ be a simple toroidal crossing space, and let $s \in \Gamma(V, \mathcal{LS}_V)$. Let $\mathcal{M}$ be the log structure, and let $q\colon \mathcal{M} \to \mathcal{P}$ be the canonical map. Let*

$$q|_X\colon \ i^*_{\log}\mathcal{M} \to \mathcal{P}|_X$$

be the map induced by restriction along $i\colon X \to V$. Then the log structure $(q|_X)^{-1}(\overline{\mathcal{F}}_{X/V})$ is canonically isomorphic to the log structure of the simple toroidal embedding $(X|\partial X)$.

Proof By definition, $\mathcal{M}$ is defined in the étale topology. Restricting and extending gives rise to a morphism of u-integral log structures $\gamma\colon (\mathcal{M}^{\mathrm{Zar}})^{\mathrm{\acute{e}t}} \to \mathcal{M}$. Since $\mathcal{P}$ is defined in the Zariski topology, γ is an isomorphism on the level of ghost sheaves. Because both log structures are u-integral, γ is an isomorphism. Therefore, $\mathcal{M}$ is defined in the Zariski topology.

Let $\mathcal{N} = (q|_X)^{-1}(\overline{\mathcal{F}}_{X/V}) \subseteq i^*_{\log}\mathcal{M}$, considered in the Zariski topology of X. As is the case for any sheaf of faces in the ghost sheaf, this is a log structure with ghost sheaf $\overline{\mathcal{F}}_{X/V}$. For any non-empty $U \subseteq X^\circ$, the map $\Gamma(U, \mathcal{P}|_X) \to \mathcal{P}_{\eta_X}$ is an isomorphism, so we have $\overline{\mathcal{F}}_{X/V}|_{X^\circ} = 0$. Thus, the universal property of the compactifying log structure $\mathcal{M}_X$ yields a canonical map $\varepsilon\colon \mathcal{N} \to \mathcal{M}_X$.

We extend it to a map of associated étale log structures. Note that $\mathcal{M}_X^{\mathrm{\acute{e}t}}$ is nothing but the compactifying log structure in the étale topology, and that restricting the étale log structures to the Zariski topology yields back the original Zariski log structures. Therefore, it is sufficient to prove that $\varepsilon\colon \mathcal{N} \to \mathcal{M}_X$ is an isomorphism étale locally.

Let $v \in X$, and choose a special chart as in Corollary 9.56, consisting of a strict étale morphism $\pi\colon U \to V$ and a strict smooth morphism $\tau\colon U \to V(\sigma)$ of log schemes (not just of spaces with a pre–ghost structure) as well as a point $u \in U$ with $\pi(u) = v$ and $\tau(u) = 0$. Let $F \subsetneq P_\sigma$ be the face with $Y := \pi^{-1}(X) = \tau^{-1}(V_F)$.

Since $\mathcal{P}$ is defined in the Zariski topology, we have $\pi^{-1}\mathcal{P} = \tau^{-1}\mathcal{P}_\sigma$ as sheaves in the Zariski topology of U. Inside $\pi^{-1}\mathcal{P}$, we have $\widehat{\mathcal{K}}_{Y/U}$ constructed in analogy with $\widehat{\mathcal{K}}_{X/V}$. When expanding the definitions, an easy argument yields that $\pi^{-1}\widehat{\mathcal{K}}_{V/X} = \widehat{\mathcal{K}}_{Y/U}$. Similarly, we have $\tau^{-1}\widehat{\mathcal{K}}_{V_F/V(\sigma)} = \widehat{\mathcal{K}}_{Y/U}$. Then, we also have $\pi^{-1}\overline{\mathcal{F}}_{X/V} = \overline{\mathcal{F}}_{Y/U} = \tau^{-1}\overline{\mathcal{F}}_{V_F/V(\sigma)}$.

Now, we find

$$\pi^*_{\log}\mathcal{N} = (q|_Y)^{-1}(\overline{\mathcal{F}}_{Y/U}) =: \mathcal{N}_{Y/U}.$$

After setting $q_\sigma \colon \mathcal{M}_{V(\sigma)} \to \mathcal{P}_\sigma$ and $\mathcal{N}_\sigma := q_\sigma^{-1}(\overline{\mathcal{F}}_{V_F/V(\sigma)})$, we also find $\mathcal{N}_{Y/U} = \tau_{\log}^* \mathcal{N}_\sigma$. Since $\tau_{\log}^* \mathcal{M}_{V_F} = \mathcal{M}_Y$ for the divisorial log structures on the strata, it is sufficient to show that $\varepsilon_\sigma \colon \mathcal{N}_\sigma \to \mathcal{M}_{V_F}$ is an isomorphism of log structures in the Zariski topology.

The log scheme $U(\sigma)$ is a simple toroidal embedding, so we have a log structure $\mathcal{F}_{V_F/U(\sigma)}$ as above on V_F. As a subsheaf of $\mathcal{M}_{V(\sigma)}|_{V_F}$, this coincides with $\mathcal{N}_\sigma$ because a section of $\mathcal{M}_{V(\sigma)}|_{V_F}$ restricts to zero in $\mathcal{P}_\sigma|_{V_F}$ at the generic point of V_F if and only if it is in the sheaf of faces in $\mathcal{M}_{V(\sigma)}|_{V_F}$ generated by F. Therefore, we have already seen that $\varepsilon_\sigma \colon \mathcal{N}_\sigma \to \mathcal{M}_{V_F}$ is an isomorphism in Lemma 9.68. $\qquad\square$

Let us continue to assume that we have a log smooth structure $s \in \Gamma(V, \mathcal{LS}_V)$. Then $\widetilde{\mathcal{K}}_{X/V} := q^{-1}(\widehat{\mathcal{K}}_{X/V}) \subseteq \mathcal{M}$ is a sheaf of ideals. We see from the local model $V_F \subset V(\sigma)$ that $\mathcal{I}_{X/V} = \alpha(\widetilde{\mathcal{K}}_{X/V}) \cdot \mathcal{O}_V$ is the ideal sheaf which cuts out $X \subset V$. Furthermore, $\widetilde{\mathcal{K}}_{X/V}$ is a coherent sheaf of ideals. Therefore, we obtain an ideally log étale map $i \colon [X/V]_s \to V_s$ of idealized log schemes, where V_s is V endowed with the log structure coming from s, and with the empty ideal sheaf. Similarly, we have an ideally log smooth map $\ell \colon [X/V]_s \to X$, where X carries the divisorial log structure of $\partial X \subset X$. As above, this yields a short exact sequence

$$0 \to \Omega^1_{X/0} \to \Omega^1_{[X/V]_s/0} \to \Omega^1_{[X/V]_s/X} \to 0$$

and an isomorphism $i^* \Omega^1_{V_s/0} \cong \Omega^1_{[X/V]_s/0}$. We have an isomorphism

$$\mathcal{O}_X \otimes_{\mathbb{Z}} (\mathcal{P}^{\mathrm{gp}}|_X)/(\overline{\mathcal{F}}^{\mathrm{gp}}_{X/V}) \to \Omega^1_{[X/V]_s/X}$$

as in Lemma 9.72.

Lemma 9.76 *The quotient* $\Phi_{X/V} := (\mathcal{P}^{\mathrm{gp}}|_X)/(\overline{\mathcal{F}}^{\mathrm{gp}}_{X/V})$ *is a constant sheaf with stalk* $\mathbb{Z}^{m+1}$, *where* $m = \operatorname{codim}(X, V)$.

Proof To compute the quotient locally in the étale topology, it is sufficient to compute it on the local model $V_F \subset V(\sigma)$. There, we have $(\mathcal{P}^{\mathrm{gp}}|_{V_F})/(\overline{\mathcal{F}}^{\mathrm{gp}}_{V_F/V(\sigma)}) = P_\sigma^{\mathrm{gp}}/F^{\mathrm{gp}}$, as we computed in the proof of Lemma 9.71. Since $\operatorname{codim}(V_F, U(\sigma)) = m+1$, this is isomorphic to $\mathbb{Z}^{m+1}$. Since X is integral and $(\mathcal{P}^{\mathrm{gp}}|_X)/(\overline{\mathcal{F}}^{\mathrm{gp}}_{X/V})$ is defined in the Zariski topology, this sheaf is constant with stalk $\mathbb{Z}^{m+1}$. $\qquad\square$

When setting $\mho_{X/V} := \bigwedge_{\mathbb{Z}}^{m+1} \Phi_{X/V}$, we find $\omega_{X/0} \otimes_{\mathbb{Z}} \mho_{X/V} \cong i^* \omega_{V_s/0}$. The left-hand side is independent of $s \in \Gamma(V, \mathcal{LS}_V)$, and indeed well-defined without choosing a log smooth structure. This is the key observation for proving Theorem 9.66.

9.9.5 Comparing Two Closed Strata in a Simple Toroidal Crossing Space

Let now $D \subset X \subset V$ be two closed strata in the simple toroidal crossing space $(V, \mathcal{P}, \bar{\rho})$. Let $m = \mathrm{codim}(X, V)$ and $c = \mathrm{codim}(D, V)$. Note that we have a canonical map $\overline{\mathcal{F}}_{D/V} \to i_{D/X}^{-1} \mathcal{F}_{X/V}$. Then, we have a short exact sequence

$$0 \to (i_{D/X}^{-1} \overline{\mathcal{F}}_{X/V}^{\mathrm{gp}})/(\overline{\mathcal{F}}_{D/V}^{\mathrm{gp}}) \to (\mathcal{P}^{\mathrm{gp}}|_D)/(\overline{\mathcal{F}}_{D/V}^{\mathrm{gp}}) \to (\mathcal{P}^{\mathrm{gp}}|_D)/(i_{D/X}^{-1} \overline{\mathcal{F}}_{X/V}^{\mathrm{gp}}) \to 0$$

of Abelian sheaves in the Zariski topology of D. We already know by Lemma 9.76 that the middle term is constant with stalk $\mathbb{Z}^{c+1}$, and that the right term is constant with stalk $\mathbb{Z}^{m+1}$. Therefore, the left term is constant with stalk $\mathbb{Z}^{c-m}$. Furthermore, we know by Lemma 9.75 that $\overline{\mathcal{F}}_{X/V}^{\mathrm{gp}} = \overline{\mathcal{M}}_X$ and $\overline{\mathcal{F}}_{D/V}^{\mathrm{gp}} = \overline{\mathcal{M}}_D$. Thus, the left-hand term is canonically isomorphic to $\Phi_{D/X} = \mathcal{M}_{[D/X]/D}^{\mathrm{gp}}$, and we obtain an isomorphism

$$\check{\chi}_{D/X/V} : \; \mho_{D/X} \otimes_{\mathbb{Z}} \left(i_{D/X}^{-1} \mho_{X/V} \right) \to \mho_{D/V}$$

from the exact sequence. Together with $\hat{\rho}_{D/X} : \omega_{X/0} \to \omega_{D/0} \otimes_{\mathbb{Z}} \mho_{D/X}$, this yields the desired map $\rho_{D/X}$ in Theorem 9.66 as the composition

$$\omega_{X/0} \otimes_{\mathbb{Z}} \mho_{X/V} \xrightarrow{\hat{\rho}_{D/X} \otimes \mathrm{id}} \omega_{D/0} \otimes_{\mathbb{Z}} \mho_{D/X} \otimes_{\mathbb{Z}} i_{D/X}^{-1} \mho_{X/V} \xrightarrow{\mathrm{id} \otimes \check{\chi}_{D/X/V}} \omega_{D/0} \otimes_{\mathbb{Z}} \mho_{D/V}.$$

Lemma 9.77 *We have* $\rho_{S/D} \circ \rho_{D/X} = \rho_{S/X}$ *for closed strata* $S \subset D \subset X$.

Proof Let $m = \mathrm{codim}(X, V)$, $c = \mathrm{codim}(D, V)$, and $e = \mathrm{codim}(S, V)$. First, we show that

$$
\begin{array}{ccc}
\mho_{S/D} \otimes_{\mathbb{Z}} \mho_{D/X}|_S \otimes_{\mathbb{Z}} \mho_{X/V}|_S & \xrightarrow{\mathrm{id} \otimes \check{\chi}_{D/X/V}} & \mho_{S/D} \otimes_{\mathbb{Z}} \mho_{D/V}|_S \\
\downarrow{\scriptstyle \hat{\chi}_{S/D/X} \otimes \mathrm{id}} & & \downarrow{\scriptstyle \check{\chi}_{S/D/V}} \\
\mho_{S/X} \otimes_{\mathbb{Z}} \mho_{X/V}|_S & \xrightarrow{\check{\chi}_{S/X/V}} & \mho_{S/V}
\end{array}
$$

commutes. Let $\alpha \in \mho_{S/D} = \bigwedge^{e-c} \Phi_{S/D}$, $\beta \in \mho_{D/X}|_S = \bigwedge^{c-m} \Phi_{D/X}|_S$, and $\gamma \in \mho_{X/V}|_S = \bigwedge^{m+1} \Phi_{X/V}|_S$. Note that we have inclusions

$$\Phi_{S/D} \subseteq \Phi_{S/X} \subseteq \Phi_{S/V}$$

and quotient maps

$$\Phi_{S/V} \to \Phi_{D/V}|_S \to \Phi_{X/V}|_S$$

as well as $\Phi_{S/X} \to \Phi_{D/X}|_S$. Let $\gamma' \in \bigwedge^{m+1} \Phi_{D/V}|_S$ be a lift of γ, and let $\gamma'' \in \bigwedge^{m+1} \Phi_{S/V}|_S$ be a lift of γ'. Let furthermore $\beta'' \in \bigwedge^{c-m} \Phi_{S/X}$ be a lift of β. Now the upper right composition is

$$\alpha \otimes \beta \otimes \gamma \mapsto \alpha \otimes (\beta \wedge \gamma') \mapsto \alpha \wedge \beta'' \wedge \gamma''$$

while the lower left composition is

$$\alpha \otimes \beta \otimes \gamma \mapsto (\alpha \wedge \beta'') \otimes \gamma \mapsto \alpha \wedge \beta'' \wedge \gamma''.$$

Therefore, the diagram commutes. The claim now follows from Lemma 9.74.   □

9.9.6   Concluding the Proof of Theorem 9.66

Let $[C_\bullet V]$ be the poset of closed strata, considered as a category. Lemma 9.77 shows that we have a finite diagram $E : [C_\bullet V]^{\mathrm{op}} \to \mathrm{Coh}(V)$ of coherent sheaves. The limit of this diagram is a sheaf of O_V-modules, which we denote by ω_V. Since pull-back along flat morphisms preserves finite limits, this limit can be computed étale locally. Furthermore, our construction of the diagram commutes with étale morphisms.

Lemma 9.78 *Let $(V, \mathcal{P}, \bar{\rho})$ be a simple toroidal crossing space, and let $\pi : U \to V$ be étale. Let $s \in \Gamma(U, \mathcal{LS})$ be a log smooth structure. Then $\omega_V|_U \cong \omega_{U_s/0}$, where U_s is U endowed with the log structure defined by s. In particular, ω_V is a line bundle.*

Proof Note that U is a simple toroidal crossing space. For notational simplicity, let us assume that $U = V$. Above, we have seen a restriction map $\rho_{X/V} : \omega_{V_s/0} \to \omega_{X/0} \otimes_{\mathbb{Z}} \mho_{X/V}$ which induces an isomorphism after pull-back to X. With a similar argument as in Lemma 9.74 and Lemma 9.77, one may show that $\rho_{D/X} \circ \rho_{X/V} = \rho_{D/V}$. Therefore, we have an induced map $\omega_{V_s/0} \to \omega_V$.

Since $S_0 \to \mathbf{0}$ can be made ideally log smooth, and since $V_s \to S_0$ is log smooth, $\Omega^1_{V_s/\mathbf{0}}$ is locally free, and $\omega_{V_s/\mathbf{0}}$ is a line bundle. After shrinking V, we can assume that we have a trivializing section $\phi_V \in \omega_{V_s/\mathbf{0}}$. Since $i^*_{X/V}\omega_{V_s/\mathbf{0}} \to \omega_{X/\mathbf{0}} \otimes_{\mathbb{Z}} \mho_{X/V}$ is an isomorphism, $\phi_X = i^*_{X/V}(\phi_V)$ trivializes $\omega_{X/\mathbf{0}} \otimes_{\mathbb{Z}} \mho_{X/V}$. In other words, the diagram $E(X) = \omega_{X/\mathbf{0}} \otimes_{\mathbb{Z}} \mho_{X/V}$ is locally isomorphic to the diagram $F(X) = O_X$. Thus, their limits as well as the comparison maps $\omega_{V_s/\mathbf{0}} \to \omega_V$ and $O_V \to \lim_X(O_X)$ are locally isomorphic. Lemma 9.57 shows that the latter comparison map is an isomorphism, so also $\omega_{V_s/\mathbf{0}} \to \omega_V$ is an isomorphism.   □

Corollary 9.79 *In the situation of Lemma 9.78, we have a canonical isomorphism $\omega_V|_U \cong \omega_{U_s/S_0}$.*

Proof The short exact sequence

$$0 \to (f_0 \circ \pi)^* \Omega^1_{S_0/\mathbf{0}} \to \Omega^1_{U_s/\mathbf{0}} \to \Omega^1_{U_s/S_0} \to 0$$

induces an isomorphism $(f_0 \circ \pi)^* \omega_{S_0/\mathbf{0}} \otimes \omega_{U_s/S_0} \cong \omega_{U_s/\mathbf{0}}$. Now $\omega_{S_0/\mathbf{0}}$ has a canonical generator $\delta(1) = i^* \frac{dt}{t}$, where $\delta \colon \mathcal{M}_{S_0} \to \Omega^1_{S_0/\mathbf{0}}$ is the log part of the universal derivation, and $i \colon S_0 \to (\mathbb{A}^1_t | \{0\})$ is the inclusion. $\square$

Corollary 9.80 *Let $(V, \mathcal{P}, \bar\rho)$ be a simple toroidal crossing space. Then, we have an isomorphism $\omega_V \cong \omega^\circ_V$ with the normalized dualizing sheaf.*

Proof Let $s \in \Gamma(U, \mathcal{LS}_V)$ be a log smooth structure. By Tsuji's comparison result in Proposition 7.70, we have an isomorphism $\vartheta_{U_s/S_0} \colon \omega_{U_s/S_0} \cong \omega^\circ_U$ which coincides with the Verdier isomorphism on the strict locus.

The strict locus is $U^\circ = \bigcup_{X \in [C_0 U]} X^\circ$. On the interior X° of a top-dimensional stratum, the log scheme $(X | \partial X)$ carries the trivial log structure, so we have an identification $\omega_{X/\mathbf{0}}|_{X^\circ} \cong \Omega^d_{\underline{X}^\circ/S_0}$. Furthermore, for $X \in [C_0 U]$, we have a canonical trivialization $\mho_{X/V} \cong \mathbb{Z}$ given by taking the generator in $\mathcal{P}|_{X^\circ} = \mathbb{N}$. This yields an identification $\omega_V|_{U^\circ} \cong \Omega^d_{\underline{U}^\circ}$. One may now show that, after restriction to U°, the isomorphism $\omega_V|_U \cong \omega_{U_s/S_0}$ fits into a commutative diagram

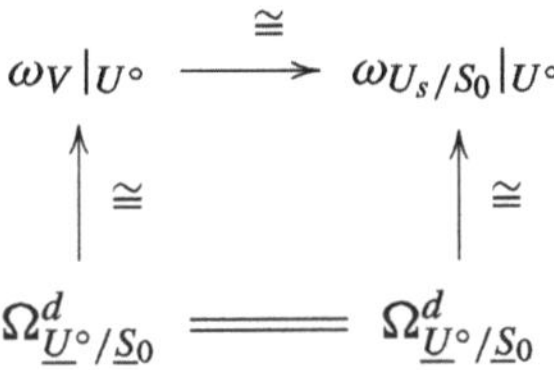

Therefore, the composition $\omega_V|_U \to \omega^\circ_U$ coincides with the Verdier isomorphism after restriction to U° once we use the left vertical identification from the diagram. In particular, this isomorphism is independent of the precise $s \in \Gamma(U, \mathcal{LS}_V)$, and they glue to a global isomorphism $\omega_V \cong \omega^\circ_V$. $\square$

9.10 Toroidal Crossing Degenerations to V

Let R be a discrete valuation $\mathbf{k}$-algebra with residue field $\mathbf{k}$, and $S = \operatorname{Spec} R$. Let $f \colon X \to S$ be a toroidal crossing degeneration. Example 9.37 shows that $V = X_0$ endowed with $\mathcal{P} := \overline{\mathcal{M}}_{X_0}$ and the corresponding global section $\bar\rho \in \mathcal{P}$ is a toroidal crossing space. Conversely, when $(V, \mathcal{P}, \bar\rho)$ is a toroidal crossing space, and $f \colon X \to S$ is a toroidal crossing degeneration, and we have an isomorphism $V \cong X_0$ of schemes, then there is at most one map $q \colon \mathcal{M}_{X_0} \to \mathcal{P}$ which turns $(X_0, \mathcal{M}_{X_0}) \to S_0$ into a section of $\mathcal{LS}_V$. Namely, any two such maps differ by an automorphism of $\mathcal{P}$, and when we analyze the supports of sections of $\mathcal{P}$ in the local models, we find that the identity is the only automorphism of $\mathcal{P}$ as a sheaf of monoids. Thus, the following definition makes sense.

Definition 9.81 (Toroidal Crossing Degenerations II) Let $(V, \mathcal{P}, \bar{\rho})$ be a toroidal crossing space. Then a *toroidal crossing degeneration to* V is a toroidal crossing degeneration $f : X \to S$ together with an isomorphism $V \cong X_0$ as schemes such that there is a map $q : \mathcal{M}_{X_0} \to \mathcal{P}$ which turns the log morphism $f_0 : X_0 \to S_0$ into a section of $\mathcal{LS}_V$.

This is equivalent to that $(\overline{\mathcal{M}}_{X_0}, f^*(1_{S_0}))$ is locally isomorphic to $(\mathcal{P}, \bar{\rho})$ as a sheaf of monoids with a distinguished section. Namely, since $(\mathcal{P}, \bar{\rho})$ has no nontrivial automorphisms, the local isomorphisms glue to a global isomorphism, and then we obtain a section of $s \in \widehat{\mathcal{LS}}_V$. The argument which shows that $\mathcal{LS}_V \subseteq \widehat{\mathcal{LS}}_V$ is well-defined now shows that in fact $s \in \mathcal{LS}_V$.

As a variant of the fact that $\mathcal{LS}_V$ classifies precisely log smooth, saturated, and vertical morphisms to S_0, we see that every section $s \in \mathcal{LS}_V$ arises locally from a toroidal crossing degeneration—at least when R is complete. By the Cohen structure theorem, we have $R = \mathbf{k}[\![t]\!]$ in this case.

Lemma 9.82 *Let* $R = \mathbf{k}[\![t]\!]$, *and let* $(V, \mathcal{P}, \bar{\rho})$ *be a toroidal crossing space. Let* $s \in \mathcal{LS}_V$ *be a section. Then every geometric point* $\bar{v} \in V$ *has an étale neighborhood* W *such that there is a toroidal crossing degeneration* $f : X \to S$ *to* $V|_W$ *which induces* $s \in \mathcal{LS}_V$.

Proof Let $f_0 : X_0 \to S_0$ be the log morphism corresponding to $s \in \mathcal{LS}_V$. Then $f_0 : X_0 \to S_0$ is log smooth, saturated, and vertical. Every geometric point $\bar{v} \in V$ admits a strict étale neighborhood $g_0 : W_0 \to X_0$ such that there is a strict étale morphism

$$h_0 : \quad W_0 \to V(\sigma) \times \mathbb{A}^{d-r} =: L_0.$$

By shrinking W_0, we can assume that it is affine. Lemma 17.15 shows that there is an étale morphism of finite type $h : W \to L$ of schemes, where $L := U_S(\sigma) \times \mathbb{A}^{d-r}$ is the base change of the local model to $S = \operatorname{Spec} \mathbb{C}[\![t]\!]$. Now $W \to S$ is a toroidal crossing degeneration, and the divisorial log structure defined by $t = 0$ is the given log structure on W_0 after restriction to the central fiber. $\qquad\square$

9.11 The Local Description of $\mathcal{LS}_V$

In [124, Thm. 3.22], Gross and Siebert give a description of the sheaf $\mathcal{LS}_V$ for the prototypes $(V(\sigma), \mathcal{P}_\sigma, \bar{\rho}_\sigma)$ of toroidal crossing spaces, which we briefly explain here. Let $\sigma \subseteq M_{\mathbb{R}}$ be a full-dimensional lattice polytope as in Construction 9.2, and let $V = (V, \mathcal{P}, \bar{\rho}) = V(\sigma) \times \mathbb{G}_m^{d-r}$ be the toroidal crossing space which we obtain from the prototype $(V(\sigma), \mathcal{P}_\sigma, \bar{\rho}_\sigma)$ by pulling back the structure along the smooth projection $V \to V(\sigma)$. If $\tau \subseteq \sigma$ is a face, then the dual face $\check{\tau} \subseteq N_{\mathbb{R}}$ defines an irreducible closed toric stratum $V_\tau \subset V$ with $V_\tau \cong \operatorname{Spec} \mathbf{k}[(\check{\tau} \cap N) \oplus \mathbb{Z}^{d-r}]$; these strata are glued along lower-dimensional strata. The irreducible components of V are $\{V_v \mid v \in \sigma \text{ is a vertex}\}$. For each face $\tau \subseteq \sigma$, we denote by $\tau^{\parallel} \subseteq M$ the intersection of the tangent space in $M_\mathbb{R}$ at τ with the lattice M. For edges $\omega \subseteq \sigma$, this tangent space is one-dimensional, i.e., $\omega^{\parallel} \cong \mathbb{Z}$; thus, we can (and do) choose primitive generators $d_\omega \in \omega^{\parallel}$. We get a choice of vertices v_ω^+ and v_ω^- of ω (such that d_ω points from v_ω^- to v_ω^+) and an orientation on the edge ω. For each two-dimensional face $\tau \subseteq \sigma$, we choose a *sign vector*

$$\epsilon_\tau : \ \{\text{edges of } \sigma\} \to \{-1, 0, 1\}$$

such that $\epsilon_\tau(\omega) = 0$ if and only if $\omega \nsubseteq \tau$, and $\sum_\omega \epsilon_\tau(\omega)\omega$ is an oriented boundary of τ. With these choices, Gross–Siebert finds:

Theorem 9.83 ([124, Thm. 3.22]) *The sheaf $\mathcal{LS}_V$ is isomorphic to the subsheaf of*

$$\bigoplus_{\dim \omega = 1} O_{V_\omega}^*$$

defined as follows. If $U \subseteq V$ is an open subset, then $\Gamma(U, \mathcal{LS}_V)$ consists of $(f_\omega)_\omega$ such that, for every two-dimensional face τ of σ, we have

$$\prod_{\dim \omega = 1} d_\omega \otimes f_\omega^{\epsilon_\tau(\omega)}|_{V_\tau} = 1 \in M \otimes_\mathbb{Z} \Gamma(U, O_{V_\tau}^*).$$

The functions $f_\omega \in O_{V_\omega}^*$ in the theorem are referred to as *slab functions*. Gross–Siebert finds in [124, Thm. 3.28] that, in their setting, the local descriptions can be glued, and the slab functions are sections of global line bundles $\mathcal{N}_\omega$ on the (normalizations of) strata of codimension 1 of $X_0(B, \mathcal{P}, s)$.

Remark 9.84 In the general case, the global nature of the line bundles $\mathcal{N}_\omega$ is not yet entirely clear. In the recent preprint [55], Corti and Ruddat define, for each closed codimension-1 stratum $\rho \in [\mathcal{S}_1 V]$ of a simple toroidal crossing space, a *wall bundle* $\mathcal{L}_\rho$ on ρ, which will be the global analog of $\mathcal{N}_\omega$ above. Then Corti and Ruddat obtain a coordinate-free global version of Theorem 9.83. This will allow us to construct sections of $\mathcal{LS}_V$ more easily in a general global setting because the $\mathcal{L}_\rho$ are coherent sheaves, as opposed to the more complicated $\mathcal{E}xt^1(\mathcal{P}^{\mathrm{gp}}/\bar{\rho}, O_V^*)$. An important application is the construction of resolutions of log singularities. Given a toroidal crossing space $(V, \mathcal{P}, \bar{\rho})$ together with a section $s \in \Gamma(V \setminus Z, \mathcal{LS}_V)$, one wishes to construct a proper birational map $\pi : Y \to V$ such that Y carries the structure of a saturated log smooth log morphism, and such that $\pi : Y \to V$ is an isomorphism over $V \setminus Z$. Given a good understanding of the behavior of $\mathcal{LS}_V$, one can construct a resolution as follows: Let us assume that $s \in \Gamma(V \setminus Z, \mathcal{LS}_V)$ is given by functions $f_\rho \in \mathcal{L}_\rho$ which vanish precisely in Z. Then one constructs a toroidal crossing space Y and a proper birational map $\pi : Y \to V$ with the following properties:

- $Y \setminus \pi^{-1}(Z) \to V \setminus Z$ is an isomorphism not only of schemes but of toroidal crossing spaces (while we have on the whole of Y only a map of schemes);
- each open stratum $S^\circ \in [\mathcal{S}_m Y]$ surjects birationally onto an open stratum of V, and this defines a one-to-one correspondence of open strata;
- to each $\rho \in [C_1 V]$ corresponds a stratum $\tilde{\rho} \in [C_1 Y]$, and we have $\rho \setminus Z = \tilde{\rho} \setminus \pi^{-1}(Z)$ as well as $\mathcal{L}_\rho|_{V \setminus Z} = \mathcal{L}_{\tilde{\rho}}|_{Y \setminus \pi^{-1}(Z)}$ via π; then the slab function $f_\rho|_{\rho \setminus Z} = f_\rho|_{\tilde{\rho} \setminus \pi^{-1}(Z)}$, which has a zero in $\mathcal{L}_\rho$ on V, extends to a nowhere vanishing section of $\mathcal{L}_{\tilde{\rho}}$ on Y.

At least in good situations, this should be possible and yields a section of $\mathcal{L}\mathcal{S}_Y$ which extends s from $V \setminus Z = Y \setminus \pi^{-1}(Z)$ to the whole of Y.

For applications in deformation theory, one sometimes wants to consider *crepant* resolutions of log singularities, often of affine log singularities, because then we have $\Theta^1_{Y/S_0} \cong \Omega^{d-1}_{Y/S_0}$. When insisting that $\pi : Y \to V$ should be crepant, examples suggest (received from Corti and Ruddat in personal communication) that it is not always possible to have that Y is a toroidal crossing space with a section of $\mathcal{L}\mathcal{S}_Y$. Instead, some singularity like $\{xy = 0\} \subset \frac{1}{r}(1, -1, a, -a)$ will persist already on the level of the underlying space of Y. This has lead Corti and Ruddat to investigate *generically* toroidal crossing spaces, which are (more or less) spaces with a pre–ghost structure $(V, \mathcal{P}, \bar{\rho})$ such that the local models of toroidal crossing spaces do not necessarily exist everywhere but only around the generic points of the strata. Here, one assumes that a stratification is a priori given, which coincides with the stratification from the toroidal crossing space structure where both are defined. Corti and Ruddat impose a compatibility condition, called *viability*, which is sufficient to construct the wall bundles $\mathcal{L}_\rho$, and to define a sheaf $\mathcal{L}\mathcal{S}_V$ which classifies some log morphisms to the standard log point S_0. Due to the possibility of coherent log singularities that are already build in into $(V, \mathcal{P}, \bar{\rho})$, this approach is far more general and flexible than the classical notion of toroidal crossing spaces which we discuss in this chapter, allowing to construct log structures on spaces that do not admit the structure of a toroidal crossing space in our sense. $\Diamond$

Since $\mathcal{U}_1 V \cap V_\omega \subseteq V_\omega$ is scheme-theoretically dense, we deduce from Theorem 9.83:

Corollary 9.85 *Let $(V, \mathcal{P}, \bar{\rho})$ be a toroidal crossing space. Then, for each (Zariski) open subset $U \subseteq V$, the restriction map $\Gamma(U, \mathcal{L}\mathcal{S}_V) \to \Gamma(U \cap \mathcal{U}_1 V, \mathcal{L}\mathcal{S}_V)$ is injective.*

9.12 The Map $\mathcal{L}\mathcal{S}_V \to \mathcal{T}^1_V$

In Chap. 9.11, we gave Gross–Siebert's local description of $\mathcal{L}\mathcal{S}_V$, whose globalization is not yet completely clear. In this chapter, we explain another approach to a global description of $\mathcal{LS}_V$, which is particularly useful for normal crossing spaces. More generally, to obtain useful information from this approach, we need the toroidal crossing space to be tame as defined above. The reference for this approach is [77]; the ideas trace back to Schröer–Siebert's work [253].

Let $(V, \mathcal{P}, \bar{\rho})$ be a toroidal crossing space, and let $U \subseteq V$ be an affine open subset. Let $(\mathcal{M}, \alpha, q, \rho)$ be a log morphism that represents a class in $\Gamma(U, \mathcal{LS}_V)$; thus, we have a log smooth morphism $U \to S_0$. We set $S_1 := \mathrm{Spec}(\mathbb{N} \to \mathbf{k}[t]/(t^2))$ where $1 \mapsto t$; it is a first order log thickening of S_0. By log smooth deformation theory, there is—up to non-unique isomorphism—a unique log smooth deformation $U_1 \to S_1$. In particular, forgetting the log structure, we get a closed embedding $i : \underline{U} \subseteq \underline{U}_1$ of schemes; it gives rise to an extension

$$0 \to O_U \to i^*\Omega^1_{\underline{U}_1} \to \Omega^1_{\underline{U}} \to 0 \qquad (9.4)$$

of sheaves on U. Here, $\Omega^1_{\underline{U}_1}$ are the *absolute* Kähler differential forms of $\underline{U}_1$; the left inclusion is given by $1 \mapsto i^*dt$. The isomorphism class of the extension is an element in $\Gamma(U, \mathcal{E}xt^1(\Omega^1_{\underline{U}}, O_U))$; this yields a map

$$\eta_V : \quad \mathcal{LS}_V \to \mathcal{E}xt^1(\Omega^1_{\underline{V}}, O_V) =: \mathcal{T}^1_V$$

of sheaves of sets. Because the log smooth deformation $U_1 \to S_1$ is only unique up to isomorphism, so is the extension; thus, a global section of $\mathcal{LS}_V$ gives rise to a section of $\mathcal{E}xt^1(\Omega^1_{\underline{V}}, O_V)$, but not to a global extension in $\mathrm{Ext}^1(\Omega^1_{\underline{V}}, O_V)$.

The target $\mathcal{T}^1_V$ is a coherent sheaf, but the source $\mathcal{LS}_V$ has, up to now, no similar structure. However, we easily endow it with an O^*_V-action by setting

$$\lambda \cdot [(\mathcal{M}, \alpha, q, \rho)] := [(\mathcal{M}, \alpha, q, \lambda^{-1}\rho)]$$

—i.e., we change $\rho \in \mathcal{M}$ to $\lambda^{-1}\rho$ without changing its image $\bar{\rho} \in \mathcal{P}$. Note that we invert the function $\lambda \in O^*_V$. The target $\mathcal{T}^1_V$ has a canonical O^*_V-action as well—it is a coherent sheaf.

Proposition 9.86 ([77, Prop. 5.2]) *The map* $\eta_V : \mathcal{LS}_V \to \mathcal{T}^1_V$ *is* O^*_V*-equivariant.*

Remark 9.87 At the level of $\bar{\iota}_{V,\bar{v}} : \widehat{\mathcal{LS}}_{V,\bar{v}} \to \mathcal{E}xt^1(\mathcal{P}^{\mathrm{gp}}/\bar{\rho}, O^*_V)_{\bar{v}}$, we see the $O^*_{V,\bar{v}}$-action as follows: If $(h_p)_p$ represents a germ $M \in \mathcal{LS}_{V,\bar{v}}$ and $\phi : P^{\mathrm{gp}} \to O^*_{V,\bar{v}}$ is a homomorphism, then $(\phi(p)h_p)_p$ represents $\phi(\bar{\rho})^{-1} \cdot M$. $\Diamond$

In order to be useful to describe sections of $\mathcal{LS}_V$, we want the map $\eta_V : \mathcal{LS}_V \to \mathcal{T}^1_V$ to be injective; however, this is not always the case.

Example 9.88 This is an example of a one-dimensional toroidal crossing space such that η_V is not injective. Let $N = \mathbb{Z}$ and $\sigma = [0, 2] \subseteq \mathbb{R} = N_{\mathbb{R}}$. We denote the associated toroidal crossing space by $(V, \mathcal{P}, \bar{\rho})$; it consists of two intersecting lines L_x and L_y. In the interiors of the lines, the stalk of $\mathcal{P}$ is $\mathbb{N}$; in the intersection point, we have $\mathcal{P}_0 = \langle(-1, 2), (1, 0)\rangle \subseteq \mathbb{Z}^2$. For every $\lambda \in \mathbf{k}^*$, the log smooth family

$$\mathrm{Spec}\,\mathbf{k}[x, y, t]/(xy - \lambda t^2) \to \mathrm{Spec}\,\mathbf{k}[t], \quad t \mapsto t,$$

induces a section $s_\lambda \in \Gamma(V, \mathcal{LS}_V)$. If $\lambda \neq \mu$, then also $s_\lambda \neq s_\mu$ However, we have

$$\eta_V(s_\lambda) = \eta_V(s_\mu) = 0$$

because the unique first order log smooth deformation has $\mathrm{Spec}\,\mathbf{k}[x, y, t]/(xy, t^2)$ as underlying space. Thus $\eta_V \colon \mathcal{LS}_V \to \mathcal{T}_V^1$ is not injective; in fact, it is the zero map because every section of $\mathcal{LS}_V$ is of the form s_λ. $\Diamond$

In order to find a criterion for injectivity of $\eta_V \colon \mathcal{LS}_V \to \mathcal{T}_V^1$, we consider (at a point $\bar{v} \in V$) the forgetful map

$$\mathcal{LS}_{V,\bar{v}} \subseteq \widehat{\mathcal{LS}}_{V,\bar{v}} \to \widehat{\mathcal{L}}_{V,\bar{v}};$$

it is invariant under the $O_{V,\bar{v}}^*$-action since this action does not affect the underlying log structure. Conversely, when two germs $M, M' \in \mathcal{LS}_{V,\bar{v}}$ have the same underlying log structure, then they differ only by an invertible function. Thus, the fibers of the forgetful map are precisely the $O_{V,\bar{v}}^*$-orbits. In general, there might be many orbits, but, for $\mathcal{P}_{\bar{v}} \cong \mathbb{N}^r$, the situation is particularly simple.

Lemma 9.89 *Let $(V, \mathcal{P}, \bar{\rho})$ be a toroidal crossing space, and let $\bar{v} \in V$ be a point with $\mathcal{P}_{\bar{v}} \cong \mathbb{N}^r$. Then $O_{V,\bar{v}}^*$ acts transitively on $\mathcal{LS}_{V,\bar{v}}$.*

Proof The statement is in [77, Lemma 5.2]. In a nutshell, this holds because $\mathbb{N}^r$ is free, so homomorphisms from $\mathbb{N}^r$ can be constructed by specifying values on a basis. $\square$

The stalk of the ghost sheaf in Example 9.88 is not free; it is in fact $\mathcal{P}_0 = \langle(-1, 2), (1, 0)\rangle \subseteq \mathbb{Z}^2$. But if instead $\mathcal{P}_{\bar{v}} \cong \mathbb{N}^r$, then the map $\eta_V \colon \mathcal{LS}_V \to \mathcal{T}_V^1$ turns out to be injective.

Lemma 9.90 ([77, Lemma 5.3]) *Let $(V, \mathcal{P}, \bar{\rho})$ be a toroidal crossing space, and let $\bar{v} \in V$ be a geometric point with $\mathcal{P}_{\bar{v}} \cong \mathbb{N}^r$. Then:*

(a) *For $M \in \mathcal{LS}_{V,\bar{v}}$, the map $\mu_M \colon O_{V,\bar{v}} \to \mathcal{T}_{V,\bar{v}}^1$ is surjective.*

(b) *The map $\eta_{V,\bar{v}} \colon \mathcal{LS}_{V,\bar{v}} \to \mathcal{T}_{V,\bar{v}}^1$ is injective.*

(c) *The image of $\eta_{V,\bar{v}}$ is $(\mathcal{T}_{V,\bar{v}}^1)^* \subseteq \mathcal{T}_{V,\bar{v}}^1$, the elements that generate $\mathcal{T}_{V,\bar{v}}^1$ as an $O_{V,\bar{v}}$-module.*

Injectivity is not restricted to toroidal crossing spaces with $\mathcal{P}_{\bar{v}} \cong \mathbb{N}^r$ for all $\bar{v} \in V$; it extends to tame toroidal crossing spaces as defined above.

Theorem 9.91 *If $(V, \mathcal{P}, \bar{\rho})$ is a* tame *toroidal crossing space, then $\eta_V \colon \mathcal{LS}_V \to \mathcal{T}_V^1$ is injective.*

Proof As in [77], this follows from Corollary 9.85. $\square$

9.13 Generically Log Smooth Families from $\mathcal{LS}_V$

Given a projective or just proper toroidal crossing space $(V, \mathcal{P}, \bar{\rho})$, there is often no global section of $\mathcal{LS}_V$ to endow it with a global log structure, cf. [124]. Even if there is one, it is often enlightening for the (flat) deformation theory of V to allow certain log singularities. In our theory, we achieve this by selecting a *log singular locus* $Z \subset V$ of codimension ≥ 2 and a section $s \in \Gamma(V \setminus Z, \mathcal{LS}_V)$. Since V is Cohen–Macaulay and thus satisfies Serre's condition (S_2), this gives rise to a generically log smooth family $f_0 \colon (X_0, U_0) \to S_0$. This family is particularly well-behaved if the following condition is satisfied. Our main motivation for this condition is that it makes the deformation theory of f_0 more tractable in the setup of unisingular deformations in Sect. 14.3.

Definition 9.92 (Well-adjusted Triples) A *well-adjusted triple* (V, Z, s) consists of a toroidal crossing space $(V, \mathcal{P}, \bar{\rho})$ which is pure of some dimension d, a closed subset $Z \subset C_1 V \subset V$ such that the intersection $Z \cap S$ with any closed stratum $S \in [C_m V]$ is properly contained in S, and a section $s \in \Gamma(V \setminus Z, \mathcal{LS}_V)$.

Remark 9.93 We imagine Z to show inadequate behavior, claiming space that it does not deserve, if this condition is violated; hence the name. $\qquad\qquad \diamond$

The codimension condition is automatic if (V, Z, s) is well-adjusted; Z may be empty or non-reduced. Note that being well-adjusted is an étale local property: When $\bigcup_i (V_i, Z_i, s_i) \to (V, Z, s)$ is an étale cover, then (V, Z, s) is well-adjusted if and only if each (V_i, Z_i, s_i) is well-adjusted. The first property which makes well-adjusted triples nice is the following.

Lemma 9.94 *Let (V, Z, s) be a well-adjusted triple, and let $j \colon V \setminus Z \to V$ be the inclusion. Then we have $\mathcal{P} = j_*(\mathcal{P}|_{V \setminus Z})$.*

Proof Let $W \to V$ be an étale open. We show that $\Gamma(W, \mathcal{P}) \to \Gamma(W \setminus Z, \mathcal{P})$ is bijective. By the definition of a toroidal crossing space, every point $w \in W$ has an étale neighborhood $\pi_w \colon W_w \to W$ with an element $u_w \in W_w$ with $\pi_w(u_w) = w$ such that we have a full-dimensional lattice polytope σ_w and a smooth map $\tau_w \colon W_w \to V(\sigma_w)$ with $\tau_w(u_w) = 0$.

The preimage of every closed stratum T of $V(\sigma_w)$ is a normal closed subscheme $\tau^{-1}(T)$ of W_w. Therefore, there is only one connected component C° of $\tau_w^{-1}(T^\circ)$ whose closure C contains u_w. Namely, as a full-dimensional closed subscheme of a normal scheme, C is a connected component of $\tau_w^{-1}(T)$, and only one connected component of $\tau_w^{-1}(T)$ contains u_w. The union of the remaining connected components of $\tau_w^{-1}(T)$ is a closed subset of V. Since $V(\sigma)$ has only finitely many strata, we can remove them for every closed stratum T of $V(\sigma_w)$ and still obtain an open subset of W_w which contains u_w. In other words, by shrinking W_w, we can assume that $\tau_w^{-1}(T)$ is always connected, and that we have a one-to-one correspondence between open strata of W_w and open strata of $V(\sigma_w)$.

On W_w, we can compute $\Gamma(W_w, \mathcal{P})$ and $\Gamma(W_w \setminus Z, \mathcal{P})$ in the Zariski topology because $\mathcal{P}$ is nothing but the étale sheaf associated with $\tau_w^{-1}\mathcal{P}_{V(\sigma_w)}$. For every open stratum T of $V(\sigma_w)$, let η_T be the generic point of the corresponding stratum of W_w. Moreover, let η be the generic point of the deepest stratum, corresponding to $\{0\} \subset V(\sigma_w)$. Since Z does not contain any stratum, we have $\eta_T \in W_w \setminus Z$ for every T, and therefore a map $\Gamma(W_w \setminus Z, \mathcal{P}) \to \bigoplus_T \mathcal{P}_{\eta_T}$. Since $\mathcal{P}$ is constant on every $\tau_w^{-1}(T^\circ)$, this map is injective. Namely, when $a \neq b$ for $a, b \in \Gamma(W_w \setminus Z, \mathcal{P})$, then there is some $y \in W_w$ with $a_y \neq b_y$, but there is some T such that $\mathcal{P}_y \to \mathcal{P}_{\eta_T}$ is an isomorphism. Similarly, $\Gamma(W_w, \mathcal{P}) \to \bigoplus_T \mathcal{P}_{\eta_T}$ is an isomorphism, and we find that $\Gamma(W_w, \mathcal{P}) \to \Gamma(W_w \setminus Z, \mathcal{P})$ is injective. Now $\Gamma(W, \mathcal{P}) \to \prod_w \Gamma(W_w, \mathcal{P})$ is injective so that $\Gamma(W, \mathcal{P}) \to \Gamma(W \setminus Z, \mathcal{P})$ must be injective as well.

Let $U \subseteq W_w$ be a Zariski open subset with $u_w \in U$. Then $0 \in \tau_w(U) \subseteq V(\sigma_w)$ is open so that $\eta_T \in U$ for all T. Therefore, already $\Gamma(U, \mathcal{P}) \to \mathcal{P}_\eta$ is injective. Since $\mathcal{P}_\eta \cong \mathcal{P}_w$ is the colimit of the $\Gamma(U, \mathcal{P})$, and since $\mathcal{P}_\eta$ is finitely generated, we can find some $u_w \in U \subseteq W_w$ such that $\Gamma(U, \mathcal{P}) \to \mathcal{P}_\eta$ is an isomorphism. Since also $\Gamma(U \setminus Z, \mathcal{P}) \to \mathcal{P}_\eta$ is injective, we find that the latter map is an isomorphism, and so is $\Gamma(U, \mathcal{P}) \to \Gamma(U \setminus Z, \mathcal{P})$. Thus, by further shrinking each W_w, we can assume that $\Gamma(W_w, \mathcal{P}) \to \Gamma(W_w \setminus Z, \mathcal{P})$ is bijective. Now, we use this open cover to compute $\Gamma(W, \mathcal{P}) \to \Gamma(W \setminus Z, \mathcal{P})$ by means of the sheaf condition. Bijectivity for each W_w and injectivity for each intersection of two opens yield that $\Gamma(W, \mathcal{P}) \to \Gamma(W \setminus Z, \mathcal{P})$ is bijective. $\qquad\square$

This allows us to reconstruct $(V, \mathcal{P}, \bar{\rho})$ from $f_0 \colon (X_0, U_0) \to S_0$ once we know that f_0 comes from a well-adjusted triple.

In the context of unisingular deformations, we will see with Lemma 14.21 another important property of well-adjusted triples. We also need the following result.

Lemma 9.95 *Let $f_0 \colon (X_0, U_0) \to S_0$ be the generically log smooth family associated with a well-adjusted triple (V, Z, s). Then $f_0 \colon (X_0, U_0) \to S_0$ is log Gorenstein.*

Proof The underlying scheme V is Gorenstein, so its normalized dualizing sheaf ω_V° is a line bundle. Since the log morphism $f_0 \colon U_0 \to S_0$ is vertical, we have $\mathcal{W}_{U_0/S_0}^d \cong \omega_V^\circ|_{U_0}$ by Proposition 7.70. Thus, $\mathcal{W}_{X_0/S_0}^d \cong \omega_V^\circ$ is a line bundle. $\qquad\square$

Part III
Global Deformation Theory

Chapter 10
Generically Log Smooth Deformations

We introduce deformations of generically log smooth families and describe their infinitesimal automorphisms. This theory works well only if the generically log smooth families under consideration have the base change property. As a variant to deal with situations where we do not have the base change property, we introduce enhanced generically log smooth families. For both the plain case and the enhanced case, we introduce systems of deformations. They are a device to fix the local deformation theory and thus simplify the deformation problem.

10.1 Definition of Generically Log Smooth Deformations

As a general framework to study deformations of generically log smooth families, we introduce the following definition.

Definition 10.1 Let $S_0 = \mathrm{Spec}(Q \to \mathbf{k})$ be a log point, and let $f_0\colon (X_0, U_0) \to S_0$ be a generically log smooth family. Let $A \in \mathbf{Art}_Q$. Then a *generically log smooth deformation* over S_A is a Cartesian diagram

$$
\begin{array}{ccc}
(X_0, U_0) & \xrightarrow{\;i_A\;} & (X_A, U_A) \\
\downarrow{\scriptstyle f_0} & & \downarrow{\scriptstyle f_A} \\
S_0 & \longrightarrow & S_A
\end{array}
$$

S. Felten, *Global Logarithmic Deformation Theory*, Lecture Notes
in Mathematics 2373, https://doi.org/10.1007/978-3-031-98751-9_10

of partial log schemes such that $f_A\colon (X_A, U_A) \to S_A$ is a generically log smooth family. In particular, we have $i^{-1}(U_A) = U_0$. A *morphism* of generically log smooth deformations is a commutative diagram

$$
\begin{array}{ccccc}
(X_0, U_0) & \xrightarrow{\ i_B\ } & (X_B, U_B) & \xrightarrow{\ c\ } & (X_{B'}, U_{B'}) \\
\downarrow{\scriptstyle f_0} & & \downarrow{\scriptstyle f_B} & & \downarrow{\scriptstyle f_{B'}} \\
S_0 & \xrightarrow{\hspace{2cm}} & S_B & \xrightarrow{\ S_\phi\ } & S_{B'}
\end{array}
$$

of partial log schemes where $S_\phi\colon S_B \to S_{B'}$ is induced by a morphism $\phi\colon B' \to B$ in $\mathbf{Art}_Q$, where the right-hand square is Cartesian, and where $c \circ i_B = i_{B'}$. Isomorphism classes of generically log smooth deformations form a functor of Artin rings

$$
\mathrm{LD}^{\mathrm{gen}}_{X_0/S_0}\colon \quad \mathbf{Art}_Q \to \mathbf{Set}.
$$

Lemma 10.2 *The functor* $\mathrm{LD}^{\mathrm{gen}}_{X_0/S_0}$ *is a neat deformation functor, i.e., it satisfies* (H_0), (H_1), (H_2), *and* (H_2^+) *defined in Appendix A.*

Proof This is completely analogous to the proof that the log smooth deformation functor LD is a deformation functor, see [160, Thm. 8.7]. As in the log smooth case, the verification of Condition (H_2^+) from Definition A.9 is analogous to the verification of Condition (H_2). $\qquad\square$

To give an impression of this functor, let us provide the following fact.

Lemma 10.3 *For every* $A \in \mathbf{Art}_Q$, *we have a Cartesian square*

$$
\begin{array}{ccc}
\mathrm{LD}^{\mathrm{gen}}_{X_0/S_0}(A) & \longrightarrow & \mathrm{LD}_{U_0/S_0}(A) \\
\downarrow & & \downarrow \\
\mathrm{Def}_{X_0}(A) & \longrightarrow & \mathrm{Def}_{U_0}(A)
\end{array}
$$

of sets with the log smooth deformation functor LD_{U_0/S_0} *and the flat deformation functor* Def_{X_0}.

Proof Let $P(A)$ be the fiber product; its elements consist of a flat deformation $X_A \to S_A$ and a log smooth deformation $U_A \to S_A$ such that there is an isomorphism $O_{X_A}|_{U_0} \cong O_{U_A}$. By choosing one such isomorphism, we see that the canonical map $\mathrm{LD}^{\mathrm{gen}}_{X_0/S_0}(A) \to P(A)$ is surjective. For injectivity, note that the canonical map

$$
\rho\colon \quad \mathrm{LD}^{\mathrm{gen}}_{X_0/S_0}(A) \to \mathrm{LD}_{U_0/S_0}(A)
$$

is injective because we can extend an isomorphism of two structure sheaves from
U_0 to X_0. $\square$

10.2 Infinitesimal Automorphisms

We generalize the treatment of infinitesimal automorphisms in [76] from log smooth
deformations to generically log smooth deformations. Let $B' \to B$ be a surjection
in $\mathbf{Art}_Q$ with kernel $I \subset B'$, write $S \to S'$ for $S_B \to S_{B'}$, and let

$$
\begin{array}{ccccc}
X_0 & \longrightarrow & X & \longrightarrow & X' \\
\downarrow{\scriptstyle f_0} & & \downarrow{\scriptstyle f} & & \downarrow{\scriptstyle f'} \\
S_0 & \longrightarrow & S & \longrightarrow & S'
\end{array}
$$

be a Cartesian diagram of generically log smooth deformations with log smooth loci
U_0, U, and U'. The ideal sheaf of the closed subscheme $X \subset X'$ is $\mathcal{I} = I \cdot \mathcal{O}_{X'}$.
An automorphism of $f' \colon X' \to S'$ over $f \colon X \to S$ consists of an automorphism
$\phi \colon \mathcal{O}_{X'} \to \mathcal{O}_{X'}$ of sheaves of rings, compatible with $f' \colon X' \to S'$ and $\mathcal{O}_{X'} \to \mathcal{O}_X$,
and an automorphism $\Phi \colon \mathcal{M}_{U'} \to \mathcal{M}_{U'}$ of sheaves of monoids, compatible with
the log part of $f' \colon X' \to S'$ and $\mathcal{M}_{U'} \to \mathcal{M}_U$; on U', they must be compatible with
$\alpha \colon \mathcal{M}_{U'} \to \mathcal{O}_{U'}$ so that (ϕ, Φ) is an automorphism of a log scheme. They form a
sheaf of groups $\mathcal{A}ut_{X'/X}$ on X'.

Lemma 10.4 *Let V be an open subset of X_0. Then the natural restriction map*

$$
\Gamma(V, \mathcal{A}ut_{X'/X}) \to \Gamma(V \cap U_0, \mathcal{A}ut_{X'/X})
$$

*is an isomorphism. Thus, $\mathcal{A}ut_{X'/X} = j_* \mathcal{A}ut_{U'/U}$ where $j \colon U' \to X'$ is the
inclusion.*

Proof Since $Z' = X' \setminus U'$ is of codimension ≥ 2 and $f' \colon X' \to S'$ is a flat
morphism with S_2-fibers, we have $\mathcal{O}_{X'} = j_* \mathcal{O}_{U'}$ for the inclusion $j \colon U' \to X'$.
Thus, given ϕ on $U' \cap V$, it can be extended in a unique way to an automorphism
of sheaves of groups on $X' \cap V$. It satisfies all compatibilities automatically. $\square$

Since $U' \to S'$ is log smooth, the results of [76] hold for $\mathcal{A}ut_{U'/U}$. Let us briefly
recall them. The sheaf of groups $\mathcal{A}ut_{U'/U}$ is isomorphic to the sheaf $\mathcal{D}er_{U'/S'}(\mathcal{I})$
of relative log derivations (D, Δ) with values in $\mathcal{I}$, i.e., $D \colon \mathcal{O}_{U'} \to \mathcal{I}$ is a derivation
and $\Delta \colon \mathcal{M}_{U'} \to \mathcal{I}$ is its log part. This is a sheaf of Lie algebras as a subalgebra of
$\Theta^1_{U'/S'}$ which is filtered by

$$
F^k := F^k \mathcal{D}er_{U'/S'}(\mathcal{I}) := \mathcal{D}er_{U'/S'}(\mathcal{I}^k) \subset \mathcal{D}er_{U'/S'}(\mathcal{I})
$$

for $k \geq 1$, the sheaf of derivations with values in $\mathcal{I}^k \subset \mathcal{I}$. We have $[F^k, F^\ell] \subset F^{k+\ell}$, so the lower central series of $\mathcal{D}er_{U'/S'}(\mathcal{I})$ is eventually 0, and it is a sheaf of nilpotent Lie algebras. In particular, the Baker–Campbell–Hausdorff formula turns $\mathcal{D}er_{U'/S'}(\mathcal{I})$ into a sheaf of groups.

More precisely, in [76], we find two explicit isomorphisms

$$\mathcal{D}er_{U'/S'}(\mathcal{I}) \overset{\mathrm{Exp}}{\underset{\mathrm{Log}}{\rightleftarrows}} \mathcal{A}ut_{U'/U}.$$

Given $(D, \Delta) \in \mathcal{D}er_{U'/S'}(\mathcal{I})$, the automorphism $(\phi, \Phi) = \mathrm{Exp}_{U'/U}(D, \Delta)$ is defined by the formulae

$$\phi:\ \mathcal{O}_{U'} \to \mathcal{O}_{U'}, \qquad \phi(a) = \sum_{n=0}^{\infty} \frac{D^n(a)}{n!}$$

$$\Phi:\ \mathcal{M}_{U'} \to \mathcal{M}_{U'}, \qquad \Phi(m) = m + \alpha^{-1}\left(\sum_{n=0}^{\infty} \frac{[\Delta(m) + D]^n(1)}{n!} \right)$$

where the sums are actually finite and the symbol $\Delta(m)$ denotes the multiplication operator with this element. Conversely, given $(\phi, \Phi) \in \mathcal{A}ut_{U'/U}$, the classical part of $\mathrm{Log}_{U'/U}(\phi, \Phi)$ is

$$D:\ \mathcal{O}_{U'} \to \mathcal{I}, \quad D(a) = \sum_{n=1}^{\infty} \frac{(-1)^{n-1}[\phi - \mathrm{Id}]^n(a)}{n},$$

and the log part is

$$\Delta:\ \mathcal{M}_{U'} \to \mathcal{I}, \quad \Delta(m) = \sum_{n=1}^{\infty}\left(\sum_{k=0}^{n} \binom{n}{k} \frac{(-1)^{k+1}\alpha(\Phi^k(m) - m)}{n} \right),$$

which is a finite sum as well.

On U', we have $\mathcal{D}er_{U'/S'}(\mathcal{I}) = \mathcal{I} \cdot \Theta^1_{U'/S'}$, which is also the kernel of the restriction map $\Theta^1_{U'/S'} \to \Theta^1_{U/S}$. Thus, we have $\mathcal{A}ut_{X'/X} \cong j_*(\mathcal{I} \cdot \Theta^1_{U'/S'})$ and moreover,

$$\mathrm{Log}:\ \mathcal{A}ut_{X'/X} \overset{\cong}{\to} \ker(\Theta^1_{X'/S'} \to \Theta^1_{X/S})$$

where the map $\Theta^1_{X'/S'} \to \Theta^1_{X/S}$ is induced via push-forward from the map on U', where we have a canonical restriction of relative log derivations, and need not be surjective if the deformation does not have the base change property of

Definition 8.47 (or possibly if it is not log Gorenstein). If $f' \colon X' \to S'$ has the base change property and is log Gorenstein, then the kernel is equal to $I \cdot \Theta^1_{X'/S'}$.

If $B' \to B$ is a *first-order* extension so that $I \cdot \mathfrak{m}_{B'} = 0,$[1] then we obtain

$$\mathcal{A}ut_{X'/X} = \Theta^1_{X_0/S_0} \otimes_{\mathbf{k}} I.$$

10.2.1   Restricting Automorphisms

Let $B' \to B$ and $B'' \to B'$ be two surjections in $\mathbf{Art}_Q$, and let $f \colon X \to S$, $f' \colon X' \to S'$, $f'' \colon X'' \to S''$ be (compatible) deformations of $f_0 \colon X_0 \to S_0$. Then there is a natural restriction map

$$\rho \colon \quad \mathcal{A}ut_{X''/X} \to \mathcal{A}ut_{X'/X},$$

which fits into a diagram

$$
\begin{array}{ccccccccc}
1 & \longrightarrow & \mathcal{A}ut_{X''/X} & \xrightarrow{\;\mathrm{Log}\;} & \Theta^1_{X''/S''} & \longrightarrow & \Theta^1_{X/S} \\
 & & \Big\downarrow{\scriptstyle \rho} & & \Big\downarrow & & \Big\| \\
1 & \longrightarrow & \mathcal{A}ut_{X'/X} & \xrightarrow{\;\mathrm{Log}\;} & \Theta^1_{X'/S'} & \longrightarrow & \Theta^1_{X/S}
\end{array}
$$

of (left) exact sequences. If $f'' \colon X'' \to S''$ has the base change property and $f_0 \colon X_0 \to S_0$ is log Gorenstein, then the middle vertical arrow is surjective.[2] In this case, also ρ is surjective. In particular, it is surjective over affine open subsets— every automorphism of $f' \colon X' \to S'$ defined on an affine open subset can be lifted to $f'' \colon X'' \to S''$.

If $f'' \colon X'' \to S''$ has *not* the property that $\Theta^1_{X''/S''} \otimes_{O_{X''}} O_{X_0} \cong \Theta^1_{X_0/S_0}$, then surjectivity may fail. The following example is similar to Example 1.158 and also occurs when endowing toroidal crossing Fano spaces with a log smooth structure.

Example 10.5   Let $S = \mathbb{A}^1_t$, let

$$X = \operatorname{Spec} \mathbf{k}[x, y, z, w, t, u]/(xy - w^3 - t^3,\ zw - tu),$$

and let $f \colon X \to S$ be the obvious map. Endow both spaces with the compactifying log structures coming from $\{t = 0\}$. This is a generically log smooth family, which

---

[1] See Appendix A for our conventions about small extensions.
[2] Since $W^d_{X''/S''}|_{X_0} \cong W^d_{X_0/S_0}$ is a line bundle and $W^d_{X''/S''}$ is flat over S'', it is a line bundle as well, cf. Remark 12.12. Thus, $f'' \colon X'' \to S''$ is log Gorenstein.

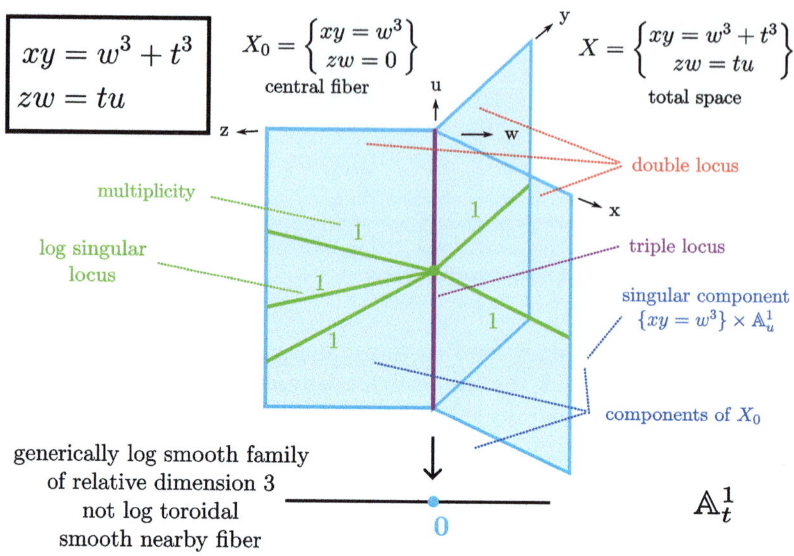

Fig. 10.1 Example 10.5. Compare also with Example 1.158. © Simon Felten 2025. All rights reserved

is depicted in Fig. 10.1. The central fiber X_0 has three irreducible components intersecting in three strata $\mathbb{A}^2_{u,z}$, $\mathbb{A}^2_{u,x}$, and $\mathbb{A}^2_{u,y}$ of codimension 1. The part of the log singular locus inside the central fiber is given by three lines $\{u^3 + z^3 = 0\}$ inside $\mathbb{A}^2_{u,z}$ and one line $\{u = 0\}$ each inside $\mathbb{A}^2_{u,x}$ and $\mathbb{A}^2_{u,y}$. The family is unisingular in the sense of Definition 14.20.

The sheaf of relative classical derivations on $f\colon X \to S$ equals the sheaf of relative log derivations on $f\colon X \to S$. This allows us to compute the restriction maps

$$\Theta^1_{X_2/S_2} \to \Theta^1_{X_1/S_1} \to \Theta^1_{X_0/S_0}$$

explicitly, and it turns out that $\mathcal{A}ut_{X_2/X_0} \to \mathcal{A}ut_{X_1/X_0}$ is *not* surjective. In fact, the cokernel of the map is a **k**-vector space of dimension 1. We have carried out the computation in Macaulay2, and an annotated script for this computation is included in Appendix F. The computation is done over $\mathbb{Q}$, but everything should commute with flat base change to **k**. $\diamond$

10.2.2 The Action of Automorphisms on the Lie–Rinehart Algebra and the Gerstenhaber Calculus

Let us assume that $f_0 \colon X_0 \to S_0$ is log Gorenstein, and that our deformations have the base change property. Let $\varphi \colon X' \to X'$ be an automorphism of the deformation $f' \colon X' \to S'$ over $f \colon X \to S$. Let us write $\varphi \colon X'_1 \to X'_2$ to distinguish the roles of source and target. As part of the definition, we have a map $\phi \colon \mathcal{W}^0_{X'_2/S'} \to \mathcal{W}^0_{X'_1/S'}$. Pull-back of differential forms gives us

$$d\varphi^* \colon \ \mathcal{W}^1_{X'_2/S'} \to \mathcal{W}^1_{X'_1/S'},$$

and the exterior powers of this map give maps $d^i\varphi^*$ on $\mathcal{W}^i_{X'/S'}$ after push-forward. Considering $\mathcal{V}^{-1}_{X'_2/S'}$ as the dual of $\mathcal{W}^1_{X'_2/S'}$ (via the universal property of the log differential forms, or equivalently, via the contraction map) gives us

$$T\varphi^* \colon \ \mathcal{V}^{-1}_{X'_2/S'} \to \mathcal{V}^{-1}_{X'_1/S'}$$

via pre-composition with $(d\varphi^*)^{-1}$ and post-composition with ϕ, i.e., if $\xi \in \mathcal{V}^{-1}_{X'_2/S'}$ and $\alpha \in \mathcal{W}^1_{X'_1/S'}$, then

$$T\varphi^*(\xi) \lrcorner\, \alpha = \phi(\xi \lrcorner (d\varphi^*)^{-1}(\alpha)).$$

Finally, we obtain a map $T^p\varphi^*$ on $\mathcal{V}^p_{X'/S'}$ via exterior powers and push-forward of $T\varphi^*$. Collectively, they form an (a priori only set-theoretic) self-map of the Gerstenhaber calculus $\mathcal{VW}^\bullet_{X'/S'}$. The following result identifies this self-map as a specific gauge transform of $\mathcal{VW}^\bullet_{X'/S'}$, and thus as an automorphism.

Lemma 10.6 *Assume that $f' \colon X' \to S'$ has the base change property and is log Gorenstein. Let $\theta = \mathrm{Log}(\varphi)$. Then the above-constructed collection of maps $\{d^i\varphi^*, T^p\varphi^*\}$ is equal to the gauge transform*

$$\exp_{-\theta} \colon \ \mathcal{V}\mathcal{W}^\bullet_{X'/S'} \to \mathcal{V}\mathcal{W}^\bullet_{X'/S'}$$

in the sense of Definition 4.3. In particular, it is an automorphism of the Gerstenhaber calculus. This induces a one-to-one correspondence between automorphisms of $f' \colon X' \to S'$ over $f \colon X \to S$ and gauge transforms induced by elements $\theta \in I \cdot \mathcal{V}^{-1}_{X'/S'}$.

Proof This is essentially [76, Lemma 3.2], but the proof there is not completely correct. First, we work on U'. Then $\phi \colon \mathcal{O}_{X'} \to \mathcal{O}_{X'}$ is equal to $\exp_{-\theta}$ by definition; namely, if $\theta = (D, \Delta)$, then $[\theta, g] = -D(g)$ for $g \in \mathcal{O}_{X'}$. From the axioms of a Gerstenhaber calculus, we deduce $\theta \lrcorner \partial g = [-\theta, g] \lrcorner 1 = D(g)$.[3] Let us write $\Box_\theta \colon \mathcal{W}^1_{U'/S'} \to \mathcal{W}^1_{U'/S'}$, $\alpha \mapsto \partial(\theta \lrcorner \alpha)$. Since $d\varphi^* \circ \partial = \partial \circ \phi$, we have

$$d\varphi^*(\partial g) = \sum_{n=0}^{\infty} \frac{\partial D^n(g)}{n!} = \sum_{n=0}^{\infty} \frac{\Box_\theta^n(\partial g)}{n!} = \sum_{n=0}^{\infty} \frac{\mathcal{L}^n_{-\theta}(\partial g)}{n!} = \exp_{-\theta}(\partial g).$$

The second last equality holds due to $\mathcal{L}_{-\theta}(\partial g) = \nabla_\theta(\partial g) + \theta \lrcorner \partial^2 g$. Since we have both $d\varphi^*(g \cdot \alpha) = \phi(g) \cdot d\varphi^*(\alpha)$ and $\exp_{-\theta}(g \cdot \alpha) = \phi(g) \cdot \exp_{-\theta}(\alpha)$, we have $d\varphi^* = \exp_{-\theta}$ on the submodule of $\mathcal{W}^1_{U'/S'}$ generated by elements of the form ∂g. On the strict locus of $U' \to S'$, the log differentials agree with the classical differentials, so this is everything. Since the strict locus is scheme-theoretically dense, we get $d\varphi^* = \exp_{-\theta}$ everywhere on U'.

There is a second, more conceptual proof of $d\varphi^* = \exp_{-\theta}$, which does not rely on the density of the strict locus.[4] In general, $\mathcal{W}^1_{U'/S'}$ is locally generated by elements of the form $\delta(m)$, where $\delta \colon \mathcal{M}_{U'} \to \mathcal{W}^1_{U'/S'}$ is the log part of the universal derivation, see [222, IV, Prop. 1.2.11]. This map satisfies $d\varphi^* \circ \delta = \delta \circ \Phi$, i.e., we have

$$d\varphi^*(\delta(m)) = \delta \left(m + \alpha^{-1} \left(\sum_{n=0}^{\infty} \frac{[\Delta(m) + D]^n(1)}{n!} \right) \right)$$

$$= \delta(m) + \left(\sum_{n=0}^{\infty} \frac{[\Delta(m) + D]^n(1)}{n!} \right)^{-1} \cdot d \left(\sum_{n=0}^{\infty} \frac{[\Delta(m) + D]^n(1)}{n!} \right)$$

where $d \colon \mathcal{O}_{U'} \to \mathcal{W}^1_{U'/S'}$ is the classical part of the universal derivation, and where $\Delta(m)$ denotes the operator which multiplies the argument with $\Delta(m) \in \mathcal{O}_{U'}$. Using the formula in the proof of [76, Lemma 2.2], we can show that

$$\left(\sum_{n=0}^{\infty} \frac{[\Delta(m) + D]^n(1)}{n!} \right)^{-1} = \sum_{n=0}^{\infty} \frac{[-\Delta(m) + D]^n(1)}{n!}.$$

---

[3] We see here a nice example of how our choice $[-, -] = -[-, -]_{\mathrm{sn}}$ interacts with our chosen axioms for a Gerstenhaber calculus. Since we do not switch signs in the contraction map, we get $D(a) = \theta \lrcorner \partial a$. Then, due to the sign in the Lie–Rinehart homotopy formula, we have $\partial D(a) = -\mathcal{L}_\theta(a)$, and this is what we need because we are using $\exp_{-\theta}$.

[4] If one works with *integral* log smooth morphisms instead of saturated ones, then the strict locus may be empty, for example in the central fiber of $\mathbb{A}^2 \to \mathbb{A}^1$, $t \mapsto x^2 y^2$. See also Chap. 15.2 for another situation which needs the more general proof.

Since $\partial\delta(m) = 0$, we have $\mathcal{L}_{-\theta}(\delta(m)) = \Box_\theta(\delta(m))$, and then we obtain inductively $\partial\Box^n_\theta(\delta(m)) = 0$ and $\mathcal{L}^n_{-\theta}(\delta(m)) = \Box^n_\theta(\delta(m))$. Thus, in order to show $d\varphi^*(\delta(m)) = \exp_{-\theta}(\delta(m))$, it is sufficient to show

$$\Box^n_\theta(\delta(m)) = \sum_{k=0}^{n}\binom{n}{k}[-\Delta(m) + D]^{n-k}(1) \cdot \partial[\Delta(m) + D]^k(1) := A_n(m).$$

Unfortunately, this seems to be surprisingly difficult. As the first step, we wish to show

$$\partial[\Delta(m) + D]^n(1) = \sum_{k=1}^{n}\binom{n}{k}[\Delta(m) + D]^{n-k}(1) \cdot \Box^k_\theta(\delta(m))$$

for $n \geq 1$ (the sum really starts at $k = 1$). This is easy to verify for the first few values of n, but a direct inductive argument does not seem to be available. Assume that we know this formula up to n. Then we compute, for $1 \leq p \leq n + 1$,

$$P_p(m):= \sum_{k=1}^{p}\binom{p}{k}\left(\partial[\Delta(m) + D]^{p-k}(1)\right) \wedge \Box^k_\theta(\delta(m)) \quad \in \mathcal{W}^2_{U'/S'}$$

$$= \sum_{\substack{k,\ell\geq 1 \\ k+\ell\leq p}} \frac{p!}{k!\ell!(p - k - \ell)!}[\Delta(m) + D]^{p-k-\ell} \wedge \Box^\ell_\theta(\delta(m)) \wedge \Box^k_\theta(\delta(m)) = 0.$$

The sum is zero since $\Box^\ell_\theta(\delta(m)) \wedge \Box^k_\theta(\delta(m)) = -\Box^k_\theta(\delta(m)) \wedge \Box^\ell_\theta(\delta(m))$, and the sum is symmetric in k and ℓ. This shows

$$\sum_{k=1}^{p}\binom{p}{k}\partial[\Delta(m) + D]^{p-k}(1) \cdot \left(\theta \lrcorner \Box^k_\theta(\delta(m))\right)$$

$$= \sum_{k=1}^{p}\binom{p}{k} D[\Delta(m) + D]^{p-k} \cdot \Box^k_\theta(\delta(m))$$

via the formula for $\theta \lrcorner (\alpha \wedge \beta)$ since $D(g) = \theta \lrcorner \partial g$ for $g \in O_{U'}$. After this preparation, the induction step is a direct computation of $\partial[\Delta(m) + D]^{n+1}(1)$. As soon as we have this formula, we can compute

$$A_n(m) = \sum_{k=0}^{n}\binom{n}{k}[-\Delta(m) + D]^{n-k}(1) \cdot \sum_{\ell=0}^{k-1}\binom{k}{\ell}[\Delta(m) + D]^\ell(1) \cdot \Box^{k-\ell}_\theta(\delta(m))$$

$$= \sum_{p=1}^{n}\sum_{q=0}^{n-p} \frac{n!}{(n - p - q)!p!q!}[-\Delta(m) + D]^q(1)$$

$$\cdot \, [\Delta(m) + D]^{n-p-q}(1) \cdot \square_\theta^p(\delta(m))$$

$$= \sum_{p=1}^{n} \binom{n}{p} [\Delta(0) + D]^{n-p}(1) \cdot \square_\theta^p(\delta(m)) = \square_\theta^n(\delta(m))$$

by setting $p = k - \ell$ and $q = n - k$.

Since both $d^\bullet \varphi^*$ and $\exp_{-\theta}$ are compatible with the $\wedge$-product, we obtain $d^i \varphi^* = \exp_{-\theta}$ on $\mathcal{W}^i_{U'/S'}$. A direct computation with the definition yields

$$T\varphi^*(\xi) \lrcorner \, \alpha = \exp_{-\theta}(\xi) \lrcorner \, \alpha$$

for $\xi \in \Theta^1_{U'/S'}$ and $\alpha \in \mathcal{W}^1_{U'/S'}$, so $T\varphi^*(\xi) = \exp_{-\theta}(\xi)$. Then both $T^p \varphi^*$ and $\exp_{-\theta}$ are compatible with the $\wedge$-product, so we have $T^p \varphi^* = \exp_{-\theta}$ as well. Since all sheaves $\mathcal{F}$ involved satisfy $j_* \mathcal{F}|_U = \mathcal{F}$, we obtain the assertion on X' from the one on U' by push-forward. Finally, infinitesimal automorphisms are in one-to-one correspondence with elements of $I \cdot \Theta^1_{X'/S'}$, and they are in turn in one-to-one correspondence with gauge transforms of the Gerstenhaber calculus by Lemma 4.6 since $\mathcal{V}^\bullet_{X_0/S_0}$ is strictly faithful. $\square$

Remark 10.7 We write $\mathrm{Exp}(\theta)$ with a capital "E" for the automorphism induced by $\theta \in I \cdot \Theta^1_{X'/S'}$, and we write $\exp_{-\theta}$ with a lower case e for the gauge transform. In forming $\mathrm{Exp}(\theta)$, we use the positive Lie bracket of $\Theta^1_{X'/S'}$, and in forming $\exp_{-\theta}$, we use the negative Lie bracket of $\mathcal{V}^{-1}_{X'/S'}$. $\Diamond$

When we restrict this construction to $\mathcal{V}^0_{X'/S'}$ and $\mathcal{V}^{-1}_{X'/S'}$, then we obtain an induced gauge transform of the Lie–Rinehart algebra $\mathcal{LR}^\bullet_{X'/S'}$. Exactly as in the case of the Gerstenhaber calculus in Lemma 10.6, automorphisms of $f' \colon X' \to S'$ are in one-to-one correspondence with gauge transforms induced by $\theta \in I \cdot \mathcal{LR}^T_{X'/S'}$.

10.3 Enhanced Generically Log Smooth Families

In Sect. 1.12 in the introduction, we argued that many interesting generically log smooth families do not have the base change property, i.e., the formation of the Gerstenhaber calculus $\mathcal{V} \mathcal{W}^\bullet_{X/S}$ does not commute with base change. For instance, the families in Example 1.152, Example 1.158, and Example 1.159 all do not have the base change property. With Example 10.5, we saw a further example of this phenomenon in this chapter. Here, we provide our theory of *enhanced generically log smooth families*, which stabilizes the Gerstenhaber calculus under base change.

Definition 10.8 An *enhanced generically log smooth family of relative dimension* $d \geq 1$ is a tuple

$$(f \colon (X, U) \to S, \, \mathcal{G}^{\bullet}_{X/S}, \, \mathcal{A}^{\bullet}_{X/S}, \, \varpi^{\bullet})$$

where:

- $f \colon (X, U) \to S$ is a generically log smooth family of relative dimension d;
- $\mathcal{W}^{d}_{X/S}$ is a line bundle, i.e., $f \colon X \to S$ is log Gorenstein;
- the pair

$$\mathcal{GC}^{\bullet}_{X/S} = (\mathcal{G}^{\bullet}_{X/S}, \mathcal{A}^{\bullet}_{X/S})$$

  is a two-sided Gerstenhaber calculus of dimension d in the context $\mathfrak{Coh}(X/S)$; in particular, its pieces are flat over S;
- $\varpi^{\bullet} \colon \mathcal{G}^{\bullet}_{X/S} \to \mathcal{V}^{\bullet}_{X/S}$ and $\varpi^{\bullet} \colon \mathcal{A}^{\bullet}_{X/S} \to \mathcal{W}^{\bullet}_{X/S}$ form a map of two-sided Gerstenhaber calculi, i.e., ϖ^{p} and ϖ^{i} are O_X-linear maps which are compatible with the two $\wedge$-products, with the left contraction $\vdash$ and the right contraction $\dashv$, with the bracket $[-, -]$, with the de Rham differential ∂, with the Lie derivative $\mathcal{L}_{-}(-)$, and with the constants;
- $\varpi^{0} \colon \mathcal{G}^{0}_{X/S} \to \mathcal{V}^{0}_{X/S}$ and $\varpi^{0} \colon \mathcal{A}^{0}_{X/S} \to \mathcal{W}^{0}_{X/S}$ are isomorphisms;
- $\varpi^{d} \colon \mathcal{A}^{d}_{X/S} \to \mathcal{W}^{d}_{X/S}$ is an isomorphism; in particular, $\mathcal{A}^{d}_{X/S}$ is a line bundle;
- the Gerstenhaber calculus $\mathcal{GC}^{\bullet}_{X/S}$ is locally Batalin–Vilkovisky, i.e., every local volume form $\omega \in \mathcal{A}^{d}_{X/S}$ gives rise to a local isomorphism $\kappa_{\omega} \colon \mathcal{G}^{p}_{X/S} \cong \mathcal{A}^{p+d}_{X/S}$;
- $\varpi^{\bullet}$ is an isomorphism on U for all $\mathcal{G}^{p}_{X/S}$ and $\mathcal{A}^{i}_{X/S}$.

Example 10.9 Let $f \colon (X, U) \to S$ be a generically log smooth family which has the base change property and is log Gorenstein. Then, we can consider it as an enhanced generically log smooth family by setting $\mathcal{A}^{\bullet}_{X/S} = \mathcal{W}^{\bullet}_{X/S}$ and $\mathcal{G}^{\bullet}_{X/S} = \mathcal{V}^{\bullet}_{X/S}$. Namely, these sheaves are flat over S by Corollary 8.50. $\Diamond$

Given a map $b \colon T \to S$, we define the fiber product $g \colon Y = X \times_S T$ in the obvious way. The underlying generically log smooth family is the fiber product $(X, U) \times_S T \to T$ of partial log schemes. On the level of coherent sheaves, the Gerstenhaber calculus is given by

$$\mathcal{G}^{p}_{Y/T} := c^{*}\mathcal{G}^{p}_{X/S} \quad \text{and} \quad \mathcal{A}^{i}_{Y/T} := c^{*}\mathcal{A}^{i}_{X/S},$$

with the obvious maps $\varpi^{\bullet}$ to $\mathcal{V}^{p}_{Y/T}$ and $\mathcal{W}^{i}_{Y/T}$. Every operation on $\mathcal{GC}^{\bullet}_{X/S}$ is a multilinear differential operator, a concept which we discuss in Sect. C in the appendix. In particular, every operation can be pulled back to an operation on $\mathcal{GC}^{\bullet}_{Y/T}$ by Proposition C.19, even if that operation is not O_X-linear. Since multi-compositions and linear combinations of multilinear differential operators are multilinear differential operators, we can consider every relation in $\mathcal{GC}^{\bullet}_{X/S}$ as a multilinear differential operator which vanishes. Now the pull-back is compatible with multi-compositions and linear combinations, so the relation also holds in $\mathcal{GC}^\bullet_{Y/T}$. Thus, $\mathcal{GC}^\bullet_{Y/T}$ is a two-sided Gerstenhaber calculus in the context $\mathfrak{Coh}(Y/T)$. The log Gorenstein and local Batalin–Vilkovisky condition as well as the isomorphy on U are preserved due to their $\mathcal{O}_X$-linear nature. Therefore, we have constructed an enhanced generically log smooth family. With this construction, any enhanced generically log smooth family satisfies the base change property when considering $\mathcal{A}^\bullet_{X/S}$ and $\mathcal{G}^\bullet_{X/S}$ instead of $\mathcal{W}^\bullet_{X/S}$ and $\mathcal{V}^\bullet_{X/S}$.

Example 10.10 Let $\Sigma = \mathrm{Spec}(\mathbb{N} \to \mathbf{k}[\![t]\!])$, and let $f\colon (X, U) \to \Sigma$ be the base change along $\Sigma \to (\mathbb{A}^1_t | \{0\})$ of any of the generically log smooth families in the examples listed at the beginning of this section, for instance Example 10.5. We set $\mathcal{A}^\bullet_{X/\Sigma} = \mathcal{W}^\bullet_{X/\Sigma}$ and $\mathcal{G}^\bullet_{X/\Sigma} = \mathcal{V}^\bullet_{X/\Sigma}$. To see that this turns $f\colon (X, U) \to \Sigma$ into an enhanced generically log smooth family, we have to verify that each $\mathcal{A}^i_{X/\Sigma}$ and each $\mathcal{G}^p_{X/\Sigma}$ is flat over Σ. This is the case because they have no t-torsion. Now we can form the base change along $S_0 \to \Sigma$ and obtain a new enhanced generically log smooth family over S_0. The comparison maps $\varpi^\bullet\colon \mathcal{A}^\bullet_{X_0/S_0} \to \mathcal{W}^\bullet_{X_0/S_0}$ are not all isomorphisms because the original family does not have the base change property. Therefore, we have constructed some truly enhanced examples of enhanced generically log smooth families. In Construction 10.27 below, we further elaborate on this way of constructing enhanced generically log smooth families. In Example 10.28, we further elaborate on the family of Example 10.5.   ◊

When $p\colon (X', U') \to (X, U)$ is strict, étale, separated, and of finite type[5] but not necessarily accurate, then we construct similarly an induced two-sided Gerstenhaber calculus $\mathcal{GC}^\bullet_{X'/S}$ which turns $f \circ p\colon (X', U') \to S$ into an enhanced generically log smooth family. This allows us to consider enhanced generically log smooth families locally in the étale topology.

These two constructions motivate the following notion of a morphism between enhanced generically log smooth families.

Definition 10.11 Let $f\colon (X, U) \to S$ and $g\colon (Y, V) \to T$ be enhanced generically log smooth families of relative dimension d. Then a *morphism* from g to f is a tuple

$$\varphi = (b, c, d^\bullet\varphi^*, T^\bullet\varphi^*)$$

where:

- $b\colon T \to S$ is a morphism of log schemes;
- $c\colon (Y, V) \to (X, U)$ is a morphism of partial log schemes with $f \circ c = b \circ g$ as morphisms of partial log schemes;
- the induced morphism $(Y, V) \to (X, U) \times_S T$ is strict and étale;

---

[5] Étale alone is not sufficient as the resulting family is required to be separated and of finite type.

- $(d^\bullet\varphi^*, T^\bullet\varphi^*)\colon c^*\mathcal{GC}^\bullet_{X/S} \to \mathcal{GC}^\bullet_{Y/T}$ is an isomorphism of two-sided Gerstenhaber calculi which is compatible with the two maps $\varpi^\bullet$ to $c^*\mathcal{VW}^\bullet_{X/S}$ and $\mathcal{VW}^\bullet_{Y/T}$; here, c^* of a two-sided Gerstenhaber calculus is formed as indicated above.

Note that we do not assume that $c\colon (Y, V) \to (X, U)$ is accurate. This allows us to consider shrinking U as a morphism of enhanced generically log smooth families. The assumption that $(Y, V) \to (X, U) \times_S T$ is strict and étale is necessary in order to make sense of the map $T^\bullet\varphi^*$ being part of a morphism of two-sided Gerstenhaber calculi.

Often, the map $\varpi^\bullet$ is injective, for example when $\mathcal{GC}^\bullet_{X/S}$ is obtained as the pull-back of $\mathcal{VW}^\bullet_{X/S}$ in a family over a one-dimensional base (such as in Example 1.152), or when $\mathcal{GC}^\bullet_{X/S}$ is the direct image from a resolution of log singularities with no contracted components. Within this monograph, we will usually assume that $\varpi^\bullet$ is injective and remains so after any base change since this simplifies the arguments—now $\mathcal{GC}^\bullet_{X/S}$ is a Gerstenhaber subcalculus of $\mathcal{VW}^\bullet_{X/S}$— and is often satisfied.

Recall that a coherent sheaf $\mathcal{F}$ is called *torsionless* if the canonical map $\mathcal{F} \to \mathcal{F}^{\vee\vee}$ into the bidual is injective. This motivates the following terminology.

Definition 10.12 We say that an enhanced generically log smooth family $f\colon X \to S$ is *torsionless* if $\mathcal{G}^p_{X/S}$ and $\mathcal{A}^i_{X/S}$ are torsionless and remain so after any base change. Equivalently, $\varpi^\bullet$ is injective and remains so after any base change. By Lemma B.4, it is sufficient to require this condition in the fibers of $f\colon X \to S$.

For generically log smooth families, we use the notation $\Theta^\bullet_{X/S}$ for the Gerstenhaber algebra of reflexive polyvector fields in positive degrees, endowed with the positive Schouten–Nijenhuis bracket. Analogously, we write $\Gamma^\bullet_{X/S}$ to denote $\mathcal{G}^\bullet_{X/S}$ in positive degrees and with the positive bracket, i.e., the negative of the bracket on $\mathcal{G}^\bullet_{X/S}$. In particular, we will use $\Gamma^1_{X/S}$ when studying infinitesimal automorphisms.

10.4 Systems of Deformations

As we have argued in Sect. 1.7, arbitrary generically log smooth deformations are too general and too pathological for our purpose of constructing deformations of degenerate schemes. In that section, we introduced two devices to restrict the generically log smooth deformations which we admit: classes of log singularities $\mathcal{C}$ and systems of deformations $\mathcal{D}$. We now discuss the latter in some detail. Unlike the situation for a general class of log singularities, generically log smooth deformations of a given type $\mathcal{D}$ are locally rigid.

Definition 10.13 Let $f_0\colon (X_0, U_0) \to S_0$ be a generically log smooth family over a log point $S_0 = \mathrm{Spec}(Q \to \mathbf{k})$. Assume that $f_0\colon (X_0, U_0) \to S_0$ is log Gorenstein. We say that a Zariski open cover $\mathcal{V} = \{V_\alpha\}_\alpha$ of X_0 is:

(1) *pre-admissible* if it consists of (only) finitely many affine open immersions $j_\alpha\colon V_\alpha \to X_0$, i.e., for every affine subset $W \subseteq X_0$, the intersection $W \cap V_\alpha$ is affine as well;

(2) *weakly admissible* if it is pre-admissible and

$$H^n(V_{\alpha_1} \cap \ldots \cap V_{\alpha_r}, \mathcal{O}_{X_0}|_{\alpha_1\ldots\alpha_r}) = 0, \quad H^n(V_{\alpha_1} \cap \ldots \cap V_{\alpha_r}, \Theta^1_{X_0/S_0}|_{\alpha_1\ldots\alpha_r}) = 0$$

for all $n \geq 1$ and indices $\alpha_1, \ldots, \alpha_r$;

(3) *admissible* if it is pre-admissible and

$$H^n(V_{\alpha_1} \cap \ldots \cap V_{\alpha_r}, \mathcal{F}|_{\alpha_1\ldots\alpha_r}) = 0$$

for $\mathcal{F} = \mathcal{G}^p_{X_0/S_0}, \mathcal{A}^i_{X_0/S_0}$, all $n \geq 1$, and indices $\alpha_1, \ldots, \alpha_r$;

(4) *affine* if every V_α is an affine scheme.

Every affine open cover is admissible, and every admissible open cover is weakly admissible.

Remark 10.14 Under these conditions, we can compute the cohomology of the respective sheaves from the Čech complex. $\qquad\qquad\qquad\qquad\qquad\qquad\qquad\qquad\diamond$

Definition 10.15 Let $f_0\colon (X_0, U_0) \to S_0$ be a generically log smooth family over a log point $S_0 = \mathrm{Spec}(Q \to \mathbf{k})$. Assume that $f_0\colon (X_0, U_0) \to S_0$ is log Gorenstein. Let $\mathcal{V} = \{V_\alpha\}_\alpha$ be a weakly admissible open cover of X_0. A *system of deformations* $\mathscr{D}$ of $f_0\colon (X_0, U_0) \to S_0$, subordinate to $\mathcal{V}$, is a tuple

$$((V_{\alpha;A}, U_{\alpha;A}) \to S_A, \ \rho_{\alpha;BB'}, \ \psi_{\alpha\beta;A})$$

where:

(a) for every $A \in \mathbf{Art}_Q$, the structure $(V_{\alpha;A}, U_{\alpha;A}) \to S_A$ is a generically log smooth deformation of $X_0|_\alpha := X_0|_{V_\alpha}$ over S_A; we assume that this deformation has the base change property, i.e., $\mathcal{W}^k_{V_{\alpha;A}/S_A}$ is flat over S_A;

(b) for every $B' \to B$ in $\mathbf{Art}_Q$, the structure $\rho_{\alpha;BB'}\colon (V_{\alpha;B}, U_{\alpha;B}) \to (V_{\alpha;B'}, U_{\alpha;B'})$ is a morphism of generically log smooth deformations; they are compatible with compositions and, by definition, with the maps from $X_0|_\alpha$ that are part of a generically log smooth deformation;

(c) for each intersection $V_\alpha \cap V_\beta$, the structure

$$\psi_{\alpha\beta;A}\colon (V_{\alpha;A}, U_{\alpha;A})|_{\alpha\beta} \xrightarrow{\cong} (V_{\beta;A}, U_{\beta;A})|_{\alpha\beta}$$

is an isomorphism of generically log smooth deformations; they commute with the restrictions $\rho_{BB'}$, and, by definition, with the maps from $X_0|_{\alpha\beta}$; we do not require any cocycle condition on triple intersections $V_\alpha \cap V_\beta \cap V_\gamma$.

We say that a generically log smooth deformation $f_A \colon (X_A, U_A) \to S_A$ of $f_0 \colon (X_0, U_0) \to S_0$ is *of type* $\mathscr{D}$ if we can find isomorphisms $\chi_\alpha \colon (X_A, U_A)|_\alpha \cong (V_{\alpha;A}, U_{\alpha;A})$ of generically log smooth deformations for each α. Isomorphism classes of generically log smooth deformations of type $\mathscr{D}$ form a functor of Artin rings

$$\mathrm{LD}^{\mathscr{D}}_{X_0/S_0} \colon \quad \mathbf{Art}_Q \to \mathbf{Set}.$$

Remark 10.16 The condition $(X_A, U_A)|_\alpha \cong (V_{\alpha;A}, U_{\alpha;A})$ is equivalent to that there is an (étale) open cover $\{U_i\}_i$ by affines subordinate to $\mathcal{V}$ such that we can find isomorphisms $(X_A, U_A)|_i \cong (V_{\alpha(i);A}, U_{\alpha;A})|_i$ of generically log smooth deformations. The proof is by induction over small extensions, where $H^1(V_\alpha, \mathscr{A}ut_{X'/X}) = H^1(V_\alpha, \Theta^1_{X_0/S_0} \otimes_{\mathbf{k}} I) = 0$, and then we have to arrange the automorphisms on U_i to be the given one on V_α over the smaller ring, which we achieve by liftability of all automorphisms on the affines U_i. $\Diamond$

Example 10.17 Let $f_0 \colon X_0 \to S_0$ be separated, log smooth, and saturated. Let $\mathcal{V} = \{V_\alpha\}_\alpha$ be a finite open affine cover. The log smooth deformations give rise to a system of deformations $\mathscr{D}$. Generically log smooth deformations of type $\mathscr{D}$ are the same as log smooth deformations. $\Diamond$

Example 10.18 Let (V, Z, s) be a well-adjusted triple in the sense of Definition 9.92, and let $f_0 \colon (X_0, U_0) \to S_0$ be the associated generically log smooth family. Assume that it is of elementary Gross–Siebert type. Then Corollary 14.37 gives a system of deformations $\mathscr{D}$. This is the most important system of deformations studied in this monograph. Namely, it is not log smooth, but we can prove that its deformation functor is unobstructed in the Calabi–Yau case. $\Diamond$

Lemma 10.19 *Let $\mathscr{D}$ be a system of deformations. Then $\mathrm{LD}^{\mathscr{D}}_{X_0/S_0}$ is a neat deformation functor, i.e., it satisfies (H_0), (H_1), (H_2), and (H_2^+) defined in Appendix A.*

Proof The condition (H_0) is clear. To check (H_1), let $A' \to A$ be arbitrary and $A'' \to A$ be a surjection in $\mathbf{Art}_Q$, and let $B := A' \times_A A''$. Let $f \colon X \to S_A$ be of type $\mathscr{D}$, and let $f' \colon X' \to S_{A'}$ and $f'' \colon X'' \to S_{A''}$ be two liftings of type $\mathscr{D}$. We show that

$$Y := (|X_0|, \, \mathcal{O}_{X'} \times_{\mathcal{O}_X} \mathcal{O}_{X''}, \, \mathcal{M}' \times_{\mathcal{M}} \mathcal{M}'')$$

is a common lifting of type $\mathscr{D}$ over S_B. After restricting to V_α, we have an isomorphism $X'|_\alpha \cong V_{\alpha;A'}$ by assumption. By restriction, it induces an isomorphism

$X|_\alpha \cong V_{\alpha;A}$. Since all automorphisms of $X|_\alpha$ can be lifted along $X''|_\alpha \to X|_\alpha$,[6] we can modify the existent isomorphism $X''|_\alpha \cong V_{\alpha;A''}$ to be compatible with the restriction $X''|_\alpha \to X|_\alpha$ and the given isomorphism $X|_\alpha \cong V_{\alpha;A}$. Then the universal property of Y yields a map $Y|_\alpha \to V_{\alpha;B}$ of generically log smooth families which is compatible with everything. The base change of this map to $S_{A'}$ is an isomorphism, so it must be a closed immersion on underlying schemes. Let $J' \subset B$ be the kernel of $B \to A'$, and let J'' be the kernel of $B \to A''$. If $g \in O_{V_{\alpha;B}}$ is in the kernel of $Y|_\alpha \to V_{\alpha;B}$, it must be in both $J' \cdot O_{V_{\alpha;B}}$ and $J'' \cdot O_{V_{\alpha;B}}$. Since $V_{\alpha;B} \to S_B$ is flat, extension of ideals from S_B commutes with intersection; we have $J' \cap J'' = 0$, hence $g = 0$. Because all other maps are strict, $Y|_\alpha \to V_{\alpha;B}$ is strict, too; in particular, it is an isomorphism of generically log smooth families. Finally, given (H_1), the conditions (H_2) and (H_2^+) are inherited from $\mathrm{LD}^{\mathrm{gen}}_{X_0/S_0}$. $\qquad\square$

Remark 10.20 Note that, as usual, the proof does not show bijectivity of the natural map $\eta\colon F(A' \times_A A'') \to F(A') \times_{F(A)} F(A'')$ in general because Y depends on the choice of the two maps $X \to X'$ and $X \to X''$, which are not encoded in the deformation functor. $\qquad\Diamond$

10.5   Enhanced Systems of Deformations

In the Definition 10.15 of a system of deformations $\mathscr{D}$, we require that every local model $V_{\alpha;A} \to S_A$ should have the base change property. This is necessary because compatibility with base change as well as flatness (cf. Lemma 8.49) is essential in the construction of the curved two-sided Gerstenhaber calculus in Chap. 13. Working with enhanced generically log smooth families allows us to use a modified Gerstenhaber calculus $\mathcal{GC}^\bullet_{X/S}$, flat over S, instead of the sometimes non-flat $\mathcal{VW}^\bullet_{X/S}$, and thus to study the deformation theory of many generically log smooth families which do not have the base change property. In this section, we define our notion of an *enhanced system of deformations*, which adapts Definition 10.15 to the setting of enhanced generically log smooth families. For simplicity, we assume that our enhanced generically log smooth families are torsionless as defined above.[7]

When it comes to infinitesimal automorphisms of enhanced generically log smooth families, then we have to distinguish between *inner automorphisms* and *outer automorphisms*. The latter are precisely the automorphisms in the sense of

---

[6] This is because $H^1(V_\alpha, I'' \cdot \Theta^1_{X''/S_{A''}}) = 0$ for I'' the kernel of $A'' \to A$.

[7] As long as $\varpi^{-1}\colon \mathcal{G}^{-1}_{X/S} \to \mathcal{V}^{-1}_{X/S}$ is injective and remains so after any base change, the theory should work with not too many modifications. However, if we drop the injectivity of ϖ^{-1}, then new complications arise due to the presence of gauge transforms which act trivially on the Gerstenhaber calculus.

Definition 10.11. The former are those outer automorphisms which are of the form $\text{Exp}(\theta)$ for some $\theta \in I \cdot \Gamma^1_{X'/S'} = I \cdot \mathcal{G}^{-1}_{X'/S'}$ under the map

$$\text{Exp:}\quad I \cdot \Gamma^1_{X'/S'} \xrightarrow{\varpi} I \cdot \Theta^1_{X'/S'} \subseteq \ker(\Theta^1_{X'/S'} \to \Theta^1_{X/S}) \xrightarrow{\text{Exp}} \mathcal{A}ut_{X'/X},$$

i.e., the gauge transform on $\mathcal{G}C^\bullet_{X'/S'} \subseteq \mathcal{V}\mathcal{W}^\bullet_{X'/S'}$ is given by $\exp_{-\theta}$. This obviously defines an (outer) automorphism in the sense of Definition 10.11 (cf. Lemma 10.6). Since Exp is injective, every inner automorphism is induced by a unique $\theta \in I \cdot \Gamma^1_{X'/S'}$. The outer automorphisms are those induced by elements $\theta \in \ker(\Theta^1_{X'/S'} \to \Theta^1_{X/S})$ with the property that $\exp_{-\theta}$ preserves $\mathcal{G}C^\bullet_{X'/S'} \subseteq \mathcal{V}\mathcal{W}^\bullet_{X'/S'}$.

Definition 10.21 Let $f_0 \colon X_0 \to S_0$ be a torsionless enhanced generically log smooth family over a log point $S_0 = \text{Spec}(Q \to \mathbf{k})$. Let $\mathcal{V} = \{V_\alpha\}_\alpha$ be a pre-admissible open cover of X_0. We write $\mathcal{V}\mathcal{W}^\bullet_0 := \mathcal{V}\mathcal{W}^\bullet_{X_0/S_0}$ as well as $\mathcal{G}C^\bullet_0 := \mathcal{G}C^\bullet_{X_0/S_0}$ for short.

(1) An *enhanced system of deformations* $\mathcal{D}$ subordinate to $\mathcal{V}$ is a tuple

$$(V_{\alpha;A} \to S_A,\ \rho_{\alpha;BB'},\ \psi_{\alpha\beta;A},\ \mathcal{G}C^\bullet_{\alpha;A})$$

where:

(a) for every $A \in \mathbf{Art}_Q$, the object $V_{\alpha;A} \to S_A$ is a generically log smooth deformation of $X_0|_\alpha := X_0|_{V_\alpha}$ over S_A; the two-sided Gerstenhaber calculus

$$\mathcal{G}C^\bullet_{\alpha;A} \subseteq \mathcal{V}\mathcal{W}^\bullet_{V_{\alpha;A}/S_A} := \mathcal{V}\mathcal{W}^\bullet_{\alpha;A}$$

in the context $\mathfrak{Coh}(V_{\alpha;A}/S_A)$ turns $V_{\alpha;A} \to S_A$ into an enhanced generically log smooth family; the restriction map $\mathcal{V}\mathcal{W}^\bullet_{\alpha;A} \to \mathcal{V}\mathcal{W}^\bullet_0|_\alpha$ maps $\mathcal{G}C^\bullet_{\alpha;A}$ to $\mathcal{G}C^\bullet_0|_\alpha$, and the induced map after base change is an isomorphism;

(b) for every $B' \to B$ in $\mathbf{Art}_Q$, we have a map $\rho_{\alpha;BB'} \colon V_{\alpha;B} \to V_{\alpha;B'}$ of generically log smooth deformations; they are compatible with compositions and, by definition, with the maps from $X_0|_\alpha$ that are part of a generically log smooth deformation; the restriction map $\mathcal{V}\mathcal{W}^\bullet_{\alpha;B'} \to \mathcal{V}\mathcal{W}^\bullet_{\alpha;B}$ maps $\mathcal{G}C^\bullet_{\alpha;B'}$ to $\mathcal{G}C^\bullet_{\alpha;B}$ and induces an isomorphism after base change;

(c) on intersections $V_\alpha \cap V_\beta$, we have isomorphisms $\psi_{\alpha\beta;A} \colon V_{\alpha;A}|_{\alpha\beta} \xrightarrow{\cong} V_{\beta;A}|_{\alpha\beta}$ of generically log smooth deformations; they commute with the restrictions $\rho_{BB'}$, and, by definition, with the maps from $X_0|_{\alpha\beta}$; the induced isomorphism

$$\psi^*_{\alpha\beta;A} \colon\ \mathcal{V}\mathcal{W}^\bullet_{\beta;A}|_{\alpha\beta} \to \mathcal{V}\mathcal{W}^\bullet_{\alpha;A}|_{\alpha\beta}$$

maps $\mathcal{G}C^\bullet_{\beta;A}|_{\alpha\beta}$ isomorphically onto $\mathcal{G}C^\bullet_{\alpha;A}|_{\alpha\beta}$;

(d) applying Lemma 10.6 on the log smooth locus yields a unique element

$$o_{\alpha\beta\gamma;A} \in \Gamma(V_\alpha \cap V_\beta \cap V_\gamma \cap U_0, \mathfrak{m}_A \cdot \mathcal{V}_{\alpha;A}^{-1})$$

such that the cocycle $\psi_{\gamma\alpha;A} \circ \psi_{\beta\gamma;A} \circ \psi_{\alpha\beta;A}$ induces $\exp_{-o_{\alpha\beta\gamma;A}}$ on the Gerstenhaber calculus $\mathcal{V}\mathcal{W}_{\alpha;A}^\bullet$ on $V_\alpha \cap V_\beta \cap V_\gamma \cap U_0$; since this Gerstenhaber calculus is reflexive, this is actually a section on $V_\alpha \cap V_\beta \cap V_\gamma$ of the kernel of $\mathcal{V}_{\alpha;A}^{-1} \to \mathcal{V}_0^{-1}|_\alpha$, and $\exp_{-o_{\alpha\beta\gamma;A}}$ is the induced action of the cocycle; we assume that

$$o_{\alpha\beta\gamma;A} \in \Gamma(V_\alpha \cap V_\beta \cap V_\gamma, \mathfrak{m}_A \cdot \mathcal{G}_{\alpha;A}^{-1});$$

in other words, the cocycles must be inner automorphisms;

(e) the open cover $\mathcal{V}$ is admissible, i.e., $H^n(V_{\alpha_1} \cap \ldots \cap V_{\alpha_r}, \mathcal{F}) = 0$ for $\mathcal{F} = \mathcal{G}_0^p, \mathcal{A}_0^i$, for all $n \geq 1$, and for all indices $\alpha_1, \ldots, \alpha_r$.

(2) An *enhanced generically log smooth deformation of type* $\mathcal{D}$ of $f_0 \colon X_0 \to S_0$ is a generically log smooth deformation $f_A \colon X_A \to S_A$ of $f_0 \colon X_0 \to S_0$ together with isomorphisms $\chi_\alpha \colon X_A|_\alpha \cong V_{\alpha;A}$ of generically log smooth deformations such that

$$\psi_{\beta\alpha;A} \circ \chi_\beta|_{\alpha\beta} \circ \chi_\alpha^{-1}|_{\alpha\beta} \colon \; V_{\alpha;A}|_{\alpha\beta} \to V_{\alpha;A}|_{\alpha\beta}$$

is an inner automorphism of the enhanced generically log smooth family $V_{\alpha;A} \to S_A$. Transporting $\mathcal{G}\mathcal{C}_{\alpha;A}^\bullet \subseteq \mathcal{V}\mathcal{W}_{\alpha;A}^\bullet$ along χ_α gives rise to a Gerstenhaber calculus $\mathcal{G}\mathcal{C}_{X_A/S_A}^\bullet \subseteq \mathcal{V}\mathcal{W}_{X_A/S_A}^\bullet$, which is independent of α on overlaps, hence well-defined, due to our condition on the automorphisms on overlaps. This turns $f_A \colon X_A \to S_A$ into an enhanced generically log smooth family.

(3) Let $B' \to B$ be a morphism in $\mathbf{Art}_\Lambda$, and let $f' \colon (X_{B'}, \{\chi_\alpha'\}) \to S_{B'}$ and $f \colon (X_B, \{\chi_\alpha\}) \to S_B$ be two enhanced generically log smooth deformations of type $\mathcal{D}$. Then a *morphism* $c \colon X_B \to X_{B'}$ is a morphism of generically log smooth deformations such that $\chi'_\alpha|_B \circ \chi_\alpha^{-1}$ is an inner automorphism of the enhanced generically log smooth family $V_{\alpha;B} \to S_B$.

(4) Two enhanced generically log smooth deformations $f \colon X_A \to S_A$ and $f' \colon X_A' \to S_A$ are *equivalent* if there is an isomorphism between them. This means that we have an isomorphism of generically log smooth deformations such that $\chi'_\alpha \circ \chi_\alpha^{-1}$ is an inner automorphism. Equivalence classes of enhanced generically log smooth deformations of type $\mathcal{D}$ form a functor of Artin rings

$$\mathrm{ELD}_{X_0/S_0}^{\mathcal{D}} \colon \; \mathbf{Art}_Q \to \mathbf{Set}.$$

Lemma 10.22 *The functor* $\mathrm{ELD}_{X_0/S_0}^{\mathcal{D}}$ *is a neat deformation functor, i.e., it satisfies* (H_0), (H_1), (H_2), *and* (H_2^+) *defined in Appendix A.*

Proof (H_0) is clear. For (H_1), set $B = A' \times_A A''$ and note that

$$V_{\alpha;A'} \times_{V_{\alpha;A}} V_{\alpha;A''} = V_{\alpha;B}, \quad \mathcal{G}C_{\alpha;A'} \times_{\mathcal{G}C_{\alpha;A}} \mathcal{G}C_{\alpha;A''} = \mathcal{G}C_{\alpha;B}$$

by the methods of Lemma 10.19 and Lemma 12.17 below. Then (H_1) follows essentially like in the proof of Lemma 10.19. To see (H_2) and (H_2^+), we use the method also used in the proof of Lemma 12.17 below. $\square$

Of course, if $\mathcal{G}C_0^\bullet = \mathcal{V}W_0^\bullet$ and $\mathcal{G}C_{\alpha;A}^\bullet = \mathcal{V}W_{\alpha;A}^\bullet$, then we have defined nothing new and just obtain generically log smooth deformations of type $\mathcal{D}$.

Example 10.23 Let $f_0 \colon (X_0, U_0) \to S_0$ be a generically log smooth family of surfaces. Assume that an enhanced structure is given Zariski locally around every point $z \in Z_0$, together with enhanced deformations over any $A \in \mathbf{Art}_Q$. Then we can construct an enhanced system of deformations $\mathcal{D}$ since the family is log smooth wherever two local patches overlap. For instance, we can apply the procedure in Example 1.152 and obtain an enhanced system of deformations already when we only know that the deformation given there exists locally. $\Diamond$

Example 10.24 In Example 1.160 in the introduction, we have seen an enhanced system of deformations which is not a plain system of deformations in the sense of Definition 10.15. We expect that similar enhanced systems of deformations can be constructed on other Fano toroidal crossing schemes. $\Diamond$

Remark 10.25 Although the local models for the deformations do not have the base change property as required in Definition 10.15, the functor $\mathrm{LD}_{X_0/S_0}^{\mathcal{D}}$ makes sense for an enhanced system of deformations as well, albeit that it is in general only a functor of Artin rings and not a deformation functor since the proof of condition (H_1) in Lemma 10.19 may fail. Since we do not require $H^n(V_\alpha, \mathcal{V}_{X_0/S_0}^{-1}) = 0$ for $n \geq 1$, and since $H^2(V_\alpha, \mathcal{V}_{X_0/S_0}^{-1})$ is no longer an obstruction space for liftings along small extensions in the situation where not all local automorphisms lift, the isomorphisms $\chi_\alpha \colon X_A|_\alpha \cong V_{\alpha;A}$ must be relaxed to X_A being locally isomorphic to $V_{\alpha;A}$, say on a cover $\mathcal{U} = \{U_i\}_i$ which refines $\mathcal{V}$. Moreover, the change of comparison map between χ_i and χ_j is no longer an inner automorphism of $V_{\alpha;A}|_{ij}$ as an enhanced generically log smooth family but only an automorphism of $V_{\alpha;A}|_{ij}$ as a generically log smooth family. In particular, these deformations do not necessarily carry a structure of enhanced generically log smooth family as dictated by $\mathcal{D}$. We obtain a comparison map

$$\mathrm{ELD}_{X_0/S_0}^{\mathcal{D}} \to \mathrm{LD}_{X_0/S_0}^{\mathcal{D}}$$

which, in general, may be neither injective nor surjective. Namely, deformations of first order can still be classified as usual, and the map is given by

$$H^1(X_0, \mathcal{G}_{X_0/S_0}^{-1}) \to H^1(X_0, \mathcal{V}_{X_0/S_0}^{-1})$$

on the level of tangent spaces. Note that $\mathbf{k} \cdot \varepsilon \otimes \mathcal{V}^{-1}_{X_0/S_0} = \ker(\mathcal{V}^{-1}_{X_\varepsilon/S_\varepsilon} \to \mathcal{V}^{-1}_{X_0/S_0})$ is the sheaf of infinitesimal first order automorphisms of $f_0 \times \mathbf{k}[\varepsilon]/(\varepsilon^2)$ as a generically log smooth family. $\diamond$

10.5.1 The Ubiquity of Enhanced Systems of Deformations

We conclude this section by showing that many generically log smooth one-parameter deformations to an integral scheme can be interpreted as arising from an enhanced system of deformations, albeit that we may not know how to construct it when we are not already given the deformation. We start with an easy lemma.

Lemma 10.26 *Let R be one of the rings $\mathbf{k}[t]$, $\mathbf{k}[t]_{(t)}$, or $\mathbf{k}[\![t]\!]$. Let $f: X \to S = \operatorname{Spec} R$ be a flat morphism whose fibers are pure of dimension d and satisfy (S_2). Let $U \subseteq X$ be an open subset such that $Z = X \setminus U$ has relative codimension ≥ 2, and let $\mathcal{F}$ be a reflexive sheaf on X which is locally free on U. Let $S_0 \subseteq S$ be the closed subscheme defined by $t = 0$, and let $f_0: X_0 \to S_0$ be the base change of $f: X \to S$. Let $\mathcal{F}_0 := \mathcal{F}|_{X_0}$. Then the natural map $\rho: \mathcal{F}_0 \to j_*\mathcal{F}_0|_{U_0}$ is injective.*

Proof Inside U, every stalk $\mathcal{F}_x$ is flat over $O_{S, f(x)}$ and hence over R. Since R is a Dedekind domain, flatness is equivalent to being torsion-free. In particular, $\mathcal{F}_x$ has no t-torsion for $x \in U$, and hence $\mathcal{F} = j_*\mathcal{F}|_U$ has no t-torsion at any stalk. Thus, the left map in the sequence

$$0 \to \mathcal{F} \xrightarrow{\cdot t} \mathcal{F} \to \mathcal{F}_0 \to 0$$

is injective, and since $X_0 \subseteq X$ is defined by the ideal (t), the sequence is exact. Let $\theta \in \mathcal{F}$ be such that $\theta|_{X_0}$ is in the kernel of $\rho: \mathcal{F}_0 \to j_*\mathcal{F}_0|_{U_0}$. Then $\theta|_U$ is in the image of the multiplication with t. Let $g \in \Gamma(U, \mathcal{F})$ be such that $tg = \theta|_U$. Since $\mathcal{F}$ is reflexive, our element g can be extended to X, and we have $tg = \theta$ for this extension. $\square$

Here is the construction of the system of deformations.

Construction 10.27 Let $\Sigma = \operatorname{Spec}(\mathbb{N} \to \mathbf{k}[\![t]\!])$, and let $f: (X, U) \to \Sigma$ be a log Gorenstein generically log smooth family of relative dimension d. Let us assume that $f: X \to \Sigma$ is vertical, which ensures that the generic fiber X_η carries the trivial log structure, hence we have a genuine deformation to a scheme. We also assume that the fibers of $f: X \to \Sigma$ are Cohen–Macaulay. Let us first study the geometry of $f: X \to \Sigma$. For simplicity, we assume that X is connected. Using [267, 055J], also the generic fiber X_η is connected. Since X_η carries the trivial log structure, it is smooth on U_η, i.e., regular in codimension 1. Since it also satisfies Serre's condition (S_2), we find that X_η is normal. Because it is connected, it is then integral as well. We may show similarly that the total space X is normal; namely, the generic points of the irreducible components of X_0 are regular. Since X is connected, it must be integral. By Lemma 8.54,[8] our log Gorenstein assumption ensures that both fibers X_0 and X_η are Gorenstein. Then also the total space X is Gorenstein.

On X, we have a locally Batalin–Vilkovisky Gerstenhaber calculus $\mathcal{VW}^\bullet_{X/\Sigma}$ from Proposition 8.5. Its pieces are locally free on U, so they are flat over $\mathbf{k}[\![t]\!]$, and hence they have no t-torsion, neither on U nor on X. Since $\mathbf{k}[\![t]\!]$ is a Dedekind domain, this implies that these sheaves are flat over Σ. Lemma 10.26 shows that we have injections $\mathcal{VW}^P_{X/\Sigma}|_{X_0} \subseteq \mathcal{VW}^P_{X_0/S_0}$, and on the fiber X_η this holds because $X_\eta \subseteq X$ is an open subset. Then Lemma B.4 shows that this map is injective for any base change $g\colon Y \to T$ of $\hat{f}\colon X \to \Sigma$.

Given a, say affine, cover $\mathcal{V} = \{\hat{V}_\alpha\}_\alpha$ of X, we obtain an enhanced system of deformations $\mathcal{D}$ for $f_0\colon X_0 \to S_0$ subordinate to $\mathcal{V}$ by setting $V_{\alpha;A} := (f \times_S S_A)|_{\hat{V}_\alpha}$ and $\mathcal{GC}^\bullet_{\alpha;A} := \mathcal{VW}^\bullet_{X/\Sigma}|_{V_{\alpha;A}}$. The restriction maps and comparison isomorphisms come from the global nature of $f \times_S S_A$. Then the cocycles have the required form because they are the identity. Obviously, $f \times_S S_A$ is an enhanced generically log smooth deformation of $f_0\colon X_0 \to S_0$ of type $\mathcal{D}$. $\Diamond$

Finally, let us illustrate the construction with the family from Example 10.5.

Example 10.28 Let

$$f\colon\quad X = \operatorname{Spec} \mathbf{k}[x, y, z, w, u, t]/(xy - t^3 - w^3,\ zw - tu) \to \operatorname{Spec}\mathbf{k}[t] = S$$

be the obvious map $t \mapsto t$. It is smooth over $t \neq 0$, and generically log smooth once we endow it with the compactifying log structure coming from $t = 0$. Outside

$$\{0\} = \{x = y = z = w = u = t = 0\},$$

it is unisingular in the sense of Definition 14.20. More precisely, it has local models of the form $xy = tz$ respectively $xy = t^3z$. Let us denote the Gerstenhaber calculus by $\mathcal{GC}^\bullet_S := \mathcal{VW}^\bullet_S = (\mathcal{V}^\bullet_{X/S}, \mathcal{W}^\bullet_{X/S})$. Then we obtain an enhanced system of deformations with a single affine open $V_\alpha = X_0$ by setting $\mathcal{GC}^\bullet_A := \mathcal{GC}^\bullet_S \otimes_S S_A$. The map $f\colon X \to S$ is flat because $\mathcal{O}_X$ is torsion-free and $\mathbf{k}[t]$ is a Dedekind domain. Similarly, $\mathcal{G}^p_S$ and $\mathcal{A}^i_S$ are flat over S. The central fiber is reduced and pure of dimension 3. It is the product of the boundary of the Gorenstein affine toric variety $\operatorname{Spec}\mathbf{k}[x, y, z, w]/(xy - w^3)$ with $\mathbb{A}^1_u$ and hence Gorenstein itself. In particular, $f\colon X \to S$ is a Gorenstein and hence Cohen–Macaulay morphism. As in Example 10.5, we can compute $\mathcal{G}^{-1}_S$ as the relative classical derivations. Then $\mathcal{G}^{-3}_S = (\bigwedge^3 \mathcal{G}^{-1}_S)^{\vee\vee} = j_*\mathcal{G}^{-3}_S|_U$ is isomorphic to $\mathcal{O}_X$, where $U \subseteq X$ is the log smooth locus—this seems to be the hardest part of the computation. Then $\mathcal{A}^3_S \cong (\mathcal{G}^{-1}_S)^\vee \cong \mathcal{O}_X$ is a line bundle, and this still holds, of course, after any base change, i.e., the family is log Gorenstein, which was to be expected in view

---

[8] It is here that we use our assumption that the fibers of $f\colon X \to \Sigma$ are Cohen–Macaulay and do not just satisfy Serre's condition (S_2).

of Lemma 8.54 since $f\colon X \to S$ is vertical. Another computation in Macaulay2 shows that

$$\mathcal{G}_S^{-1} \otimes_S S_0 = \mathcal{G}_0^{-1} \to j_*\mathcal{G}_0^{-1}|_{U_0} \tag{10.1}$$

is injective, showing purity at G^{-1} (in the sense of Definition 12.1). Similarly, one can show purity at G^{-2}. Of course, we already know this by Lemma 10.26. Since the map (10.1) is not surjective, not every infinitesimal automorphism of generically log smooth families corresponds to a gauge transform for $\mathcal{G}_0^{-1}$. This is possible because $(\mathcal{G}_0^\bullet, \mathcal{A}_0^\bullet)$ is not equal to the reflexive Gerstenhaber calculus $(\mathcal{V}_{X_0/S_0}^\bullet, \mathcal{W}_{X_0/S_0}^\bullet)$ and its variants over infinitesimal deformations who show up in Lemma 10.6.   ◊

Chapter 11
Deformations with a Vector Bundle

A generically log smooth family with a vector bundle $f\colon (X, U, \mathcal{E}) \to S$ is, as the name says, a generically log smooth family $f\colon (X, U) \to S$ together with a vector bundle $\mathcal{E}$ on X.[1] Let us assume that $\mathcal{E}$ is of constant rank $r \geq 1$. A morphism from a generically log smooth family with a vector bundle $g\colon (Y, V, \mathcal{E}_Y) \to T$ to another one $f\colon (X, U, \mathcal{E}_X) \to S$ is a morphism of generically log smooth families—consisting of two maps $c\colon Y \to X$ and $b\colon T \to S$ with $V \subseteq c^{-1}(U)$—together with a homomorphism $\mathcal{E}_X \to c_*\mathcal{E}_Y$ of O_X-modules which induces an isomorphism $c^*\mathcal{E}_X \cong \mathcal{E}_Y$.

Similarly, an enhanced generically log smooth family with a vector bundle $f\colon (X, U, \mathcal{E}) \to S$ consists of an enhanced generically log smooth family $f\colon (X, U) \to S$ and a vector bundle $\mathcal{E}$ on X, which we assume to be of some constant rank $r \geq 1$. A morphism is a morphism of enhanced generically log smooth families together with a compatible map $\mathcal{E}_X \to c_*\mathcal{E}_Y$ which induces an isomorphism $c^*\mathcal{E}_X \cong \mathcal{E}_Y$.

In this chapter, we study the deformation theory of these objects. Our main result is the unobstructedness theorem for line bundles Theorem 16.2. Our original intention had been to apply the theory developed in Chap. 13 directly to the Lie–Rinehart pair constructed below, and to analyze the characteristic Lie–Rinehart pair that we obtain from it. This is the strategy employed by Iacono and Manetti

---

[1] We will be mainly interested in the case of line bundles because then the deformation functor is unobstructed in the Calabi–Yau case. However, the basic theory works for vector bundles as well. We restrict to vector bundles as opposed to more general modules because they have an obvious local deformation once we have a deformation of the underlying space.

S. Felten, *Global Logarithmic Deformation Theory*, Lecture Notes
in Mathematics 2373, https://doi.org/10.1007/978-3-031-98751-9_11

in [154].[2] In our proof of Theorem 16.2, we have however adopted a simpler and more geometric strategy, which reduces deformations of line bundles to deformations of a $\mathbb{P}^1$-bundle which is log Calabi–Yau. It is still open to investigate the relation between the characteristic Gerstenhaber calculus of that $\mathbb{P}^1$-bundle and the characteristic Lie–Rinehart pair.

11.1   Generically Log Smooth Deformations with a Vector Bundle

Let $S_0 = \mathrm{Spec}(Q \to \mathbf{k})$ for some sharp toric monoid Q, and let $f_0\colon (X_0, U_0, \mathcal{E}_0) \to S_0$ be a generically log smooth family of relative dimension d with a vector bundle $\mathcal{E}_0$ of rank r.

Definition 11.1 Let $A \in \mathbf{Art}_Q$. Then a *generically log smooth deformation with a vector bundle* is a map of generically log smooth families with a vector bundle

$$
\begin{array}{ccc}
(X_0, U_0, \mathcal{E}_0) & \xrightarrow{\;\;i\;\;} & (X_A, U_A, \mathcal{E}_A) \\
\downarrow{\scriptstyle f_0} & & \downarrow{\scriptstyle f_A} \\
S_0 & \longrightarrow & S_A
\end{array}
$$

such that the underlying diagram of partial log schemes is Cartesian. In particular, $i^{-1}(U_A) = U_0$. Morphisms and isomorphisms of generically log smooth deformations with a vector bundle are defined in analogy with the case of generically log smooth deformations, in particular, they are compatible with the map from the central fiber. Isomorphism classes of generically log smooth deformations with a vector bundle form a functor of Artin rings

$$
\mathrm{LD}^{\mathrm{gen}}_{X_0/S_0}(\mathcal{E}_0)\colon \quad \mathbf{Art}_Q \to \mathbf{Set}. \tag{11.1}
$$

Lemma 11.2 *The functor of Artin rings* (11.1) *is a neat deformation functor, i.e., it satisfies* (H_0), (H_1), (H_2), *and* (H_2^+) *defined in Appendix A.*

Proof The condition (H_0) is clear. For (H_1), let $B = A' \times_A A''$, let $S = S_A$, $S' = S_{A'}$, $S'' = S_{A''}$, $T = S_B$, and let $f'\colon (X', \mathcal{E}') \to S'$ and $f''\colon (X'', \mathcal{E}'') \to S''$ be two

<hr>

deformations which restrict to $f: (X, \mathcal{E}) \to S$ via a choice of the restriction map. For the lift $g: Y \to T$, we take as underlying generically log smooth deformation

$$Y := (|X_0|, \, \mathcal{O}_{X'} \times_{\mathcal{O}_X} \mathcal{O}_{X''}, \, \mathcal{M}_{U'} \times_{\mathcal{M}_U} \mathcal{M}_{U''}).$$

Then we set $\mathcal{F} := \mathcal{E}' \times_{\mathcal{E}} \mathcal{E}''$ along the chosen maps $\mathcal{E}' \to \mathcal{E}$ and $\mathcal{E}'' \to \mathcal{E}$; this is an $\mathcal{O}_Y$-module because the restrictions on the level of the vector bundles are compatible with the restrictions on the level of the structure sheaves, and it is quasi-coherent because the construction commutes with localization (in local sections $f \in \mathcal{O}_Y$ over affines). Then $\mathcal{F}$ is locally free by [80, Thm. 2.2].

For (H_2) and (H_2^+), let $A = \mathbf{k}$. Let $g: (Z, \mathcal{H}) \to T$ be a simultaneous lift of $f': (X', \mathcal{E}') \to S'$ and $f'': (X'', \mathcal{E}'') \to S''$. Let $g: (Y, \mathcal{F}) \to T$ be the lift constructed above. The universal property of the fiber products gives a map $Z \to Y$ of generically log smooth families together with a compatible map $\mathcal{F} \to \mathcal{H}$ of modules. Since generically log smooth deformations form a neat deformation functor, $Z \to Y$ is an isomorphism (cf. [160, Lemma 9.2]). But then also $\mathcal{F} \to \mathcal{H}$ is an isomorphism because both coherent sheaves are flat over T, and the pull-back of the map to S_0 is an isomorphism. $\qquad\square$

11.2 Derivations of $f: (X, \mathcal{E}) \to S$

We have seen in Chap. 10.2 that infinitesimal automorphisms of deformations of generically log smooth families are controlled by log derivations. Here, we introduce the analogous notion of derivations for generically log smooth families with a vector bundle $f: (X, \mathcal{E}) \to S$.

Definition 11.3 A *derivation* of $f: (X, \mathcal{E}) \to S$ on an open subset $V \subseteq X$ is a triple (D, Δ, u) where $D: \mathcal{O}_X|_V \to \mathcal{O}_X|_V$ is a relative derivation of $f: X \to S$, the map

$$\Delta: \quad \mathcal{M}_U|_{V \cap U} \to \mathcal{O}_U|_{V \cap U}$$

is a homomorphism of sheaves of monoids which turns (D, Δ) into a relative log derivation of $f: U \cap V \to S$, and $u: \mathcal{E}|_V \to \mathcal{E}|_V$ is a homomorphism of Abelian sheaves such that

$$u(a \cdot e) = D(a) \cdot e + a \cdot u(e)$$

for $a \in \mathcal{O}_X, e \in \mathcal{E}$. In particular, u is $f^{-1}(\mathcal{O}_S)$-linear. Derivations of $f: (X, \mathcal{E}) \to S$ form a sheaf $\Theta^1_{X/S}(\mathcal{E})$ of $\mathcal{O}_X$-modules.

At this point, we do not yet know that $\Theta^1_{X/S}(\mathcal{E})$ is coherent. However, we have an O_X-linear forgetful map $\Theta^1_{X/S}(\mathcal{E}) \to \Theta^1_{X/S}$ forgetting u, and we have an O_X-linear inclusion

$$\mathcal{E}nd(\mathcal{E}) \to \Theta^1_{X/S}(\mathcal{E}), \quad A \mapsto (D(a) = 0, \ \Delta(m) = 0, \ u(e) = A(e)).$$

This gives rise to a sequence

$$0 \to \mathcal{E}nd(\mathcal{E}) \to \Theta^1_{X/S}(\mathcal{E}) \to \Theta^1_{X/S} \to 0 \qquad \text{(AE)}$$

of O_X-modules, which we call the *Atiyah extension* in analogy with the classical deformation theory of smooth manifolds with a line bundle.

Lemma 11.4 *The sequence* (AE) *is exact and Zariski locally split. In particular, $\Theta^1_{X/S}(\mathcal{E})$ is a coherent reflexive sheaf, locally free of rank $d + r^2$ on U.*

Proof As is easily checked, this sequence is exact on the left and in the middle for the sections over any open subset. If $V \subseteq X$ is an open on which $\mathcal{L}|_V \cong O_X|_V$ via the local basis $e_1, \ldots, e_r$, and if $(D, \Delta) \in \Gamma(V, \Theta^1_{X/S})$, then we form a splitting by setting $u(\sum_i a_i e_i) := \sum_i D(a_i)e_i$. $\square$

We turn $\Theta^1_{X/S}(\mathcal{E})$ into a sheaf of Lie algebras by setting

$$[(D_1, \Delta_1, u_1), (D_2, \Delta_2, u_2)]$$
$$:= (D_1 \circ D_2 - D_2 \circ D_1, \ D_1 \circ \Delta_2 - D_2 \circ \Delta_1, \ u_1 \circ u_2 - u_2 \circ u_1).$$

This is a natural extension of the Lie algebra structure on $\Theta^1_{X/S}$; the Lie bracket is $f^{-1}(O_S)$-linear but not O_X-linear. The sheaf $\Theta^1_{X/S}(\mathcal{E})$ also fits into a Lie–Rinehart pair

$$\mathcal{LRP}^\bullet_{X/S}(\mathcal{E})$$

as follows: We set

$$\mathcal{LRP}^F = O_X, \quad \mathcal{LRP}^T = \Theta^1_{X/S}(\mathcal{E}), \quad \mathcal{LRP}^E = \mathcal{E}$$

as modules over the ring $O_X = \mathcal{LRP}^F$. Then we set $\nabla^F_{(D,\Delta,u)}(a) := -D(a)$ and $\nabla^E_{(D,\Delta,u)}(e) := -u(e)$ as well as

$$\nabla^T_{(D_1, \Delta_1, u_1)}(D_2, \Delta_2, u_2) := -[(D_1, \Delta_1, u_1), (D_2, \Delta_2, u_2)],$$

where the $(-)$-sign is in accordance with our convention of taking the negative of the Schouten–Nijenhuis bracket.

Lemma 11.5 *We have:*

(1) *If $\Theta^1_{X/S}$ is flat over S, then $\mathcal{LRP}^\bullet_{X/S}(\mathcal{E})$ is a Lie–Rinehart pair in $\mathfrak{Coh}(X/S)$.*
(2) *If $f\colon X \to S$ has the base change property and is log Gorenstein,[3] then the formation of $\mathcal{LRP}^\bullet_{X/S}(\mathcal{E})$ commutes with base change.*
(3) *$\mathcal{LRP}^\bullet_{X/S}(\mathcal{E})$ is Z-faithful in the sense of Definition 3.40 for $Z = \{F, E\}$ in accordance with the discussion after Definition 3.37.*

Proof Part (a) is mostly computational. Flatness is required because of our definition of the context $\mathfrak{Coh}(X/S)$. For part (b), let $b\colon T \to S$, let $c\colon Y = X \times_S T \to X$, and let $g\colon (Y, c^*\mathcal{E}) \to T$ be the generically log smooth family with a vector bundle obtained by base change. Then we have obvious maps $O_X \to c_*O_Y$ and $\mathcal{E} \to c_*c^*\mathcal{E}$, which induce isomorphisms on Y. To construct the map

$$\Theta^1_{X/S}(\mathcal{E}) \to c_*\Theta^1_{Y/T}(c^*\mathcal{E}),$$

let $\theta = (D, \Delta, u) \in \Theta^1_{X/S}(\mathcal{E})$. Then (D, Δ) corresponds to a homomorphism $k\colon \mathcal{W}^1_{X/S} \to O_X$, which can be pulled back to $\mathcal{W}^1_{Y/T} \to O_Y$. The map $u\colon \mathcal{E} \to \mathcal{E}$ is a differential operator of order 1, so there is a pull-back $c^*\mathcal{E} \to c^*\mathcal{E}$ as well. This gives the desired section of $c_*\Theta^1_{Y/T}(c^*\mathcal{E})$. Since this map is compatible with the two Atiyah extensions and $c^*\Theta^1_{X/S} \cong \Theta^1_{Y/T}$ by the base change property and log Gorenstein assumption, we obtain an isomorphism on Y. It remains to show that the maps ∇^P on $\mathcal{LRP}^\bullet_{Y/T}(c^*\mathcal{E})$ are the ones induced via pull-back of multilinear differential operators from $\mathcal{LRP}^\bullet_{X/S}(\mathcal{E})$. We can assume that X, S, T, Y are all affine. Then, on global sections of Y, the claim follows from the compatibility of the maps on X and O_T-(multi-)linearity. On open subsets of Y, the claim follows from the unique extendability of multilinear differential operators to localizations. For part (c), let

$$\theta = (D, \Delta, u) \in \Gamma(V, \Theta^1_{X/S}(\mathcal{E}))$$

be such that $D(a) = 0$ for all $a \in \Gamma(V', O_X)$ and $u(e) = 0$ for all $e \in \Gamma(V', \mathcal{E})$ for $V' \subseteq V$. Since $\Theta^1_{X/S}$ is strictly faithful, we have $(D, \Delta) = 0$. Since $u = 0$, we find $\theta = 0$, so $\mathcal{LRP}^\bullet_{X/S}(\mathcal{E})$ is Z-faithful. $\qquad\square$

Remark 11.6 Note that $\mathcal{LRP}^\bullet_{X/S}(\mathcal{E})$ may not be Z-faithful if we would choose $Z = \{F\}$. Faithfulness is only achieved by considering the action on O_X and on $\mathcal{E}$ at once. $\qquad\Diamond$

---

[3] In this case, $\Theta^1_{X/S}$ is flat over S by Proposition 8.5.

11.3   Infinitesimal Automorphisms

Let $f_0 \colon (X_0, \mathcal{E}_0) \to S_0$ be a generically log smooth family with a vector bundle. Let $B' \to B$ be a surjection in $\mathbf{Art}_Q$ with kernel $I \subset B'$, and let $f \colon (X, \mathcal{E}) \to S = S_B$ and $f' \colon (X', \mathcal{E}') \to S' = S_{B'}$ be two generically log smooth deformations with a vector bundle. Assume we have a morphism $f \to f'$. We denote the sheaf of automorphisms of $f' \colon (X', \mathcal{E}') \to S'$ over $f \colon (X, \mathcal{E}) \to S$ by $\mathcal{A}ut_{(X',\mathcal{E}')/(X,\mathcal{E})}$. A section over $V \subseteq X'$ consists of an automorphism $\phi \colon O_{X'} \to O_{X'}$ of sheaves of rings on V, a map $\Phi \colon M_{X'} \to M_{X'}$ on $V \cap U'$ that yields an automorphism of log schemes, and a map $\psi \colon \mathcal{E}' \to \mathcal{E}'$ on V which is compatible with ϕ. As in Lemma 10.4, for every open $V \subseteq X_0$, the restriction

$$\Gamma(V, \mathcal{A}ut_{(X',\mathcal{E}')/(X,\mathcal{E})}) \to \Gamma(V \cap U_0, \mathcal{A}ut_{(X',\mathcal{E}')/(X,\mathcal{E})})$$

is an isomorphism. Let

$$\Theta^1_{X'/S'}(\mathcal{E}', I)$$

be the sheaf of those derivations (D, Δ, u) of $f \colon (X', \mathcal{E}') \to S'$ with D and Δ mapping into $I \cdot O_{X'}$ and u mapping into $I \cdot \mathcal{E}'$.

Lemma 11.7 *We have two inverse isomorphisms*

$$\Theta^1_{X'/S'}(\mathcal{E}', I) \; \underset{\mathrm{Log}}{\overset{\mathrm{Exp}}{\rightleftarrows}} \; \mathcal{A}ut_{(X',\mathcal{E}')/(X,\mathcal{E})}$$

of sheaves of groups, where the left-hand side is endowed with the Baker–Campbell–Hausdorff product induced from the original[4] Lie bracket $[-, -] = -\nabla^T$. If $\theta = (D, \Delta, u) \in \Theta^1_{X'/S'}(\mathcal{E}', I)$, then $(\phi, \Phi, \psi) = \mathrm{Exp}(\theta)$ is given by the formulae in Chap. 10.2 for ϕ, Φ, and by

$$\psi \colon \; \mathcal{E}' \to \mathcal{E}', \quad e \mapsto \sum_{n=0}^{\infty} \frac{u^n(e)}{n!} = e + u(e) + \frac{1}{2}u^2(e) + \dots;$$

if $(\phi, \Phi, \psi) \in \mathcal{A}ut_{(X',\mathcal{E}')/(X,\mathcal{E})}$, then $(D, \Delta, u) = \mathrm{Log}(\phi, \Phi, \psi)$ is given by the formulae in Chap. 10.2, and by

$$u \colon \; \mathcal{E}' \to \mathcal{E}', \quad e \mapsto \sum_{n=1}^{\infty} \frac{(-1)^{n-1}[\psi - \mathrm{Id}]^n(e)}{n}.$$

---

[4] As in the discussion of automorphisms of generically log smooth families above, we use the original (not negative) Lie bracket on $\Theta^1(X/S, \mathcal{E})$ when describing automorphisms.

Proof First, let us work on U_0. Given $(D, \Delta, u) \in \Theta^1_I(X'/S', \mathcal{E}')$, a rather easy direct computation shows that $(\phi, \Phi, \psi) = \mathrm{Exp}(D, \Delta, u)$ is an automorphism. For the converse, the hardest part is to show that $u = \mathrm{Log}(\psi)$ satisfies $u(ae) = D(a)e + au(e)$. This can be achieved by an easy adaptation of the corresponding part in the proof of [76, Lemma 2.3]. Because $\mathrm{Exp}(u)$ and $\mathrm{Log}(\psi)$ are given by the formulae for the exponential respectively the logarithm, they are inverse to each other. Thus, we have two isomorphisms of sheaves of *sets* on U_0. Now let $Z_0 = X_0 \setminus U_0$. Since $O_{X'}$ and $\mathcal{E}'$ are Z_0-closed, i.e., $j_* O_{U'} = O_{X'}$ and $j_* \mathcal{E}'|_{U'} = \mathcal{E}'$, it is easy to show that the automorphism sheaf is Z_0-closed. Similarly, $\Theta^1_{X'/S'}(\mathcal{E}', I)$ is Z_0-closed; in fact, both $I \cdot O_{X'}$ and $I \cdot \mathcal{E}'$ are Z_0-closed as well because they are the kernels of $O_{X'} \to O_X$ respectively $\mathcal{E}' \to \mathcal{E}$, and both $O_{X'}$ and $\mathcal{E}'$ are flat over S'. Thus, we have two isomorphisms of sheaves of sets on X'. The proof that Exp is a group homomorphism for the Baker–Campbell–Hausdorff formula is similar to the proof of Lemma 4.4. Then its inverse Log is a group homomorphism as well. $\square$

Remark 11.8 If $\Theta^1_{X'/S'}$ is flat over S', then the canonical map

$$I \cdot \Theta^1_{X'/S'}(\mathcal{E}') \to \Theta^1_{X'/S'}(\mathcal{E}', I)$$

is an isomorphism. Namely, the right side is the kernel of $\Theta^1_{X'/S'}(\mathcal{E}') \to \Theta^1_{X/S}(\mathcal{E})$, which can be computed as the left side if $\Theta^1_{X'/S'}$, hence $\Theta^1_{X'/S'}(\mathcal{E}')$, is flat over S'. $\Diamond$

11.3.1 *Automorphisms of the Lie–Rinehart Pair*

Let us assume that $f_0 \colon (X_0, \mathcal{E}_0) \to S_0$ is log Gorenstein, and that every deformation which we consider has the base change property. Let $f' \colon (X', \mathcal{E}') \to S'$ be a deformation over S' with base change $f \colon (X, \mathcal{E}) \to S$, and let $\varphi = (\phi, \Phi, \psi)$ be an automorphism of f' over f. We wish to construct an induced action on $\mathcal{LRP}^\bullet_{X'/S'}(\mathcal{E}')$. On $\mathcal{LRP}^F_{X'/S'}(\mathcal{E}') = O_{X'}$ and $\mathcal{LRP}^E_{X'/S'}(\mathcal{E}') = \mathcal{E}'$, the action should be given by ϕ respectively ψ. As we have seen above, we have $\phi = \exp_{-\theta}$ and $\psi = \exp_{-\theta}$ for $\theta = \mathrm{Log}(\varphi)$. Then we just *define* the induced action of φ on $\mathcal{LRP}^T_{X'/S'}(\mathcal{E}')$ by $\exp_{-\theta}$. Namely, this is automatically a gauge transform and hence an automorphism of Lie–Rinehart pairs, it induces automatically the identity on $f \colon (X, \mathcal{E}) \to S$, and it is automatically compatible with $T\varphi^*$ along the projection in the Atiyah extension (AE) since this projection is a morphism of Lie algebras and $T\varphi^*$ is equal to $\exp_{-\theta}$ in $\mathcal{LR}^T_{X'/S'}$. This construction yields a one-to-one correspondence between automorphisms in $\mathcal{A}ut_{(X',\mathcal{E}')/(X,\mathcal{E})}$ and gauge transforms of $\mathcal{LRP}^\bullet_{X'/S'}(\mathcal{E}')$ induced from $\theta \in I \cdot \mathcal{LRP}^T_{X'/S'}(\mathcal{E}')$ because $\mathcal{LRP}^\bullet_{X'/S'}(\mathcal{E}')$ is Z-faithful by Lemma 11.5.

11.4   Systems of Deformations

Similar to the case of plain generically log smooth deformations, we wish to devise a method to define a subfunctor of $\mathrm{LD}^{\mathrm{gen}}_{X_0/S_0}(\mathcal{E}_0)$ which classifies specific generically log smooth deformations with a vector bundle which are locally unique. Let us fix an open cover of X_0 which is weakly admissible for $f_0 \colon X_0 \to S_0$ in the sense of Definition 10.13. This specifies how deformations of the underlying generically log smooth family should look like locally. If we had allowed arbitrary coherent sheaves $\mathcal{E}_0$, then we would also need to specify the local deformations of $\mathcal{E}_0$, but since we insist that $\mathcal{E}_0$ is a vector bundle, we just require the deformation to be a vector bundle as well. In fact, every flat deformation $\mathcal{E}$ of $\mathcal{E}_0$, i.e., $f \colon X \to S$ is flat and $\mathcal{E}$ is flat over S, is locally free. Namely, a local trivialization can be lifted to $\mathcal{E}$.

Definition 11.9 Let $\mathcal{V} = \{V_\alpha\}_\alpha$ be an open cover of X_0 by finitely many open immersions. Then it is *weakly $\mathcal{E}_0$-admissible* if it is weakly admissible and

$$\mathcal{E}_0|_\alpha \cong O_{X_0}^{\oplus r}|_\alpha$$

for every α, where we write $\mathcal{S}|_\alpha := \mathcal{S}|_{V_\alpha}$ for any sheaf $\mathcal{S}$ on X_0.

In this case, a flat deformation $\mathcal{E}$ of $\mathcal{E}_0$ is not only locally trivial but trivial on each V_α. Namely, since $H^1(V_\alpha, O_{X_0}|_\alpha) = 0$, we can lift the trivialization of $\mathcal{E}_0|_\alpha$ to $\mathcal{E}|_\alpha$.

Definition 11.10 Let $f_0 \colon X_0 \to S_0$ be a log Gorenstein generically log smooth family of relative dimension d, and let $\mathcal{E}_0$ be a vector bundle of rank r on X_0. Let $\mathcal{V} = \{V_\alpha\}_\alpha$ be a weakly $\mathcal{E}_0$-admissible open cover of X_0. Then a *system of deformations* $\mathscr{D}$ for $f_0 \colon (X_0, \mathcal{E}_0) \to S_0$ subordinate to $\mathcal{V}$ is a system of deformations for $f_0 \colon X_0 \to S_0$ subordinate to $\mathcal{V}$. A generically log smooth deformation with a vector bundle $f_A \colon (X_A, \mathcal{E}_A) \to S_A$ is *of type $\mathscr{D}$* if $f \colon X_A \to S_A$ is of type $\mathscr{D}$. Isomorphism classes of these deformations form a functor of Artin rings

$$\mathrm{LD}^{\mathscr{D}}_{X_0/S_0}(\mathcal{E}_0) \colon \quad \mathbf{Art}_Q \to \mathbf{Set}. \tag{11.2}$$

Deformations of type $\mathscr{D}$ are now such deformations which are, on V_α, isomorphic to $(V_{\alpha;A}, O_{V_{\alpha;A}}^{\oplus r}) \to S_A$.

Lemma 11.11 *The functor of Artin rings* (11.2) *is a neat deformation functor, i.e., it satisfies* (H_0), (H_1), (H_2), *and* (H_2^+) *defined in Appendix A.*

Proof This is a combination of the proofs of Lemma 10.19 and Lemma 11.2. $\square$

11.5 Enhanced Systems of Deformations

Now we present the version for enhanced generically log smooth families. As usual, we restrict to the simpler torsionless case.

Definition 11.12 Let $f_0\colon X_0 \to S_0$ be a torsionless enhanced generically log smooth family of relative dimension d, and let $\mathcal{E}_0$ be a vector bundle on X_0 of constant rank $r \geq 1$. Let $\mathcal{V} = \{V_\alpha\}_\alpha$ be a weakly $\mathcal{E}_0$-admissible open cover of X_0. Then an *enhanced system of deformations* $\mathscr{D}$ for $f_0\colon (X_0, \mathcal{E}_0) \to S_0$ subordinate to $\mathcal{V}$ is an enhanced system of deformations $\mathscr{D}$ subordinate to $\mathcal{V}$. Enhanced deformations of type $\mathscr{D}$ are given by enhanced generically log smooth deformations $f\colon X_A \to S_A$ of type $\mathscr{D}$, i.e., generically log smooth deformations $f_A\colon X_A \to S_A$ together with isomorphisms $\chi_\alpha\colon X_A|_\alpha \cong V_{\alpha;A}$, together with a vector bundle $\mathcal{E}_A$ on X_A of constant rank r. Two enhanced generically log smooth deformations with a vector bundle *of type $\mathscr{D}$* are *equivalent* if there is an isomorphism between them together with a compatible isomorphism of the vector bundles. Equivalence classes of enhanced deformations form a deformation functor

$$\mathrm{ELD}^{\mathscr{D}}_{X_0/S_0}(\mathcal{E}_0)\colon \quad \mathbf{Art}_Q \to \mathbf{Set}.$$

Since we require $H^1(V_\alpha, O_{X_0}) = 0$, we have $\mathcal{E}_A|_\alpha \cong O^{\oplus r}_{V_\alpha;A}$.

Let $f\colon (X_A, \mathcal{E}_A) \to S_A$ be an enhanced generically log smooth deformation with a vector bundle of type $\mathscr{D}$. Through the comparison isomorphisms $\chi_\alpha\colon X_A|_\alpha \cong V_{\alpha;A}$, we have a distinguished subsheaf $\Gamma^1_{X_A/S_A} \subseteq \Theta^1_{X_A/S_A}$. We obtain an Atiyah extension

$$0 \to \mathcal{E}nd(\mathcal{E}_A) \to \Gamma^1_{X_A/S_A}(\mathcal{E}_A) \to \Gamma^1_{X_A/S_A} \to 0$$

by considering triples (D, Δ, u) with $(D, \Delta) \in \Gamma^1_{X_A/S_A}$ and an additive map $u\colon \mathcal{E}_A \to \mathcal{E}_A$ satisfying $u(a \cdot e) = D(a) \cdot e + a \cdot u(e)$. A Lie bracket is given on $\Gamma^1_{X_A/S_A}(\mathcal{E}_A)$ with the same formula as for $\Theta^1_{X_A/S_A}(\mathcal{E}_A)$. The kernel of

$$\Gamma^1_{X_{B'}/S_{B'}}(\mathcal{E}_{B'}) \to \Gamma^1_{X_B/S_B}(\mathcal{E}_B)$$

is given by $I \cdot \Gamma^1_{X_{B'}/S_{B'}}(\mathcal{E}_{B'})$ (since the sheaves are flat over the base), and every $\theta = (D, \Delta, u) \in I \cdot \Gamma^1_{X_{B'}/S_{B'}}(\mathcal{E}_{B'})$ induces an (inner) automorphism $\mathrm{Exp}(\theta)$ of the enhanced deformation which acts as $\exp_{-(D,\Delta)}$ on the two-sided Gerstenhaber calculus, and by

$$\psi(e) = \sum_{n=0}^{\infty} \frac{u^n(e)}{n!} = e + u(e) + \frac{1}{2}u^2(e) + \dots$$

on the vector bundle $\mathcal{E}_{B'}$. Every automorphism of the enhanced deformation is of this form because it induces an inner automorphism of the underlying enhanced generically log smooth family by definition.

We form a Lie–Rinehart pair consisting of $\mathcal{F} = O_{X_A}$, $\mathcal{T} = \Gamma^1_{X_A/S_A}(\mathcal{E}_A)$, and $\mathcal{E} = \mathcal{E}_A$ by endowing it with the negative brackets, i.e.,

$$\nabla^F_\theta(a) = -D(a), \quad \nabla^T_\theta(\xi) = -[\theta, \xi]_{\mathrm{Lie}}, \quad \nabla^E_\theta(e) = -u(e);$$

we denote it by $\mathcal{LRP}^\bullet_{X_A/S_A}(\mathcal{E}_A)$.

We shall occasionally also write $\mathcal{G}^{-1}_{X/S}(\mathcal{E})$ for $\Gamma^1_{X/S}(\mathcal{E})$ when considered endowed with the negative Lie bracket and the negative action on $\mathcal{E}$.

11.6   From a Vector Bundle to Its Determinant

In the above situation, we can form the determinant $\det(\mathcal{E}_A)$ of $\mathcal{E}_A$, and we can forget $\mathcal{E}_A$. This gives rise to a diagram

$$
\begin{array}{ccc}
D_r := \mathrm{ELD}^{\mathcal{D}}_{X_0/S_0}(\mathcal{E}_0) & \xrightarrow{\quad\pi\quad} & D_0 := \mathrm{ELD}^{\mathcal{D}}_{X_0/S_0} \\
& \searrow{\rho} \qquad\qquad {\sigma}\nearrow & \\
& D_1 := \mathrm{ELD}^{\mathcal{D}}_{X_0/S_0}(\det(\mathcal{E}_0)) &
\end{array}
$$

of deformation functors. In this section, we examine these maps briefly.

The main results of this monograph show that both D_0 and D_1 are smooth deformation functors if $\mathcal{A}^d_0 \cong O_{X_0}$ and certain technical conditions are met. Since the obstructions to extending a line bundle $\mathcal{L}_B$ from X_B to a given thickening $X_{B'}$ lie in $H^2(X_0, O_{X_0}) \otimes_{\mathbf{k}} I$, the map $\sigma : D_1 \Rightarrow D_0$ is smooth if $H^2(X_0, O_{X_0}) = 0$. However, if $H^2(X_0, O_{X_0}) \neq 0$, then the map is not always smooth.

Example 11.13 This example is taken from [139, Exer. 6.7(c)-(d)]. Let $X_0 \subset \mathbb{P}^3$ be the smooth quartic surface defined by $f_0 = x^4 + y^4 + xz^3 + yw^3$, and let $\mathcal{L}_0$ be the line bundle associated with the divisor $Y_0 = \{x = y = 0\}$. Since X_0 is a K3 surface, we have $H^2(X_0, O_{X_0}) \cong \mathbf{k} \neq 0$. Consider the deformation X of X_0 over $\mathbf{k}[t]/(t^2)$ defined by $f = f_0 + tz^2w^2$. Then there is no line bundle $\mathcal{L}$ on X with $\mathcal{L}|_0 \cong \mathcal{L}_0$. Thus, the map $\sigma : D_1 \Rightarrow D_0$ is not a smooth morphism of deformation functors. $\diamond$

Similarly, if $H^2(X_0, \mathcal{E}nd(\mathcal{E}_0)) = \mathrm{Ext}^2(\mathcal{E}_0, \mathcal{E}_0) = 0$, then the map $\pi : D_r \Rightarrow D_0$ is smooth by [139, Thm. 7.1].

Example 11.14 This example is due to R. Thomas. Let $X_0 \subseteq \mathbb{P}^2 \times \mathbb{P}^2$ be the $(3, 3)$-divisor of [268],[5] and let $\mathcal{E}_0$ be the vector bundle of rank 2 of [268, Thm. 1.1]. Let $X_1 = X_0 \times \operatorname{Spec} \mathbb{C}[t]/(t^2)$, and let $X_2 = X_0 \times \operatorname{Spec} \mathbb{C}[t]/(t^3)$. Let $\mathcal{E}_1$ be a lift of $\mathcal{E}_0$ to X_1 which is not trivial. Then $\mathcal{E}_1$ cannot be lifted to a vector bundle $\mathcal{E}_2$ on X_2. Thus, the map $\pi : D_2 \Rightarrow D_0$ of deformation functors is not smooth. $\Diamond$

Remark 11.15 It is currently an open question if, in this example, the deformation functor D_2 itself is smooth. Related to this, if D_2 is smooth here, it is open to find some Calabi–Yau space (say smooth projective) X_0 together with a vector bundle $\mathcal{E}_0$ of rank $r \geq 2$ such that D_r is not smooth. $\Diamond$

The trace map in linear algebra gives rise to an O_{X_0}-linear trace map

$$\operatorname{tr}: \quad \mathcal{E}nd(\mathcal{E}_0) \to O_{X_0}.$$

It satisfies $\operatorname{tr}(\mathrm{Id}) = r$, i.e., the composition with the diagonal embedding

$$L: \quad O_{X_0} \to \mathcal{E}nd(\mathcal{E}_0), \quad \lambda \mapsto (e \mapsto \lambda \cdot e),$$

is the multiplication with r. Since $\operatorname{char}(\mathbf{k}) = 0$, the trace map is split by $\frac{1}{r}L$. Thus, the induced map

$$H^2(\operatorname{tr}): \quad \operatorname{Ext}^2(\mathcal{E}_0, \mathcal{E}_0) \to H^2(X_0, O_{X_0})$$

is split and surjective; we denote its kernel by $\operatorname{Ext}^2(\mathcal{E}_0, \mathcal{E}_0)_0$.

Proposition 11.16 *Assume that* $\operatorname{Ext}^2(\mathcal{E}_0, \mathcal{E}_0)_0 = 0$. *Then* $\rho : D_r \Rightarrow D_1$ *is a smooth map of deformation functors.*

Proof Let $B' \to B$ be a first-order extension in $\mathbf{Art}_\Lambda$ with kernel $I \subset B'$, let $f : (X, \mathcal{E}) \to S_B$ be an enhanced deformation of generically log smooth families with a vector bundle, let $\mathcal{L} = \det(\mathcal{E})$, and let $f' : (X', \mathcal{L}') \to S_{B'}$ be a lift to $S_{B'}$. We have to show that there is a vector bundle $\mathcal{E}'$ on X' which lifts $\mathcal{E}$ and satisfies $\det(\mathcal{E}') \cong \mathcal{L}'$ in a compatible way. By [139, Thm. 7.1], the obstruction for lifting $\mathcal{E}$ lies in $H^2(X_0, I \otimes_{\mathbf{k}} \mathcal{E}nd(\mathcal{E}_0))$. Similarly, the obstruction for lifting $\mathcal{L}$ to X' lies in $H^2(X_0, I \otimes_{\mathbf{k}} \mathcal{E}nd(\mathcal{L}_0))$, where $\mathcal{L}_0 := \det(\mathcal{E}_0)$. Given some local lift $\mathcal{E}'$ of $\mathcal{E}$ and a local automorphism of $\mathcal{E}'$ over $\mathcal{E}$ determined by $\theta \in I \otimes_{\mathbf{k}} \mathcal{E}nd(\mathcal{E}_0)$, the induced local automorphism of $\mathcal{L}' = \det(\mathcal{E}')$ is determined by $\operatorname{tr}(\theta) \in I \otimes_{\mathbf{k}} \mathcal{E}nd(\mathcal{L}_0)$. Thus, if $o \in I \otimes_{\mathbf{k}} \operatorname{Ext}^2(\mathcal{E}_0, \mathcal{E}_0)$ is the obstruction to lifting $\mathcal{E}$ to X', then

$$H^2(\operatorname{tr})(o) \in I \otimes_{\mathbf{k}} H^2(X_0, O_{X_0})$$

---

[5] The published version of this article contains a mistake which is corrected in a newer version on the arXiv.

is the obstruction to lifting $\mathcal{L}$ to X'. However, we have $H^2(\mathrm{tr})(o) = 0$ in our situation, and $H^2(\mathrm{tr})$ is injective by assumption. Thus, $o = 0$, and a lift $\mathcal{E}'$ exists. $\quad\square$

Example 11.17 We continue Example 11.14. Since $H^2(X_0, O_{X_0}) = 0$, the map $\sigma: D_1 \Rightarrow D_0$ is smooth. Thus, given the non-trivial $\mathcal{E}_1$, we can find a lift $\mathcal{L}_2$ of $\mathcal{L}_1 := \det(\mathcal{E}_1)$ to X_2. However, a lift $\mathcal{E}_2$ still does not exist, so $\rho: D_2 \Rightarrow D_1$ is not smooth. $\quad\diamond$

Chapter 12
Geometric Families of $\mathcal{P}$-Algebras

When we start with a log Gorenstein generically log smooth family $f_0\colon X_0 \to S_0$, we can form a sheaf of Gerstenhaber calculi $(\mathcal{V}^\bullet_{X_0/S_0}, \mathcal{W}^\bullet_{X_0/S_0})$ on X_0 by Proposition 8.5. When $f_A\colon X_A \to S_A$ is a deformation of f_0 with the base change property, then its sheaf of Gerstenhaber calculi is a deformation of $(\mathcal{V}^\bullet_{X_0/S_0}, \mathcal{W}^\bullet_{X_0/S_0})$.

In this chapter, we define and study sheaves of $\mathcal{P}$-algebras for algebraic structures $\mathcal{P}$ carrying a Cartan structure, as well as their infinitesimal deformations. When we want to capture as much information as possible, we can work with two-sided Gerstenhaber calculi. When we want only a minimal structure that is enough for basic deformation theory, we can work with Lie–Rinehart algebras. In the case of deformations with a vector bundle, we work with Lie–Rinehart pairs.

While we are currently mostly interested in families which come from log geometry, this theory is of independent interest as it can also be applied beyond log geometry.

Definition 12.1 Let $\mathcal{P}$ be an algebraic structure carrying a Cartan structure in the sense of Definition 3.37. Let S be a locally Noetherian scheme. A *geometric family of $\mathcal{P}$-algebras* of dimension d is a tuple

$$(f\colon X \to S,\ U^\circ \subseteq U \subseteq X,\ \mathcal{E}^\bullet)$$

where:

- $f\colon X \to S$ is a separated and flat morphism of finite type of locally Noetherian schemes whose geometric fibers (over the algebraic closures of the residue fields) are reduced, satisfy Serre's property (S_2), and are pure of dimension d;
- $U \subseteq X$ is a Zariski open subset satisfying the codimension condition;

S. Felten, *Global Logarithmic Deformation Theory*, Lecture Notes
in Mathematics 2373, https://doi.org/10.1007/978-3-031-98751-9_12

- $U^\circ \subseteq U$ is a Zariski open subset such that $U^\circ \to S$ is smooth, and $U_s^\circ \subseteq X_s$ is scheme-theoretically dense in the fiber for every $s \in S$;
- $\mathcal{E}^\bullet$ is a Cartanian $\mathcal{P}$-algebra in $\mathfrak{Coh}(X/S)$;
- the map $O_X \to \mathcal{F} := \mathcal{E}^F$, $1 \mapsto 1_F$, is an isomorphism of O_X-modules;
- for a morphism $b \colon T \to S$ of locally Noetherian schemes, let $c \colon Y := X \times_S T \to X$ be the induced map via pull-back; then for all such $b \colon T \to S$, we assume that the $\mathcal{P}$-algebra $c^*\mathcal{E}^\bullet$ in $\mathfrak{Coh}(Y/T)$ is Z-faithful (in the sense of Definition 3.40).

When no confusion is likely, we write a geometric family of $\mathcal{P}$-algebras as $f \colon (X, \mathcal{E}^\bullet) \to S$ for short. A *morphism* of geometric families of $\mathcal{P}$-algebras from $g \colon (Y, V, V^\circ, \mathcal{E}_Y^\bullet) \to T$ to $f \colon (X, U, U^\circ, \mathcal{E}_X^\bullet) \to S$ consists of two morphisms $b \colon T \to S$ and $c \colon Y \to X$ of schemes, satisfying $c^{-1}(U) = V$ and $c^{-1}(U^\circ) = V^\circ$ and giving a Cartesian diagram, and maps

$$\mathcal{E}_X^P \to c_*\mathcal{E}_Y^P$$

of O_X-modules that induce isomorphisms $c^*\mathcal{E}_X^P \cong \mathcal{E}_Y^P$ and are compatible with all constants in $\mathcal{P}(P)$ and all operations in $\mathcal{P}(P_1, \ldots, P_n; Q)$. A geometric family of $\mathcal{P}$-algebras $f \colon (X, \mathcal{E}^\bullet) \to S$ is called:

- *strictly faithful* if $c^*\mathcal{E}^\bullet$ is strictly faithful for all $b \colon T \to S$;
- *tame at* $P \in D(\mathcal{P})$ if $\mathcal{E}^P|_U$ is locally free;
- *pure at* $P \in D(\mathcal{P})$ if $c^*\mathcal{E}^P \to j_*(c^*\mathcal{E}^P)|_V$ is injective after every base change $b \colon T \to S$;
- *closed at* $P \in D(\mathcal{P})$ if $c^*\mathcal{E}^P \to j_*(c^*\mathcal{E}^P)|_V$ is an isomorphism after every base change $b \colon T \to S$;[1]
- *reflexive at* $P \in D(\mathcal{P})$ if it is tame at P and closed at P;
- *locally free at* $P \in D(\mathcal{P})$ if $\mathcal{E}^P$ is a locally free sheaf.

Remark 12.2 Several remarks are in order.

(1) Concretely, $\mathcal{E}^P$ consists of the following structures: For each $P \in D(\mathcal{P})$, we have a coherent sheaf $\mathcal{E}^P$ of O_X-modules, flat over S; for each $\gamma \in \mathcal{P}(P)$, we have a global section $\gamma \in \mathcal{E}^P$; for each $\mu \in \mathcal{P}(P; Q)$, we have an $f^{-1}(O_S)$-linear map $\mu \colon \mathcal{E}^P \to \mathcal{E}^Q$; for each $\mu \in \mathcal{P}(P, Q; R)$, we have an $f^{-1}(O_S)$-bilinear map $\mu \colon \mathcal{E}^P \times \mathcal{E}^Q \to \mathcal{E}^R$. They must satisfy all relations required by the algebraic structure $\mathcal{P}$—we do not just assume that $\mathcal{E}^P$ is a pre-algebra.

---

[1] The names "pure" and "closed" refer to the terminology in [131, Defn. 5.9.9].

(2) We write $\mathcal{F} := \mathcal{E}^F$ and $\mathcal{T} := \mathcal{E}^T$. Since $O_X \to \mathcal{F}$ is an isomorphism and $*^P : \mathcal{F} \times \mathcal{E}^P \to \mathcal{E}^P$ is O_X-bilinear, the $\mathcal{F}$-module structure coincides with the O_X-module structure on every $\mathcal{E}^P$, and in particular, the product on $\mathcal{F}$ coincides with the product on O_X. Thus, every $\mu \in \mathcal{P}_N(P_1, \ldots, P_n; Q)$ is a differential operator of order N with respect to $O_X / f^{-1}(O_S)$.

(3) Let $b \colon T \to S$ be a morphism of locally Noetherian schemes with base change $c \colon Y \to X$. Then each $c^*\mathcal{E}^P$ is a coherent sheaf on Y, flat over T. By Proposition C.19, every multilinear differential operator $\mu \colon \mathcal{E}^{P_1} \times \ldots \times \mathcal{E}^{P_n} \to \mathcal{E}^Q$ pulls back to a well-defined multilinear differential operator $c^*\mu \colon c^*\mathcal{E}^{P_1} \times \ldots \times c^*\mathcal{E}^{P_n} \to c^*\mathcal{E}^Q$. Since this construction preserves compositions and the identity on objects $\mathcal{E}^P$, and is compatible with constants, the relations of $\mathcal{P}$ hold in $c^*\mathcal{E}^\bullet$ as well. Thus, $c^*\mathcal{E}^\bullet$ is a Cartanian $\mathcal{P}$-algebra in $\mathfrak{Coh}(Y/T)$.

(4) Since $c^*\mathcal{E}^\bullet$ is a Cartanian $\mathcal{P}$-algebra in $\mathfrak{Coh}(Y/T)$, the requirement that $c^*\mathcal{E}^\bullet$ should be Z-faithful is defined. A fortiori, if $\mathcal{E}^\bullet$ remains Z-faithful after any base change, so does $c^*\mathcal{E}^\bullet$. Thus, if $(f \colon X \to S, U \subseteq X, \mathcal{E}^\bullet)$ is a geometric family of $\mathcal{P}$-algebras, so is $(g \colon Y \to T, V \subseteq Y, c^*\mathcal{E}^\bullet)$. Concretely, Z-faithfulness means that, if $\theta \in c^*\mathcal{T}$ is a local section with $\nabla^P_\theta = 0$ as a map $\nabla^P_\theta \colon c^*\mathcal{E}^P \to c^*\mathcal{E}^P$ for all $P \in Z$, then $\theta = 0$.

(5) Since $U \subseteq X$ satisfies the codimension condition and the fibers of $f \colon X \to S$ have Serre's property (S_2), the natural map $O_X \to j_*O_U$ is an isomorphism by [141, Prop. 3.5]. This holds after any base change. Thus, every geometric family of $\mathcal{P}$-algebras is closed at F.

(6) If $\mathcal{E}^\bullet$ is strictly faithful, then the natural map $\mathcal{T} \to j_*\mathcal{T}|_U$ is injective. Namely, if $\theta|_U = 0$, then $\nabla^F_\theta|_U = 0$, but $O_X = j_*O_U$, so $\nabla^F_\theta = 0$. Conversely, if $\mathcal{E}^\bullet$ is pure at T and $\mathcal{E}^\bullet|_U$ is strictly faithful, then $\mathcal{E}^\bullet$ is strictly faithful.

(7) Let $f \colon (X, \mathcal{E}^\bullet) \to S$ be pure or closed at $P \in D(\mathcal{P})$. Then, for every $s \in S$, the natural map $\mathcal{E}^P_s \to j_*\mathcal{E}^P_s|_{U_s}$ is injective respectively bijective. Conversely, if this holds, then it holds after any base change to the spectrum of a field, and then $f \colon (X, \mathcal{E}^\bullet) \to S$ is pure respectively closed at $P \in D(\mathcal{P})$ by Lemmas B.4 and B.5.

(8) In practice, we usually know that $\mathcal{E}^P|_U$ is locally free, that $\mathcal{E}^\bullet|_{U^\circ}$ is Z-faithful or even strictly faithful after any base change, and that $\mathcal{T}_s \to j_*\mathcal{T}_s|_{U_s}$ is injective for $s \in S$. Then we can conclude[2] that the family is tame at every $P \in D(\mathcal{P})$, pure at T, and Z-faithful respectively strictly faithful. This is one reason why we carry the two open subsets U and U° as part of the structure. $\Diamond$

Example 12.3 Here is an example of a geometric family of Gerstenhaber calculi which arises from a classical construction. Let

$$f \colon \mathbb{A}^4 \to \mathbb{A}^1_t, \quad t \mapsto x^2 + y^2 + z^2 + w^3.$$

---

[2] The map $\mathcal{E}^P|_U \to j^\circ_* \mathcal{E}^P|_{U^\circ}$ is injective if $\mathcal{E}^P$ is locally free on U because $U^\circ \to U$ is scheme-theoretically dense in every fiber and thus after any base change.

Then the central fiber is the normal A_2-threefold singularity, and f is a smoothing. We take $\Lambda = \mathbf{k}[\![t]\!]$ and obtain a Gerstenhaber calculus as the base change of the classical reflexive relative Gerstenhaber calculus of $f \colon \mathbb{A}^4 \to \mathbb{A}_t^1$. Since $f \colon \mathbb{A}^4 \to \mathbb{A}_t^1$ is smooth outside the isolated singularity in the central fiber, $\mathcal{E}^P|_U$ is locally free when we set $U := \mathbb{A}^4 \setminus \{0\}$. A direct calculation using Macaulay2 shows that $\mathcal{T}_0 \to j_* \mathcal{T}_0|_{U_0}$ is injective. Since $\mathcal{E}^\bullet$ is strictly faithful on U, we find that $\mathcal{E}_A^\bullet$ is tame and strictly faithful for every $A \in \mathbf{Art}_\Lambda$. $\diamond$

For practical applications, let us show the following lemma. The proof is a little more involved than one may expect because of the non-$\mathcal{O}_X$-linear nature of the involved maps.

Lemma 12.4 *Assume that $Z \subseteq D(\mathcal{P})$ is finite. Then, to show that $c^* \mathcal{E}^\bullet$ is Z-faithful (respectively strictly faithful) for all $b \colon T \to S$, it is sufficient to show that the fiber $\mathcal{E}_s^\bullet$ is Z-faithful (respectively strictly faithful) for every point $s \in S$.*

Proof Let $f \colon (X, \mathcal{E}^\bullet) \to S$ be a candidate for a geometric family of $\mathcal{P}$-algebras which satisfies everything except possibly the last condition on Z-faithfulness. Assume furthermore that $\mathcal{E}_s^\bullet$ is Z-faithful (respectively strictly faithful) for every point $s \in S$. First, let K be a field, and let $\mathrm{Spec}\, K \to S$ be a morphism. Let $s \in S$ be the image. Then we have a factorization

$$\mathrm{Spec}\, K \to \mathrm{Spec}\, \kappa(s) \to S.$$

Let $\{\lambda_i\}$ be a $\kappa(s)$-basis of K. Let $V \subseteq X_s$ be an affine open subset, and assume $\theta = \sum_i \theta(i) \otimes \lambda_i \in \Gamma(V_K, \mathcal{T}_K)$ with $\theta(i) \in \Gamma(V, \mathcal{T}_{\kappa(s)})$ satisfies $\nabla_\theta^M = 0$ for all $M \in Z$. Then, for $m \in \mathcal{E}_{\kappa(s)}^M$, we have $\nabla_\theta^M(m) = \sum_i \nabla_{\theta(i)}^M(m) \otimes \lambda_i = 0$, so $\nabla_{\theta(i)}^M(m) = 0$ and hence $\theta(i) = 0$ by assumption. Thus, $\theta = 0$.

Next, let $W \subseteq V_K$ be a standard open subset defined by a function g, and let $\theta \in \Gamma(W, \mathcal{T}_K)$ with $\nabla_\theta^M = 0$ for all $M \in Z$. We have $g^n \nabla_\theta^M = \nabla_{g^n \theta}^M$. Thus, we can assume without loss of generality that θ is the restriction of a section $\theta \in \Gamma(V_K, \mathcal{T}_K)$. Let $W' \subseteq V_K$ be another standard open, and let $m \in \Gamma(W', \mathcal{E}_K^M)$ be a section. Then $\nabla_\theta^M(m)|_{W \cap W'} = 0$, so $g^n \cdot \nabla_\theta^M(m) = 0$ in $\Gamma(W', \mathcal{E}_K^M)$ for some $n \geq 0$. The coherent subsheaves $\mathcal{Z}^M(n) := \{m \in \mathcal{E}_K^M \mid g^n \cdot m = 0\}$ for an ascending sequence in the coherent sheaf $\mathcal{E}_K^M$; since X_K is Noetherian, the sequence becomes stationary. Thus, there is some n with $g^n \cdot \nabla_\theta^M(m) = 0$ for all local sections $m \in \mathcal{E}_K^M$. Since Z is finite, we can find an n which works for all $M \in Z$ simultaneously. Then we find $g^n \theta = 0$ because we already know Z-faithfulness (respectively strict faithfulness) for sections in $\Gamma(V_K, \mathcal{T}_K)$; in particular, $\theta|_W = 0$ so that $\mathcal{E}_K^\bullet$ is Z-faithful (respectively strictly faithful).

For the next step, let A be an Artinian local ring with residue field K, and let $\mathrm{Spec}\, A \to S$ be a morphism. Then the proof of Lemma 4.6 shows that $\mathcal{E}_A^\bullet$ is Z-faithful (respectively strictly faithful) if $\mathcal{E}_K^\bullet$ is.

Finally, let $T = \mathrm{Spec}\, \mathcal{O}_T$ be Noetherian affine, and let $b \colon T \to S$ be a morphism. Let $\theta \in \Gamma(V, \mathcal{T}_T)$ be a section over an affine open $V \subseteq Y$ with $\nabla_\theta^M = 0$ for $M \in Z$. Then, for a map $\mathcal{O}_T \to A$ to an Artinian local ring A, the induced differential operator ∇_θ^M on $\mathcal{E}_A^M$ is zero as well. Since $\mathcal{E}_A^\bullet$ is Z-faithful (respectively strictly faithful), we find $\theta|_A = 0$. Then $\theta = 0$ by Lemma B.3. $\square$

12.1 Notions of Good Geometric Families of $\mathcal{P}$-Algebras

In Definition 12.1, we have given a general definition to capture geometric families of $\mathcal{P}$-algebras for all algebraic structures $\mathcal{P}$ carrying a Cartan structure at once. However, for specific algebraic structures $\mathcal{P}$, we wish to impose additional hypotheses unique to that $\mathcal{P}$. It is here that U° comes into play to tie the family of $\mathcal{P}$-algebras to what happens classically on the smooth locus. These definitions are not essential but rather illustrate the concept of geometric families of $\mathcal{P}$-algebras.

Let $\mathcal{T}_{X/S}$ be the sheaf of classical relative derivations, i.e., the dual of $\underline{\Omega}_{X/S}^1$. For every geometric family of $\mathcal{P}$-algebras, we have a map

$$a\colon \ \mathcal{T} \to \mathcal{T}_{X/S}, \quad \theta \mapsto (g \mapsto -\nabla_\theta^F(g)),$$

of Lie algebras (when endowing the right-hand side with the negative Lie bracket)[3] and O_X-modules, which corresponds to the *anchor map* of a Lie algebroid.

Definition 12.5 A geometric family of Lie–Rinehart algebras $f\colon (X, \mathcal{F}, \mathcal{T}) \to S$ is *good* if it is strictly faithful, tame at F and T, pure at T, and if the anchor map $a\colon \mathcal{T}|_{U^\circ} \to \mathcal{T}_{U^\circ/S}$ is an isomorphism.

If we have a Lie–Rinehart pair $(\mathcal{F}, \mathcal{T}, \mathcal{E})$, then we have an extended anchor map

$$a^E\colon \ \mathcal{T} \to \mathcal{T}_{X/S}(\mathcal{E}), \quad \theta \mapsto (g \mapsto -\nabla_\theta^F(g),\ e \mapsto -\nabla_\theta^E(e)),$$

where $\mathcal{T}_{X/S}(\mathcal{E})$ denotes derivations of $f\colon (X, \mathcal{E}) \to S$ in the sense of Definition 11.3.

Definition 12.6 A geometric family of Lie–Rinehart pairs $f\colon (X, \mathcal{F}, \mathcal{T}, \mathcal{E}) \to S$ is *good* if it is Z-faithful, tame and pure at T, if $\mathcal{E}$ is a vector bundle of constant rank r, and if the extended anchor map $a^E\colon \mathcal{T}|_{U^\circ} \to \mathcal{T}_{U^\circ/S}(\mathcal{E})$ is an isomorphism.

Definition 12.7 A geometric family of Gerstenhaber algebras $f\colon (X, \mathcal{G}^\bullet) \to S$ is *good* if:

- the relative dimension d of $f\colon X \to S$ equals the dimension d of the Gerstenhaber algebra $\mathcal{G}^\bullet$, which is in degrees $-d \le p \le 0$;
- it is strictly faithful as well as tame and pure at all $\mathcal{G}^p$;
- the anchor map $a\colon \mathcal{T}|_{U^\circ} \to \mathcal{T}_{U^\circ/S}$ is an isomorphism;
- $\mathcal{G}^\bullet|_U$ is isomorphic to the exterior algebra of $\mathcal{G}^{-1}|_U$ via the $\wedge$-product;

---

[3] With the negative sign we account for our construction of the bracket on $\mathcal{T} = \mathcal{G}^{-1}$.

- $\mathcal{G}^{-d}$ is a line bundle.

Definition 12.8   A geometric family of two-sided Gerstenhaber calculi

$$f:\ (X, \mathcal{G}^{\bullet}, \mathcal{A}^{\bullet}) \to S$$

is *good* if:

- the relative dimension d of $f: X \to S$ equals the dimension d of the two-sided Gerstenhaber calculus $(\mathcal{G}^{\bullet}, \mathcal{A}^{\bullet})$;
- it is strictly faithful as well as tame and pure at all $\mathcal{G}^p$ and $\mathcal{A}^i$;
- it is *locally Batalin–Vilkovisky*, i.e., $\mathcal{A}^d$ is a line bundle, and a local generator $\omega \in \mathcal{A}^d$ induces a local isomorphism $\mathcal{G}^p \cong \mathcal{A}^{p+d}$ for every $-d \le p \le 0$; if this property holds for one local generator, it holds for every local generator (cf. Proposition 3.28);
- the anchor map $a: \mathcal{G}^{-1}|_{U^{\circ}} \to \mathcal{T}_{U^{\circ}/S}$ is an isomorphism;
- $\mathcal{G}^{\bullet}|_U$ is isomorphic to the exterior algebra of $\mathcal{G}^{-1}|_U$ via the $\wedge$-product;
- $\mathcal{A}^{\bullet}|_U$ is isomorphic to the exterior algebra of $\mathcal{A}^1|_U$ via the $\wedge$-product;
- the induced map

$$a^{\vee}:\ \mathcal{A}^1|_{U^{\circ}} \to \mathcal{H}om(\mathcal{T}_{U^{\circ}/S}, O_{U^{\circ}}) \cong \Omega^1_{\underline{U}^{\circ}/\underline{S}}, \quad \alpha \mapsto (a^{-1}(\theta) \mapsto (\theta \lrcorner \alpha)),$$

  is an isomorphism of O_X-modules.

Remark 12.9   If $f: (X, \mathcal{G}^{\bullet}) \to S$ is a good geometric family of Gerstenhaber algebras, then $\mathcal{G}^{\bullet}|_{U^{\circ}}$ is isomorphic to the Gerstenhaber algebra $\mathcal{G}^{\bullet}_{U^{\circ}/S}$ of the strict log smooth family $f: U^{\circ} \to S$. Similarly, if $f: (X, \mathcal{G}^{\bullet}, \mathcal{A}^{\bullet}) \to S$ is a good geometric family of two-sided Gerstenhaber calculi, then $(\mathcal{G}^{\bullet}, \mathcal{A}^{\bullet})|_{U^{\circ}}$ is isomorphic to the Gerstenhaber calculus of $f: U^{\circ} \to S$. For example, one can show that

$$a^{\vee} \circ \partial = \partial_{\mathrm{dR}}:\ O_{U^{\circ}} \to \Omega^1_{\underline{U}^{\circ}/\underline{S}},$$

where ∂_{dR} is the usual de Rham differential, by exploiting the definition of the anchor map a and the axioms of a Gerstenhaber calculus. $\Diamond$

Example 12.10   Let $f: X \to S$ be a log Gorenstein generically log smooth family. Then its Gerstenhaber calculus in the sense of Proposition 8.5 forms a good geometric family of two-sided Gerstenhaber calculi when we take for U° the strict locus of $f: U \to S$, which is scheme-theoretically dense in every fiber by [76, Prop. 10.1]. $\Diamond$

12.2 Geometric Deformations of $\mathcal{P}$-Algebras

Let $\mathcal{P}$ be an algebraic structure carrying a Cartan structure. Let Λ be a complete local Noetherian $\mathbf{k}$-algebra[4] with residue field $\mathbf{k}$, and let

$$f_0: \quad (X_0, U_0, U_0^\circ, \mathcal{E}_0^\bullet) \to S_0 = \operatorname{Spec} \mathbf{k}$$

be a geometric family of $\mathcal{P}$-algebras.

Definition 12.11 Let $f_0: X_0 \to S_0$ be as above, and let $A \in \mathbf{Art}_\Lambda$. Then a *geometric deformation of $\mathcal{P}$-algebras* of $f_0: X_0 \to S_0$ over A is a geometric family of $\mathcal{P}$-algebras

$$f_A: \quad (X_A, U_A, U_A^\circ, \mathcal{E}_A^\bullet) \to S_A$$

together with a morphism from $f_0: X_0 \to S_0$ (which, by definition, induces an isomorphism after base change). If $f_0: X_0 \to S_0$ is a good geometric family of $\mathcal{P}$-algebras, then we say that a geometric deformation $f_A: X_A \to S_A$ is *good* if it is good as a geometric family of $\mathcal{P}$-algebras.

Remark 12.12 When we already know that $f_0: (X_0, \mathcal{E}_0^\bullet) \to S_0$ is a geometric family of $\mathcal{P}$-algebras, checking the conditions for $f_A: (X_A, \mathcal{E}_A^\bullet) \to S_A$ becomes easier. Assume that $f_A: X_A \to S_A$ is a separated and flat thickening of $f_0: X_0 \to S_0$ of finite type, and let $\mathcal{E}_A^\bullet$ be a Cartanian $\mathcal{P}$-algebra in $\mathfrak{Coh}(X_A/S_A)$ together with a map $\mathcal{E}_A^\bullet \to \mathcal{E}_0^\bullet$ of O_{X_A}-modules which is compatible with all constants and operations and induces an isomorphism after base change to X_0. This is enough to conclude that $f_A: (X_A, \mathcal{E}_A^\bullet) \to S_A$ is a geometric deformation of $f_0: (X_0, \mathcal{E}_0^\bullet) \to S_0$ when we take U_A and U_A° such that $U_A = U_0$ and $U_A^\circ = U_0^\circ$ on underlying schemes. The most interesting part of the proof is Z-faithfulness, which follows from Lemma 12.4, and that $O_{X_A} \to \mathcal{F}_A$ is an isomorphism. The latter morphism is surjective because its pull-back to X_0 is surjective. The target $\mathcal{F}_A$ is flat over A so that the embedding of the kernel is universally injective (over A). Then the pull-back to X_0 is the kernel of $O_{X_0} \to \mathcal{F}_0$, hence 0. Thus, the kernel vanishes itself, and $O_{X_A} \to \mathcal{F}_A$ is an isomorphism.

Similarly, if $f_0: (X_0, \mathcal{E}_0^\bullet)$ is tame or locally free at $P \in D(\mathcal{P})$, then f_A is tame respectively locally free at $P \in D(\mathcal{P})$. Since $\mathcal{E}_A^P$ is flat over S_A, it follows from Lemma B.4 respectively Lemma B.5 that f_A is pure respectively closed at $P \in D(\mathcal{P})$ if f_0 is. Then, if f_0 is reflexive at $P \in D(\mathcal{P})$, the deformation f_A is reflexive at $P \in D(\mathcal{P})$ as well. If f_0 is strictly faithful, so is f_A by Lemma 12.4. $\Diamond$

Remark 12.13 When we already know that $f_0: (X_0, \mathcal{E}_0^\bullet) \to S_0$ is a *good* geometric family of Gerstenhaber calculi, a deformation $f_A: (X_A, \mathcal{E}_A^\bullet) \to S_A$ is good as well. A local generator of $\mathcal{A}_0^d$ can be lifted to a local section of $\mathcal{A}_A^d$, which

---

[4] We do *not* assume that Λ must be of the form $\mathbf{k}[\![Q]\!]$.

induces a local isomorphism $O_{X_A} \cong \mathcal{A}_A^d$ similarly to the case of $\mathcal{F}_A$ above. Then, the local generator gives locally a homomorphism $\kappa_\omega \colon \mathcal{G}_A^p \to \mathcal{A}_A^{p+d}$ whose pullback to X_0 is an isomorphism; since $\mathcal{A}_A^{p+d}$ is flat over A, we find that this κ_ω is an isomorphism. Similar arguments complete the proof. $\Diamond$

This definition is only intended to be language. Be aware of the curious fact that, if $A, A' \in \mathbf{Art}_\Lambda$ are isomorphic as $\mathbf{k}$-algebras but not as Λ-algebras, then there is no real difference in the notions of deformations over A or A' but nonetheless we distinguish them.

We do not wish to form isomorphism classes and a functor of Artin rings at this point as we do not yet have defined a useful notion of isomorphism of geometric deformations of $\mathcal{P}$-algebras. To this end, we need gauge transforms in the sense of Definition 4.3.

Let $B' \to B$ be a surjection in $\mathbf{Art}_\Lambda$ with kernel $I \subset B'$; let $f' \colon X_{B'} \to S_{B'}$ be a geometric deformation of $\mathcal{P}$-algebras of $f_0 \colon X_0 \to S_0$ over B, and let $f \colon X_B \to S_B$ be the base change to B. For $\theta \in \Gamma(X_0, I \cdot \mathcal{T}_{B'})$, the *gauge transform* $\exp_\theta$ consists of sheaf maps

$$\exp_\theta \colon \qquad \mathcal{E}_{B'}^P \to \mathcal{E}_{B'}^P, \qquad p \mapsto \sum_{n=0}^{\infty} \frac{(\nabla_\theta^P)^n(p)}{n!},$$

which give rise to an automorphism $\exp_\theta \colon X_{B'} \to X_{B'}$ of geometric deformations $\mathcal{P}$-algebras.[5] Let us denote by $\mathcal{G}au_{X_{B'}/X_B}$ the sheaf of gauge transforms, considered as a sheaf of groups which acts by automorphisms of $f' \colon X_{B'} \to S_{B'}$. If $B' \to B$ and $B'' \to B'$ are two surjections in $\mathbf{Art}_\Lambda$, then we obtain a restriction map

$$\mathcal{G}au_{X_{B''}/X_B} \to \mathcal{G}au_{X_{B'}/X_B},$$

which coincides with the restriction map $I_{B''/B} \cdot \mathcal{G}_{B''}^{-1} \to I_{B'/B} \cdot \mathcal{G}_{B'}^{-1}$; it is surjective[6] because our deformations satisfy, by definition, $\mathcal{T}_{B'} \otimes_{B'} B = \mathcal{T}_B$.

If $\theta, \xi \in \Gamma(X_0, I \cdot \mathcal{G}_{B'}^{-1})$ with $\exp_\theta = \exp_\xi$, then $\theta = \xi$ by induction over small extensions, using Z-faithfulness on each small extension. Thus, gauge transforms form a subsheaf

$$\mathcal{G}au_{X_{B'}/X_B} \subseteq \mathcal{A}ut_{X_{B'}/X_B}$$

of all automorphisms of geometric deformations of $\mathcal{P}$-algebras.[7]

---

[5] Use $O_{X_{B'}} = \mathcal{F}_{B'}$ for the automorphism of underlying schemes.

[6] Recall from Example 10.5 that a generically log smooth deformation does not need to have this property if we take all automorphisms into account.

[7] This is the reason why we require Z-faithfulness; if we do not have it, then there are "ghost" gauge transforms which act by the identity.

12.3   Systems of Deformations

We define a device to control geometric deformations of $\mathcal{P}$-algebras locally, analogous to the case of generically log smooth families. Let Λ and $f_0 \colon (X_0, \mathcal{E}_0^\bullet) \to S_0$ be as above.

Definition 12.14 A pre-admissible covering $\mathcal{V} = \{V_\alpha\}_\alpha$ of X_0 is *admissible* with respect to $\mathcal{E}_0^\bullet$ if, for every $P \in D(\mathcal{P})$, we have

$$H^i(V_{\alpha_1} \cap \ldots \cap V_{\alpha_r}, \mathcal{E}_0^P|_{\alpha_1 \ldots \alpha_r}) = 0$$

for all $i \geq 1$ and indices $\alpha_1, \ldots, \alpha_r$, where we write $\mathcal{F}|_{\alpha_1 \ldots \alpha_r} := \mathcal{F}|_{V_{\alpha_1} \cap \ldots \cap V_{\alpha_r}}$.

We require this condition to compute the cohomology of $\mathcal{E}_0^P$ from the Čech complex of $\{V_\alpha\}_\alpha$, and to make deformations on V_α unique up to isomorphism. The following definition essentially goes back to [38, §2]. The key point is that the cocycles must be gauge transforms.

Definition 12.15 Let $\mathcal{P}$ be an algebraic structure carrying a Cartan structure, let $f_0 \colon (X_0, U_0, U_0^\circ, \mathcal{E}_0^\bullet) \to S_0$ be a geometric family of $\mathcal{P}$-algebras, and let $\{V_\alpha\}_\alpha$ be an admissible open cover.

(1) A *system of deformations* $\mathscr{D}$ for $f_0 \colon (X_0, \mathcal{E}_0^\bullet) \to S_0$ is a tuple

$$\mathscr{D} = (f_{\alpha;A} \colon (V_{\alpha;A}, \mathcal{E}_{\alpha;A}^\bullet) \to S_A, \ \rho_{\alpha;BB'}, \ \psi_{\alpha\beta;A}, \ o_{\alpha\beta\gamma;A})$$

where:

- for every $A \in \mathbf{Art}_\Lambda$, the object $f_{\alpha;A} \colon (V_{\alpha;A}, \mathcal{E}_{\alpha;A}^\bullet) \to S_A$ is a geometric deformation of $\mathcal{P}$-algebras of $f_0 \colon (X_0, \mathcal{E}_0^\bullet) \to S_0$; in particular, $(V_{\alpha;0}, \mathcal{E}_{\alpha;A}^\bullet) = (X_0, \mathcal{E}_0^\bullet)|_\alpha$;
- for every map $B' \to B$ in $\mathbf{Art}_\Lambda$, the object

$$\rho_{\alpha;BB'} \colon \ (V_{\alpha;B}, \mathcal{E}_{\alpha;B}^\bullet) \to (V_{\alpha;B'}, \mathcal{E}_{\alpha;B'}^\bullet)$$

  is a map of geometric families of $\mathcal{P}$-algebras, compatible with the map from $V_{\alpha;0}$; they induce isomorphisms $\mathcal{E}_{\alpha;B'}^P \otimes_{B'} B \cong \mathcal{E}_{\alpha;B}^P$ by the definition of a morphism of geometric families of $\mathcal{P}$-algebras;
- for every $A \in \mathbf{Art}_\Lambda$, the object

$$\psi_{\alpha\beta;A} \colon \ (V_{\alpha;A}, \mathcal{E}_{\alpha;A}^\bullet)|_{\alpha\beta} \to (V_{\beta;A}, \mathcal{E}_{\beta;A}^\bullet)|_{\alpha\beta}$$

  is an isomorphism of geometric deformations of $\mathcal{P}$-algebras; we assume them to be compatible with the restriction maps $\rho_{\alpha;BB'}$; in particular, $\psi_{\alpha\beta;0}$ is the identity; they satisfy $\psi_{\alpha\beta;A}^{-1} = \psi_{\beta\alpha;A}$;

- the cocycles $\psi_{\gamma\alpha;A} \circ \psi_{\beta\gamma;A} \circ \psi_{\alpha\beta;A}$ are equal to (unique) gauge transforms $\exp_{o_{\alpha\beta\gamma;A}}$ for $o_{\alpha\beta\gamma;A} \in \Gamma(V_\alpha \cap V_\beta \cap V_\gamma, \mathfrak{m}_A \cdot \mathcal{T}_{\alpha;A})$.

(2) A *geometric deformation of $\mathcal{P}$-algebras of type $\mathscr{D}$* is a geometric deformation $f_A \colon (X_A, \mathcal{E}_A^\bullet) \to S_A$ of $\mathcal{P}$-algebras of $f_0 \colon (X_0, \mathcal{E}_0^\bullet) \to S_0$ together with isomorphisms

$$\chi_\alpha \colon \ (X_A, \mathcal{E}_A^\bullet)|_\alpha \cong (V_{\alpha;A}, \mathcal{E}_{\alpha;A}^\bullet)$$

of geometric deformations of $\mathcal{P}$-algebras such that

$$\psi_{\beta\alpha;A} \circ \chi_\beta|_{\alpha\beta} \circ \chi_\alpha^{-1}|_{\alpha\beta}$$

is a gauge transform of $(V_{\alpha;A}, \mathcal{E}_{\alpha;A}^\bullet)|_{\alpha\beta}$ for all indices α, β.

(3) If $B' \to B$ is a map in $\mathbf{Art}_\Lambda$, and if

$$f \colon \ (X_B, \mathcal{E}_B^\bullet) \to S_B \quad \text{and} \quad f' \colon \ (X_{B'}, \mathcal{E}_{B'}^\bullet) \to S_{B'}$$

are two geometric deformations of $\mathcal{P}$-algebras of type $\mathscr{D}$ with local isomorphisms χ_α and χ'_α, then a *morphism* is a morphism of geometric deformations of $\mathcal{P}$-algebras—consisting of maps $O_{X_{B'}} \to O_{X_B}$ and $\mathcal{E}_{B'}^\bullet \to \mathcal{E}_B^\bullet$ inducing an isomorphism in pull-backs, and compatible with the maps to the central fiber—with the following property: The local isomorphism χ'_α induces, via the Cartesian squares, a local isomorphism $\chi'_\alpha|_B \colon (X_B, \mathcal{E}_B^\bullet)|_\alpha \cong (V_{\alpha;B}, \mathcal{E}_{\alpha;B}^\bullet)$. Then $\chi'_\alpha|_B \circ \chi_\alpha^{-1}$ must be a gauge transform of $(V_{\alpha;B}, \mathcal{E}_{\alpha;B}^\bullet)$.

(4) Two geometric deformations of type $\mathscr{D}$ are *equivalent* if there is an isomorphism $\sigma \colon (X_A, \mathcal{E}_A^\bullet) \to (X'_A, \mathcal{E}_A'^\bullet)$ of geometric deformations of $\mathcal{P}$-algebras such that $\chi'_\alpha \circ \sigma|_\alpha \circ \chi_\alpha^{-1}$ is a gauge transform of $(V_{\alpha;A}, \mathcal{E}_{\alpha;A}^\bullet)$. This is precisely an isomorphism of geometric deformations of $\mathcal{P}$-algebras of type $\mathscr{D}$ because we have $\chi'_\alpha \circ \sigma_\alpha = \chi'_\alpha|_A$ by the construction of $\chi'_\alpha|_A$. Equivalence classes of geometric deformations of $\mathcal{P}$-algebras of type $\mathscr{D}$ form a functor of Artin rings

$$\mathrm{GDef}^{\mathscr{D}}(\mathcal{E}_0^\bullet, -) \colon \ \mathbf{Art}_\Lambda \to \mathbf{Set}.$$

Remark 12.16 At this point, we see why it makes sense to work over $\mathbf{Art}_\Lambda$. Namely, for $A, A' \in \mathbf{Art}_\Lambda$, we may choose completely different local models $V_{\alpha;A}$ even if $A \cong A'$ as $\mathbf{k}$-algebras. For example, in the case of log smooth deformations over $S_0 = \mathrm{Spec}(\mathbb{N} \to \mathbf{k})$, we obtain different local models over $A = \mathbf{k}[t]/(t^2)$ and $A' = \mathbf{k}[s]/(s^2)$ with $1 \mapsto t$ respectively $1 \mapsto 0$. $\Diamond$

Lemma 12.17 *The functor* $\mathrm{GDef}^{\mathscr{D}}(\mathcal{E}_0^\bullet, -)$ *is a neat deformation functor, i.e., it satisfies* (H_0), (H_1), (H_2), *and* (H_2^+) *defined in Appendix A.*

Proof Condition (H_0) is clear. For condition (H_1), let $A' \to A$ be arbitrary and $A'' \to A$ be surjective in $\mathbf{Art}_\Lambda$, and let $B := A' \times_A A''$. In the case of the local models $V_{\alpha;C}$, we form

$$Y := (|V_\alpha|, \ O_{V_{\alpha;A'}} \times_{O_{V_{\alpha;A}}} O_{V_{\alpha;A''}}, \ \mathcal{E}^P_{\alpha;A'} \times_{\mathcal{E}^P_{\alpha;A}} \mathcal{E}^P_{\alpha;A''});$$

due to the universal property, we obtain a map $Y \to V_{\alpha;B}$. As in the case of $\mathrm{LD}^{\mathscr{D}}_{X_0/S_0}$ in Lemma 10.19, we first obtain that $O_Y \cong O_{V_{\alpha;B}}$ is an isomorphism—for the surjectivity, we use $O_Y \otimes_B A' \cong O_{V_{\alpha;A'}}$ and that we can apply [267, 09ZW]. With the same injectivity argument as for O_Y, we find that $\mathcal{E}^P_{\alpha;B} \to \mathcal{E}^P_Y$ is injective. However, it is also surjective because $\mathcal{E}^P_Y \otimes_B A' \cong \mathcal{E}^P_{\alpha;A'}$ by [80, Thm. 2.2]. Thus, $Y = V_{\alpha;B}$. The general case is then very similar to the case of $\mathrm{LD}^{\mathscr{D}}_{X_0/S_0}$ in Lemma 10.19 because we can lift gauge transforms from $X_A|_\alpha$ to $X_{A''}|_\alpha$. The proof of (H_2) and (H_2^+) is analogous to the log smooth case in [160, Thm. 8.7], using the above-constructed push-out of geometric deformations of $\mathcal{P}$-algebras. $\qquad\square$

If $B' \to B$ is a surjection and $f : (X_B, \mathcal{E}^\bullet_B) \to S_B$ and $f' : (X_{B'}, \mathcal{E}^\bullet_{B'}) \to S_{B'}$ are two compatible geometric deformations of $\mathcal{P}$-algebras of type $\mathscr{D}$, then we have the sheaf

$$\mathcal{A}ut^{\mathscr{D}}_{X_{B'}/X_B}$$

of relative automorphisms. It consists precisely of the gauge transforms, i.e.,

$$\mathcal{A}ut^{\mathscr{D}}_{X_{B'}/X_B} = \mathcal{G}au_{X_{B'}/X_B}$$

since gauge transforms correspond to gauge transforms under isomorphisms of $\mathcal{P}$-algebras—and since gauge transforms inject into all automorphisms. In particular, if $B' \to B$ is a first-order extension with kernel $I \subset B'$, then we get

$$\mathcal{A}ut^{\mathscr{D}}_{X_{B'}/X_B} = \mathcal{T}_0 \otimes_{\mathbf{k}} I.$$

Now let $f : (X_B, \mathcal{E}^\bullet_B) \to S_B$ a geometric deformation of $\mathcal{P}$-algebras of type $\mathscr{D}$. On V_α, the local models allow a lift to $S_{B'}$. Since all automorphisms can be lifted, we can compare them with maps that become the identity over S_B. This yields the following standard result.

Proposition 12.18 *Let $f_0 : X_0 \to S_0$ be a geometric family of $\mathcal{P}$-algebras, and let $\mathscr{D}$ be a system of deformations over Λ. Let $B' \to B$ be a first-order extension with kernel I, and assume that we have a geometric deformation $f : X_B \to S_B$ of type $\mathscr{D}$. Then:*

(1) The automorphisms of a given lifting $f' : X_{B'} \to S_{B'}$ lie in $H^0(X_0, \mathcal{T}_0) \otimes_{\mathbf{k}} I$.
(2) Given one lifting, the set of all liftings is a torsor under $H^1(X_0, \mathcal{T}_0) \otimes_{\mathbf{k}} I$.
(3) The obstruction to the existence of a lifting is in $H^2(X_0, \mathcal{T}_0) \otimes_{\mathbf{k}} I$.

12.4   Systems of Deformations from Geometry

We construct systems of deformations of geometric families of $\mathcal{P}$-algebras from the log-geometric deformation problems we are interested in.

12.4.1   Systems of Deformations for Generically Log Smooth Families

Let $f_0\colon X_0 \to S_0$ be a log Gorenstein generically log smooth family over $S_0 = \mathrm{Spec}(Q \to \mathbf{k})$, and let $\mathcal{V} = \{V_\alpha\}_\alpha$ be a weakly admissible open cover. When we have a system of deformations $\mathscr{D}$ for f_0 subordinate to $\mathcal{V}$, then we can form an associated system of good geometric deformations of Lie–Rinehart algebras $\mathscr{D}^{\mathrm{lr}}$ such that geometric deformations of Lie–Rinehart algebras of type $\mathscr{D}^{\mathrm{lr}}$ correspond one-to-one to generically log smooth deformations of type $\mathscr{D}$. On the central fiber, we take for U_0° the strict locus U_0^{str} of $f\colon U_0 \to S_0$, which is smooth because $f_0\colon U_0 \to S_0$ is log smooth, and scheme-theoretically dense by [76, Prop. 10.1]. We have a geometric family of Lie–Rinehart algebras $\mathcal{LR}^\bullet_{X_0/S_0}$. This is a *good* geometric family of Lie–Rinehart algebras essentially by construction. Since $\mathcal{V}$ is weakly admissible with respect to $f_0\colon X_0 \to S_0$, it is also admissible with respect to $\mathcal{LR}^\bullet_{X_0/S_0}$. The local model $V_{\alpha;A} \to S_A$ from $\mathscr{D}$ gives rise to a geometric deformation of Lie–Rinehart algebras $(V_{\alpha;A}, \mathcal{LR}^\bullet_{V_{\alpha;A}/S_A})$; the pieces are flat because we assume each local deformation to have the base change property. The $(V_{\alpha;A}, \mathcal{LR}^\bullet_{V_{\alpha;A}/S_A})$ come with restriction maps and comparison isomorphisms induced from those of $\mathscr{D}$. The cocycles are gauge transforms by Lemma 10.6.

Analogously, we can also form a good geometric family of Gerstenhaber algebras $\mathcal{V}^\bullet_{X_0/S_0}$ on X_0, or a good geometric family $\mathcal{V}\mathcal{W}^\bullet_{X_0/S_0}$ of two-sided Gerstenhaber calculi. If $\mathcal{V}$ is not only weakly admissible but admissible, then we can form an analogous system of geometric deformations of Gerstenhaber algebras $\mathscr{D}^{\mathrm{g}}$ respectively a system of geometric deformations of Gerstenhaber calculi $\mathscr{D}^{\mathrm{gc}}$. Again, the log Gorenstein assumption and the base change property are crucial to make $\mathcal{V}\mathcal{W}^\bullet_{X_A/S_A}$ flat over S_A and compatible with base change. Again, Lemma 10.6 is crucial to make the cocycles gauge transforms.

In all three cases, we have the following result, which we formulate only for the most important case of two-sided Gerstenhaber calculi.

Proposition 12.19 *Let $\Lambda = \mathbf{k}[\![Q]\!]$, let $f_0\colon X_0 \to S_0$ be a generically log smooth family, and let $\mathscr{D}$ be a system of generically log smooth deformations subordinate to an admissible cover $\mathcal{V} = \{V_\alpha\}_\alpha$. Let $\mathscr{D}^{\mathrm{gc}}$ be the associated system of geometric deformations of two-sided Gerstenhaber calculi. Then the natural map*

$$\mathrm{LD}^{\mathscr{D}}_{X_0/S_0} \Rightarrow \mathrm{GDef}^{\mathscr{D}^{\mathrm{gc}}}(\mathcal{V}\mathcal{W}^\bullet_{X_0/S_0}, -),$$

$$(f\colon X_A \to S_A) \mapsto (X_A, U_A, U_A^{\mathrm{str}}, \mathcal{V}^\bullet_{X_A/S_A}, \mathcal{W}^\bullet_{X_A/S_A}),$$

is an isomorphism of functors of Artin rings.

Proof Let $f_A \colon X_A \to S_A$ be a generically log smooth deformation of type $\mathcal{D}$. Write $(\mathcal{V}^\bullet, \mathcal{W}^\bullet)$ for its two-sided Gerstenhaber calculus. First note that the image of the map is in fact a (good) geometric family of Gerstenhaber calculi. Next, note that we have a geometric *deformation* of two-sided Gerstenhaber calculi because $U_A^{\mathrm{str}}|_{X_0} = U_0^{\mathrm{str}}$ and $f_A \colon X_A \to S_A$ has the base change property. It is of type $\mathcal{D}^{\mathrm{gc}}$ because the (existent but not fixed) isomorphisms $(X_A, \mathcal{V}^\bullet, \mathcal{W}^\bullet)|_\alpha \cong (V_{\alpha;A}, \mathcal{V}^\bullet_{\alpha;A}, \mathcal{W}^\bullet_{\alpha;A})$ of generically log smooth families give rise to a choice of the desired isomorphisms of two-sided Gerstenhaber calculi. On overlaps, the comparison maps must be gauge transforms by Lemma 10.6. If we choose other isomorphisms, then they differ, again by Lemma 10.6, by a gauge transform so that the resulting geometric deformations of two-sided Gerstenhaber calculi of type $\mathcal{D}^{\mathrm{gc}}$ are equivalent. That the map is an isomorphism follows from tracking gluings of the pieces on V_α as in the proof of [76, Prop. 4.2]. $\square$

12.4.2   Systems of Deformations for Enhanced Generically Log Smooth Families

Let $f_0 \colon X_0 \to S_0$ be a torsionless enhanced generically log smooth family with an enhanced system of deformations $\mathcal{D}$ subordinate to $\mathcal{V} = \{V_\alpha\}_\alpha$. Then we obtain a system of geometric deformations of Gerstenhaber calculi $\mathcal{D}^{\mathrm{gc}}$ by forgetting the log structures and only keeping the underlying schemes as well as the Gerstenhaber calculi. Strict faithfulness of the Gerstenhaber calculus is guaranteed by assuming the enhanced generically log smooth family to be torsionless. The following result is clear from the definitions.

Proposition 12.20 *Let $f_0 \colon X_0 \to S_0$ be a torsionless enhanced generically log smooth family, and let $\mathcal{D}$ be an enhanced system of deformations subordinate to $\mathcal{V}$. Then we have an isomorphism*

$$\mathrm{ELD}^{\mathcal{D}}_{X_0/S_0} \Rightarrow \mathrm{GDef}^{\mathcal{D}^{\mathrm{gc}}}(\mathcal{G}_0^\bullet, \mathcal{A}_0^\bullet, -)$$

of functors of Artin rings.

12.4.3   Systems of Deformations for Families with Vector Bundles

Let $f_0 \colon (X_0, \mathcal{E}_0) \to S_0$ be a torsionless enhanced generically log smooth family with a vector bundle $\mathcal{E}_0$ of rank r, and let $\mathcal{V} = \{V_\alpha\}_\alpha$ be an $\mathcal{E}_0$-admissible open cover of X_0. Let $\mathcal{D}$ be a system of deformations subordinate to $\mathcal{V}$. Now $\mathcal{LRP}^\bullet_0 := \mathcal{LRP}^\bullet_{X_0/S_0}(\mathcal{E}_0)$ is a good geometric family of Lie–Rinehart pairs when we take for U_0° the strict locus of $f_0 \colon U_0 \to S_0$ as above. We construct a system of geometric deformations of Lie–Rinehart pairs $\mathscr{D}^{\mathrm{lrp}}$ as follows: When $V_{\alpha;A} \to S_A$ is a local model from $\mathscr{D}$, then we take

$$\mathcal{LRP}^{\bullet}_{\alpha;A} = \mathcal{LRP}^{\bullet}_{V_{\alpha;A}/S_A}(O^{\oplus r}_{V_{\alpha;A}}).$$

It makes sense to define $\mathcal{LRP}^E_{\alpha;A}$ as above because $\mathcal{E}_0$ is trivial on V_α by assumption. We take the restriction maps $\rho_{\alpha;BB'}$ induced from $\mathscr{D}$. Over S_0, we have to *choose* a trivialization $\mathcal{E}_0|_\alpha \cong O^{\oplus r}_{V_{\alpha;\mathbf{k}}}$ to obtain $\mathcal{LRP}^E_0|_\alpha \cong \mathcal{LRP}^E_{\alpha;\mathbf{k}}$. The definition of the comparison isomorphisms $\psi_{\alpha\beta;A}$ needs special care. Namely, if we just took the comparison isomorphisms from $O_{V_{\alpha;A}}$ for $O^{\oplus r}_{V_{\alpha;A}}$, then they would not give the correct map over S_0. Instead, we have to lift the isomorphism

$$O^{\oplus r}_{V_{\alpha;\mathbf{k}}}|_{\alpha\beta} \cong \mathcal{E}_0|_{\alpha\beta} \cong O^{\oplus r}_{V_{\beta;\mathbf{k}}}|_{\alpha\beta}$$

order by order along $A_k = \mathbf{k}[Q]/\mathrm{m}^{k+1}_Q$ and then take the base change to an arbitrary A. The cocycles are gauge transforms because we have gauge transforms on the level of the underlying enhanced generically log smooth family, and any compatible automorphism of the vector bundle then gives rise to a gauge transform in $\mathcal{G}^{-1}(\mathcal{E})$.

Proposition 12.21 *Let $f_0 \colon X_0 \to S_0$ be a torsionless enhanced generically log smooth family, let $\mathcal{E}_0$ be a vector bundle on X_0, and let $\mathscr{D}$ be a system of deformations subordinate to an $\mathcal{E}_0$-admissible open cover $\mathcal{V} = \{V_\alpha\}_\alpha$ of X_0. Let $\mathscr{D}^{\mathrm{lrp}}$ be the associated system of geometric deformations of Lie–Rinehart pairs. Then the natural map*

$$\mathrm{ELD}^{\mathscr{D}}_{X_0/S_0}(\mathcal{E}_0) \Rightarrow \mathrm{GDef}^{\mathscr{D}^{\mathrm{lrp}}}(\mathcal{LRP}^{\bullet}_0, -),$$

$$(f_A \colon (X_A, \mathcal{E}_A) \to S_A) \mapsto \mathcal{LRP}^{\bullet}_{X_A/S_A}(\mathcal{E}_A),$$

is an isomorphism of functors of Artin rings.

Proof The map $f_A \colon X_A \to S_A$ together with $|U_A| = |U_0|$ and $|U^\circ_A| = |U^\circ_0|$ as topological spaces, and with $\mathcal{LRP}^{\bullet}_{X_A/S_A}(\mathcal{E}_A)$ is a good geometric family of Lie–Rinehart pairs, essentially by Lemma 11.5. The map $f_0 \to f_A$ induces a map $\mathcal{LRP}^{\bullet}_{X_A/S_A}(\mathcal{E}_A) \to \mathcal{LRP}^{\bullet}(X_0/S_0, \mathcal{E}_0)$ which exhibits $\mathcal{LRP}^{\bullet}_{X_A/S_A}(\mathcal{E}_A)$ as a geometric *deformation* of Lie–Rinehart pairs. Now let $\chi_\alpha \colon X_A|_\alpha \cong V_{\alpha;A}$ be the local isomorphism which turns f_A into a deformation of type $\mathscr{D}$. By assumption, we have $\mathcal{E}_0|_\alpha \cong O^{\oplus r}_{X_0}|_\alpha$. This isomorphism can be lifted to an isomorphism $\mathcal{E}_A|_\alpha \cong O^{\oplus r}_{X_A}|_\alpha$ because $H^0(V_\alpha, O_{X_A}) \to H^0(V_\alpha, O_{X_0})$ and hence $H^0(V_\alpha, \mathcal{E}_A) \to H^0(V_\alpha, \mathcal{E}_0)$ is surjective. This induces an isomorphism $\mathcal{LRP}^{\bullet}_{X_A/S_A}(\mathcal{E}_A)|_\alpha \cong \mathcal{LRP}^{\bullet}_{\alpha;A}$ of geometric deformations of Lie–Rinehart pairs. The comparison maps between χ_α and χ_β are gauge transforms because they are induced from comparisons of enhanced generically log smooth deformations with a vector bundle. Thus we have an element in $\mathrm{GDef}^{\mathscr{D}^{\mathrm{lrp}}}(\mathcal{LRP}^{\bullet}_0, A)$. When we choose different trivializations of $\mathcal{E}_0$, different lifts of the trivialization to A, or different isomorphisms $\chi_\alpha \colon X_A|_\alpha \cong V_{\alpha;A}$, then the geometric deformations of Lie–Rinehart pairs of type $\mathscr{D}^{\mathrm{lrp}}$ are all equivalent. Similarly, if $f_A \colon (X_A, \mathcal{E}_A) \to S_A$ is isomorphic to $f'_A \colon (X'_A, \mathcal{E}'_A) \to S_A$, we obtain isomorphic geometric deformations of Lie–Rinehart pairs of type $\mathscr{D}^{\mathrm{lrp}}$. Thus, our map is well-defined on the level of objects. It is then easy to see that it is in fact a natural transformation. The proof of bijectivity is analogous to [76, Prop. 4.2]. $\square$

Chapter 13
The Characteristic Algebra

We explain the construction of the curved Gerstenhaber calculus associated with a deformation problem of Gerstenhaber calculi, and thus associated with a logarithmic deformation problem. This curved Gerstenhaber calculus is called the "characteristic Gerstenhaber calculus." We prove that deformations correspond with gauge equivalence classes of solutions of the extended Maurer–Cartan equation. In this chapter, we work in the following situation.

Situation 13.1 We fix a complete local Noetherian $\mathbf{k}$-algebra Λ with residue field $\mathbf{k}$. We have an algebraic structure $\mathcal{P}$ carrying a Cartan structure in the sense of Definition 3.37, and $f_0 \colon (X_0, \mathcal{E}_0^\bullet) \to S_0$ is a geometric family of $\mathcal{P}$-algebras in the sense of Definition 12.1. We have an open cover $\mathcal{V} = \{V_\alpha\}_\alpha$ of X_0 which is admissible with respect to $\mathcal{E}_0^\bullet$, and $\mathcal{D}$ is a system of deformations for f_0 subordinate to $\mathcal{V}$ over Λ. We have another open cover $\mathcal{U} = \{U_i\}_i$ of X_0 such that the union $\mathcal{U} \cup \mathcal{V}$ is admissible with respect to $\mathcal{E}_0^\bullet$.

In this situation, we construct a Λ-linear (Cartanian) $\mathcal{P}^{\mathrm{crv}}$-pre-algebra (in the sense of Definition 5.1)

$$E_{X_0/\Lambda}^{\bullet,\bullet} := E^{\bullet,\bullet}(X_0/S_0, \mathcal{E}_0^\bullet, \mathcal{V}, \mathcal{U}, \mathcal{D}),$$

the *characteristic algebra* of $f_0 \colon (X_0, \mathcal{E}_0^\bullet) \to S_0$ with respect to $\mathcal{V}, \mathcal{U}, \mathcal{D}$.[1] It depends on further choices which are suppressed in the notation, see Remark 13.34. Then

$$L_{X_0/\Lambda}^{\bullet} := E_{X_0/\Lambda}^{T,\bullet}$$

---

[1] We think of $E_{X_0/\Lambda}^{\bullet,\bullet}$ as characterizing both the deformation problem and the geometry of the central fiber $f_0 \colon X_0 \to S_0$. The deformation functor can always be extracted from the characteristic algebra, and for example in the case of the two-sided Gerstenhaber calculus of a generically log smooth family, the characteristic algebra also recovers the Hodge diamond.

S. Felten, *Global Logarithmic Deformation Theory*, Lecture Notes in Mathematics 2373, https://doi.org/10.1007/978-3-031-98751-9_13

is a Λ-linear curved Lie algebra controlling the deformation functor $\mathrm{GDef}^{\mathscr{D}}(\mathcal{E}_0^{\bullet}, -)$, as we will see in Lemma 13.36. Here, T is the index of the piece of the $\mathcal{P}$-algebra which controls gauge transforms; such an index is required as part of a Cartan calculus, cf. Definition 3.37. If $f_A \colon (X_A, \mathcal{E}_A^{\bullet}) \to S_A$ is the deformation corresponding to $\phi \in \mathfrak{m}_A \cdot (L_{X_0/\Lambda}^1 \otimes_\Lambda A)$, then

$$H^k(X_A, \mathcal{E}_A^P) = H^k(E_{X_0/\Lambda}^{P,\bullet} \otimes_\Lambda A, \ \bar{\partial} + \nabla_\phi^P)$$

for every domain (= index) $P \in D(\mathcal{P})$ of $\mathcal{P}$-algebras. The most important special case of Situation 13.1 is the case of (two-sided) Gerstenhaber calculi.

Situation 13.2 We fix a complete local Noetherian $\mathbf{k}$-algebra Λ with residue field $\mathbf{k}$. We have a good geometric family $f_0 \colon (X_0, \mathcal{G}_0^{\bullet}, \mathcal{A}_0^{\bullet}) \to S_0$ of two-sided Gerstenhaber calculi of relative dimension $d \geq 1$, see Definition 12.8. We have an admissible open cover $\mathcal{V} = \{V_\alpha\}_\alpha$ of X_0, and $\mathscr{D}$ is a system of deformations for f_0 subordinate to $\mathcal{V}$. Each deformation in $\mathscr{D}$ is good by Remark 12.13. We have another open cover $\mathcal{U} = \{U_i\}_i$ of X_0 such that $\mathcal{U} \cup \mathcal{V}$ is admissible with respect to $(\mathcal{G}_0^{\bullet}, \mathcal{A}_0^{\bullet})$.

In this situation, the characteristic algebra is actually a Λ-linear curved two-sided Gerstenhaber calculus; we denote it by

$$(PV_{X_0/\Lambda}^{\bullet,\bullet}, DR_{X_0/\Lambda}^{\bullet,\bullet}). \tag{13.1}$$

In Situation 13.2, $\mathcal{A}_0^d$ is a line bundle. If $\mathcal{A}_0^d \cong O_{X_0}$, then we say f_0 is *Calabi–Yau*. In this case, (13.1) is a Λ-linear curved two-sided Batalin–Vilkovisky calculus by Lemma 13.40. Consequently, if (13.1) is quasi-perfect, then $\mathrm{GDef}^{\mathscr{D}}(\mathcal{G}_0^{\bullet}, \mathcal{A}_0^{\bullet}, -)$ is unobstructed by Theorem 5.24. In particular, this applies in our situation of main interest.

Situation 13.3 We fix a sharp toric monoid Q and set $\Lambda := \mathbf{k}[\![Q]\!]$; we set $S_0 := \mathrm{Spec}(Q \to \mathbf{k})$. We have a torsionless enhanced generically log smooth family $f_0 \colon X_0 \to S_0$, an admissible open cover $\mathcal{V} = \{V_\alpha\}_\alpha$ of X_0 with admissibility as specified in Definition 10.21, and a system of deformations $\mathscr{D}$ subordinate to $\mathcal{V}$ in the sense of Definition 10.21. Furthermore, we have another open cover $\mathcal{U} = \{U_i\}_i$ of X_0 such that $\mathcal{U} \cup \mathcal{V}$ is admissible. Similar to Example 12.10, when taking the associated geometric families of two-sided Gerstenhaber calculi $\mathcal{GC}_{X_A/S_A}^{\bullet}$ and the system of deformations $\mathscr{D}^{\mathrm{gc}}$, we are in Situation 13.2. We have an isomorphism

$$\mathrm{ELD}_{X_0/S_0}^{\mathscr{D}} \cong \mathrm{GDef}^{\mathscr{D}^{\mathrm{gc}}}(\mathcal{G}_0^{\bullet}, \mathcal{A}_0^{\bullet}, -)$$

of functors of Artin rings by Proposition 12.20.

In this situation, if $f_0 \colon X_0 \to S_0$ is log Calabi–Yau, then (13.1) is a Λ-linear curved two-sided Batalin–Vilkovisky calculus which controls the deformation functor $\mathrm{ELD}_{X_0/S_0}^{\mathscr{D}}$.

In another direction, we can also construct a Λ-linear curved Lie–Rinehart pair which controls enhanced generically log smooth deformations with a vector bundle.

Situation 13.4 We fix a sharp toric monoid Q and set $\Lambda := \mathbf{k}[\![Q]\!]$; we set $S_0 := \mathrm{Spec}(Q \to \mathbf{k})$. We have a torsionless enhanced generically log smooth family $f_0 \colon X_0 \to S_0$, a vector bundle $\mathcal{E}_0$ on X_0 of rank r, an $\mathcal{E}_0$-admissible open cover $\mathcal{V} = \{V_\alpha\}_\alpha$ of X_0,[2] and a system of deformations $\mathscr{D}$ subordinate to $\mathcal{V}$ in the sense of Definition 11.12. Furthermore, we have another open cover $\mathcal{U} = \{U_i\}_i$ of X_0 such that $\mathcal{U} \cup \mathcal{V}$ is $\mathcal{E}_0$-admissible. When taking the Lie–Rinehart pairs $\mathcal{LRP}^\bullet_{X_A/S_A}(\mathcal{E}_A)$ and the system of deformations $\mathscr{D}^{\mathrm{lrp}}$, we are in Situation 13.1 since the cohomology of $\mathcal{G}^{-1}_{X_0/S_0}(\mathcal{E}_0)$ vanishes on each $V_{\alpha_1} \cap \ldots \cap V_{\alpha_s}$. We have an isomorphism

$$\mathrm{ELD}^{\mathscr{D}}_{X_0/S_0}(\mathcal{E}_0) \cong \mathrm{GDef}^{\mathscr{D}^{\mathrm{lrp}}}(\mathcal{LRP}^\bullet_0, -)$$

of functors of Artin rings by Proposition 12.21.

The main part of this chapter is an adaptation of [76, §§5–8], generalizing it from the case of Gerstenhaber algebras to $\mathcal{P}$-algebras, in particular two-sided Gerstenhaber calculi.

13.1 The Thom–Whitney Resolution

The Thom–Whitney resolution is an acyclic resolution of complexes which is well-suited to preserve additional algebraic structures on these complexes—exactly what we want to have when going from a sheaf of $\mathcal{P}$-algebras to a curved $\mathcal{P}$-algebra. In its present form, the Thom–Whitney resolution goes back to [217]; it seems that its first use in algebraic infinitesimal deformation theory is in Iacono–Manetti's algebraic proof of the Bogomolov–Tian–Todorov theorem in [152], where the reader can find extensive references to their prior use in topology. Of course, the Thom–Whitney resolution features prominently in Chan–Leung–Ma's method of deforming log spaces [38]. In [76], we gave a recollection of its basic properties, which we summarize again for convenience and to fix notations.

We denote the category of sets $[n] = \{0, \ldots, n\}$ with order-preserving injections as morphisms by Δ_{mon}; then a *semicosimplicial* object in a category C is a covariant functor $\Delta_{\mathrm{mon}} \to C$. Explicitly, a semicosimplicial object is a list of objects $A_0, A_1, \ldots$ together with $n + 1$ maps $\partial_{k,n} \colon A_{n-1} \to A_n$ for each $n \geq 1$ satisfying $\partial_{\ell,n+1}\partial_{k,n} = \partial_{k+1,n+1}\partial_{\ell,n}$.

A *semisimplicial* object is a contravariant functor $\Delta_{\mathrm{mon}} \to C$. For us, the most important semisimplicial object is the semisimplicial differential graded

---

[2] Here, $\mathcal{E}_0$-admissibility means that, on the one hand, $\mathcal{V}$ is weakly $\mathcal{E}_0$-admissible as used in Definition 11.12, and, on the other hand, $\mathcal{V}$ is admissible as specified in Definition 10.21.

commutative algebra $(A_{PL})_n$ formed by the dgca of (global algebraic) differential forms on $H_n = \{t_0 + \ldots + t_n = 1\} \subset \mathbb{A}_{\mathbf{k}}^{n+1}$ together with maps $\delta^{k,n} : (A_{PL})_n \to (A_{PL})_{n-1}$ induced by pull-back along inclusions of coordinate hyperplanes in $\mathbb{A}^{n+1}$.

If V^Δ is a semicosimplicial complex of vector spaces—here, each V_n is a complex $(V_n^\bullet, d)$ of vector spaces—then we can form the *Thom–Whitney bicomplex*

$$C_{TW}^{i,j}(V^\Delta) = \left\{ (x_n)_n \in \prod_{n \in \mathbb{N}} (A_{PL})_n^i \otimes_{\mathbf{k}} V_n^j \;\middle|\; \forall k, n : \right.$$

$$\left. (\delta^{k,n} \otimes \mathrm{Id})(x_n) = (\mathrm{Id} \otimes \partial_{k,n})(x_{n-1}) \right\}.$$

It comes with two natural differentials

$$\delta_1((a_n \otimes v_n)_n) = (da_n \otimes v_n)_n \quad \text{for} \quad (a_n \otimes v_n)_n \in C_{TW}^{i,j}(V^\Delta),$$

induced from the differential of $(A_{PL})_n$, and

$$\delta_2((a_n \otimes v_n)_n) = (-1)^i (a_n \otimes dv_n)_n \quad \text{for} \quad (a_n \otimes v_n)_n \in C_{TW}^{i,j}(V^\Delta),$$

induced from the differential of $V_n^\bullet$. The Thom–Whitney bicomplex is an exact functor, i.e., if

$$0 \to U^\Delta \to V^\Delta \to W^\Delta \to 0$$

is exact, then

$$0 \to C_{TW}^{i,j}(U^\Delta) \to C_{TW}^{i,j}(V^\Delta) \to C_{TW}^{i,j}(W^\Delta) \to 0$$

is exact as well for all i, j. We shall also need the following result.

Lemma 13.5 *Let V_m^Δ be a directed system of semicosimplicial complexes of vector spaces, and let V^Δ be the colimit in the sense that each V_n^j is the colimit of the $V_{m;n}^j$. Assume that the system V_m^Δ is bounded in the sense that there is $N > 0$ such that $V_{m;n}^j = 0$ for all $n > N$ and all m and j. Then the canonical map*

$$\varinjlim_m C_{TW}^{i,j}(V_m^\Delta) \to C_{TW}^{i,j}(V^\Delta)$$

is an isomorphism of complexes of vector spaces.

Proof It is easy to show that V^Δ is a semicosimplicial complex of vector spaces as well, so the statement makes sense. The proof of the main statement is straightforward once we use that each element either in $C_{TW}^{i,j}(V_m^\Delta)$ or $C_{TW}^{i,j}(V^\Delta)$

is represented by a *finite* sequence $(x_n)_n$ with $x_n \in (A_{\mathrm{PL}})_n^i \otimes V_{m;n}^j$ respectively $(A_{\mathrm{PL}})_n^i \otimes V_n^j$, and that $(A_{\mathrm{PL}})_n^i \otimes V_n^j$ is the colimit of the system $(A_{\mathrm{PL}})_n^i \otimes V_{m;n}^j$.   □

Let $S/\mathbf{k}$ be a Noetherian scheme defined over $\mathbf{k}$, and let $f \colon X \to S$ be a morphism of Noetherian schemes. Let $\mathcal{U} = \{U_i\}_i$ be a pre-admissible open cover of X in the sense of Definition 10.13, and let $\mathcal{F}$ be a quasi-coherent sheaf on X. Then we can form the *Čech semicosimplicial sheaf* $\mathcal{F}(\mathcal{U})$ as in [76, Ex. 5.1]—its objects are the quasi-coherent sheaves

$$\mathcal{F}(\mathcal{U})_n = \prod_{i_0 < \ldots < i_n} j_*(\mathcal{F}|_{U_{i_0} \cap \ldots \cap U_{i_n}})$$

where $j \colon U_{i_0} \cap \ldots \cap U_{i_n} \to X$ is the open immersion, and the maps are the usual Čech maps. This semicosimplicial object is *bounded* in the sense that $\mathcal{F}(\mathcal{U})_n = 0$ for $n \gg 0$. Since S is a $\mathbf{k}$-scheme, the sections $\Gamma(V, \mathcal{F}(\mathcal{U})_n)$ form a $\mathbf{k}$-vector space on each open $V \subseteq X$. By considering $\mathcal{F}(\mathcal{U})$ as a semicosimplicial complex concentrated in degree 0, we can form its Thom–Whitney bicomplex $C_{\mathrm{TW}}^{\bullet,\bullet}(\Gamma(V, \mathcal{F}(\mathcal{U})))$. They form a presheaf of $\mathbf{k}$-vector spaces on X which is actually a sheaf, as explained e.g. in [76, Constr. 5.3]. Moreover, we have a natural O_X-action which turns each $C_{\mathrm{TW}}^{i,0}(\mathcal{F}(\mathcal{U}))$ into a quasi-coherent sheaf. The following fact is not discussed in [76].

Lemma 13.6 *If $\mathcal{F}$ is flat over S, then $C_{\mathrm{TW}}^{i,0}(\mathcal{F}(\mathcal{U}))$ is flat over S as well.*

Proof Assume that $S = \mathrm{Spec}(R)$ is affine. Using that $U_i \to X$ is an affine morphism, we find that each $\mathcal{F}(\mathcal{U})_n$ is flat over S. Then, if $V \subseteq X$ is affine, we get that the (actually finite) product

$$\prod_{n \in \mathbb{N}} (A_{\mathrm{PL}})_n^i \otimes_{\mathbf{k}} \Gamma(V, \mathcal{F}(\mathcal{U})_n)$$

is flat over S. Similar to the proof of [76, Lemma 5.2], we find

$$\Gamma(V, C_{\mathrm{TW}}^{i,0}(\mathcal{F}(\mathcal{U}))) \otimes_R Q = \Gamma(V, C_{\mathrm{TW}}^{i,0}((\mathcal{F} \otimes_R Q)(\mathcal{U})))$$

for every finitely presented R-module Q. This allows us to conclude that the inclusion

$$\Gamma(V, C_{\mathrm{TW}}^{i,0}(\mathcal{F}(\mathcal{U}))) \to \prod_{n \in \mathbb{N}} (A_{\mathrm{PL}})_n^i \otimes_{\mathbf{k}} \Gamma(V, \mathcal{F}(\mathcal{U})_n)$$

is universally injective so that the left term is flat over R.   □

The first differential

$$\delta_1: \quad C_{\mathrm{TW}}^{i,0}(\mathcal{F}(\mathcal{U})) \to C_{\mathrm{TW}}^{i+1,0}(\mathcal{F}(\mathcal{U}))$$

is O_X-linear, and this, together with the canonical map

$$\mathcal{F} \to C_{\mathrm{TW}}^{0,0}(\mathcal{F}(\mathcal{U})), \quad f \mapsto (1 \otimes (f|_{U_{i_0} \cap \ldots \cap U_{i_n}})_{i_0 < \ldots < i_n})_n,$$

turns $C_{\mathrm{TW}}^{\bullet,0}(\mathcal{F}(\mathcal{U}))$ into a resolution of $\mathcal{F}$. Namely, this complex is quasi-isomorphic (on the level of sections over any open subset) to the Čech complex $\check{C}^{\bullet}(\mathcal{U}, \mathcal{F})$ by [217, Thm. 2.14].

Definition 13.7 Fix a pre-admissible cover $\mathcal{U}$ of X. Then the *Thom–Whitney resolution* of a quasi-coherent sheaf $\mathcal{F}$ is the complex $\mathrm{TW}^{\bullet}(\mathcal{F}) := C_{\mathrm{TW}}^{\bullet,0}(\mathcal{F}(\mathcal{U}))$ of quasi-coherent sheaves with O_X-linear differentials, which is quasi-isomorphic to $\mathcal{F}$ via the canonical map $\mathcal{F} \to \mathrm{TW}^0(\mathcal{F})$. This defines a functor.

Remark 13.8 The Thom–Whitney resolution is actually defined on the level of sheaves of $\mathbb{C}$-vector spaces. The O_X-module structure comes in addition to that; the Thom–Whitney resolution commutes with the forgetful functor from (quasi-coherent) O_X-modules to Abelian sheaves. $\diamond$

Remark 13.9 When applying the Thom–Whitney resolution to a complex $(\mathcal{F}^{\bullet}, \partial)$ of sheaves, we have to apply our above sign convention: The map $\mathrm{TW}^i(\partial)$ acquires a sign $(-1)^i$. Similarly, if $h: \mathcal{F}^{\bullet} \to \mathcal{G}^{\bullet}$ is a map of degree $|h|$ between complexes of sheaves, then our convention will be that $\mathrm{TW}^i(h)$ acquires a sign $(-1)^{|h| \cdot i}$. $\diamond$

Lemma 13.10 *The functor* $\mathrm{TW}^i(-)$ *is exact on quasi-coherent sheaves*[3] *for all* $i \geq 0$.

Proof Each embedding $U_{i_0} \cap \ldots \cap U_{i_n} \to X$ is an affine open immersion, so the direct image is an exact functor on quasi-coherent sheaves. Thus, a short exact sequence

$$0 \to \mathcal{E} \to \mathcal{F} \to \mathcal{G} \to 0$$

of quasi-coherent sheaves gives rise to a short exact sequence

$$0 \to \mathcal{E}(\mathcal{U})_n \to \mathcal{F}(\mathcal{U})_n \to \mathcal{G}(\mathcal{U})_n \to 0,$$

so we have a short exact sequence of semicosimplicial O_X-modules. Then, on each affine open subset of X, the above mentioned exactness of the Thom–Whitney

---

[3] The author believes that the lemma is not true in general for arbitrary Abelian sheaves.

construction on semicosimplicial vector spaces yields that

$$0 \to \mathrm{TW}^i(\mathcal{E}) \to \mathrm{TW}^i(\mathcal{F}) \to \mathrm{TW}^i(\mathcal{G}) \to 0$$

is exact. $\qquad\qquad\qquad\qquad\qquad\qquad\qquad\qquad\qquad\qquad\qquad\qquad\square$

Lemma 13.11 *The functor* $\mathrm{TW}^i(-)$ *preserves filtered colimits of quasi-coherent sheaves for all* $i \geq 0$.

Proof On a Noetherian scheme, a filtered colimit $\mathcal{F} = \varinjlim_m \mathcal{F}_m$ of quasi-coherent sheaves has the property that $\varinjlim_m \Gamma(W, \mathcal{F}_m) = \Gamma(W, \mathcal{F})$ for all open subsets $W \subseteq X$. Since $\mathcal{U}$ is a finite open cover, there is an N such that $\mathcal{F}_m(\mathcal{U})_n = 0$ for all $n > N$ and all m. Then we have $\varinjlim_m \Gamma(W, \mathrm{TW}^i(\mathcal{F}_m)) = \Gamma(W, \mathrm{TW}^i(\mathcal{F}))$ for all open subsets $W \subseteq X$ by Lemma 13.5. In particular, $\varinjlim_m \mathrm{TW}^i(\mathcal{F}_m) = \mathrm{TW}^i(\mathcal{F})$ as sheaves. $\qquad\qquad\qquad\qquad\qquad\qquad\qquad\qquad\qquad\qquad\qquad\qquad\square$

Let us say that an open cover $\mathcal{U} = \{U_i\}_i$ of X by finitely many affine open immersions $j_i \colon U_i \to X$ is *acyclic* with respect to a family $\{\mathcal{F}_k\}_k$ of quasi-coherent sheaves if $H^\ell(U_{i_0} \cap \ldots \cap U_{i_n}, \mathcal{F}_k) = 0$ for all $\ell \geq 1$, all indices $i_0, \ldots, i_n$, and all $\mathcal{F}_k$. Then, if $\mathcal{U}$ is acyclic with respect to $\mathcal{F}$, the complex $\mathrm{TW}^\bullet(\mathcal{F})$ computes the cohomology of $\mathcal{F}$ by [267, 0FLH]. Also, in this case, each $\mathrm{TW}^i(\mathcal{F})$ is acyclic for $\Gamma(X, -)$. Unfortunately, the proof of acyclicity in [76, Lemma 5.6] cannot be easily generalized to a non-affine open cover $\mathcal{U}$; instead, we have the following more conceptual proof.

Lemma 13.12 *Assume that* $\mathcal{U}$ *is an acyclic open cover of* X *with respect to* $\mathcal{F}$*. Let* $V \subseteq X$ *be an open subset with* $H^\ell(V \cap U_{i_0} \cap .. \cap U_{i_n}, \mathcal{F}) = 0$ *for all indices* $i_0, \ldots, i_n$*. Then* $H^\ell(V, \mathrm{TW}^k(\mathcal{F})) = 0$ *for* $\ell \geq 1$*; hence, we have*

$$H^\ell(V, \mathcal{F}) = H^\ell\left(\Gamma\big(V, \mathrm{TW}^\bullet(\mathcal{F})\big), \delta_1\right).$$

In particular, this holds for $V = X$ *and* $V = U_{j_0} \cap \ldots \cap U_{j_m}$.

Proof Since X is (locally) Noetherian, the injective objects of $\mathrm{QCoh}(X)$ are precisely those injective objects of $\mathrm{Mod}(X)$ which are quasi-coherent. This is a consequence of [137, Ch. II, Thm. 7.18], as pointed out by A. Preygel on math*overflow*. Now let $\mathcal{F} \to \mathcal{I}^\bullet$ be a quasi-coherent injective resolution of $\mathcal{F}$. Then each

$$\mathcal{I}^m(\mathcal{U})_n = \prod_{i_0 < \ldots < i_n} j_*(\mathcal{I}^m|_{U_{i_0} \cap \ldots \cap U_{i_n}})$$

is a quasi-coherent and injective sheaf as well; indeed, the restriction is injective, and then the direct image is injective by [267, 02N5]. In particular, each $\mathcal{I}^m(\mathcal{U})_n$ is a flasque sheaf by [138, Lemma 2.4]. Since the Thom–Whitney resolution is an exact functor, this implies that $\mathrm{TW}^k(I^m)$ is flasque as well because it is constructed by application of the functor to each open subset. Since $\mathrm{TW}^k(-)$ is an exact functor on quasi-coherent sheaves, we have an exact sequence

$$0 \to \mathrm{TW}^k(\mathcal{F}) \to \mathrm{TW}^k(I^0) \to \mathrm{TW}^k(I^1) \to \ldots$$

of quasi-coherent sheaves. Thus, $\mathrm{TW}^k(I^\bullet)$ computes the cohomology of $\mathrm{TW}^k(\mathcal{F})$. Since each $U_i \to X$ is an affine morphism, we have

$$H^\ell(V, \mathcal{F}(\mathcal{U})_n) = \bigoplus_{i_0 < \ldots < i_n} H^\ell(V \cap U_{i_0} \cap \ldots \cap U_{i_n}, \mathcal{F}) = 0 \qquad (13.2)$$

for $\ell \geq 1$. Since $I^\bullet(\mathcal{U})_n$ computes the cohomology of $\mathcal{F}(\mathcal{U})_n$, this shows that

$$0 \to H^0(V, \mathcal{F}(\mathcal{U})_n) \to H^0(V, I^0(\mathcal{U})_n) \to H^0(V, I^1(\mathcal{U})_n) \to \ldots$$

is exact. But then

$$0 \to H^0(V, \mathrm{TW}^k(\mathcal{F})) \to H^0(V, \mathrm{TW}^k(I^0)) \to H^0(V, \mathrm{TW}^k(I^1)) \to \ldots$$

is exact as well due to the exactness of the Thom–Whitney construction on semicosimplicial vector spaces, and we find $H^\ell(V, \mathrm{TW}^k(\mathcal{F})) = 0$ for $\ell \geq 1$. $\square$

Remark 13.13 This is not true for the cohomology on a general open subset $V \subseteq X$, even if each U_i is affine. $\Diamond$

If $b\colon T \to S$ is an affine morphism of finite type, and if $c\colon Y = X \times_T S \to X$, then $c^{-1}\mathcal{U}$ is an open cover of Y, and we have a natural isomorphism

$$c^* C_{\mathrm{TW}}^{i,0}(\mathcal{F}(\mathcal{U})) \cong C_{\mathrm{TW}}^{i,0}((c^*\mathcal{F})(c^{-1}\mathcal{U})) \qquad (13.3)$$

of quasi-coherent sheaves by [76, Lemma 5.2] and [267, 02KG]. However, in general, $c^{-1}(\mathcal{U})$ may not be acyclic with respect to $c^*\mathcal{F}$ so that $c^* C_{\mathrm{TW}}^{\bullet,0}(\mathcal{F}(\mathcal{U}))$ may not compute the cohomology of $c^*\mathcal{F}$. Of course, if $\mathcal{U}$ is an affine cover, then $c^{-1}\mathcal{U}$ is affine as well and hence acyclic. The following case is also of interest for us.

Lemma 13.14 *Assume that $S = \mathrm{Spec}\, A$ is the spectrum of an Artinian local $\mathbf{k}$-algebra A with residue field $\mathbf{k}$, and let $b\colon S_0 \to S$ be the inclusion induced by $A \to \mathbf{k}$. Let $c\colon X_0 \to X$ be the inclusion of the central fiber. Assume that $f\colon X \to S$ is flat and separated, and assume that $\mathcal{F}$ is flat over S. Then an open cover $\mathcal{U}$ of X by finitely many affine open immersions is acyclic with respect to $\mathcal{F}$ if and only if $c^{-1}\mathcal{U}$ is acyclic with respect to $\mathcal{F}_0 := c^*\mathcal{F}$.*

Proof We compare $H^\ell(U_{i_0} \cap \ldots \cap U_{i_n}, \mathcal{F})$ with $H^\ell(U_{i_0} \cap \ldots \cap U_{i_n}, \mathcal{F}_0)$ by means of [279, Thm. 0.4]. $\square$

13.1.1 The Thom–Whitney Resolution of a Geometric Deformation of $\mathcal{P}$-Algebras

Let $\mathcal{P}$ be an algebraic structure carrying a Cartan structure, and let $f_0\colon (X_0, \mathcal{E}_0^\bullet) \to S_0$ be a geometric family of $\mathcal{P}$-algebras. Let $\mathcal{U} = \{U_i\}_i$ be an open cover of X_0 which is admissible with respect to $\mathcal{E}_0^\bullet$. Let $W_0 \subseteq X_0$ be some Zariski open subset, and let $f\colon (W, \mathcal{E}^\bullet) \to S$ be a geometric deformation of $f_0|_{W_0}$ over $S = \operatorname{Spec} A$ for $A \in \mathbf{Art}_\Lambda$. We define

$$\mathrm{TW}^{P,q}(\mathcal{E}^\bullet) := C_{\mathrm{TW}}^{q,0}(\mathcal{E}^P(\mathcal{U}))$$

for $q \geq 0$ and $P \in D(\mathcal{P})$; this is a quasi-coherent sheaf on X, flat over S. The natural map

$$\mathcal{E}^P \to \mathrm{TW}^{P,\bullet}(\mathcal{E}^\bullet)$$

is a resolution of $\mathcal{E}^P$, but it is not necessarily acyclic and does not necessarily compute the cohomology of $\mathcal{E}^P$ unless $W_0 = X_0$.[4] We denote the differential by

$$\bar{\partial}\colon\ \mathrm{TW}^{P,q}(\mathcal{E}^\bullet) \to \mathrm{TW}^{P,q+1}(\mathcal{E}^\bullet);$$

it is $\mathcal{O}_W$-linear. Every constant $\gamma \in \mathcal{P}(P)$ gives rise to a constant $\gamma \in \mathrm{TW}^{P,0}(\mathcal{E}^\bullet)$ via $\mathcal{E}^P \to \mathrm{TW}^{P,0}(\mathcal{E}^\bullet)$. Every unary operation $\mu \in \mathcal{P}_N(P;\, Q)$ gives rise to an unary operation

$$\mu\colon\ \mathrm{TW}^{P,q}(\mathcal{E}^\bullet) \to \mathrm{TW}^{Q,q}(\mathcal{E}^\bullet), \quad (a_n \otimes p_n)_n \mapsto (-1)^{q\cdot|\mu|} \cdot (a_n \otimes \mu(p_n))_n,$$

with $|\mu| := |Q| - |P|$. The choice of the sign is compatible with our prior convention for the sign of the second differential in the Thom–Whitney bicomplex of a semicosimplicial complex of vector spaces. It can be interpreted as swapping a_n and μ. Every binary operation $\mu \in \mathcal{P}_N(P,\, Q;\, R)$ gives rise to a binary operation

$$\mu\colon\ \mathrm{TW}^{P,q}(\mathcal{E}^\bullet) \times \mathrm{TW}^{Q,q'}(\mathcal{E}^\bullet) \to \mathrm{TW}^{R,q+q'}(\mathcal{E}^\bullet),$$

$$((a_n \otimes p_n)_n,\ (b_n \otimes q_n)_n) \mapsto (-1)^{(|P|+|\mu|)\cdot q'} \cdot ((a_n \wedge b_n) \otimes \mu(p_n, q_n))_n\ ,$$

with $|\mu| := |R| - |Q| - |P|$. This formula can be interpreted as the sign which we obtain from swapping b_n and the operator $\mu(p_n, -)$, which is of degree $|R|-|Q|$. The sign convention is abstracted from [38, Defn. 3.9] and the subsequent discussion.

---

[4] This is because we may have $H^\ell(W \cap U_{i_0} \cap \ldots \cap U_{i_n}, \mathcal{E}^P) \neq 0$.

Remark 13.15 Unfortunately, we do not know a reasonable sign convention for operations of higher arity. In a Gerstenhaber algebra, different compositions of $\wedge$ and $[-,-]$ suggest different rules for ternary operations. This is the main reason why we require $\mathcal{P}(P_1, \ldots, P_n; Q) = \emptyset$ for $n \geq 3$ in the definition of a Cartan structure. $\Diamond$

Proposition 13.16 $\mathrm{TW}^{\bullet,\bullet}(\mathcal{E}^\bullet)$ *is a Z-faithful bounded Cartanian* $\mathcal{P}^{\mathrm{crv}}$*-pre-algebra in the context* $\mathfrak{QCoh}(W/S)$ *with* $\ell = 0$. *Its formation commutes with base change in* $\mathbf{Art}_\Lambda$.

Proof Since each $\mathrm{TW}^{P,q}(\mathcal{E}^\bullet)$ is a quasi-coherent sheaf, flat over S, and each operation is A-multilinear, we have indeed a $\mathcal{P}^{\mathrm{crv}}$-pre-algebra when setting $\ell = 0$. Since $\mathcal{F}(\mathcal{U})_n = 0$ for n exceeding the number M of opens in $\mathcal{U}$, we have $\mathrm{TW}^{P,q}(\mathcal{E}^\bullet) = 0$ for $q > M$ since then $(A_{\mathrm{PL}})_n^q = 0$ for all $n \leq M$; thus, $\mathrm{TW}^{\bullet,\bullet}(\mathcal{E}^\bullet)$ is (uniformly) bounded.

If $\mu \in \mathcal{P}_0(P; Q)$ or $\mu \in \mathcal{P}_0(P, Q; R)$, then the induced map is O_W-(bi-)linear; in particular, the induced products

$$*^P: \quad \mathrm{TW}^{F,q}(\mathcal{E}^\bullet) \times \mathrm{TW}^{P,q'}(\mathcal{E}^\bullet) \to \mathrm{TW}^{P,q+q'}(\mathcal{E}^\bullet)$$

are O_W-bilinear. Graded commutativity (for $P = F$) and unitality of this product are easy; it is also associative because the signs of the two ways to evaluate

$$(a_n \otimes f_n)_n *^F (b_n \otimes g_n)_n *^P (c_n \otimes p_n)_n$$

turn out to agree. If $\mu \in \mathcal{P}_N(P; P)$, then, for $a = (a_n \otimes g_n)_n \in \mathrm{TW}^{F,0}(\mathcal{E}^\bullet)$, we find

$$D_{1;a}((b_n \otimes p_n)_n) = (-1)^{|b||\mu|}\big((a_n \wedge b_n) \otimes (D_{1;g_n}(p_n))\big)_n;$$

repeated application of this formula shows that, on $\mathrm{TW}^{\bullet,\bullet}(\mathcal{E}^\bullet)$, μ is a differential operator of order N with respect to $\mathrm{TW}^{F,0}(\mathcal{E}^\bullet)/A$. Similar formulae hold in case $\mu \in \mathcal{P}_N(P, Q; R)$ with the factor $(-1)^{(|P|+|\mu|)|c|}$ for μ applied to $(b_n \otimes p_n)_n$ and $(c_n \otimes q_n)_n$ for both $D_{1;a}$ and $D_{2;a}$. Thus, μ is a bilinear differential operator with respect to $\mathrm{TW}^{F,0}(\mathcal{E}^\bullet)/A$ in this case. In particular, $\mathrm{TW}^{\bullet,\bullet}(\mathcal{E}^\bullet)$ is semi-Cartanian as a $\mathcal{P}^{\mathrm{bg}}$-pre-algebra in the sense of Definition 3.42. That it is a Cartanian $\mathcal{P}^{\mathrm{bg}}$-pre-algebra (in the sense of Definition 3.42) follows from an easy but tedious computation comparing signs.

For $\gamma \in \mathcal{P}(P)$, we have $\bar{\partial}(1 \otimes \gamma) = d(1) \otimes \gamma = 0$. The two formulae for $\bar{\partial}\mu(p)$ and $\bar{\partial}\mu(p, q)$ are straightforward. Finally, we have $\bar{\partial}^2(p) = \nabla_\ell^P(p)$ and $\bar{\partial}(\ell) = 0$ since $\bar{\partial}^2 = 0$ and $\ell = 0$. Thus, $\mathrm{TW}^{\bullet,\bullet}(\mathcal{E}^\bullet)$ is a Cartanian $\mathcal{P}^{\mathrm{crv}}$-pre-algebra in the sense of Definition 3.44.

The proof of Z-faithfulness is very similar to [76, Lemma 6.3]; we just have to choose instead of a function $f \in O(V \cap U_{i_0} \cap \ldots \cap U_{i_n})$ an appropriate section of some $\mathcal{E}^M$ for $M \in Z$.

That the formation of the quasi-coherent sheaf $\mathrm{TW}^{p,q}(\mathcal{E}^{\bullet})$ commutes with base change along $B' \to B$ in $\mathbf{Art}_\Lambda$ follows from (13.3). Obviously, the formation of constants commutes with base change. Finally, on affine opens, the formation of an operation μ commutes with base change; since μ is a multilinear differential operator over both B' and B, the operation over B is the induced one from the one over B' via our base change construction. $\qquad\qquad\qquad\qquad\qquad\square$

Remark 13.17 In special cases, additional relations are satisfied.

(1) If $\mathcal{E}^{\bullet}$ is a Lie–Rinehart algebra (respectively pair), then $\mathrm{TW}^{\bullet,\bullet}(\mathcal{E}^{\bullet})$ is a differential bigraded Lie–Rinehart algebra (respectively pair).
(2) If $\mathcal{E}^{\bullet}$ is a Gerstenhaber algebra (respectively calculus), then $\mathrm{TW}^{\bullet,\bullet}(\mathcal{E}^{\bullet})$ is a differential bigraded Gerstenhaber algebra (respectively calculus). In this situation, we also write $\mathrm{TW}^{p,q}(\mathcal{G}^{\bullet})$ and $\mathrm{TW}^{i,j}(\mathcal{A}^{\bullet})$ for the pieces of $\mathrm{TW}^{\bullet,\bullet}(\mathcal{E}^{\bullet})$. Explicitly, the bigraded $\wedge$-product and the bigraded Lie bracket on $\mathrm{TW}^{\bullet,\bullet}(\mathcal{G}^{\bullet})$ are given by

$$(a_n \otimes \theta_n)_n \wedge (b_n \otimes \xi_n)_n = (-1)^{|b||\theta|}((a_n \wedge b_n) \otimes (\theta_n \wedge \xi_n))_n$$

and

$$[(a_n \otimes \theta_n)_n, (b_n \otimes \xi_n)_n] = (-1)^{(|\theta|+1)|b|}((a_n \wedge b_n) \otimes [\theta_n, \xi_n])_n.$$

The bigraded $\wedge$-product on $\mathrm{TW}^{\bullet,\bullet}(\mathcal{A}^{\bullet})$ is given by

$$(a_n \otimes \alpha_n)_n \wedge (b_n \otimes \beta_n)_n = (-1)^{|b||\alpha|}((a_n \wedge b_n) \otimes (\alpha_n \wedge \beta_n))_n;$$

the $\mathcal{O}_W$-bilinear contraction map is

$$(a_n \otimes \theta_n)_n \lrcorner (b_n \otimes \alpha_n)_n = (-1)^{|\theta||b|}((a_n \wedge b_n) \otimes (\theta_n \lrcorner \alpha_n))_n,$$

and the $f^{-1}(\mathcal{O}_S)$-bilinear Lie derivative is

$$\mathcal{L}_{(a_n \otimes \theta_n)_n}((b_n \otimes \alpha_n)_n) = (-1)^{(|\theta|+1)|b|}((a_n \wedge b_n) \otimes \mathcal{L}_{\theta_n}(\alpha_n))_n.$$

(3) If $\mathcal{E}^{\bullet}$ is a two-sided Gerstenhaber calculus, then $\mathrm{TW}^{\bullet,\bullet}(\mathcal{E}^{\bullet})$ is a differential bigraded two-sided Gerstenhaber calculus. The bigraded left contraction on $\mathrm{TW}^{\bullet,\bullet}(\mathcal{A}^{\bullet})$ is given by

$$(a_n \otimes \theta_n) \vdash (b_n \otimes \alpha_n)_n = (-1)^{|\theta||b|}((a_n \wedge b_n) \otimes (\theta_n \vdash \alpha_n))_n.$$

A direct computation shows that all the relations required in Definition 3.22 are satisfied.

(4) If $\mathcal{E}^{\bullet} = (\mathcal{G}^{\bullet}, \mathcal{A}^{\bullet})$ is a Gerstenhaber calculus, and $\omega \in \Gamma(W, \mathcal{A}^d)$ is a global section which turns it into a Batalin–Vilkovisky calculus by Proposition 3.28,

then $(1 \otimes \omega)_n \in \mathrm{TW}^{d,0}(\mathcal{A}^\bullet)$ turns $\mathrm{TW}^{\bullet,\bullet}(\mathcal{E}^\bullet)$ into a differential bigraded Batalin–Vilkovisky calculus. The operations κ, v, and Δ on $\mathrm{TW}^{\bullet,\bullet}(\mathcal{E}^\bullet)$ are induced from these operations on $\mathcal{E}^\bullet$. Obviously, if $\mathcal{E}^\bullet$ is two-sided, so is $\mathrm{TW}^{\bullet,\bullet}(\mathcal{E}^\bullet)$ since this holds for the Gerstenhaber calculi. $\diamond$

13.2 Bigraded Thom–Whitney Deformations of $\mathcal{P}$-Algebras

In this section, we study geometric deformations of $f_0 \colon (X_0, \mathcal{E}_0^\bullet) \to S_0$ via their Thom–Whitney resolutions. So let us assume that we are in Situation 13.1.[5] First, we can apply the Thom–Whitney resolution (with respect to $\mathcal{U}$) to $\mathcal{E}_0^\bullet$ and obtain

$$\mathrm{TW}_0^{\bullet,\bullet} := \mathrm{TW}^{\bullet,\bullet}(\mathcal{E}_0^\bullet)$$

as a Cartanian $\mathcal{P}^{\mathrm{crv}}$-pre-algebra on the central fiber. Next, we apply the Thom–Whitney resolution to each local model $\mathcal{E}_{\alpha;A}^\bullet$ and obtain the Cartanian $\mathcal{P}^{\mathrm{crv}}$-pre-algebra

$$\mathrm{TW}_{\alpha;A}^{\bullet,\bullet} := \mathrm{TW}^{\bullet,\bullet}(\mathcal{E}_{\alpha;A}^\bullet)$$

on $V_{\alpha;A}$. They come with restriction maps

$$\mathrm{TW}(\rho_{\alpha;BB'})^* \colon \quad \mathrm{TW}_{\alpha;B'}^{\bullet,\bullet} \to \mathrm{TW}_{\alpha;B}^{\bullet,\bullet}$$

and comparison isomorphisms

$$\mathrm{TW}(\psi_{\alpha\beta;A})^* \colon \quad \mathrm{TW}_{\beta;A}^{\bullet,\bullet}|_{\alpha\beta} \xrightarrow{\cong} \mathrm{TW}_{\alpha;A}^{\bullet,\bullet}|_{\alpha\beta}$$

on overlaps. The symbol $*$ indicates that now the arrow direction is reversed compared with the original geometric arrow directions of these maps.

The Thom–Whitney resolution preserves gauge transforms. If $\theta \in I_{B'/B} \cdot \mathcal{E}_{B'}^T$ induces a gauge transform of some $\mathcal{P}$-algebra $\mathcal{E}_{B'}^\bullet$ over B', then the induced automorphism of the Thom–Whitney resolution $\mathrm{TW}^{\bullet,\bullet}(\mathcal{E}_{B'}^\bullet)$ is the gauge transform of

$$(1 \otimes \theta)_n \in I_{B'/B} \cdot \mathrm{TW}^{T,0}(\mathcal{E}_{B'}^\bullet).$$

---

[5] Under Lemma 13.12, the admissibility of the *union* $\mathcal{U} \cup \mathcal{V}$ ensures that a Thom–Whitney resolution with respect to $\mathcal{U}$ computes the cohomology on V_α. If both covers consist of affines, this is automatic.

Compared to a general gauge transform of $\mathrm{TW}^{\bullet,\bullet}(\mathcal{E}_{B'}^{\bullet})$, this gauge transform is special because it is compatible with $\bar{\partial}$ due to $\bar{\partial}((1 \otimes \theta)_n) = 0$; a general gauge transform interacts with $\bar{\partial}$ according to Lemma 5.3. We find that the cocycle

$$\mathrm{TW}(\psi_{\gamma\alpha;A})^* \circ \mathrm{TW}(\psi_{\beta\gamma;A})^* \circ \mathrm{TW}(\psi_{\alpha\beta;A})^*$$

is a gauge transform and moreover compatible with $\bar{\partial}$.

In analogy with [76, Defn. 6.4], we define *bigraded Thom–Whitney deformations* of $f_0\colon (X_0, \mathcal{E}_0^{\bullet}) \to S_0$. This gives a notion of the underlying deformation of *bigraded $\mathcal{P}$-algebras* when we forget the horizontal differential $\bar{\partial}$ of $\mathrm{TW}^{\bullet,\bullet}(\mathcal{E}^{\bullet})$, i.e., the differential which increases the second index. Studying this object is easier than studying the actual Thom–Whitney resolutions and gives additional insight into their structures—in fact, while there may be many geometric deformations of Gerstenhaber calculi, their Thom–Whitney resolutions all have the same underlying deformation of bigraded $\mathcal{P}$-algebras once we forget the horizontal differential $\bar{\partial}$. In [76], we called such a deformation a *Thom–Whitney–Gerstenhaber deformation*.

Definition 13.18 In the above situation, a *bigraded Thom–Whitney deformation* over $A \in \mathbf{Art}_\Lambda$ is a Cartanian $\mathcal{P}^{\mathrm{bg}}$-pre-algebra $\mathcal{H}^{\bullet,\bullet}$ in the context $\mathfrak{Flat}(|X_0|; A)$ together with a map

$$\rho^*\colon \ \mathcal{H}^{\bullet,\bullet} \to \mathrm{TW}^{\bullet,\bullet}(\mathcal{E}_0^{\bullet})$$

of $\mathcal{P}^{\mathrm{bg}}$-pre-algebras and isomorphisms

$$\chi_\alpha^*\colon \ \mathrm{TW}_{\alpha;A}^{\bullet,\bullet} \xrightarrow{\cong} \mathcal{H}^{\bullet,\bullet}|_\alpha$$

of $\mathcal{P}^{\mathrm{bg}}$-pre-algebras in $\mathfrak{Flat}(|V_\alpha|; A)$ which are compatible with the two restrictions $\rho^*|_\alpha$ and $\mathrm{TW}(\rho_{\alpha;A\mathbf{k}})^*$. The composition

$$\mathrm{TW}(\psi_{\beta\alpha;A})^* \circ (\chi_\alpha^*)^{-1} \circ \chi_\beta^*\colon \ \mathrm{TW}_{\beta;A}^{\bullet,\bullet}|_{\alpha\beta} \to \mathrm{TW}_{\beta;A}^{\bullet,\bullet}|_{\alpha\beta}$$

must be a gauge transform of $\mathrm{TW}_{\beta;A}^{\bullet,\bullet}|_{\alpha\beta}$ by some element $\theta \in I_{B'} \cdot \mathrm{TW}_{\beta;A}^{T,0}|_{\alpha\beta}$ which *may not* satisfy $\bar{\partial}\theta = 0$.

If $B' \to B$ is a map in $\mathbf{Art}_\Lambda$, and $(\mathcal{H}_B^{\bullet,\bullet}, \rho^*, (\chi_\alpha^*)_\alpha)$ and $(\mathcal{H}_{B'}^{\bullet,\bullet}, \rho'^*, (\chi'^*_\alpha)_\alpha)$ are two bigraded Thom–Whitney deformations, then a *morphism* is a map

$$r^*\colon \ \mathcal{H}_{B'}^{\bullet,\bullet} \to \mathcal{H}_B^{\bullet,\bullet}$$

of $\mathcal{P}^{\mathrm{bg}}$-pre-algebras which is compatible with ρ^* and ρ'^*, which induces an isomorphism $\mathcal{H}_{B'}^{P,q} \otimes_{B'} B \cong \mathcal{H}_B^{P,q}$ of sheaves of B-modules, and such that

$$(\chi'^*_\alpha|_B)^{-1} \circ \chi_\alpha^*$$

is a gauge transform of $\mathrm{TW}^{\bullet,\bullet}_{\alpha;A}$. Every morphism over the identity $A \to A$ is an isomorphism.

Remark 13.19 There is no horizontal (= increasing the second index) differential $\bar{\partial}$ on $\mathcal{H}^{\bullet,\bullet}$. Correspondingly, we do not require the comparison gauge transforms on overlaps to be compatible with $\bar{\partial}$. Also, this definition does not fix a scheme structure of a deformation of X_0 over S_A. The closest that we get to a structure sheaf is $\mathcal{H}^{0,0}$, but this is, of course, not locally isomorphic to O_{V_α}. $\Diamond$

Example 13.20 Let $f : (X_A, \mathcal{E}^\bullet_A) \to S_A$ be a geometric deformation of $\mathcal{P}$-algebras of type $\mathscr{D}$. Then $\mathrm{TW}^{\bullet,\bullet}(\mathcal{E}^\bullet_A)$ has a canonical structure of a bigraded Thom–Whitney deformation induced by the comparison isomorphisms χ_α that are part of a geometric deformation of type $\mathscr{D}$. $\Diamond$

Lemma 13.21 *Let* $\mathcal{H}^{\bullet,\bullet}$ *be a bigraded Thom–Whitney deformation over A. Then*

$$(\mathfrak{m}_A \cdot \mathcal{H}^{T,0}, \odot) \to (\mathcal{A}ut(\mathcal{H}^{\bullet,\bullet}), \circ), \quad \theta \mapsto \exp_\theta,$$

is an isomorphism of sheaves of groups, i.e., the automorphisms of $\mathcal{H}^{\bullet,\bullet}$ *as a bigraded Thom–Whitney deformation are precisely the gauge transforms.*

Proof Every gauge transform is indeed an automorphism of $\mathcal{H}^{\bullet,\bullet}$, so the map is well-defined. It is injective by Z-faithfulness. Locally on V_α, the map is an isomorphism due to our definition of an isomorphism of bigraded Thom–Whitney deformations, which requires that the induced map on $\mathrm{TW}^{\bullet,\bullet}_{\alpha;A}$ is a gauge transform. $\square$

Corollary 13.22 *Let* $\mathcal{H}^{\bullet,\bullet}_{B'}$ *be a bigraded Thom–Whitney deformation over B', with base change* $\mathcal{H}^{\bullet,\bullet}_B$ *to B. Then every global automorphism of* $\mathcal{H}^{\bullet,\bullet}_B$ *can be lifted to* $\mathcal{H}^{\bullet,\bullet}_{B'}$.

Proof It is sufficient to assume that $B' \to B$ is a small extension. Then we have an exact sequence

$$0 \to \mathrm{TW}^{T,0}_0 \otimes_{\mathbf{k}} I \to \mathfrak{m}_{B'} \cdot \mathcal{H}^{T,0}_{B'} \to \mathfrak{m}_B \cdot \mathcal{H}^{T,0}_B \to 0.$$

By Lemma 13.12, the right-hand map is surjective on global sections. $\square$

As in [76, Lemma 6.6], we have the following result.

Lemma 13.23 *Let* $B' \to B$ *be a first-order extension in* $\mathbf{Art}_\Lambda$ *with kernel* $I \subset B'$. *Let* $\mathcal{H}^{\bullet,\bullet}$ *be a bigraded Thom–Whitney deformation over B.*

(1) *Given a lifting* $\mathcal{H}'^{\bullet,\bullet}$ *to B', the relative automorphisms are in*

$$H^0(X_0, \mathrm{TW}^{T,0}_0) \otimes_{\mathbf{k}} I.$$

(2) *Given a lifting $\mathcal{H}'^{\bullet,\bullet}$ to B', the isomorphism classes of liftings are in*

$$H^1(X_0, \mathrm{TW}_0^{T,0}) \otimes_{\mathbf{k}} I.$$

(3) *The obstructions to the existence of a lifting are in*

$$H^2(X_0, \mathrm{TW}_0^{T,0}) \otimes_{\mathbf{k}} I.$$

Since $H^\ell(X_0, \mathrm{TW}_0^{T,0}) = 0$ for $\ell \geq 1$ by Lemma 13.12—we are now working on the whole space X_0, not on an open subspace where this may fail—the obstructions vanish, and there is a single isomorphism class of a lifting.

Definition 13.24 For $A = A_0 = \mathbf{k}$, we define the *characteristic bigraded sheaf* $\mathcal{E}_A^{\bullet,\bullet}$ as the Cartanian $\mathcal{P}^{\mathrm{bg}}$-pre-algebra

$$\mathcal{E}_0^{\bullet,\bullet} := \mathcal{E}_{A_0}^{\bullet,\bullet} := \mathrm{TW}_0^{\bullet,\bullet}$$

(by forgetting the differential $\bar{\partial}$). For $A = A_k = \Lambda/\mathfrak{m}_\Lambda^{k+1}$, we define $\mathcal{E}_A^{\bullet,\bullet}$ by inductively choosing a lift $\mathcal{E}_{k+1}^{\bullet,\bullet} := \mathcal{E}_{A_{k+1}}^{\bullet,\bullet}$ of $\mathcal{E}_k^{\bullet,\bullet} = \mathcal{E}_{A_k}^{\bullet,\bullet}$. Then, for a general A, we define

$$\mathcal{E}_A^{\bullet,\bullet} := \mathcal{E}_{A_k}^{\bullet,\bullet} \otimes_{A_k} A$$

as the tensor product along the canonical map $A_k \to A$ for an arbitrary $k \geq 0$ with $\mathfrak{m}_A^{k+1} = 0$.

Remark 13.25

(1) Since $\mathcal{E}_{k+1}^{\bullet,\bullet}$ is defined as a lift of $\mathcal{E}_k^{\bullet,\bullet}$, we have isomorphisms

$$\mathcal{E}_{k+1}^{\bullet,\bullet} \otimes_{A_{k+1}} A_k \cong \mathcal{E}_k^{\bullet,\bullet}$$

as part of the choice of a lift. Then, for a general A, we have an induced isomorphism

$$\mathcal{E}_{k+1}^{\bullet,\bullet} \otimes_{A_{k+1}} A \cong \mathcal{E}_k^{\bullet,\bullet} \otimes_{A_k} A$$

which does not depend on any further choice. In this sense, $\mathcal{E}_A^{\bullet,\bullet}$ is well-defined.

(2) If $\mathcal{E}_k^{\bullet,\bullet}$ and $\tilde{\mathcal{E}}_k^{\bullet,\bullet}$ are two systems of chosen lifts, then we can lift the isomorphism $\Phi_0 \colon \mathcal{E}_0^{\bullet,\bullet} = \tilde{\mathcal{E}}_0^{\bullet,\bullet}$ in a non-canonical and non-unique way order by order to obtain a compatible system of isomorphisms $\Phi_k \colon \mathcal{E}_k^{\bullet,\bullet} \cong \tilde{\mathcal{E}}_k^{\bullet,\bullet}$. After this choice, the induced isomorphisms $\Phi_A \colon \mathcal{E}_A^{\bullet,\bullet} \cong \tilde{\mathcal{E}}_A^{\bullet,\bullet}$ are unique (separately for every possible choice of $A_k \to A$).

(3) When $f_A \colon (X_A, \mathcal{E}_A^\bullet) \to S_A$ is a geometric deformation of $\mathcal{P}$-algebras, then $\mathrm{TW}^{\bullet,\bullet}(\mathcal{E}_A^\bullet)$ is non-uniquely isomorphic to $\mathcal{E}_A^{\bullet,\bullet}$. The isomorphism can be constructed by lifting along the steps of a decomposition of $A \to \mathbf{k}$ into small extensions.

(4) In the case of a Gerstenhaber algebra, we write $\mathcal{PV}_A^{\bullet,\bullet}$, and in the case of a Gerstenhaber calculus, we write additionally $\mathcal{DR}_A^{\bullet,\bullet}$. $\Diamond$

13.3   Predifferentials on Bigraded Thom–Whitney Deformations

In this section, we generalize [76, §7] from Gerstenhaber algebras to $\mathcal{P}$-algebras. The local model $\mathrm{TW}_{\alpha;A}^{\bullet,\bullet}$ of a bigraded Thom–Whitney deformation comes with a differential

$$\bar{\partial}_{\alpha;A} \colon \ \mathrm{TW}_{\alpha;A}^{P,q} \to \mathrm{TW}_{\alpha;A}^{P,q+1}.$$

Every $\phi \in \mathfrak{m}_A \cdot \mathrm{TW}_{\alpha;A}^{T,1}$ defines a modification $\bar{\partial}_{\alpha;A} + \nabla_\phi^P(-)$, which turns $\mathrm{TW}_{\alpha;A}^{\bullet,\bullet}$ into another Cartanian $\mathcal{P}^{\mathrm{crv}}$-pre-algebra with $\ell_{\alpha;A}(\phi) = \bar{\partial}_{\alpha;A}(\phi) + \frac{1}{2}\nabla_\phi^T(\phi)$. If $\theta \in \mathfrak{m}_A \cdot \mathrm{TW}_{\alpha;A}^{T,0}$, then Lemma 5.3 shows that

$$\exp_\theta(\bar{\partial}_{\alpha;A}(p) + \nabla_\phi^P(p)) = \bar{\partial}_{\alpha;A}(\exp_\theta(p)) + \nabla_{\exp_\theta * \phi}^P(\exp_\theta(p)),$$

i.e., maps of the form $\bar{\partial}_{\alpha;A} + \nabla_\phi^P(-)$ remain of this form under gauge transforms, albeit with a new value of ϕ.

Definition 13.26 Let $\mathcal{H}^{\bullet,\bullet}$ be a bigraded Thom–Whitney deformation over $A \in \mathbf{Art}_\Lambda$. Then a *predifferential* on $\mathcal{H}^{\bullet,\bullet}$ consists of (global) maps $\bar{\partial} \colon \mathcal{H}^{P,q} \to \mathcal{H}^{P,q+1}$, one for each domain $P \in D(\mathcal{P})$ and each $q \geq 0$, with

$$(\chi_\alpha^*)^{-1} \circ \bar{\partial}|_\alpha \circ \chi_\alpha^* = \bar{\partial}_{\alpha;A} + \nabla_{\phi_\alpha}^P(-)$$

for some $\phi_\alpha \in \mathfrak{m}_A \cdot \mathrm{TW}_{\alpha;A}^{T,0}$ (the same for all P and q). Two predifferentials $\bar{\partial}_1$ and $\bar{\partial}_2$ are *gauge equivalent* if there is an automorphism $\psi^* \colon \mathcal{H}^{\bullet,\bullet} \to \mathcal{H}^{\bullet,\bullet}$ with $(\psi^*)^{-1} \circ \bar{\partial}_1 \circ \psi^* = \bar{\partial}_2$.

A predifferential turns $\mathcal{H}^{\bullet,\bullet}$ into a Cartanian $\mathcal{P}^{\mathrm{crv}}$-pre-algebra.

Lemma 13.27 *Let $\mathcal{H}^{\bullet,\bullet}$ be a bigraded Thom–Whitney deformation over $A \in \mathbf{Art}_\Lambda$, and let $\bar{\partial}$ be a predifferential. Then, there is a unique global section $\ell \in \mathfrak{m}_A \cdot \mathcal{H}^{T,2}$ with $\bar{\partial}^2 = \nabla_\ell^P(-)$ for every $P \in D(\mathcal{P})$. This turns the triple $(\mathcal{H}^{\bullet,\bullet}, \bar{\partial}, \ell)$ into a global Cartanian $\mathcal{P}^{\mathrm{crv}}$-pre-algebra.*

Proof The element ϕ_α in the definition is unique since $\mathrm{TW}^{\bullet;\bullet}_{\alpha;A}$ is Z-faithful. Since

$$(\bar\partial_{\alpha;A} + \nabla^{P}_{\phi_\alpha}(-))^2 = \nabla^{P}_{\ell_{\alpha;A}(\phi_\alpha)}(-),$$

we also have $\bar\partial^2 = \nabla^{P}_{\ell_\alpha}(-)$ for

$$\ell_\alpha := \chi^*_\alpha(\ell_{\alpha;A}(\phi_\alpha)) \in \mathfrak{m}_A \cdot \mathcal{H}^{T,2}|_\alpha.$$

By Z-faithfulness, they glue to a global section $\ell \in \mathfrak{m}_A \cdot \mathcal{H}^{T,2}$ with $\bar\partial^2 = \nabla^{P}_\ell(-)$. We have $\bar\partial(\ell) = 0$ since $(\bar\partial_{\alpha;A} + \nabla^{T}_{\phi_\alpha})(\ell_{\alpha;A}(\phi_\alpha)) = 0$. The remaining conditions from Definition 3.44 for a Cartanian $\mathcal{P}^{\mathrm{crv}}$-pre-algebra hold because they are invariant under adding $\nabla^{P}_{\phi_\alpha}$ to $\bar\partial_{\alpha;A}$. $\qquad\square$

A predifferential is a *differential* if $\bar\partial^2 = 0$. This is the case if and only if the corresponding ℓ vanishes, due to Z-faithfulness.

We say that two pairs $(\mathcal{H}^{\bullet,\bullet}_1, \bar\partial_1)$ and $(\mathcal{H}^{\bullet,\bullet}_2, \bar\partial_2)$ consisting of a bigraded Thom–Whitney deformation together with a predifferential are *isomorphic* if there is an isomorphism $r^* : \mathcal{H}^{\bullet,\bullet}_2 \cong \mathcal{H}^{\bullet,\bullet}_1$ of bigraded Thom–Whitney deformations with $\bar\partial_2 = (r^*)^{-1} \circ \bar\partial_1 \circ r^*$. We collect a few obvious results which ensure that the functor of isomorphism classes of pairs $(\mathcal{H}^{\bullet,\bullet}, \bar\partial)$ is well-behaved.

Lemma 13.28 *The following hold:*

(1) *If $(\mathcal{H}^{\bullet,\bullet}_1, \bar\partial_1)$ and $(\mathcal{H}^{\bullet,\bullet}_2, \bar\partial_2)$ are isomorphic, then $\bar\partial_1$ is a differential if and only if $\bar\partial_2$ is a differential.*

(2) *On the same $\mathcal{H}^{\bullet,\bullet}$, the two pairs $(\mathcal{H}^{\bullet,\bullet}, \bar\partial_1)$ and $(\mathcal{H}^{\bullet,\bullet}, \bar\partial_2)$ are isomorphic if and only if $\bar\partial_1$ and $\bar\partial_2$ are gauge equivalent.*

(3) *If $\mathcal{H}^{\bullet,\bullet}_1$ and $\mathcal{H}^{\bullet,\bullet}_2$ are two isomorphic bigraded Thom–Whitney deformations, and if $\bar\partial_1$ is a predifferential on $\mathcal{H}^{\bullet,\bullet}_1$, then there is some predifferential $\bar\partial_2$ on $\mathcal{H}^{\bullet,\bullet}_2$ such that the pairs become isomorphic.*

(4) *If $r^* : \mathcal{H}^{\bullet,\bullet}_{B'} \to \mathcal{H}^{\bullet,\bullet}_B$ is a morphism of bigraded Thom–Whitney deformations over $B' \to B$, and $\bar\partial'$ is a predifferential on $\mathcal{H}^{\bullet,\bullet}_{B'}$, then there is a unique induced predifferential $\bar\partial$ on $\mathcal{H}^{\bullet,\bullet}_B$ which commutes with r^*. If $\bar\partial'$ is a differential, so is $\bar\partial$.[6]*

(5) *If $(\mathcal{H}^{\bullet,\bullet}_1, \bar\partial_1)$ and $(\mathcal{H}^{\bullet,\bullet}_2, \bar\partial_2)$ are isomorphic pairs over B', so are $\mathcal{H}^{\bullet,\bullet}_1 \otimes_{B'} B$ and $\mathcal{H}^{\bullet,\bullet}_2 \otimes_{B'} B$ with their induced predifferentials.*

Definition 13.29 Isomorphism classes of pairs $(\mathcal{H}^{\bullet,\bullet}, \bar\partial)$ consisting of a bigraded Thom–Whitney deformation $\mathcal{H}^{\bullet,\bullet}$ and a differential (not just a predifferential) $\bar\partial$ form a functor of Artin rings

$$\mathrm{TWD}^{\mathscr{D}}(\mathcal{E}^\bullet_0, -) : \quad \mathbf{Art}_\Lambda \to \mathbf{Set}.$$

---

[6] The converse may fail. Also note that $B' \to B$ does not need to be a surjection.

Example 13.20 yields a natural transformation

$$\mathbf{tw}\colon\quad \mathrm{GDef}^{\mathscr{D}}(\mathcal{E}_0^{\bullet}, -) \Rightarrow \mathrm{TWD}^{\mathscr{D}}(\mathcal{E}_0^{\bullet}, -)$$

of functors of Artin rings. Conversely, when we have a pair $(\mathcal{H}^{\bullet,\bullet}, \bar{\partial})$ with $\bar{\partial}^2 = 0$, then we can take the cohomology sheaf $\mathcal{E}_A^P := H^0(\mathcal{H}^{P,\bullet}, \bar{\partial})$ for each $P \in D(\mathcal{P})$. All constants and operations descend to $\mathcal{E}_A^{\bullet}$, so that it is a $\mathcal{P}$-pre-algebra (in some context, say sheaves of A-modules on $|X_0|$). Since [76, Lemma 7.3] holds here as well—namely, using our admissibility assumption with respect to $\mathcal{U} \cup \mathcal{V}$ and Lemma 13.12, we have $H^1(V_\alpha, \mathcal{E}_{\alpha;A}^T) = 0$, and this cohomology is computed by $(\mathrm{TW}_{\alpha;A}^{T,\bullet}, \bar{\partial})$—the $\mathcal{P}$-pre-algebra $\mathcal{E}_A^{\bullet}$ is locally isomorphic to the local model $\mathcal{E}_{\alpha;A}^{\bullet}$. Comparing two of these isomorphisms on the level of bigraded Thom–Whitney deformations with a differential shows that the isomorphisms with $(\mathrm{TW}_{\alpha;A}^{\bullet,\bullet}, \bar{\partial}_{\alpha;A})$ respectively $(\mathrm{TW}_{\beta;A}^{\bullet,\bullet}, \bar{\partial}_{\beta;A})$ differ from $\mathrm{TW}(\psi_{\alpha\beta;A})^*$ by a gauge transform $\exp_\theta$ which respects differentials. However, such a gauge transform must satisfy $\bar{\partial}\theta = 0$, i.e., it comes from a gauge transform of $\mathcal{E}_{\alpha;A}^{\bullet}$. Thus, $O_{X_A} := \mathcal{E}_A^F$ turns $X_A \to S_A$ into a flat deformation of $X_0 \to S_0$, and $\mathcal{E}_A^{\bullet}$ is a geometric deformation of $\mathcal{P}$-algebras of $\mathcal{E}_0^{\bullet}$. This defines a natural transformation

$$\mathbf{h}\colon\quad \mathrm{TWD}^{\mathscr{D}}(\mathcal{E}_0^{\bullet}, -) \Rightarrow \mathrm{GDef}^{\mathscr{D}}(\mathcal{E}_0^{\bullet}, -)$$

of functors of Artin rings. As in [76, Lemma 7.4], we have the following result:

Proposition 13.30 *The two natural transformations* $\mathbf{tw}$ *and* $\mathbf{h}$ *are inverse to each other, so that* $\mathrm{GDef}^{\mathscr{D}}(\mathcal{E}_0^{\bullet}, -) \cong \mathrm{TWD}^{\mathscr{D}}(\mathcal{E}_0^{\bullet}, -)$.

13.4   Construction of $E_{X_0/\Lambda}^{\bullet,\bullet}$

We generalize [76, §8] to the case of $\mathcal{P}$-algebras. Recall from Definition 13.24 that we have a system $\mathcal{E}_A^{\bullet,\bullet}$ of bigraded Thom–Whitney deformations over $A \in \mathbf{Art}_\Lambda$ together with restriction maps between them along $B' \to B$. We will construct a compatible system of predifferentials on $\mathcal{E}_A^{\bullet,\bullet}$. Then $E_{X_0/\Lambda}^{\bullet,\bullet}$ is a Cartanian $\mathcal{P}^{\mathrm{crv}}$-pre-algebra in the context $\mathfrak{Comp}(\Lambda)$ obtained as a limit of global sections of $\mathcal{E}_A^{\bullet,\bullet}$.

Let $B' \to B$ be a map in $\mathbf{Art}_\Lambda$. Every predifferential (locally) on $\mathcal{E}_{B'}^{\bullet,\bullet}$ restricts uniquely to $\mathcal{E}_B^{\bullet,\bullet}$ as we have seen in Lemma 13.28. Conversely, if $B' \to B$ is a surjection, and if we have a predifferential $\bar{\partial}_B$ on $\mathcal{E}_B^{\bullet,\bullet}$, it is always the restriction of some predifferential $\bar{\partial}_{B'}$ on $\mathcal{E}_{B'}^{\bullet,\bullet}$. To see this, it is sufficient to assume that $B' \to B$ is a small extension with kernel $I \subset B'$. Then, as in [76, §8], the predifferentials on $\mathcal{E}_{B'}^{\bullet,\bullet}$ which restrict to $\bar{\partial}_B$ form a torsor under $I \cdot \mathcal{E}_{B'}^{T,1} \cong \mathcal{E}_0^{T,1} \otimes_{\mathbf{k}} I$. Since $H^1(X_0, \mathcal{E}_0^{T,1}) = 0$ by Lemma 13.12, every torsor is trivial, so it has a global section $\bar{\partial}_{B'}$.

Definition 13.31 The *characteristic curved sheaf* $(\mathcal{E}_A^{\bullet,\bullet}, \bar{\partial}_A)$ is obtained from the characteristic bigraded sheaf $\mathcal{E}_A^{\bullet,\bullet}$ as follows:

(1) Over $A_0 = \mathbf{k}$, we have the differential $\bar{\partial}_0$ of $\mathcal{E}_0^{\bullet,\bullet}$.

(2) Assume that we have $\bar{\partial}_k$ over $A_k = \Lambda/\mathfrak{m}_\Lambda^{k+1}$. We have just seen that, along any surjection $B' \to B$, the restriction map of predifferentials is surjective. Therefore, we can choose a lift $\bar{\partial}_{k+1}$ of $\bar{\partial}_k$ over A_{k+1}.

(3) Let $A \in \mathbf{Art}_\Lambda$ be general. When $\mathfrak{m}_A^{k+1} = 0$, there is a canonical map $A_k \to A$ induced by $\Lambda \to A$. Then, we set $\bar{\partial}_A = \bar{\partial}_k|_A$. This predifferential is independent of our choice of k. By construction, we have $\bar{\partial}_{B'}|_B = \bar{\partial}_B$ for every map $B' \to B$ in $\mathbf{Art}_\Lambda$.

(4) We obtain a unique element $\ell_A \in H^0(X_0, \mathfrak{m}_A \cdot \mathcal{E}_A^{T,2})$ with $\bar{\partial}_A^2 = \nabla_\ell(-)$ and $\bar{\partial}_A(\ell_A) = 0$ by Lemma 13.27; these sections satisfy $\ell_{B'}|_B = \ell_B$ for every $B' \to B$ in $\mathbf{Art}_\Lambda$.

The *characteristic algebra* over $\Lambda \in \mathbf{Art}_\Lambda$ is then given by global sections:

$$E_{X_0/A}^{\bullet,\bullet} := H^0(X_0, (\mathcal{E}_A^{\bullet,\bullet}, \bar{\partial}_A, \ell_A)).$$

Lemma 13.32 *We have the following:*

(1) $\mathcal{E}_A^{P,q}$ *is acyclic, i.e.,* $H^\ell(X_0, \mathcal{E}_A^{P,q}) = 0$ *for* $\ell \geq 1$.

(2) $E_{X_0/A}^{P,q}$ *is a flat A-module.*[7]

(3) $E_{X_0/A}^{\bullet,\bullet}$ *is a Cartanian $\mathcal{P}^{\mathrm{crv}}$-pre-algebra in the context $\mathfrak{Comp}(A)$.*

(4) *For each $B' \to B$, we have a restriction map $E_{X_0/B'}^{P,q} \to E_{X_0/B}^{P,q}$ which induces an isomorphism $E_{X_0/B'}^{P,q} \otimes_{B'} B \cong E_{X_0/B}^{P,q}$.*

Proof If $B' \to B$ is a small extension, then we have an exact sequence

$$0 \to I \otimes_{\mathbf{k}} \mathcal{E}_0^{P,q} \to \mathcal{E}_{B'}^{P,q} \to \mathcal{E}_B^{P,q} \to 0.$$

The left-hand term is acyclic by Lemma 13.12. Thus, if the right-hand term is acyclic, the middle term is acyclic as well, so the statement follows from induction over small extensions.

For flatness of $E_{X_0/A}^{P,q}$, let $\check{C}^\bullet(\mathcal{W}, \mathcal{E}_{X_0/A}^{P,q})$ be the Čech resolution of $\mathcal{E}_{X_0/A}^{P,q}$ in sheaves for some finite affine open cover $\mathcal{W}$ of X_0. We can assume that every open subset in $\mathcal{W}$ is contained in some V_α. This computes the cohomology of $\mathcal{E}_{X_0/A}^{P,q}$, so we have an exact sequence

$$0 \to E_{X_0/A}^{P,q} \to H^0(X_0, \check{C}^0(\mathcal{W}, \mathcal{E}_{X_0/A}^{P,q}))$$

$$\to \ldots \to H^0(X_0, \check{C}^{N-1}(\mathcal{W}, \mathcal{E}_{X_0/A}^{P,q})) \to 0$$

---

[7] The question of flatness of the global sections is not discussed in [76].

where N is the number of open subsets in $\mathcal{W}$. From the second term on, each term is a flat A-module because $\mathcal{E}^{P,q}_{X_0/A}|_\alpha \cong \mathrm{TW}^{P,q}_{\alpha;A}$ is an $O_{V_{\alpha;A}}$-module which is flat over S_A by Lemma 13.6, so sections over an intersection of opens in $\mathcal{W}$ form a flat A-module. Then $E^{P,q}_{X_0/A}$ is a flat A-module by repeated application of [279, Lemma 0.2.1].

Once we know flatness, $E^{\bullet,\bullet}_{X_0/A}$ is a Cartanian $\mathcal{P}^{\mathrm{crv}}$-pre-algebra in $\mathfrak{Comp}(A)$ because $\mathcal{E}^{\bullet,\bullet}_A$ is one in $\mathfrak{Flat}(|X_0|; A)$. The last statement is similar to [76, Cor. 5.7]. $\qquad\qquad\Box$

Definition 13.33 The *characteristic algebra* is the limit[8]

$$E^{\bullet,\bullet}_{X_0/\Lambda} := \varprojlim_{k\to\infty} E^{\bullet,\bullet}_{X_0/A_k}.$$

Each $E^{P,q}_{X_0/\Lambda}$ is a flat Λ-module by [267, 0912], and it is complete by construction. Thus, $E^{\bullet,\bullet}_{X_0/\Lambda}$ is a Cartanian $\mathcal{P}^{\mathrm{crv}}$-pre-algebra in the context $\mathfrak{Comp}(\Lambda)$. For each $A \in \mathbf{Art}_\Lambda$, we have $E^{\bullet,\bullet}_{X_0/A} \cong E^{\bullet,\bullet}_{X_0/\Lambda} \otimes_\Lambda A$.

Remark 13.34 We collect the choices on which $E^{\bullet,\bullet}_{X_0/\Lambda}$ depends:

(1) on the open cover $\mathcal{V}$ and the system of deformations $\mathscr{D}$;
(2) on the open cover $\mathcal{U}$;
(3) on the liftings $\mathcal{E}^{\bullet,\bullet}_{k+1}$ of the bigraded Thom–Whitney deformation $\mathcal{E}^{\bullet,\bullet}_k$;
(4) on the liftings of the predifferentials along $\mathcal{E}^{\bullet,\bullet}_{k+1} \to \mathcal{E}^{\bullet,\bullet}_k$.

Therefore, the characteristic algebra is not canonical, but any choice is sufficient for our purposes. Furthermore, the author expects that any two choices are homotopy equivalent in an appropriate sense, but this question is beyond the scope of this monograph. $\qquad\qquad\Diamond$

Remark 13.35 In the case of Gerstenhaber algebras, we write $PV^{\bullet,\bullet}_{X_0/\Lambda}$ for $E^{\bullet,\bullet}_{X_0/\Lambda}$. It is a Λ-linear curved Gerstenhaber algebra in this case. In the case of (two-sided) Gerstenhaber calculi, we write additionally $DR^{\bullet,\bullet}_{X_0/\Lambda}$ for the part corresponding to the de Rham complex. Then $(PV^{\bullet,\bullet}_{X_0/\Lambda}, DR^{\bullet,\bullet}_{X_0/\Lambda})$ is a Λ-linear curved (two-sided) Gerstenhaber calculus. $\qquad\qquad\Diamond$

13.5 Maurer–Cartan Elements in $E^{\bullet,\bullet}_{X_0/\Lambda}$

The key reason to study $E^{\bullet,\bullet}_{X_0/\Lambda}$ is because it controls $\mathrm{TWD}^{\mathscr{D}}(\mathcal{E}^\bullet_0, -)$ and thus $\mathrm{GDef}^{\mathscr{D}}(\mathcal{E}^\bullet_0, -)$. In fact, the Λ-linear curved Lie algebra $L^\bullet := E^{T,\bullet}_{X_0/\Lambda}$ is sufficient

---

[8] Recall that the formation of inverse limits of sheaves commutes with taking sections. Thus, we obtain the same characteristic algebra either by taking first global sections of $\mathcal{E}^{P,q}_A$ and then the limit, or vice versa. This also means that the characteristic algebra behaves well with respect to gluing the characteristic algebras of open subsets.

for this purpose. If $\phi_A \in \mathfrak{m}_A \cdot L_A^1$ is a solution of the extended Maurer–Cartan equation

$$\bar{\partial}_A \phi_A + \frac{1}{2}[\phi_A, \phi_A] + \ell_A = 0 \in L_A^2,$$

then $\bar{\partial}_A + \nabla_{\phi_A}(-)$ is a differential on $\mathcal{E}_A^{\bullet,\bullet}$, so the pair $(\mathcal{E}_A^{\bullet,\bullet}, \bar{\partial}_A + \nabla_{\phi_A}(-))$ gives an element of $\mathrm{TWD}^{\mathscr{D}}(\mathcal{E}_0^\bullet, A)$, defining a natural transformation

$$\mathbf{mc}\colon \quad \mathrm{Def}(L^\bullet, -) \Rightarrow \mathrm{TWD}^{\mathscr{D}}(\mathcal{E}_0^\bullet, -)$$

of functors of Artin rings from the deformation functor of $L^\bullet$, introduced in Chap. 5.1.1.

Lemma 13.36 *The map $\mathbf{mc}$ is a well-defined natural isomorphism. In particular, $L^\bullet$ controls the deformation functor $\mathrm{GDef}^{\mathscr{D}}(\mathcal{E}_0^\bullet, -)$.*

Proof If ϕ and ϕ' are gauge equivalent Maurer–Cartan elements in $\mathfrak{m}_A \cdot L_A^1$, then an element $\theta \in \mathfrak{m}_A \cdot L_A^0$ with $\exp_\theta * \phi = \phi'$ defines a gauge transform of $\mathcal{E}_A^{\bullet,\bullet}$ which transforms $\bar{\partial}$ into $\bar{\partial}'$. Thus, $\mathbf{mc}$ is well-defined. For injectivity, let ϕ and ϕ' be two Maurer–Cartan elements with $\mathbf{mc}(\phi) = \mathbf{mc}(\phi')$. Then there is some automorphism of $\mathcal{E}_A^{\bullet,\bullet}$ which transforms $\bar{\partial}$ into $\bar{\partial}'$. By Lemma 13.21, this automorphism must be a gauge transform $\exp_\theta$ for some $\theta \in \mathfrak{m}_A \cdot L_A^0 = H^0(X_0, \mathfrak{m}_A \cdot \mathcal{E}_A^{T,0})$. By Z-faithfulness, we find $\phi' = \exp_\theta * \phi$ from Lemma 5.3 (its variant for $\mathcal{P}$-algebras), so ϕ and ϕ' are gauge equivalent as elements of $\mathfrak{m}_A \cdot L_A^1$ and thus define the same element in $\mathrm{Def}(L^\bullet, A)$. For surjectivity, we can assume that $\mathcal{H}^{\bullet,\bullet} = \mathcal{E}_A^{\bullet,\bullet}$ in any pair $(\mathcal{H}^{\bullet,\bullet}, \bar{\partial})$ by Lemma 13.28. Since predifferentials form a torsor under $\mathfrak{m}_A \cdot \mathcal{E}_A^{T,1}$, we have an element $\phi \in \mathfrak{m}_A \cdot H^0(X_0, \mathcal{E}_A^{T,1}) = \mathfrak{m}_A \cdot L_A^1$ with $\bar{\partial} = \bar{\partial}_A + \nabla_\phi(-)$. Since $\bar{\partial}$ is a differential by assumption, ϕ must satisfy the extended Maurer–Cartan equation due to Z-faithfulness. But then $(\mathcal{H}^{\bullet,\bullet}, \bar{\partial})$ is the image of ϕ under $\mathbf{mc}$. $\qquad\square$

Corollary 13.37 *In Situation 13.2, we have a Λ-linear curved two-sided Gerstenhaber calculus*

$$(PV_{X_0/\Lambda}^{\bullet,\bullet}, DR_{X_0/\Lambda}^{\bullet,\bullet})$$

with $\mathrm{GDef}^{\mathscr{D}}(\mathcal{G}_0^\bullet, \mathcal{A}_0^\bullet, -) \cong \mathrm{Def}(L^\bullet, -)$ for $L^\bullet = PV_{X_0/\Lambda}^{-1,\bullet}$.

Corollary 13.38 *In Situation 13.3, we have a Λ-linear curved two-sided Gerstenhaber calculus*

$$(PV_{X_0/\Lambda}^{\bullet,\bullet}, DR_{X_0/\Lambda}^{\bullet,\bullet})$$

with controls enhanced generically log smooth deformations of $f_0\colon X_0 \to S_0$, i.e., $\mathrm{ELD}_{X_0/S_0}^{\mathscr{D}} \cong \mathrm{Def}(L^\bullet, -)$ for $L^\bullet = PV_{X_0/\Lambda}^{-1,\bullet}$.

To obtain the second open cover $\mathcal{U}$ in Situation 13.3, we may choose any affine open cover $\mathcal{U}$. To see this, we have to show admissibility with respect to $\mathcal{U} \cup \mathcal{V}$. Since $V_{\alpha_0} \cap \ldots \cap V_{\alpha_m} \to X_0$ is an affine morphism and $U_{i_0} \cap \ldots \cap U_{i_n}$ is affine, their intersection is affine as well, so all coherent sheaves are acyclic on this intersection.

Corollary 13.39 *In Situation 13.4, we have a Λ-linear curved Lie–Rinehart pair*

$$LRP^{\bullet,\bullet}_{X_0/\Lambda}(\mathcal{E}_0)$$

which controls enhanced generically log smooth deformations of $f_0 \colon X_0 \to S_0$ together with a vector bundle, i.e., we have $\mathrm{ELD}^{\mathscr{D}}_{X_0/S_0}(\mathcal{E}_0) \cong \mathrm{Def}(L^\bullet, -)$ for $L^\bullet = LRP^{T,\bullet}_{X_0/\Lambda}(\mathcal{E}_0)$.

As above, an affine open cover $\mathcal{U}$ satisfies the condition that $\mathcal{U} \cup \mathcal{V}$ must be $\mathcal{E}_0$-admissible.

13.6   The Calabi–Yau Case

Suppose we are in Situation 13.2, and assume that f_0 is Calabi–Yau, i.e., $\mathcal{A}^d_0 \cong \mathcal{O}_{X_0}$ via the choice of an appropriate global section $\omega_0 \in \Gamma(X_0, \mathcal{A}^d_0)$. Let $(\mathcal{PV}^{\bullet,\bullet}_k, \mathcal{DR}^{\bullet,\bullet}_k)$ be the characteristic sheaf over $A_k = \Lambda/\mathfrak{m}^{k+1}_\Lambda$. By the canonical map $\mathcal{A}^d_0 \to \mathcal{DR}^{d,0}_0$, we consider ω_0 as an element $\hat{\omega}_0$ of $\mathcal{DR}^{d,0}_0$. By assumption,

$$\kappa_0 \colon \quad \mathcal{G}^p_0 \to \mathcal{A}^{p+d}_0, \quad \theta \mapsto (\theta \lrcorner \omega_0),$$

is an isomorphism of $\mathcal{O}_{X_0}$-modules. Applying the Thom–Whitney functor $\mathrm{TW}^q(-)$ to this map yields the contraction $\hat{\kappa}_0$ with $\hat{\omega}_0$ (when we use positive signs instead of $(-1)^q$). Thus, each $\hat{\kappa}_0 \colon \mathcal{PV}^{p,q}_0 \to \mathcal{DR}^{p+d,q}_0$ is an isomorphism. Since each $\mathcal{DR}^{d,0}_{k+1} \to \mathcal{DR}^{d,0}_k$ is surjective, we can lift $\hat{\omega}_0$ order by order and obtain a compatible system of elements $\hat{\omega}_k \in \mathcal{DR}^{d,0}_k$. It is now easy to see that the contraction maps

$$\hat{\kappa}_k \colon \quad \mathcal{PV}^{p,q}_k \to \mathcal{DR}^{p+d,q}_k, \quad \theta \mapsto (\theta \lrcorner \hat{\omega}_k),$$

are isomorphisms of sheaves of A_k-modules as well, using the description of the kernel as $I \otimes_k \mathcal{DR}^{p,q}_0$. Now Proposition 3.28 turns each $(\mathcal{PV}^{\bullet,\bullet}_k, \mathcal{DR}^{\bullet,\bullet}_k)$ into a sheaf of curved two-sided Batalin–Vilkovisky calculi. This shows the following result.

Lemma 13.40 *In Situation 13.2, assume furthermore that f_0 is Calabi–Yau. Then*

$$(PV^{\bullet,\bullet}_{X_0/\Lambda}, DR^{\bullet,\bullet}_{X_0/\Lambda})$$

can be endowed with a (non-canonical) structure of Λ-linear curved two-sided Batalin–Vilkovisky calculus.

Proof Take global sections and then the inverse limit along the A_k. The structure is not canonical because it depends both on the choice of ω_0 and on the choice of the liftings $\hat{\omega}_k$ of $\hat{\omega}_0$. $\qquad\square$

Corollary 13.41 *In Situation 13.3, if $f_0\colon X_0 \to S_0$ is log Calabi–Yau, then*

$$(PV^{\bullet,\bullet}_{X_0/\Lambda},\, DR^{\bullet,\bullet}_{X_0/\Lambda})$$

can be endowed with a (non-canonical) structure of Λ-linear curved two-sided Batalin–Vilkovisky calculus.

13.7 Summary for Generically Log Smooth Families

For the reader's convenience, we summarize the theory in the case of a generically log smooth family. Let Q be a sharp toric monoid, let $\Lambda = \mathbf{k}[\![Q]\!]$, and let $f_0\colon X_0 \to S_0$ be a log Gorenstein generically log smooth family. Let $\mathcal{V}$ be an admissible open cover of X_0 in the sense of Definition 10.13, and let $\mathcal{D}$ be a system of deformations subordinate to $\mathcal{V}$. In particular, all local deformations $V_{\alpha;A} \to S_A$ have the base change property. Let $\mathcal{U}$ be another admissible open cover of X_0 with the property that $\mathcal{U} \cup \mathcal{V}$ is admissible as well; for example this is satisfied if $\mathcal{U}$ consists of affine open subsets. We have a Λ-linear curved two-sided Gerstenhaber calculus

$$(PV^{\bullet,\bullet}_{X_0/\Lambda},\, DR^{\bullet,\bullet}_{X_0/\Lambda}) \tag{13.4}$$

as characteristic algebra, and with the Λ-linear curved Lie algebra

$$L^{\bullet}_{X_0/\Lambda} := PV^{-1,\bullet}_{X_0/\Lambda},$$

we have a sequence

$$\mathrm{LD}^{\mathcal{D}}_{X_0/S_0} \cong \mathrm{GDef}^{\mathcal{D}}(\mathcal{E}^{\bullet}_0, -) \cong \mathrm{TWD}^{\mathcal{D}}(\mathcal{E}^{\bullet}_0, -) \cong \mathrm{Def}(L^{\bullet}_{X_0/\Lambda}, -)$$

of isomorphisms of deformation functors, i.e., $L^{\bullet}_{X_0/\Lambda}$ controls the deformation functor $\mathrm{LD}^{\mathcal{D}}_{X_0/S_0}$. If $\phi \in \mathfrak{m}_A \cdot L^1_A$ is a Maurer–Cartan element corresponding to a generically log smooth family $f_A\colon X_A \to S_A$ of type $\mathcal{D}$, then we have acyclic resolutions

$$\mathcal{V}^p_{X_A/S_A} \to (\mathcal{PV}^{p,\bullet}_A, \bar{\partial}_A + \nabla_\phi(-)), \qquad \mathcal{W}^i_{X_A/S_A} \to (\mathcal{DR}^{i,\bullet}_A, \bar{\partial}_A + \nabla_\phi(-))$$

so that we find

$$H^q(X_A, \Theta^p_{X_A/S_A}) = H^q(X_A, \mathcal{V}^p_{X_A/S_A}) \cong H^q(PV^{p,\bullet}_{X_0/A}, \bar\partial_A + \nabla_\phi(-))$$

and

$$H^j(X_A, \mathcal{W}^i_{X_A/S_A}) \cong H^j(DR^{i,\bullet}_{X_0/A}, \bar\partial_A + \nabla_\phi(-)).$$

Similarly, we have a resolution

$$(\mathcal{W}^\bullet_{X_A/S_A}, \partial) \to \Big(\bigoplus_{i+j=\bullet} \mathcal{DR}^{i,j}_A, \; \partial + \bar\partial_A + \nabla_\phi(-) \Big)$$

with acyclic pieces so that we have

$$\mathbb{H}^k(X_A, \mathcal{W}^\bullet_{X_A/S_A}) \cong H^k\Big(\bigoplus_{i+j=\bullet} DR^{i,j}_{X_0/A}, \; \partial + \bar\partial_A + \nabla_\phi(-) \Big).$$

By the computation in Lemma 5.28, we have an isomorphism

$$H^k\Big(\bigoplus_{i+j=\bullet} DR^{i,j}_{X_0/A}, \; \partial + \bar\partial_A + \nabla_\phi(-) \Big) \cong H^k\Big(\bigoplus_{i+j=\bullet} DR^{i,j}_{X_0/A}, \; \partial + \bar\partial_A + \ell_A \lrcorner(-) \Big)$$

so that the hypercohomology is, to a certain extent, independent of the Maurer–Cartan element ϕ.

If (13.4) is quasi-perfect, then the cohomologies $H^k(X_A, \mathcal{W}^i_{X_A/S_A})$ and the hypercohomology $\mathbb{H}^k(X_A, \mathcal{W}^\bullet_{X_A/S_A})$ are flat A-modules, and their formation commutes with base change. The Hodge–de Rham spectral sequence

$$E_1^{pq} = H^q(X_A, \mathcal{W}^p_{X_A/S_A}) \Rightarrow \mathbb{H}^{p+q}(X_A, \mathcal{W}^\bullet_{X_A/S_A})$$

degenerates at E_1.

If (13.4) is quasi-perfect, and $f_0 \colon X_0 \to S_0$ is log Calabi–Yau, then also the cohomology $H^k(X_A, \mathcal{V}^p_{X_A/S_A})$ is a flat A-module, and its formation commutes with base change. Furthermore, the deformation functor $\mathrm{LD}^{\mathscr{D}}_{X_0/S_0}$ is unobstructed.

If $f_0 \colon X_0 \to S_0$ is proper, then all cohomologies and hypercohomologies are finitely generated A-modules.

Part IV
Applications

Chapter 14
Log Toroidal Families of Gross–Siebert Type

In Chap. 8, we defined a log toroidal family as a generically log smooth family which has certain local models $f: A_{P,\mathcal{F}} \to A_Q$. A generically log smooth family $f: X_A \to S_A$ for $A \in \mathbf{Art}_\mathbb{N}$ is a *log toroidal family of Gross–Siebert type* if it admits more special local models of the form discussed below. Families of *elementary* or *standard* Gross–Siebert type are then defined by imposing additional conditions on the local models. These local models have first been investigated by Gross and Siebert in [125]. There, they observe that toric log Calabi–Yau spaces, i.e., certain generically log smooth families arising from a canonical construction in [124], have local models of this form. One of the key insights of [125] is that, once certain mild conditions are met, infinitesimal generically log smooth deformations which have local models of elementary Gross–Siebert type admit a well-behaved deformation theory analogous to the deformation theory of log smooth morphisms to the standard log point S_0. In our language, this means that we can construct a system of deformations in the sense of Definition 10.15 in this situation. This is one of the main motivations for our concept of a system of deformations, giving a large class of examples which are not log smooth.

In this chapter, we first review the local models considered by Gross and Siebert. Then, we define carefully what we mean by a generically log smooth family having these local models. Finally, we explain the deformation theory of these objects as developed in [125], slightly reformulated in our language. If the reader has not yet done so, it is advisable to read Chap. 9 on toroidal crossing spaces first.

14.1 Local Models of Gross–Siebert Type

They are specific elt data which arise from the Gross–Siebert program. Their central fibers can be considered as well-adjusted triples (V, Z, s) in the sense of Definition 9.92.

© The Author(s), under exclusive license to Springer Nature Switzerland AG 2025

S. Felten, *Global Logarithmic Deformation Theory*, Lecture Notes in Mathematics 2373, https://doi.org/10.1007/978-3-031-98751-9_14

14.1.1 The Construction of the Local Model

The reference for the construction is [125, Constr. 2.1]. The input datum is as follows. It builds upon the construction of the local model $(V(\sigma), \mathcal{P}_\sigma, \bar{\rho}_\sigma)$ for toroidal crossing spaces, which we saw in Construction 9.2.

Definition 14.1 A *weak Gross–Siebert local model datum* is a tuple

$$(M', N', \tau, \Delta_1, \ldots, \Delta_q)$$

where $M' \cong \mathbb{Z}^d$ is a lattice, N' is its dual lattice, $\tau \subseteq M'_\mathbb{R}$ is a convex lattice polytope (= bounded polyhedron) of full dimension d, and $\Delta_1, \ldots, \Delta_q \subseteq M'_\mathbb{R}$ are convex lattice polytopes which are allowed to be not of full dimension. A *(strong) Gross–Siebert local model datum* is a weak Gross–Siebert local model datum where the normal fan $\check{\Sigma}_0$ of τ is a subdivision of the normal fan $\check{\Sigma}_i$ of Δ_i for each $1 \leq i \leq q$.

Here, $q = 0$ is allowed, i.e., there may not be any Δ_i. For convenience, we also set $\Delta_0 := \tau$. Examples can be found below and in the introduction, for instance Example 1.79. From the input datum, we construct a generically log smooth family $f : \mathbb{A}_P \to \mathbb{A}^1_t$, the *local model of Gross–Siebert type*. The more relevant case (and the only one considered by Gross and Siebert) is the strong case while some of the constructions also work for weak Gross–Siebert local model data, providing some additional interesting examples of log singularities.

Each Δ_i gives rise to a piecewise linear function

$$\check{\psi}_i(n) := -\inf\{\langle m, n \rangle \mid m \in \Delta_i\}$$

on $N'_\mathbb{R}$. On each cone $\sigma \in \check{\Sigma}_i$, the function $\check{\psi}_i$ is linear. If the input datum is strong, then the function $\check{\psi}_i$ is linear on each $\sigma \in \check{\Sigma}_0$ since $\check{\Sigma}_0$ is a subdivision of each $\check{\Sigma}_i$.

Gross–Siebert defines a sharp toric monoid

$$P' = P'(M', N', \tau) = \{n + a_0 e_0^* \in N' \oplus \mathbb{Z} \mid a_0 \geq \check{\psi}_0(n)\},$$

which is exactly what we get when we apply Construction 9.2 to τ, as we have seen in Lemma 9.10. The Gorenstein degree is $\rho' = (0, 1) \in N' \oplus \mathbb{Z}$ here; it induces an injective and saturated homomorphism $\theta' : \mathbb{N} \to P'$, $1 \mapsto \rho'$, of sharp toric monoids.

There is a unique subset $E' \subset P'$ such that every $p' \in P'$ has a unique decomposition $p' = e' + k\rho'$ for $e' \in E$ and $k \in \mathbb{N}$. This is called the *essential locus*. Explicitly, we have

$$E' = \{n + a_0 e_0^* \mid a_0 = \check{\psi}_0(n)\}.$$

A face $F \subseteq P'$ is called *essential* (or *critical* in the language of [222]) if $F \subseteq E'$; then E' is the union of the essential faces. Under the bijection

$$\phi': \ N' \to E', \quad n \mapsto (n, \check{\psi}_0(n)),$$

the essential faces in P' correspond exactly to the toric monoids in

$$\mathcal{E}_0 := \{N' \cap \sigma \mid \sigma \in \check{\Sigma}_0\},$$

the set of intersections between cones in $\check{\Sigma}_0$ and the lattice N'. Since ρ' generates the interior of P', every face in P' is essential.

As in Construction 9.2, we obtain a log smooth and saturated morphism $f': (\mathbb{A}_{P'} | D_{P'}) \to (\mathbb{A}_t^1 | \{0\})$. Its central fiber defines a toroidal crossing space $V(\tau)$, which is stratified by toric strata of the form $V_{F'} := \operatorname{Spec} \mathbf{k}[F'] \subset \mathbb{A}_{P'}$ for essential faces $F' \subseteq E'$. If the input datum is strong, the central fiber of the local model to be constructed will be a well-adjusted triple (V, Z, s) (in the sense of Definition 9.92) on the toroidal crossing space $V := V(\tau) \times \mathbb{A}^q$.

To construct the local model, we use the remaining polytopes $\Delta_1, \ldots, \Delta_q$ to construct a sharp toric monoid P.

Definition 14.2 Let $(M', N', \tau, \Delta_1, \ldots, \Delta_q)$ be a weak Gross–Siebert local model datum. Then we set

$$P = P(M', N', \tau, \Delta_1, \ldots, \Delta_q)$$

$$= \left\{ n + a_0 e_0^* + \sum_{i=1}^{q} a_i e_i^* \in N' \oplus \mathbb{Z} \oplus \mathbb{Z}^q \ \middle| \ \forall i : a_i \geq \check{\psi}_i(n) \right\}.$$

An injective and saturated homomorphism $\theta : \mathbb{N} \to P$ is given by $\theta(1) = \rho = e_0^*$.

The monoid P is Gorenstein with Gorenstein degree $s = \sum_{i=0}^{q} e_i^*$. However, this element is not the element used for the map $\theta : \mathbb{N} \to P$, which is simply $\rho = e_0^*$. Then we obtain a log morphism

$$f : \ (\mathbb{A}_P | \{t = 0\}) \to (\mathbb{A}_t^1 | \{0\}),$$

where $\mathbb{A}_P = \operatorname{Spec} \mathbf{k}[P]$. At this point, we could consider f as an elementary log toroidal family with log smooth locus $U_{P/\mathbb{N}}$ constructed in Sect. 8.3, but more about the log singular locus below.

Over $\mathbb{A}_t^1 \setminus \{0\}$, the log structure is trivial. The family is trivial as well, with fiber $\operatorname{Spec} \mathbf{k}[P_\rho/\mathbb{Z}\rho]$ where

$$P_\rho/\mathbb{Z}\rho = \left\{ n + \sum_{i=1}^{q} a_i e_i^* \in N' \oplus \mathbb{Z}^q \ \middle|\ a_i \geq \check{\psi}_i(n) \right\}.$$

After setting

$$\Delta_+ := \operatorname{Conv}\left(\bigcup_{i=1}^{q} \Delta_i \times \{e_i\} \right) \subseteq M_\mathbb{R}' \oplus \mathbb{R}^q$$

and $K_+ := \operatorname{Cone}(\Delta_+) \subseteq M_\mathbb{R}' \oplus \mathbb{R}^q$ the cone spanned by Δ_+, we have

$$P_\rho/\mathbb{Z}\rho = K_+^\vee \cap (N' \oplus \mathbb{Z}^q).$$

In general, the family f is not smooth over $\mathbb{A}_t^1 \setminus \{0\}$. Let us denote its singular locus by Z^*. We will later impose additional conditions on Δ_+ that bound the "badness" of the singularities in the general fiber.

From now on until the end of the section, we assume that the input datum is strong, i.e., that $\check{\Sigma}_0$ is a subdivision of each $\check{\Sigma}_i$. To describe the central fiber $f^{-1}(0)$ in this case, we use the unique set

$$E = \left\{ n + \sum_{i=0}^{q} a_i e_i^* \ \middle|\ a_0 = \check{\psi}_0(n) \right\}$$

such that every $p \in P$ has a unique decomposition $p = e + k\rho$ with $e \in E$ and $k \geq 0$, i.e., the essential locus of $\theta \colon \mathbb{N} \to P$. Then $f^{-1}(0)$ is stratified by toric strata of the form $V_F := \operatorname{Spec} \mathbf{k}[F] \subset \mathbb{A}_P$ for *essential faces* $F \subseteq E$. The bijection

$$\phi \colon \ N' \oplus \mathbb{N}^q \to E, \quad (n, b_1, \ldots, b_q) \mapsto n + \check{\psi}_0(n) e_0^* + \sum_{i=1}^{q} (b_i + \check{\psi}_i(n)) e_i^*,$$

allows us to describe the essential faces: If $F' \in \mathcal{E}_0$ and $G \subseteq \mathbb{N}^q$ is a face, then $\phi(F' + G)$ is an essential face of P. Every essential face of P is of this form for unique $F' \in \mathcal{E}_0$ and $G \subseteq \mathbb{N}^q$. We denote the set of essential faces by $\mathcal{E}_0 \oplus \mathbb{N}^q$.

The bijection ϕ induces an isomorphism of underlying schemes between the central fiber of $f^{-1}(0)$ and $V(\tau) \times \mathbb{A}^q$. In particular, it has a stratification coming from $V(\tau)$. If ρ is a face of τ and $F' = \sigma_\rho \cap N'$ for the cone $\sigma_\rho \in \check{\Sigma}_0$ associated with ρ, then we write $V_\rho := \operatorname{Spec} \mathbf{k}[F'] \times \mathbb{A}^q$ for the corresponding stratum.

We investigate the log singular locus inside the central fiber. We denote the set of edges in τ by $\Omega(\tau)$. For $1 \leq i \leq q$,

$$\Omega_i := \{\omega \mid \sigma_\omega \in \check{\Sigma}_i\} \subseteq \Omega(\tau)$$

is the subset of those edges whose corresponding cones σ_ω are contained in $\check{\Sigma}_i$; this is the case if and only if the function $\check{\psi}_i$ bends along σ_ω, i.e., $\check{\psi}_i$ is not linear on the union of the two adjacent maximal cones of $\check{\Sigma}_0$. Let us write $F'_\omega = N' \cap \sigma_\omega$ for the associated toric monoid in N', and let

$$F_{\omega;i} := F'_\omega \oplus \{(a_1, \ldots, a_q) \in \mathbb{N}^q \mid a_i = 0\} \in \mathcal{E}_0 \oplus \mathbb{N}^q.$$

Via $\phi\colon N' \oplus \mathbb{N}^q \cong E$, this defines a closed subset $\operatorname{Spec} \mathbf{k}[F_{\omega;i}]$ of the stratum V_ω of codimension 1. Gross–Siebert defines

$$Z_i := \bigcup_{\omega \in \Omega_i} \operatorname{Spec} \mathbf{k}[F_{\omega;i}] \quad \text{and} \quad Z^\| := \bigcup_{i=1}^q Z_i.$$

Then $Z^\|$ is pure of dimension $d + q - 2$, and each component $\operatorname{Spec} \mathbf{k}[F_{\omega;i}]$ is affine toric.[1] Since the second summand in F'_ω is properly contained in $\mathbb{N}^q$, the intersection $Z \cap V_\rho$ is a proper subset of the stratum $V_\rho = \operatorname{Spec} \mathbf{k}[F'_\rho \oplus \mathbb{N}^q]$ for every face $\rho \subset \tau$. We now set $Z := Z^\| \cup Z^*$, where $Z^* = \operatorname{Sing}(f^{-1}(\{t \neq 0\}))$.

The complement $\mathbb{A}_P \setminus Z^\|$ is a union of open subsets of the form $U_F = \operatorname{Spec} \mathbf{k}[P_F]$ for faces $F \subseteq P$ and the monoid localization $P_F = P + F^{\mathrm{gp}}$. We have to take those faces F which either contain ρ or which are essential and satisfy $F \not\subseteq F_{\omega;i}$ for all $1 \leq i \leq q$ and $\omega \in \Omega_i$. Let us denote the latter set of faces by $\mathcal{F}$ for short.

If $\rho \in F$, then $U_F \subseteq f^{-1}(\mathbb{A}^1_t \setminus \{0\})$ does not contain any point of the central fiber. Thus, $U_F \setminus Z^* \to \mathbb{A}^1_t \setminus \{0\}$ is smooth. If $F \in \mathcal{F}$, decompose $F = \phi(F' + G)$ with $F' \in \mathcal{E}_0$ and $G \subseteq \mathbb{N}^q$ a face. A careful computation yields an isomorphism $P_F \cong P'_{F'} \oplus \mathbb{N}^q$ compatibly with ρ and ρ' so that $f\colon (U_F|\{t = 0\}) \to (\mathbb{A}^1_t|\{0\})$ is log smooth. In particular, $U_F \setminus \{t = 0\}$ is smooth, and $U_F \cap Z^* = \varnothing$. We see that $Z = Z^\| \cup Z^*$ is closed and that $f\colon \mathbb{A}_P \to \mathbb{A}^1_t$ is log smooth on $U = \mathbb{A}_P \setminus Z$. This choice of log smooth locus turns f into a generically log smooth family.

Definition 14.3 A *local model of Gross–Siebert type* is a generically log smooth family

$$f\colon \quad (\mathbb{A}_P, \mathbb{A}_P \setminus (Z^\| \cup Z^*)|\{t = 0\}) \to (\mathbb{A}^1_t|\{0\})$$

---

[1] The log singular locus Z constructed here is always contained in the log singular locus $Z_{P/\mathbb{N}}$ of Sect. 8.3, but it may be smaller than its restriction to the central fiber. The reason for this is that our construction here does take into account only some of the essential faces of rank $d - 2$ while the construction in Sect. 8.3 includes all of them. For an example, see Example 14.13 below.

arising from the above construction for a Gross–Siebert local model datum $(M', N', \tau, \Delta_1, \ldots, \Delta_q)$.

We emphasize that the local model datum is not allowed to be weak in this definition. We now assemble these local models in a class of log singularities in the sense of Definition 1.84.

Definition 14.4 The class of log singularities of *Gross–Siebert type* $\mathscr{C}^{GS}$ is the set of all local models of Gross–Siebert type.

Recall that, in order to define a section of $\mathcal{LS}_V$ on the toroidal crossing space $V = V(\tau) \times \mathbb{A}^q$, we must exhibit an isomorphism between the ghost sheaves of $\mathbb{A}_P$ and $\mathbb{A}_{P'} \times \mathbb{A}^q$ on the two central fibers—identified via ϕ—which is compatible with the maps to the base $\mathbb{A}_t^1$, and which turns the two log structures into log structures of the same type. Both is achieved by $P_F \cong P'_{F'} \oplus \mathbb{N}^q$ as follows: We have a commutative diagram

$$
\begin{array}{ccccc}
U_F & \longrightarrow & \mathbb{A}_P & \overset{f}{\longrightarrow} & \mathbb{A}_t^1 \\
\downarrow {\scriptstyle \cong} & & & & \| \\
U_{F'} \times \mathbb{A}^q & \longrightarrow & \mathbb{A}_{P'} \times \mathbb{A}^q & \longrightarrow & \mathbb{A}_t^1
\end{array}
$$

of log schemes. It induces an isomorphism $f^{-1}(0) \cap U_F \cong V \cap U_{F'}$ on the central fiber; the underlying morphism of schemes is *not* the restriction of $f^{-1}(0) \cong V(\tau) \times \mathbb{A}^q$ via ϕ but differs from it by an automorphism compatible with the pre-ghost structure on V. Taking this automorphism into account, we obtain—only on the central fiber, not the whole of U_F—the correct identifications of the ghost sheaf, and it also implies that both log structures are of the same type because this is always the case if two log structures are obtained from the same one along two different étale maps that are compatible with each other on the level of the ghost sheaves in a suitable sense.

14.1.2   The Local Model in Codimension 1

Let us first study an example of the construction.

Example 14.5 Let $\ell \geq 1$, and let $m_i \geq 0$ for $1 \leq i \leq q$ for some $q \geq 1$. We take $M' = \mathbb{Z}$ and $\tau = [0, \ell]$, $\Delta_i = [0, m_i]$. The functions are $\check{\psi}_0(n) = -\inf\{\ell n, 0\}$ and $\check{\psi}_i(n) = -\inf\{m_i n, 0\}$. Let $f = 1$ be the generator of N'. Now

$$
\bar{x} := f, \quad \bar{y} := -f + \ell e_0^* + \sum_{i=1}^{q} m_i e_i^*, \quad \bar{t} := e_0^*, \quad \bar{z}_i := e_i^* \text{ for } i = 1, \ldots, q
$$

are elements of $P = P(M', N', \tau, \Delta_1, \ldots, \Delta_q)$, and they generate it as a monoid. With $x = z^{\bar{x}}$ etc., we obtain

$$\mathbf{k}[P] = \mathbf{k}[x, y, t, z_1, \ldots, z_q] \Big/ \left(xy - t^\ell \prod_{i=1}^q z_i^{m_i} \right).$$

The central fiber is $\mathbf{k}[x, y, z_1, \ldots, z_q]$, and the log singular locus in the central fiber is given, with a natural possibly non-reduced scheme structure, by

$$Z^{\|} = \left\{ \prod_{i=1}^q z_i^{m_i} = 0, \ x = 0, \ y = 0, \ t = 0 \right\}.$$

We denote this family (with a possible additional factor of $\mathbb{A}^s$) by

$$L(\ell; m_1, \ldots, m_q; s) := L(\ell; m_1, \ldots, m_q) \times \mathbb{A}^s := \operatorname{Spec} \mathbf{k}[P] \times \mathbb{A}^s \to \mathbb{A}^1_t.$$

$$\Diamond$$

Let us go back to a general local model of Gross–Siebert type. For every $\omega \in \Omega(\tau)$, we have an open subset $U_\omega := U_{F_\omega} = \operatorname{Spec} \mathbf{k}[P_{F_\omega}]$ with $F_\omega := \phi(F'_\omega)$. Together, they cover the locus $\mathcal{U}_1 V$ on the central fiber where at most two components intersect. With

$$\check{\psi}_{i;\omega}(n) := \sup\{\check{\psi}_i(n + f) - \check{\psi}_i(f) \mid f \in \phi(F'_\omega)\},$$

we have

$$P_\omega := P_{F_\omega} = \left\{ n + \sum_{i=0}^q a_i e_i^* \ \middle| \ \forall i : a_i \geq \check{\psi}_{i;\omega}(n) \right\}.$$

Let σ^+ and σ^- be the two maximal cones which are adjacent to σ_ω in $\check{\Sigma}_0$, and let $F^+ = N' \cap \sigma^+$ and $F^- = N' \cap \sigma^-$. Let $H^+ = F^+ + (F'_\omega)^{\mathrm{gp}}$ and $H^- = F^- + (F'_\omega)^{\mathrm{gp}}$. Then $\check{\psi}_{i;\omega}$ is linear on both H^+ and H^- for all $0 \leq i \leq q$. If $h \in H^+$ is an element with $H^+ = (F'_\omega)^{\mathrm{gp}} \oplus h \cdot \mathbb{N}$ and $H^- = (F'_\omega)^{\mathrm{gp}} \oplus (-h) \cdot \mathbb{N}$—such an element exists— then we find an isomorphism

$$Q \oplus (F'_\omega)^{\mathrm{gp}} \xrightarrow{\cong} P_\omega,$$

$$\bar{x} \mapsto h + \sum_{i=0}^q \check{\psi}_{i;\omega}(h) e_i^*, \quad \bar{y} \mapsto -h + \sum_{i=0}^q \check{\psi}_{i;\omega}(-h) e_i^*, \quad \bar{t} \mapsto e_0^*, \quad \bar{z}_i \mapsto e_i^*,$$

where Q is the monoid P of Example 14.5 with $\ell = \check{\psi}_{0;\omega}(h) + \check{\psi}_{0;\omega}(-h) \geq 1$ and $m_i = \check{\psi}_{i;\omega}(h) + \check{\psi}_{i;\omega}(-h) \geq 0$. We have $m_i \geq 1$ if and only if $\omega \in \Omega_i$. Summarizing the discussion, we find that $f \colon \mathbb{A}_P \to \mathbb{A}_t^1$ is (locally) isomorphic to $L(\ell; m_1, \ldots, m_q; d-1) \to \mathbb{A}_t^1$ on U_ω. In particular, the central fiber is locally isomorphic to $L_0(\ell; m_1, \ldots, m_q; d-1)$ on $\mathcal{U}_1 V$ where $L_0(-)$ has the obvious meaning.

14.1.3 The Case Where Δ_+ is an Elementary Simplex or a Standard Simplex

In [125], local models where Δ_+ is an elementary simplex or a standard simplex in $M'_\mathbb{R} \oplus \mathbb{R}^q$ (of some dimension; it cannot have full dimension) plays a special role because then $f \colon \mathbb{A}_P \to \mathbb{A}_t^1$ is particularly well-behaved. To avoid any confusion, we fix the following definition.

Definition 14.6 Let M be a lattice, and $P \subseteq M_\mathbb{R}$ be a lattice polytope, not necessarily of full dimension.

(1) We say that P is a *simplex* if the number of vertices of P is equal to $\dim(P)+1$, the minimal number possible.
(2) We say that P is an *elementary simplex* if it is a simplex, and it contains no other lattice points than its vertices.
(3) We say that P is a *standard simplex* if it is an elementary simplex, and if, for some (equivalently any) numbering $v_0, \ldots, v_d$ of its vertices, we can extend $v_1 - v_0, \ldots, v_d - v_0$ to a basis of the lattice M. This is equivalent to that $v_1 - v_0, \ldots, v_d - v_0$ is a basis of the monoid $A_\mathbb{Z}(P) - v_0$ where $A(P)$ is the real affine space spanned by P, and $A_\mathbb{Z}(P) := A(P) \cap M$.[2]

Example 14.7 $\mathrm{Conv}((0, 1, 0), (1, 1, 0), (1, 0, 0), (0, 0, 2)) \subseteq \mathbb{R}^3$ is an elementary simplex which is not a standard simplex. $\diamond$

We study the standard case first. The following result also holds for weak Gross–Siebert local model data.

Lemma 14.8 ([125, Prop. 2.2]) *The general fiber of $f \colon \mathbb{A}_P \to \mathbb{A}_t^1$ is smooth if and only if Δ_+ is a standard simplex.*

Proof The general fiber is $\mathrm{Spec}\,\mathbf{k}[P_\rho/\mathbb{Z}\rho]$ with $P_\rho/\mathbb{Z}\rho = K_+^\vee \cap (N' \oplus \mathbb{Z}^q)$. Thus, the general fiber is smooth if and only if K_+ is a regular cone. The primitive ray generators of K_+ are precisely the vertices $v_0, \ldots, v_d$ of Δ_+; they form part of a

---

[2] To see the equivalence of the two formulations, consider the torsion in the quotient of M respectively $A_\mathbb{Z}(P)$ by the sublattice generated by $v_1 - v_0, \ldots, v_d - v_0$.

basis of $M' \oplus \mathbb{Z}^q$ if and only if Δ_+ is a standard simplex. For the non-trivial direction, note that $M' \oplus \mathbb{Z}^q = \mathbb{Z}v_0 \oplus \{(m, a_1, \ldots, a_q) \mid \sum_{i=1}^{q} a_i = 1\}$.   □

When Δ_+ is an elementary simplex, the singularities of the general fiber are very mild. Again, the result also holds for weak local model data.

Lemma 14.9 ([125, Prop. 2.2]) *Assume that Δ_+ is an elementary simplex. Then the general fiber of $f \colon \mathbb{A}_P \to \mathbb{A}_t^1$ is normal, Gorenstein, terminal, $\mathbb{Q}$-factorial, smooth in codimension 3, and has Abelian quotient singularities.*

Proof As a toric variety, the general fiber $\operatorname{Spec} \mathbf{k}[P_\rho/\mathbb{Z}\rho]$ is normal and Cohen–Macaulay. We obtain from the criterion at the beginning of [6, §6] that it is Gorenstein because the primitive ray generators of K_+ are precisely the vertices of Δ_+. Then [56, Prop. 11.4.12] shows that the general fiber has terminal singularities. A dimension count yields that $\operatorname{Spec} \mathbf{k}[P_\rho/\mathbb{Z}\rho]$ is a simplicial toric variety, and hence it is $\mathbb{Q}$-factorial and has Abelian quotient singularities. Then [56, Prop. 11.4.22] yields that $\operatorname{Spec} \mathbf{k}[P_\rho/\mathbb{Z}\rho]$ is, as a simplicial Gorenstein terminal toric variety, smooth in codimension 3.[3]   □

In particular, we can apply the following well-known theorem of Altmann, which is proven in [6, §6]:

Theorem 14.10 *Let Y be a $\mathbb{Q}$-Gorenstein affine toric variety. Assume that Y is smooth in codimension 3. Then Y is rigid, i.e., $\mathcal{T}_Y^1 = 0$.*

Corollary 14.11 *Assume that Δ_+ is an elementary simplex. Then the general fiber of the morphism $f \colon \mathbb{A}_P \to \mathbb{A}_t^1$ is rigid.*

Thus, the general fiber does not admit any non-trivial local deformation.

We now define subclasses of $\mathscr{C}^{\mathrm{GS}}$.

Definition 14.12 The class of log singularities of *elementary Gross–Siebert type*

$$\mathscr{C}^{\mathrm{eGS}} \subset \mathscr{C}^{\mathrm{GS}}$$

consists of all *local models of elementary Gross–Siebert type*, i.e., local models of Gross–Siebert type where Δ_+ is an elementary simplex. The class of log singularities of *standard Gross–Siebert type* $\mathscr{C}^{\mathrm{sGS}} \subset \mathscr{C}^{\mathrm{eGS}}$ consists of all *local models of standard Gross–Siebert type*, i.e., local models of Gross–Siebert type where Δ_+ is a standard simplex.

Next, we investigate the local model in codimension 1 in the elementary case. Assume that Δ_+ is an elementary simplex. Since, for $1 \leq i \leq q$, each Δ_i is a face

---

[3] $\mathbb{Q}$-factoriality is the only one of these properties which is only Zariski local but not étale local. In particular, we cannot conclude that a space which has the general fiber as an étale local model is $\mathbb{Q}$-factorial.

of Δ_+, they must be elementary simplices as well. Since each vertex of Δ_i shows up in Δ_+, we find $\sum_{i=1}^{q} \dim(\Delta_i) \le d = \dim(\tau)$. Let $\omega \in \Omega(\tau)$ be an edge, and set

$$m_{i;\omega} := \check{\psi}_{i;\omega}(h) + \check{\psi}_{i;\omega}(-h), \quad m_\omega := \sum_{i=1}^{q} m_{i;\omega},$$

where we maintain the notations from the previous section. As stated above, we have $m_{i;\omega} \ne 0$ if and only if $\omega \in \Omega_i$. In this case, Δ_i has an edge ω_i which is parallel to ω; the lattice length of this edge is $m_{i;\omega}$, i.e., the edge contains $m_{i;\omega} + 1$ lattice points, including the two vertices. Since each Δ_i is an elementary simplex, we have $m_{i;\omega} \le 1$. Now assume that we have two indices $i \ne j$ with both $m_{i;\omega} \ne 0$ and $m_{j;\omega} \ne 0$. Let $\omega_k^{\pm}$ be the two vertices of the edge ω_k with $k = i, j$. Then the four vertices $\omega_i^{\pm} \times e_i$ and $\omega_j^{\pm} \times e_j$ of Δ_+ lie in a common plane, contradicting the assumption that Δ_+ is a simplex. Thus, $m_\omega \le 1$. This means that the central fiber of $f \colon \mathbb{A}_P \to \mathbb{A}_t^1$ is particularly simple on $\mathcal{U}_1 V$ because it is locally isomorphic to the central fiber of

$$\operatorname{Spec} \mathbb{C}[x, y, t, z_1, \ldots, z_q, u_1, \ldots, u_s]/(xy - t^{\ell} z_i) \to \mathbb{A}_t^1$$

for some $1 \le i \le q$ and $s = d - 1$. The equation $xy = t^{\ell} \prod_{i=1}^{q} z_i^{m_i}$ simplifies to $xy = t^{\ell} z_i$.

14.1.4 Some Additional Examples

With Examples 1.79, 1.83, 1.156, 1.157, we have already seen some examples of the Gross–Siebert local-model construction. Here are some additional examples. In particular, we will see that there are local models of elementary Gross–Siebert type which are not of standard Gross–Siebert type, a phenomenon that appears only in relative dimension ≥ 4.

Example 14.13 (xy = zw = t) We set $\tau = [0, 1] \times [0, 1] \subseteq \mathbb{R}^2$ and $q = 0$. Then

$$\check{\psi}_0(n_1, n_2) = -\inf\{0, n_1, n_2, n_1 + n_2\}.$$

The monoid

$$P = \{(n_1, n_2, a_0) \mid a_0 \ge \check{\psi}_0(n_1, n_2)\}$$

is generated by

$$\bar{x} = (1, 0, 0), \quad \bar{y} = (-1, 0, 1), \quad \bar{z} = (0, 1, 0), \quad \bar{w} = (0, -1, 1).$$

We also set $\bar{t} = (0, 0, 1)$. Thus, we have an isomorphism

$$\mathbf{k}[P] \cong \mathbf{k}[x, y, z, w]/(xy - zw).$$

Indeed, the surjection must be an isomorphism because both sides are integral of the same dimension 3. This is the same monoid as in Example 1.79, but now the map is given by $t = xy = zw$. This family is log smooth and vertical. The central fiber consists of four copies of $\mathbb{A}^2$. Each of them intersects two other components in a copy of $\mathbb{A}^1$, and the third other component in a single point $\{0\}$. The intersection of all four components is this single point $\{0\}$. Obviously, the central fiber is not a normal crossing space, but it is a toroidal crossing space. The log singular locus $Z = \bigcup_{i=1}^{q} Z_i$ constructed above is empty because $q = 0$. However, the log singular locus $Z = Z_{P/\mathbb{N}}$ defined in Sect. 8.3, whose complement we denote by $U_{P/\mathbb{N}}$, is non-empty even when restricted to the central fiber. Namely, $F = \{(0, 0, 0)\}$ is an essential face of rank $d - 2 = 0$, so $\{0\} = V_F \subset Z_{P/\mathbb{N}}$. For an illustration, see Fig. 14.1. $\Diamond$

Example 14.14 ($xy = tu$, $zw = tu$) We take again $\tau = [0, 1] \times [0, 1] \subseteq \mathbb{R}^2$, but this time, we set $q = 1$ and $\Delta_1 = \tau$. Then we have

$$\check{\psi}_0(n_1, n_2) = -\inf\{0, n_1, n_2, n_1 + n_2\} = \check{\psi}_1(n_1, n_2).$$

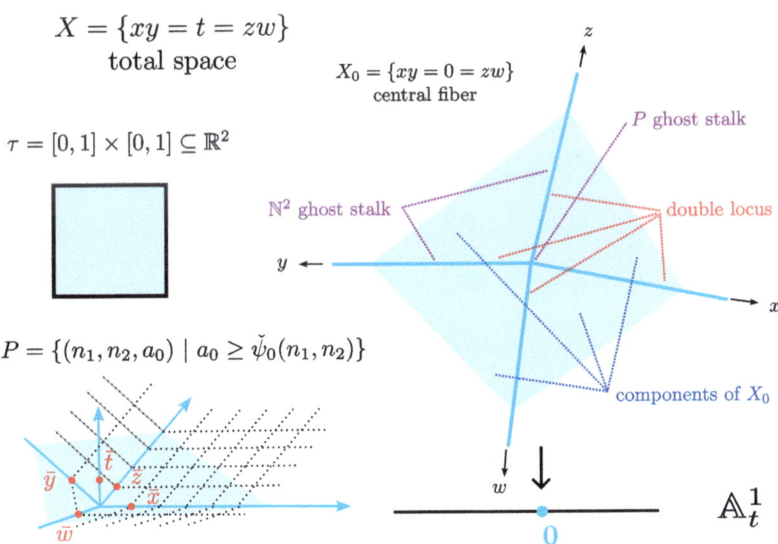

Fig. 14.1 Example 14.13. In the lower left corner, we have the monoid P with its four generators $\bar{x}$, $\bar{y}$, $\bar{z}$, and $\bar{w}$. We also see the Gorenstein degree $\bar{t} = \bar{x} + \bar{y} = \bar{z} + \bar{w}$. On the right, we have the central fiber of the associated family $\mathbb{A}_P \to \mathbb{A}^1_t$. It consists of four copies of $\mathbb{A}^2$, each intersecting in a line, and all four intersecting in a single point. The family is log smooth, saturated, and vertical.

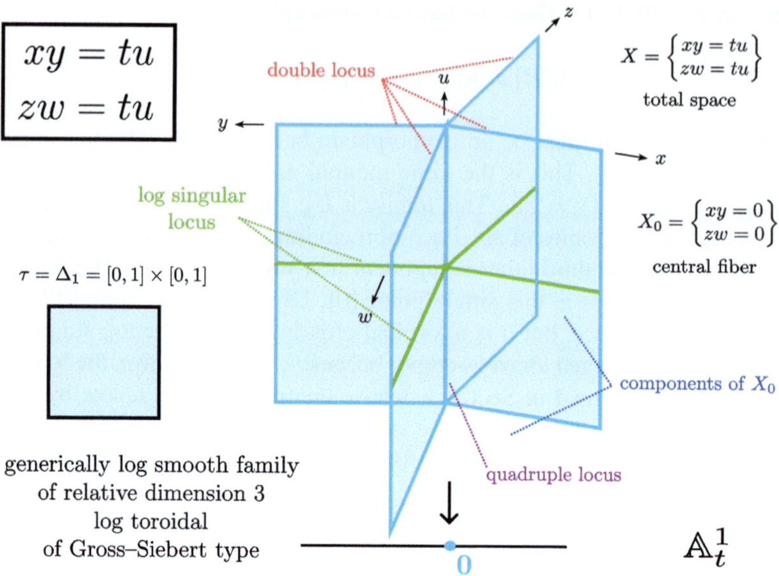

Fig. 14.2 Example 14.14. The graphic shows in light blue the double locus of the central fiber X_0 of the local model of Gross–Siebert type from Example 14.14. The four irreducible components of X_0 are not drawn and lie between the four components of the double locus. © Simon Felten 2025. All rights reserved

The monoid

$$P = \{(n_1, n_2, a_0, a_1) \mid a_i \geq \check{\psi}_i(n_1, n_2), \; i = 0, 1\}$$

is generated by

$$\bar{x} = (1, 0, 0, 0), \quad \bar{y} = (-1, 0, 1, 1), \quad \bar{t} = (0, 0, 1, 0),$$
$$\bar{z} = (0, 1, 0, 0), \quad \bar{w} = (0, -1, 1, 1), \quad \bar{u} = (0, 0, 0, 1).$$

Thus, we have $\mathbf{k}[x, y, z, w, t, u]/(xy - tu, zw - tu) \cong \mathbf{k}[P]$. The nearby fiber has an A_1-threefold singularity so that the family is not log toroidal of standard Gross–Siebert type. In fact, $\Delta_+ \cong [0, 1] \times [0, 1]$ is not even a simplex. In the central fiber, the log singular locus is given by $u = 0$ inside the double locus. It consists of four lines, one in each component of X_0, which meet at 0 inside the quadruple locus (Fig. 14.2).   ◊

Example 14.15 (xy = tu, zw = t) We take again $\tau = [0, 1] \times [0, 1] \subseteq \mathbb{R}^2$ and $q = 1$ as well, but this time, we set $\Delta_1 = [0, 1] \times \{0\} \subseteq \mathbb{R}^2$. Then we have

$$\check{\psi}_0(n_1, n_2) = -\inf\{0, n_1, n_2, n_1 + n_2\}, \quad \check{\psi}_1(n_1, n_2) = -\inf\{0, n_1\}.$$

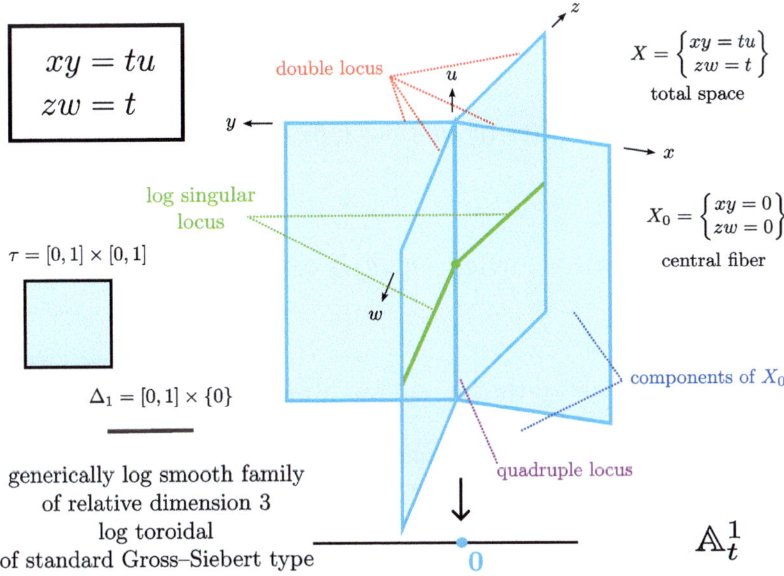

Fig. 14.3 Example 14.15. As in Fig. 14.2, the graphic only shows the double locus of the central fiber but not the irreducible components of the central fiber. This time, we have a local model of *standard* Gross–Siebert type. © Simon Felten 2025. All rights reserved

The monoid

$$P = \{(n_1, n_2, a_0, a_1) \mid a_i \geq \check{\psi}_i(n_1, n_2), \; i = 0, 1\}$$

is generated by

$$\begin{aligned}
\bar{x} &= (1, 0, 0, 0), & \bar{y} &= (-1, 0, 1, 1), & \bar{t} &= (0, 0, 1, 0), \\
\bar{z} &= (0, 1, 0, 0), & \bar{w} &= (0, -1, 1, 0), & \bar{u} &= (0, 0, 0, 1).
\end{aligned}$$

The equations are slightly different now: We obtain $xy = tu$ and $zw = t$. Now the nearby fiber is smooth so that $f \colon X := \mathbb{A}_P \to \mathbb{A}^1_t$ is log toroidal of standard Gross–Siebert type. In fact, we have $\Delta_+ \cong [0, 1]$, a standard simplex. As a scheme, the central fiber is the as in the previous Example 14.14. However, in this example, the log singular locus Z consists only of two lines in diametrically opposed components of the double locus, i.e., the other two components of the double locus contain only the intersection point 0 of the two lines in the quadruple locus as log singular locus (see Fig. 14.3). When taking the equations $xy = t(u + 1)$ and $zw = t(u - 1)$, then we can even pull the four branches of the previous Example 14.14 apart into two times two branches.   ◇

Example 14.16 This is an example of a log toroidal family which is of elementary Gross–Siebert type but not of standard Gross–Siebert type. We start with the quadrangle

$$\tau = \mathrm{Conv}((1, 0), (0, 1), (-1, 0), (0, -1)) \subseteq \mathbb{R}^2,$$

so we have

$$\check{\psi}_0(n_1, n_2) = -\inf\{n_1, -n_1, n_2, -n_2\} = \sup\{|n_1|, |n_2|\}.$$

Thus,

$$P' = \{(n_1, n_2, a_0) \mid a_0 \geq \check{\psi}_0(n_1, n_2)\}$$

is the cone over a square with edge length 2. This monoid has nine generators and 20 relations (according to a computation in Macaulay2), so we will not describe the family in terms of generators and relations. To obtain our example of a family which is of elementary but not standard Gross–Siebert type, we set $q = 2$ and take the two diagonal intervals $\Delta_1 = \mathrm{Conv}((0, 0), (1, 1))$ and $\Delta_2 = \mathrm{Conv}((0, 0), (1, -1))$. Then

$$\check{\psi}_1(n_1, n_2) = -\inf\{0, n_1 + n_2\}, \quad \check{\psi}_2(n_1, n_2) = -\inf\{0, n_1 - n_2\},$$

and the family is given by the monoid

$$P = \{(n_1, n_2, a_0, a_1, a_2) \mid a_i \geq \check{\psi}_i(n_1, n_2), \ i = 0, 1, 2\}.$$

By definition, we have

$$\Delta_+ = \mathrm{Conv}\left(\bigcup_{q=1}^{2} \Delta_i \times \{e_i\} \right)$$

$$= \mathrm{Conv}((0, 0, 1, 0), (1, 1, 1, 0), (0, 0, 0, 1), (1, -1, 0, 1)) \subseteq \mathbb{R}^4.$$

Thus, Δ_+ is contained in the affine space

$$(0, 0, 1, 0) + \mathbb{R} \cdot (1, 0, 0, 0) + \mathbb{R} \cdot (0, 1, 0, 0) + \mathbb{R} \cdot (0, 0, -1, 1).$$

After the affine transformation given by these vectors, we have

$$\Delta_+ \cong \mathrm{Conv}((0, 0, 0), (1, 1, 0), (0, 0, 1), (1, -1, 1)) \subseteq \mathbb{R}^3.$$

This is an elementary simplex, but the differences of these points do not form a basis of $\mathbb{Z}^3 \subseteq \mathbb{R}^3$, i.e., this is not a standard simplex. $\Diamond$

14.1.5 Some Examples of Weak Gross–Siebert Type

We give examples of the local-model construction when $(M', N', \tau, \Delta_1, \ldots, \Delta_q)$ is only a weak Gross–Siebert local model datum. In this case, our description of the central fiber as $V(\tau) \times \mathbb{A}^q$ breaks down, and the central fiber has no longer an underlying toroidal crossing space.

Example 14.17 ($xy = su$, $uv = t$) We take the triangle

$$\tau = \mathrm{Conv}((0,0), (1,0), (0,1)) \subseteq \mathbb{R}^2$$

and the interval $\Delta_1 = [0,1] \times \{0\} \subseteq \mathbb{R}^2$. The log smooth family $U(\tau) \to \mathbb{A}_t^1$ is given by $xyz = t$. We have

$$\check{\psi}_0(n_1, n_2) = -\inf\{0, n_1, n_2\} \quad \text{and} \quad \check{\psi}_1(n_1, n_2) = -\inf\{0, n_1\}.$$

A computation in Macaulay2 yields

$$(\bar{x}\ \bar{y}\ \bar{v}\ \bar{u}\ \bar{s}) = \begin{pmatrix} 1 & -1 & 0 & 0 & 0 \\ 0 & -1 & 1 & -1 & 0 \\ 0 & 1 & 0 & 1 & 0 \\ 0 & 1 & 0 & 0 & 1 \end{pmatrix}$$

as a Hilbert basis of P. As usual, we have $\bar{t} = (0,0,1,0)^\mathsf{T}$ so that $\bar{t} = \bar{u} + \bar{v}$. Then the associated generically log smooth family is given by

$$X := \mathrm{Spec}\, \mathbf{k}[x, y, u, v, s]/(xy - su) \to \mathbb{A}_t^1, \quad t \mapsto uv,$$

which is illustrated in Fig. 14.4. The central fiber is given by

$$\mathrm{Spec}\, \mathbf{k}[x, y, u, v, s]/(xy - su, uv);$$

it has three irreducible components $\mathbb{A}^3_{x,v,s}$, $\mathbb{A}^3_{y,v,s}$, and $\mathrm{Spec}\, \mathbf{k}[x, y, u, s]/(xy - su)$. In particular, the central fiber is not isomorphic to $V(\tau) \times \mathbb{A}^1$ already as a scheme.

On $\{s \neq 0\}$, the family is given by $xyv = ts$ and hence log smooth. Similarly, the family is log smooth on $\{x \neq 0\}$, $\{y \neq 0\}$, and $\{u \neq 0\}$. It remains $\{x = y = u = s = 0\} = \mathbb{A}_v^1$ as potential log singular locus. On $\{v \neq 0\}$, the family is also given by $xyv = ts$, which exhibits $x_0 y_0 = t_0 z_0$ as a local model for the log singularity. In particular, $X_0 \setminus \{x = y = s = u = 0\}$ is the maximal open log smooth subset, and $Z = \{x = y = s = u = 0\}$ is the log singular locus.

In the log singular point $\{0\}$, there is no structure of a toroidal crossing space on X_0. Namely, if it would come from a monoid of rank ≤ 3, then we would have a smooth factor, but the singularity in the component $\mathrm{Spec}\, \mathbf{k}[x, y, u, s]/(xy - su)$ is

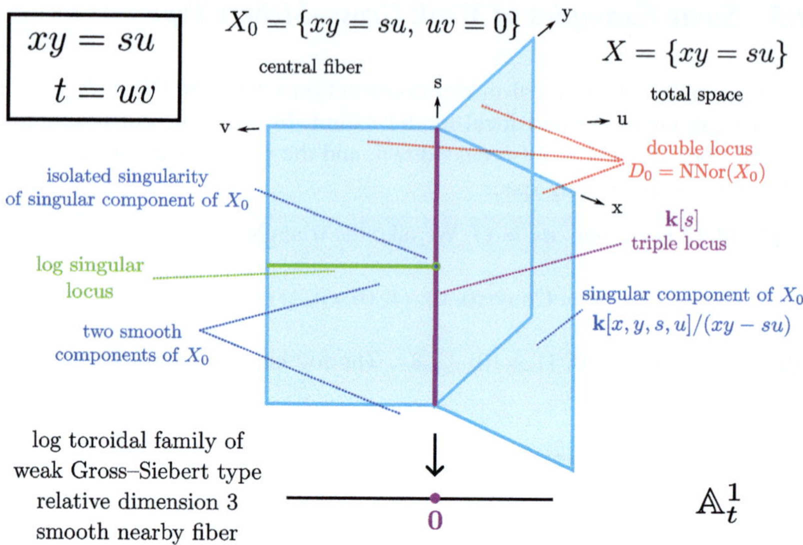

Fig. 14.4 Example 14.17. © Simon Felten 2025. All rights reserved

isolated. If it would come from a monoid of rank ≥ 4, then there would be at least four (local) irreducible components meeting at $\{0\}$, which is not the case. $\Diamond$

Example 14.18 ($xy = su$, $uvw = t$) We start with the 3-simplex

$$\tau = \mathrm{Conv}((0, 0, 0), (1, 0, 0), (0, 1, 0), (0, 0, 1)) \subseteq \mathbb{R}^3$$

and the interval

$$\Delta_1 = \mathrm{Conv}((0, 0, 0), (0, 0, 1)).$$

Then a Hilbert basis of $P = P(\mathbb{Z}^2, \mathbb{Z}^2, \tau, \Delta_1)$ is computed in Macaulay2 as

$$\left(\bar{x}\ \bar{y}\ \bar{u}\ \bar{v}\ \bar{w}\ \bar{s} \right) = \begin{pmatrix} 0 & -1 & -1 & 1 & 0 & 0 \\ 0 & -1 & -1 & 0 & 1 & 0 \\ 1 & -1 & 0 & 0 & 0 & 0 \\ 0 & 1 & 1 & 0 & 0 & 0 \\ 0 & 1 & 0 & 0 & 0 & 1 \end{pmatrix}.$$

The associated generically log smooth family is

$$X := \mathrm{Spec}\,\mathbf{k}[x, y, u, v, w, s]/(xy - su) \to \mathbb{A}^1_t, \quad t \mapsto uvw.$$

On the open subsets $\{x \neq 0\}$, $\{y \neq 0\}$, $\{s \neq 0\}$, and $\{u \neq 0\}$, the family is log smooth. It remains $Z = \{x = y = s = u = 0\} \cong \mathbb{A}^1_{v,w}$ as potential log singular locus. On $\{vw \neq 0\}$, the family is given by $xyvw = ts$, which shows that the log singularities there are modeled on $xy = tz$ and confirms that Z is the log singular locus.

The central fiber $V = X_0$ has four irreducible components:

$$V_w = \{t = w = 0\} \cong \mathrm{Spec}\ \mathbf{k}[x, y, u, v, s]/(xy - su)$$

$$V_v = \{t = v = 0\} \cong \mathrm{Spec}\ \mathbf{k}[x, y, u, w, s]/(xy - su)$$

$$V_y = \{t = u = y = 0\} \cong \mathrm{Spec}\ \mathbf{k}[x, v, w, s]$$

$$V_x = \{t = u = x = 0\} \cong \mathrm{Spec}\ \mathbf{k}[y, v, w, s]$$

Then, we have $Z = V_x \cap V_y \cap \{s = 0\}$, so the log singular locus is fully contained in one component of the double locus of V. The part of the log singular locus where the local model is not $xy = tz$ is $(V_x \cap V_y \cap V_w \cap \{s = 0\}) \cup (V_x \cap V_y \cap V_v \cap \{s = 0\})$, which is exactly the part where the log singular locus is also contained in a third irreducible component of V. $\diamond$

14.2 Log Toroidal Families of Gross–Siebert Type

A log toroidal family of (elementary/standard) Gross–Siebert type is a log toroidal family whose local models arise from the construction in the previous section, with Δ_+ an elementary/standard simplex in the elementary/standard case. Unlike in Definition 8.27, where we have defined a general log toroidal family, we require that a log toroidal family of Gross–Siebert type has precisely the log singular locus Z constructed above.[4] In other words, it is a log toroidal family accurately of class $\mathscr{C}^{GS}$, $\mathscr{C}^{eGS}$, respectively $\mathscr{C}^{sGS}$.

Definition 14.19 Let $S_0 = \mathrm{Spec}(\mathbb{N} \to \mathbf{k})$ be the standard log point. A generically log smooth family $f_0 \colon (X_0, U_0) \to S_0$ is *log toroidal of Gross–Siebert type* if it is accurately of class $\mathscr{C}^{GS}$. Similarly, for $A \in \mathbf{Art}_\mathbb{N}$, a generically log smooth deformation $f \colon (X_A, U_A) \to S_A$ of f_0 is *log toroidal of Gross–Siebert type* if it is accurately of class $\mathscr{C}^{GS}$—in other words, if we can find finitely many commutative diagrams

---

[4] The examples of local models of Gross–Siebert type show that the log singular locus $Z_{P/Q}$ of an elementary log toroidal family $f \colon A_{P,\mathcal{F}} \to A_Q$, although it is technically small enough to turn $f \colon A_{P,\mathcal{F}} \to A_Q$ into a generically log smooth family, is often unnaturally large. In particular, this justifies the flexibility that we have chosen in Definition 8.27.

$$
\begin{array}{ccccccc}
(X_A, U_A) & \xleftarrow{\;p_i\;} & (W_i, U_i') & \xrightarrow{\;h_i\;} & (L_{i;A}, U_{i;A}) & \longrightarrow & (\mathbb{A}_{P_i}, U_i) \\
\downarrow{\scriptstyle f_A} & & & & \downarrow & & \downarrow{\scriptstyle f_i} \\
S_A & =\!=\!=\!=\!=\!=\!=\!=\!=\!=\!=\!=\!=\!= & & & S_A & \longrightarrow & \mathbb{A}_t^1
\end{array}
$$

of partial log schemes where $f_i \colon (\mathbb{A}_{P_i}, U_i) \to \mathbb{A}_t^1$ are the local models associated with data $(M_i', N_i', \tau_i, \Delta_{1;i}, \ldots, \Delta_{q_i;i})$, the map $S_A \to \mathbb{A}_t^1$ is the canonical morphism, $(L_{i;A}, U_{i;A})$ is the fiber product of partial log schemes, $h_i \colon (W_i, U_i') \to (L_{i;A}, U_{i;A})$ is étale, accurate, and strict, $p_i \colon (W_i, U_i') \to (X_A, U_A)$ is étale, accurate, and strict, and $X_A = \bigcup_i p_i(W_i)$ is an open covering. We say it is *log toroidal of elementary Gross–Siebert type* if all local models can be chosen such that $\Delta_{+;i}$ is an elementary simplex. We say it is *log toroidal of standard Gross–Siebert type* if all local models can be chosen such that $\Delta_{+;i}$ is a standard simplex.

The local models of the deformation are not required to be the same as the ones exhibited for the central fiber. So, if one has two different local models of (elementary/standard) Gross–Siebert type, and they happen to define the same central fiber, then both define valid and different log toroidal deformations of (elementary/standard) Gross–Siebert type. However, we will see later that, in the elementary case at least, these deformations are locally unique up to non-unique isomorphism, in the same way as log smooth deformations are.

When $f_0 \colon (X_0, U_0) \to S_0$ is a log toroidal family of Gross–Siebert type, then we turn X_0 into a toroidal crossing space by setting $\mathcal{P} := j_* \overline{\mathcal{M}}_{U_0}$. This is a toroidal crossing space by Lemma 9.94 since the central fibers of the local models of Gross–Siebert type are well-adjusted triples $(V(\tau) \times \mathbb{A}^q, Z^{\|}, s)$. In particular, (X_0, Z_0, s) is a well-adjusted triple.

14.3   Unisingular Deformations

In [125], Gross and Siebert show that toric log Calabi–Yau spaces admit a well-behaved deformation theory. This deformation theory works so well because toric log Calabi–Yau spaces admit local models of elementary Gross–Siebert type. Namely, in this case, log toroidal deformations controlled by these models are locally rigid. We guide the reader through the argument for local rigidity and formulate the results in their natural generality of log toroidal families of elementary Gross–Siebert type. To do so, we introduce the auxiliary notion of a *unisingular* generically log smooth family and unisingular deformations thereof. Log toroidal families of elementary Gross–Siebert type are unisingular, and their unisingular deformations are always log toroidal of elementary Gross–Siebert type as well. Gross and Siebert call these deformations *divisorial deformations* in [125]. Note however that our concept of unisingular deformations is more general, and that, conversely, one may apply the term "divisorial deformations" to log toroidal deformations which are not unisingular.

In Appendix C of [243], Ruddat and Siebert studied the divisorial deformation functor of a (simple) toric log Calabi–Yau space and showed that it has a hull by verifying Schlessinger's conditions. Leaving finite-dimensionality of the tangent space aside, we generalize this in two directions: On the one hand, we show that isomorphism classes of unisingular deformations form a deformation functor, and on the other hand, we have already seen in Lemma 10.19 that isomorphism classes of deformations of type $\mathscr{D}$, for any system of deformations $\mathscr{D}$, form a deformation functor. In the case of divisorial deformations of a log toroidal family of elementary Gross–Siebert type, such as for example a toric log Calabi–Yau space, the two functors are of course isomorphic.

The theory of unisingular deformations is of interest for more general log singularities than log toroidal ones of elementary Gross–Siebert type, but then deformations will be, in general, no longer locally unique. We discuss an instance of this in Example 14.41.

14.3.1 Unisingular Families over the Standard Log Point

As a special case of $L(\ell; m_1, \ldots, m_q; s)$ from Example 14.5, we write

$$L(\ell) := \operatorname{Spec} \mathbf{k}[x, y, z, t]/(xy - t^\ell z) \to \mathbb{A}^1_t$$

with the divisorial log structures induced by $t = 0$. It becomes generically log smooth once we remove the point $Z(\ell) = \{p_0\}$ in the central fiber $L_0(\ell) = \operatorname{Spec} \mathbf{k}[x, y, z]/(xy)$. As a variant in higher dimensions, we write $L(\ell; s) := L(\ell) \times \mathbb{A}^s$. We denote the log smooth locus of $L(\ell; s)$ by $U(\ell; s)$, and the log singular locus by $Z(\ell; s)$.

Let $f \colon X = \mathbb{A}_P \to \mathbb{A}^1_t$ be a local model of Gross–Siebert type with central fiber $V = V(\tau) \times \mathbb{A}^q$. We have seen above that $f_0 \colon X_0 \to S_0$ is of the form $xy = t^\ell \prod_{i=1}^q z_i^{m_i}$ on the open subset $\mathcal{U}_1 V \subseteq V$. When Δ_+ is an elementary simplex, we have seen that this simplifies considerably to $xy = t^\ell z$. In other words, $f_0 \colon X_0 \to S_0$ is isomorphic to $L(\ell; s)_0 \to S_0$ on $\mathcal{U}_1 V$ in this case. We say that $f_0 \colon X_0 \to S_0$ is *unisingular* because z has exponent one in this equation—the log singular locus has multiplicity one on $\mathcal{U}_1 V$.[5] More generally, when $f_0 \colon X_0 \to S_0$ is a log toroidal family of elementary Gross–Siebert type, then f_0 has local models of the form $L_0(\ell; s) \to S_0$ on $\mathcal{U}_1 X_0$ for the induced structure of a toroidal crossing space on X_0.

Log toroidal families of elementary Gross–Siebert type are not the only families which are modeled on $L(\ell; s) \to \mathbb{A}^1_t$ outside of codimension 3. For instance, the

---

[5] The prefix "uni-" is derived from Latin "ūnus, ūna, ūnum," meaning "one."

families in Examples 1.158 and 10.5 have this property, as well as the local model of weak Gross–Siebert type in Example 14.17. We capture this phenomenon with the following definition.

Definition 14.20 Let $S_0 = \mathrm{Spec}(\mathbb{N} \to \mathbf{k})$ be the standard log point.

(1) A *pre-unisingular generically log smooth family* is a vertical generically log smooth family $f_0 \colon (X_0, U_0) \to S_0$ together with an open subset

$$U_0 \subseteq \widetilde{U}_0 \subseteq X_0$$

such that every point $\bar{x} \in Z_0 \cap \widetilde{U}_0$ admits an accurate strict étale neighborhood $\pi \colon (W_0, U_0') \to (X_0, U_0)$ and an accurate strict étale morphism

$$\tau \colon (W_0, U_0') \to (L_0(\ell; s), U_0(\ell; s))$$

such that the two maps $(W_0, U_0') \to S_0$ coincide.
(2) A pre-unisingular generically log smooth family is *unisingular* if the following conditions hold:

   (a) the complement $\widetilde{Z}_0 = X_0 \setminus \widetilde{U}_0$ has codimension ≥ 3;
   (b) for every $x \in \widetilde{Z}_0$, we have

$$\mathrm{depth}(\mathcal{T}^1_{X_0, x}) \geq 1,$$

   where $\mathcal{T}^1_{X_0} = \mathcal{E}xt^1(\Omega^1_{\underline{X}_0}, \mathcal{O}_{X_0})$ since X_0 is reduced.

(3) A well-adjusted triple (V, Z, s) is *unisingular* if its associated generically log smooth family $f_0 \colon (X_0, U_0) \to S_0$ together with $\widetilde{U}_0 := U_0 \cup \mathcal{U}_1 V$ is pre-unisingular.

The additional Condition (a) is used in Lemma 14.34, and the additional Condition (b) is used in Proposition 14.29. The usefulness of the well-adjustedness condition on a triple (V, Z, s) is precisely that it ensures Condition (b).

Lemma 14.21 *Let (V, Z, s) be a unisingular well-adjusted triple. Then the associated generically log smooth family $f_0 \colon (X_0, U_0) \to S_0$ together with $\widetilde{U}_0 := U_0 \cup \mathcal{U}_1 V$ is unisingular.*

Proof We clearly have $\mathrm{codim}(\widetilde{Z}_0, X_0) \geq 3$, so we have to show $\mathrm{depth}(\mathcal{T}^1_{X_0, x}) \geq 1$ for every $x \in \widetilde{Z}_0$. By shrinking V around x in the étale topology, we can assume that we have a smooth map $\tau \colon V \to V(\sigma)$ with $\tau(x) = 0$ for some polytope σ, as in the definition of a toroidal crossing space. Let $e = \dim(\sigma)$ so that $\mathcal{S}_e V(\sigma) = \{0\}$ for the stratification. Thus, $x \in \mathcal{S}_e V$. Since Z does not contain any stratum, we have $e < d = \dim(V)$, which implies that the relative dimension of τ is $d - e \geq 1$. Since τ is smooth, one may show that $\mathcal{T}^1_V \cong \tau^* \mathcal{T}^1_{V(\sigma)}$. Therefore, the depth formula [267, 0338] shows $\mathrm{depth}(\mathcal{T}^1_{V, x}) = \mathrm{depth}(\mathcal{T}^1_{V(\sigma), 0}) + d - e \geq 1$. $\square$

Corollary 14.22 *Let $f_0\colon (X_0, U_0) \to S_0$ be a generically log smooth family of elementary Gross–Siebert type. Consider X_0 as a toroidal crossing space in the canonical way. Then $\widetilde{U}_0 := U_0 \cup \mathcal{U}_1 X_0$ turns f_0 unisingular.*

14.3.2 The Unisingular Deformation Functor

We now define the deformation theory of unisingular families.

Definition 14.23 Let $f_0\colon (X_0, U_0) \to S_0$ be a pre-unisingular generically log smooth family. Then a *unisingular deformation* of $f_0\colon (X_0, U_0) \to S_0$ is a generically log smooth deformation $f_A\colon (X_A, U_A) \to S_A$ such that for every $x \in Z_0 \setminus \widetilde{Z}_0$, there is an accurate strict étale neighborhood $\pi\colon (W_A, U'_A) \to (X_A, U_A)$ and an accurate strict étale morphism $\tau\colon (W_A, U'_A) \to (L_A(\ell; s), U_A(\ell; s))$ such that the two maps $(W_A, U'_A) \to S_A$ coincide. We denote the functor of isomorphism classes of unisingular deformations by

$$\mathrm{LD}^{\mathrm{uni}}_{X_0/S_0}\colon \quad \mathbf{Art}_{\mathbb{N}} \to \mathbf{Set}.$$

Example 14.24 Let $f_0\colon (X_0, U_0) \to S_0$ be a log toroidal family of elementary Gross–Siebert type. Then every generically log smooth deformation $f_A\colon (X_A, U_A) \to S_A$ which is a log toroidal family of elementary Gross–Siebert type is a unisingular deformation. $\Diamond$

Lemma 14.25 *The functor*

$$\mathrm{LD}^{\mathrm{uni}}_{X_0/S_0}\colon \quad \mathbf{Art}_{\mathbb{N}} \to \mathbf{Set}$$

of unisingular deformations is a neat deformation functor, i.e., it satisfies (H_0), (H_1), (H_2), and (H_2^+) defined in Appendix A.

Proof The condition (H_0) is clear. For (H_1), let $B = A'' \times_A A'$ for some morphism $A' \to A$ and a surjective morphism $A'' \to A$, let $X_A \to S_A$ be a unisingular deformation, and let $X_{A'} \to S_{A'}$, $X_{A''} \to S_{A''}$ be unisingular liftings (they come with a *choice* of map $X_A \to X_{A'}$, $X_A \to X_{A''}$, which is not encoded in the deformation functor). We have to show that the generically log smooth family

$$X_{A'} \sqcup_{X_A} X_{A''} := (|X_0|,\ \mathcal{O}_{X_{A'}} \times_{\mathcal{O}_{X_A}} \mathcal{O}_{X_{A''}},\ \mathcal{M}' \times_{\mathcal{M}} \mathcal{M}'')$$

is unisingular. After appropriate shrinking of the étale neighborhoods of the point $x \in Z_0 \setminus \widetilde{Z}_0$, this amounts to showing that the canonical map

$$L(n)_{A'} \sqcup_{L(n)_A} L(n)_{A''} \to L(n)_B$$

is an isomorphism where $L(n)_B := L(n) \times_{\mathbb{A}^1} S_B$ etc. This is true on the level of O because both are flat thickenings of $L(n)_{\underline{A'}}$, cf. also [251, Cor. 3.6]; it is true on the level of M because it holds for O^* and $\overline{M}$. Finally, (H_2) and (H_2^+) follow from (H_1) because $\mathrm{LD}^{\mathrm{gen}}_{X_0/S_0}$ satisfies (H_2) and (H_2^+). $\square$

14.3.3   Unisingular Deformations on $\widetilde{U}_0 \subseteq X_0$

The power of unisingular deformations relies on the fact that they are locally rigid on $\widetilde{U}_0$. We now review the argument from [125] for this fact. On U_0, local rigidity follows from log smoothness, so we focus on geometric points $\bar{v} \in \widetilde{U}_0 \setminus U_0$.

First, we revisit the local models $L(\ell; s) \to \mathbb{A}^1_t$. They are local models of Gross–Siebert type associated with $M' = \mathbb{Z} \cdot f$, $\tau = [0, \ell]$, $\Delta_1 = [0, 1]$, and $\Delta_i = \{0\}$ for $2 \leq i \leq q$ with $q = s + 1$. Let $P = P(M', N', \tau, \Delta_1, \ldots, \Delta_q)$. Let

$$\sigma \subseteq \mathbb{R} \cdot f \oplus \mathbb{R} \cdot e_0 \oplus \bigoplus_{i=1}^{q} \mathbb{R} \cdot e_i$$

be the cone generated by $\ell \cdot f + e_0$, $f + e_1$, and e_i for $0 \leq i \leq q$. Then, we have $P = \sigma^\vee \cap (N' \oplus \mathbb{Z}^{q+1})$. We subdivide σ into a cone σ_1 generated by $f + e_1$ and e_i for $0 \leq i \leq q$ and a cone σ_2 generated by e_0, $f + e_1$, $\ell f + e_0$, and e_i for $2 \leq i \leq q$. This subdivision gives rise to a birational toric morphism $\pi \colon \widetilde{L}(\ell; s) \to L(\ell; s)$. Explicitly, $\widetilde{L}(\ell; s)$ has two open subsets given by

$$\widetilde{O}(1) = \mathrm{Spec}\, \mathbf{k}[x, t, \hat{z}, w_2, \ldots, w_q]$$

with $\hat{z} \hat{=} z/x$ and

$$\widetilde{O}(2) = \mathrm{Spec}\, \mathbf{k}[\hat{x}, y, z, t, w_2, \ldots, w_q]/(\hat{x}y - t^\ell)$$

with $\hat{x} \hat{=} x/z$. The overlap is given by inverting $\hat{z}$ respectively $\hat{x}$. From the explicit description, we see that π is an isomorphism over

$$U(\ell; s) = L(\ell; s) \setminus \{x = y = z = t = 0\}.$$

In particular, it is an isomorphism over $\{t \neq 0\}$. The central fiber $\widetilde{L}_0(\ell; s)$ has two irreducible components. One of them maps isomorphically onto the irreducible component of $L_0(\ell; s)$ given by $x = 0$ while the other irreducible component of $\widetilde{L}_0(\ell; s)$ maps birationally onto the irreducible component of $L_0(\ell; s)$ given by $y = 0$. In particular, $\pi_0 \colon \widetilde{L}_0(\ell; s) \to L_0(\ell; s)$ is birational. The following observation of Gross–Siebert makes the resolution π useful to study deformations.

Lemma 14.26 ([125, Lemma 2.12]) *In the above situation,*

$$\left(\widetilde{L}(\ell; s) \,\middle|\, \widetilde{L}_0(\ell; s)\right) \to \left(\mathbb{A}_t^1 \,\middle|\, \{0\}\right)$$

is log smooth and saturated. We have

$$\pi_* \Theta^1_{\widetilde{L}_0(\ell;s)/S_0} = \Theta^1_{L_0(\ell;s)/S_0} \quad \text{and} \quad R^q \pi_* \Theta^1_{\widetilde{L}_0(\ell;s)/S_0} = 0$$

for $q \geq 1$.

As a consequence, we show the following local rigidity result for unisingular deformations.

Lemma 14.27 (cf. [125, Lemma 2.15]) *Let $f_0 \colon (X_0, U_0) \to S_0$ be a pre-unisingular generically log smooth family, and let $\bar{v} \to Z_0 \setminus \widetilde{Z}_0$ be a geometric point. Let W_0 be an étale neighborhood of $\bar{v}$ inside $\widetilde{U}_0$, and let $W_A \to S_A$ be a unisingular deformation. Let $B \to A$ be a surjection, and let $W_B \to S_B$ and $W_B' \to S_B$ be two lifts. Then, after shrinking W_0 around $\bar{v}$, we have $W_B \cong W_B'$ as generically log smooth lifts of the deformation $W_A \to S_A$.*

Proof The values ℓ and s of the local model at $\bar{v}$ are uniquely determined. Namely, ℓ is determined by the ghost sheaf near $\bar{v}$, and s is determined by the dimension of X_0. Since both $W_B \to S_B$ and $W_B' \to S_B$ are unisingular deformations, we have accurate strict étale morphisms $p_B \colon \hat{W}_B \to L_B(\ell; s)$ and $p_B' \colon \hat{W}_B' \to L_B(\ell; s)$ for étale neighborhoods $\hat{W}_B$ and $\hat{W}_B'$ of $\bar{v}$. By replacing W_0 with the fiber product $\hat{W}_0 \times_{W_0} \hat{W}_0'$, we can assume that we actually have accurate strict étale maps $p_B \colon W_B \to L_B(\ell; s)$ and $p_B' \colon W_B' \to L_B(\ell; s)$. On the central fiber, they induce two accurate strict étale maps $p_0, p_0' \colon W_0 \to L_0(\ell; s)$ between the same partial log schemes. When arguing as in the proof of Lemma 9.94, we see that we can shrink W_0 such that it has only two irreducible components which are both smooth. Then, by swapping the variables x and y of one copy of $L(\ell; s)$ if necessary, we can assume that p_0 and p_0' map each irreducible component of W_0 to the same irreducible component of $L_0(\ell; s)$.

Next, we form the two fiber products of p_0 respectively p_0' with $\widetilde{L}_0(\ell; s) \to L_0(\ell; s)$. This gives us two maps $\widetilde{W}_0 \to W_0$ and $\widetilde{W}_0' \to W_0$. They are both birational and surjective, they induce both a one-to-one correspondence between the irreducible components, they are both isomorphic over the same irreducible component, and since the formation of the blow-up commutes with flat base change, they are both the blow-up of the same ideal sheaf on the other irreducible component. Because $\widetilde{W}_0$ and $\widetilde{W}_0'$ are obtained by closed gluing of their irreducible components, we find that they are isomorphic as schemes over W_0. Furthermore, their log smooth structures are isomorphic over $U_0 \cap W_0$. Now, $\mathcal{LS}_{\widetilde{W}_0}$ has injective restrictions to open subsets which are dense on the double locus. Therefore, we find $\widetilde{W}_0 \cong \widetilde{W}_0'$ as log schemes over S_0.

By further shrinking W_0 if necessary, we can assume that W_0 is affine. Now, $\widetilde{W}_A$ and $\widetilde{W}'_A$ are log smooth deformations of $\widetilde{W}_0$, so they are isomorphic by Lemma 14.26. When we restrict this isomorphism to $U_A \cap \widetilde{W}_A$ respectively $U_A \cap \widetilde{W}'_A$, then we obtain an automorphism of W_A as a generically log smooth deformation. Again by Lemma 14.26, this can be lifted to an automorphism of $\widetilde{W}_A$. Therefore, we can assume that we have $\widetilde{W}_A \cong \widetilde{W}'_A$ extending the identity on $U_A \cap W_A$.

Similarly, we have $\widetilde{W}_B \cong \widetilde{W}'_B$ as log smooth lifts of $\widetilde{W}_A \to S_A$. We restrict the isomorphism to $U_B \cap \widetilde{W}_B$ respectively $U'_B \cap \widetilde{W}'_B$. This then extends to an isomorphism of generically log smooth lifts of $W_A \to S_A$ because $j_* O_{U_B \cap W_B} = O_{W_B}$ and $j_* O_{U'_B \cap W'_B} = O_{W'_B}$. $\qquad\square$

This implies the following.

Proposition 14.28 *Let* $f_0 \colon (X_0, U_0) \to S_0$ *be a pre-unisingular generically log smooth family, and let* $W_0 \subseteq \widetilde{U}_0$ *be an affine open subset. Then, for every* $A \in \mathbf{Art}_\mathbb{N}$, *there is a unisingular deformation* $W_A \to S_A$ *of* $W_0 \to S_0$, *which is unique up to non-unique isomorphism.*

Moreover, when $S_A \to S_B$ *is a small thickening by an ideal* $I \subset B$, *and when a unisingular deformation* $\widetilde{U}_A \to S_A$ *of* $\widetilde{U}_0 \to S_0$ *is given, then:*

(1) *the automorphisms of a lifting* $\widetilde{U}_B \to S_B$ *are in* $H^0(\widetilde{U}_0, \Theta^1_{\widetilde{U}_0/S_0} \otimes I)$;

(2) *the different liftings are classified by* $H^1(\widetilde{U}_0, \Theta^1_{\widetilde{U}_0/S_0} \otimes I)$;

(3) *the obstruction to the existence of a lifting is in* $H^2(\widetilde{U}_0, \Theta^1_{\widetilde{U}_0/S_0} \otimes I)$.

Proof Because automorphisms of a lifting (as a generically log smooth deformation) are classified by $\Theta^1_{\widetilde{U}_0/S_0} \otimes I$, the assertion follows from the same arguments as in the classical theory of smooth deformations. $\qquad\square$

14.3.4　Unisingular Deformations of Affine Spaces

We study unisingular deformations of an affine unisingular generically log smooth family. In particular, we compute the tangent space of the unisingular deformation functor.

Proposition 14.29 ([125, Lemma 2.16]) *Let* $f_0 \colon (X_0, U_0) \to S_0$ *be a unisingular generically log smooth family, and assume that* X_0 *is affine. Let* $f_A \colon (X_A, U_A) \to S_A$ *be a unisingular deformation of* f_0, *and let* $S_A \to S_B$ *be a first-order extension by an ideal* $I \subset B$. *Assume that a lifting* $f_B \colon (X_B, U_B) \to S_B$ *is given. Then all liftings over* S_B *have the same underlying scheme* X_B.

Proof Let $(Y_B, V_B) \to S_B$ be another lifting. By the previous proposition, after restriction to $\widetilde{U}_0$, the difference to $(X_B, U_B) \to S_B$ is captured by an element

$$\theta \in H^1(\widetilde{U}_0, \Theta^1_{X_0/S_0} \otimes I) \subseteq \mathrm{Ext}^1_{O_{\widetilde{U}_0}}(\mathcal{W}^1_{\widetilde{U}_0/S_0} \otimes I, O_{\widetilde{U}_0}).$$

Similarly, the difference between the underlying flat deformations of X_0 is captured by an element

$$\underline{\theta} \in \mathrm{Ext}^1_{X_0}(\Omega^1_{\underline{X_0}} \otimes I, \mathcal{O}_{\widetilde{U}_0}),$$

where we use the classical differential forms. Let us now write

$$\Theta^1 = \Theta^1_{X_0/S_0}, \quad \mathcal{W}^1 = \mathcal{W}^1_{X_0/S_0}, \quad \mathcal{T} = \mathcal{T}_{X_0/S_0}, \quad \Omega^1 = \Omega^1_{\underline{X_0}}, \quad \mathcal{O} = \mathcal{O}_{X_0}$$

for brevity, where $\mathcal{T}_{X_0/S_0}$ are the classical derivations. In [125, Lemma 2.16], Gross and Siebert consider the following commutative diagram:

$$
\begin{array}{ccccc}
H^1(\widetilde{U}_0, \Theta^1 \otimes I) & \longrightarrow & \mathrm{Ext}^1_{\widetilde{U}_0}(\mathcal{W}^1 \otimes I, \mathcal{O}) & & \\[2mm]
\Big\downarrow{\scriptstyle \kappa} & {\scriptstyle \psi} \searrow & \Big\downarrow & & \\[4mm]
H^1(\widetilde{U}_0, \mathcal{T} \otimes I) & \longrightarrow & \mathrm{Ext}^1_{\widetilde{U}_0}(\Omega^1 \otimes I, \mathcal{O}) & \overset{\pi}{\longrightarrow} & H^0(\widetilde{U}_0, \mathcal{E}xt^1(\Omega^1 \otimes I, \mathcal{O})) \\[2mm]
\Big\uparrow & & \Big\uparrow{\scriptstyle \chi} & & \Big\uparrow{\scriptstyle \rho} \\[4mm]
H^1(X_0, \mathcal{T} \otimes I) & \longrightarrow & \mathrm{Ext}^1_{X_0}(\Omega^1 \otimes I, \mathcal{O}) & \overset{\varpi}{\longrightarrow} & H^0(X_0, \mathcal{E}xt^1(\Omega^1 \otimes I, \mathcal{O}))
\end{array}
$$

The left horizontal maps are injective, and the lower two rows are exact in the middle. We have $\psi(\theta) = \chi(\underline{\theta})$.

Since f_0 is unisingular, we have $\mathrm{depth}(\mathcal{T}^1_{X_0,x}) \geq 1$ for every $x \in \widetilde{Z}_0$, where $\mathcal{T}^1_{X_0} = \mathcal{E}xt^1(\Omega^1, \mathcal{O})$. Since X_0 is affine, this implies that ρ is injective. By exactness in the middle row, we have $\pi \circ \psi(\theta) = 0$, and hence we find $\varpi(\underline{\theta}) = 0$. However, affinity of X_0 also implies $H^1(X_0, \mathcal{T} \otimes I) = 0$, so ϖ is injective, and we have $\underline{\theta} = 0$. Therefore, the underlying flat deformations of Y_B and X_B are the same. $\square$

The proof shows a little bit more: θ lies in the kernel of the map

$$\kappa: \quad H^1(\widetilde{U}_0, \Theta^1_{X_0/S_0} \otimes I) \to H^1(\widetilde{U}_0, \mathcal{T}_{X_0/S_0} \otimes I)$$

in the upper left corner of the above diagram because the left horizontal maps are injective. Let us now define $\mathcal{B}_{X_0/S_0}$ as the cokernel in the short exact sequence

$$0 \to \Theta^1_{X_0/S_0} \to \mathcal{T}_{X_0/S_0} \to \mathcal{B}_{X_0/S_0} \to 0.$$

Then the associated long exact sequence shows that θ lies in $T(X_0/S_0) \otimes I$ for

$$T(X_0/S_0) := \mathrm{coker}\left(H^0(\widetilde{U}_0, \mathcal{T}_{X_0/S_0}) \to H^0(\widetilde{U}_0, \mathcal{B}_{X_0/S_0})\right).$$

Conversely, still assuming that one lifting to S_B is given, every $\theta \in T(X_0/S_0) \otimes I$ gives rise to a unisingular lifting of $X_A \to S_A$ to S_B because the extension of the structure sheaf to the whole space is already given.

Corollary 14.30 *Under the assumptions of Proposition 14.29, the set of all unisingular liftings is a torsor under $T(X_0/S_0) \otimes I$. In particular, if $T(X_0/S_0)$ is of finite dimension, then $\mathrm{LD}^{\mathrm{uni}}_{X_0/S_0}$ has a hull.*

We give a more convenient description of $T(X_0/S_0)$. Since $\mathcal{T}_{X_0/S_0}$ is reflexive, the restriction map $H^0(X_0, \mathcal{T}_{X_0/S_0}) \to H^0(\widetilde{U}_0, \mathcal{T}_{X_0/S_0})$ is bijective. Since X_0 is affine, the map $H^0(X_0, \mathcal{T}_{X_0/S_0}) \to H^0(X_0, \mathcal{B}_{X_0/S_0})$ is surjective. Thus, we find

$$T(X_0/S_0) = \mathrm{coker}\left(H^0(X_0, \mathcal{B}_{X_0/S_0}) \to H^0(\widetilde{U}_0, \mathcal{B}_{X_0/S_0}) \right).$$

Let $j\colon \widetilde{U}_0 \to X_0$ be the inclusion, and let C_{X_0/S_0} be the cokernel in

$$\mathcal{B}_{X_0/S_0} \to j_*(\mathcal{B}_{X_0/S_0}|_{\widetilde{U}_0}) \to C_{X_0/S_0} \to 0.$$

Then, we have $T(X_0/S_0) = H^0(X_0, C_{X_0/S_0})$.

14.3.5   Coherence of C_{X_0/S_0}

Let $f_0\colon (X_0, U_0) \to S_0$ be a unisingular generically log smooth family. In this section, we consider C_{X_0/S_0} and show that it is coherent. This justifies the remaining assumption $\mathrm{codim}(\widetilde{Z}_0, X_0) \geq 3$ in our definition of unisingularity, which we have not yet used.

Note that the formation of both $\mathcal{B}_{X_0/S_0}$ and C_{X_0/S_0} commute with accurate strict étale morphisms that also preserve $\widetilde{U}_0$. Therefore, we can compute either of them on local models.

We first analyze $\mathcal{B}_{X_0/S_0}$, and we start with our local model $xy = t^\ell z$ of unisingularity.

Example 14.31 Let $X = L(\ell)$ and $S = \mathbb{A}^1_t$. We obtain $\Theta^1_{X/S} = \mathcal{T}_{X/S}$ as the kernel in the exact sequence

$$0 \to \mathcal{T}_{X/S} \to O_X \cdot \partial_x \oplus O_X \cdot \partial_y \oplus O_X \cdot \partial_z \xrightarrow{\left(y\ x\ -t^\ell \right)} O_X.$$

This kernel is generated by

$$x\partial_x - y\partial_y, \quad y\partial_y + z\partial_z, \quad t^\ell\partial_x + y\partial_z, \quad t^\ell\partial_y + x\partial_z.$$

On the central fiber, $\mathcal{T}_{X_0/S_0}$ is the kernel of

$$0 \to \mathcal{T}_{X_0/S_0} \to O_{X_0} \cdot \partial_x \oplus O_{X_0} \cdot \partial_y \oplus O_{X_0} \cdot \partial_z \xrightarrow{\begin{pmatrix} y & x & 0 \end{pmatrix}} O_{X_0},$$

which is generated by $x\partial_x$, $y\partial_y$, and ∂_z. The image of $\Theta^1_{X_0/S_0}$ inside $\mathcal{T}_{X_0/S_0}$ is then generated by $x\partial_x - y\partial_y$, $y\partial_y + z\partial_z$, $y\partial_z$, and $x\partial_z$. Thus, we find

$$O_{X_0}/(x, y) \cong \mathcal{B}_{X_0/S_0}, \quad 1 \mapsto \partial_z.$$

Therefore, $\mathcal{B}_{X_0/S_0}$ is a line bundle on $\mathrm{Sing}(X_0)$ with its reduced scheme structure.
$\Diamond$

The following characterization of $\mathcal{B}_{X_0/S_0}$ only needs a pre-unisingular generically log smooth family (so we could even take $\widetilde{U}_0 = U_0$).

Lemma 14.32 *Let $f_0\colon (X_0, U_0) \to S_0$ be a pre-unisingular generically log smooth family. Then $\mathrm{Supp}(\mathcal{B}_{X_0/S_0})$ is given by the non-normal locus $\mathrm{NNor}(X_0)$ with its reduced scheme structure. When we set $D_0 = \mathrm{Supp}(\mathcal{B}_{X_0/S_0})$, then the map*

$$\delta\colon \ \mathcal{B}_{X_0/S_0} \to \mathcal{H}om(\mathcal{H}om(\mathcal{B}_{X_0/S_0}, O_{D_0}), O_{D_0}) := \mathcal{B}^{\vee\vee}_{X_0/S_0}$$

is injective.

Proof Since U_0 can be considered as a toroidal crossing space, $\mathrm{Reg}(U_0)$ is the strict locus of $f_0\colon U_0 \to S_0$. Therefore, we have $\mathcal{B}_{X_0/S_0}|_{\mathrm{Reg}(U_0)} = 0$. On $\mathrm{Sing}(U_0) = \mathrm{NNor}(U_0)$, we have that $\Theta^1_{U_0/S_0}$ is locally free while $\mathcal{T}_{U_0/S_0}$ is not locally free since U_0 has étale locally the form $\mathrm{Spec}\,\mathbf{k}[x, y, z_1, \ldots, z_k]/(xy)$ in codimension 1; thus, $\mathcal{B}_{X_0/S_0,x} \neq 0$ for every $x \in \mathrm{NNor}(U_0)$. Around a point $x \in Z_0 \setminus \widetilde{Z}_0$, we have already seen that $\mathcal{B}_{X_0/S_0}$ is a line bundle on $\mathrm{Sing}(X_0)$ with its reduced scheme structure. Thus, we have

$$|\mathrm{Supp}(\mathcal{B}_{X_0/S_0})|_{\widetilde{U}_0} = \mathrm{NNor}(\widetilde{U}_0)$$

as sets.

Let now $x \in \mathrm{Nor}(X_0)$ be a point where X_0 is normal, and let $x \in W_0 \subseteq X_0$ be a small open neighborhood. Then $W_0 \cap U_0$ is smooth, and $W_0 \cap U_0 \to S_0$ is strict so that $\Theta^1_{X_0/S_0} = \mathcal{T}_{X_0/S_0}$ on $W_0 \cap U_0$. Since both sheaves are reflexive, the equality holds on W_0, and we have $\mathcal{B}_{X_0/S_0} = 0$ on W_0. Therefore, we have

$$|\mathrm{Supp}(\mathcal{B}_{X_0/S_0})| \subset \mathrm{NNor}(X_0)$$

as closed subsets of $|X_0|$.

Assume that $\mathrm{NNor}(\widetilde{U}_0) \subseteq \mathrm{NNor}(X_0)$ would not be dense. Then we can find a point $x \in \widetilde{Z}_0 \cap \mathrm{NNor}(X_0)$ which admits an open neighborhood $x \in W_0 \subseteq X_0$ such that $W_0 \cap \mathrm{NNor}(\widetilde{U}_0) = \varnothing$. Therefore, $\mathrm{NNor}(W_0) \subset \widetilde{Z}_0$ so that $\dim(\mathrm{NNor}(W_0)) \leq$

$d - 2$. In particular, W_0 satisfies the condition (R_1), and it satisfies (S_2) since f_0 is a generically log smooth family. In other words, W_0 is normal, contradicting our assumption that $x \in \mathrm{NNor}(W_0)$. Thus, $\mathrm{NNor}(\widetilde{U}_0) \subseteq \mathrm{NNor}(X_0)$ is dense, and we find

$$|\mathrm{Supp}(\mathcal{B}_{X_0/S_0})| = \mathrm{NNor}(X_0)$$

as sets.

On the log smooth locus U_0, the family is given by $xy = t^\ell$ in codimension 1, i.e., there is an open subset $U_0' \subseteq U_0$ such that f_0 has locally this form (or is smooth), and such that $X_0 \setminus U_0'$ has codimension ≥ 2. As a consequence of our computation for $xy = t^\ell z$, the sheaf $\mathcal{B}_{X_0/S_0}$ is a line bundle on $\mathrm{NNor}(U_0')$ with its reduced scheme structure. Therefore, $D_0|_{U_0'} = \mathrm{Supp}(\mathcal{B}_{X_0/S_0})|_{U_0'}$ is reduced.

Let η be the generic point of an irreducible component of $\mathrm{NNor}(X_0)$. Arguing as above, we see that $\dim(\overline{\{\eta\}}) = d - 1$. Thus, $\eta \in U_0'$, and $\mathcal{O}_{D_0,\eta}$ is a field, i.e., D_0 satisfies condition (R_0).

For a point $x \in U_0$, the sheaf $\Theta^1_{X_0/S_0}$ is locally free. Since U_0 is Cohen–Macaulay, we find $\mathrm{depth}(\Theta^1_{X_0/S_0,x}) = \dim(\mathcal{O}_{X_0,x})$. Since $\mathcal{T}_{X_0/S_0}$ is reflexive, we have $\mathrm{depth}(\mathcal{T}_{X_0/S_0}) \geq \min\{\dim(\mathcal{O}_{X_0,x}), 2\}$. Therefore, we find

$$\mathrm{depth}(\mathcal{B}_{X_0/S_0,x}) \geq \min\{\dim(\mathcal{O}_{X_0/S_0,x}) - 1, 2\}.$$

In particular, if $x \in \mathrm{NNor}(X_0)$ but not a generic point of an irreducible component of $\mathrm{NNor}(X_0)$, then the depth is ≥ 1.

Let $c \in \mathrm{NNor}(X_0)$ be a point which is not a generic point of an irreducible component, and let $C = \overline{\{c\}}$ so that $\mathrm{depth}(\mathcal{B}_{X_0/S_0,x}) \geq 1$ for every $x \in C$. Thus, the restriction map

$$H^0(W_0, \mathcal{B}_{X_0/S_0}) \to H^0(W_0 \setminus C, \mathcal{B}_{X_0/S_0})$$

is injective for every open subset $W_0 \subseteq X_0$ by Grothendieck [132, Prop. 5.10.2]. Now let W_0 be affine, and let $a \in H^0(W_0, \mathcal{O}_{D_0})$ be a function with $a|_{W_0 \setminus C} = 0$. Assume that $a \neq 0$. By the definition of D_0 as the support of $\mathcal{B}_{X_0/S_0}$, there is some $s \in H^0(W_0, \mathcal{B}_{X_0/S_0})$ with $a \cdot s \neq 0$. Then $(a \cdot s)|_{W_0 \setminus C} \neq 0$ by injectivity, contradicting the fact that $a|_{W_0 \setminus C} = 0$. Therefore,

$$H^0(W_0, \mathcal{O}_{D_0}) \to H^0(W_0 \setminus C, \mathcal{O}_{D_0})$$

is injective, and [132, Prop. 5.10.2] shows that $\mathrm{depth}(\mathcal{O}_{D_0,x}) \geq 1$ for all $x \in C$. In other words, D_0 is reduced.

On the open subset U_0' which we considered earlier, $\mathcal{B}_{X_0/S_0}$ is a line bundle on D_0. Therefore, $\delta \colon \mathcal{B}_{X_0/S_0} \to \mathcal{B}_{X_0/S_0}^{\vee\vee}$ is an isomorphism on U_0'. Furthermore, δ is an isomorphism on $\mathrm{Nor}(X_0)$ since source and target are zero there. After setting $U_0'' = U_0' \cup \mathrm{Nor}(X_0)$, we argue as above and obtain that

$$H^0(W_0, \mathcal{B}_{X_0/S_0}) \to H^0(W_0 \cap U_0'', \mathcal{B}_{X_0/S_0})$$

is injective for all open subsets $W_0 \subseteq X_0$. But then δ must be injective. $\qquad\square$

While δ is an isomorphism for $L(\ell) \to \mathbb{A}_t^1$, this is not the case for arbitrary vertical saturated log smooth morphisms.

Example 14.33 We consider $X = \mathrm{Spec}\,\mathbf{k}[x, y, z, w]/(xy - zw)$ together with the map to $\mathbb{A}_t^1$ given by $t \mapsto xy$ as in Example 14.13. Then one can check in Macaulay2 that $\delta\colon \mathcal{B}_{X_0/S_0} \to \mathcal{B}_{X_0/S_0}^{\vee\vee}$ is not surjective. $\qquad\Diamond$

In particular, we have $j_*(\mathcal{B}_{X_0/S_0}|_{\widetilde{U}_0}) \neq \mathcal{B}_{X_0/S_0}^{\vee\vee}$ in general. Furthermore, reflexivity on the right-hand side does not guarantee that the restriction maps of $\mathcal{B}_{X_0/S_0}^{\vee\vee}$ are isomorphisms when we remove a subset of codimension ≥ 2 because we do not know if D_0 satisfies Serre's condition (S_2). Nonetheless, in practice, δ often is an isomorphism on $\widetilde{U}_0$, and one can often show that D_0 satisfies Serre's condition (S_2). In this case, $j_*(\mathcal{B}_{X_0/S_0}|_{\widetilde{U}_0})$ can be computed as $\mathcal{B}_{X_0/S_0}^{\vee\vee}$; in particular, it is coherent in this case. Below, we will see several examples where $j_*(\mathcal{B}_{X_0/S_0}|_{\widetilde{U}_0})$ can be computed like this, for instance Example 14.41.

We are now ready to show coherence of C_{X_0/S_0} under the more general assumptions of our Definition 14.20 of unisingularity. The following lemma finally uses our assumption on the codimension of $\widetilde{Z}_0$.

Lemma 14.34 *Let $f_0\colon (X_0, U_0) \to S_0$ be a unisingular generically log smooth family. Then C_{X_0/S_0} is coherent.*

Proof The sheaf is clearly quasi-coherent, so the point is finite generatedness. We show that $j_*(\mathcal{B}_{X_0/S_0}|_{\widetilde{U}_0})$ is finitely generated, and we can assume that X_0 and hence also D_0 is affine.

Let $\mathfrak{p} \subset O_{D_0}$ be an associated prime of $\mathcal{B}_{X_0/S_0}$, and assume that $\mathfrak{p} \in \widetilde{U}_0$. Since $\delta\colon \mathcal{B}_{X_0/S_0} \to \mathcal{B}_{X_0/S_0}^{\vee\vee}$ is injective, $\mathfrak{p}$ is also an associated prime of $\mathcal{B}_{X_0/S_0}^{\vee\vee}$.

Let $a \in O_{D_0}$ be a non-zero divisor. Then a is also a non-zero divisor on $\mathcal{B}_{X_0/S_0}^{\vee\vee}$, and hence a is not contained in any associated prime of $\mathcal{B}_{X_0/S_0}^{\vee\vee}$. Now assume that $\mathfrak{q}$ is an associated prime of $\mathcal{B}_{X_0/S_0}^{\vee\vee}$ which is not a minimal prime in O_{D_0}. By prime avoidance, there must be an element $a \in \mathfrak{q}$ which is not in the union of the minimal primes of O_{D_0}. Thus, a is a non-zero divisor, and hence not contained in any associated prime of $\mathcal{B}_{X_0/S_0}^{\vee\vee}$. This shows that all associated primes of $\mathcal{B}_{X_0/S_0}^{\vee\vee}$, and hence all associated primes of $\mathcal{B}_{X_0/S_0}$, are minimal primes of O_{D_0}.

Let again $\mathfrak{p} \in D_0$ be an associated prime. Then $\mathfrak{p}$ is the generic point of an irreducible component of D_0, and hence we have $\dim(\overline{\{\mathfrak{p}\}}) = d - 1$. Since $\dim(\widetilde{Z}_0) \leq d - 3$, we find $\mathrm{codim}(\overline{\{\mathfrak{p}\}} \cap \widetilde{Z}_0, \overline{\{\mathfrak{p}\}}) \geq 2$, and the coherence criterion [132, Cor. 5.11.4] yields the coherence of $j_*(\mathcal{B}_{X_0/S_0}|_{\widetilde{U}_0})$ since D_0, as an affine scheme of finite type over $\mathbf{k}$, can be (locally) embedded into a regular scheme. $\qquad\square$

14.3.6 Unisingular Deformations of Families of Elementary Gross–Siebert Type

We first look at the local models of elementary Gross–Siebert type.

Example 14.35 Let $f_0 \colon X_0 \to S_0$ be the central fiber of a local model $f \colon \mathbb{A}_P \to \mathbb{A}_t^1$ of elementary Gross–Siebert type, i.e., Δ_+ is an elementary simplex. In this situation, Gross and Siebert computed in [125, Lemma 2.14] the sheaf $\mathcal{B}_{X_0/S_0}$ explicitly as a subsheaf of $\bigoplus_\omega \mathcal{L}_\omega$, where each $\mathcal{L}_\omega$ is a line bundle on a closed codimension-1 stratum $V_\omega \in [\mathcal{S}_1 X_0]$. As pointed out at the end of the proof of [125, Prop. 2.16], their description shows that the restriction map

$$H^0(X_0, \mathcal{B}_{X_0/S_0}) \to H^0(\widetilde{U}_0, \mathcal{B}_{X_0/S_0})$$

is bijective. Namely, the complement of $\widetilde{U}_0 \cap V_\omega$ in V_ω is of codimension ≥ 2 so that $H^0(X_0, \mathcal{L}_\omega) \to H^0(\widetilde{U}_0, \mathcal{L}_\omega)$ is an isomorphism for every ω. Now the equations that cut out $\mathcal{B}_{X_0/S_0}$ as a subsheaf of the direct sum do not change under the restriction $X_0 \rightsquigarrow \widetilde{U}_0$ either. Therefore, we have $T(X_0/S_0) = 0$, and unisingular liftings are unique if they exist. Conversely, the local model provides a unisingular deformation over each $A \in \mathbf{Art}_\mathbb{N}$, so that there is at least one unisingular deformation for each $A \in \mathbf{Art}_\mathbb{N}$. By induction over small extensions, every unisingular deformation over A is isomorphic to that given one. In particular, $\mathrm{LD}^{\mathrm{uni}}_{X_0/S_0} = \{*\}$. The hull is $\mathbf{k}[\![t]\!]$, and the miniversal, in fact universal, family is given by pull-back of $f \colon \mathbb{A}_P \to \mathbb{A}_t^1$ to $\mathbf{k}[\![t]\!]$. ◇

Let $f_0 \colon (X_0, U_0) \to S_0$ be a log toroidal family of elementary Gross–Siebert type with $\widetilde{U}_0 = U_0 \cup \mathcal{U}_1 X_0$ as above. Since the formation of C_{X_0/S_0} commutes with accurate strict étale morphisms preserving $\widetilde{U}_0$, the example shows that $C_{X_0/S_0} = 0$. Thus, unisingular deformations are locally rigid if they exist.

Every unisingular deformation $f_A \colon (X_A, U_A) \to S_A$ is in fact a log toroidal family of elementary Gross–Siebert type. This is proven via induction over small extensions in $\mathbf{Art}_\mathbb{N}$. If $f_B \colon (X_B, U_B) \to S_B$ is a unisingular deformation, then its pull-back to S_A along a small extension $B \to A$ must admit local models of elementary Gross–Siebert type. These local models can be lifted to S_B, and the lifts must be locally isomorphic to $f_B \colon (X_B, U_B) \to S_B$. Thus, the latter family admits local models as well.

If $f_A \colon (X_A, U_A) \to S_A$ is a unisingular deformation, then the local models show that X_A can be covered by étale opens where the deformation can be lifted to S_B. This is not yet the existence of a unique deformation over every affine open, but we can get this from the standard theory of [134, Exposé III], applied to our context.

Proposition 14.36 *Let $f_0 \colon (X_0, U_0) \to S_0$ be a log toroidal family of elementary Gross–Siebert type. Let $B \to A$ be a first-order extension with kernel $I \subset B$, and assume that we have a unisingular deformation $f_A \colon (X_A, U_A) \to S_A$. Then:*

(1) *The automorphisms of any lifting $f_B\colon X_B \to S_B$ lie in $H^0(X_0, \Theta^1_{X_0/S_0}) \otimes_{\mathbf{k}} I$.*
(2) *Given one lifting to S_B, the set of all liftings is a torsor under $H^1(X_0, \Theta^1_{X_0/S_0}) \otimes_{\mathbf{k}} I$.*
(3) *The obstruction to the existence of a lifting to S_B is in $H^2(X_0, \Theta^1_{X_0/S_0}) \otimes_{\mathbf{k}} I$.*

Proof Let $j\colon U_0 \to X_0$ be the inclusion. Using Lemma 10.4, we find that the automorphisms of a lifting lie in $j_*(\Theta^1_{U_0/S_0} \otimes_{\mathbf{k}} I)$. This is equal to $\Theta^1_{X_0/S_0} \otimes_{\mathbf{k}} I$. The rest follows from general theory (use a Čech cover in the étale topology). $\square$

Corollary 14.37 *Let $f_0\colon (X_0, U_0) \to S_0$ be a log toroidal family of elementary Gross–Siebert type. Then it admits a system of deformations $\mathscr{D}$ in the sense of Definition 10.15. Every local model in $\mathscr{D}$ is log toroidal of elementary Gross–Siebert type.*

Proof The central fiber $f_0\colon X_0 \to S_0$ is log Gorenstein because this is the case for the local models of elementary Gross–Siebert type, whose central fibers come equipped with the structure of a toroidal crossing space. Let $\{V_\alpha\}_\alpha$ be an affine cover. For each k, we can choose inductively a unisingular deformation $V_{\alpha;k}$ over S_k which is compatible with restrictions along $S_k \to S_{k+1}$. They have the base change property by Theorem 8.63. Given α, β, we have $V_{\alpha;0}|_{\alpha\beta} = V_{\beta;0}|_{\alpha\beta}$. Then, by induction on k and by using the uniqueness of unisingular deformations an the affine open $V_\alpha \cap V_\beta$, we can lift this to an isomorphism $\psi_{\alpha\beta;k}\colon V_{\alpha;k}|_{\alpha\beta} \cong V_{\beta;k}|_{\alpha\beta}$ which is compatible with the restrictions along $S_k \to S_{k+1}$. Every general $A \in \mathbf{Art}_{\mathbb{N}}$ admits a canonical map $S_A \to S_k$ for all $k \gg 0$; they are compatible with $S_k \to S_{k+1}$ and with every map $S_B \to S_{B'}$. Then we can define $V_{\alpha;A}$, $\rho_{\alpha;BB'}$, and $\psi_{\alpha\beta;A}$ by pull-back. $\square$

14.3.7   More Examples of C_{X_0/S_0}

Considering only unisingular deformations simplifies the deformation problem not only for log toroidal families of elementary Gross–Siebert type. Here are some additional examples. Let us start with a local model of non-elementary Gross–Siebert type, namely the one with non-rigid deformations which we have encountered in Example 1.83.

Example 14.38 We consider

$$f\colon \ (X|X_0) = (\mathrm{Spec}\,\mathbf{k}[x, y, z, w, t]/(xy - tzw)|\{t = 0\}) \to (\mathbb{A}^1_t|\{0\})$$

with a log singular locus that restricts to $Z_0 = \{x = y = zw = 0\}$ in the central fiber. This is clearly unisingular with $\widetilde{Z}_0 = \{0\}$. An explicit computation in Macaulay2 shows that $\mathcal{B}_{X_0/S_0}$ is supported on $\{x = y = 0\} \cong \mathbb{A}^2_{z,w}$, which is Cohen–Macaulay. The canonical map $\delta\colon \mathcal{B}_{X_0/S_0} \to \mathcal{B}^{\vee\vee}_{X_0/S_0}$ is injective, and its cokernel is the skyscraper sheaf with stalk $\mathbf{k}$ in 0. This implies $C_{X_0/S_0} \cong \mathrm{coker}(\delta)$, and therefore, we have $\dim_{\mathbf{k}} T(X_0/S_0) = 1$. Contrary to what one might expect at first glance after looking at Example 1.83, the miniversal unisingular family is not given by the equation $xy = tzw + t^2 s$ but by the equation $xy = tzw + ts$ over $B = \mathrm{Spec}(\mathbb{N} \to \mathbf{k}[t, s], \ 1 \mapsto t)$. The proof of miniversality needs the Kodaira–Spencer map of unisingular families and lies beyond the scope of this monograph. We hope to elaborate on this elsewhere. $\diamond$

Let us now turn to our two examples of local models of weak Gross–Siebert type.

Example 14.39 We continue Example 14.17 given by the equations $xy = su$ and $uv = t$. Let $f\colon (X, U) \to \mathbb{A}^1_t$ be the family in that example with its log singular locus $Z = \{x = y = s = u = 0\}$. Let $\widetilde{Z}_0 = \{0\}$. We have already seen that $\widetilde{U}_0 = X_0 \setminus \widetilde{Z}_0$ turns f_0 into a pre-unisingular family. Furthermore, we have $\mathrm{codim}(\widetilde{Z}_0, X_0) = 3$. A computation in Macaulay2 shows that the support of $\mathcal{T}^1_{X_0}$ is given by $D_0 = \mathrm{Sing}(X_0) = \mathrm{NNor}(X_0) \cong \mathrm{Spec}\,\mathbf{k}[x, y, v, s]/(xy, yv, xv)$. As a product of a reduced scheme with a smooth scheme, this is (S_2). Another computation shows that $\mathcal{T}^1_{X_0} \to \mathcal{H}om(\mathcal{H}om(\mathcal{T}^1_{X_0}, \mathcal{O}_{D_0}), \mathcal{O}_{D_0})$ is an isomorphism. From this, we see easily that $\mathrm{depth}(\mathcal{T}^1_{X_0,0}) \geq 2$. In particular, $f_0\colon (X_0, U_0) \to S_0$ is unisingular.

Also $\mathcal{B}_{X_0/S_0}$ is supported on D_0, and a computation yields that $\mathcal{B}_{X_0/S_0} \to \mathcal{B}^{\vee\vee}_{X_0/S_0}$ is an isomorphism. Since D_0 is (S_2), we find that

$$\mathcal{B}^{\vee\vee}_{X_0/S_0} \to j_*(\mathcal{B}^{\vee\vee}_{X_0/S_0}|_{D_0 \setminus \widetilde{Z}_0})$$

is an isomorphism, and therefore $C_{X_0/S_0} = 0$. Thus, unisingular deformations of $f_0\colon (X_0, U_0) \to S_0$ are rigid. $\diamond$

Example 14.40 We continue Example 14.18 given by the equations $xy = su$ and $uvw = t$. Let $f\colon (X, U) \to \mathbb{A}^1_t$ be the family from that example with its log singular locus $Z = \{x = y = s = u = 0\}$. We have already seen that the family is pre-unisingular on $\{vw \neq 0\}$. It remains

$$\widetilde{Z} = \widetilde{Z}_0 = \{x = y = s = u = vw = 0\} \cong \mathrm{Spec}\,\mathbf{k}[v, w]/(vw),$$

which has codimension 3. The support D_0 of $\mathcal{T}^1_{X_0}$ is the singular locus of X_0 with its reduced scheme structure. Furthermore, $\mathcal{T}^1_{X_0} \to \mathcal{H}om(\mathcal{H}om(\mathcal{T}^1_{X_0}, \mathcal{O}_{D_0}), \mathcal{O}_{D_0})$ is an isomorphism. Therefore, $\mathcal{T}^1_{X_0}$ has injective restrictions to dense open subsets in D_0, and so we find $\mathrm{depth}(\mathcal{T}^1_{X_0,z}) \geq 1$ for every $z \in \widetilde{Z}_0$. In particular, $f_0\colon X_0 \to S_0$ is unisingular.

Explicitly, we have

$$D_0 \cong \mathrm{Spec}\,\mathbf{k}[x, y, u, v, w, s]/(uv, uw, xy - su, xvw, yvw).$$

One may check that this satisfies Serre's condition (S_2) on $\{x \neq 0\}$, $\{y \neq 0\}$, $\{u \neq 0\}$, $\{v \neq 0\}$, and $\{w \neq 0\}$. The complement of these loci is one-dimensional while $\dim(D_0) = 3$. Therefore, D_0 satisfies Serre's condition (S_2).

When computing $\mathcal{B}_{X_0/S_0}$ explicitly in Macaulay2, we obtain that the support of $\mathcal{B}_{X_0/S_0}$ is precisely D_0 with its reduced scheme structure, and that the canonical map

$$\delta: \quad \mathcal{B}_{X_0/S_0} \to \mathcal{B}_{X_0/S_0}^{\vee\vee}$$

is an isomorphism. Since D_0 is (S_2), this implies $j_*(\mathcal{B}_{X_0/S_0}|_{\tilde{U}_0}) = \mathcal{B}_{X_0/S_0}^{\vee\vee}$ and therefore $C_{X_0/S_0} = 0$. $\Diamond$

Let us now consider the family given by $xy = w^2 + t^2$ and $zw = tu$. Like in the example $xy = tzw$, unisingular deformations are not rigid here.

Example 14.41 Let $\tau \subseteq \mathbb{R}^2$ be the $(1, 1, 2)$-triangle as considered in Example 1.157. It defines a log smooth morphism $U(\tau) \to \mathbb{A}_t^1$ by Construction 9.2, which is precisely the family given by $xy = w^2$ and $zw = t$, considered in Example 1.66. Its central fiber $V(\tau)$ can be considered as a toroidal crossing space, and we obtain another toroidal crossing space $V = V(\tau) \times \mathbb{A}_u^1$.

Let

$$X = \operatorname{Spec} \mathbf{k}[x, y, z, w, u, t]/(xy - w^2 - t^2, \, zw - tu),$$

and consider the map $f: X \to \mathbb{A}_t^1$ as in Example 1.158. Let $Z = Z_0$ be the log singular locus as defined in that example. The central fiber X_0 is isomorphic to $V = V(\tau) \times \mathbb{A}_u^1$ as a scheme, and f induces a section s of $\mathcal{LS}_V$ on $V \setminus Z$. Then (V, Z, s) is a unisingular well-adjusted triple. Namely, we saw already in Example 1.158 that f has local models of the form $xy = t^2z$ or $xy = tz$ around log singular points in $Z_0 \setminus \{0\}$.

An explicit computation in Macaulay2 shows that $\mathcal{B}_{X_0/S_0}$ is supported on $D_0 = \operatorname{Sing}(X_0)$ with its reduced scheme structure, and that

$$D_0 \cong \operatorname{Spec} \mathbf{k}[x, y, z, u]/(xy, yz, xz).$$

Therefore, D_0 satisfies Serre's condition (S_2).

Another explicit computation shows that the canonical map $\delta: \mathcal{B}_{X_0/S_0} \to \mathcal{B}_{X_0/S_0}^{\vee\vee}$ is injective, and that its cokernel is supported in $\tilde{Z}_0 = \{0\}$. Therefore, we have $j_*(\mathcal{B}_{X_0/S_0}|_{\tilde{U}_0}) = \mathcal{B}_{X_0/S_0}^{\vee\vee}$, and C_{X_0/S_0} can be computed as the cokernel of δ. Then, we obtain

$$\dim_{\mathbf{k}} T(X_0/S_0) = \dim_{\mathbf{k}} H^0(X_0, C_{X_0/S_0}) = 1.$$

In particular, unisingular deformations are not rigid, but $\mathrm{LD}^{\mathrm{uni}}_{X_0/S_0}$ has a hull and thus a miniversal family.[6] $\diamond$

More generally, we see that $T(X_0/S_0)$ is finite-dimensional if $\dim(X_0) = 3$. Namely, in this case, the complement of $\tilde{U}_0$, and therefore the support of C_{X_0/S_0}, is zero-dimensional. Conversely, if $\dim(X_0) \geq 4$, then we expect in general $\mathrm{Supp}(C_{X_0/S_0})$ to have dimension ≥ 1, and therefore, we expect $T(X_0/S_0)$ to be infinite-dimensional.

14.3.8   The Interpretation of $\mathcal{B}_{X_0/S_0}$

We provide an intrinsic interpretation of the role of $\mathcal{B} := \mathcal{B}_{X_0/S_0}$, which came out of a long exact sequence in the previous section. This material seems new; it is not contained in the source [125] of the rest of the material. Let $f_0 \colon (X_0, U_0) \to S_0$ be the family associated to a unisingular well-adjusted triple, let $f_A \colon (X_A, U_A) \to S_A$ be a unisingular thickening, let $S_A \to S_B$ be a small thickening by $I \subseteq B$, and assume that we have a unisingular lifting $f_B \colon (X_B, U_B) \to S_B$. On $\tilde{U}_0$, we introduce a *presheaf* $\mathcal{D}$ of generically log smooth deformations which are locally isomorphic to the given deformation $f_B \colon (X_B, U_B) \to S_B$:

Definition 14.42 Let $W_0 \subseteq \tilde{U}_0$ be open, and let W_B be the corresponding open subset of X_B. Then a section in $\Gamma(W_0, \mathcal{D})$ is an isomorphism class of quadruples $(\mathcal{M}, \alpha, \pi, \rho)$ where $\alpha \colon \mathcal{M} \to \mathcal{O}_{X_B}|_{W_B \cap U_B}$ is a log structure on the scheme $W_B \cap U_B$, the homomorphism of sheaves of monoids $\pi \colon \mathcal{M} \to \mathcal{M}_{X_A}|_{W_A \cap U_A}$ defines a strict closed embedding $(W_A \cap U_A, \mathcal{M}_{X_A}) \to (W_B \cap U_B, \mathcal{M})$, and $\rho \in \Gamma(W_0 \cap U_0, \mathcal{M})$ satisfies $\pi(\rho) = f_A^{\flat}(1)$ for $1 \in \mathcal{M}_{S_A}$ and $\alpha(\rho) = f_B^{\sharp}(t)$. Every geometric point $\bar{x} \in W_0$ must have an étale neighborhood $V_0 \to W_0$ such that the above data fit into a commutative diagram

$$\begin{array}{ccc}
\mathcal{M}|_{V_B \cap U_B} & \xrightarrow[\cong]{\;\;\Phi\;\;} & \mathcal{M}_{X_B}|_{V_B \cap U_B} \\
\;\;\downarrow{\alpha} & & \downarrow \\
\mathcal{O}_{X_B}|_{V_B \cap U_B} & \xrightarrow[\cong]{\;\;\phi\;\;} & \mathcal{O}_{X_B}|_{V_B \cap U_B}
\end{array}$$

which defines an isomorphism of local liftings of $f_A \colon (X_A, U_A) \to S_A$. An isomorphism between $(\mathcal{M}, \alpha, \pi, \rho)$ and $(\mathcal{M}', \alpha', \pi', \rho')$ is an isomorphism $\Phi \colon \mathcal{M} \to \mathcal{M}'$ of sheaves of monoids such that $\alpha' \circ \Phi = \alpha$, $\pi' \circ \Phi = \pi$, and $\Phi(\rho) = \rho'$.

---

[6] In joint work in progress with Andrés David Gómez Villegas, the author studies the miniversal family of this example. We hope to report the results elsewhere.

Lemma 14.43 *The presheaf $\mathcal{D}$ is a sheaf on $\widetilde{U}_0$.*

Proof It suffices to show that $(\mathcal{M}, \alpha, \pi, \rho)$ has no non-trivial automorphisms. Every automorphism is in particular an automorphism of the log smooth lifting $(X_{A'}|_{W_0 \cap U_0}, \mathcal{M}) \to S_{A'}$, i.e., it is given by a log derivation

$$\theta \in H^0(U_0 \cap W_0, \Theta^1_{X_0/S_0} \otimes I) = H^0(W_0, \Theta^1_{X_0/S_0} \otimes I)$$

as is discussed in [125, Lemma 2.10] and more generally in [76]. On the structure sheaf, the automorphism is trivial, so θ maps to 0 under the forgetful map $\Theta^1_{X_0/S_0} \otimes I \to \mathcal{T}_{X_0/S_0} \otimes I$; the latter is injective, so $\theta = 0$. $\qquad\square$

Let $\mathcal{I} = I \cdot \mathcal{O}_{X_B}$ be the kernel of the map $\mathcal{O}_{X_B} \to \mathcal{O}_{X_A}$. Let $\mathcal{A}ut$ be the sheaf of automorphisms of X_B as a *scheme* which induce the identity on X_A and are compatible with $f_B \colon X_B \to S_B$. If $\phi \colon \mathcal{O}_{X_B} \to \mathcal{O}_{X_B}$ is such an automorphism, then we obtain a derivation

$$\delta_\phi = \phi - \mathrm{id} \colon \ \mathcal{O}_{X_0} \to \mathcal{I} \cong \mathcal{O}_{X_0} \otimes_k I;$$

this defines an isomorphism $\mathcal{A}ut \cong \mathcal{T}_{X_0/S_0} \otimes_k I$ of sheaves of groups. Now $\mathcal{A}ut$ acts on $\mathcal{D}$ via composition:

$$\mathcal{A}ut \times \mathcal{D} \to \mathcal{D}, \quad (\phi, (\mathcal{M}, \alpha, \pi, \rho)) \mapsto (\mathcal{M}, \phi \circ \alpha, \pi, \rho).$$

Lemma 14.44 *The action $\mathcal{A}ut \times \mathcal{D} \to \mathcal{D}$ induces an action $(\mathcal{B} \otimes_k I) \times \mathcal{D} \to \mathcal{D}$ via the sequence*

$$0 \to \Theta^1_{X_0/S_0} \otimes_k I \to \mathcal{T}_{X_0/S_0} \otimes_k I \to \mathcal{B} \otimes_k I \to 0.$$

Under this action, $\mathcal{D}$ is a trivial $(\mathcal{B} \otimes_k I)$-torsor.

Proof First, we show that the action descends to $\mathcal{B} \otimes_k I$. Let $W_0 \subseteq \widetilde{U}_0$ be an open, let $f' \colon X' \to S_B$ be a deformation representing a class $\gamma \in \Gamma(W_0, \mathcal{D})$, and assume that $\phi \in \Gamma(W_0, \mathcal{A}ut)$ is in the image of $\Theta^1_{X_0/S_0} \otimes_k I \to \mathcal{T}_{X_0/S_0} \otimes_k I$; this means that we can find $\Phi \colon \mathcal{M}|_{W_0 \cap U_0} \to \mathcal{M}|_{W_0 \cap U_0}$ such that (ϕ, Φ) is an automorphism of X', cf. [125, Lemma 2.10]. Now the diagram

$$
\begin{array}{ccc}
\mathcal{M} & \xrightarrow{\ \Phi\ } & \mathcal{M} \\
{\scriptstyle \phi \circ \alpha}\big\downarrow & & \big\downarrow{\scriptstyle \alpha} \\
\mathcal{O}_{X_{A'}} & =\!=\!= & \mathcal{O}_{X_{A'}}
\end{array}
$$

shows that $\phi \cdot \gamma = \gamma$ in $\Gamma(W_0, \mathcal{D})$, so the action descends. The arguments that $\mathcal{D}$ is a torsor under $\mathcal{B} \otimes_k I$ are similar. It is trivial because the given lifting $f_B \colon X_B \to S_B$ defines a global section in $\Gamma(\widetilde{U}_0, \mathcal{D})$. $\qquad\square$

Chapter 15
The Gerstenhaber Calculus of Log Toroidal Families

Let Q be a sharp toric monoid, let $\Lambda = \mathbb{C}[\![Q]\!]$, and let $f_0\colon X_0 \to S_0$ be a generically log smooth family. Let $A_Q = \mathrm{Spec}(Q \to \mathbb{C}[Q])$, and let $a_A\colon S_A \to A_Q$ be the canonical chart of S_A for $A \in \mathbf{Art}_\Lambda$. In Chap. 8, we have defined what it means for $f_A\colon X_A \to S_A$ to be log toroidal with respect to $a\colon S_A \to A_Q$. Namely, it admits certain local models $A_{P,\mathcal{F}} \to A_Q$, which are elementary log toroidal families.

In this chapter, we show that the characteristic Gerstenhaber calculus of a proper log toroidal family is perfect.

When studying this chapter in detail, we recommend that the reader keeps [38, 75, 77] at hand. References are to these articles, even when we have included the result in Chap. 8. We work in the following situation:

Situation 15.1 We have a proper generically log smooth family $f_0\colon X_0 \to S_0$ of relative dimension $d \geq 1$ which is log toroidal with respect to $a_0\colon S_0 \to A_Q$ and log Gorenstein. We fix an admissible open cover $\mathcal{V} = \{V_\alpha\}_\alpha$ of X_0 in the sense of Definition 10.13, and $\mathscr{D}$ is a system of deformations subordinate to $\mathcal{V}$. We assume that all deformations in $\mathscr{D}$ are log toroidal with respect to $a_A\colon S_A \to A_Q$.[1] In particular, they have automatically the base change property by Felten et al. [77, Cor. 7.9]. We fix another open cover $\mathcal{U} = \{U_i\}_i$ of X_0 such that $\mathcal{U} \cup \mathcal{V}$ is admissible.

Here is the main result of this chapter:[2]

---

[1] Neither do we require that at every point, there is a unique local model, nor do we require that every local model on some deformation extends to a thickening of the deformation. We just require that around every point on each deformation, there is some local model. We also do not require the log smooth loci to coincide as we did for log toroidal families of Gross–Siebert type.

[2] This section has been written before we became aware of the left contraction $\vdash$ that characterizes a two-sided Gerstenhaber calculus. The left contraction $\vdash$ is not relevant for this chapter, so all Gerstenhaber calculi are one-sided. Also, chapter was written before it was planned to include Chap. 8, which essentially only gives an exposition of a theory already developed elsewhere.

© The Author(s), under exclusive license to Springer Nature Switzerland AG 2025
S. Felten, *Global Logarithmic Deformation Theory*, Lecture Notes in Mathematics 2373, https://doi.org/10.1007/978-3-031-98751-9_15

Theorem 15.2 *In the above situation, the characteristic Λ-linear curved Gerstenhaber calculus $(PV_{X_0/\Lambda}^{\bullet,\bullet}, DR_{X_0/\Lambda}^{\bullet,\bullet})$ is perfect. In particular, if $f_0\colon X_0 \to S_0$ is log Calabi–Yau, then $\mathrm{LD}_{X_0/S_0}^{\mathscr{D}}$ is unobstructed.*

Proof The degeneration of the Hodge–de Rham spectral sequence of $f_0\colon X_0 \to S_0$ is [77, Thm. 1.9]. The (hyper-)cohomology $H^k(DR_{X_0/\mathbb{C}}^{\bullet}, d) \cong \mathbb{H}^k(X_0, \mathcal{W}_{X_0/S_0}^{\bullet})$ is finitely generated because X_0 is proper. If we assume for a moment that

$$H^k(DR_{X_0/A}^{\bullet}, d) \to H^k(DR_{X_0/\mathbb{C}}^{\bullet}, d) \tag{15.1}$$

is surjective, then Proposition D.3 yields that it is an isomorphism; thus, Nakayama's Lemma, which holds for modules over the Artinian local ring A without any finiteness assumption,[3] yields that $H^k(DR_{X_0/A}^{\bullet}, d)$ is finitely generated—just take a finitely generated submodule which surjects onto $H^k(DR_{X_0/\mathbb{C}}^{\bullet}, d)$. The difficult part is to show the surjectivity of (15.1). This is the content of Sect. 15.3 below. For the unobstructedness, we use Corollary 13.41. $\square$

Example 15.3 Let (V, Z, s) be a well-adjusted triple such that the associated generically log smooth family $f_0\colon X_0 \to S_0$ is a log toroidal family of elementary Gross–Siebert type in the sense of Definition 14.19. Let $\mathcal{V}$ and $\mathcal{U}$ be affine open covers. Let $\mathscr{D}$ be a system of unisingular deformations obtained from Corollary 14.37. Assume that $f_0\colon X_0 \to S_0$ is log Calabi–Yau. Then $\mathrm{LD}_{X_0/S_0}^{\mathscr{D}} = \mathrm{LD}_{X_0/S_0}^{\mathrm{uni}}$ is unobstructed. $\Diamond$

In the remainder of this chapter, we prove the surjectivity of the map (15.1) in the above proof. If a Maurer–Cartan solution exists, this is a consequence of [77, Thm. 1.10], respectively a variant thereof. However, in the non-Calabi–Yau case, the statement is true even if no Maurer–Cartan solution exists, and in the Calabi–Yau case, we need to know the surjectivity before we can conclude that a Maurer–Cartan solution exists. Essentially, this result is [38, Lemma 4.18]. Our proof will be essentially the same, slightly more conceptual; the point is that our construction of $(PV_{X_0/\Lambda}^{\bullet,\bullet}, DR_{X_0/\Lambda}^{\bullet,\bullet})$ is very similar to the one in [38] but technically not quite the same, and we want to make sure that our algebraic version is actually perfect in the sense of Definition 5.18, allowing a proof of unobstructedness in the Calabi–Yau case.

A *monoid ideal* $K \subseteq Q$ is a subset such that $k \in K$ and $q \in Q$ implies $k+q \in K$. If $K \subseteq Q$ is a proper[4] monoid ideal with $Q \setminus K$ finite, then $A_K := \mathbb{C}[Q]/\mathbb{C}[K]$

---

[3] See Appendix D.2.

[4] Proper means $K \neq Q$.

is in $\mathbf{Art}_\Lambda$. Because the Artinian local rings $A_k = \mathbb{C}[\![Q]\!]/\mathfrak{m}_Q^{k+1}$ are of this form for $(Q \setminus \{0\})^{k+1}$, and every $A \in \mathbf{Art}_\Lambda$ admits a map $A_k \to A$ for some $k \geq 0$, it is sufficient to prove the surjectivity in (15.1) for $A = A_k$ and in particular for $A = A_K$.

15.1 Absolute Differential Forms

Let $\mathbf{0} := \mathrm{Spec}(0 \to \mathbb{C})$. Then $A_Q \to \mathbf{0}$ is a saturated log smooth morphism of relative dimension $r := \mathrm{rk}(Q^{\mathrm{gp}})$. The log de Rham complex is

$$\Omega^\bullet_{A_Q/\mathbf{0}} = \bigoplus_{k=0}^{r} \mathbb{C}[Q] \otimes_{\mathbb{C}} \bigwedge_{\mathbb{C}}^{k} (Q^{\mathrm{gp}} \otimes_{\mathbb{Z}} \mathbb{C})$$

with de Rham differential $\partial(z^q \otimes \alpha) = z^q \otimes (q \wedge \alpha)$ on global sections. For every $A \in \mathbf{Art}_\Lambda$, we have a log de Rham complex $\Omega^\bullet_{S_A/\mathbf{0}}$ of the composite $S_A \to A_Q \to \mathbf{0}$ as well. When $K \subseteq Q$ is a monoid ideal with $Q \setminus K$ finite and $A = A_K :=$ $\mathbb{C}[Q]/\mathbb{C}[K]$, then this complex has a particularly simple form. Namely, in this case, [222, IV, Cor. 2.3.3] yields $\Omega^1_{A_Q/\mathbf{0}} \otimes_{\mathbb{C}[Q]} A_K \cong \Omega^1_{S_K/\mathbf{0}}$, where $S_K := \mathrm{Spec}(Q \to A_K)$, i.e., we have

$$\Omega^\bullet_{S_K/\mathbf{0}} = \bigoplus_{k=0}^{r} A_K \otimes_{\mathbb{C}} \bigwedge_{\mathbb{C}}^{k} (Q^{\mathrm{gp}} \otimes_{\mathbb{Z}} \mathbb{C})$$

as graded A_K-module. Since the de Rham differential must be compatible, we find $\partial(z^q \otimes \alpha) = z^q \otimes (q \wedge \alpha)$; now S_K is a punctual scheme, so this actually describes the full de Rham complex of sheaves.[5] We say that $\Omega^\bullet_{S_K/\mathbf{0}}$ is the *absolute de Rham complex* on S_K.

Let $f \colon X_K \to S_K$ be a generically log smooth deformation of type $\mathscr{D}$. On U_K, we have the de Rham complex $\Omega^\bullet_{U_K/\mathbf{0}}$ of the composite $X_K \to S_K \to \mathbf{0}$. Then we define the *absolute de Rham complex* of $X_K \to S_K$ by

$$\mathcal{W}^\bullet_{X_K/\mathbf{0}} := j_* \Omega^\bullet_{U_K/\mathbf{0}}.$$

---

[5] The example $Q = 0$ and $A_\varepsilon = \mathbb{C}[\varepsilon]/(\varepsilon^2)$ shows that this description of $\Omega^\bullet_{S_A/\mathbf{0}}$ is not true for a general A which is not of the form $\mathbb{C}[Q]/\mathbb{C}[K]$. Namely, in this case, we have $\Omega^1_{A_Q/\mathbf{0}} = 0$ but $\Omega^1_{S_\varepsilon/\mathbf{0}} = \mathbb{C} \cdot \partial\varepsilon \neq 0$. In particular, $\Omega^1_{S_A/\mathbf{0}}$ is not locally free in general.

Of course, the de Rham differential ∂ is $\mathbb{C}$-linear but not A_K-linear. On U_K, we have the sequence

$$0 \to f^*\Omega^1_{S_K/0} \to \Omega^1_{U_K/0} \to \Omega^1_{U_K/S_K} \to 0$$

which is exact and locally split by Ogus [222, IV, Thm. 3.2.3]. In particular, $\Omega^1_{U_K/0}$ is locally free of rank $d + r$. Taking exterior powers and the direct image under $j\colon U_K \to X_K$, we obtain an exact sequence

$$0 \to f^*\Omega^1_{S_K/0} \wedge \mathcal{W}^{k-1}_{X_K/0} \to \mathcal{W}^k_{X_K/0} \to \mathcal{W}^k_{X_K/S_K} \to 0,$$

where the right-hand map is surjective by the local computation in [77, Cor. 7.11, Cor. 7.12], and where the left-hand side is *defined* as

$$j_*\mathrm{im}\left(f^*\Omega^1_{S_K/0} \otimes \Omega^{k-1}_{U_K/0} \to \Omega^k_{U_K/0}, \ \ f^*\mu \otimes \alpha \mapsto f^*\mu \wedge \alpha\right) \subseteq \mathcal{W}^k_{X_K/0}.$$

A careful local computation using [77, Cor. 7.11, Cor. 7.12] shows that this is actually equal to the image of

$$f^*\Omega^1_{S_K/0} \otimes \mathcal{W}^{k-1}_{X_K/0} \to \mathcal{W}^k_{X_K/0}.$$

Then we define a decreasing filtration of $\mathcal{W}^\bullet_{X_K/0}$ by

$$F^s\mathcal{W}^k_{X_K/0} := \mathrm{im}\left(f^*\Omega^s_{S_K/0} \otimes \mathcal{W}^{k-s}_{X_K/0} \to \mathcal{W}^k_{X_K/0}\right).$$

We have $F^0\mathcal{W}^\bullet_{X_K/0} = \mathcal{W}^\bullet_{X_K/0}$ and $F^{r+1}\mathcal{W}^\bullet_{X_K/0} = 0$. Each $F^s\mathcal{W}^\bullet_{X_K/0}$ is closed under the de Rham differential ∂.

When $\mathrm{Gr}^s\mathcal{W}^k_{X_K/0} := F^s\mathcal{W}^k_{X_K/0}/F^{s+1}\mathcal{W}^k_{X_K/0}$ is the quotient of the filtration, then we obtain a commutative diagram

$$
\begin{array}{ccccccccc}
0 & \to & f^*\Omega^s_{S_K/0} \otimes F^1\mathcal{W}^{k-s}_{X_K/0} & \to & f^*\Omega^s_{S_K/0} \otimes \mathcal{W}^{k-s}_{X_K/0} & \to & f^*\Omega^s_{S_K/0} \otimes \mathcal{W}^{k-s}_{X_K/S_K} & \to & 0 \\
 & & \downarrow & & \downarrow & & \downarrow{\scriptstyle \tau^k_s} & & \\
0 & \longrightarrow & F^{s+1}\mathcal{W}^k_{X_K/0} & \longrightarrow & F^s\mathcal{W}^k_{X_K/0} & \longrightarrow & \mathrm{Gr}^s\mathcal{W}^k_{X_K/0} & \longrightarrow & 0.
\end{array}
$$

Using the local direct sum decomposition of $\mathcal{W}^k_{X_K/0}$ on U_K, we can show that τ^k_s (the vertical map on the right-hand side) is an isomorphism on U_K, and then it must be an isomorphism on X_K because the source is Z-closed. Then also the target $\mathrm{Gr}^s\mathcal{W}^k_{X_K/0}$ must be Z-closed. Furthermore, a diagram chase shows that

$$(-1)^s \cdot \tau^{k+1}_s(f^*\mu \otimes \partial(\alpha)) = \partial\,\tau^k_s(f^*\mu \otimes \alpha)$$

because the second summand of $\partial(f^*\mu \wedge \alpha)$ is inside $F^{s+1}\mathcal{W}^k_{X_K/\mathbf{0}}$. Thus,

$$\tau_s: \quad f^{-1}\Omega^s_{S_K/\mathbf{0}} \otimes_{A_K} \mathcal{W}^\bullet_{X_K/S_K}[-s] \to \mathrm{Gr}^s\mathcal{W}^\bullet_{X_K/\mathbf{0}}$$

is an isomorphism of complexes, where the convention for $[-s]$ is that the differential acquires a factor $(-1)^{-s}$.

The log morphism $U_K \to \mathbf{0}$ has an O_{X_K}-bilinear contraction map

$$\lrcorner: \quad \Theta^p_{U_K/\mathbf{0}} \times \Omega^i_{U_K/\mathbf{0}} \to \Omega^{i-p}_{U_K/\mathbf{0}}.$$

Via the inclusion $\Theta^p_{U_K/S_K} \subseteq \Theta^p_{U_K/\mathbf{0}}$, this gives rise to the contraction map

$$\lrcorner: \quad \mathcal{V}^p_{X_K/S_K} \times \mathcal{W}^i_{X_K/\mathbf{0}} \to \mathcal{W}^{p+i}_{X_K/\mathbf{0}}.$$

Then we can define a Lie derivative

$$\mathcal{L}: \quad \mathcal{V}^p_{X_K/S_K} \times \mathcal{W}^i_{X_K/\mathbf{0}} \to \mathcal{W}^{p+i+1}_{X_K/\mathbf{0}}$$

via the Lie–Rinehart homotopy formula. Together with the obvious $\wedge$-product, they turn $(\mathcal{V}^\bullet_{X_K/S_K}, \mathcal{W}^\bullet_{X_K/\mathbf{0}})$ into an *absolute Gerstenhaber calculus* in the sense of the following definition. We include directly also the bigraded and curved versions in this definition, which we use below. Beyond the scope of this monograph, absolute Gerstenhaber calculi are useful for a deeper Hodge-theoretic study of the spaces carrying them, for example via logarithmic semi-infinite variations of Hodge structures as in [38, §6].

Definition 15.4 Let Q be a sharp toric monoid with $r := \mathrm{rk}(Q^{\mathrm{gp}})$, and let $K \subseteq Q$ be a proper monoid ideal with $Q \setminus K$ finite. Let $(S, C, O, \mathbf{C})$ be a context where C is the constant sheaf on the site S with stalk $A_K = \mathbb{C}[Q]/\mathbb{C}[K]$.

(1) An *absolute Gerstenhaber calculus* of dimension $d \geq 1$

$$(\mathcal{G}^\bullet, \wedge, 1_G, [-,-], \mathcal{K}^\bullet, \wedge, 1_K, \partial, \lrcorner, \mathcal{L}, \epsilon, F^\bullet\mathcal{K}^\bullet)$$

consists of:

- a Gerstenhaber algebra $(\mathcal{G}^\bullet, \wedge, 1_G, [-,-])$ of dimension d;
- an O-module $\mathcal{K}^i \in \mathbf{C}$ of degree i for every $0 \leq i \leq d + r$; for i outside this range, we set $\mathcal{K}^i = 0$ whenever necessary;
- an O-bilinear product $-\wedge-: \mathcal{K}^i \times \mathcal{K}^{i'} \to \mathcal{K}^{i+i'}$ and a global section $1_K \in \mathcal{K}^0$ such that

$$(\alpha \wedge \beta) \wedge \gamma = \alpha \wedge (\beta \wedge \gamma); \quad \alpha \wedge \beta = (-1)^{|\alpha||\beta|}\beta \wedge \alpha; \quad 1_K \wedge \alpha = \alpha;$$

- a $\mathbb{C}$-linear[6] map $\partial\colon \mathcal{K}^i \to \mathcal{K}^{i+1}$ with $\partial^2 = 0$, satisfying the derivation rule

$$\partial(\alpha \wedge \beta) = \partial(\alpha) \wedge \beta + (-1)^{|\alpha|}\alpha \wedge \partial(\beta),$$

the *absolute de Rham differential*;
- an O-bilinear map $\lrcorner\colon \mathcal{G}^p \times \mathcal{K}^i \to \mathcal{K}^{p+i}$ such that

$$1_G \lrcorner \alpha = \alpha \quad \text{and} \quad (\theta \wedge \xi) \lrcorner \alpha = \theta \lrcorner (\xi \lrcorner \alpha);$$

it is called the *absolute contraction map*;
- a $\mathbb{C}$-bilinear map $\mathcal{L}_-(-)\colon \mathcal{G}^p \times \mathcal{K}^i \to \mathcal{K}^{p+i+1}$ such that

$$\mathcal{L}_{[\theta,\xi]}(\alpha) = \mathcal{L}_\theta(\mathcal{L}_\xi(\alpha)) - (-1)^{(|\theta|+1)(|\xi|+1)}\mathcal{L}_\xi(\mathcal{L}_\theta(\alpha));$$

it is called the *absolute Lie derivative*;
- an A_K-linear injective map $\epsilon\colon f^{-1}\Omega^\bullet_{S_K/0} \to (\mathcal{K}^\bullet, 1_K, \wedge, \partial)$ of differential graded algebras with $\epsilon(1) = 1_K$; here, $f^{-1}\Omega^i_{S_K/0}$ is the constant sheaf on $\mathcal{S}$ with stalk $\Omega^i_{S_K/0}$; we assume that $f^{-1}\Omega^i_{S_K/0} \otimes_{A_K} O \to \mathcal{K}^i$ is injective as well and consider $f^{-1}\Omega^i_{S_K/0} \otimes_{A_K} O$ as a subsheaf of $\mathcal{K}^i$ via this map; we write $f^*\Omega^i_{S_K/0} := f^{-1}\Omega^i_{S_K/0} \otimes_{A_K} O$ for short; these notations are justified by considering $f\colon (\mathcal{S}, C) \to (\{*\}, A_K)$ as a map of ringed sites;
- a decreasing filtration $F^s\mathcal{K}^\bullet$ of $\mathcal{K}^\bullet$ by O-submodules, where $F^s\mathcal{K}^i$ is defined as the image of $f^*\Omega^s_{S_K/0} \otimes \mathcal{K}^{i-s} \to \mathcal{K}^i$, $f^*\mu \otimes \alpha \mapsto f^*\mu \wedge \alpha$.

They need to have the following properties:

- the mixed Leibniz rule, the Lie–Rinehart homotopy formula, and the formulae for contraction with $\theta \in \mathcal{G}^0$ respectively $\theta \in \mathcal{G}^{-1}$ hold as stated in Definition 3.13; the map $\lambda\colon \mathcal{G}^0 \to \mathcal{K}^0$, $\theta \mapsto \theta \lrcorner 1_K$, is an isomorphism of O-modules with $\lambda(\theta \wedge \xi) = \lambda(\theta) \wedge \lambda(\xi)$;
- for $\theta \in \mathcal{G}^p$, $\mu \in \Omega^s_{S_K/0}$, and $\alpha \in \mathcal{K}^i$, we have the commutation relation

$$[\theta\lrcorner,\ \epsilon(\mu) \wedge (-)](\alpha) := \theta \lrcorner (\epsilon(\mu) \wedge \alpha) - (-1)^{|\theta||\mu|}\epsilon(\mu) \wedge (\theta \lrcorner \alpha) = 0; \tag{15.2}$$

in particular, $\mathcal{L}$ is A_K-linear in the second variable but not necessarily in the first one;
- the filtration $F^s\mathcal{K}^\bullet$ is compatible with $\wedge$, ∂, $\lrcorner$, and $\mathcal{L}$ in that we have induced maps

$$\wedge\colon\ F^s\mathcal{K}^i \times F^{s'}\mathcal{K}^{i'} \to F^{s+s'}\mathcal{K}^{i+i'}, \qquad \partial\colon\ F^s\mathcal{K}^i \to F^s\mathcal{K}^{i+1},$$

$$\lrcorner\colon\ \mathcal{G}^p \times F^s\mathcal{K}^i \to F^s\mathcal{K}^{p+i}, \qquad\qquad \mathcal{L}\colon\ \mathcal{G}^p \times F^s\mathcal{K}^i \to F^s\mathcal{K}^{p+i+1};$$

---

[6] It is not necessarily A_K-linear.

- let $\mathcal{A}^\bullet := F^0\mathcal{K}^\bullet/F^1\mathcal{K}^\bullet$; we assume that $\mathcal{A}^\bullet$ is concentrated in degrees $[0, d]$, and that $\mathcal{A}^i \in \mathbf{C}$; for $a \in A_K$, we have automatically $\partial\,\epsilon(a) \in F^1\mathcal{K}^1$; thus, the differential on $\mathcal{A}^\bullet$ is A_K-linear;
- the induced map

$$\tau_s:\quad \Omega^s_{S_K/\mathbf{0}} \otimes_{A_K} \mathcal{A}^\bullet[-s] \to \mathrm{Gr}^s\mathcal{K}^\bullet$$

  is an isomorphism of complexes for all s.

The contraction map and the Lie derivative descend to $\mathcal{A}^\bullet$; since all necessary relations are satisfied, the pair $(\mathcal{G}^\bullet, \mathcal{A}^\bullet)$ then becomes an ordinary Gerstenhaber calculus of dimension d.

(2)   A *bigraded absolute Gerstenhaber calculus* of dimension $d \geq 1$

$$(\mathcal{G}^{\bullet,\bullet}, \wedge, 1_G, [-,-],\ \mathcal{K}^{\bullet,\bullet}, \wedge, 1_K, \partial, \lrcorner, \mathcal{L}, \epsilon, F^\bullet\mathcal{K}^{\bullet,\bullet})$$

consists of:

- a bigraded Gerstenhaber algebra $(\mathcal{G}^{\bullet,\bullet}, \wedge, 1_G, [-,-])$ of dimension d in $(\mathcal{S}, \mathcal{C}, \mathcal{O}, \mathbf{C})$;
- for every $0 \leq i \leq d + r$ and every $j \geq 0$, an $\mathcal{O}$-module $\mathcal{K}^{i,j} \in \mathbf{C}$ of bidegree (i, j) and total degree $i + j$; for (i, j) outside this range, we set $\mathcal{K}^{i,j} = 0$;
- an A_K-bilinear product $-\wedge-: \mathcal{K}^{i,j} \times \mathcal{K}^{i',j'} \to \mathcal{K}^{i+i',j+j'}$ and an element $1_K \in \mathcal{K}^{0,0}$ such that

$$(\alpha \wedge \beta) \wedge \gamma = \alpha \wedge (\beta \wedge \gamma); \quad \alpha \wedge \beta = (-1)^{|\alpha||\beta|}\beta \wedge \alpha; \quad 1_K \wedge \alpha = \alpha;$$

- a $\mathbb{C}$-linear map $\partial: \mathcal{K}^{i,j} \to \mathcal{K}^{i+1,j}$ with $\partial^2 = 0$ and satisfying the derivation rule

$$\partial(\alpha \wedge \beta) = \partial(\alpha) \wedge \beta + (-1)^{|\alpha|}\alpha \wedge \partial(\beta),$$

  the *absolute de Rham differential*;
- an injective degree-preserving A_K-linear homomorphism

$$\epsilon:\quad f^{-1}\Omega^\bullet_{S_K/\mathbf{0}} \to \mathcal{K}^{\bullet,0}$$

  of differential graded-commutative algebras (with $\epsilon(1) = 1$); furthermore, we assume that the induced map $f^*\Omega^i_{S_K/\mathbf{0}} \to \mathcal{K}^{i,0}$ is injective;
- an $\mathcal{O}$-bilinear map $\lrcorner: \mathcal{G}^{p,q} \times \mathcal{K}^{i,j} \to \mathcal{K}^{p+i,q+j}$ such that $1 \lrcorner \alpha = \alpha$ and $(\theta \wedge \xi) \lrcorner \alpha = \theta \lrcorner (\xi \lrcorner \alpha)$, the *absolute contraction map*;
- an A_K-bilinear map $\mathcal{L}_-(-): \mathcal{G}^{p,q} \times \mathcal{K}^{i,j} \to \mathcal{K}^{p+i+1,q+j}$ such that

$$\mathcal{L}_{[\theta,\xi]}(\alpha) = \mathcal{L}_\theta(\mathcal{L}_\xi(\alpha)) - (-1)^{(|\theta|+1)(|\xi|+1)}\mathcal{L}_\xi(\mathcal{L}_\theta(\alpha)),$$

the *absolute Lie derivative*;

- a decreasing filtration $F^s \mathcal{K}^{\bullet,\bullet}$ of $\mathcal{K}^{\bullet,\bullet}$ by O-submodules, where $F^s \mathcal{K}^{i,j}$ is the image of $f^* \Omega^s_{S_K/0} \otimes \mathcal{K}^{i-s,j} \to \mathcal{K}^{i,j}$, $f^*\mu \otimes \alpha \mapsto f^*\mu \wedge \alpha$.

They need to have the following properties:

- the mixed Leibniz rule, the Lie–Rinehart homotopy formula, and the formulae for contraction with $\theta \in G^{0,q}$ and $\theta \in G^{-1,q}$ hold as stated in Definition 3.13; the map $\lambda \colon G^{0,q} \to \mathcal{K}^{0,q}, \theta \mapsto \theta \lrcorner 1_K$, is an O-linear isomorphism with $\lambda(\theta \wedge \xi) = \lambda(\theta) \wedge \lambda(\xi)$;
- for $\theta \in G^{p,q}$, $\mu \in \Omega^s_{S_K/0}$, and $\alpha \in \mathcal{K}^{i,j}$, we have the commutation relation

$$[\theta\lrcorner,\ \epsilon(\mu) \wedge (-)](\alpha) := \theta \lrcorner (\epsilon(\mu) \wedge \alpha) - (-1)^{|\theta||\mu|}\epsilon(\mu) \wedge (\theta \lrcorner \alpha) = 0;$$

$$(15.3)$$

- the filtration $F^s \mathcal{K}^{\bullet,\bullet}$ is compatible with $\wedge$, ∂, $\lrcorner$, and $\mathcal{L}$ in that we have induced maps

$$\wedge \colon\ F^s \mathcal{K}^{i,j} \times F^{s'} \mathcal{K}^{i',j'} \to F^{s+s'} \mathcal{K}^{i+i',j+j'}, \quad \partial \colon F^s \mathcal{K}^{i,j} \to F^s \mathcal{K}^{i+1,j},$$

$$\lrcorner \colon\ G^{p,q} \times F^s \mathcal{K}^{i,j} \to F^s \mathcal{K}^{p+i,q+j}, \quad \mathcal{L} \colon G^{p,q} \times F^s \mathcal{K}^{i,j} \to F^s \mathcal{K}^{p+i+1,q+j};$$

- let $\mathcal{A}^{\bullet,\bullet} := F^0 \mathcal{K}^{\bullet,\bullet}/F^1 \mathcal{K}^{\bullet,\bullet}$; we assume that $\mathcal{A}^{\bullet,\bullet}$ is concentrated in degrees $[0, d]$ for the first variable, and that $\mathcal{A}^{i,j} \in \mathbf{C}$; for $a \in A_K$, we have automatically $\partial \epsilon(a) \in F^1 \mathcal{K}^{1,0}$; thus, the differential on $\mathcal{A}^{\bullet,\bullet}$ is A_K-linear;
- the induced map

$$\tau_{s,j} \colon\ f^{-1} \Omega^s_{S_K/0} \otimes_{A_K} \mathcal{A}^{\bullet,j}[-s] \to \mathrm{Gr}^s \mathcal{K}^{\bullet,j}$$

is an isomorphism of complexes for all s.

It is called *bounded* if $G^{p,q} = 0$ for $q \gg 0$ and $\mathcal{K}^{i,j} = 0$ for $j \gg 0$.

(3) A *curved absolute Gerstenhaber calculus* of dimension $d \geq 1$ is a bigraded absolute Gerstenhaber calculus together with an A_K-linear map $\bar{\partial} \colon G^{p,q} \to G^{p,q+1}$, an element $\ell \in I_K \cdot G^{-1,2}$, and an A_K-linear map $\bar{\partial} \colon \mathcal{K}^{i,j} \to \mathcal{K}^{i,j+1}$ such that $(G^{\bullet,\bullet}, \bar{\partial}, \ell)$ is a curved Gerstenhaber algebra and the map $\bar{\partial} \colon \mathcal{K}^{i,j} \to \mathcal{K}^{i,j+1}$ satisfies the derivation rules

$$\bar{\partial}(\alpha \wedge \beta) = \bar{\partial}\alpha \wedge \beta + (-1)^{|\alpha|}\alpha \wedge \bar{\partial}\beta, \quad \bar{\partial}(\theta \lrcorner \alpha) = (\bar{\partial}\theta) \lrcorner \alpha + (-1)^{|\theta|}\theta \lrcorner \bar{\partial}\alpha$$

as well as $\bar{\partial}^2(\alpha) = \mathcal{L}_\ell(\alpha)$ and $\partial\bar{\partial} + \bar{\partial}\partial = 0$. Moreover, we assume that $\bar{\partial}\epsilon(\mu) = 0$ for $\mu \in f^{-1}\Omega^s_{S_K/0}$.

(4) A *differential bigraded absolute Gerstenhaber calculus* is a curved absolute Gerstenhaber calculus with $\ell = 0$.

In the above situation, we write $\mathcal{K}^i_{X_K/S_K} := \mathcal{W}^i_{X_K/0}$.

Lemma 15.5 *The pair* $(\mathcal{V}^{\bullet}_{X_K/S_K}, \mathcal{K}^{\bullet}_{X_K/S_K})$ *is an absolute Gerstenhaber calculus in the context* $\mathfrak{Coh}(X_K/S_K)$.

Proof Since the absolute differential forms $\Omega^1_{U_K/0}$ on U_K are locally free of rank $d + r$, the pieces $\mathcal{K}^i_{X_K/S_K}$ are in the range $0 \leq i \leq d + r$. From [77, Cor. 7.12], we find that $\mathcal{W}^i_{X_K/0}$ is flat over S_K because, in the local model, it is a free A_K-module.[7] The $\wedge$-product and the de Rham differential on $\mathcal{K}^{\bullet}_{X_K/S_K}$ have the required properties because $\mathcal{K}^{\bullet}_{X_K/S_K}$ is the direct image of the de Rham complex of $U_K \to \mathbf{0}$. Similarly, the properties of the contraction $\lrcorner$ and the Lie derivative $\mathcal{L}$ are inherited from $U_K \to \mathbf{0}$. The map $\epsilon \colon f^{-1}\Omega^{\bullet}_{S_K/0} \to \mathcal{K}^{\bullet}_{X_K/S_K}$ comes via the direct image from U_K from the adjunction of $\Omega^{\bullet}_{S_K/0} \to f_* j_* \Omega^{\bullet}_{U_K/0}$. Here, we use that $j_*(f^{-1}\Omega^i_{S_K/0})|_{U_K} = f^{-1}\Omega^i_{S_K/0}$. Since $f^*\Omega^1_{S_K/0} \to \Omega^1_{U_K/0}$ is locally split injective, $f^*\Omega^i_{S_K/0} \to \mathcal{W}^i_{X_K/0}$ is injective as well. The inclusion $A_K \to \mathcal{O}_{X_K}$ is injective because every stalk of $\mathcal{O}_{X_K}$ is an A_K-algebra which is flat and hence free as an A_K-module. Since $f^{-1}\Omega^i_{S_K/0}$ is a free A_K-module, also the induced map $f^{-1}\Omega^i_{S_K/0} \to f^*\Omega^i_{S_K/0}$ is injective; hence, ϵ is injective. The filtration $F^s\mathcal{K}^{\bullet}_{X_K/S_K}$ is defined as specified in the definition. The mixed Leibniz rule and the formulae for the contraction with $\theta \in \mathcal{V}^0_{X_K/S_K}$ and $\theta \in \mathcal{V}^{-1}_{X_K/S_K}$ are inherited from the Gerstenhaber calculus of $U_K \to \mathbf{0}$. The Lie–Rinehart homotopy formula holds by the definition of $\mathcal{L}$. The map $\lambda \colon \mathcal{V}^0_{X_K/S_K} \to \mathcal{K}^0_{X_K/S_K}$ is obviously an isomorphism of $\mathcal{O}_{X_K}$-algebras because both are equal to $\mathcal{O}_{X_K}$. To show the commutation relation (15.2), first note that it holds for $p = 0$. For $p = -1$ and $s = 0$, we have $\theta \lrcorner \epsilon(\mu) = 0$ by degree reasons. For $s = 1$, we have $\theta \lrcorner \epsilon(\mu) = 0$ because θ is a relative derivation of $X_K \to S_K$. Since $\Omega^{\bullet}_{S_K/0}$ is generated in degree 1, the relation $\theta \lrcorner \epsilon(\mu) = 0$ follows for all s from the formula for $\theta \lrcorner (\alpha \wedge \beta)$. Then the same formula yields the commutation relation (15.2) for $p = -1$ and all s. Then, on U_K, the commutation relation follows for all p from $(\theta \wedge \xi) \lrcorner \alpha = \theta \lrcorner (\xi \lrcorner \alpha)$ because, there, $\mathcal{V}^p_{X_K/S_K}$ is generated by $\mathcal{V}^{-1}_{X_K/S_K}$. Finally, on X_K, the relation is inherited from U_K via direct image. Every element of $F^s\mathcal{K}^i_{X_K/S_K}$ is locally a sum of elements of the form $\epsilon(\mu) \wedge \alpha$ for $\mu \in \Omega^s_{S_K/0}$ and $\alpha \in \mathcal{K}^{i-s}_{X_K/S_K}$. Then direct computations show that $F^s\mathcal{K}^{\bullet}_{X_K/S_K}$ is closed under ∂, $\lrcorner$, and $\mathcal{L}$. We have already seen above that the remaining statements hold. $\qquad\square$

---

[7] Note that, in [77, Cor. 7.12], we have $E_K = E + (Q \setminus K)$, and that $e \in H$ if and only if $e + q \in H$ for $e \in E$ and $q \in Q \setminus K$.

15.2   The Characteristic Absolute Gerstenhaber Calculus

In this section, we study Thom–Whitney resolutions of absolute Gerstenhaber calculi, and we construct an analog of the characteristic sheaf of curved Gerstenhaber calculi $(\mathcal{PV}_A^{\bullet,\bullet}, \mathcal{DR}_A^{\bullet,\bullet})$ in absolute Gerstenhaber calculi.

Let $f\colon X_K \to S_K$ be a generically log smooth deformation of $f_0\colon X_0 \to S_0$ of type $\mathcal{D}$. Since the Thom–Whitney resolution is a $\mathbb{C}$-linear construction, we can apply it to the absolute Gerstenhaber calculus $(\mathcal{V}_{X_K/S_K}^\bullet, \mathcal{K}_{X_K/S_K}^\bullet)$ as well.

Lemma 15.6 *Let* $(\mathcal{V}_{X_K/S_K}^{\bullet,\bullet}, \mathcal{K}_{X_K/S_K}^{\bullet,\bullet})$ *be the Thom–Whitney resolution of the absolute Gerstenhaber calculus* $(\mathcal{V}_{X_K/S_K}^\bullet, \mathcal{K}_{X_K/S_K}^\bullet)$ *with operations according to the sign conventions in Sect. 13.1.*

(1) $(\mathcal{V}_{X_K/S_K}^{\bullet,\bullet}, \mathcal{K}_{X_K/S_K}^{\bullet,\bullet})$ *is a differential bg absolute Gerstenhaber calculus in* $\mathfrak{Coh}(X_K/S_K)$.
(2) $F^s\mathcal{K}_{X_K/S_K}^{\bullet,\bullet}$ *is the Thom–Whitney resolution of* $F^s\mathcal{K}_{X_K/S_K}^\bullet$.
(3) $\mathcal{W}_{X_K/S_K}^{\bullet,\bullet} := F^0\mathcal{K}_{X_K/S_K}^{\bullet,\bullet}/F^1\mathcal{K}_{X_K/S_K}^{\bullet,\bullet}$ *is the Thom–Whitney resolution of* $\mathcal{W}_{X_K/S_K}^\bullet$.

Proof We just give some brief indication. The map $\epsilon\colon f^{-1}\Omega_{S_K/0}^\bullet \to \mathcal{K}_{X_K/S_K}^{\bullet,0}$ is the composition of the original ϵ with the injection $\mathcal{K}_{X_K/S_K}^\bullet \to \mathcal{K}_{X_K/S_K}^{\bullet,0}$. The filtered piece $F^s\mathcal{K}^i$ is the image of the map

$$\bigwedge^s (Q^{\mathrm{gp}} \otimes_{\mathbb{Z}} \mathbb{C}) \otimes_{\mathbb{C}} \mathcal{K}^{i-s} \to \mathcal{K}^i, \quad \mu \otimes \alpha \mapsto \epsilon(\mu) \wedge \alpha.$$

On the one hand, we can consider this as an O_{X_K}-linear map between two (plain, non-graded) coherent sheaves; then we can apply the Thom–Whitney resolution, which is an exact functor for each j, and obtain that the Thom–Whitney resolution $\mathrm{TW}^j(F^s\mathcal{K}^i) := C_{\mathrm{TW}}^{j,0}(F^s\mathcal{K}^i(\mathcal{U}))$ is the image of the map

$$\bigwedge^s (Q^{\mathrm{gp}} \otimes_{\mathbb{Z}} \mathbb{C}) \otimes_{\mathbb{C}} \mathrm{TW}^j(\mathcal{K}^{i-s}) \to \mathrm{TW}^j(\mathcal{K}^i),$$

$$\mu \otimes (a_n \otimes \alpha_n)_n \mapsto (a_n \otimes (\epsilon(\mu) \wedge \alpha_n))_n.$$

On the other hand, $F^s\mathcal{K}^{i,j}$ is, by definition, the image of the map

$$\bigwedge^s (Q^{\mathrm{gp}} \otimes_{\mathbb{Z}} \mathbb{C}) \otimes_{\mathbb{C}} \mathcal{K}^{i-s,j} \to \mathcal{K}^{i,j},$$

$$\mu \otimes (a_n \otimes \alpha_n)_n$$

$$\mapsto (1 \otimes \epsilon(\mu))_n \wedge (a_n \otimes \alpha_n)_n = (-1)^{js}((1 \wedge a_n) \otimes (\epsilon(\mu) \wedge \alpha_n))_n,$$

where the $\wedge$-product on the right has to be formed according to the conventions for bilinear maps, as they hold for the construction of the $\wedge$-product on $\mathcal{K}^{\bullet,\bullet}_{X_K/S_K}$. Although the signs differ, the two maps have the same image, so $F^s\mathcal{K}^{i,j}_{X_K/S_K}$ is the (j-th piece of the) Thom–Whitney resolution of $F^s\mathcal{K}^i_{X_K/S_K}$, embedded via $\mathrm{TW}^j(-)$ applied to the embedding $F^s\mathcal{K}^i \to \mathcal{K}^i$ sitting in degree i. Then, by exactness, $\mathrm{TW}^j(\mathcal{W}^i) \cong F^0\mathcal{K}^{i,j}/F^1\mathcal{K}^{i,j}$. This isomorphism is compatible with all operations (that are compatible with $\mathcal{K}^i \to \mathcal{W}^i$) because $\mathrm{TW}^j(\mathcal{W}^i)$ and $\mathrm{TW}^j(\mathcal{K}^i)$ sit in the same degrees and hence have the same sign conventions. With this preparation, we find that the map $\tau_{s,j}$ in the definition of a bg absolute Gerstenhaber calculus is an isomorphism of complexes because of the analogous condition on $\mathcal{K}^\bullet_{X_K/S_K}$. We define $\ell = 0$. The remaining conditions are more or less straightforward. $\square$

Every infinitesimal automorphism φ of $f : X_K \to S_K$ induces an automorphism $(T^\bullet_\varphi, d^\bullet_\varphi)$ of the absolute Gerstenhaber calculus $(\mathcal{V}^\bullet_{X_K/S_K}, \mathcal{K}^\bullet_{X_K/S_K})$. Since $\mathcal{V}^\bullet_{X_K/S_K}$ is the usual Gerstenhaber algebra of $f : X_K \to S_K$, the induced automorphism there is $T^\bullet_\varphi = \exp_{-\theta}$ for a unique $\theta \in I_K \cdot \mathcal{V}^{-1}_{X_K/S_K}$, where $I_K \subseteq A_K$ is the kernel of $A_K \to \mathbb{C}$. By the second proof of Lemma 10.6, we find that $d^\bullet\varphi = \exp_{-\theta}$ as well, where $\exp_{-\theta}$ is constructed via the absolute Lie derivative on $\mathcal{K}^\bullet_{X_K/S_K}$.[8]

A gauge transform $\exp_{-\theta}$ of $(\mathcal{V}^\bullet_{X_K/S_K}, \mathcal{K}^\bullet_{X_K/S_K})$ induces a gauge transform of the Thom–Whitney resolution $(\mathcal{V}^{\bullet,\bullet}_{X_K/S_K}, \mathcal{K}^{\bullet,\bullet}_{X_K/S_K})$ given by $(1 \otimes (-\theta))_n \in I_K \cdot \mathcal{V}^{-1,0}_{X_K/S_K}$.

In Situation 15.1, we can apply this construction, using the open cover $\mathcal{U}$ for the Thom–Whitney resolution, to the local deformations $V_{\alpha;A_K} \to S_K$. Since this will be sufficient, we restrict to the powers of the maximal ideal $K = (Q \setminus \{0\})^k$ and write A_k, S_k, and $V_{\alpha;k}$. We obtain differential bigraded absolute Gerstenhaber calculi $(\mathcal{V}^{\bullet,\bullet}_{\alpha;k}, \mathcal{K}^{\bullet,\bullet}_{\alpha;k})$ together with comparison isomorphisms

$$\mathrm{TW}(\psi_{\alpha\beta;k})^* : \quad \mathcal{V}^{\bullet,\bullet}_{\beta;k}|_{\alpha\beta} \to \mathcal{V}^{\bullet,\bullet}_{\alpha;k}|_{\alpha\beta}, \quad \mathcal{K}^{\bullet,\bullet}_{\beta;k}|_{\alpha\beta} \to \mathcal{K}^{\bullet,\bullet}_{\alpha;k}|_{\alpha\beta}$$

as well as restriction maps as in Sect. 13.2. The cocycles of $\mathrm{TW}(\psi_{\alpha\beta;k})^*$ are gauge transforms.

We have a characteristic curved Gerstenhaber algebra $\mathcal{P}\mathcal{V}^{\bullet,\bullet}_k$ in the sense of Definition 13.31. By definition, it comes with local isomorphisms

$$\chi^*_{\alpha;k} : \quad \mathcal{V}^{\bullet,\bullet}_{\alpha;k} \to \mathcal{P}\mathcal{V}^{\bullet,\bullet}_k|_\alpha$$

of bigraded Gerstenhaber algebras, compatible with restriction maps. On overlaps, we have that

$$\mathrm{TW}(\psi_{\beta\alpha;k})^* \circ (\chi^*_{\alpha;k})^{-1} \circ \chi^*_{\beta;k} : \quad \mathcal{V}^{\bullet,\bullet}_{\beta;k}|_{\alpha\beta} \to \mathcal{V}^{\bullet,\bullet}_{\beta;k}|_{\alpha\beta}$$

---

[8] The first proof of Lemma 10.6 is not applicable here because $\mathcal{K}^i_{X_K/S_K}$ is even on the strict and smooth locus locally not generated by elements of the form ∂g for $g \in \mathcal{O}_{X_K}$.

is the gauge transform $\exp_{\theta_{\alpha\beta;k}}$ for a—since we have strict faithfulness—unique element $\theta_{\alpha\beta;k} \in I_K \cdot V_{\beta;k}^{-1,0}|_{\alpha\beta}$. The predifferential $\bar{\partial}_k$ on $\mathcal{PV}_k^{\bullet,\bullet}$ satisfies

$$(\chi_{\alpha;k}^*)^{-1} \circ \bar{\partial}_k|_\alpha \circ \chi_{\alpha;k}^* = \bar{\partial}_{\alpha;k} + \nabla_{\phi_{\alpha;k}}(-)$$

for unique elements $\phi_{\alpha;k} \in I_K \cdot V_{\alpha;k}^{-1,1}$. All these data are compatible with restriction maps because they are constructed via an order-by-order lift along $A_{k+1} \to A_k$.

We use these data to construct the characteristic curved sheaf of absolute Gerstenhaber calculi.

Definition 15.7 In the above situation, the sheaf $\mathcal{DA}_k^{i,j}$ is given by

$$\Gamma(W, \mathcal{DA}_k^{i,j})$$

$$= \{s_\alpha \in \Gamma(W \cap V_\alpha, \mathcal{K}_{\alpha;k}^{i,j}) \mid \mathrm{TW}(\psi_{\alpha\beta;k})^*(s_\alpha|_{\alpha\beta}) = \exp_{\theta_{\alpha\beta;k}}(s_\beta|_{\alpha\beta})\},$$

i.e., by the gluing of the pieces $\mathcal{K}_{\alpha;k}^{i,j}$, identified along $\mathrm{TW}(\psi_{\alpha\beta;k})^* \circ \exp_{\theta_{\alpha\beta;k}}^{-1}$.

The gluings satisfy the cocycle condition because, after transporting the gauge transforms along the maps $\mathrm{TW}(\psi_{\alpha\beta;k})^*$, which commute with gauge transforms, the questions comes down to comparing the $\theta_{\alpha\beta;k}$ with the elements $o_{\alpha\beta\gamma;k}$ which control the cocycles of $\mathrm{TW}(\psi_{\alpha\beta;k})^*$. Now the cocycle condition is satisfied for $\mathcal{PV}_k^{\bullet,\bullet}$, so it must be satisfied for $\mathcal{DA}_k^{i,j}$ as well. We also have induced restriction maps $\mathcal{DA}_{k+1}^{i,j} \to \mathcal{DA}_k^{i,j}$.

We endow $\mathcal{DA}_k^{\bullet,\bullet}$ with the (bigraded) operations $\wedge$, $\lrcorner$, ∂, and $\mathcal{L}$. They are defined locally and commute with both $\mathrm{TW}(\psi_{\alpha\beta;k})^*$ and gauge transforms, so they are well-defined on $\mathcal{DA}_k^{\bullet,\bullet}$. We also endow $\mathcal{DA}_k^{\bullet,\bullet}$ with a predifferential $\bar{\partial}_k$ of bidegree $(0,1)$, which is given on V_α by $\bar{\partial}_{\alpha;k} + \nabla_{\phi_{\alpha;k}}(-)$. This is well-defined because it is so on the Gerstenhaber algebra $\mathcal{PV}_k^{\bullet,\bullet}$. Namely, $\mathrm{TW}(\psi_{\alpha\beta;k})^*$ commutes with all predifferentials, and for the gauge transform part of the comparison, the equations for the transform of $\phi_{\alpha;k}$ into $\phi_{\beta;k}$ (in Lemma 5.14) are the same on $\mathcal{DA}_k^{\bullet,\bullet}$ and $\mathcal{PV}_k^{\bullet,\bullet}$. Moreover, all operations are compatible with restriction maps. This shows:

Lemma 15.8 *The pair* $(\mathcal{PV}_k^{\bullet,\bullet}, \mathcal{DA}_k^{\bullet,\bullet})$ *forms a sheaf of curved absolute Gerstenhaber calculi.*

Proof As discussed above, we have globally defined sheaves, operations, and constants. For $\mu \in \Omega_{S_K/\mathbf{0}}^s$, the commutation relation (15.3) shows that $\theta \lrcorner \epsilon(\mu) = 0$ on $\mathcal{K}_{\alpha;k}^{s,1}$ for $\theta \in V_{\alpha;k}^{-1,1}$. Thus, we have $\mathcal{L}_\theta(\epsilon(\mu)) = 0$ by the Lie–Rinehart homotopy formula, and this implies $\epsilon(\mu) \in \mathcal{DA}_k^{s,0}$, giving the map ϵ. The constant ℓ already exists globally as part of $\mathcal{PV}_k^{\bullet,\bullet}$. Now we have all the data required by Definition 15.4.

Each $(\mathcal{V}_{\alpha;k}^{\bullet,\bullet}, \mathcal{K}_{\alpha;k}^{\bullet,\bullet})$ is a curved absolute Gerstenhaber calculus when endowed with the predifferential $\bar{\partial}_{\alpha;k} + \nabla_{\phi_{\alpha;k}}(-)$ and the constant

$$\ell_{\alpha;k} := \bar{\partial}_{\alpha;k}(\phi_{\alpha;k}) + \frac{1}{2}[\phi_{\alpha;k}, \phi_{\alpha;k}].$$

The condition $\bar{\partial}\epsilon(\mu) = 0$ is preserved under changing the predifferential by $\nabla_{\phi}(-)$ because of the commutation relation (15.3). Since $(\mathcal{PV}_k^{\bullet,\bullet}, \mathcal{DA}_k^{\bullet,\bullet})$ is locally isomorphic to this curved absolute Gerstenhaber calculus, and since all properties required in Definition 15.4 are local, it is a curved absolute Gerstenhaber calculus as well. $\qquad\square$

15.3 The Proof of Surjectivity

In this section, we show that the map in (15.1) is surjective for $A = A_k$ when we are in Situation 15.1. For $Q = 0$, nothing is to show. For $Q = \mathbb{N}$, we employ an old trick which Steenbrink ascribes in [261] to Katz, referring to [256]. It also has been used in [125, Thm. 4.1], and subsequently in [38, Ass. 4.15] and in [77, Thm. 1.10], where we prove the surjectivity in the case of an actual log toroidal deformation. Finally, the case of a general Q can be reduced to $Q = \mathbb{N}$.

So let us first assume that $Q = \mathbb{N}$. By Lemma 13.12, the sheaves $\mathcal{DR}_0^{i,j}$ are Γ-acyclic since $\mathcal{U}$ is admissible. Then each $\mathcal{DR}_k^{i,j}$ must be Γ-acyclic as well because of the exact sequence

$$0 \to I \otimes_{\mathbb{C}} \mathcal{DR}_0^{i,j} \to \mathcal{DR}_{k+1}^{i,j} \to \mathcal{DR}_k^{i,j} \to 0. \tag{15.4}$$

Since $DR_{X_0/A_k}^{i,j} = \Gamma(X_0, \mathcal{DR}_k^{i,j})$, we have $H^n(DR_{X_0/A_k}^\bullet, d) = \mathbb{H}^n(X_0, (\mathcal{DR}_k^\bullet, d))$, so it is sufficient to prove the surjectivity in (actual) hypercohomology. Next, we form a complex

$$\mathcal{DA}_k^n[u] := \bigoplus_{s=0}^{\infty} \mathcal{DA}_k^n \cdot u^s$$

with differential $e_d(\alpha \cdot u^s) := d(\alpha) \cdot u^s + s\bar{\rho} \wedge \alpha \cdot u^{s-1}$. Here, $\bar{\rho} \in \Gamma(X_0, \mathcal{DA}_k^1)$ is the following element: Since $Q = \mathbb{N}$, we have an element 1 in the sheaf of monoids $\mathcal{M}_{S_k}$ on S_k. Applying the universal log derivation, we obtain

$$[1] \in \Omega_{S_k/\mathbf{0}}^1 = \mathbb{C}[\mathbb{N}] \otimes_{\mathbb{C}} (\mathbb{N}^{\mathrm{gp}} \otimes_{\mathbb{Z}} \mathbb{C}),$$

and then $\bar{\rho} := \epsilon([1])$. The operator d in the formula is $d = \partial + \bar{\partial} + \ell \lrcorner (-)$. Then we consider the composition

$$\pi: \quad (\mathcal{DA}_k^\bullet[u], e_d) \to (\mathcal{DA}_k^\bullet, d) \to (\mathcal{DR}_k^\bullet, d) \to (\mathcal{DR}_0^\bullet, d).$$

Each of them is compatible with the differential, and each of them is surjective (the first map is the projection onto the u^0-summand). Thus, we can define a complex $(\mathcal{R}^\bullet, e_d)$ by the exact sequence

$$0 \to (\mathcal{R}^\bullet, e_d) \to (\mathcal{DA}_k^\bullet[u], e_d) \to (\mathcal{DR}_0^\bullet, d) \to 0.$$

It is then sufficient to prove $\mathbb{H}^n(X_0, \mathcal{R}^\bullet) = 0$ to conclude the surjectivity in (15.1). This is the main goal of the remainder of this section. We have to overcome three main difficulties: First, the main argument works only for complex analytic spaces, not schemes, secondly, we have no global deformation X_k for which $\mathcal{DA}_k^\bullet[u]$ is quasi-coherent, so analytification needs special care, and thirdly, we have to analytify the differential operator e_d, so we cannot just take a pull-back along $\varphi: X^{\mathrm{an}} \to X$, even locally.

15.3.1   The Local Model

Let $(\mathbb{N} \subset P, \mathcal{F})$ be an elt datum; it gives rise to a log morphism $f: A_{P,\mathcal{F}} \to A_{\mathbb{N}}$, which serves as an étale local model for a generically log smooth deformation of type $\mathscr{D}$ in Situation 15.1. Let $k \geq 0$, and let $f: L_k \to S_k$ be the base change of the local model to $S_k = \mathrm{Spec}(\mathbb{N} \to \mathbb{C}[t]/(t^{k+1}))$. Let $K = (\mathbb{N}^+)^{k+1}$ be the corresponding monoid ideal.

Let $(\mathcal{V}_k^\bullet, \mathcal{K}_k^\bullet)$ be the absolute Gerstenhaber calculus of $f: L_k \to S_k$, with relative de Rham complex $\mathcal{W}_k^\bullet$. Then we define a complex $(\mathcal{K}_k^\bullet[u], e)$ with

$$\mathcal{K}_k^n[u] := \bigoplus_{s=0}^{\infty} \mathcal{K}_k^n \cdot u^s \quad \text{and} \quad e(\alpha \cdot u^s) := \partial(\alpha) \cdot u^s + s\bar{\rho} \wedge \alpha \cdot u^{s-1}$$

similar to the above one with $\bar{\rho} = \epsilon([1])$. We define $(\mathcal{R}_k^\bullet, e)$ as the kernel in the sequence

$$0 \to (\mathcal{R}_k^\bullet, e) \to (\mathcal{K}_k^\bullet[u], e) \to (\mathcal{W}_0^\bullet, \partial) \to 0. \tag{15.5}$$

Let $\tilde{f}: \tilde{L}_k \to \tilde{S}_k$ be the analytification of $f: L_k \to S_k$, where we denote the analytification with $\tilde{}$ in order to not load our notation even more with the heavy $(-)^{\mathrm{an}}$. The pieces of the complexes in (15.5) are (quasi-)coherent sheaves, and the maps between them are $\mathcal{O}_{L_k}$-linear. Thus, we can analytify them. The de Rham differential ∂ on $\mathcal{W}_0^\bullet$ is a first-order differential operator, so we can analytify it as well in a unique way by Proposition E.7 and obtain $\tilde{\partial}$. But e is a first-order differential operator as well, so we find a unique analytification $\tilde{e}$. The uniqueness properties in Proposition E.7 show that

$$0 \to (\tilde{\mathcal{R}}_k^{\bullet}, \tilde{e}) \to (\tilde{\mathcal{K}}_k^{\bullet}[u], \tilde{e}) \to (\tilde{\mathcal{W}}_0^{\bullet}, \tilde{\partial}) \to 0$$

is a short exact sequence of complexes whose pieces are quasi-coherent analytic sheaves, and whose differentials are analytic differential operators of first order.

Proposition 15.9 *The complex* $(\tilde{\mathcal{R}}_k^{\bullet}, \tilde{e})$ *has no cohomology at the stalk at* $0 \in \tilde{L}_k$.

Proof In principle, this comes down to a calculation which is already in [125, Thm. 4.1]. There, the computation is done on the global sections on L_k of the algebraic counterparts, and it is wrongly concluded that this would yield Γ-acyclicity of the global algebraic complex. In [77, Lemma 12.1], we have remedied this situation by working on the stalks of the analytification, which needs some additional consideration of convergence questions. However, strictly speaking, the proof of [77, Lemma 12.1] is still incomplete because we made an implicit assumption about the form of $\tilde{\partial}$ and $\tilde{e}$ on the analytic stalks, which we did not prove back then. So here comes what is still missing.

The remaining gap can be closed by using the sequence topology on analytic stalks, which we review in Appendix E.1. In the discussion after [77, Lemma 7.13], we have shown that the local analytic ring at $0 \in \tilde{L}_k$ is

$$\mathcal{O}_{\tilde{L}_k,0} = \left\{ \sum_{e \in E_K} \alpha_e z^e \in \mathbb{C}[\![E_K]\!] \;\middle|\; \sup_{e \in E_K \setminus 0} \left\{ \frac{\log|\alpha_e|}{h(e)} \right\} < \infty \right\}$$

with the notations from there, i.e., $E_K = P \setminus (P + K)$ and $h \colon P \to \mathbb{N}$ is a local homomorphism of sharp toric monoids. It is now quite easy to show that any sequence of finite truncations of $\alpha = \sum_{e \in E_K} \alpha_e z^e$ which exhausts E_K actually converges to α in the sequence topology. Namely, this can be shown directly for the stalk of $(\operatorname{Spec} \mathbb{C}[\mathbb{N}^r])^{\mathrm{an}}$ when described as in [77, Lemma 7.13]; then this property descends along a local surjection $\phi \colon \mathbb{N}^r \to P$ to our sharp toric monoid P and the stalk of $\tilde{A}_P$; and finally, this property descends along the closed embedding $\tilde{L}_k \subset \tilde{A}_P$ of complex analytic spaces. Furthermore, for a stalk of a coherent analytic sheaf as in [77, Lemma 7.14], the same is true. Namely, each such stalk is, in the sequence topology, a closed subset of $\mathcal{O}_{\tilde{L}_k,0}^{\oplus r}$ carrying the product topology. Then

it is also true for the quasi-coherent analytic sheaf $\tilde{\mathcal{K}}_k^n[u]$. By Proposition E.7 and Lemma E.6, the analytifications $\tilde{\partial}$ and $\tilde{e}$ are continuous for the sequence topologies; thus, we can evaluate them on the stalks by evaluating them on finite truncations, and then taking the limit. However, the finite truncations come from sections of the algebraic quasi-coherent sheaves on L_k; thus, the analytic differential operators $\tilde{\partial}$ and $\tilde{e}$ actually have the form used in our proof of [77, Lemma 12.1].   □

15.3.2   The Transfer to $f : V_{\alpha;k} \to S_k$

Consider $f : V_{\alpha;k} \to S_k$, and let $(\mathcal{V}^\bullet_{\alpha;k}, \mathcal{K}^\bullet_{\alpha;k})$ be the absolute Gerstenhaber calculus. As for the local model, we define a complex $(\mathcal{K}^\bullet_{\alpha;k}[u], e)$ with

$$\mathcal{K}^n_{\alpha;k}[u] := \bigoplus_{s=0}^{\infty} \mathcal{K}^n_{\alpha;k} \cdot u^s \quad \text{and} \quad e(\alpha \cdot u^s) := \partial(\alpha) \cdot u^s + s\bar{\rho} \wedge \alpha \cdot u^{s-1}.$$

We define $(\mathcal{R}^\bullet_{\alpha;k}, e)$ as the kernel of $(\mathcal{K}^\bullet_{\alpha;k}[u], e) \to (\mathcal{W}^\bullet_{\alpha;0}, \partial)$. Let $\tilde{f} : \tilde{V}_{\alpha;k} \to \tilde{S}_k$ be the analytification of $f : V_{\alpha;k} \to S_k$. As above, we obtain an analytification

$$0 \to (\tilde{\mathcal{R}}^\bullet_{\alpha;k}, \tilde{e}) \to (\tilde{\mathcal{K}}^\bullet_{\alpha;k}[u], \tilde{e}) \to (\tilde{\mathcal{W}}^\bullet_{\alpha;0}, \tilde{\partial}) \to 0. \tag{15.6}$$

By assumption, $f : V_{\alpha;k} \to S_k$ is, locally in the étale topology, isomorphic to $f : L_k \to S_k$ for some elementary log toroidal datum $(\mathbb{N} \subset P, \mathcal{F})$, which, in general, depends on the point $v \in V_{\alpha;k}$. By Felten [75, Cor. 4.16], if $v \in V_{\alpha;k}(\mathbb{C})$ is a $\mathbb{C}$-valued point, we can arrange the étale roof such that $v = 0$ in $L_k(\mathbb{C})$.

If $f : X_k \to S_k$ is a generically log smooth family, and $g : Y_k \to X_k$ is a strict étale morphism, then we have

$$g^* \mathcal{W}^i_{X_k/S_k} \cong \mathcal{W}^i_{Y_k/S_k}, \quad g^* \mathcal{V}^p_{X_k/S_k} \cong \mathcal{V}^p_{Y_k/S_k}, \quad g^* \mathcal{K}^i_{X_k/S_k} \cong \mathcal{K}^i_{Y_k/S_k}$$

canonically. These maps are compatible with g^{-1} applied to all operations in the (absolute or relative) Gerstenhaber calculus. Using this together with Proposition E.7[9] and the fact that $\tilde{g} : \tilde{Y}_k \to \tilde{X}_k$ is a local isomorphism in the Euclidean topology, we find that the stalk of the exact sequence (15.6) for $Y_k \to S_k$ at $y \in \tilde{Y}_k$ is the same as the stalk of the exact sequence (15.6) for $X_k \to S_k$ at $\tilde{g}(y) \in \tilde{X}_k$. Applying this along the étale roof of the local model at $v \in \tilde{V}_{\alpha;k}$, we find that $(\tilde{\mathcal{R}}^\bullet_{\alpha;k}, \tilde{e})$ is an acyclic complex.[10]

Applying the Thom–Whitney resolution (with respect to $\mathcal{U}$) to the original algebraic version of (15.6) yields a short exact sequence

$$0 \to (\mathrm{TW}^\bullet(\mathcal{R}^\bullet_{\alpha;k}), e, \bar{\partial}) \to (\mathrm{TW}^\bullet(\mathcal{K}^\bullet_{\alpha;k}[u]), e, \bar{\partial}) \to (\mathrm{TW}^\bullet(\mathcal{W}^\bullet_{\alpha;0}), \partial, \bar{\partial}) \to 0 \tag{15.7}$$

---

[9] The point is that the two analytifications of ∂ and e either on $\tilde{X}_k$ or on $\tilde{Y}_k$ agree via $\tilde{g}$.

[10] Note that it does not matter that we may change the log smooth open subset of the generically log smooth family in comparing $V_{\alpha;k}$ with its local models, as is allowed in [77, Defn. 4.1]. Namely, the absolute and relative Gerstenhaber calculi are independent of the choice of open subset of log smoothness.

of sheaves of double complexes, where we apply our sign conventions of Chap. 13.1 in the formation of ∂ and e on the resolution. We can analytify this short exact sequence as well because it consists of quasi-coherent sheaves with first-order differential operators between them.[11] Each horizontal row

$$\mathcal{R}^i_{\alpha;k} \to \mathrm{TW}^\bullet(\mathcal{R}^i_{\alpha;k}, \bar{\partial})$$

is a resolution in the category of quasi-coherent sheaves. Thus, also the analytification

$$\tilde{\mathcal{R}}^i_{\alpha;k} \to \widetilde{\mathrm{TW}}^\bullet(\mathcal{R}^i_{\alpha;k}, \tilde{\bar{\partial}})$$

is a resolution in the category of quasi-coherent analytic sheaves. In the vertical direction, we have analytic first-order differential operators $\tilde{e}$ such that $\tilde{\bar{\partial}}\tilde{e} + \tilde{e}\tilde{\bar{\partial}} = 0$. In particular, we have a quasi-isomorphism

$$(\tilde{\mathcal{R}}^\bullet_{\alpha;k}, \tilde{e}) \to (\mathrm{Tot}^\bullet\widetilde{\mathrm{TW}}^\bullet(\mathcal{R}^\bullet_{\alpha;k}), \tilde{e} + \tilde{\bar{\partial}}) := \widetilde{\mathrm{Tot}}^\bullet(\mathcal{R}^\bullet_{\alpha;k})$$

of complexes of sheaves so that the latter complex is acyclic.

When $\phi \in \mathfrak{m}_k \cdot \mathcal{V}^{-1,1}_{\alpha;k}$, then we can modify the (pre-)differential $\bar{\partial}$ to obtain a predifferential $\bar{\partial}_\phi := \bar{\partial} + \mathcal{L}_\phi(-)$. In the short exact sequence (15.7), only the horizontal differentials change from $\bar{\partial}$ to $\bar{\partial}_\phi$. Setting $E := e^\phi \in \mathrm{Tot}^0(\mathrm{TW}^\bullet(\mathcal{V}^\bullet_{\alpha;k}))$, we obtain a diagram of exact sequences of total complexes

$$
\begin{array}{ccccc}
(\mathrm{Tot}^\bullet\mathrm{TW}^\bullet(\mathcal{R}^\bullet_{\alpha;k}), e + \bar{\partial}) & \hookrightarrow & (\mathrm{Tot}^\bullet\mathrm{TW}^\bullet(\mathcal{K}^\bullet_{\alpha;k}[u]), e + \bar{\partial}) & \twoheadrightarrow & (\mathrm{Tot}^\bullet\mathrm{TW}^\bullet(\mathcal{W}^\bullet_{\alpha;0}), \partial + \bar{\partial}) \\[4pt]
\Phi_E \uparrow \ \cong & & \Phi_E \uparrow \ \cong & & \Big\| \\[4pt]
\begin{array}{c}(\mathrm{Tot}^\bullet\mathrm{TW}^\bullet(\mathcal{R}^\bullet_{\alpha;k}),\\ e + \bar{\partial}_\phi + \ell_\phi \lrcorner (-))\end{array} & \hookrightarrow & \begin{array}{c}(\mathrm{Tot}^\bullet\mathrm{TW}^\bullet(\mathcal{K}^\bullet_{\alpha;k}[u]),\\ e + \bar{\partial}_\phi + \ell_\phi \lrcorner (-))\end{array} & \twoheadrightarrow & (\mathrm{Tot}^\bullet\mathrm{TW}^\bullet(\mathcal{W}^\bullet_{\alpha;0}), \partial + \bar{\partial})
\end{array}
$$

similar to Lemma 5.28. Thus, the analytification of the complex on the lower left side is acyclic. Furthermore, the restriction of our complex $(\mathcal{R}^\bullet, e_d) \subseteq (\mathcal{D}\mathcal{A}^\bullet_k[u], e_d)$ above to $V_{\alpha;k}$ is isomorphic to the lower left complex once we take $\phi = \phi_{\alpha;k}$.

---

[11] The Thom–Whitney construction does not commute with analytification, i.e., when we apply the analytic analog of the Thom–Whitney construction to $\mathcal{F}^{\mathrm{an}}$, this is in general not the same as the analytification of $\mathrm{TW}^\bullet(\mathcal{F})$. The reason for this is that direct images from open subsets do not commute with analytification. Consider for example the structure sheaf on $\mathbb{A}^1 \subseteq \mathbb{P}^1$.

15.3.3   From Local to Global Analytification

We wish to find a global analytification $(\tilde{\mathcal{R}}^{\bullet}, \tilde{e}_d)$ of $(\mathcal{R}^{\bullet}, e_d)$ which is locally the one coming from $\tilde{V}_{\alpha;k} \to V_{\alpha;k}$. Then $(\tilde{\mathcal{R}}^{\bullet}, \tilde{e}_d)$ must be Γ-acyclic because it is acyclic, and we wish to conclude from this that $(\mathcal{R}^{\bullet}, e_d)$ is Γ-acyclic as well. The first part is quite easy. On underlying spaces, we have a global analytification map $\varphi \colon |\tilde{X}_0| \to |X_0|$. Then, on each V_α, we can analytify separately and obtain a map

$$\varphi^{-1}(\mathcal{R}^{\bullet}, e_d)|_\alpha \to (\tilde{\mathcal{R}}_\alpha^{\bullet}, \tilde{e}_d),$$

where the right-hand side is the analytification of $(\mathcal{R}^{\bullet}, e_d)|_\alpha$ along $\tilde{V}_{\alpha;k} \to V_{\alpha;k}$ by definition. On overlaps $V_\alpha \cap V_\beta$, we obtain comparison isomorphisms which are compatible with the differentials. Since these comparison isomorphisms come from the tensor product representation, they fit into a commutative diagram

$$
\begin{array}{ccc}
\varphi^{-1}(\mathcal{R}^{\bullet}, e_d)|_{\alpha\beta} & \longrightarrow & \varphi^{-1}(\mathcal{R}^{\bullet}, e_d)|_{\alpha\beta} \otimes_{\varphi^{-1}O_{\alpha;k}|_{\alpha\beta}} \varphi^{-1}\tilde{O}_{\alpha;k}|_{\alpha\beta} \\
\Big\| & & \Big\downarrow \cong \\
\varphi^{-1}(\mathcal{R}^{\bullet}, e_d)|_{\alpha\beta} & \longrightarrow & \varphi^{-1}(\mathcal{R}^{\bullet}, e_d)|_{\alpha\beta} \otimes_{\varphi^{-1}O_{\beta;k}|_{\alpha\beta}} \varphi^{-1}\tilde{O}_{\beta;k}|_{\alpha\beta}
\end{array}
$$

with the identity on the left. Thus, also the cocycles of the comparison isomorphism fit into such a diagram with the identity on the left. However, on the analytic stalks, the cocycles are locally bounded and hence continuous for the sequence topology; since the image of $\varphi^{-1}\mathcal{R}^{\bullet}|_\alpha$ is dense in $\tilde{\mathcal{R}}_\alpha^{\bullet}$, this implies that the cocycles are the identity because finitely generated subsheaves of $\mathcal{R}_\alpha^{\bullet}$ are coherent, and the sequence topology on stalks of coherent sheaves is Hausdorff, cf. the proof of Lemma E.8.[12] Thus, we have a global analytification $(\tilde{\mathcal{R}}^{\bullet}, \tilde{e}_d)$ of $(\mathcal{R}^{\bullet}, e_d)$, and $(\tilde{\mathcal{R}}^{\bullet}, \tilde{e}_d)$ is acyclic, hence also Γ-acyclic.

15.3.4   Comparison of Cohomologies

If we have a global scheme structure, i.e., a gluing of the pieces $V_{\alpha;k}$, then we can conclude that $(\mathcal{R}^{\bullet}, e_d)$ is acyclic by comparing the two spectral sequences of the stupid filtrations and applying the classical GAGA comparison of sheaf cohomologies of proper schemes X, which holds not only for coherent but also for quasi-coherent sheaves $\mathcal{F}$ by Lemma E.3. Since we do not have a global scheme

---

[12] Note that the cocycles are not $\tilde{O}_{k;\alpha}|_{\alpha\beta\gamma}$-linear because the cocycles are, in general, not the identity on the structure sheaf. Nonetheless, they are locally bounded operators in the sense of Appendix E.1, and they are continuous on finitely generated subsheaves.

structure a priori, more care is needed. Nonetheless, we have a map of spectral sequences[13]

$$(E_1^{p,q} := H^q(X_0, \mathcal{R}^p) \Rightarrow \mathbb{H}^{p+q}(X_0, \mathcal{R}^\bullet))$$

$$\to (\tilde{E}_r^{p,q} := H^q(\tilde{X}_0, \tilde{\mathcal{R}}^p) \Rightarrow \mathbb{H}^{p+q}(\tilde{X}_0, \tilde{\mathcal{R}}^\bullet)),$$

and we know already that the abutment on the right-hand side is 0.

First, by Lemma 13.32, we know that $\mathcal{DR}_k^{i,j}$ is Γ-acyclic. Now $\mathcal{DA}_k^{i,j}$ has a decreasing filtration $F^s \mathcal{DA}_k^{i,j}$ with subquotients isomorphic to $f^{-1}\Omega_{S_k/0}^s \otimes_{A_k} \mathcal{DR}_k^{i,j}$. Thus, $\mathcal{DA}_k^{i,j}$ is Γ-acyclic as well. Since X_0 is Noetherian, direct limits commute with cohomology; thus, $\mathcal{DA}_k^{i,j}[u]$ is Γ-acyclic, and then $\mathcal{R}^n$ must be Γ-acyclic as well. For $\ell \geq 2$, we have directly $H^\ell(X_0, \mathcal{R}^n) = 0$, and for $\ell = 1$, this follows from the surjectivity of

$$H^0(X_0, \mathcal{DA}_k^{i,j}[u]) \to H^0(X_0, \mathcal{DR}_0^{i,j}).$$

Analogously to $\tilde{\mathcal{R}}^\bullet$, we can also analytify $\mathcal{DR}_k^{\bullet,\bullet}$ and $\mathcal{DA}_k^{\bullet,\bullet}$ globally, yielding $\widetilde{\mathcal{DR}}_k^{\bullet,\bullet}$ and $\widetilde{\mathcal{DA}}_k^{\bullet,\bullet}$. The sheaf $\widetilde{\mathcal{DR}}_0^{i,j}$ is the usual analytification of the Γ-acyclic quasi-coherent sheaf $\mathcal{DR}_0^{i,j}$ on X_0, so it is Γ-acyclic by Lemma E.3. Similar to the algebraic case, this implies that first $\widetilde{\mathcal{DR}}_k^{i,j}$ and then also $\widetilde{\mathcal{DA}}_k^{i,j}$ is Γ-acyclic. Since cohomology commutes with filtered colimits on the compact Hausdorff space $\tilde{X}_0$ by Godement [108, Thm. 4.12.1], also $\widetilde{\mathcal{DA}}_k^{i,j}[u]$ is Γ-acyclic. Then also $\tilde{\mathcal{R}}^p$ must be Γ-acyclic because $\widetilde{\mathcal{DA}}_k^p[u] \to \widetilde{\mathcal{DR}}_0^p$ is surjective on the level of global sections. This shows that the maps $E_1^{p,q} \to \tilde{E}_1^{p,q}$ of the above map of spectral sequences is an isomorphism for $q \geq 1$. For the case $q = 0$, first observe that

$$H^0(X_0, \mathcal{DR}_0^{i,j}) \to H^0(\tilde{X}_0, \widetilde{\mathcal{DR}}_0^{i,j})$$

is an isomorphism by Lemma E.3. Using (15.4) and its analytic counterpart, we can show that

$$H^0(X_0, \mathcal{DR}_k^{i,j}) \to H^0(\tilde{X}_0, \widetilde{\mathcal{DR}}_k^{i,j})$$

is an isomorphism by induction on k. Using the filtration, we can show similarly that

$$H^0(X_0, \mathcal{DA}_k^{i,j}) \to H^0(\tilde{X}_0, \widetilde{\mathcal{DA}}_k^{i,j})$$

---

[13] Cf. also [75, Lemma 2.46] for a discussion of the construction.

is an isomorphism, and then $E_1^{p,q} \to \tilde{E}_1^{p,q}$ must be an isomorphism for $q = 0$ as well, where we use again that, on both spaces X_0 and $\tilde{X}_0$, cohomology commutes with filtered colimits, in order to obtain the isomorphism on the level of $\mathcal{DA}_k^p[u]$. This shows $\mathbb{H}^n(X_0, (\mathcal{R}^\bullet, e_d)) = 0$ for all $n \geq 0$, concluding the proof in the case $Q = \mathbb{N}$.

15.3.5   The Case of a General Q

To reduce this case to $Q = \mathbb{N}$, we use an idea of Chan–Leung–Ma, see [38, §4.3.2], which we also used in the case of a globally given log toroidal family in [75].

We have to show the surjectivity of $\mathbb{H}^n(X_0, (\mathcal{DR}_k^\bullet, d)) \to \mathbb{H}^n(X_0, (\mathcal{DR}_0^\bullet, d))$. Let $K = (Q^+)^{k+1}$ be the monoid ideal corresponding to A_k. Then we can find a finite decreasing filtration

$$Q^+ = I_0 \supset I_1 \supset \ldots \supset K$$

as in [75, Lemma 8.11], i.e., each I_n is a monoid ideal, we have

$$\dim_{\mathbb{C}}(\mathbb{C}[I_n]/\mathbb{C}[I_{n+1}]) = 1,$$

and for each n, there is a monoid homomorphism $h_n\colon Q \to \mathbb{N}$ with $I_n = h_n^{-1}((i, \infty))$ and $I_{n+1} = h_n^{-1}((i + 1, \infty))$ for some $i \geq 0$. Then it is sufficient to show the surjectivity for the restriction along $\mathbb{C}[Q]/\mathbb{C}[I_{n+1}] \to \mathbb{C}[Q]/\mathbb{C}[I_n]$. Let us use the following notations:

$I = I_{n+1}$	$A' = \mathbb{C}[Q]/\mathbb{C}[I]$	$A = \mathbb{C}[Q]/\mathbb{C}[J]$
$J = I_n$	$S' = \mathrm{Spec}(Q \to A')$	$S = \mathrm{Spec}(Q \to A)$
$h = h_n$	$B' = \mathbb{C}[t]/(t^{i+1})$	$B = \mathbb{C}[t]/(t^i)$
$T_0 = \mathrm{Spec}(\mathbb{N} \to \mathbb{C})$	$T' = \mathrm{Spec}(\mathbb{N} \to \mathbb{C}[t]/(t^{i+1}))$	$T = \mathrm{Spec}(\mathbb{N} \to \mathbb{C}[t]/(t^i))$

Then h induces a commutative diagram

$$
\begin{array}{ccccc}
T_0 & \longrightarrow & T & \longrightarrow & T' \\
\downarrow{\scriptstyle b_0} & & \downarrow{\scriptstyle b} & & \downarrow{\scriptstyle b'} \\
S_0 & \longrightarrow & S & \longrightarrow & S'
\end{array}
$$

of log schemes. By Felten [75, Prop. 4.7], the base change $g_0\colon Y_0 \to T_0$ of $f_0\colon X_0 \to S_0$ along $T_0 \to S_0$ is a log toroidal family with respect to $T_0 \to A_{\mathbb{N}}$. These two log toroidal families have the same underlying space, and the induced maps $\mathcal{W}^\bullet_{X_0/S_0} \to \mathcal{W}^\bullet_{Y_0/T_0}$ and $\mathcal{V}^\bullet_{X_0/S_0} \to \mathcal{V}^\bullet_{Y_0/T_0}$ (via pull-back of maps $\mathcal{W}^1_{X_0/S_0} \to O_{X_0}$) are isomorphisms.

Let $V_{\alpha;I}$ and $V_{\alpha;J}$ be the local models over S' respectively S of the system of deformations $\mathscr{D}$. They induce a kind of system of deformations on T and T' via pull-back along b and b'. We denote the pull-backs by $V_{\alpha;i-1}$ and $V_{\alpha;i}$. Again by Felten [75, Prop. 4.7], they are log toroidal families. The induced maps $\mathcal{W}^\bullet_{\alpha;J} \otimes_A B \to \mathcal{W}^\bullet_{\alpha;i-1}$ and $\mathcal{W}^\bullet_{\alpha;I} \otimes_{A'} B' \to \mathcal{W}^\bullet_{\alpha;i}$ are isomorphisms by applying Lemma B.5 to the sources. Then we also have induced isomorphisms $\mathcal{V}^\bullet_{\alpha;J} \otimes_A B \to \mathcal{V}^\bullet_{\alpha;i-1}$ and $\mathcal{V}^\bullet_{\alpha;I} \otimes_{A'} B' \to \mathcal{V}^\bullet_{\alpha;i}$. Let $(\mathcal{PV}^{\bullet;\bullet}_I, \mathcal{DR}^{\bullet;\bullet}_I)$ be a choice of a characteristic sheaf of curved Gerstenhaber calculi over S'. Then

$$(\mathcal{PV}^{\bullet;\bullet}_i, \mathcal{DR}^{\bullet;\bullet}_i) := (\mathcal{PV}^{\bullet;\bullet}_I \otimes_{A'} B', \mathcal{DR}^{\bullet;\bullet}_I \otimes_{A'} B')$$

is a sheaf of curved Gerstenhaber calculi as well, and we define $(\mathcal{PV}^{\bullet;\bullet}_{i-1}, \mathcal{DR}^{\bullet;\bullet}_{i-1})$ similarly as the base change along $b \colon T \to S$. When denoting the Thom–Whitney resolution of $(\mathcal{V}^\bullet_{\alpha;i}, \mathcal{W}^\bullet_{\alpha;i})$ by $(\mathcal{V}^{\bullet;\bullet}_{\alpha;i}, \mathcal{W}^{\bullet;\bullet}_{\alpha;i})$ as usual, we have isomorphisms

$$(\mathcal{V}^{\bullet;\bullet}_{\alpha;i}, \mathcal{W}^{\bullet;\bullet}_{\alpha;i}) \cong (\mathcal{V}^{\bullet;\bullet}_{\alpha;I}, \mathcal{W}^{\bullet;\bullet}_{\alpha;I}) \otimes_{A'} B' \xrightarrow{\chi^*_{\alpha;I} \otimes_{A'} B'} (\mathcal{PV}^{\bullet;\bullet}_i, \mathcal{DR}^{\bullet;\bullet}_i)|_\alpha$$

of *bigraded* Gerstenhaber calculi since the Thom–Whitney resolution commutes with affine base change. When ignoring the induced predifferential $\bar{\partial}_i$, the above $(\mathcal{PV}^{\bullet;\bullet}_i, \mathcal{DR}^{\bullet;\bullet}_i)$ becomes a bigraded Thom–Whitney deformation in the sense of Definition 13.18. However, $\bar{\partial}_i$ is also a predifferential on it in the sense of Definition 13.26. The same is true for $(\mathcal{PV}^{\bullet;\bullet}_{i-1}, \mathcal{DR}^{\bullet;\bullet}_{i-1})$, which is a bigraded Thom–Whitney deformation with predifferential $\bar{\partial}_{i-1}$. Furthermore, we have a restriction map between them which is compatible with all data.

The above discussion for $Q = \mathbb{N}$ applies to $(\mathcal{PV}^{\bullet;\bullet}_i, \mathcal{DR}^{\bullet;\bullet}_i)$. This shows that the map

$$\mathbb{H}^n(X_0, (\mathcal{DR}^\bullet_i, d)) \to \mathbb{H}^n(X_0, (\mathcal{DR}^\bullet_0, d))$$

is surjective, where $d = \partial + \bar{\partial} + \ell \lrcorner (-)$, and where $\mathcal{DR}^\bullet_0$ is the total complex of the Thom–Whitney resolution of either $\mathcal{W}^\bullet_{X_0/S_0}$ or $\mathcal{W}^\bullet_{Y_0/T_0}$, which coincide. Then also

$$\mathbb{H}^n(X_0, (\mathcal{DR}^\bullet_i, d)) \to \mathbb{H}^n(X_0, (\mathcal{DR}^\bullet_{i-1}, d))$$

must be surjective by Nakayama's lemma since both are free B'- respectively B-modules with the same base change to $\mathbb{C}$. After this preparation, a diagram chase as in the proof of [75, Lemma 8.9] completes the proof of surjectivity along the base change $\mathbb{C}[Q]/\mathbb{C}[I] \to \mathbb{C}[Q]/\mathbb{C}[J]$, and thus the proof of Theorem 15.2.

Chapter 16
Deformations of Line Bundles

We explore the relation between deformations of a generically log smooth family and deformations of the generically log smooth family together with a line bundle. The main result which we prove in this chapter is that in the log toroidal log Calabi–Yau case, these deformations are unobstructed. This gives the version of the logarithmic Bogomolov–Tian–Todorov theorem for pairs consisting of a log Calabi–Yau log toroidal family and a line bundle. Suppose we are in the following situation:

Situation 16.1 We have a sharp toric monoid Q, and we set $\Lambda = \mathbf{k}[\![Q]\!]$. We have a torsionless enhanced generically log smooth family $f_0\colon X_0 \to S_0$ of relative dimension $d \geq 1$ together with a line bundle $\mathcal{L}_0$ on X_0, and $\mathcal{V} = \{V_\alpha\}_\alpha$ is an admissible open cover such that $\mathcal{L}_0|_\alpha$ is isomorphic to $O_{X_0}|_\alpha$ via a trivializing section $s_{0;\alpha} \in \mathcal{L}_0|_\alpha$. Then we have coordinate transformation functions $u_{0;\alpha\beta} \in \Gamma(V_\alpha \cap V_\beta, O^*_{X_0})$ with $s_{0;\alpha} = u_{0;\alpha\beta} \cdot s_{0;\beta}$. We fix an enhanced system of deformations $\mathscr{D}$ for f_0 subordinate to $\mathcal{V}$.

Below, we construct a $\mathbb{P}^1$-bundle $p_0\colon P_0(\mathcal{L}_0) \to X_0$ together with an enhanced system of deformations $\mathscr{D}(\mathcal{L}_0)$ of the enhanced generically log smooth family $g_0\colon P_0(\mathcal{L}_0) \to S_0$ subordinate to $\mathcal{V}(\mathcal{L}_0) := \{p_0^{-1}(V_\alpha)\}_\alpha$ and an isomorphism

$$\mathrm{ELD}^{\mathscr{D}}_{X_0/S_0}(\mathcal{L}_0) \Rightarrow \mathrm{ELD}^{\mathscr{D}(\mathcal{L}_0)}_{P_0(\mathcal{L}_0)/S_0}$$

of deformation functors. The map $p_0\colon P_0(\mathcal{L}_0) \to X_0$ is obviously projective, and over $U_0 \subseteq X_0$, it is log smooth and satisfies $\Omega^1_{P_0(\mathcal{L}_0)/X_0} \cong O_{P_0(\mathcal{L}_0)}$. Thus, if $f_0\colon X_0 \to S_0$ is proper respectively log Calabi–Yau, then $g_0\colon P_0(\mathcal{L}_0) \to S_0$ is proper respectively log Calabi–Yau. Moreover, we will see in Lemma 16.10 below that $g_0\colon P_0(\mathcal{L}_0) \to S_0$ as well as its deformations are log toroidal families if this is the case for $f_0\colon X_0 \to S_0$ and its deformations.

© The Author(s), under exclusive license to Springer Nature Switzerland AG 2025 481
S. Felten, *Global Logarithmic Deformation Theory*, Lecture Notes
in Mathematics 2373, https://doi.org/10.1007/978-3-031-98751-9_16

Theorem 16.2 *In Situation 16.1, assume that* $\mathbf{k} = \mathbb{C}$, *that* $f_0 \colon X_0 \to S_0$ *is proper and log Calabi–Yau, that* $f_0 \colon X_0 \to S_0$ *is log toroidal with respect to* $a_0 \colon S_0 \to A_Q$, *and that each local model* $V_{\alpha;A} \to S_A$ *is log toroidal with respect to* $a_A \colon S_A \to A_Q$. *Then the deformation functor* $\mathrm{LD}^{\mathscr{D}}_{X_0/S_0}(\mathcal{L}_0)$ *is unobstructed.*

Proof We apply Theorem 15.2 to the log toroidal family $g_0 \colon P_0(\mathcal{L}_0) \to S_0$.   □

Remark 16.3 The idea for the construction of $P_0(\mathcal{L}_0) \to S_0$ is taken from [154], where the classical case of a smooth algebraic variety X_0 is treated. While unobstructedness of pairs was already known in the classical case, the primary goal of that article is to show that the dg Lie algebra controlling the deformations of the pair $(X_0, \mathcal{L}_0)$ is homotopy Abelian, a stronger result which implies the unobstructedness. Here, we content ourselves with proving the unobstructedness.   ◇

Remark 16.4 Deformations of pairs $(X_0, \mathcal{L}_0)$ for $Q = \mathbb{N}$ have been also studied in [40]. Our study of the curved Lie–Rinehart pair controlling the deformations of $(X_0, \mathcal{L}_0)$ has been inspired by and is essentially contained in that work. However, this work does not study the $\mathbb{P}^1$-bundle $P_0(\mathcal{L}_0)$, and hence they do not obtain the above unobstructedness result. Instead, their main result [40, Thm. 1.1] is a relation between the deformation theory of a pair $(X_0, \mathcal{F}_0^{\bullet})$ with a vector bundle $\mathcal{F}_0$, and the deformation theory of $(X_0, \det(\mathcal{F}_0))$. With Theorem 16.2, we can remove their condition that $(X_0, \det(\mathcal{F}_0))$ must be (more or less) unobstructed in [40, Thm. 1.1] because we know it holds. Presumably, this also simplifies [41], where they apply [40, Thm. 1.1].   ◇

16.1   The Construction of $P_A(\mathcal{L}_A)$

In Situation 16.1, let $Y_0 \subseteq X_0$ be an open subset of the form $V_{\alpha_1} \cap \ldots \cap V_{\alpha_r}$ (possibly $Y_0 = X_0$), and let $f_A \colon (Y_A, \mathcal{L}_A) \to S_A$ be an enhanced generically log smooth deformation with a line bundle of type $\mathscr{D}$. We have a $\mathbb{P}^1$-bundle

$$\mathbb{P}(O_{Y_A} \oplus \mathcal{L}_A) := \mathrm{Proj}\,\mathrm{Sym}^{\bullet}(O_{Y_A} \oplus \mathcal{L}_A)$$

over Y_A. On $Y_{A;\alpha} := Y_A \cap V_{\alpha}$, we choose a lift $s_{A;\alpha}$ of $s_{0;\alpha}$, which is possible due to our assumption on the form of Y_0. This lift trivializes $\mathcal{L}_A|_{\alpha}$, and we have coordinate transformation functions $u_{A;\alpha\beta}$ satisfying $s_{A;\alpha} = u_{A;\alpha\beta} \cdot s_{A;\beta}$, which are a lift of $u_{0;\alpha\beta}$. In particular, we have

$$\mathrm{Sym}^{\bullet}(O_{Y_A} \oplus \mathcal{L}_A)|_{\alpha} \cong O_{Y_{A;\alpha}}[R_{\alpha}, S_{\alpha}]$$

where R_{α} corresponds to $1 \in O_{Y_A}$ and S_{α} corresponds to $s_{A;\alpha} \in \mathcal{L}_A$. Thus,

$$W_{\alpha} := \mathbb{P}(O_{Y_A} \oplus \mathcal{L}_A)|_{\alpha} \cong Y_{A;\alpha} \times \mathbb{P}^1,$$

and we decompose $W_\alpha = W'_\alpha \cup W''_\alpha$ into two affine patches $W'_\alpha = \{S_\alpha \neq 0\}$ with coordinate $x_\alpha = R_\alpha/S_\alpha$ and $W''_\alpha = \{R_\alpha \neq 0\}$ with coordinate $y_\alpha = S_\alpha/R_\alpha$. The coordinate transformation on $W'_\alpha \cap W'_\beta$ is given by $x_\alpha = u_{A;\alpha\beta}^{-1} \cdot x_\beta$, and the coordinate transformation on $W''_\alpha \cap W''_\beta$ is given by $y_\alpha = u_{A;\alpha\beta} \cdot y_\beta$; since $W'_\alpha \cap W_\beta = W'_\beta$ and $W''_\alpha \cap W_\beta = W''_\beta$, this together with $y_\alpha = x_\alpha^{-1}$ on $W'_\alpha \cap W''_\alpha$ is enough to describe all coordinate transformations.

We see from these coordinate transformations that

$$\Delta_{0,A} := \{x_\alpha = 0\}, \quad \Delta_{\infty,A} := \{y_\alpha = 0\}$$

are well-defined closed subschemes of $\mathbb{P}(O_{Y_A} \oplus \mathcal{L}_A)$; they are sections of the projection, and they are the complements of W''_α respectively W'_α. We denote their ideal sheaves by $\mathcal{I}_{0,A}$ and $\mathcal{I}_{\infty,A}$. They are line bundles, so they give rise to a Deligne–Faltings log structure $\gamma_1 \colon \mathcal{I}_{0,A} \to O_\mathbb{P}$, $\gamma_2 \colon \mathcal{I}_{\infty,A} \to O_\mathbb{P}$ in the sense of [222, III, Defn. 1.7.1]. This gives rise to a log scheme over $\underline{Y}_A$, the underlying scheme of Y_A endowed with the trivial log structure, which we denote by

$$p'_A \colon \ P'_A(\mathcal{L}_A) \to \underline{Y}_A.$$

Lemma 16.5 *The morphism p'_A is smooth and log smooth of relative dimension 1, and we have a global isomorphism $\Omega^1_{P'_A(\mathcal{L}_A)/\underline{Y}_A} \cong O_{P'_A(\mathcal{L}_A)}$.*

Proof The formation of the associated log structure from a Deligne–Faltings log structure commutes with pull-back of both types of log structures. Thus, for each α, we have a Cartesian diagram

$$
\begin{array}{ccc}
P'_A(\mathcal{L}_A)|_\alpha & \xrightarrow{\ \pi\ } & \mathbb{P}^1(0+\infty) \\
\downarrow & & \downarrow \\
\underline{Y}_{A;\alpha} & \longrightarrow & \operatorname{Spec} \mathbf{k}
\end{array}
$$

of log schemes, where $\mathbb{P}^1(0+\infty)$ denotes $\mathbb{P}^1$ endowed with the log structure coming from the Deligne–Faltings log structure of the ideals of the two points 0 and ∞. By [222, III, Prop. 1.7.3], the latter log schemes is equal to $\mathbb{P}^1$ endowed with the divisorial log structure of the two points. Thus, the morphism on the right is smooth and log smooth, and so is its (local) base change p'_A.

Let x and y be the two coordinates on $\mathbb{P}^1$. Then we have $\pi^* x = x_\alpha$ and $\pi^* y = y_\alpha$. On $\mathbb{A}^1_x(0) \subseteq \mathbb{P}^1(0+\infty)$, we have an element $\hat{x} \in \mathcal{M}$ in the monoid sheaf mapping to $x \in O$. When $\delta_{\mathbb{P}^1}$ is the log part of the universal derivation, we have $\Omega^1_{\mathbb{A}^1_x(0)/\mathbf{k}} \cong O \cdot \delta_P(\hat{x})$. In particular, when denoting by $\hat{x}_\alpha$ the element in $\mathcal{M}_{P'_A(\mathcal{L}_A)|W'_\alpha}$ induced from $\hat{x}$, we have

$$\Omega^1_{P'_A(\mathcal{L}_A)/\underline{Y}_A}|W'_\alpha \cong O_{W'_\alpha} \cdot \delta_{P/Y}(\hat{x}_\alpha).$$

Similarly, we have

$$\Omega^1_{P'_A(\mathcal{L}_A)/\underline{Y}_A}|_{W''_\alpha} \cong \mathcal{O}_{W''_\alpha} \cdot \delta_{P/Y}(\hat{y}_\alpha).$$

Since $y = x^{-1}$ on $\mathbb{A}^1_x(0) \cap \mathbb{A}^1_y(\infty)$, we have $\hat{y}_\alpha = \hat{x}_\alpha^{-1}$ on $W'_\alpha \cap W''_\alpha$; thus, $\delta_{P/Y}(\hat{x}_\alpha) = -\delta_{P/Y}(\hat{y}_\alpha)$. On overlaps $W'_\alpha \cap W'_\beta$, we have $x_\alpha = u^{-1}_{A;\alpha\beta} \cdot x_\beta$; it follows from the construction of the log structure associated with a Deligne–Faltings log structure that we have $\hat{x}_\alpha = u^{-1}_{A;\alpha\beta} \cdot \hat{x}_\beta$ as well. Thus, $\delta_{P/Y}(\hat{x}_\alpha) = \delta_{P/Y}(u^{-1}_{A;\alpha\beta}) + \delta_{P/Y}(\hat{x}_\beta) = \delta_{P/Y}(\hat{x}_\beta)$; the latter equality follows from $(p'_A)^{-1}\mathcal{O}^*_{Y_A}$-linearity of the universal relative derivation. Similarly, we have $\delta_{P/Y}(\hat{y}_\alpha) = \delta_{P/Y}(\hat{y}_\beta)$. In particular, there is a global section δ of $\Omega^1_{P'_A(\mathcal{L}_A)/\underline{Y}_A}$ with $\delta|_{W'_\alpha} = \delta_{P/Y}(\hat{x}_\alpha)$ and $\delta|_{W''_\alpha} = -\delta_{P/Y}(\hat{y}_\alpha)$. This is the trivialization in the last assertion of the lemma. Note also that δ is independent of the choice of $s_{A;\alpha}$. $\qquad\square$

We define the map $p_A \colon P_A(\mathcal{L}_A) \to Y_A$ of (non-enhanced) generically log smooth families by the Cartesian diagram

$$
\begin{array}{ccc}
P_A(\mathcal{L}_A) & \longrightarrow & P'_A(\mathcal{L}_A) \\
\downarrow{\scriptstyle p_A} & & \downarrow{\scriptstyle p'_A} \\
Y_A & \longrightarrow & \underline{Y}_A.
\end{array}
$$

The space $P_A(\mathcal{L}_A)$ carries a log structure on $p_A^{-1}(U_A)$; there, the map p_A is smooth and log smooth. The composition with $f_A \colon Y_A \to S_A$ turns $g_A \colon P_A(\mathcal{L}_A) \to S_A$ into a generically log smooth family over S_A. We write $\Omega^1_{P_A(\mathcal{L}_A)/Y_A}$ for the relative differential forms, rather than $\mathcal{W}^1_{P_A(\mathcal{L}_A)/Y_A}$, since p_A is, as a base change of p'_A, in some sense log smooth. We denote the global generator constructed in Lemma 16.5 by $\gamma = \delta_{P/Y}(\hat{x}_\alpha) = -\delta_{P/Y}(\hat{y}_\alpha)$.

The direct image from $p_A^{-1}(U_A)$ gives an exact sequence

$$0 \to p_A^* \mathcal{W}^1_{Y_A/S_A} \to \mathcal{W}^1_{P_A(\mathcal{L}_A)/S_A} \to \Omega^1_{P_A(\mathcal{L}_A)/Y_A} \to 0. \tag{16.1}$$

Namely, p_A is flat, so $p_A^* \mathcal{W}^1_{Y_A/S_A}$ is already Z-closed. Except for surjectivity, the sequence is exact because j_* is left exact. The map on the right is surjective because $\delta_{P/S}(\hat{x}_\alpha)$ is a preimage of $\gamma = \delta_{P/Y}(\hat{x}_\alpha)$ on $W'_\alpha \cap p_A^{-1}(U_A)$, and similarly for W''_α. We also find

$$\mathcal{W}^{d+1}_{P_A(\mathcal{L}_A)/S_A} \cong p_A^* \mathcal{W}^d_{Y_A/S_A}$$

because this holds on $p_A^{-1}(U_A)$ since $\Omega^1_{P_A(\mathcal{L}_A)/Y_A} \cong \mathcal{O}_{P_A(\mathcal{L}_A)}$, and both sides are Z-closed; in particular, due to our log Gorenstein assumption, they are line bundles.

Next, we turn $g_A \colon P_A(\mathcal{L}_A) \to S_A$ into an *enhanced* generically log smooth family. The exact sequence (16.1) is locally split, and any two splittings differ by a section of

$$\mathcal{H}om(\Omega^1_{P_A(\mathcal{L}_A)/Y_A}, p_A^* \mathcal{W}^1_{Y_A/S_A}).$$

Inside $p_A^* \mathcal{W}^1_{Y_A/S_A}$, we have $p_A^* \mathcal{A}^1_{Y_A/S_A}$, and the local splittings constructed above and given by $\gamma \mapsto \delta_{P/S}(\hat{x}_\alpha) = -\delta_{P/S}(\hat{y}_\alpha)$ all differ by a section of

$$\mathcal{H}om(\Omega^1_{P_A(\mathcal{L}_A)/Y_A}, p_A^* \mathcal{A}^1_{Y_A/S_A}).$$

Thus, we have a class of *distinguished splittings*, which differ by a section in that $\mathcal{O}_{P_A(\mathcal{L}_A)}$-module. Given a local distinguished splitting

$$B_1 \colon \quad \Omega^1_{P_A(\mathcal{L}_A)/Y_A} \to \mathcal{W}^1_{P_A(\mathcal{L}_A)/S_A},$$

we define

$$\mathcal{A}^i_{P_A(\mathcal{L}_A)/S_A} := p_A^* \mathcal{A}^i_{Y_A/S_A} + p_A^* \mathcal{A}^{i-1}_{Y_A/S_A} \wedge B_1(\Omega^1_{P_A(\mathcal{L}_A)/Y_A}) \subseteq \mathcal{W}^i_{P_A(\mathcal{L}_A)/S_A}.$$

This is independent of the choice of distinguished splitting B_1. At this point, the discussion is analogous to the one given later in Chap. 18.1 in more detail, so we just summarize quickly the situation. We have a diagram

$$
\begin{array}{ccccccccc}
0 & \longrightarrow & p_A^* \mathcal{W}^i_{Y_A/S_A} & \overset{E_i}{\longrightarrow} & \mathcal{W}^i_{P_A(\mathcal{L}_A)/S_A} & \overset{Q_i}{\longrightarrow} & p_A^* \mathcal{W}^{i-1}_{Y_A/S_A} \otimes \Omega^1_{P_A(\mathcal{L}_A)/Y_A} & \longrightarrow & 0 \\
& & \uparrow & & \uparrow & & \uparrow & & \\
0 & \longrightarrow & p_A^* \mathcal{A}^i_{Y_A/S_A} & \overset{E_i}{\longrightarrow} & \mathcal{A}^i_{P_A(\mathcal{L}_A)/S_A} & \overset{Q_i}{\longrightarrow} & p_A^* \mathcal{A}^{i-1}_{Y_A/S_A} \otimes \Omega^1_{P_A(\mathcal{L}_A)/Y_A} & \longrightarrow & 0
\end{array}
$$

where both rows are locally split exact sequences. A distinguished local splitting B_1 induces a splitting

$$T_i \colon \quad \mathcal{A}^i_{P_A(\mathcal{L}_A)/S_A} \to p_A^* \mathcal{A}^i_{Y_A/S_A}$$

which satisfies $T_i(\alpha) \wedge T_j(\beta) = T_{i+j}(\alpha \wedge \beta)$. Dualizing the diagram, we obtain a locally split exact sequence

$$0 \to p_A^* \mathcal{V}^{p+1} \otimes \Theta^1_{P_A(\mathcal{L}_A)/Y_A} \overset{I_p}{\to} \mathcal{V}^p_{P_A(\mathcal{L}_A)/S_A} \overset{F_p}{\to} p_A^* \mathcal{V}^p_{Y_A/S_A} \to 0;$$

when dualizing the splitting T_{-p} of E_{-p}, we obtain a splitting

$$S_p \colon \quad p_A^* \mathcal{V}^p_{Y_A/S_A} \to \mathcal{V}^p_{P_A(\mathcal{L}_A)/S_A}$$

of F_p. These functions satisfy a large number of identities given in Lemma 18.5. Then we define

$$\mathcal{G}^p_{P_A(\mathcal{L}_A)/S_A} := I_p(p_A^* \mathcal{G}^{p+1}_{Y_A/S_A} \otimes \Theta^1_{P_A(\mathcal{L}_A)/Y_A}) + S_p(p_A^* \mathcal{G}^p_{Y_A/S_A}) \subseteq \mathcal{V}^p_{P_A(\mathcal{L}_A)/S_A},$$

which is independent of the original choice of distinguished splitting B_1. The proof of Proposition 18.6 shows that we obtain an enhanced generically log smooth family by setting

$$\mathcal{G}C^\bullet_{P_A(\mathcal{L}_A)/S_A} := (\mathcal{G}^\bullet_{P_A(\mathcal{L}_A)/S_A}, \mathcal{A}^\bullet_{P_A(\mathcal{L}_A)/S_A}).$$

The construction of $p_A \colon P_A(\mathcal{L}_A) \to Y_A$ commutes with base change along $B' \to B$. Namely, the formation of the $\mathbb{P}^1$-bundle $\mathbb{P}(O_{Y_A} \oplus \mathcal{L}_A)$ commutes with base change, the formation of the ideals $I_{0,A}$ and $I_{\infty,A}$ commutes with base change, the formation of the Deligne–Faltings log structure and its associated log structure commutes with base change, and the pull-back along $Y_A \to \underline{Y}_A$ commutes with base change. From the above diagram with E_i and Q_i, we see that the formation of $\mathcal{G}C^\bullet_{P_A(\mathcal{L}_A)/S_A}$ commutes with base change as well. Thus, the formation of $g_A \colon P_A(\mathcal{L}_A) \to S_A$ as an enhanced generically log smooth family commutes with base change.

In particular, we have an enhanced generically log smooth family $g_0 \colon P_0(\mathcal{L}_0) \to S_0$, and $g_A \colon P_A(\mathcal{L}_A) \to S_A$ is a deformation thereof over $p_0^{-1}(Y_0) \subseteq P_0(\mathcal{L}_0)$.

When $\mathcal{G}C^\bullet_{Y_A/S_A} = \mathcal{V}\mathcal{W}^\bullet_{Y_A/S_A}$, then $\mathcal{G}C^\bullet_{P_A(\mathcal{L}_A)/S_A} = \mathcal{V}\mathcal{W}^\bullet_{P_A(\mathcal{L}_A)/S_A}$ as well so that we do not leave the setting of generically log smooth families in the case where we have started in this setting.

16.2 Isomorphisms and Automorphisms

Let $f \colon (Y_A, \mathcal{L}_A) \to S_A$ and $\tilde{f} \colon (\tilde{Y}_A, \tilde{\mathcal{L}}_A) \to S_A$ be two enhanced generically log smooth deformations with a line bundle of type $\mathcal{D}$, and assume that $\varphi = (\phi, \Phi, \psi) \colon (Y_A, \mathcal{L}_A) \to (\tilde{Y}_A, \tilde{\mathcal{L}}_A)$ is an isomorphism over $f_0 \colon (Y_0, \mathcal{L}_0) \to S_0$. Then we have an induced isomorphism

$$P_A(\varphi) \colon \ P_A(\mathcal{L}_A) \xrightarrow{\cong} \tilde{P}_A(\tilde{\mathcal{L}}_A)$$

of enhanced generically log smooth families which is compatible with $(\phi, \Phi) \colon Y_A \cong \tilde{Y}_A$ and with the identity on $(Y_0, \mathcal{L}_0)$. This construction is functorial for compositions of isomorphisms, and it is compatible with base change along $B' \to B$.

More generally, we also have this construction for isomorphisms of the underlying (non-enhanced) generically log smooth families. Before going to the enhanced case, we study automorphisms in the non-enhanced case. So let $\varphi = (\phi, \Phi, \psi)$ be an automorphism of $f : (Y_A, \mathcal{L}_A) \to S_A$ as a non-enhanced generically log smooth family with a line bundle. Let us write $\psi(s_{A;\alpha}) = v_\alpha \cdot s_{A;\alpha}$. This determines $v_\alpha \in \Gamma(Y_{A;\alpha}, O_{Y_A}^*)$ uniquely, and we have $v_\alpha|_0 = 1$. First, φ induces an automorphism of $\mathrm{Sym}^\bullet(O_{Y_A} \oplus \mathcal{L}_A)$, and this in turn induces an automorphism of $\mathbb{P}(O_{Y_A} \oplus \mathcal{L}_A)$, which we denote by $\hat{\phi}$. We have $\hat{\phi}(x_\alpha) = v_\alpha^{-1} \cdot x_\alpha$ on W'_α and $\hat{\phi}(y_\alpha) = v_\alpha \cdot y_\alpha$ on W''_α. We also have an induced automorphism $\hat{\phi}_0$ of the ideal $\mathcal{I}_{0,A}$ and an induced automorphism $\hat{\phi}_\infty$ of the ideal $\mathcal{I}_{\infty,A}$. Together, they define an automorphism of the Deligne–Faltings log structure, and hence an automorphism $(\hat{\phi}, \hat{\Phi}')$ of the log scheme $P'_A(\mathcal{L}_A)$. By construction, we have a commutative diagram

$$
\begin{array}{ccccc}
\mathcal{I}_{0,A}^* & \longrightarrow & M_{P'_A(\mathcal{L}_A)} & \longrightarrow & O_{P'_A(\mathcal{L}_A)} \\
\downarrow {\scriptstyle \hat{\phi}_0} & & \downarrow {\scriptstyle \hat{\Phi}'} & & \downarrow {\scriptstyle \hat{\phi}} \\
\mathcal{I}_{0,A}^* & \longrightarrow & M_{P'_A(\mathcal{L}_A)} & \longrightarrow & O_{P'_A(\mathcal{L}_A)}
\end{array}
$$

and a similar one for $\mathcal{I}_{\infty,A}^*$. On W'_α, the element x_α is a generator of $\mathcal{I}_{0,A}$, so we have $x_\alpha \in \mathcal{I}_{0,A}^*$. Thus, we find $\hat{\Phi}'(\hat{x}_\alpha) = v_\alpha^{-1} + \hat{x}_\alpha$ on W'_α, and similarly $\hat{\Phi}'(\hat{y}_\alpha) = v_\alpha + \hat{y}_\alpha$ on W''_α. On $p_A^{-1}(U_0) \cap W'_\alpha$, we obtain $\hat{\Phi}(\hat{x}_\alpha) = v_\alpha^{-1} + \hat{x}_\alpha$, and on $p_A^{-1}(U_0) \cap W''_\alpha$, we obtain $\hat{\Phi}(\hat{y}_\alpha) = v_\alpha + \hat{y}_\alpha$, both after base change along $Y_A \to \underline{Y}_A$, where $\hat{\Phi}$ is the log part of the induced automorphism on $P_A(\mathcal{L}_A)$.

We have a canonical isomorphism

$$
p_A^* \mathcal{L}_A \xrightarrow{\cong} \mathcal{I}_{\infty,A} \otimes \mathcal{I}_{0,A}^{-1}, \qquad s_{A;\alpha} \mapsto \begin{cases} 1 \otimes x_\alpha^{-1}, & W'_\alpha \\ y_\alpha \otimes 1, & W''_\alpha \end{cases}.
$$

When we apply $\hat{\phi}_\infty \otimes \hat{\phi}_0^{-1}$ on the right, then we obtain an induced automorphism $\hat{\psi} : p_A^* \mathcal{L}_A \to p_A^* \mathcal{L}_A$ with $\hat{\psi}(a \cdot s) = \hat{\phi}(a) \cdot \hat{\psi}(s)$ and $\hat{\psi}(s_{A;\alpha}) = v_\alpha \cdot s_{A;\alpha}$. We can reconstruct $\hat{\psi}$ from $(\hat{\phi}, \hat{\Phi})$ since $\hat{\Phi}(\hat{y}_\alpha) = v_\alpha + \hat{y}_\alpha$.

We construct a map

$$
\rho : \ (p_A)_* \mathcal{A}ut_{P_A(\mathcal{L}_A)/P_0(\mathcal{L}_0)} \to \mathcal{A}ut_{(Y_A,\mathcal{L}_A)/(Y_0,\mathcal{L}_0)}
$$

in the other direction. Let $(\hat{\phi}, \hat{\Phi})$ be an automorphism of $g_A : P_A(\mathcal{L}_A) \to S_A$. Since $(p_A)_* O_{P_A(\mathcal{L}_A)} = O_{Y_A}$, we can define $\phi := (p_A)_* \hat{\phi}$. There is a derivation $(\hat{D}, \hat{\Delta})$ such that $(\hat{\phi}, \hat{\Phi}) = \exp(\hat{D}, \hat{\Delta})$. Thus, $\hat{\Phi}$ is of the form

$$
\hat{\Phi}(m) = m + \alpha^{-1} \left(\sum_{n=0}^{\infty} \frac{[\hat{\Delta}(m) + \hat{D}]^n (1)}{n!} \right).
$$

In particular, this also holds for

$$(p_A)_* \hat{\Phi}: \quad (p_A)_* \mathcal{M}_{P_A(\mathcal{L}_A)} \to (p_A)_* \mathcal{M}_{P_A(\mathcal{L}_A)},$$

and when we use the same formula with $\hat{\Delta}$ replaced by

$$\mathcal{M}_{Y_A} \to (p_A)_* \mathcal{M}_{P_A(\mathcal{L}_A)} \to \mathcal{O}_{Y_A}$$

and $\hat{D}$ replaced by $(p_A)_* \hat{D}$, we obtain a map $\Phi: \mathcal{M}_{Y_A} \to \mathcal{M}_{Y_A}$ which is compatible with $(p_A)_* \hat{\Phi}$ along $\mathcal{M}_{Y_A} \to (p_A)_* \mathcal{M}_{P_A(\mathcal{L}_A)}$. In particular, the pair (ϕ, Φ) is a log morphism, and it is in fact an automorphism because we can apply the same construction to the inverse of $(\hat{\phi}, \hat{\Phi})$, and because this construction preserves compositions and the identity.

Given $(\hat{\phi}, \hat{\Phi})$, there is a unique $v_\alpha \in \Gamma(W_\alpha'' \cap p_A^{-1}(U_0), \mathcal{O}^*_{P_A(\mathcal{L}_A)})$ with $\hat{\Phi}(\hat{y}_\alpha) = \hat{y}_\alpha + v_\alpha$. It can in fact be extended to $W_\alpha \cap p_A^{-1}(U_A)$ such that $\hat{\Phi}(\hat{x}_\alpha) = \hat{x}_\alpha + v_\alpha^{-1}$ on W_α'. Since $(p_A)_* \mathcal{O}_{P_A(\mathcal{L}_A)} = \mathcal{O}_{Y_A}$, we have indeed $v_\alpha \in \Gamma(Y_{A;\alpha}, \mathcal{O}^*_{Y_A})$. A direct computation based on the transformation behavior of $\hat{y}_\alpha$ yields $v_\alpha = \hat{\phi}(u_{A;\alpha\beta}) \cdot u_{A;\alpha\beta}^{-1} \cdot v_\beta$. This shows that

$$\hat{\psi}: \quad p_A^* \mathcal{L}_A \to p_A^* \mathcal{L}_A, \quad a \cdot s_{A;\alpha} \mapsto \hat{\phi}(a) \cdot v_\alpha \cdot s_{A;\alpha},$$

is well-defined. Moreover, we have $\hat{\psi}(a \cdot s) = \hat{\phi}(a) \cdot \hat{\psi}(s)$. As a consequence of the projection formula, we have $(p_A)_* p_A^* \mathcal{L}_A = \mathcal{L}_A$. Then we set $\psi := (p_A)_* \hat{\psi}$. Since this ψ forms, together with ϕ and Φ, in fact an automorphism of $f_A: (Y_A, \mathcal{L}_A) \to S_A$ over $f_0: (Y_0, \mathcal{L}_0) \to S_0$, this completes the construction of ρ.

We observe that $\rho(P_A(\phi, \Phi, \psi)) = (\phi, \Phi, \psi)$. For ϕ and ψ, this is straightforward. For Φ, the easiest way to see this is by noting that the forgetful map $(\phi, \Phi) \to \phi$ is injective because of the density of the strict locus; namely, on the strict locus, the forgetful map is bijective, and the automorphism sheaf has injective restrictions to the strict locus. This shows that

$$P_A: \quad \mathcal{A}ut_{(Y_A, \mathcal{L}_A)/(Y_0, \mathcal{L}_0)} \to (p_A)_* \mathcal{A}ut_{P_A(\mathcal{L}_A)/P_0(\mathcal{L}_0)}$$

is injective, and that ρ is surjective.

Proposition 16.6 *The maps $P_A(-)$ and ρ are inverse isomorphisms of groups.*

In order to prove this, we study derivations first. The exact sequence (16.1) above is locally split. Thus, its dual sequence

$$0 \to \mathcal{O}_{P_A(\mathcal{L}_A)} \to \Theta^1_{P_A(\mathcal{L}_A)/S_A} \to \mathcal{H}om(p_A^* W^1_{Y_A/S_A}, \mathcal{O}_{P_A(\mathcal{L}_A)}) \to 0$$

is locally split exact as well. The right-hand side is canonically isomorphic to $p_A^* \Theta^1_{Y_A/S_A}$.

In the $\mathbb{P}^1$-bundle $p_A \colon P_A(\mathcal{L}_A) \to Y_A$, we have

$$(p_A)_* O_{P_A(\mathcal{L}_A)} = O_{Y_A}, \quad R^q (p_A)_* O_{P_A(\mathcal{L}_A)} = 0 \text{ for } q \geq 1.$$

Furthermore, for *every* coherent sheaf $\mathcal{F}$ on Y_A, we have the projection formula

$$(p_A)_* p_A^* \mathcal{F} = \mathcal{F}, \quad R^q (p_A)_* p_A^* \mathcal{F} = 0 \text{ for } q \geq 1.$$

This is a consequence of the very general projection formula [190, Prop. 3.9.4]. Thus, we obtain an exact sequence

$$0 \to O_{Y_A} \xrightarrow{\iota} (p_A)_* \Theta^1_{P_A(\mathcal{L}_A)/S_A} \xrightarrow{\pi} \Theta^1_{Y_A/S_A} \to 0.$$

Proposition 16.7 *This exact sequence is isomorphic to the Atiyah sequence* (AE) *before Lemma 11.4.*

Proof Let $\mathbb{U} \subseteq Y_A$ be an open subset, and let $(D, \Delta) \in \Gamma(p_A^{-1}(\mathbb{U}), \Theta^1_{P_A(\mathcal{L}_A)/S_A})$ be a derivation. Then $\pi(D, \Delta)$ is given by the adjoint maps of $p_A^{-1} O_{Y_A} \to O_{P_A}$ and $p_A^{-1} \mathcal{M}_{Y_A} \to O_{P_A}$, using that $(p_A)_* O_{P_A} = O_{Y_A}$. Given such (D, Δ), we can furthermore form a map $E \colon p_A^* \mathcal{L}_A \to p_A^* \mathcal{L}_A$ given by

$$E(a \cdot s_{A;\alpha}) := D(a) \cdot s_{A;\alpha} + a \cdot \Delta(\hat{y}_\alpha) \cdot s_{A;\alpha}$$

on $p_A^{-1}(\mathbb{U}) \cap W_\alpha$. Although $\hat{y}_\alpha$ is only defined on W_α'', the element $\Delta(\hat{y}_\alpha)$ is well-defined on W_α because we have $\Delta(\hat{y}_\alpha) = -\Delta(\hat{x}_\alpha)$ on $W_\alpha' \cap W_\alpha''$. The map E is well-defined because it turns out that

$$E(a \cdot s_{A;\alpha}) = E(a \cdot u_{A;\alpha\beta} \cdot s_{A;\beta})$$

using the formula for W_α on the left and the one for W_β on the right. Furthermore, for $s \in p_A^* \mathcal{L}_A$ and $a \in O_{P_A(\mathcal{L}_A)}$, we have $E(a \cdot s) = D(a) \cdot s + a \cdot E(s)$. Since $(p_A)_* p_A^* \mathcal{L}_A = \mathcal{L}_A$, we find that sections on $p_A^{-1}(\mathbb{U})$ give rise to a derivation of $f \colon (Y_A, \mathcal{L}_A) \to S_A$. In other words, we have defined a map

$$\eta \colon (p_A)_* \Theta^1_{P_A(\mathcal{L}_A)/S_A} \to \Theta^1_{Y_A/S_A}(\mathcal{L}_A).$$

It is easy to see that this map is O_{Y_A}-linear. Furthermore, a direct computation shows that the diagram

$$
\begin{array}{ccccccccc}
0 & \longrightarrow & O_{Y_A} & \longrightarrow & (p_A)_* \Theta^1_{P_A(\mathcal{L}_A)/S_A} & \longrightarrow & \Theta^1_{Y_A/S_A} & \longrightarrow & 0 \\
& & \| & & \downarrow{\scriptstyle \eta} & & \| & & \\
0 & \longrightarrow & O_{Y_A} & \longrightarrow & \Theta^1_{Y_A/S_A}(\mathcal{L}_A) & \longrightarrow & \Theta^1_{Y_A/S_A} & \longrightarrow & 0
\end{array}
$$

is not only commutative on the right but also commutative on the left. Hence η is an isomorphism. $\qquad\square$

Proof of Proposition 16.6 It is quite easy to show that

$$\mathfrak{m}_A \cdot (p_A)_* \Theta^1_{P_A(\mathcal{L}_A)/S_A} = (p_A)_*(\mathfrak{m}_A \cdot \Theta^1_{P_A(\mathcal{L}_A)/S_A}).$$

Thus, we have a diagram

$$
\begin{array}{ccc}
(p_A)_* \mathscr{A}ut_{P_A(\mathcal{L}_A)/P_0(\mathcal{L}_0)} & \xrightarrow{\ \rho\ } & \mathscr{A}ut_{(Y_A,\mathcal{L}_A)/(Y_0,\mathcal{L}_0)} \\
{\scriptstyle (p_A)_* \exp}\Big\uparrow{\scriptstyle \cong} & & {\scriptstyle \exp}\Big\uparrow{\scriptstyle \cong} \\
\mathfrak{m}_A \cdot (p_A)_* \Theta^1_{P_A(\mathcal{L}_A)/S_A} & \xrightarrow[\ \cong\ ]{\ \eta\ } & \mathfrak{m}_A \cdot \Theta^1_{Y_A/S_A}(\mathcal{L}_A).
\end{array}
$$

This diagram is commutative. To see this, we have to show that

$$(\phi_1, \Phi_1, \psi_1) := \rho(\exp(\hat{D}, \hat{\Delta})) = \exp(\eta(\hat{D}, \hat{\Delta})) -: (\phi_2, \Phi_2, \psi_2).$$

The equality $\phi_1 = \phi_2$ is straightforward. Again, by the density of the strict locus, we also find $\Phi_1 = \Phi_2$. For $\psi_1 = \psi_2$, it suffices to show $\exp(\hat{E}) = \hat{\psi}$, where $\hat{E}$ is defined by the formula in the proof of Proposition 16.7, and $\hat{\psi}$ is constructed from $(\hat{\phi}, \hat{\Phi})$ as described above. Thus, we have to show $\exp(\hat{E})(s_{A;\alpha}) = v_\alpha \cdot s_{A;\alpha}$. On the one hand, we have

$$v_\alpha = \sum_{n=0}^{\infty} \frac{[\hat{\Delta}(\hat{y}_\alpha) + \hat{D}]^n(1)}{n!}.$$

On the other hand, we have

$$\hat{E}(a \cdot s_{A;\alpha}) = (\hat{\Delta}(\hat{y}_\alpha) \cdot a + D(a)) \cdot s_{A;\alpha},$$

proving the claim by induction. It is clear that the two vertical arrows are isomorphisms. Since the formation of η commutes with base change along $A \to \mathbf{k}$, also the lower horizontal map is an isomorphism. Thus, ρ is an isomorphism, and so is $P_A(-)$ since $\rho \circ P_A(-) = \mathrm{id}$. Finally, to see that the maps are group homomorphisms, it is sufficient to note that $P_A(-)$ is a group homomorphism by the functorial nature of its construction. $\qquad\square$

Now we turn to the enhanced case. From the definitions, we find that

$$\Gamma^1_{P_A(\mathcal{L}_A)/S_A} = (F_{-1})^{-1}(p_A^* \Gamma^1_{Y_A/S_A})$$

so that we have

$$(p_A)_* \Gamma^1_{P_A(\mathcal{L}_A)/S_A} = \pi^{-1}(\Gamma^1_{Y_A/S_A}).$$

Similarly, we have

$$\Gamma^1_{Y_A/S_A}(\mathcal{L}_A) = q^{-1}(\Gamma^1_{Y_A/S_A})$$

where $q : \Theta^1_{Y_A/S_A}(\mathcal{L}_A) \to \Theta^1_{Y_A/S_A}$ is the surjection in the Atiyah extension. Thus, we obtain an induced isomorphism

$$\eta : \ (p_A)_* \Gamma^1_{P_A(\mathcal{L}_A)/S_A} \xrightarrow{\cong} \Gamma^1_{Y_A/S_A}(\mathcal{L}_A).$$

From the proof of Proposition 16.6, we see that ρ gives rise to a one-to-one correspondence between gauge transforms of $g_A : P_A(\mathcal{L}_A) \to S_A$ and gauge transforms of $f_A : (Y_A, \mathcal{L}_A) \to S_A$. In particular, if φ is a gauge transform, then $P_A(\varphi)$ is a gauge transform as well.

16.3   The Enhanced System of Deformations $\mathscr{D}(\mathcal{L}_0)$

We use the construction of $P_A(\mathcal{L}_A)$ to form an enhanced system of deformations $\mathscr{D}(\mathcal{L}_0)$ for $g_0 : P_0(\mathcal{L}_0) \to S_0$, subordinate to $\mathcal{V}(\mathcal{L}_0) := \{p_0^{-1}(V_\alpha)\}_\alpha$. We endow each local model $V_{\alpha;A} \to S_A$ of $\mathscr{D}$ with the line bundle $\mathcal{L}_{\alpha;A} := O_{V_{\alpha;A}}$. For the restriction maps along some $B' \to B$, we take the obvious map $\mathcal{L}_{\alpha;B'} \to \mathcal{L}_{\alpha;B}$. For the restriction to $B = \mathbf{k}$, we use the identification $\mathcal{L}_0|_\alpha \cong O_{X_0}|_\alpha$ via $s_{0;\alpha} \in \mathcal{L}_0|_\alpha$. For the comparison maps, we start from $\mathcal{L}_{\alpha;0}|_{\alpha\beta} \cong \mathcal{L}_0|_{\alpha\beta} \cong \mathcal{L}_{\beta;0}|_{\alpha\beta}$ and lift this order by order to an isomorphism $\mathcal{L}_{\alpha;k}|_{\alpha\beta} \cong \mathcal{L}_{\beta;k}|_{\alpha\beta}$.[1] For a general A, we take an appropriate pull-back from some A_k. Now we form $P_{\alpha;A} := P_A(\mathcal{L}_{\alpha;A})$. They come with restriction maps along $B' \to B$, and on overlaps, we have isomorphisms $P_{\alpha;A}|_{\alpha\beta} \cong P_{\beta;A}|_{\alpha\beta}$. The cocycles on overlaps are inner automorphisms because $P_A(\varphi)$ is a gauge transform whenever φ is a gauge transform. The open cover $\mathcal{V}(\mathcal{L}_0)$ can be seen to be admissible as specified in Definition 10.21 by using that a distinguished splitting B_1 exists on every $p_0^{-1}(V_\alpha)$, using the general projection formula mentioned above, and using the appropriate Leray spectral sequence. This gives the enhanced system of deformations $\mathscr{D}(\mathcal{L}_0)$.

If $f : (Y_A, \mathcal{L}_A) \to S_A$ is an enhanced deformation of type $\mathscr{D}$, its restriction to $Y_0 \cap V_\alpha$ is isomorphic to $(V_{\alpha;A}, \mathcal{L}_{\alpha;A}) \to S_A$ via a specified isomorphism which differ by an inner automorphism on $V_\alpha \cap V_\beta$. Thus, we have an isomorphism from $P_A(\mathcal{L}_A)|_\alpha$ to $P_{\alpha;A}$, differing by an inner automorphism on $V_\alpha \cap V_\beta$. In other

---

[1] In general, the identification $\mathcal{L}_{\alpha;k}|_{\alpha\beta} = O_{V_{\alpha;k}}|_{\alpha\beta} \cong O_{V_{\beta;k}}|_{\alpha\beta} = \mathcal{L}_{\beta;k}|_{\alpha\beta}$ does not work because it is not compatible with our choice on $k = 0$.

words, $P_A(\mathcal{L}_A) \to S_A$ is an enhanced generically log smooth deformation of $g_0 \colon P_0(\mathcal{L}_0) \to S_0$ of type $\mathscr{D}(\mathcal{L}_0)$. This defines a natural transformation

$$\mathrm{ELD}^{\mathscr{D}}_{X_0/S_0}(\mathcal{L}_0) \Rightarrow \mathrm{ELD}^{\mathscr{D}(\mathcal{L}_0)}_{P_0(\mathcal{L}_0)/S_0}.$$

Lemma 16.8 *This is an isomorphism of deformation functors.*

Proof Since $P_A(-)$ induces a bijection on automorphisms, this follows from tracking the gluing like in the proof of [76, Prop. 4.2]. $\qquad\qquad\square$

An important special case is when both $f_0 \colon X_0 \to S_0$ is log toroidal with respect to $a_0 \colon S_0 \to A_Q$, and each local model $V_{\alpha;A} \to S_A$ is log toroidal with respect to $a_A \colon S_A \to A_Q$. In this situation, also $g_A \colon P_A(\mathcal{L}_A) \to S_A$ is log toroidal with respect to $a_A \colon S_A \to A_Q$. To see this, first let $(Q \subset P, \mathcal{F})$ be an elt datum in the sense of Definition 8.8. Let $\hat{P} := P \times \mathbb{N}$, and let $\hat{\mathcal{F}} := \{F \times \mathbb{N} \mid F \in \mathcal{F}\} \cup \{P \times 0\}$. Mimicking the construction of $P_A(\mathcal{L}_A)$ for $A_{P,\mathcal{F}}$, we form a commutative diagram

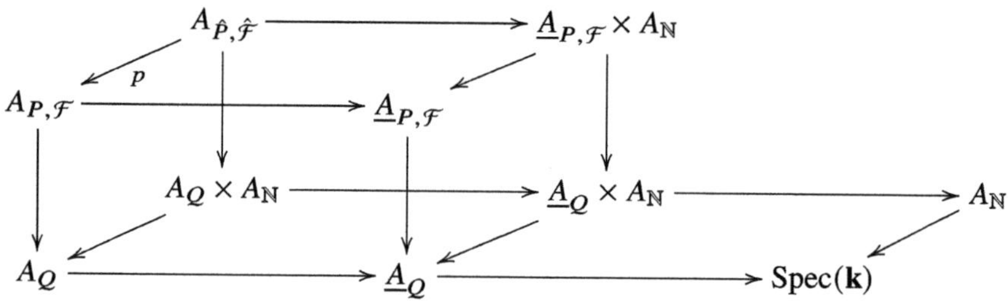

of log schemes. In this diagram, the log structure is defined everywhere on every entry, but it is not necessarily coherent. Since $A_Q \times A_\mathbb{N} = A_{Q \oplus \mathbb{N}}$, the lower two horizontal squares are Cartesian. Also the middle vertical square is Cartesian. Let x be the coordinate of $A_\mathbb{N}$ on the very right. By [222, III, Prop. 1.7.3], $\underline{A}_Q \times A_\mathbb{N}$ is the log scheme with divisorial log structure defined by $\{x = 0\}$, and $\underline{A}_{P,\mathcal{F}} \times A_\mathbb{N}$ carries also the divisorial log structure defined by $\{x = 0\}$. This shows that the arrows emanating from $A_{\hat{P},\hat{\mathcal{F}}}$ are well-defined and give rise to a commutative diagram. On $A_{P,\mathcal{F}}$, we have a log smooth locus $U_{P/Q} = U_1 \cup U_2$ such that $A_{P,\mathcal{F}}|_{U_1} = A_P|_{U_1}$, and $A_{P,\mathcal{F}}|_{U_2} \to A_Q$ is strict and smooth. One can then show that the left vertical square is Cartesian on $U_{P/Q}$ by analyzing U_1 and U_2 separately. For U_1, one uses that $A_{P_1 \oplus P_2} \cong A_{P_1} \times A_{P_2}$, and for U_2 one uses Lemma 16.9 below, a result which seems to be missing in [222] although similar results are discussed. Now also the upper horizontal square is Cartesian on $U_{P/Q}$. Finally, the log structure on $\underline{A}_{P,\mathcal{F}} \times A_\mathbb{N}$ is the one associated with the Deligne–Faltings log structure coming from the ideal (x) because the map to $A_\mathbb{N}$ is strict. Thus, on $p^{-1}(U_{P/Q})$, the map $p \colon A_{\hat{P},\hat{\mathcal{F}}} \to A_{P,\mathcal{F}}$ is precisely the analogue of either W'_α or W''_α in the construction of $P_A(\mathcal{L}_A)$.

Lemma 16.9 *Let Q be a sharp toric monoid, and let $p \colon X \to A_Q$ be a strict and smooth map. Then X carries the divisorial log structure defined by $p^{-1}(D_Q)$.*

Proof First assume that we have a factorization via an étale map $r\colon X \to A_Q \times \mathbb{A}^r$ and the projection $q\colon A_Q \times \mathbb{A}^r \to A_Q$. Following the proof of [222, III, Thm. 1.9.4], it is sufficient to show that

$$q^{-1}\underline{\Gamma}_{D_Q}(Div^+_{A_Q}) \to \underline{\Gamma}_{D_Q \times \mathbb{A}^r}(Div^+_{A_Q \times \mathbb{A}^r})$$

is an isomorphism in order to show that $A_Q \times \mathbb{A}^r$ carries the divisorial log structure defined by $D_Q \times \mathbb{A}^r$. Here, $\underline{\Gamma}_Y(Div^+_X)$ denotes the sheaf of effective Cartier divisors with support in $Y \subset X$. Now the argument in the proof of [222, III, Lemma 1.6.7] shows that this is the case. Since every irreducible component of $D_Q \times \mathbb{A}^r$ is pure of codimension 1 and normal, hence geometrically unibranch, we can apply [222, III, Prop. 1.6.5] to the étale map $r\colon X \to A_Q \times \mathbb{A}^r$ and obtain, after untangling the definitions, that X carries the divisorial log structure defined by $Y := r^{-1}(D_Q \times \mathbb{A}^r)$. Because every smooth map $p\colon X \to A_Q$ can, locally in the étale topology, be factorized as above, X carries the divisorial log structure defined by $p^{-1}(D_Q)$ in the étale topology. Another application of [222, III, Prop. 1.6.5] yields that this is the same as the divisorial log structure in the Zariski topology. $\qquad\square$

With this preparation, we show:

Lemma 16.10 *In Situation 16.1, assume that $f_0\colon X_0 \to S_0$ is log toroidal with respect to $a_0\colon S_0 \to A_Q$, and that each local model $V_{\alpha;A} \to S_A$ is log toroidal with respect to $a_A\colon S_A \to A_Q$. Let $f_A\colon (Y_A, \mathcal{L}_A) \to S_A$ be a generically log smooth deformation with a line bundle of type $\mathscr{D}$. Then $g_A\colon P_A(\mathcal{L}_A) \to S_A$ is log toroidal with respect to $a_A\colon S_A \to A_Q$.*

Proof If $A_{P,\mathcal{F}} \to A_Q$ is a local model of $V_{\alpha;A} \to S_A$, then $A_{\hat{P},\hat{\mathcal{F}}} \to A_Q$ is a local model of $P_A(O_{V_{\alpha;A}})$ constructed over $V_{\alpha;A}$. Thus, $P_A(\mathcal{L}_A)$ is log toroidal with respect to $a_A\colon S_A \to A_Q$. $\qquad\square$

Chapter 17
Algebraic Deformations

In the previous chapters, we showed the existence of infinitesimal and formal deformations. Already in the introduction in Sects. 1.1 and 1.9, we discussed how to pass from a formal deformation to an analytic one-parameter deformation or an algebraic one-parameter deformation. In this chapter, we complement our treatment from the introduction with additional considerations. As in the introduction, we restrict our attention to the standard log point with $Q = \mathbb{N}$ and mostly work over $\mathbf{k} = \mathbb{C}$. We work in the following situation.

Situation 17.1 Let $\Lambda = \mathbb{C}[\![t]\!]$, and let $f_0 \colon X_0 \to S_0$ be a torsionless enhanced generically log smooth family of relative dimension d. For simplicity, assume that X_0 is Cohen–Macaulay. Assume that $f_0 \colon X_0 \to S_0$ is projective, and let $\mathcal{L}_0$ be an ample line bundle on X_0. Assume that X_0 is connected. Let $\mathcal{V} = \{V_\alpha\}_\alpha$ be an $\mathcal{L}_0$-admissible open cover of X_0, and let $\mathscr{D}$ be an enhanced system of deformations subordinate to $\mathscr{D}$. Assume that $\mathrm{ELD}^{\mathscr{D}}_{X_0/S_0}(\mathcal{L}_0)$ is unobstructed. The family $f_0 \colon X_0 \to S_0$ may or may not be log Calabi–Yau.

Example 17.2 Theorem 16.2 shows that we are in Situation 17.1 when the enhanced generically log smooth family is an actual generically log smooth family, all local models are log toroidal with respect to $a_A \colon S_A \to A_{\mathbb{N}}$, and $f_0 \colon X_0 \to S_0$ is log Calabi–Yau. $\Diamond$

We recall and expand our discussion of the existence of the algebraic deformation in Situation 17.1, cf. Sect. 1.9. Let us fix a deformation $f_k \colon (X_k, \mathcal{L}_k) \to S_k$ for every $k \geq 0$. Let $N > 0$ be such that $\mathcal{L}_0^{\otimes N}$ is very ample and $H^n(X_0, \mathcal{L}_0^{\otimes N}) = 0$ for $n \geq 1$, and let $\mathcal{N}_k := \mathcal{L}_k^{\otimes N}$. Then $\mathcal{N}_0$ defines a closed embedding $\Phi_0 \colon X_0 \to \mathbb{P}^s$ for $s = h^0(X_0, \mathcal{M}_0) - 1$. By lifting the generators of $\mathcal{N}_0$, we obtain a surjection $O_{X_k}^{\oplus(s+1)} \to \mathcal{N}_k$, which defines a morphism $\Phi_k \colon X_k \to \mathbb{P}^s_{S_k}$. The formation of this morphism is compatible with base change so that $\Phi_k \times_{S_k} S_0 = \Phi_0$. This shows that Φ_k is a closed immersion, and that $\mathcal{N}_k$ is very ample.

© The Author(s), under exclusive license to Springer Nature Switzerland AG 2025

S. Felten, *Global Logarithmic Deformation Theory*, Lecture Notes in Mathematics 2373, https://doi.org/10.1007/978-3-031-98751-9_17

The limit of the flat deformations $f_k \colon X_k \to S_k$ is a formal scheme $f_\infty \colon \mathfrak{X} \to \mathfrak{S}$ over $\mathfrak{S} = \operatorname{Spf} \mathbb{C}[\![t]\!]$. Let us recall the proof that $\mathfrak{X}$ is a Noetherian formal scheme, as given in [139, Prop. 21.1]: Let $W \subseteq X_0$ be affine, and let $B_k = \Gamma(W, O_{X_k})$. By [267, 09B8], the limit $B_\infty = \varprojlim_k B_k$ is t-adically complete, and $(W, O_{\mathfrak{X}}|_W)$ is the formal completion of $\operatorname{Spec} B_\infty$ in $\operatorname{Spec} B_0$. It thus remains to show that B_∞ is Noetherian. Since $f_0 \colon X_0 \to S_0$ is of finite type, we can find a surjection $A_0[\underline{x}] \to B_0$. We can lift the images of the variables to any order so that we obtain maps $P_k := A_k[\underline{x}] \to B_k$. By [267, 09ZW], this map is surjective. Thus, we have a surjective homomorphism $P_\infty := \varprojlim_k A_k[\underline{x}] \to B_\infty$. Let $A = \mathbb{C}[\![t]\!]$. Then P_∞ is the t-adic completion of $A[\underline{x}]$. In particular, P_∞ is Noetherian, and so is its quotient B_∞.[1]

The closed immersions Φ_k give rise to a closed immersion $\Phi_\infty \colon \mathfrak{X} \to \mathfrak{P}^s$ of formal schemes over $\mathfrak{S}$, as can be seen by using [129, Lemma 10.14.4]. Here, $\mathfrak{P}^s$ is the limit of $\mathbb{P}^s_{S_k} \to S_k$, and it is also equal to the formal completion of $\mathbb{P}^s_\Sigma \to \Sigma$ in $t \cdot O_{\mathbb{P}^s_\Sigma}$, where $\Sigma = \operatorname{Spec} \mathbb{C}[\![t]\!]$. Now the Grothendieck existence theorem [130, Cor. 5.1.8] guarantees that closed subschemes of $\mathbb{P}^s_\Sigma$ are in one-to-one correspondence with closed formal subschemes of $\mathfrak{P}^s$. Thus, there is a closed subscheme $\Phi \colon X \to \mathbb{P}^s_\Sigma$ whose completion in $X_0 = X \cap \mathbb{P}^s_0$ is $\mathfrak{X}$; it induces the closed immersions $\Phi_k \colon X_k \to \mathbb{P}^s_{S_k}$ over S_k. This gives rise to the map $f \colon X \to \Sigma$, which is of finite type and proper, and which induces $f_k \colon X_k \to S_k$ after base change. It follows from [267, 0523] that X is flat over Σ at all points in X_0. Since the flat locus is open and $f \colon X \to \Sigma$ is a closed morphism, this implies that $f \colon X \to \Sigma$ is flat. By [267, 045U], the Cohen–Macaulay locus of $f \colon X \to \Sigma$ is open, and it contains X_0, so we find that $f \colon X \to \Sigma$ is a Cohen–Macaulay morphism, and X is Cohen–Macaulay since the base Σ is so. Using [267, 02NM], we can show that $f \colon X \to \Sigma$ is of relative dimension d. Since t is a non-zero divisor due to flatness, the Krull intersection theorem together with the reducedness of X_0 shows that X is reduced at points in X_0. Then X is reduced because the support of any section is closed. The generic fiber X_η is reduced because it is the localization of X in t. It follows from [267, 055J] that X_η is connected since X_0 is connected. Thus, we have shown:

Lemma 17.3 *In Situation 17.1, there is a morphism $f \colon X \to \Sigma$ of finite type over $\Sigma = \operatorname{Spec} \mathbb{C}[\![t]\!]$ which induces, via base change, the underlying morphism of schemes of some enhanced generically log smooth deformation $f_k \colon X_k \to S_k$ of type $\mathscr{D}$. The morphism $f \colon X \to \Sigma$ is separated, flat, projective, Cohen–Macaulay,*

---

[1] An earlier version of the manuscript has a mistake at this point. When denoting by I_k the kernel of $A_k[\underline{x}] \to B_k$, it is true that $I_{k+1} \otimes_{A_{k+1}} A_k = I_k$. However, it may not be true that $I = \{f \in A[\underline{x}] \mid \forall k \colon f|_k \in I_k\}$ is an ideal with $I \otimes_A A_k = I_k$—the ideal I might not surject onto I_k. In this case, it is no longer true that we would have $(A[\underline{x}]/I) \otimes_A A_k = B_k$. It is plausible that there are indeed affine formal families $(f_k \colon X_k \to S_k)_k$ which are not the completion of any scheme of finite type over $\operatorname{Spec} \mathbf{k}[\![t]\!]$, but we currently do not have a proven counterexample.

of relative dimension d, and has reduced connected fibers. The total space X is connected, reduced, and Cohen–Macaulay.

17.1   Properties of the Generic Fiber X_η

First, we investigate smoothness, integrality, and normality of the generic fiber X_η. We already introduced the following notion of *formal smoothings* in Definition 1.23. The notion seems to appear first in Tziolas' article [273, Defn. 11.6], although the concept is probably much older, cf. also [267, 0E7T]. In [139, p. 202], Hartshorne defines the closely related notion of being *formally smoothable*, and on page 214, he claims that this definition is new.

Definition 17.4 Let $S_0 = \operatorname{Spec} \mathbf{k}$, and let $f_0 \colon X_0 \to S_0$ be a separated scheme of finite type, reduced, connected, Cohen–Macaulay of dimension $d \geq 1$. Let $S_k = \operatorname{Spec} \mathbf{k}[t]/(t^{k+1})$, and let $(f_k \colon X_k \to S_k)_k$ be a compatible family of flat deformations of $f_0 \colon X_0 \to S_0$. Then $(f_k)_k$ is a *formal smoothing* of $f_0 \colon X_0 \to S_0$ if there is some $m \geq 0$ such that t^m is in the d-th Fitting ideal $\operatorname{Fitt}_d(\Omega^1_{X_m/S_m})$ of the classical Kähler differentials $\Omega^1_{X_m/S_m}$ of $f_m \colon X_m \to S_m$. In this case, the property holds for all $k \geq m$.

Note that the notion of a formal smoothing is étale local, i.e., when we have a system of compatible surjective étale maps $g_k \colon Y_k \to X_k$, then $(X_k \to S_k)_k$ is a formal smoothing if and only if $(Y_k \to S_k)_k$ is a formal smoothing. We have stated the following result in the analytic case already in Lemma 1.21 in the introduction.

Lemma 17.5 *Let $\Sigma = \operatorname{Spec} \mathbf{k}[\![t]\!]$. In the situation of Definition 17.4, let $f \colon X \to \Sigma$ be a flat and separated morphism of finite type inducing the formal smoothing $(f_k)_k$. Then there is an open subset $W \subseteq X$ with $W \cap X_0 = X_0$ such that W_η is smooth.*

Proof This is a variant of [267, 0E7T], cf. also [273, Prop. 11.8].   □

The converse holds as well, i.e., if $f \colon X \to \Sigma$ is a separated morphism of finite type, and if X_η is smooth, then the induced system $(f_k \colon X_k \to S_k)_k$ is a formal smoothing.

The point now is that saturated log smooth maps give rise to formal smoothings in codimension 1. In the spirit of Definition 1.20, let us say that a log smooth and saturated morphism $f_0 \colon X_0 \to S_0$ is *juxta-smooth* if the locally unique log smooth deformations constitute a formal smoothing.

Lemma 17.6 *Let $f_0 \colon U_0 \to S_0$ be a separated log smooth and saturated morphism over the standard log point $S_0 = \operatorname{Spec}(\mathbb{N} \to \mathbb{C})$. Then there is an open subset $\widetilde{U}_0 \subseteq U_0$ which is juxta-smooth, and such that $U_0 \setminus \widetilde{U}_0$ has codimension ≥ 2 in U_0.*

Proof A saturated and log smooth morphism $f_0 \colon U_0 \to S_0$ has local models $A_P \to A_\mathbb{N}$ for saturated injections $h \colon \mathbb{N} \to P$. Let $\rho = h(1)$, and let $R \subseteq P$ be the face generated by ρ. Then we have a factorization $A_P \to A_R \to A_\mathbb{N}$. The fibers of

$A_R \to A_{\mathbb{N}}$ over $t \neq 0$ are smooth because $\mathbb{N} \to R$ is vertical. The fibers over $z^\rho \neq 0$ of $A_P \to A_R$ are given by $A_{P/R}$, a normal scheme. When we remove the closure of the singular locus in $A_{P/R} \times A_R^*$ from A_P, then the complement is a deformation to a smooth general fiber, and moreover the removed locus has the right codimension. $\qquad\square$

Corollary 17.7 *In Situation 17.1, there is a juxta-smooth open subset $\widetilde{U}_0 \subseteq U_0 \subseteq X_0$ with $\mathrm{codim}(X_0 \setminus \widetilde{U}_0, X_0) \geq 2$. Moreover, there is an open subset $\widetilde{U} \subseteq X$ with $\widetilde{U} \cap X_0 = \widetilde{U}_0$ and such that $\widetilde{U}_\eta$ is smooth. In particular, both X and X_η are normal.*

Proof The family $(f_k \colon \widetilde{U}_k \to S_k)_k$ is a formal smoothing so that we can find an open subset $\widetilde{U} \subseteq X$ such that $\widetilde{U} \cap X_0 = \widetilde{U}_0$ and $\widetilde{U}_\eta$ is smooth. Recall that X is Cohen–Macaulay. To show that X is normal, let $x \in X$ be such that $\dim(O_{X,x}) = 1$. We have to show that $O_{X,x}$ is regular. If $x \in X_0$, then $O_{X_0,x}$ is reduced and of dimension 0 because $f \colon X \to \Sigma$ is flat, hence $O_{X_0,x}$ is a field, and $t \in O_{X,x}$ generates the maximal ideal. Using the Krull intersection theorem, we find that $O_{X,x}$ is integral, and then it is a discrete valuation ring, hence regular. In the case $x \in X_\eta$, let $C = \overline{\{x\}}$ be the closure in X. Then C is irreducible of dimension d. Since $f \colon X \to \Sigma$ is closed, we find $C \cap X_0 \neq \emptyset$. Let $c \in C \cap X_0$ be a closed point. The formula in [267, 02JS] shows that $\dim_c(C \cap X_0) \geq d - 1$. Thus, we have $C \cap \widetilde{U}_0 \neq \emptyset$, and hence $x \in \widetilde{U}$. This shows that $O_{X,x}$ is regular, and hence that X is normal. Since X_η is a localization of X, it is normal as well. $\qquad\square$

Another consequence of Lemma 17.5 and its converse is that, in Situation 17.1, the smoothness of X_η depends only on the system of deformations $\mathscr{D}$ if $f_0 \colon X_0 \to S_0$ is proper.

Definition 17.8 A system of deformations $\mathscr{D}$ is *juxta-smooth* if $\{V_{\alpha;k}\}_k$ is a formal smoothing of $V_{\alpha;0}$.

In Situation 17.1, if $f_0 \colon X_0 \to S_0$ is proper, and $\mathscr{D}$ is juxta-smooth, then X_η is smooth.

Corollary 17.9 *In Situation 17.1, assume additionally that $f_0 \colon X_0 \to S_0$ is proper, and that both $f_0 \colon X_0 \to S_0$ and all deformations in $\mathscr{D}$ are log toroidal of standard Gross–Siebert type. Then the generic fiber X_η is smooth.*

Proof Lemma 14.8 shows that $\mathscr{D}$ is juxta-smooth. $\qquad\square$

We can also strengthen the formal and analytic smoothing result [77, Thm. 1.1] for normal crossing spaces to an algebraic version.

Corollary 17.10 *Let V be a connected projective normal crossing space, assume that $\mathcal{T}_V^1$ is globally generated, and that V is Calabi–Yau, i.e., $\omega_V^\circ \cong O_V$. Then there is a flat projective morphism $f \colon X \to \Sigma$ with $f^{-1}(0) = V$ and a connected smooth projective generic fiber X_η.*

Proof We will see in Theorem 18.44 that we can endow V with the structure of a log toroidal family $f_0\colon X_0 \to S_0$ of standard Gross–Siebert type, cf. also [77, Prop. 6.9]. $\square$

Remark 17.11 In view of Chap. 18, this remains true when ω_V^{-1} admits a log regular section s_0 with respect to the log structure $f_0\colon X_0 \to S_0$. Indeed, the system of deformations $\mathscr{D}(s_0)$ remains juxta-smooth although the log structure changes. An example of this is [77, Thm. 1.1], which we have worked out in a variant in Theorem 18.44. $\Diamond$

In a sense, the singularities of the generic fiber are always determined by $\mathscr{D}$, not only in the juxta-smooth case. The key insight is the following variant of Artin's approximation.

Lemma 17.12 *Let $R = \mathbb{C}[\![t]\!]$ and $\Sigma = \operatorname{Spec} R$. Let $f\colon X \to \Sigma$ and $f'\colon X' \to \Sigma$ be two morphisms of finite type, and assume that we have a system of compatible isomorphisms $\varphi_k\colon X_k \overset{\cong}{\to} X'_k$. Let $x \in X_0 \subset X$ be a point in the central fiber, and let $x' = \varphi_0(x) \in X'_0 \subset X'$. Let $c \geq 0$. Then there is a scheme V of finite type over Σ together with a point $v \in V$ and two étale Σ-morphisms $g\colon V \to X$ and $h\colon V \to X'$ with the following properties: $g(v) = x$ and we have an isomorphism $\kappa(x) \cong \kappa(v)$ of residue fields; $h(v) = x'$ and we have an isomorphism $\kappa(x') \cong \kappa(v)$ of residue fields; we have $\varphi_c \circ g_c = h_c$ for the base change $g_c\colon V_c \to X_c$ and $h_c\colon V_c \to X'_c$.*

Proof The base ring R is an excellent discrete valuation ring, so we can apply [15, Thm. 1.10]. We may assume that X is affine; let $B := \Gamma(X, O_X) = R[y_1, \ldots, y_n]/(f_1, \ldots, f_m)$. We set

$$A := O^h_{X,x}, \quad A' := O^h_{X',x'},$$

the Henselian local rings, i.e., the local rings in the *Nisnevich* topology. Note that for $A_k := A \otimes_R R/(t^{k+1})$ and A'_k analogously, we have $A_k = O^h_{X_k,x}$ and $A'_k = O^h_{X'_k,x'}$. The isomorphisms φ_k induce compatible isomorphisms $\phi_k\colon A_k \cong A'_k$. In particular, we obtain composite homomorphisms $\psi_k\colon B \to A \to A_k \to A'_k$. In other words, we have a formal solution of the system of equations $f_j(\hat{y}_1, \ldots, \hat{y}_n) = 0$ with values in the t-completion $\hat{A}' = \varprojlim A'_k$ of A'. By [15, Thm. 1.10], we can find $y_1, \ldots, y_n \in A'$ which satisfy $f_j(y_1, \ldots, y_n) = 0$ and $y_i \equiv \hat{y}_i \mod t^{c+1}$. Thus, we have a homomorphism $\psi\colon B \to A'$ which coincides with ψ_k after mapping to A'_c (but not necessarily for $k > c$). Now we can find an étale map $h\colon U' \to X'$ from an affine Σ-scheme U' together with a point $u' \in U'$ such that $h(u') = x'$ and we have an isomorphism $\kappa(x') \cong \kappa(u')$, and such that we have a map $\chi\colon B \to \Gamma(U', O_{U'}) =: B'$ which coincides with $\psi\colon B \to A'$ after composition with $B' \to A'$. By shrinking U' in the Zariski topology, we may assume that $B' \to A'$ is injective. Since $\chi\colon B \to B'$ induces an isomorphism of residue fields and of completions at x and u', we can assume that $\chi\colon B \to B'$ is étale after further shrinking U' in the Zariski topology, by [133, Prop 17.6.3]. It remains to show that $\varphi_c \circ g_c = h_c$, or more precisely, that $\chi_c \colon B_c \to B'_c$ equals the composition

$$B_c \xrightarrow{\varphi_{c*}} O_{X'_c} \xrightarrow{h^*_c} B'_c.$$

By construction, the two maps are equal after composing with $B'_c \to A'_c$. By shrinking U'_c in the Zariski topology, we can assume that $B'_c \to A'_c$ is injective. Then we choose V as an open subset of U' which, after base change to S_c, gives rise to the shrunken U'_c. $\qquad\square$

Remark 17.13 This result is different from the similar [15, Cor. 2.6] in Artin's original article in that the latter does not show that we recover the original isomorphism over S_c. $\qquad\Diamond$

Corollary 17.14 *In Situation 17.1, assume that we have, for each open V_α, morphisms $U_\alpha \to \Sigma$ of finite type with $U_\alpha \times_\Sigma S_k = V_\alpha$ as schemes. Then the singularities of the generic fiber X_η of $f\colon X \to \Sigma$ are étale equivalent to the singularities of $U_{\alpha;\eta}$.*

Proof Use that $f_0\colon X_0 \to S_0$ is proper to show that the result holds globally on X_η. $\qquad\square$

This is particularly interesting for log toroidal families of elementary Gross–Siebert type since, in this case, we have complete control over the generic fiber of the local models, as explained in Lemma 14.9 and Corollary 14.11. We need the following preparation.

Lemma 17.15 *Let $\Sigma = \operatorname{Spec} \mathbb{C}[\![t]\!]$, and let $f\colon X \to \Sigma$ be a morphism of finite type between affine schemes. Let $g_0\colon Y_0 \to X_0$ be an étale morphism of finite type between affine schemes. Then there is an étale morphism $g\colon Y \to X$ of finite type between affine schemes such that g induces $g_0\colon Y_0 \to X_0$ over S_0.*

Proof Let $B = \Gamma(X, O_X)$, let $B_0 = \Gamma(X_0, O_{X_0})$, and let $C_0 = \Gamma(Y_0, O_{Y_0})$. Since $g_0\colon Y_0 \to X_0$ is étale, we have $C_0 = B_0[x_1, \ldots, x_n]/(f_1, \ldots, f_n)$ with

$$\Delta := \det\left(\frac{\partial f_j}{\partial x_i}\right)$$

invertible in C_0. Let $f'_1, \ldots, f'_n \in B[x_1, \ldots, x_n]$ be arbitrary lifts of $f_1, \ldots, f_n$, let $\Delta' := \det\left(\frac{\partial f'_j}{\partial x_i}\right)$, and set

$$C := \left(B[x_1, \ldots, x_n]/(f'_1, \ldots, f'_n)\right)_{\Delta'}.$$

Then $B \to C$ is étale and induces $B_0 \to C_0$ after base change. $\qquad\square$

Corollary 17.16 *In Situation 17.1, assume that $f_0 \colon X_0 \to S_0$ is proper, and assume that both $f_0 \colon X_0 \to S_0$ and all deformations in $\mathcal{D}$ are log toroidal of elementary Gross–Siebert type. Then the generic fiber X_η has toroidal singularities which are smooth in codimension 3, locally rigid, terminal, Gorenstein, and Abelian quotient singularities.*[2]

Proof Let $f \colon X \to \Sigma$ be the algebraic deformation. By subdividing $\mathcal{V}$, we may assume that $\mathcal{V}$ is an affine cover of X_0. Let $W_\alpha \subseteq X$ be an affine open subset with $W_\alpha \cap X_0 = V_\alpha$. Around a point $x \in V_\alpha$, we may find an étale roof $Y_0 \to V_\alpha$, $Y_0 \to L_0$ where L_0 is the central fiber of a local model of elementary Gross–Siebert type. Let $Y \to W_\alpha$ be an étale morphism obtained from $Y_0 \to W_\alpha$ via Lemma 17.15, and let $Y' \to L$ be one obtained from $Y_0 \to L_0$. Here, $L \to S$ is the local model of elementary Gross–Siebert type. Since infinitesimal log toroidal deformations of elementary Gross–Siebert type are locally unique, we have $Y_k \cong Y'_k$ for all $k \geq 0$. Now Lemma 17.12 shows that W_α and L are étale locally isomorphic after possibly shrinking W_α. Then the generic fiber has the stated properties on the new W_α. Since $f_0 \colon X_0 \to S_0$ is proper, the new W_α collectively cover X and thus X_η. □

In other words, the generic fiber X_η has very mild singularities, and furthermore, they cannot be improved by a flat deformation of X_η.

Remark 17.17 In the preprint [242], Ruddat gives a variant of Lemma 17.12 in which the isomorphism φ_k does not need to be given for all $k \geq 0$. Instead, one fixes the first family $f \colon X \to \Sigma$, and then there is a number N, depending on $f \colon X \to \Sigma$, such that the statement of Lemma 17.12 holds when φ_k is given for $k \leq 4N + 1$, for any flat $f' \colon X' \to \Sigma$ of finite type. In order for this statement to hold, one needs the crucial assumption that the generic fiber of the first given family $f \colon X \to \Sigma$ is *locally rigid*, i.e., $\mathcal{T}^1_{X_\eta/\eta} = 0$ for the first tangent sheaf in the sense of Lichtenbaum–Schlessinger. As we have just seen, this assumption is satisfied in the case of elementary Gross–Siebert type. The relevance of Ruddat's version is that, by Corollary 1.19, one can find, for every $c \geq 0$, an *analytic* family over a small disk Δ which coincides with the analytification of $f_c \colon X_c \to S_c$ after base change to S_c. Then the general fiber of this family must have the same singularities as the general fiber of the analytification of the local model of elementary Gross–Siebert type. The advantage of this picture is that we do not need to have an ample line bundle on X_0. However, for algebraic degenerations arising from Situation 17.1, the weaker Lemma 17.12 is enough.   ◊

---

[2] $\mathbb{Q}$-factoriality does not translate to X_η because this property is not étale local.

17.2 Comparing the Two Log Structures

Suppose again we are in Situation 17.1, and assume for simplicity that $f_0 \colon X_0 \to S_0$ is vertical. Let $f \colon X \to \Sigma$ be the algebraic deformation whose existence is guaranteed by Lemma 17.3. The base Σ carries two log structures. On the one hand, we have the log structure defined by the chart, i.e., $\Sigma = \mathrm{Spec}(\mathbb{N} \to \mathbb{C}[\![t]\!])$. On the other hand, the divisor $\{t = 0\}$ defines a log structure on Σ. The two log structures are canonically isomorphic via $\mathbb{N} \to \mathcal{M}_{(\Sigma,0)}$, $1 \mapsto t$.

Similarly, we can endow the total space X with the divisorial log structure defined by the divisor $X_0 = \{t = 0\}$ and obtain a log scheme $(X|X_0)$. Via pull-back, this turns each $f_k \colon X_k \to S_k$ into a morphism of log schemes. Let us denote it by $g_k \colon Y_k \to S_k$ to distinguish it from the structure of a generically log smooth family that is already given on $f_k \colon X_k \to S_k$. In this section, we show that $g_k \colon Y_k \to S_k$ is, after restriction to $U_0 \subseteq X_0$, canonically isomorphic to $f_k \colon U_k \to S_k$. We denote the restriction of $g_k \colon Y_k \to S_k$ by $g_k \colon V_k \to S_k$. First, we show that $f_0 \colon U_0 \to S_0$ is isomorphic to $g_0 \colon V_0 \to S_0$.

Lemma 17.18 *In Situation 17.1, assume that $f_0 \colon U_0 \to S_0$ is vertical. Let $g_0 \colon V_0 \to S_0$ be defined as above. Then there is a unique isomorphism $\psi_0 \colon U_0 \xrightarrow{\;\cong\;} V_0$ of log schemes over S_0 which is the identity on underlying schemes.*

Proof Let $W \subseteq X$ be an affine open subset such that $W_0 := W \cap X_0 \subseteq U_0$. Let $\gamma_0 \colon N_0 \to W_0$ be a strict étale morphism with N_0 affine, and assume that we have another strict étale morphism $\eta_0 \colon N_0 \to V(\sigma) \times \mathbb{A}^{d-r} =: L_0$. Here, we use the notation from Construction 9.2. When we apply Lemma 17.15 to g_0, we obtain an étale morphism $\gamma \colon N \to W$, and when we apply the lemma to η_0, then we obtain an étale morphism $\eta \colon N' \to U_\Sigma(\sigma) \times_\Sigma \mathbb{A}^{d-r}_\Sigma =: L$. We apply Lemma 17.12 to $N' \to L \to \Sigma$. Since both $N_k \to S_k$ and $N'_k \to S_k$ are log smooth deformations of the affine log smooth log scheme $N_0 \to S_0$, we have compatible isomorphisms $N_k \cong N'_k$ over S_k up to arbitrary order. Then we can find étale neighborhoods $\bar{N} \to N$ and $\bar{N}' \to N'$ of any given point $v \in N_0$ together with an isomorphism $\phi \colon \bar{N} \cong \bar{N}'$ over Σ which is compatible with $N_0 = N'_0$ on the central fiber.

We analyze the log structure is this picture. On W, we have the divisorial log structure defined by W_0. Let us denote this log scheme by M. Via base change, we have a log morphism $M_0 \to S_0$. Since $\bar{N} \to N \to W$ is étale, this map becomes strict as a morphism of log schemes $(\bar{N}|\bar{N}_0) \to M$. Since $\bar{N} \cong \bar{N}'$, we also have $(\bar{N}|\bar{N}_0) \cong (\bar{N}|\bar{N}'_0)$. On L, the log structure is canonically isomorphic to the divisorial log structure defined by $L_0 \subset L$. Since also $\bar{N}' \to L$ is étale, we obtain a strict étale morphism $(\bar{N}'|\bar{N}'_0) \to (L|L_0)$. In particular, we have a strict étale morphism $\bar{N}'_0 \to L_0$, where the original log structures are the same as the one pulled back from the divisorial ones. Moreover, on $\bar{N}_0$, the log structure pulled back from the divisorial one on $\bar{N}$ coincides with the one pulled back from W_0. However, also $\bar{N}_0 \to M_0$ is strict étale by construction. In other words, the log structures on W_0 and M_0 are étale locally isomorphic.

On U_0, we have the structure of a toroidal crossing space by taking the ghost sheaf of the log structure. Thus, we obtain a section $s \in \mathcal{LS}_{U_0}$. Locally, the ghost sheaf of the log structure coming from $X_0 \subseteq X$ is isomorphic to $\mathcal{P}$. Since $\mathcal{P}$ has no non-trivial automorphisms, the two ghost sheaves coincide. Thus, $X_0 \subseteq X$ defines a section of $\mathcal{LS}_{U_0}$ as well. The two sections must be locally the same, so they are in fact globally the same. In other words, the two log structures coincide on U_0.

Essentially with the same argument, we show the uniqueness of $\psi_0 \colon U_0 \cong V_0$ over S_0. Namely, let $\Psi \colon U_0 \to U_0$ be an automorphism over S_0 which is the identity on underlying schemes, given by $\varphi \colon \mathcal{M}_{U_0} \to \mathcal{M}_{U_0}$. Then $\bar{\varphi} \colon \overline{\mathcal{M}}_{U_0} \to \overline{\mathcal{M}}_{U_0}$ is an automorphism of the ghost sheaf, i.e., an automorphism of $\mathcal{P}$. In particular, it must be trivial. But then $\varphi \colon \mathcal{M}_{U_0} \to \mathcal{M}_{U_0}$ must be the identity because sections of $\mathcal{LS}_{U_0}$ do not have any non-trivial automorphisms. $\square$

The proof of the lemma also shows the following result:

Corollary 17.19 $g_k \colon V_k \to S_k$ *is a log smooth deformation of* $g_0 \colon V_0 \to S_0$.

Proof We keep the notations from the previous proof. Endow $\bar{N}_k$ with the pull-back of the divisorial log structure on $\bar{N}$. Then $\bar{N}_k \to M_k$ is strict étale. Since also $\bar{N}_k \cong \bar{N}'_k \to L_k$ is strict étale, we find that $\bar{N}_k \to S_k$ is log smooth. Since log smoothness is an étale local property, we find that $M_k \to S_k$ and hence $g_k \colon V_k \to S_k$ is log smooth. $\square$

This shows that $f_k \colon U_k \to S_k$ and $g_k \colon V_k \to S_k$ are locally isomorphic, but it does not yet show that they are globally isomorphic. The proof that a global isomorphism exists is surprisingly intricate.

The reason why the proof is so intricate is as follows: Consider the forgetful map

$$\mathcal{A}ut_k(U_{k+1}) \to \mathcal{A}ut_k(\underline{U}_{k+1})$$

from the sheaf of automorphisms of the log scheme U_{k+1} inducing the identity on U_k to the sheaf of automorphisms of the underlying scheme $\underline{U}_{k+1}$ inducing the identity on $\underline{U}_k$. This map is always injective since $\Theta^1_{U_0/S_0} \otimes_{\mathbf{k}} I$ has injective restriction maps to the strict locus U_0° of $U_0 \to S_0$, and on U_0°, the two sheaves are the same. However, this restriction map is not surjective. Thus, there a fewer logarithmic automorphisms than classical automorphisms. As a result, two log smooth deformations $f_{k+1} \colon U_{k+1} \to S_{k+1}$ and $g_{k+1} \colon V_{k+1} \to S_{k+1}$ may be non-isomorphic even if the underlying schemes are isomorphic, and f_k and g_k are isomorphic as log morphisms. The key result to circumvent this problem in the above situation is the following.

Proposition 17.20 *Let* $f_0 \colon U_0 \to S_0$ *be a saturated, log smooth, and vertical morphism over the standard log point* $S_0 = \mathrm{Spec}(\mathbb{N} \to \mathbf{k})$. *Then there is a number* $\mu \geq 1$, *depending only on* $f_0 \colon U_0 \to S_0$, *with the following property: Let* $k \geq 0$ *and* $\ell \geq k + \mu$, *and let* $f_\ell \colon U_\ell \to S_\ell$ *be a log smooth deformation. Let* $\phi_\ell \colon \underline{U}_\ell \to \underline{U}_\ell$ *be an automorphism of schemes over* S_ℓ. *Then there is a unique automorphism*

$\Phi_k : U_k \to U_k$ *of log schemes over S_k whose underlying automorphism of schemes* $\underline{\Phi}_k : \underline{U}_k \to \underline{U}_k$ *coincides with the restriction ϕ_k of ϕ_ℓ to $\underline{U}_k$.*

Proof Let us denote by $\mathcal{T}_k = \mathcal{T}_{U_k/S_k}$ the sheaf of relative classical derivations. Then restriction of derivations yields a homomorphism $\mathcal{T}_k \to \mathcal{T}_0$ of sheaves. More precisely, this homomorphism is given by $D(\bar{a}) = \overline{D(a)}$ for $a \in O_{U_k} =: O_k$, where $\bar{a} \in O_0$ is its restriction. Let $\mathcal{B}_k$ be the kernel of this homomorphism. Then the formulas in Sect. 10.2 produce an isomorphism $\mathrm{Exp} : \mathcal{B}_k \to \mathcal{A}ut(U_k/U_0)$. Similarly, by the results of Sect. 10.2, we have an isomorphism $\mathrm{Exp} : \mathcal{A}_k \to \mathcal{A}ut(U_k/U_0)$, where $\mathcal{A}_k$ is the kernel of the restriction map $\Theta_k^1 := \Theta_{U_k/S_k}^1 \to \Theta_{U_0/S_0}^1 =: \Theta_0^1$. We have already seen above that $\mathcal{A}_k \to \mathcal{B}_k$ is injective, showing in particular the uniqueness of Φ_k. In other words, we have to show that, given $\theta_\ell \in \mathcal{B}_\ell$, we have $\theta_\ell|_k \in \mathcal{A}_k \subseteq \mathcal{B}_k$ if $\ell \geq k + \mu$. Since all these sheaves are Z-closed for any closed subset $Z \subset U_0$ of codimension ≥ 2, it is sufficient to show this on an open subset $W_0 \subseteq U_0$ whose complement has codimension ≥ 2. Furthermore, the statement is étale local, so it is sufficient to prove it for $U_0 \to S_0$ of the following form: $U_0 \to S_0$ is the central fiber of

$$\mathrm{Spec}\,\mathbf{k}[x, y, z_1, \ldots, z_q, t]/(xy - t^m) \to \mathrm{Spec}\,\mathbf{k}[t].$$

The m in this equation is the μ from the statement. More precisely, the μ in the statement is the maximum over all m that occur in local models of the given $f_0 : U_0 \to S_0$ in codimension 1. We achieve this through a direct computation. For simplicity, we can assume $q = 0$; in the other cases, the computation is the same. Let $R = \mathbf{k}[x, y, t]/(xy - t^m)$. Then

$$\mathcal{T} := \mathcal{T}_{\mathrm{Spec}\,R/\mathbb{A}^1} = \ker\left(R^2 \to R,\ (a, b) \mapsto ya + xb\right).$$

When writing $R_k = R/(t^{k+1})$, we find similarly

$$\mathcal{T}_k = \ker\left(R_k^2 \to R_k,\ (a, b) \mapsto ya + xb\right).$$

The image of $\Theta_k^1 \to \mathcal{T}_k$ is represented by $a, b \in \mathbf{k}[x, y, t]$ with $ya + xb \in (xy - t^m)$. It is now sufficient to show the following: If $\theta_\ell \in \mathcal{T}_\ell$ with $\ell \geq k + m$, then $\theta_\ell|_k \in \Theta_k^1$. If $\theta_\ell \in \mathcal{B}_\ell$, then $\theta_\ell|_k \in \mathcal{A}_k$ is automatic provided that $\theta_\ell|_k \in \Theta_k^1$. Elements of $\mathcal{T}_\ell$ are represented by $c, d \in \mathbf{k}[x, y, t]$ such that $yc + xd \in (xy - t^m, t^{\ell+1})$. It suffices to show the following: Let c, d be such that $yc + xd \in (xy - t^m, t^{\ell+1})$. Then there are a, b with $a \equiv c \mod t^{k+1}$ and $b \equiv d \mod t^{k+1}$ such that $ya + xb \in (xy - t^m)$. Let $yc + xd = f(xy - t^m) + gt^{\ell+1}$. Then we achieve this by setting $a := c$ and $d := c - ygt^{\ell+1-m}$. In fact, we have

$$ya + xb = yc + x(c - ygt^{\ell+1-m}) = f(xy - t^m) + gt^{\ell+1} - xygt^{\ell+1-m}$$

$$= (f - gt^{\ell+1-m})(xy - t^m).$$

This is possible because we assumed $\ell \geq k + m$. $\qquad\square$

Corollary 17.21 *There are unique isomorphisms $\psi_k \colon U_k \cong V_k$ of log schemes over S_k which are the identity on underlying schemes and induce ψ_0 on the central fiber. In particular, the family of isomorphisms $(\psi_k)_k$ is compatible with base change.*

Proof Uniqueness follows from injectivity of $\mathcal{A}_k \to \mathcal{B}_k$. Now fix an open affine cover $\mathcal{U} = \{U_\alpha\}_\alpha$ of U_0, and let $\phi_{k;\alpha} \colon \underline{U}_{k;\alpha} \to \underline{V}_{k;\alpha}$ be the identity. Choose isomorphisms of log schemes $\Phi_{k;\alpha} \colon U_{k;\alpha} \to V_{k;\alpha}$ compatible with restrictions. This is possible because both are log smooth. Note that $\underline{\Phi}_{k;\alpha}$ does not need to be the identity. Now let $\Psi_{k;\alpha} = \phi_{k;\alpha}^{-1} \circ \Phi_{k;\alpha} \colon \underline{U}_{k;\alpha} \to \underline{U}_{k;\alpha}$. Since $\Psi_{k;\alpha}$ extends to arbitrarily high order as an automorphism of schemes, it can be extended to a unique automorphism of log schemes $\Psi_{k;\alpha} \colon U_{k;\alpha} \to U_{k;\alpha}$. Then $\psi_k = (\Phi_{k;\alpha} \circ \Psi_{k;\alpha}^{-1})_\alpha$ is the desired isomorphism. $\qquad\square$

Chapter 18
Modifications of the Log Structure

The main unobstructedness results apply to log Calabi–Yau spaces $f_0 \colon X_0 \to S_0$. If the anti-canonical bundle $\mathcal{A}^d_{X_0/S_0}$ is only assumed to be effective, we can (in some cases) use a section s_0 of it to modify the log structure on X_0 and obtain a log Calabi–Yau space $f_0 \circ h_0 \colon X_0(s_0) \to S_0$ whose deformation theory is closely related to the deformation theory of $f_0 \colon X_0 \to S_0$. Then, we can often infer unobstructedness of the deformations of f_0 from unobstructedness of the deformations of $f_0 \circ h_0$. In this final chapter, we explain precisely how to modify the log structure and establish the basic properties of this construction.

The main application of modifications is Theorem 18.44, a smoothing result for normal crossing spaces V with the property that $(\omega^\circ_V)^\vee$ is globally generated (instead of trivial). With [77, Thm. 1.1], we gave a very similar result. However, the assumptions in [77] are somewhat stronger, and the proof is quite sketchy in the non-Calabi–Yau case. Here, we give a full account of the necessary considerations.

We work in the generality of enhanced generically log smooth families. While this situation is considerably more subtle than the case of plain generically log smooth families, we believe that this generality is necessary for some applications such as in the situations described in Sect. 1.12. The most general setup for this chapter is the following.

Situation 18.1 Let $f \colon X \to S$ be an enhanced generically log smooth family of relative dimension d with Gerstenhaber calculus $\mathcal{GC}^\bullet_{X/S}$, endowed with maps

$$\varpi^\bullet \colon \mathcal{G}^\bullet_{X/S} \to \mathcal{V}^\bullet_{X/S} \quad \text{and} \quad \varpi^\bullet \colon \mathcal{A}^\bullet_{X/S} \to \mathcal{W}^\bullet_{X/S}.$$

Recall that $\varpi^\bullet$ must be an isomorphism on U, and that each $\mathcal{A}^i_{X/S}$ and $\mathcal{G}^p_{X/S}$ must be flat over S. By definition, $f \colon X \to S$ is log Gorenstein, and we have $\mathcal{A}^d_{X/S} = \mathcal{W}^d_{X/S}$. We fix a line bundle $\mathcal{L}$ on X and a global section $s \in \mathcal{L}$.

© The Author(s), under exclusive license to Springer Nature Switzerland AG 2025
S. Felten, *Global Logarithmic Deformation Theory*, Lecture Notes
in Mathematics 2373, https://doi.org/10.1007/978-3-031-98751-9_18

18.1 Log Pre-regular Sections and $f \circ p \colon L(X) \to S$

In this section, we study log pre-regular sections $s \in \mathcal{L}$, and we investigate the generically log smooth family $f \circ p \colon L(X) \to S$ used to define the modification $f \circ h \colon X(s) \to S$. We turn $f \circ p \colon L(X) \to S$ into an enhanced generically log smooth family.

Definition 18.2 (Log Pre-regular Sections)   In Situation 18.1, we say that s is a *log pre-regular section* if:

(a) the induced map $s \colon O \to \mathcal{L}$ is injective;
(b) the effective Cartier divisor $H = \mathrm{div}(s)$ is flat over S; we write $i \colon H \to X$ for the embedding;
(c) the intersection $Z \cap H \subseteq H$ has codimension ≥ 2 in every fiber of $f \circ i \colon H \to S$, and furthermore, $\mathrm{depth}(O_{H_s,z}) \geq 2$ for every $z \in Z_s \cap H_s$ and every $s \in S$.[1]

In this case, we denote the underlying scheme $\underline{X}$ of X, endowed with the Deligne–Faltings log structure defined by $s^\vee \colon \mathcal{L}^\vee \to O_X$, by $\underline{X}(s)$. We write $X(s)$ for the fiber product[2] of $X \to \underline{X} \leftarrow \underline{X}(s)$; this space carries a log structure on $U \subseteq X$. We denote the induced map by $h \colon X(s) \to X$; this is a map of schemes everywhere, and a map of log schemes on U.

The notion of a log pre-regular section $s \in \mathcal{L}$ is stable under base change; namely, since $O_H = \mathrm{coker}(\mathcal{L}^\vee \to O_X)$ is flat over S, injectivity of $\mathcal{L}^\vee \to O_X$ is preserved under base change. Similarly, the construction of $h \colon X(s) \to X$ commutes with base change. We have a second, equivalent construction of $h \colon X(s) \to X$, which is more useful in studying $f \circ h \colon X(s) \to S$. For the sake of explicit computations, let us introduce a finite affine open cover $\mathcal{V} = \{V_\alpha\}_\alpha$ of X such that $\mathcal{L}|_\alpha$ is trivial. Let $e_\alpha \in \mathcal{L}|_\alpha$ be a trivializing section, and let $s = s_\alpha \cdot e_\alpha$ with $s_\alpha \in O_X|_\alpha$. We denote the coordinate transforms by $\gamma_{\alpha\beta} \in O_X^*|_{\alpha\beta}$, i.e., $e_\alpha = \gamma_{\alpha\beta} e_\beta$. We denote the dual trivializing sections by $e_\alpha^\vee \in \mathcal{L}^\vee|_\alpha$; then the coordinate transform on $\mathcal{L}^\vee$ is $e_\alpha^\vee = \gamma_{\alpha\beta}^{-1} e_\alpha^\vee$. We have $s^\vee(e_\alpha^\vee) = s_\alpha$ under the map $s^\vee \colon \mathcal{L}^\vee \to O_X$. The function s_α transforms as $s_\alpha = \gamma_{\alpha\beta}^{-1} s_\beta$. Let us denote the coordinate rings by $R_\alpha := \Gamma(V_\alpha, O_X)$.
  Let

$$L = \mathrm{Spec}_{O_X} \bigoplus_{\ell \geq 0} (\mathcal{L}^\vee)^{\otimes \ell}$$

---

[1] This ensures $j_* O_{U \cap H} = O_H$. Since we have required $f \colon X \to S$ only to have (S_2)-fibers, the condition on the codimension is not sufficient. Note that the fibers of $f \circ i \colon H \to S$ satisfy Serre's condition (S_2) because H is cut out by a non-zero divisor, and $U \to S$ has Cohen–Macaulay fibers.
[2] Formed in the category of all log schemes. The log structure on $X(s)$ is fine and saturated; thus, $\underline{X}(s) \to \underline{X}$ is a saturated morphism, and $U(s)$ is fine and saturated as well.

be the total space of $\mathcal{L}$, and let $p \colon L \to X$ be the projection. On $p^{-1}(U)$, we endow L with the log structure pulled back from $U \subseteq X$. The coordinate ring of L on $W_\alpha := p^{-1}(V_\alpha)$ is $R_\alpha[x_\alpha]$ with x_α corresponding to $e_\alpha^\vee \in \mathcal{L}|_\alpha$. On overlaps, we have $x_\alpha = \gamma_{\alpha\beta}^{-1} x_\beta$. The projection $p \colon L \to X$ is given by $R_\alpha \to R_\alpha[x_\alpha]$. Let $o \colon X \to L$ be the map corresponding to $0 \in \mathcal{L}$, which is given by $R_\alpha[x_\alpha] \to R_\alpha$, $x_\alpha \mapsto 0$. The ideal sheaf of $o \colon X \to L$ is $\varepsilon_0 \colon p^*\mathcal{L}^\vee \hookrightarrow \mathcal{O}_L$ with $p^*e_\alpha^\vee \mapsto x_\alpha$. This gives rise to a Deligne–Faltings log structure on the underlying space $\underline{L}$ of L, which we denote by $\underline{L}(X)$. As above, we can form the fiber product $L(X)$ of $L \to \underline{L} \leftarrow \underline{L}(X)$, which is a log scheme on $p^{-1}(U)$. By abuse of notation, we will write $p \colon L(X) \to X$ as well.

Let $s \colon X \to L$ be the map corresponding to $s \in \mathcal{L}$; it is given by $R_\alpha[x_\alpha] \to R_\alpha$, $x_\alpha \mapsto s_\alpha$. The ideal sheaf of $s \colon X \to L$ is $\varepsilon_s \colon p^*\mathcal{L}^\vee \hookrightarrow \mathcal{O}_L$ with $p^*e_\alpha^\vee \mapsto x_\alpha - s_\alpha$. Then we define a log scheme $X'(s)$ by pulling back the log structure from $L(X)$ along $s \colon X \to L$. This log structure is defined on $s^{-1}(p^{-1}(U)) = U$. Since $s^*\varepsilon_0 \colon \mathcal{L}^\vee \to \mathcal{O}_X$ is precisely $s^\vee \colon \mathcal{L}^\vee \to \mathcal{O}_X$, and since the pull-back of Deligne–Faltings structures is the same as the pull-back of the associated log structures, it is easy to see that this second construction coincides with the first one in Definition 18.2.

Lemma 18.3 *In the above situation, we have:*

(1) *The map $p \colon L \to X$ is strict and smooth of relative dimension* 1. *We have* $\Omega^1_{L/X} \cong p^*\mathcal{L}^\vee$.
(2) *The map $p \colon L(X) \to X$ is smooth and log smooth of relative dimension* 1. *We have* $\Omega^1_{L(X)/X} = \mathcal{O}_L \cdot \gamma_s$, *where we have* $\gamma_s|_{U \cap V_\alpha} = \delta_{L(X)/X}(m_\alpha)$ *in a local chart.*

Proof It is clear that $p \colon L \to X$ is strict and smooth. The notation $\Omega^1_{L/X}$ means the classical relative differentials. They are, on W_α, generated by dx_α. Since we have $dx_\alpha = \gamma_{\alpha\beta}^{-1} \cdot dx_\beta$ on overlaps, we find $\Omega^1_{L/X} \cong p^*\mathcal{L}^\vee$. Since $\underline{p} \colon \underline{L}(X) \to \underline{X}$ is locally a base change of $A_\mathbb{N} \to \operatorname{Spec} \mathbb{Z}$, it is smooth and log smooth. Let us denote by $\underline{m}_\alpha \in \mathcal{M}_{\underline{L}(X)}(W_\alpha)$ the section induced by $x_\alpha \in \varepsilon_0(p^*\mathcal{L}^\vee)$ via the construction of the associated log structure of a Deligne–Faltings structure. Then $\Omega^1_{\underline{L}(X)/\underline{X}}$ is locally free of rank 1 and, on W_α, generated by $\delta(\underline{m}_\alpha)$, where δ denotes the log part of the universal derivation. Due to $\underline{m}_\alpha = \gamma_{\alpha\beta}^{-1} \underline{m}_\beta$, we find $\delta(\underline{m}_\alpha) = \delta(\underline{m}_\beta)$ on overlaps, so $\Omega^1_{\underline{L}(X)/\underline{X}} \cong \mathcal{O}_L$. Now $p \colon L(X) \to X$ is a base change of $\underline{p} \colon \underline{L}(X) \to \underline{X}$, so we have $\Omega^1_{L(X)/X} \cong \Omega^1_{\underline{L}(X)/\underline{X}}$ where the former sheaf is defined, and we consider $\Omega^1_{L(X)/X}$ to be defined globally via this isomorphism. This is obviously the same as taking the direct image from $p^{-1}(U)$. $\qquad\square$

The composition $f \circ p \colon L(X) \to S$ is clearly a generically log smooth family. We now turn to the problem of endowing it with a Gerstenhaber calculus $(\mathcal{G}^\bullet_{L(X)/S}, \mathcal{A}^\bullet_{L(X)/S})$ which turns it into an enhanced generically log smooth family, compatibly with $f \colon X \to S$. We set $\mathcal{A}^0_{L(X)/S} := \mathcal{O}_L$. Consider the exact sequence

$$0 \to p^* \mathcal{W}^1_{X/S} \to \mathcal{W}^1_{L(X)/S} \to \Omega^1_{L(X)/X} \to 0. \tag{18.1}$$

It is clearly an exact sequence on $p^{-1}(U)$. Since p is flat, $p^* \mathcal{W}^1_{X/S}$ is reflexive and hence $p^{-1}(Z)$-closed. Since $\Omega^1_{L(X)/X}$ is $p^{-1}(Z)$-closed as well, we can define the sequence as the direct image from $p^{-1}(U)$. If $m_\alpha \in \Gamma(p^{-1}(U) \cap W_\alpha, \mathcal{M}_{L(X)})$ is the image of $\underline{m}_\alpha$ under $L(X) \to \underline{L}(X)$, then $\delta(m_\alpha) \in \Gamma(p^{-1}(U) \cap W_\alpha, \mathcal{W}^1_{L(X)/S})$ is a preimage of the generator of $\Omega^1_{L(X)/X}$, so the map is surjective. Since $\Omega^1_{L(X)/X}$ is free, the exact sequence is locally split. Any two splittings differ by a map in $\mathcal{H}om(\Omega^1_{L(X)/X}, p^* \mathcal{W}^1_{X/S})$. There is, however, a set of distinguished splittings such that any two distinguished splittings differ by a map in $\mathcal{H}om(\Omega^1_{L(X)/X}, p^* \mathcal{A}^1_{X/S})$. One distinguished splitting is given by

$$\gamma_s = \delta_{L(X)/X}(m_\alpha) \mapsto \delta_\alpha := \delta_{L(X)/S}(m_\alpha).$$

On overlaps, we have

$$\delta_{L(X)/S}(\gamma_{\alpha\beta}^{-1} \cdot m_\beta) = \gamma_{\alpha\beta} \cdot d_{L(X)/S}(\gamma_{\alpha\beta}^{-1}) + \delta_{L(X)/S}(m_\beta).$$

Because of $d_{L(X)/S}(\gamma_{\alpha\beta}^{-1}) \in p^* \mathcal{A}^1_{X/S}$, the difference is given by a map to $p^* \mathcal{A}^1_{X/S} \subseteq p^* \mathcal{W}^1_{X/S}$. Thus, the class of distinguished splittings is globally consistent. If we modify the trivializing section e_α by a unit u_α, then a similar computation shows that the class of distinguished splittings is independent of the choice of e_α. If $B_1 \colon \Omega^1_{L(X)/X} \to \mathcal{W}^1_{L(X)/S}$ is some distinguished splitting, then we set

$$\mathcal{A}^1_{L(X)/S} := p^* \mathcal{A}^1_{X/S} \oplus B_1(\Omega^1_{L(X)/X}) \subseteq \mathcal{W}^1_{L(X)/S}.$$

This is independent of the choice of B_1, and thus globally well-defined; it fits into a split exact sequence

$$0 \to p^* \mathcal{A}^1_{X/S} \to \mathcal{A}^1_{L(X)/S} \to \Omega^1_{L(X)/X} \to 0.$$

For $i \geq 2$, we set

$$\mathcal{A}^i_{L(X)/S} := p^* \mathcal{A}^i_{X/S} + p^* \mathcal{A}^{i-1}_{X/S} \wedge B_1(\Omega^1_{L(X)/X}) \subseteq \mathcal{W}^i_{L(X)/X},$$

where, again, $B_1 \colon \Omega^1_{L(X)/X} \to \mathcal{W}^1_{L(X)/S}$ is some local distinguished splitting. This is independent of the choice of B_1 and hence globally well-defined. We have a diagram

$$
\begin{array}{ccccccccc}
0 & \longrightarrow & p^*\mathcal{W}^i_{X/S} & \xrightarrow{\;E_i\;} & \mathcal{W}^i_{L(X)/S} & \xrightarrow{\;Q_i\;} & p^*\mathcal{W}^{i-1}_{X/S} \otimes \Omega^1_{L(X)/X} & \longrightarrow & 0 \\[2mm]
 & & \Big\uparrow & & \Big\uparrow & & \Big\uparrow & & \\[2mm]
0 & \longrightarrow & p^*\mathcal{A}^i_{X/S} & \xrightarrow{\;E_i\;} & \mathcal{A}^i_{L(X)/S} & \xrightarrow{\;Q_i\;} & p^*\mathcal{A}^{i-1}_{X/S} \otimes \Omega^1_{L(X)/X} & \longrightarrow & 0.
\end{array}
$$

First, we obtain both rows as exact sequences on $p^{-1}(U)$ via the canonical construction. The upper row is then obtained as the direct image from $p^{-1}(U)$; it is exact on the right because we can construct explicit preimages via a splitting B_1 from above. It follows from the definition of $\mathcal{A}^i_{L(X)/S}$ that the lower row is well-defined as a sequence of subsheaves of the upper row, and it is easy to see that it is exact. A distinguished splitting B_1 as above gives rise to a simultaneous (local) splitting of both rows. We denote the induced splitting on the left-hand side by

$$
T_i \colon \quad \mathcal{W}^i_{L(X)/S} \to p^*\mathcal{W}^i_{X/S};
$$

it restricts to a map $T_i \colon \mathcal{A}^i_{L(X)/S} \to p^*\mathcal{A}^i_{X/S}$.

Lemma 18.4 *We have $T_i(\alpha) \wedge T_j(\beta) = T_{i+j}(\alpha \wedge \beta)$.*

Proof Let $\delta := B_1(\gamma_s)$, depending on our choice of B_1. Then, locally, every element $\alpha \in \mathcal{A}^i_{L(X)/S}$ can be written uniquely as $\alpha = E_i(\alpha_1) + E_{i-1}(\alpha_2) \wedge \delta$ with $\alpha_1 \in p^*\mathcal{A}^i_{X/S}$ and $\alpha_2 \in p^*\mathcal{A}^{i-1}_{X/S}$. Then $T_i(E_i(\alpha_1) + E_{i-1}(\alpha_2) \wedge \delta) = \alpha_1$. We have

$$
(E_i(\alpha_1) + E_{i-1}(\alpha_2) \wedge \delta) \wedge (E_j(\beta_1) + E_{j-1}(\beta_2) \wedge \delta)
$$
$$
= E_{i+j}(\alpha_1 \wedge \beta_1) + E_{i+j-1}(\alpha_1 \wedge \beta_2 + \alpha_2 \wedge \beta_1) \wedge \delta,
$$

and hence $T_i(\alpha) \wedge T_j(\beta) = T_{i+j}(\alpha \wedge \beta)$. $\square$

The maps E_i and Q_i are independent of B_1. To capture the dependence of T_i from B_1, let $\delta = B_1(\gamma_s)$ and $\delta' = B_1'(\gamma_s)$. Then there is a unique $\varepsilon \in p^*\mathcal{A}^1_{X/S}$ with $\delta' = \delta + E_1(\varepsilon)$. Now a direct computation yields $T_i'(\alpha) = T_i(\alpha) - Q_i(\alpha) \wedge \varepsilon$, where we consider $Q_i(\alpha) \in p^*\mathcal{W}^{i-1}_{X/S}$ by identifying the generator γ_s of $\Omega^1_{L(X)/X}$ with 1.

We write $\mathcal{V}^\bullet_{L(X)/S}$ for the reflexive Gerstenhaber algebra of polyvector fields on $f \circ p \colon L(X) \to S$. By dualizing the split exact sequence above, we get a split exact sequence

$$
0 \to p^*\mathcal{V}^{p+1}_{X/S} \otimes \Theta^1_{L(X)/X} \xrightarrow{\;I_p\;} \mathcal{V}^p_{L(X)/S} \xrightarrow{\;F_p\;} p^*\mathcal{V}^p_{X/S} \to 0 \tag{18.2}
$$

of reflexive sheaves. In the middle and on the right, we identify these sheaves with the duals via the contraction map. We have $\Theta^1_{L(X)/X} = O_L \cdot \gamma_s^\vee$ for the formal dual of γ_s, and we identify on the left via

$$p^* \mathcal{V}^{p+1}_{X/S} \otimes \Theta^1_{L(X)/X} \to \mathcal{H}om(p^* \mathcal{W}^{-p+1}_{X/S} \otimes \Omega^1_{L(X)/X}, O_L),$$

$$\theta \otimes \gamma_s^\vee \mapsto (\alpha \otimes \gamma_s \mapsto (-1)^{p+1}\theta \lrcorner\, \alpha).$$

We define an element $\delta^\vee \in \mathcal{V}^{-1}_{L(X)/S}$, a priori depending on the choice of B_1, by setting $\delta^\vee \lrcorner\, \delta = 1$ and $\delta^\vee \lrcorner\, E_1(\alpha) = 0$. However, we also have $\delta^\vee \lrcorner\, (\delta + E_1(\varepsilon)) = 1$, so $\delta^\vee$ is actually independent of the choice of B_1. Dualizing the splitting T_{-p} of E_{-p} gives a splitting

$$S_p: \quad p^* \mathcal{V}^p_{X/S} \to \mathcal{V}^p_{L(X)/S}$$

of F_p.

Lemma 18.5 *The following formulae hold for any distinguished splitting B_1:*

(1) $\delta^\vee \lrcorner\, E_i(\alpha) = 0$ *for* $\alpha \in p^* \mathcal{W}^i_{X/S}$;

(2) $S_p(\theta) \vdash \delta = 0$ *for* $\theta \in p^* \mathcal{V}^p_{X/S}$;

(3) $\theta \lrcorner\, E_i(\alpha) = F_{-i}(\theta) \lrcorner\, \alpha$ *for* $\theta \in \mathcal{V}^{-i}_{L(X)/S}$, $\alpha \in p^* \mathcal{W}^i_{X/S}$;

(4) $I_p(F_{p+1}(\theta) \otimes \gamma_s^\vee) = \theta \wedge \delta^\vee$ *for* $\theta \in \mathcal{V}^{p+1}_{L(X)/S}$;

(5) $F_p(\theta) \wedge F_q(\xi) = F_{p+q}(\theta \wedge \xi)$ *for* $\theta \in \mathcal{V}^p_{L(X)/S}$, $\xi \in \mathcal{V}^q_{L(X)/S}$;

(6) $\theta \lrcorner\, E_i(\alpha) = E_{p+i}(F_p(\theta) \lrcorner\, \alpha)$ *for* $\theta \in \mathcal{V}^p_{L(X)/S}$, $\alpha \in p^* \mathcal{W}^i_{X/S}$;

(7) $F_{p+i}(\theta \vdash E_i(\alpha)) = F_p(\theta) \vdash \alpha$ *for* $\theta \in \mathcal{V}^p_{L(X)/S}$, $\alpha \in p^* \mathcal{W}^i_{X/S}$;

(8) $\theta \lrcorner\, T_i(\alpha) = S_{-i}(\theta) \lrcorner\, \alpha$ *for* $\theta \in p^* \mathcal{V}^{-i}_{X/S}$, $\alpha \in \mathcal{W}^i_{L(X)/S}$;

(9) $S_p(\theta) \wedge S_q(\xi) = S_{p+q}(\theta \wedge \xi)$ *for* $\theta \in p^* \mathcal{V}^p_{X/S}$, $\xi \in p^* \mathcal{V}^q_{X/S}$;

(10) $\theta \lrcorner\, T_i(\alpha) = T_{p+i}(S_p(\theta) \lrcorner\, \alpha)$ *for* $\theta \in p^* \mathcal{V}^p_{X/S}$, $\alpha \in \mathcal{W}^i_{L(X)/S}$;

(11) $S_p(\theta) \lrcorner\, E_i(\alpha) = E_{p+i}(\theta \lrcorner\, \alpha)$ *for* $\theta \in p^* \mathcal{V}^p_{X/S}$, $\alpha \in p^* \mathcal{W}^i_{X/S}$;

(12) $S_p(\theta) \lrcorner\, (E_{i-1}(\beta) \wedge \delta) = E_{p+i-1}(\theta \lrcorner\, \beta) \wedge \delta$ *for* $\theta \in p^* \mathcal{V}^p_{X/S}$, $\beta \in p^* \mathcal{W}^{i-1}_{X/S}$;

(13) $S_p(\theta) \vdash E_i(\alpha) = S_{p+i}(\theta \vdash \alpha)$ *for* $\theta \in p^* \mathcal{V}^p_{X/S}$, $\alpha \in p^* \mathcal{W}^i_{X/S}$;

(14) $(S_{p+1}(\xi) \wedge \delta^\vee) \vdash E_i(\alpha) = (-1)^i S_{p+1}(\xi \vdash \alpha) \wedge \delta^\vee$ *for* $\xi \in p^* \mathcal{V}^{p+1}_{X/S}$, $\alpha \in p^* \mathcal{W}^i_{X/S}$.

Proof We start with (1). If $i = 0$, then we have $\delta^\vee \lrcorner\, E_i(\alpha) = 0$ for degree reasons. If $i = 1$, then we have $\delta^\vee \lrcorner\, E_i(\alpha) = 0$ by definition. Since $|\delta^\vee| = -1$, we have a formula for computing $\delta^\vee \lrcorner\, E_{i+j}(\alpha \wedge \beta)$ in terms of $\delta^\vee \lrcorner\, E_i(\alpha)$ and $\delta^\vee \lrcorner\, E_j(\beta)$, which yields $\delta^\vee \lrcorner\, E_i(\alpha) = 0$ for every α which is locally a $\wedge$-product of elements in degree 1. This holds everywhere on $p^{-1}(U)$; since $p^* \mathcal{W}^i_{X/S}$ has injective restrictions to $p^{-1}(U)$, the claim follows. The proof of (2) is similar, starting with $T_1(\delta) = 0$ and using (8).

The formula in (3) holds by the definition of F_{-i} as the dual of E_i.

To see (4), contract both sides with $\alpha = E_{-p}(\alpha_1) + E_{-p+1}(\alpha_2) \wedge \delta$, using the formula of (3) and our identification above.

We show (5) at a stalk at $x \in p^{-1}(U)$. On the one hand, we have

$$F_{-1}(\theta_r) \wedge \ldots \wedge F_{-1}(\theta_1) \lrcorner (\alpha_1 \wedge \ldots \wedge \alpha_r) = \sum_{\sigma \in S_r} (-1)^{\operatorname{sgn}(\sigma)} \prod_{j=1}^{r} F_{-1}(\theta_j) \lrcorner \alpha_{\sigma(j)};$$

on the other hand, we have

$$F_{-r}(\theta_r \wedge \ldots \wedge \theta_1) \lrcorner (\alpha_1 \wedge \ldots \wedge \alpha_r) = (\theta_r \wedge \ldots \wedge \theta_1) \lrcorner E_1(\alpha_1) \wedge \ldots \wedge E_1(\alpha_r)$$

$$= \sum_{\sigma \in S_r} (-1)^{\operatorname{sgn}(\sigma)} \prod_{j=1}^{r} \theta_j \lrcorner E_1(\alpha_{\sigma(j)}).$$

Since $p^* W^r_{X/S}$ is locally free at $x \in p^{-1}(U)$, this implies

$$F_{-1}(\theta_1) \wedge \ldots \wedge F_{-1}(\theta_r) = F_{-r}(\theta_1 \wedge \ldots \wedge \theta_r),$$

and since $p^* V^{-1}_{X/S}$ is locally free at $x \in p^{-1}(U)$, the claim follows.

It suffices to show the formula in (6) on $p^{-1}(U)$. The formula is obvious for $p = 0$; for $p = -1$ and $i = 0$, it is trivial, and for $p = -1$ and $i = 1$, it is the very definition of the map F_{-1}. We obtain the formula for $p = -1$ and arbitrary i via the product formula for right contractions. Using (5), we generalize to $p \leq -2$. The proof of (7) is analogous.

The formula in (8) is nothing but the definition of S_{-i} as the dual of T_i. Then (9) is analogous to (5), and (10) is analogous to (6).

The remaining four formulae are now easy once we use $T_1(\delta) = 0$. $\square$

We set

$$\mathcal{G}^p_{L(X)/S} := I_p(p^* \mathcal{G}^{p+1}_{X/S} \otimes \mathcal{V}^{-1}_{L(X)/X}) + S_p(p^* \mathcal{G}^p_{X/S}) \subseteq \mathcal{V}^p_{L(X)/S}.$$

A careful computation shows that, under a change of distinguished splitting B_1, we have $S'_p(\theta) = S_p(\theta) - I_p((\theta \vdash \varepsilon) \otimes \gamma_s^\vee)$, where again $\varepsilon \in p^* \mathcal{A}^1_{X/S}$ is such that $E_1(\varepsilon) = B'_1(\gamma_s) - B_1(\gamma_s)$. Since $\theta \vdash \varepsilon \in p^* \mathcal{G}^{p+1}_{X/S}$ if $\theta \in p^* \mathcal{G}^p_{X/S}$, we find that $\mathcal{G}^p_{L(X)/S}$ is independent of the choice of B_1, and hence globally well-defined.

Every element $\theta \in \mathcal{G}^p_{L(X)/S}$ can be uniquely written as

$$\theta = S_{p+1}(\theta_1) \wedge \delta^\vee + S_p(\theta_2)$$

with $\theta_1 \in p^* \mathcal{G}^{p+1}_{X/S}$ and $\theta_2 \in p^* \mathcal{G}^p_{X/S}$ once a local distinguished splitting B_1 is chosen.

Proposition 18.6 *The family $f \circ p \colon L(X) \to S$, endowed with*

$$\mathcal{GC}^{\bullet}_{L(X)/S} = (\mathcal{G}^{\bullet}_{L(X)/S}, \mathcal{A}^{\bullet}_{L(X)/S}),$$

is an enhanced generically log smooth family and log Gorenstein.

Proof By construction, we have $\mathcal{A}^{0}_{L(X)/S} = \mathcal{W}^{0}_{L(X)/S} = \mathcal{O}_L$. By assumption, we have $\mathcal{A}^{d}_{X/S} = \mathcal{W}^{d}_{X/S}$, and this is a line bundle. Since

$$\mathcal{W}^{d+1}_{L(X)/S} = p^* \mathcal{W}^{d}_{X/S} \otimes \Omega^{1}_{L(X)/X} \quad \text{and} \quad \mathcal{A}^{d+1}_{L(X)/S} = p^* \mathcal{A}^{d}_{L(X)/S} \otimes \Omega^{1}_{L(X)/S},$$

we find $\mathcal{A}^{d+1}_{L(X)/S} = \mathcal{W}^{d+1}_{L(X)/S}$, and this is a line bundle.

We show that $\mathcal{A}^{\bullet}_{L(X)/S} \subseteq \mathcal{W}^{\bullet}_{L(X)/S}$ is a differential graded subalgebra. Since

$$E_i \colon \quad p^* \mathcal{W}^{i}_{X/S} \to \mathcal{W}^{i}_{L(X)/S}$$

is compatible with the $\wedge$-product, we see that $\mathcal{A}^{\bullet}_{L(X)/S} \subseteq \mathcal{W}^{\bullet}_{L(X)/S}$ is closed under the $\wedge$-product. On W_α, we have

$$d_{L(X)/S}(x_\alpha) = x_\alpha \cdot \delta_{L(X)/S}(m_\alpha) \in B_1(\Omega^{1}_{L(X)/X})$$

for the original distinguished splitting B_1, which we have constructed first. Thus, we have $\partial(\mathcal{O}_L) \subseteq \mathcal{A}^{1}_{L(X)/S} \subseteq \mathcal{W}^{1}_{L(X)/S}$; another direct computation yields that $\partial(\mathcal{A}^{i}_{L(X)/S}) \subseteq \mathcal{A}^{i+1}_{L(X)/S}$, so $\mathcal{A}^{\bullet}_{L(X)/S}$ is a complex.

$\mathcal{G}^{\bullet}_{L(X)/S}$ is closed under the $\wedge$-product. Namely, let $\theta = S_{p+1}(\theta_1) \wedge \delta^{\vee} + S_p(\theta_2)$ and $\xi = S_{q+1}(\xi_1) \wedge \delta^{\vee} + S_q(\xi_2)$; then we have

$$(S_{p+1}(\theta_1) \wedge \delta^{\vee} + S_p(\theta_2)) \wedge (S_{q+1}(\xi_1) \wedge S_q(\theta_2))$$

$$= S_{p+q+1}(\theta_1 \wedge \xi_2 + \xi_1 \wedge \theta_2) \wedge \delta^{\vee} + S_{p+q}(\theta_2 \wedge \xi_2)$$

so that $\theta \wedge \xi \in \mathcal{G}^{p+q}_{L(X)/S}$. Similarly, we have

$$(S_{p+1}(\theta_1) \wedge \delta^{\vee} + S_p(\theta_2)) \mathbin{\lrcorner} (E_i(\alpha_1) + E_{i-1}(\alpha_2) \wedge \delta)$$

$$= E_{p+i}((\theta_2 \mathbin{\lrcorner} \alpha_1) + (-1)^{i-1}(\theta_1 \mathbin{\lrcorner} \alpha_2)) + E_{p+i-1}(\theta_2 \mathbin{\lrcorner} \alpha_2) \wedge \delta, \qquad (18.3)$$

showing that $\theta \mathbin{\lrcorner} \alpha \in \mathcal{A}^{p+i}_{L(X)/S}$ for $\theta \in \mathcal{G}^{p}_{L(X)/S}$ and $\alpha \in \mathcal{A}^{i}_{L(X)/S}$. By the Lie–Rinehart homotopy formula, $\mathcal{GC}^{\bullet}_{L(X)/S}$ is closed under the Lie derivative $\mathcal{L}_-(-)$. From (18.3), we also see that $\mathcal{GC}^{\bullet}_{L(X)/S}$ is locally Batalin–Vilkovisky. Namely, if

$\omega \in \mathcal{A}^d_{X/S}$ is a local volume form on X, then $E_d(p^*\omega) \wedge \delta \in \mathcal{A}^{d+1}_{L(X)/S}$ is a local volume form on $L(X)$. Contracting with this form yields

$$(S_{p+1}(\theta_1) \wedge \delta^\vee + S_p(\theta_2)) \lrcorner\, E_d(p^*\omega) \wedge \delta$$

$$= E_{p+d+1}((-1)^d \theta_1 \lrcorner\, p^*\omega) + E_{p+d}(\theta_2 \lrcorner\, p^*\omega) \wedge \delta,$$

so this induces an isomorphism $\mathcal{G}^p_{L(X)/S} \cong \mathcal{A}^{p+d+1}_{L(X)/S}$. Now the mixed Leibniz rule yields that $\mathcal{G}^\bullet_{L(X)/S}$ is closed under $[-, -]$. For, we have $[\theta, \xi] \in \mathcal{G}^{p+q+1}_{L(X)/S}$ if and only if

$$[\theta, \xi] \lrcorner\, (E_d(p^*\omega) \wedge \delta) \in \mathcal{A}^{p+q+d+2}_{L(X)/S},$$

and the latter is the case by evaluating the term with the mixed Leibniz rule, and using closedness under $\mathcal{L}_-(-)$. We have

$$(S_{p+1}(\theta_1) \wedge \delta^\vee + S_p(\theta_2)) \vdash (E_i(\alpha_1) + E_{i-1}(\alpha_2) \wedge \delta)$$

$$= (-1)^i S_{p+i+1}(\theta_1 \vdash \alpha_1) \wedge \delta^\vee + S_{p+i}(\theta_2 \vdash \alpha_1 + (-1)^{i-1}\theta_1 \vdash \alpha_2),$$

so $\mathcal{GC}^\bullet_{L(X)/S}$ is closed under the left contraction $\vdash$.

Each term $\mathcal{G}^p_{L(X)/S}$ and $\mathcal{A}^i_{L(X)/S}$ is flat over S because of the locally split exact sequences these terms fit in. On $p^{-1}(U)$, the Gerstenhaber calculus $\mathcal{GC}^\bullet_{L(X)/S}$ coincides with the reflexive Gerstenhaber calculus by construction. Finally, $\mathcal{G}^p_{L(X)/S}$ and $\mathcal{A}^i_{L(X)/S}$ are $p^{-1}(Z)$-pure in every fiber since they decompose locally as a direct sum of terms of the form $p^*\mathcal{G}^p_{X/S}$ and $p^*\mathcal{A}^i_{X/S}$, and they are $p^{-1}(Z)$-pure in every fiber by assumption. $\qquad\square$

18.2 Log Quasi-regular and Log Regular Sections

In this section, we introduce the notions of a log quasi-regular section $s \in \mathcal{L}$ and a log regular section $s \in \mathcal{L}$. The latter will be the correct notion to study modifications $f \circ h\colon X(s) \to S$ in a deformation-theoretic setting. If the enhanced generically log smooth family is a plain generically log smooth family, i.e., its Gerstenhaber calculus is reflexive, then the two notions coincide.[3]

Let $s \in \mathcal{L}$ be a log pre-regular section, and let

$$X(s) \xrightarrow{e} N(s) \xrightarrow{k} L(X)$$

---

[3] We do not need the base change property for this to hold.

be the first infinitesimal neighborhood, i.e., it is the closed subscheme of $L(X)$ defined by the square of the ideal sheaf $\varepsilon_s \colon p^*\mathcal{L}^\vee \to O_L$, $p^*e_\alpha^\vee \mapsto x_\alpha - s_\alpha$, of $X(s) \subset L(X)$, and on $N(s) \cap p^{-1}(U)$, it is endowed with the inverse image of the log structure on $L(X)$. As topological spaces, we have $|X(s)| = |N(s)|$ and $|N(s) \cap p^{-1}(U)| = |U|$. We have an exact sequence

$$0 \to \mathcal{L}^\vee \xrightarrow{e_\alpha^\vee \mapsto x_\alpha - s_\alpha} O_{N(s)} \to O_{X(s)} \to 0,$$

which exhibits $\mathcal{L}^\vee$ as the ideal sheaf of $X(s) \subset N(s)$. Let us write $\varepsilon_N \colon \mathcal{L}^\vee \to O_{N(s)}$ for the corresponding map. A *splitting* of $e \colon X(s) \to N(s)$ is a morphism $r \colon N(s) \to X(s)$ over S with $r \circ e = \mathrm{id}_{X(s)}$; more precisely, $r \colon N(s) \to X(s)$ is supposed to be a morphism of schemes everywhere, and a morphism of log schemes on U. Local splittings form a sheaf on the topological space X, which we denote by

$$\mathrm{Sp}_{X(s)/N(s)}.$$

Explicitly, a splitting consists of two maps

$$r^* \colon \ O_{X(s)} \to O_{N(s)} \quad \text{and} \quad r^* \colon \ M_{X(s)}|_U \to M_{N(s)}|_U.$$

There is an action of $\mathcal{V}^{-1}_{X(s)/S} \otimes \mathcal{L}^\vee = \mathcal{D}er_{X(s)/S}(\mathcal{L}^\vee)$ on $\mathrm{Sp}_{X(s)/N(s)}$, which is, when $\theta = (D, \Delta)$ is a log derivation with values in $\mathcal{L}^\vee$, explicitly given by

$$(\theta \odot r)^*(a) = r^*(a) + \varepsilon_N \circ D(a), \quad (\theta \odot r)^*(m) = (1 + \varepsilon_N \circ \Delta(m)) \cdot r^*(m).$$

Here, we consider $\mathcal{V}^\bullet_{X(s)/S} := j_*\Theta^{-\bullet}_{U(s)/S}$ as the complex of Z-*closed* polyvector fields, as $U(s) \to S$ may not be log smooth. First, this action is given on U, but since both $\mathcal{V}^{-1}_{X(s)/S} \otimes \mathcal{L}^\vee$ and $\mathrm{Sp}_{X(s)/N(s)}$ are Z-closed,[4] we obtain the action on X as the direct image from U. By [222, IV, Thm. 2.2.2], $\mathrm{Sp}_{X(s)/N(s)}|_U$ is a pseudo-torsor over $\mathcal{V}^{-1}_{U(s)/S} \otimes \mathcal{L}^\vee$. Then $\mathrm{Sp}_{X(s)/N(s)}$ is automatically a pseudo-torsor over $\mathcal{V}^{-1}_{X(s)/S} \otimes \mathcal{L}^\vee$.

Definition 18.7 (Log Quasi-regular Sections) In Situation 18.1, let $s \in \mathcal{L}$ be a log pre-regular section. Then s is a *log quasi-regular section* if $\mathrm{Sp}_{X(s)/N(s)}$ is a torsor over $\mathcal{V}^{-1}_{X(s)/S} \otimes \mathcal{L}^\vee$, i.e., if local splittings exist everywhere.

It is clear that every base change of a log quasi-regular section is again a log quasi-regular section. The proof of [222, IV, Thm. 3.2.2] shows that $U(s) \to S$ is log smooth if and only if $s|_U$ is a log quasi-regular section. In particular, if $s \in \mathcal{L}$

---

[4] If we have a splitting on $W \cap U$ for some open subset $W \subseteq X$, then we can extend the classical part of the morphism to W since we have $O_{X(s)}|_W = j_*O_{X(s)}|_{W \cap U}$ and $O_{N(s)}|_W = j_*O_{N(s)}|_{W \cap U}$; the latter holds because $O_{N(s)}$ is an extension of $O_{X(s)}$ by $\mathcal{L}^\vee$.

is a log quasi-regular section, then $f \circ h \colon X(s) \to S$ is a generically log smooth family. We have the following criterion for log quasi-regularity:

Lemma 18.8 *Let $s \in \mathcal{L}$ be a log pre-regular section. Then $s^*\mathcal{W}^1_{L(X)/S}$ is reflexive, and the canonical exact sequence of the embedding $X(s) \subset L(X)$ extends to a (not necessarily exact) sequence*

$$0 \to \mathcal{L}^\vee \xrightarrow{G_1} s^*\mathcal{W}^1_{L(X)/S} \xrightarrow{R_1} \mathcal{W}^1_{X(s)/S} \to 0$$

with $\mathcal{W}^1_{X(s)/S} := j_*\Omega^1_{U(s)/S}$, even if $U(s) \to S$ is not log smooth. The section $s \in \mathcal{L}$ is log quasi-regular if and only if this sequence is exact and locally split.

Proof We consider the diagram

$$
\begin{array}{ccccccccc}
0 & \longrightarrow & \mathcal{L}^\vee & \xrightarrow{\ \bar{d}_{X(s)}\ } & s^*\mathcal{W}^1_{L(X)/S} & \xrightarrow{\ ds^*\ } & \mathcal{W}^1_{X(s)/S} & \longrightarrow & 0 \\
& & \big\uparrow{\scriptstyle 0} & & \big\uparrow{\scriptstyle \pi} & {\scriptstyle\phi}\nwarrow\ \ & \big\uparrow{\scriptstyle de^*}\ \ \vdots\ {\scriptstyle dr^*} & & \\
& & k^*\mathcal{I}_{N(s)} & \xrightarrow{\ \bar{d}_{N(s)}\ } & k^*\mathcal{W}^1_{L(X)/S} & \longrightarrow & \mathcal{W}^1_{N(s)/S} & \longrightarrow & 0
\end{array}
$$

of sheaves on the topological space X. On U, the upper row is the canonical exact sequence of the strict closed immersion $U(s) \subset L(X)|_{p^{-1}(U)}$. It is exact on U in the middle and on the right by [222, IV, Prop. 2.3.2]. From the locally split exact sequence (18.1), it follows that $s^*\mathcal{W}^1_{L(X)/S}$ is reflexive. Hence we can define the upper row on X as its direct image from U. The sheaf $\mathcal{I}_{N(s)}$ is the ideal sheaf of $N(s) \subset L(X)$; on U, the lower row is the canonical exact sequence of this embedding. It is also possible to show that $k^*p^*\mathcal{W}^1_{X/S}$ is Z-closed, hence $k^*\mathcal{W}^1_{L(X)/S}$ is Z-closed, and we can define the lower row on X as the direct image from U. A direct computation shows that $\pi \circ \bar{d}_{N(s)} = 0$; hence, on U, π descends to a map $\phi \colon \mathcal{W}^1_{N(s)/S} \to s^*\mathcal{W}^1_{L(X)/S}$. Since both source and target are Z-closed, this map can be extended to be defined on X.[5]

First assume that $s \in \mathcal{L}$ is log quasi-regular. Then, on U, the upper row is exact and locally split by [222, IV, Prop. 2.3.2]. In particular, on X, it is exact on the left and in the middle. For a local splitting $r \in \mathrm{Sp}_{X(s)/N(s)}$, we have $ds^* \circ \phi \circ dr^* = \mathrm{id}$, first on U, and then on X. Thus, ds^* is surjective on X, and the upper row is locally split exact everywhere.

Conversely, assume that the sequence is exact and locally split. Let Sp' be the sheaf of splittings of this exact sequence. Under the assumption, this is a $\mathcal{H}om(\mathcal{W}^1_{X(s)/S}, \mathcal{L}^\vee)$-torsor. The above construction yields a map

$$\eta \colon \ \mathrm{Sp}_{X(s)/N(s)} \to \mathrm{Sp}'$$

---

[5] Note that we have not shown surjectivity of the lower row on the right.

of sheaves. A careful computation (using that elements of the form $\delta(m)$ generate $\Omega^1_{U(s)/S}$) shows that η is equivariant for the $\mathcal{V}^{-1}_{X(s)/S}\otimes\mathcal{L}^\vee$-pseudo-torsor structure on the left and the $\mathcal{H}om(\mathcal{W}^1_{X(s)/S}, \mathcal{L}^\vee)$-torsor structure on the right, with the obvious identification. By [222, IV, Thm. 3.2.2], $\mathrm{Sp}_{X(s)/N(s)}$ is a torsor on U. Thus, η is an isomorphism on U. If $\rho \in \mathrm{Sp}'$ is a local section on an open subset $W \subseteq X$, then $\eta^{-1}(\rho|_{W\cap U})$ can be canonically extended to W; thus, $s \in \mathcal{L}$ is a log quasi-regular section. $\square$

Let $s \in \mathcal{L}$ be a log quasi-regular section. From (18.2), we find that $s^*\mathcal{V}^{-1}_{X/S}$ is locally free on U and reflexive; thus

$$s^*\mathcal{V}^{-1}_{L(X)/S} \cong \mathcal{H}om(s^*\mathcal{W}^1_{L(X)/S}, O_X),$$

and we obtain a locally split exact sequence

$$0 \to \mathcal{V}^{-1}_{X(s)/S} \xrightarrow{J_{-1}} s^*\mathcal{V}^{-1}_{L(X)/S} \xrightarrow{H_{-1}} \mathcal{L} \to 0$$

by dualizing the exact sequence of Lemma 18.8. The map $s^*\mathcal{G}^{-1}_{L(X)/S} \to s^*\mathcal{V}^{-1}_{L(X)/S}$ is injective, and after setting

$$\mathcal{G}^{-1}_{X(s)/S} := (J_{-1})^{-1}(s^*\mathcal{G}^{-1}_{L(X)/S}) \subseteq \mathcal{V}^{-1}_{X(s)/S},$$

we have an exact sequence

$$0 \to \mathcal{G}^{-1}_{X(s)/S} \to s^*\mathcal{G}^{-1}_{L(X)/S} \to \mathcal{L}.$$

It is natural to ask for this map to be surjective, and, as it turns out, this is a very useful condition in many respects.

Definition 18.9 (Log Regular Sections)   In Situation 18.1, let $s \in \mathcal{L}$ be a log quasi-regular section. Then s is a *log regular section* if the canonical map

$$s^*\mathcal{G}^{-1}_{L(X)/S} \subseteq s^*\mathcal{V}^{-1}_{L(X)/S} \to \mathcal{L}$$

is surjective.

Since the formation of the map $s^*\mathcal{G}^{-1}_{L(X)/S} \to \mathcal{L}$ commutes with base change, the condition of being a log regular section is stable under base change. If $s \in \mathcal{L}$ is a log regular section, then the exact sequence is locally split because $\mathcal{L}$ is a line bundle. If $r \in \mathrm{Sp}_{X(s)/N(s)}$ is a splitting, let us denote the induced splitting of $R_1\colon s^*\mathcal{W}^1_{L(X)/S} \to \mathcal{W}^1_{X(s)/S}$ by $C_1[r]$, and the induced splitting of $G_1\colon \mathcal{L}^\vee \to s^*\mathcal{W}^1_{L(X)/S}$ by $U_1[r]$. With this notation, we construct a map

$$\psi\colon\ \mathrm{Sp}_{X(s)/N(s)} \to \mathcal{H}om(\mathcal{W}^1_{X/S}, \mathcal{L}^\vee)$$

as

$$\psi(r) = U_1[r] \circ E_1 \colon \quad \mathcal{W}^1_{X/S} \xrightarrow{\; E_1 \;} s^* \mathcal{W}^1_{L(X)/S} \xrightarrow{\; U_1[r] \;} \mathcal{L}^\vee .$$

For $r' = \theta \odot r$ with $\theta \in \mathcal{V}^{-1}_{X(s)/S} \otimes \mathcal{L}^\vee = \mathcal{H}om(\mathcal{W}^1_{X(s)/S}, \mathcal{L}^\vee)$, we find

$$U_1[r'](\alpha) - U_1[r](\alpha) = -\theta(R_1(\alpha))$$

for $\alpha \in s^* \mathcal{W}^1_{L(X)/S}$. Thus,

$$\psi(r')(\beta) - \psi(r)(\beta) = -\theta(R_1 \circ E_1(\beta))$$

for $\beta \in \mathcal{W}^1_{X/S}$, so ψ is $\mathcal{V}^{-1}_{X(s)/S} \otimes \mathcal{L}^\vee$-equivariant when it acts on $\mathcal{H}om(\mathcal{W}^1_{X/S}, \mathcal{L}^\vee)$ via the negative of the embedding via the canonical map

$$dh^* = R_1 \circ E_1 \colon \quad \mathcal{W}^1_{X/S} \to \mathcal{W}^1_{X(s)/S},$$

i.e., via $-Th_* \otimes \mathcal{L}^\vee \colon \mathcal{V}^{-1}_{X(s)/S} \otimes \mathcal{L}^\vee \to \mathcal{V}^{-1}_{X/S} \otimes \mathcal{L}^\vee$. This map is injective since it is an isomorphism on $X \setminus H$, which is dense due to our log pre-regularity assumption.

Lemma 18.10 *Let $s \in \mathcal{L}$ be a log quasi-regular section. Then the following statements are equivalent:*

(i) *s is a log regular section, i.e., $s^* \mathcal{G}^{-1}_{L(X)/S} \to \mathcal{L}$ is surjective;*
(ii) *everywhere on X, there is locally a splitting $r \in \mathrm{Sp}_{X(s)/N(s)}$ such that*

$$\psi(r) \in \mathcal{G}^{-1}_{X/S} \otimes \mathcal{L}^\vee \subseteq \mathcal{V}^{-1}_{X/S} \otimes \mathcal{L}^\vee .$$

Proof First, assume that $s \in \mathcal{L}$ is a log regular section. Then we can find locally an element $\theta_\alpha \in s^* \mathcal{G}^{-1}_{L(X)/S}$ with $H_{-1}(\theta_\alpha) = e_\alpha$. In particular, this defines a splitting of H_{-1} with $e_\alpha \mapsto \theta_\alpha$, and due to the isomorphism $\eta \colon \mathrm{Sp}_{X(s)/N(s)} \to \mathrm{Sp}'$, we can find a splitting $r \in \mathrm{Sp}_{X(s)/N(s)}$ such that $e_\alpha \mapsto \theta_\alpha$ is given by $V_{-1}[r]$, the dual of $U_1[r]$. Since $F_{-1}(\theta_\alpha) \in \mathcal{G}^{-1}_{X/S}$, we have

$$F_{-1} \circ V_{-1}[r] \colon \quad \mathcal{L} \to \mathcal{G}^{-1}_{X/S}.$$

This implies $\psi(r) = U_1[r] \circ E_1 \in \mathcal{G}^{-1}_{X/S} \otimes \mathcal{L}^\vee$, as desired. Conversely, if $\psi(r) \in \mathcal{G}^{-1}_{X/S} \otimes \mathcal{L}^\vee$, then $F_{-1} \circ V_{-1}[r](e_\alpha) \in \mathcal{G}^{-1}_{X/S}$, so $\theta_\alpha := V_{-1}[r] \in s^* \mathcal{G}^{-1}_{L(X)/S}$ by the variant of (18.2) for $\mathcal{G}^\bullet$. Then $H_{-1}(\theta_\alpha) = e_\alpha$, so $s^* \mathcal{G}^{-1}_{L(X)/S} \to \mathcal{L}$ is surjective, and s is a log regular section. $\square$

Let

$$\psi^{-1}(\mathcal{V}_{X/S}^{-1} \otimes \mathcal{L}^\vee) =: \widetilde{\mathrm{Sp}} \subseteq \mathrm{Sp}_{X(s)/N(s)}$$

be the subsheaf of those splittings $r \in \mathrm{Sp}_{X(s)/N(s)}$ with $\psi(r) \in \mathcal{G}_{X/S}^{-1} \otimes \mathcal{L}^\vee$. We have an induced map

$$Th_* : \quad \mathcal{G}_{X(s)/S}^{-1} \xrightarrow{J_{-1}} s^* \mathcal{G}_{L(X)/S}^{-1} \xrightarrow{F_{-1}} \mathcal{G}_{X/S}^{-1},$$

and under this map, we have

$$\psi(\theta \odot r) = \psi(r) - (Th_* \otimes \mathcal{L}^\vee)(\theta)$$

for $\theta \in \mathcal{G}_{X(s)/S}^{-1} \otimes \mathcal{L}^\vee$, i.e., $\mathcal{G}_{X(s)/S}^{-1} \otimes \mathcal{L}^\vee$ acts on $\widetilde{\mathrm{Sp}}$. If $r', r \in \widetilde{\mathrm{Sp}}$ are two local sections, then there is some $\theta \in \mathcal{V}_{X(s)/S}^{-1} \otimes \mathcal{L}^\vee$ with $r' = \theta \odot r$. Since

$$\psi(\theta \odot r) = \psi(r) - (Th_* \otimes \mathcal{L}^\vee)(\theta),$$

we find $(Th_* \otimes \mathcal{L}^\vee)(\theta) \in \mathcal{G}_{X/S}^{-1} \otimes \mathcal{L}^\vee$. Then, $(J_{-1} \otimes \mathcal{L}^\vee)(\theta) \in s^* \mathcal{G}_{L(X)/S}^{-1} \otimes \mathcal{L}^\vee$, and hence $\theta \in \mathcal{G}_{X(s)/S}^{-1} \otimes \mathcal{L}^\vee$ more or less by definition. Thus, $\widetilde{\mathrm{Sp}}$ is a $\mathcal{G}_{X(s)/S}^{-1} \otimes \mathcal{L}^\vee$-torsor. We summarize this as follows:

Proposition 18.11 *Let $s \in \mathcal{L}$ be a log regular section. Then there is a distinguished submodule $\mathcal{G}_{X(s)/S}^{-1} \subseteq \mathcal{V}_{X(s)/S}^{-1}$ of log derivations on $f \circ h : X(s) \to S$, depending only on $s \in \mathcal{L}$ and the enhanced generically log smooth family $f : X \to S$, together with a canonical $\mathcal{G}_{X(s)/S}^{-1} \otimes \mathcal{L}^\vee$-torsor $\widetilde{\mathrm{Sp}}$ of special splittings of $e : X(s) \to N(s)$.*

Using $\widetilde{\mathrm{Sp}}$, we give yet another equivalent characterization of log regularity.

Lemma 18.12 *Let $s \in \mathcal{L}$ be a log quasi-regular section. Then s is log regular if and only if, locally on each V_α, there is a log derivation $\theta = (D, \Delta) \in \mathcal{G}_{X/S}^{-1}$ with $D(s_\alpha) = 1 + a \cdot s_\alpha$ for some (locally defined) function $a \in \mathcal{O}_X$.*

Proof First assume that s is log regular. Recall that $i : H \to X$ is the inclusion. We consider the diagram

$$
\begin{array}{ccccccccc}
0 & \longrightarrow & i^* \mathcal{W}_{X/S}^1 & \xrightarrow{i^* E_1} & i^* s^* \mathcal{W}_{L(X)/S}^1 & \xrightarrow{i^* Q_1} & i^* s^* \Omega_{L(X)/X}^1 & \longrightarrow & 0 \\
& & \uparrow{\scriptstyle M} & & \| & & & & \\
0 & \longrightarrow & i^* \mathcal{L}^\vee & \xrightarrow{i^* G_1} & i^* s^* \mathcal{W}_{L(X)/S}^1 & \xrightarrow{i^* R_1} & i^* s^* \mathcal{W}_{X(s)/S}^1 & \longrightarrow & 0
\end{array}
$$

with two locally split exact rows. Since $i^* Q_1 \circ i^* G_1 (i^* e_\alpha^\vee) = 0$, there is a map $M : i^* \mathcal{L}^\vee \to i^* \mathcal{W}_{X/S}^1$ as indicated. For any splitting $U_1[r]$ of G_1, we have $i^* U_1[r] \circ$

$M = \mathrm{id}_{i_* \mathcal{L}^\vee}$. Now let $r \in \widetilde{\mathrm{Sp}}$ be a local splitting, and let $(D, \Delta) \in \mathcal{G}_{X/S}^{-1}$ be such that $\psi(r) = (D, \Delta) \otimes e_\alpha^\vee$. Then we have $D(s_\alpha) \otimes e_\alpha^\vee = U_1[r](ds_\alpha)$, and since $M(i^* e_\alpha^\vee) = -i^* ds_\alpha$, we have

$$i^* D(s_\alpha) \otimes i^* e_\alpha^\vee = -i^* U_1[r] \circ M(i^* e_\alpha^\vee) = -i^* e_\alpha^\vee.$$

Thus, $1 + D(s_\alpha) \in \mathrm{Ann}(i^* \mathcal{L}^\vee) = \mathcal{I}_H = (s_\alpha)$, and hence we can write $D(s_\alpha) = -1 - a \cdot s_\alpha$ for some $a \in O_X$. Replacing (D, Δ) with $(-D, -\Delta)$ yields the desired log derivation.

Conversely, let $\theta = (D, \Delta)$ be a log derivation with $D(s_\alpha) = 1 + a \cdot s_\alpha$. Let

$$\xi := -a \cdot s^* \delta^\vee - S_{-1}(\theta)$$

where S_{-1} is the splitting associated with the original distinguished splitting $B_1(\gamma_s) = \delta_\alpha$. We have $\xi \in s^* \mathcal{G}_{L(X)/S}^{-1}$ due to $\theta \in \mathcal{G}_{X/S}^{-1}$. Since $H_{-1} \colon s^* \mathcal{V}_{L(X)/S}^{-1} \to \mathcal{L}$ is the dual of $G_1 \colon \mathcal{L}^\vee \to s^* \mathcal{W}_{L(X)/S}^1$, and since $G_1(e_\alpha^\vee) = s_\alpha \cdot s^* \delta^\vee - ds_\alpha$, we have

$$H_{-1}(\xi)(e_\alpha^\vee) = \xi \lrcorner\, G_1(e_\alpha^\vee) = 1.$$

Thus, $H_{-1}(\xi) = e_\alpha$, and $s^* \mathcal{G}_{L(X)/S}^{-1} \to \mathcal{L}$ is surjective. $\square$

Remark 18.13 Note that this criterion is independent of the choice of s_α: For another choice $u_\alpha \cdot s_\alpha$ with u_α invertible, we find $D(u_\alpha \cdot s_\alpha) = 1 + (u_\alpha \cdot a + D(u_\alpha)) \cdot s_\alpha$. $\Diamond$

18.3 The Gerstenhaber Calculus on $f \circ h \colon X(s) \to S$

Let $s \in \mathcal{L}$ be a log regular section. The exact sequence of Lemma 18.8 yields a locally split exact sequence

$$0 \to \mathcal{W}_{X(s)/S}^{i-1} \otimes \mathcal{L}^\vee \xrightarrow{G_i} s^* \mathcal{W}_{L(X)/S}^i \xrightarrow{R_i} \mathcal{W}_{X(s)/S}^i \to 0.$$

Namely, on U, we obtain this sequence as usual; since $s^* \mathcal{W}_{L(X)/S}^i$ is reflexive, we can extend it via the direct image to X; and finally, the right-hand side is surjective because

$$s^* \Omega_{L(X)/S}^i |_U \to \Omega_{U(s)/S}^i$$

is locally split on X (rather than merely locally on U). Note that $s^* \mathcal{A}^i_{L(X)/S} \subseteq s^* \mathcal{W}^i_{L(X)/S}$. We define

$$\mathcal{A}^i_{X(s)/S} := R_i(s^* \mathcal{A}^i_{L(X)/S}) \subseteq \mathcal{W}^i_{X(s)/S}.$$

Dually, we have a locally split exact sequence

$$0 \to \mathcal{V}^p_{X(s)/S} \xrightarrow{J_p} s^* \mathcal{V}^p_{L(X)/S} \xrightarrow{H_p} \mathcal{V}^{p+1}_{X(s)/S} \otimes \mathcal{L} \to 0$$

since $s^* \mathcal{V}^p_{L(X)/S} \cong \mathcal{H}om(s^* \mathcal{W}^{-p}_{L(X)/S}, O_X)$. Note that $s^* \mathcal{G}^p_{L(X)/S} \subseteq s^* \mathcal{V}^p_{L(X)/S}$. We define

$$\mathcal{G}^p_{L(X)/S} := J_p^{-1}(s^* \mathcal{G}^p_{L(X)/S}) \subseteq \mathcal{V}^p_{X(s)/S},$$

extending our above definition of $\mathcal{G}^{-1}_{X(s)/S}$. These definitions make sense for log quasi-regular sections $s \in \mathcal{L}$, but we need log regularity in order for them to be well-behaved.

Obviously, we have well-defined comparison maps

$$dh^* = R_i \circ E_i : \ \mathcal{A}^i_{X/S} \to \mathcal{A}^i_{X(s)/S}, \quad Th_* = F_p \circ J_p : \ \mathcal{G}^p_{X(s)/S} \to \mathcal{G}^p_{X/S},$$

which are associated with $h \colon X(s) \to X$. Recall that $o \colon X \to L$ is the zero section. Since $L \setminus o(X) = L(X) \setminus o(X)$, the morphism $h \colon X(s) \to X$ is an isomorphism on $X \setminus H$, so dh^* and Th_* are isomorphisms on $X \setminus H$ on the level of $\mathcal{V}^\bullet$ and $\mathcal{W}^\bullet$. This remains true for $\mathcal{G}^\bullet$ and $\mathcal{A}^\bullet$ under the definition we have given:

Lemma 18.14 *The maps $dh^* \colon \mathcal{A}^i_{X/S} \to \mathcal{A}^i_{X(s)/S}$ and $Th_* \colon \mathcal{G}^p_{X(s)/S} \to \mathcal{G}^p_{X/S}$ are isomorphisms on $X \setminus H$.*

Proof Both maps are injective because they are isomorphisms on the level of $\mathcal{V}^\bullet$ and $\mathcal{W}^\bullet$, so it suffices to show surjectivity. If $\alpha \in \mathcal{A}^i_{X(s)/S}$ inside $X \setminus H$, then, locally, we can find $\beta = E_i(\beta_1) + E_{i-1}(\beta_2) \wedge \delta \in s^* \mathcal{A}^i_{L(X)/S}$ with $R_i(\beta) = \alpha$. Now

$$R_i(\beta) = R_i \circ E_i(\beta_1) + R_{i-1} \circ E_{i-1}(\beta_2) \wedge R_1(\delta)$$

with $R_1(\delta) = s_\alpha^{-1} \cdot R_1 \circ E_1(ds_\alpha),^{6,7}$ so we have

$$\alpha = R_i(\beta) = R_i \circ E_i(\beta_1 + s_\alpha^{-1} \cdot \beta_2 \wedge ds_\alpha),$$

---

^{6} Here we take $\delta = B_1(\gamma_s)$ for our original distinguished splitting B_1 used to define the class of distinguished splittings.

^{7} In the computation, we use $R_i(\alpha) \wedge R_j(\beta) = R_{i+j}(\alpha \wedge \beta)$, which holds by definition of R_i.

i.e., dh^* is surjective. For the surjectivity of Th_*, let $\theta \in \mathcal{G}^p_{X/S}$. Since $Th_* \colon \mathcal{V}^p_{X(s)/S} \to \mathcal{V}^p_{X/S}$ is an isomorphism on $X \setminus H$, we can find $\theta' \in \mathcal{V}^p_{X(s)/S}$ with $Th_*(\theta') = \theta$. Let

$$S_p \colon \ \mathcal{V}^p_{X/S} \to s^* \mathcal{V}^p_{L(X)/S}$$

be the splitting of F_p associated with the original distinguished splitting B_1. Let $i = -p$ and $\alpha \in s^* \mathcal{W}^i_{L(X)/S}$. On the one hand, we have

$$J_p(\theta') \lrcorner (E_i(\alpha_1) + E_{i-1}(\alpha_2) \wedge \delta) = \theta \lrcorner \alpha_1 + \theta \lrcorner (s_\alpha^{-1} \cdot \alpha_2 \wedge ds_\alpha).$$

On the other hand, we have

$$S_p(\theta) \lrcorner (E_i(\alpha_1) + E_{i-1}(\alpha_2) \wedge \delta) = \theta \lrcorner \alpha_1.$$

Thus, we have

$$(J_p(\theta') - S_p(\theta)) \lrcorner (E_i(\alpha_1) + E_{i-1}(\alpha_2) \wedge \delta) = \theta \lrcorner (s_\alpha^{-1} \cdot \alpha_2 \wedge ds_\alpha)$$
$$= (-1)^{i-1} s_\alpha^{-1} \cdot (\theta \vdash ds_\alpha) \lrcorner \alpha_2 = s_\alpha^{-1} \cdot I_p((\theta \vdash ds_\alpha) \otimes \gamma_s^\vee) \lrcorner \alpha.$$

Since $S_p(\theta) \in s^* \mathcal{G}^p_{L(X)/S}$ and $s_\alpha^{-1} \cdot I_p((\theta \vdash ds_\alpha) \otimes \gamma_s^\vee) \in s^* \mathcal{G}^p_{L(X)/S}$, we have $J_p(\theta') \in s^* \mathcal{G}^p_{L(X)/S}$, and hence $\theta' \in \mathcal{G}^p_{X(s)/S}$. $\square$

We work again on the whole of X. We have

$$G_1(1 \otimes e_\alpha^\vee) = s_\alpha \cdot s^* \delta - d_{X/S}(s_\alpha),$$

i.e., $G_1(1 \otimes e_\alpha^\vee) \in s^* \mathcal{A}^1_{L(X)/S}$, so we find $G_i(\mathcal{A}^{i-1}_{X(s)/S} \otimes \mathcal{L}^\vee) \subseteq s^* \mathcal{A}^i_{L(X)/S}$ since $s^* \mathcal{A}^\bullet_{L(X)/S} \subseteq s^* \mathcal{W}^\bullet_{L(X)/S}$ is closed under the $\wedge$-product. Thus, we have a sequence

$$0 \to \mathcal{A}^{i-1}_{X(s)/S} \otimes \mathcal{L}^\vee \xrightarrow{G_i} s^* \mathcal{A}^i_{L(X)/S} \xrightarrow{R_i} \mathcal{A}^i_{X(s)/S} \to 0. \tag{18.4}$$

It is exact on the left and on the right by construction.

Lemma 18.15 *If $s \in \mathcal{L}$ is a log regular section, then the sequence (18.4) is exact in the middle and locally split.*

Proof We start with exactness in the middle. By construction, we have $R_i \circ G_i = 0$. From Lemma 18.14, it is not so hard to see that (18.4) is exact in the middle on $X \setminus H$. So let us work with the stalks at some $x \in H$. Let $\alpha \in s^* \mathcal{A}^i_{L(X)/S}$ be such that $R_i(\alpha) = 0$. From the exactness of the corresponding sequence for $\mathcal{W}^\bullet$, we find that there is some $\beta \in s^* \mathcal{W}^{i-1}_{L(X)/S}$ with $\beta \wedge G_1(1 \otimes e^\vee_\alpha) = \alpha$. After decomposing $\beta = E_{i-1}(\beta_1) + E_{i-2}(\beta_2) \wedge \delta$ with $\beta_1 \in \mathcal{W}^{i-1}_{X/S}$ and $\beta_2 \in \mathcal{W}^{i-2}_{X/S}$, we have

$$\alpha = -E_i(\beta_1 \wedge ds_\alpha) + E_{i-1}(s_\alpha \cdot \beta_1 + \beta_2 \wedge ds_\alpha) \wedge s^* \delta.$$

This shows $\beta_1 \wedge ds_\alpha \in \mathcal{A}^i_{X/S}$ and $s_\alpha \cdot \beta_1 + \beta_2 \wedge ds_\alpha \in \mathcal{A}^{i-1}_{X/S}$. Let $\theta \in \mathcal{G}^{-1}_{X/S}$ be a log derivation as obtained from Lemma 18.12, i.e., $\theta \rightharpoonup ds_\alpha = 1 + a \cdot s_\alpha =: u$. Since we are at the stalk at $x \in H$, the function $u \in O_{X,x}$ is invertible. We set

$$\tilde\beta_1 := (-1)^{i-1} u^{-1} \cdot \theta \rightharpoonup (\beta_1 \wedge ds_\alpha) \in \mathcal{A}^{i-1}_{X/S};$$

then we have $\tilde\beta_1 \wedge ds_\alpha = \beta_1 \wedge ds_\alpha$. Furthermore, we have

$$((-1)^i u^{-1} \cdot s_\alpha \cdot \theta \rightharpoonup (\beta_1 - \tilde\beta_1)) \wedge ds_\alpha = s_\alpha \cdot (\beta_1 - \tilde\beta_1),$$

in other words, if $\tilde\gamma \in \mathcal{W}^{i-1}_{X/S}$ is the left-hand side without the ds_α-term, then we have $(\tilde\gamma + \beta_2) \wedge ds_\alpha \in \mathcal{A}^{i-1}_{X/S}$. After setting

$$\tilde\beta_2 = (-1)^{i-2} u^{-1} \cdot \theta \rightharpoonup ((\tilde\gamma + \beta_2) \wedge ds_\alpha) \in \mathcal{A}^{i-2}_{X/S},$$

we have $s_\alpha \cdot \tilde\beta_1 + \tilde\beta_2 \wedge ds_\alpha = s_\alpha \cdot \beta_1 + \beta_2 \wedge ds_\alpha$. Thus,

$$\tilde\beta := E_{i-1}(\tilde\beta_1) + E_{i-2}(\tilde\beta_2) \wedge s^* \delta \in s^* \mathcal{A}^{i-1}_{L(X)/S},$$

and $\tilde\beta \wedge G_1(1 \otimes e^\vee_\alpha) = \alpha$. In other words, $G_i(R_{i-1}(\tilde\beta) \otimes e^\vee_\alpha) = \alpha$, and (18.4) is exact in the middle.

To show that the sequence is locally split, let $\theta = (D, \Delta) \in \mathcal{G}^{-1}_{X/S}$ be such that $D(s_\alpha) = 1 + a \cdot s_\alpha$, and let $\xi = -a \cdot s^* \delta^\vee - S_{-1}(\theta) \in s^* \mathcal{G}^{-1}_{L(X)/S}$. Then we obtain a splitting of G_i as

$$U_i : \ s^* \mathcal{A}^i_{L(X)/S} \rightarrow \mathcal{A}^{i-1}_{X(s)/S} \otimes \mathcal{L}^\vee, \quad \alpha \mapsto (-1)^{i-1} R_{i-1}(\xi \rightharpoonup \alpha) \otimes e^\vee_\alpha.$$

If $\beta \in \mathcal{A}^{i-1}_{X(s)/S}$, then $G_i(\beta \otimes e^\vee_\alpha) = \tilde\beta \wedge G_1(e^\vee_\alpha)$ for any lift $\tilde\beta \in s^* \mathcal{A}^{i-1}_{L(X)/S}$ with $R_{i-1}(\tilde\beta) = \beta$, so

$$U_i \circ G_i(\beta \otimes e^\vee_\alpha) = R_{i-1}((-1)^{i-1}(\xi \rightharpoonup \tilde\beta) \wedge G_1(e^\vee_\alpha) + \tilde\beta \wedge (\xi \rightharpoonup G_1(e^\vee_\alpha))) \otimes e^\vee_\alpha = \beta \otimes e^\vee_\alpha$$

since $\xi \rightharpoonup G_1(e^\vee_\alpha) = 1$. $\hfill\square$

Next, we study the dual of the sequence (18.4). If $\theta \in s^* \mathcal{V}^p_{L(X)/S}$ is such that $H_p(\theta) = \xi \otimes e_\alpha$ for some $\xi \in \mathcal{V}^{p+1}_{X(s)/S}$, then we have

$$\xi \lrcorner R_{i-1}(\beta) = (\theta \vdash G_1(e_\alpha^\vee)) \lrcorner \beta$$

for every $\beta \in s^* \mathcal{W}^{i-1}_{L(X)/S}$ for $i = -p$. In other words, $J_{p+1}(\xi) = \theta \vdash G_1(e_\alpha^\vee)$. If $\theta \in s^* \mathcal{G}^p_{L(X)/S}$, then $\theta \vdash G_1(e_\alpha^\vee) \in s^* \mathcal{G}^{p+1}_{L(X)/S}$, and hence $\xi \in \mathcal{G}^{p+1}_{X(s)/S}$ by definition. Thus, we have a well-defined sequence

$$0 \to \mathcal{G}^p_{X(s)/S} \xrightarrow{J_p} s^* \mathcal{G}^p_{L(X)/S} \xrightarrow{H_p} \mathcal{G}^{p+1}_{X(s)/S} \otimes \mathcal{L} \to 0, \qquad (18.5)$$

which is exact on the left and in the middle.

Lemma 18.16 *The sequence (18.5) is exact on the right and locally split.*

Proof We construct a local splitting of H_p. Let $\theta_\alpha = (D_\alpha, \Delta_\alpha) \in \mathcal{G}^{-1}_{X/S}$ be such that $D_\alpha(s_\alpha) = 1 + a \cdot s_\alpha$, and let $\xi_\alpha := -a \cdot s^* \delta^\vee - S_{-1}(\theta_\alpha) \in s^* \mathcal{G}^{-1}_{L(X)/S}$.[8] Let $\theta \in \mathcal{G}^{p+1}_{X(s)/S}$. Since H_p is surjective for $\mathcal{V}^\bullet$, there is some $\xi \in \mathcal{V}^p_{L(X)/S}$ with $H_p(\xi) = \theta \otimes e_\alpha$, and hence $J_{p+1}(\theta) = \xi \vdash G_1(e_\alpha^\vee)$. We define the splitting by

$$V_p \colon \ \mathcal{G}^{p+1}_{X(s)/S} \otimes \mathcal{L} \to s^* \mathcal{G}^p_{L(X)/S}, \qquad \theta \otimes e_\alpha \mapsto J_{p+1}(\theta) \wedge \xi_\alpha.$$

Then $H_p \circ V_p(\theta \otimes e_\alpha) = \theta' \otimes e_\alpha$ for some θ' with

$$
\begin{aligned}
J_{p+1}(\theta') &= (J_{p+1}(\theta) \wedge \xi_\alpha) \vdash G_1(e_\alpha^\vee) \\
&= -(\xi \vdash G_1(e_\alpha^\vee)) \vdash G_1(e_\alpha^\vee) + J_{p+1}(\theta) \wedge (\xi_\alpha \vdash G_1(e_\alpha^\vee)),
\end{aligned}
$$

i.e., $\theta' = \theta$ since $\xi_\alpha \vdash G_1(e_\alpha^\vee) = 1$. Thus V_p is a splitting of H_p. $\qquad\square$

With this preparation, we show:

Proposition 18.17 *Let $s \in \mathcal{L}$ be a log regular section. Assume that the pieces of $\mathcal{G}^p_{X/S}$ and $\mathcal{A}^i_{X/S}$ are torsionless in every fiber. Then $f \circ h \colon X(s) \to S$ is a log Gorenstein generically log smooth family. The relative log canonical bundle is*

$$\mathcal{W}^d_{X(s)/S} \cong \mathcal{W}^d_{X/S} \otimes \mathcal{L}.$$

When we endow $f \circ h \colon X(s) \to S$ with $(\mathcal{G}^\bullet_{X(s)/S}, \mathcal{A}^\bullet_{X(s)/S})$, this is an enhanced generically log smooth family whose pieces are torsionless in every fiber. The formation of the Gerstenhaber calculus commutes with base change in S.

---

[8] Note that this does not only depend on α.

Proof Considering the exact sequence (18.4) for $i = d + 1$ together with its variant for $\mathcal{W}^\bullet$ and the embedding, we find that

$$\varpi^{d+1} \otimes \mathcal{L}^\vee : \quad \mathcal{A}^d_{X(s)/S} \otimes \mathcal{L}^\vee \to \mathcal{W}^d_{X(s)/S} \otimes \mathcal{L}^\vee$$

is an isomorphism, and that both sides are isomorphic to the line bundle

$$s^* \mathcal{W}^{d+1}_{L(X)/S} \cong \mathcal{W}^d_{X/S} \otimes s^* \Omega^1_{L(X)/X} \cong \mathcal{W}^d_{X/S}.$$

The exact sequences (18.4) and (18.5) show that each piece $\mathcal{G}^p_{X(s)/S}$ and $\mathcal{A}^i_{X(s)/S}$ is flat over S, and that they are torsionless in every fiber since this holds for $\mathcal{G}^p_{X/S}$ and $\mathcal{A}^i_{X/S}$, and hence for $s^* \mathcal{G}^p_{L(X)/S}$ and $s^* \mathcal{A}^i_{L(X)/S}$.

Because J_0 and R_0 are isomorphisms, we have $\mathcal{A}^0_{X(s)/S} = O_X$ and $\mathcal{G}^0_{X(s)/S} = O_X$. Since R_i is compatible with the $\wedge$-product, $\mathcal{A}^\bullet_{X(s)/S} \subseteq \mathcal{W}^\bullet_{X(s)/S}$ is closed under the $\wedge$-product. To see that $\mathcal{A}^\bullet_{X(s)/S}$ is a complex, we consider the diagram

$$
\begin{array}{ccccc}
\mathcal{A}^i_{L(X)/S} & \longrightarrow & s_* s^* \mathcal{A}^i_{L(X)/S} & \longrightarrow & s_* \mathcal{A}^i_{X(s)/S} \\
\downarrow & & \downarrow & & \downarrow \\
\mathcal{W}^i_{L(X)/S} & \longrightarrow & s_* s^* \mathcal{W}^i_{L(X)/S} & \longrightarrow & s_* \mathcal{W}^i_{X(s)/S} = \!\!= j_* s_* \Omega^i_{U(s)/S}.
\end{array}
$$

The lower composition is compatible with the de Rham differential ∂ since it is so on $p^{-1}(U)$. The upper row remains surjective. Then the claim follows from the fact that $\mathcal{A}^\bullet_{L(X)/S} \subseteq \mathcal{W}^\bullet_{L(X)/S}$ is a subcomplex.

The graded subsheaf $\mathcal{G}^\bullet_{X(s)/S} \subseteq \mathcal{V}^\bullet_{X(s)/S}$ is closed under the $\wedge$-product because J_p is compatible with the $\wedge$-product. Similar to the proof of Lemma 18.5, we can show that

$$R_{p+i}(J_p(\theta) \lrcorner \, \alpha) = \theta \lrcorner R_i(\alpha)$$

for $\theta \in \mathcal{V}^p_{X(s)/S}$ and $\alpha \in s^* \mathcal{W}^i_{L(X)/S}$. Using this formula, we see that $\mathcal{G}C^\bullet_{X(s)/S}$ is closed under the right contraction $\lrcorner$. Similarly, we have[9]

$$J_p(\theta) \vdash \alpha = J_{p+i}(\theta \vdash R_i(\alpha)),$$

so $\mathcal{G}C^\bullet_{X(s)/S}$ is closed under the left contraction $\vdash$.

Next, we show that $\mathcal{G}C^\bullet_{X(s)/S}$ is locally Batalin–Vilkovisky. Let $\omega \in \mathcal{A}^d_{X(s)/S}$ be a local volume form, and let $\tilde{\omega} \in s^* \mathcal{A}^d_{L(X)/S}$ be such that $R_d(\tilde{\omega}) = \omega$. A local

---

[9] Note that this holds for arbitrary $\alpha \in s^* \mathcal{W}^i_{L(X)/S}$—the left contraction with an element in the image of J_p is always in the image of J_{p+i}.

volume form of $s^*\mathcal{A}^{d+1}_{L(X)/S}$ is now given by $\hat{\omega} = \tilde{\omega} \wedge G_1(e_\alpha^\vee)$. Given ω, this form is independent of the chosen lift $\tilde{\omega}$. We have to show that

$$\kappa_\omega\colon\ \mathcal{G}^p_{X(s)/S} \to \mathcal{A}^{p+d}_{X(s)/S}, \qquad \theta \mapsto \theta \lrcorner\, \omega,$$

is an isomorphism. By construction, the corresponding map on the level of $\mathcal{V}^p_{X(s)/S}$ and $\mathcal{W}^{p+d}_{X(s)/S}$ is an isomorphism. In particular, our κ_ω is injective, and for $\alpha \in \mathcal{A}^{p+d}_{X(s)/S}$, we can find $\theta \in \mathcal{V}^p_{X(s)/S}$ with $\theta \lrcorner\, \omega = \alpha$. We have to show that $\theta \in \mathcal{G}^p_{X(s)/S} \subseteq \mathcal{V}^p_{X(s)/S}$, i.e., $J_p(\theta) \in s^*\mathcal{G}^p_{L(X)/S}$. Since $\mathcal{G}C^\bullet_{L(X)/S}$ is locally Batalin–Vilkovisky, it is sufficient to show $J_p(\theta) \lrcorner\, \hat{\omega} \in s^*\mathcal{A}^{p+d+1}_{L(X)/S}$. If $\tilde{\alpha} \in s^*\mathcal{A}^{p+d}_{L(X)/S}$ is a lift of α, then we find $R_{p+d}(J_p(\theta) \lrcorner\, \tilde{\omega}) = R_{p+d}(\tilde{\alpha})$ from the above formulae. In particular, there is some $\beta \in \mathcal{W}^{p+d-1}_{X(s)/S} \otimes \mathcal{L}^\vee$ with $J_p(\theta) \lrcorner\, \tilde{\omega} = G_{p+d}(\beta) + \tilde{\alpha}$. By induction on p, we can show that

$$J_p(\theta) \lrcorner\, \hat{\omega} = (J_p(\theta) \lrcorner\, \tilde{\omega}) \wedge G_1(e_\alpha^\vee)$$

although there is in general no easy formula for the action of the right contraction on a $\wedge$-product on the right. Thus,

$$J_p(\theta) \lrcorner\, \hat{\omega} = \tilde{\alpha} \wedge G_1(e_\alpha^\vee) \in s^*\mathcal{A}^{p+d+1}_{L(X)/S},$$

as desired.

Now we obtain that $\mathcal{G}C^\bullet_{X(s)/S}$ is closed under the Lie derivative $\mathcal{L}_-(-)$ from the Lie–Rinehart homotopy formula, and that $\mathcal{G}C^\bullet_{X(s)/S}$ is closed under the bracket $[-,-]$ from the mixed Leibniz rule.

Considering the exact sequence (18.4), we see by induction on i that the formation of $\mathcal{A}^i_{X(s)/S}$ commutes with base change because the formation of $s^*\mathcal{A}^i_{L(X)/S}$ commutes with base change. Since $\mathcal{G}C^\bullet_{X(s)/S}$ is locally Batalin–Vilkovisky, then the same is true for $\mathcal{G}^p_{X(s)/S}$. $\hfill\square$

18.4 Thickenings of Log Regular Sections

In this section, we show that a thickening $s \in \mathcal{L}$ of a log regular section $s_0 \in \mathcal{L}_0$ is again a log regular section. In other words, the condition of log regularity does not impose additional restrictions on infinitesimal deformations once we require it on the central fiber. We work in the following situation:

Situation 18.18 Let $S_0 = \mathrm{Spec}(Q \to \mathbf{k})$, and let $f_0\colon X_0 \to S_0$ be an enhanced generically log smooth family of relative dimension d with Gerstenhaber calculus

$$\varpi^\bullet\colon\ (\mathcal{G}^\bullet_{X_0/S_0}, \mathcal{A}^\bullet_{X_0/S_0}) \subseteq (\mathcal{V}^\bullet_{X_0/S_0}, \mathcal{W}^\bullet_{X_0/S_0}).$$

Let $\mathcal{L}_0$ be a line bundle on X_0, and let s_0 be a global section. We consider infinitesimal deformations $f_A\colon X_A \to S_A$ of $f_0\colon X_0 \to S_0$ with Gerstenhaber calculus

$$\varpi^{\bullet}\colon\ (\mathcal{G}^{\bullet}_{X_A/S_A}, \mathcal{A}^{\bullet}_{X_A/S_A}) \subseteq (\mathcal{V}^{\bullet}_{X_A/S_A}, \mathcal{W}^{\bullet}_{X_A/S_A}).$$

Then $\mathcal{L}$ is a line bundle on X_A with a chosen isomorphism $\mathcal{L}|_0 = \mathcal{L}_0$, and $s \in \mathcal{L}$ is a global section with $s|_0 = s_0$.

First, we prove our claim for log pre-regular sections.

Lemma 18.19 *Assume that $s_0 \in \mathcal{L}_0$ is a log pre-regular section. Then $s \in \mathcal{L}$ is a log pre-regular section.*

Proof It is clear that $\mathrm{codim}(Z \cap H, H) \geq 2$ in every fiber since $f\colon X \to S$ and $f_0\colon X_0 \to S_0$ have the same fibers; this also applies to the condition on the depth. The open subset $X_0 \setminus H_0 \subseteq X_0$ is scheme-theoretically dense, so $X_A \setminus H_A \subseteq X_A$ is dense in every fiber, and by Remark 8.37, $O_{X_A} \to j_* O_{X_A \setminus H_A}$ is injective. Thus, $s^{\vee}\colon \mathcal{L}^{\vee} \to O_{X_A}$ is injective. Now let $\mathbf{M}$ be the class of all finitely generated A-modules M such that $\mathcal{L}^{\vee} \otimes_A M \to O_{X_A} \otimes_A M$ remains injective. By assumption, we have $A/\mathfrak{m}_A \in \mathbf{M}$, and the same argument that shows $\mathcal{L}^{\vee} \to O_{X_A}$ injective also shows $A/I \in \mathbf{M}$ for all ideals $I \subseteq \mathfrak{m}_A \subseteq A$. Since both $\mathcal{L}^{\vee}$ and O_{X_A} are flat over S_A, we find that every extension M of two modules $M', M'' \in \mathbf{M}$ is in $\mathbf{M}$ as well. Since every finitely generated module M has a filtration by submodules such that each quotient is of the form A/I for some ideal $I \subseteq A$, we find that $\mathbf{M}$ is the class of all finitely generated A-modules. Then the sequence

$$0 \to \mathcal{L}^{\vee} \to O_{X_A} \to O_{H_A} \to 0$$

is universally exact over A, and since O_{X_A} is flat over A, we find that O_{H_A} is flat over A by [267, 058P]. $\qquad\square$

On the log smooth locus U_A, it is easy to see that s is log regular if s_0 is log regular. First, there is a splitting in $\mathrm{Sp}_{X_A(s)/N_A(s)}$ on $U_A \setminus H_A$, given by $p\colon L_A \to X_A$, so $G_1\colon \mathcal{L}^{\vee} \to s^* W^1_{L_A(X_A)/S_A}$ is injective. Since $\Omega^1_{U_0(s_0)/S_0}$ is locally free, the same is true for $\Omega^1_{U_A(s)/S_A}$, so the exact sequence in Lemma 18.8 is locally split on U_A. On the log singular locus, this result is considerably more difficult to prove. We start with a lemma.

Lemma 18.20 *In Situation 18.1, let $s \in \mathcal{L}$ be a log quasi-regular section. Let Q be the cokernel in the exact sequence*

$$0 \to \mathcal{V}^{-1}_{X(s)/S} \xrightarrow{Th_*} \mathcal{V}^{-1}_{X/S} \xrightarrow{q} Q \to 0.$$

Let $V \subseteq X$ be an affine open subset. Then the induced map

$$\Gamma(V \cap U, \mathcal{V}_{X/S}^{-1}) \to \Gamma(V \cap U, \mathcal{Q})$$

is surjective.

Proof First, we consider the following diagram:

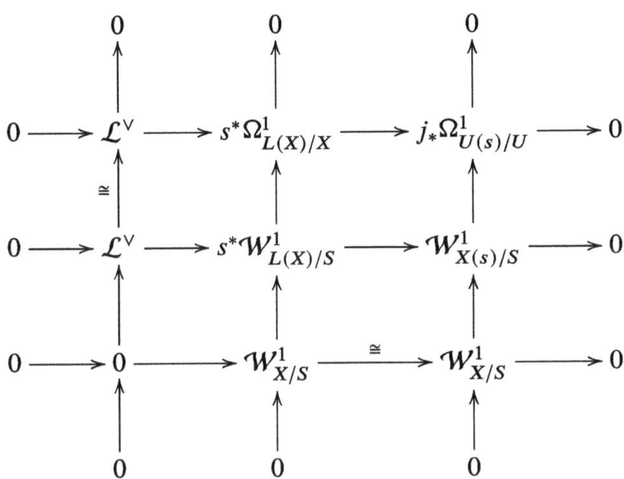

The middle row is exact by Lemma 18.8 since $s \in \mathcal{L}$ is a log quasi-regular section. The upper left horizontal map is an isomorphism on $X \setminus H$ and hence injective since both the source and the target are line bundles. On U, the upper row is exact in the middle and on the right by construction since $U(s) \to L(U) := L(X)|_{p^{-1}(U)}$ is a strict closed immersion. Since the upper left horizontal map is equal to $s^\vee \colon \mathcal{L}^\vee \to O_X$ under $O_X \cong s^* \Omega^1_{L(X)/X}$, $1 \mapsto s^* \gamma_s$, its cokernel is O_H. Since $j_* O_{U \cap H} = O_H$, the map $O_H \to j_* \Omega^1_{U(s)/U}$ induced from the upper row is not only an isomorphism on U but on X; in particular, the upper row is exact. Similar arguments show that the right column is exact. We obtain the surjectivity on the top since we already know that the middle column is exact and locally split, i.e., the upper left composition in the upper right square is surjective, so the right map must be surjective as well. Let us write O_H for $j_* \Omega^1_{U(s)/U}$ from now on.

By dualizing the above diagram with $\mathcal{H}om(-, O_X)$, we obtain the diagram in Fig. 18.1. The spaces in the diagram at the places of 0_1 and 0_2 may not actually be 0, but the horizontal map to 0_1 and the vertical map to 0_2 are surjective because they come from dualizing a locally split exact sequence. Thus, we can put a 0 in their places in the diagram. From the upper row, we find $\mathcal{E}xt^1(O_H, O_X) \cong O_H \otimes \mathcal{L}$. This sheaf satisfies $j_*(O_H \otimes \mathcal{L})|_U = O_H \otimes \mathcal{L}$, so the upper horizontal left map is surjective for sections on $V \cap U$. Similarly, the map $s^* \mathcal{V}_{L(X)/S}^{-1} \to \mathcal{L}$ is surjective for sections on $V \cap U$, so the map A is surjective for sections on $V \cap U$. The maps A and B have the same source and target, and both have the same kernel

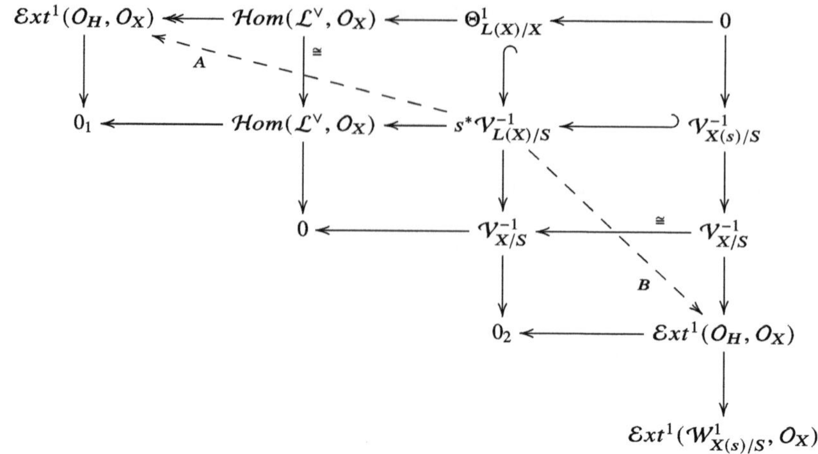

Fig. 18.1 The dual of the diagram in the proof of Proposition 18.20

$\Theta^1_{L(X)/X} + \mathcal{V}^{-1}_{X(s)/S}$. The map A is surjective, and the map B is surjective over U since $\mathcal{W}^1_{X(s)/S}$ is locally free there. Thus, on U, there is a unique automorphism γ of $\mathcal{E}xt^1(O_H, O_X)$ with $B = \gamma \circ A$; in particular, the map B is surjective for sections on $V \cap U$. Since the canonical map $Q \to \mathcal{E}xt^1(O_H, O_X)$ is an isomorphism on U, we find that $\mathcal{V}^{-1}_{X/S} \to Q$ is surjective for sections on $V \cap U$. $\square$

Corollary 18.21 *We have $Q \cong \mathcal{E}xt^1(O_H, O_X) \cong O_H \otimes \mathcal{L}$.*

Proof The map $\Gamma(V, Q) \to \Gamma(V \cap U, Q)$ is both injective because Q is a quotient of a reflexive sheaf by a reflexive subsheaf, and it is surjective as an easy application of Lemma 18.20. Thus, $Q \to \mathcal{E}xt^1(O_H, O_X)$ is an isomorphism. $\square$

With this preparation, we can prove that, in Situation 18.18, an infinitesimal thickening s of a log regular section s_0 is again a log regular section. Note that the proof below is not correct if we assume only log quasi-regular because the map $\mathcal{V}^{-1}_{X_{A'}/S_{A'}} \to \mathcal{V}^{-1}_{X_A/S_A}$ need not be surjective.

Proposition 18.22 *In Situation 18.18, if $s_0 \in \mathcal{L}_0$ is a log regular section, then $s \in \mathcal{L}$ is a log regular section.*

Proof We prove the result by induction over small extensions $A' \to A$ with kernel $I \subset A'$. So let $f' \colon X_{A'} \to S_{A'}$ be a deformation of $f \colon X_A \to S_A$ as an enhanced generically log smooth family, and assume that we have a deformation $\mathcal{L}'$ of $\mathcal{L}$ and a section $s' \in \mathcal{L}'$ with $s'|_A = s$. We consider the diagram in Fig. 18.2.

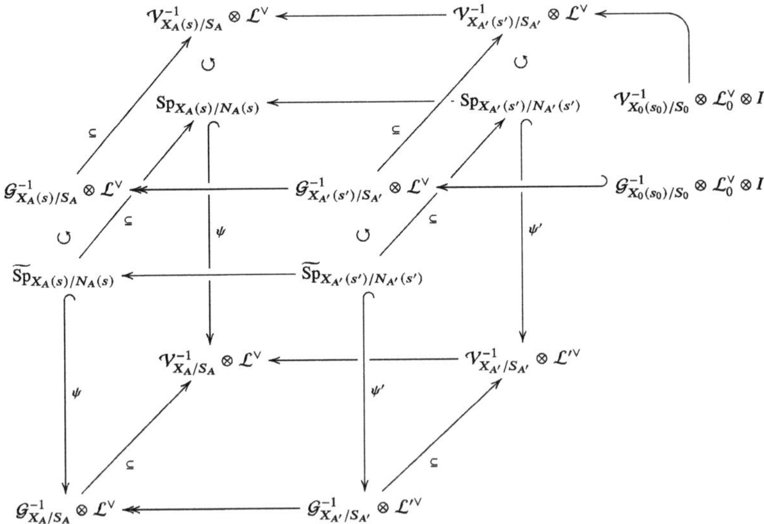

Fig. 18.2 The sheaves in the proof of Proposition 18.22

Let $V \subseteq X$ be a (small) affine open subset. Since $s \in \mathcal{L}$ is log regular, we have a section $r \in \widetilde{\mathrm{Sp}}_{X_A(s)/N_A(s)}$ over V. Let $\mathcal{U} = \{U_i\}_i$ be an affine open cover of $V \cap U$. Since the restriction map

$$\widetilde{\mathrm{Sp}}_{X_{A'}(s')/N_{A'}(s')} \to \widetilde{\mathrm{Sp}}_{X_A(s)/N_A(s)}$$

is surjective on U,[10] we can find lifts $r_i' \in \widetilde{\mathrm{Sp}}_{X_{A'}(s')/N_{A'}(s')}$ of r on U_i if the U_i are small enough. On the other hand, we can find an element $g' \in \mathcal{G}^{-1}_{X_{A'}/S_{A'}} \otimes \mathcal{L}'^{\vee}$ over V with $g'|_A = \psi(r)$. Since $r_i'|_A = r_j'|_A$, we have, due to the torsor structure, an element $\theta_{ij} \in \mathcal{G}^{-1}_{X_0(s_0)/S_0} \otimes \mathcal{L}_0^{\vee} \otimes I$ over $U_i \cap U_j$ with $r_j' = \theta_{ij} \odot r_i'$. Then

$$\psi'(r_j') - \psi'(r_i') = Th_*(\theta_{ij}) \in \mathcal{G}^{-1}_{X_0/S_0} \otimes \mathcal{L}_0^{\vee} \otimes I.$$

On the other hand, we also have $g_i' := \psi'(r_i') - g' \in \mathcal{G}^{-1}_{X_0/S_0} \otimes \mathcal{L}_0^{\vee} \otimes I$, and $g_j' - g_i' = \psi'(r_j') - \psi'(r_i')$. When we denote the cohomology class represented by θ_{ij} as

$$[\theta] \in H^1(U \cap V, \, \mathcal{G}^{-1}_{X_0(s_0)/S_0} \otimes \mathcal{L}_0^{\vee} \otimes I),$$

---

[10] This is because of the torsor structure that we have; we do, however, not yet know the surjectivity on X because, there, the source might be empty.

this shows $Th_*[\theta] = 0 \in H^1(U \cap V, \mathcal{G}^{-1}_{X_0/S_0} \otimes \mathcal{L}^\vee_0 \otimes I)$. Lemma 18.20 implies that Th_* is injective as a map on $H^1(U \cap V, -)$, i.e., $[\theta] = 0$. Then there are sections $\theta_i \in \mathcal{G}^{-1}_{X_0(s_0)/S_0} \otimes \mathcal{L}^\vee_0 \otimes I$ over U_i with $\theta_j - \theta_i = \theta_{ij}$. Then the local splittings $\tilde{r}'_i := (-\theta_i) \odot r'_i$ glue to a splitting r' on $V \cap U$ with $r'|_A = r$, i.e., $s' \in \mathcal{L}'$ is a log quasi-regular section. Since the formation of $s'^* \mathcal{G}^{-1}_{L_{A'}(X_{A'})/S_{A'}} \to \mathcal{L}'$ commutes with base change, s' is in fact a log regular section. $\square$

18.5   Infinitesimal Automorphisms and Isomorphisms

We study infinitesimal automorphisms and isomorphisms in the following situation:

Situation 18.23 Let $S_0 = \mathrm{Spec}(Q \to \mathbf{k})$, and let $f_0 \colon X_0 \to S_0$ be an enhanced generically log smooth family of relative dimension d with Gerstenhaber calculus

$$\varpi^\bullet \colon \ (\mathcal{G}^\bullet_{X_0/S_0}, \mathcal{A}^\bullet_{X_0/S_0}) \subseteq (\mathcal{V}^\bullet_{X_0/S_0}, \mathcal{W}^\bullet_{X_0/S_0}).$$

Note that we assume $\varpi^\bullet$ to be injective. Let $\mathcal{L}_0$ be a line bundle on X_0, and let s_0 be a *log regular* global section. We fix an infinitesimal deformation $f \colon X \to S$ of $f_0 \colon X_0 \to S_0$ over $S = S_A$ with Gerstenhaber calculus

$$\varpi^\bullet \colon \ (\mathcal{G}^\bullet_{X/S}, \mathcal{A}^\bullet_{X/S}) \subseteq (\mathcal{V}^\bullet_{X/S}, \mathcal{W}^\bullet_{X/S}).$$

We have a line bundle $\mathcal{L}$ on X with a chosen isomorphism $\mathcal{L}|_0 = \mathcal{L}_0$, and $s \in \mathcal{L}$ is a global section with $s|_0 = s_0$. Furthermore, we have a not necessarily small extension $A' \to A$ with kernel $I \subset A'$, and we consider a thickening $f' \colon X' \to S'$ of $f \colon X \to S$ over $S' = S_{A'}$ as an enhanced generically log smooth family. We have a line bundle $\mathcal{L}'$ on X' with chosen isomorphism $\mathcal{L}'|_A = \mathcal{L}$, and we consider global sections $s' \in \mathcal{L}'$ with $s'|_A = s$.

Let $f'_1 \colon X'_1 \to S'$ and $f'_2 \colon X'_2 \to S'$ be two thickenings of $f \colon X \to S$ as an enhanced generically log smooth family, and assume that we have line bundles $\mathcal{L}'_1$ and $\mathcal{L}'_2$ with sections s'_1 and s'_2, both being thickenings of $s \in \mathcal{L}$. If $\varphi \colon X'_1 \xrightarrow{\cong} X'_2$ is an isomorphism of enhanced generically log smooth families over $f \colon X \to S$, i.e., $(d\varphi^*, T\varphi_*)$ is compatible with the chosen Gerstenhaber calculi $\mathcal{GC}^\bullet_{X'_i/S'}$ inside $(\mathcal{V}^\bullet_{X'_i/S'}, \mathcal{W}^\bullet_{X'_i/S'})$, and if we are given an isomorphism $\psi \colon \mathcal{L}'_2 \cong \varphi_* \mathcal{L}'_1$ with $\psi(s'_2) = s'_1$, then we have a commutative diagram

$$
\begin{array}{ccccccc}
X'_1(s'_1) & \longrightarrow & L'_1(X'_1) & \longrightarrow & L'_1 & \longrightarrow & X'_1 \\
\cong \downarrow \varphi_s & & \cong \downarrow \psi & & \cong \downarrow \psi & & \cong \downarrow \varphi \\
X'_2(s'_2) & \longrightarrow & L'_2(X'_2) & \longrightarrow & L'_2 & \longrightarrow & X'_2
\end{array}
$$

of isomorphisms which induce the identity after restriction to S. Note that the isomorphism φ_s on the left-hand side is an isomorphism of enhanced generically log smooth families, i.e., compatible with the Gerstenhaber calculus $\mathcal{GC}^{\bullet}_{X'_i(s'_i)/S'}$.

We compute these maps in the case of an inner automorphism of $f'\colon X' \to S'$ over $f\colon X \to S$, with the same $\mathcal{L}'$ but allowing two different sections s'_1 and s'_2. Recall that the automorphisms of $f'\colon X' \to S'$ as a non-enhanced generically log smooth family are classified by log derivations in the kernel $I \cdot \mathcal{V}^{-1}_{X'/S'}$ of the possibly non-surjective map $\mathcal{V}^{-1}_{X'/S'} \to \mathcal{V}^{-1}_{X/S}$. The induced automorphism of the reflexive Gerstenhaber calculus is $\exp_{-\theta}$ where $\theta = (D, \Delta)$ is the relative log derivation. If $\theta \in I \cdot \mathcal{G}^{-1}_{X'/S'} \subseteq I \cdot \mathcal{V}^{-1}_{X'/S'}$, which is the kernel of the surjective map $\mathcal{G}^{-1}_{X'/S'} \to \mathcal{G}^{-1}_{X/S}$, then we say that θ induces an *inner automorphism* of the enhanced generically log smooth family; in this case, $\mathcal{GC}^{\bullet}_{X/S}$ is invariant under the gauge transform $\exp_{-\theta}$.

When we want to take the map $\psi\colon \mathcal{L}' \to \mathcal{L}'$ into account, then we need, in addition to (D, Δ), an $f'^{-1}(O_{S'})$-linear map $u\colon \mathcal{L}' \to \mathcal{L}'$ with $u(a \cdot e) = D(a) \cdot e + a \cdot u(e)$. The classifying space of such triples (D, Δ, u) fits into the Atiyah extension

$$0 \to O_{X'} \to \mathcal{V}^{-1}_{X'/S'}(\mathcal{L}') \to \mathcal{V}^{-1}_{X'/S'} \to 0,$$

which is locally split exact. Those triples (D, Δ, u) in $I \cdot \mathcal{V}^{-1}_{X'/S'}(\mathcal{L}')$ give the infinitesimal automorphisms of $f'\colon (X', \mathcal{L}') \to S'$ as a non-enhanced generically log smooth family with a line bundle. When we consider only such (D, Δ, u) with $(D, \Delta) \in \mathcal{G}^{-1}_{X'/S'}$, then we obtain the modified Atiyah extension

$$0 \to O_{X'} \to \mathcal{G}^{-1}_{X'/S'}(\mathcal{L}') \to \mathcal{G}^{-1}_{X'/S'} \to 0,$$

which is locally split exact as well. The triples (D, Δ, u) in $I \cdot \mathcal{G}^{-1}_{X'/S'}(\mathcal{L}')$ are precisely those with $(D, \Delta) \in I \cdot \mathcal{G}^{-1}_{X'/S'}$, and they give rise to an (inner) automorphism of $f'\colon (X', \mathcal{L}') \to S'$ as an enhanced generically log smooth family with a line bundle.

For $(D, \Delta, u) \in \mathcal{G}^{-1}_{X'/S'}(\mathcal{L}')$, we have, by definition,

$$\phi\colon \ O_{X'} \to O_{X'}, \quad a \mapsto \sum_{n=0}^{\infty} \frac{D^n(a)}{n!} = a + D(a) + \frac{1}{2}D^2(a) + \dots,$$

as well as $\Phi\colon \mathcal{M}_{U'} \to \mathcal{M}_{U'}$ defined by the more complicated formula in Chap. 10.2. On $\mathcal{L}'$, we have

$$\psi\colon \ \mathcal{L}' \to \mathcal{L}', \quad e \mapsto \sum_{n=0}^{\infty} \frac{u^n(e)}{n!} = e + u(e) + \frac{1}{2}u^2(e) + \dots.$$

We use the notations from Chap. 18.1 on $f' \colon X' \to S'$, i.e., e_α is a local trivializing section of $\mathcal{L}'$ etc. When we define $w_\alpha \in \mathcal{O}_{X'}$ by $u(e_\alpha) = w_\alpha \cdot e_\alpha$, then we have $\psi(e_\alpha) = v_\alpha \cdot e_\alpha$ with

$$v_\alpha = 1 + \sum_{n=1}^{\infty} \frac{[D + w_\alpha]^{n-1}(w_\alpha)}{n!} = 1 + w_\alpha + \frac{D(w_\alpha) + w_\alpha^2}{2} + \dots .$$

On the level of $\mathcal{O}_{L'}$, we have $\psi \colon R_\alpha[x_\alpha] \to R_\alpha[x_\alpha]$ with $\psi(a) = \phi(a)$ for $a \in R_\alpha$ and $\psi(x_\alpha) = v_\alpha^{-1} \cdot x_\alpha$. For the embedding $\varepsilon_0 \colon p'^* \mathcal{L}' \to \mathcal{O}_{L'}$, $p'^* e_\alpha^\vee \mapsto x_\alpha$, we find that it is compatible with ψ when we map

$$p'^* \mathcal{L}' \to p'^* \mathcal{L}', \quad p'^* e_\alpha^\vee \mapsto v_\alpha^{-1} \cdot p'^* e_\alpha^\vee,$$

as is natural. Thus, we have an isomorphism of Deligne–Faltings structures, and hence an isomorphism $\psi \colon \underline{L}'(X') \to \underline{L}'(X')$, yielding the isomorphism $\psi \colon L'(X') \cong L'(X')$ over $\psi \colon L' \to L'$.

Let us write $s_1' = s_{\alpha;1} \cdot e_\alpha$ and $s_2' = s_{\alpha;2} \cdot e_\alpha$. Since $\psi(s_2') = s_1'$, we find $\phi(s_{\alpha;2}) = v_\alpha^{-1} \cdot s_{\alpha;1}$. This confirms that we have a commutative diagram

$$
\begin{array}{ccc}
X'(s_1') & \xrightarrow{\ s_{\alpha;1} \leftarrow\!\shortmid x_\alpha\ } & L'(X') \\[2pt]
{\scriptstyle v_\alpha^{-1} \cdot s_{\alpha;1} \leftarrow\!\shortmid s_{\alpha;2}} \big\downarrow & & \big\downarrow {\scriptstyle v_\alpha^{-1} \cdot x_\alpha \leftarrow\!\shortmid x_\alpha} \\[2pt]
X'(s_2') & \xrightarrow{\ s_{\alpha;2} \leftarrow\!\shortmid x_\alpha\ } & L'(X'),
\end{array}
$$

and, because the log structure is defined as the inverse image from $L'(X')$, a compatible isomorphism $\varphi_s \colon X'(s_1') \cong X'(s_2')$ of log schemes (and in fact enhanced generically log smooth families since our construction is functorial on that level). This allows us to construct both isomorphisms between different choices of s' and automorphisms of each $X'(s')$.

First, we show that, locally, $X'(s_1') \cong X'(s_2')$ for any two choices s_1' and s_2' with $s_i'|_A = s$. In other words, up to (non-unique local) isomorphism, $X'(s')$ is independent of the choice of s'.

Lemma 18.24 *Let s_1' and s_2' be two sections of $\mathcal{L}'$ with $s_i'|_A = s$. Then there is locally an element $\theta = (D, \Delta, u) \in I \cdot \mathcal{G}_{X'/S'}^{-1}(\mathcal{L}')$ with $\psi(s_2') = s_1'$.*

Proof First, we show the statement for a small extension, so assume that $I^2 = 0$, and more precisely that $I = (i)$ with $i^2 = 0$. We want to have $\psi(s_2') = s_1'$, which is equivalent to

$$s_{\alpha;2} + D(s_{\alpha;2}) = (1 - w_\alpha) \cdot s_{\alpha;1},$$

where $u(e_\alpha) = w_\alpha \cdot e_\alpha$. Let $s_{\alpha;2} = s_{\alpha;1} + i \cdot x$, and let $(D', \Delta') \in I \cdot \mathcal{G}^{-1}_{X'/S'}$ with $D'(s_{\alpha;2}) = 1 + y \cdot s_{\alpha;2}$, which exists by Lemma 18.12. After setting $(D, \Delta) := -i \cdot x \cdot (D', \Delta')$, the above equation becomes

$$i \cdot x \cdot y \cdot s_{\alpha;2} = w_\alpha \cdot s_{\alpha;2}$$

since $w_\alpha \cdot s_{\alpha;1} = w_\alpha \cdot s_{\alpha;2}$. Let u' be arbitrary such that $(D, \Delta, u') \in I \cdot \mathcal{G}^{-1}_{X'/S'}(\mathcal{L}')$, and let w'_α be such that $u'(e_\alpha) = w'_\alpha \cdot e_\alpha$. Then setting $u := u' + (i \cdot x \cdot y - w'_\alpha)$ yields the desired element $(D, \Delta, u) \in I \cdot \mathcal{G}^{-1}_{X'/S'}(\mathcal{L}')$. In the general case, we apply an inductive argument over small extensions. When $A' = A_n \to \ldots \to A_0 = A$ is a decomposition into small extensions, then we use the flatness and compatibility with base change of $\mathcal{G}^{-1}_{X'/S'}$ to lift the constructed isomorphism from A_k to A_{k+1}, thus producing the situation that $s'_1|_{A_k} = s'_2|_{A_k}$. $\qquad\square$

Remark 18.25 Note that this proof does not work for $\mathcal{V}^{-1}_{X'/S'}$ since the base change property may be violated, i.e., we may have constructed an isomorphism in a step $A_{k+1} \to A_k$ which does not lift to A'. $\qquad\Diamond$

After we have established that there is, up to an isomorphism which can be explicitly constructed, locally only one $X'(s')$, we study the automorphisms of $f': X'(s') \to S'$.

Lemma 18.26 *Let $\theta = (D, \Delta, u) \in I \cdot \mathcal{G}^{-1}_{X'/S'}(\mathcal{L}')$. Then we have $\psi(s') = s'$ if and only if $u(s') = 0$.*

Proof We have $\psi(s') = \sum_{n=0}^\infty \frac{u^n(s')}{n!}$, so $\psi(s') = s'$ is equivalent to $\sum_{n=1}^\infty \frac{u^n(s')}{n!} = 0$. If $u(s') = 0$, then obviously $\psi(s') = s'$. Conversely, assume that $u(s') \neq 0$ but $\psi(s') = s'$. Then there is some $m > 0$ with $u(s') \in I^m \cdot \mathcal{L}'$ but $u(s') \notin I^{m+1} \cdot \mathcal{L}'$. However, $u^2(s') \in I^{m+1} \cdot \mathcal{L}'$ because $(D, \Delta, u) \in I \cdot \mathcal{G}^{-1}_{X'/S'}(\mathcal{L}')$. Now $\sum_{n=1}^\infty \frac{u^n(s')}{n!} = 0$ implies $u(s') \in I^{m+1} \cdot \mathcal{L}'$, a contradiction. $\qquad\square$

In view of Lemma 18.26, we write

$$\mathcal{G}^{-1}_{X'/S'}(\mathcal{L}', s') := \{(D, \Delta, u) \in \mathcal{G}^{-1}_{X'/S'}(\mathcal{L}') \mid u(s') = 0\},$$

and analogously for $\mathcal{V}^{-1}_{X'/S'}(\mathcal{L}', s')$. Thus, the automorphisms of $f': X'(s') \to S'$ which come from X' are in $I \cdot \mathcal{G}^{-1}_{X'/S'}(\mathcal{L}') \cap \mathcal{G}^{-1}_{X'/S'}(\mathcal{L}', s')$.[11] A priori, they are *outer* automorphisms of $f': X'(s') \to S'$, but we will show now that they are in fact inner automorphisms, i.e., we construct a map

$$\mu: \ I \cdot \mathcal{G}^{-1}_{X'/S'}(\mathcal{L}') \cap \mathcal{G}^{-1}_{X'/S'}(\mathcal{L}', s') \to I \cdot \mathcal{G}^{-1}_{X'(s')/S'}$$

---

[11] At this point, we do not yet know that $I \cdot \mathcal{G}^{-1}_{X'/S'}(\mathcal{L}') \cap \mathcal{G}^{-1}_{X'/S'}(\mathcal{L}', s') = I \cdot \mathcal{G}^{-1}_{X'/S'}(\mathcal{L}', s')$; however, we will see this below. The problem is that we do not know here that the injectivity of $\mathcal{G}^{-1}_{X'/S'}(\mathcal{L}', s') \to \mathcal{G}^{-1}_{X'/S'}(\mathcal{L}')$ is preserved under base change.

such that $\psi_s : X'(s') \cong X'(s')$ is the gauge transform induced by $\mu(D, \Delta, u)$.

First, we have to compute the gauge transform on $(\mathcal{V}^\bullet_{L'(X')/S'}, \mathcal{W}^\bullet_{L'(X')/S'})$ induced by $\psi : L'(X') \cong L'(X')$. In order to do so, we start by comparing the Atiyah extension with the exact sequence (18.2). Unfortunately, we have not found a direct functorial comparison map, and we have to construct it explicitly on charts.

Lemma 18.27 *We have an isomorphism*

$$
\begin{array}{ccccccccc}
0 & \longrightarrow & p'^*\mathcal{O}_{X'} & \longrightarrow & p'^*\mathcal{V}^{-1}_{X'/S'}(\mathcal{L}') & \longrightarrow & p'^*\mathcal{V}^{-1}_{X'/S'} & \longrightarrow & 0 \\
& & \downarrow{\scriptstyle 1 \mapsto -\gamma_s^\vee} & & \downarrow{\scriptstyle c} & & \| & & \\
0 & \longrightarrow & \Theta^1_{L'(X')/X'} & \longrightarrow & \mathcal{V}^{-1}_{L'(X')/S'} & \longrightarrow & p'^*\mathcal{V}^{-1}_{X'/S'} & \longrightarrow & 0
\end{array}
$$

of short exact sequences.[12] This isomorphism is compatible with $\mathcal{G}^{-1}$, i.e., when replacing $\mathcal{V}^{-1}$ with $\mathcal{G}^{-1}$, we still have an isomorphism of short exact sequences.

Proof Given $(D, \Delta) \in \mathcal{V}^{-1}_{X'/S'}$ on V_α, we can find a unique u_α such that $u_\alpha(e_\alpha) = 0$ and $(D, \Delta, u_\alpha) \in \mathcal{V}^{-1}_{X'/S'}(\mathcal{L}')$. Namely, start with an arbitrary u, and then set $u_\alpha(e) := u(e) - w_\alpha \cdot e$, where $u(e_\alpha) = w_\alpha \cdot e_\alpha$. This defines a local splitting

$$
K: \quad \mathcal{V}^{-1}_{X'/S'} \to \mathcal{V}^{-1}_{X'/S'}(\mathcal{L}')
$$

of the Atiyah extension. Obviously, it splits the Atiyah extension for $\mathcal{G}^{-1}$ instead of $\mathcal{V}^{-1}$ as well. If $K' : \mathcal{V}^{-1}_{X'/S'} \to \mathcal{V}^{-1}_{X'/S'}(\mathcal{L}')$ is the analogous splitting on V_β, then we have

$$
K'(D, \Delta) - K(D, \Delta) = \Delta(\gamma_{\alpha\beta})
$$

on overlaps $V_\alpha \cap V_\beta$, where $\gamma_{\alpha\beta}$ is the transition function with $e_\alpha = \gamma_{\alpha\beta} e_\beta$, and we consider $\Delta(\gamma_{\alpha\beta})$ as the multiplication operator on $\mathcal{L}'$, i.e., embedded via $\mathcal{O}_{X'} \to \mathcal{V}^{-1}_{X'/S'}(\mathcal{L}')$.

On the other hand, we also have our original distinguished splitting B_1 of (18.1). Let B_1 be the splitting on V_α, and let B_1' be the splitting on V_β. Let

$$
S_{-1}, \; S'_{-1} : \quad p'^*\mathcal{V}^{-1}_{X'/S'} \to \mathcal{V}^{-1}_{L'(X')/S'}
$$

be the two induced splittings of (18.2). Then a slightly tedious direct computation shows that

$$
S'_{-1}(p'^*\theta) - S_{-1}(p'^*\theta) = -\Delta(\gamma_{\alpha\beta}) \cdot \delta^\vee
$$

---

[12] The map $1 \mapsto -\gamma_s^\vee$ rather than $1 \mapsto \gamma_s^\vee$ is, though somewhat surprising, not a mistake.

for $\theta = (D, \Delta) \in \mathcal{V}^{-1}_{X'/S'}$. Now we define the comparison map

$$c: \quad p'^* \mathcal{V}^{-1}_{X'/S'}(\mathcal{L}') \to \mathcal{V}^{-1}_{L'(X')/S'}, \quad a + b \cdot p'^* K(D, \Delta)$$

$$\mapsto -a \cdot \delta^{\vee} + b \cdot S_{-1}(p'^*(D, \Delta)).$$

The above formulae for K and K' respectively S_{-1} and S'_{-1} show that c is well-defined. The same computation also shows that c is independent of the choice of e_α as a local trivialization of $\mathcal{L}'$, and hence canonical. It is obvious that the diagram in the statement above commutes with this definition of c. Since the maps on the left and on the right are isomorphisms, the map in the middle is an isomorphism as well. By construction, c maps $p'^* \mathcal{G}^{-1}_{X'/S'}(\mathcal{L}')$ into $\mathcal{G}^{-1}_{L'(X')/S'}$. Since we still have a morphism of short exact sequences with isomorphisms on the sides, c restricted to $p'^* \mathcal{G}^{-1}_{X'/S'}(\mathcal{L}')$ is an isomorphism as well, mapping onto $\mathcal{G}^{-1}_{L'(X')/S'}$. $\qquad \square$

With this comparison map c, we get the expected gauge transform.

Lemma 18.28 *Let* $\theta = (D, \Delta, u) \in I \cdot \mathcal{V}^{-1}_{X'/S'}(\mathcal{L}')$. *Then the gauge transform on* $\mathcal{V}\mathcal{W}^{\bullet}_{L'(X')/S'}$ *induced by* $\psi: L'(X') \to L'(X')$ *is given by* $\exp_{-c(p'^*\theta)}$. *In particular, if* $\theta \in I \cdot \mathcal{G}^{-1}_{X'/S'}(\mathcal{L}')$, *then the gauge transform preserves the Gerstenhaber subcalculus* $\mathcal{GC}^{\bullet}_{L'(X')/S'}$.

Proof We write $\psi: L'(X') \to L'(X')$ as above for the morphism induced by (D, Δ, u). First, we show that $\psi^*(b) = \exp_{-c(p'^*\theta)}(b)$ for functions $b \in \mathcal{O}_{L'}$. For a function $a \in \mathcal{O}_{X'}$, we have $[-c(p'^*\theta), p'^*a] = p'^* D(a)$. Thus, we have

$$\psi^*(p'^*a) = p'^* \varphi^*(a) = \sum_{n=0}^{\infty} \frac{p'^* D^n(a)}{n!} = \exp_{-c(p'^*\theta)}(p'^*a).$$

Since $[S_{-1}(p'^*(D, \Delta)), x_\alpha] = -S_{-1}(p'^*(D, \Delta)) \lrcorner (x_\alpha \cdot \delta_\alpha) = 0$, we find

$$[-c(p'^*\theta), x_\alpha] = -p'^*(w_\alpha) \cdot x_\alpha$$

where w_α is such that $u(e_\alpha) = w_\alpha \cdot e_\alpha$. This shows

$$\exp_{-c(p'^*\theta)}(x_\alpha) = \left(1 + \sum_{n=1}^{\infty}(-1) \cdot \frac{[D - p'^* w_\alpha]^{n-1}(p'^* w_\alpha)}{n!}\right) \cdot x_\alpha = v_\alpha^{-1} \cdot x_\alpha,$$

where the second equality is as in the proof of Lemma 10.6, i.e., $\psi^*(x_\alpha) = \exp_{-c(p'^*\theta)}(x_\alpha)$. Since $\mathcal{O}_{L'}$ is, as a ring, locally generated by $\mathcal{O}_{X'}$ and x_α, we obtain our claim on $\mathcal{O}_{L'}$.

From here, we proceed as in the proof of Lemma 10.6. First, we obtain

$$d\psi^*(\partial a) = \exp_{-c(p'^*\theta)}(\partial a)$$

for functions $a \in O_{L'}$, and then we have $d^1 \psi^* = \exp_{-c(p'^*\theta)}$ first on the strict locus of $f' \colon L'(X') \to S'$ and then everywhere. Both sides are compatible with the $\wedge$-product, so we get the equality in all degrees. Afterward, we get $T^1 \psi^* = \exp_{-c(p'^*\theta)}$ from the formula $T^1 \psi^*(\xi) \smile \alpha = \exp_{-c(p'^*\theta)}(\xi) \smile \alpha$, which is obtained from the definition of $T^1 \psi^*$ by applying $\exp_{-c(p'^*\theta)}$ in the cases where we already know our claimed formula. Finally, we obtain all degrees $T^p \psi^*$ similarly to $d^i \psi^*$. $\qquad\square$

Since

$$H_{-1}(s'^* c(p'^*\theta)) = (s'^* c(p'^*\theta) \smile G_1(e_\alpha^\vee)) \cdot e_\alpha$$

under the map $H_{-1} \colon s'^* \mathcal{V}^{-1}_{L'(X')/S'} \to \mathcal{L}'$, a direct computation shows that

$$H_{-1}(s'^* c(p'^*\theta)) = -(w_\alpha \cdot s_\alpha + D(s_\alpha)) \cdot e_\alpha = -u(s),$$

where $\theta = (D, \Delta, u)$, and w_α is such that $u(e_\alpha) = w_\alpha \cdot e_\alpha$. Thus, we have $H_{-1}(s'^* c(p'^*\theta)) = 0$ if and only if $u(s) = 0$, i.e., if and only if $\theta \in \mathcal{V}^{-1}_{X'/S'}(\mathcal{L}', s')$. If in fact $\theta \in \mathcal{G}^{-1}_{X'/S'}(\mathcal{L}')$, then we obtain $s'^* c(p'^*\theta) \in \mathcal{G}^{-1}_{X'(s')/S'}$. Finally, by considering the maps to the variants for $f \colon X \to S$, we find $s'^* c(p'^*\theta) \in I \cdot \mathcal{G}^{-1}_{X'(s')/S'}$. This defines our map

$$\mu \colon \ I \cdot \mathcal{G}^{-1}_{X'/S'}(\mathcal{L}') \cap \mathcal{G}^{-1}_{X'/S'}(\mathcal{L}', s') \to I \cdot \mathcal{G}^{-1}_{X'(s')/S'}, \quad \theta \mapsto s'^* c(p'^*\theta).$$

Lemma 18.29 *Let $\theta = (D, \Delta, u) \in I \cdot \mathcal{V}^{-1}_{X'/S'}(\mathcal{L}') \cap \mathcal{V}^{-1}_{X'/S'}(\mathcal{L}', s')$. Then the gauge transform on $\mathcal{V}\mathcal{W}^\bullet_{X'(s')/S'}$ induced by $\varphi_s \colon X'(s') \cong X'(s')$ is given by $\exp_{-\mu(\theta)}$. If*

$$\theta \in I \cdot \mathcal{G}^{-1}_{X'/S'}(\mathcal{L}') \cap \mathcal{G}^{-1}_{X'/S'}(\mathcal{L}', s'),$$

then the gauge transform preserves the Gerstenhaber subcalculus $\mathcal{GC}^\bullet_{X'(s')/S'}$.

Proof Instead of tracking the actions through the construction, we apply the following trick: Recall that $h \colon X'(s') \to X'$ is an isomorphism over $X' \setminus H'$. In particular, we have a commutative diagram $h \circ \varphi_s = \varphi \circ h$ of isomorphisms on $X' \setminus H'$. Under the identification given by h, we have $(d\varphi_s^*, T\varphi_s^*) = (d\varphi^*, T\varphi^*)$; the latter is the gauge transform $\exp_{-\theta}$. Expanding our definition of μ, we find $Th_* \mu(D, \Delta, u) = (D, \Delta)$, so, on $X' \setminus H'$, the automorphism $(d\varphi_s^*, T\varphi_s^*)$ is given by the gauge transform $\exp_{-Th_*(D,\Delta)} = \exp_{-\mu(\theta)}$. Now $(d\varphi_s^*, T\varphi_s^*)$ and $\exp_{-\mu(\theta)}$ are two automorphisms of $\mathcal{V}\mathcal{W}^\bullet_{X'(s')/S'}$ which coincide on the scheme-theoretically dense open subset $X' \setminus H'$, so they must be equal due to local freeness on U' and reflexivity of $\mathcal{V}^p$ and $\mathcal{W}^i$. $\qquad\square$

As in the proof of Lemma 18.29, we consider now the map

$$Th_*\colon\ I\cdot\mathcal{G}^{-1}_{X'(s')/S'} \to I\cdot\mathcal{G}^{-1}_{X'/S'}.$$

By construction, we have $Th_* \circ \mu(D, \Delta, u) = (D, \Delta)$. To exploit this map, we establish the following basic facts about $\mathcal{V}^{-1}_{X'/S'}(\mathcal{L}', s') \to \mathcal{V}^{-1}_{X'/S'}$. Parts (3) and (4) are not needed for our argument, but we found them worth mentioning here.

Lemma 18.30 *The following hold:*

(1) *the map $\mathcal{V}^{-1}_{X'/S'}(\mathcal{L}', s') \to \mathcal{V}^{-1}_{X'/S'}$ is injective;*

(2) *$(D, \Delta) \in \mathcal{V}^{-1}_{X'/S'}$ is in the image of the map if and only if $D(s_\alpha) = a \cdot s_\alpha$ for some local function $a \in \mathcal{O}_{X'}$, i.e., if and only if D preserves the ideal $\mathcal{L}'^\vee \subseteq \mathcal{O}_{X'}$ of $H' \subset X'$.*

(3) *if $(D, \Delta, u) \in \mathcal{V}^{-1}_{X'/S'}(\mathcal{L}', s')$, then $s' \circ D = u \circ s'$ where $s'\colon \mathcal{O}_{X'} \to \mathcal{L}'$ is the map induced by the section s';*

(4) *if $(D, \Delta, u) \in \mathcal{V}^{-1}_{X'/S'}(\mathcal{L}', s')$, then there is a unique map $u^\vee\colon \mathcal{L}'^\vee \to \mathcal{L}'^\vee$ with $s'^\vee \circ u^\vee = D \circ s'^\vee$, where $s'^\vee\colon \mathcal{L}'^\vee \to \mathcal{O}_{X'}$ is the canonical map associated with the section s'; this map $u^\vee$ is $f'^{-1}(\mathcal{O}_{S'})$-linear and satisfies $u^\vee(a \cdot e^\vee) = D(a) \cdot e^\vee + a \cdot u^\vee(e^\vee)$.*

Proof First, we show (1). Let (D, Δ, u) and (D, Δ, u') be in $\mathcal{V}^{-1}_{X'/S'}(\mathcal{L}', s')$. Then there is some $a \in \mathcal{O}_{X'}$ with $v(e) = u(e) + a \cdot e$. Since $u(s) = v(s) = 0$, we obtain $a \cdot s = 0$; since each s_α is a non-zero divisor, we get $a = 0$. For (2), first observe that, if $(D, \Delta, u) \in \mathcal{V}^{-1}_{X'/S'}(\mathcal{L}', s')$, then

$$0 = u(s_\alpha \cdot e_\alpha) = D(s_\alpha) \cdot e_\alpha + s_\alpha \cdot u(e_\alpha),$$

so $D(s_\alpha) = a \cdot s_\alpha$ for some $a \in \mathcal{O}_{X'}$. Conversely, if $(D, \Delta) \in \mathcal{V}^{-1}_{X'/S'}$ with $D(s_\alpha) = a \cdot s_\alpha$, we can find first some locally defined $u\colon \mathcal{L}' \to \mathcal{L}'$ with $(D, \Delta, u) \in \mathcal{V}^{-1}_{X'/S'}(\mathcal{L}')$. Let $x \in \mathcal{O}_{X'}$ be such that $u(e_\alpha) = x \cdot e_\alpha$. Then $\hat{u}(e) := u(e) - (a + x) \cdot e$ defines a preimage of (D, Δ) since $\hat{u}(s) = 0$. For (3), note that $u(b \cdot s) = D(b) \cdot s + b \cdot u(s) = D(b) \cdot s$. For (4), a map $u^\vee$ with the desired property must satisfy

$$s_0^\vee(u^\vee(b \cdot e_\alpha^\vee)) = b \cdot D(s_\alpha) + s_\alpha \cdot D(b) = s_\alpha \cdot (b \cdot a + D(b)),$$

where again $D(s_\alpha) = a \cdot s_\alpha$. Thus, we must have $u^\vee(b \cdot e_\alpha^\vee) = (b \cdot a + D(b)) \cdot e_\alpha^\vee$. A direct computation shows that this is independent of the choice of e_α and hence well-defined on overlaps; another direct computation shows the claimed relations. $\square$

From part (1) of Lemma 18.30 it follows that μ is injective. Now let $\theta' = (D', \Delta') \in I \cdot \mathcal{G}^{-1}_{X'(s')/S'}$, and let $(D, \Delta) = Th_*(D', \Delta')$. Then $D = D'$ as a derivation on $\mathcal{O}_{X'}$, and we have $D(s_\alpha) = D'(s_\alpha) = s_\alpha \cdot \Delta'(m_\alpha)$, i.e., we have some $(D, \Delta, u) \in \mathcal{V}^{-1}_{X'/S'}(\mathcal{L}', s')$ lifting (D, Δ) by part (2) of Lemma 18.30. When we look into the actual construction, then we find $(D, \Delta, u) \in I \cdot \mathcal{G}^{-1}_{X'/S'}(\mathcal{L}') \cap \mathcal{G}^{-1}_{X'/S'}(\mathcal{L}', s')$, i.e., (D, Δ, u) is in the domain of μ, and we have $\mu(D, \Delta, u) = \theta'$ since Th_* is injective. Thus, μ is an isomorphism.

In the same way as μ, we can also define

$$\tilde{\mu} : \ \mathcal{G}^{-1}_{X'/S'}(\mathcal{L}', s') \to \mathcal{G}^{-1}_{X'(s')/S'}, \quad \theta \mapsto s'^* c(p'^* \theta).$$

Just as μ, also $\tilde{\mu}$ is an isomorphism. Because the formation of $\mathcal{G}^{-1}_{X'(s')/S'}$ commutes with base change, so does the formation of $\mathcal{G}^{-1}_{X'/S'}(\mathcal{L}', s')$. This implies that

$$I \cdot \mathcal{G}^{-1}_{X'/S'}(\mathcal{L}') \cap \mathcal{G}^{-1}_{X'/S'}(\mathcal{L}', s') = I \cdot \mathcal{G}^{-1}_{X'/S'}(\mathcal{L}', s').$$

Corollary 18.31 *Every element* $\theta = (D, \Delta, u) \in I \cdot \mathcal{G}^{-1}_{X'/S'}(\mathcal{L}', s')$ *induces a (geometric) automorphism of* $f' \circ h' : X'(s') \to S'$ *whose action on the Gerstenhaber calculus is the gauge transform* $\exp_{-\mu(\theta)}$, *and every inner automorphism of* $f' \circ h' : X'(s') \to S'$ *is of this form for a unique* θ.

18.6   Enhanced Systems of Deformations

Suppose we are in the following situation:

Situation 18.32 Let $f_0 : X_0 \to S_0$ be a torsionless enhanced generically log smooth family of relative dimension d over $S_0 = \mathrm{Spec}(Q \to \mathbf{k})$. Let $\mathcal{D}$ be an enhanced system of deformations of $f_0 : X_0 \to S_0$, subordinate to an affine open cover $\mathcal{V} = \{V_\alpha\}$. Let $\mathcal{L}_0$ be a line bundle on X_0, and let $s_0 \in \mathcal{L}_0$ be a log regular global section. Assume that $\mathcal{L}_0|_\alpha$ is trivial.

We construct an enhanced system of deformations $\mathcal{D}(s_0)$ for $f_0 \circ h_0 : X_0(s_0) \to S_0$. Let us write $O_{\alpha;A}$ for the structure sheaf on $V_{\alpha;A}$. We choose

(1) a trivialization $e_{\alpha;0}$ of $\mathcal{L}_0|_\alpha$.

Then we choose, for every α,

(2) sections $s_{\alpha;A} \in O_{\alpha;A}$

which are compatible with restrictions, and such that $s_{\alpha;0} \cdot e_{\alpha;0} = s|_\alpha$. We set

$$\tilde{V}_{\alpha;A} := V_{\alpha;A}(s_{\alpha;A}),$$

the modification of the local model $V_{\alpha;A}$ by the section $s_{\alpha;A} \in O_{\alpha;A}$. This gives a collection of generically log smooth deformations of $X_0(s_0)|_\alpha$ which is compatible with base change. Next, for every $\alpha \neq \beta$, we choose a collection of

(3)   isomorphisms $\gamma^*_{\alpha\beta;A} \colon O_{\beta;A}|_{\alpha\beta} \cong O_{\alpha;A}|_{\alpha\beta}$

which is compatible with restrictions, and which induces the isomorphism[13]

$$\gamma^*_{\alpha\beta;0} \colon \quad (O_{X_0}|_\beta)|_{\alpha\beta} \cong (\mathcal{L}_0|_\beta)|_{\alpha\beta} = (\mathcal{L}_0|_\alpha)|_{\alpha\beta} \cong (O_{X_0}|_\alpha)_{\alpha\beta}$$

—which comes from our choice of $e_{\alpha;0}$—on the central fiber. These isomorphisms must be O-linear for $\psi_{\alpha\beta;A} \colon V_{\alpha;A}|_{\alpha\beta} \cong V_{\beta;A}|_{\alpha\beta}$. Additionally, we assume that $(\gamma^*_{\alpha\beta;A})^{-1} = \gamma^*_{\beta\alpha;A}$. By applying Lemma 18.24 inductively along A_k, we can find

(4) $\theta_{\alpha\beta;A} = (D_{\alpha\beta;A}, \Delta_{\alpha\beta;A}, u_{\alpha\beta;A}) \in \mathfrak{m}_A \cdot \mathcal{G}^{-1}_{\alpha;A}(O_{\alpha;A})|_{\alpha\beta}$

with $\exp_{-\theta_{\alpha\beta;A}}(\gamma^*_{\alpha\beta;A}(s_{\beta;A})) = s_{\alpha;A}$ and $\theta_{\alpha\beta;A} = -\theta_{\beta\alpha;A}$ under the identification given by $(\psi^*_{\alpha\beta;A}, \gamma^*_{\alpha\beta;A})$, and which are compatible with restrictions. Then we replace $\psi_{\alpha\beta;A}$ with $\hat{\psi}_{\alpha\beta;A} := \psi_{\alpha\beta;A} \circ \exp(-\theta_{\alpha\beta;A})$, where $\exp(-\theta)$ is the automorphism of enhanced generically log smooth families which acts via $\exp_{-\theta}$ on the Gerstenhaber calculus. The new cocycles are gauge transforms in $\mathfrak{m}_A \cdot \mathcal{G}^{-1}_{\alpha;A}|_{\alpha\beta\gamma}$ as well, so we still have an enhanced system of deformations. The new $\hat{\psi}^*_{\alpha\beta;A}$ together with $\exp_{-\theta_{\alpha\beta;A}} \circ \gamma^*_{\alpha\beta;A}$ yields the comparison isomorphism

$$\tilde{\psi}_{\alpha\beta;A} \colon \quad \tilde{V}_{\alpha;A}|_{\alpha\beta} \cong \tilde{V}_{\beta;A}|_{\alpha\beta}$$

for the enhanced system of deformations $\mathcal{D}(s_0)$. These isomorphisms are compatible with restrictions. By construction, they also preserve the Gerstenhaber calculi. Also by construction, the cocycles are in $\mathfrak{m}_A \cdot \mathcal{G}^{-1}_{V_{\alpha;A}(s_{\alpha;A})/S_A}|_{\alpha\beta\gamma}$. Thus, we have indeed constructed an enhanced system of deformations $\mathcal{D}(s_0)$ for $f_0 \circ h_0 \colon X_0(s_0) \to S_0$. It gives rise to a deformation functor

$$\mathrm{ELD}^{\mathcal{D}(s_0)}_{X_0(s_0)/S_0} \colon \quad \mathbf{Art}_Q \to \mathbf{Set}.$$

Lemma 18.33 *This deformation functor is, up to canonical isomorphism, independent of the choices that we made in the construction of $\mathcal{D}(s_0)$.*

Proof Let $\mathcal{D}(s_0)$ and $\mathcal{D}'(s_0)$ be two choices of enhanced systems of deformations which come out of the above construction for different choices of the parameters. Let us denote the local models of the two systems by $\tilde{V}_{\alpha;A}$ respectively $\tilde{V}'_{\alpha;A}$. Then there is a (non-unique) isomorphism $\bar{\chi}_\alpha \colon \tilde{V}_{\alpha;A} \to \tilde{V}'_{\alpha;A}$ which comes from

---

[13] Now we have $e_\beta = \gamma_{\alpha\beta} e_\alpha$ rather than our previous formula $e_\alpha = \gamma_{\alpha\beta} e_\beta$, which does not fit in well with our conventions about enhanced systems of deformations.

Lemma 18.24. An enhanced generically log smooth deformation $\tilde{X}_A \to S_A$ of $f_0 \colon X_0(s_0) \to S_0$ of type $\mathscr{D}(s_0)$ comes with isomorphisms $\chi_\alpha \colon \tilde{X}_A|_\alpha \cong \tilde{V}_{\alpha;A}$. We define the image of our comparison isomorphism between the deformation functors by taking the enhanced generically log smooth family $\tilde{X}_A \to S_A$ together with the comparison isomorphisms $\bar{\chi}_\alpha \circ \chi_\alpha \colon \tilde{X}_A|_\alpha \cong \tilde{V}'_{\alpha;A}$. This is again an enhanced generically log smooth deformation because the two comparison isomorphisms $\tilde{\psi}_{\alpha\beta;A}$ and $\tilde{\psi}'_{\alpha\beta;A}$ differ by gauge transforms. Any two choices of $\bar{\chi}_\alpha$ differ by a gauge transform, so the map is independent of the choice of $\bar{\chi}_\alpha$ on the level of the deformation functor since the images are equivalent as enhanced generically log smooth deformations. The map is an isomorphism since we can define the analogous map in the opposite direction, and the composition is the identity. $\qquad\square$

We can define a similar deformation functor

$$\mathrm{ELD}^{\mathscr{D}}_{X_0/S_0}(\mathcal{L}_0, s_0) \colon \quad \mathbf{Art}_Q \to \mathbf{Set}$$

which classifies enhanced generically log smooth deformations of $f_0 \colon X_0 \to S_0$ of type $\mathscr{D}$ together with a line bundle $\mathcal{L}$ deforming $\mathcal{L}_0$, and a section $s \in \mathcal{L}$ with $s|_0 = s_0$. We have the obvious map

$$\mathrm{ELD}^{\mathscr{D}}_{X_0/S_0}(\mathcal{L}_0, s_0) \Rightarrow \mathrm{ELD}^{\mathscr{D}(s_0)}_{X_0(s_0)/S_0}$$

of deformation functors. By Lemma 18.26 and Corollary 18.31, the objects classified by the two deformation functors have the same automorphisms. Then the map is an isomorphism of deformation functors with the argument employed, for example, in Proposition 12.19. On the other hand, we have also a forgetful morphism

$$\mathrm{ELD}^{\mathscr{D}}_{X_0/S_0}(\mathcal{L}_0, s_0) \overset{\Phi}{\Rightarrow} \mathrm{ELD}^{\mathscr{D}}_{X_0/S_0}(\mathcal{L}_0) \overset{\Psi}{\Rightarrow} \mathrm{ELD}^{\mathscr{D}}_{X_0/S_0}$$

of deformation functors. Here, the middle term denotes enhanced generically log smooth deformations with a line bundle.

Lemma 18.34 *The following statements hold:*

(1) *If $H^1(X_0, \mathcal{L}_0) = 0$, then Φ is smooth and, in particular, surjective.*
(2) *If $H^2(X_0, O_{X_0}) = 0$, then Ψ is smooth and, in particular, surjective.*
(3) *If $H^1(X_0, O_{X_0}) = 0$, then Ψ is injective.*
(4) *If $\mathcal{L}_0 = (\mathcal{W}^d_{X_0/S_0})^\vee$ (respectively $\mathcal{W}^d_{X_0/S_0}$), then Ψ admits a splitting*

$$\Sigma \colon \quad \mathrm{ELD}^{\mathscr{D}}_{X_0/S_0} \Rightarrow \mathrm{ELD}^{\mathscr{D}}_{X_0/S_0}(\mathcal{L}_0)$$

given by $\mathcal{L}_A = (\mathcal{W}^d_{X_A/S_A})^\vee$ (respectively $\mathcal{W}^d_{X_A/S_A}$). In particular, Ψ is surjective.

For us, the most interesting case is

$$\mathcal{L}_0 = (\mathcal{W}^d_{X_0/S_0})^\vee$$

since, then, $f_0 \circ h_0 \colon X_0(s_0) \to S_0$ is log Calabi–Yau by Proposition 18.17.

Corollary 18.35 *Assume that we are in Situation 18.32 with $\mathcal{L}_0 = (\mathcal{W}^d_{X_0/S_0})^\vee$.*

(1) *If $H^1(X_0, \mathcal{L}_0) = 0$, then unobstructedness of $\mathrm{ELD}^{\mathscr{D}(s_0)}_{X_0(s_0)/S_0}$ implies unobstructedness of $\mathrm{ELD}^{\mathscr{D}}_{X_0/S_0}$.*

(2) *If additionally either $H^1(X_0, O_{X_0}) = 0$ or $H^2(X_0, O_{X_0}) = 0$, then $\mathrm{ELD}^{\mathscr{D}}_{X_0/S_0}$ is unobstructed if and only if $\mathrm{ELD}^{\mathscr{D}(s_0)}_{X_0(s_0)/S_0}$ is unobstructed.*

Example 18.36 Let $f_0 \colon X_0 \to S_0$ be a vertical enhanced generically log smooth family which is log Fano, i.e., $\omega^\vee_{X_0/S_0} = (\mathcal{W}^d_{X_0/S_0})^\vee$ is an ample line bundle. If the Gorenstein scheme X_0 satisfies Kodaira vanishing for the canonical bundle, then the conditions of the corollary are satisfied. This is, for example, the case if X_0 has semi–log canonical singularities, cf. [181, Cor. 6.6] and the dual statement in the sense of [138, III, Thm. 7.6].[14] In particular, this is the case for a toroidal crossing space $(V, \mathcal{P}, \bar{\rho})$ by Lemma 9.8.   ◊

18.7 Log Regular Sections in Practice

To obtain unconditional unobstructedness results from Corollary 18.35, we need to construct log regular sections. We do not yet know how to do this in general. In this section, we present some considerations in this direction. We start with an elementary result that helps us to study log regularity of sections $s \in \mathcal{L}$ on local models.

Lemma 18.37 *Let $f \colon X \to S$ be an enhanced generically log smooth family, and let $g \colon X' \to X$ be an étale cover. Then $s \in \mathcal{L}$ is log regular if and only if $g^*s \in g^*\mathcal{L}$ is log regular.*

Proof Injectivity of $\mathcal{L}^\vee \to O_X$, flatness of O_H over the base ([267, 0584]), and having codimension and depth ≥ 2 for $H \cap Z \subseteq H$ are all properties which are equivalent on X and X' under an étale cover $g \colon X' \to X$, so $s \in \mathcal{L}$ is log pre-regular if and only if $g^*s \in g^*\mathcal{L}$ is log pre-regular. The formation of the sequence in Lemma 18.8 commutes with the faithfully flat map $g \colon X' \to X$, so it is locally split exact on X if and only if it is locally split exact on X'. Thus, $s \in \mathcal{L}$ is log

---

[14] While Kodaira vanishing in the sense of $H^q(X, L^{-1}) = 0$ for $0 \leq q \leq d - 1$ and an ample line bundle L needs that X is Cohen–Macaulay, Fujino has given a generalization of the dual statement $H^q(X, \omega_X \otimes L) = 0$ for $q \geq 1$ without the Cohen–Macaulay assumption on X, see [98, Thm. 1.8] and [99, Thm. 1.3].

quasi-regular if and only if $g^*s \in g^*\mathcal{L}$ is log quasi-regular. Since also the formation of the map $s^*\mathcal{G}^{-1}_{L(X)/S} \to \mathcal{L}$ commutes with $g\colon X' \to X$, the claim follows. $\square$

Next, we show that the construction of [77, Lemma 6.10] yields log regular sections in our sense, so our theory is applicable in that setting. Let $g\colon H \to S$ be a torsionless enhanced generically log smooth family. We obtain an enhanced generically log smooth family $f\colon X \to S$ by setting $X = H \times \mathbb{A}^1$; when $q\colon H \times \mathbb{A}^1 \to H$ is the projection, then the two-sided Gerstenhaber calculus is given by

$$\mathcal{A}^i_{X/S} = q^*\mathcal{A}^i_{H/S} \oplus q^*\mathcal{A}^{i-1}_{H/S} \wedge dx \quad \text{and} \quad \mathcal{G}^p_{X/S} = q^*\mathcal{G}^p_{H/S} \oplus q^*\mathcal{G}^{p+1}_{H/S} \wedge \partial_x,$$

where x is the variable in $\mathbb{A}^1$. Now we set $\mathcal{L} = O_X$ and $s = x$.

Lemma 18.38 *Consider the enhanced generically log smooth family $f\colon H \times \mathbb{A}^1_x \to S$. Then $x \in O_X =: \mathcal{L}$ is a log regular section.*

Proof This section is obviously log pre-regular, and we have $X(s) = H \times A_{\mathbb{N}}$. The first exact sequence for $s^*\mathcal{W}^1_{L(X)/S}$ becomes

$$0 \to q^*\mathcal{W}^1_{H/S} \oplus O_X \cdot dx \to q^*\mathcal{W}^1_{H/S} \oplus O_X \cdot dx \oplus O_X \cdot s^*\delta_\alpha \to O_X \cdot s^*\gamma_s \to 0.$$

Then the second sequence is

$$0 \to \mathcal{L}^\vee \xrightarrow{e_\alpha^\vee \mapsto x \cdot s^*\delta_\alpha - dx} q^*\mathcal{W}^1_{H/S} \oplus O_X \cdot dx \oplus O_X \cdot s^*\delta_\alpha \xrightarrow{R_1} q^*\mathcal{W}^1_{H/S} \oplus O_X \cdot \frac{dx}{x} \to 0.$$

The map on the left is obviously injective, so we already know that the sequence is exact on the left and in the middle from its construction. We have $R_1(dx) = x \cdot \frac{dx}{x}$ and $R_1(s^*\delta_\alpha) = \frac{dx}{x}$, and on $q^*\mathcal{W}^1_{H/S}$, the map is the identity; in particular, R_1 is surjective, the sequence is locally split, and $s \in \mathcal{L}$ is log quasi-regular. The dual exact sequence is

$$0 \to q^*\mathcal{V}^{-1}_{H/S} \oplus O_X \cdot x\partial_x \to q^*\mathcal{V}^{-1}_{H/S} \oplus O_X \cdot \partial_x \oplus O_X \cdot s^*\delta^\vee \xrightarrow{H_{-1}} \mathcal{L} \to 0.$$

Here, $H_{-1}(\partial_x) = e_\alpha$ and $H_{-1}(s^*\delta^\vee) = x \cdot e_\alpha$. Thus, H_{-1} remains surjective even after restricting to $s^*\mathcal{G}^{-1}_{L(X)/S} \subseteq s^*\mathcal{V}^{-1}_{L(X)/S}$, so that $s \in \mathcal{L}$ is indeed log regular. $\square$

At least in the log smooth case, a converse is true: When $f\colon X \to S$ is log smooth, and $s \in \mathcal{L}$ a log regular section, then the zero locus H of s is log smooth as well; then both X and $H \times \mathbb{A}^1$ have the same local models around points $x \in H$, so $f\colon X \to S$ together with $s \in \mathcal{L}$ arises étale locally from this construction. We do not know if the converse is true in general.

Definition 18.39 (Log Transversal Sections)   In Situation 18.1, we say that a log regular section $s \in \mathcal{L}$ is *log transversal* if $f \colon X \to S$ together with $s \in \mathcal{L}$ arises from the above construction étale locally around every point $x \in H$.[15]

In the log transversal case, if $H \to S$ is log toroidal, then both $f \colon X \to S$ and $f \circ h \colon X(s) \to S$ are log toroidal.[16]

Open Problem 18.40   Find an example of $f \colon X \to S$ with a log regular section $s \in \mathcal{L}$ which is not log transversal, or show that log regularity implies log transversality.

Here are two examples which are not log regular.

Example 18.41   Let $f_0 \colon X_0 \to S_0$ be the central fiber of $xy = tz$. Let $\mathcal{L}_0 = \mathcal{O}_{X_0}$, and let $s = z$. Then $s \in \mathcal{L}$ is not even log pre-regular since $H \cap Z$ has codimension 1 in H. Also the criterion in Lemma 18.8 is violated. Namely, we have

$$s^* \mathcal{W}^1_{\mathcal{L}_0(X_0)/S_0} = \mathcal{W}^1_{X_0/S_0} \oplus \mathcal{O}_{X_0} \cdot s^* \delta_\alpha.$$

The map

$$G_1 \colon \ \mathcal{L}^\vee \to s^* \mathcal{W}^1_{\mathcal{L}_0(X_0)/S_0}, \quad e_\alpha^\vee \mapsto z \cdot s^* \delta_\alpha - dz,$$

can be computed explicitly since we know $\mathcal{W}^1_{X_0/S_0}$ explicitly by Proposition 8.57. It turns out that G_1 is injective, but the cokernel is not reflexive, so the map R_1 in Lemma 18.8 is not surjective. $\diamond$

Example 18.42   Let $f_0 \colon X_0 \to S_0$ be the central fiber of $xy = tzw$ (see also Example 1.83), let $\mathcal{L} = \mathcal{O}_{X_0}$, and $s = z + w$. The family $f_0 \colon X_0 \to S_0$ is log toroidal. We have $H_0 = \operatorname{Spec} \mathbf{k}[x, y, z, w]/(xy, z + w)$. Since $H_0 \cap Z = \{0\}$, the section $s \in \mathcal{L}$ is log pre-regular. The derivations of the family $f \colon X \to S$ given by $xy = tzw$ can be computed as the kernel $\mathcal{T}_{X/S} = \Theta^1_{X/S}$ of the matrix

$$R = \begin{pmatrix} y & x & -tz & -tw \end{pmatrix} \colon \ \mathcal{O}_X^{\oplus 4} \to \mathcal{O}_X.$$

When $\theta \in \mathcal{T}_{X/S}$, then $\theta \lrcorner (dz + dw)$ is the image of the matrix

$$M = \begin{pmatrix} 0 & 0 & 1 & 1 \end{pmatrix} \colon \ \mathcal{T}_{X/S} \to \mathcal{O}_X.$$

---

[15] We leave it to the reader to construct the two-sided Gerstenhaber calculus on $H \subset X$ to turn it into an enhanced generically log smooth family so that the comparison between X and $H \times \mathbb{A}^1$ makes sense in that case.

[16] It is probably also true that, if $f \colon X \to S$ is log toroidal in this case, then also $H \to S$ is log toroidal.

Thus, $dz + dw = M^{\vee}(1)$ for the dual matrix $M^{\vee}\colon O_X \to \mathcal{W}^1_{X/S} = \mathcal{H}om(\mathcal{T}_{X/S}, O_X)$. This allows us to compute $dz + dw \in \mathcal{W}^1_{X_0/S_0}$ as the image $M_0^{\vee}(1)$, and then we can compute the map

$$G_1\colon \mathcal{L}^{\vee} \to s^*\mathcal{W}^1_{L_0(X_0)/S_0} = O_{X_0}\cdot s^*\delta_\alpha \oplus \mathcal{W}^1_{X_0/S_0}, \quad e_\alpha^{\vee} \mapsto (z+w)\cdot s^*\delta_\alpha - d(z+w),$$

explicitly. It turns out that G_1 is injective, and the cokernel is reflexive. Thus, the sequence in Lemma 18.8 is exact. However, the exact sequence is not locally split. Namely, if this were the case, then the dual $G_1^{\vee}$ would be surjective, which is not the case. Thus, $s = z + w \in O_{X_0}$ is not log (quasi-)regular. $\qquad \Diamond$

Finally, we do not yet know in general under which conditions global log regular sections exist.

Open Problem 18.43 Give criteria on $f_0\colon X_0 \to S_0$ for the existence of a log regular section $s_0 \in \mathcal{L}_0$. For example, is it true that a global log regular section $s_0 \in \mathcal{L}_0$ exists if $f_0\colon X_0 \to S_0$ is log smooth, and $\mathcal{L}_0$ is very ample? How to deal with the ample case when $\mathcal{L}_0$ is not necessarily globally generated?

18.8   Smoothing Normal Crossing Spaces

In this final section, we conclude the monograph with an application of the theory developed in this chapter and indeed of the theory developed in the whole monograph. We discuss a variant of [77, Thm. 1.1] in our joint work with Filip and Ruddat, which gives sufficient conditions for a normal crossing space V to be smoothable. Proving this theorem was our original motivation for developing the theory of modifying the log structure, and in the proof, all major themes of this monograph come together.

Recall that a normal crossing space V carries a canonical structure of a toroidal crossing space $(V, \mathcal{P}, \bar{\rho})$, that we have a natural injection $\eta\colon \mathcal{LS}_V \to \mathcal{T}^1_V$ into the first tangent sheaf $\mathcal{T}^1_V = \mathcal{E}xt^1(\Omega^1_V, O_V)$, and that the latter is a line bundle on the double locus $D = \mathrm{Sing}(V)$. The normal crossing space V admits a stratification according to the rank of $\mathcal{P}$, which is discussed in Sect. 9.6. We write ω_V° for the normalized dualizing sheaf.

Theorem 18.44 (Smoothing Normal Crossing Spaces) *Let $V/\mathbb{C}$ be a proper normal crossing space such that every open stratum $S^{\circ} \in [S_k V]$ is quasi-projective for every $k \geq 0$. Assume that both $(\omega_V^{\circ})^{\vee}$ and $\mathcal{T}^1_V$ are globally generated. Then, there is a closed subset $Z \subset V$, a section $s \in \Gamma(V \setminus Z, \mathcal{LS}_V)$, and a section $e_0 \in \Gamma(V, (\omega_V^{\circ})^{\vee})$ such that:*

(a) *(V, Z, s) is a well-adjusted triple in the sense of Definition 9.92;*
(b) *the associated generically log smooth family $f_0\colon (X_0, U_0) \to S_0$ is log toroidal of standard Gross–Siebert type;*

(c) $e_0 \in (\omega_V^\circ)^\vee$ *is a log regular section;*
(d) *the modification* $g_0 \colon X_0(e_0) \to S_0$ *is log Calabi–Yau and log toroidal, and every deformation in* $\mathcal{D}(e_0)$ *is log toroidal as well, where* $\mathcal{D}$ *is a system of log toroidal deformations of standard Gross–Siebert type for* $f_0 \colon X_0 \to S_0$, *and* $\mathcal{D}(e_0)$ *is its modification.*

In particular, V admits a formal algebraic one-parameter smoothing. If V is projective, then V admits an algebraic smoothing $f \colon X \to \Sigma$ *over* $\Sigma = \mathrm{Spec}\, \mathbb{C}[\![t]\!]$.

To the author's knowledge, this is the strongest known smoothing theorem for normal crossing spaces (maybe together with its variant [77, Thm. 1.1]). The proof of this theorem occupies the rest of this section. Example 1.111 in the introduction illustrates the theorem. The author believes that a similar result holds for more general toroidal crossing spaces, but we do not yet have the necessary techniques to construct sections of $\mathcal{LS}_V$ in a generality comparable with the theorem.

Remark 18.45 In the projective case of Theorem 18.44, the algebraic deformation $f \colon X \to \Sigma$ over $\Sigma = \mathrm{Spec}\, \mathbb{C}[\![t]\!]$ is a priori only a morphism of schemes but not of log schemes. Nonetheless, we can endow X with the divisorial log structure from the central fiber, and this turns $f \colon X \to \Sigma$ into a log toroidal family of standard Gross–Siebert type, as follows from the method of Lemma 17.18 and the discussion before it. Thus, we have a well-behaved log de Rham complex $\mathcal{W}^\bullet_{X/\Sigma}$ on $f \colon X \to \Sigma$, whose Hodge–de Rham spectral sequence degenerates at E_1 by Theorem 8.65. Since the central fiber of this log morphism $f \colon X \to \Sigma$ coincides with $f_0 \colon X_0 \to S_0$ by Lemma 17.18, we can compute the Hodge numbers of the smoothing X_η as the log Hodge numbers of $f_0 \colon X_0 \to S_0$. $\Diamond$

The Proof of Theorem 18.44

The central fiber $f_0 \colon (X_0, U_0) \to S_0$ will be accurately of class $\mathscr{C}^{\mathrm{anc}}$ for the class of log singularities $\mathscr{C}^{\mathrm{anc}}$ from Example 1.86. After having discussed the local models of Gross–Siebert type in Sect. 14.1, we revisit the class $\mathscr{C}^{\mathrm{anc}}$.

Construction 18.46 We elaborate on the class of log singularities $\mathscr{C}^{\mathrm{anc}}$ from Example 1.86. First, $\mathscr{C}^{\mathrm{anc}}$ contains the semistable local models $\mu_{d;r}^{\mathrm{nc}} \colon (\mathbb{A}^{d+1} | H_{d;r}^{\mathrm{nc}}) \to (\mathbb{A}_t^1 | \{0\})$ for all $d \geq 0$ and $0 \leq r \leq d$. These local models are log smooth. To see that they are of standard Gross–Siebert type, we consider the Gross–Siebert local model datum (in the sense of Definition 14.1) given by $q = d - r \geq 0$, the lattice $M' = \mathbb{Z}^r$, the standard simplex

$$\tau = \mathrm{Conv}(0, f_1, \ldots, f_r) \subseteq M'_\mathbb{R} = \mathbb{R}^r$$

where $f_1, \ldots, f_r$ is the standard basis of $M' = \mathbb{Z}^r$, and $\Delta_1 = \ldots = \Delta_q = \{0\}$. We have $\Delta_+ = \{0\}$ so that Δ_+ is a standard simplex. In particular, $Z^* = \varnothing$. Now, we have

$$\check{\psi}_0(n) = -\inf\{0, n_1, \ldots, n_r\}$$

and $\check{\psi}_1(n) = \ldots = \check{\psi}_q(n) = 0$. Since each $\check{\psi}_i(n)$ for $1 \leq i \leq q$ is constant, we have $\Omega_i = \varnothing$ so that $Z^{\parallel} = \varnothing$. Let us write $f_1^*, \ldots, f_r^*$ for the basis of N' which is dual to $f_1, \ldots, f_r$. The monoid $P = P(M', N', \tau, \Delta_1, \ldots, \Delta_q)$ from Definition 14.2 is generated by

$$\bar{x}_0 = e_0^* - \sum_{i=1}^{r} f_i^*, \qquad \bar{x}_i = f_i^* \quad \text{for } 1 \leq i \leq r, \qquad \bar{u}_i = e_i^* \quad \text{for } 1 \leq i \leq q.$$

Therefore, we have $\mathbb{C}[P] \cong \mathbb{C}[x_0, x_1, \ldots, x_r, u_1, \ldots, u_q]$, and the map $\mathbb{C}[t] \to \mathbb{C}[P]$ is given by $t \mapsto x_0 \cdot \ldots \cdot x_r$ since $e_0^* = \sum_{i=0}^{r} \bar{x}_i$. In other words, $\mu_{d;r}^{\text{nc}}$ is a local model of standard Gross–Siebert type.

Secondly, $\mathscr{C}^{\text{anc}}$ contains local models $\mu_{d;r}^{\text{anc}}$ for $d \geq 2$ and $1 \leq r \leq d - 1$. Their total space can be written as

$$M(r; q) = \operatorname{Spec} \mathbb{C}[x_0, x_1, \ldots, x_r, t, u_1, \ldots, u_q]/(x_0 \cdot \ldots \cdot x_r - t \cdot u_1)$$

with $q = d - r \geq 1$. We have also seen these local models in Example 8.24, where we studied them as a preparation for showing in Example 8.44 that certain pencils of normal crossing divisors are log toroidal families. To recognize the local models as local models of standard Gross–Siebert type, we consider the Gross–Siebert local model datum given by $M' = \mathbb{Z}^r$, the standard simplex $\tau = \operatorname{Conv}(0, f_1, \ldots, f_r) \subseteq M_{\mathbb{R}}'$ from above, $\Delta_1 = \tau$, and $\Delta_2 = \ldots = \Delta_q = 0$. We have $\Delta_+ = \tau \times \{e_1\} \subseteq M_{\mathbb{R}}' \oplus \sum_{i=1}^{q} \mathbb{R} \cdot e_i$ so that Δ_+ is a standard simplex. In particular, $Z^* = \varnothing$. Next, we have

$$\check{\psi}_0(n) = \check{\psi}_1(n) = -\inf\{0, n_1, \ldots, n_r\}$$

and $\check{\psi}_2 = \ldots = \check{\psi}_q = 0$. Thus, the monoid $P = P(M', N', \tau, \Delta_1, \ldots, \Delta_q)$ is generated by

$$\bar{x}_0 = e_0^* + e_1^* - \sum_{i=1}^{r} f_i^*, \qquad\qquad \bar{x}_i = f_i^* \quad \text{for } 1 \leq i \leq r,$$

$$\bar{t} = e_0^*, \qquad\qquad \bar{u}_i = e_i^* \quad \text{for } 1 \leq i \leq q,$$

and we have $\mathbb{C}[P] \cong M(r; q)$, compatibly with the two maps to $\mathbb{A}_t^1$. Since $\check{\psi}_i$ is constant for $i \geq 2$, we have $\Omega_i = \varnothing$ and consequently $Z_i = \varnothing$ for $i \geq 2$. Since $\check{\psi}_0 = \check{\psi}_1$, every edge of τ corresponds to a cone in $\check{\Sigma}_0$ along which $\check{\psi}_1$ bends, so Ω_1 is the set of all edges of τ. One may now show that $Z_1 = \operatorname{Sing}(M_0) \cap \{u_1 = 0\}$, where M_0 is the central fiber of $f \colon M(r; q) \to \mathbb{A}_t^1$. In particular, we have $Z = \operatorname{Sing}(M_0) \cap \{u_1 = 0\}$. This turns $\mu_{d;r}^{\text{anc}}$ into a generically log smooth family. It is now clear from the construction that we have $\mathscr{C}^{\text{anc}} \subset \mathscr{C}^{\text{sGS}}$. $\Diamond$

Let $M_0 = M_0(r; q) = M(r; q) \times_{\mathbb{A}^1} \operatorname{Spec} \mathbb{C}$ be the central fiber. The exact sequence

$$0 \to O_{M_0} \xrightarrow{1 \mapsto dt} O_{M_0}\langle dx_0, \ldots, dx_r \rangle \to \Omega^1_{\underline{M_0}} \to 0$$

of the deformation $x_0 \cdot \ldots \cdot x_r = t$ (which is log smooth but considered as a flat deformation here) defines a surjection $O_{M_0} \to \mathcal{T}^1_{M_0}$. The image is the reference section $s_0 \in \mathcal{LS}_{M_0} \subseteq \mathcal{T}^1_{M_0}$. Now a direct computation shows that the extension class of $x_0 \cdot \ldots \cdot x_r = tu_1$ is $s = u_1 \cdot s_0$, so outside the zero locus of s, this is also the class of the induced log structure.

Using the reference log structure on M_0, we get

$$\mathcal{W}^{r+q}_{M_0^{\mathrm{ref}}/S_0} = O_{M_0} \cdot \frac{dx_1}{x_1} \wedge \ldots \wedge \frac{dx_r}{x_r} \wedge du_1 \wedge \ldots \wedge du_q =: O_{M_0} \cdot \varepsilon_0$$

Since the log canonical bundle ω_{M_0/S_0} is independent of the chosen log structure (and equal to the dualizing sheaf $\omega^\circ_{M_0}$ of the underlying Gorenstein scheme), we find $\omega_{M_0/S_0} = \omega_{M_0^{\mathrm{ref}}/S_0} = O_{M_0} \cdot \varepsilon_0$, and thus $(\omega_{M_0/S_0})^\vee = O_{M_0} \cdot \varepsilon_0^\vee$. From Lemma 18.38, we see that $u_2 \cdot \varepsilon_0^\vee$ is a log regular (indeed log transversal) section of $\omega^\vee_{M_0/S_0}$ (of course assuming $q \geq 2$ now). This log regular section will give rise to the local model for the modified and log Calabi–Yau family $g_0 \colon X_0(e_0) \to S_0$.

Lemma 18.47 *The normal crossing space $M_0 = M_0(r; q)$ together with $s = u_1 \cdot s_0 \in \mathcal{T}^1_{M_0}$ and $e_0 = u_2 \cdot \varepsilon_0^\vee \in \omega^\vee_{M_0/S_0}$ has the following properties:*

(i) *on every open stratum $S^\circ \in [\mathcal{S}_k M_0]$ for $k \geq 0$, the intersection $E \cap S^\circ \subset S^\circ$ is a smooth effective Cartier divisor for $E = \operatorname{div}(e_0)$;*
(ii) *on every open stratum $S^\circ \in [\mathcal{S}_k M_0]$ for $k \geq 1$, the intersection $Z \cap S^\circ \subset S^\circ$ is a smooth effective Cartier divisor for $Z = \operatorname{div}(s)$;*
(iii) *for every open stratum $S^\circ \in [\mathcal{S}_k M_0]$ for $k \geq 1$, the intersection $Z \cap E \cap S^\circ$ is a smooth effective Cartier divisor both on $Z \cap S^\circ$ and on $E \cap S^\circ$.*

Proof All statements are clear. $\square$

Definition 18.48 (Well-formed Triples) We say that a triple (V, s, e_0) with a normal crossing space V, a section $s \in \mathcal{T}^1_V$, and a section $e_0 \in (\omega^\circ_V)^\vee$ is *well-formed* if the statements of Lemma 18.47 hold.

In the situation of Theorem 18.44, we construct a well-formed triple (V, s, e_0). Note the following version of Bertini's theorem, which follows from the method of the proof of [138, III, Cor. 10.9].

Theorem 18.49 (Bertini) *Let $X/\mathbb{C}$ be a smooth quasi-projective variety, let $\mathcal{L}$ be a line bundle on X, and let $\mathfrak{d} \subseteq H^0(X, \mathcal{L})$ be a finite-dimensional linear system without base points, i.e., $\mathfrak{d} \otimes_{\mathbb{C}} O_X \to \mathcal{L}$ is surjective. Then there is a dense Zariski open subset $\mathfrak{u} \subseteq \mathfrak{d}$ such that $\operatorname{div}(e) \subset X$ is a smooth effective Cartier divisor for every $e \in \mathfrak{u}$, possibly empty or with several connected components.*

Note that the statement is also true for a homomorphism $\mathfrak{b} \to H^0(X, \mathcal{L})$ which need not be injective.

Corollary 18.50 *In the situation of Theorem 18.44, there are global sections $s \in \mathcal{T}_V^1$ and $e_0 \in (\omega_V^\circ)^\vee$ such that (V, s, e_0) is a well-formed triple.*

Proof Each open stratum $S^\circ \in [\mathcal{S}_k V]$ is a smooth quasi-projective variety. We apply Bertini's theorem to all open strata S° separately. First, we construct $e_0 \in (\omega_V^\circ)^\vee$ with the desired property by intersecting the relevant open subsets of $H^0(V, (\omega_V^\circ)^\vee)$ for all open strata S°. Then we can construct a dense open subset $\mathfrak{u} \subseteq H^0(D, \mathcal{T}_V^1)$ such that the condition on $Z \cap S^\circ \subset S^\circ$ is satisfied. Since also $E \cap S^\circ$ decomposes into smooth quasi-projective varieties, we achieve that $Z \cap E \cap S^\circ \subset E \cap S^\circ$ is a smooth effective Cartier divisor by possibly shrinking $\mathfrak{u}$. Then $Z \cap E \cap S^\circ \subset Z \cap S^\circ$ must be a smooth effective Cartier divisor as well. $\square$

Remark 18.51 In [77, Thm. 1.1], it was assumed instead that $D = \mathrm{Sing}(V)$ itself is projective (while not assuming that V is projective), and that an appropriate section $e_0 \in (\omega_V^\circ)^\vee$ is already given.[17] Then we do not need the quasi-projectivity of the strata in $[\mathcal{S}_0 V]$, and for $S^\circ \in [\mathcal{S}_k V]$ for $k \geq 1$, it is automatic. $\Diamond$

We investigate the local structure of well-formed triples (V, s, e_0). Recall that, given a local ring $R = O_{X,x}$ of a smooth scheme $X/\mathbb{C}$ at a $\mathbb{C}$-valued point $x \in X$, we have

$$\Omega_{R/\mathbb{C}}^1 \cong \bigoplus_{k=1}^{n} R \cdot dy_k$$

for any regular system of parameters $y_1, \ldots, y_n \in \mathfrak{m}_R$. Thus, for every R-module M and every sequence $m_1, \ldots, m_n \in M$, there is a unique $\mathbb{C}$-linear derivation $D \colon R \to M$ with $D(y_k) = m_k$. We write ∂_i for the unique derivation with $\partial_i(y_k) = \delta_{ik}$. Recall the following basic fact:

Lemma 18.52 *Let $R = O_{X,x}$ be a regular local ring of dimension $n \geq 1$, essentially of finite type over $\mathbb{C}$, and let $y_1, \ldots, y_n \in \mathfrak{m}_R$ be a regular system of parameters. For an element $f \in \mathfrak{m}_R$, the following statements are equivalent:*

(i) *$R/(f)$ is a regular local ring of dimension $n - 1$;*
(ii) *$f \notin \mathfrak{m}_R^2$;*
(iii) *there is a $\mathbb{C}$-linear derivation $D \colon R \to R$ with $D(f) \notin \mathfrak{m}_R$;*
(iv) *there is an index $1 \leq i \leq n$ with $\partial_i(f) \notin \mathfrak{m}_R$.*

Proof This essentially follows from the discussion in [267, 07PD]. $\square$

---

[17] This is the meaning of the term *effective* anti-canonical bundle in [77, Thm. 1.1].

With this preparation, we also get:

Lemma 18.53 *Let $R = O_{X,x}$ be a regular local ring of dimension $n \geq 2$, essentially of finite type over $\mathbb{C}$, and let $y_1, \ldots, y_n \in \mathfrak{m}_R$ be a regular system of parameters. For elements $f, g \in \mathfrak{m}_R$, the following statements are equivalent:*

(i) *both $R/(f)$ and $R/(g)$ are regular local rings of dimension $n-1$, and $R/(f,g)$ is a regular local ring of dimension $n-2$;*
(ii) *there are indices $1 \leq i, j \leq n$ such that $\partial_i(f) \notin \mathfrak{m}_R$, $\partial_j(g) \notin \mathfrak{m}_R$, and $\partial_i(f)\partial_j(g) - \partial_i(g)\partial_j(f) \notin \mathfrak{m}_R$.*

Proof Let $f_k := [\partial_k(f)] \in R/\mathfrak{m}_R = \mathbf{k}$ and $g_k := [\partial_k(g)] \in R/\mathfrak{m}_R = \mathbf{k}$. Then we have $[f] = \sum_{k=1}^n f_k[y_k]$ and $[g] = \sum_{k=1}^n g_k[y_k]$ in $\mathfrak{m}_R/\mathfrak{m}_R^2$. First suppose that we have indices i, j with $f_i \neq 0$, $g_j \neq 0$, and $f_i g_j - f_j g_i \neq 0$. In particular, we have $i \neq j$, and also $f \neq 0$, $g \neq 0$. By the above lemma, $R/(f)$ and $R/(g)$ are regular local rings of dimension $n - 1$. For notational simplicity, suppose that $i = 1$ and $j = 2$. Note that $f, y_2, \ldots, y_n$ is a regular system of parameters for $R/(f)$, so $\bar{y}_2, \ldots, \bar{y}_n \in \mathfrak{m}_{R/(f)}$ is a regular system of parameters for $R/(f)$. Now

$$[\bar{g}] = \sum_{k=1}^n g_k[\bar{y}_k] = \sum_{k=2}^n (g_k - g_1 f_k/f_1)[\bar{y}_k]$$

because $0 = [\bar{f}] = \sum_{k=1}^n f_k[\bar{y}_k]$ and $f_1 \neq 0$. Since $g_2 f_1 - g_1 f_2 \neq 0$, we have $[\bar{g}] \neq 0 \in \mathfrak{m}_{R(f)}/\mathfrak{m}_{R/(f)}^2$, so $\bar{g} \notin \mathfrak{m}_{R/(f)}^2$, and hence $R/(f,g)$ is regular of dimension $n-2$.

Conversely, assume the regularity statement. Suppose that there is some $f_i \neq 0$ such that $g_i = 0$. Because $R/(g)$ is regular of dimension $n-1$, there is some $g_j \neq 0$ with $j \neq i$. Then $f_i g_j - f_j g_i = f_i g_j \neq 0$, so we are finished. Similarly, when there is some $g_j \neq 0$ with $f_j = 0$, our claim follows. Thus, we may assume that $f_i = 0$ if and only if $g_i = 0$, and that there is some index j with $f_j, g_j \neq 0$. Now suppose that the claim is wrong. In particular, for each i, we have $f_i = 0$ or $f_i g_j = f_j g_i$, i.e., $f_i/f_j = g_i/g_j$. Since $f_i = 0$ implies $g_i = 0$, we always have $f_i/f_j = g_i/g_j$. But then $[f]/f_j = [g]/g_j$ in $\mathfrak{m}_R/\mathfrak{m}_R^2$. This shows $\bar{g} \in \mathfrak{m}_{R/(f)}^2$, contradicting the regularity of $R/(f,g)$ (under the assumption that this ring has dimension $n-2$). $\qquad\square$

Let $p: Y \to X$ be a morphism between two schemes of finite type over $\mathbb{C}$. Then $p: Y \to X$ is étale if and only if the induced map $p^*: \widehat{O}_{X,p(y)} \to \widehat{O}_{Y,y}$ of complete local rings is an isomorphism for all $\mathbb{C}$-valued points $y \in Y$. Namely, by [267, 0C4G], the map $p^*: O_{X,p(y)} \to O_{Y,y}$ is flat, and then $p: Y \to X$ is étale at y by [267, 02GU, (6)]. This allows us to show the following structure result for well-formed triples. The proof essentially follows a part of [125, Thm. 2.6].

Proposition 18.54 *Let (V, s, e_0) be a well-formed triple, and let $v \in D = \mathrm{Sing}(V) \subset V$ be a $\mathbb{C}$-valued point.*

(1) *Assume that $s(v) = 0$ and $e_0(v) = 0$. Then there is an étale neighborhood $\chi\colon (W, w) \to (V, v)$ of v together with an étale morphism $\phi\colon W \to M_0(r; q)$ to*

$$M_0(r; q) = \operatorname{Spec}\mathbb{C}[x_0, \ldots, x_r, u_1, \ldots, u_q]/(x_0 \cdot \ldots \cdot x_r)$$

for $q \geq 2$ with $\phi(w) = 0$ such that $\chi^*(s) = \phi^*(u_1 \cdot s_0) \in \mathcal{T}_W^1$ and $\chi^*(e_0) = v \cdot \phi^*(u_2 \cdot \varepsilon_0^\vee) \in (\omega_W^\circ)^\vee$ for some invertible function $v \in O_W^*$, where $s_0 \in \mathcal{T}_{M_0}^1$ and $\varepsilon_0^\vee \in (\omega_{M_0}^\circ)^\vee$ are as above.

(2) *Assume that $s(v) = 0$ and $e_0(v) \neq 0$. Then there is an étale neighborhood $\chi\colon (W, w) \to (V, v)$ of v together with an étale morphism $\phi\colon W \to M_0(r; q)$ for $q \geq 1$ with $\phi(w) = 0$ such that $\chi^*(s) = \phi^*(u_1 \cdot s_0) \in \mathcal{T}_W^1$.*

Proof Note that the statement makes sense because $p^*\mathcal{T}_X^1 = \mathcal{T}_Y^1$ and $p^*(\omega_X^\circ)^\vee = (\omega_Y^\circ)^\vee$ for an étale morphism $p\colon Y \to X$. We show the first statement, and the proof of the second statement is similar. Let $T^\circ \in [\mathcal{S}_k V]$ be the open stratum with $v \in T^\circ$. Then we set $r = k$ and $q = d - k$, where $d = \dim(V)$. Since V is a normal crossing space, we can find an étale neighborhood $\chi\colon (W, w) \to (V, v)$ together with an étale morphism $\psi\colon W \to M_0(r; q) =: M_0$ with $\psi(w) = 0$. By shrinking W, we can assume that W is affine. Since s_0 trivializes $\mathcal{T}_{M_0}^1$, and $\varepsilon_0^\vee$ trivializes $(\omega_{M_0}^\circ)^\vee$, we can find functions $f, g \in O_W$ with $\chi^*s = f \cdot \psi^*(s_0)$ and $\chi^*e_0 = g \cdot \psi^*(\varepsilon_0^\vee)$. We have $f(w) = 0$ and $g(w) = 0$. Let us denote the stratum of W by T° as well; by shrinking W, we can assume that $T^\circ \subset W$ is a (smooth) closed subscheme. Let us denote the images of f, g in $\bar{R} := O_{T^\circ}$ by $\bar{f}, \bar{g}$. By the well-formedness assumption, $\bar{R}/\bar{f}$, $\bar{R}/\bar{g}$, and $\bar{R}/(\bar{f}, \bar{g})$ are regular local rings of dimension $q - 1$ respectively $q - 2$. Now $y_1 := \psi^*u_1, \ldots, y_q := \psi^*u_q$ forms a regular system of parameters of $\bar{R}$. Thus, we can find indices $1 \leq i, j \leq q$ with $\partial_i(\bar{f}) \notin \mathfrak{m}_{\bar{R}}$, $\partial_j(\bar{g}) \notin \mathfrak{m}_{\bar{R}}$, and $\partial_i(\bar{f})\partial_j(\bar{g}) - \partial_i(\bar{g})\partial_j(\bar{f}) \notin \mathfrak{m}_{\bar{R}}$. For simplicity of notation, let us assume $i = 1$ and $j = 2$. Now we define a morphism $\phi\colon W \to M_0(r; q)$ by $\phi^*(u_1) = f$, $\phi^*(u_2) = g$, and $\phi^*(u_k) = \psi^*(u_k)$ for $k \geq 3$ as well as $\phi^*(x_k) = \psi^*(x_k)$. Note that $\phi^*(u_k) \in \mathfrak{m}_{W,w}$ for all $1 \leq k \leq q$ as well as $\phi^*(x_k) \in \mathfrak{m}_{W,w}$, so we have $\phi(w) = 0$. Since $f, g, y_3, \ldots, y_q$ is a system of parameters for $\bar{R}$—here we use the condition on the derivations—we find that $\phi^*(\mathfrak{m}_0)$ generates $\mathfrak{m}_w \subseteq O_{W,w}$. Thus, the induced map $\phi^*\colon \widehat{O}_{M_0,0} \to \widehat{O}_{W,w}$ of complete local rings is surjective. However, we also have an isomorphism $\psi^*\colon \widehat{O}_{M_0,0} \cong \widehat{O}_{W,w}$ from our first étale morphism ψ, so $\phi^* \circ (\psi^*)^{-1}$ is a surjective endomorphism of a Noetherian ring, hence an isomorphism. Since $\phi\colon W \to M_0(r; q)$ is of finite type, it is étale at w, hence étale after further shrinking W.

Now let $\pi\colon M_0(r; q) \to M_0(r; 0)$ be the projection. On the one hand, we have $s_0 = \pi^*(\tilde{s}_0)$ for the corresponding section on $M_0(r; 0)$. On the other hand, we have

$\pi \circ \phi = \pi \circ \psi$. Thus, we have $\chi^*(s) = f \cdot \psi^*(s_0) = \phi^*(u_1) \cdot \phi^*(s_0)$. The case of $\varepsilon_0^\vee$ is more complicated. We have $\varepsilon_0 = \pi^* \tilde{\varepsilon}_0 \otimes du_1 \wedge \ldots \wedge du_q$, so

$$
\begin{aligned}
\phi^*(\varepsilon_0) &= (\pi \circ \phi)^* \tilde{\varepsilon}_0 \otimes df \wedge dg \wedge d\psi^*(u_3) \wedge \ldots \wedge d\psi^*(u_q) \\
&= (\pi \circ \psi)^* \tilde{\varepsilon}_0 \otimes (\partial_1(f)\partial_2(g) - \partial_2(f)\partial_1(g)) d\psi^*(u_1) \wedge \ldots \wedge d\psi^*(u_q) \\
&= (\partial_1(f)\partial_2(g) - \partial_2(f)\partial_1(g)) \cdot \psi^*(\varepsilon_0)
\end{aligned}
$$

Here, $\partial_k \colon O_W \to O_W$ is the induced derivation from $\partial_{u_k} \colon O_{M_0} \to O_{M_0}$. These derivations restrict to T°, so after shrinking W, we know that the factor is invertible on W. Then we find $\chi^*(e_0) = g \cdot \psi^*(\varepsilon_0^\vee) = v \cdot \phi^*(u_2) \cdot \phi^*(\varepsilon_0^\vee)$ for an invertible function v. $\square$

Proof of Theorem 18.44 (V, Z, s) is obviously a well-adjusted triple. From the local model of Proposition 18.54, we see that the log structure on $f_0 \colon X_0 \to S_0$ is log toroidal of standard Gross–Siebert type. Namely, the map $\eta \colon \mathcal{LS}_V \to \mathcal{T}_V^1$ is compatible with étale pull-back. Outside the double locus D, the section $e_0 \in (\omega_V^\circ)^\vee$ is log regular because it is the strict embedding of a smooth divisor in a strict log smooth space. Inside D, Proposition 18.54 shows that e_0 is log regular up to multiplication with a unit, but then $e_0 \in (\omega_V^\circ)^\vee$ itself is log regular because this notion is invariant under isomorphisms of the triple $(V, \mathcal{L}, e_0)$. The modification $g_0 \colon X_0(e_0) \to S_0$ is log toroidal because the pull-back from $M_0(r; q)$ is log toroidal, and because this pull-back is isomorphic to the pull-back from V since the two sections of $(\omega_V^\circ)^\vee$ differ only by a unit locally. Then $g_0 \colon X_0(e_0) \to S_0$ is log Calabi–Yau because the chosen line bundle is the anti-canonical bundle $(\omega_V^\circ)^\vee$. The thickenings of u_2 on the infinitesimal log toroidal deformations of $M_0(r; q)$ are log transversal so that the modification in u_2 is log toroidal as well. When pulling these deformations back to W along $\phi \colon W \to M_0(r; q)$, we obtain local log toroidal deformations of $g_0 \colon X_0(e_0) \to S_0$ which coincide with the modifications of the log toroidal deformations of $f_0 \colon X_0 \to S_0$. Thus, the deformations of type $\mathcal{D}(e_0)$ are log toroidal. The formal smoothing is obtained from the unobstructedness of $\mathrm{LD}_{X_0(e_0)/S_0}^{\mathcal{D}(e_0)}$ by Theorem 15.2 via the map $\Psi \circ \Phi$ to $\mathrm{LD}_{X_0/S_0}^{\mathcal{D}}$. The algebraic smoothing in the projective case is obtained from the unobstructedness of $\mathrm{LD}_{X_0(e_0)/S_0}^{\mathcal{D}(e_0)}(\mathcal{L}_0)$ for a very ample line bundle $\mathcal{L}_0$ as in Chap. 17. $\square$

Appendix A
Deformation Functors

We review the notion of a deformation functor and briefly discuss the existence of a universal obstruction theory.

A.1 Artinian Local Rings

Let Λ be a complete local Noetherian ring with residue field $\mathbf{k}$, and let $p_\Lambda \colon \Lambda \to \mathbf{k}$ be the projection. We denote the maximal ideal by $\mathfrak{m}_\Lambda \subset \Lambda$. In the main text, we assume that the residue field $\mathbf{k}$ is of characteristic 0, and that a ring homomorphism $e \colon \mathbf{k} \to \Lambda$ with $p_\Lambda \circ e = \mathrm{id}_\mathbf{k}$ is given, but this assumption is not necessary for the general discussion of deformation functors.

Definition A.1 ($\mathbf{Art}_\Lambda$) An *Artinian local Λ-algebra with residue field* $\mathbf{k}$ is a triple (A, e_A, p_A), where A is an Artinian local ring, $e_A \colon \Lambda \to A$ is a local ring homomorphism, and $p_A \colon A \to \mathbf{k}$ is a surjective local ring homomorphism such that $p_A \circ e_A = p_\Lambda \colon \Lambda \to \mathbf{k}$. Together with homomorphisms $\varphi \colon B \to A$ such that $e_A = \varphi \circ e_B$ and $p_B = p_A \circ \varphi$, they form a category $\mathbf{Art}_\Lambda$.

Note that a homomorphism $\varphi \colon B \to A$ in $\mathbf{Art}_\Lambda$ is necessarily local. We define *complete local Noetherian Λ-algebras with residue field* $\mathbf{k}$ analogously. Together with compatible local homomorphisms (locality now not being automatic), they form a category $\mathbf{Comp}_\Lambda$.

Let $A \in \mathbf{Art}_\Lambda$. Then we have a finite filtration $A \supseteq \mathfrak{m}_A \supseteq \mathfrak{m}_A^2 \supseteq \ldots \supseteq \mathfrak{m}_A^{N+1} = 0$. Each quotient $\mathfrak{m}_A^k / \mathfrak{m}_A^{k+1}$ is a finite-dimensional $\mathbf{k}$-vector space and thus a finitely generated Λ-module. In particular, A is finitely generated as a Λ-module.

Let $\varphi \colon B \to A$ and $\psi \colon A' \to A$ be homomorphisms in $\mathbf{Art}_\Lambda$. Then the set-theoretic fiber product $B' := A' \times_A B$—not to be confused with the tensor product—has a canonical structure of a local Λ-algebra with a unique prime ideal $\mathfrak{m}_{B'}$ and residue field $\mathbf{k}$. Since A' and B are finitely generated Λ-modules, so is

S. Felten, *Global Logarithmic Deformation Theory*, Lecture Notes in Mathematics 2373, https://doi.org/10.1007/978-3-031-98751-9

B'. In particular, B' is a finitely generated Λ-algebra, thus Noetherian and hence Artinian. It is the fiber product in $\mathbf{Art}_\Lambda$.

When Λ is understood, we write

$$A_k = \Lambda/\mathfrak{m}_\Lambda^{k+1}$$

for integers $k \geq 0$. In particular, we set $A_0 = \mathbf{k}$. We write furthermore

$$A_\varepsilon = \mathbf{k}[\varepsilon]/(\varepsilon^2)$$

with the Λ-algebra structure induced by $\Lambda \to A_0 \to A_\varepsilon$. More generally, we can consider any $A \in \mathbf{Art}_\mathbf{k}$ as an element of $\mathbf{Art}_\Lambda$ via $\Lambda \to A_0 \to A$, and this defines a fully faithful embedding $\mathbf{Art}_\mathbf{k} \to \mathbf{Art}_\Lambda$.

For a sharp toric monoid Q, we obtain a complete local Noetherian ring $\mathbf{k}[\![Q]\!]$ by completing the monoid ring $\mathbf{k}[Q]$ in the maximal ideal $\mathbf{k}[Q^+]$, where $Q^+ = Q \setminus \{0\}$ is the set of non-units. In this case, we write $\mathbf{Art}_Q$ for $\mathbf{Art}_{\mathbf{k}[\![Q]\!]}$, and similarly $\mathbf{Comp}_Q$ for $\mathbf{Comp}_{\mathbf{k}[\![Q]\!]}$.

A *first-order extension* is an exact sequence

$$e: \quad 0 \to I \to B \overset{\varphi}{\to} A \to 0$$

where $\varphi\colon B \to A$ is a surjection in $\mathbf{Art}_\Lambda$ such that the kernel $I \subset B$ satisfies $I \cdot \mathfrak{m}_B = 0$. A *small extension* is a first-order extension such that I is a principal ideal; in this case, I is a one-dimensional $\mathbf{k}$-vector space. Every surjection in $\mathbf{Art}_\Lambda$ factors into a sequence of small extensions.[1]

A.2   Deformation Functors

Definition A.2 (Functors of Artin Rings)   A *functor of Artin rings*[2] is a covariant functor $F\colon \mathbf{Art}_\Lambda \to \mathbf{Set}$ with the following property:

(H_0)   The set $F(A_0)$ consists of a single element.

Together with natural transformations, they form a category $\mathbf{FunArt}_\Lambda$.

The category $\mathbf{FunArt}_\Lambda$ has all (small) limits $\lim_i F_i$: they are constructed as the limits of $F_i(A)$ over every $A \in \mathbf{Art}_\Lambda$. In particular, $\mathbf{FunArt}_\Lambda$ has a terminal object T given by $T(A) = \{*\}$ for every $A \in \mathbf{Art}_\Lambda$, where $\{*\}$ is an arbitrary one-element

---

[1] Our convention for the definition of a small extension is compatible with most of the literature. In [199] and related works, what we call a "first-order extension" is called a "small extension."

[2] The name "functor of Artin rings" rather than "functor of Artinian rings" is well established, and we follow this convention although we otherwise use "Artinian" in analogy with "Noetherian."

set. By abuse of notation, we denote this terminal functor of Artin rings by $\{*\}$ as well.

A natural transformation $\xi\colon F \Rightarrow G$ of functors of Artin rings is *smooth* if, for every surjection $\varphi\colon B \to A$ in $\mathbf{Art}_\Lambda$, the induced map

$$F(B) \to F(A) \times_{G(A)} G(B)$$

is surjective. A smooth map $\xi\colon F \Rightarrow G$ is *surjective* in the sense that $F(A) \to G(A)$ is surjective for every $A \in \mathbf{Art}_\Lambda$. If the unique natural transformation $F \Rightarrow \{*\}$ is smooth, then F is called *unobstructed*. A functor of Artin rings F is unobstructed if and only if $F(\varphi)\colon F(B) \to F(A)$ is surjective for every surjection $\varphi\colon B \to A$ in $\mathbf{Art}_\Lambda$.

When $\xi\colon F \Rightarrow G$ and $\theta\colon G \Rightarrow H$ are two smooth morphisms, then $\theta \circ \xi\colon F \Rightarrow H$ is smooth as well. As a partial converse, if $\theta \circ \xi\colon F \Rightarrow H$ is smooth and $\xi\colon F \Rightarrow G$ is surjective, then $\theta\colon G \Rightarrow H$ is smooth.

Let F be a functor of Artin rings. Then the set

$$\mathrm{T}_F := F(A_\varepsilon)$$

is called the *tangent space* of F.

We study the geometry of functors of Artin rings. For every $R \in \mathbf{Comp}_\Lambda$, we have a functor of Artin rings

$$h_R\colon \quad \mathbf{Art}_\Lambda \to \mathbf{Set}, \quad A \mapsto \mathrm{Hom}_{\mathbf{Comp}_\Lambda}(R, A).$$

This defines a fully faithful contravariant functor $h\colon \mathbf{Comp}_\Lambda \to \mathbf{FunArt}_\Lambda$. A functor of Artin rings is *pro-representable* if it is in the essential image.

Being pro-representable is the ideal situation from the perspective of describing the geometry of F, but most functors of Artin rings which arise in practice are not pro-representable. Having a *hull* is often a good approximation of being pro-representable. In the terminology of [251], a *pro-couple* is a pair (R, ξ) with $R \in \mathbf{Comp}_\Lambda$ and $\xi\colon h_R \Rightarrow F$ a natural transformation.

Definition A.3 (Versal Formal Families) Let F be a functor of Artin rings, and let (R, ξ) be a pro-couple.

(a) (R, ξ) is a *versal formal family* if $\xi\colon h_R \Rightarrow F$ is smooth.
(b) (R, ξ) is a *miniversal formal family* if $\xi\colon h_R \Rightarrow F$ is smooth, and the induced map $\mathrm{T}_\xi\colon \mathrm{T}_{h_R} \to \mathrm{T}_F$ is bijective.
(b) (R, ξ) is a *universal formal family* if $\xi\colon h_R \Rightarrow F$ is an isomorphism.

A *hull* of F is a miniversal formal family (R, ξ).

A pro-couple (R, ξ) can be interpreted as a sequence of elements $\xi_k \in F(R_k)$, where $R_k = R/\mathfrak{m}_R^{k+1}$. If $\xi\colon h_R \Rightarrow F$ is a versal formal family, then every $\alpha \in F(A)$

is obtained as $F(\psi)(\xi_k)$ for some $k \geq 0$ and some $\psi: R_k \to A$. Furthermore, if $\varphi: B \to A$ is a surjection, and if $\beta \in F(B)$ is such that $F(\varphi)(\beta) = \alpha$, then there is some $d \geq 0$ and some $\psi': R_{k+d} \to B$ such that $\beta = F(\psi')(\xi_{k+d})$ and such that $\varphi \circ \psi' = \psi \circ \pi$, where $\pi: R_{k+d} \to R_k$ is the quotient map. Intuitively, one may say that we know all deformations described by F if we know a versal formal family $(\xi_k)_k$. A hull can be understood as a versal formal family which cannot be made smaller, a *minimal* one—thus explaining the name "miniversal" formal family.

Before we recall Schlessinger's criterion for having a hull, we recall the notion of a *deformation functor*. Let $\varphi: B \to A$ be surjective, and let $\psi: A' \to A$ be an arbitrary homomorphism in $\mathbf{Art}_\Lambda$. Then we have an induced map

$$\eta: \quad F(A' \times_A B) \to F(A') \times_{F(A)} F(B).$$

Definition A.4 (Deformation Functors) A *deformation functor* is a functor of Artin rings $F: \mathbf{Art}_\Lambda \to \mathbf{Set}$ such that the following two conditions hold:

(H_1) The natural map η is surjective for every surjection $\varphi: B \to A$.
(H_2) The natural map η is bijective for $B = A_\varepsilon$ and $A = A_0$.

Together with natural transformations, deformation functors form a full subcategory $\mathbf{DefoFun}_\Lambda$ of $\mathbf{FunArt}_\Lambda$.

Remark A.5 The fiber product of deformation functors, formed in the category of functors of Artin rings, is not in general a deformation functor, see [199, Exer. 3.9.4]. $\Diamond$

This definition is justified by the fact that the tangent space $\mathrm{T}_F = F(A_\varepsilon)$ carries a natural structure of a $\mathbf{k}$-vector space—for a general functor of Artin rings, T_F is just a set. Furthermore, if $\varphi: B \to A$ is a small extension with kernel $I \subset B$, and if $\beta \in F(B)$ with $\alpha := F(\varphi)(\beta) \in F(A)$, then there is a natural transitive action of $\mathrm{T}_F \otimes_{\mathbf{k}} I$ on $F(\varphi)^{-1}(\alpha)$. If $\xi: F \Rightarrow G$ is a natural transformation, then $\mathrm{T}_\xi: \mathrm{T}_F \to \mathrm{T}_G$ is $\mathbf{k}$-linear, and for a small extension $\varphi: B \to A$, the $\mathrm{T} \otimes_{\mathbf{k}} I$-actions are compatible.

We can now state Schlessinger's famous criterion of [251] for having a hull.

Theorem A.6 *Let $F: \mathbf{Art}_\Lambda \to \mathbf{Set}$ be a functor of Artin rings. Then F has a hull if and only if it is a deformation functor and the following condition is satisfied:*

(H_3) *The $\mathbf{k}$-vector space T_F is of finite dimension.*

If a hull exists, then it is unique up to non-unique isomorphism.

Schlessinger also characterized pro-representable functors of Artin rings in [251]. The first criterion below is Schlessinger's original formulation, while the second criterion is the one chosen by Hartshorne in [139].

Theorem A.7 *Let F be a deformation functor satisfying (H_3). Then F is pro-representable if and only if the following condition holds:*

(H_4) *Let $\varphi\colon B \to A$ be a small extension. Then*

$$F(B \times_A B) \to F(B) \times_{F(A)} F(B)$$

is bijective.

Furthermore, F is pro-representable if and only if the following alternative condition holds:

(H_4') *If $\varphi\colon B \to A$ is a small extension with kernel I, and $\beta \in F(B)$ with $\alpha :=$ $F(\varphi)(\beta)$, then the natural action of $T_F \otimes_{\mathbf{k}} I$ on $F(\varphi)^{-1}(\alpha)$ is bijective.*

Following [199], we say:

Definition A.8 (Homogeneous Functors) A functor of Artin rings F is *homogeneous* if the natural map

$$\eta\colon \ F(A' \times_A B) \to F(A') \times_{F(A)} F(B)$$

is bijective whenever $\varphi\colon B \to A$ is surjective

Thus, a homogeneous functor of Artin rings is a deformation functor, and any functor of Artin rings F is pro-representable if and only if it is homogeneous and T_F is of finite dimension.

A.3 Obstruction Theories

We discuss *obstruction theories*. Given a deformation functor $F\colon \mathbf{Art}_\Lambda \to \mathbf{Set}$, an obstruction theory is a tool to answer the question if, given $A \in \mathbf{Art}_\Lambda$, $\alpha \in F(A)$, and a first-order extension $\varphi\colon B \to A$, there is some $\beta \in F(B)$ with $F(\varphi)(\beta) = \alpha$. Obstruction theories for a given deformation functor form a category. Precise definitions can be found below.

When a concrete deformation problem is given, it is often possible to construct some obstruction theory explicitly. More intrinsically, for many deformation functors F, the category of obstruction theories has an initial object. This *universal obstruction theory* can be constructed purely out of F, without using the original deformation problem which defines F. However, to construct a universal obstruction theory from F, it is not sufficient to assume that F is a deformation functor. Instead, we have to impose some additional condition. The following condition is both convenient to check and sufficient to construct a universal obstruction theory.

Definition A.9 (Neatness) A deformation functor $F\colon \mathbf{Art}_\Lambda \to \mathbf{Set}$ is *neat* if it satisfies the following condition:

(H_2^+) The natural map

$$\eta\colon\quad F(A' \times_A B) \to F(A') \times_{F(A)} F(B)$$

is bijective for $A = A_0$.

The author learned this condition from Manetti's monograph [199], where Manetti imposes (H_2^+) as an additional condition in the definition of a deformation functor, precisely due to its role in constructing a universal obstruction theory. This condition is absent in Schlessinger's original definition because it is not necessary to endow the tangent space T_F with the structure of a $\mathbf{k}$-vector space, and also not necessary to show the existence of a hull as in Theorem A.6. Respecting Schlessinger's original definition, which is widely accepted in the literature, we have opted to consider (H_2^+) as a property of deformation functors rather than as an amendment of the definition.

We never use the universal obstruction theory in this monograph, but we anticipate to use it in future works. Since the construction of the universal obstruction theory needs an additional condition which we wish to check for the deformation functors considered in this monograph, we have included the present section. In the remainder of the section, we give some indication of why condition (H_2^+) is sufficient. Namely, in [199], Manetti considers only the case $\Lambda = \mathbf{k}$, so we establish that our condition (H_2^+) as stated above is correct for general Λ.

A.3.1   The Definition of an Obstruction Theory

We have introduced first-order extensions already above. Here, we use the following slight variation of their definition. We write $\mathbf{Vect}$ for the category of $\mathbf{k}$-vector spaces and $\mathbf{vect}$ for the full subcategory of finite-dimensional $\mathbf{k}$-vector spaces.

Definition A.10 (First-Order Extensions) The *category of first-order extensions* $\mathbf{Ex}$ is defined as follows. An object is a tuple (A, I, B, φ, μ), where $\varphi\colon B \to A$ is a surjection in $\mathbf{Art}_\Lambda$, where I is a finite-dimensional $\mathbf{k}$-vector space, and where $\mu\colon I \to B$ is a group homomorphism such that $\mu(p_B(b) \cdot i) = b \cdot \mu(i)$ for every $b \in B$ and $i \in I$, such that

$$0 \to I \xrightarrow{\mu} B \xrightarrow{\varphi} A \to 0$$

is a short exact sequence of Abelian groups. A *homomorphism* between first-order extensions $e_2 = (A_2, I_2, B_2, \varphi_2, \mu_2)$ and $e_1 = (A_1, I_1, B_1, \varphi_1, \mu_1)$ is a commutative diagram

$$
\begin{array}{ccccccccc}
e_2: & 0 & \longrightarrow & I_2 & \longrightarrow & B_2 & \xrightarrow{\varphi_2} & A_2 & \longrightarrow & 0 \\
 & & & \downarrow{f} & & \downarrow{\varpi} & & \downarrow{\pi} & & \\
e_1: & 0 & \longrightarrow & I_1 & \longrightarrow & B_1 & \xrightarrow{\varphi_1} & A_1 & \longrightarrow & 0
\end{array}
$$

of Abelian groups where $\pi: A_2 \to A_1$ and $\varpi: B_2 \to B_1$ are in $\mathbf{Art}_\Lambda$, and where $f: I_2 \to I_1$ is $\mathbf{k}$-linear.

Note that we suppress the dependency on Λ in our notation $\mathbf{Ex}$. For $A \in \mathbf{Art}_\Lambda$ and $I \in \mathbf{vect}$ a finite-dimensional $\mathbf{k}$-vector space, we write $\mathbf{Ex}(A, I) \subseteq \mathbf{Ex}$ for the full subcategory where A and I are fixed, and where f and π are the identity. This is a groupoid.

With this preparation, we give the formal definition of an obstruction theory.

Definition A.11 (Obstruction Theories) Let F be a functor of Artin rings. Then an *obstruction theory* is a pair (V, v) where:

(a) V is a $\mathbf{k}$-vector space;
(b) v is a collection of maps of sets $v_e: F(A) \to V \otimes_{\mathbf{k}} I$, one for each $A \in \mathbf{Art}_\Lambda$, $I \in \mathbf{vect}$, and $e \in \mathbf{Ex}(A, I)$.

These data are such that, for every map $(f, \varpi, \pi): e_2 \to e_1$ of first-order extensions and every $\alpha_2 \in F(A_2)$, we have

$$(\mathrm{id}_V \otimes f)(v_{e_2}(\alpha_2)) = v_{e_1}(F(\pi)(\alpha_2)).$$

A *homomorphism* of obstruction theories $(V, v) \to (W, w)$ is a homomorphism $\theta: V \to W$ of $\mathbf{k}$-vector spaces such that $w_e(\alpha) = (\theta \otimes \mathrm{id}_I)(v_e(\alpha))$ for every first-order extension e and every $\alpha \in F(A)$.

If there is some $\beta \in F(B)$ with $F(\varphi)(\beta) = \alpha$, then $v_e(\alpha) = 0$ for any obstruction theory (V, v). Conversely, if an obstruction theory (V, v) has the property that $v_e(\alpha) = 0$ implies the existence of some $\beta \in F(B)$ with $F(\varphi)(\beta) = \alpha$, then (V, v) is called *complete*.

A.3.2 *Existence of a Universal Obstruction Theory*

We discuss how to adapt the proof of the existence of a universal obstruction theory given by Manetti in [199, § 3.8] from $\Lambda = \mathbf{k}$ to a general Λ. For $A \in \mathbf{Art}_\Lambda$ and

$I \in \mathbf{vect}$, we have seen the groupoid $\mathbf{Ex}(A, I)$ of first-order extensions above. We write $\mathrm{Ex}(A, I)$ for the set of isomorphism classes in $\mathbf{Ex}(A, I)$.

For $e \in \mathbf{Ex}(A, I)$ and $f \colon I \to J$, we obtain a push-forward

$$f_*(e) \colon \qquad 0 \to J \to J \oplus_I B \to A \to 0$$

in $\mathbf{Ex}(A, J)$ as in [199, § 3.8]. Similarly, for $\pi \colon C \to A$ in $\mathbf{Art}_\Lambda$, we obtain a pull-back

$$\pi^*(e) \colon \qquad 0 \to I \to B \times_A C \to C \to 0$$

in $\mathbf{Ex}(C, I)$ as in [199, § 3.8]. The proof that $f_*(e)$ and $\pi^*(e)$ are first-order extensions in $\mathbf{Ex}$ is slightly tedious but straightforward. One may check that, on the level of isomorphism classes, we obtain a functor

$$\mathrm{Ex}(-, -) \colon \quad \mathbf{Art}_\Lambda^{\mathrm{op}} \times \mathbf{vect} \to \mathbf{Set}.$$

The same construction as in [199, § 3.8] turns $\mathrm{Ex}(A, I)$ into a $\mathbf{k}$-vector space, and one may check that we obtain a functor

$$\mathrm{Ex}(-, -) \colon \quad \mathbf{Art}_\Lambda^{\mathrm{op}} \times \mathbf{vect} \to \mathbf{Vect},$$

i.e., that f_* and π^* are $\mathbf{k}$-linear. Furthermore, the induced map

$$\mathrm{Hom}(I, J) \to \mathrm{Hom}(\mathrm{Ex}(A, I), \mathrm{Ex}(A, J)), \quad f \mapsto f_*,$$

is $\mathbf{k}$-linear.

Theorem A.12 *Let $F \colon \mathbf{Art}_\Lambda \to \mathbf{Set}$ be a neat deformation functor. Then F admits a universal obstruction theory (O_F, ob), i.e., an obstruction theory such that for any obstruction theory (V, v), there is a unique morphism of obstruction theories $(O_F, \mathrm{ob}) \to (V, v)$. The universal obstruction theory (O_F, ob) is complete.*

Proof As in [199, § 3.8], for every $\alpha \in F(A)$, the subset $F(A, I, \alpha) \subseteq \mathrm{Ex}(A, I)$ of first-order extensions such that there is some $\beta \in F(B)$ with $F(\varphi)(\beta) = \alpha$ is a $\mathbf{k}$-vector subspace. Thus, we can form the quotient

$$E_F(A, \alpha) := \mathrm{Ex}(A, \mathbf{k})/F(A, \mathbf{k}, \alpha).$$

For any $\pi \colon C \to A$ and $\gamma \in F(C)$ with $F(\pi)(\gamma) = \alpha$, we obtain a well-defined and injective $\mathbf{k}$-linear map $\pi^* \colon E_F(A, \alpha) \to E_F(C, \gamma)$. The proof of [199, Lemma 3.8.4] carries over to our setting and shows that $\pi_1^* = \pi_2^*$ whenever we have two maps $\pi_1, \pi_2 \colon C \to A$ in $\mathbf{Art}_\Lambda$ with $F(\pi_1)(\gamma) = \alpha = F(\pi_2)(\gamma)$. This is the step where we use (H_2^+). Now we obtain a diagram of $\mathbf{k}$-vector spaces indexed by pairs (A, α) with $A \in \mathbf{Art}_\Lambda$ and $\alpha \in F(A)$. There is a single homomorphism

$E_F(A, \alpha) \to E_F(C, \gamma)$ if and only if there is a $\pi : C \to A$ as before, and no homomorphism otherwise. We define O_F as the colimit of this diagram.

Let $e \in \mathbf{Ex}(A, I)$ and $\alpha \in F(A)$. When $w_1, \ldots, w_n$ is a basis of I, then we set

$$\mathrm{ob}_e(\alpha) := \sum_i (w_i^\vee)_*(e) \otimes w_i \in E_F(A, \alpha) \otimes_{\mathbf{k}} I,$$

where $w_i^\vee : I \to \mathbf{k}$ is the $\mathbf{k}$-vector space homomorphism characterized by $w_i^\vee(w_j) = \delta_{ij}$ with $\delta_{ij} = 1$ if and only if $i = j$ and $\delta_{ij} = 0$ otherwise. As in [199, § 3.8], one may check that this is independent of the basis and defines an obstruction theory (O_F, ob). The proof of universality is similar to the one in [199, § 3.8].

Since [199, Lemma 3.8.3] carries over to our setting, we see that (O_F, ob) is complete. $\qquad\square$

Appendix B
From the Fibers to the Total Space

Let $f\colon X = \operatorname{Spec} R \to \operatorname{Spec} \Lambda = S$ be a flat family, let M be an R-module, and let $\mathcal{F}$ be the associated coherent sheaf on X. In this chapter, we collect technical results about the injectivity and bijectivity of the restriction maps from larger to smaller open subsets, and we compare this between $\mathcal{F}$ and its restrictions to the fibers X_s of $f\colon X \to S$.

Lemma B.1 *Assume that $\mathcal{F}_s = 0$ on X_s for all points $s \in S$. Then $\mathcal{F} = 0$. If $f\colon X \to S$ maps closed points to closed points (for example if S is Jacobson), then it is sufficient to check closed points $s \in S$.*

Proof Let $x \in X$ be a point with image $s = f(x) \in S$. Then we have a factorization

$$x = \operatorname{Spec} \kappa(x) \xrightarrow{i_{x,s}} X_s \xrightarrow{i_s} X$$

of $i_x\colon x \to X$. Thus, we have $i_x^* \mathcal{F} = i_{x,s}^* \mathcal{F}_s = 0$, so $\mathcal{F}_x/(\mathfrak{m}_x \cdot \mathcal{F}_x) = 0$. Since $\mathcal{F}_x$ is a finitely generated $\mathcal{O}_{X,x}$-module, Nakayama's lemma implies $\mathcal{F}_x = 0$. Then $\mathcal{F} = 0$ follows from [16, Prop. 3.8]. The second claim is then clear. $\qquad\square$

Remark B.2 If $\mathcal{F}$ is not finitely generated, the statement is false, at least the part about the closed points. Let $\Lambda = R = \mathbb{C}[x]$, and let M be the field of fractions of $\mathbb{C}[x]$. Then $i_x^* M = 0$ for all closed points $x \in X = \mathbb{A}^1$. $\qquad\Diamond$

Lemma B.3 *Let $m \in M = \Gamma(X, \mathcal{F})$, and assume that $m|_{X_A} = 0$ for all fibers $X_A \to \operatorname{Spec} A$ over Artinian local rings $A = \mathcal{O}_{S,s}/\mathfrak{m}_s^{k+1}$ for $k \geq 0$ and points $s \in S$. Then $m = 0$.*

Proof Assume that $m \neq 0$. Then there is a point $x \in X$ with $m \neq 0$ in $\mathcal{F}_x$. Since $\mathcal{F}_x$ is a finitely generated $\mathcal{O}_{X,x}$-module and $\mathcal{O}_{X,x}$ is a Noetherian local ring, the Krull intersection theorem implies that we can find $k + 1 \geq 1$ such that $m \notin \mathfrak{m}_x^{k+1} \cdot \mathcal{F}_x$. Let $B = \mathcal{O}_{X,x}/\mathfrak{m}_x^{k+1}$; then $m \neq 0$ in $\mathcal{F} \otimes_R B$. Since the inclusion $\operatorname{Spec} B \to X$ factors through X_A with $A = \mathcal{O}_{S,f(x)}/\mathfrak{m}_{f(x)}^{k+1}$, this is a contradiction to our assumption. $\qquad\square$

© The Author(s), under exclusive license to Springer Nature Switzerland AG 2025
S. Felten, *Global Logarithmic Deformation Theory*, Lecture Notes
in Mathematics 2373, https://doi.org/10.1007/978-3-031-98751-9

Let $Z \subseteq X$ be a closed subset. Following [131, Ch. IV, §5.9], a coherent sheaf $\mathcal{F}$ on X is *Z-pure* if $H^0(X, \mathcal{F}) \rightarrow H^0(X \setminus Z, \mathcal{F})$ is injective, and *Z-closed* if $H^0(X, \mathcal{F}) \rightarrow H^0(X \setminus Z, \mathcal{F})$ is bijective. By Grothendieck [131, Ch. IV, Prop. 5.10.2], a coherent sheaf $\mathcal{F}$ is Z-pure if and only if $\mathrm{depth}_{O_{X,z}}(\mathcal{F}_z) \geq 1$ for every $z \in Z$, and Z-closed if and only if $\mathrm{depth}_{O_{X,z}}(\mathcal{F}_z) \geq 2$ for every $z \in Z$. In demonstrating how to go from the fibers to the total space with these properties, we use that

$$\mathrm{depth}_B(N) = \mathrm{depth}_A(A) + \mathrm{depth}_{B \otimes_A k}(N \otimes_A k)$$

for a local homomorphism $A \rightarrow B$ of Noetherian local rings, the residue field k of A, and a finitely generated B-module N which is flat over A. This statement is a special case of [131, Ch. IV, Prop. 6.3.1].

Lemma B.4 *Let $\mathcal{F}$ be flat over S. Let $U \subseteq X$ be an open subset, and assume that $H^0(X_s, \mathcal{F}_s) \rightarrow H^0(U_s, \mathcal{F}_s)$ is injective for all points $s \in S$. Then $H^0(X, \mathcal{F}) \rightarrow H^0(U, \mathcal{F})$ is injective.*

Proof Let $Z := X \setminus U$. We have to show $\mathrm{depth}_{O_{X,z}}(\mathcal{F}_z) \geq 1$ for all $z \in Z$. For a point $z \in Z$, let $s := f(z)$, and let X_s be the fiber over s. Then we have $z \in X_s$ and $O_{X_s,z} = O_{X,z} \otimes_{O_{S,s}} \kappa(s)$. By assumption, we have $\mathrm{depth}_{O_{X_s,z}}((\mathcal{F}_s)_z) \geq 1$. Since $\mathrm{depth}(O_{S,s}) \geq 0$, we find $\mathrm{depth}_{O_{X,z}}(\mathcal{F}_z) \geq 1$. $\qquad\square$

Lemma B.5 *Let $\mathcal{F}$ be flat over S. Let $U \subseteq X$ be an open subset, and assume that $H^0(X_s, \mathcal{F}_s) \rightarrow H^0(U_s, \mathcal{F}_s)$ is bijective for all points $s \in S$. Then $H^0(X, \mathcal{F}) \rightarrow H^0(U, \mathcal{F})$ is bijective.*

Proof The proof is the same as for the previous lemma, with 1 replaced by 2. $\qquad\square$

Appendix C
Multilinear Differential Operators

We review the theory of multilinear differential operators in algebraic geometry. In particular, we construct universal multilinear differential operators.

Let $f\colon X \to S$ be a morphism of schemes, and let $\mathcal{F}$ and $\mathcal{G}$ be quasi-coherent O_X-modules. Following [267, 0G3P] or [133, §16.8], a *differential operator of order k* is, inductively, an $f^{-1}(O_S)$-linear map of sheaves $D\colon \mathcal{F} \to \mathcal{G}$ such that $D_r(f) := D(rf) - rD(f)\colon \mathcal{F} \to \mathcal{G}$ is a differential operator of order $k-1$ for every $r \in O_X$. For $k = 0$, a differential operator of order 0 is, by definition, nothing but an O_X-linear homomorphism. Here, we give a generalization for multilinear operators

$$\mu\colon \quad \mathcal{F}_1 \times \ldots \times \mathcal{F}_n \to \mathcal{G},$$

the cases of most interest for us being the Lie bracket $[-, -]$ and the Lie derivative $\mathcal{L}$ in a Gerstenhaber calculus. Due to lack of reference, we have to develop everything from scratch, although the theory is very similar to the classical case $n = 1$. The theory works in great generality, without any Noetherianity or other finiteness assumptions. It comes in two very similar flavors, one for a total degree $k \geq 0$, and one for a tuple $(k_1, \ldots, k_n)$ as degree.

C.1 Multilinear Differential Operators in Commutative Algebra

Let $\phi\colon A \to R$ be a ring homomorphism, and let $F_1, \ldots, F_n$ as well as G be R-modules.

Definition C.1 Let

$$\mu\colon \quad F_1 \times \ldots \times F_n \to G$$

© The Author(s), under exclusive license to Springer Nature Switzerland AG 2025

S. Felten, *Global Logarithmic Deformation Theory*, Lecture Notes in Mathematics 2373, https://doi.org/10.1007/978-3-031-98751-9

be an A-multilinear map. Then μ is a *multilinear differential operator of total order* 0 if it is R-multilinear. For $1 \leq \ell \leq n$ and $r \in R$, we set $L_r : G \to G$, $g \mapsto rg$, and

$$L_{r;\ell} : \quad F_1 \times \ldots \times F_n \to F_1 \times \ldots \times F_n, \quad (f_1, \ldots, f_n) \mapsto (f_1, \ldots, rf_\ell, \ldots, f_n).$$

For $r \in R$ and $1 \leq \ell \leq n$, we define

$$\Psi_{r;\ell}(\mu) := \mu \circ L_{r;\ell} - L_r \circ \mu : \quad F_1 \times \ldots \times F_n \to G,$$

$$(f_1, \ldots, f_n) \mapsto \mu(f_1, \ldots, rf_\ell, \ldots, f_n) - r\mu(f_1, \ldots, f_n).$$

These operations commute, i.e., $\Psi_{r;\ell} \circ \Psi_{s;m}(\mu) = \Psi_{s;m} \circ \Psi_{r;\ell}(\mu)$. Then, for $k \geq 1$, we say that μ is a *multilinear differential operator of total order k* if $\Psi_{r;\ell}(\mu)$ is a multilinear differential operator of total order $k - 1$ for every $r \in R$ and every $1 \leq \ell \leq n$. We denote the set of multilinear differential operators of order k by $\mathrm{Diff}^k_{R/A}(F_1, \ldots, F_n; G)$. We consider it as an R-module via $r \cdot \mu := L_r \circ \mu$.

Similarly, we say that μ is a *multilinear differential operator of order* $(0, \ldots, 0)$ if it is R-linear. Then, for a tuple $\underline{k} = (k_1, \ldots, k_n)$ with $k_\ell \geq 0$ for all $1 \leq \ell \leq n$, we say that μ is a *multilinear differential operator of order $\underline{k}$* if $\Psi_{r;\ell}(\mu)$ is a multilinear differential operator of order $(k_1, \ldots, k_\ell - 1, \ldots, k_n)$ for all $r \in R$ and all $1 \leq \ell \leq n$ with $k_\ell \geq 1$, and if it is R-linear in the ℓ-th entry for ℓ with $k_\ell = 0$. We denote the set of multilinear differential operators of order $\underline{k}$ by $\mathrm{Diff}^{\underline{k}}_{R/A}(F_1, \ldots, F_n; G)$. It is an R-module via $r \cdot \mu := L_r \circ \mu$.

Multilinear differential operators of total order k and order $\underline{k}$ are closely related. If μ is a multilinear differential operator of total order k, then it is also of order $(k, \ldots, k)$; if μ is of order $\underline{k} = (k_1, \ldots, k_n)$, then it is also of total order $|\underline{k}| := \sum_{\ell=1}^n k_\ell$. In the following, we work out the theory for total order k, and then we state without proof the variant for order $\underline{k}$.

When

$$\nu : \quad G_1 \times \ldots \times G_n \to H$$

is a multilinear differential operator of total order k_0, and

$$\mu_i : \quad F_{i1} \times \ldots \times F_{im_i} \to G_i$$

are multilinear differential operators of total order k_i, then

$$\nu \circ (\mu_1, \ldots, \mu_n) : \quad F_{11} \times \ldots \times F_{nm_n} \to H$$

is a multilinear differential operator of total order $k = k_0 + \sum_{i=1}^n k_i$. Namely, this is true for $k_0 = k_i = 0$, and we have

$$\Psi_{r;i\ell}(\nu \circ (\mu_1, \ldots, \mu_n)) = \Psi_{r;i}(\nu) \circ (\mu_1, \ldots, \mu_n) + \nu \circ (\mu_1, \ldots, \Psi_{r;\ell}(\mu_i), \ldots, \mu_n).$$

Here, $\Psi_{r;i\ell}$ is applied to the $F_{i\ell}$-entry. We call $\nu \circ (\mu_1, \ldots, \mu_n)$ the *multi-composition*. Multi-composition yields an A-multilinear map

$$\Gamma_{R/A}: \ \mathrm{Diff}^{k_0}_{R/A}(G_1, \ldots, G_n; H) \times \mathrm{Diff}^{k_1}_{R/A}(F_{11}, \ldots, F_{1m_1}; G_1) \times \ldots$$

$$\ldots \times \mathrm{Diff}^{k_n}_{R/A}(F_{n1}, \ldots, F_{nm_n}; G_n) \to \mathrm{Diff}^{k_0+k_1+\ldots+k_n}_{R/A}(F_{11}, \ldots, F_{nm_n}; H).$$

C.1.1 An Analog of the Module of Principal Parts

Let $\mu \in \mathrm{Diff}^k_{R/A}(F_1, \ldots, F_n; G)$. Then for all sequences $r_1, \ldots, r_k \in R$ and $\ell_1, \ldots, \ell_k \in \{1, \ldots, n\}$, we have

$$\sum_{I \subseteq \{1,\ldots,k\}} \prod_{i \notin I}(-r_i)\mu\left(f_1 \cdot \prod_{i \in I, \ell_i=1} r_i, \ldots, f_n \cdot \prod_{i \in I, \ell_i=n} r_i\right) = 0 \qquad \text{(C.1)}$$

for all $f_1 \in F_1, \ldots, f_n \in F_n$, where the sum is over all subsets $I \subseteq \{1, \ldots, k\}$. Conversely, if an A-multilinear map

$$\mu: \ F_1 \times \ldots \times F_n \to G$$

satisfies (C.1) for all $r_1, \ldots, r_k$ and $\ell_1, \ldots, \ell_k$, then $\mu \in \mathrm{Diff}^k_{R/A}(F_1, \ldots, F_n; G)$.

Our first aim now is to construct an analog of the module of principal parts. Let us consider the free R-module

$$\mathrm{Free}_R(F_1, \ldots, F_n) := \bigoplus_{(f_1,\ldots,f_n)\in F_1\times\ldots\times F_n} R \cdot [f_1, \ldots, f_n]$$

on generators $[f_1, \ldots, f_n]$; inside it, let $\mathrm{I}^k_{R/A}(F_1, \ldots, F_n)$ be the R-submodule generated by

$$[f_1, \ldots, f_\ell + f'_\ell, \ldots, f_n] - [f_1, \ldots, f_\ell, \ldots, f_n] - [f_1, \ldots, f'_\ell, \ldots, f_n]$$

and

$$[f_1, \ldots, a \cdot f_\ell, \ldots, f_n] - a \cdot [f_1, \ldots, f_\ell, \ldots, f_n]$$

for $a \in A$ as well as

$$\sum_{I \subseteq \{0,\ldots,k\}} \prod_{i \notin I}(-r_i)\left[f_1 \cdot \prod_{i \in I, \ell_i=1} r_i, \ldots, f_n \cdot \prod_{i \in I, \ell_i=n} r_i\right] \qquad \text{(HPF)}$$

for $r_0, \ldots, r_k \in R$, $1 \leq \ell_0, \ldots, \ell_k \leq n$, and $f_1 \in F_1, \ldots, f_n \in F_n$. Here, (HPF) stands for "higher product formula." Let

$$P^k_{R/A}(F_1, \ldots, F_n) := \mathrm{Free}_R(F_1, \ldots, F_n)/I^k_{R/A}(F_1, \ldots, F_n)$$

be the quotient.

Lemma C.2 *The map*

$$\mu^u_{R/A}: \quad F_1 \times \ldots \times F_n \to P^k_{R/A}(F_1, \ldots, F_n), \quad (f_1, \ldots, f_n) \mapsto [f_1, \ldots, f_n],$$

is an A-multilinear differential operator of total order k. It is universal in the sense that every A-multilinear differential operator of total order k

$$\mu: \quad F_1 \times \ldots \times F_n \to G$$

factors uniquely through an R-linear homomorphism

$$h: \quad P^k_{R/A}(F_1, \ldots, F_n) \to G$$

as $\mu = h \circ \mu^u_{R/A}$. In particular,

$$\mathrm{Diff}^k_{R/A}(F_1, \ldots, F_n; G) = \mathrm{Hom}_R(P^k_{R/A}(F_1, \ldots, F_n), G).$$

Proof μ_u is an A-multilinear differential operator of total order k because all required relations are satisfied by construction. These relations are preserved when postcomposing with R-linear homomorphisms so that $h \circ \mu^u_{R/A}$ is always an A-multilinear differential operator of total order k. This defines a map

$$\mathrm{Hom}_R(P^k_{R/A}(F_1, \ldots, F_n), G) \to \mathrm{Diff}^k_{R/A}(F_1, \ldots, F_n; G).$$

When μ is an element on the right, then we can set $h_\mu[f_1, \ldots, f_n] := \mu(f_1, \ldots, f_n)$; due to the relations for μ, this is well-defined, and we have $h_\mu \circ \mu^u_{R/A} = \mu$. If $h \circ \mu^u_{R/A} = h' \circ \mu^u_{R/A}$, then $h[f_1, \ldots, f_n] = h'[f_1, \ldots, f_n]$, so $h = h'$ since this is an R-generating system of the module $P^k_{R/A}(F_1, \ldots, F_n)$.   □

Variant C.3 *There is a universal A-multilinear differential operator of order $\underline{k}$*

$$\mu^u_{R/A}: \quad F_1 \times \ldots \times F_n \to P^{\underline{k}}_{R/A}(F_1, \ldots, F_n), \quad (f_1, \ldots, f_n) \mapsto [f_1, \ldots, f_n].$$

In particular, $\mathrm{Diff}^{\underline{k}}_{R/A}(F_1, \ldots, F_n; G) = \mathrm{Hom}_R(P^{\underline{k}}_{R/A}(F_1, \ldots, F_n), G).$

Next, we exhibit a canonical generating system for $P_{R/A}^k(F_1, \ldots, F_n)$; this allows us to conclude that $P_{R/A}^k(F_1, \ldots, F_n)$ is finitely generated in good cases. Let $X \subset R$ be a generating system of R as an A-*algebra*, and let

$$M(X) := \left\{ \prod_{i=1}^{\ell} x_i \;\middle|\; x_1, \ldots, x_\ell \in X \right\} \cup \{1\}$$

be the set of monomial expressions in X. If E_i is a system of generators of F_i as an R-module, then $P_{R/A}^k(F_1, \ldots, F_n)$ will be generated, as an R-module, by elements $[p_1 e_1, \ldots, p_n e_n]$ with $e_i \in E_i$ and $p_i \in M(X)$. However, a subset of these generators is already sufficient:

Lemma C.4 *Let $M^n(k; X)$ be the set of vectors $(p_1, \ldots, p_n) \in M(X)^n$ with $\sum \deg(p_i) \le k$. Then*

$$P_{R/A}^k(F_1, \ldots, F_n)$$

is generated, as an R-module, by $[p_1 e_1, \ldots, p_n e_n]$ with $(p_1, \ldots, p_n) \in M^n(k; X)$ and $e_i \in E_i$. In particular, if $\phi \colon A \to R$ is of finite type, and each F_i is a finitely generated R-module, then $P_{R/A}^k(F_1, \ldots, F_n)$ is a finitely generated R-module as well.

Proof For $k = 0$, we have $M(0; X) = \{1\}$ and $P_{R/A}^0(F_1, \ldots, F_n) = F_1 \otimes_R \ldots \otimes_R F_n$, so the statement is true. For $k \ge 1$, let $G(k) \subseteq P_{R/A}^k(F_1, \ldots, F_n)$ be the submodule generated by the specified elements. Then, by definition, $[p_1 e_1, \ldots, p_n e_n] \in G(k)$ for $(p_i)_i \in M^n(k; X)$ and $e_i \in E_i$. We show by induction that, for all $\ell \ge k$, we have $[p_1 e_1, \ldots, p_n e_n] \in G(k)$ for $(p_i)_i \in M(\ell; X)$ and $e_i \in E_i$. We can assume that $\prod_{i=1}^n p_i$ has degree $\ge k + 1$; then we can find $(p_i')_i \in M^n(\ell - k - 1; X)$ and monomials q_i such that $\prod_{i=1}^n q_i$ has degree $k + 1$ and $p_i' q_i = p_i$. By using formula (HPF), we can express $[p_1 e_1, \ldots, p_n e_n]$ in terms of the form $Q \cdot [P_1 e_1, \ldots, P_n e_n]$ with $Q \in M(X)$ and $(P_i)_i \in M(\ell - 1; X)$. By induction hypothesis, these terms are contained in $G(k)$, so $[p_1 e_1, \ldots, p_n e_n] \in G(k)$ as well. Since $P_{R/A}^k(F_1, \ldots, F_n)$ is generated, as an R-module, by $[p_1 e_1, \ldots, p_n e_n]$ with $p_i \in M(X)$ and $e_i \in E_i$ for all k, our claim follows. $\qquad\square$

Variant C.5 *Let $M^n(\underline{k}; X)$ be the set of vectors $(p_1, \ldots, p_n) \in M(X)^n$ with $\deg(p_\ell) \le k_\ell$. Then*

$$P_{R/A}^k(F_1, \ldots, F_n)$$

is generated, as an R-module, by $[p_1 e_1, \ldots, p_n e_n]$ with $(p_1, \ldots, p_n) \in M^n(\underline{k}; X)$ and $e_i \in E_i$. In particular, if $\phi \colon A \to R$ is of finite type, and each F_i is a finitely generated R-module, then $P_{R/A}^k(F_1, \ldots, F_n)$ is a finitely generated R-module as well.

C.1.2 Localization on R

In order to study the behavior of multilinear differential operators on sheaves, we study localization, so let $S \subset R$ be a multiplicative system. We start with a rather strong uniqueness result of extensions from R to $S^{-1}R$, which is enforced by (HPF).

Lemma C.6 *Let $F_1, \ldots, F_n$ be R-modules, and let G_S be an $S^{-1}R$-module. Let*

$$\mu, \nu: \quad S^{-1}F_1 \times \ldots \times S^{-1}F_n \to G_S$$

be two A-multilinear differential operators of total order k, and assume that

$$\mu(f_1, \ldots, f_n) = \nu(f_1, \ldots, f_n)$$

for $f_i \in F_i$. Then $\mu = \nu$.

Proof We prove the statement by induction on k. For $k = 0$, the statement follows from $S^{-1}R$-linearity. So assume that we know the statement for some k, and let μ, ν be A-multilinear differential operators of total order $k + 1$ which satisfy $\mu(f_1, \ldots, f_n) = \nu(f_1, \ldots, f_n)$ for $f_i \in F_i$. For $1 \leq \ell \leq n$ and $s \in S$, both $\Psi_{s;\ell}(\mu)$ and $\Psi_{s;\ell}(\nu)$ are A-multilinear differential operators of total order k; by induction hypothesis, we have $\Psi_{s;\ell}(\mu) = \Psi_{s;\ell}(\nu)$ since they are equal for $f_i \in F_i$. For $f_i \in F_i$ and $s_1 \in S$, we have

$$
\begin{aligned}
\mu(f_1/s_1, f_2, \ldots, f_n) &= s_1^{-1} \cdot (\mu(f_1, f_2, \ldots, f_n) - \Psi_{s;\ell}(\mu)(f_1/s_1, f_2, \ldots, f_n)) \\
&= s_1^{-1} \cdot (\nu(f_1, f_2, \ldots, f_n) - \Psi_{s;\ell}(\nu)(f_1/s_1, f_2, \ldots, f_n)) \\
&= \nu(f_1/s_1, f_2, \ldots, f_n).
\end{aligned}
$$

Repeating this argument inductively over ℓ, we show that

$$\mu(f_1/s_1, \ldots, f_\ell/s_\ell, f_{\ell+1}, \ldots, f_n) = \nu(f_1/s_1, \ldots, f_\ell/s_\ell, f_{\ell+1}, \ldots, f_n)$$

for all ℓ and hence $\mu = \nu$. $\qquad\square$

If $\mu \in \mathrm{Diff}^k_{R/A}(F_1, \ldots, F_n; G)$ is given, then an extension over $S^{-1}R$ is not only unique if it exists but actually exists. This is the computationally most intricate part of the basic theory of multilinear differential operators. Our proof closely follows [267, 0G36], which is the same statement for $n = 1$.

Lemma C.7 *Let $S \subset R$ be a multiplicative system. Then, for every $k \geq 0$, there is a unique map*

$$\rho_k: \quad \mathrm{Diff}^k_{R/A}(F_1, \ldots, F_n; G) \to \mathrm{Diff}^k_{S^{-1}R/A}(S^{-1}F_1, \ldots, S^{-1}F_n; S^{-1}G)$$

such that

$$\begin{array}{ccc}
F_1 \times \ldots \times F_n & \xrightarrow{\ \ \mu\ \ } & G \\
\downarrow & & \downarrow \\
S^{-1}F_1 \times \ldots \times S^{-1}F_n & \xrightarrow{\ \rho_k(\mu)\ } & S^{-1}G
\end{array}$$

commutes. The map is compatible among all k in the sense that $\rho_k(\mu) = \rho_{k+1}(\mu)$ if both are defined. The map is compatible with the multiplication operators in that, for $r \in R$ and $1 \le \ell \le n$, we have $L_r \circ \rho_k(\mu) = \rho_k(L_r \circ \mu)$ and $\rho_k(\mu) \circ L_{r;\ell} = \rho_k(\mu \circ L_{r;\ell})$; in particular, we have $\rho_{k-1}(\Psi_{r;\ell}(\mu)) = \Psi_{r;\ell}(\rho_k(\mu))$.

Proof In this proof, we use the shorter notation $\mu_{r;\ell} := \Psi_{r;\ell}(\mu)$. The uniqueness follows from Lemma C.6. We show existence by induction on k. For $k = 0$, we obtain $\rho_0(\mu)$ as the localization $S^{-1}\mu$. For the induction step, we construct $\hat{\mu} = \rho_k(\mu)$ in several steps. Let $\hat{F}_i$ be the image of F_i in $S^{-1}F_i$. First, we define

$$\hat{\mu}_0 \colon \ \hat{F}_1 \times \ldots \times \hat{F}_n \to S^{-1}G \quad \text{via} \quad \hat{\mu}_0(f_1, \ldots, f_n) := \mu(f_1, \ldots, f_n) \in S^{-1}G.$$

We have to show that this is well-defined. So assume that $f_\ell = f'_\ell \in S^{-1}F_\ell$, i.e., $sf_\ell = sf'_\ell$ for some $s \in S$. Then we have

$$\begin{aligned}
ss'\mu(f_1, \ldots, f_\ell, \ldots, f_n) &= s'\mu(f_1, \ldots, sf_\ell, \ldots, f_n) - s'\mu_{s;\ell}(f_1, \ldots, f_\ell, \ldots, f_n) \\
&= s'\mu(f_1, \ldots, sf'_\ell, \ldots, f_n) - s'\mu_{s;\ell}(f_1, \ldots, f'_\ell, \ldots, f_n) \\
&= ss'\mu(f_1, \ldots, f'_\ell, \ldots, f_n)
\end{aligned}$$

for some $s' \in S$ with $s'\mu_{s;\ell}(f_1, \ldots, f_\ell, \ldots, f_n) = s'\mu_{s;\ell}(f_1, \ldots, f'_\ell, \ldots, f_n)$, which exists by induction hypothesis. Thus, $\hat{\mu}_0(f_1, \ldots, f_n)$ is well-defined. It is easy to see that $\hat{\mu}_0$ is A-multilinear.

For the next step, let $\hat{\mu}_{r;1} := \rho_{k-1}(\mu_{r;1})$ for $r \in R$, which exists by the induction hypothesis. Note that

$$\hat{\mu}_{r;1}(f_1, \ldots, f_n) = \hat{\mu}_0(rf_1, \ldots, f_n) - r\hat{\mu}_0(f_1, \ldots, f_n)$$

for $f_i \in \hat{F}_i$. Then we define a map

$$\hat{\mu}_1 \colon \ S^{-1}F_1 \times \hat{F}_2 \times \ldots \times \hat{F}_n \to S^{-1}G$$

with the formula

$$\hat{\mu}_1(f_1/s_1, f_2, \ldots, f_n) := s_1^{-1} \cdot (\hat{\mu}_0(f_1, f_2, \ldots, f_n) - \hat{\mu}_{s_1;1}(f_1/s_1, f_2, \ldots, f_n)).$$

To show that it is well-defined, assume $f_1/s_1 = f_1'/s_1'$. Without loss of generality, we can assume that $f_1' = f_1 t$ and $s_1' = s_1 t$ for some $t \in S$. We have

$$\mu_{s_1 t;1} \circ L_{s_1;1} - L_{s_1} \circ \mu_{s_1 t;1} = \mu_{s_1;1} \circ L_{s_1 t;1} - L_{s_1 t} \circ \mu_{s_1;1}$$

as operators in $\mathrm{Diff}_{R/A}^{k-1}(F_1, \ldots, F_n; G)$ by direct computation. Applying ρ_{k-1}, the same equation holds for μ replaced with $\hat{\mu}$. We also have

$$s_1 t \hat{\mu}_0(f_1, \ldots, f_n) - s_1 \hat{\mu}_0(t f_1, \ldots, f_n) = \hat{\mu}_{s_1;1}(t f_1, \ldots, f_n) - \hat{\mu}_{s_1 t;1}(f_1, \ldots, f_n)$$

by direct computation. From this, we obtain

$$\hat{\mu}_1(f_1/s_1, f_2, \ldots, f_n) = \hat{\mu}_1(f_1'/s_1', f_2, \ldots, f_n)$$

with the above definition applied to either side, and hence $\hat{\mu}_1(f_1/s_1, f_2, \ldots, f_n)$ is well-defined in $S^{-1}G$. It is easy to see that $\hat{\mu}_1$ is A-multilinear.

Now let $r \in R$. Then we have $(\mu_{r;1})_{s_1;1} = (\mu_{s_1;1})_{r;1}$; consequently, we have $(\hat{\mu}_{r;1})_{s_1;1} = (\hat{\mu}_{s_1;1})_{r;1}$. Then we obtain

$$(\hat{\mu}_1)_{r;1}(f_1/s_1, \ldots, f_n) := \hat{\mu}_1(r f_1/s_1, \ldots, f_n) - r \hat{\mu}_1(f_1/s_1, \ldots, f_n)$$

$$= \frac{1}{s_1}(\hat{\mu}_0)_{r;1}(f_1, \ldots, f_n) - \frac{1}{s_1}(\hat{\mu}_{s_1;1})_{r;1}(f_1/s_1, \ldots, f_n)$$

$$= \frac{1}{s_1}\hat{\mu}_{r;1}(f_1, \ldots, f_n) - \frac{1}{s_1}(\hat{\mu}_{r;1})_{s_1;1}(f_1/s_1, \ldots, f_n)$$

$$= \hat{\mu}_{r;1}(f_1/s_1, \ldots, f_n).$$

For $s \in S$, we have

$$(\hat{\mu}_1)_{1/s;1}(f_1/s_1, \ldots, f_n) := \hat{\mu}_1(f_1/(s_1 s), \ldots, f_n) - \frac{1}{s}\hat{\mu}_1(f_1/s_1, \ldots, f_n)$$

$$= -\frac{1}{s s_1}\hat{\mu}_{s s_1;1}(f_1/(s_1 s), \ldots, f_n)$$

$$+ \frac{1}{s s_1}\hat{\mu}_{s_1;1}(f_1/s_1, \ldots, f_n)$$

$$= -(L_{1/s} \circ \hat{\mu}_{s;1} \circ L_{1/s;1})(f_1/s_1, \ldots, f_n)$$

by applying $\hat{\mu}_{s_1 s;1} = \hat{\mu}_{s_1;1} \circ L_{s;1} + L_{s_1} \circ \hat{\mu}_{s;1}$, which holds since it holds over R, and then we apply ρ_{k-1}.

In the next step, we construct inductively a map

$$\hat{\mu}_p\colon\ S^{-1}F_1 \times \ldots \times S^{-1}F_p \times \hat{F}_{p+1} \times \ldots \times \hat{F}_n \to S^{-1}G$$

by the formula

$$\hat{\mu}_p(f_1/s_1, \ldots, f_p/s_p, \ldots, f_n)$$
$$:= s_p^{-1} \cdot (\hat{\mu}_{p-1}(f_1/s_1, \ldots, f_p, \ldots, f_n) - \hat{\mu}_{s_p;p}(f_1/s_1, \ldots, f_p/s_p, \ldots, f_n)),$$

where, as above, $\hat{\mu}_{s_p;p} := \rho_{k-1}(\mu_{s_p;p})$. Similar to the case $\hat{\mu}_1$, we obtain that $\hat{\mu}_p$ is well-defined and A-multilinear. Also similar to the case $\hat{\mu}_1$, we can show that

$$(\hat{\mu}_p)_{r;p} = \hat{\mu}_{r;p} \quad \text{and} \quad (\hat{\mu}_p)_{1/s;p} = -L_{1/s} \circ \hat{\mu}_{s;p} \circ L_{1/s;p}.$$

We wish to show these equations as well for p replaced with $1 \leq \ell \leq p - 1$, i.e.,

$$(\hat{\mu}_p)_{r;\ell} = \hat{\mu}_{r;\ell} \quad \text{and} \quad (\hat{\mu}_p)_{1/s;\ell} = -L_{1/s} \circ \hat{\mu}_{s;\ell} \circ L_{1/s;\ell},$$

where again $\hat{\mu}_{r;\ell} := \rho_{k-1}(\mu_{r;\ell})$. They are not yet shown because $\hat{\mu}_p$ has a larger domain than $\hat{\mu}_\ell$. By induction on p, we can assume that we already know these equations for $\hat{\mu}_{p-1}$. Then the claim follows from two straightforward computations.

We set $\hat{\mu} := \hat{\mu}_n$; it is a well-defined A-multilinear map which makes the diagram in the statement of the Lemma commute. For $r \in R$, we have $(\hat{\mu})_{r;\ell} = \rho_{k-1}(\mu_{r;\ell})$, so this is an A-multilinear differential operator of total order $k - 1$ by the induction hypothesis. Similarly, $(\hat{\mu})_{1/s;\ell}$ is an A-multilinear differential operator of total order $k - 1$ for $s \in S$. Then

$$(\hat{\mu})_{r/s;\ell} = (\hat{\mu})_{r;\ell} \circ L_{1/s;\ell} + L_r \circ (\hat{\mu})_{1/s;\ell}$$

is an A-multilinear differential operator of total order $k - 1$ as well. Thus, $\hat{\mu}$ is an A-multilinear differential operator of total order k. The compatibility claims all follow from Lemma C.6.   □

Remark C.8 Localization commutes with multi-composition. Namely, the diagram in Lemma C.7 commutes with $\nu \circ (\mu_1, \ldots, \mu_n)$ and $\rho(\nu) \circ (\rho(\mu_1), \ldots, \rho(\mu_n))$, so uniqueness shows that

$$\rho(\nu \circ (\mu_1, \ldots, \mu_n)) = \rho(\nu) \circ (\rho(\mu_1), \ldots, \rho(\mu_n)). \qquad \diamond$$

Variant C.9 *In the situation of Lemma C.7, if μ is a multilinear differential operator of order $\underline{k} = (k_1, \ldots, k_n)$, then $\rho_k(\mu)$ is a multilinear differential operator of order k, where $k = \sum_{\ell=1}^n k_\ell$.*

With this preparation, we can show that the formation of $\mathrm{P}^k_{R/A}(F_1, \ldots, F_n)$ commutes with localization.

Lemma C.10 *We have a canonical isomorphism*

$$\Phi_k: \quad S^{-1}\mathrm{P}^k_{R/A}(F_1, \ldots, F_n) \to \mathrm{P}^k_{S^{-1}R/A}(S^{-1}F_1, \ldots, S^{-1}F_n)$$

of $S^{-1}R$-modules.

Proof First note that $S^{-1}\mathrm{Free}_R(F_1, \ldots, F_n) = \mathrm{Free}_{S^{-1}R}(F_1, \ldots, F_n)$. Then we get an $S^{-1}R$-homomorphism

$$\tilde{\Phi}_k: \quad \mathrm{Free}_{S^{-1}R}(F_1, \ldots, F_n) \to \mathrm{P}^k_{S^{-1}R/A}(S^{-1}F_1, \ldots, S^{-1}F_n)$$

by $\tilde{\Phi}_k[f_1, \ldots, f_n] := [f_1, \ldots, f_n]$. It descends to a map Φ_k from $S^{-1}\mathrm{P}^k_{R/A}(F_1, \ldots, F_n)$ since every relation in $S^{-1}\mathrm{I}^k_{R/A}(F_1, \ldots, F_n)$ is also in $\mathrm{I}^k_{S^{-1}R/A}(S^{-1}F_1, \ldots, S^{-1}F_n)$.

Next, Lemma C.7 gives us an A-multilinear differential operator

$$v_k := \rho_k(\mu^u_{R/A}): \quad S^{-1}F_1 \times \ldots \times S^{-1}F_n \to S^{-1}\mathrm{P}^k_{R/A}(F_1, \ldots, F_n)$$

of total order k. Thus, we have a unique $S^{-1}R$-homomorphism

$$h_k: \quad \mathrm{P}^k_{S^{-1}R/A}(S^{-1}F_1, \ldots, S^{-1}F_n) \to S^{-1}\mathrm{P}^k_{R/A}(F_1, \ldots, F_n)$$

with $h_k \circ \mu^u_{S^{-1}R/A} = v_k$. On the other hand, we also have $\Phi_k \circ v_k = \mu^u_{S^{-1}R/A}$ by Lemma C.6 since they are equal on $F_1 \times \ldots \times F_n$. Then the universal property of $\mu^u_{S^{-1}R/A}$ shows that $\Phi_k \circ h_k = \mathrm{Id}$; in particular, h_k is injective. Now $S^{-1}\mathrm{P}^k_{R/A}(F_1, \ldots, F_n)$ is, as an $S^{-1}R$-module, generated by $v_k(f_1, \ldots, f_k)$ for $f_1 \in F_1, \ldots, f_n \in F_n$. Thus, h_k is also surjective. But then Φ_k must be an isomorphism as well. $\qquad\square$

Variant C.11 *For $\underline{k} = (k_1, \ldots, k_n)$, we have a canonical isomorphism*

$$\Phi_{\underline{k}}: \quad S^{-1}\mathrm{P}^{\underline{k}}_{R/A}(F_1, \ldots, F_n) \to \mathrm{P}^{\underline{k}}_{S^{-1}R/A}(S^{-1}F_1, \ldots, S^{-1}F_n)$$

of $S^{-1}R$-modules.

C.1.3   Base Change

Let $\phi: A \to R$ and $\psi: A \to B$ be ring homomorphisms, and set $T := R \otimes_A B$. Given an A-multilinear differential operator $\mu \in \mathrm{Diff}^k_{R/A}(F_1, \ldots, F_n; G)$, we can form the A-multilinear map

$$\mu \otimes_A B: \quad F_1 \otimes_A B \times \ldots \times F_n \otimes_A B \to G \otimes_A B,$$

which is in fact a multilinear differential operator for T/B. It is clear that this construction commutes with multi-compositions. We have the analog of the well-known base change result for $\Omega^1_{R/A}$.

Lemma C.12 *In the above situation, we have a canonical isomorphism*

$$C^k_{B/A}: \quad \mathrm{P}^k_{R/A}(F_1, \ldots, F_n) \otimes_A B \to \mathrm{P}^k_{T/B}(F_1 \otimes_A B, \ldots, F_n \otimes_A B)$$

of T-modules.

Proof The proof is very similar to that of Lemma C.10. We have

$$\mathrm{Free}_R(F_1, \ldots, F_n) \otimes_A B = \mathrm{Free}_T(F_1, \ldots, F_n),$$

and then we obtain the map $C^k_{B/A}$ by preservation of relations. On the other hand, we have a B-multilinear differential operator

$$\mu^u_{R/A} \otimes_A B: \quad F_1 \otimes_A B \times \ldots \times F_n \otimes_A B \to \mathrm{P}^k_{R/A}(F_1, \ldots, F_n) \otimes_A B$$

which gives rise to a map

$$h_k: \quad \mathrm{P}^k_{T/B}(F_1 \otimes_A B, \ldots, F_n \otimes_A B) \to \mathrm{P}^k_{R/A}(F_1, \ldots, F_n) \otimes_A B$$

with $h_k \circ \mu^u_{T/B} = \mu^u_{R/A} \otimes_A B$. By the universal property of $\mu^u_{T/B}$, we have $C^k_{B/A} \circ h_k = \mathrm{Id}$, so h_k is injective. However, as a T-module, the target is generated by $(\mu^u_{R/A} \otimes_A B)(f_1, \ldots, f_n)$ for $f_i \in F_i$, so h_k is surjective as well. Then $C^k_{B/A}$ is an isomorphism. $\square$

Variant C.13 *Let $\psi: A \to B$ be a ring homomorphism, and let $T := R \otimes_A B$. Let $\underline{k} = (k_1, \ldots, k_n)$. Then we have a canonical isomorphism*

$$C^{\underline{k}}_{B/A}: \quad \mathrm{P}^{\underline{k}}_{R/A}(F_1, \ldots, F_n) \otimes_A B \to \mathrm{P}^{\underline{k}}_{T/B}(F_1 \otimes_A B, \ldots, F_n \otimes_A B)$$

of T-modules.

C.1.4 Localization on A

For the construction of a sheaf, we also need the following basic compatibility. Recall that $\mathrm{Spec}(A) \to \mathrm{Spec}(A')$ is an open immersion if and only if $A' \to A$ is flat, of finite presentation, and an epimorphism of rings. This is more general than being a localization in a multiplicative system $T \subset A'$.

Lemma C.14 *Let $\psi \colon A' \to A$ be flat, of finite presentation, and an epimorphism of rings. Then*

$$\mathrm{P}^k_{R/A'}(F_1, \ldots, F_n) = \mathrm{P}^k_{R/A}(F_1, \ldots, F_n).$$

Proof Let $\phi' := \phi \circ \psi$. Then, using [267, 04VN], we obtain a Cartesian diagram

$$
\begin{array}{ccccc}
A & \overset{\cong}{\longrightarrow} & A \otimes_{A'} A & \longrightarrow & R \otimes_{A'} A \\
\uparrow & & \cong \uparrow & & \cong \uparrow \\
A' & \longrightarrow & A & \longrightarrow & R
\end{array}
$$

of rings. Then we have

$$\mathrm{P}^k_{R/A}(F_1, \ldots, F_n) = \mathrm{P}^k_{R/A}(F_1 \otimes_{A'} A, \ldots, F_n \otimes_{A'} A) = \mathrm{P}^k_{R/A'}(F_1, \ldots, F_n) \otimes_{A'} A$$

$$= \mathrm{P}^k_{R/A'}(F_1, \ldots, F_n) \otimes_R R = \mathrm{P}^k_{R/A'}(F_1, \ldots, F_n)$$

by Lemma C.12. $\qquad\square$

Variant C.15 *In the setting of Lemma C.14, we have*

$$\mathrm{P}^{\underline{k}}_{R/A'}(F_1, \ldots, F_n) = \mathrm{P}^{\underline{k}}_{R/A}(F_1, \ldots, F_n)$$

for $\underline{k} = (k_1, \ldots, k_n)$.

C.1.5 *Étale Localization of Multilinear Differential Operators*

First, we given an alternative, more geometric description of $\mathrm{P}^k_{R/A}(F_1, \ldots, F_n)$ in the case of a total degree k. Let us consider the exact sequence

$$0 \to I_\Delta \to R \otimes_A R \otimes_A \ldots \otimes_A R \xrightarrow{r_0 \otimes \ldots \otimes r_n \mapsto r_0 \cdot \ldots \cdot r_n} R \to 0 \tag{C.2}$$

which corresponds to the diagonal embedding. The kernel I_Δ is generated by elements of the form

$$1 \otimes 1 \otimes \ldots \otimes r \otimes \ldots \otimes 1 - r \otimes 1 \otimes \ldots \otimes 1 \otimes \ldots \otimes 1.$$

In particular, for $k \geq 0$, the power I_Δ^{k+1} is generated by elements of the form

$$\sum_{I \subseteq \{0,\ldots,k\}} \left(\prod_{i \notin I} (-r_i) \right) \otimes \left(\prod_{i \in I, \ell_i = 1} r_i \right) \otimes \ldots \otimes \left(\prod_{i \in I, \ell_i = n} r_i \right)$$

for $r_0, \ldots, r_k \in R$ and $1 \leq \ell_0, \ldots, \ell_k \leq n$. Taking the tensor product of the exact sequence (C.2) over $T = R \otimes_A R \otimes_A \ldots \otimes_A R$ with $B = R \otimes_A F_1 \otimes \ldots \otimes F_n$, we obtain

$$I_\Delta^{k+1} \otimes_T B \to R \otimes_A F_1 \otimes_A \ldots \otimes_A F_n \to (T/I_\Delta^{k+1}) \otimes_T B \to 0.$$

Comparing with the definition of $P_{R/A}^k(F_1, \ldots, F_n)$, we find a canonical isomorphism

$$(T/I_\Delta^{k+1}) \otimes_T (R \otimes_A F_1 \otimes_A \ldots \otimes_A F_n) \cong P_{R/A}^k(F_1, \ldots, F_n).$$

Now let $\psi : R \to R'$ be a ring map. It is easy to see that we have a canonical induced homomorphism of R'-modules

$$R' \otimes_R P_{R/A}^k(F_1, \ldots, F_n) \to P_{R'/A}^k(R' \otimes_R F_1, \ldots, R' \otimes_R F_n),$$

$$[f_1, \ldots, f_n] \mapsto [1 \otimes f_1, \ldots, 1 \otimes f_n].$$

This map fits into a commutative diagram

$$
\begin{array}{ccccc}
F_1 \times \ldots \times F_n & \longrightarrow & P_{R/A}^k(F_1, \ldots, F_n) & \longrightarrow & R' \otimes_R P_{R/A}^k(F_1, \ldots, F_n) \\
\downarrow & & & & \downarrow \\
R' \otimes_R F_1 \times \ldots \times R' \otimes_R F_n & & \longrightarrow & & P_{R'/A}^k(R' \otimes_R F_1, \ldots, R' \otimes_R F_n)
\end{array}
$$

Lemma C.16 *Assume that the right vertical map is an isomorphism. Let G' be an R'-module, and let G be the R'-module G' considered as an R-module via $\psi : R \to R'$. Then, for every $\mu \in \mathrm{Diff}_{R/A}^k(F_1, \ldots, F_n; G)$, there is a unique $\nu \in \mathrm{Diff}_{R'/A}^k(R' \otimes_R F_1, \ldots, R' \otimes_R F_n; G')$ such that*

$$
\begin{array}{ccc}
F_1 \times \ldots \times F_n & \xrightarrow{\;\;\mu\;\;} & G \\
\downarrow & & \| \\
R' \otimes_R F_1 \times \ldots \times R' \otimes_R F_n & \xrightarrow{\;\;\nu\;\;} & G'
\end{array}
$$

commutes.

Proof This follows from

$$\mathrm{Hom}_R(E, G) = \mathrm{Hom}_{R'}(R' \otimes_R E, G')$$

for $E = \mathrm{P}^k_{R/A}(F_1, \ldots, F_n)$. $\square$

Now the conditions of Lemma C.16 are satisfied if $\psi: R \to R'$ is étale.

Proposition C.17 *Assume that* $\psi: R \to R'$ *is étale, in particular of finite presentation. Then*

$$R' \otimes_R \mathrm{P}^k(F_1, \ldots, F_n) \to \mathrm{P}^k_{R'/A}(R' \otimes_R F_1, \ldots, R' \otimes_R F_n)$$

is an isomorphism.

Proof We work geometrically. Let $S = \mathrm{Spec}\, A$, $X = \mathrm{Spec}\, R$, and $X' = \mathrm{Spec}\, R'$. Let $g: X' \to X$ and $f: X \to S$ be the induced maps. Let $M = X \times_S X \times_S \ldots \times_S X$, and we define M' analogously. We consider the commutative diagram

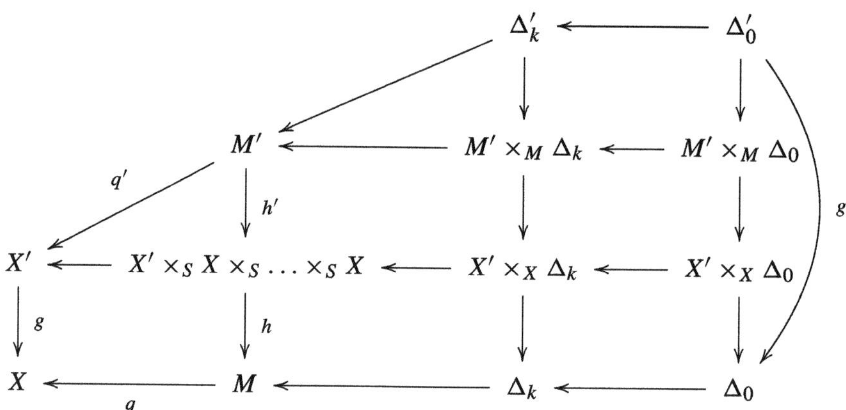

where $q: M \to X$ and $q': M' \to X'$ are the projections onto the first factors, $\Delta_0 \to M$ and $\Delta'_0 \to M'$ are the diagonal embeddings, the lower and middle squares are Cartesian, and $\Delta_k \to M$ as well as $\Delta'_k \to M'$ are closed immersions defined by $\mathcal{I}^{k+1}_\Delta$ respectively $\mathcal{I}^{k+1}_{\Delta'}$. The lower and middle vertical arrows are étale. Since g is étale as well, we find that $\Delta'_0 \to M' \times_M \Delta_0$ is étale. Since this map is also a closed immersion because $\Delta'_0 \to X'$ is an isomorphism, we find that $\Delta'_0 \to M' \times_M \Delta_0$ is the embedding of a connected component. Note that $M' \times_M \Delta_0 \to M'$ is defined by $\mathcal{O}_{M'} \cdot \mathcal{I}_\Delta$, and $\Delta'_0 \to M'$ is defined by $\mathcal{I}_{\Delta'}$. Thus, these two ideals coincide on the open subset $U' := M' \setminus (M' \times_M \Delta_0 \setminus \Delta'_0)$, and hence $\mathcal{O}_{M'} \cdot \mathcal{I}^{k+1}_\Delta$ coincides with $\mathcal{I}^{k+1}_{\Delta'}$ on U'. Thus, $\Delta'_k \to M' \times_M \Delta_k$ is also the inclusion of a connected component; in particular, this map is an étale closed immersion. Then the base change $P \to M' \times_M \Delta_0$ of this map to $M' \times_M \Delta_0$ is an étale closed immersion, and $\Delta'_0 \to P$ is étale. This map is also a closed immersion because $\Delta'_0 \to X' \times_X \Delta_0$

is an isomorphism. It is then easy to see that the upper right square is Cartesian as well. Now $\Delta'_k \to X' \times_X \Delta_k$ is a closed immersion because it is the thickening of a closed immersion, see [267, 09ZW], and then it must be an isomorphism because it is étale and a homeomorphism. In a nutshell, the outer square in

$$
\begin{array}{ccccc}
X' & \longleftarrow & M' & \longleftarrow & \Delta'_k \\
\downarrow{\scriptstyle g} & {\scriptstyle q'} & \downarrow{\scriptstyle p} & {\scriptstyle i'} & \downarrow{\scriptstyle p_k} \\
X & \longleftarrow & M & \longleftarrow & \Delta_k. \\
& {\scriptstyle q} & & {\scriptstyle i} &
\end{array}
$$

is Cartesian. After setting $B = R \otimes_A F_1 \otimes_A \ldots \otimes_A F_n$ as a module on M and $B' = R' \otimes_A (R' \otimes_R F_1) \otimes_A \ldots \otimes_A (R' \otimes_R F_n)$ as a module on M', we have

$$
g^* \mathrm{P}^k_{R/A}(F_1, \ldots, F_n) \cong g^*(q \circ i)_* i^* B \cong (q' \circ i')_* p_k^* i^* B
$$

$$
\cong (q' \circ i')_* (i')^* B' \cong \mathrm{P}^k_{R'/A}(R' \otimes_R F_1, \ldots, R' \otimes_R F_n)
$$

because g is étale,[1] and this isomorphism coincides with the canonical isomorphism.
$\qquad\square$

Étale localization is compatible with multi-compositions and with base change $A \to B$.

C.2   Multilinear Differential Operators on Schemes

Let $f : X \to S$ be a morphism of schemes, and let $\mathcal{F}_1, \ldots, \mathcal{F}_n$ as well as $\mathcal{G}$ be quasi-coherent O_X-modules.

Definition C.18 Let

$$
\mu : \ \mathcal{F}_1 \times \ldots \times \mathcal{F}_n \to \mathcal{G}
$$

be an $f^{-1}(O_S)$-multilinear map. We say that it is a *multilinear differential operator of total order* 0 if it is O_X-multilinear. Then we say inductively that it is a *multilinear differential operator of total order* k if, for every open $V \subseteq X$, every $1 \le \ell \le n$, and every $r \in \Gamma(V, O_X)$, the map

$$
\Psi_{r;\ell}(\mu) := \mu \circ L_{r;\ell} - L_r \circ \mu : \ \mathcal{F}_1|_V \times \ldots \times \mathcal{F}_n|_V \to \mathcal{G}|_V
$$

---

[1] Note that flat base change holds when the other morphism is qcqs; since $\Delta_k \to X$ is a morphism of affine schemes, this is always true.

is a multilinear differential operator of total order $k - 1$. Multilinear differential operators of total order k form a sheaf $\mathcal{D}\!\mathit{iff}^k_{X/S}(\mathcal{F}_1, \ldots, \mathcal{F}_n; \mathcal{G})$ of O_X-modules via $r \cdot \mu := L_r \circ \mu$.

Similarly, we say that μ is a *multilinear differential operator of order* $(0, \ldots, 0)$ if it is O_X-multilinear. Then, for a tuple $\underline{k} = (k_1, \ldots, k_n)$ with $k_\ell \geq 0$ for all $1 \leq \ell \leq n$, we say that μ is a *multilinear differential operator of order* $\underline{k}$ if $\Psi_{r;\ell}(\mu)$ is a multilinear differential operator of order $(k_1, \ldots, k_\ell - 1, \ldots, k_n)$ for all open $V \subseteq X$, all $r \in \Gamma(V, O_X)$, and all $1 \leq \ell \leq n$ with $k_\ell \geq 1$, and if it is O_X-linear in the ℓ-th entry for ℓ with $k_\ell = 0$. We denote the sheaf of multilinear differential operators of order $\underline{k}$ by $\mathcal{D}\!\mathit{iff}^{\underline{k}}_{X/S}(\mathcal{F}_1, \ldots, \mathcal{F}_n; \mathcal{G})$. It is an O_X-module via $r \cdot \mu := L_r \circ \mu$.

Proposition C.19 *In the above situation, there is a quasi-coherent sheaf*

$$\mathcal{P}^k_{X/S}(\mathcal{F}_1, \ldots, \mathcal{F}_n)$$

on X together with an $f^{-1}(O_S)$-multilinear differential operator

$$\mu^u_{X/S}: \ \mathcal{F}_1 \times \ldots \times \mathcal{F}_n \to \mathcal{P}^k_{X/S}(\mathcal{F}_1, \ldots, \mathcal{F}_n)$$

which induces an isomorphism

$$\mathcal{H}om_{O_X}(\mathcal{P}^k_{X/S}(\mathcal{F}_1, \ldots, \mathcal{F}_n), \mathcal{G}) \cong \mathcal{D}\!\mathit{iff}^k_{X/S}(\mathcal{F}_1, \ldots, \mathcal{F}_n; \mathcal{G})$$

for every quasi-coherent sheaf $\mathcal{G}$ via $h \mapsto h \circ \mu^u_{X/S}$. If $b: T \to S$ is a morphism of schemes and $c: Y := X \times_S T \to X$ is the fiber product, then we have a canonical isomorphism

$$C^k_{T/S}: \ c^* \mathcal{P}^k_{X/S}(\mathcal{F}_1, \ldots, \mathcal{F}_n) \cong \mathcal{P}^k_{Y/T}(c^*\mathcal{F}_1, \ldots, c^*\mathcal{F}_n)$$

which induces a pull-back map

$$c^*: \ \mathcal{D}\!\mathit{iff}^k_{X/S}(\mathcal{F}_1, \ldots, \mathcal{F}_n; \mathcal{G}) \to c_* \mathcal{D}\!\mathit{iff}^k_{Y/T}(c^*\mathcal{F}_1, \ldots, c^*\mathcal{F}_n; c^*\mathcal{G})$$

for every quasi-coherent sheaf $\mathcal{G}$.

Proof First, we fix an affine open $U \subseteq X$ such that there is some affine open $V \subseteq S$ with $U \subseteq f^{-1}(V)$. Let $A = \Gamma(V, O_S)$ and $R = \Gamma(U, O_X)$. If $S \subset R$ is a multiplicative system, let $U_S := \mathrm{Spec}(S^{-1}R) \subseteq U$. We set

$$\Gamma(U_S, \mathcal{P}^k_{U/S}(\mathcal{F}_1, \ldots, \mathcal{F}_n)) := \mathrm{P}^k_{S^{-1}R/A}(S^{-1}F_1, \ldots, S^{-1}F_n)$$

where $F_i := \Gamma(U, \mathcal{F}_i)$. By Lemma C.14, this is independent of the choice of V. By Lemma C.10, we can set $\mathcal{P}^k_{U/S}(\mathcal{F}_1, \ldots, \mathcal{F}_n)$ to be the quasi-coherent sheaf associated with $\mathrm{P}^k_{R/A}(F_1, \ldots, F_n)$. By Lemma C.7, we obtain an A-multilinear differential operator

$$\mu^u_{U/S}: \quad \mathcal{F}_1|_U \times \ldots \times \mathcal{F}_n|_U \to \mathcal{P}^k_{U/S}(\mathcal{F}_1, \ldots \mathcal{F}_n)$$

first on every U_S, and then, by the sheaf property, on every open subset of U. Another application of Lemma C.14 together with a careful topological argument shows that $\mu^u_{U/S}$ is $f^{-1}(O_S)$-multilinear. By induction on k, we show that a map is a multilinear differential operator of total order k on the level of sheaves if and only if it is on all affine opens U_S, so $\mu^u_{U/S}$ is indeed an $f^{-1}(O_S)$-multilinear differential operator of total order k. This also shows

$$\Gamma(U_S, \mathcal{D}iff^k_{X/S}(\mathcal{F}_1, \ldots, \mathcal{F}_n; \mathcal{G})) = \mathrm{Diff}^k_{S^{-1}R/A}(S^{-1}F_1, \ldots, S^{-1}F_n; S^{-1}G).$$

In particular, the homomorphism

$$\mathcal{H}om_{O_U}(\mathcal{P}^k_{U/S}(\mathcal{F}_1, \ldots, \mathcal{F}_n), \mathcal{G}) \to \mathcal{D}iff^k_{U/S}(\mathcal{F}_1|_U, \ldots, \mathcal{F}_n|_U; \mathcal{G}|_U);$$

induced by $\mu^u_{U/S}$ is an isomorphism.

If $U_1, U_2 \subseteq X$ are two affine open subsets as above, then, by the universal property, $\mathcal{P}^k_{U_1/S}(\mathcal{F}_1, \ldots, \mathcal{F}_n; \mathcal{G})$ and $\mathcal{P}^k_{U_2/S}(\mathcal{F}_1, \ldots, \mathcal{F}_n; \mathcal{G})$ are canonically isomorphic on the overlap $U_1 \cap U_2$; in particular, there is a unique gluing $\mathcal{P}^k_{X/S}(\mathcal{F}_1, \ldots, \mathcal{F}_n; \mathcal{G})$ to a quasi-coherent sheaf on X together with a multilinear differential operator $\mu^u_{X/S}$ of total order k into it; $\mu^u_{X/S}$ has the desired universal property.

Consider a base change along $b: T \to S$ as in the statement of the proposition. Then we obtain a multilinear differential operator of total order k

$$\nu: \quad \mathcal{F}_1 \times \ldots \times \mathcal{F}_n \to c_* c^* \mathcal{F}_1 \times \ldots \times c_* c^* \mathcal{F}_n \to c_* \mathcal{P}^k_{Y/T}(c^* \mathcal{F}_1, \ldots, c^* \mathcal{F}_n)$$

relative to X/S. Thus, we have unique homomorphism

$$h: \quad \mathcal{P}^k_{X/S}(\mathcal{F}_1, \ldots, \mathcal{F}_n) \to c_* \mathcal{P}^k_{Y/T}(c^* \mathcal{F}_1, \ldots, c^* \mathcal{F}_n)$$

with $h \circ \mu^u_{X/S} = \nu$. The adjoint map is the map $C^k_{B/A}$ of Lemma C.12 on appropriate affine open subsets, so it is an isomorphism; this is the map $C^k_{T/S}$ above. To form the pull-back map c^* on the level of multilinear differential operators, we use the natural map

$$c^* \mathcal{H}om(\mathcal{P}^k_{X/S}(\mathcal{F}_1, \ldots, \mathcal{F}_n), \mathcal{G}) \to \mathcal{H}om(c^* \mathcal{P}^k(\mathcal{F}_1, \ldots, \mathcal{F}_n), c^* \mathcal{G}),$$

which need not be an isomorphism.[2] $\square$

---

[2] Already the example of derivations with values in O_X shows that c^* does not necessarily induce an isomorphism $c^* \mathcal{D}iff^k_{X/S}(\mathcal{F}_1, \ldots, \mathcal{F}_n; \mathcal{G}) \to \mathcal{D}iff^k_{Y/T}(c^* \mathcal{F}_1, \ldots, c^* \mathcal{F}_n; c^* \mathcal{G})$.

Remark C.20 The pull-back map c^* is compatible with multi-compositions. This is because c^* is given by $(-) \otimes_A B$ on affine open subsets $\operatorname{Spec} R \subseteq X$, which is compatible with multi-compositions. $\diamond$

Variant C.21 *Proposition C.19 holds with k replaced by $\underline{k} = (k_1, \dots, k_n)$ as well.*

Lemma C.22 *Assume that $f\colon X \to S$ is locally of finite type, and that $\mathcal{F}_1, \dots, \mathcal{F}_n$ are quasi-coherent sheaves of finite type. Then $\mathcal{P}^k_{X/S}(\mathcal{F}_1, \dots, \mathcal{F}_n)$ and $\mathcal{P}^{\underline{k}}_{X/S}(\mathcal{F}_1, \dots, \mathcal{F}_n)$ are quasi-coherent sheaves of finite type.*

Proof Let $U \subseteq X$ be an affine open as in the proof of Proposition C.19. By further shrinking U and V, we can assume that R is a finitely generated A-algebra. By Authors [267, 01PB], each R-module F_i is finitely generated. Thus, by Lemma C.4, $\mathrm{P}^k_{R/A}(F_1, \dots, F_n)$ is a finitely generated R-module, and then $\mathcal{P}^k_{X/S}(\mathcal{F}_1, \dots, \mathcal{F}_n)$ is of finite type. $\square$

We also have étale localization on the level of schemes.

Lemma C.23 *In the above situation, let $g\colon X' \to X$ be étale, in particular locally of finite presentation. Then we have a canonical isomorphism*

$$g^* \mathcal{P}^k_{X/S}(\mathcal{F}_1, \dots, \mathcal{F}_n) \to \mathcal{P}^k_{X'/S}(g^* \mathcal{F}_1, \dots, g^* \mathcal{F}_n)$$

which induces a pull-back map

$$g^*\colon \ \mathcal{D}\mathit{iff}^k_{X/S}(\mathcal{F}_1, \dots, \mathcal{F}_n; \mathcal{G}) \to g_* \mathcal{D}\mathit{iff}^k_{X'/S}(g^* \mathcal{F}_1, \dots, g^* \mathcal{F}_n; g^* \mathcal{G})$$

for every quasi-coherent sheaf $\mathcal{G}$. This pull-back map is compatible with multi-compositions and base change.

Appendix D
Spectral Sequences

We review a criterion for a spectral sequence over a field to degenerate at the first page. Furthermore, we discuss various basic properties of spectral sequences over Artinian local rings.

D.1   A Criterion for E_1-Degeneration over k

Here, we give a criterion for a spectral sequence associated with a double complex $(B^{\bullet,\bullet}, \partial, \bar{\partial})$ to degenerate at the page E_1. We use this criterion in the proof of Lemma 6.12, which is a crucial step in the proof of Theorem 6.8, the unobstructedness of the quantum extended Maurer–Cartan functor. The criterion is that the first spectral sequence associated with $B^{\bullet,\bullet}$ degenerates at the page E_1 if and only if

$$H^\bullet(B^\bullet[\![\hbar]\!], \bar{\partial} + \hbar\partial)$$

is a flat $\mathbf{k}[\![\hbar]\!]$-module. The criterion goes back at least to [171, Defn. 4.13], where it is only implicit; it is also claimed explicitly just after [38, Ass. 5.4], but none of the two references give a proof. Therefore, we provide the proof here. In fact, we give a more precise criterion which also applies to some form of partial degeneration which we specify in Definition D.1 below.

Let $(B^{\bullet,\bullet}, \partial, \bar{\partial})$ be a double complex of $\mathbf{k}$-vector spaces, bounded below in both indices. Recall that our convention is that $\partial\bar{\partial} + \bar{\partial}\partial = 0$. Let us first review the construction of the spectral sequence. We have inclusions

$$I_r^{p,q} \subseteq K_r^{p,q} \subseteq B^{p,q}$$

S. Felten, *Global Logarithmic Deformation Theory*, Lecture Notes in Mathematics 2373, https://doi.org/10.1007/978-3-031-98751-9

such that $E_{r+1}^{p,q} = K_r^{p,q}/I_r^{p,q}$, starting with

$$K_0^{p,q} = \ker(\bar{\partial}\colon B^{p,q} \to B^{p,q+1}), \quad I_0^{p,q} = \mathrm{im}(\bar{\partial}\colon B^{p,q-1} \to B^{p,q}).$$

For $r \geq 1$, an r-*zigzag* is a sequence $\alpha_\bullet = (\alpha_1, \ldots, \alpha_r)$ with $\alpha_i \in B^{p-1+i,q+1-i}$ and $\bar{\partial}\alpha_1 = 0$ as well as $\bar{\partial}\alpha_{i+1} + \partial\alpha_i = 0$ for $1 \leq i \leq r - 1$. Then we have, for $r \geq 0$, that $\alpha \in K_r^{p,q}$ if and only if there is an $(r + 1)$-zigzag $\alpha_\bullet$ with $\alpha_1 = \alpha$. The differential is given by

$$d_{r+1}[\alpha] = [\partial\alpha_{r+1}] \in E_{r+1}^{p+r+1,q-r} = K_r^{p+r+1,q-r}/I_r^{p+r+1,q-r}$$

for any $(r + 1)$-zigzag $\alpha_\bullet$ with $\alpha_1 = \alpha$. These claims can be proven by induction on r, using that $\partial\alpha_{r+1} \in I_r^{p+r+1,q-r} \subseteq I_{r+1}^{p+r+1,q-r}$ if and only if there is an element β^0 and a sequence of i-zigzags

$$\beta_\bullet^1 = (\beta_1^1), \quad \beta_\bullet^2 = (\beta_1^2, \beta_2^2), \quad \ldots, \quad \beta_\bullet^r = (\beta_1^r, \ldots, \beta_r^r)$$

such that $0 = \partial\alpha_{r+1} + \bar{\partial}\beta^0 + \sum_i \partial\beta_i^i$.

Definition D.1 We say that the first spectral sequence of $(B^{\bullet,\bullet}, \partial, \bar{\partial})$ degenerates at E_1 *in* $[k - 1, k]$ if $d_r^{p,q} = 0$ for $p + q = k - 1$ and $r \geq 1$. This is equivalent to $K_0^{p,q} = K_1^{p,q} = \ldots = K_\infty^{p,q}$ for $p + q = k - 1$. We say that the spectral sequence degenerates *at level k* if it degenerates in $[k - 1, k]$ and in $[k, k + 1]$. This is equivalent to $E_1^{p,q} = E_2^{p,q} = \ldots = E_\infty^{p,q}$.

For the criterion, we form a new double complex $(B^{\bullet,\bullet}[\![\hbar]\!], \hbar\partial, \bar{\partial})$ of $\mathbf{k}[\![\hbar]\!]$-modules with

$$B^{p,q}[\![\hbar]\!] := \left\{ \sum_{i=0}^{\infty} b_i \hbar^i \;\middle|\; b_i \in B^{p,q} \right\},$$

i.e., in general $B^{p,q}[\![\hbar]\!] \neq B^{p,q} \otimes_{\mathbf{k}} \mathbf{k}[\![\hbar]\!]$. They are only equal if $B^{p,q}$ has finite dimension, which we do not assume. However, concerning the total complexes, we have an equality

$$\bigoplus_{p+q=k} B^{p,q}[\![\hbar]\!] = B^k[\![\hbar]\!] := \left\{ \sum_{i=0}^{\infty} b_i \hbar^i \;\middle|\; b_i \in B^k = \bigoplus_{p+q=k} B^{p,q} \right\}$$

because the direct sum is finite. Note that we have modified ∂ to $\hbar\partial$.

Since $\mathbf{k}[\![\hbar]\!]$ is a discrete valuation ring, a $\mathbf{k}[\![\hbar]\!]$-module is flat if and only if it is torsion-free.

Proposition D.2 *In the above situation, the first spectral sequence of $(B^{\bullet,\bullet}, \partial, \bar{\partial})$ degenerates at E_1 in $[k - 1, k]$ for $k \in \mathbb{Z}$ if and only if $H^k(B^\bullet[\![\hbar]\!], \bar{\partial} + \hbar\partial)$ is a flat*

$\mathbf{k}[\![\hbar]\!]$-*module. In particular, the spectral sequence degenerates at* E_1 *if and only if* $H^k(B^\bullet[\![\hbar]\!], \bar{\partial} + \hbar\partial)$ *is a flat* $\mathbf{k}[\![\hbar]\!]$-*module for all* k.

Proof First assume that the first spectral sequence of $(B^{\bullet,\bullet}, \partial, \bar{\partial})$ degenerates at E_1 in $[k-1, k]$, i.e., we have $K_0^{p,q} = \ldots = K_\infty^{p,q}$ for $p + q = k - 1$. Consider the double complex $\tilde{B}^{\bullet,\bullet} := (B^{\bullet,\bullet}[\![\hbar]\!], \partial, \bar{\partial})$ with the unaltered differentials ∂ and $\bar{\partial}$. Then we have $\tilde{K}_r^{p,q} = K_r^{p,q}[\![\hbar]\!]$ and $\tilde{I}_r^{p,q} = I_r^{p,q}[\![\hbar]\!]$.

Next, we consider the double complex $\check{B}^{\bullet,\bullet} = (B^{\bullet,\bullet}[\![\hbar]\!], \hbar\partial, \bar{\partial})$ with the modified differential. We have $\check{d}_0 = \tilde{d}_0$, so we find $\check{K}_0^{p,q} = \tilde{K}_0^{p,q}$. We already know that

$$\tilde{K}_0^{p,q} = \tilde{K}_1^{p,q} = \ldots = \tilde{K}_\infty^{p,q}$$

for $p + q = k - 1$. If $\tilde{\alpha} \in \tilde{K}_r^{p,q}$, then we can find an $(r+1)$-zigzag $\tilde{\alpha}_\bullet$ with $\tilde{\alpha}_1 = \tilde{\alpha}$. Setting $\check{\alpha}_i = \hbar^{i-1}\tilde{\alpha}_i$, we obtain an $(r+1)$-zigzag $\check{\alpha}_\bullet$ for $\check{\alpha} := \tilde{\alpha}$ with respect to the modified double complex $\check{B}^{\bullet,\bullet}$. Thus, we have $\check{\alpha} \in \check{K}_r^{p,q}$, and hence $\check{K}_r^{p,q} = \tilde{K}_r^{p,q} = \check{K}_0^{p,q}$. This shows $\check{K}_0^{p,q} = \ldots = \check{K}_\infty^{p,q}$ for $p + q = k - 1$, hence

$$\check{I}_1^{p,q} = \check{I}_2^{p,q} = \ldots = \check{I}_\infty^{p,q}$$

for $p + q = k$. In particular, $\check{E}_\infty^{p,q}$ is a $\mathbf{k}[\![\hbar]\!]$-submodule of $\check{E}_1^{p,q} = \tilde{E}_1^{p,q} = E_1^{p,q}[\![\hbar]\!]$ so that $\check{E}_\infty^{p,q}$ is torsion-free and hence flat. The abutment $H^k(B^\bullet[\![\hbar]\!], \bar{\partial} + \hbar\partial)$ is a flat $\mathbf{k}[\![\hbar]\!]$-module since it has a filtration by the flat $\mathbf{k}[\![\hbar]\!]$-modules $\check{E}_\infty^{p,q}$.

Conversely, assume that the first spectral sequence of $(B^{\bullet,\bullet}, \partial, \bar{\partial})$ does *not* degenerate at E_1 in $[k-1, k]$. Let $\alpha \in K_{r-1}^{p,q}$ be such that $\alpha \notin K_r^{p,q}$ for some $r \geq 1$ and $p + q = k - 1$. Then there is an r-zigzag $\alpha_\bullet$ with $\alpha_1 = \alpha$ but no such $(r+1)$-zigzag. In particular, there is no $\gamma \in B^{p+r,q-r}$ such that $\bar{\partial}\gamma = \beta := -\partial\alpha_r$. Note that $\partial\beta = 0$ and $\bar{\partial}\beta = 0$, so also $(\bar{\partial} + \hbar\partial)(\beta) = 0$, i.e., β defines a class $[\beta]_{\text{tot}} \in H^k(B^\bullet[\![\hbar]\!], \bar{\partial} + \hbar\partial)$. We have $[\beta]_{\text{tot}} \neq 0$ because otherwise there would be a γ with $\bar{\partial}\gamma = \beta$ as above. Now let

$$\alpha' := \sum_{i=1}^{r} \hbar^i \alpha_i.$$

A direct computation shows that $(\bar{\partial} + \hbar\partial)(\alpha') = \hbar^{r+1}\beta$. In other words, we have $\hbar^{r+1}[\beta]_{\text{tot}} = 0$ so that $H^k(B^\bullet[\![\hbar]\!], \bar{\partial} + \hbar\partial)$ has $\hbar$-torsion and is not flat. $\qquad\square$

D.2   Spectral Sequences over Artinian Local Rings

We prove Proposition D.4 below about the behavior of spectral sequences over Artinian local rings under base change. In this section, A is an Artinian local $\mathbf{k}$-algebra with residue field $\mathbf{k}$ and maximal ideal $\mathfrak{m}_A \subset A$.

As pointed out in [279], Nakayama's lemma holds for a general A-module M, not just for finitely generated ones. More precisely, if M is an A-module, and $N \subseteq M$ is a submodule such that the induced map $N \to M_0 := M \otimes_A \mathbf{k}$ is surjective, then $N = M$.

By Authors [267, 051G], an A-module M is flat if and only if it is projective if and only if it is free. By Wahl [279, Lemma 0.2.1], in an exact sequence

$$0 \to M' \to M \to M'' \to 0$$

of A-modules, if two of them are flat, so is the third. In particular, if we have an exact sequence of finite length, and all modules but one are flat, the remaining one is flat as well. The following result is [279, Thm. A.1].

Proposition D.3 *Let $C^\bullet$ be a complex of flat A-modules (not necessarily bounded), and let*

$$\phi_A^q: \quad H^q(C^\bullet) \otimes_A \mathbf{k} \to H^q(C^\bullet \otimes_A \mathbf{k})$$

be the canonical map. Then:

(1) *If ϕ_A^q is surjective, then it is an isomorphism.*
(2) *If $H^q(C^\bullet)$ is flat over A, then ϕ_A^q is injective.*
(3) *If ϕ_A^q is injective, and ϕ_A^{q-1} is an isomorphism, then ϕ_A^q is an isomorphism.*
(4) *Any two of the following statements together imply the third:*

    (i) *ϕ_A^q is an isomorphism.*
    (ii) *ϕ_A^{q-1} is an isomorphism.*
    (iii) *$H^q(C^\bullet)$ is flat over A.*

This allows us to prove the following result for spectral sequences. Recall that we assume $\partial\bar{\partial} + \bar{\partial}\partial = 0$ in a double complex.

Proposition D.4 *Let $(C^{\bullet,\bullet}, \partial, \bar{\partial})$ be a double complex of flat A-modules, bounded from below in both variables. Let $(C^\bullet, \partial + \bar{\partial})$ be the associated total complex. Let $C_0^{\bullet,\bullet} := C^{\bullet,\bullet} \otimes_A \mathbf{k}$, and assume that the first spectral sequence*

$$'E_{0;1}^{p,q} = H^q(C_0^{p,\bullet}, \bar{\partial}) \Rightarrow H_0^{p+q} := H^{p+q}(C_0^\bullet, \partial + \bar{\partial})$$

associated with $C_0^{\bullet,\bullet}$ degenerates at E_1. Assume furthermore that the restriction map

$$H_A^n := H^n(C^\bullet, \partial + \bar{\partial}) \to H_0^n := H^n(C_0^\bullet, \partial + \bar{\partial})$$

is surjective for all n. Let

$$'E_{A;1}^{p,q} = H^q(C^{\bullet,\bullet}, \bar{\partial}) \Rightarrow H_A^{p+q} = H^{p+q}(C^\bullet, \partial + \bar{\partial})$$

be the first spectral sequence associated with $C^{\bullet,\bullet}$. We have an induced map $'E_{A;r} \to {}'E_{0;r}$ of spectral sequences. Then the following hold:

(1) *The filtered pieces $F^p H_A^n$ of $'E_{A;r}$ are flat A-modules, and the canonical maps*

$$\rho^p: \quad F^p H_A^n \to F^p H_0^n$$

are surjective, inducing an isomorphism $F^p H_A^n \otimes_A \mathbf{k} \cong F^p H_0^n$.

(2) *Each $'E_{A;r}^{p,q}$ is a flat A-module, and the canonical map $'E_{A;r}^{p,q} \to {}'E_{0;r}^{p,q}$ is surjective, inducing an isomorphism $'E_{A;r}^{p,q} \otimes_A \mathbf{k} \cong {}'E_{0;r}^{p,q}$. In particular, $H^q(C^{p,\bullet}, \bar{\partial})$ is a flat A-module, and $H^q(C^{p,\bullet}, \bar{\partial}) \to H^q(C_0^{p,\bullet}, \bar{\partial})$ is surjective, inducing an isomorphism $H^q(C^{p,\bullet}, \bar{\partial}) \otimes_A \mathbf{k} \cong H^q(C_0^{p,\bullet}, \bar{\partial})$.*

(3) *The spectral sequence $'E_{A;r}$ degenerates at E_1.*

Proof Since $A \to \mathbf{k}$ can be split into a succession of small extensions, we may prove the result by induction on the length of A. So let $A \to \bar{A}$ be a small extension, and assume that the result holds for $\bar{A}$. Note that $\bar{C}^{\bullet,\bullet} := C^{\bullet,\bullet} \otimes_A \bar{A}$ satisfies the assumptions. We obtain an exact sequence

$$0 \to C_0^{\bullet,\bullet} \otimes_{\mathbf{k}} I \to C^{\bullet,\bullet} \to \bar{C}^{\bullet,\bullet} \to 0$$

of double complexes, where $I \subset A$ is the kernel of $A \to \bar{A}$. First, we show that

$$\rho^p: \quad F^p H_A^n \xrightarrow{r^p} F^p H_{\bar{A}}^n \to F^p H_0^n$$

is surjective. By the induction hypothesis, it is sufficient to show that r^p is surjective. Let $d = \partial + \bar{\partial}$ for short. For $p \ll 0$, we have $F^p H^n = H^n$, so in this case r^p is surjective. To show that in fact all restrictions r^p are surjective, assume the contrary and let p be the lowest value such that r^p is not surjective. Let $\bar{c} = (\bar{c}^{i,j})_{i \geq p}$ with $\bar{c}^{i,j} \in \bar{C}^{i,j}$ and $d(\bar{c}) = 0$ be a representative of a class $[\bar{c}] \in F^p H^n(\bar{C}^\bullet, d)$ which is not in the image of r^p. By assumption, r^{p-1} is surjective. Thus, we can find a $(\tilde{c}^{i,j})_{i \geq p-1}$ with $\tilde{c}^{i,j} \in C^{i,j}$ and $d(\tilde{c}) = 0$ such that $[\tilde{c}]|_{\bar{A}} = [\bar{c}]$. This means that $\tilde{c}|_{\bar{A}} + d(\bar{b}) = \bar{c}$ for some $\bar{b} \in \bar{C}^{n-1}$. Since $C^{n-1} \to \bar{C}^{n-1}$ is surjective, we can find a lift $b \in C^{n-1}$ of $\bar{b}$. When we set $c := \tilde{c} + db \in C^n$, then this element satisfies $d(c) = 0$ and $c|_{\bar{A}} = \bar{c}$. In particular, for $i < p$, we have $c^{i,j} \in I \cdot C^{i,j} = C_0^{i,j} \otimes_{\mathbf{k}} I$. We construct a new element $e \in I \cdot C^n$. For $i < p$, we set $e^{i,j} = c^{i,j}$. Let m be the smallest index with $e^{m,n-m} \neq 0$. By the construction of the differentials d_r of the spectral sequence $'E_{0;r}$,

$$\pm \bar{\partial} e^{m+r,n-m-r} = \mp \partial e^{m+r-1,n-m-r+1}$$

represents $d_r[e^{m,n-m}]$ until we reach $e^{p-1,n-p+1}$. However, by assumption, $'E_{0;r}$ degenerates at E_1 so that $d_r[e^{m,n-m}] = 0$. Thus, we can construct all further $e^{i,j}$ for $i \geq p$ by choosing a preimage in $C_0^{i,j} \otimes_{\mathbf{k}} I$ of $-\partial e^{i-1,j+1}$ under $\bar{\partial}$, which exists because $[-\partial e^{i-1,j+1}] = \pm d_r[e^{m,n-m}]$ is the zero class. This element $e \in C_0^n \otimes_{\mathbf{k}} I$ satisfies $d(e) = 0$, and we have $(c - e)^{i,j} = 0$ for $i < p$. Thus, $d(c - e) = 0$, and $c - e$ defines a class in $F^p H_A^n$ with $r^p[c - e] = [\bar{c}]$. Hence, r^p is surjective.

Now since $F^p H_A^n \to F^p H_0^n$ is surjective, the induced map

$$'E_{A;\infty}^{p,q} = F^p H_A^{p+q}/F^{p+1} H_A^{p+q} \to F^p H_0^{p+q}/F^{p+1} H_0^{p+q} = 'E_{0;\infty}^{p,q}$$

is surjective as well. By assumption, we have $'E_{0;\infty}^{p,q} = 'E_{0;1}^{p,q}$, and $'E_{A;\infty}^{p,q} = Z_{A;\infty}^{p,q}/B_{A;\infty}^{p,q}$ is a subquotient of $'E_{A;1}^{p,q}$ for A-submodules $B_{A;\infty}^{p,q} \subseteq Z_{A;\infty}^{p,q} \subseteq 'E_{A;1}^{p,q}$. In particular, $Z_{A;\infty}^{p,q} \to 'E_{0;1}^{p,q}$ is surjective, and Nakayama's lemma (as discussed at the beginning of the section) shows that $Z_{A;\infty}^{p,q} = 'E_{A;1}^{p,q}$. Then $d_1 : 'E_{A;1}^{p,q} \to 'E_{A;1}^{p+1,q}$ must be the zero map, for otherwise $Z_{A;\infty}^{p,q} \subsetneq 'E_{A;1}^{p,q}$. Since this holds for all (p, q), we have $d_1 = 0$. Repeating this argument step by step, we find $d_r = 0$ for all $r \geq 1$; in particular, $'E_{A;r}$ degenerates at E_1.

Now we have $'E_{A;\infty}^{p,q} = 'E_{A;1}^{p,q}$ due to degeneration at E_1. Thus, the induced map $'E_{A;1}^{p,q} \to 'E_{0;1}^{p,q}$ is surjective. Proposition D.3 applied to $(C^{p,\bullet}, \bar{\partial})$ yields that $'E_{A;1}^{p,q} \otimes_A \mathbf{k} \cong 'E_{0;1}^{p,q}$ is an isomorphism, and that $'E_{A;1}^{p,q}$ is a flat A-module. Since $'E_{A;r}^{p,q} = 'E_{A;1}^{p,q}$, they are also flat and surject to $'E_{0;r}^{p,q}$, inducing isomorphisms.

Since $'E_{A;\infty}^{p,q} \cong 'E_{A;1}^{p,q}$ is flat, and since H_A^n is flat as well by Proposition D.3 applied to $(C^\bullet, d)$, the filtered pieces $F^p H_A^n$ must be flat. Finally, we show $F^p H_A^n \otimes_A \mathbf{k} \cong F^p H_0^n$ by induction on p. For $p \ll 0$, we have $F^p H_A^n = H_A^n$, so it follows from Proposition D.3 applied to $(C^\bullet, d)$. Then we have an exact sequence

$$0 \to F^{p+1} H_A^n \to F^p H_A^n \to 'E_{A;\infty}^{p,n-p} \to 0$$

with only flat modules so that it remains exact after applying $(-) \otimes_A \mathbf{k}$. This shows that $F^{p+1} H_A^n \otimes_A \mathbf{k} \to F^{p+1} H_0^n$ is injective, and surjectivity follows from surjectivity of $F^{p+1} H_A^n \to F^{p+1} H_0^n$.   □

Corollary D.5 *In the above situation, let $A \to B$ be a homomorphism of Artinian local $\mathbf{k}$-algebras with residue field $\mathbf{k}$. Then the induced maps*

$$H^q(C^{p,\bullet}, \bar{\partial}) \otimes_A B \to H^q(C^{p,\bullet} \otimes_A B, \bar{\partial})$$

and $F^p H_A^n \otimes_A B \to F^p H_B^n$ are isomorphisms of B-modules.

Proof Let M_A be any of the modules over A in the statement, and let M_B respectively M_0 be their variant over B respectively $\mathbf{k}$. Since $M_A \to M_0$ is surjective, the induced map $M_A \otimes_A B \to M_0$ is surjective as well; then the image of $M_A \otimes_A B \to M_B$ surjects onto $M_0 \cong M_B \otimes_B \mathbf{k}$, so $M_A \otimes_A B \to M_B$ is surjective by Nakayama's lemma. Since M_B is a flat B-module, the inclusion of the kernel K of $M_A \otimes_A B \to M_B$ is universally injective. Thus $K \otimes_B \mathbf{k} = 0$ and hence $K = 0$.   □

Appendix E
Analytification

Let $X/\mathbb{C}$ be a scheme locally of finite type. By Serre's famous article [255] and subsequent work, we have a complex analytic space X^{an} together with a map $\varphi\colon X^{\mathrm{an}} \to X$ of locally $\mathbb{C}$-ringed spaces which is universal among all morphisms of locally $\mathbb{C}$-ringed spaces from a complex analytic space $\mathcal{Z}$ to X; see [134, XII] for a nice exposition. In this chapter, we prove the existence of a unique analytification of first-order differential operators between quasi-coherent sheaves in Proposition E.7. Furthermore, we extend the classical comparison of the algebraic and analytic cohomologies on a proper scheme X from coherent sheaves to quasi-coherent sheaves in Lemma E.3. For neither of the two results we could find a reference. Both results are needed to show Theorem 15.2 in Chap. 15.

We write O_X^{an} for the structure sheaf of X^{an}. This notation is less heavy than $O_{X^{\mathrm{an}}}$, and it is correct in that $O_{X^{\mathrm{an}}} \cong \varphi^* O_X$.

E.1 Preliminaries

The Sequence Topology As described in [113], every ring of convergent power series $K_n := \mathbb{C}\{x_1, \ldots, x_n\}$ carries a special topology induced from an exhaustion by Banach algebras, the *sequence topology*. More precisely, for every $t = (t_1, \ldots, t_n) \in \mathbb{R}_+^n$, we have a map

$$\| \cdot \|_t\colon \quad \mathbb{C}[\![x_1, \ldots, x_n]\!] \to \mathbb{R}_{\geq 0} \cup \{\infty\}, \quad \sum_\nu a_\nu x^\nu \mapsto \sum_\nu |a_\nu| \cdot t^\nu;$$

then $B_t\{x_1, \ldots, x_n\} := \{f \in \mathbb{C}[\![x_1, \ldots, x_n]\!] \mid \|f\|_t < \infty\}$ is a Banach algebra, and K_n is the union of these for all $t \in \mathbb{R}_+^n$.

© The Author(s), under exclusive license to Springer Nature Switzerland AG 2025
S. Felten, *Global Logarithmic Deformation Theory*, Lecture Notes
in Mathematics 2373, https://doi.org/10.1007/978-3-031-98751-9

The sequence topology can be extended in a canonical way to all analytic $\mathbb{C}$-algebras A such that every ideal $I \subseteq A$ is closed, and in every surjection $\phi: A \to A/I$ of analytic $\mathbb{C}$-algebras, the sequence topology on A/I is the quotient topology of the one on A. Every homomorphism $\phi: A \to B$ of analytic $\mathbb{C}$-algebras is continuous. Furthermore, it can be canonically extended to all finitely generated A-modules M such that $A^{\oplus n}$ carries the product topology, in every surjection $h: M \to N$, the sequence topology on N is the quotient of the one on M, and all homomorphisms $h: M \to N$ of finitely generated A-modules are continuous. Every injection $h: M \to N$ exhibits M as a closed subset of N. The sequence topology is always Hausdorff.

We extend the sequence topology to arbitrary A-modules as follows: Every A-module M is the direct limit of the system M_α of its finitely generated submodules. Then we endow M with the colimit topology of this directed system. This construction has the following properties:

Lemma E.1 *Let A be an analytic $\mathbb{C}$-algebra, and let M be an A-module.*

(1) *Let $(M_i)_{i \in I}$ be a directed system of finitely generated A-modules with colimit M. Then the colimit topology of the system $(M_i)_i$ coincides with the colimit topology of the system $(M_\alpha)_\alpha$.*

(2) *Let $h: M \to N$ be A-linear. Then h is continuous for the sequence topologies on M and N.*

(3) *If $p: M \to N$ is A-linear and surjective, then the sequence topology on N is the quotient of the sequence topology on M.*

(4) *If $i: M \to N$ is A-linear and injective, then M is a closed subset of N, and the sequence topology on M is the induced topology from the sequence topology on N.*

Proof All statements are more or less straightforward. Let $M_{\alpha(i)}$ be the image of M_i under the map $f_i: M_i \to M$ of the directed system $(M_i)_i$, and let $\bar{f}_i: M_i \to M_{\alpha(i)}$ be the induced map. Let $U \subseteq M$ be a subset. Then $f_i^{-1}(U) = \bar{f}_i^{-1}(U \cap M_{\alpha(i)})$. Since the sequence topology on $M_{\alpha(i)}$ is the quotient topology of the sequence topology on M_i, we have that $U \cap M_{\alpha(i)}$ is open in $M_{\alpha(i)}$ if and only if $f_i^{-1}(U)$ is open in M_i. Now if U is open in the sequence topology on M, then $U \cap M_{\alpha(i)}$ is open in $M_{\alpha(i)}$, thus $f_i^{-1}(U)$ is open in M_i, and hence U is open in the colimit topology of $(M_i)_i$. Conversely, if U is open in the colimit topology of $(M_i)_i$, then $U \cap M_{\alpha(i)}$ is open in $M_{\alpha(i)}$ for all i. Since M is the colimit of $(M_i)_i$, for every M_α, there is some i with $M_\alpha \subseteq M_{\alpha(i)}$. Then $U \cap M_\alpha$ is open in M_α because $M_\alpha \to M_{\alpha(i)}$ is continuous, so U is open in the sequence topology on M. The proofs of the other statements are similar. $\square$

Remark E.2 Every open in the product topology on $M \times N = M \oplus N$ is open in the sequence topology. However, the converse is not clear.[1] Correspondingly, it is not clear if the sequence topology on M is Hausdorff if M is not finitely generated.   ◊

Quasi-Coherent Analytic Sheaves Let X be a complex analytic space. By definition, a sheaf $\mathcal{F}$ of O_X-modules is coherent if it is locally of finite presentation. We say that it is *quasi-coherent* if it admits locally some presentation, and it is *locally generated by sections* if we can find an open cover $\{\mathcal{U}_i\}_i$ of X such that $H^0(\mathcal{U}_i, \mathcal{F}) \otimes_{\mathbb{C}} O_{\mathcal{U}_i} \to \mathcal{F}|_{\mathcal{U}_i}$ is surjective.

Following [159, Defn. 1.1.6], we say that a sheaf $\mathcal{F}$ of O_X-modules is *pseudo-coherent*[2] if, for every open subset $\mathcal{U} \subseteq X$, every O_X-submodule $\mathcal{E} \subseteq \mathcal{F}|_{\mathcal{U}}$ which is locally of finite type is actually coherent.

Renaming for a moment the concept of quasi-coherence in [52, Defn. 2.1.1], we say that a sheaf $\mathcal{F}$ of O_X-modules is *locally ind-coherent* if we can find an open cover $\{\mathcal{U}_i\}_i$ of X such that each $\mathcal{F}|_{\mathcal{U}_i}$ is a (small filtered) direct limit of a system $\{\mathcal{F}_{i,\alpha}\}$ of coherent $O_{\mathcal{U}_i}$-modules. It follows from the proof of [52, Lemma 2.1.8] that the image of an O_X-linear map $h \colon \mathcal{E} \to \mathcal{F}$ from a coherent sheaf $\mathcal{E}$ to a locally ind-coherent sheaf $\mathcal{F}$ is coherent. Thus, every locally ind-coherent sheaf is pseudo-coherent.

Cartan's Theorem A states that $H^0(X, \mathcal{F}) \otimes_{\mathbb{C}} O_X \to \mathcal{F}$ is surjective if X is a Stein space and $\mathcal{F}$ a coherent analytic sheaf.[3] Thus, after refining the open cover $\{\mathcal{U}_i\}$ in the definition of a locally ind-coherent sheaf $\mathcal{F}$ such that each $\mathcal{U}_i$ is a Stein space, we find that a locally ind-coherent sheaf $\mathcal{F}$ is also locally generated by sections. Conversely, if a sheaf $\mathcal{F}$ of O_X-modules is both pseudo-coherent and locally generated by sections, then it is locally ind-coherent.

By Conrad [52, Lemma 2.1.9], kernels, cokernels, extensions, and tensor products of locally ind-coherent sheaves are locally ind-coherent. In particular, a locally ind-coherent sheaf is quasi-coherent in the usual sense because the kernel of the surjection $H^0(\mathcal{U}_i, \mathcal{F}) \otimes_{\mathbb{C}} O_{\mathcal{U}_i} \to \mathcal{F}|_{\mathcal{U}_i}$ is locally generated by sections. Conversely, a quasi-coherent analytic sheaf $\mathcal{F}$ is locally the cokernel of a map between locally ind-coherent sheaves, so every quasi-coherent sheaf is locally ind-coherent. Thus, the two notions are equivalent.

---

[1] In general, filtered colimits do not commute with finite products in the category of topological spaces. This holds for compactly generated Hausdorff spaces (maybe under some additional hypotheses on the filtered system and the maps between the spaces), but, in general, the sequence topology even on a finitely generated A-module M is not compactly generated.

[2] Compare this also with the usage of the term in [29]. In [267], the term is used for a different concept.

[3] One might be tempted to believe that a finite number of global sections in $H^0(X, \mathcal{F})$ would be sufficient to generate $\mathcal{F}$. However, this is not true. For an example, take a sheaf of ideals $\mathcal{I}$ in $O_{\mathbb{C}}$ generated by a sequence of functions as in the example in [143, p. 181]. The ideal sheaf $\mathcal{I}$ is coherent because, on every bounded domain $U \subseteq \mathbb{C}$, finitely many of the generators are sufficient. But globally, there is no finite number of sections generating $\mathcal{I}$ because they would also generate $\mathcal{I}(\mathbb{C}) \subseteq O_{\mathbb{C}}(\mathbb{C})$, which is not the case because this ideal is not finitely generated.

If $\mathcal{F}$ is a quasi-coherent analytic sheaf on X, then, for every $x \in X$, we endow the stalk $\mathcal{F}_x$ with the sequence topology constructed above. If $h \colon \mathcal{F} \to \mathcal{G}$ is an $\mathcal{O}_X$-linear map of quasi-coherent analytic sheaves, then the induced map $h \colon \mathcal{F}_x \to \mathcal{G}_x$ is continuous.

If $X/\mathbb{C}$ is a scheme locally of finite type and $\mathcal{F}$ is a quasi-coherent sheaf, then $\mathcal{F}^{\mathrm{an}}$ is a quasi-coherent analytic sheaf.

A GAGA Result for Quasi-Coherent Sheaves Let $X/\mathbb{C}$ be a scheme of finite type with analytification X^{an}. For every $\mathcal{O}_X$-module $\mathcal{F}$, there is a canonical and functorial map

$$H^p(X, \mathcal{F}) \to H^p(X^{\mathrm{an}}, \mathcal{F}^{\mathrm{an}}). \qquad (\mathrm{E}.1)$$

As is well-known, this map is an isomorphism if $X/\mathbb{C}$ is proper and $\mathcal{F}$ is coherent. However, it is true for quasi-coherent sheaves as well.[4]

Lemma E.3 *Let $X/\mathbb{C}$ be proper, and let $\mathcal{F}$ be a quasi-coherent $\mathcal{O}_X$-module. Then the map in* (E.1) *is an isomorphism of $\mathbb{C}$-vector spaces.*

Proof Since X is Noetherian, $\mathcal{F}$ is the direct limit of its coherent subsheaves. Thus, we have a directed system $\{\mathcal{F}_\alpha\}$ of coherent sheaves with colimit $\mathcal{F}$. Since $(-)^{\mathrm{an}}$ is left adjoint to φ_*, it preserves colimits. Thus, $\mathcal{F}^{\mathrm{an}}$ is the colimit of the directed system $\{\mathcal{F}_\alpha^{\mathrm{an}}\}$ of coherent analytic sheaves. Since the map in (E.1) is functorial, we get a commutative diagram

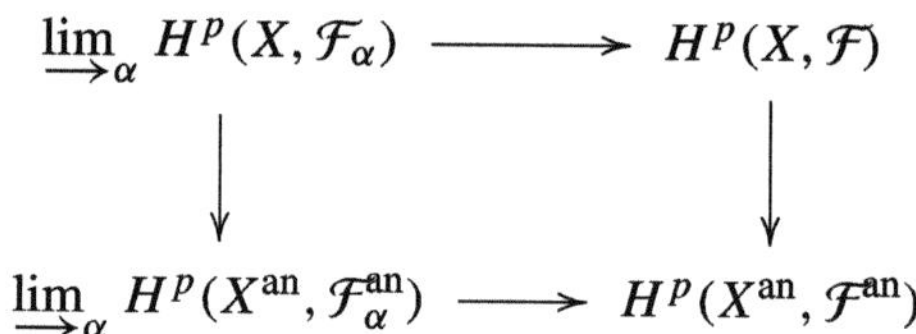

of cohomology $\mathbb{C}$-vector spaces. The upper horizontal arrow is an isomorphism by Authors [267, 01FF] because X is a Noetherian topological space. The left vertical arrow is an isomorphism because (E.1) is an isomorphism for each coherent sheaf $\mathcal{F}_\alpha$. Since X^{an} is a compact Hausdorff space, the lower horizontal arrow is an isomorphism by Godement [108, Thm. 4.12.1]; namely, cohomology with compact support is the same as cohomology on a compact space. $\square$

Density Let X be an affine scheme of finite type over $\mathbb{C}$, with analytification X^{an}. Let $\mathcal{F}$ be a quasi-coherent sheaf on X, and let $x \in X^{\mathrm{an}}$ be a point. Then we have an induced map $q \colon H^0(X, \mathcal{F}) \to \mathcal{F}_{\varphi(x)} \to \mathcal{F}_x^{\mathrm{an}}$.

---

[4] Although Lemma E.3 is in the preliminaries section, it is not needed in the proof of Proposition E.7 below. Instead, we need it directly in the proof of Theorem 15.2.

Lemma E.4 *The image of $q\colon H^0(X, \mathcal{F}) \to \mathcal{F}_x^{\mathrm{an}}$ is dense in $\mathcal{F}_x^{\mathrm{an}}$ for the sequence topology.*

Proof First, consider the map $q\colon P_n := \mathbb{C}[x_1, \ldots, x_n] \to \mathbb{C}\{x_1, \ldots, x_n\} -: K_n$. Let $A \subseteq K_n$ be a closed subset which contains the image of q. Then, for every $t \in \mathbb{R}_+^n$, the intersection $A \cap B_t\{x_1, \ldots, x_n\}$ is a closed subset of $B_{t;n} := B_t\{x_1, \ldots, x_n\}$ which contains the image of q as well. If $f \in B_{t;n}$, set $f_k := \sum_{|v|\le k} a_v x^v \in \mathbb{C}[x_1, \ldots, x_n] \subseteq A \cap B_{t;n}$, where $|v| := v_1 + \ldots + v_n$. Then $(f_k)_k$ converges to f, so $f \in A \cap B_{t;n}$ because A is closed, and thus $A \cap B_{t;n} = B_{t;n}$. This implies $A = K_n$.

Next, let $R = \Gamma(X, O_X)$ be the coordinate ring of X. Then we can find a presentation $R = \mathbb{C}[x_1, \ldots, x_n]/I$. After a coordinate transformation, we can assume that $x \in X$ corresponds to $\{x_1 = \ldots = x_n = 0\}$. The analytic local ring of X^{an} at x is $K_n/I \cdot K_n$, and we have a commutative diagram

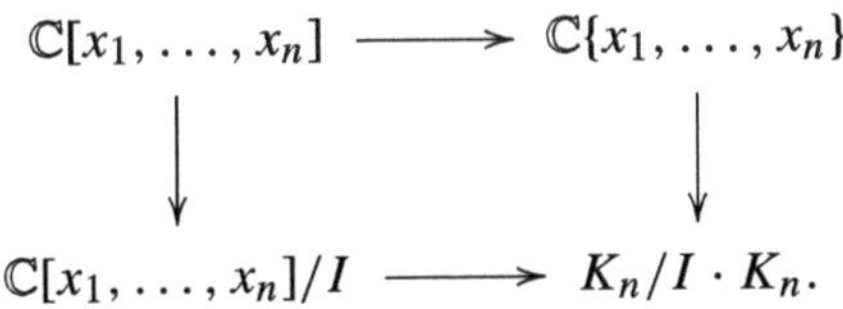

Since $K_n \to K_n/I \cdot K_n$ is continuous and surjective, we find that the image of $\mathbb{C}[x_1, \ldots, x_n]/I$ is dense.

Considering the product topology on $O_{X,x}^{\mathrm{an}, \oplus k}$, we obtain the claim for finite free O_X-modules. Then, we obtain it for coherent sheaves $\mathcal{F}$ from some surjection $O_X^{\oplus k} \to \mathcal{F}$ since $O_{X,x}^{\mathrm{an}, \oplus k} \to \mathcal{F}_x^{\mathrm{an}}$ is continuous and surjective. Finally, if $\mathcal{F}$ is quasi-coherent, we have $\mathcal{F} = \varinjlim \mathcal{F}_\alpha$ for the coherent subsheaves $\mathcal{F}_\alpha$ of $\mathcal{F}$, and at the same time, we have $\mathcal{F}_x^{\mathrm{an}} = \varinjlim \mathcal{F}_{\alpha,x}^{\mathrm{an}}$. Since the image of $H^0(X, \mathcal{F}_\alpha) \to \mathcal{F}_{\alpha,x}^{\mathrm{an}}$ is dense, we obtain the claim for $\mathcal{F}$ because the sequence topology on $\mathcal{F}_x^{\mathrm{an}}$ is the colimit topology of the sequence topologies on the $\mathcal{F}_{\alpha,x}^{\mathrm{an}}$. $\qquad\square$

Locally Bounded Operators For technical reasons, we need to introduce the following notion. Let $\mathcal{F}$ and $\mathcal{G}$ be two pseudo-coherent analytic sheaves on X. Then a map $h\colon \mathcal{F} \to \mathcal{G}$ of sheaves of sets, not necessarily O_X-linear, is called *locally bounded* if the following holds: For every open subset $\mathcal{U} \subseteq X$ and every coherent analytic subsheaf $\mathcal{E}$ of $\mathcal{F}|_{\mathcal{U}}$, there is an open cover $\{\mathcal{U}_i\}$ of $\mathcal{U}$ and a coherent analytic subsheaf $\mathcal{E}_i'$ of $\mathcal{G}|_{\mathcal{U}_i}$ such that h maps $\mathcal{E}|_{\mathcal{U}_i}$ into $\mathcal{E}_i' \subseteq \mathcal{G}|_{\mathcal{U}_i}$.

Derivations and Analytification If $f\colon X \to S$ is a morphism of complex analytic spaces, then there is a coherent analytic sheaf $\Omega_{X/S}^1$ together with a canonical $f^{-1}(O_S)$-linear derivation

$$d^{\mathrm{an}}\colon \ O_X \to \Omega_{X/S}^1.$$

Unlike the simpler algebraic case, it is only universal among $f^{-1}(O_S)$-linear derivations from O_X into a coherent analytic sheaf $\mathcal{E}$, but (most likely) not among derivations into an arbitrary O_X-module. According to [250, Exp. 14], which unfortunately does not discuss the universal property, the canonical derivation can be constructed as follows: Let

$$\Delta\colon\ X \to \mathcal{P} := X \times_S X$$

be the diagonal, and let $\mathcal{I}$ be its ideal sheaf as a closed analytic subspace. The two projections give rise to two maps $\mathrm{pr}_1^*, \mathrm{pr}_2^*\colon O_X \to O_\mathcal{P}$, whose composition with $O_\mathcal{P} \to O_\Delta$ is the same. Then $d^{\mathrm{an}} = \mathrm{pr}_2^* - \mathrm{pr}_1^*$ is their difference, considered as a map $O_X \to \mathcal{I}/\mathcal{I}^2$. The latter is supported on $\Delta \subseteq \mathcal{P}$, so it can be considered as a sheaf on X.

Now let $f\colon X \to S$ be a morphism of finite type between schemes X and S which are of finite type over $\mathbb{C}$. Let $f^{\mathrm{an}}\colon X^{\mathrm{an}} \to S^{\mathrm{an}}$ be the analytification. The (now for all O_X-modules) universal derivation

$$d\colon\ O_X \to \Omega^1_{X/S}$$

arises from the same construction via the algebraic diagonal $\Delta\colon X \to P := X \times_S X$. Thus, we have a commutative diagram

$$
\begin{array}{ccc}
\varphi_* O_X^{\mathrm{an}} & \xrightarrow{\ \varphi_* d^{\mathrm{an}}\ } & \varphi_* \Omega^1_{X^{\mathrm{an}}/S^{\mathrm{an}}} \\[2mm]
\Big\uparrow & & \Big\uparrow \\[2mm]
O_X & \xrightarrow{\ \ d\ \ } & \Omega^1_{X/S}
\end{array}
$$

and an isomorphism $\varphi^* \Omega^1_{X/S} \cong \Omega^1_{X^{\mathrm{an}}/S^{\mathrm{an}}}$.

First-Order Differential Operators Let $f\colon X \to S$ be a morphism of complex analytic spaces. Let $\mathcal{F}$ and $\mathcal{G}$ be two quasi-coherent sheaves on X. Then a *differential operator of first order relative to X/S* is an $f^{-1}(O_S)$-linear map $D\colon \mathcal{F} \to \mathcal{G}$ such that, for every local section $a \in O_X$, the map

$$D_a\colon\ \mathcal{F} \to \mathcal{G},\ f \mapsto D(af) - aD(f),$$

is O_X-linear. These operators form a sheaf $\mathcal{D}\!\mathit{iff}^1_{X/S}(\mathcal{F}, \mathcal{G})$ of O_X-modules under the O_X-action given by $(a \cdot D)(f) := a \cdot D(f)$.

Lemma E.5 *Let $\mathcal{F}$ and $\mathcal{G}$ be coherent analytic sheaves. Then $\mathcal{D}\!\mathit{iff}^1_{X/S}(\mathcal{F}, \mathcal{G})$ is coherent.*

Proof Let $\tilde{\Omega}^1_{X/S}$ be the sheaf of Kähler differential forms of the morphism $f\colon X \to S$ of locally ringed spaces. Its universal property gives rise to a canonical O_X-linear homomorphism $\chi\colon \tilde{\Omega}^1_{X/S} \to \Omega^1_{X/S}$ to the usual analytic differential forms which induces an isomorphism

$$\mathcal{H}om(\Omega^1_{X/S}, \mathcal{G}) \to \mathcal{H}om(\tilde{\Omega}^1_{X/S}, \mathcal{G})$$

for every coherent analytic sheaf $\mathcal{G}$ by the universal property of $\Omega^1_{X/S}$. By Authors [267, 0G3V], applied to the morphism $f\colon X \to S$ of locally ringed spaces, we obtain a short exact sequence

$$0 \to \tilde{\Omega}^1_{X/S} \otimes \mathcal{F} \to \tilde{\mathcal{P}}^1_{X/S}(\mathcal{F}) \to \mathcal{F} \to 0$$

of O_X-modules with the sheaf of principal parts in the middle. Applying $\mathcal{H}om(-, \mathcal{G})$, we find a long exact sequence

$$0 \to \mathcal{H}om(\mathcal{F}, \mathcal{G}) \xrightarrow{a} \mathcal{D}iff^1_{X/S}(\mathcal{F}, \mathcal{G}) \xrightarrow{b} \mathcal{H}om(\tilde{\Omega}^1_{X/S} \otimes \mathcal{F}, \mathcal{G}) \xrightarrow{c} \mathcal{E}xt^1(\mathcal{F}, \mathcal{G}).$$

From the tensor-hom adjunction, we find that the third sheaf is coherent. Since the last one is coherent as well, so is the kernel of the map c between them. But then the image of b is coherent, and thus the sheaf of first-order differential operators is an extension of coherent sheaves, so coherent itself. □

We use our notion of locally bounded operators in the following result.

Lemma E.6 *Let $D\colon \mathcal{F} \to \mathcal{G}$ be a locally bounded first-order differential operator relative to X/S between quasi-coherent analytic sheaves. Then, for every $x \in X$, the induced map $D\colon \mathcal{F}_x \to \mathcal{G}_x$ is continuous for the sequence topologies on $\mathcal{F}_x$ and $\mathcal{G}_x$.*

Proof We start with the case where both $\mathcal{F}$ and $\mathcal{G}$ are coherent. In this case, every operator is locally bounded, so this condition is empty. By Grauert and Remmert [113, III, §4], a derivation $D\colon O_{X,x} \to \mathcal{G}_x$ is continuous for the sequence topologies if $\mathcal{G}$ is coherent. Every first-order differential operator $D\colon O_{X,x} \to \mathcal{G}_x$ can be decomposed as $D = D_0 + D_1$ with an $O_{X,x}$-linear map $D_0(a) := aD(1)$, and a derivation $D_1(a) := D(a) - D_0(a)$. Since both D_0 and D_1 as well as the summation map $\mathcal{G}_x \times \mathcal{G}_x \to \mathcal{G}_x$ are continuous, we find that every first-order differential operator $D\colon O_{X,x} \to \mathcal{G}_x$ is continuous. Now if $\mathcal{F}$ is coherent, we can find a surjection $\pi\colon O^{\oplus n}_{X,x} \to \mathcal{F}_x$. The sequence topology on $\mathcal{F}_x$ is the quotient topology along π, so D is continuous if and only if $D \circ \pi$ is continuous. The sum $O^{\oplus n}_{X,x}$ carries the product topology, so D is continuous if and only if the maps $D \circ \pi_i$, with $\pi_i\colon O_{X,x} \to O^{\oplus n}_{X,x} \to \mathcal{F}_x$ the n summands, are continuous. However, they are first-order differential operators, so they are continuous. In the more general quasi-coherent case, we write $\mathcal{F}_x = \varinjlim \mathcal{F}_{\alpha,x}$. Let $U \subseteq \mathcal{G}_x$ be an open subset. By our definition of the sequence topology on $\mathcal{F}_x$, we have to show that $D^{-1}(U) \cap \mathcal{F}_{\alpha,x}$ is open for all α. Since D is locally bounded by assumption, there is a finitely generated submodule $\mathcal{G}_{\alpha,x}$ of $\mathcal{G}_x$ which contains $D(\mathcal{F}_{\alpha,x})$. Then $D^{-1}(U) \cap \mathcal{F}_{\alpha,x} = D^{-1}(\mathcal{G}_{\alpha,x} \cap U) \cap \mathcal{F}_{\alpha,x} = D_\alpha^{-1}(\mathcal{G}_{\alpha,x} \cap U)$ for $D_\alpha : \mathcal{F}_{\alpha,x} \to \mathcal{G}_{\alpha,x}$. Since both D_α and $\mathcal{G}_{\alpha,x} \to \mathcal{G}_x$ are continuous, the claim follows. $\square$

E.2   Analytification of First-Order Differential Operators

Let $f : X \to S$ be a morphism of finite type between schemes X and S which are of finite type over $\mathbb{C}$.[5] Let $\mathcal{F}$ and $\mathcal{G}$ be two quasi-coherent sheaves on X. Then a differential operator of first order relative to X/S is an $f^{-1}(O_S)$-linear map $D : \mathcal{F} \to \mathcal{G}$ such that, for every local section $a \in O_X$, the map $D_a : \mathcal{F} \to \mathcal{G}$, $f \mapsto D(af) - aD(f)$, is O_X-linear. In this section, we prove the following result.

Proposition E.7   *In the above situation, we have the following:*

(1) *There is a unique locally bounded differential operator*

$$D^{\mathrm{an}} : \quad \mathcal{F}^{\mathrm{an}} \to \mathcal{G}^{\mathrm{an}}$$

of first order relative to $X^{\mathrm{an}}/S^{\mathrm{an}}$ such that

$$\varphi_* D^{\mathrm{an}} \circ c_{\mathcal{F}} = c_{\mathcal{G}} \circ D : \quad \mathcal{F} \to \varphi_* \mathcal{G}^{\mathrm{an}}$$

where $c_{\mathcal{F}} : \mathcal{F} \to \varphi_* \mathcal{F}^{\mathrm{an}}$ and $c_{\mathcal{G}} : \mathcal{G} \to \varphi_* \mathcal{G}^{\mathrm{an}}$ are the adjunction maps.
(2) *Equivalently, we can characterize D^{an} as the unique locally bounded differential operator $D^{\mathrm{an}} : \mathcal{F}^{\mathrm{an}} \to \mathcal{G}^{\mathrm{an}}$ of first order relative to $X^{\mathrm{an}}/S^{\mathrm{an}}$ such that*

$$D^{\mathrm{an}} \circ c'_{\mathcal{F}} = c'_{\mathcal{G}} \circ \varphi^{-1} D : \quad \varphi^{-1}\mathcal{F} \to \mathcal{G}^{\mathrm{an}}$$

where $c'_{\mathcal{F}} : \varphi^{-1}\mathcal{F} \to \mathcal{F}^{\mathrm{an}}$ and $c'_{\mathcal{G}} : \varphi^{-1}\mathcal{G} \to \mathcal{G}^{\mathrm{an}}$ are the adjunction maps.
(3) *If $\mathcal{E}$ is quasi-coherent and $h : \mathcal{E} \to \mathcal{F}$ is O_X-linear, then $D^{\mathrm{an}} \circ h^{\mathrm{an}} = (D \circ h)^{\mathrm{an}}$.*
(4) *If $\mathcal{H}$ is quasi-coherent and $h : \mathcal{G} \to \mathcal{H}$ is O_X-linear, then $h^{\mathrm{an}} \circ D^{\mathrm{an}} = (h \circ D)^{\mathrm{an}}$.*
(5) *If $E : \mathcal{G} \to \mathcal{H}$ is another differential operator of first order relative to X/S between quasi-coherent sheaves, then $E^{\mathrm{an}} \circ D^{\mathrm{an}} = 0$ if and only if $E \circ D = 0$.*

We will obtain this as a consequence of a number of preliminary results.

Lemma E.8   *Let $\mathcal{F}$ and $\mathcal{G}$ be two quasi-coherent sheaves of modules on X, and let $h_1, h_2 : \mathcal{F}^{\mathrm{an}} \to \mathcal{G}^{\mathrm{an}}$ be two locally bounded maps of sheaves of sets such that $h_1, h_2 : \mathcal{F}_x^{\mathrm{an}} \to \mathcal{G}_x^{\mathrm{an}}$ are continuous for the sequence topologies. Assume that $\varphi_* h_1 \circ$*

---

[5] Although some of the theory works for locally Noetherian schemes as well, we restrict here to the Noetherian case as we shall only be interested in Noetherian schemes anyway. Then we can use that every quasi-coherent sheaf on a scheme X is the colimit of its coherent subsheaves.

$c_{\mathcal{F}} = \varphi_* h_2 \circ c_{\mathcal{F}}$ *as maps* $\mathcal{F} \to \varphi_* \mathcal{F}^{\mathrm{an}} \to \varphi_* \mathcal{G}^{\mathrm{an}}$. *Then* $h_1 = h_2$. *In particular, if* $h: \mathcal{F} \to \mathcal{G}$ *is* O_X-*linear, then* $h^{\mathrm{an}}: \mathcal{F}^{\mathrm{an}} \to \mathcal{G}^{\mathrm{an}}$ *is the unique* O_X^{an}-*linear map with* $\varphi_* h^{\mathrm{an}} \circ c_{\mathcal{F}} = c_{\mathcal{G}} \circ h$.

Proof It suffices to show that $h_1 = h_2$ at the stalks $\mathcal{F}_x^{\mathrm{an}} \to \mathcal{G}_x^{\mathrm{an}}$. Let $U \subseteq X$ be an affine open subset with $x \in U$. Since $\varphi_* h_1 \circ c_{\mathcal{F}} = \varphi_* h_2 \circ c_{\mathcal{F}}$, the compositions of the two maps at the stalks with $H^0(U, \mathcal{F}) \to \mathcal{F}_x^{\mathrm{an}}$ are equal. Now let $A_x \subseteq \mathcal{F}_x^{\mathrm{an}}$ be the subset where h_1 and h_2 are equal, i.e., we have just seen that the image of $q: H^0(U, \mathcal{F}) \to \mathcal{F}_x^{\mathrm{an}}$ is contained in A_x. Let $M \subseteq \mathcal{F}_x^{\mathrm{an}}$ be a finitely generated $O_{X^{\mathrm{an}},x}$-submodule.[6] Since $\mathcal{F}$ is pseudo-coherent, we can find, locally around x, a coherent subsheaf $\mathcal{E} \subseteq \mathcal{F}$ with $\mathcal{E}_x = M$. Since both h_1 and h_2 are locally bounded, we can find, again locally around x, another coherent subsheaf $\mathcal{E}' \subseteq \mathcal{G}$ such that both h_1 and h_2 map $\mathcal{E}$ into $\mathcal{E}'$. Let $N \subseteq \mathcal{G}_x^{\mathrm{an}}$ be the stalk of $\mathcal{E}'$ at x. Then we have two induced continuous maps $h_1, h_2: M \to N$; continuity of these maps follows from the fact that the sequence topology on N is the induced subspace topology from $\mathcal{G}_x^{\mathrm{an}}$. Since N is finitely generated, and hence its sequence topology is Hausdorff, the set where h_1 and h_2 agree on M is closed, i.e., $A_x \cap M$ is closed. Then $A_x \subseteq \mathcal{F}_x^{\mathrm{an}}$ is closed by the definition of the sequence topology on $\mathcal{F}_x^{\mathrm{an}}$. Since A_x contains a dense subset of $\mathcal{F}_x^{\mathrm{an}}$, we have $A_x = \mathcal{F}_x^{\mathrm{an}}$, and hence $h_1 = h_2$. $\qquad\square$

Proposition E.9 *Let* $D: O_X \to \mathcal{F}$ *be a derivation relative to* X/S *with values in a coherent sheaf* $\mathcal{F}$. *Then there is a unique derivation* $D^{\mathrm{an}}: O_X^{\mathrm{an}} \to \mathcal{F}^{\mathrm{an}}$ *relative to* $X^{\mathrm{an}}/S^{\mathrm{an}}$ *satisfying* $\varphi_* D^{\mathrm{an}} \circ c_O = c_{\mathcal{F}} \circ D$ *for the adjunction maps* $c_O: O \to \varphi_* O_X^{\mathrm{an}}$ *and* $c_{\mathcal{F}}: \mathcal{F} \to \varphi_* \mathcal{F}^{\mathrm{an}}$.

Proof Consider the diagram

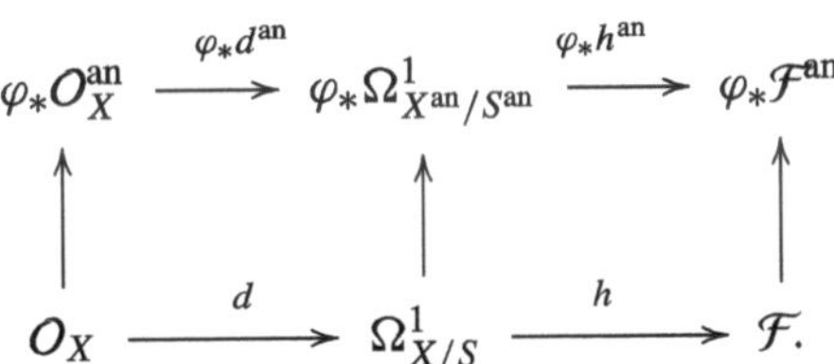

We define $D^{\mathrm{an}} := h^{\mathrm{an}} \circ d^{\mathrm{an}}$. This is a derivation over $X^{\mathrm{an}}/S^{\mathrm{an}}$ which makes the diagram commutative. If we have another derivation D' with this property, then it factors as $D' = h' \circ d^{\mathrm{an}}$; namely, $\mathcal{F}^{\mathrm{an}}$ is a coherent analytic sheaf so that the universal property of d^{an} applies to it. Using that the algebraic $d: O_X \to \Omega_{X/S}^1$ is universal also for derivations with values in $\varphi_* \mathcal{F}^{\mathrm{an}}$, we can show that $h' = h^{\mathrm{an}}$. Then we have $D' = D^{\mathrm{an}}$. $\qquad\square$

We now obtain Proposition E.7 in the special case where both $\mathcal{F}$ and $\mathcal{G}$ are coherent.

---

[6] The proof is so subtle because we do not know if the sequence topology on $\mathcal{G}_x^{\mathrm{an}}$ is Hausdorff.

Proposition E.10 *Let $D: \mathcal{F} \to \mathcal{G}$ be a differential operator of first order relative to X/S between coherent sheaves. Then there is a unique differential operator $D^{\mathrm{an}}: \mathcal{F}^{\mathrm{an}} \to \mathcal{G}^{\mathrm{an}}$ of first order relative to $X^{\mathrm{an}}/S^{\mathrm{an}}$ satisfying $\varphi_* D^{\mathrm{an}} \circ c_{\mathcal{F}} = c_{\mathcal{G}} \circ D$.*

Proof By definition, for every $a \in O_X$, the map $H_a(f) := D(af) - aD(f)$ is O_X-linear. This defines a sheaf homomorphism

$$H: \quad O_X \to \mathcal{H}om(\mathcal{F}, \mathcal{G}).$$

By direct computation, this is a derivation relative to X/S. Thus, we have an analytification

$$H^{\mathrm{an}}: \quad O_X^{\mathrm{an}} \to \mathcal{H}om(\mathcal{F}, \mathcal{G})^{\mathrm{an}} \xrightarrow{\cong} \mathcal{H}om(\mathcal{F}^{\mathrm{an}}, \mathcal{G}^{\mathrm{an}}),$$

where the right-hand isomorphism comes from the fact that $\varphi: X^{\mathrm{an}} \to X$ is flat and $\mathcal{F}$ is finitely presented (otherwise, it is just a map). Then we define

$$D^{\mathrm{an}}: \quad \varphi^{-1}\mathcal{F} \otimes_{\varphi^{-1}O_X} O_X^{\mathrm{an}} \to \varphi^{-1}\mathcal{G} \otimes_{\varphi^{-1}O_X} O_X^{\mathrm{an}}, \quad f \otimes a \mapsto D(f) \otimes a + H_a(f \otimes 1).$$

If $b \in \varphi^{-1}O_X$, then $D^{\mathrm{an}}(f \otimes ba) = D^{\mathrm{an}}(fb \otimes a)$, showing that the map is well-defined. It is also a differential operator of order 1 relative to $X^{\mathrm{an}}/S^{\mathrm{an}}$. Furthermore, it satisfies $\varphi_* D^{\mathrm{an}} \circ c_{\mathcal{F}} = c_{\mathcal{G}} \circ D$. Then, one may show uniqueness by using the uniqueness of H and of the composition $\varphi^{-1}\mathcal{F} \to \mathcal{F}^{\mathrm{an}} \to \mathcal{G}^{\mathrm{an}}$. $\qquad \square$

Before we can prove Proposition E.7, we need a final preparation.

Lemma E.11 *Let $D: \mathcal{F} \to \mathcal{G}$ be a differential operator of first order relative to X/S between quasi-coherent sheaves. Then D is globally bounded, i.e., for every coherent subsheaf $\mathcal{E} \subseteq \mathcal{F}$, the image $D(\mathcal{E})$ is contained in a coherent subsheaf $\mathcal{E}' \subseteq \mathcal{G}$.*

Proof As a consequence of [267, 01PF], it is sufficient to work on an affine open subset $U \subseteq X$. Let $e_1, \ldots, e_n$ be a set of generators of $\mathcal{E}$, and let $a_1, \ldots, a_m$ be a set of generators of $O_X(U)$ as a $\mathbb{C}$-algebra. Let $\mathcal{E}' \subseteq \mathcal{G}$ be the submodule generated by $D(e_j)$ for $1 \leq j \leq n$ and $D(a_i e_j)$ for $1 \leq i \leq m$, $1 \leq j \leq n$. Since D is $\mathbb{C}$-linear and satisfies $D(abf) - aD(bf) - bD(af) + abD(f) = 0$ for $a, b \in O_X$, $f \in \mathcal{F}$, we find that $D(\mathcal{E}(U))$ is contained in $\mathcal{E}'(U)$. Then there is a unique first-order differential operator $E: \mathcal{E} \to \mathcal{E}'$ with $\Gamma(U, E) = \Gamma(U, D)$ by Proposition C.19. Applying Lemma C.6 to standard open subsets of U and, if necessary, enlarging $\mathcal{E}'$ for each standard open individually, we find that $D|_U = E$. Thus, $D(\mathcal{E}) \subseteq \mathcal{E}'$. $\qquad \square$

Proof of Proposition E.7 We start with (1). Let $\{\mathcal{F}_\alpha\}_\alpha$ be a directed system of coherent subsheaves of $\mathcal{F}$ with colimit $\mathcal{F}$. For each α, let $\mathcal{G}_\alpha$ be the intersection of all (quasi-)coherent subsheaves of $\mathcal{G}$ which contain the image of $D_\alpha := D|_{\mathcal{F}_\alpha}: \mathcal{F}_\alpha \to \mathcal{G}$. Then $\mathcal{G}_\alpha$ is a coherent subsheaf of $\mathcal{G}$ by Lemma E.11, and for $\alpha \leq \beta$, we have $\mathcal{G}_\alpha \subseteq \mathcal{G}_\beta$. Proposition E.10 yields a differential operator $D_\alpha^{\mathrm{an}}: \mathcal{F}_\alpha^{\mathrm{an}} \to \mathcal{G}_\alpha^{\mathrm{an}}$ for each α. For $\alpha \leq \beta$, let us write $i_{\alpha\beta}: \mathcal{F}_\alpha \to \mathcal{F}_\beta$ respectively $j_{\alpha\beta}: \mathcal{G}_\alpha \to \mathcal{G}_\beta$ for the inclusions. Then we have $D_\beta^{an} \circ i_{\alpha\beta}^{an} = j_{\alpha\beta}^{an} \circ D_\alpha^{an}$ because both are the unique analytification of the differential operator $D_\beta \circ i_{\alpha\beta} = j_{\alpha\beta} \circ D_\alpha$. Now $\mathcal{F}^{an}$ is the direct limit of $\{\mathcal{F}_\alpha^{an}\}_\alpha$ in the category of Abelian sheaves on X^{an}, so there is an induced map $D^{an}\colon \mathcal{F}^{an} \to \mathcal{G}^{an}$ coming from the compositions $\mathcal{F}_\alpha^{an} \to \mathcal{G}_\alpha^{an} \to \mathcal{G}^{an}$. It is a differential operator of first order relative to X^{an}/S^{an} because its restriction to each $\mathcal{F}_\alpha^{an} \subseteq \mathcal{F}^{an}$ is. It satisfies $\varphi_* D^{an} \circ c_\mathcal{F} = c_\mathcal{G} \circ D$ because the compositions of the two maps with $i_{\alpha\beta}\colon \mathcal{F}_\alpha \to \mathcal{F}$ are the same for all α, and $\mathcal{F}$ is the direct limit of $\{\mathcal{F}_\alpha\}_\alpha$ in the category of Abelian sheaves on X.

To show that D^{an} is locally bounded, let $\mathcal{U} \subseteq X^{an}$ be an open subset, and let $\mathcal{E} \subseteq \mathcal{F}^{an}|_\mathcal{U}$ be a coherent analytic subsheaf. Let $\{\mathcal{U}_i\}_i$ be an open cover of $\mathcal{U}$ such that each $\mathcal{E}|_{\mathcal{U}_i}$ is finitely generated. By refining $\{\mathcal{U}_i\}_i$ if necessary, we can assume that we have, for each index i, an index α_i with $\mathcal{E}|_{\mathcal{U}_i} \subseteq \mathcal{F}_{\alpha_i}^{an}|_{\mathcal{U}_i}$. Namely, at every point $x \in \mathcal{U}_i$, the germs of the finitely many generators of $\mathcal{E}|_{\mathcal{U}_i}$ must all be contained in the stalk $\mathcal{F}_{\alpha,x}^{an}$ for some α. Then D^{an} maps $\mathcal{E}|_{\mathcal{U}_i}$ into the coherent analytic subsheaf $\mathcal{G}_{\alpha_i}^{an}|_{\mathcal{U}_i}$ of $\mathcal{G}^{an}|_{\mathcal{U}_i}$. Thus, D^{an} is locally bounded. Uniqueness follows from Lemma E.8, using Lemma E.6.

For (3) and (4), first note that the statement holds if all involved modules are coherent. Then, to see (3), use a representation of $\mathcal{E}$ as a direct limit of coherent subsheaves $\mathcal{E}_\alpha$, let $\mathcal{F}_\alpha := h(\mathcal{E}_\alpha)$, and let $\mathcal{G}_\alpha$ be constructed from this $\mathcal{F}_\alpha$ like in the construction of D^{an} above. To see (4), start with a representation of $\mathcal{F}$ as a direct limit of coherent subsheaves $\mathcal{F}_\alpha$, let $\mathcal{G}_\alpha$ be as above, and let $\mathcal{H}_\alpha := h(\mathcal{G}_\alpha)$. Then use that this $\mathcal{H}_\alpha$ is the smallest submodule of $\mathcal{H}$ which contains $h \circ D(\mathcal{F}_\alpha)$.

For (5), note that $E^{an} \circ D^{an}$ is locally bounded and induces continuous maps $\mathcal{F}_x^{an} \to \mathcal{H}_x^{an}$ on the stalks. If $E \circ D = 0$, then $\varphi_* E^{an} \circ \varphi_* D^{an} \circ c_\mathcal{F} = 0$, so $E^{an} \circ D^{an} = 0$ by Lemma E.8. For the converse, first note that $c_\mathcal{H}\colon \mathcal{H} \to \varphi_* \mathcal{H}^{an}$ is injective. Namely, using the Krull intersection theorem, we find that $\mathcal{E}_{\varphi(x)} \to \mathcal{E}_x^{an}$ is injective for $\mathcal{E}$ coherent; then, however, it is easy to generalize to quasi-coherent $\mathcal{E}$ using the direct limit description. Now the injectivity of $c_\mathcal{H}$ implies $E \circ D = 0$. $\square$

Appendix F
The Script for Example 10.5

We provide an annotated Macaulay2 script which we used to establish that not all
infinitesimal automorphisms lift in Example 10.5.

```
-- ------------------------------------------------

-- this is a script to show that for a general generically
-- log smooth deformation, automorphisms over affine opens
-- do not always lift

-- the example that we give is given by the equations
-- xy = w^3 + t^3 and zw = tu


-- ------------------------------------------------


-- we will use the method reflexify of the following package
-- to compute the map into the bidual

needsPackage("Divisor")

-- ------------------------------------------------

-- definition of the family over A^1_t

OX = QQ[x,y,z,w,t,u]/(x*y - w^3 - t^3, z*w - t*u)

-- classical relative tangent space T_X/S;
-- it is canonically isomorphic to the relative
-- log tangent space once we use the compactifying
-- log structures coming from t = 0

relativeTangentMap = matrix{{y,x,0,-3*w^2,0},{0,0,w,z,-t}}
```


S. Felten, *Global Logarithmic Deformation Theory*, Lecture Notes
in Mathematics 2373, https://doi.org/10.1007/978-3-031-98751-9

```
TXS = kernel(relativeTangentMap)


-- --------------------------------------------------

-- we compute the base change to the
-- Artinian ring Q[t]/(t^3)

OX2 = OX/ideal(t^3)

preThetaXS2 = TXS ** OX2
-- this is isomorphic to the relative log tangent sheaf
-- on the log smooth locus

ThetaXS2 = target(reflexify(preThetaXS2,ReturnMap=>true))
-- this is the reflexive hull; it is isomorphic to the
-- relative log tangent sheaf


-- --------------------------------------------------

-- we compute further base change to Q[t]/(t^2)

OX1 = OX2/ideal(t^2)

r21 = map(OX1,OX2)

pThetaXS21 = tensor(r21,ThetaXS2)
-- this is isomorphic to the relative log tangent sheaf
-- on the log smooth locus

c21 = reflexify(pThetaXS21,ReturnMap=>true)

ThetaXS1 = target(c21)
-- the reflexive hull is isomorphic to
-- the relative log tangent sheaf

k21 = map(ThetaXS1,ThetaXS2,r21,
                matrix c21 * matrix coverMap(pThetaXS21))
-- this is the restriction map from the
-- relative log tangent sheaf over Q[t]/(t^3)
-- to the one over Q[t]/(t^2)


-- ------------------

-- we compute further base change
-- to S0 = Spec (N -> Q[t]/(t))

OX0 = OX1/ideal(t)

r10 = map(OX0,OX1)

pThetaXS10 = tensor(r10,ThetaXS1)
```

```
c10 = reflexify(pThetaXS10,ReturnMap=>true)

ThetaXS0 = target(c10)

k10 = map(ThetaXS0,ThetaXS1,r10,
                matrix c10 * matrix coverMap(pThetaXS10))

-- this is the restriction map from ThetaXS1 to ThetaXS0

k20 = k10 * k21
-- the restriction from ThetaXS2 to ThetaXS0

-- note that using k20, k21, and k10 is difficult
-- from a programming perspective because Macaulay2
-- does not easily work with maps
-- between modules over different rings


-- ---------------------------------------------------


K20 = kernel k20
-- one of the few operations supported for maps
-- between modules over different rings

-- unfortunately, K20 is not given as a
-- submodule of ThetaXS2 but as a
-- submodule of the generators module;
-- we fix this with the following map

embK20 = inducedMap(ThetaXS2,K20,gens ThetaXS2)

Aut20 = image embK20

K10 = kernel k10

embK10 = inducedMap(ThetaXS1,K10,gens ThetaXS1)

Aut10 = image embK10

-- in principle, we now have the desired restriction map
-- k21: Aut20 -> Aut10 that we wish to show not surjective;
-- unfortunately, studying this map is pretty complicated
-- in Macaulay2---we cannot just command
-- image(Aut20 -> ThetaXS2 -> ThetaXS1)
-- and then compare it to Aut10


-- ----------------------------------------


-- we compute k21(Aut10) as a subset of ThetaXS1

G = source coverMap(K20)

p21 = k21 * embK20 * coverMap(K20)
-- the image of p21 in ThetaXS1 is k21(Aut20);
-- the source G of this map is a free module
```

```
r = (rank G) - 1
-- the rank of G - 1 = number of generators in this module

ze = p21(G_0)
-- the image of the first generator of G in ThetaXS1
-- happens to be 0; thus, this object is the
-- zero object in ThetaXS1

-- because OX2 -> OX1 is surjective, the image of p21
-- is the OX1-submodule spanned by all p21(G_i);
-- we make a list that contains precisely
-- the non-zero image vectors

L = {} -- empty list

for i from 0 to r do
     if (p21(G_i) != ze) then L = append(L,p21(G_i));

-- next, we wish to create the submodule in ThetaXS1
-- that is generated by the vectors in the list;
-- since Macaulay2 still remembers them
-- as coming from another ring,
-- it cannot easily form this submodule;
-- instead we extract the symbols of this list
-- and use them to define a new matrix over OX1

st = toString(L)

"L-saving-file.txt" <<
        "use OX1" << endl << "ML = matrix" << st << close

load "L-saving-file.txt"
-- it is here that we had to remove all zero entries
-- because they would not be interpreted over OX1
-- but over ZZ -- at least if I make only the list
-- in the file and try to make
-- the matrix on the M2 prompt

-- unfortunately, now the degrees of ML and of ThetaXS1
-- do not match; we need the following comparison map,
-- which seems to me can only be defined by explicitly
-- spelling out the matrix
-- (it has been generated with an id command,
-- then printed to a file,
-- and from there copied into this script);


P = inducedMap(ThetaXS1, image ML, matrix
{{1, 0_OX1, 0,0,0,0,0,0,0,0}, {0,1,0,0,0,0,0,0,0,0},
{0,0,1,0,0,0,0,0,0,0}, {0,0,0,1,0,0,0,0,0,0},
{0,0,0,0,1,0,0,0,0,0}, {0,0,0,0,0,1,0,0,0,0},
{0,0,0,0,0,0,1,0,0,0}, {0,0,0,0,0,0,0,1,0,0},
{0,0,0,0,0,0,0,0,1,0}, {0,0,0,0,0,0,0,0,0,1}})
```

```
-- now image P is the image of Aut20 under k21 in ThetaXS1

-- a sanity check:

assert(isSubset(image P,Aut10))
-- this is what we expect

-- the result:

assert(not isSubset(Aut10, image P))

-- hence image P is a proper subset of Aut10,
-- and there are morphisms in Aut10 that do not lift to Aut20


-- -------------------------------------

-- we compute how much the two differ

Q = Aut10 / image P

Qs = Q ** OX1/ann(Q)
-- Q and Qs are the same as QQ-vector spaces

Qsp = prune Qs

-- the ring of Qsp, i.e., OX1/ann(Q),
-- happens to be isomorphic to QQ

assert(isFreeModule(Qsp))

assert(rank Qsp == 1)

-- thus Qsp, and hence Q, is a QQ-vector space of dimension 1
```

References

1. Abramovich, D., Chen, Q.: Stable logarithmic maps to Deligne-Faltings pairs II. Asian J. Math. **18**(3), 465–488 (2014). https://doi.org/10.4310/AJM.2014.v18.n3.a5
2. Abramovich, D., Chen, Q., Gross, M., Siebert, B.: Decomposition of degenerate Gromov-Witten invariants. Compos. Math. **156**(10), 2020–2075 (2020). https://doi.org/10.1112/s0010437x20007393
3. Abramovich, D., Karu, K.: Weak semistable reduction in characteristic 0. Invent. Math. **139**(2), 241–273 (2000). https://doi.org/10.1007/s002229900024
4. Abramovich, D., Wise, J.: Birational invariance in logarithmic Gromov-Witten theory. Compos. Math. **154**(3), 595–620 (2018). https://doi.org/10.1112/S0010437X17007667
5. Altmann, K.: Computation of the vector space T^1 for affine toric varieties. J. Pure Appl. Algebra **95**(3), 239–259 (1994). https://doi.org/10.1016/0022-4049(94)90060-4
6. Altmann, K.: Minkowski sums and homogeneous deformations of toric varieties. Tohoku Math. J. (2) **47**(2), 151–184 (1995). https://doi.org/10.2748/tmj/1178225590
7. Altmann, K.: Infinitesimal deformations and obstructions for toric singularities. J. Pure Appl. Algebra **119**(3), 211–235 (1997). https://doi.org/10.1016/S0022-4049(96)00029-1
8. Altmann, K.: The versal deformation of an isolated toric Gorenstein singularity. Invent. Math. **128**(3), 443–479 (1997). https://doi.org/10.1007/s002220050148
9. Altmann, K.: One parameter families containing three-dimensional toric-Gorenstein singularities. In: Explicit Birational Geometry of 3-folds, *London Mathematical Society Lecture Note Series*, vol. 281, pp. 21–50. Cambridge University Press, Cambridge (2000)
10. Ambro, F.: Cyclic covers and toroidal embeddings. Eur. J. Math. **2**(1), 9–44 (2016). https://doi.org/10.1007/s40879-015-0084-y
11. Anderson, D.: Okounkov bodies and toric degenerations. Math. Ann. **356**(3), 1183–1202 (2013). https://doi.org/10.1007/s00208-012-0880-3
12. Argüz, H.: Real loci in (log) Calabi-Yau manifolds via Kato-Nakayama spaces of toric degenerations. Eur. J. Math. **7**(3), 869–930 (2021). https://doi.org/10.1007/s40879-021-00454-z
13. Artebani, M., Dolgachev, I.: The Hesse pencil of plane cubic curves. Enseign. Math. (2) **55**(3-4), 235–273 (2009). https://doi.org/10.4171/LEM/55-3-3
14. Artin, M.: On the solutions of analytic equations. Invent. Math. **5**, 277–291 (1968). https://doi.org/10.1007/BF01389777
15. Artin, M.: Algebraic approximation of structures over complete local rings. Inst. Hautes Études Sci. Publ. Math. **36**, 23–58 (1969). http://www.numdam.org/item?id=PMIHES_1969__36__23_0

16. Atiyah, M.F., Macdonald, I.G.: Introduction to Commutative Algebra. Addison-Wesley Publishing Co., Reading (1969)
17. Barannikov, S.: Quantum periods. I. Semi-infinite variations of Hodge structures. Internat. Math. Res. Notices **23**, 1243–1264 (2001). https://doi.org/10.1155/S1073792801000599
18. Barannikov, S.: Non-commutative periods and mirror symmetry in higher dimensions. Comm. Math. Phys. **228**(2), 281–325 (2002). https://doi.org/10.1007/s002200200656
19. Barannikov, S., Kontsevich, M.: Frobenius manifolds and formality of Lie algebras of polyvector fields. Internat. Math. Res. Notices **4**, 201–215 (1998). https://doi.org/10.1155/S1073792898000166
20. Barannikov, S.A.: Extended Moduli Spaces and Mirror Symmetry in Dimensions $n > 3$. ProQuest LLC, Ann Arbor, MI (1999). http://gateway.proquest.com/openurl?url_ver=Z39.88-2004&rft_val_fmt=info:ofi/fmt:kev:mtx:dissertation&res_dat=xri:pqdiss&rft_dat=xri:pqdiss:9931178. Thesis (Ph.D.)–University of California, Berkeley
21. Bashkirov, D., Voronov, A.A.: The BV formalism for L_∞-algebras. J. Homotopy Relat. Struct. **12**(2), 305–327 (2017). https://doi.org/10.1007/s40062-016-0129-z
22. Batyrev, V.V.: Dual polyhedra and mirror symmetry for Calabi-Yau hypersurfaces in toric varieties. J. Algebraic Geom. **3**(3), 493–535 (1994)
23. Batyrev, V.V., Borisov, L.A.: On Calabi-Yau complete intersections in toric varieties. In: Higher-Dimensional Complex Varieties (Trento, 1994), pp. 39–65. de Gruyter, Berlin (1996)
24. Bellier-Millès, J., Drummond-Cole, G.C.: Homotopy theory of curved operads and curved algebras (2020)
25. Berkovich, V.G.: Smooth p-adic analytic spaces are locally contractible. Invent. Math. **137**(1), 1–84 (1999). https://doi.org/10.1007/s002220050323
26. Castaño Bernard, R., Matessi, D.: Conifold transitions via affine geometry and mirror symmetry. Geom. Topol. **18**(3), 1769–1863 (2014). https://doi.org/10.2140/gt.2014.18.1769
27. Bingener, J.: Offenheit der Versalität in der analytischen Geometrie. Math. Z. **173**(3), 241–281 (1980). https://doi.org/10.1007/BF01159663
28. Bogomolov, F.A.: Hamiltonian Kählerian manifolds. Dokl. Akad. Nauk SSSR **243**(5), 1101–1104 (1978)
29. Bourbaki, N.: Éléments de mathématique. Fascicule XXVII. Algèbre commutative. Chapitre 1: Modules plats. Chapitre 2: Localisation. Actualités Scientifiques et Industrielles [Current Scientific and Industrial Topics], No. 1290. Hermann, Paris (1961)
30. Bousseau, P.: Tropical refined curve counting from higher genera and lambda classes. Invent. Math. **215**(1), 1–79 (2019). https://doi.org/10.1007/s00222-018-0823-z
31. Bousseau, P.: The quantum tropical vertex. Geom. Topol. **24**(3), 1297–1379 (2020). https://doi.org/10.2140/gt.2020.24.1297
32. Braun, C., Lazarev, A.: Homotopy BV algebras in Poisson geometry. Trans. Moscow Math. Soc. **74**, 217–227 (2013). https://doi.org/10.1090/s0077-1554-2014-00216-8
33. Calaque, D., Grivaux, J.: Formal moduli problems and formal derived stacks. In: Derived Algebraic Geometry, *Panorama Synthèses*, vol. 55, pp. 85–145. Soc. Math. France, Paris ([2021] ©2021)
34. Candelas, P., de la Ossa, X.C., Green, P.S., Parkes, L.: A pair of Calabi-Yau manifolds as an exactly soluble superconformal theory. Nuclear Phys. B **359**(1), 21–74 (1991). https://doi.org/10.1016/0550-3213(91)90292-6
35. Casas, J.M., Ladra, M., Pirashvili, T.: Crossed modules for Lie-Rinehart algebras. J. Algebra **274**(1), 192–201 (2004). https://doi.org/10.1016/j.jalgebra.2003.10.001
36. Chan, K., Leung, N.C., Ma, Z.N.: Scattering diagrams from asymptotic analysis on Maurer-Cartan equations. J. Eur. Math. Soc. (JEMS) **24**(3), 773–849 (2022). https://doi.org/10.4171/JEMS/1100
37. Chan, K., Leung, N.C., Ma, Z.N.: Smoothing, scattering, and a conjecture of Fukaya (2022)
38. Chan, K., Leung, N.C., Ma, Z.N.: Geometry of the Maurer-Cartan equation near degenerate Calabi-Yau varieties. J. Differential Geom. **125**(1), 1–84 (2023). https://doi.org/10.4310/jdg/1695236591

39. Chan, K., Ma, Z.N.: Tropical counting from asymptotic analysis on Maurer-Cartan equations. Trans. Am. Math. Soc. **373**(9), 6411–6450 (2020). https://doi.org/10.1090/tran/8128

40. Chan, K., Ma, Z.N.: Smoothing pairs over degenerate Calabi-Yau varieties. Int. Math. Res. Not. IMRN **4**, 2582–2614 (2022). https://doi.org/10.1093/imrn/rnaa212

41. Chan, K., Ma, Z.N., Suen, Y.H.: Tropical Lagrangian multi-sections and smoothing of locally free sheaves over degenerate Calabi-Yau surfaces. Adv. Math. **401**, Paper No. 108280, 37 (2022). https://doi.org/10.1016/j.aim.2022.108280

42. Chen, Q.: The degeneration formula for logarithmic expanded degenerations. J. Algebraic Geom. **23**(2), 341–392 (2014). https://doi.org/10.1090/S1056-3911-2013-00614-1

43. Chen, Q.: Stable logarithmic maps to Deligne-Faltings pairs I. Ann. of Math. (2) **180**(2), 455–521 (2014). https://doi.org/10.4007/annals.2014.180.2.2

44. Chuang, J., Lazarev, A., Mannan, W.H.: Cocommutative coalgebras: homotopy theory and Koszul duality. Homology Homotopy Appl. **18**(2), 303–336 (2016). https://doi.org/10.4310/HHA.2016.v18.n2.a17

45. Cirici, J., Horel, G.: Formality of hypercommutative algebras of kähler and calabi-yau manifolds (2023)

46. Clemens, H.: Homological equivalence, modulo algebraic equivalence, is not finitely generated. Inst. Hautes Études Sci. Publ. Math. **58**, 19–38 (1983). http://www.numdam.org/item?id=PMIHES_1983__58__19_0

47. Coates, T., Corti, A., Galkin, S., Golyshev, V., Kasprzyk, A.: Mirror symmetry and Fano manifolds. In: European Congress of Mathematics, pp. 285–300. European Mathematical Society, Zürich (2013)

48. Coates, T., Corti, A., Galkin, S., Kasprzyk, A.: Quantum periods for 3-dimensional Fano manifolds. Geom. Topol. **20**(1), 103–256 (2016). https://doi.org/10.2140/gt.2016.20.103

49. Coates, T., Corti, A., da Silva Jr., G.: On the topology of Fano smoothings. In: Interactions with lattice polytopes, *Springer Proceedings in Mathematics & Statistics*, vol. 386, pp. 135–156. Springer, Cham ([2022] ©2022). https://doi.org/10.1007/978-3-030-98327-7_6

50. Coates, T., Galkin, S., Kasprzyk, A., Strangeway, A.: Quantum periods for certain four-dimensional Fano manifolds. Exp. Math. **29**(2), 183–221 (2020). https://doi.org/10.1080/10586458.2018.1448018

51. Conrad, B.: Grothendieck duality and base change. In: Lecture Notes in Mathematics, vol. 1750. Springer-Verlag, Berlin (2000). https://doi.org/10.1007/b75857

52. Conrad, B.: Relative ampleness in rigid geometry. Ann. Inst. Fourier (Grenoble) **56**(4), 1049–1126 (2006). http://aif.cedram.org/item?id=AIF_2006__56_4_1049_0

53. Corti, A., Filip, M., Petracci, A.: Mirror symmetry and smoothing Gorenstein toric affine 3-folds. In: Facets of algebraic geometry, vol. I. London Mathematical Society Lecture Note Series, vol. 472, pp. 132–163. Cambridge University Press, Cambridge (2022)

54. Corti, A., Hacking, P., Petracci, A.: Smoothing gorenstein toric fano 3-folds (2024). https://arxiv.org/abs/2412.06500

55. Corti, A., Ruddat, H.: How to make log structures (2024)

56. Cox, D.A., Little, J.B., Schenck, H.K.: Toric varieties. In: Graduate Studies in Mathematics, vol. 124. American Mathematical Society, Providence, RI (2011). https://doi.org/10.1090/gsm/124

57. Danilov, V.I.: The geometry of toric varieties. Akademiya Nauk SSSR i Moskovskoe Matematicheskoe Obshchestvo. Uspekhi Matematicheskikh Nauk **33**(2(200)), 85–134, 247 (1978)

58. Deligne, P., Griffiths, P., Morgan, J., Sullivan, D.: Real homotopy theory of Kähler manifolds. Invent. Math. **29**(3), 245–274 (1975). https://doi.org/10.1007/BF01389853

59. Deligne, P., Illusie, L.: Relèvements modulop 2 et décomposition du complexe de de rham. Inventiones mathematicae **89**(2), 247–270 (1987)

60. Deligne, P., Mumford, D.: The irreducibility of the space of curves of given genus. Inst. Hautes Études Sci. Publ. Math. **36**, 75–109 (1969). http://www.numdam.org/item?id=PMIHES_1969__36__75_0

61. Denef, J.: Some remarks on toroidal morphisms. arXiv preprint arXiv:1303.4999 (2013)

62. Doi, M., Yotsutani, N.: Differential geometric global smoothings of simple normal crossing complex surfaces with trivial canonical bundle. Complex Manifolds **10**(1), 1–38 (2023). https://doi.org/10.1515/coma-2022-0143

63. Dotsenko, V., Shadrin, S., Vallette, B.: De Rham cohomology and homotopy Frobenius manifolds. J. Eur. Math. Soc. (JEMS) **17**(3), 535–547 (2015). https://doi.org/10.4171/JEMS/510

64. Dotsenko, V., Shadrin, S., Vallette, B.: Maurer-Cartan methods in deformation theory—the twisting procedure. In: London Mathematical Society Lecture Note Series, vol. 488. Cambridge University Press, Cambridge (2024)

65. Douady, A.: Le problème des modules locaux pour les espaces C-analytiques compacts. Ann. Sci. École Norm. Sup. (4) **7**, 569–602 (1974). http://www.numdam.org/item?id=ASENS_1974_4_7_4_569_0

66. Drummond-Cole, G.C.: Formal formality of the hypercommutative algebras of low dimensional Calabi-Yau varieties. Comm. Math. Phys. **327**(2), 433–441 (2014). https://doi.org/10.1007/s00220-014-2018-9

67. Drummond-Cole, G.C., Vallette, B.: The minimal model for the Batalin-Vilkovisky operad. Selecta Math. (N.S.) **19**(1), 1–47 (2013). https://doi.org/10.1007/s00029-012-0098-y

68. Eisenbud, D., Harris, J.: The geometry of schemes. In: Graduate Texts in Mathematics, vol. 197. Springer, New York (2000)

69. Faltings, G.: F-isocrystals on open varieties: results and conjectures. In: The Grothendieck Festschrift, Vol. II. Progress in Mathematics, vol. 87, pp. 219–248. Birkhäuser Boston, Boston, MA (1990)

70. Fantechi, B., Franciosi, M., Pardini, R.: Deformations of semi-smooth varieties (2021)

71. Fantechi, B., Franciosi, M., Pardini, R.: Smoothing semi-smooth stable Godeaux surfaces. Algebr. Geom. **9**(4), 502–512 (2022). https://doi.org/10.14231/ag-2022-015

72. Fantechi, B., Manetti, M.: On the T^1-lifting theorem. J. Algebraic Geom. **8**(1), 31–39 (1999)

73. Felten, S.: Good differential forms for a family of schemes. Master Thesis (2018)

74. Felten, S.: Log smooth deformation theory via Gerstenhaber algebras. arXiv preprint:2001.02995 (2020)

75. Felten, S.: Log toroidal families (2021). https://doi.org/10.25358/openscience-5658. PhD Thesis

76. Felten, S.: Log smooth deformation theory via Gerstenhaber algebras. Manuscripta Math. **167**(1-2), 1–35 (2022). https://doi.org/10.1007/s00229-020-01255-6

77. Felten, S., Filip, M., Ruddat, H.: Smoothing toroidal crossing spaces. Forum Math. Pi **9**, Paper No. e7, 36 (2021). https://doi.org/10.1017/fmp.2021.8

78. Felten, S., Petracci, A.: The logarithmic Bogomolov-Tian-Todorov theorem. Bull. Lond. Math. Soc. **54**(3), 1051–1066 (2022)

79. Felten, S., Petracci, A., Robins, S.: Deformations of log Calabi-Yau pairs can be obstructed (2022)

80. Ferrand, D.: Conducteur, descente et pincement. Bull. Soc. Math. France **131**(4), 553–585 (2003). https://doi.org/10.24033/bsmf.2455

81. Filippini, S.A., Stoppa, J.: Block-Göttsche invariants from wall-crossing. Compos. Math. **151**(8), 1543–1567 (2015). https://doi.org/10.1112/S0010437X14007994

82. Fiorenza, D., Manetti, M.: L_∞ structures on mapping cones. Algebra Number Theory **1**(3), 301–330 (2007). https://doi.org/10.2140/ant.2007.1.301

83. Fiorenza, D., Manetti, M.: A period map for generalized deformations. J. Noncommut. Geom. **3**(4), 579–597 (2009). https://doi.org/10.4171/JNCG/47

84. Fiorenza, D., Manetti, M., Martinengo, E.: Cosimplicial DGLAs in deformation theory. Comm. Algebra **40**(6), 2243–2260 (2012). https://doi.org/10.1080/00927872.2011.577479

85. Fischer, G.: Complex analytic geometry. In: Lecture Notes in Mathematics, vol. 538. Springer, Berlin-New York (1976)

86. Forster, O., Knorr, K.: Konstruktion verseller Familien kompakter komplexer Räume. In: Lecture Notes in Mathematics, vol. 705. Springer, Berlin (1979)

87. Friedman, R.: Global smoothings of varieties with normal crossings. Ann. of Math. (2) **118**(1), 75–114 (1983). https://doi.org/10.2307/2006955
88. Friedman, R.: Simultaneous resolution of threefold double points. Math. Ann. **274**(4), 671–689 (1986). https://doi.org/10.1007/BF01458602
89. Friedman, R.: On threefolds with trivial canonical bundle. In: Complex geometry and Lie theory (Sundance, UT, 1989). In: Proceedings of Symposia in Pure Mathematics, vol. 53, pp. 103–134. American Mathematical Society, Providence, RI (1991). https://doi.org/10.1090/pspum/053/1141199
90. Friedman, R., Laza, R.: Deformations of Calabi-Yau varieties with isolated log canonical singularities (2023)
91. Friedman, R., Laza, R.: Deformations of Calabi-Yau varieties with k-liminal singularities (2023)
92. Friedman, R., Laza, R.: Deformations of singular Fano and Calabi-Yau varieties (2023)
93. Friedman, R., Laza, R.: Deformations of some local Calabi-Yau manifolds (2023)
94. Friedman, R., Laza, R.: The higher Du Bois and higher rational properties for isolated singularities (2023)
95. Friedman, R., Laza, R.: Higher Du Bois and higher rational singularities (2023)
96. Friedman, R., Laza, R.: Higher Du Bois and higher rational singularities. Duke Math. J. **173**(10), 1839–1881 (2024). https://doi.org/10.1215/00127094-2023-0051. Appendix by Morihiko Saito
97. Friedman, R., Morrison, D.R. (eds.): The birational geometry of degenerations. In: Progress in Mathematics, vol. 29. Birkhäuser, Boston, MA (1983). Based on papers presented at the Summer Algebraic Geometry Seminar held at Harvard University, Cambridge, Mass (1981)
98. Fujino, O.: Fundamental theorems for semi log canonical pairs. Algebr. Geom. **1**(2), 194–228 (2014). https://doi.org/10.14231/AG-2014-011
99. Fujino, O.: Kodaira vanishing theorem for log-canonical and semi-log-canonical pairs (2015)
100. Fujisawa, T., Nakayama, C.: Geometric log Hodge structures on the standard log point. Hiroshima Math. J. **45**(3), 231–266 (2015). http://projecteuclid.org/euclid.hmj/1448323766
101. Fujisawa, T., Nakayama, C.: Geometric polarized log Hodge structures with a base of log rank one. Kodai Math. J. **43**(1), 57–83 (2020). https://doi.org/10.2996/kmj/1584345688
102. Fujita, T.: On del Pezzo fibrations over curves. Osaka J. Math. **27**(2), 229–245 (1990). http://projecteuclid.org/euclid.ojm/1200782303
103. Gálvez-Carrillo, I., Tonks, A., Vallette, B.: Homotopy Batalin-Vilkovisky algebras. J. Noncommut. Geom. **6**(3), 539–602 (2012). https://doi.org/10.4171/JNCG/99
104. Gerstenhaber, M.: The cohomology structure of an associative ring. Ann. of Math. (2) **78**, 267–288 (1963). https://doi.org/10.2307/1970343
105. Getzler, E.: Operads and moduli spaces of genus 0 Riemann surfaces. In: The Moduli Space of Curves (Texel Island, 1994). Progress in Mathematics, vol. 129, pp. 199–230. Birkhäuser Boston, Boston, MA (1995). https://doi.org/10.1007/978-1-4612-4264-2_8
106. Getzler, E.: Lie theory for nilpotent L_∞-algebras. Ann. of Math. (2) **170**(1), 271–301 (2009). https://doi.org/10.4007/annals.2009.170.271
107. Gillam, W.D.: Logarithmic stacks and minimality. Internat. J. Math. **23**(7), 1250069, 38 (2012). https://doi.org/10.1142/S0129167X12500693
108. Godement, R.: Topologie algébrique et théorie des faisceaux. Publications de l'Institut de Mathématique de l'Université de Strasbourg, XIII. Hermann, Paris (1973). Troisième édition revue et corrigée
109. Goldman, W.M., Millson, J.J.: The homotopy invariance of the Kuranishi space. Illinois J. Math. **34**(2), 337–367 (1990). http://projecteuclid.org/euclid.ijm/1255988270
110. Gonciulea, N., Lakshmibai, V.: Degenerations of flag and Schubert varieties to toric varieties. Transform. Groups **1**(3), 215–248 (1996). https://doi.org/10.1007/BF02549207
111. Graefnitz, T.: Tropical correspondence for smooth del Pezzo log Calabi-Yau pairs. J. Algebraic Geom. **31**(4), 687–749 (2022)
112. Grauert, H.: Der Satz von Kuranishi für kompakte komplexe Räume. Invent. Math. **25**, 107–142 (1974). https://doi.org/10.1007/BF01390171

113. Grauert, H., Remmert, R.: Analytische Stellenalgebren. Die Grundlehren der mathematischen Wissenschaften, Band 176. Springer, Berlin (1971). Unter Mitarbeit von O. Riemenschneider

114. Grauert, H., Remmert, R.: Theory of Stein spaces. In: Grundlehren der Mathematischen Wissenschaften, vol. 236. Springer, Berlin (1979). Translated from the German by Alan Huckleberry

115. Grauert, H., Remmert, R.: Coherent analytic sheaves. In: Grundlehren der mathematischen Wissenschaften [Fundamental Principles of Mathematical Sciences], vol. 265. Springer, Berlin (1984). https://doi.org/10.1007/978-3-642-69582-7

116. Greuel, G.M., Lossen, C., Shustin, E.: Introduction to singularities and deformations. In: Springer Monographs in Mathematics. Springer, Berlin (2007)

117. Grignou, B.L., i Lucio, V.R.: A new approach to formal moduli problems (2023)

118. Gross, M.: Deforming Calabi-Yau threefolds. Math. Ann. **308**(2), 187–220 (1997). https://doi.org/10.1007/s002080050072

119. Gross, M.: Toric degenerations and Batyrev-Borisov duality. Math. Ann. **333**(3), 645–688 (2005). https://doi.org/10.1007/s00208-005-0686-7

120. Gross, M., Hacking, P., Keel, S.: Mirror symmetry for log Calabi-Yau surfaces I. Publ. Math. Inst. Hautes Études Sci. **122**, 65–168 (2015). https://doi.org/10.1007/s10240-015-0073-1

121. Gross, M., Hacking, P., Keel, S., Siebert, B.: The mirror of the cubic surface. In: Recent Developments in Algebraic Geometry—to Miles Reid for his 70th Birthday. London Mathematical Society Lecture Note Series, vol. 478, pp. 150–182. Cambridge University Press, Cambridge (2022)

122. Gross, M., Hacking, P., Siebert, B.: Theta functions on varieties with effective anti-canonical class. Mem. Am. Math. Soc. **278**(1367), xii+103 (2022). https://doi.org/10.1090/memo/1367

123. Gross, M., Pandharipande, R., Siebert, B.: The tropical vertex. Duke Math. J. **153**(2), 297–362 (2010). https://doi.org/10.1215/00127094-2010-025

124. Gross, M., Siebert, B.: Mirror symmetry via logarithmic degeneration data. I. J. Differ. Geom. **72**(2), 169–338 (2006)

125. Gross, M., Siebert, B.: Mirror symmetry via logarithmic degeneration data, II. J. Algebraic Geometry **19**(4), 679–780 (2010). https://doi.org/10.1090/S1056-3911-2010-00555-3

126. Gross, M., Siebert, B.: From real affine geometry to complex geometry. In: Annals of Mathematics, pp. 1301–1428 (2011)

127. Gross, M., Siebert, B.: Logarithmic Gromov-Witten invariants. J. Am. Math. Soc. **26**(2), 451–510 (2013). https://doi.org/10.1090/S0894-0347-2012-00757-7

128. Gross, M., Siebert, B.: Intrinsic mirror symmetry and punctured Gromov-Witten invariants. In: Algebraic Geometry: Salt Lake City 2015. Proceedings of Symposia in Pure Mathematics, vol. 97.2, pp. 199–230. American Mathematical Society, Providence, RI (2018). https://doi.org/10.1090/pspum/097.2/01705

129. Grothendieck, A.: Éléments de géométrie algébrique. I. Le langage des schémas. Inst. Hautes Études Sci. Publ. Math. (4), 228 (1960). http://www.numdam.org/item?id=PMIHES_1960__4__228_0

130. Grothendieck, A.: Éléments de géométrie algébrique. III. Étude cohomologique des faisceaux cohérents. I. Inst. Hautes Études Sci. Publ. Math. (11), 167 (1961). http://www.numdam.org/item?id=PMIHES_1961__11__167_0

131. Grothendieck, A.: Éléments de géométrie algébrique. IV. Étude locale des schémas et des morphismes de schémas. II. Inst. Hautes Études Sci. Publ. Math. (24), 231 (1965). http://www.numdam.org/item?id=PMIHES_1965__24__231_0

132. Grothendieck, A.: éléments de géométrie algébrique. IV. étude locale des schémas et des morphismes de schémas. II. Institut des Hautes Études Scientifiques. Publications Mathématiques (24), 231 (1965)

133. Grothendieck, A.: Éléments de géométrie algébrique. IV. Étude locale des schémas et des morphismes de schémas IV. Inst. Hautes Études Sci. Publ. Math. (32), 361 (1967). http://www.numdam.org/item?id=PMIHES_1967__32__361_0

134. Grothendieck, A. (ed.): Séminaire de géométrie algébrique du Bois Marie 1960-61. Revêtements étales et groupe fondamental (SGA 1). Un séminaire dirigé par Alexander Grothendieck. Augmenté de deux exposés de M. Raynaud., édition recomposée et annotée du original publié en 1971 par springer edn. Société Mathématique de France, Paris (2003)

135. Grothendieck, A., Dieudonné, J.A.: Eléments de géométrie algébrique. I. Grundlehren der Mathematischen Wissenschaften [Fundamental Principles of Mathematical Sciences], vol. 166. Springer, Berlin (1971)

136. Hartshorne, R.: Complete intersections and connectedness. Am. J. Math. **84**, 497–508 (1962). https://doi.org/10.2307/2372986

137. Hartshorne, R.: Residues and duality. Lecture Notes in Mathematics, No. 20. Springer, Berlin (1966). Lecture notes of a seminar on the work of A. Grothendieck, given at Harvard 1963/64, With an appendix by P. Deligne

138. Hartshorne, R.: Algebraic geometry. Springer, New York (1977). Graduate Texts in Mathematics, No. 52

139. Hartshorne, R.: Deformation theory. In: Graduate Texts in Mathematics, vol. 257. Springer, New York (2010). https://doi.org/10.1007/978-1-4419-1596-2

140. Hashimoto, K., Sano, T.: Examples of non-Kähler Calabi-Yau 3-folds with arbitrarily large b_2. Geom. Topol. **27**(1), 131–152 (2023). https://doi.org/10.2140/gt.2023.27.131

141. Hassett, B., Kovács, S.J.: Reflexive pull-backs and base extension. J. Algebraic Geometry **13**(2), 233–247 (2004). https://doi.org/10.1090/S1056-3911-03-00331-X

142. He, Y.H.: The Calabi-Yau landscape. From geometry, to physics, to machine learning. Lecture Notes in Mathematics, vol. 2293. Springer, Cham (2021). https://doi.org/10.1007/978-3-030-77562-9

143. Henriksen, M.: On the ideal structure of the ring of entire functions. Pacific J. Math. **2**, 179–184 (1952). http://projecteuclid.org/euclid.pjm/1103051864

144. Hinich, V.: Descent of Deligne groupoids. Internat. Math. Res. Notices **5**, 223–239 (1997). https://doi.org/10.1155/S1073792897000160

145. Hinich, V., Schechtman, V.: Deformation theory and Lie algebra homology. I. Algebra Colloq. **4**(2), 213–240 (1997)

146. Hinich, V., Schechtman, V.: Deformation theory and Lie algebra homology. II. Algebra Colloq. **4**(3), 291–316 (1997)

147. Hinich, V.A., Schechtman, V.V.: On homotopy limit of homotopy algebras. In: K-theory, Arithmetic and Geometry (Moscow, 1984–1986), Lecture Notes in Mathematics, vol. 1289, pp. 240–264. Springer, Berlin (1987). https://doi.org/10.1007/BFb0078370

148. Huebschmann, J.: Poisson cohomology and quantization. J. Reine Angew. Math. **408**, 57–113 (1990). https://doi.org/10.1515/crll.1990.408.57

149. Huebschmann, J.: Lie-Rinehart algebras, Gerstenhaber algebras and Batalin-Vilkovisky algebras. Ann. Inst. Fourier (Grenoble) **48**(2), 425–440 (1998). http://www.numdam.org/item?id=AIF_1998__48_2_425_0

150. Iacono, D.: Deformations and obstructions of pairs (X, D). Int. Math. Res. Not. IMRN **19**, 9660–9695 (2015). https://doi.org/10.1093/imrn/rnu242

151. Iacono, D.: On the abstract Bogomolov-Tian-Todorov theorem. Rend. Mat. Appl. (7) **38**(2), 175–198 (2017)

152. Iacono, D., Manetti, M.: An algebraic proof of Bogomolov-Tian-Todorov theorem. In: Deformation spaces, Aspects of Mathematics, E40, pp. 113–133. Vieweg + Teubner, Wiesbaden (2010). https://doi.org/10.1007/978-3-8348-9680-3_5

153. Iacono, D., Manetti, M.: On deformations of pairs (manifold, coherent sheaf). Canad. J. Math. **71**(5), 1209–1241 (2019). https://doi.org/10.4153/cjm-2018-027-8

154. Iacono, D., Manetti, M.: Homotopy abelianity of the DG-Lie algebra controlling deformations of pairs (variety with trivial canonical bundle, line bundle). Internat. J. Math. **32**(11), Paper No. 2150086, 7 (2021). https://doi.org/10.1142/S0129167X21500865

155. Illusie, L., Kato, K., Nakayama, C.: Quasi-unipotent logarithmic riemann-hilbert correspondences. J. Math. Sci. Univ. Tokyo **12**(1), 1–66 (2005)

156. Jahnke, P., Radloff, I.: Terminal Fano threefolds and their smoothings. Math. Z. **269**(3-4), 1129–1136 (2011). https://doi.org/10.1007/s00209-010-0780-8

157. Kachi, Y.: Global smoothings of degenerate Del Pezzo surfaces with normal crossings. J. Algebra **307**(1), 249–253 (2007). https://doi.org/10.1016/j.jalgebra.2006.03.029

158. Kadeishvili, T.V.: The algebraic structure in the homology of an $A(\infty)$-algebra. Soobshch. Akad. Nauk Gruzin. SSR **108**(2), 249–252 (1982)

159. Kashiwara, M., Kawai, T.: On holonomic systems of microdifferential equations. III. Systems with regular singularities. Publ. Res. Inst. Math. Sci. **17**(3), 813–979 (1981). https://doi.org/10.2977/prims/1195184396

160. Kato, F.: Log smooth deformation theory. Tohoku Math. J. Second Series **48**(3), 317–354 (1996)

161. Kato, F.: Log smooth deformation and moduli of log smooth curves. Int. J. Math. **11**(02), 215–232 (2000)

162. Kato, K.: Logarithmic structures of Fontaine-Illusie. In: Algebraic Analysis, Geometry, and Number Theory (Baltimore, MD, 1988), pp. 191–224. Johns Hopkins University Press, Baltimore, MD (1989)

163. Kato, K.: Toric singularities. Am. J. Math. **116**(5), 1073–1099 (1994)

164. Kato, K., Matsubara, T., Nakayama, C.: Log C^∞-functions and degenerations of Hodge structures. In: Algebraic Geometry 2000, Azumino (Hotaka). Advanced Studies in Pure Mathematics, vol. 36, pp. 269–320. Mathematical Society of Japan, Tokyo (2002). https://doi.org/10.2969/aspm/03610269

165. Kato, K., Nakayama, C., Usui, S.: Classifying spaces of degenerating mixed Hodge structures. I. Borel-Serre spaces. In: Algebraic Analysis and Around. Advanced Studies in Pure Mathematics, vol. 54, pp. 187–222. Mathematical Society of Japan, Tokyo (2009). https://doi.org/10.2969/aspm/05410187

166. Kato, K., Nakayama, C., Usui, S.: Moduli of log mixed Hodge structures. Proc. Japan Acad. Ser. A Math. Sci. **86**(7), 107–112 (2010). https://doi.org/10.3792/pjaa.86.107

167. Kato, K., Nakayama, C., Usui, S.: Classifying spaces of degenerating mixed Hodge structures, II: spaces of SL(2)-orbits. Kyoto J. Math. **51**(1), 149–261 (2011). https://doi.org/10.1215/0023608X-2010-023

168. Kato, K., Nakayama, C., Usui, S.: Classifying spaces of degenerating mixed Hodge structures, III: spaces of nilpotent orbits. J. Algebraic Geom. **22**(4), 671–772 (2013). https://doi.org/10.1090/S1056-3911-2013-00629-3

169. Kato, K., Usui, S.: Logarithmic Hodge structures and classifying spaces. In: The Arithmetic and Geometry of Algebraic Cycles (Banff, AB, 1998). CRM Proceedings & Lecture Notes, vol. 24, pp. 115–130. American Mathematical Society, Providence, RI (2000). https://doi.org/10.1090/crmp/024/06

170. Kato, K., Usui, S.: Classifying spaces of degenerating polarized Hodge structures. In: Annals of Mathematics Studies, vol. 169. Princeton University Press, Princeton, NJ (2009). https://doi.org/10.1515/9781400837113

171. Katzarkov, L., Kontsevich, M., Pantev, T.: Hodge theoretic aspects of mirror symmetry. In: From Hodge Theory to Integrability and TQFT tt*-geometry. Proceedings of Symposia in Pure Mathematics, vol. 78, pp. 87–174. American Mathematical Society, Providence, RI (2008). https://doi.org/10.1090/pspum/078/2483750

172. Kawamata, Y.: Unobstructed deformations. A remark on a paper of Z. Ran: "Deformations of manifolds with torsion or negative canonical bundle" [J. Algebraic Geom. **1**(2), 279–291 (1992); MR1144440 (93e:14015)]. J. Algebraic Geom. **1**(2), 183–190 (1992)

173. Kawamata, Y., Namikawa, Y.: Logarithmic deformations of normal crossing varieties and smoothing of degenerate Calabi-Yau varieties. Invent. Math. **118**(3), 395–409 (1994). https://doi.org/10.1007/BF01231538

174. Kempf, G., Knudsen, F.F., Mumford, D., Saint-Donat, B.: Toroidal embeddings. I. Lecture Notes in Mathematics, vol. 339. Springer, Berlin (1973)

175. Khoroshkin, A., Markarian, N., Shadrin, S.: Hypercommutative operad as a homotopy quotient of BV. Comm. Math. Phys. **322**(3), 697–729 (2013). https://doi.org/10.1007/s00220-013-1737-7

176. Kollár, J.: Singularities of the minimal model program. In: Cambridge Tracts in Mathematics, vol. 200. Cambridge University Press, Cambridge (2013). https://doi.org/10.1017/CBO9781139547895. With a collaboration of Sándor Kovács

177. Kollár, J., Miyaoka, Y., Mori, S.: Rational connectedness and boundedness of Fano manifolds. J. Differential Geom. **36**(3), 765–779 (1992). http://projecteuclid.org/euclid.jdg/1214453188

178. Kontsevich, M.: Homological algebra of mirror symmetry. In: Proceedings of the International Congress of Mathematicians, vol. 1, 2 (Zürich, 1994), pp. 120–139. Birkhäuser, Basel (1995)

179. Kontsevich, M.: Deformation quantization of Poisson manifolds. Lett. Math. Phys. **66**(3), 157–216 (2003). https://doi.org/10.1023/B:MATH.0000027508.00421.bf

180. Kosmann-Schwarzbach, Y.: Exact Gerstenhaber algebras and Lie bialgebroids, pp. 153–165 (1995). https://doi.org/10.1007/BF00996111. Geometric and algebraic structures in differential equations

181. Kovács, S.J., Schwede, K., Smith, K.E.: The canonical sheaf of Du Bois singularities. Adv. Math. **224**(4), 1618–1640 (2010). https://doi.org/10.1016/j.aim.2010.01.020

182. Kravchenko, O.: Deformations of Batalin-Vilkovisky algebras. In: Poisson Geometry (Warsaw, 1998). Banach Center Publications, vol. 51, pp. 131–139. Polish Academy of Sciences Institute of Mathematics, Warsaw (2000)

183. Kulikov, V.S.: Degenerations of $K3$ surfaces and Enriques surfaces. Uspehi Mat. Nauk **32**(3(195)), 167–168 (1977)

184. Kuranishi, M.: On the locally complete families of complex analytic structures. Ann. of Math. (2) **75**, 536–577 (1962). https://doi.org/10.2307/1970211

185. Kříž, I., May, J.P.: Operads, algebras, modules and motives. Astérisque (233), iv+145 (1995)

186. Lee, N.H.: Calabi-Yau construction by smoothing normal crossing varieties. Internat. J. Math. **21**(6), 701–725 (2010). https://doi.org/10.1142/S0129167X10006173

187. Lee, N.H.: d-semistable Calabi-Yau threefolds of type III. Manuscripta Math. **161**(1-2), 257–281 (2020). https://doi.org/10.1007/s00229-018-1097-x

188. Leung, N.C., Ma, Z.N., Young, M.B.: Refined scattering diagrams and theta functions from asymptotic analysis of Maurer-Cartan equations. Int. Math. Res. Not. IMRN **5**, 3389–3437 (2021). https://doi.org/10.1093/imrn/rnz220

189. Li, T.: Log structures on generalized semi-stable varieties. Acta Math. Sin. (Engl. Ser.) **23**(7), 1217–1232 (2007). https://doi.org/10.1007/s10114-005-0854-4

190. Lipman, J.: Notes on derived functors and Grothendieck duality. In: Foundations of Grothendieck Duality for Diagrams of Schemes. Lecture Notes in Mathematics, vol. 1960, pp. 1–259. Springer, Berlin (2009). https://doi.org/10.1007/978-3-540-85420-3

191. Loday, J.L., Vallette, B.: Algebraic operads. In: Grundlehren der mathematischen Wissenschaften [Fundamental Principles of Mathematical Sciences], vol. 346. Springer, Heidelberg (2012). https://doi.org/10.1007/978-3-642-30362-3

192. Losev, A., Shadrin, S.: From Zwiebach invariants to Getzler relation. Comm. Math. Phys. **271**(3), 649–679 (2007). https://doi.org/10.1007/s00220-007-0217-3

193. Lu, P., Tian, G.: The complex structure on a connected sum of $S^3 \times S^3$ with trivial canonical bundle. Math. Ann. **298**(4), 761–764 (1994). https://doi.org/10.1007/BF01459760

194. i Lucio, V.R.: Curved operadic calculus (2022)

195. Lurie, J.: Derived algebraic geometry X: formal moduli problems (2011). https://people.math.harvard.edu/~lurie/papers/DAG-X.pdf

196. Mandel, T., Ruddat, H.: Tropical quantum field theory, mirror polyvector fields, and multiplicities of tropical curves. Int. Math. Res. Not. IMRN **4**, 3249–3304 (2023). https://doi.org/10.1093/imrn/rnab332

197. Manetti, M.: Lectures on deformations of complex manifolds (deformations from differential graded viewpoint). Rend. Mat. Appl. (7) **24**(1), 1–183 (2004)

198. Manetti, M.: Differential graded Lie algebras and formal deformation theory. In: Algebraic Geometry—Seattle 2005. Part 2. Proceedings of Symposia in Pure Mathematics, vol. 80, Part 2, pp. 785–810. American Mathematical Society, Providence, RI (2009). https://doi.org/10.1090/pspum/080.2/2483955

199. Manetti, M.: Lie methods in deformation theory. In: Springer Monographs in Mathematics. Springer, Singapore ([2022] ©2022). https://doi.org/10.1007/978-981-19-1185-9

200. Manetti, M., Meazzini, F.: Formal deformation theory in left-proper model categories. New York J. Math. **25**, 1259–1311 (2019)

201. Manetti, M., Meazzini, F.: Deformations of algebraic schemes via Reedy-Palamodov cofibrant resolutions. Indag. Math. (N.S.) **31**(1), 7–32 (2020). https://doi.org/10.1016/j.indag.2019.08.007

202. Manin, Y.I.: Frobenius manifolds, quantum cohomology, and moduli spaces. In: American Mathematical Society Colloquium Publications, vol. 47. American Mathematical Society, Providence, RI (1999). https://doi.org/10.1090/coll/047

203. Markl, M., Voronov, A.A.: The MV formalism for IBL_∞- and BV_∞-algebras. Lett. Math. Phys. **107**(8), 1515–1543 (2017). https://doi.org/10.1007/s11005-017-0954-y

204. Matsubara, T.: On log Hodge structures of higher direct images. Kodai Math. J. **21**(2), 81–101 (1998). https://doi.org/10.2996/kmj/1138043866

205. Maulik, D., Ranganathan, D.: Logarithmic Donaldson-Thomas theory. Forum Math. Pi **12**, Paper No. e9, 63 (2024). https://doi.org/10.1017/fmp.2024.1

206. Maunder, J.: Koszul duality and homotopy theory of curved Lie algebras. Homology Homotopy Appl. **19**(1), 319–340 (2017). https://doi.org/10.4310/HHA.2017.v19.n1.a16

207. Merkulov, S.A.: Strong homotopy algebras of a Kähler manifold. Internat. Math. Res. Notices **1999**(3), 153–164 (1999). https://doi.org/10.1155/S1073792899000070

208. Merkulov, S.A.: Frobenius$_\infty$ invariants of homotopy Gerstenhaber algebras. I. Duke Math. J. **105**(3), 411–461 (2000). https://doi.org/10.1215/S0012-7094-00-10533-9

209. Minagawa, T.: Deformations of weak Fano 3-folds with only terminal singularities. Osaka J. Math. **38**(3), 533–540 (2001). http://projecteuclid.org/euclid.ojm/1153492511

210. Minagawa, T.: Global smoothing of singular weak Fano 3-folds. J. Math. Soc. Japan **55**(3), 695–711 (2003). https://doi.org/10.2969/jmsj/1191418998

211. Morrison, D.R.: Mirror symmetry and rational curves on quintic threefolds: a guide for mathematicians. J. Am. Math. Soc. **6**(1), 223–247 (1993). https://doi.org/10.2307/2152798

212. Nakayama, C., Ogus, A.: Relative rounding in toric and logarithmic geometry. Geom. Topol **14**(4), 2189–2241 (2010). https://doi.org/10.2140/gt.2010.14.2189

213. Namikawa, Y.: On deformations of Calabi-Yau 3-folds with terminal singularities. Topology **33**(3), 429–446 (1994). https://doi.org/10.1016/0040-9383(94)90021-3

214. Namikawa, Y.: Deformation theory of Calabi-Yau threefolds and certain invariants of singularities. J. Algebraic Geom. **6**(4), 753–776 (1997)

215. Namikawa, Y.: Smoothing Fano 3-folds. J. Algebraic Geom. **6**(2), 307–324 (1997)

216. Namikawa, Y.: Stratified local moduli of Calabi-Yau threefolds. Topology **41**(6), 1219–1237 (2002). https://doi.org/10.1016/S0040-9383(01)00038-6

217. Navarro Aznar, V.: Sur la théorie de Hodge-Deligne. Invent. Math. **90**(1), 11–76 (1987). https://doi.org/10.1007/BF01389031

218. Neeman, A.: An improvement on the base-change theorem and the functor $f^!$. Bull. Iranian Math. Soc. **49**(3), Paper No. 25, 163 (2023). https://doi.org/10.1007/s41980-023-00768-6

219. Nishinou, T., Siebert, B.: Toric degenerations of toric varieties and tropical curves. Duke Math. J. **135**(1), 1–51 (2006). https://doi.org/10.1215/S0012-7094-06-13511-1

220. Nobile, A.: From formal smoothings to geometric smoothings. Rend. Mat. Appl. (7) **44**(3), 181–210 (2023)

221. Oda, T.: Convex bodies and algebraic geometry. In: Ergebnisse der Mathematik und ihrer Grenzgebiete (3) [Results in Mathematics and Related Areas (3)], vol. 15. Springer, Berlin (1988). An introduction to the theory of toric varieties, Translated from the Japanese

222. Ogus, A.: Lectures on Logarithmic Algebraic Geometry. In: Cambridge Studies in Advanced Mathematics, vol. 178. Cambridge University Press, Cambridge (2018). https://doi.org/10.1017/9781316941614

223. Olsson, M.C.: Log Algebraic Stacks and Moduli of Log Schemes. ProQuest LLC, Ann Arbor, MI (2001). http://gateway.proquest.com/openurl?url_ver=Z39.88-2004&rft_val_fmt=info:ofi/fmt:kev:mtx:dissertation&res_dat=xri:pqdiss&rft_dat=xri:pqdiss:3019758. Thesis (Ph.D.)–University of California, Berkeley

224. Olsson, M.C.: Universal log structures on semi-stable varieties. Tohoku Math. J. (2) **55**(3), 397–438 (2003). http://projecteuclid.org/euclid.tmj/1113247481

225. Olsson, M.C.: (Log) twisted curves. Compos. Math. **143**(2), 476–494 (2007). https://doi.org/10.1112/S0010437X06002442

226. Park, J.S.: Semi-classical quantum field theories and Frobenius manifolds. Lett. Math. Phys. **81**(1), 41–59 (2007). https://doi.org/10.1007/s11005-007-0165-z

227. Persson, U.: On degenerations of algebraic surfaces. Mem. Am. Math. Soc. **11**(189), xv+144 (1977). https://doi.org/10.1090/memo/0189

228. Persson, U., Pinkham, H.: Degeneration of surfaces with trivial canonical bundle. Ann. Math. (2) **113**(1), 45–66 (1981). https://doi.org/10.2307/1971133

229. Persson, U., Pinkham, H.: Some examples of nonsmoothable varieties with normal crossings. Duke Math. J. **50**(2), 477–486 (1983). http://projecteuclid.org/euclid.dmj/1077303204

230. Peters, C.A.M., Steenbrink, J.H.M.: Mixed Hodge structures. In: Ergebnisse der Mathematik und ihrer Grenzgebiete. 3. Folge. A Series of Modern Surveys in Mathematics [Results in Mathematics and Related Areas. 3rd Series. A Series of Modern Surveys in Mathematics], vol. 52. Springer, Berlin (2008)

231. Petracci, A.: Some examples of non-smoothable Gorenstein Fano toric threefolds. Math. Z. **295**(1-2), 751–760 (2020). https://doi.org/10.1007/s00209-019-02369-8

232. Petracci, A.: On deformations of toric Fano varieties. In: Interactions with Lattice Polytopes. Springer Proceedings in Mathematics & Statistics, vol. 386, pp. 287–314. Springer, Cham ([2022] ©2022). https://doi.org/10.1007/978-3-030-98327-7_14

233. Positselski, L.: Two kinds of derived categories, Koszul duality, and comodule-contramodule correspondence. Mem. Am. Math. Soc. **212**(996), vi+133 (2011). https://doi.org/10.1090/S0065-9266-2010-00631-8

234. Pridham, J.P.: Unifying derived deformation theories. Adv. Math. **224**(3), 772–826 (2010). https://doi.org/10.1016/j.aim.2009.12.009

235. Prince, T.: Smoothing toric Fano surfaces using the Gross-Siebert algorithm. Proc. Lond. Math. Soc. (3) **117**(3), 617–660 (2018). https://doi.org/10.1112/plms.12153

236. Prince, T.: Smoothing Calabi-Yau toric hypersurfaces using the Gross-Siebert algorithm. Compos. Math. **157**(7), 1441–1491 (2021). https://doi.org/10.1112/s0010437x21007132

237. Ran, Z.: Deformations of Calabi-Yau Kleinfolds. In: Essays on Mirror Manifolds, pp. 451–457. International Press, Hong Kong (1992)

238. Ran, Z.: Deformations of manifolds with torsion or negative canonical bundle. J. Algebraic Geom. **1**(2), 279–291 (1992)

239. Rinehart, G.S.: Differential forms on general commutative algebras. Trans. Am. Math. Soc. **108**, 195–222 (1963). https://doi.org/10.2307/1993603

240. Robert-Nicoud, D., Vallette, B.: Higher lie theory (2020)

241. Ruddat, H.: Log Hodge groups on a toric Calabi-Yau degeneration. In: Mirror Symmetry and Tropical Geometry. In: Contemporary Mathematics, vol. 527, pp. 113–164. American Mathematical Society, Providence, RI (2010). https://doi.org/10.1090/conm/527/10402

242. Ruddat, H.: Local uniqueness of approximations and finite determinacy of log morphisms (2019)

243. Ruddat, H., Siebert, B.: Period integrals from wall structures via tropical cycles, canonical coordinates in mirror symmetry and analyticity of toric degenerations. Publ. Math. Inst. Hautes Études Sci. **132**, 1–82 (2020). https://doi.org/10.1007/s10240-020-00116-y

244. Sano, T.: On deformations of $\mathbb{Q}$-Fano 3-folds. J. Algebraic Geom. **25**(1), 141–176 (2016). https://doi.org/10.1090/jag/672

245. Sano, T.: Deforming elephants of $\mathbb{Q}$-Fano 3-folds. J. Lond. Math. Soc. (2) **95**(1), 23–51 (2017). https://doi.org/10.1112/jlms.12000
246. Sano, T.: On deformations of $\mathbb{Q}$-Fano threefolds II. J. Reine Angew. Math. **730**, 251–261 (2017). https://doi.org/10.1515/crelle-2014-0125
247. Sano, T.: Deformations of weak $\mathbb{Q}$-Fano 3-folds. Internat. J. Math. **29**(7), 1850049, 23 (2018). https://doi.org/10.1142/S0129167X18500490
248. Sano, T.: Examples of non-Kähler Calabi-Yau manifolds with arbitrarily large b_2. J. Topol. **14**(4), 1448–1460 (2021). https://doi.org/10.1112/topo.12212
249. Sastry, P.: Base change and Grothendieck duality for Cohen-Macaulay maps. Compos. Math. **140**(3), 729–777 (2004). https://doi.org/10.1112/S0010437X03000654
250. Séminaire Henri Cartan, 13ième année: 1960/61. Familles d'espaces complexes et fondements de la géométrie analytique. Fasc. 1 et 2: Exp. 1–21. Secrétariat Mathématique, École Normale Supérieure, Paris (1962). 2ième édition, corrigée
251. Schlessinger, M.: Functors of Artin rings. Trans. Am. Math. Soc. **130**, 208–222 (1968). https://doi.org/10.2307/1994967
252. Schmid, W.: Variation of Hodge structure: the singularities of the period mapping. Invent. Math. **22**, 211–319 (1973). https://doi.org/10.1007/BF01389674
253. Schröer, S., Siebert, B.: Toroidal crossings and logarithmic structures. Adv. Math. **202**(1), 189–231 (2006). https://doi.org/10.1016/j.aim.2005.03.006
254. Sernesi, E.: Deformations of algebraic schemes. In: Grundlehren der mathematischen Wissenschaften [Fundamental Principles of Mathematical Sciences], vol. 334. Springer, Berlin (2006)
255. Serre, J.P.: Géométrie algébrique et géométrie analytique. Ann. Inst. Fourier (Grenoble) **6**, 1–42 (1955/56). http://aif.cedram.org/item?id=AIF_1955__6__1_0
256. Groupes de monodromie en géométrie algébrique. II. In: Lecture Notes in Mathematics, vol. 340. Springer, Berlin (1973). Séminaire de Géométrie Algébrique du Bois-Marie 1967–1969 (SGA 7 II), Dirigé par P. Deligne et N. Katz
257. Shentu, J.: Smoothing of semistable Fano varieties. Osaka J. Math. **57**(3), 617–645 (2020). https://projecteuclid.org/euclid.ojm/1594627219
258. Shentu, J., Wang, D.: Notes on algebraic log stack. Internat. J. Math. **27**(10), 1650081, 23 (2016). https://doi.org/10.1142/S0129167X16500816
259. Soukhanov, L.: A Surface Degeneration with Non-Collapsible Contractible Dual Complex. Preprint, arXiv:2012.10995 [math.AG] (2020) (2020). https://arxiv.org/abs/2012.10995
260. Stasheff, J.: Deformation theory and the Batalin-Vilkovisky master equation. In: Deformation Theory and Symplectic Geometry (Ascona, 1996). Mathematical Physics Studies, vol. 20, pp. 271–284. Kluwer Academic Publishers, Dordrecht (1997)
261. Steenbrink, J.: Limits of Hodge structures. Invent. Math. **31**(3), 229–257 (1975/76). https://doi.org/10.1007/BF01403146
262. Steenbrink, J.H.M.: Logarithmic embeddings of varieties with normal crossings and mixed Hodge structures. Math. Ann. **301**(1), 105–118 (1995). https://doi.org/10.1007/BF01446621
263. Stevens, J.: Semistable K3-surfaces with icosahedral symmetry. Enseign. Math. (2) **48**(1-2), 91–126 (2002)
264. Talpo, M., Vistoli, A.: Deformation theory from the point of view of fibered categories. In: Handbook of moduli, vol. III, pp. 281–397. International Press, Somerville, MA; Higher Education Press, Beijing (2013)
265. Talpo, M., Vistoli, A.: Infinite root stacks and quasi-coherent sheaves on logarithmic schemes. Proc. Lond. Math. Soc. (3) **116**(5), 1187–1243 (2018). https://doi.org/10.1112/plms.12109
266. Terilla, J.: Smoothness theorem for differential BV algebras. J. Topol. **1**(3), 693–702 (2008). https://doi.org/10.1112/jtopol/jtn019
267. The Stacks Project Authors: Stacks project (2025). http://stacks.math.columbia.edu/
268. Thomas, R.P.: An obstructed bundle on a Calabi-Yau 3-fold. Adv. Theor. Math. Phys. **3**(3), 567–576 (1999). https://doi.org/10.4310/ATMP.1999.v3.n3.a4

269. Tian, G.: Smoothness of the universal deformation space of compact Calabi-Yau manifolds and its Petersson-Weil metric. In: Mathematical Aspects of String Theory (San Diego, California, 1986). Advanced Series in Mathematical Physics, vol. 1, pp. 629–646. World Scientific Publishing, Singapore (1987)
270. Todorov, A.N.: The Weil-Petersson geometry of the moduli space of SU($n \geq 3$) (Calabi-Yau) manifolds. I. Comm. Math. Phys. **126**(2), 325–346 (1989). http://projecteuclid.org/euclid.cmp/1104179854
271. Tsuji, T.: p-adic étale cohomology and crystalline cohomology in the semi-stable reduction case. Inventiones Mathematicae **137**(2), 233–411 (1999)
272. Tsuji, T.: Poincaré duality for logarithmic crystalline cohomology. Compositio Math. **118**(1), 11–41 (1999). https://doi.org/10.1023/A:1001020809306
273. Tziolas, N.: Smoothings of schemes with nonisolated singularities. Michigan Math. J. **59**(1), 25–84 (2010). https://doi.org/10.1307/mmj/1272376026
274. Tziolas, N.: First order deformations of schemes with normal crossing singularities. Manuscripta Math. **136**(3-4), 345–363 (2011). https://doi.org/10.1007/s00229-011-0444-y
275. Tziolas, N.: Smoothings of Fano varieties with normal crossing singularities. Proc. Edinb. Math. Soc. (2) **58**(3), 787–806 (2015). https://doi.org/10.1017/S0013091515000024
276. Usui, S.: Log Hodge theoretic formulation of mirror symmetry for Calabi-Yau threefolds. Vietnam J. Math. **42**(3), 345–363 (2014). https://doi.org/10.1007/s10013-014-0085-z
277. Verdier, J.L.: Base change for twisted inverse image of coherent sheaves. In: Algebraic Geometry (Internat. Colloq., Tata Inst. Fund. Res., Bombay, 1968). Tata Institute of Fundamental Research Studies in Mathematics, vol. 4, pp. 393–408. Tata Institute of Fundamental Research, Bombay (1969)
278. Voronov, A.A.: Quantizing deformation theory II. Pure Appl. Math. Q. **16**(1), 125–152 (2020). https://doi.org/10.4310/PAMQ.2020.v16.n1.a3
279. Wahl, J.M.: Equisingular deformations of normal surface singularities. I. Ann. of Math. (2) **104**(2), 325–356 (1976). https://doi.org/10.2307/1971049

Index

LECTURE NOTES IN MATHEMATICS Springer

Editors in Chief: J.-M. Morel, B. Teissier;

Editorial Policy

1. Lecture Notes aim to report new developments in all areas of mathematics and their applications – quickly, informally and at a high level. Mathematical texts analysing new developments in modelling and numerical simulation are welcome.

   Manuscripts should be reasonably self-contained and rounded off. Thus they may, and often will, present not only results of the author but also related work by other people. They may be based on specialised lecture courses. Furthermore, the manuscripts should provide sufficient motivation, examples and applications. This clearly distinguishes Lecture Notes from journal articles or technical reports which normally are very concise. Articles intended for a journal but too long to be accepted by most journals, usually do not have this "lecture notes" character. For similar reasons it is unusual for doctoral theses to be accepted for the Lecture Notes series, though habilitation theses may be appropriate.

2. Besides monographs, multi-author manuscripts resulting from SUMMER SCHOOLS or similar INTENSIVE COURSES are welcome, provided their objective was held to present an active mathematical topic to an audience at the beginning or intermediate graduate level (a list of participants should be provided).

   The resulting manuscript should not be just a collection of course notes, but should require advance planning and coordination among the main lecturers. The subject matter should dictate the structure of the book. This structure should be motivated and explained in a scientific introduction, and the notation, references, index and formulation of results should be, if possible, unified by the editors. Each contribution should have an abstract and an introduction referring to the other contributions. In other words, more preparatory work must go into a multi-authored volume than simply assembling a disparate collection of papers, communicated at the event.

3. Manuscripts should be submitted either online at www.editorialmanager.com/lnm to Springer's mathematics editorial in Heidelberg, or electronically to one of the series editors. Authors should be aware that incomplete or insufficiently close-to-final manuscripts almost always result in longer refereeing times and nevertheless unclear referees' recommendations, making further refereeing of a final draft necessary. The strict minimum amount of material that will be considered should include a detailed outline describing the planned contents of each chapter, a bibliography and several sample chapters. Parallel submission of a manuscript to another publisher while under consideration for LNM is not acceptable and can lead to rejection.

4. In general, **monographs** will be sent out to at least 2 external referees for evaluation.

   A final decision to publish can be made only on the basis of the complete manuscript, however a refereeing process leading to a preliminary decision can be based on a pre-final or incomplete manuscript.

   Volume Editors of **multi-author works** are expected to arrange for the refereeing, to the usual scientific standards, of the individual contributions. If the resulting reports can be forwarded to the LNM Editorial Board, this is very helpful. If no reports are forwarded or if other questions remain unclear in respect of homogeneity etc, the series editors may wish to consult external referees for an overall evaluation of the volume.

5. Manuscripts should in general be submitted in English. Final manuscripts should contain at least 100 pages of mathematical text and should always include

   – a table of contents;
   – an informative introduction, with adequate motivation and perhaps some historical remarks: it should be accessible to a reader not intimately familiar with the topic treated;
   – a subject index: as a rule this is genuinely helpful for the reader.
   – For evaluation purposes, manuscripts should be submitted as pdf files.

6. Careful preparation of the manuscripts will help keep production time short besides ensuring satisfactory appearance of the finished book in print and online. After acceptance of the manuscript authors will be asked to prepare the final LaTeX source files (see LaTeX templates online: https://www.springer.com/gb/authors-editors/book-authors-editors/manuscriptpreparation/5636) plus the corresponding pdf- or zipped ps-file. The LaTeX source files are essential for producing the full-text online version of the book, see http://link.springer.com/bookseries/304 for the existing online volumes of LNM). The technical production of a Lecture Notes volume takes approximately 12 weeks. Additional instructions, if necessary, are available on request from lnm@springer.com.

7. Authors receive a total of 30 free copies of their volume and free access to their book on SpringerLink, but no royalties. They are entitled to a discount of 33.3 % on the price of Springer books purchased for their personal use, if ordering directly from Springer.

8. Commitment to publish is made by a *Publishing Agreement*; contributing authors of multiauthor books are requested to sign a *Consent to Publish form*. Springer-Verlag registers the copyright for each volume. Authors are free to reuse material contained in their LNM volumes in later publications: a brief written (or e-mail) request for formal permission is sufficient.

Addresses:
Professor Jean-Michel Morel, CMLA, École Normale Supérieure de Cachan, France
E-mail: moreljeanmichel@gmail.com

Professor Bernard Teissier, Equipe Géométrie et Dynamique,
Institut de Mathématiques de Jussieu – Paris Rive Gauche, Paris, France
E-mail: bernard.teissier@imj-prg.fr

Springer: Ute McCrory, Mathematics, Heidelberg, Germany,
E-mail: lnm@springer.com